Handbook of

Applications of
Chaos Theory

HANDBOOK OF

APPLICATIONS OF CHAOS THEORY

EDITED BY

CHRISTOS H. SKIADAS
CHARILAOS SKIADAS

CRC Press
Taylor & Francis Group
Boca Raton London New York

CRC Press is an imprint of the
Taylor & Francis Group, an **informa** business

A CHAPMAN & HALL BOOK

CRC Press
Taylor & Francis Group
6000 Broken Sound Parkway NW, Suite 300
Boca Raton, FL 33487-2742

First issued in paperback 2020

ISBN-13: 978-1-4665-9043-4 (hbk)
ISBN-13: 978-0-367-73704-7 (pbk)

Library of Congress Cataloging-in-Publication Data

Names: Skiadas, Christos H. | Skiadas, Charilaos.
Title: Handbook of applications of chaos theory / [edited by] Christos H
Skiadas, Charilaos Skiadas.
Description: Boca Raton : Taylor & Francis, 2016. | "A CRC title."
Identifiers: LCCN 2015040349 | ISBN 9781466590434 (alk. paper)
Subjects: LCSH: Mathematical analysis. | Chaotic behavior in systems.
Classification: LCC QA320 .H35 2016 | DDC 003/.857--dc23
LC record available at http://lccn.loc.gov/2015040349

Contents

SECTION III   Chaotic Data Analysis, Equations, and Applications

SECTION IV   Chaos in Plasma

SECTION V   Chaos in Flows and Turbulence

SECTION VI Chaos and Quantum Theory

SECTION VII Optics and Chaos

SECTION VIII Chaos Theory in Biology and Medicine

SECTION IX Chaos in Mechanical Sciences

SECTION X   Chaotic Pattern Recognition

SECTION XI   Chaos in Socioeconomic and Human Sciences

SECTION XII   Chaos In Music

Preface

We are happy to introduce the *Handbook of Applications of Chaos Theory*. Chaos and related theory have expanded to cover almost every scientific field. More than explaining and modeling unexplored phenomena in nature and society, chaos uses vital parts of nonlinear dynamical systems theory along with the already established chaotic theory to open new frontiers and fields of study.

This book serves as a valuable reference for scientists, engineers, and practitioners in the fields of science, technology, and society where chaos theory is used to model unexplored cases and give rise to new applications. The need for a handbook which includes the main parts of chaos theory in various fields, along with related applications, is obvious. The book is designed to be informative and useful to people from a variety of fields with different educations and specific needs.

Section I on Chaos and Nonlinear Dynamics includes six important chapters focused on the intermittency route to chaos, evolutionary dynamics and deterministic chaos, and transition to phase synchronization chaos. Furthermore, the Kolmogorov–Taylor law of turbulence is presented along with the presentation of the nonlinear dynamics of the two-dimensional chaos map and the fractal set for the snow crystal. This section includes also a chapter on the fractional Chen oscillators.

Section II includes important contributions on strange attractors, self-exciting and hidden attractors, stability theory, and Lyapunov exponents. Furthermore, the bifurcation analysis of simple and chaos oscillators along with synchronization of coupled systems and capture of particle into resonance is presented.

The chaotic analysis and the related equations and applications are presented in Section III. Integral equations and large-time behavior of solutions to evolution equations are presented along with characterization of time-series data and examining empirical wavelet coefficients and denoising of chaotic data in the phase space. The geometry of the local instability in Hamiltonian dynamics is analyzed along with the chaos analysis of dynamics in microscopy imaging. The robust stability of discrete-time hybrid systems is presented along with the cohomological theory of stochastic dynamics and the demystification of self-organized criticality.

Section IV includes two important chapters presenting the state of the art of chaos in plasma physics and plasma harmonic and overtone coupling. Plasma and related studies and applications have been fast growing over the last decade. It is perhaps the ideal field of chaos science as chaos is a crucial part of plasma systems.

Flows and turbulence are presented in Section V including wave turbulence in vibrating plates, nonlinear dynamics of oceanic flows, the suspensions of maps to flows, and the Lagrangian coherent structures at the onset of hyperchaos in two-dimensional flows.

Chaotic interference versus decoherence regarding external noise, state mixing, and quantum–classical correspondence is analyzed and an application of microwave networks to simulation of quantum graphs is presented in Section VI.

Section VII on Optics and Chaos includes a detailed presentation of the chaotic, rogue, and noisy optical dissipative solitons along with parhelic-like circle and chaotic light scattering and interesting forms of the hyperbolic prism, the Poincaré disc, and foams.

Applications of chaos theory in biology and medicine are presented in the chapters of Section VIII including applications of extreme value theory for dynamical systems to the analysis of blood pressure data, the Comb models for transport along spiny dendrites, and applications of chaos theory methods in clinical digital pathology.

Section IX on Chaos in Mechanical Sciences includes the system augmentation for detection and sensing with theory and applications, unveiling complexity of church bells dynamics using an experimentally validated hybrid dynamical model, and the multiple Duffing problems based on hill-top bifurcation theory on MFM models.

The science and art of chaotic pattern recognition is presented in Section X. Pattern recognition and encryption systems based on chaos theory are a part of the fast growing chaos literature with many interesting applications.

Section XI on Chaos in Socioeconomic and Human Sciences includes chapters on chaos in economics, on human fuzzy rationality as a novel mechanism of emergent phenomena, and on chaos in monolingual and bilingual speech.

Section XII is devoted to chaos and music, and especially on how composers approach chaos including a survey of applications of chaos theory in musical arts and research.

We are indebted to the authors of this book for their contributions, the participants of the Chaotic Modeling and Simulation Conference series providing material for the book, and many colleagues advising and proposing related topics. Special thanks to the staff of CRC/Taylor & Francis for their support and especially Sunil Nair, Alexander Edwards, Sarfraz Khan, and Jill Jurgensen.

Christos H. Skiadas
Athens, Greece

Charilaos Skiadas
Hanover, Indiana

Editors

Christos H. Skiadas, PhD, was the founder and director of the Data Analysis and Forecasting Laboratory in the Technical University of Crete. Former vice-rector of the Technical University of Crete and chairman of the Department of Production Engineering and Management, he participated in many committees and served as a consultant in various firms while directing several scientific and applied projects. His main scientific and research interests are innovation and innovation diffusion modeling and forecasting, life table data modeling, and healthy life expectancy estimates, including applications in finance and insurance. He also conducted research in deterministic, stochastic, and chaotic modeling, conducted postdoctoral research at the University of Exeter and was a visiting fellow at Université Libre de Bruxelles (ULB) in Brussels. He has published more than 80 papers, 3 monographs, 12 books, and many reports, and has organized and chaired several national and international conferences.

He is chair of the Chaotic Modeling and Simulation Conference series and coauthor of the book *Chaotic Modelling and Simulation* CRC/Taylor & Francis, 2009.

Charilaos Skiadas, PhD, is an assistant professor in mathematics and computer science at Hanover College, Indiana. After earning his PhD in mathematics from the University of Chicago, and being an avid computer programmer all his life, he has developed interests in a wide array of mathematical and computing topics, ranging from algebraic geometry to statistics and programming languages and, lately, data science.

He is coauthor of the book *Chaotic Modelling and Simulation* CRC/Taylor & Francis, 2009.

Contributors

Maricel Agop
Department of Physics
"Gheorghe Asachi" Technical University
Iasi, Romania

Nguyen H. Tuan Anh
Vietnam National University
University of Science
Ho Chi Minh City, Vietnam

Ichiro Ario
Department of Civil and Environmental
    Engineering
Hiroshima University, Higashi-hiroshima
Hiroshima Prefecture, Japan

Elena Babatsouli
Institute of Monolingual and Bilingual
    Speech
Chania, Greece

Szymon Bauch
Institute of Physics
Polish Academy of Sciences
Warsaw, Poland

Piotr Brzeski
Technical University of Łódź
Łódź, Poland

M.V. Budyansky
Laboratory of Nonlinear
    Dynamical Systems
Pacific Oceanological Institute
    of the Russian Academy
    of Sciences
Vladivostok, Russia

Olivier Cadot
Institute of Mechanical Sciences and
    Industrial Applications (IMSIA)
ENSTA-ParisTech, CNRS, EDF, CEA Université
    Paris-Saclay
Palaiseau, France

Dragos Calitoiu
Carleton University
Ottawa, Ontario, Canada

Ana R.M. Carvalho
Faculty of Sciences
University of Porto
Porto, Portugal

Abraham C.-L. Chian
National Institute for Space Research (INPE)
and
Institute of Aeronautical Technology
São Paulo, Brazil
and
School of Mathematical Sciences
University of Adelaide
Adelaide, Australia

Kiran D'Souza
Mechanical and Aerospace Engineering
    Department
The Ohio State University
Columbus, Ohio

Ezequiel del Rio
Department of Applied Physics
E.T.S.I. Aeronautica y del espacio
Universidad Politécnica de Madrid
Madrid, Spain

Dan-Gheorghe Dimitriu
Faculty of Physics
"Alexandru Ioan Cuza" University
Iasi, Romania

Michele Ducceschi
Acoustics and Audio Group
University of Edinburgh
Edinburgh, United Kingdom

Sergio Elaskar
Departamento de Aeronáutica
Universidad Nacional de Córdoba and CONICET
Córdoba, Argentina

Bogdan I. Epureanu
Department of Mechanical Engineering
University of Michigan
Ann Arbor, Michigan

Marisa Faggini
Dipartimento di Scienze Economiche
  e Statistiche
University of Salerno
Fisciano, Italy

Davide Faranda
Laboratoire des Sciences du Climat et de
  l'Environnement
Université Paris-Saclay
Gif-sur-Yvette, France

Matthieu Garcin
Université Paris 1 Panthéon-Sorbonne
Natixis Asset Management
Paris, France

L. Horwitz
Physics Department
Ariel University
Ariel, Israel

and

School of Physics
Tel Aviv University
Tel Aviv, Israel

and

Department of Physics
Bar-Ilan University
Ramat Gan, Israel

Thomas Humbert
Service de Physique de l'Etat Condensé
CEA Saclay
Université Paris-Saclay
Gif-sur-Yvette, France

Alexander Iomin
Technion—Israel Institute of Technology
Haifa, Israel

Christophe Josserand
Sorbonne Universités, UPMC Université Paris
Institut Jean Le Rond d'Alembert
Paris, France

Vladimir L. Kalashnikov
Aston Institute of Photonic Technologies
Aston University
Birmingham, United Kingdom

Tomasz Kapitaniak
Faculty of Mechanical Engineering
Technical University of Łódź
Łódź, Poland

Shunji Kawamoto
Osaka Prefecture University
Osaka, Japan

Oleg Mikhailovich Kiselev
Institute of Mathematics
Ufa Science Centre of Russian Academy
  of Science
Ufa, Russia

Wlodzmierz Klonowski
Nalecz Institute of Biocybernetics
  and Biomedical Engineering
Polish Academy of Sciences
Warsaw, Poland

Christopher W. Kulp
Department of Astronomy and Physics
Lycoming College
Williamsport, Pennsylvania

Nikolay Kuznetsov
Faculty of Mathematics and Mechanics
St. Petersburg State University
St. Petersburg, Russia

and

Department of Mathematical Information
  Technology
University of Jyväskylä
Jyväskylä, Finland

Jakob Lund Laugesen
Department of Physics
Technical University of Denmark
Lyngby, Denmark

Victor J. Law
Department of Chemical Engineering
University of Patras
Rio, Greece

and

School of Mechanical Engineering
University College Dublin
Dublin, Ireland

Michał Ławniczak
Institute of Physics
Polish Academy of Sciences
Warsaw, Poland

Gennady Leonov
Faculty of Mathematics and Mechanics
St. Petersburg State University
and
Institute for Problems in
  Mechanical Engineering
Russian Academy of Sciences
St. Petersburg, Russia

J. Levitan
Physics Department
Ariel University
Ariel, Israel

and

Department of Physics
Technical University of Denmark
Lyngby, Denmark

Roger Lewandowski
IRMAR, UMR 6625
University of Rennes 1, Campus Beaulieu
Rennes, France

M. Lewkowicz
Physics Department
Ariel University
Ariel, Israel

Ihor Lubashevsky
The University of Aizu
Fukushima, Japan

Scott Mc Laughlin
University of Leeds
West Yorkshire, United Kingdom

Vicenç Méndez
Departament de Física
Universitat Autònoma de Barcelona
Cerdanyola del Vallès, Spain

Benjamin Miquel
Department of Applied Mathematics
University of Colorado
Boulder, Colorado

Rodrigo A. Miranda
Faculty UnB-Gama
and
Institute of Physics
University of Brasília
Brasília, Brazil

Nicolas Mordant
Laboratoire des Écoulements Géophysiques
  et Industriels (LEGI)
Université de Grenoble Alpes and CNRS
Grenoble, France

Erik Mosekilde
Department of Physics
The Technical University of Denmark
Lyngby, Denmark

Yusuke Nishiuchi
National Institute of Technology
Kochi College
Kochi, Japan

Brandon J. Niskala
Department of Astronomy and Physics
Lycoming College
Williamsport, Pennsylvania

B. John Oommen
Carleton University
Ottawa, Ontario, Canada

and

University of Agder
Kristiansand, Norway

Igor V. Ovchinnikov
Electrical Engineering Department
University of California at Los Angeles
Los Angeles, California

Anna Parziale
Dipartimento di Scienze Economiche e Statistiche
University of Salerno
Fisciano (Salerno), Italy

Przemyslaw Perlikowski
Technical University of Łódź
Łódź, Poland

and

National University of Singapore
Singapore

Tuan D. Pham
Department of Biomedical Engineering
Linkoping University
Linkoping, Sweden

C.M.A. Pinto
Department of Mathematics
Polytechnic of Porto
Porto, Portugal

S.V. Prants
Laboratory of Nonlinear Dynamical Systems
Pacific Oceanological Institute of the Russian
    Academy of Sciences
Vladivostok, Russia

Ke Qin
University of Electronic Science and Technology
    of China
Sichuan, China

Grienggrai Rajchakit
Department of Mathematics
Maejo University
Chiang Mai, Thailand

Alexander G. Ramm
Mathematics Department
Kansas State University
Manhattan, Kansas

Erico L. Rempel
Institute of Aeronautical Technology
and
National Institute for Space Research
São Paulo, Brazil

Adriane B. Schelin
Institute of Physics
University of Brasília
Brasília, Brazil

Leszek Sirko
Institute of Physics
Polish Academy of Sciences
Warsaw, Poland

Valentin V. Sokolov
Budker Institute of Nuclear Physics
and
Novosibirsk State Technical University
Novosibirsk, Russia

John Starrett
New Mexico Institute of
    Mining and Technology
Socorro, New Mexico

Cyril Touzé
Institute of Mechanical Sciences
    and Industrial Applications (IMSIA)
ENSTA-ParisTech, CNRS, EDF, CEA Université
    Paris-Saclay
Palaiseau, France

Adriana Pedrosa Biscaia Tufaile
Soft Matter Laboratory, Escola de Artes
Ciências e Humanidades
Universidade de São Paulo
São Paulo, Brazil

Alberto Tufaile
Soft Matter Laboratory, Escola de Artes
Ciências e Humanidades
Universidade de São Paulo
São Paulo, Brazil

Tetsushi Ueta
Center for Admin. Info. Tech.
The University of Tokushima
Tokushima, Japan

M.Yu. Uleysky
Laboratory of Nonlinear Dynamical Systems
Pacific Oceanological Institute of the
    Russian Academy of Sciences
Vladivostok, Russia

Dang Van Liet
Vietnam National University
University of Science
Ho Chi Minh City, Vietnam

Ikuo Wada
Department of Cell Science
Institute of Biomedical Sciences
Fukushima Medical University
Fukushima, Japan

Ivan Zelinka
Department of Computer Science
VŠB-Technical University of
    Ostrava
Ostrava-Poruba, Czech Republic

Oleg V. Zhirov
Budker Institute of Nuclear Physics
and
Novosibirsk State University
Novosibirsk, Russia

Y. Ben Zion
Department of Physics
Bar-Ilan University
Ramat Gan, Israel

Zhanybai T. Zhusubaliyev
Department of Computer Science
South West State University
Kursk, Russia

I

Chaos and Nonlinear Dynamics

The Intermittency Route to Chaos

Ezequiel del Rio and
Sergio Elaskar

1.1 Introduction

Intermittency is a particular route to the deterministic chaos characterized by spontaneous transitions between laminar and chaotic dynamics. For the first time, this concept has been introduced by Pomeau and Maneville in the context of the Lorenz system [1,2]. Later, intermittency has been found in a variety of different systems, including, for example, periodically forced nonlinear oscillators, Rayleigh–Bénard convection, derivative nonlinear Schrödinger (DNLS) equation, and development of turbulence in hydrodynamics (see, e.g., References 3–7).

Besides this, there are other types of intermittencies such as type V, X, on–off, eyelet, and ring [8–13]. A more general case of on–off intermittency is the so-called in–out intermittency. An interesting review of on–off and in-out intermittencies can be found in Reference 14.

Proper qualitative and quantitative characterizations of intermittency based on experimental data are especially useful for studying problems with partial or complete lack of knowledge on exact governing equations, as it frequently happens, e.g., in economics, biology, and medicine (see, e.g., References 15 and 16).

It is interesting to note that most of the above-cited references are devoted to systems that have more than one dimension. Inspite of this, they can be described by a one-dimensional map. This phenomenon is typical of systems that contract volume in phase space [17].

All cases of Pomeau and Maneville intermittency have been classified into three types: I, II, and III [18]. The local laminar dynamics of type-I intermittency evolves in a narrow channel, whereas the laminar behavior of type-II and type-III intermittencies develops around a fixed point of its generalized Poincaré maps.

Another characteristic attribute of intermittency is the *global reinjection mechanism* that maps trajectories of the system from the chaotic region back into the local laminar phase. The reinjection mechanism from the chaotic phase into the laminar region is dependent on the chaotic phase behavior, so it is a global property, hence the probability density of reinjection (RPD) of the system back from chaotic burst into points in laminar zone is determined by the dynamics in the chaotic region. Only in a few cases it is possible to an analytical expression for RPD, let us say $\phi(x)$. It is also difficult to obtain RPD experimentally or numerically because of the large amount of data needed to cover each small subset of length Δx, which belong to the reinjection zone. Because of all this, different approximations have been used in the literature to study the intermittency phenomenon. The most common approximation is to consider RPD uniform and thus independent of the reinjection point [3,18–25]. In different investigations, it is assumed an artificial approximation and consider that the reinjection is in a fixed point [22,26].

Here, we present some results of novel view of a recent theory on the intermittency phenomenon based on a new two-parameter class of RPDs appearing in many maps with intermittency. For specific values of the parameters, we recover the classical theory developed for uniform RPD. For this review, we mainly follow References 27 through 31.

First, let us briefly describe the theoretical framework that accounts for a wide class of dynamical systems exhibiting intermittency. We consider a general 1-D map

$$x_{n+1} = G(x_n), \qquad G : \mathbb{R} \to \mathbb{R} \tag{1.1}$$

which exhibits intermittency. Note that map (1.1) can come, for instance, from a Poincaré map of a continuous dynamical system. Let us introduce the dynamics corresponding to the three types of intermittencies around the unstable fixed point. The local laminar dynamics of type-I intermittency is determined by the Poincaré map in the form

$$x_{n+1} = \varepsilon + x_n + ax_n^p \tag{1.2}$$

where $a > 0$ accounts for the weight of the nonlinear component and ε is a controlling parameter ($\varepsilon \ll 1$). The laminar behavior of type-II and type-III intermittencies develops around a fixed point of generalized Poincare Poincaré maps:

$$x_{n+1} = (1 + \varepsilon)x_n + ax_n^p \qquad \text{Type-II} \tag{1.3}$$

$$x_{n+1} = -(1 + \varepsilon)x_n - ax_n^p \qquad \text{Type-III} \tag{1.4}$$

where $a > 0$ accounts for the weight of the nonlinear component and ε is a controlling parameter ($|\varepsilon| \ll 1$). For $\varepsilon \gtrsim 0$, the fixed point $x_0 = 0$ becomes unstable, and hence trajectories slowly escape from the origin, preserving and reversing orientation for type-II and type-III intermittencies, respectively (see Figure 1.1b and c). In some pioneer papers devoted to type-I intermittency, the nonlinear component in Equation 1.2 is quadratic (i.e., $p = 2$) and cubic for type-II and type-III, i.e., $p = 3$ in Equations 1.3 and 1.4 but actually this restriction is not necessary. In any case, for $\varepsilon > 0$, there is an unstable fixed point at $x = 0$ for type-II and type-III and there is not a fixed point at $x = 0$ for type-I, and hence, the trajectories slowly move along the narrow channel formed with the bisecting line as illustrated in Figure 1.1a.

Figure 1.1b illustrates a map having type-II intermittency given by

$$x_{n+1} = G(x_n) \equiv \begin{cases} F(x_n) & x_n \leq x_r \\ (F(x_n) - 1)^{\gamma} & x_n > x_r \end{cases} \tag{1.5}$$

here $F(x) = (1 + \varepsilon)x + ax^p$ with $a = 1 - \varepsilon$ and x_r is the root of the equation $F(x_r) = 1$. Note that map (1.5) is a generalization of the map used by Manneville in his pioneering paper Reference 19, that is, for $\gamma = 1$ map (1.5) can be written as $x_{n+1} = (F(x_n) \bmod 1)$ and if $p = 2$ we recover the Manneville map.

For $\varepsilon > 0$, an iterated point x_n of a starting point x_0 close to the origin, increases in a process driven by parameters ε and p as indicated in Figure 1.1b. When x_n becomes larger than x_r, a chaotic burst occurs

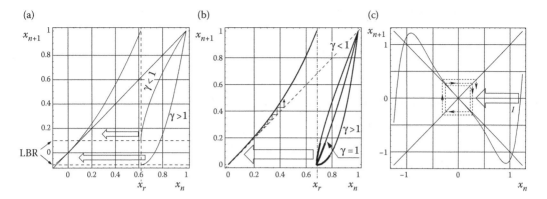

FIGURE 1.1 Maps illustrating type-I, II, and III intermittencies. Also it is indicated the reinjection mechanism into the laminar region by empty arrows. (a) Map having type-I intermittency. There are indicated two LBR corresponding with two reinjected mechanisms according with the values of γ of Equation 1.6. (b) Map having type-II intermittency. Three reinjected mechanisms are also indicated according with Equation 1.5. (c) Typical map shows type-III intermittency. Dashed arrow illustrates the trajectory inside of laminar region.

that will be interrupted when x_n is again mapped into the laminar region, from the region labelled with heavy black segments. This reinjection process is indicated by a big arrow in Figure 1.1b.

The next modification of map 1.5 illustrates the type-I intermittency (see Figure 1.1a)

$$x_{n+1} = G(x_n) = \begin{cases} \varepsilon + x_n + a|x_n|^p & \text{if } x_n < x_r \\ (1 - \hat{x}) \left(\frac{x_n - x_r}{1 - x_r} \right)^{\gamma} + \hat{x} & \text{otherwise} \end{cases} \tag{1.6}$$

where x_r is the root of the equation $\varepsilon + x_n + x_n^p = 1$ and the parameter $\gamma > 1$ driven the nonlinear term of the reinjection mechanism. The parameter $\hat{x}$ corresponds with the so-called lower boundary reinjection (LBR) point and it indicates the limit value for the reinjection that forms the chaotic region into the laminar one. Recently, it has been demonstrated that the statistical behavior of the intermittency can be strongly affected by the value of the LBR [32].

Note that ε and p modified the duration of the laminar phase where the dynamics of the system looks periodic and x_n is less than some value, lets say c. Note that the function RPD will strongly depend on parameter γ that determines the curvature of the map in region marked by a heavy black segment in Figure 1.1b. Only points in that region will be mapped inside of the laminar region. Note that when γ increases, the number of points that will be mapped around the unstable fixed point $x = 0$ also increases, hence we expect that the classical hypothesis of uniform RPD used to develop the classical intermittency theory does not work. In the next section, we will study a more general RPD.

1.2 The Reinjection Probability Distribution

In general, it is very difficult to get $\phi(x)$ analytically, however, for a map like model (1.5) we can use a simple analysis to guess the behavior of $\phi(x)$ near $x = 0$, as the parameter γ changes. To do this, note that all points falling close to the point $x = 0$, are coming from the point close to $x = x_r$ indicated by heavy lines in Figure 1.1b, so $\phi(x)$ is related to the invariant density $\rho(x')$ of map (1.5), where x' refers to the preceding interaction, that is $x' = G_2^{-1}(x)$, where $G_2^{-1}(x)$ is the inverse function of $G(x)$ considering only the definition for $x > x_r$. We also need to rescale $\rho(x')$ taking into account the slope of the function G for points closed to x_r and laying in the right side of x_r. Hence, we get points close to $x = 0$

$$\phi(x) = \rho(x')\frac{b}{\left.\frac{dG(\tau)}{d\tau}\right|_{\tau=x'}} \tag{1.7}$$

where b is a normalization constant. Note that in Equation 1.7, the slope given by $\lim_{\tau\to x_r^+}\left.\frac{dG(\tau)}{d\tau}\right|_{\tau=x'}$ is zero for values of γ bigger than 1 and we have $\lim_{\tau\to x_r^+}\left.\frac{dG(\tau)}{d\tau}\right|_{\tau=x'} = \infty$ if $\gamma < 1$. Hence, we expect that for $\gamma < 1$ the RPD vanished near $x = 0$ and, on the other hand, if $\gamma > 1$ we expect $\lim_{x\to 0^+}\phi(x) = \infty$. For map 1.5, expression 1.7 gives

$$\phi(x) = \frac{b\,\rho(x')}{\gamma\,G'(x')}x^{\frac{1}{\gamma}-1}, \tag{1.8}$$

where G' indicates the derivative of the function G. In the linear approximation of $F(x)$ in the interval $(x_r, G_2^{-1}(c))$, we can consider F' as a constant. Note that the interval $(x_r, G_2^{-1}(c))$ will be mapped into the whole laminar region $(0, c)$. Now, if the density $\rho(x')$ is uniform, we get the RPD

$$\phi(x) = bx^\alpha \quad \text{where } \alpha = \frac{1}{\gamma} - 1. \tag{1.9}$$

that was verified in Reference 27.

Whereas in map (1.5) in Figure 1.1b, the LBR is taken as zero it is important to emphasize that a positive LBR produces a gap around the unstable point in the Poincaré map, as has been experimentally observed from early times [3,4,33]. The lower bound of the reinjection appears in the function $\phi(x) = bx^\alpha$ as a positive shift in x, hence the RPD of Equation 1.9 becomes

$$\phi(x) = b(x - \hat{x})^\alpha \quad \text{where } \alpha = \frac{1}{\gamma} - 1. \tag{1.10}$$

that corresponds with the RPD found in map (1.6) in Figure 1.1a. It is interesting to note that in the case of $\hat{x} > 0$ we have an upper cutoff for the number of iterations in the laminar region, l, because there are no reinjected points arbitrarily close to the fixed point.

From an experimental or numerical point of view, we can observe a gap in the case of $\hat{x} = 0$ if at the same time $0 < \alpha$ holds and we do not have a large number of points. This is because $\phi(x)$ approaches zero as x tends to the unstable point [27]. Note that in this case, the gap length decreases as the number of points increases and finally the gap disappears for a large enough number of points.

With reference to what the map shows in Figure 1.1c, it is clear that to reinject into the laminar region, a point lying in the vicinity of the maximum or minimum, there are necessarily two interactions of the map, whoever, the RPD still can be represented by a power law like Equation 1.10 but in this case, it is necessary to determine the value of α which is not given by Equation 1.10. Moreover, if $\hat{x} < 0$ is due to the symmetry of the map, the reinjection probability function must also be symmetric and it can be described by the two overlapping functions, each one having the form given by Equation 1.10. So that, for $\hat{x} < 0$ we have

$$\phi(x) = \begin{cases} b\left[(|x_i| + x)^\alpha + (|x_i| - x)^\alpha\right] & \text{if } |x| \leqslant |x_i| \\ b\left(|x_i| + x\right)^\alpha & \text{if } |x_i| < x \leqslant c \\ b\left(|x_i| - x\right)^\alpha & \text{if } -c < x \leqslant -|x_i| \end{cases} \tag{1.11}$$

where $b > 0$ is again obtained by the standard normalization condition.

It is interesting to note that the power law (1.10) is already verified in a wide class of 1D-maps even in some classical "pathological" cases that deviate significantly from the classical predictions as in Section 1.5. Regarding the classical hypothesis of uniform RPD, it holds for map (1.5) and (1.6), only in the case of $\gamma = 1$, where x_r is not an extreme point, however, it is false for $\gamma \neq 1$ where the RPD is given by Equation 1.10. This means that whenever the hypothesis of uniform reinjection does not work in general, it usually works for $\rho(x')$ when it is generated in no extreme points.

From the mathematical RPD shape, it is possible to analytically estimate the fundamental characteristic of the intermittency, that is the probability density of the length of laminar phase $\psi(l)$, depending on l, that approximates the number of iterations in the laminar region, i.e., the length of the laminar phase. Note that the function $\psi(l)$ can be estimated from time series, as it is usual to characterize the intermittency type. The characteristic exponent β, depending on $\psi(l)$, defined through the relation $\bar{l} \to \varepsilon^{-\beta}$, is also a good indicator of the intermittency behavior.

The next section is devoted to evaluate the RPD, that is the key point to determine the rest of the properties associated with a specific intermittency.

1.3 Assessment of RPD Function from Numerical and Experimental Data

The RPD function determines the statistical distribution of trajectories leaving the chaotic region. As the RPD is in general a power law of Equation 1.10, the key point to solve the problem of model fitting is to introduce the following integral characteristic:

$$M(x) = \begin{cases} \dfrac{\int_{x_s}^{x} \tau\, \phi(\tau)\, d\tau}{\int_{x_s}^{x} \phi(\tau)\, d\tau} & \text{if } \int_{x_s}^{x} \phi(\tau)d\tau \neq 0 \\ 0 & \text{otherwise} \end{cases} \tag{1.12}$$

where x_s is some "starting" point. The interesting property of the function $M(x)$ is that it is a linear function for an RPD given by Equation 1.10, hence the function $M(x)$ is a useful tool to find the parameters determining the RPD. Setting a constant $c > 0$ that limits the laminar region, we define the domain of M, i.e., $M : [x_0 - c, x_0 + c] \to \mathbb{R}$, where x_0 is the fixed point of maps of the Figure 1.1b and c, inside of the laminar phase of intermittency. In the mentioned case, we have $x_0 = 0$.

As $M(x)$ is an integral characteristic, its numerical estimation is more robust than direct evaluation of $\phi(x)$. This allows reducing statistical fluctuations even for a relatively small data set or data with a high level of noise.

1.3.1 Fitting Linear Model to Experimental Data

To approximate numerically $M(x)$, we notice that it is an average over reinjection points in the interval (x_s, x), hence we can write

$$M(x) \approx M_j \equiv \frac{1}{j} \sum_{k=1}^{j} x_k, \qquad x_{j-1} < x \leq x_j \tag{1.13}$$

where the data set (N reinjection points) $\{x_j\}_{j=1}^{N}$ has been previously ordered, i.e., $x_j \leq x_{j+1}$.

For a wide class of maps exhibiting type-I, type-II, or type-III intermittency, the numerical and experimental data show that $M(x)$ follows the linear law

$$M(x) = \begin{cases} m(x - \hat{x}) + \hat{x} & \text{if } x \geq \hat{x} \\ 0 & \text{otherwise} \end{cases} \tag{1.14}$$

where $m \in (0, 1)$ is a free parameter and $\hat{x}$ is the LBR, i.e., $\hat{x} \approx \inf\{x_j\}$. Then using Equation 1.12, we obtain the corresponding RPD:

$$\phi(x) = b(\alpha)(x - \hat{x})^{\alpha}, \quad \text{with} \quad \alpha = \frac{2m - 1}{1 - m} \tag{1.15}$$

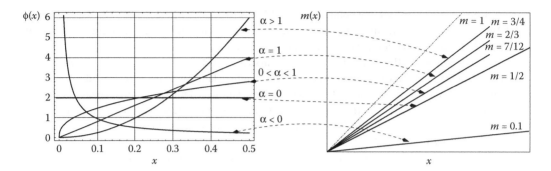

FIGURE 1.2 RPD for different values of α showing decreasing and nondecreasing functions. It is sketched the corresponding slope for the function $M(x)$. Dashes line represents the limit value $m = 1$.

where $b(\alpha)$ is a constant chosen to satisfy $\int_{-\infty}^{\infty} \phi(x)\, dx = 1$. At this point, we note that the linear approximation (1.14) for the numerical or experimental data determines the RPD given by Equation 1.15. Figure 1.2 displays different RPD depending on the exponent α for $\hat{x} = 0$ and $c = 0.5$. It is also shown how the free parameter α depends on the slope m according to Equation 1.15. For $m = 1/2$, we recover the most common approach with uniform RPD, i.e., $\phi(x) = \text{cnst}$, widely considered in the literature. For $m < 1/2$, we have $\alpha < 0$ and the RPD increases without bound for $x \to 0$ as shown in Figure 1.2. In the opposite case $m > 1/2$, we have $\phi(0) = 0$. In this last case, the two possibilities for the RPD, concave or convex are separated by the slope $m = 2/3$ (see Figure 1.2). The RPD (1.15) has two limit cases:

$$\phi_0(x) = \lim_{m \to 0} \phi(x) = \delta(x - \hat{x}) \tag{1.16}$$

$$\phi_1(x) = \lim_{m \to 1} \phi(x) = \delta(x - c) \tag{1.17}$$

Note that $b(\alpha) \to 0$ in these cases.

1.3.2 How to Deal with Short Data Sets

For relatively big data sets (thousands of points), Equation 1.13 provides faithful description of the RPD. However, for small data sets (usually available in experiments) it may lead to a bias in estimation of the parameters. Ordinary least-squares fitting using Equations 1.13 and 1.14 tends to underestimate the value of m.

Let $\{x_j\}_{j=1}^{N}$ be an available properly ordered (say, $x_j \geq x_{j+1}$) small data set of experimental points. Assuming that the exact values m_{exc} and $\hat{x}_{\text{exc}}$ are the exact values of m and $\hat{x}$, we can evaluate the error provided by Equations 1.13 and 1.14:

$$
\begin{aligned}
\varepsilon_1 &= (1 - m_{\text{exc}})(\hat{x}_{\text{exc}} - x_1) \\
\varepsilon_j &\equiv M(x_j) - M_j = \varepsilon_1 + \sum_{k=1}^{j-1} \left(\frac{k}{j} - m_{\text{exc}} \right)(x_k - x_{k+1}), \\
j &= 2, 3, \dots, N
\end{aligned}
\tag{1.18}
$$

where M_j are provided by Equation 1.13 applied over the data set $\{x_j\}_{j=1}^{N}$. We note that the straight line (1.14) intersects the bisector line in the point $(\hat{x}_{\text{exc}},\ M(\hat{x}_{\text{exc}}))$. Thus, ε_1 quantifies the deviation of the first data point from this value. Since $x_1 \leq \hat{x}_{\text{exc}}$ for any data set, the error at the first data point is always nonnegative, $\varepsilon_1 \geq 0$, which leads to a systematic error. Moreover, ε_1 propagates the other errors, which causes significant bias in the least-squares fitting of small data sets.

Since $\varepsilon_1 \propto (\hat{x}_{exc} - x_1)$, to reduce the effect of ε_1, the data set must have a point close to $\hat{x}_{exc}$. This is usually the case for intermittencies with $0 < m < 1/2$, because the $\lim_{x \to \hat{x}_{exc}} \phi(x) = \infty$, i.e., the probability to find $x_1 \approx x_{exc}$ is high enough.

In view of the earlier mentioned, we modify the fitting procedure. The main idea on how to reduce the bias is to introduce an "extra point," z, to the data. This extra point satisfies $z > x_1$. Then we adjust its location in such a way that the newly obtained values of M_j would not have a significant bias.

Before proceeding, we introduce the following notation. Given two vectors $u, v \in \mathbb{R}^N$ we define their mean and covariance:

$$\bar{u} = \frac{1}{N} \sum_{j=1}^{N} u_j, \quad S_{uv} = \frac{1}{N} \sum_{j=1}^{N} (u_j - \bar{u})(v_j - \bar{v}) \tag{1.19}$$

then S_{uu} and S_{vv} are the variances of u and v, respectively.

Let us now introduce three vectors $w, h, y \in \mathbb{R}^N$:

$$w_j = \frac{1}{j+1}, \quad h_j = jM_jw_j, \quad y_j = h_j + zw_j \tag{1.20}$$

The vectors w, h, and y define weights, weighted values of M_j, and ordinates of new data points, respectively. Then, we can apply the standard-least squares fitting to $\{(x_j, \; y_j(z))\}_{j=1}^{N}$, which gives $(y = mx + p)$:

$$m(z) = \frac{S_{xy}}{S_{xx}}, \quad p(z) = \bar{y} - \bar{x}\frac{S_{xy}}{S_{xx}} \tag{1.21}$$

Simple but tedious calculations provide the variance of residuals:

$$S_{rr}(z) = S_{ww}z^2 + 2S_{wh}z + S_{hh} - \frac{(S_{xh} + zS_{xw})^2}{S_{xx}} \tag{1.22}$$

We then select z by minimizing $S_{rr}(z)$:

$$z = \frac{S_{xh}S_{xw} - S_{wh}S_{xx}}{S_{xx}S_{ww} - S_{xw}^2} \tag{1.23}$$

Finally, we estimate the optimal value of m by

$$m_{opt} = \frac{S_{xh} + zS_{xw}}{S_{xx}} \tag{1.24}$$

where z is given by Equation 1.23. The interested reader can find in Reference 30, an application of this method to study a classical map with $m > 1/2$. In the following sections, we will show how the new RPD (1.15) produces quite different results from which we know form the classical theory.

1.4 Length of Laminar Phase and Characteristic Exponent

The probability of finding a laminar phase of length between l and $l + dl$ is $dl\psi(l)$, where the $\psi(l)$ is the duration probability density of the laminar phase. It is to characterize the type of intermittency to compare the analytical prediction for $\psi(l)$ with numerical or experimental evaluation of it. In this chapter, we explain how the RPD of Equation 1.15 can modify the classical result about $\psi(l)$. The method used is similar for the three types of intermittencies here, however, whereas for type-II and type-III, it is possible to find the analytical solution, for type-I it is not possible in the general case.

First we study type-II following Reference 28. To do this, we introduce the next continuous differential equation to approximate the dynamics of the local map (1.3) in the laminar region

$$\frac{dx}{dl} = \varepsilon x + ax^p \tag{1.25}$$

where l approximates the number of iterations in the laminar region, i.e., the length of the laminar phase. Note that l depends on the reinjection point x and it is given by

$$l(x, c) = \int_x^c \frac{d\tau}{a\tau^p + \varepsilon\tau} \tag{1.26}$$

After integration it yields

$$l(x, c) = \frac{1}{\varepsilon}\left[\ln\left(\frac{c}{x}\right) - \frac{1}{p-1}\ln\left(\frac{ac^{(p-1)} + \varepsilon}{ax^{(p-1)} + \varepsilon}\right)\right]. \tag{1.27}$$

Note that Equation 1.27 refers to a local behavior of the map in the laminar region and it determines the length of laminar period, however, the length statistic of the laminar phases, $\psi(l)$, is also affected by the density $\phi(x)$, which is a global property as

$$\psi(l) = \phi(X(l))\left|\frac{dX(l)}{dl}\right| = \phi(X(l))\left|\varepsilon X(l) + aX(l)^p\right| \tag{1.28}$$

where $X(l)$ is the inverse function of $l(x, c)$ and we have used Equation 1.25. Note that $\psi(l)$ depends on one side on the local parameters ε and p, and on the other side, on the global parameters α and $\hat{x}$ determined by the linear function $M(x)$ according to Equation 1.15. The reader can find a detailed analysis on this field in References 27, 28, 31.

Regarding type-III intermittency, in the laminar region the sign x_n changes in each mapping. However, $|x|$ can be approximated by Equation 1.25, consequently the previous results on type-II intermittency can be also applied in the case of type-III intermittency.

Let us consider now the case of type-I intermittency. In this case, the equivalent to Equation 1.25 for type-I is

$$\frac{dx}{dl} = \varepsilon + ax^p, \tag{1.29}$$

from which we obtain $l = L(x, c)$ as a function of x

$$L(x, c) = \frac{c}{\varepsilon 2}F_1\left(\frac{1}{p}, 1; 1 + \frac{1}{p}; -\frac{ac}{\varepsilon}\right) - \frac{x}{\varepsilon 2}F_1\left(\frac{1}{p}, 1; 1 + \frac{1}{p}; -\frac{ax}{\varepsilon}\right) \tag{1.30}$$

in terms of the Gauss hypergeometric function $_2F_1(a, b; c; z)$ [34]. In the case of $p = 2$, $L(x, c)$ can be given by

$$L(x, c) = \frac{1}{\sqrt{a\varepsilon}}\left[\tan^{-1}\left(\sqrt{\frac{a}{\varepsilon}}c\right) - \tan^{-1}\left(\sqrt{\frac{a}{\varepsilon}}x\right)\right]. \tag{1.31}$$

In the case of type-I intermittency, Equation 1.28 transforms

$$\psi(l) = \phi(X(l, c))\left|\frac{dX(l, c)}{dl}\right| = \phi(X(l, c))\left|aX(l, c)^p + \varepsilon\right| \tag{1.32}$$

Whereas, the explicit expression for $X(l, c)$ can be obtained only for a few cases, however, ψ can be plotted in all cases by using the parametrization

$$(L(x, c), \psi'(x)) = \left(L(x, c), \phi(x)\left|\varepsilon + ax^p\right|\right). \tag{1.33}$$

TABLE 1.1 Type-I Intermittency

		$L(x_{r1})$	$L(x_{r2})$	$\lim_{l \to l_{max}} \psi(l)$
$\alpha > 0$	$\hat{x} > 0$	$\nexists$	$\nexists$	0
$\alpha > 0$	$\hat{x} < 0$	min	MAX	0
$\alpha < 0$	$\hat{x} > 0$	$\nexists$	min	∞
$\alpha < 0$	$\hat{x} < 0$	min	$\nexists$	∞
$\alpha = 0$	$\hat{x} < 0$	min	$\nexists$	$\dfrac{\varepsilon + a\hat{x}^p}{\lvert\hat{x} + c\rvert}$
$\alpha = 0$	$\hat{x} > 0$	$\nexists$	$\nexists$	$\dfrac{\varepsilon + a\hat{x}^p}{\lvert\hat{x} + c\rvert}$

Note: Classification of the $\psi(l)$ local extrema types, minimum (min) or maximum (MAX), at $L(x_{r1})$ and $L(x_{r2})$, according to α and $\hat{x}$ values in the RPD. The limits $\lim_{l \to l_{max}} \psi(l)$, depending on α, are also given.

where instead of l, we have taken the coordinate of the reinjected points x as the free parameter. According to Equation 1.33, the value of the function $\psi(l_{max})$ for the maximum length l_{max} is given by

$$\lim_{l \to l_{max}} \psi(l) = \lim_{x \to \hat{x}} \psi'(x) = \begin{cases} 0 & \text{if } \alpha > 0 \\ b(\varepsilon + a\hat{x}^p) & \text{if } \alpha = 0 \\ \infty & \text{if } \alpha < 0 \end{cases} \qquad (1.34)$$

which depends on α. It is interesting to observe that if $\alpha > 0$ we have $\psi(l_{max}) = 0$ and the graph for this function is very different from the one obtained for the classical $\psi(l)$ that can be seen in References 18 and 35, for instance. The reader can find all possible shapes for the $\psi(l)$ in Reference 31.

The extreme points of the function $\psi(l)$ can be obtained, taking into account Equations 1.29 and 1.32, from

$$\frac{d\psi(l)}{dl} = \left((\varepsilon + ax^p)\frac{d\phi(x)}{dx} + ap\phi(x)\,x^{p-1} \right) \left| \frac{dX(l)}{dl} \right| = 0 \qquad (1.35)$$

Since Equation 1.25 imposes $dX(l)/dl \neq 0$ for $\varepsilon \neq 0$, the expression between the square brackets in Equation 1.35 must be zero for $x \in (\hat{x}, c)$ and $\varepsilon \approx 0$, the roots can be finally approximated as

$$x_{r1} \approx 0 \quad \text{and} \quad x_{r2} \approx \frac{p\hat{x}}{\alpha + p}. \qquad (1.36)$$

These estimated values show that the density $\psi(l)$ extrema can occur at two points, $L(x_{r1})$ and $L(x_{r2})$, provided that x_{r1} and x_{r2} lie in $(\hat{x}, c)$. Table 1.1 shows all possibilities matching this restriction for both roots. In the second line of table 1.1, we have $\psi(l_{max}) = 0$, but now $\psi(l)$ has a local maximum, what is a remarkable characteristic is not given by the classical theory on type-I intermittency.

The slope of $\psi(l)$ at the final interval point is determined by the factor $(x - \hat{x})^{(\alpha-1)}$, thus, for $\alpha > 0$ we have

$$\lim_{l \to l_{max}} \frac{d\psi(l)}{dl} = \begin{cases} \infty & \text{if } \alpha < 1 \\ (\varepsilon + a\hat{x}^p)^2 & \text{if } \alpha = 1 \\ 0 & \text{if } \alpha > 1 \end{cases} \qquad (1.37)$$

For the particular case, $\alpha = 0$ corresponding with the uniform reinjection the exact roots $x_{r1} = 0$ and $x_{r2} = \hat{x}$ hold for any value of ε. Observe that x_{r2} lies on the lower limit of the laminar interval whereas x_{r1} yields to a minimum only if $\hat{x} < 0$.

1.4.1   Characteristic Relations

Let us describe how the characteristic exponent is affected by the RPD of Equation 1.15. This exponent, β, defined by the characteristic relation

$$\bar{l} \propto \frac{1}{\varepsilon^{\beta}} \tag{1.38}$$

describes, for small values of ε, how fast the length of the laminar phase grows whereas ε decreases. Traditionally, a single value is admitted depending on the intermittencies type [18]. The mean value of l is defined by

$$\bar{l} = \int_{0}^{\infty} s\psi(s)\,\mathrm{d}s. \tag{1.39}$$

Taking into account the function ψ, depending on the parameter $\hat{x}$ and α, we found that the characteristic exponent β is not a single value as is usually established. According to Equation 1.39, intermittencies type-II and type-III have the same characteristic exponent. We separate the following cases:

- Case A: $\hat{x} \approx x_0$
  A1: $m \in (0, 1 - \frac{1}{p})$ or equivalent $\alpha \in (-1, p-2)$. Equations 1.15, 1.28, and 1.39 give

$$\beta = \frac{\alpha + 2 - p}{1 - p} = \frac{1 + p(m-1)}{(1-p)(1-m)} \tag{1.40}$$

  Particularly $\lim_{m \to 0} \beta = 1$ and $\lim_{m \to 1 - \frac{1}{p}} \beta = 0$.
  A2: $m \in \left(1 - \frac{1}{p}, 1\right)$ or equivalent $p - 2 < \alpha$. We have in this case

$$\beta = 0 \tag{1.41}$$

- Case B: $\hat{x} > x_0$. There is an upper cutoff for l and in the limit $\varepsilon \to 0$ the value $\bar{l}$ practically does not change, hence

$$\beta = 0 \tag{1.42}$$

- Case C: $\hat{x} < x_0$.

$$\beta = \frac{p-2}{p-1}. \tag{1.43}$$

  as in the uniform reinjection.

In a similar way, for type-I intermittency, we find the following cases:

- Case D: $\hat{x} \approx x_0$
  D1: $m \in \left(0, 1 - \frac{1}{p}\right)$ or equivalent $\alpha \in (-1, p-2)$. Equations 1.15, 1.32, and 1.39 give

$$\beta = \frac{p - \alpha - 2}{p} = 1 - \frac{1}{(1-m)p}. \tag{1.44}$$

  Particularly $\lim_{m \to 0} \beta = 1 - \frac{1}{p}$ and $\lim_{m \to 1 - \frac{1}{p}} \beta = 0$.
  D2: $m \in \left(1 - \frac{1}{p}, 1\right)$ or equivalent $p - 2 < \alpha$. Now we have

$$\beta = 0 \tag{1.45}$$

- Case E: $\hat{x} > x_0$.

$$\beta = 0 \tag{1.46}$$

- Case F: $\hat{x} < x_0$.

F1: $\alpha > 0$

$$\beta = \frac{p-2}{p}. \tag{1.47}$$

F2: $\alpha < 0$

$$\beta = \frac{p-1}{p}. \tag{1.48}$$

We would like to emphasize that the value of β in Case C can be obtained by setting $m = 1/2$ in the expression of Case A1, getting the same characteristic exponent that in the uniform reinjection case, even in the no uniform scenario. In fact, if the next two conditions are true: (i) $\phi(x_0) \neq 0$ and (ii) $\frac{d\phi(x)}{dx}\big|_{x=x_0}$ is bounded, then the function $M(x)$ can be approximated close to $x = x_0$ as $M(x) = (x - \hat{x})/2 + \hat{x}$, as in the case of uniform RPD [28].

To prove this, note that for $\phi(x_0) \neq 0$, we have

$$\lim_{x \to x_0} M(x) = \lim_{x \to x_0} \frac{x\,\phi(x)}{\phi(x)} = 0 \tag{1.49}$$

and the slope of the function $M(x)$ for x close to $\hat{x}$ in given, after using the l'Hôpital theorem, by

$$\lim_{x \to \hat{x}} \frac{dM(x)}{dx} = 1 - \frac{1}{2} \lim_{x \to \hat{x}} \left[\frac{\phi'(x) \int_0^x \tau \phi(\tau)\, d\tau}{\phi(x) \int_0^x \phi(\tau)\, d\tau} \right] - \frac{1}{2} \lim_{x \to \hat{x}} \left[\frac{x\,\phi(x)}{\int_0^x \phi(\tau)\, d\tau} \right] \tag{1.50}$$

Now, because of $\frac{d\phi(x)}{dx}\big|_{x=x_0}$ is bounded, we obtain, with Equation 1.49 the value $1/2$ for the limit (1.50). As a consequence, the uniform reinjection is not needed to obtain the classical values of the characteristic exponent, what is more, the requirements on the RPD are only provided by the two assumptions previously imposed in this discussion. In Case C with $\hat{x} < x_0$, the RPD given by Equation 1.11 also satisfies these properties, consequently, the value of β corresponds with the value given by substituting $m = 1/2$ in the Case A1.

This argument can be applied to the type-I intermittency, however, the scenario in the laminar phase is different to the type-II and III, hence only Case F1 can be obtained by setting $m = 1/2$ in the expression for β corresponding to Case D1. A more complete discussion on this topic can be found in Reference 31.

As we will show here, in certain situations the described limit values of β cannot be attained numerically because it requires prohibitively small values of ε. Particularly in Case C if $\hat{x} \lesssim x_0$ the characteristic relation matches Case A ($\hat{x} = x_0$) for small enough values of ε.

1.5 Unification of Anomalous and Standard Intermittencies under a Single Framework

In this section, we show how the described method in the preceding sections can also be applied to pathological cases of intermittency described in the literature [20,36]. We use "pathological" in the sense that they are characterized by a significant deviation of main characteristics from the values predicted on the basis of the uniform distribution. We select two celebrated cases of anomalous intermittency, one with very low value of the parameter m and other one with a very high value of m. Hence, in both cases m is far from $m = 0.5$.

1.5.1 Laugesen Type-III Intermittency

The anomalous cases described by Laugesen and colleagues [36], the authors proposed the next map

$$G(x) = -x(1 + \varepsilon + x^2)e^{-dx^2} \tag{1.51}$$

where the reinjection strongly compresses trajectories in such a way that the RPD becomes similar to δ-function. From a local point of view, the dynamics corresponds with a type-III intermittency but the

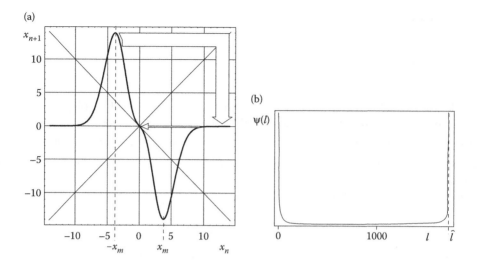

FIGURE 1.3   (a) Sketch of the map (1.1), (1.51) exhibiting anomalous type-III intermittency. Thick arrow illustrates mapping of points from the chaotic region (around the maximum of $G(x)$) into the region with practically zero tangent of $G(x)$. Then thin arrow indicates the following reinjection of these points into the laminar region. (b) Typical duration probability density of the laminar phase. In dashed line is represented the cutoff limit $\hat{l}$.

probability density function of the laminar length deviates significantly from the prediction made by the classical theory. In the original paper, the authors argued that the observed deviation is due to strongly nonuniform reinjection as shown in Figure 1.3a. In spite of this, our approach still accurately describes the intermittent behavior.

To estimate the function $M(x)$, we numerically evaluate the expression (1.13). Owing to the symmetry of map (1.51), we considered only reinjected points coming from one side of the map. As expected, the data obtained fit well to the linear law (1.14). Thus, we can conclude that the power law (1.15) generated by trajectories passing around the maximum and minimum of $G(x)$ is robust against strong compression in the reinjection mechanism indicates in Figure 1.3. Least-squares fit of the numerical data gives $m \approx 0.1$, that differs significantly from $m = 1/2$ corresponding to the classical uniform RPD. Substituting the found value into Equation 1.15, we determine the exponent $\alpha \approx -0.9$, obtaining a value near minus one that corresponds to δ RPD.

Concerning the length of the laminar phase, the linear approximation (1.14) was found $\hat{x} > 0$, so we conclude that there is a gap around zero without reinjected points and according to our classification, we are in Case B described in Section 1.4.1, hence there exists an upper cutoff for l. The cutoff length, $\hat{l}$, is given by

$$X(\hat{l}) = \hat{x}$$

In this case, the probability density function $\phi(X(l))$ grows to infinity ($\alpha < 0$) as $l \to \hat{l}$ and in accordance with Equation 1.28 $\psi \to \infty$. It is worth noting that the presence of a cutoff is not a sufficient condition for unbounded growth of ψ as $l \to \hat{l}$. Besides, it is also necessary that $m \in (0, 1/2)$. In the next section, we shall show a counterexample. The cutoff value $\hat{l}$ increases as ε decreases, and in the limit

$$\hat{l}_0 = \lim_{\varepsilon \to 0} \hat{l}(\varepsilon) = \frac{1}{2a} \left(\frac{1}{\hat{x}^2} - \frac{1}{c^2} \right) \qquad (1.52)$$

which also corresponds to the characteristic exponent $\beta = 0$. For $d = 0.1$ we have $\hat{x} > 0$ but very close to zero giving a very large cutoff length $\bar{l} \approx 10^{12}$, hence we have $\bar{l} \ll \hat{l}$. In this scenario, Case A1 ($\hat{x} = 0$) can provide reasonable approximation for the characteristic exponent β [30].

In the original paper on this anomalous intermittency [36], the two separate analytical arguments to estimate the behavior of $\psi(l)$ in opposite limits ($l \to 0$ and $l \to \hat{l}$) have been proposed. We note that our approach provides approximation of $\psi(l)$ in a single shot.

1.5.2 Pikovsky Intermittency

Another classical example of nonstandard intermittency can be observed in the Pikovsky map [20]:

$$x_{n+1} = G(x_n) = \begin{cases} f(x_n) & x_n \geq 0 \\ -f(-x_n) & x_n < 0 \end{cases} \qquad (1.53)$$

where $f(x) = x^q + hx - 1$ ($q, h > 0$). Map (1.53) has no fixed points and to facilitate the study of its dynamics, it is convenient to introduce the second iteration, i.e., to consider Equation 1.1 with $G(x) = f^2(x) = f(f(x))$. In what follows, we shall deal with this new map.

Figure 1.4 illustrates the map where two unstable fixed points in the typical laminar regions with type-II intermittency. Since the map is symmetrical, we shall describe the upper fixed point only, i.e., $x_0 > 0$. We define two reinjection intervals $I_l = [h - c, h]$ and $I_r = [G(-1), G(-1) + c]$, where c is a constant defining the extension of the laminar region. Points are mapped into the interval I_l from the branch of $G(x)$ with the end point at $(0, h)$, whereas the interval I_r receives trajectories from the branch starting at $(-1, G(-1))$ (Figure 1.4, dashed arrow). If $G(-1) > h$ then there is a gap between these intervals (Figure 1.4), whereas in the opposite case the intervals overlap.

In the nonoverlapping case, there exist two chaotic attractors. Their basins of attraction depend on the controlling parameter q and by playing with this, we can merge them thus obtaining a single chaotic attractor. In the latter case, trajectories can stay for a long time either in the region $|x| < x_0$ or in $|x| > x_0$ and then "jump" between these parts of the attractor. Laminar phases alternate the chaotic dynamics.

Here, we explain the properties of the map in the nonoverlapping case. For a more complete analysis of this map using the method above explicated, the reader can follow Reference 30.

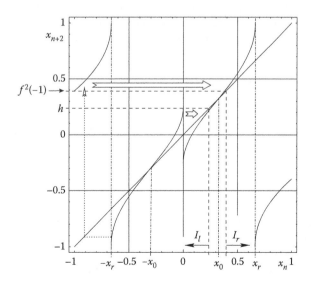

FIGURE 1.4 Second iteration of the map (1.53) demonstrating the Pikovsky type-II intermittency in the nonoverlapping case with a gap between two reinjection intervals. Empty arrows show reinjections into two disjoint intervals I_l and I_r for the upper laminar region. For the interval I_r, dashed arrow shows the routes of reinjection from the infinity tangent region.

Let us assume that $I_l \cap I_r = \varnothing$ (Figure 1.4), then the map has two attractors and consequently two independent chaotic behaviors with intermittency selected by initial conditions. Therefore, the integral characteristic $M(x)$ has two independent branches.

To evaluate $M(x)$, we set the starting point in Equation 1.12 to $x_s^r = x_0 - c$ and $x_s^l = x_0 + c$ for the intervals I_r and I_l, respectively. We notice that $\hat{x}_r = \inf_{x_j \in I_r}\{x_j\} \approx F^2(-x_r^+)$, whereas $\hat{x}_l = \sup_{x_j \in I_l}\{x_j\} \approx F(0^-)$. Thus, to adapt the numerical approximation (1.13) to the interval I_l, we sort the reinjection points in reverse order, i.e., $x_j \geq x_{j+1}$.

For both branches of $M(x)$, the slope is significantly higher than 0.5 due to the infinite tangent generating the power law (1.15). In Figure 1.4, this corresponds to the short arrow indicating reinjection into the interval I_l from the region $x \lesssim x_0$ with near infinite tangent of $G(x)$ at $x = 0$. Other singular point is $-x_r$. We notice that points $x \gtrsim -x_r$ are mapped to the region near $G(-1)$ (see dashed trajectory in Figure 1.4) and finally, after the second iteration, they enter in the laminar interval I_r (long arrow).

Since in this case $(x_0 - \hat{x}_l) > 0$ and $(\hat{x}_r - x_0) > 0$ there is a gap that determines the corresponding cutoff lengths $\hat{l}_l$ and $\hat{l}_r$. Therefore, the length of the laminar phase is bounded. However, in this case we have $m_l, m_r > 0.5$, and hence $\alpha_l, \alpha_r > 0$ and then $\psi(\hat{l}_l) = \psi(\hat{l}_r) = 0$. Thus, the asymptotic behavior of the probability density of the length of laminar phase at $l \to \hat{l}$ is opposite to the blow up observed in the previous case.

Note that the least-squares fitting provide faithful description of the $M(x)$ for relatively big data sets (thousands of points). However, for small data sets (usually available in experiments) it may lead to a bias (Section 1.3.2). This bias can be significant especially in the case like the actual one because the RPD and its first and second derivative are equal to zero at $x = \hat{x}_{\mathrm{exc}}$. Then, we expect a large gap between $\hat{x}_{\mathrm{exc}}$ and the first point in the data set. Moreover, the value of m estimated by using Equation 1.13 and the least-squares fitting depends on the size of data set, and for data sets smaller than 1000 points, the estimated value is consistently below the exact one.

In this scenario, it is better to implement the modified method adapted to fit short data sets (Equations 1.20, 1.23, and 1.24). In the case of the map shown in Figure 1.4, even with short data sets the modified method provides acceptable results [30].

In the original Pikovky paper about map 1.53 [20], the author considered only a set of parameters in the overlapping scenario, that is, $(x_0 - \hat{x}_l) < 0$ and $(\hat{x}_r - x_0) < 0$. Moreover, the conditions for $m \approx 1/2$ referred in Section 1.4.1 hold, hence, they found the simplest scenario and the characteristic exponent corresponding with the uniform reinjection, even if the RPD is not constant.

1.6 Effect of Noise on the RPD

In the previous section, we used the function $M(x)$ as an useful tool to study the RPD. In the noisy case, we also use this function to investigate the new noisy RPD, let say noisy reinjection probability density (NRPD), in systems with intermittency. Figure 1.5 shows the noise effect on a point near the maximun for the next map having type-III intermittency,

$$x_{n+1} = -(1 + \varepsilon)\, x_n - a x_n^3 + d x_n^6\, \sin(x_n) + \sigma \xi_n, \tag{1.54}$$

where $-(1 + \varepsilon)\, x_n - a x_n^3$ $(a > 0)$ is the standard local map for type-III intermittency, whereas the term $d x_n^6 \sin(x_n)$ $(d > 0)$ provides the reinjection mechanism into the laminar region around the critical point $x_0 = 0$. In map (1.54) ξ_n is a noise with $<\xi_m, \xi_n> = \delta(m - n)$ and $<\xi_n> = 0$ and σ is the noise strength. As Figure 1.5 illustrates, the RPD corresponding to the noiseless map is generated around the maximum and minimum of the map by a mechanism that is robust against noise. Following this argument, we can obtain the NRPD, let us say $\Phi(x)$, from the noiseless RPD according to the convolution

$$\Phi(x) = \int \phi(y) g(x - y, \sigma) dy, \tag{1.55}$$

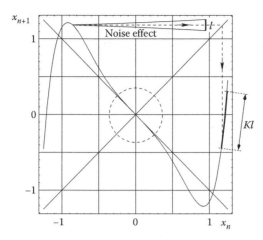

FIGURE 1.5 Noisy map with type-III intermittency. Dashed line between the two solid lines indicate the effect of the noiseless map on a point near the maximum. These solid lines indicate the effect of the noisy map on the same point, that will be mapped on the region shows by a heavy line on the graph of the map. The dashed circle with radius c indicates the laminar region.

where $g(x, \sigma)$ is the probability density of the noise term $\sigma \xi_n$ in Equation (1.54). The reader can find a more complete information in Reference 29.

In the case of uniform distributed noise, after some algebraic manipulation, we get the NRPD as

$$\Phi(x) = \frac{1}{c^{1+\alpha}} \frac{(|x| + K\sigma)^{1+\alpha} - S(|x| - K\sigma)||x| - K\sigma|^{1+\alpha}}{2K\sigma}. \tag{1.56}$$

where we denote by $S(x)$ the sign function that extracts the sign from its argument. In Equation 1.56, the factor K is due to the length amplification indicated in Figure 1.5 where the interval of length equal to l is mapped into a new interval of length Kl. We emphasize that, according to Equation 1.56, the factor K produces an amplification of the effect of the noise. Note that K should be equal to 1 in the case on direct reinjection from the maximum to minimum point, as in the case on type-I and II shown in Figure 1.1a and b.

Some consequences can be derived from the NRPD of Equation 1.56. First, for $|x| >> K\sigma$ the NRPD approaches to the noiseless RPD and second, for $x \approx x_0$ (note that in this example we set $x_0 = 0$) we have a constant function, that is uniform reinjection. The described consequences of Equation 1.56 for the NRPD can be better investigated by using the $M(x)$, because now, the noisy $M(x)$ looks like a piece linear function with two slopes. The first one corresponding to the noiseless RPD is observed far from the x_0, that is, on the right side a given value χ in Figure 1.6b. The second slope approaches to $1/2$ corresponding to uniform reinjection and is observed on the left side of χ. This means that, by the analysis of the noisy data, we can predict the RPD function for the noiseless case. To do this, we proceed like in the noiseless case as explained in the previous sections, but considering only the data on the right side of χ in Figure 1.6b. That is, by least mean square analysis, we can calculate the slope m in Equation 1.15, that determines the reinjection function in the noiseless case. Note that now, $K\sigma$ is the single free parameter in Equation 1.56.

It is important to note that when the noise is applied to the whole map, the function $M(x)$ evidences that, on the right side of χ, the reinjection function is robust against the noise but on the left side of χ, the noise changes the RPD approaching it to the uniform reinjection, at least locally around $x = 0$.

Regarding the uniform RPD, note that the piece linear function approximation of $M(x)$ shown in Figure 1.6b becomes a linear approximation because the two slopes meet in a single one. This means that the effect of noise on the RPD is not too important for uniform reinjection. Owing to this fact, much research devoted to the noise on the local Poincaré map has been published so far, there are only a few studies focused on the effect of noise on the RPD.

We will find a similar scenario type-II intermittencies. The case of type-I can be investigated in a similar way, but this type of intermittency presents a different behavior [37].

1.7 Conclusions and Discussion

In this chapter, an overview of type-I, II, and III intermittencies and a recent method to investigate them are reported.

The main point to describe the intermittency behavior is to determine the RPD. Through the use of M studied in Section 1.3, we have set a way to obtain an analytical description for the RPD, the density of laminar length and the characteristic relations.

The quantity $M(x)$ has more reliable numerical and experimental access than $\phi(x)$. In a number of cases, the linear approximation $M(x) \approx m(x - \hat{x}) + \hat{x}$ fits very well the numerical or experimental data. According to this approximation, we have $\phi(x) = b(x - \hat{x})^{\alpha}$, hence we have found a rich variety of possible profiles for the function $\psi(l)$. Note that the new RPD is a generalization of the usual uniform reinjection approximation which corresponds to $\alpha = 0$ or $m = 1/2$.

Because the probability density of the length of laminar phase $\psi(l)$ depends on the RPD, the $\psi(l)$ shapes are qualitatively different from the classical one.

Also, it extended the characteristic relation for type-I, II, and III intermittencies. Now, the critical exponent β is determined, through the quantities m, $\hat{x}$, and p as reported in Section 1.4.1, hence very different RPDs can lead to the same characteristic exponent β. In the case of positive LBR, that is, $\hat{x} > 0$, a cutoff value for the laminar phase length appears. In this case, the characteristic exponent β is zero. By contrast, for $\hat{x} \approx 0$, the characteristic exponent is a function depending on m and p.

It is worthy to recall that for $m = 0.5$, the classical uniform reinjection is recovered, together with its corresponding characteristic relation. However, it is important to emphasize that this characteristic relation can also emerge from a special kind of RPD which is very different from the one corresponding to the uniform reinjection case. In this sense, in Section 1.4.1 are referred the general requirements to be satisfied by an RPD in order to exhibit the same characteristic relation corresponding to the uniform reinjection.

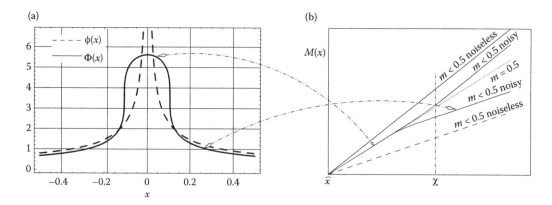

FIGURE 1.6   (a) Dashed lines indicates a typical noiseless RPD (with $\alpha < 0$) for map of Equation 1.54 with $\sigma = 0$. Solid line corresponds with noisy case according with Equation 1.56. (b) Typical shapes of $M(x)$ for noiseless and noisy cases as indicates. Dashed line correspond with dashed RPD of the sub-figure (a). Dots line shows the uniform reinjection with $m = 1/2$.

The described framework also includes the limit cases ($m \approx 0$ and $m \approx 1$) describing anomalous intermittencies published in the literature with a very different behavior given by $m = 0.5$ corresponding with the classical intermittency with uniform RPD.

Even though, there are certainly many papers devoted to the analysis of the effect of noise on the laminar region, the effect of noise on the RPD has not been fully considered. Note that the noise effect on the uniform RPD can be neglected if it does not change the uniform distribution, however this is not the case for a more general RPD.

In Section 1.6, we propose an analytical description of the NRPD valid for type-I, II, and type-III intermittency. We start making a numerical evaluation of the function $M(x)$. From this knowledge, we obtain the RPD corresponding to the noiseless map, that is generated around the maximum and minimum of the map. We find that this mechanism is robust against noise, hence we can use the RPD to obtain an analytical description of the NRPD. It is also important to note that from the RPD, obtained from noisy data, we have a complete description of the noiseless system.

References

1. P. Manneville and Y. Pomeau. Intermittency and the Lorenz model. *Phys. Lett. A*, 75:1–2, 1979.
2. Y. Pomeau and P. Manneville. Intermittent transition to turbulence in dissipative dynamical systems. *Commun. Math. Phys.*, 74(2):189–197, 1980.
3. M. Dubois, M. Rubio, and P. Berge. Experimental evidence of intermittencies associated with a subharmonic bifurcation. *Phys. Rev. Lett.*, 51:1446–1449, 1983.
4. E. del Rio, M.G. Velarde, and A. Rodriguez-Lozano. Long time data series and difficulties with the characterization of chaotic attractors: A case with intermittency III. *Chaos Soliton Fract.*, 4:2169–2179, 1994.
5. S. G. Stavrinides, A. N. Miliou, Th. Laopoulos, and A. N. Anagnostopoulos. The intermittency route to chaos of an electronic digital oscillator. *Int. J. Bifurcat. Chaos*, 18:1561–1566, 2008.
6. G. Krause, S. Elaskar, and E. del Rio. Type-I intermittency with discontinuous reinjection probability density in a truncation model of the derivative nonlinear schrodinger equation. *Nonlinear Dynam.*, 77:455–466, 2014.
7. G. Sanchez-Arriaga, J. Sanmartin, and S. Elaskar. Damping models in the truncated derivative nonlinear schrodinger equation. *Phys. Plasmas*, 14:082108, 2007.
8. H. Kaplan. Return to type-I intermittency. *Phys. Rev. Lett.*, 68:553–557, 1992.
9. T. Price and P. Mullin. An experimental observation of a new type of intermittency. *Physica D*, 48:29–52, 1991.
10. N. Platt, E. Spiegel, and C. Tresser. On-off intermittency: A mechanism for bursting. *Phys. Rev. Lett.*, 70:279–282, 1993.
11. A. Pikovsky, G. Osipov, M. Rosenblum, M. Zaks, and J. Kurths. Attractor-repeller collision and eyelet intermittency at the transition to phase synchronization. *Phys. Rev. Lett.*, 79:47–50, 1997.
12. K. Lee, Y. Kwak, and T. Lim. Phase jumps near a phase synchronization transition in systems of two coupled chaotic oscillators. *Phys. Rev. Lett.*, 81:321–324, 1998.
13. A. Hramov, A. Koronovskii, M. Kurovskaya, and S. Boccaletti. Ring intermittency in coupled chaotic oscillators at the boundary of phase synchronization. *Phys. Rev. Lett.*, 97:114101, 2006.
14. S.G. Stavrinides and A.N. Anagnostopoulos. The route from synchronization to desynchronization of chaotic operating circuits and systems. In S. Banerjee and L. Rondoni, editors, *Applications of Chaos and Nonlinear Dynamics in Science and Engineering*. Springer-Verlag, Berlin, 2013.
15. J. Zebrowski and R. Baranowski. Type I intermittency in nonstationary systems models and human heart rate variability. *Physica A*, 336:74–83, 2004.
16. A. Chian. Complex systems approach to economic dynamics. In *Lecture Notes in Economics and Mathematical Systems*, 39–50. Springer, Berlin, 2007.
17. E. Ott. *Chaos in Dynamical Systems*. Cambridge University Press, Cambridge, 2008.

18. H. Schuster and W. Just. *Deterministic Chaos. An Introduction.* Wiley-VCH Verlag GmbH & Co. KGaA, Weinheim, Germany, 2005.

19. P. Manneville. Intermittency, self-similarity and 1/f spectrum in dissipative dynamical systems. *J. Phys,* 41:1235–1243, 1980.

20. A.S. Pikovsky. A new type of intermittent transition to chaos. *J. Phys A,* 16:L109–L112, 1983.

21. C.M. Kim, O.J. Kwon, E.K. Lee, and H. Lee. New characteristic relations in type-I intermittency. *Phys. Rev. Lett.,* 73:525–528, 1994.

22. C.M. Kim, G.S. Yim, J.W. Ryu, and Y.J. Park. Characteristic relations of type-III intermittency in an electronic circuit. *Phys. Rev. Lett.,* 80:5317–5320, 1998.

23. C.M. Kim, G.S. Yim, Y.S. Kim, J.M. Kim, and H.W. Lee. Experimental evidence of characteristic relations of type-I intermittency in anelectronic circuit. *Phys. Rev. E,* 56:2573–2577, 1997.

24. J.H. Cho, M.S. Ko, Y.J. Park, and C.M. Kim. Experimental observation of the characteristic relations of type-I intermittency in the presence of noise. *Phys. Rev. E,* 65:036222, 2002.

25. W.H. Kye, S. Rim, and C.M. Kim. Experimental observation of characteristic relations of type-III intermittency in the presence of noise in a simple electronic circuit. *Phys. Rev. E,* 68:036203, 2003.

26. W.H. Kye and C.M. Kim. Characteristic relations of type-I intermittency in the presence of noise. *Phys. Rev. E,* 62:6304–6307, 2000.

27. E. del Rio and S. Elaskar. New characteristic relations in type-II intermittency. *Int. J. Bifurcat Chaos,* 20:1185–1191, 2010.

28. S. Elaskar, E. del Rio, and J.M. Donoso. Reinjection probability density in type-III intermittency. *Physica A,* 390:2759–2768, 2011.

29. E. del Rio, M.A.F. Sanjuán, and S. Elaskar. Effect of noise on the reinjection probability density in intermittency. *Commun. Nonlinear Sci. Numer. Simulat.,* 17:3587–3596, 2012.

30. E. del Rio, S. Elaskar, and V.A. Makarov. Theory of intermittency applied to classical pathological cases. *Chaos,* 23:033112, 2013.

31. E. del Rio, S. Elaskar, and J.M. Donoso. Laminar length and characteristic relation in type-I intermittency. *Commun. Nonlinear Sci. Numer. Simulat.,* 19:967–976, 2014.

32. G. Krause, S. Elaskar, and E. del Rio. Type-I intermittency with discontinuous reinjection probability density in a truncation model of the derivative nonlinear schrodinger equation. *Nonlinear Dynam.,* 77(3):455–466, 2014.

33. Y. Ono, K. Fukushima, and T. Yazaki. Critical behavior for the onset of type-III intermittency observed in an electronic circuit. *Phys. Rev.,* 52:4520–4522, 1995.

34. M. Abramowitz and I. A. Stegun. *Handbook of Mathematical Functions.* Dover, USA, 1970.

35. J. E. Hirsch, B. A. Huberman, and D.J. Scalapino. Theory of intermittency. *Phys. Rev. A,* 25:519–532, 1982.

36. J. Laugesen, N. Carlsson, E. Mosekilde, and Bountis. Anomalous statistics for type-III intermittency. *Open Syst. Inf. Dynam.,* 4:393–405, 1997.

37. G. Krause, S. Elaskar, and E. del Rio. Noise effect on statistical properties of type-I intermittency. *Physica A,* 402:318–329, 2014.

2

Deterministic Chaos and Evolutionary Dynamics: Mutual Relations

Ivan Zelinka

2.1 Introduction

Deterministic chaos and evolutionary algorithms (EAs) seem to be two different areas of research that are not joined together, but this is not in fact true. Both areas are tightly joined, as is discussed in this chapter. Because the discussed topics are wide and it is not possible to discuss all in the limited space of this chapter, only the main ideas are reported here. For detailed study, it is recommended to use references provided in the chapter. To better understand the background and consequently discussed mutual relations between chaos and evolutionary dynamics, that is in the fact discrete dynamical system, a brief overview of EAs is done here. For more detailed text about evolutionary techniques, it is recommended to read [1–4].

The chapter is organized as follows. In the beginning, a brief overview of EA is introduced that, followed by a discussion on mutual relations between chaos and evolution. After that, how chaos can be used to randomize (i.e., mutate) evolutionary dynamic and the fact that chaos can be observed inside EAs is explained. In the second part, it is reported how EAs can be used to control/synthesize/identify chaotic systems. At the end, there is a conclusion that summarizes the main ideas of this chapter and proposes a few open questions for future research. In the beginning, let us briefly explain the core idea of EAs in order to better understand their iterative and discrete nature. In recent years, a broad class of algorithms has been developed for heuristic optimization such as genetic algorithms [5], simulated annealing [6,7] and differential evolution (DE) [3], self-organizing migrating algorithms (SOMA) [2], bee algorithm, firefly algorithm, and many more. Most engineering problems can be defined as optimization problems and solutions to such problems are usually difficult to find as their parameters usually include variables of different types such as floating point or integer variables. EAs such as the genetic algorithms, particle swarm [8], ant colony optimization [9], scatter search [10], DE, etc. have been successfully used in the past for these engineering problems. They can offer solutions to almost any problem in a simplified manner: they are able to handle optimizing tasks with mixed variables, including the appropriate constraints, and they do not rely on the existence of derivatives or auxiliary information about the system, for example its transfer function.

EAs are based on ideas of the Darwin and Mendel theory of evolution [4]. The evolutionary principles are transferred into computational methods in a simplified form that will be outlined now.

The evolutionary principles follow iterated procedures:

1. Setting of the algorithm parameters: for each algorithm, parameters must be defined that control the run of the algorithm or terminate it regularly, if the termination criteria defined in advance are fulfilled. Part of this point is the definition of the cost function (objective function) or, as the case may be, what is called fitness—a modified return value of the objective function). The objective function is usually a mathematical model of the problem, whose minimization or maximization leads to the solution of the problem. This function with possible limiting conditions is some kind of "environmental equivalent" in which the quality of current individuals is assessed.

2. Generation of the initial population (generally $N \times M$ matrix, where N is the number of parameters of an individual—D and M is the number of individuals in the population). An individual is a vector of numbers having such a number of components as the number of optimized parameters of the objective function. These components are set randomly and each individual thus represents one possible specific solution of the problem. The set of individuals is called a population.

3. All the individuals are evaluated through a defined objective (or cost) function and to each of them is assigned value of the return objective function.

4. Now parents are selected according to their quality and according to other criterions.

5. Descendants are created by crossbreeding the parents. The process of crossbreeding is different for each algorithm. Parts of parents are changed in classic genetic algorithms, in a DE, crossbreeding is a certain vector operation, etc.

6. Every descendant is mutated. In other words, a new individual is changed by means of a suitable *random* process.

7. Every new individual is evaluated in the same manner as in step 3.

8. The best individuals are selected.

9. The best individuals fill a new population.

10. The old population is eliminated and is replaced by a new population; step 4 represents further continuation.

Steps 4–10 are iteratively repeated until the number of generations specified by the user is reached.

Simply said, EAs are simple iterative process of numerical operations, that combine individuals (candidate solutions) in order to get a better one. EAs are usually very powerful and can be successfully used to solve various complex problems. On the other hand their dynamics can show interesting patterns and structures that clearly point to the existence of deterministic chaos in it. For wider introduction about evolutionary dynamics, it is recommended to read References 1 and 4.

Various relations and similarities can be seen amongst EAs (i.e., their dynamics). Here, we discuss three different point of views on deterministic chaos and EAs. The first part discusses use of deterministic chaos on EAs dynamics (randomization of evolutionary dynamics and existence of chaos inside evolutionary dynamics) and in the second part, there is discussed how can evolution be used in deterministic chaos control (DCC), synthesis, and identification. At the end, in the third part of this chapter, we outline interdisciplinary approach that joins evolutions, chaos, complex networks, and their control into one unified scheme.

Results referred in this chapter are based on the research of wide scientific community. This chapter is an introduction into topic that can be beneficial to the interdisciplinary thinking research communities.

2.2 Deterministic Chaos and Evolutionary Dynamics

Deterministic chaos is observable in many nonlinear and complex systems. As typical examples, we can list weather, bio-systems, electronic systems, physics, and more. Usually, it is taken as a product of nonlinearity and is eliminated or controlled by means of different algorithms. However, it can be used as a tool to improve behavior of selected algorithms and also can be observed inside algorithm dynamics. This is discussed here.

2.2.1 Randomization of Evolutionary Dynamics

The term chaos covers a rather broad class of phenomena whose behavior may seem erratic, chaotic at first glance. Often, this term is used to denote phenomena which are of a purely stochastic nature such as the motion of molecules in a vessel with gas and so. The discovery of the phenomenon of deterministic chaos brought the need to identify manifestations of this phenomenon also in experimental data. Deterministically, chaotic systems are necessarily nonlinear, and conventional statistical procedures, which are mostly linear, are insufficient for their analysis. If the output of a deterministically chaotic system is subjected to linear methods, such a signal will appear as the result of a pseudorandom process.

The general idea, presented in many research papers up to now, is to replace the default pseudorandom number generator (PRNG) with the discrete chaotic map. As the discrete chaotic map is a set of equations with a static start position, we created a random start positions of the map, in order to have different start positions for different experiments. This random position is initialized with the default PRNG, as a one–off randomizer. Once the start position of the chaotic map has been obtained, the map generates the next sequence using its current position.

Till now, chaos was observed in many of various systems (including evolutionary ones) and in the last few years, it was also used to replace PRNGs in EAs. Let us mention, for example, research papers like [11] (a comprehensive overview of mutual intersection between EAs and chaos is discussed here), one of the first uses of chaos inside EAs [12–17] discussing the use of deterministic chaos inside particle swarm algorithm instead of PRNGs, [18–21] investigating relations between chaos and randomness, or the latest one [22–24] using chaos with EAs in applications, amongst the others.

The mutual intersection between evolutionary techniques and chaotic systems has become relatively hot topic in the last few years. PRNGs, many interesting papers have been published about the use of chaotic dynamics instead of PRNGs. This is represented by the usage of evolutionary techniques for the optimization of chaos control [11,25–27] which is summarized in Reference 4. Also, another research on the use of chaos to improve evolutionary performance was done in References 12–17. The idea of using chaotic systems instead of pseudorandom processes has been presented in several research fields and in many applications with promising results [28,29]. Recent research in chaos driven heuristics has been done with the predisposition that unlike stochastic approaches, a chaotic approach is able to bypass local optima stagnation. A chaotic approach generally uses the chaotic map in the place of a PRNG [30]. This causes the heuristic to map unique regions, since the chaotic map iterates to new regions. The task is then to select a very good chaotic map (and its adaptation) as the PRNG. The chaotic systems of interest are usually discrete dissipative chaotic systems. The initial concept of embedding chaotic dynamics into the EAs is given in Reference 12. Later, the initial study [31] was focused on the simple embedding of chaotic systems in the form of chaos pseudorandom number generator (CPRNG) for DE and SOMA in the task of optimal PID tuning. Several papers have been recently focused on the connection of heuristic and chaotic dynamics either in the form of hybridizing of DE with chaotic searching algorithm [32] or in the form of chaotic mutation factor and dynamically changing weighting and crossover factor in self-adaptive chaos differential evolution [33]. Also, the particle swarm optimization (PSO) algorithm with elements of chaos was introduced as CPSO [34] or CPSO combined with chaotic local search [35]. The focus of our research is the embedding of chaotic systems in the form of CPRNG for EAs. The primary aim of this work is not to develop a new type of PRNG, which should pass many statistical tests, but to try to use and test the implementation of natural chaotic dynamics into EA as a CPRNG. This research was later extended with the successful experiments with chaos driven DE [36,37] with simple test functions in low dimensions and in the task of chemical reactor geometry optimization [38]. The concept of chaos DE proved itself to be a powerful heuristic also in the combinatorial problems domain [39,40]. At the same time, the chaos embedded PSO with inertia weigh strategy was closely investigated and more experiments were performed with the concept of chaos driven DE in higher dimensions and with more complex benchmark functions [41]. The interconnection between PSO algorithm and pure CPRNGs was intensively studied within the PID controller optimization issue [42], followed by the introduction of a PSO strategy driven alternately by two chaotic systems [43] and novel chaotic multi choice PSO strategy (chaos MC-PSO) [44]. Recently, the chaotic firefly algorithm was also introduced [45,46] and finally, it was proven that EAs do not require pseudorandom processes at all and work well and even better with simple deterministic sequences [47]. Another research joining deterministic chaos and PRNG has been done for example in Reference 18. The possibility of the generation of random or pseudorandom numbers by use of the ultra weak multidimensional coupling of p one-dimensional (1D) dynamical systems is discussed there. Another paper [48] deeply investigates logistic map as a PRNG and is compared with contemporary PRNGs. A comparison of logistic map results is made with conventional methods of generating pseudorandom numbers. The approach was used to determine the number, delay, and period of the orbits of the logistic map at varying degrees of reported precision (3–23 bits). A logistic map was also used in Reference 19 like a chaos-based true random number generator embedded in reconfigurable specialized hardware. Paper [19] proposed an algorithm of generating PRNG, which is called (couple map lattice based on discrete chaotic iteration) and combines the couple map lattice and chaotic iteration. In Reference 20, authors exploit interesting properties of chaotic systems to design a random bit generator, called CCCBG, in which two chaotic systems are cross-coupled with each other. For evaluation of the bit streams generated by the CCCBG, the four basic tests are performed there including the NIST suite tests.

2.2.2 Deterministic Chaos inside Evolutionary Dynamics

Deterministic chaos has been also observed, mathematically proved, and numerically demonstrated in EAs, especially in genetic algorithms as reported in Reference 49.

In that original research paper, dynamical system models of genetic algorithms modeling the expected behavior or the algorithm as the population size goes to infinity are considered. Whose work is based on research of References 50 through 52. An elegant theory of simple genetic algorithms is based on random heuristic search on the idea of a heuristic map G there. An important point of the research in Reference 49 is that the map, reported in this paper as G, includes all of the dynamics of the simple genetic algorithm, based on specific equations that are based on truncation selection and mutation heuristic function. Sample bifurcation diagrams are depicted in Figures 2.1 through 2.4. Ideas about chaos in simple genetic algorithms are explained in detail there and it is also numerically demonstrated, that chaos can be observed in heuristic algorithms. A novel method joining evolutions, complex networks, and coupled map lattices (CML) systems is described there. Chaotic regimes inside evolutionary dynamics can be very important and certainly have impact on their performance and thus, this is open field for future research.

2.3 EAs and Deterministic Chaos

Since now, it has been outlined how chaos is or can be used inside EAs. The opposite side of the topic discussed above is the use of evolutions on chaotic systems. Few important selected

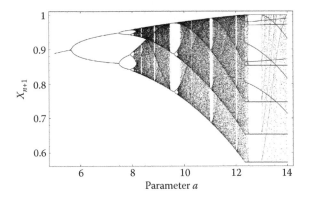

FIGURE 2.1 Bifurcation diagram of simple genetic algorithm for $a \in [54,68]$, $b = 1$, and $T = 7/8$.

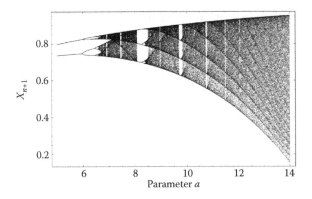

FIGURE 2.2 Bifurcation diagram of simple genetic algorithm for $a \in [54,68]$, $b = 7$, and $T = 7/8$.

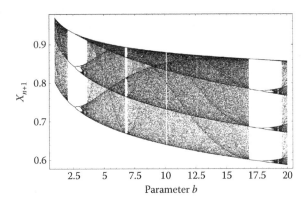

FIGURE 2.3 Bifurcation diagram of simple genetic algorithm for $a = 9$, $b \in [51,73]$, and $T = 7/8$.

topics are discussed here such as the evolutionary control of deterministic chaos, evolutionary control of CML systems, evolutionary synthesis of chaotic systems, evolutionary identification, and synchronization of chaotic systems. At the end, EAs as discrete systems with complex dynamics are discussed.

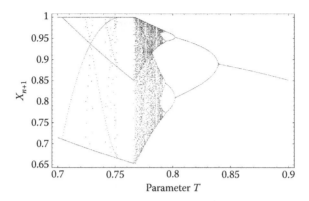

FIGURE 2.4   Bifurcation diagram of simple genetic algorithm for $a = 4$, $b = 1$, and $T \in [0.7, 0.9]$.

2.3.1 Evolutionary Control of Deterministic Chaos

Generally said, EAs are known as powerful tools for almost any difficult and complex optimization problem. But the quality of optimization process results mostly depends on proper design of used cost function, especially when the EAs are used for optimization of chaos control. The results of numerous simulations lend weight to the argument that deterministic chaos in general and also any technique to control chaos are sensitive to proper parameters set up, initial conditions, and in the case of optimization, they are also extremely sensitive to the construction of used cost function.

The main aim of chaos control by EAs is focused on EA implementation to methods for chaos control for the purpose of obtaining better results, which means faster reaching of desired stable state and superior stabilization, which could be robust and effective to optimize difficult problems in the real world. In other words, use EAs on control deals with an investigation on the optimization of the control of chaos by means of EA and constructing of the cost function securing the improvement of system behavior and faster stabilization to desired periodic orbits. The control law can be based on various methods as, for example, on two Pyragas methods [4,11,27,53,54]: delay feedback control—TDAS and extended delay feedback control—ETDAS. As models of deterministic chaotic systems, 1D logistic equation and two-dimensional (2D) Henon map were used. The SOMA and DE were used. Also, the comparison with classical control technique—OGY is presented.

Some research in this field has been recently done using EA for optimization of local control of chaos based on a Lyapunov approach [55,56]. But the approach described here is unique, novel, and up to date, and it was not used or mentioned anywhere. We use EA to search for optimal setting of adjustable parameters of arbitrary control method to reach desired state or behavior of chaotic system. The complexity of such a process is visible in Figure 2.5 in which space of all possible solutions (x, y) and their quality (z) are depicted. As an example of successful chaos control is in Figure 2.6, see, for example, [27]. In this case a chaotic system (Henon) in a four periodic orbit has been stabilized. In the figure is captured 100 repeated experiments and as it is visible, EAs successfully stabilized the system after 1400 iterations (in the worst case). Usually, the solution discovered by EAs has been faster than classical control techniques [4].

2.3.2 Evolutionary Control of CML Systems

Deterministic chaos, discovered by Lorenz [57] is a fairly active area of research in the last few decades. The Lorenz system produces one of the well-known canonical chaotic attractors in a simple three-dimensional autonomous system of ordinary differential equations [57,58]. For discrete chaos, there is another famous chaotic system, called logistic equation [59]. Logistic equation is based on a predator–prey model showing chaotic behavior. This simple model is widely used in the study of chaos, where other similar models exist

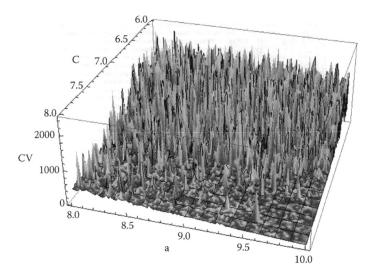

FIGURE 2.5 Cost-function surface representing problem of synchronization.

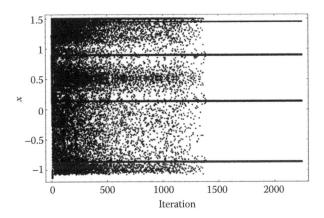

FIGURE 2.6 Best individual solution, p-4 orbit.

(canonical logistic equation [60] and 1D or 2D coupled map lattices [61]) based on it. Since then, a large set of nonlinear systems that can produce chaotic behavior have been observed and analyzed. Chaotic systems thus have become a vitally important part of science and engineering in theoretical as well as practical levels of research. The most interesting and applicable notions are, for example, chaos control and chaos synchronization related to secure communication, amongst others. Recently, the study of chaos is focused not only along the traditional trends but also on understanding and analyzing principles, with the new intention of controlling and utilizing chaos as demonstrated in References 62 and 63. The term chaos control was first coined by Ott, Grebogi, and Yorke in 1990. It represents a process in which a control law is derived and used so that the original chaotic behavior can be stabilized on a constant level of output value or a n-periodic cycle. Since the first experiment of chaos control, many control methods have been developed and some are based on the first approach [64], including pole placement [65,66] and delay feedback [67,68]. Another research has been done on CML control by Reference 69, and special feedback methods for controlling spatio-temporal on-off intermittency has been used there. In Reference 61 the

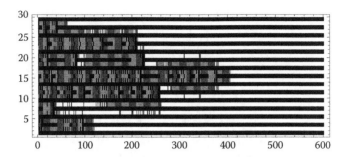

FIGURE 2.7   CML T_1S_2 in configuration 30×600—stabilization after 400 iterations is visible.

control of CML systems while in References 4, 25, and 26 the use of EAs on CML control are summarized. Many methods were adapted for the so-called spatiotemporal chaos represented by CML. Control laws derived for CML are usually based on existing system structures [61], or using an external observer [70]. The evolutionary approach for control was also successfully developed in, for example, [25,55,56,70]. Many published methods of DCC (originally developed for classic DCC) were adapted for the so-called spatiotemporal chaos represented by CML, given by Equation 2.1. Models of this kind are based on a set of spatiotemporal (for 1D, Figure 2.7 and its fitness landscape Figure 2.8) or spatial cells which represents the appropriate state of system elements. Typical example is CML based on the so-called logistic equation, [59,70,71] which is used to simulate behavior of a system which consists of n mutually joined cells via nonlinear coupling, usually noted like ε. The mathematical description of CML system is given by Equation 2.1. Function, which is represented by $f(x_n(i))$ is "arbitrary" discrete system — in this case study, logistic equations have been selected to substitute $f(x_n(i))$. CML description based on Equation 2.1.

$$x_{n+1}(i) = (1 - \varepsilon)f(x_n(i)) + \frac{\varepsilon}{2}(f(x_n(i-1)) + f(x_n(i+1)))   \qquad (2.1)$$

The main question in that research was if EAs are able to control and stabilize chaotic systems like CML, and if they are capable of controlling CML like a black box system, i.e., when the structure of controlled system is unknown. All experiments here were designed to check this idea, and confirm or cancel this idea. Comparison has been done with control based on analysis of CML system [61,72] and analytic derivation of control law for CML. Behavior of controlled CML is as demonstrated on Figure 2.7. It is clearly visible that it is fully random on the beginning, however, later on, it is stabilized. More details about classical

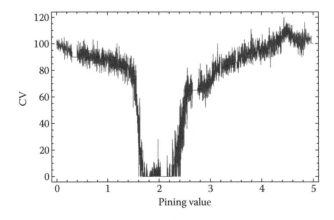

FIGURE 2.8   CML landscape for T_1S_2 in configuration 30×600. Comparing with another landscapes like with T_1S_1 is this much more complex.

control of CML systems can be found in Reference 61 while in References 4, 25, and 26 the principles of evolutionary control of CML systems are explained and demonstrated.

2.3.3 Evolutionary Synthesis of Chaotic Systems

In recent years, interest in softcomputing methods is increasing, including EAs in particular. These algorithms are based on similar principles of biological evolution in the real world as already mentioned. One of the typical uses of EAs is to solve computationally hard problems which are too complex to be solved by conventional methods. In their canonical form, EAs can be used only for numerical estimation of parameters (usually, arguments of a given cost function). Together with EAs in the canonical form, another modification allows the use of EAs as symbolic "constructors", i.e., a processor, for synthesizing complex structures in a symbolic way, based on some predefined simple elements (mathematical operators or electronic elements like diode, transistor, etc.). The term "symbolic way" specifies that mathematical structures and equations, electronic systems, etc., are generated from those simple elements just mentioned.

Given the above background in References 73 through 75, the main motivation of investigation was the question *"Is it possible to synthesize the mathematical description of a new chaotic system, based on simple and elementary mathematical objects, by means of evolutionary computation?"* This question was also based partially on the fact that in engineering applications, it is very often vitally important to know not only when chaos can be generated but also how to generate it [76,77]. This is extremely important in cryptography, for example, where chaotic systems are often used in the design. From a mathematical point of view, it is quite clear that there are some classes of chaotic systems which can be represented by one canonical form (one class — one canonical form) [78]. However, generally speaking, it is not so easy to exactly synthesize a chaotic system with specified features by means of classical mathematical methods. A positive answer to the question mentioned above would open possibilities to synthesize not only a set of not-yet-described chaotic systems, but also some chaotic systems with predefined features. It is believed that such possibilities would have an important impact on the engineering design of various complex nonlinear systems, especially chaotic systems. The main idea of evolutionary synthesis is that in the evolutionary process basic building blocks (i.e., mathematical objects like $+, -, /, x, \ldots$) that are used in chaotic systems are used like individuals. Evolution then synthesizes systems that exhibit chaos under user defined criteria. Selected examples of evolutionary synthesized chaotic systems from Reference 4 are depicted in Figures 2.9 and 2.10.

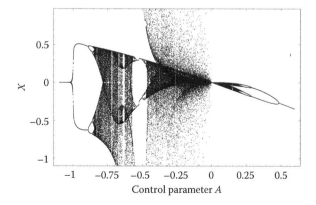

FIGURE 2.9 Bifurcation diagram of $x_{n+1} = \dfrac{x}{\frac{x^3}{2A^3(x-A)(A+x)}+A}$.

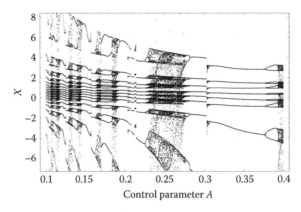

FIGURE 2.10 Bifurcation diagram of $x_{n+1} = \frac{\frac{A^2(A-x)}{x(A(-x)+A+2x)} - 2A + x - 1}{x}$.

2.3.4 Evolutionary Identification and Synchronization of Chaotic Systems

Another relation-question is if EAs are able to synchronize simple chaotic systems. Main attention has been paid to continuous chaotic systems, i.e., to the Rössler and Lorenz systems. All experiments in References 4 and 79 were designed to check this idea and confirm or cancel this idea and were designed as simply as possible, to show the methodology of EAs use.

Synchronization is a dynamical process during which one system is synchronized, slaved remotely by another synchronizing, master system so that the synchronized system is in a certain manner following the behavior of the master system. The word "synchronization" comes from the Greek word "synchronos" (συνχρονος) in which συν (syn) means the same (common,. . .) and χρονος (chronos) means the "time". Synchronization can be divided into the following classes [80], [81], or [61]:

- *Identical synchronization.* This synchronization may occur when two identical chaotic oscillators are mutually coupled (unidirectional or bidirectional coupling), or when one of them drives the other, which is the case of numerical study A (Lorenz–Lorenz). Basically, if $\{x_1, x_2, \ldots, x_n\}$ is a set of state (dynamical) variables of the master system as well as $\{x'_1, x'_2, \ldots, x'_n\}$ of the slave system, then both systems are synchronized under certain initial conditions and $t \to \infty$ is true that $|x_1 - x'_1| \to 0$. This says nothing more than for time large enough is dynamics of both systems in a good approximation. This kind of synchronization is usually called identical synchronization.

- *Generalized synchronization* differs from the previous case by the fact that coupled chaotic oscillators are different and that the dynamical state of one of the oscillators is completely determined by the state of the other.

- *Phase synchronization* is another case of synchronization which occurs when the oscillators coupled are not totally identical and the amplitudes of the oscillator remain not synchronized, while only the oscillator phases evolve in synchrony. There is a geometrical interpretation of this case of synchronization. It is possible to find a so-called plane in phase space in which the projection of the trajectories of the oscillator follows a rotation around a well-defined center. The phase is defined by the angle $\varphi(t)$, described by the segment which joins the center of rotation and the projection of the trajectory point onto the plane.

- *Anticipated and lag synchronization.* Let us say that we have a synchronizing system with state variables $\{x_1, x_2, \ldots, x_n\}$ and a synchronized system with state variables $\{x'_1, x'_2, \ldots, x'_n\}$. Anticipated and lag synchronization occurs when is true that $x'_1(t) = x_1(t + \tau)$. This relation, in fact, says that the dynamics of one of the systems follows, or anticipates, the dynamics of the other and whose dynamics is described by delay differential equations.

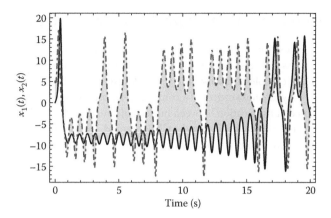

FIGURE 2.11 Principle of synchronization—difference (the gray area) is minimized.

- *Amplitude envelop synchronization* is the kind of synchronization that may appear between two weakly coupled chaotic oscillators. Comparing with another the cases of synchronization, there is no correlation between phases or amplitudes. One can observe a periodic envelop that has the same frequency in the two systems. The magnitude of that envelop has the same order as the difference between the average frequencies of oscillation of both systems. It is important to note that phase synchronization can develop from amplitude one, when the strength of the coupling force between two amplitude envelop synchronized oscillators increase in time.

The main principle of synchronization by EAs is depicted in Figure 2.11, see also Reference 4. A rich amount of literature working with synchronization exists. As a representative literature, we can recommend [80], [81], or [61], all three books are well written and highly readable. Another research works are [82,83] (synchronization based on time-series analysis), [84] (robustness of synchronized systems is studied there), and many others. As a very good starting reference, the above-mentioned books can be used [80], [81], or [61].

EA can be also used for identification (reconstruction) of mathematical description of chaotic systems. This topic is widely discussed in References 4 and 79. In this chapter we discuss the possibility of using EA for the reconstruction of chaotic systems. The main aim of this chapter is to show that EAs are capable of the reconstruction of chaotic systems without any partial knowledge of internal structure, i.e., based only on measured data and predefined set of basic mathematical objects. Algorithms such as SOMA [2] and DE [3] were used in reported experiments here. The systems selected for numerical experiments in Reference 79 are the well-known Lorenz system, simplest quadratic flow, double scroll, damped driven pendulum, and Nose-Hoover oscillator. According to obtained results, it can be stated that evolutionary reconstruction is an alternative and promising way to identify chaotic systems. Identification of various dynamical systems is vitally important in the case of practical applications as well as in theory. A rich set of various methods for dynamical system identification has been developed in the past. In the case of chaotic dynamics, it is, for example, the well-known reconstruction of a chaotic attractor based on research of Reference 85 which has shown that, after the transients have died out, one can reconstruct the trajectory of the attractor from the measurement of a single component. Since the entire trajectory contains a large amount of information, the series of papers by References 86 and 87 is introduced to show a set of averaged coordinate invariant numbers (generalized dimensions, entropies, and scaling indices) by which different strange attractors can be distinguished. The method presented in this research is based on EAs, see Reference 1, which allows the reconstruction not only of chaotic attractors as a geometrical object, but also their mathematical description, based on methods of symbolic regression [73–75]. It is recommended to study those papers.

2.4 EAs as Discrete Systems with Complex Dynamics

Comprehensive and complex topics joining EAs, complex systems, and chaos are discussed in References 88 through 91. In these papers is presented a novel method for visualizing the dynamics of EAs consequently in the form of complex networks and then CML systems. The analogy between individuals in populations in an arbitrary EA and vertices of a complex network is discussed, as well as between edges in a complex network and communication between individuals in a population. The possibility of visualizing the dynamics of a complex network using the CML method, see Figure 2.12, and control by means of chaos control techniques are also discussed. The scheme of this topic is depicted in Figure 2.13. This

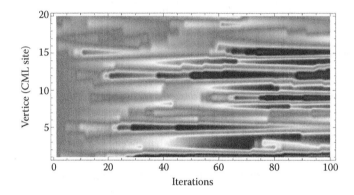

FIGURE 2.12 CML behavior based on evolutionary dynamics. One of many examples.

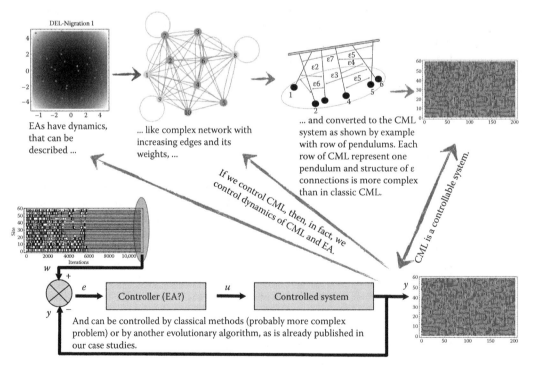

FIGURE 2.13 Mutual relations between chaos, evolution complex dynamics and CML systems as reported in References 88 through 91.

method allows also the study of the presence of the chaos inside complex network dynamics as well as of EAs in more natural way as already done in References 4 and 49. The research reported in References 88 through 91 is focused on mutual intersection of few interesting fields of research whose core topics are about EAs in general. It discusses recent progress in EAs that can be considered as a discrete dynamical complex system with inherent nonlinear dynamics. As already reported in many research papers and books, this dynamics can generate a different kind of behavior including a chaotic one and can be visualized as a complex geometrical structure. In References 88 through 91 relations between evolutionary dynamics, their visualization as a complex network, and novel methods of its control as well as its possible analysis are explained. Control is based on controlling the CML system, that is used to model spatiotemporal behavior (including the chaotic one) and its relation to complex networks (i.e., to evolutionary dynamics) is explained and is based on the classical feedback loop control scheme. On the contrary with classical school books that use classical CML systems and classical control methods, in this case, due to high complexity of our CML system structure that can vary in time, evolutionary methods are the only suitable candidates to control these kind of systems, as reported for example in References 4, 25, 55, and 56.

2.5 Conclusion

In this chapter, we have discussed topics and re-opened questions about mutual relations between chaotic systems and EAs. We show that deterministic chaos can be effectively used as a randomization of EAs, but can also be observed inside evolutionary dynamics. Vice versa, EAs can be easily used to control, synchronize, synthesize, and identify chaotic systems, as reported in numerous research papers. It is clear that EAs are nothing other than a kind of discrete dynamical system exhibiting complex structures and dynamics. Despite the fact that the initial research has been done in this way, it is clear that there are a lot of unanswered questions like (for example) *What is impact of specific dynamic regimes in evolutionary dynamics on its performance? Can control of evolutionary dynamics significantly improve its performance? What kind of chaotic systems can be controlled/synthesized/identified by EAs?* Also very important issues rise about so called hidden attractors, as reported in References 92 through 94. The hidden attractors are a special set of points that reflect the dynamics of observed systems as reported in References 92 through 94. In general and from a computational point of view, attractors can be regarded as self-excited and hidden attractors. Self-excited attractors can be localized numerically by a standard computational procedure, in which after a transient process a trajectory, starting from a point of unstable manifold in a neighborhood of equilibrium, reaches a state of oscillation, therefore, one can easily identify it. In contrast, for a hidden attractor, a basin of attraction does not intersect with small neighborhoods of equilibrium. While classical attractors are self-excited, attractors can, therefore, be obtained numerically by standard computational procedure. For localization of hidden attractors, *it is necessary* to develop special procedures, since there are no similar transient processes leading to such attractors. If the hidden attractor is present in the system dynamics and if coincidentally reached, then device (airplane, el. circuit, etc...) starts to show quasi-cyclic behavior, that can, based on the kind of device, cause real disasters. As an example, we can take the Gripen jet fighter crash[*] or F-22 raptor crash landing[†] caused by computer malfunction that led into oscillations (called also wind-up in control theory). Hidden attractors as a part of deterministic chaos, can be studied in References 92 through 94. Its identification in black-box systems is one of another open questions for future research.

Acknowledgment

This work was supported by the grant no. GACR P103/15/06700S of the Grant Agency of Czech Republic.

---

[*] https://www.youtube.com/watch?v=jP-QMmzGL5I
[†] https://www.youtube.com/watch?v=faB5bIdksi8

References

1. Back T., Fogel D. B., Michalewicz Z. 1997. *Handbook of Evolutionary Computation*, Institute of Physics, London.
2. Zelinka I. 2004. *SOMA— Self organizing migrating algorithm*, In Babu B. V., Onwubolu G. (eds), New Optimization Techniques in Engineering, Chapter 7, 33 p, Springer-Verlag, Heidelberg, ISBN 3-540-20167X.
3. Price K. 1999. *An introduction to differential evolution*, In Corne D., Dorigo M., Glover F. (eds), New Ideas in Optimization, McGraw-Hill, London, UK, 79108, ISBN 007-709506-5.
4. Zelinka I., Celikovsky S., Richter H., Chen G. 2010. *Evolutionary Algorithms and Chaotic Systems*, Springer-Verlag.
5. Holland J. 1992. Genetic algorithms, *Scientific American*, July 44–50.
6. Kirkpatrick S., Gelatt C. Jr., Vecchi M. 1983. Optimization by simulated annealing. *Science*, 220(4598), 671–680.
7. Cerny V. 1985. Thermodynamical approach to the traveling salesman problem: An efficient simulation algorithm, *Journal of Optimization Theory and Applications*, 45(1), 41–51.
8. Kennedy J., Eberhart R. 1995. Particle swarm optimization, *Proceedings of IEEE International Conference on Neural Networks IV*, Perth, WA, 1942–1948.
9. Dorigo M., Stutzle T. 2004. *Ant Colony Optimization*. MIT Press, Cambridge.
10. Laguna M., Marti R. 2003. Scatter Search—Methodology and Implementations in C., Springer, Heidelberg.
11. Senkerik R., Zelinka I., Davendra D., Oplatkova Z. 2010. Utilization of SOMA and differential evolution for robust stabilization of chaotic logistic equation, *Computers and Mathematics with Applications*, 60(4), 1026–1037.
12. Caponetto R., Fortuna L., Fazzino S., Xibilia M. 2003. Chaotic sequences to improve the performance of evolutionary algorithms, *IEEE Transactions on Evolutionary Computation*, 7(3), 289–304.
13. Pluhacek M., Senkerik R., Davendra D., Kominkova Oplatkova Z. 2013. On the behaviour and performance of chaos driven PSO algorithm with inertia weight, *Computers and Mathematics with Applications*, 66(2), 122–134.
14. Pluhacek M., Budikova V., Senkerik R., Oplatkova Z., Zelinka I. 2012. Extended initial study on the performance of enhanced PSO algorithm with lozi chaotic map, In *Proceedings of Nostradamus 2012: International Conference on Prediction*, Modeling and Analysis of Complex Systems, Springer Series: Advances in Intelligent Systems and Computing, Vol. 192, 167–178, ISBN 978-3-642-33226-5.
15. Pluhacek M., Senkerik R., Zelinka I. 2012. Impact of various chaotic maps on the performance of chaos enhanced PSO algorithm with inertia weight–an initial study, In *Proceedings of Nostradamus 2012: International Conference on Prediction*, Modeling and Analysis of Complex Systems, Springer Series: Advances in Intelligent Systems and Computing, Vol. 192, 153–166, ISBN 978-3-642-33226-5.
16. Pluhacek M., Senkerik R., Davendra D., Zelinka I. 2012. PID controller design for 4th order system by means of enhanced PSO algorithm with lozi chaotic map, In *Proceedings of 18th International Conference on Soft Computing—MENDEL 2012*, 35–39, ISBN 978-80- 214-4540-6.
17. Pluhacek M., Budikova V., Senkerik R., Oplatkova Z., Zelinka I. 2012. On the performance of enhanced PSO algorithm with lozi chaotic map a initial study, In *Proceedings of 18th International Conference on Soft Computing—MENDEL 2012*, 40–45, ISBN 978-80- 214-4540-6.
18. Lozi R. 2012. Emergence of randomness from chaos, International Journal of Bifurcation and Chaos, 22(2) 1250021, World Scientific Publishing Company, DOI 10.1142/S0218127412500216.
19. Wang X. Y., Qin X. 2012. A new pseudo-random number generator based on CML and chaotic iteration, Nonlinear Dynamics An International Journal of Nonlinear Dynamics and Chaos in Engineering Systems, *Nonlinear Dynamics*, 70(2) 1589–1592, ISSN 0924-090X, DOI 10.1007/s11071-012-0558-0.

20. Pareek N. K., Patidar V., Sud K. K. 2010. A random bit generator using chaotic maps, *International Journal of Network Security*, 10(1), 3238.

21. Xing-Yuan W., Lei Y. 2012. Design of pseudo-random bit generator based on chaotic maps, *International Journal of Modern Physics B*, 26(32), 1250208 (9 pages), World Scientific Publishing Company, DOI 10.1142/S0217979212502086.

22. Zhang S. Y., Xingsheng L. G. 2010. A hybrid co-evolutionary cultural algorithm based on particle swarm optimization for solving global optimization problems, In *International Conference on Life System Modeling and Simulation/International Conference on Intelligent Computing for Sustainable Energy and Environment (LSMS-ICSEE)*, Wuxi, People's Republic of China, September 17–20.

23. Hong W. -C., Dong Y., Zhang W. Y., Chen L. -Y., Panigrahi B. K. 2013. Cyclic electric load forecasting by seasonal SVR with chaotic genetic algorithm, *International Journal of Electrical Power and Energy Systems*, 44(1), 604–614, DOI 10.1016/j.ijepes.2012.08.010.

24. Senkerik R., Davendra D., Zelinka I., Oplatkova Z., Pluhacek M. 2012. Optimization of the batch reactor by means of chaos driven differential evolution, In *Proceedings of SOCO 2012: Soft Computing Models in Industrial and Environmental Applications*, Springer Series: Advances in Intelligent Systems and Computing, Vol. 188, 93–102, ISBN 978-3-642-32922-7.

25. Zelinka I. 2006. Investigation on realtime deterministic chaos control by means of evolutionary algorithms, *Proceedings of the 1st IFAC Conference on Analysis and Control of Chaotic Systems*, Reims, France, 211–217.

26. Zelinka I. 2009. Real-time deterministic chaos control by means of selected evolutionary algorithms, *Engineering Applications of Artificial Intelligence*, 22(2), 283–297, DOI 10.1016/j.engappai.2008.07.008.

27. Zelinka I., Senkerik R., Navratil E. 2009. Investigation on evolutionary optimitazion of chaos control, *Chaos, Solitons and Fractals*, 40(1), 111–129, DOI 10.1016/j.chaos.2007.07.045.

28. Lee J. S., Chang K. S. 1996. Applications of chaos and fractals in process systems engineering, *Journal of Process Control*, 6(23), 71–87.

29. Wu J., Lu J., Wang J. 2009. Application of chaos and fractal models to water quality time series prediction, *Environmental Modelling and Software*, 24(5), 2009, 632–636.

30. Aydin I., Karakose M., Akin E. 2010. Chaotic-based hybrid negative selection algorithm and its applications in fault and anomaly detection, *Expert Systems with Applications*, 37(7), 2010, 5285–5294.

31. Davendra D., Zelinka I., Senkerik R. 2010. Chaos driven evolutionary algorithms for the task of PID control, *Computers and Mathematics with Applications*, 60(4), 1088–1104, ISSN 0898-1221.

32. Liang W., Zhang L., Wang M. 2011. The chaos differential evolution optimization algorithm and its application to support vector regression machine, *Journal of Software*, 6(7), pp. 1297–1304.

33. Zhenyu G., Bo Ch., Min Z., Binggang C. 2006. Self-adaptive chaos differential evolution, *Lecture Notes in Computer Science*, 4221, 972–975.

34. Coelho L. S., Mariani V. C. 2009. A novel chaotic particle swarm optimization approach using Hénon map and implicit filtering local search for economic load dispatch, *Chaos, Solitons and Fractals*, 39(2), 510–518.

35. Hong W. Ch. 2009. Chaotic particle swarm optimization algorithm in a support vector regression electric load forecasting model, *Energy Conversion and Management*, 50(1), 105–117.

36. Senkerik R., Davendra D., Zelinka I., Pluhacek M., Oplatkova Z. 2012. An investigation on the chaos driven differential evolution: An initial study, In *Proceedings of 5th International Conference on Bioinspired Optimization Methods and Their Applications*, BIOMA 2012, Slovenia, 185–194, ISBN 978-961-264-043-9.

37. Senkerik R., Davendra D., Zelinka I., Pluhacek M., Oplatkova Z. 2012. An investigation on the differential evolution driven by selected discrete chaotic systems, In *Proceedings of the 18th International Conference on Soft Computing–MENDEL 2012*, Czech Republic, 157–162, ISBN 978-80-214-4540-6.

38. Senkerik R., Pluhacek M., Oplatkova Z. K., Davendra D., Zelinka I. 2013. Investigation on the differential evolution driven by selected six chaotic systems in the task of reactor geometry optimization, *2013 IEEE Congress on Evolutionary Computation (CEC)*, Mexico, 3087–3094, ISBN 978-1-4799-0453-2.

39. Davendra D., Zelinka I., Senkerik R., Bialic M. L. 2010. Chaos driven evolutionary algorithm for the traveling salesman problem, In *Traveling Salesman Problem, Traveling Sales-man Problem Theory and Applications*, InTech, 55–70, ISBN: 978-953-307-426-9.

40. Davendra D., Bialic-Davendra M., Senkerik R. 2013. Scheduling the lot-streaming flowshop scheduling problem with setup time with the chaos-induced enhanced differential evolution, *2013 IEEE Symposium on Differential Evolution (SDE)*, Singapore, 119–126.

41. Senkerik R., Pluhacek M., Zelinka I., Kominkova Oplatkova Z., Vala R., Jasek R. 2014. Performance of chaos driven differential evolution on shifted benchmark functions Set, International Joint Conference SOCO13-CISIS13-ICEUTE13, *Advances in Intelligent Systems and Computing*, Czech Republic, Vol. 239, 41–50.

42. Pluhacek M., Senkerik R., Davendra D., Zelinka I. 2013. Designing PID controller for DC motor by means of enhanced PSO algorithm with dissipative chaotic map, Soft Computing Models in Industrial and Environmental Applications, *Advances in Intelligent Systems and Computing*, Nostradamus 2013, Czech Republic, Vol. 188, 475–483.

43. Pluhacek M., Senkerik R., Zelinka I., Davendra D. 2013. Chaos PSO algorithm driven alternately by two different chaotic maps—An initial study, *2013 IEEE Congress on Evolutionary Computation (CEC)*, Mexico, 2444–2449, ISBN 978-1-4799-0453-2.

44. Pluhacek M., Senkerik R., Zelinka I. 2014. Multiple choice strategy based PSO algorithm with chaotic decision making a preliminary study, International Joint Conference SOCO13-CISIS13-ICEUTE13, *Advances in Intelligent Systems and Computing*, Czech Republic, Vol. 239, 2014, 21–30.

45. Coelho L. S., Mariani V. C. 2012. Firefly algorithm approach based on chaotic Tinkerbell map applied to multivariable PID controller tuning, Computers and Mathematics with Applications, 64(8), 2371–2382.

46. Gandomi A. H., Yang X.-S., Talatahari S., Alavi A. H. 2013. Firefly algorithm with chaos, *Communications in Nonlinear Science and Numerical Simulation*, 18(1), 89–98.

47. Zelinka I., Senkerik R., Pluhacek M. 2013. Do evolutionary algorithms indeed require randomness? *2013 IEEE Congress on Evolutionary Computation (CEC)*, Mexico, 2283–2289, ISBN 978-1-4799-0453-2.

48. Persohn K. J., Povinelli R. J. 2012. Analyzing logistic map pseudorandom number generators for periodicity induced by finite precision floating-point representation, *Chaos, Solitons and Fractals*, 45, 238–245.

49. Wright A., Agapie A. 2001. Cyclic and chaotic behavior in genetic algorithms, In *Proceedings of Genetic and Evolutionary Computation Conference*, GECCO, San Francisco, CA, July 7–11.

50. Vose M., Liepins G. 1991. Punctuated equilibria in genetic search, *Complex Systems*, 5, 31–44.

51. Vose M., Wright A. 1994. Simple genetic algorithms with linear fitness, *Evolutionary Computation*, 4(2), 347–368.

52. Vose M. 1999. *The Simple Genetic Algorithm: Foundations and Theory*, MIT Press, Cambridge, MA.

53. Pyragas K. 1995. Control of chaos via extended delay feedback, *Physics Letters*, 206, 323–330.

54. Pyragas K. 1992. Continuous control of chaos by self-controlling feedback, *Physics Letters*, 170, 421–428.

55. Richter H., Reinschke K. J. 2000. Optimization of local control of chaos by an evolutionary algorithm, *Physica D*, 144, 309–334.

56. Richter H. 2002. *An Evolutionary Algorithm for Controlling Chaos: The Use of Multi-objective Fitness Functions*, In Merelo Guervs J. J., Panagiotis A., Beyer H. G., Fernndez Villacanas J. L., Schwefel H. P. (eds), Parallel Problem Solving from Nature-PPSNVII, Lecture Notes in Computer Science, Vol. 2439, Springer-Verlag, Berlin Heidelberg New York, 308–317.

57. Lorenz E. N. 1963. Deterministic nonperiodic flow, *Journal of the Atmospheric Sciences*, 20(2), 130–141.

58. Stewart I. 2000. The Lorenz attractor exists, *Nature*, 406, 948–949.

59. May R. 1976. Simple mathematical model with very complicated dynamics, *Nature*, 261, 45–67.

60. Gilmore R., Lefranc M. 2002. *The Topology of Chaos: Alice in Stretch and Squeezeland*, Wiley-Interscience, New York.

61. Schuster H. G. 1999. *Handbook of Chaos Control*, Wiley-VCH, New York.

62. Chen G., Dong X. 1998. From Chaos to Order: Methodologies, Perspectives and Applications, World Scientific, Singapore.

63. Wang X., Chen G. 2000. Chaotification via arbitrarily small feedback controls: Theory, method, and applications, *International Journal of Bifurcation and Chaos*, 10, 549–570.

64. Ott E., Grebogi C., Yorke J. A. 1990. Controlling chaos, *Physical Review Letters*, 64, 1196–1199.

65. Grebogi C., Lai Y. C. 1999. Controlling chaos, In Schuster H. G. (ed.), *Handbook of Chaos Control*, 1–20, Wiley-VCH, New York.

66. Zou Y. -L., Luo X. -S., Chen G. 2006. Pole placement method of controlling chaos in DC–DC buck converters, *Chinese Physics*, 15, 1719–1724.

67. Just W. 1999. Principles of time delayed feedback control, In Schuster H. G. (ed.), *Handbook of Chaos Control*, 21–42, Wiley-VCH, New York.

68. Just W., Benner H., Reibold E. 2003. Theoretical and experimental aspects of chaos control by time-delayed feedback, *Chaos*, 13, 259–266.

69. Deilami M. Z., Rahmani Ch. Z., Motlagh M. R. J. 2007. Control of spatio-temporal on–off intermittency in random driving diffusively coupled map lattices, *Chaos, Solitons, Fractals*, Available online December 21.

70. Chen G. 2000. *Controlling Chaos and Bifurcations in Engineering Systems*, CRC Press, Boca Raton, FL.

71. Hilborn R. C. 1994. *Chaos and Nonlinear Dynamics*, Oxford University Press, New York, ISBN 0-19-508816-8.

72. Hu G., Xie F., Xiao J., Yang J., Qu Z. 1999. Control of patterns and spatiotemporal chaos and its application, In Schuster H. G. (ed.), *Handbook of Chaos Control*, 43–86, Wiley VCH, New York.

73. Zelinka I., Davendra D., Senkerik R., Oplatkova Z., Jasek R. 2011. Analytical programming a novel approach for evolutionary synthesis of symbolic structures, In Kita E. (ed.), *Evolutionary Algorithms*, InTech, ISBN 978-953-307-171-8.

74. Koza J. R., Bennet F. H., Andre D., Keane M. 1999. *Genetic Programming III*, Morgan Kaufmann Publishers, ISBN 1-55860-543-6.

75. O'Neill M., Ryan C. 2002. *Grammatical Evolution: Evolutionary Automatic Programming in an Arbitrary Language*, Kluwer Academic Publishers, ISBN 1402074441.

76. Chen G., Dong X. 1998. *From Chaos to Order: Methodologies, Perspectives and Applications*, World Scientific, Singapore.

77. Perruquetti W., Barbot J. P. 2005. *Chaos in Automatic Control*, CRC, Bota Raton, FL.

78. Gilmore R., Lefranc M. 2002. *The Topology of Chaos: Alice in Stretch and Squeezeland*, Wiley-Interscience, New York.

79. Zelinka I., Chadli M., Davendra D., Senkerik R., Jasek R. An investigation on evolutionary reconstruction of continuous chaotic systems, *Mathematical and Computer Modeling*, 57(1–2), 2–15, DOI 10.1016/j.mcm.2011.06.034.

80. Pikovsky A., Rosemblum M., Kurths J. 2001. *Synchronization: A Universal Concept in Nonlinear Sciences*, Cambridge University Press, Cambridge, ISBN 0-521-53352-X.

81. Gonzalez-Miranda J. M. 2004. *Synchronization and Control of Chaos: An Introduction for Scientists and Engineers*, Imperial College Press, London. ISBN 1-86094-488-4.

82. Sushchik M. M., Rulkov N. F., Tsimring L. S., Abarbanel H. D. I. Generalized synchronization of chaos in directionally coupled chaotic systems, In *Proceedings of 1995 International Symposium on Nonlinear Theory and its Applications*, IEEE, 1995, vol. 2, 949–952.

83. Brown R., Rulkov N. F., Tracy E. R. 1994. Modeling and synchronization chaotic system from time-series data. *Physical Review E*, 49, 3784.
84. Rulkov N. F., Sushchik M. M. 1997. Robustness of synchronized chaotic oscillations. *International Journal of Bifurcation and Chaos*, 7, 625.
85. Takens F. 1981. *Detecting strange attractors in turbulence*, Lecture Notes in Mathematics, 366–381.
86. Halsey T. C., Jensen M. H., Kadanoff L. P., Pro-caccia I., Schraiman B. I. 1986. Fractal measures and their singularities: The characterization of strange sets, *Physical Review* 33 A, 1141.
87. Eckmann J. P., Procaccia I. 1986. Fluctuation of dynamical scaling indices in non-linear systems, *Physical Review* 34 A, 659.
88. Zelinka I., Davendra D., Snasel V., Jasek R., Senkerik R., Oplatkova Z. 2010. *Preliminary investigation on relations between complex networks and evolutionary algorithms dynamics*, CIMSIM 2010, Krakow, Poland, 148–153.
89. Zelinka I., Davendra D., Chadli M., Senkerik R., Dao T.T., Skanderova L. 2012. Evolutionary dynamics and complex networks, In *Handbook of Optimization*, Springer Series on Intelligent Systems.
90. Zelinka I., Skanderova L., Saloun P., Senkerik R., Pluhacek M. 2013. *Hidden complexity of evolutionary dynamics analysis*, ISCS 2013, Prague, Czech Republic.
91. Zelinka I. Davendra D., Senkerik R., Jasek R. 2012. Do evolutionary algorithm dynamics create complex network structures? *Complex Systems*, 20(2), 127–140, ISSN 0891-2513.
92. Leonov G. A., Kuznetsov N. V., Prediction of hidden oscillations existence in nonlinear dynamical systems: Analytics and simulation, *Advances in Intelligent Systems and Computing*, Vol. 210 AISC, Springer, pp. 5–13.
93. Leonov G. A., Kuznetsov N. V. 2013. Hidden attractors in dynamical systems: From hidden oscillations in Hilbert–Kolmogorov, Aizerman, and Kalman problems to hidden chaotic attractor in Chua circuits, *International Journal of Bifurcation and Chaos*, 23(1), 1–69, art. no. 1330002.
94. Kuznetsov N., Kuznetsova O., Leonov G., Vagaitsev V. 2013. Analytical-numerical localization of hidden attractor in electrical Chua's circuit, *Lecture Notes in Electrical Engineering*, Vol. 174 LNEE, Springer, Heidelberg, Germany, 149–158.

3

On the Transition to Phase Synchronized Chaos

Erik Mosekilde,
Jakob Lund Laugesen,
and Zhanybai T.
Zhusubaliyev

3.1 Introduction

Chaotic phase synchronization [3,11,22] denotes an interesting form of synchronization in which a chaotic attractor adjusts the frequencies of its internal dynamics to the rhythm of an external forcing signal, or to the dynamics of another chaotic oscillator, while the oscillator's amplitude continues to vary in an essentially uncorrelated manner. This form of chaotic synchronization is clearly distinguished from the complete or full synchronization one can observe in the interaction between two identical chaotic oscillators [30,34].

Complete synchronization has attracted significant attention both because of its potential application in areas such as chaos control [42,46] and secure communication [4,44] and because of the broad range of interesting nonlinear dynamic phenomena that can be observed as the synchronized state loses its stability [27]. Besides the so-called locally and globally riddled basins of attraction [2,21], these phenomena also include on–off intermittency [36], attractor bubbling, and blow-out bifurcations [29]. All these phenomena are derived from a situation in which, although the trajectories on the average are attracted to the synchronized state, any neighborhood of this state may contain a dense set of trajectories that are repelled from it. Under these conditions, the stability criterion takes a rather unusual form: all periodic orbits embedded in the synchronized chaotic state must be transversely stable [13]. The application of this condition for a system of two coupled identical Rössler oscillators has been demonstrated, for instance, by Yanchuck et al. [48].

In recent years, the interest in application of chaos synchronization for secure communication has led many investigators to consider different forms of the so-called projective synchronization by which the requirement of complete identity of the interacting oscillators can be reformulated to allow for different forms of scaling [25]. At the same time, a significant effort to study the synchronization properties of different fractional order chaotic oscillators has been initiated [31,47].

Phase synchronized chaos, as described in this chapter is characteristic for interacting nonidentical chaotic oscillators. This form of synchronization occurs in a wide variety of different physical, chemical, and biological systems, including coupled electronic oscillators [3], arrays of interacting chaotic lasers [9], plasma discharge tubes paced by a low-amplitude wave generator [40], coupled electrochemical reactors [50], sub-threshold behaviors of interacting nerve cells [26], and oscillatory dynamics in the blood flow regulation of neighboring functional units of the kidney [14].

In the animate world, the complex coordination of oscillatory modes associated with chaotic phase synchronization is likely to translate into structure that can adapt to a wide range of functional states and rapidly switch from un-synchronized to almost coherent dynamics. It is well-established, for instance, that the activation of the smooth muscle cells in the arteriolar wall involves a transition from a state of seemingly random fluctuations of intracellular calcium concentrations into a state of nearly coherent cellular oscillations [12]. Detailed bifurcation analyses of chaotic phase synchronization in multidimensional biological systems have been performed, for instances, for chains of microbiological reactors [27], for interacting functional units of the kidney [23], and for scale-free networks of bursting neurons [7].

In several experiments with a single periodically driven chaotic oscillator, one can observe [6,27,33] how the frequency spectrum remains relatively unaffected by the variation of the forcing frequency until the system enters the region of chaotic phase synchronization where the mean spectral frequency locks to the forcing frequency. The width of the synchronization interval typically increases with the forcing amplitude and, as the system leaves this interval, the mean frequency of the chaotic oscillator again starts to slide away from the forcing frequency. This type of behavior is associated with the fact that, while the amplitude of the endogenous oscillations is controlled by a balance between the inherent instability of the system on one side and nonlinear restraining mechanisms on the other, there is no preferred value for the phase of an unforced oscillator. The application of a forcing signal breaks this time translational symmetry and, even for relatively small forcing amplitudes, the chaotic oscillator will adjust its frequencies and phases [33].

From its first observation, the phenomenon of chaotic phase synchronization has attracted significant theoretical interest [35,41,42], and the concepts and the methods developed through works in this area have been used to interpret experimental time series from many different sources [15]. Apart from the above-mentioned changes in the spectral distribution, the transition between phase synchronized and nonsynchronized chaos is reflected in the specific variation of the Lyapunov exponents, in characteristic changes of the spectral distribution of the oscillations, and through changes in the size and form of the Poincaré section associated with the attracting state [45].

For spiral-type chaos, it is known that the edge of the synchronization zone consists of a dense set of saddle–node bifurcations that delineate the range of existence for the different unstable periodic cycles, which constitute the chaotic state. Drawing on the analogy with the tangent bifurcation along the edge of the synchronization zone for a forced limit cycle oscillator, Pikovsky et al. [32] suggested that a large number of attractor–repellor collisions take place at the transition to chaotic phase synchronization. For systems of coupled period-doubling oscillators, Postnov et al. [37] have described a nested structure of phase synchronized regions for different attractor families associated with the transition to phase synchronization, and Vadivasova et al. [45] have provided a preliminary overview of the regions of existence for the main behavioral modes of the harmonically forced Rössler system. Moreover, by applying a lift to the phase variable (such that phase points separated by 2π no longer are considered the same), Rosa et al. [39] have described the transition to chaotic phase synchronization as a boundary crises mediated by an unstable–unstable pair bifurcation on a branched manifold. So far, however, the details of how the saddle–node bifurcations arise, how they are organized, and how the resonance tori that exist in the synchronization regime relate to the period-doubled ergodic tori that are known to exist outside the resonance zone [5,43], have remained largely unexplored.

An important contribution toward the development of a more detailed picture is the theory of cyclic (also called C-type) criticality developed by Kuznetsov et al. [16–18]. In particular, this work has demonstrated that the period-doubling transition to chaos along the edge of a resonance tongue displays an unusual scaling behavior, involving subsequent pairs of period-doubling bifurcations. Inspired by these results, we have recently applied continuation techniques [10,19] to perform a more complete analysis of the bifurcation structure involved [24,51]. In this way, we have demonstrated how the dynamics of the periodically forced Rössler system develop through cascades of period-doubling bifurcations of both the node and saddle cycles in a direction transverse to the original period-1 resonance torus, thereby producing a system of the so-called multilayered resonance tori [52–54].

After each period-doubling bifurcation, new saddle–node bifurcation curves are born on both sides of the synchronized zone in order to delineate the range of existence for the emerging resonance cycles. For a particular side of the synchronized zone, these saddle–node bifurcation curves emanate alternately from the stable and the unstable branches of the period-doubling curve and thereby causing the characteristic cyclic structure of the transition where the scaling relates to pairs of subsequent period-doubling transitions. Moreover, the new saddle–node bifurcation curves arise close to, but not in the so-called fold-flip bifurcation points [20] where the period-doubling curves are tangent to the saddle–node bifurcation curves produced in the preceding period-doubling bifurcation. Additional (local and/or global) bifurcations are, therefore, required to close the holes between the saddle–node bifurcation curves [24].

By following the development of the phase portrait for the various stable and unstable resonance cycles in a periodically forced Rössler system through the period-doubling cascade the present chapter illustrates the formation, reconstruction, and final breakdown of increasingly complicated multilayered resonance tori. This chapter also explains how the saddle–node bifurcation curves along the edge of the resonance zone are organized and illustrates the transition from multi-layered resonance torus to period-doubled ergodic torus.

3.2 Main Bifurcation Structure of the 1:1 Synchronization Zone

Let us consider the periodically forced Rössler system

$$\dot{x} = -y - z + A\sin(\omega t); \quad \dot{y} = x + ay; \quad \dot{z} = b + z(x - c) \tag{3.1}$$

that has also served as a vehicle for the investigation of chaotic phase synchronization in many previous studies [35,41,42]. Here x, y, and z are the dynamic variables of the unforced oscillator, and $A\sin(\omega t)$ represents the external forcing. The parameters a and b and the forcing amplitude A will be kept constant at the values $a = b = 0.2$ and $A = 0.1$, whereas the nonlinearity parameter c and the forcing frequency ω are used as bifurcation parameters. For the above parameter values, the unforced Rössler system undergoes a Hopf bifurcation at $c = 0.4$ and, for increasing values of c, the system exhibits a Feigenbaum cascade of period-doubling bifurcations. When an external periodic forcing is applied in the regime of simple periodic oscillations, the Rössler system displays regions of two-mode quasiperiodic (ergodic) dynamics interrupted by a dense set of resonance zones where the internally generated periodic oscillations synchronize with the external forcing. The 1:1 resonance generally produces the most prominent synchronization regime, and the purpose of the current analysis is to examine how the structure of this region develops as the parameter c increases through the range, where the unforced Rössler system undergoes its period-doubling transition to chaos.

To illustrate some of the characteristic features of the transition to phase synchronized chaos, Figure 3.1a shows a scan across the main region of 1:1 phase synchronization [6,27,33]. Here, we have plotted the ratio between the mean spectral frequency ω_1 of the forced Rössler system in Equation 3.1 and the frequency ω of the applied forcing as a function of the forcing frequency. The synchronization zone clearly stands out as an interval of forcing frequencies in which ω_1 coincides with ω. For forcing frequencies below this interval,

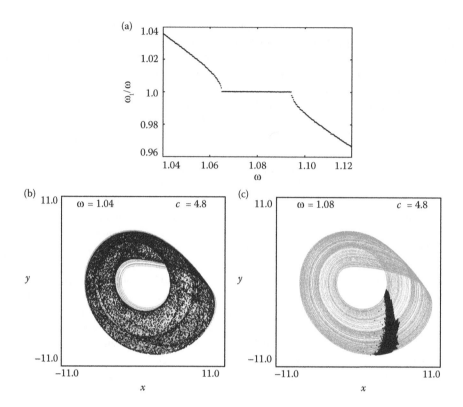

FIGURE 3.1 Main features of the transition to phase synchronized chaos. (a) Synchronization characteristic: Ratio of the mean spectral frequency ω_1 to the forcing frequency ω as a function of the forcing frequency. Distribution of stroboscopically sampled phase space positions for the forced Rössler system (b) outside and (c) inside the range of chaotic phase synchronization. Projections of the forced Rössler oscillator are shown as background for the stroboscopic points in (b) and (c).

the mean spectral frequency exceeds the forcing frequency and, when the forcing frequency crosses the upper edge of the synchronization interval, ω_1 no longer follows the variation of ω.

Figure 3.1b and c provide an alternative illustration of the transition from unsynchronized to phase synchronized chaos in response to a shift in the forcing frequency. Here we have superimposed a projection of the phase space trajectory for the periodically forced Rössler system with a distribution of points obtained by stroboscopic marking of the position of the system after each completed period of the forcing signal. Figure 3.1b shows the typical point distribution observed outside the region of phase synchronization, and Figure 3.1c shows an example of the characteristic point distribution in the range of phase synchronized chaos. Each distribution involves the order of 1000 points. Note, how the point distribution for the unsynchronized state spreads nearly uniformly across the whole attractor, while the spread of the points is significantly smaller in the phase synchronized state. It is also interesting to note that, even for the unsynchronized attractor, the density of stroboscopic points is very low in a particular region. This is likely to be the region where the oscillator rapidly approaches its equilibrium point along the stable manifold (i.e., the z-axis).

To better understand what happens in the transition to phase synchronized chaos let us start by examining the transitions that occur for lower values of the nonlinearity parameter c as the unforced Rössler system takes its first steps of the period-doubling cascade. The two-dimensional bifurcation diagram in Figure 3.2 provides an overview of the first four period-doubling bifurcations in the 1:1 resonance tongue [24].

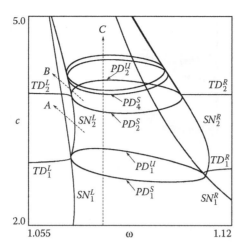

FIGURE 3.2 Two-dimensional bifurcation diagram for the 1:1 resonance zone of the periodically forced Rössler oscillator in (1). Bifurcation parameters are the forcing frequency ω and the nonlinearity parameter c of the unforced Rössler system. To delineate the range of existence of the emerging resonance cycles, each period-doubling bifurcation gives rise to a new pair of saddle–node bifurcation curves along the edge of the resonance tongue. (The figure was constructed by means of conventional continuation methods [19] using software made available by E. J. Doedel et al. Auto 2000: Continuation and bifurcation software for ordinary differential equations (with homcont). Technical report, California Institute of Technology, 2002.)

As mentioned above, the bifurcation parameters are the forcing frequency ω and the nonlinearity parameter c for the Rössler oscillator. PD^S and PD^U denote period-doubling bifurcation curves for stable (node) and unstable (saddle) resonance cycles, respectively, and SN refers to the saddle–node bifurcation curves observed along the sides of the resonance zone. TD locates torus-doubling bifurcations for the ergodic tori that exist outside the resonance zone, and the arrows $A \rightarrow A$ and $B \rightarrow B$ relate to one-dimensional bifurcation scans to be discussed in the following.

Below the first period-doubling bifurcation curve PD_1^S, the resonance zone is delineated to the left and right by the saddle–node bifurcation curves SN_1^L and SN_1^R, respectively. In this region, the forced Rössler system displays a stable, synchronized period-1 cycle N_1 and a corresponding saddle cycle S_1, born together in the saddle–node bifurcation SN_1^L (or SN_1^R) and both situated on a closed invariant curve that represents the resonance torus. Along the lower branch PD_1^S of the first period-doubling curve, the stable period-1 cycle undergoes its first period-doubling bifurcation while the corresponding saddle cycle period doubles at PD_1^U. At the edge of the resonance zone, the two period-1 cycles merge, and the period-doubling bifurcations occur simultaneously. Above the curve PD_1^U in Figure 3.2, the forced Rössler system displays a pair of saddle and doubly unstable saddle period-1 cycles together with a pair of saddle and stable node period-2 cycles.

The saddle–node bifurcation curves SN_1^L and SN_1^R continue up along the edge of the resonance zone to delineate the region of resonant period-1 dynamics. However, the synchronization zone for the new period-2 cycles is not the same as that for the period-1 cycles, and a new set of saddle–node bifurcation curves SN_2^L and SN_2^R are required to delineate the region of period-2 dynamics. These new saddle–node bifurcation curves originate from points on the period-doubling curve PD_1 (PD_1^S or PD_1^U) near the so-called fold-flip bifurcation points [20] where the period-doubling curve is tangent to the saddle–node bifurcation curves SN_1^L and SN_1^R. However, as closer inspection shows [24], the new saddle–node bifurcation curves do not emanate precisely from the fold-flip bifurcation points. There is a gap between the saddle–node bifurcation curves and, as we shall see, a number of additional local and global bifurcations are required to complete the border of the resonance zone for the period-2 cycles.

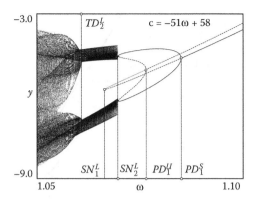

FIGURE 3.3 One-dimensional scan along the direction A in Figure 3.2. The pair of saddle and doubly unstable 1:1 resonance cycles merge and disappear in SN_1^L. The pair of saddle and stable 2:2 cycles merge in SN_2^L, leading to a period-2 invariant torus. This torus subsequently undergoes a torus-doubling at TD_2^L.

The stable period-2 solution undergoes a new period doubling at PD_2^S, and the saddle period-2 solution period doubles at PD_2^U. The saddle–node bifurcation curves SN_2^L and SN_2^R are tangent to these period-doubling curves, and close to the points of tangency a new pair of saddle–node bifurcation curves SN_4^L and SN_4^R are born to delineate the range of existence for the period-4 resonance dynamics (not visible in the scale of the figure). As the value of c continues to increase, the same process repeats over and over again until the system undergoes a transition to phase synchronized chaos. Inspection of Figure 3.2 shows that, at least in the middle of the resonance zone, this transition occurs before the saddle solution has undergone its second period-doubling bifurcation. Inspection of Figure 3.2 also suggests that the torus-doubling bifurcations TD^L and TD^R that take place in the quasiperiodic regime outside the resonance tongue are coupled with the period-doubling bifurcations of the resonance cycles in the tongue.

As an illustration to the two-dimensional bifurcation diagram in Figure 3.2, the one-dimensional bifurcation diagram in Figure 3.3 shows the transitions that take place along the direction A in Figure 3.2. Full curves represent stable periodic cycles, dashed curves the saddle cycles, and dotted curves the doubly unstable node solutions. Notice how the stable 1:1 solution that exists in the upper right corner of the figure undergoes a period-doubling at PD_1^S while the corresponding 1:1 saddle solution suffers its first period doubling at PD_1^U. From here we can follow the two solutions (now as a saddle cycle and a doubly unstable node) to the saddle–node bifurcation SN_1^L to the left in the figure. This saddle–node bifurcation defines the zone edge for the period-1 cycles. The saddle and stable node 2:2 resonance cycles merge at SN_2^L to give birth to a period-doubled ergodic torus. Finally, when crossing the torus-doubling bifurcation curve TD_2^L, the ergodic torus undergoes a new period-doubling transition.

Figure 3.4 shows a similar one-dimensional bifurcation diagram for the direction B in Figure 3.1. After the first period-doubling bifurcations for the 1:1 resonance node and saddle cycles at PD_1^S and PD_1^U, the interconnected period-doubling processes continue to the left in the figure. Here we can locate the period-doubling bifurcation PD_2^S for the stable period-2 cycle and the bifurcation PD_2^U for the corresponding saddle cycle. Each pair of saddle and doubly unstable node cycles born in these bifurcations can subsequently be followed to the saddle–node bifurcation that demarcates their synchronization zone.

In Figure 3.4, the boundary of the resonance zone consists of saddle–node bifurcations for the period-1, period-2, and period-4 cycles, but only the period-4 node is stable, and both the 1:1 and the 2:2 resonance tori have been destroyed. Hence, we observe that the period-4 resonance torus ends in a saddle–node bifurcation in which an ergodic period-4 torus is born. The period-1 saddle and doubly unstable node cycles continue to exist into the region of the stable period-4 ergodic torus. As the system moves further away from the resonance zone, the ergodic torus starts to fold and it finally undergoes torus destruction at the point ETD where its different layers begin to mix.

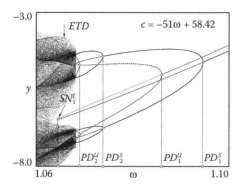

FIGURE 3.4 One-dimensional scan along the direction B in Figure 3.2. As the forcing frequency ω is reduced, we can follow two subsequent period-doubling bifurcations of both the synchronized period-1 node and saddle cycles. The resonant period-4 torus ends in a saddle–node bifurcation that gives birth to a period-4 ergodic torus. The torus-doubling process only occurs in a restricted region on the sides of the resonance tongue. At the point ETD, the ergodic torus is destroyed, and to the left of ETD the system displays nonsynchronous chaos.

3.3 Bifurcation Structure Near a Fold-Flip Bifurcation

To understand the above transitions in more detail we need to focus on the bifurcation structure close to the fold-flip bifurcation points where the saddle–node bifurcation is tangent to the corresponding period-doubling bifurcation curve. As explained above, the new saddle-node bifurcation curves do not emerge from the fold-flip bifurcation points, but from points on the period-doubling curve close to these points. The distance to the fold-flip bifurcation points decreases with increase in forcing amplitude, but never vanishes. The generic situation is that the new saddle–node bifurcation curves emerge from the stable branch of the period-doubling curve in one side of the resonance zone and from the unstable branch in the other side. Moreover, for a given side, the emergence of the new saddle–node bifurcation curve alternately takes place from the stable and the unstable branch of the period-doubling curves.

Figure 3.5a shows a slightly redrawn version of Figure 3.2 in which the regions of existence for the stable 1:1, 2:2, 4:4, etc., resonance cycles are displayed in different shades and the region of quasiperiodic (or nonsynchronized chaotic) dynamics outside the resonance zone is dark gray. Figure 3.5b and c shows magnifications of the regions around the two fold-flip bifurcation points for the period-1 cycles. Figure 3.5b considers the situation at the left-hand side of the resonance tongue where the new saddle–node bifurcation curve SN_1^L emerges from the point Q_2 on the unstable branch PD_1^U of the period-doubling curve, and Figure 3.5c presents the situation near the right-hand side of the tongue where SN_2^R emerges from a point on the stable branch PD_1^S of the period-doubling curve.

To close the gap between the saddle–node bifurcation curves SN_1^L and SN_2^L (and avoid that the period-2 cycle produced at PD_1^S can escape from the resonance zone) the system makes use of a subcritical torus-birth (or Andronov–Hopf) bifurcation T_2 in conjunction with a complex set of global bifurcations G_2 through which the unstable two-branched torus generated in T_2 is transformed into a stable period-doubled ergodic torus [24]. The alternative case where the saddle–node bifurcation curve emerges from the stable branch of PD_1 does not leave a gap through which the 2:2 resonance cycles can escape. However, the section of PD_1^S that connects the point at which SN_2^R is born with the point of tangency with SN_1^R must be subcritical. This allows saddle–node bifurcations to take place along that part of SN_2^R that falls below its point of intersection with SN_1^R. We also note that SN_2^R is born to the left of the already existing tongue edge. Hence, SN_2^R must intersect SN_1^R and thereafter, at least in beginning, proceed to the right of this saddle–node bifurcation curve.

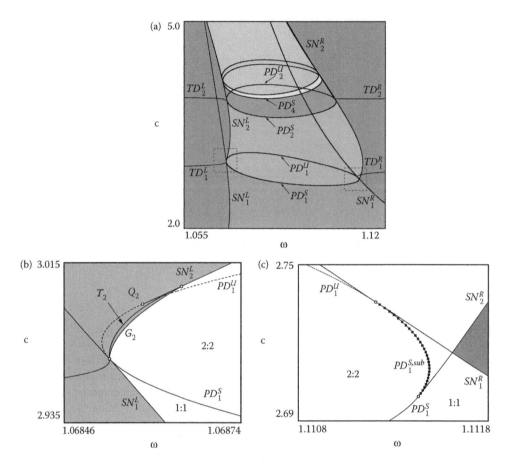

FIGURE 3.5 Bifurcation structure for the 1:1 resonance tongue in the periodically forced Rössler system. (a) Overview of the first four period-doubling bifurcations. Only the first couple of saddle–node bifurcations along the zone edge can be resolved in the applied scale. (b) Enlargement of the region around PD_1 and SN_1^L. The gap between the saddle–node bifurcations SN_1^L and SN_2^L is closed by the torus-birth bifurcation T_2 and the global bifurcations that take place near G_2. (c) Enlargement of the corresponding region to the right in the resonance zone. Here the new saddle-node bifurcation curve emerges from the stable branch of the period-doubling curve. Part of this branch (marked by small black squares) then has to be subcritical.

This completes our discussion of the detailed bifurcation structure near the fold-flip bifurcation point. Similar structures arise near the fold-flip bifurcation points of the subsequent period-doubling bifurcations and, as we have already indicated, the two different scenarios alternate in opposite phase along the sides of the resonance zone. This is the basis for the characteristic cyclic character of the period-doubling transition along the edge of a resonance tongue [16–18]. At the same time we can conclude that the above systematic leads the new saddle–node bifurcation curves to proceed alternately inside and outside the resonance zone delineated by the immediately preceding saddle–node bifurcation curves.

3.4 Period Doubling of the Ergodic Torus

Let us consider the above transitions once more, but now with a direct focus on the processes that lead to the birth and transformation of the large period-2 ergodic torus that arise in the torus-birth bifurcation T_2. In connection with Figure 3.5b we have already discussed how the saddle–node bifurcation curve SN_2^L emanates from the point Q_2 on the unstable branch of the period-doubling curve PD_1^U. This point does not coincide with the point at which PD_1 is tangent to the edge of the synchronization region for period-1 dynamics, thus leaving a gap in the boundary of the resonance zone. However, as described in Section 3.3, additional local and global bifurcations are in place to ensure that the 2:2 cycles do not escape the resonance area through the gap left by the saddle–node bifurcation curves. Inspection of Figure 3.6 (which is a slightly redrawn version of Figure 3.5b immediately allows us to locate the saddle–node bifurcation curves SN_1^L and SN_2^L, the two branches of the period-doubling curve PD_1, and the point Q_2 where SN_2^L is born. The torus bifurcation curve T_2 is seen to bridge the gap between the two saddle–node bifurcation curves. At this torus bifurcation, the stable period-2 cycle produced at the lower branch of the period-doubling curve PD_1^S, and now transformed into a stable focus, loses its stability in a subcritical torus-birth bifurcation. As previously noted, the curve G_2 represents a set of closely situated local and global bifurcations that serve to stabilize the ergodic period-2 torus and thereby establish the required boundary for the synchronization zone. Finally, the torus-doubling curve TD_1^L represents the processes by which the period-2 ergodic torus transforms into the ergodic period-1 torus that exists outside the resonance tongue for values of the nonlinearity parameter c below TD_1^L.

In the presence of these additional bifurcations, the brute force bifurcation diagram calculated along the elliptic curve denoted C_a in Figure 3.6 takes the form illustrated in Figure 3.7. At both ends of this diagram we observe the ergodic period-1 torus that exists below TD_1^L and to the left of the resonance zone. As we follow the transitions from left to right through the intermediate range of Figure 3.7 we first meet the saddle–node bifurcation SN_1^L at the edge of the resonance tongue where the 1:1 node cycle is born. At PD_1^S, this is followed by the period-doubling bifurcation in which the 1:1 node is transformed into a 1:1 saddle cycle while producing a stable 2:2 cycle transverse to the torus surface. Hereafter follows the sequence G_2 of closely situated local and global bifurcations that give birth to both the large period-2 torus that dominates most of the right-hand side of the diagram and to an unstable two-branch torus around the 2:2 focus cycle. The unstable two-branch torus again disappears in the subcritical torus-birth bifurcation T_2 while the large period-2 torus continues to exist until it undergoes the aforementioned reverse torus-doubling bifurcation at the point TD_1^L.

The sketches in Figure 3.8 give a clearer account of the structure of local and global bifurcations observed in the region around the birth of the saddle–node bifurcation curve SN_2^L. The curves along which these bifurcation diagrams are thought to be drawn are indicated in Figure 3.6 by the letters C_a, C_b, and C_c.

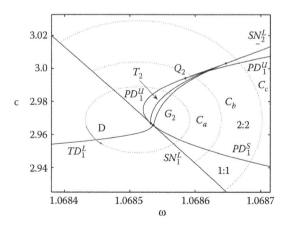

FIGURE 3.6 Magnification of part of the bifurcation diagram near the fold-flip bifurcation point. PD_1^S and PD_1^U are the stable and unstable branches of the period-doubling bifurcation curve and Q_2 denotes the point in which the saddle–node bifurcation curve SN_2^L starts. G_2 represents the sequence of bifurcations that give rise to the large period-2 ergodic torus seen in Figure 3.7, and TD_1^L is the torus-doubling bifurcation in which this torus is transformed into an ordinary period-1 ergodic torus. Note that some of the substructure disappears when the forcing amplitude becomes sufficiently large.

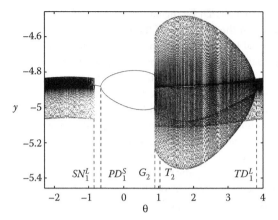

FIGURE 3.7 Brute force bifurcation diagram calculated along the elliptic curve C_a in Figure 3.6. θ is a measure of the position along C_a with 0 (and 2π) representing the outmost right point and π the outmost left point. For increasing values of θ the diagram first shows the saddle–node bifurcation SN_1^L that occurs when the system enters the 1:1 resonance zone. Hereafter follows first the period-doubling PD_1^S that produces the stable 2:2 resonant node and the subcritical torus-birth bifurcation T_2 in which the 2:2 cycle loses its stability. G_2 represents a sequence of bifurcations that give birth to the large ergodic period-2 torus and TD_1^L denotes the torus-doubling bifurcation in which this torus transforms into an ordinary period-1 ergodic torus.

Starting from the bottom, panel C_a first shows the saddle–node bifurcation through which the 1:1 node and saddle cycles are born as the system enters the resonance domain. Hereafter follows the period-doubling bifurcation PD_1^S on the 1:1 node. At T_2, the 2:2 cycle generated in this bifurcation (now a stable focus) undergoes a subcritical torus-birth bifurcation and turns into an unstable focus. In accordance with the brute force diagram in Figure 3.7, the unstable two-branch torus produced in this bifurcation disappears in a global bifurcation close to the point G_2. As we continue the scan, the unstable 2:2 focus cycle (now a doubly unstable saddle) undergoes a reverse period-doubling bifurcation in PD_1^U while the 1:1 saddle destabilizes into a doubly unstable cycle. This latter cycle finally disappears in a saddle–node bifurcation

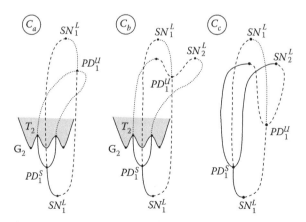

FIGURE 3.8 Bifurcation diagrams drawn along parts of the curves C_a, C_b, and C_c in Figure 3.6 and extended in both ends to the saddle–node bifurcation curve SN_1^L. Note how the 2:2 cycles in all cases are captured before they can escape from the resonance zone. Full lines represent stable node or focus solutions, dashed lines saddle solutions, and dotted lines doubly unstable node or unstable focus solutions. A more detailed description of the transformations that take place at G_2 may be found in our recent work [28].

at the tongue edge SN_1^L. We shall refer to the sequence of bifurcations that occur at G_2 as a torus-fold transition. A more detailed description of the processes associated with this transition has been presented in previous work [28].

If we denote the period-doubling bifurcation PD_1^U that ends the life of the 2:2 cycle in the upper end of panel C_a as subcritical, the corresponding period-doubling bifurcation in panel C_b has become supercritical as the saddle–node bifurcation curve SN_2^L has transformed the doubly unstable 2:2 saddle into a 2:2 saddle cycle with a single unstable direction. In this way the torus-birth bifurcation T_2 serves to degrade the stability of the 2:2 cycle so that it can annihilate with the 2:2 saddle cycle at the upper branch of the period-doubling curve. Finally, in panel C_c the torus bifurcation on the 2:2 cycle no longer occurs, the saddle–node bifurcation SN_2^L has overtaken the role of delineating the edge of the resonance tongue for the 2:2 cycles, and we recover the scenario discussed in Section 3.3.

In order to provide a clear impression the torus-doubling bifurcation that occurs along TD_1^L, Figure 3.9 presents a series of Poincaré sections of the ergodic torus observed in the one-dimensional brute force bifurcation diagram. As defined in the caption to Figure 3.7, the parameter θ is the angle along the curve C_a in Figure 3.6. For $\theta = 3.8$, the Poincaré section shows an ordinary (i.e., period-1) ergodic torus. Note, however, that there is an uneven distribution of points along the periphery, indicating a "hesitation" of the system near the top of the section. This is the well-known indication that the system approaches a resonance zone, in this case it is the 1:1 resonance.

As θ is reduced, one can observe how the invariant curve starts to split into two different windings, as the quasiperiodic oscillator alternately chooses one route over the other. This is another indication of the fact that the resonance mode from which the torus originates has undergone a period-doubling transition in the direction transverse to the periphery of the closed invariant curve. As θ is further reduced, the separation between the two windings is seen to continue to grow, and for larger values of the parameter c, one can observe how the two windings of the period-2 torus move apart.

This completes our discussion of the main bifurcation structure near the fold-flip bifurcation point for the case where the new saddle–node bifurcation curve emerges from the unstable branch of the period-doubling curve. The alternative situation where the saddle–node curve emerges from the stable branch of PD_1^S, i.e., below the point of tangency, is found at the right-hand edge of the resonance zone. In this case the birth of SN_2^R does not leave a gap between resonant and ergodic dynamics. This can easily be checked by trying to find a path from the 2:2 resonant region to ergodicity without crossing SN_2^R. Hence, no additional bifurcation is required. However, the period-doubling curve that connects the point where SN_2^R is born to the point of tangency with SN_1^R must be subcritical in order to account for that part of SN_2^R that exists below PD_1^U. Moreover, SN_2^R must cross SN_1^R and at least at the beginning proceed along the right side of SN_1^R.

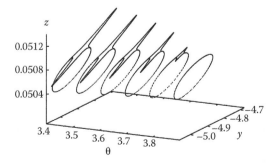

FIGURE 3.9 Illustration of the transition that occurs as the forced Rössler system crosses the torus doubling bifurcation curve TD_1^L in the direction indicated by the arrow D in Figure 3.6. The figure presents a series of Poincaré sections of the ergodic torus for different positions θ along the curve C_a. Compare with the bifurcation diagram in Figure 3.7.

Similar bifurcation structures are observed at the points of tangency at the next period-doubling bifurcation, except that here the subcritical case is found at the left edge and the supercritical with the additional set of torus and global bifurcations at the right-hand edge.

3.5 Formation and Reconstruction of Multi-Layered Resonance Tori

After the above discussion of the bifurcation structure for a periodically forced period-doubling oscillator, the next problem is to describe the internal organization of the many different resonance cycles that arise as increasing values of the nonlinearity parameter c takes the oscillator up through the period-doubling cascade. This leads us to study the formation and reconstruction of the so-called multilayered resonance tori [51] and to examine how these structures communicate with the ergodic tori that exist outside the resonance zone. In this discussion we shall again refer to the 2D bifurcation diagram in Figure 3.2.

Below the first period-doubling curve PD_1^S, the forced Rössler system displays a stable synchronized period-1 cycle N_1 and a corresponding saddle cycle S_1, both situated on the closed invariant curve that represents the resonance torus. As shown in Figure 3.10a, this corresponds to the normal resonance structure for a forced limit-cycle oscillator. Along the curve PD_1^S, the stable period-1 cycle undergoes its first period-doubling bifurcation. This gives rise to the formation of a stable period-2 resonance cycle N_2 while

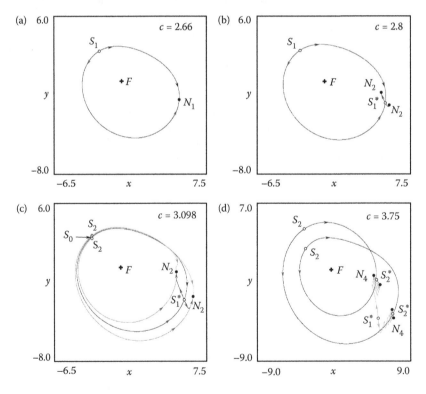

FIGURE 3.10 Phase portraits of the resonance structure at different values of the nonlinearity parameter along the scan line C in Figure 3.2. (a) Original 1:1 resonance torus with its stable node N_1 and saddle cycle S_1. (b) Phase portrait after the first period-doubling of the node cycle. (c) Phase portrait after the transverse period doubling of the original saddle cycle S_1. The system now displays a stable double-layered resonance torus consisting of the node N_2, the saddle S_2, and the unstable manifold of the saddle. (d) Phase portrait after the second period-doubling of the original resonance node cycle. $\omega = 1.08$.

the original resonance node is transformed into a period-1 saddle cycle S_1^*. As illustrated in Figure 3.10b, this period doubling takes place in a direction transverse to the resonance torus. The stable period-2 cycle is situated away from the resonance torus, and the unstable manifold of S_1^* is transverse to this torus.

With further increase of the nonlinearity parameter c, the system crosses the curve PD_1^U. Here, the saddle solution S_1 undergoes a period-doubling bifurcation thus giving birth to the doubly unstable period-1 saddle cycle S_0 and to a new period-2 saddle S_2. As shown in Figure 3.10c, this transition again takes place in a direction transverse to the resonance torus. Note, however, how the unstable manifold of the new period-2 saddle cycle S_2 connects to the period-2 node N_2. The original period-1 resonance torus has now lost its transverse stability and a new, the so-called double-layered resonance torus [51,54], has been born. This torus consists of the period-2 node N_2 together with the corresponding saddle cycle S_2 and the unstable manifold of this cycle. With further increase of the nonlinearity parameter, the period-2 cycle N_2 undergoes a new period-doubling bifurcation in the direction transverse to the original resonance torus. As shown in Figure 3.10d, this bifurcation gives birth to the stable period-4 resonance cycle N_4 and to the period-2 saddle cycle S_2^*.

Inspection of Figure 3.10d reveals an interesting structure of the unstable manifold from the period-1 saddle S_1^* as it approaches the stable period-4 cycle N_4. In order to visualize this structure in more detail, Figure 3.11 shows a magnification of the region around S_1^* and N_4. Note, how the unstable manifold from S_1^* approaches the stable period-4 cycle N_4 in an oscillatory manner that appears to be controlled by the eigenvalues of the saddle cycle S_2^*. These eigenvalues are positive and numerically less than unity in the direction of the stable manifold and negative and numerically larger than unity in direction of the unstable manifold.

With further increase of the nonlinearity parameter c, the stable resonance node continues its transverse period-doubling process with the formation of period-8 and period-16 resonance nodes. This is illustrated in Figure 3.12a and c for $c = 3.936$ and $c = 3.947$, respectively. Figure 3.12b demonstrates how the folded approach from the unstable period-1 resonance cycle S_1^* continues to develop finer and finer details as the periodicity of the stable node cycle increases. With even further increase in c, the system crosses into a regime where the original resonance node N_1 has completed its transverse period-doubling cascade whereas the original saddle cycle S_1 has period doubled only once. As illustrated in Figure 3.13, the dynamics around the original node cycle is now chaotic while the resonance saddle cycle continues its transverse period-doubling cascade with the formation of saddle cycles of period-4 Figure 3.13a, period-8 Figure 3.13b and period-16 Figure 3.13c.

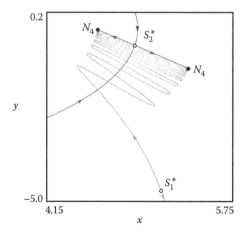

FIGURE 3.11 Magnification of part of the phase portrait in Figure 3.10d around the saddle cycle S_1^* and the stable period-4 node N_4. Note, how the unstable manifold from S_1^* oscillates as it approaches N_4. A similar structure is observed in Figure 3.6 for higher levels of the period-doubling cascade.

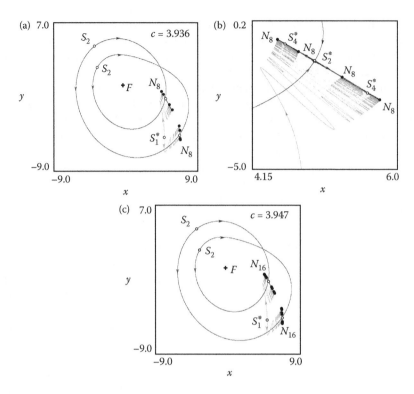

FIGURE 3.12   Phase portraits after the third (a) and the fourth (c) transverse period-doubling of the original period-1 resonance node. (b) Magnification of part of the phase portrait in Figure 3.6a between the saddle cycle S_1^* and the stable period-8 node N_8. While developing finer and finer structures, the folded approach of the unstable manifold from S_1^* to the stable node cycle continues to occur as the period-doubling process proceeds. $\omega = 1.08$.

The above description has presented some of the main structures that one can observe in a vertical scan of the 2D-bifurcation diagram in Figure 3.2. Let us now consider the phenomena that arise in a scan along the direction $A \to A$ in Figure 3.2, i.e., along the line $c = -51\omega + 58$. Starting at a point between the period-doubling curves PD_1^U and PD_2^S, this scan takes the system across the saddle–node bifurcation curves SN_2^L and SN_1^L and into the region of quasiperiodicity. A view on the bifurcation structure to the left of the resonance zone in Figure 3.2 shows that the scan leads the system into a region where the ergodic torus has undergone a first period-doubling bifurcation [5,43]. The idea of the scan is, therefore, to examine how the double-layered resonance torus in the synchronization zone transforms into the period-doubled ergodic torus that exists to the left of this zone.

Figures 3.14a is similar to Figure 3.10c and shows the organization of the resonance cycles after the first period-doubling transitions of the period-1 node and saddle cycles (N_1 and S_1) transversely to the period-1 resonance torus. These transitions have given birth to the formation of a double-layered resonance torus, and the original period-1 resonance torus is now unstable. As shown in Figure 3.14b, breakdown of the period-1 torus involves a splitting of the unstable manifold of S_0 into two branches that are both attracted by the double-layered resonance cycle. This implies that a trajectory starting close to S_0 may approach the stable double-layered torus in an alternating manner. As the system hereafter crosses the saddle–node bifurcation curve SN_2^L, the resonance cycles N_2 and S_2 merge and disappear leaving a period-doubled ergodic torus together with a pair of singly and doubly unstable saddle cycles S_1^* and S_0 (Figure 3.14c). In this way, communication across the zone boundary from resonant period-2 dynamics to period-doubled quasiperiodicity takes the form of a transformation of a double-layered resonance torus into a period-doubled ergodic torus of similar shape and size. Finally, S_1^* and S_0 also merge and disappear

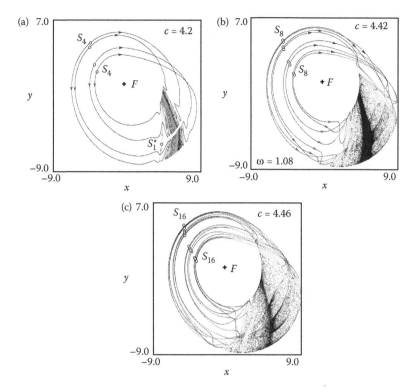

FIGURE 3.13 Phase portraits after completion of the transverse period-doubling for the original resonance node. (a) In the region around this node, the dynamics is now chaotic, and the unstable manifold of S_4 serves as an inset to the chaotic region. As the nonlinearity parameter is further increased, the resonance saddle cycle continues its period-doubling process through the formation of (b) period-8 and (c) period-16 saddle cycles. $\omega = 1.08$.

in the saddle–node bifurcation SN_1^L (Figure 3.14d). However, this happens only after the stable dynamics of the system has turned ergodic.

3.6 Bifurcation Structure in the Period-3 Window

Let us complete our bifurcation analysis for the region of period-doubling bifurcations in the forced Rössler system by demonstrating that similar bifurcation phenomena take place in the periodic windows that exist in the chaotic regime for higher values of the nonlinearity parameter c. Figure 3.15 shows the main bifurcation structure of the period-3 window. Here, SN_3^B denotes the saddle–node bifurcation in which the 3:3 resonant node and saddle cycles are born. This curve continues up along the two sides of the resonance tongue, now denoted SN_3^L and SN_3^R, respectively. The closed curve PD_3 represents the first period-doubling bifurcation. The resonant 3:3 node cycle period doubles along the lower branch of this curve, and the 3:3 saddle cycle doubles its period along the upper branch. Similarly, the closed curve PD_6 represents the second period doubling with the node cycle period doubling its period at the lower branch and the saddle cycle at the upper branch. With its pronounced cusp structure, the bifurcation curve SN_3^C represents a couple of saddle–node bifurcations in which the 3:3 saddle cycle generated in the period-doubling bifurcation PD_3 first loses and subsequently regains stability in a secondary direction. Finally, T_3 is a torus-birth bifurcation curve that serves a purpose similar to that of the torus bifurcation curve T_2 in Figure 3.5c. This curve closes a hole in the edge of the resonance tongue between SN_3^L and the saddle–node bifurcation curve that delineates the sides of the resonance zone for the 6:6 cycles.

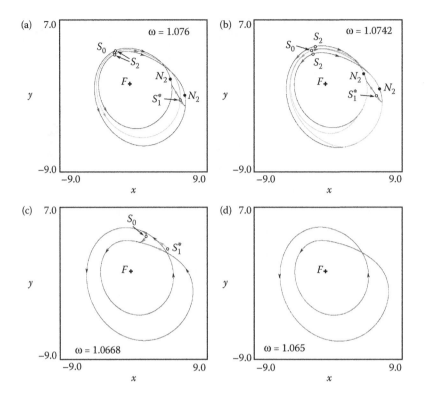

FIGURE 3.14 Stages in the transition from double-layered resonance torus to period-doubled ergodic torus in a scan along the direction $A \to A$ in Figure 3.2. (a) Double-layered resonance torus produced through period doubling of the original period-1 node and saddle cycles. The corresponding period-1 resonance torus is unstable. (b) Destruction of the period-1 resonance torus. (c) Disappearance of the period-2 resonance cycles N_2 and S_2 in the saddle–node bifurcation SN_2^L. This marks the transition to ergodic dynamics. (d) Disappearance of the saddle cycles S_0 (doubly unstable) and S_1^* in the saddle–node bifurcation curve SN_1^L. This final bifurcation takes place in the quasiperiodic regime outside the resonance zone.

In close accordance with the bifurcation structure observed for the 1:1 resonance regime, examination of the bifurcation structure in the 3:3 region confirms that

1. The period-doubling cascades for the node and saddle solutions are interconnected. At the edge of the synchronization tongue, the two bifurcations are simultaneous, but away from the tongue edge the node solution (with the applied forcing type) bifurcates before the saddle solution.
2. Each pair of period-doubling bifurcations generates a new pair of saddle–node bifurcations that define the edges of the resonance zone at the next level in the cascade.
3. Additional torus and global bifurcations serve to close the gap between the new and the previous borders of the resonance zone.

3.7 Accumulating Set of Saddle–Node Bifurcations

Let us finally focus on the problems that relate to the accumulation of the saddle–node bifurcation curves along the edges of the resonance zone and to the continuation of these curves into the range of phase synchronized chaos. Except, perhaps, for the first two to three bifurcations that are model dependent,

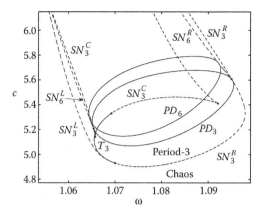

FIGURE 3.15 Main bifurcation structure in the period-3 window. The lower saddle–node bifurcation curve marks the onset of the period-3 resonance solutions. For increasing values of the nonlinearity parameter c, the same saddle–node curve extends up along both tongue edges. PD_3 and PD_6 represent the first and second period doubling with the node solution bifurcating at the lower branch and the saddle solution at the upper branch.

scaling analyses performed by Kuznetsov et al. [16] have demonstrated that the alternation of the bifurcation processes between the two sides of the synchronization zone can be observed for many generations of period-doubling bifurcations, at least up to period-256. This represents the basis for their formulation of a special scaling theory for C-type criticality [16–18].

Based on the bifurcation analyses presented in Section 3.2, we have argued that, at least immediately after their formation, saddle–node bifurcation curves emerging from the stable branch of a period-doubling curve cross out the zone boundary as defined by the preceding saddle–node bifurcation curve along the same edge. On the other hand, saddle–node bifurcation curves that emerge from the unstable branch proceed inside the resonance zone defined by the former saddle–node bifurcation curve. Hence, we can represent the characteristic cyclic behavior associated with the generation of saddle–node bifurcation curves in the period-doubling cascade on the form outlined in Figure 3.16.

The sketch in Figure 3.17 addresses the same problem from a more global point of view. Use of continuation techniques [10,19] has allowed us to follow the saddle–node bifurcations that delineate the left-hand

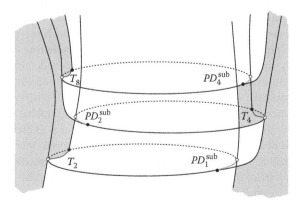

FIGURE 3.16 Sketch to illustrate the alternation of two different scenarios for the birth of new saddle–node bifurcation curves along the edge of the resonance zone. Segments denoted T_2, T_4, and T_8 represent torus-birth bifurcations and segments denoted PD_1, PD_2, and PD_4 represent subcritical period-doubling bifurcations. Note, how new saddle–node bifurcation curves along a given side alternately move to the right and the left of the preceding saddle–node bifurcation curve.

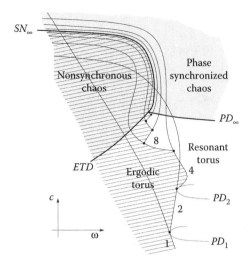

FIGURE 3.17 Sketch illustrating the cascade of saddle–node bifurcations along the left-hand side of the 1:1 reso-nance zone for the periodically forced Rössler oscillator. While distorting the scales, the sketch preserves the observed systematic of the structure. The curves denoted 1, 2, 4, 8, etc., represent the saddle–node bifurcation curves SN_1^L, SN_2^L, SN_4^L, SN_8^L, etc. The gray zone represents the region of phase synchronized chaos, and the hatched gray zone is the region of nonsynchronous chaos.

side of the resonance zone through seven to eight generations of period-doubling bifurcations and thus to outline the structure of the accumulating set of saddle–node bifurcations. The curves denoted 1, 2, 4, 8, etc., represent the saddle–node bifurcation curves SN_1^L, SN_2^L, SN_4^L, SN_8^L, etc., and PD_1, PD_2 locate the first period-doubling curves. SN_∞ denotes the numerically determined accumulation curve for the saddle-node bifurcation curves, PD_∞ similarly represents the accumulation curve for the period-doubling cas-cade, and ETD represents the curve of ergodic torus destruction. The resonance zone is presented with a white background, the region in which ergodic tori (of different periodicity) exist is hatched, the region of phase synchronized chaos (neglecting the occurrence of periodic windows) is gray, and the region of nonsynchronous chaos is hatched with a gray background. As before, the bifurcation parameters are the forcing frequency ω and the nonlinearity parameter c of the Rössler oscillator.

Closer inspection of Figure 3.17 reveals that although the first few saddle–node bifurcation curves fol-low their own courses, a systematic structure soon materializes in which the saddle–node bifurcation curves alternate around a final accumulation curve SN_∞. It is also interesting to note that the hatched region of ergodic torus dynamics, unaffected by several of the saddle–node bifurcation curves, consis-tently penetrates the structure until it finds the saddle–node bifurcation curve of the highest periodicity for the given value of c. This is also the saddle–node bifurcation that provides the boundary of existence for the stable node cycle. As mentioned above, this implies that the period-doubled ergodic torus that exists for a given value of c communicates with a multi-layered resonance torus of the same periodicity, form, and size. To complete this description it is important to note that the well-organized structure of alter-nating saddle–node bifurcation curves tends to dissolve as the system progresses deeper into the chaotic regime. Hence, we cannot provide a description of the organization of the saddle–node bifurcation curves for the fully developed phase synchronized chaos.

3.8 Conclusions

Chaotic phase synchronization is an extremely interesting area of research with obvious applications to many different problems in physics, chemistry, biology, and other fields of science and technology

[1,23,40]. However, although work on chaotic phase synchronization has already contributed many useful results [6,27,33], our understanding of this area is still far from complete. In particular, many questions related to the synchronization of multi-mode oscillations remain unsolved. Already the definition of the phase for a given chaotic oscillator can present a complicated problem, and in many situations it may not even be possible to find a satisfactory solution [8,33]. This is the case, for instance, for the Lorenz attractor where the dynamics shifts, in an apparently random manner, between rotation around one or the other of the two symmetric equilibrium points [49].

Other chaotic systems display more or less random shifts between small loops and loops with much longer transition times, and it may not be possible, in a consistent manner, to decide which loops constitute a cycle [33]. This situation arises even for the Rössler system if the parameters are chosen in the regime of the so-called funnel attractor. However, problems of this type do not arise for the parameter values we have used in the present study. As illustrated in Figure 3.1b, the projection of the Rössler attractor onto the *xy*-plane for these parameter values produces a regular rotation of the phase point around origo. Hence, the dynamics is sufficiently simple to allow for a consistent definition of the phase variable. In practice, the angle between any fixed direction in the *xy*-plane and the direction from origo to the instantaneous phase-point is often sufficient [33]. More theoretical studies may also involve determination of the neutrally stable direction, i.e., the direction of vanishing Lyapunov exponent [8].

In spite of these complications it is clear that many chaotic systems react to an external forcing in a manner that in many ways resembles the reaction observed for the simple Rössler system. Moreover, one can often use some of the diagnostic tools described in the beginning of this chapter without actually introducing the concept of a phase [33]. The mean frequency, for instance, typically displays a variation similar to the synchronization characteristics shown in Figure 3.1a. This means that one can delineate a restricted range of forcing frequencies where the mean frequency of the forced chaotic system approximately follows the variation of the forcing frequency. The lack of precise frequency synchronization may then be revealed by the rounding of the otherwise sharp edges of the synchronization characteristic at the synchronization thresholds. In experimental studies, a similar rounding phenomenon may be associated with the presence of noise.

When the stroboscopic projection technique illustrated in Figure 3.1b and c is applied, one might find that the distribution of dots across the forced state from time to time turns unstable and undergoes a major reorganization to again return the same basic distribution. In this connection the distribution of return times of the different loops in the unforced chaotic oscillator may also play a role. If this distribution is sufficiently inhomogeneous, the distribution of stroboscopic points may break up into several groups.

In view, primarily, of analyzing the phase relationships for experimental time series, Pikovsky et al. [35] have suggested the use of a Hilbert transformation approach to determine the degree of phase synchronization between two interacting chaotic oscillators. Provided the series is sufficiently long, the Hilbert transformation allows one to define the instantaneous phase (and amplitude) of an experimentally observed time series. With similar definitions for other locking ratios, 1:1 phase synchronization between two interacting, chaotic oscillators may then be said to occur if, with a sufficiently small margin μ, the difference between the two phases continues to fall near some constant value. Theoretically, phase synchronization may be said to occur if the small margin $\mu \ll 2\pi$. In experimental studies of interacting chaotic oscillators one often finds that the phase difference remains constant within a relatively small margin μ for a certain period of time to then jump one or more 2π steps as the oscillators for a moment loose synchronization. The question whether phase synchronization occurs or not then becomes a matter of the frequency of such 2π steps. We have successfully used the Hilbert transformation approach to demonstrate, for instance, the occurrence of phase synchronization between the proximal tubular pressure oscillations for a pair of neighboring nephrons of the kidney [14].

The purpose of this chapter has been to present the results of a relatively detailed bifurcation analysis of the transition to chaotic phase synchronization in a periodically forced Rössler oscillator. With this purpose, we have applied a standard continuation method [10] to follow the bifurcations that occur within and along the edges of the 1:1 synchronization zone as increasing values of the so-called nonlinearity parameter c take the system up through the cascade of period-doubling bifurcations.

This approach has allowed us to examine the interplay between a significant number of interesting phenomena, including (i) the relation between the interconnected period-doubling bifurcations for the node and saddle cycles that occur in the resonance zone, (ii) the formation, transformation, and elimination of different forms of multi-layered resonance tori, (iii) the communication between the resonance tori and the quasiperiodic oscillations that exist outside the resonance zone, (iv) the torus doubling bifurcations that occur along the edges of the resonance zone, and (v) the alternating involvement of torus-birth bifurcations and subcritical period-doubling bifurcations in restricting the stable resonance modes from escaping from their designated regions of parameter space. It is precisely this alternation of different bifurcations that produce the special scaling structure (cyclic criticality) that characterizes the period-doubling bifurcation along the edge of a resonance tongue [17,18].

If one extends the investigation to consider the transition to chaotic phase synchronization for a pair of coupled, nearly identical period-doubling systems, the first thing to notice is that the presence of a coupling splits the synchronized modes into symmetric and anti-symmetric states [38]. If synchronization takes place into a symmetric state, the two oscillators move in phase, the explored part of phase space remains relatively narrow, and the subsequent bifurcations in general maintain the original symmetry. If, on the other hand, the synchronization occurs into an anti-symmetric state, the two oscillators move with opposite phases, excursions are made to a much larger part of phase space, and the subsequent bifurcations typically involve torus-birth processes and homoclinic bifurcations.

The bifurcation structure one can observe along the edge of a synchronization tongue that leads to symmetric states is essentially the same as the structure we have described for the periodically forced Rössler attractor.

However, along the other edge of the synchronization zone, where the phase locking typically leads to anti-symmetric states, one can observe a very different bifurcation structure [23]. Here, the edge of the resonance zone involves cascades of bifurcations of different type. A cascade of saddle–node bifurcations serves to delineate the range of existence of the resonance cycles. Like in the bifurcation diagram (Figure 3.2) for the periodically forced Rössler system, the saddle–node bifurcations curves emerge from points on the corresponding period-doubling curve and continue up along the edge of the resonance zone. Moreover, this process still occurs in an alternating manner to the left and to the right of the former saddle-node bifurcation curve. However, the saddle–node bifurcation curves no longer mark the transition from stable periodic dynamics to quasiperiodicity.

This function has been taken over by a cascade of torus bifurcation curves that stretch up along the saddle–node bifurcation curves, but inside the resonance zone. In this way, a region is formed where stable quasiperiodic oscillations coexist with doubly and triply unstable resonance modes. The torus bifurcation curves emerge from the points of intersection between the corresponding period-doubling curve and the former torus bifurcation curve. They are supported, close to the transition to chaos, by points on the corresponding saddle–node bifurcation curves.

From each of these points, a curve of homoclinic bifurcations stretches between the torus bifurcation curve and the corresponding saddle–node bifurcation curve all the way down to the point where the torus bifurcation curve was born.

The torus bifurcation curves produce a set of period doubled quasiperiodic states, and the overall result of this reconstruction of the bifurcation diagram for the anti-symmetric modes is that part of the resonance zone is invaded by different quasiperiodic modes. This includes the occurrence of phase-modulated quasiperiodicity. Let us finally note that a recent study of the transition to chaotic phase synchronization for two coupled, nearly identical two-oscillator systems (nephrons) [23] has demonstrated essentially the same bifurcation structures for synchronization into symmetric, respectively anti-symmetric states. This result applies, provided that the 1:5 or 1:4 synchronization for the internal modes of the individual oscillator remained unaffected by the coupling.

Acknowledgments

We thank S.P. Kuznetsov for valuable discussions about the bifurcation structure associated with C-type criticality.

References

1. A. Ahlborn and U. Parlitz. Experimental observation of chaotic phase synchronization of a periodically modulated frequency-doubled ND:YAG laser. *Optics Lett.*, 34:2754–2756, 2009.
2. J. C. Alexander, J. A. Yorke, Z. You, and I. Kan. Riddled basins. *Int. J. Bif. Chaos*, 2:795–813, 1992.
3. V. S. Anishchenko, T. E. Vadivasova, D. E. Postnov, and M. A. Safonova. Synchronization of chaos. *Int. J. Bifurcat. Chaos*, 2:633–644, 1992.
4. A. Argyris, D. Syvridis, and L. Lager et al. Chaos-based communication using commercial fibre optic links. *Nature*, 438:343–346, 2005.
5. A. Arnéodo, P. H. Coullet, and E. A. Spiegel. Cascade of period doublings of tori. *Phys. Lett. A*, 94:1–6, 1983.
6. A. Balanov, N. Janson, D. Postnov, and O. Sosnovtseva. *Synchronization: From Simple to Complex*. Springer-Verlag, Berlin, 2009.
7. C. A. S. Batista, A. M. Batista, J. A. C. de Pontes, R. L. Viana, and S. R. Lopes. Chaotic phase synchronization in scale-free networks of bursting neurons. *Phys. Rev. E*, 76:016218, 2007.
8. S. Boccaletti, J. Kurths, G. Osipov, D. L. Valladares, and C. S. Zhu. The synchronization of chaotic systems. *Phys. Rep.*, 366:1–101, 2002.
9. D. J. DeShazer, R. Breban, E. Ott, and R. Roy. Detecting phase synchronization in a chaotic laser array. *Phys. Rev. Lett.*, 87:044101, 2001.
10. E. J. Doedel, R. C. Paffenroth, A. R. Champneys, T. F. Fairgrieve, Yu. A. Kuznetsov, B. E. Oldeman, B. Sandstede, and X. Wang. Auto 2000: Continuation and bifurcation software for ordinary differential equations (with homcont). Technical report, California Institute of Technology, 2002.
11. G. I. Dykman, P. S. Landa, and Y. I. Neymark. Synchronizing the chaotic oscillations by external force. *Chaos Soli. Fract.*, 1:339–353, 1991.
12. H. Gustafsson, A. Bülow, and H. Nilsson. Rhythmic contractions of isolated, pressurized small arteries from rat. *Acta Physiol. Scand.*, 152:145–152, 1994.
13. J. F. Heagy, T. L. Carroll, and L. M. Pecora. Desynchronization by periodic orbits. *Phys. Rev. E*, 52:R1253–R1256, 1995.
14. N.-H. Holstein-Rathlou, K.-P. Yip, O. V. Sosnovtseva, and E. Mosekilde. Synchronization phenomena in nephron–nephron interaction. *Chaos*, 11:417–426, 2001.
15. H. Kantz and T. Schreiber. *Nonlinear Time Series Analysis*. Cambridge University Press, UK, 1997.
16. A. P. Kuznetsov, S. P. Kuznetsov, and I. R. Sataev. A variety of period-doubling universality classes in multi-parameter analysis of transition to chaos. *Physica D*, 109:91–112, 1997.
17. S. P. Kuznetsov, A. P. Kuznetsov, and I. R. Sataev. Multiparameter critical situations, universality and scaling in two-dimensional period-doubling maps. *J. Stat. Phys.*, 121:697–748, 2005.
18. S. P. Kuznetsov and I. R. Sataev. Universality and scaling for the breakup of phase synchronization at the onset of chaos in a periodically driven Rössler oscillator. *Phys. Rev. E*, 64:046214, 2001.
19. Yu. A. Kuznetsov. *Elements of Applied Bifurcation Theory*. Springer-Verlag, Berlin, 2004.
20. Yu. A. Kuznetsov, H. G. E. Meijer, and L. Van Veen. The fold-flip bifurcation. *Int. J. Bifurcat. Chaos*, 14:2253–2282, 2004.
21. Y.-C. Lai, C. Grebogi, J. A. Yorke, and S. C. Venkataramani. Riddling bifurcation in chaotic dynamical systems. *Phys. Rev. Lett.*, 77:55–58, 1996.
22. P. S. Landa and M. G. Rosenblum. Synchronization and chaotization of oscillations in coupled self-oscillating systems. *Appl. Mech. Rev.*, 46:414–426, 1993.

23. J. L. Laugesen, E. Mosekilde, and N.-H. Holstein-Rathlou. Synchronization of period-doubling oscillations in vascular coupled nephrons. *Chaos*, 21:033128, 2011.

24. J. L. Laugesen, E. Mosekilde, and Zh. T. Zhusubaliyev. Bifurcation structure of the C-type period-doubling transition. *Physica D*, 241:488–496, 2012.

25. V. L. Maistrenko, Y. L. Maistrenko, and E. Mosekilde. Chaotic synchronization and antisynchronization in coupled sine maps. *Int. J. Bif. Chaos*, 15:2161–2177, 2007.

26. V. Makarenko and R. Llinás. Experimentally determined chaotic phase synchronization in a neuronal system. *Proc. Natl. Acad. Sci.*, 95:15747–15752, 1998.

27. E. Mosekilde, Yu. Maistrenko, and D. Postnov. *Chaotic Synchronization: Applications to Living Systems.* World Scientific, Singapore, 2002.

28. E. Mosekilde, Zh. T. Zhusubaliyev, J. L. Laugesen, and O. Yanochkina. On the structure of phase synchronized chaos. *Chaos Soliton. Fract.*, 46:28–37, 2013.

29. E. Ott and J. C. Sommerer. Blowout bifurcations: The occurrence of riddled basins and on–off intermittency. *Phys. Lett. A*, 188:39–47, 1994.

30. L. M. Pecora and T. L. Carroll. Synchronization in chaotic systems. *Phys. Rev. Lett.*, 64:821–824, 1991.

31. G. Peng, Y. Jiang, and F. Chen. Generalized projective synchronization of fractional order chaotic systems. *Physica A*, 387:3738–3746, 2008.

32. A. Pikovsky, G. Osipov, M. Rosenblum, M. Zaks, and J. Kurths. Attractor–repeller collision and eyelet intermittency at the transition to phase synchronization. *Phys. Rev. Lett.*, 79:47–50, 1997.

33. A. Pikovsky, M. Rosenblum, and J. Kurths. *Synchronization: A Universal Concept in Nonlinear Science.* Cambridge University Press, UK, 2001.

34. A. S. Pikovsky. On the interaction of strange attractors. *Z. Phys. B*, 55:149–152, 1984.

35. A. S. Pikovsky, M. G. Rosenblum, G. V. Osipov, and J. Kurths. Phase synchronization of chaotic oscillators by external driving. *Physica D*, 104:219–238, 1997.

36. N. Platt, E. A. Spiegel, and C. Tresser. On–off intermittency: A mechanism for bursting. *Phys. Rev. Lett.*, 70:279–282, 1993.

37. D. E. Postnov, T. E. Vadivasova, O. V. Sosnovtseva, A. G. Balanov, V. S. Anishchenko, and E. Mosekilde. Role of multistability in the transition to chaotic phase synchronization. *Chaos*, 9:227–232, 1999.

38. C. Reick and E. Mosekilde. Emergence of quasiperiodicity in symmetrically coupled, identical period-doubling systems. *Phys. Rev. E*, 52:1418–1435, 1995.

39. E. Rosa, E. Ott, and M. H. Hess. Transition to phase synchronization of chaos. *Phys. Rev. Lett.*, 80:1642–1645, 1998.

40. E. Rosa, W. B. Pardo, C. M. Ticos, J. A. Walkenstein, and M. Monti. Phase synchronization of chaos in a plasma discharge tube. *Int. J. Bifurcat. Chaos*, 10:2551–2563, 2000.

41. M. G. Rosenblum, A. S. Pikovsky, and J. Kurths. Phase synchronization of chaotic oscillators. *Phys. Rev. Lett.*, 76:1804–1807, 1996.

42. N. F. Rulkov. Images of synchronized chaos: Experiments with circuits. *Chaos*, 6:262–279, 1996.

43. M. Sekikawa, N. Inaba, T. Yoshinaga, and T. Tsubouchi. Bifurcation structure of successive torus doubling. *Phys. Lett. A*, 348:187–194, 2006.

44. A. Uchida, K. Amano, and M. Inoue et al. Fast physical random bit generation with chaotic semiconductor lasers. *Nature Photon.*, 2:728–732, 2008.

45. T. E. Vadivasova, A. G. Balanov, O. V. Sosnovtseva, D. E. Postnov, and E. Mosekilde. Synchronization in driven chaotic systems: Diagnostics and bifurcations. *Phys. Lett. A*, 253:66–74, 1999.

46. C. W. Wu and L. O. Chua. A unified framework for synchronization and control of dynamical systems. *Int. J. Bif. Chaos*, 4:979–998, 1994.

47. X. Wu and Y. Lu. Generalized projective synchronization of the fractional order Chen hyperchaotic system. *Nonl. Dyn.*, 57:25–35, 2009.

48. S. Yanchuck, Yu. Maistrenko, and E. Mosekilde. Loss of synchronization in coupled Rössler systems. *Physica D*, 154:26–42, 2001.

49. L. Zhao, Y.-C. Lai, R. Wang, and J.-Y. Gao. Limits to chaotic phase synchronization. *Eur. Phys. Lett.*, 66:324, 2004.

50. C. Zhou, J. Kurths, I. Z. Kiss, and J. L. Hudson. Noise-enhanced phase synchronization of chaotic oscillators. *Phys. Rev. Lett.*, 89:014101, 2002.

51. Zh. T. Zhusubaliyev, J. L. Laugesen, and E. Mosekilde. From multi-layered resonance tori to period-doubled ergodic tori. *Phys. Lett. A*, 374:2534–2538, 2010.

52. Zh. T. Zhusubaliyev and E. Mosekilde. Formation and destruction of multilayered tori in coupled map systems. *Chaos*, 18:037124, 2008.

53. Zh. T. Zhusubaliyev and E. Mosekilde. Multilayered tori in a system of two coupled logistic maps. *Phys. Lett. A*, 373:946–951, 2009.

54. Zh. T. Zhusubaliyev and E. Mosekilde. Novel routes to chaos through torus breakdown in non-invertible maps. *Physica D*, 238:589–602, 2009.

4

The Kolmogorov–Taylor Law of Turbulence: What Can Rigorously Be Proved?

Roger Lewandowski

4.1 Introduction

Most of realistic flows are turbulent, looking chaotic, disordered, and unpredictable. The physical processes that govern turbulence are far to be all understood, although it attracted a lot of attention in the literature since the nineteenth century. Therefore, turbulence remains one of the main challenges of modern science, which has an impact from the human, social, and economic standpoints, especially in the understanding of climate change and ecological issues.

On the basis of works carried out by Boussinesq [1] and Reynolds [27], it has been soon recognized that the velocity field of a turbulent flow can be decomposed as the sum of a mean field and a fluctuation. Initially, the mean velocity field was specified by its long time averaged. Later, Taylor [29] introduced the notion of statistical means, where the velocity and the pressure of the flow are considered as abstract random fields. The mean velocity is then its mathematical expectation, which also allowed Taylor to define the notion of isotropic turbulence, through algebraic properties satisfied by correlation tensors.

In Reference 21, Kolmogorov mainly revisited Taylor's work. He generalized the notion of isotropy by introducing local isotropy, stating somehow that all turbulent flows are locally isotropic. In this context, he retrieved the law of the 2/3, already found in Reference 29. However, the main impact of Kolmogorov's paper lies in a footnote, in which he explains that turbulence can be depicted by a structure of eddies of different scales that interact with each other, the scales being distributed in a continuous range. The large-scale eddies transfer energy to smaller ones, and then they transfer energy to more smaller, and so on until a final scale that dissipates energy as heat.

Kolmogorov's paper is written from a physical viewpoint, without considering the Navier-Stokes equations (NSE). The aim of the present work is to define a mathematical framework in which notions such as homogeneity and isotropy can be set up, the correlation tensors of the flow can be properly defined, and in which we can find the right mathematical hypotheses that allow to rigorously describe the structure of the covariance matrix in order to prove standard laws of the turbulence, such as the law of the 2/3. To do so, we are led to make connections between the physical concepts of References 21 and 29 and modern mathematical results about the NSE.

We start with a short overview about the NSE, considering throughout this text the incompressible case. Then, we define the length–time bases that provide a framework for performing the dimensional analysis, which yields to introduce the notion of generalized Reynolds number and to give the dimensionless form of the NSE. We state the Reynolds similarity principle in this context and seek for solutions of the NSE that verify this similarity principle. We next define general homogeneous and isotropic tensors. We prove algebraic results in order to fully characterize isotropic tensors of orders 1 and 2.

Once we are done with this technical background, we recall the suitable probabilistic framework, which was initially introduced in Reference 5 by considering smooth solutions of the NSE. The underlying probabilistic space is a set of initial data, over which we are able to construct a probability measure. This construction should be compared with the notion of statistical solutions, developed for instance in Reference 16.

The correlation tensors can therefore be specified, yielding the definition of homogeneous and isotropic turbulence. We also introduce by this way the Reynolds stress, which plays a central role in the turbulence models.

We then consider the correlation tensor of order 2 in a ball of fluid δV, centered at a point $\mathbf{x}_0$, and over a short time interval. From the probability viewpoint, this tensor, denoted here by $B(\mathbf{x}, \mathbf{x}_0)$, is the covariant matrix of the velocities at $\mathbf{x}_0$ and any other point $\mathbf{x}$ of the ball of fluid. Therefore, $B(\mathbf{x}, \mathbf{x}_0)$ measures how the two random velocity vectors at $\mathbf{x}_0$ and $\mathbf{x}$ are dependent. We consider a homogeneous, isotropic, and stationary turbulence, in the case of smooth solutions of the NSE. We then perform an expansion of $B(\mathbf{x}, \mathbf{x}_0)$ when $r = |\mathbf{x} - \mathbf{x}_0|$ goes to zero, and we show that the main term in this expansion is entirely characterized by a function $E = E(r)$. We show next that in some inertial range $[r_1, r_2]$, $E(r) \sim (\mathcal{E} r)^{\frac{2}{3}}$, $\mathcal{E}$ being the mean dissipation at $\mathbf{x}_0$, which is the law of the 2/3. To do so, we carefully establish the mathematical concept of similarity principle.

So far we know, this work is the first attempt to give a rigorous proof of the 2/3 law. Nevertheless, the well-known law of the $-5/3$ about the velocity spectrum, and everything which relates, has attracted a lot of attention in the mathematical literature (see, for instance, References 5, 8, 10–12, 15, and 18).

4.2 About the NSE

Let $\Omega \subset \mathbb{R}^3$ be a smooth domain, $\Gamma = \partial\Omega$ its boundary, $T > 0$ a given time, $Q = [0, T] \times \Omega$. In the following $\mathbf{v} = \mathbf{v}(t, \mathbf{x})$, $p = p(t, \mathbf{x})$, denote the Euler velocity and the pressure of the fluid at any $(t, \mathbf{x}) \in Q$.

A divergence free field $\mathbf{v}_0$ being given, the incompressible NSE set in Q with the no-slip boundary condition and $\mathbf{v}_0$ as initial data are the following:

$$\begin{cases} \partial_t \mathbf{v} + (\mathbf{v} \cdot \nabla)\,\mathbf{v} - \nabla \cdot (2\nu D\,\mathbf{v}) + \nabla p = \mathbf{f} & \text{in } Q, \quad \text{(a)} \\ \nabla \cdot \mathbf{v} = 0 & \text{in } Q, \quad \text{(b)} \\ \mathbf{v} = 0 & \text{on } \Gamma, \quad \text{(c)} \\ \mathbf{v} = \mathbf{v}_0 & \text{at } t = 0, \quad \text{(d)} \end{cases} \qquad (4.1)$$

where $\nu > 0$ is the kinematic viscosity, usually a function of the temperature,

$$D\mathbf{v} = \frac{1}{2}(\nabla\mathbf{v} + \nabla\mathbf{v}^t),$$

is the deformation tensor, and $\mathbf{f}$ is any external force (gravity, electromagnetic force, etc.).

Equation 4.1a is the momentum equation. Equation 4.1b is the mass conservation equation that expresses in this case the incompressibility of the fluid. Equation 4.1c is the boundary condition, namely, the no-slip condition, whereas (4.1d) is the initial data. Notice that when ν is constant, we deduce from the incompressibility condition the identity

$$\nabla \cdot (2\nu D\mathbf{v}) = \nu\Delta\mathbf{v}.$$

By writing $\mathbf{v} = (v_1, v_2, v_3)$, the momentum equation (4.1a) becomes component-by-component

$$\partial_t v_i + v_j \frac{\partial v_i}{\partial x_j} - \frac{\partial}{\partial x_j}\left(\nu\left(\frac{\partial v_i}{\partial x_j} + \frac{\partial v_j}{\partial x_i}\right)\right) + \frac{\partial p}{\partial x_i} = f_i \tag{4.2}$$

by using the Einstein summation convention. Moreover, the incompressibility condition (4.1b) reads

$$\frac{\partial v_j}{\partial x_j} = 0. \tag{4.3}$$

It is convenient to notice that the incompressibility condition yields

$$v_j\frac{\partial v_i}{\partial x_j} = \frac{\partial(v_i v_j)}{\partial x_j},$$

hence,

$$(\mathbf{v}\cdot\nabla)\mathbf{v} = \nabla\cdot(\mathbf{v}\otimes\mathbf{v}),$$

where $\mathbf{v}\otimes\mathbf{v} = (v_i v_j)_{1\leq i,j\leq 3}$.

We distinguish two types of solutions to the NSE:

1. The "strong" solutions over a small time interval $[0, T_{\max}[$ Fujita and Kato [17]
2. The "weak" (also turbulent) solutions, global in time, Leray [19,22]
3. Strong solutions are essentially $C^{1,\alpha}$ over $[0, T_{\max}[\times\Omega$, as long as the data are smooth enough. A given $\mathbf{f}$ being fixed, the time $T_{\max}$ only depends on ν and $\|\mathbf{v}_0\|$ for a suitable norm, and whatever the choice of $\mathbf{v}_0$ smooth enough, the corresponding solution is unique, yielding the writing

$$\mathbf{v} = \mathbf{v}(t, \mathbf{x}, \mathbf{v}_0), \quad p = p(t, \mathbf{x}, \mathbf{v}_0).$$

Strong solutions might be globally defined in time over $[0, \infty[$ when the initial data $\mathbf{v}_0$ is "small enough" or the viscosity ν is "large" enough, which means that the flow is rather laminar (see also References 3, 4, 6, and 20).

4. Weak solutions are such that the velocity, denoted here by $\mathbf{v} = \mathbf{v}(t)$, is considered as a trajectory into a suitable Hilbert space $H \subset L^2(\Omega)^3$, and is weakly continuous from $[0, T]$ into H, for any $T > 0$. This means that any $\eta \in H$ being given, the function $t \to \langle\mathbf{v}(t), \eta\rangle$ is a continuous function of t, where $\langle\cdot,\cdot\rangle$ denotes the scalar product in H (see References 9, 14, 24, 25, and 30).

Whatever the type of solution one considers, it is not known if it develops a singularity in finite time [13], which means that for some $(t_0, \mathbf{x}_0)$,

$$\lim_{\substack{t\to t_0 \\ \mathbf{x}\to\mathbf{x}_0}} |\mathbf{v}(t, \mathbf{x})| = \infty,$$

a question studied References 2, 7, 23, and 28. Moreover, it is not known if the Leray–Hopf solution is unique or not.

4.3 Similarity and Reynolds Number

4.3.1 Dimensions

Each physical field $\psi = \psi(t, \mathbf{x})$ involved in incompressible flows, can be decomposed as

$$\psi = \ell^{d_\ell(\psi)} \tau^{d_\tau(\psi)}, \tag{4.4}$$

where $\tau = \tau(t, \mathbf{x})$ is a time field (expressed in seconds) and $\ell = \ell(t, \mathbf{x})$ is a length field (expressed in meters). Because of incompressibility, the mass does not play any role in the dimensional analysis carried out in the following. In the expression above,

$$\mathbb{D}(\psi) = (d_\ell(\psi), d_\tau(\psi)) \in \mathbb{Q}^2 \tag{4.5}$$

is the dimension of ψ. Notice that in particular, $\mathbb{D}(\mathbf{x}) = (1, 0)$, $\mathbb{D}(t) = (0, 1)$. Any field ψ such that $\mathbb{D}(\psi) = (0, 0)$ is said to be dimensionless. We also use the standard notation

$$[\psi] = \mathcal{L}^{d_\ell(\psi)} \mathcal{T}^{d_\tau(\psi)}, \tag{4.6}$$

which is useful in practical calculations. The following table gives the dimension of the main scalar fields involved in turbulent flows.

Scalar Field	Dimension $\mathbb{D}$
Kinematic viscosity ν	$(2, -1)$
Scalar velocity u	$(1, -1)$
Pressure per mass density p	$(2, -2)$
Kinetic energy per mass density $E = (1/2)|\mathbf{v}|^2$	$(2, -2)$
Dissipation per mass density $\varepsilon = 2\nu|D\mathbf{v}|^2$	$(2, -3)$

The following table gives the main dimension of the vector and tensor fields involved in turbulent flows.

Vector and Tensor Field	Dimension $\mathbb{D}$
Velocity $\mathbf{v}$	$(1, -1)$
Deformation tensor $D\mathbf{v}$	$(0, -1)$
Vorticity ω	$(0, -1)$
Source term $\mathbf{f}$, force per mass unit	$(1, -2)$
Stress tensor per mass density $(1/\rho)\sigma$	$(2, -2)$

4.3.2 Length–Time Bases and Generalized Reynolds Numbers

Definition 4.1

A length–time basis is a couple

$$b = (\lambda, \tau), \tag{4.7}$$

where λ a given constant length and τ a constant time.

Definition 4.2

Let $\psi = \psi(t, \mathbf{x})$ (constant, scalar, vector, tensor, etc.,) be any given field defined on a cylinder $Q = [0, T] \times \Omega$. Let ψ_b be the field defined by

$$\psi_b(t', \mathbf{x}') = \lambda^{-d_\ell(\psi)} \tau^{-d_\tau(\psi)} \psi(\tau t', \lambda \mathbf{x}'), \quad (t', \mathbf{x}') \in Q_b = \left[0, \frac{T}{\tau}\right] \times \frac{1}{\lambda}\Omega, \tag{4.8}$$

where t' and $\mathbf{x}'$ are dimensionless. It is easily checked that ψ_b is dimensionless. We say that $\psi_b = \psi_b(t', \mathbf{x}')$ is the *b*-dimensionless field deduced from ψ.

Let $b = (\lambda, \tau)$ be the length–time basis related to a given scale, and

$$V = \lambda \tau^{-1}, \tag{4.9}$$

be the convective associated velocity. We assume in what follows that the kinematic viscosity ν is constant. Therefore, according to formula (4.8), the b-dimensionless field ν_b deduced from ν is expressed as

$$\nu_b = \lambda^{-2}\tau\nu = \frac{\nu}{V\lambda}, \tag{4.10}$$

by involving the associated convective velocity V given by Equation 4.9. Let $Re(b)$ be the dimensionless number defined by

$$Re(b) = \frac{1}{\nu_b} = \frac{V\lambda}{\nu}. \tag{4.11}$$

We observe that $Re(b)$ is of the same form as the Reynolds number used in classical fluid dynamics, defined as the quotient of the convective forces intensity by the viscous forces intensity, which is why we call $Re(b)$ a generalized Reynolds number.

4.3.3 Dimensionless Form of the NSE

Lemma 4.1

Let $b = (\lambda, \tau)$ be any length–time basis, and $(\mathbf{v}_b, p_b)$ be the *b*-dimensionless field deduced from $(\mathbf{v}, p)$. Then $(\mathbf{v}_b, p_b)$ satisfies the following dimensionless NSE:

$$\begin{cases} \partial_{t'}\mathbf{v}_b + (\mathbf{v}_b \cdot \nabla')\mathbf{v}_b - \nu_b\Delta'\mathbf{v}_b + \nabla'p_b = \mathbf{f}_b & \text{in } Q_b, \\ \nabla' \cdot \mathbf{v}_b = 0 & \text{in } Q_b, \\ \mathbf{v}_b = 0 & \text{on } \Gamma_b, \\ \mathbf{v}_b = (\mathbf{v}_0)_b & \text{at } t = 0. \end{cases} \tag{4.12}$$

Proof. It is easily checked that for any field $\psi = \psi(t, \mathbf{x})$ of class C^1 in time, of class C^2 in space,

$$\partial_t\psi(t, \mathbf{x}) = \lambda^{d_\ell(\psi)}\tau^{d_\tau(\psi)-1}\partial_{t'}\psi_b(t', \mathbf{x}'), \tag{4.13}$$

$$\nabla\psi(t, \mathbf{x}) = \lambda^{d_\ell(\psi)-1}\tau^{d_\tau(\psi)}\nabla'\psi_b(t', \mathbf{x}'), \tag{4.14}$$

$$\Delta\psi(t, \mathbf{x}) = \lambda^{d_\ell(\psi)-2}\tau^{d_\tau(\psi)}\nabla'\psi_b(t', \mathbf{x}'), \tag{4.15}$$

where $(t, \mathbf{x}) = (\tau t', \lambda \mathbf{x}')$. As

$$d_\ell(\mathbf{v}) = 1, \quad d_\tau(\mathbf{v}) = -1, \quad d_\ell(p) = 2, \quad d_\tau(p) = 1,$$

we get from Equation 4.8

$$\mathbf{v}(t, \mathbf{x}) = \lambda\tau^{-1}\mathbf{v}_b(t', \mathbf{x}'), \quad p(t, \mathbf{x}) = \lambda^2\tau^{-2}p_b(t', \mathbf{x}'). \tag{4.16}$$

Then Equation 4.12 results from Equations 4.13 through 4.15 applied to $\mathbf{v}$ and p, combined with Equation 4.16. ∎

4.3.4 Similarity

The pressure in the NSE is defined up to a constant. Therefore, it naturally belongs to quotient spaces. We denote by $\widetilde{p}$ the class of any p in a suitable quotient space, which does not need to be specified here. For $i = 1, 2$, let us consider

1. $Q^{(i)} = [0, T^{(i)}] \times \Omega^{(i)}$ two cylinders
2. $\nu^{(i)}$ two kinematical viscosities
3. $\mathbf{f}^{(i)}$ two forces per mass unit
4. $\mathbf{v}_0^{(i)} = \mathbf{v}_0^{(i)}(\mathbf{x})$ two velocity fields defined in $\Omega^{(i)}$ $(i = 1, 2)$

Definition 4.3

Let $(\mathbf{v}^{(i)}, \widetilde{p}^{(i)})$ be two flows in $Q^{(i)}$, $i = 1, 2$. We say that these two flows are similar if there exist two length–time bases b_1 and b_2, such that

$$Q_{b_1}^{(1)} = Q_{b_2}^{(2)}, \quad (\mathbf{v}_{b_1}^{(1)}, \widetilde{p}_{b_1}^{(1)}) = (\mathbf{v}_{b_2}^{(2)}, \widetilde{p}_{b_2}^{(2)}). \tag{4.17}$$

Let us consider the NSE equations, for $i = 1, 2$,

$$\begin{cases} \partial_t \mathbf{v}^{(i)} + (\mathbf{v}^{(i)} \cdot \nabla)\mathbf{v}^{(i)} - \nu^{(i)} \Delta \mathbf{v}^{(i)} + \nabla p^{(i)} = \mathbf{f}^{(i)} & \text{in } Q^{(i)}, \\ \nabla \cdot \mathbf{v}^{(i)} = 0 & \text{in } Q^{(i)}, \\ \mathbf{v}^{(i)} = 0 & \text{on } \Gamma^{(i)}, \\ \mathbf{v}^{(i)} = \mathbf{v}_0^{(i)} & \text{at } t = 0. \end{cases} \tag{4.18}$$

We assume that the data are such that each of these two NSE has a sufficiently smooth solution $(\mathbf{v}^{(i)}, \widetilde{p}^{(i)})$. The similarity hypothesis is stated as follows:

Similarity hypothesis: If there exist two length–time bases b_1 and b_2 such that

$$Q_{b_1}^{(1)} = Q_{b_2}^{(2)}, \quad (\mathbf{v}_0^{(1)})_{b_1} = (\mathbf{v}_0^{(1)})_{b_2}, \quad \mathbf{f}^{(1)}_{b_1} = \mathbf{f}^{(2)}_{b_2}, \quad \nu_{b_1} = \nu_{b_2},$$

then the two flows $(\mathbf{v}^{(i)}, \widetilde{p}^{(i)})$ are similar.

It results from the foregoing that the similarity hypothesis is satisfied if and only if the dimensionless form (4.12) of the NSE has a unique solution, which can be made sure for local time solutions "à la Fujita-Kato." Global time similarity remains a mathematical open problem, although it is intensively used in many engineering and environmental applications.

4.4 Homogeneous and Isotropic Tensor Fields

We assume that Ω is a convex set and its boundary (if any) is of class C^2. Moreover, we also assume that $\overset{\circ}{\Omega} \neq \emptyset$. We denote by $\mathcal{B}_0 = (\mathbf{e}_1, \mathbf{e}_2, \mathbf{e}_3)$, the canonical basis of $\mathbb{R}^3$.

4.4.1 Homogeneity

Definition 4.4

Let $n \geq 1$ and $p \geq 1$ integers, $\mathcal{T}_p$ the set of all tensor of order p in $\mathbb{R}^3$. We say that T is an (n, p)-order tensor field over Ω if it defines a map from Ω^{n+1} into $\mathcal{T}_p$,

$$T : \begin{cases} \Omega^{n+1} & \longrightarrow & \mathcal{T}_p, \\ (\mathbf{x}_0, \ldots, \mathbf{x}_n) & \longrightarrow & T(\mathbf{x}_0, \ldots, \mathbf{x}_n) = (T_{i_1 \ldots i_p}(\mathbf{x}_0, \ldots, \mathbf{x}_n))_{1 \leq i_1, \ldots, i_p \leq 3}. \end{cases}$$

Definition 4.5

The (n, p)-order tensor field T is said to be homogeneous if and only if it is invariant with respect to the translations, which means that $\forall \mathbf{r} \in \mathbb{R}^3$ such that $\forall j = 0, \ldots, n$, $\mathbf{x}_j + \mathbf{r} \in \Omega$, then $T(\mathbf{x}_0 + \mathbf{r}, \ldots, \mathbf{x}_n + \mathbf{r}) = T(\mathbf{x}_0, \ldots, \mathbf{x}_n)$.

The following result is straightforward.

Proposition 4.1

Let T be a homogeneous (n, p)-order tensor. Then it only depends on the n vectors $\mathbf{r}_j = \mathbf{x}_j - \mathbf{x}_0$, $j = 1, \ldots, n$.

The proof of Proposition 4.1 is straightforward. According to it, we shall denote any homogeneous tensor T by $T = T(\mathbf{r}_1, \ldots, \mathbf{r}_n)$.

It is worth noting that all this makes sense when the $\mathbf{r}_j$'s belong to the set

$$\eta_n(\Omega) = \{P_n = (\mathbf{r}_1, \ldots, \mathbf{r}_n) \in \mathbb{R}^{3n}, \exists \mathbf{x}_0 \in \Omega \text{ such that } \forall j = 1, \ldots, n, \mathbf{x}_0 + \mathbf{r}_j \in \Omega\}, \tag{4.19}$$

which is not an empty set. Indeed, as $\overset{\circ}{\Omega} \neq \emptyset$, for some $\mathbf{x}_0 \in \Omega$ and $r_0 > 0$, $B(\mathbf{x}_0, r_0) \subset \overset{\circ}{\Omega} \subset \Omega$. Therefore,

$$\{P_n = (\mathbf{r}_1, \ldots, \mathbf{r}_n) \in \mathbb{R}^{3n}, \ |\mathbf{r}_j| < r_0, \ j = 1, \ldots, n\} \subset \eta_n(\Omega).$$

From now on, we shall deal with $\eta_n(\Omega)$. For example, a homogeneous $(1, 1)$-order tensor $T = T(\mathbf{r}) = (T_1(\mathbf{r}), T_2(\mathbf{r}), T_3(\mathbf{r}))$ is a vector field defined over $\eta_1(\Omega)$. Furthermore, a homogeneous $(1, 2)$-order tensor can be viewed as a matrix

$$T = T(\mathbf{r}) = \begin{pmatrix} T_{11}(\mathbf{r}) & T_{12}(\mathbf{r}) \\ T_{21}(\mathbf{r}) & T_{22}(\mathbf{r}) \end{pmatrix}.$$

Throughout the following, we shall write $P_n = (\mathbf{r}_1, \ldots, \mathbf{r}_n) \in \eta_n(\Omega)$.

4.4.2 Dual Action

Any (n, p)-order homogeneous tensor field $T = T(P_n)$ performs a dual action over $\mathbb{R}^{3p}$ as follows. Let $A_p = (\mathbf{a}_1, \mathbf{a}_2, \ldots, \mathbf{a}_p) \in \mathbb{R}^{3p} = \mathbb{R}^3 \times \cdots \times \mathbb{R}^3$. We set $\mathbf{a}_i = (a_{i1}, a_{i2}, a_{i3})$. It is therefore natural to define the dual action of $T(P_n)$ at each $P_n \in \eta_n(\Omega)$ over $\mathbb{R}^{3p}$ by the expression

$$[T(P_n), A_p] = T_{i_1 \ldots i_p}(P_n) a_{i_1 1} \ldots a_{i_p p}, \tag{4.20}$$

that we could also write

$$[T(P_n), A_p] = T(P_n) : \mathbf{a}_1 \otimes \mathbf{a}_2 \otimes \cdots \otimes \mathbf{a}_n, \tag{4.21}$$

where ":" stands for the contracted tensor product and "$\otimes$" the tensor product.

For example, when $T = T(\mathbf{r}) = (T_1(\mathbf{r}), T_2(\mathbf{r}), T_3(\mathbf{r}))$ is a $(1, 1)$-order tensor, this dual action is the standard scalar product on $\mathbb{R}^3$,

$$[T(\mathbf{r}), \mathbf{a}] = (T(P_n), \mathbf{a}) = T_i(P_n) a_i.$$

When $T(\mathbf{r}) = (T_{ij}(\mathbf{r}))_{1 \leq ij \leq 3}$ is a $(1, 2)$-order tensor, then

$$[T(\mathbf{r}), (\mathbf{a}, \mathbf{b})] = T_{ij}(\mathbf{r}) a_i b_j = (T(\mathbf{r}) \cdot \mathbf{b}, \mathbf{a}),$$

where $T(\mathbf{r}) \cdot \mathbf{b}$ denotes the product of the matrix $T(\mathbf{r})$ with the vector $\mathbf{b}$.

4.4.3 Isotropy

Usually, a tensor field T is said to be isotropic if it exerts the same action regardless of the direction, a notion that remains to be rigorously defined. To do so, let us introduce $O_3(\mathbb{R})$, the orthogonal group, characterized by

$$Q \in O_3(\mathbb{R}) \quad \text{if and only if} \quad QQ^t = I_3,$$

where I_3 denotes the identity of $\mathbb{R}^3$, and Q^t the transpose of Q. To well define the notion of isotropy, we also will need the set

$$\iota_n(\Omega) = \{P_n = (\mathbf{r}_1, \ldots, \mathbf{r}_n) \in \eta_n(\Omega), \forall Q \in O_3(\mathbb{R}), QP_n = (Q\mathbf{r}_1, \ldots, Q\mathbf{r}_n) \in \eta_n(\Omega)\}. \tag{4.22}$$

It is easily checked that $\iota_n(\Omega) \neq \emptyset$. Finally, let $A_p = (\mathbf{a}_1, \mathbf{a}_2, \ldots, \mathbf{a}_p) \in \mathbb{R}^{3p}$, $Q \in O_3(\mathbb{R})$, and let QA_p be specified by

$$QA_p = (Q\mathbf{a}_1, Q\mathbf{a}_2, \ldots, Q\mathbf{a}_p) \in \mathbb{R}^{3p}.$$

Definition 4.6

Let T be a homogeneous (n, p)-order tensor field over Ω. The tensor T is said to be isotropic if and only if

$$\forall P_n \in \iota_n(\Omega), \quad \forall A_p \in \mathbb{R}^{3p}, \quad \forall Q \in O_3(\mathbb{R}), \quad [T(QP_n), QA_p] = [T(P_n), A_p]. \tag{4.23}$$

4.4.4 (1-1)-Order Isotropic Tensor Fields

We aim in this section to characterize (1-1)-order isotropic tensors. The main result is that the unique (1-1)-order isotropic tensor, which is smooth and has a free divergence, is the null tensor. In the following, any $\mathbf{r} = (r_1, r_2, r_3) \in \mathbb{R}^3$ being given, we set $r = |\mathbf{r}| = (r_1^2 + r_2^2 + r_3^2)^{\frac{1}{2}}$.

Theorem 4.1

Let $\mathbf{w} = \mathbf{w}(\mathbf{r}) = (w_1(\mathbf{r}), w_2(\mathbf{r}), w_3(\mathbf{r}))$ be a (1-1)-order isotropic tensor field. Then there exists a function $a = a(r)$ such that

$$\forall \mathbf{r} \in \iota_1(\Omega), \quad \mathbf{w}(\mathbf{r}) = a(r)\frac{\mathbf{r}}{r}. \tag{4.24}$$

Assume in addition that $\mathbf{w}$ is differentiable over $\iota_1(\Omega) \setminus B(0, r_0)$ for some $r_0 > 0$, and is incompressible with respect to $\mathbf{r}$. Then if $\mathbf{w}$ is not identically equal to zero, there exists a constant K such that

$$\forall \mathbf{r} \in \iota_1(\Omega), \quad \mathbf{w}(\mathbf{r}) = K\frac{\mathbf{r}}{r}, \tag{4.25}$$

for all $\mathbf{r} \in \iota_1(\Omega) \setminus B(0, r_0)$.

Proof. Let $\mathbf{a} \in \mathbb{R}^3$, $Q \in O_3(\mathbb{R})$. Remember that $[\mathbf{w}(\mathbf{r}), \mathbf{a}] = (\mathbf{w}(\mathbf{r}), \mathbf{a})$ is the standard scalar product in $\mathbb{R}^3$. From the isotropy assumption and the relation $Q^t = Q^{-1}$ because $Q \in O_3(\mathbb{R})$, we get the equalities

$$[\mathbf{w}(Q\mathbf{r}), Q\mathbf{a}] = (\mathbf{w}(Q\mathbf{r}), Q\mathbf{a}) = (Q^{-1}\mathbf{w}(Q\mathbf{r}), \mathbf{a}) = (\mathbf{w}(\mathbf{r}), \mathbf{a}) = [\mathbf{w}(\mathbf{r}), \mathbf{a}].$$

As this equality holds $\forall \mathbf{a} \in \mathbb{R}^3$, the isotropy of $\mathbf{w}$ yields

$$\forall Q \in O_3(\mathbb{R}), \quad \forall \mathbf{r} \in \iota_1(\Omega), \quad Q^{-1}\mathbf{w}(Q\mathbf{r}) = \mathbf{w}(\mathbf{r}). \tag{4.26}$$

Let $\mathbf{r} = r\mathbf{e}_1$, with $r \in [0, r_0[$ such that $r\mathbf{e}_1 \in \iota_1(\Omega)$. Let $Q \in O_3(\mathbb{R})$ partitioned into four blocks of the form

$$Q = \begin{pmatrix} 1 & 0 \\ 0 & P \end{pmatrix}, \quad P \in O_2(\mathbb{R}). \tag{4.27}$$

It is easily checked that $Q(\mathbf{e}_1) = \mathbf{e}_1$. Therefore, we deduce from Equation 4.26

$$Q^{-1}\mathbf{w}(r\mathbf{e}_1) = \mathbf{w}(r\mathbf{e}_1). \tag{4.28}$$

Let us write $\mathbf{w}(r\mathbf{e}_1) = (a(r), b(r), c(r))^t$, that we insert in Equation 4.28, which yields

$$Q\mathbf{w}(r\mathbf{e}_1) = \begin{pmatrix} 1 & 0 \\ 0 & P \end{pmatrix} \begin{pmatrix} a(r) \\ b(r) \\ c(r) \end{pmatrix} = \begin{pmatrix} a(r) \\ P\begin{pmatrix} b(r) \\ c(r) \end{pmatrix} \end{pmatrix} = \begin{pmatrix} a(r) \\ b(r) \\ c(r) \end{pmatrix}, \tag{4.29}$$

leading to

$$P\begin{pmatrix} b(r) \\ c(r) \end{pmatrix} = \begin{pmatrix} b(r) \\ c(r) \end{pmatrix},$$

which holds $\forall P \in O_2(\mathbb{R})$. Therefore, $b(r) = c(r) = 0$. Consequently,

$$\mathbf{w}(r\mathbf{e}_1) = a(r)\mathbf{e}_1.$$

Let $\mathbf{r} \in \iota_1(\Omega)$. There exists $Q \in O_3(\mathbb{R})$ such that

$$\frac{\mathbf{r}}{r} = Q(\mathbf{e}_1).$$

We deduce from Equation 4.26 that

$$\mathbf{w}(\mathbf{r}) = Q\mathbf{w}(r\mathbf{e}_1) = a(r)\frac{\mathbf{r}}{r},$$

which concludes the proof of Equation 4.24.

We are left with proving Equation 4.25. Assume that a is of class C^1 over $[r_0, r_1[$ for some $0 < r_0 < r_1$, and that $\mathbf{w}$ is incompressible over $\iota_1(\Omega) \setminus B(0, r_0)$. For simplicity, we put $\alpha(r) = a(r)/r$. In this formalism, $w_i = \alpha(r)r_i$. Since $\partial_i r = r_i/r$, we deduce from the incompressibility assumption, the differential equation

$$\nabla \cdot \mathbf{w} = r\alpha'(r) + \alpha(r) = 0. \tag{4.30}$$

Integrating Equation 4.30 over $[r_0, r_1[$ yields

$$\alpha(r) = \frac{K}{r},$$

hence the result. ∎

The following corollary is straightforward.

Corollary 4.1

Let $\mathbf{w} = \mathbf{w}(\mathbf{r}) = (w_1(\mathbf{r}), w_2(\mathbf{r}), w_3(\mathbf{r}))$ be a (1-1)-order isotropic tensor field of class C^1 over ι_1, and incompressible over ι_1. Then $\mathbf{w}$ is identically equal to 0.

4.4.5 (1-2)-Order Isotropic Tensor Fields

This section is devoted to the characterization of (1-2)-order isotropic tensor fields.

Theorem 4.2

Let $B(\mathbf{r}) = (B_{ij}(\mathbf{r}))_{1 \leq i,j \leq 3}$ be a (1-2)-order isotropic tensor field. Then there exists a function $B_d = B_d(r)$ and a function $B_n = B_n(r)$ such that $\forall \mathbf{r} \in \iota_1(\Omega)$,

$$B(\mathbf{r}) = (B_d(r) - B_n(r))\frac{\mathbf{r} \otimes \mathbf{r}}{r^2} + B_n(r)I_3. \tag{4.31}$$

Proof. Let $(\mathbf{a}, \mathbf{b}) \in (\mathbb{R}^3)^2$ be any vector. We already know that

$$[B(\mathbf{r}), (\mathbf{a}, \mathbf{b})] = B_{ij}(\mathbf{r})a_i b_j = (B(\mathbf{r}) \cdot \mathbf{b}, \mathbf{a}). \qquad (4.32)$$

The isotropy assumption yields

$$\forall Q \in O_3(\mathbb{R}), \quad (B(Q\mathbf{r}) \cdot Q\mathbf{b}, Q\mathbf{a}) = (B(\mathbf{r}) \cdot \mathbf{b}, \mathbf{a}) = (Q^{-1}B(Q\mathbf{r})Q\mathbf{b}, \mathbf{a}), \qquad (4.33)$$

where we have used $Q^t = Q^{-1}$. Since this relation holds for all $(\mathbf{a}, \mathbf{b}) \in \mathbb{R}^3 \times \mathbb{R}^3$, the isotropy assumption leads to

$$\forall \mathbf{r} \in \iota_1(\Omega), \quad Q^{-1}B(Q\mathbf{r})Q = B(\mathbf{r}), \qquad (4.34)$$

or alternatively

$$\forall \mathbf{r} \in \iota_1(\Omega), \quad B(Q\mathbf{r})Q = BR(\mathbf{r}). \qquad (4.35)$$

We consider again $\mathbf{r} = r\mathbf{e}_1$ and Q as Equation 4.27, so that $Q(\mathbf{e}_1) = \mathbf{e}_1$. We write

$$B(r\mathbf{e}_1) = \begin{pmatrix} B_d(r) & \mathbf{w}^t \\ \mathbf{v} & H \end{pmatrix}, \qquad (4.36)$$

where $B_d(r)$ is a scalar function, $\mathbf{v}$ and $\mathbf{w}$ are vectors in $\mathbb{R}^2$, and H is a 2×2 matrix, all depending on r. We deduce from Equation 4.35,

$$QB(r\mathbf{e}_1) = \begin{pmatrix} B_d(r) & \mathbf{w}^t \\ P\mathbf{v} & PH \end{pmatrix} = B(r\mathbf{e}_1)Q = \begin{pmatrix} B_d(r) & \mathbf{w}^t P \\ \mathbf{v} & HP \end{pmatrix}. \qquad (4.37)$$

Therefore,

$$\forall P \in O_2(\mathbb{R}), \quad \forall (\mathbf{v}, \mathbf{w}) \in \mathbb{R}^2 \times \mathbb{R}^2, \quad P\mathbf{v} = \mathbf{v}, \quad \mathbf{w}^t P = \mathbf{w}^t, \quad PH = HP, \qquad (4.38)$$

which yields $\mathbf{v} = \mathbf{w} = 0$. Moreover, we know from standard algebra that only scalar matrices commute with all matrices in $O_2(\mathbb{R})$, which leads to $H = B_n(r)I_2$, for some scalar function $B_n(r)$. Summarizing, we have

$$B(r\mathbf{e}_1) = \begin{pmatrix} B_d(r) & 0 & 0 \\ 0 & B_n(r) & 0 \\ 0 & 0 & B_n(r) \end{pmatrix} = (B_d(r) - B_n(r))\,\mathbf{e}_1 \otimes \mathbf{e}_1 + B_n(r)I_3. \qquad (4.39)$$

Formula (4.31) results from Equation 4.34 combined with Equation 4.39. Indeed, let $\mathbf{r} \in \iota_1(\Omega)$. There exists $Q \in O_3(\mathbb{R})$ such that $\mathbf{r} = rQ(\mathbf{e}_1)$, and we notice that

$$Q(\mathbf{e}_1 \otimes \mathbf{e}_1)Q^{-1} = (Q\mathbf{e}_1) \otimes (Q\mathbf{e}_1) = \frac{\mathbf{r} \otimes \mathbf{r}}{r^2},$$

because $Q^t = Q^{-1}$.   ∎

Remark 4.1

By writing $\mathbf{r} = (r_1, r_2, r_3)$, $B = (B_{ij})_{1 \leq ij \leq 3}$, we deduce from Equation 4.39 the relations

$$\begin{aligned} B_{11}(r, 0, 0) &= B_d(r), \\ B_{22}(r, 0, 0) &= B_{33}(r, 0, 0) = B_n(r), \\ B_{ij}(r, 0, 0) &= 0, \quad \forall i \neq j. \end{aligned} \qquad (4.40)$$

4.5 Homogeneous and Isotropic Turbulence

We will define what is a homogeneous and isotropic turbulence by the end of this section.

4.5.1 Statistics and Reynolds Stress

Following Reference 5, Chapter 4, we consider a compact set $\mathbb{K}$ of $C^{2,\alpha}$ divergence free vector field on Ω. We also know from Reference 5 that there exists a probability measure μ defined on $\mathbb{K}$. Moreover, there exists $\delta T > 0$ such that for all $\mathbf{v}_0 \in \mathbb{K}$, the NSE have a unique strong solution $\mathbf{v} = \mathbf{v}(t, \mathbf{x})$ for $t \in [0, \delta T]$, such that $\mathbf{v}(0, \mathbf{x}) = \mathbf{v}_0(\mathbf{x})$. By setting $\mathbf{v} = \mathbf{v}(t, \mathbf{x}, \mathbf{v}_0)$, the velocity becomes a random variable, which allows to consider the mean field

$$\overline{\mathbf{v}}(t, \mathbf{x}) = E_\mu(\mathbf{v}) = \int_{\mathbb{K}} \mathbf{v}(t, \mathbf{x}, \mathbf{v}_0) d\mu(\mathbf{v}_0), \tag{4.41}$$

where in addition

$$\overline{\mathbf{v}}(0, \mathbf{x}) = \overline{\mathbf{v}_0}(\mathbf{x}) = \int_{\mathbb{K}} \mathbf{v}_0(\mathbf{x}) d\mu(\mathbf{v}_0). \tag{4.42}$$

We will also consider the mean pressure:

$$\overline{p}(t, \mathbf{x}) = \int_{\mathbb{K}} p(t, \mathbf{x}, \mathbf{v}_0) d\mu(\mathbf{v}_0). \tag{4.43}$$

We can decompose $(\mathbf{v}, p)$ as follows:

$$\mathbf{v} = \overline{\mathbf{v}} + \mathbf{v}', \quad p = \overline{p} + p', \tag{4.44}$$

which is known as the Reynolds decomposition. The fields $\mathbf{v}'$ and p' are the fluctuations. We deduce from standard results in analysis the following Reynolds rules:

$$\overline{\partial_t \mathbf{v}(t, \mathbf{x}, \mathbf{v}_0)} = \partial_t \overline{\mathbf{v}}(t, \mathbf{x}), \tag{4.45}$$

$$\overline{\nabla \mathbf{v}(t, \mathbf{x}, \mathbf{v}_0)} = \nabla \overline{\mathbf{v}}(t, \mathbf{x}), \tag{4.46}$$

$$\overline{\nabla p(t, \mathbf{x}, \mathbf{v}_0)} = \nabla \overline{p}(t, \mathbf{x}). \tag{4.47}$$

Moreover, by noting that $\overline{\overline{\mathbf{v}}} = \overline{\mathbf{v}}$ and $\overline{\overline{p}} = \overline{p}$, it easily checked that:

Lemma 4.2

The fluctuation's mean vanishes, i.e.,

$$\forall (t, \mathbf{x}) \in Q_m, \quad \overline{\mathbf{v}'(t, \mathbf{x}, \mathbf{v}_0)} = 0, \quad \overline{p'(t, \mathbf{x}, \mathbf{v}_0)} = 0.$$

Therefore, applying Lemma 4.2 and taking the mean of the NSE yields the PDE system

$$\begin{cases} \partial_t \overline{\mathbf{v}} + (\overline{\mathbf{v}} \cdot \nabla)\overline{\mathbf{v}} - \nu\Delta\overline{\mathbf{v}} + \nabla\overline{p} = -\nabla \cdot \sigma^{(\mathrm{R})} + \mathbf{f} & \text{in } Q_m, \\ \nabla \cdot \overline{\mathbf{v}} = 0 & \text{in } Q_m, \\ \overline{\mathbf{v}} = 0 & \text{on } \Gamma, \\ \overline{\mathbf{v}} = \overline{\mathbf{v}_0} & \text{at } t = 0, \end{cases} \tag{4.48}$$

where $Q_m = [0, \delta T] \times \Omega$, and by assuming that $\mathbf{f}$ is a given constant field. In the equation above,

$$\sigma^{(\mathrm{R})} = \overline{\mathbf{v}' \otimes \mathbf{v}'} \tag{4.49}$$

is the Reynolds stress. A thorough presentation of this process can be found in Reference 5.

Remark 4.2

Any field ψ related to the flow still satisfies the Reynolds rules (4.45) and (4.46).

Remark 4.3

The average process introduced above does not affect the dimension, in the sense that

$$\mathbb{D}(\psi) = \mathbb{D}(\overline{\psi}),$$

for each field ψ related to the flow.

4.5.2 Correlation Tensors, Main Assumptions

Following Prandtl [26], we consider a ball of fluid $\delta V \subset \Omega$, whose diameter is equal to ℓ and which is centered at $\mathbf{x}_0$, ℓ being the Prandtl mixing length at $\mathbf{x}_0$. Following Kolmogorov [21], we consider for $(t, \mathbf{x}) \in [0, \delta T] \times \delta V$,

$$\mathbf{w}(t, \mathbf{x}) = \mathbf{v}(t, \mathbf{x}) - \mathbf{v}(t, \mathbf{x}_0), \quad \mathbf{w} = (w_1, w_2, w_3). \tag{4.50}$$

The general n-order correlation tensor $T^{(n)} = T^{(n)}(t, \mathbf{x}_0, \mathbf{r}_1, \dots, \mathbf{r}_n)$ is specified component by component by the expression

$$T^{(n)}_{i_1 \dots i_n}(t, \mathbf{x}_0, \mathbf{r}_1, \dots, \mathbf{r}_n) = \overline{w_{i_1}(t, \mathbf{x}_0 + \mathbf{r}_1) \dots w_{i_n}(t, \mathbf{x}_0 + \mathbf{r}_n)},$$

We assume that in δV the turbulence is (1) stationary and (2) homogeneous, which is reflected by

1. The correlation tensors are invariant under time translation, which yields in this case that they do not depend on t.
2. The correlation tensors are invariant under spatial translations, in the sense that

$$\forall \mathbf{r}, \quad T^{(n)}(t, \mathbf{x}_0, \mathbf{r}_1 + \mathbf{r}, \dots, \mathbf{r}_n + \mathbf{r}) = T^{(n)}(t, \mathbf{x}_0, \mathbf{r}_1, \dots, \mathbf{r}_n),$$

   so far the quantities above are well-defined.

Therefore, the correlation tensor neither depends on t nor on $\mathbf{x}_0$, that might have been chosen anywhere in δV. Consequently, we can write $T^{(n)} = T^{(n)}(\mathbf{r}_1, \dots, \mathbf{r}_n)$, for $(\mathbf{r}_1, \dots, \mathbf{r}_n) \in \eta_n(\delta V)$ (see Definition 4.1).

Remark 4.4

We could also have set

$$T^{(n)}_{i_1 \dots i_n}(t_1, \dots, t_n, \mathbf{x}_0, \mathbf{r}_1, \dots, \mathbf{r}_n) = \overline{w_{i_1}(t_1, \mathbf{x}_0 + \mathbf{r}_1) \dots w_{i_n}(t_n, \mathbf{x}_0 + \mathbf{r}_n)},$$

which is the most general correlation tensor that can be considered, the study of which is beyond the scope of this text.

Finally, we assume that the turbulence is isotropic in δV, meaning that all correlation tensors are isotropic (see Definition 4.6).

Remark 4.5

Notice that $\overline{\mathbf{w}} = \overline{\mathbf{w}}(\mathbf{r})$ is a (1-1)-order divergence free tensor, which is of class $C^{1,\alpha}$ over $[0, \delta T] \times \delta V$ and isotropic. Therefore, according to Corollary 4.1, $\overline{\mathbf{w}} = 0$. The main consequence is that the mean velocity field $\overline{\mathbf{v}}$ is constant in $[0, \delta T] \times \delta V$. Consequently, the NSE (4.48) reduces to

$$\nabla \cdot \sigma^{(R)} + \nabla \overline{p} = \mathbf{f}. \tag{4.51}$$

Definition 4.7

We say that the turbulence is homogeneous and isotropic in δV if all the correlation tensors are homogeneous and isotropic (see Definition 4.6).

Remark 4.6

It is implicitly assumed in Reference 21 that the velocity is a random vector field that has a distribution function, although the probabilistic space is not specified. This crucial point is far to be obvious, which remains an open issue in our framework. However, if this claim is true, then the correlation tensor can also be considered as the momentum of this distribution function.

Throughout the rest of this text, we shall assume that the turbulence is homogeneous and isotropic. It must be stressed that this assumption may depend on the choice of the initial data set $\mathbb{K}$ and the probability measure μ, and we do not know whether there exists $\mathbb{K}$ and μ such that homogeneity and isotropy are held.

4.6 Covariance Matrix and the Law of the 2/3

4.6.1 Asymptotic Expansion of the Covariance Matrix

We focus in this subsection on the (1-2)-order energy tensor, as first introduced in References 21 and 29:

$$B(\mathbf{r}) = T^{(2)}(\mathbf{r}, \mathbf{r}),\tag{4.52}$$

which we write for simplicity $B_{ij}(\mathbf{r}) = \overline{w_i(\mathbf{r})w_j(\mathbf{r})}$. This matrix, or tensor, is the covariance matrix of the vector fields $\mathbf{v}$ at $\mathbf{x}_0$ and at any next point $\mathbf{x}$. Roughly speaking, it measures how the velocity at a given point is correlated to the velocity at a next point. In the case the velocities follow Gaussian laws, the velocities are independent random fields if and only if B is the null matrix, which may not happen in the non-Gaussian case.

The main result of this section is Theorem 4.3, which somehow proves relation (4.25) stated at the end of Reference 21.

As we have assumed that the turbulence is homogeneous and isotropic in δV, we deduce from Theorem 4.2 that there exists two scalar functions $B_d(r)$ and $B_n(r)$ such that

$$B(\mathbf{r}) = (B_d(r) - B_n(r))\frac{\mathbf{r} \otimes \mathbf{r}}{r^2} + B_n(r)\mathrm{I}_3.\tag{4.53}$$

Moreover, we also know from Remark 4.1 that

$$B_d(r) = \overline{|w_1(r,0,0)|^2},\tag{4.54}$$

$$B_n(r) = \overline{|w_2(r,0,0)|^2} = \overline{|w_3(r,0,0)|^2},\tag{4.55}$$

$$\forall i \neq j, \quad \overline{w_i(r,0,0)w_j(r,0,0)} = 0.\tag{4.56}$$

Lemma 4.3

The following holds:

$$B_d(0) = B_n(0) = 0,\tag{4.57}$$

$$B_d'(0) = B_n'(0) = 0,\tag{4.58}$$

Proof. Equality (4.57) results from Equations 4.54 and 4.55 combined with $\mathbf{w}(0,0,0) = 0$. To check Equation 4.58, we use the Reynolds rule (4.57), and we get

$$B_d'(r) = 2\overline{\frac{\partial w_1(r,0,0)}{\partial r_1}w_1(r,0,0)},$$

hence the equality $B_d'(0)$ because $\mathbf{w}(0,0,0) = 0$. The proof of $B_n'(0) = 0$ is similar. ∎

Theorem 4.3

Assume that the mean pressure gradient is constant inside δV. Then there exists a C^1 scalar function $E = E(r)$ such that $E(0) = E'(0) = 0$ and such that

$$B(\mathbf{r}) = E(r)\frac{\mathbf{r} \otimes \mathbf{r}}{r^2} - \frac{3}{2}E(r)\mathrm{I}_3 + o(r^3). \tag{4.59}$$

Proof. We denote by $\nabla_{\mathbf{x}}$ and $\nabla_{\mathbf{r}}$ the gradient with respect to $\mathbf{x}$ and $\mathbf{r}$, respectively. The proof is divided into three steps. In the first step, we expand $\nabla_{\mathbf{r}} \cdot B(\mathbf{r})$ from Equation 4.53 using Lemma 4.3. In the second step, we expand $\nabla_{\mathbf{r}} \cdot B(\mathbf{r})$ from the NSE. In the last step, we combine the two expansions and obtain the conclusion.

Step 1. We deduce from Equation 4.53 and a standard calculation the relation

$$\nabla_{\mathbf{r}} \cdot B(\mathbf{r}) = \frac{\mathbf{r}}{r^2}\left(rB_d'(r) + (B_d(r) - B_n(r))\right). \tag{4.60}$$

Identities (4.57) and (4.58) yield the Taylor expansion

$$\begin{aligned}
B_d(r) &= \alpha_d r^2 + r^3\varepsilon_d(r), \\
B_n(r) &= \alpha_n r^2 + r^3\varepsilon_n(r),
\end{aligned} \tag{4.61}$$

where α_n and α_d are constant coefficients, and ε_d and ε_n are smooth bounded functions, the derivative of which are bounded. Moreover, by Equations 4.54 and 4.55, we get $\alpha_n \geq 0$ and $\alpha_d \geq 0$. Therefore, equality (4.60) becomes

$$\nabla_{\mathbf{r}} \cdot B(\mathbf{r}) = (3\alpha_d - \alpha_n)\mathbf{r} + r^2\varepsilon(\mathbf{r}); \tag{4.62}$$

for some smooth bounded function ε that does not need to be specified.

Step 2. We start from $B_{ij}(\mathbf{r}) = \overline{w_i(\mathbf{r})w_j(\mathbf{r})}$ combined with $w_i(\mathbf{r}) = v_i(\mathbf{x}_0 + \mathbf{r}) - v_i(\mathbf{x}_0)$, which yields the decomposition

$$B_{ij}(\mathbf{r}) = M_{ij}(\mathbf{r}) - N_{ij}(\mathbf{r}) - P_{ij}(\mathbf{r}) + Q_{ij}(\mathbf{r}), \tag{4.63}$$

where

$$\begin{cases}
M_{ij}(\mathbf{r}) = \overline{v_i(\mathbf{x}_0 + \mathbf{r})v_j(\mathbf{x}_0 + \mathbf{r})}, \\
N_{ij}(\mathbf{r}) = \overline{v_i(\mathbf{x}_0)v_j(\mathbf{x}_0 + \mathbf{r})}, \\
P_{ij}(\mathbf{r}) = \overline{v_i(\mathbf{x}_0 + \mathbf{r})v_j(\mathbf{x}_0)}, \\
Q_{ij}(\mathbf{r}) = \overline{v_i(\mathbf{x}_0)v_j(\mathbf{x}_0)}.
\end{cases} \tag{4.64}$$

In the following, we set

$$\mathbf{M} = (M_{ij})_{1\leq ij\leq 3}, \quad \mathbf{N} = (N_{ij})_{1\leq ij\leq 3}, \quad \mathbf{P} = (P_{ij})_{1\leq ij\leq 3}, \quad \mathbf{Q} = (Q_{ij})_{1\leq ij\leq 3}, \tag{4.65}$$

so that decomposition (4.63) becomes

$$B = \mathbf{M} - \mathbf{N} - \mathbf{P} + \mathbf{Q}. \tag{4.66}$$

We will now use the Reynolds decomposition $\mathbf{v} = \bar{\mathbf{v}} + \bar{\mathbf{v}}'$, and study each term in the right-hand side (r.h.s) of Equation 4.63 one after each other. We recall that $\mathbf{\sigma}^{(\mathrm{R})} = (\sigma_{ij})_{1\leq ij\leq 3}$ is the Reynolds stress.

1. We easily obtain

$$M_{ij}(\mathbf{r}) = \overline{v_i(\mathbf{x}_0 + \mathbf{r}) \cdot v_j(\mathbf{x}_0 + \mathbf{r})} + \overline{v_i'(\mathbf{x}_0 + \mathbf{r})v_j(\mathbf{x}_0 + \mathbf{r})} + \overline{v_i(\mathbf{x}_0 + \mathbf{r})v_j'(\mathbf{x}_0 + \mathbf{r})} + \sigma_{ij}(\mathbf{x}_0 + \mathbf{r}). \quad (4.67)$$

As $\overline{\mathbf{v}'} = 0$, the second and third terms in the r.h.s. of this equality vanish, hence

$$M_{ij}(\mathbf{r}) = \overline{v_i(\mathbf{x}_0 + \mathbf{r}) \cdot v_j(\mathbf{x}_0 + \mathbf{r})} + \sigma_{ij}(\mathbf{x}_0 + \mathbf{r}). \quad (4.68)$$

We already know that $\overline{\mathbf{v}}$ is constant inside δV, so that

$$\nabla_{\mathbf{r}} \cdot \mathbf{M}(\mathbf{r}) = \nabla_{\mathbf{r}} \cdot \boldsymbol{\sigma}^{(\mathrm{R})}(\mathbf{x}_0 + \mathbf{r}) = \nabla_{\mathbf{x}} \cdot \boldsymbol{\sigma}^{(\mathrm{R})}(\mathbf{x}), \quad (4.69)$$

where $\mathbf{x} = \mathbf{x}_0 + \mathbf{r}$. Therefore, by the averaged NSE (4.62), we get the relation

$$\nabla_{\mathbf{r}} \cdot \mathbf{M}(\mathbf{r}) = \mathbf{f} - \nabla_{\mathbf{x}}\overline{p}(\mathbf{x}). \quad (4.70)$$

2. The incompressibility condition yields

$$\nabla_{\mathbf{r}} \cdot \mathbf{N}(\mathbf{r}) = 0. \quad (4.71)$$

3. A similar argumentation as in point (1) allows to write

$$P_{ij}(\mathbf{r}) = \overline{v_i(\mathbf{x}_0 + \mathbf{r}) \cdot v_j(\mathbf{x}_0)} + \overline{v_i'(\mathbf{x}_0 + \mathbf{r})v_j'(\mathbf{x}_0)}. \quad (4.72)$$

As $\overline{\mathbf{v}}$ is constant in δV, we get $\nabla_{\mathbf{r}} \cdot [\overline{v_i(\mathbf{x}_0 + \mathbf{r}) \cdot v_j(\mathbf{x}_0)}] = 0$. The difficult term to deal with is the second one in the r.h.s of Equation 4.72. A Taylor expansion yields

$$v_i'(\mathbf{x}_0 + \mathbf{r}) = v_i'(\mathbf{x}_0) + \nabla_{\mathbf{x}} v_i'(\mathbf{x}_0) \cdot \mathbf{r} + (H(v_i')(\mathbf{x}_0)\mathbf{r}, \mathbf{r}) + r^3 \eta_i(\mathbf{r}), \quad (4.73)$$

where η_i is some bounded function, and $H(v_i')(\mathbf{x}_0)$ the Hessian matrix of v_i' at $\mathbf{x}_0$, which means

$$(H(v_i')(\mathbf{x}_0)\mathbf{r}, \mathbf{r}) = \frac{\partial^2 v_i'}{\partial x_p \partial x_q}(\mathbf{x}_0)r_p r_q.$$

On the one hand, $\nabla_{\mathbf{r}} \cdot \overline{v_i'(\mathbf{x}_0)v_j'(\mathbf{x}_0)} = 0$. On the other hand, a rather straightforward calculation yields

$$\frac{\partial}{\partial r_j}\left(\overline{\frac{\partial v_i'}{\partial x_q}(\mathbf{x}_0)r_q v_j'(\mathbf{x}_0)}\right) = \overline{\frac{\partial v_i'}{\partial x_j}(\mathbf{x}_0)v_j'(\mathbf{x}_0)} = \frac{\partial}{\partial x_j}\left(\overline{v_i'(\mathbf{x}_0)v_j'(\mathbf{x}_0)}\right) = \frac{\partial \sigma_{ij}}{\partial x_j}(\mathbf{x}_0), \quad (4.74)$$

where we have used the incompressibility condition combined with the Reynolds rule (4.57). Finally, let $\boldsymbol{\tau}^{(\mathrm{R})}$ denote the tensor

$$\boldsymbol{\tau}^{(\mathrm{R})} = \overline{\nabla_{\mathbf{x}}\mathbf{v}' \otimes \mathbf{v}'} = \left(\overline{\frac{\partial v_i'}{\partial x_k}v_j'}\right)_{1 \leq ijk \leq 3} = (\tau_{ijk})_{1 \leq ijk \leq 3}. \quad (4.75)$$

Using the incompressibility condition again and the Reynolds rule (4.57), we get

$$\frac{\partial}{\partial r_j}\overline{(H(v_i')(\mathbf{x}_0)\mathbf{r}, \mathbf{r})v_j'(\mathbf{x}_0)} = \frac{\partial \tau_{ijk}}{\partial x_j}r_k. \quad (4.76)$$

Therefore, these calculations lead to

$$\nabla_{\mathbf{r}} \cdot \mathbf{P} = \nabla_{\mathbf{x}} \cdot \boldsymbol{\sigma}^{(\mathrm{R})}(\mathbf{x}_0) + \nabla_{\mathbf{x}} \cdot \boldsymbol{\tau}^{(\mathrm{R})}(\mathbf{x}_0) \cdot \mathbf{r} + r^2 \eta(\mathbf{r}), \quad (4.77)$$

which yields, by the averaged NSE (4.62),

$$\nabla_{\mathbf{r}} \cdot \mathbf{P} = \mathbf{f} - \nabla_{\mathbf{x}}\overline{p}(\mathbf{x}_0) + \nabla_{\mathbf{x}} \cdot \boldsymbol{\tau}^{(\mathrm{R})}(\mathbf{x}_0) \cdot \mathbf{r} + r^2 \eta(\mathbf{r}). \quad (4.78)$$

4. As $\mathbf{Q}$ does not depend on $\mathbf{r}$, we obviously have

$$\nabla_{\mathbf{r}} \cdot \mathbf{Q} = 0. \tag{4.79}$$

Step 3. By combining Equations 4.66, 4.70, 4.71, 4.78, and 4.79, we obtain the asymptotic expansion

$$\nabla_{\mathbf{r}} \cdot B = -\nabla_{\mathbf{x}} \overline{q(\mathbf{r})} - \nabla_{\mathbf{x}} \cdot \tau^{(\text{R})}(\mathbf{x}_0) \cdot \mathbf{r} - r^2 \eta(\mathbf{r}), \tag{4.80}$$

by setting

$$q(\mathbf{r}) = q(\mathbf{x}_0 + \mathbf{r}) - q(\mathbf{x}_0). \tag{4.81}$$

As we have assumed that the mean pressure gradient is constant inside δV, the first term in the r.h.s. of Equation 4.80 vanishes, so that

$$\nabla_{\mathbf{r}} \cdot B = -\nabla_{\mathbf{x}} \cdot \tau^{(\text{R})}(\mathbf{x}_0) \cdot \mathbf{r} - r^2 \eta(\mathbf{r}), \tag{4.82}$$

that we combine with Equation 4.62, which leads to

$$\left[(3\alpha_d - \alpha_n)\text{Id} + \nabla_{\mathbf{x}} \cdot \tau^{(\text{R})}(\mathbf{x}_0) \right] \cdot \mathbf{r} + r^2 \gamma(\mathbf{r}) = 0, \tag{4.83}$$

for some bounded function γ. Since this equality holds regardless of $\mathbf{r}$, whose norm is small, we deduce

$$(3\alpha_d - \alpha_n)\text{Id} + \nabla_{\mathbf{x}} \cdot \tau^{(\text{R})}(\mathbf{x}_0) = 0. \tag{4.84}$$

We observe that the diagonal coefficients of the matrix $\nabla_{\mathbf{x}} \cdot \tau^{(\text{R})}(\mathbf{x}_0)$ satisfy

$$\frac{\partial \tau_{iji}}{\partial x_j} = \frac{\partial}{\partial x_j} \left(\overline{\frac{\partial v_i'}{\partial x_i} v_j'} \right) = 0,$$

because of the incompressibility condition. Consequently, the following holds:

$$3\alpha_d = \alpha_n, \tag{4.85}$$

$$\nabla_{\mathbf{x}} \cdot \tau^{(\text{R})}(\mathbf{x}_0) = 0, \tag{4.86}$$

in particular

$$B_n(r) = 3B_d(r) + o(r^3), \tag{4.87}$$

and hence formula (4.59) by setting $E(r) = -2B_d(r)$.   ∎

Remark 4.7

The assumption that the gradient of the mean pressure is constant is not too restrictive. Experiments suggest that this is indeed verified by flows in pipes or in boundary layers. However, the general case remains an open problem.

Remark 4.8

In Reference 21 the general two-order correlation tensor is considered as a bilinear form, which directly provides the formula

$$B(\mathbf{r}) = E(r)\frac{\mathbf{r} \otimes \mathbf{r}}{r^2} - \frac{3}{2}E(r)I_3,$$

without using the NSE. We do not know how to prove this bilinearity property.

4.6.2 Derivation of the Law of the 2/3

We prove in this section the law of the 2/3, specified by Equation 4.102 below, under Assumption 4.1. Section 4.6.1 states that,

$$B(\mathbf{r}) \sim E(r)\frac{\mathbf{r} \otimes \mathbf{r}}{r^2} - \frac{3}{2}E(r)I_3, \tag{4.88}$$

and near 0

$$E(r) \approx r^2. \tag{4.89}$$

The function $E(r)$ is obviously defined in $[0, \ell]$. The question is the behavior of $E(r)$ when r differs from 0. According to Kolmogorov [21], we assume that E is entirely driven in δV by the kinematic viscosity ν and the mean dissipation at $\mathbf{x}_0$, specified by

$$\mathcal{E} = \overline{2\nu|D\mathbf{v}(\mathbf{x}_0)|^2}.$$

It is easily checked that ν and $\mathcal{E}$ are dimensionally independent. We deduce from these quantities the length–time basis $b_0 = (\lambda_0, \tau_0)$, where

$$\lambda_0 = \nu^{\frac{3}{4}}\mathcal{E}^{-\frac{1}{4}}, \quad \tau_0 = \nu^{\frac{1}{2}}\mathcal{E}^{-\frac{1}{2}}. \tag{4.90}$$

The length scale λ_0 is known as the "Kolmogorov scale." Following Definition 4.2 and by the table in section 4.3.1, we get

$$\forall r' \in [0, \frac{\ell}{\lambda_0}[, \quad E(\lambda_0 r') = (\nu\mathcal{E})^{\frac{1}{2}}E_{b_0}(r'). \tag{4.91}$$

We first assume that

$$\lambda_0 << \ell, \tag{4.92}$$

so that we can consider this as a first approximation for the simplicity,

$$\left[0, \frac{\ell}{\lambda_0}\right[\sim \mathbb{R}_+.$$

The main Kolmogorov assumption can be translated as follows: there is a range $[r_1, r_2]$, satisfying

$$\lambda_0 << r_1 << r_2 << \ell, \tag{4.93}$$

and such that in the range $[r_1, r_2]$, E is uniquely determined by $\mathcal{E}$. This last sentence can lead to confusion, and the concept needs to be made more specific and rigorous by a definition similar to Definition 4.3, based on a similarity statement. Therefore, we shall assume the following:

Assumption 4.1

For all length-times bases $b_1 = (\lambda_1, \tau_1)$ and $b_2 = (\lambda_2, \tau_2)$,

$$\mathcal{E}_{b_1} = \mathcal{E}_{b_2} \quad \Rightarrow \quad \forall r' \in \left[\frac{r_1}{\lambda_1}, \frac{r_2}{\lambda_1}\right] \cap \left[\frac{r_1}{\lambda_2}, \frac{r_2}{\lambda_2}\right], \quad E_{b_1}(r') = E_{b_2}(r'). \tag{4.94}$$

In the following, we set

$$r'_{1,0} = \frac{r_1}{\lambda_0}, \quad r'_{2,0} = \frac{r_2}{\lambda_0}.$$

Proposition 4.2

If Assumption 4.1 holds, then there exists a constant C such that

$$\forall r' \in [r'_{1,0}, r'_{2,0}], \quad E_{b_0}(r') = C(r')^{\frac{2}{3}}. \tag{4.95}$$

Proof. The proof is based on the determination of a functional equation satisfied by E_{b_0}. Let $\alpha > 0$, and $b^{(\alpha)} = (\alpha^3 \lambda_0, \alpha^2 \tau_0)$. This choice is motivated by the equality

$$\forall \alpha > 0, \quad \mathcal{E}_{b^{(\alpha)}} = \mathcal{E}_{b_0}, \tag{4.96}$$

which is easily checked. Therefore,

$$\forall r' \in I_\alpha = [r'_{1,0}, r'_{2,0}] \cap \left[\frac{r'_{1,0}}{\alpha^3}, \frac{r'_{2,0}}{\alpha^3} \right], \quad E_{b^{(\alpha)}}(r') = E_{b_0}(r'), \tag{4.97}$$

which is consistent so far

$$\frac{r_1}{r_2} < \alpha < \frac{r_2}{r_1}.$$

According to Equation 4.93, statement (4.97) makes sense for a large range of $\alpha \in \mathbb{R}_+^\star$. By Definition 4.2, we get

$$E(\lambda_0 \alpha^3 r') = \alpha^2 (\nu \mathcal{E})^{\frac{1}{2}} E_{b^{(\alpha)}}(r'), \tag{4.98}$$

hence, $\forall r' \in I_\alpha$,

$$E_{b^{(\alpha)}}(r') = \frac{1}{\alpha^2 (\nu \mathcal{E})^{\frac{1}{2}}} E(\lambda_0 \alpha^3 r') = E_{b_0}(r'), \tag{4.99}$$

where we have used Equation 4.97. We combine Equation 4.91 with Equation 4.99, which yields

$$E_{b_0}(r') = \frac{1}{\alpha^2} E_{b_0}(\alpha^3 r'). \tag{4.100}$$

We deduce from Equation 4.100 by standard calculations

$$\forall r' \in [r'_{1,0}, r'_{2,0}], \quad E_{b_0}(r') = \left(\frac{E_{b_0}(r'_1)}{(r'_1)^{\frac{2}{3}}} \right) (r')^{\frac{2}{3}}, \tag{4.101}$$

which concludes the proof.   ∎

In conclusion, combining Equations 4.90, 4.91, and 4.101 by writing $r' = r/\lambda_0$, we get

$$\forall r \in [r_1, r_2], \quad E(r) = C(\mathcal{E} r)^{\frac{2}{3}}, \tag{4.102}$$

which is indeed the law of the $2/3$ as initially found in References 21 and 29.

Acknowledgments

The author thanks Benoit Pinier for the careful proofreading of this text. He also thanks Antoine Chambert-Loir for interesting discussions about the algebraic part of this work.

References

1. J. Boussinesq. Theorie de l'écoulement tourbillant. *Mém. prés par div. savants á la Acad. Sci.*, 23:46–50, 1877.
2. L. Caffarelli, R. Kohn, and L. Nirenberg. Partial regularity of suitable weak solutions of the Navier-Stokes equations. *Comm. Pure Appl. Math.*, 35(6):771–831, 1982.
3. M. Cannone. Harmonic analysis tools for solving the incompressible Navier-Stokes equations. In *Handbook of Mathematical Fluid Dynamics*. North-Holland, Amsterdam, Vol. 3, pages 161–244, 2004.
4. M. Cannone and Y. Meyer. Littlewood-Paley decomposition and Navier-Stokes equations. *Methods Appl. Anal.*, 2(3):307–319, 1995.
5. T. Chacòn-Rebollo and R. Lewandowski. *Mathematical and Numerical Foundations of Turbulence Models and Applications.* Birkäuser, Springer, New York, 2014.

6. J-Y. Chemin. About weak-strong uniqueness for the 3D incompressible Navier-Stokes system. *Comm. Pure Appl. Math.*, 64(12):1587–1598, 2011.

7. H.-J. Choe and J.-L. Lewis. On the singular set in the Navier-Stokes equations. *J. Funct. Anal.*, 175(2):348–369, 2000.

8. P. Constantin, C.R. Doering, and E.S. Titi. Rigorous estimates of small scales in turbulent flows. *J. Math. Phys.*, 37(12):6152–6156, 1996.

9. P. Constantin and C. Foias. *Navier-Stokes Equations*. Chicago Lectures in Mathematics. University of Chicago Press, Chicago, IL, 1988.

10. C.R. Doering and C. Foias. Energy dissipation in body-forced turbulence. *J. Fluid Mech.*, 467:289–306, 2002.

11. C.R. Doering and J.D. Gibbon. *Applied Analysis of the Navier-Stokes Equations*. Cambridge Texts in Applied Mathematics. Cambridge University Press, Cambridge, 1995.

12. C.R. Doering and E.S. Titi. Exponential decay rate of the power spectrum for solutions of the Navier-Stokes equations. *Phys. Fluids*, 7(6):1384–1390, 1995.

13. C.L. Fefferman. Existence and smoothness of the Navier-Stokes equation. In *The Millennium Prize Problems*. Clay Math. Inst., Cambridge, MA, pages 57–67, 2006.

14. E. Feireisl. *Dynamics of Viscous Incompressible Fluids*. Oxford University Press, 2004.

15. C. Foias, M.S. Jolly, O.P. Manley, R. Rosa, and R. Temam. Kolmogorov theory via finite-time averages. *Phys. D*, 212(3–4):245–270, 2005.

16. C. Foias, O. Manley, R. Rosa, and R. Temam. *Navier-Stokes Equations and Turbulence*, volume 83 of *Encyclopedia of Mathematics and Its Applications*. Cambridge University Press, Cambridge, 2001.

17. H. Fujita and T. Kato. On the Navier-Stokes initial value problem. *Arch. Ration. Mech. Anal.*, 16:269–315, 1964.

18. J.D. Gibbon and C.R. Doering. Intermittency and regularity issues in 3D Navier-Stokes turbulence. *Arch. Ration. Mech. Anal.*, 177(1):115–150, 2005.

19. E. Hopf. Über die Anfangswertaufgabe für die hydrodynamischen Grundgleichungen. *Math. Nachr.*, 4:213–231, 1951.

20. H. Koch and D. Tataru. Well-posedness for the Navier-Stokes equations. *Adv. Math.*, 157(1):22–35, 2001.

21. A.N. Kolmogorov. The local structure of turbulence in incompressible viscous fluids for very large Reynolds number. *Dokl. Akad. Nauk SSR*, 30:9–13, 1941.

22. J. Leray. Sur le mouvement d'un liquide visqueux emplissant l'espace. *Acta Math.*, 63(1):193–248, 1934.

23. F. Lin. A new proof of the Caffarelli-Kohn-Nirenberg theorem. *Comm. Pure Appl. Math.*, 51(3):241–257, 1998.

24. J.-L. Lions. *Quelques méthodes de résolution des problèmes aux limites non linéaires*. Dunod; Gauthier-Villars, Paris, 1969.

25. P-L. Lions. *Mathematical Topics in Fluid Mechanics*. Vol. 1, volume 3 of *Oxford Lecture Series in Mathematics and Its Applications*. The Clarendon Press, Oxford University Press, New York, 1996. Incompressible models, Oxford Science Publications.

26. L. Prandtl. über die ausgebildeten turbulenz. *Zeitschrift für angewandte Mathematik und Mechanik*, 5:136–139, 1925.

27. O. Reynolds. An experimental investigation of the circumstances which determine whether the motion of water shall be direct or sinuous, and of the law of resistance in parallel channels. *Philos. Trans. R. Soc.*, 174:935–982, 1883.

28. V. Scheffer. Hausdorff measure and the Navier-Stokes equations. *Comm. Math. Phys.*, 55(2):97–112, 1977.

29. G.I. Taylor. Statistical theory of turbulence, Part I–IV. *Proc. Roy. Soc. A*, 151:421–478, 1935.

30. R. Temam. *Navier-Stokes Equations*. AMS Chelsea Publishing, Providence, RI, 2001. Theory and numerical analysis, Reprint of the 1984 edition.

5

Nonlinear Dynamics of Two-Dimensional Chaotic Maps and Fractal Sets for Snow Crystals

Nguyen H. Tuan Anh,
Dang Van Liet, and
Shunji Kawamoto

5.1 Introduction

For the study of nonlinear phenomena, it is known that the simplest nonlinear difference equations have arisen in the field of biological, economic, and social sciences, and possess a rich spectrum of dynamical behavior as chaos in many respects [1–3]. A population growth is modeled as a special example, and has been afforded by the nonlinear difference equation called the logistic map. Particularly, for one-dimensional (1D) chaotic maps, a bifurcation diagram of the two-parameter quadratic family has been observed [4], and the self-adjusting logistic map with a slowly changing parameter in time has been considered [5]. Moreover, the logistic map with a periodically modulated parameter has been presented [6]. In the meantime, various chaotic sequences have been proposed for the generation of pseudorandom numbers and for the application to cryptosystems [7–9].

At the same time, a family of shapes and many other irregular patterns in nature called fractals has been discussed for geometric representation as an irregular set consisting of parts similar to the whole [10–12]. However, since the Mandelbrot map is defined as a complex map, it has been pointed out that the physics of fractals is a research subject to be born [13]. In addition, chaotic and fractal dynamics have been expanded to experimental observations with the mathematical models [14], and fractal compression has been presented to compress images using fractals [15]. Recently, a construction method of three-dimensional (3D) chaotic maps has been proposed and the fractal sets with physical analog have been shown numerically [16].

In this chapter, we derive a generalized logistic map from a generalized logistic function for population growth and discuss the dynamical behavior of the map in Section 5.2. Then, by introducing the 1D exact chaos solution, we construct two-dimensional (2D) and 3D chaotic maps including the Mandelbrot map and the Julia map in terms of real variables, and 2D maps related to the Henon map, the Lorenz map, and the Helleman map are obtained in Section 5.3. Finally, a 2D chaotic map and the fractal set are considered for the physical analog of snow crystal, and the nonlinear dynamics on the fractal set are discussed by iterating the 2D map in Section 5.4. The last section concludes the chapter.

5.2 Generalized Logistic Map

First, we introduce a generalized logistic function $P(t)$:

$$P(t) = \frac{a}{b + e^{-ct}} + d \tag{5.1}$$

with the time $t > 0$, real constants $\{a \neq 0, b > 0, c \neq 0, d \geq 0\}$, and a constant population growth term d. By differentiating 5.1, we have the first-order differential equation

$$\frac{dP}{dt} = \left(\frac{bc}{a}\right) P \left[\left(\frac{a + 2bd}{b}\right) - P\right] - \left(\frac{cd}{a}\right)(a + bd) \tag{5.2}$$

and by a variable transformation,

$$X(t) \equiv P(t) / \frac{(a + 2bd)}{b} \tag{5.3}$$

and by the difference method

$$\frac{dX}{dt} \approx \frac{X_{n+1} - X_n}{\Delta t}, n = 0, 1, 2, \ldots \tag{5.4}$$

with $X_n \equiv X(t)$ and the time step $\Delta t > 0$, we find

$$\frac{X_{n+1} - X_n}{\Delta t} \equiv \left(\frac{bc}{d}\right)\left(\frac{a + 2bd}{b}\right) X_n(1 - X_n) - \left(\frac{bcd}{a}\right)\left(\frac{a + bd}{a + 2bd}\right). \tag{5.5}$$

Then, by variable transformation,

$$x_n \equiv (B_0 / A) X_n \tag{5.6}$$

with $A \equiv 1 + \Delta t(c/a)(a + 2bd)$ and $B_0 \equiv \Delta t(c/a)(a + 2bd)$, we arrive at a 1D generalized logistic map [16]:

$$x_{n+1} = Ax_n(1 - x_n) + B. \tag{5.7}$$

Here

$$A \equiv 1 + \frac{c(a + 2bd)}{a}(\Delta t), \tag{5.8}$$

$$B \equiv \frac{-bc^2 d(a + bd)(\Delta t)^2}{a[a + c(a + 2bd)(\Delta t)]}, \tag{5.9}$$

which gives a discrete nonlinear system. If $d = 0$ in Equations 5.8 and 5.9, then Equation 5.7 yields the logistic map $x_{n+1} = Ax_n(1 - x_n)$, and the map at $A = 4.0$ has an exact chaos solution $x_n = \sin^2(C2^n)$ with a real constant $C \neq \pm m\pi/2^l$ and finite positive integers $\{l, m\}$. We call the map $x_{n+1} = 4x_n(1 - x_n)$ the kernel chaotic map of Equation 5.7. Therefore, the constant A in Equation 5.7 denotes a coefficient of the nonlinear term and B corresponds to the constant population growth term d of Equation 5.1.

It is interesting to note that the logistic function has been introduced for the population growth of city in a discussion of the discrete numerical data [17], and has found an application to fields such as biology, ecology, biomathematics, economics, probability, and statistics. Therefore, the function (5.1) has originally a discrete property for population growth, that is, a discrete nonlinear dynamics.

5.3 2D and 3D Chaotic Maps

We have the following three cases to find 2D and 3D chaotic maps from a 1D exact chaos solution:

Case 1

From an exact chaos solution

$$x_n = \sin^2(C2^n) \tag{5.10}$$

with $C \neq \pm m\pi/2^l$ and finite positive integers $\{l, m\}$ to the logistic map $x_{n+1} = 4x_n(1 - x_n)$, we have, by introducing a real parameter $\alpha \neq 0$:

$$
\begin{aligned}
x_{n+1} &= 4\cos^2(C2^n)\sin^2(C2^n) \\
&= 4(1-\alpha)\cos^2(C2^n)\sin^2(C2^n) + 4\alpha\cos^2(C2^n)\sin^2(C2^n) \\
&= 4(1-\alpha)x_n(1-x_n) + 4\alpha(x_n - \sin^4(C2^n)) \\
&\equiv 4x_n - 4(1-\alpha)x_n^2 - 4\alpha y_n
\end{aligned} \tag{5.11}
$$

with

$$y_n \equiv \sin^4(C2^n). \tag{5.12}$$

Therefore, we obtain a 2D kernel chaotic map from Equations 5.10 through 5.12:

$$x_{n+1} = 4x_n - 4(1-\alpha)x_n^2 - 4\alpha y_n, \tag{5.13}$$

$$y_{n+1} = 16(1-x_n)^2 y_n \tag{5.14}$$

and a generalized 2D chaotic map, according to the construction method [16]:

$$x_{n+1} = a_1(x_n - (1-\alpha)x_n^2 - \alpha y_n) + b_1, \tag{5.15}$$

$$y_{n+1} = a_2(1-x_n)^2 y_n + b_2 \tag{5.16}$$

with real coefficients and constants $\{a_1, a_2, b_1, b_2\}$. Here, the first equation (Equation 5.15) has the same form as the Helleman map:

$$x_{n+1} = 2ax_n + 2x_n^2 - y_n, \tag{5.17}$$

$$y_{n+1} = x_n \tag{5.18}$$

with a real coefficient a, which has been obtained from the motion of a proton in a storage ring with periodic impulses [18].

Moreover, from the exact chaos solution (5.10), we find the following 3D map:

$$
\begin{aligned}
x_{n+1} &= (2\cos(C2^n)\sin(C2^n))^2 \\
&\equiv (2y_n z_n)^2
\end{aligned} \tag{5.19}
$$

with

$$y_n \equiv \cos(C2^n), \tag{5.20}$$

$$z_n \equiv \sin(C2^n) \tag{5.21}$$

and have a 3D kernel chaotic map from Equations 5.19 through 5.21:

$$x_{n+1} = 4x_n y_n^2, \tag{5.22}$$

$$y_{n+1} = y_n^2 - x_n, \tag{5.23}$$

$$z_{n+1} = 2y_n z_n. \tag{5.24}$$

Therefore, we obtain a generalized 3D chaotic map:

$$x_{n+1} = a_1 x_n y_n^2 + b_1, \tag{5.25}$$

$$y_{n+1} = a_2 \left(y_n^2 - x_n \right) + b_2, \tag{5.26}$$

$$z_{n+1} = a_3 y_n z_n + b_3, \tag{5.27}$$

which has been discussed in Reference 16, where $\{a_1, a_2, a_3, b_1, b_2, b_3\}$ are real coefficients and constants.

Case 2

For an exact chaos solution

$$x_n = \cos(C2^n), \tag{5.28}$$

we have the following derivation by introducing a real parameter $\alpha \neq 0$:

$$
\begin{aligned}
x_{n+1} &= \cos^2(C2^n) - \sin^2(C2^n) \\
&= \cos^2(C2^n) - (1-\alpha)\sin^2(C2^n) - \alpha\sin^2(C2^n) \\
&\equiv -\alpha + (1+\alpha)\cos^2(C2^n) - (1-\alpha)y_n
\end{aligned} \tag{5.29}
$$

with

$$y_n \equiv \sin^2(C2^n). \tag{5.30}$$

Then, from Equations 5.28 through 5.30, we obtain the kernel chaotic map as

$$x_{n+1} = -\alpha + (1+\alpha)x_n^2 - (1-\alpha)y_n, \tag{5.31}$$

$$y_{n+1} = 4x_n^2 y_n \tag{5.32}$$

and a generalized 2D chaotic map:

$$x_{n+1} = a_1(-\alpha + (1+\alpha)x_n^2 - (1-\alpha)y_n) + b_1, \tag{5.33}$$

$$y_{n+1} = a_2 x_n^2 y_n + b_2 \tag{5.34}$$

with real coefficients and constants $\{a_1, a_2, b_1, b_2\}$, where the first equation (Equation 5.33) has the same form as the Henon map [19]:

$$x_{n+1} = 1 - ax_n^2 + y_n, \tag{5.35}$$

$$y_{n+1} = bx_n \tag{5.36}$$

with real coefficients $\{a, b\}$, which has been introduced as a simplified model of the Poincaré section of the Lorenz model, and is known as one of the most studied maps for dynamical systems.

Here, it is interesting to note that if we define $y_n \equiv \sin(C2^n)$ with $\alpha = 0$ in Equation 5.29, we find a generalized 2D chaotic map:

$$x_{n+1} = a_1 \left(x_n^2 - y_n^2 \right) + b_1, \tag{5.37}$$

$$y_{n+1} = a_2 x_n y_n + b_2, \tag{5.38}$$

where the case of $(a_1, a_2, b_1, b_2) = (1.0, 2.0, x_0, y_0)$ or $(1.0, 2.0, k_1, k_2)$ with initial values $\{x_0, y_0\}$ and real parameters $\{k_1, k_2\}$ corresponds to the Mandelbrot map or the Julia map in terms of real variables, respectively [16].

Case 3

Similarly, for another exact chaos solution,

$$x_n = \sin(C2^n), \tag{5.39}$$

we have the following derivation:

$$x_{n+1} = 2\cos(C2^n)\sin(C2^n),$$
$$x_{n+2} = 2(\cos^2(C2^n) - \sin^2(C2^n))x_{n+1} \tag{5.40}$$
$$= 2x_{n+1}\left(1 - 2x_n^2\right),$$
$$x_{n+1} \equiv 2x_n(1 - 2y_n) \tag{5.41}$$

with

$$y_n \equiv x_{n-1}^2. \tag{5.42}$$

Then, we find a 2D kernel chaotic map from Equations 5.41 and 5.42:

$$x_{n+1} = 2x_n(1 - 2y_n), \tag{5.43}$$
$$y_{n+1} = x_n^2 \tag{5.44}$$

and a generalized 2D chaotic map

$$x_{n+1} = a_1(x_n - 2x_n y_n) + b_1, \tag{5.45}$$
$$y_{n+1} = a_2 x_n^2 + b_2 \tag{5.46}$$

with real coefficients and constants $\{a_1, a_2, b_1, b_2\}$. It is interesting to note that the first equation (Equation 5.45) has the same form as the 2D Lorenz map [20,21]:

$$x_{n+1} = (1 + ab)x_n - bx_n y_n, \tag{5.47}$$
$$y_{n+1} = bx_n^2 + (1 - b)y_n \tag{5.48}$$

with real coefficients $\{a, b\}$, which is known to have chaotic dynamics.

Thus, it is found that the 2D chaotic maps derived from 1D exact chaos solutions (5.10), (5.28), and (5.39) include the Mandelbrot map and the Julia map, and are related to the Helleman map, the Henon map, and the 2D Lorenz map, which give chaotic behaviors and nonlinear dynamics.

5.4 Nonlinear Dynamics for Snow Crystal

According to the approach presented in Section 5.3, we introduce a 1D exact chaos solution

$$x_n = \cos(C6^n) \tag{5.49}$$

to the kernel chaotic map,

$$x_{n+1} \equiv f(x_n, y_n) \tag{5.50}$$

with

$$y_n \equiv \sin(C6^n) \tag{5.51}$$
$$x_n^2 + y_n^2 = 1, \tag{5.52}$$

and find a generalized 2D chaotic map x_n

$$x_{n+1} = a_1\left(x_n^6 - 15x_n^4 y_n^2 + 15x_n^2 y_n^4 - y_n^6\right) + k_1, \tag{5.53}$$
$$y_{n+1} = a_2\left(6x_n^5 y_n - 20x_n^3 y_n^3 + 6x_n y_n^5\right) + k_2 \tag{5.54}$$

with real coefficients and constants $\{a_1, a_2, k_1, k_2\}$.

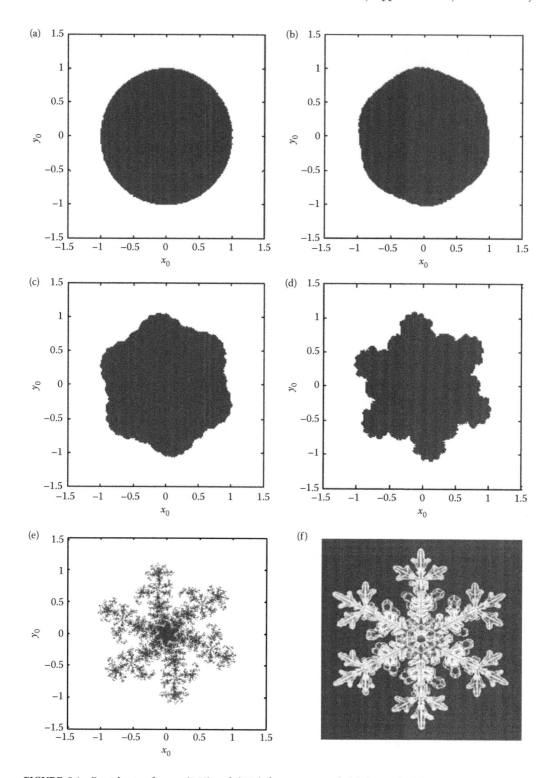

FIGURE 5.1 Fractal sets of maps (5.53) and (5.54) for snow crystal: (a) $(a_1, a_2, k_1, k_2) = (1.0, 1.0, 0.0, 0.0)$; (b) $(a_1, a_2, k_1, k_2) = (1.0, 1.0, 0.1, 0.1)$; (c) $(a_1, a_2, k_1, k_2) = (1.0, 1.0, 0.3, 0.3)$; (d) $(a_1, a_2, k_1, k_2) = (1.0, 1.0, 0.5, 0.5)$; (e) $(a_1, a_2, k_1, k_2) = (1.0, 1.0, 0.58158, 0.58158)$; (f) natural snow crystal. (From http://www.snowcrystals.net/gallary1sm/index.htm)

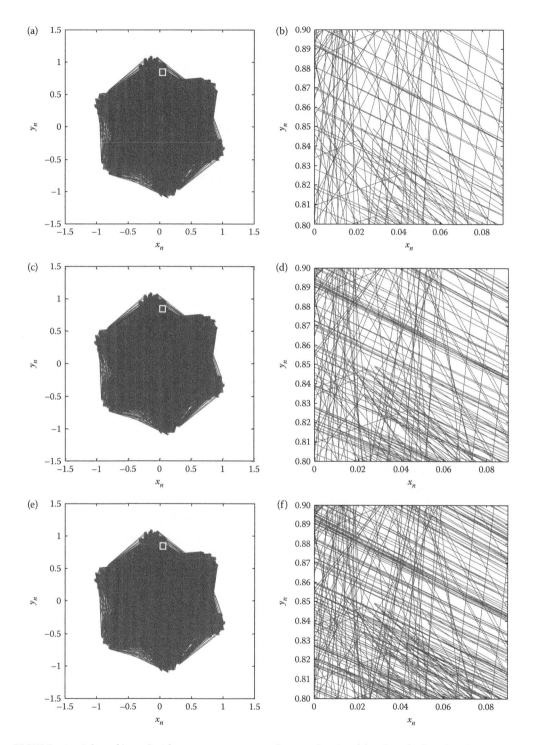

FIGURE 5.2   Orbits of (x_n, y_n) with $n = 0, 1, 2, 3, 4, 5$ given by maps (5.53) and (5.54) on the fractal set (Figure 5.1e). (a) Orbits for $n = 0, 1$ and (b) of a small framed region. (c) Orbits for $n = 0, 1, 2, 3$ and (d) of a small framed region. (e) Orbits for $n = 0, 1, 2, 3, 4, 5$ and (f) of a small framed region.

Then, the fractal set is defined by

$$M = \left\{ x_0, y_0 \in \mathbf{R} \;\middle|\; \lim_{n \to \infty} x_n, y_n < \infty \right\}, \tag{5.55}$$

where $\{x_0, y_0\}$ are initial values, and the fractal sets are illustrated in Figure 5.1, which depend on the constant parameters $\{k_1, k_2\}$. The fractal set (a) of Figure 5.1 gives a circle under condition (5.52), and (b–e) show how the fractal set (a) grows as a physical analog toward a natural snow crystal (f), which is a six-cornered dendrite-type depending on the temperature and the saturation in environment [22,23]. Here, for calculating the fractal set M, we introduce an iteration number $n = 300$ to obtain each element of M under the convergence condition $x_n^2 + y_n^2 < 4.0$ for maps (5.53) and (5.54), and the numerical calculation software MATLAB®.

Each fractal set illustrated in Figure 5.1 is a set of initial value point (x_0, y_0) defined by Equation 5.55 under the condition $x_n^2 + y_n^2 < 4.0$. For nonlinear dynamics of the 2D chaotic maps (5.53) and (5.54), the orbits of (x_n, y_n) governed by the map are calculated and shown on the fractal set of initial values in Figure 5.2, where (a) $n = 0, 1$ illustrates orbits from each initial point (x_0, y_0) to the (x_1, y_1); (b) $n = 0, 1, 2, 3$ from (x_0, y_0) to (x_3, y_3); and (c) $n = 0, 1, \ldots, 5$ from (x_0, y_0) to (x_5, y_5), for all the initial points. It is found that the orbits are complex and seem like colliding of water molecules. Here, the orbits show that we have (x_1, y_1) as the case of $n = 0$ from Equation 5.53: $x_1 = f(x_0, y_0) + k_1$ and Equation 5.54: $y_1 = g(x_0, y_0) + k_2$, (x_2, y_2) from $x_2 = f(x_1, y_1) + k_1$ and $y_1 = g(x_0, y_0) + k_2, \ldots$, and (x_{300}, y_{300}) from $x_{300} = f(x_{299}, y_{299}) + k_1$ and $y_{300} = g(x_{299}, y_{299}) + k_2$, under the condition $x_n^2 + y_n^2 < 4.0$. Then, we obtain one element of the fractal set and find that $\{(x_1, y_1), (x_2, y_2), \ldots, (x_{300}, y_{300})\}$ are other initial value points satisfying the condition $x_n^2 + y_n^2 < 4.0$ for the fractal set.

Thus, if the orbits shown in Figure 5.2 correspond to the dynamics of water molecules colliding with other ones in natural snow crystal, maps (5.53) and (5.54) may present the discrete nonlinear dynamics.

5.5 Conclusions

We have derived first the 1D generalized logistic map and have discussed that the map has originally a discrete numerical property for the population growth of a city. Then, from the 1D chaotic solution, 2D maps related to the Henon map, the 2D Lorenz map, and the Helleman map, which have chaotic dynamics, have been derived. Furthermore, the 2D chaotic maps (5.53) and (5.54) gives the fractal set for snow crystal and orbits of the map on the fractal set have been calculated numerically. As a result, it is found that the 2D chaotic maps derived from 1D exact chaotic solutions have discrete nonlinear dynamics and may express physical analogs with chaotic properties.

Acknowledgments

The authors thank Professor C. V. Tao for his encouragement and N. A. Hao of the University of Science, Ho Chi Minh City, for helping with numerical calculations.

References

1. R. M. May. Biological populations with non-overlapping generations: Stable points, stable cycles, and chaos. *Science* 15: 645–646, 1974.
2. T. Y. Li and J. A. Yorke. Period three implies chaos. *Am. Math. Mon.* 82: 985–992, 1975.
3. R. M. May. Simple mathematical models with very complicated dynamics. *Nature* 261: 459–467, 1976.
4. E. Barreto, B. R. Hunt, C. Grebogi, and J. A. Yorke. From high dimensional chaos to stable periodic orbits: The structure of parameter space. *Phys. Rev. Lett.* 78:4561–4564, 1997.

5. P. Melby, J. Kaidel, N. Weber, and A. Hubler. Adaptation to the edge of chaos in the self-adjusting logistic map. *Phys. Rev. Lett.* 84: 5991–5993, 2000.

6. T. U. Singh, A. Nandi, and R. Ramaswamy. Coexisting attractors in periodically modulated logistic maps. *Phys. Rev.* E77: 066217, 2008.

7. L. M. Pecora and T. L. Carroll. Synchronization in chaos systems. *Phys. Rev. Lett.* 64: 821–824, 1990.

8. G. Perez and H. A. Cerdeira. Extracting message masked by chaos. *Phys. Rev. Lett.* 74: 1970–1973, 1995.

9. G. D. V. Wiggeren and R. Roy. Optical communication with chaotic waveforms. *Phys. Rev. Lett.* 81: 3547–3550, 1998.

10. B. B. Mandelbrot. *The Fractal Geometry of Nature.* Freeman, San Francisco, 1982.

11. H. Peitgen and P. Richter. *The Beauty of Fractals.* Springer, New York, 1986.

12. H. Peitgen, H. Jurgens, and D. Saupe. *Chaos and Fractals – New Frontiers of Science.* Springer, New York, 1992.

13. L. P. Kadanoff. Where's the physics? *Phys. Today* 39: 6–7, 1986.

14. F. C. Moon. *Chaotic and Fractal Dynamics.* Wiley, New York, 1992.

15. M. F. Barnsley. *Fractals Everywhere.* Academic Press, New York, 1993.

16. N. T. Nhien, D. V. Liet, and S. Kawamoto. Three-dimensional chaos maps and fractal sets with physical analogue. *Proc. of CHAOS 2014*: 337–347, Lisbon, Portugal, June 7–10, 2014.

17. P. F. Verhulst. Mathematical researches into the law of population growth increase. *Nouveaux Memoires de l'Academie Royale des Sciences et Belles-Lettres de Bruxelles* 18: 1–42, 1845.

18. R. H. G. Helleman. Self-generated chaotic behavior in nonlinear mechanics. *Fundamental Problems in Statistical Mechanics* 5: 165–233, North-Holland, Amsterdam, 1980.

19. M. Henon. A two-dimensional mapping with a strange attractor. *Commun. Math. Phys.* 50: 69–77, 1976.

20. H. P. Fang and B. L. Hao. Symbolic dynamics of the Lorenz equations. *Chaos Solitons Fract.* 7: 217–246, 1996.

21. D. N. Deleanu. On the selective synchronization of some dynamical systems that exhibit chaos. *Proc. of CHAOS 2014*: 81–90, Lisbon, Portugal, June 7–10, 2014.

22. C. A. Reiter. A local cellular model for snow crystal growth. *Chaos Solitons Fract.* 23: 1111–1119, 2005.

23. http://www.snowcrystals.net/gallary1sm/index.htm

6

Fractional Chen Oscillators

C.M.A. Pinto and
Ana R.M. Carvalho

6.1 Introduction

The theory of networks of coupled cells has made a major breakthrough in the last few years, mainly due to the work of Golubitsky, Stewart, and coauthors [3–5]. These networks appear in many areas of science, such as biology, economy, ecology, neuroscience, computation, and physics [1,6,11,14]. Particular attention has been given to patterns of synchrony [9], phase-locking modes, resonance, and quasiperiodicity [1,10].

Networks of coupled cells are schematically identified with directed graphs, where the nodes (cells) represent dynamical systems, and the arrows indicate the couplings between them.

In this chapter, we analyze a fractional-order model of a network of one ring of three cells coupled to a "buffer" cell (Figure 6.1). By a cell, we mean a nonlinear system of ordinary differential equations. The full network has $\mathbf{Z}_3$ symmetry group. We consider the Chen chaotic oscillator to model the cells' internal dynamics. We observe interesting dynamical patterns, such as steady states, rotating waves, and chaos, for distinct values of the parameter c of the Chen oscillator and the derivative of fractional order $\alpha \in (0,1]$. The different patterns seem to appear through a sequence of Hopf and period-doubling (PD) bifurcations. Possible explanations for the peculiar patterns are the symmetry of the network, the dynamical characteristics of the Chen oscillator, used to model the cells' internal dynamics, and the order of the fractional derivative.

The chapter is outlined as follows. In the next subsection, we briefly review some concepts of fractional calculus (FC). In Section 6.2, we highlight important notions of the theory of coupled cell networks, for symmetric dynamical systems. In Section 6.3, we simulate the coupled cell system associated with the network in Figure 6.1. In Section 6.4, we state the main conclusions of this chapter and sketch some future research.

6.1.1 Summary of FC

The theory of FC had its start in 1695, when Leibniz exchanged letters with L'Hôpital about $D^{\frac{1}{2}}f(x)$. In FC, the concept of the derivative operator $D^{\alpha}f(x)$ is generalized to fractional values of α, the order of the

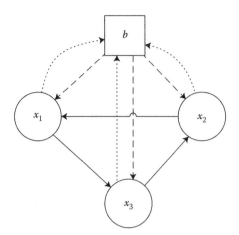

FIGURE 6.1   Network of one ring of cells coupled to a "buffer" cell with $\mathbf{Z}_3$ exact symmetry. Each node represents a cell or a dynamical system. The arrows indicate the couplings between them.

derivative. FC development is due to relevant contributions of mathematicians, such as Euler, Liouville, Riemann, and Letnikov [8,12]. In the fields of physics and engineering, FC is commonly associated with long-term memory effects [2,7].

There are several definitions of a fractional derivative of order α. The most-adopted definitions are the Riemann–Liouville, Grünwald–Letnikov (GL), and Caputo formulations. GL is defined as

$$
{}^{GL}_{a}D^{\alpha}_{t}f\left(t\right) = \lim_{h \to 0} \frac{1}{h^{\alpha}} \sum_{k=0}^{\left[\frac{t-a}{h}\right]} \left(-1\right)^{k} \left(\begin{array}{c} \alpha \\ k \end{array}\right) f\left(t - kh\right), \quad t > a, \quad \alpha > 0 \tag{6.1}
$$

where $\Gamma\left(\cdot\right)$ is Euler's γ-function, $[x]$ means the integer part of x, and h is the step time increment.

The fractional derivatives capture the history of the past dynamics, as opposed to the integer counterpart that is a "local" operator.

The GL definition inspired a discrete-time calculation algorithm, based on the approximation of the time increment h by means of the sampling period T, yielding the equation in the z domain

$$
\frac{\mathcal{Z}\left\{D^{\alpha}f\left(t\right)\right\}}{\mathcal{Z}\left\{f\left(t\right)\right\}} = \frac{1}{T^{\alpha}} \sum_{k=0}^{\infty} \frac{\left(-1\right)^{k} \Gamma\left(\alpha + 1\right)}{k!\,\Gamma\left(\alpha - k + 1\right)} z^{-k} = \left(\frac{1 - z^{-1}}{T}\right)^{\alpha} \tag{6.2}
$$

where $\mathcal{Z}$ denotes the Z-transform operator.

To apply the previous equation (6.2), it is considered an r-term truncated series:

$$
\frac{\mathcal{Z}\left\{D^{\alpha}f\left(t\right)\right\}}{\mathcal{Z}\left\{f\left(t\right)\right\}} = \frac{1}{T^{\alpha}} \sum_{k=0}^{r} \frac{\left(-1\right)^{k} \Gamma\left(\alpha + 1\right)}{k!\,\Gamma\left(\alpha - k + 1\right)} z^{-k} \tag{6.3}
$$

where, to have good approximations, a large r and a small value of T is required. This procedure is commonly known as power series expansion (PSE).

Expression (6.3) represents the Euler, or first backward difference, approximation in the so-called $s \to z$ conversion scheme. Another possibility, consists of the Tustin conversion rule. The most often-adopted generalization of the generalized derivative operator consists of $\alpha \in \mathbf{R}$.

6.2 Network

A network of cells is represented as a directed graph, where the nodes represent the cells and the arrows the couplings between them. Cells and arrows are classified according to certain types [5]. Cells of the same type have the same internal dynamics, and arrows with the same label identify equal couplings. Each cell is a dynamical system. The input set of a cell is the set of edges directed to that cell. Figure 6.1 depicts a coupled cell network, where the nodes are drawn as circles (cells in the rings) and squares ("buffer" cell). There are three different types of coupling (three distinct arrow types).

Coupled cell systems are dynamical systems consistent with the architecture or topology of the graph representing the network. Each cell c_j of the network has an internal phase space P_j. The total phase space of the network being the direct product of internal phase spaces of each cell, $P = \prod_{i=1}^{n} P_i$. Coordinates on P_j are denoted by x_j and coordinates on P are denoted by $(x_1, \ldots, x_n)$. The state of the system at time t is $(x_1(t), \ldots, x_n(t))$, where $x_j(t) \in P_j$ is the state of cell c_j at time t.

A vector field f on P is called *admissible*, for a given network, if it satisfies two conditions [4]: (i) the domain condition—each component f_j corresponding to a cell c_j is a function of the variables associated with the cells c_k that have edges directed to c_j; (ii) the pull-back condition—two components f_j and f_k corresponding to cells c_j and c_k with isomorphic input sets are identical up to a suitable permutation of the relevant variables.

In this chapter, we consider an important class of networks, namely, the ones that possess a group of symmetries. A symmetry of a coupled cell system is the group of permutations of the cells (and arrows) that preserve the network structure (including cell labels and arrow labels) and its action on P is by permutation of cell coordinates. Thus, it is a transformation of the phase space that sends solutions to solutions. The network in Figure 6.1 is an example of a network with $\mathbf{Z}_3$ symmetry.

6.3 Numerical Results

In this section, we simulate the coupled cell system associated with the network depicted in Figure 6.1. We consider the Chen oscillator to model the internal dynamics of each cell in the three ring and a unidimensional phase space for the "buffer" cell. The total phase space is thus 10th dimensional. The dynamics of a singular ring cell is given by [13]

$$
\begin{aligned}
\dot{u} &= a(v - u) \\
\dot{v} &= (c - a)u - uw + cv \\
\dot{w} &= uv - b_1 w
\end{aligned}
\tag{6.4}
$$

where $a = 35$, $b_1 = 3$, and c are real parameters. The unidimensional dynamics of the "buffer" cell is given by [10]

$$
f(u) = \mu u - \frac{1}{10} u^2 - u^3
\tag{6.5}
$$

where $\mu = -1.0$ is a real parameter.

The fractional coupled cell system of equations associated with the network in Figure 6.1 is given by

$$
\begin{aligned}
\frac{d^\alpha x_j}{dt^\alpha} &= g(x_j) + c_1(x_j - x_{j+1}) + db \quad j = 1, \ldots, 3 \\
\frac{d^\alpha b}{dt^\alpha} &= f(b)
\end{aligned}
\tag{6.6}
$$

where $g(u)$ represents the dynamics of each Chen oscillator, b is the "buffer" cell, $c_1 = -5$, $d = 0.2$, and the indexing assumes $x_4 \equiv x_1$. We consider that the coupling between all cells is linear and is done only in the first variable of each Chen oscillator.

We start with $c = 15$ and increase c till $c = 24.5$. Figure 6.2 depicts the patterns of the network for three values of the fractional derivative α and $c = 15$. We observe an equilibria for all values of α.

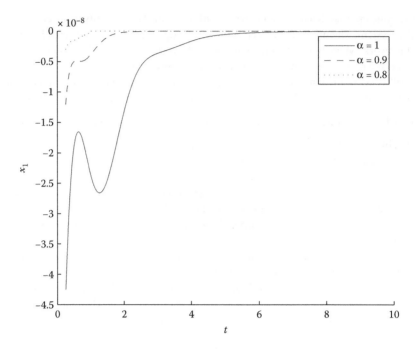

FIGURE 6.2 Dynamics of the coupled cell system (6.6) for $c = 15$ and three values of the fractional derivative $\alpha = \{0.8, 0.9, 1.0\}$.

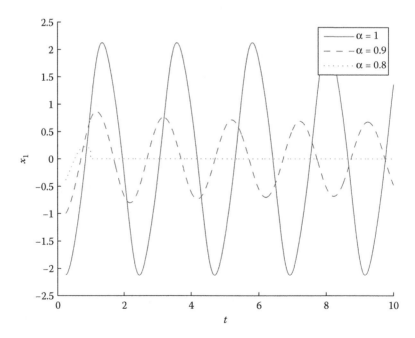

FIGURE 6.3 Dynamics of the coupled cell system (6.6) for $c = 16$ and three values of the fractional derivative $\alpha = \{0.8, 0.9, 1.0\}$.

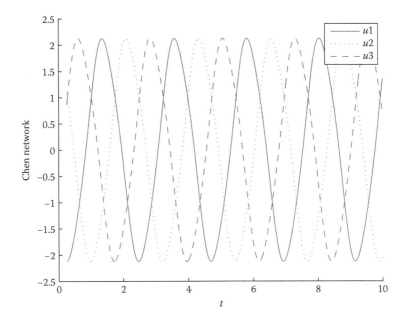

FIGURE 6.4 Rotating wave of the coupled cell system (6.6) for $c = 16$ and $\alpha = 1.0$. A similar wave is observed for $\alpha = 0.9$ with a smaller amplitude.

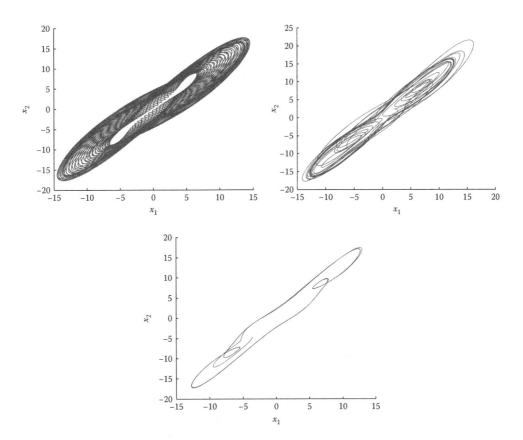

FIGURE 6.5 Phase plot of a Chen oscillator of the coupled cell system (6.6) for $c = 23$ and $\alpha = 1.0$ (left) and $\alpha = 0.9$ (right).

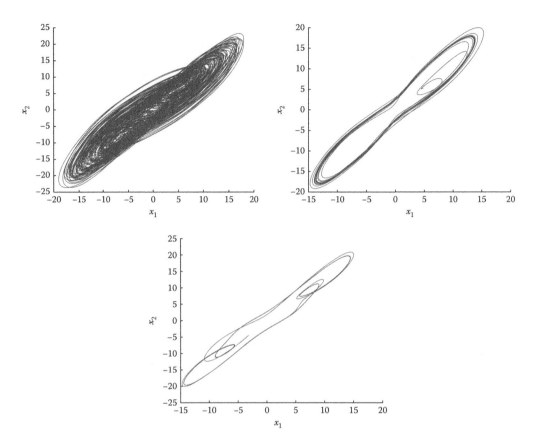

FIGURE 6.6 Phase plot of a Chen oscillator of the coupled cell system (6.6) for $c = 24.5$ and $\alpha = 1.0$ (up, left), $\alpha = 0.9$ (up, right), and $\alpha = 0.8$ (bottom).

In Figure 6.2, the dynamics of the fractional coupled system (6.6) for $c = 16$, and $\alpha \in \{0.8, 0.9, 1.0\}$ are shown. A Hopf bifurcation of the system is observed as c is increased from $c = 15$ to $c = 16$, and for $\alpha = 1.0$. The model depicts a rotating-wave state, where the cells in the three ring are $1/3$ of the period out of phase (Figure 6.4). This rotating-wave state is explained by the symmetry of the network [1]. Moreover, one can distinguish another Hopf bifurcation as the fractional derivative α is decreased from $\alpha = 0.9$ to $\alpha = 0.8$. The periodic orbit is lost and a stable equilibrium appears. The periodic orbit at $\alpha = 0.9$ has the same period as the one for $\alpha = 1.0$ but smaller amplitude.

We increase c another time to $c = 23$. Figure 6.5 shows the dynamical features of the system for $\alpha \in \{0.8, 0.9, 1.0\}$. The motion is quasiperiodic for $\alpha = 1.0$. For $\alpha = 0.9$, the dynamics of system (6.6) is still quasiperiodic but is "simpler" than for $\alpha = 1.0$. For $\alpha = 0.8$, system (6.6) exhibits a periodic orbit.

In Figure 6.6, we depict the motions of the fractional-coupled cell system (6.6) for $c = 24.5$ and $\alpha = \{0.8, 0.9, 1.0\}$. For $\alpha = 1.0$ the system is chaotic. Observing the figures for $\alpha = 1.0$ as c is increased, we note that the system has undergone a cascade of PD bifurcations that led to chaos. At $\alpha = 0.9$ it seems that the system is at a (Figures 6.4 and 6.5) periodic orbit of a long period. At $\alpha = 0.8$ the orbit is periodic and simpler than the dynamics seen for the other values of α. As c is increasing, a more complex dynamics is observed for all α.

Increasing the value of the parameter c till $c = 28$, provides other interesting patterns of the system, for distinct values of α. For $\alpha = 1$ the model again depicts a rotating-wave state [10]. We believe the system undergoes a sequence of period-halving (PH) bifurcations. On the contrary, for $\alpha = 0.9$ and $\alpha = 0.8$ the

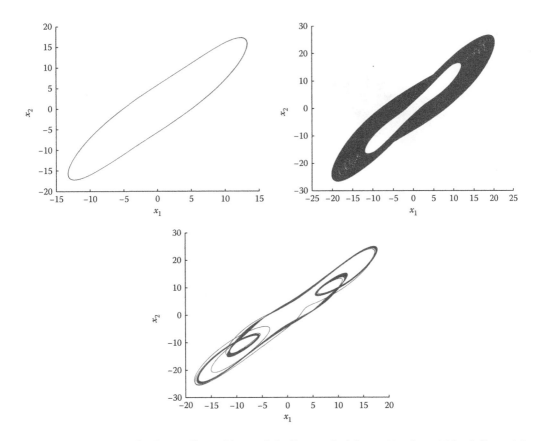

FIGURE 6.7 Phase plot of a Chen oscillator of the coupled cell system (6.6) for $c = 28$ and $\alpha = 1.0$ (up, left), $\alpha = 0.9$ (up, right), and $\alpha = 0.8$ (bottom).

system is still away from the PH bifurcation points. It appears that PD bifurcations are taking place (see Figure 6.7).

For $c = 29$, there is another cascade of PD bifurcations for $\alpha = 1.0$, again leading to a chaotic state. For $\alpha = 0.8$, as c is increased from $c = 28$, the system exhibits the first cascade of PD bifurcations, and the dynamics is chaotic. For $\alpha = 0.9$, the system seems to have undergone the full PD bifurcation cascade and has started the PH bifurcations (see Figure 6.8). This should be further verified.

From the observation of the figures in this section, one can conclude that there is a variety of curious phenomena exhibited by system (6.6). These features are attributed to the variation of parameter c of the Chen oscillator and to the order of the fractional derivative α. It seems that the system undergoes Hopf bifurcations, PD bifurcations, and PH bifurcations for all values of α as the parameter c is varied, but the bifurcation points are distinct for the three values of α. Analytical proof of the exotic behaviors seen in this section will be further studied in future work.

6.4 Conclusions

We analyze curious patterns arising in a fractional-order network of one ring of three cells coupled to a "buffer" cell. We observe a broad range of dynamical features for increasing c (Figure 6.8) and as α is decreased from 1. The exotic behaviors are explained by the symmetry of the network, the characteristics of the Chen oscillator, used to model the cells' internal dynamics, and the fractional-order derivative. Future

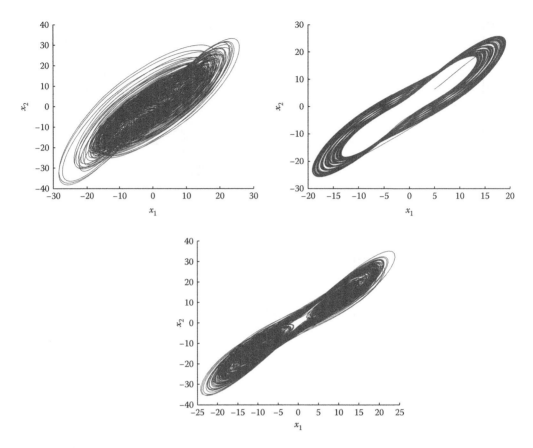

FIGURE 6.8 Phase plot of a Chen oscillator of the coupled cell system (6.6) for $c = 29$ and $\alpha = 1.0$ (up, left), $\alpha = 0.9$ (up, right), and $\alpha = 0.8$ (bottom).

work will focus on theoretically analyzing the role of the derivative of fractional order α as a bifurcation parameter of the network.

Acknowledgments

The author Pinto was partially funded by the European Regional Development Fund through the program COMPETE and by the Portuguese Government through the FCT—Fundação para a Ciência e a Tecnologia under the project PEst-C/MAT/UI0144/2013. The research of Ana Carvalho was supported by an FCT grant with reference SFRH/BD/96816/2013.

References

1. F. Antoneli, A.P.S. Dias, and C.M.A. Pinto Pinto. Quasi-periodic states in coupled rings of cells. *Communications in Nonlinear Science and Numerical Simulations*, 15:1048–1062, 2010.
2. D. Baleanu, K. Diethelm, E. Scalas, and J.J. Trujillo. *Fractional Calculus: Models and Numerical Methods*. World Scientific Publishing Company, Singapore, 2012.
3. M. Golubitsky, M. Nicol, and I. Stewart. Some curious phenomena in coupled cell systems. *Journal of Nonlinear Science*, 14:207–236, 2004.

4. M. Golubitsky and I. Stewart. Nonlinear dynamics of networks: The groupoid formalism. *Bulletin of the American Mathematical Society*, 43:305–364, 2006.

5. M. Golubitsky, I. Stewart, and A. Török. Patterns of synchrony in coupled cell networks with multiple arrows. *SIAM Journal on Applied Dynamical Systems*, 4(1):78–100, 2005.

6. X. Gong, D. Liu, and B. Wang. Chaotic system synchronization with tridiagonal structure and its initial investigation in complex power systems. *Journal of Vibration and Control*, 3(20):447–457, 2014.

7. C.M. Ionescu. *The Human Respiratory System: An Analysis of the Interplay between Anatomy, Structure, Breathing and Fractal Dynamics*. Springer-Verlag, London, 2013.

8. K.B. Oldham and J. Spanier. *The Fractional Calculus: Theory and Application of Differentiation and Integration to Arbitrary Order*. Academic Press, New York, 1974.

9. A. Pikovsky, M. Rosenblum, and J. Kurths. *Synchronization, a Universal Concept in Nonlinear Sciences*. Cambridge University Press, Cambridge, 2001.

10. C.M.A. Pinto and A.R.M. Carvalho. Strange patterns in one ring of Chen oscillators coupled to a "buffer" cell. *Journal of Vibration and Control*. Published online in December 2014. DOI 10.1177/1077546314561486.

11. C.M.A. Pinto and M. Golubitsky. Central pattern generators for bipedal locomotion. *Journal of Mathematical Biology*, 53:474–489, 2006.

12. S.G. Samko, A.A. Kilbas, and O.I. Marichev. *Fractional Integrals and Derivatives: Theory and Applications*. Gordon and Breach Science Publishers, London, 1993.

13. T. Ueta and G. Chen. Bifurcation analysis of Chen's attractor. *International Journal of Bifurcation and Chaos*, 10:1917–1931, 2000.

14. J. Zhu, J. Lu, and X. Yu. Flocking of multi-agent nonholonomic systems with proximity graphs. *IEEE Transactions on Circuits and Systems*, 60(1):199–210, 2013.

II

Strange Attractors, Bifuracation, and Related Theory

II

7

Strange Attractors and Classical Stability Theory: Stability, Instability, Lyapunov Exponents and Chaos

Nikolay Kuznetsov and
Gennady Leonov

7.1 Attractors of Dynamical Systems

Consider the dynamical systems generated by the differential equations

$$\frac{dx}{dt} = f(x), \quad t \in \mathbb{R}^1, \, x \in \mathbb{R}^n \tag{7.1}$$

and by the difference equations

$$x(t+1) = f(x(t)), \quad t \in \mathbb{Z}, \, x \in \mathbb{R}^n. \tag{7.2}$$

Here, $\mathbb{R}^n$ is a Euclidean space, $\mathbb{Z}$ is the set of integers, and $f(x)$ is a vector-function: $\mathbb{R}^n \to \mathbb{R}^n$.

Definition 7.1

We say that Equation 7.1 or 7.2 *generates a dynamical system* if for any initial data $x_0 \in \mathbb{R}^n$ the trajectory $x(t) = x(t, x_0)$ is uniquely determined for $t \geq 0$. Here, $x(0, x_0) = x_0$.

It is well known that the solutions of a dynamical system satisfy the semigroup property

$$x(t + s, x_0) = x\left(t, x(s, x_0)\right) \tag{7.3}$$

for all $t \geq 0, s \geq 0$.

For Equation 7.1, there are many existence and uniqueness theorems (see, e.g., [25]) that can be used for determining the corresponding dynamical system with the phase space $\mathbb{R}^n$. For Equation 7.2, it is readily shown that in all cases, the trajectory, defined for all $t = 0, 1, 2, \ldots$, satisfying Equation 7.3, and having initial condition x_0, is unique. Thus Equation 7.2 always generates a dynamical system with phase space $\mathbb{R}^n$.

A dynamical system generated by Equation 7.1 is called *continuous*. Equation 7.2 generates a *discrete dynamical system*.

The definitions of attractors are, as a rule, due to References 10, 47, 55, 56, and 62.

Definition 7.2

A set K is said to be *invariant* for a dynamical system if $x(t, K) = K \; \forall t \geq 0$. Here

$$x(t, K) = \{x(t, x_0) \mid x_0 \in K\}.$$

Property 1. Invariant set K is said to be *locally attractive* for a dynamical system if for a certain ε-neighborhood $K(\varepsilon)$ of the set K the following relation

$$\lim_{t \to +\infty} \rho(K, x(t, x_0)) = 0, \quad \forall x_0 \in K(\varepsilon)$$

is satisfied. Here, $\rho(K, x)$ is a distance from the point x to the set K, defined by the formula

$$\rho(K, x) = \inf_{z \in K} |z - x|,$$

$K(\varepsilon)$ is a set of points x for which $\rho(K, x) < \varepsilon$.

Property 2. Invariant set K is said to be *globally attractive* for a dynamical system if

$$\lim_{t \to +\infty} \rho(K, x(t, x_0)) = 0, \quad \forall x_0 \in \mathbb{R}^n.$$

Property 3. Invariant set K is said to be *uniformly locally attractive* for a dynamical system if for a certain ε-neighborhood $K(\varepsilon)$, any number $\delta > 0$, and any bounded set B, there exists a number $t(\delta, B) > 0$ such that

$$x(t, B \cap K(\varepsilon)) \subset K(\delta), \quad \forall t \geq t(\delta, B).$$

Here

$$x(t, B \cap K(\varepsilon)) = \{x(t, x_0) \mid x_0 \in B \cap K(\varepsilon)\}.$$

Property 4. Invariant set K is said to be *uniformly globally attractive* for a dynamical system if for any number $\delta > 0$ and any bounded set $B \subset \mathbb{R}^n$, there exists a number $t(\delta, B) > 0$ such that

$$x(t, B) \subset K(\delta), \quad \forall t \geq t(\delta, B).$$

Definition 7.3

For a dynamical system, an invariant bounded* closed set K is said to be

1. *Attractor* if it is a locally attractive set (i.e., it satisfies property 1)
2. *Global attractor* if it is a globally attractive set (i.e., it satisfies property 2)
3. *B-attractor* if it is a uniformly locally attractive set (i.e., it satisfies property 3)
4. *Global B-attractor* if it is a uniformly globally attractive set (i.e., it satisfies property 4)

Remark 7.1

This definition implies that a global B-attractor is also a global attractor. Consequently, it is rational to introduce a notion of *minimal global attractor* (and *minimal attractor*, respectively) [14], namely, the smallest bounded closed invariant set, possessing property 2 (property 1). Further, the attractors (global attractors) will be interpreted as minimal attractors (minimal global attractors).

From a computational point of view, the numerical check of property 1 for all initial states of the phase space of a dynamical system is not a feasible task. The natural generalization of the notion of an attractor is weaker requirements of attraction, almost everywhere or on the set of the positive Lebesgue measure (see, e.g., [66]).

Definition 7.4

For an attractor K, its *basin of attraction* is a set $B(K) \subset \mathbb{R}^n$ such that

$$\lim_{t \to +\infty} \rho(K, \mathrm{x}(t, \mathrm{x}_0)) = 0, \quad \forall\, \mathrm{x}_0 \in B(K).$$

Usually in numerical experiments, there is observed an attractor (or global attractor). The notion of B-attractor is mostly used in the theory of dimension, where the coverings of invariant set by balls are considered. The requirement of uniform attraction in property 3 implies that a global B-attractor involves a set of stationary points S and the corresponding unstable manifolds $W^u(S) = \{\mathrm{x}_0 \in \mathbb{R}^n \mid \lim_{t \to -\infty} \rho(S, \mathrm{x}(t, \mathrm{x}_0)) = 0\}$ (see, e.g., [14]). The same is true for B-attractor, if in property 4 the considered neighborhood $K(\varepsilon)$ contains some of the stationary points S.

Note that minimal global attractor involves the set S and its basin of attraction involves the set $W^u(S)$.

From a computational point of view, the numerical check of property 3 by numerical methods is also a difficult problem; therefore, if the basin of attraction involves unstable manifolds of equilibria, then computation of minimal attractor and unstable manifolds attracted to it may be regarded as an approximation of B-attractor (see, for example, the visualization of Lorenz attractor from a neighborhood of zero saddle equilibria).

Similar to the autonomous systems, for the analysis and visualization of a perturbed dynamical system, one can consider extended phase space and introduce various notions of attractors (see, e.g., [12,33]), or, for example, regard time t as a phase space variable, obeying the equation $\dot{t} = 1$; for the systems periodic in time, it is also possible to introduce cylindrical phase space and consider the behavior of trajectories on a Poincaré section.

7.2 Strange Attractors and the Classical Definitions of Instability

One of the basic characteristics of a strange attractor is the sensitivity of its trajectories to the initial data. We consider the correlation of such "sensitivity" with a classical notion of instability. We recall first the basic definitions of stability [55,56].

---

* Sometimes, in the definition of an attractor, the property of boundedness is omitted. For example, unbounded attractors are considered for nonautonomous systems in the extended phase space.

Consider a nonautonomous system

$$\frac{dx}{dt} = F(x,t), \quad t \in \mathbb{R}^1, \quad x \in \mathbb{R}^n, \tag{7.4}$$

where $F(x,t)$ is a continuous vector-function, and

$$x(t+1) = F(x(t),t), \quad t \in \mathbb{Z}, \quad x \in \mathbb{R}^n. \tag{7.5}$$

Denote by $x(t) = x(t,t_0,x_0)$ the solution of Equation 7.4 or 7.5 with initial data t_0, x_0: $x(t_0,t_0,x_0) = x_0$, and suppose the solution is given on the interval $t_0 < t < +\infty$.

Definition 7.5

The solution $x(t,t_0,x_0)$ is said to be *Lyapunov stable* if for any $\varepsilon > 0$ and $t_0 \geq 0$ there exists $\delta(\varepsilon,t_0)$ such that

1. All the solutions $x(t,t_0,y_0)$, satisfying the condition

$$|x_0 - y_0| \leq \delta(\varepsilon,t_0)$$

   are defined for $t \geq t_0$
2. For these solutions, the inequality

$$|x(t,t_0,x_0) - x(t,t_0,y_0)| \leq \varepsilon, \quad \forall t \geq t_0$$

   is valid

If $\delta(\varepsilon,t_0)$ is independent of t_0, the Lyapunov stability is called *uniform*.

Definition 7.6

The solution $x(t,t_0,x_0)$ is said to be *asymptotically Lyapunov stable* if it is Lyapunov stable and for any $t_0 \geq 0$ there exists $\Delta(t_0) > 0$ such that the solution $x(t,t_0,y_0)$, satisfying the condition $|x_0 - y_0| \leq \Delta(t_0)$, has the following property:

$$\lim_{t \to +\infty} |x(t,t_0,x_0) - x(t,t_0,y_0)| = 0.$$

Definition 7.7

The solution $x(t,t_0,x_0)$ is said to be *Krasovsky stable* if there exist positive numbers $\delta(t_0)$ and $R(t_0)$ such that for any y_0, satisfying the condition

$$|x_0 - y_0| \leq \delta(t_0),$$

the solution $x(t, t_0, y_0)$ is defined for $t \geq t_0$ and satisfies

$$|x(t, t_0, x_0) - x(t, t_0, y_0)| \leq R(t_0)|x_0 - y_0|, \quad \forall t \geq t_0.$$

If δ and R are independent of t_0, then Krasovsky stability is called *uniform*.

Definition 7.8

The solution $x(t, t_0, x_0)$ is said to be *exponentially stable* if there exist the positive numbers $\delta(t_0)$, $R(t_0)$, and $\alpha(t_0)$ such that for any y_0, satisfying the condition

$$|x_0 - y_0| \leq \delta(t_0),$$

the solution $x(t, t_0, y_0)$ is defined for all $t \geq t_0$ and satisfies

$$|x(t, t_0, x_0) - x(t, t_0, y_0)| \leq R(t_0) \exp(-\alpha(t_0)(t - t_0))|x_0 - y_0|, \quad \forall t \geq t_0.$$

If δ, R, and α are independent of t_0, then exponential stability is called *uniform*.

Consider now dynamical systems (see Equations 7.1 and 7.2) and introduce the following notation:

$$L^+(x_0) = \{x(t, x_0) \mid t \in [0, +\infty)\}.$$

Definition 7.9

The trajectory $x(t, x_0)$ of a dynamical system is said to be *Poincaré stable* (or *orbitally stable*) if for any $\varepsilon > 0$ there exists $\delta(\varepsilon) > 0$ such that for all y_0, satisfying the inequality $|x_0 - y_0| \leq \delta(\varepsilon)$, the relation

$$\rho(L^+(x_0), x(t, y_0)) \leq \varepsilon, \quad \forall t \geq 0$$

is satisfied. If, in addition, for a certain number δ_0 and for all y_0, satisfying the inequality $|x_0 - y_0| \leq \delta_0$, the relation

$$\lim_{t \to +\infty} \rho(L^+(x_0), x(t, y_0)) = 0$$

holds, then the trajectory $x(t, x_0)$ is said to be *asymptotically Poincaré stable* (or *asymptotically orbitally stable*).

Note that for continuous dynamical systems we have $t \in \mathbb{R}^1$, and for a discrete dynamical systems we have $t \in \mathbb{Z}$.

We now introduce the definition of Zhukovsky stability for continuous dynamical systems. For this purpose, we must consider the following set of homeomorphisms:

$$\text{Hom} = \{\tau(\cdot) \mid \tau : [0, +\infty) \to [0, +\infty), \tau(0) = 0\}.$$

The functions $\tau(t)$ from the set Hom play the role of the reparametrization of time for the trajectories of system (7.1).

Definition 7.10

The trajectory $x(t, x_0)$ of system (7.1) is said to be *Zhukovsky stable* if for any $\varepsilon > 0$ there exists $\delta(\varepsilon) > 0$ such that for any vector y_0, satisfying the inequality $|x_0 - y_0| \leq \delta(\varepsilon)$, the function $\tau(\cdot) \in \text{Hom}$ can be found such that the inequality [59,74,75]

$$|x(t, x_0) - x(\tau(t), y_0)| \leq \varepsilon, \quad \forall t \geq 0$$

is valid. If, in addition, for a certain number $\delta_0 > 0$ and any y_0 from the ball $\{y | |x_0 - y| \leq \delta_0\}$ the function $\tau(\cdot) \in$ Hom can be found such that the relation

$$\lim_{t \to +\infty} |x(t, x_0) - x(\tau(t), y_0)| = 0$$

holds, then the trajectory $x(t, x_0)$ is *asymptotically stable in the sense of Zhukovsky*.

This means that Zhukovsky stability is Lyapunov stability for the suitable reparametrization of each of the perturbed trajectories.

Recall that, by definition, Lyapunov instability is the negation of Lyapunov stability. Analogous statements hold for Krasovsky, Poincaré, and Zhukovsky instability.

The following obvious assertions can be formulated.

Proposition 7.1

1. For continuous dynamical systems, Lyapunov stability implies Zhukovsky stability, and Zhukovsky stability implies Poincaré stability.
2. For discrete dynamical systems, Lyapunov stability implies Poincaré stability.
3. For equilibria, all the above definitions due to Lyapunov, Zhukovsky, and Poincaré are equivalent.
4. For periodic trajectories of discrete dynamical systems with continuous $f(x)$, the definitions of Lyapunov and Poincaré stability are equivalent.
5. For the periodic trajectories of continuous dynamical systems with differentiable $f(x)$, the definitions of Poincaré and Zhukovsky stability are equivalent.

Also, there are well-known examples of periodic trajectories of continuous systems that happen to be Lyapunov unstable but Poincaré stable.

Now we proceed to compare the definitions given above with the effect of trajectory sensitivity to the initial data for strange attractors.

Lyapunov instability cannot characterize the "mutual repulsion" of continuous trajectories due to small variations in initial data. Neither can Poincaré instability characterize this repulsion. In this case, the perturbed solution can leave the ε-neighborhood of a certain segment of the unperturbed trajectory (the effect of repulsion) while simultaneously entering the ε-neighborhood of another segment (the property of Poincaré stability). Thus, mutually repulsive trajectories can be Poincaré stable. Let us consider these effects in more detail.

In computer experiments, it often happens that the trajectories, situated on the unstable manifold of a saddle singular point, everywhere densely fill the B-attractor (or that portion of it consisting of the bounded trajectories). This can be observed on the B-attractor of the Lorenz system (see Figure 7.1).

Example 7.1

Consider the following linearized equations of two decoupled pendula:

$$\dot{x}_1 = y_1, \quad \dot{y}_1 = -\omega_1^2 x_1,$$
$$\dot{x}_2 = y_2, \quad \dot{y}_2 = -\omega_2^2 x_3. \tag{7.6}$$

The solutions are

$$x_1(t) = A \sin(\omega_1 t + \varphi_1(0)),$$
$$y_1(t) = A\omega_1 \cos(\omega_1 t + \varphi_1(0)),$$
$$x_2(t) = B \sin(\omega_2 t + \varphi_2(0)),$$
$$y_2(t) = B\omega_2 \cos(\omega_2 t + \varphi_2(0)).$$

For fixed A and B, the trajectories of the system are situated on two-dimensional torus

$$\omega_1^2 x_1^2 + y_1^2 = A^2, \qquad \omega_2^2 x_2^2 + y_2^2 = B^3.$$

When ω_1/ω_2 is irrational, the trajectories are everywhere densely situated on the torus for any initial data $\varphi_1(0)$ and $\varphi_2(0)$.

This implies asymptotic Poincaré stability of the trajectories of the dynamical system on tori. However, the motion of the points $x(t, x_0)$ and $x(t, y_0)$ along the trajectories occurs in such a way that they do not tend toward each other as $t \to +\infty$. Neither are the trajectories "pressed" toward each other. Hence, the intuitive conception of asymptotic stability as a convergence of objects toward each other is in contrast to the formal definition of Poincaré.

It is clear that a similar effect is lacking for the notion of Zhukovsky stability: in the case under consideration, asymptotic Zhukovsky stability does not occur.

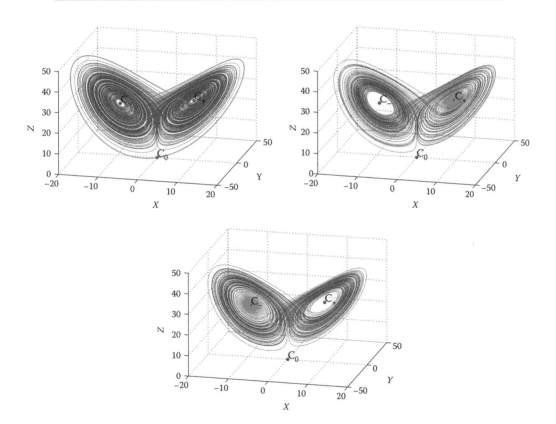

FIGURE 7.1 The Lorenz system with the classical parameters $\dot{x} = 10(y-x), \dot{y} = 28x - y - xz, \dot{z} = -8/3z + xy$. Numerical visualization of self-excited classical Lorenz attractor by trajectories from small neighborhoods of unstable equilibria.

Example 7.2

We reconsider the dynamical system (7.6) with ω_1/ω_2 irrational. Change the flow of trajectories on the torus as follows. Cut the toroidal surface along a certain segment of the fixed trajectory from the point z_1 to the point z_2. Then the surface is stretched diffeomorphically along the torus so that a cut is mapped into the circle with the fixed points z_1 and z_2 (Figure 7.2). Denote by H the interior of the circle.

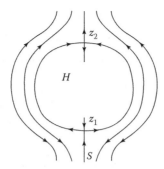

FIGURE 7.2   Trajectories of dynamical system (7.6).

Change the dynamical system so that z_1 and z_2 are saddle stationary points and the semicircles connecting z_1 and z_2 are heteroclinic trajectories, tending as $t \to +\infty$ and $t \to -\infty$ to z_2 and z_1, respectively (Figure 7.2).

Outside the "hole" H, after the diffeomorphic stretching, the disposition of trajectories on the torus is the same.

Consider the behavior of the system trajectories from the Poincaré and Zhukovsky points of view.

Outside the hole H, the trajectories are everywhere dense on torus. They are therefore, as before, asymptotically Poincaré stable.

Now we consider a certain δ-neighborhood of the point z_0, situated on the torus and outside the set H. The trajectory leaving z_0 is either dense everywhere or coincides with the separatrix S of the saddle z_1, tending to z_1 as $t \to +\infty$ (Figure 7.2). Then there exists a time t such that some trajectories, leaving the δ-neighborhood of z_0, are situated in a small neighborhood of z_1 to the right of the separatrix S. At time t, the remaining trajectories, leaving this neighborhood of z_0, are situated to the left of S. It is clear that in this case the trajectories, situated to the right and to the left of S, envelop the hole H on the right and left, respectively. It is also clear that these trajectories are repelled from each other; hence, the trajectory leaving z_0 is Zhukovsky unstable.

Thus, a trajectory can be asymptotically Poincaré stable and Zhukovsky unstable.

This example shows that the trajectories are sensitive to the initial data and can diverge considerably after some time. The notion of Zhukovsky instability is adequate to such a sensitivity.

Note that the set of such sensitive trajectories is situated on the smooth manifold, named "a torus minus the hole H." Thus, the bounded invariant set of trajectories, which are sensitive to the initial data, do not always have a noninteger Hausdorff dimension or the structure of the Cantor set.

Hence, from among the classical notions of instability for studying strange attractors, the most adequate ones are Zhukovsky instability (in the continuous case) and Lyapunov instability (in the discrete case).

7.3 Lyapunov Characteristic Exponents, Lyapunov Exponents, and Regular Systems

The linearization technique is widely used in the study of strange attractors and chaos in nonlinear dynamical systems. As a rule, the authors ignore the justification of linearization procedure and use various characteristics of a linearized system to construct various numerical characteristics of attractors of the original nonlinear systems (Lyapunov dimensions, entropies, and so on). However, the approach, based on linearizations along the nonstationary trajectories on the strange attractors, requires a rigorous justification [40,41,46,60,61].

The problem of the investigation of the solution $x(t)$ can be reduced to the problem of the stability of the trivial solution $y(t) \equiv 0$ by transformation $x = y + x(t)$. Then one can consider systems (7.4) and (7.5) with a marked linear part. In the continuous case, one has

$$\frac{dx}{dt} = A(t)x + f(t, x), \; x \in \mathbb{R}^n, \tag{7.7}$$

where $A(t)$ is a continuous $(n \times n)$-matrix and $f(\cdot, \cdot)$ is a continuous vector-function.

In the discrete case, one has

$$x(t+1) = A(t)x(t) + f\big(t, x(t)\big), \; x(t) \in \mathbb{R}^n. \tag{7.8}$$

Suppose, in a certain neighborhood $\Omega(0)$ of the point $x = 0$, the nonlinear parts of systems (7.7) and (7.8) satisfy the following condition:

$$|f(t, x)| \leq \kappa |x|^\nu \; \forall t \geq 0, \quad \forall x \in \Omega(0), \quad \kappa > 0, \nu > 1. \tag{7.9}$$

We shall say that the first approximation system for Equation 7.7 is the following linear system

$$\frac{dx}{dt} = A(t)x \tag{7.10}$$

and that for discrete system (7.8) is the linear system

$$x(t+1) = A(t)x(t). \tag{7.11}$$

Consider a fundamental matrix $X(t) = X(t, x_0) = \big(x_1(t), ..., x_n(t)\big)$, which consists of the linearly independent solutions $\{x_i(t)\}_1^n$ of the first approximation system. The fundamental matrix is often assumed to satisfy the following condition: $X(0, x_0) = I_n$, where I_n is a unit $(n \times n)$-matrix.

For the determinant of the fundamental matrix, one has the Ostrogradsky–Liouville formula, which in the continuous case is as follows:

$$\det X(t) = \det X(0) \exp \left(\int_0^t \mathrm{Tr} A(\tau) d\tau \right) \tag{7.12}$$

and in the discrete one takes the form

$$\det X(t) = \det X(0) \prod_{j=0}^{t-1} \det A(j). \tag{7.13}$$

For convenience, introduce a function $\mathcal{X}(\cdot) = \frac{1}{t} \ln |\cdot|$, where $|\cdot|$ is the Euclidian norm.

Definition 7.11

The Lyapunov characteristic exponents (LCEs) of matrix $X(t, x_0)$ are the numbers (or symbols $\pm\infty$) [63]:

$$\mathrm{LCE}_i(x_0) = \limsup_{t \to +\infty} \mathcal{X}(x_i(t)), \quad i = 1, ..., n.$$

Consider singular values of the matrix $X(t, x_0)$: $\sigma_i(t, x_0) = \sigma_i(X(t, x_0))$, which are defined as the square roots of the eigenvalues of matrix $X(t, x_0)^* X(t, x_0)$.

Definition 7.12

The Lyapunov exponents (LEs) of matrix $X(t, x_0)$ are the numbers (or symbols $\pm\infty$) [67]

$$\mathrm{LE}_i(x_0) = \limsup_{t \to +\infty} \mathcal{X}(\sigma_i(t)), \quad i = 1, ..., n.$$

Sometimes, the values $\limsup_{t\to+\infty} \mathcal{X}(x(t))$ and $\liminf_{t\to+\infty} \mathcal{X}(x(t))$ are called upper and lower LCE or LE of a solution $x(t)$.

If in the above definitions a finite limit exists (i.e., $\lim_{t\to+\infty} \mathcal{X}(\cdot)$), then the corresponding value is called *exact*. Often it is natural to consider the ordered sets LCEs and LEs. For this purpose, considering the decreasing sequences $\mathrm{LCE}_i(t', x_0) = \mathcal{X}(x_i(t'))$ and $\mathrm{LE}_i(t', x_0) = \mathcal{X}(\sigma_i(t'))$ for each $t = t'$, one first obtains the ordered (for all considered t) sets of functions

$$
\begin{aligned}
\mathrm{LCE}_1^o(t, x_0) &\geq \mathrm{LCE}_2^o(t, x_0) \geq ... \geq \mathrm{LCE}_n^o(t, x_0), \\
\mathrm{LE}_1^o(t, x_0) &\geq \mathrm{LE}_2^o(t, x_0) \geq ... \geq \mathrm{LE}_n^o(t, x_0)
\end{aligned}
\tag{7.14}
$$

and then considers the corresponding limit values

$$
\begin{aligned}
\mathrm{LCE}_1^o(x_0) &\geq \mathrm{LCE}_2^o(x_0) \geq ... \geq \mathrm{LCE}_n^o(x_0), \\
\mathrm{LE}_1^o(x_0) &\geq \mathrm{LE}_2^o(x_0) \geq ... \geq \mathrm{LE}_n^o(x_0).
\end{aligned}
\tag{7.15}
$$

Often, the definitions of LCE and LE give the same values, but for a particular system, they may be different [41].[*] Using, e.g., Courant–Fischer theorem [27], it is possible to show that the ordered LCE^o's majorizes the ordered LE^o's: $\mathrm{LE}_i^o(x_0) \leq \mathrm{LCE}_i^o(x_0)$ and the largest LCE^o is equal to the largest LE^o: $\mathrm{LCE}_1^o(x_0) = \mathrm{LE}_1^o(x_0)$.

A fundamental matrix $x(t, x_0)$ is said to be normal [63] if the sum of corresponding LCEs is minimal in comparison with other fundamental matrices; in any normal fundamental systems of solutions, the number of solutions with equal characteristic exponents is the same. For any fundamental matrix $X(t, x_0)$, there exists a nonsingular linear transformation Q such that the fundamental matrix $X(t, x_0)Q$ is a normal fundamental matrix. In contrast, LEs are independent of the choice of fundamental matrix (see, e.g., a rigorous proof in Reference 41).

Note that there are various essential generalizations of LCEs or LEs (see, e.g., [6,11,16,30,69]).

For numerical computation of LCEs and LEs, various algorithms have been developed (see, e.g., [7,20, 46]), based on QR and singular value decompositions (SVD) of fundamental matrix have been developed. However, such algorithms perform poorly in the case of coincidence or closeness of two or more LCE$_i$ or LE$_i$ and in the case of nonregular linearizations. Various methods (see [1,26,71,73] and others) have also been developed for the estimation of LEs from an experimental time series produced by some unknown system. However, there are known examples in which the results of such computations differ substantially from the analytical results [3].

The existence of different definitions of exponential growth rate, computational methods, and related assumptions led to the appeal: "*Whatever you call your exponents, please state clearly how are they being computed*" [15].

7.3.1 Regular Systems

Consider a normal fundamental systems of solutions $X(t)$. Let $\Sigma = \sum_1^n \mathrm{LCE}_i$.

Definition 7.13

A linear system is said to be *regular* [63] if for the sum of its characteristic exponents σ the following

---

[*] For example matrix $R(t) = \begin{pmatrix} 1 & g(t) - \dfrac{1}{g(t)} \\ 0 & 1 \end{pmatrix}$ allows to construct various interesting examples. Here, $\mathrm{LCE}_1^o = \max\left(\limsup_{t\to+\infty} \mathcal{X}[g(t)], \limsup_{t\to+\infty} \mathcal{X}[g^{-1}(t)]\right)$, $\mathrm{LCE}_2^o = 0$; $\mathrm{LE}_{1,2}^o = \max, \min\left(\limsup_{t\to+\infty} \mathcal{X}[g(t)], \limsup_{t\to+\infty} \mathcal{X}[g^{-1}(t)]\right)$.

relation holds:

$$\Sigma = \lim_{t \to +\infty} \inf \frac{1}{t} \ln |\det X(t)|.$$

Taking into account formula (7.12), in the continuous case, one obtains a classical definition [2,17] of the regularity of system

$$\Sigma = \lim_{t \to +\infty} \inf \frac{1}{t} \int_0^t \mathrm{Tr} A(\tau) \, d\tau.$$

Similarly, formula (7.13) gives a definition of regularity [18] in the discrete case

$$\Sigma = \lim_{t \to +\infty} \inf \frac{1}{t} \ln \prod_{j=0}^{t-1} |\det A(j)|.$$

Definition 7.14

The number

$$\Gamma = \Sigma - \lim_{t \to +\infty} \inf \frac{1}{t} \ln |\det X(t)|$$

is called an irregularity coefficient of linear system.

As was shown in Reference 17, the systems with constant and periodic coefficients are regular. For continuous [17] and discrete systems [18,23], the following is well known.

Lemma 7.1

(Lyapunov inequality) Let all LCEs of linear system be $< +\infty$ (or be $> -\infty$). Then, for any fundamental system of solutions $X(t)$, the following inequality holds

$$\lim_{t \to +\infty} \sup \frac{1}{t} \ln |\det X(t)| \le \Sigma. \tag{7.16}$$

Therefore, from the condition of regularity, it follows that exact LCEs exist. The opposite may be not valid (see [11,60]).

Example 7.3

Consider an example of a nonregular system, all LCEs of which are exact [11]. Consider system (7.10) with the matrix

$$A(t) = \begin{pmatrix} 0 & 1 \\ 0 & (\cos \ln t - \sin \ln t - 1) \end{pmatrix}, \quad t \ge 1 \tag{7.17}$$

and its fundamental matrix $X(t)$

$$X(t) = \left(x_1(t), x_2(t) \right) = \begin{pmatrix} 1 & \int_1^t e^{\gamma(\tau)} \, d\tau \\ 0 & e^{\gamma(t)} \end{pmatrix},$$

where $\gamma(t) = t(\cos \ln t - 1)$. In this case for the determinant of fundamental matrix, the following relation

$$\lim_{t \to +\infty} \inf \frac{1}{t} \ln |\det X(t)| = -2 \tag{7.18}$$

is satisfied. Consider characteristic exponents of solutions. For $x_1(t)$, one has

$$\lim_{t\to+\infty} \sup \frac{1}{t} \ln |x_1(t)| = \lim_{t\to+\infty} \inf \frac{1}{t} \ln |x_1(t)| = 0. \qquad (7.19)$$

Since $e^{\gamma(t)} \le 1$ for $t \ge 1$, one concludes that the LCE of $x_2(t)$ is less than or equal to zero

$$\lim_{t\to+\infty} \sup \frac{1}{t} \ln |x_2(t)| \le 0.$$

On the other hand, since the integral of $e^{\gamma(\tau)}$ is divergent, namely

$$\int_1^{+\infty} e^{\gamma(\tau)} \, d\tau = +\infty \qquad (7.20)$$

for $x_2(t)$, one has the following estimate:

$$\lim_{t\to+\infty} \inf \frac{1}{t} \ln |x_2(t)| \ge 0.$$

This implies that

$$\lim_{t\to+\infty} \sup \frac{1}{t} \ln |x_2(t)| = \lim_{t\to+\infty} \inf \frac{1}{t} \ln |x_2(t)| = 0. \qquad (7.21)$$

Thus, by Equations 7.18, 7.19, and 7.21, the linear system with matrix (7.17) has exact characteristic exponents but is nonregular:

$$\Gamma = 2.$$

7.4 Perron Effects

In 1930, O. Perron [68] showed that the negativeness of the largest LCE (or LE) of the first approximation system does not always result in the stability of zero solution of the original system. Furthermore, in an arbitrary small neighborhood of zero, the solutions of the original system with positive LCE can be found. The effect of sign reversal of LCE of the solutions of the first approximation system and of the original system with the same initial data is called the Perron effect [60].

We now present the outstanding result of Perron [68] and its discrete analog [23,42,45] (see also [4,6, 19,31,34–38,60,61]).

Consider the following system:

$$\begin{aligned} \frac{dx_1}{dt} &= -ax_1 \\ \frac{dx_2}{dt} &= (\sin(\ln(t+1)) + \cos(\ln(t+1)) - 2a)x_2 + x_1^2 \end{aligned} \qquad (7.22)$$

and its discrete analog

$$\begin{aligned} x_1(t+1) &= \exp(-a)x_1(t) \\ x_2(t+1) &= \frac{\exp((t+2)\sin\ln(t+2) - 2a(t+1))}{\exp((t+1)\sin\ln(t+1) - 2at)} x_2(t) + x_1(t)^2. \end{aligned} \qquad (7.23)$$

Here, a is a number satisfying the following inequalities:

$$1 < 2a < 1 + \frac{1}{2}\exp(-\pi). \qquad (7.24)$$

The solution of the first approximation system for systems (7.22) and (7.23) takes the form

$$x_1(t) = \exp(-at)x_1(0)$$
$$x_2(t) = \exp\big((t+1)\sin(\ln(t+1)) - 2at\big)x_2(0).$$

It is obvious that by condition (7.24) for the solution of the first approximation system for $x_1(0) \neq 0, x_2(0) \neq 0$, one has

$$\limsup_{t\to+\infty} \mathcal{X}(x_1(t)) = -a, \ \limsup_{t\to+\infty} \mathcal{X}(x_2(t)) = 1 - 2a < 0.$$

This implies that a zero solution of a linear system of the first approximation is Lyapunov stable.

Consider the solution of system (7.22)

$$x_1(t) = \exp(-at)x_1(0),$$
$$x_2(t) = \exp((t+1)\sin(\ln(t+1)) - 2at)$$
$$\times \left(x_2(0) + x_1(0)^2 \int_0^t \exp(-(\tau+1)\sin(\ln(\tau+1)))d\tau \right). \tag{7.25}$$

Assuming $t = t_k = \exp\left(\left(2k+\frac{1}{2}\right)\pi\right) - 1$, where k is an integer, one obtains

$$\exp\big((t+1)\sin(\ln(t+1)) - 2at\big) = \exp\big((1-2a)t + 1\big), \quad (1+t)e^{-\pi} - 1 > 0,$$

$$\int_0^t \exp\big(-(\tau+1)\sin(\ln(\tau+1))\big)d\tau$$

$$> \int_{f(k)}^{g(k)} \exp\big(-(\tau+1)\sin(\ln(\tau+1))\big)\,d\tau$$

$$> \int_{f(k)}^{g(k)} \exp\left(\frac{1}{2}(\tau+1)\right)d\tau > \int_{f(k)}^{g(k)} \exp\left(\frac{1}{2}(\tau+1)\exp(-\pi)\right)d\tau$$

$$= \exp\left(\frac{1}{2}(t+1)\exp(-\pi)\right)(t+1)\left(\exp\left(-\frac{2\pi}{3}\right) - \exp(-\pi)\right),$$

where

$$f(k) = (1+t)\exp(-\pi) - 1,$$
$$g(k) = (1+t)\exp\left(-\frac{2\pi}{3}\right) - 1.$$

Hence, one has the following estimate:

$$\exp\big((t+1)\sin(\ln(t+1)) - 2at\big)\int_0^t \exp\big(-(\tau+1)\sin(\ln(\tau+1))\big)d\tau$$

$$> \exp\left(\frac{1}{2}(2+\exp(-\pi))\right)\left(\exp\left(-\frac{2\pi}{3}\right) - \exp(-\pi)\right)(t+1)$$

$$\times \exp\left(\left(1 - 2a + \frac{1}{2}\exp(-\pi)\right)t\right). \tag{7.26}$$

From the last inequality and condition (7.24), it follows that for $x_1(0) \neq 0$, one of the characteristic exponents of solutions of system (7.22) is positive:

$$\limsup_{t \to +\infty} \mathcal{X}(x_1(t)) = -a, \ \limsup_{t \to +\infty} \mathcal{X}(x_2(t)) \geq 1 - 2a + e^{-\pi}/2 > 0. \tag{7.27}$$

Thus, one obtains that all characteristic exponents of the first approximation system are negative but almost all solutions of the original system (7.22) tend exponentially to infinity as $t_k \to +\infty$.

Consider now the solution of discrete system (7.23)

$$\begin{aligned}
x_1(t) &= x_1(0)e^{-at} \\
x_2(t) &= \exp\left((t+1)\sin\ln(t+1) - 2at\right) \\
&\quad \times \left(x_2(0) + x_1(0)^2 \sum_{k=0}^{t-1} \exp\left(-(k+2)\sin\ln(k+2) + 2a\right)\right)
\end{aligned} \tag{7.28}$$

and show that for this system inequalities (7.27) are also satisfied. For this purpose, one obtains the estimate similar to estimate (7.26) in the discrete case.

Obviously, for any $N > 0$ and $\delta > 0$, there exists a natural number $(t' = t'(N, \delta), \ t' > N)$ such that

$$\sin\ln(t' + 1) > 1 - \delta.$$

Then

$$\exp\left((t'+1)\sin\ln(t'+1) - 2at'\right) \geq \exp\left((1 - \delta - 2a)t' + 1 - \delta\right). \tag{7.29}$$

Estimate from below the second multiplier in the expression for $x_2(t)$. For sufficiently large t' there exists a natural number m

$$m \in \left(\frac{t'+1}{e^\pi} - 2, t'\right)$$

such that

$$\sin\ln(m+2) \leq -\frac{1}{2}.$$

Then, one has

$$-(m+2)\sin\ln(m+2) + 2a \geq \frac{t'+1}{2e^\pi}.$$

This implies the following estimate:

$$\sum_{k=0}^{t'-1} \exp\left(-(k+2)\sin\ln(k+2) + 2a\right) \geq \exp\left((t'+1)\frac{1}{2}e^{-\pi}\right). \tag{7.30}$$

From Equations 7.29 and 7.30 and condition (7.24), it follows that for $x_1(0) \neq 0$, one of the characteristic exponents of solutions (7.28) of system (7.23) is positive and inequalities (7.27) are satisfied.

Consider an example that shows the possibility of the sign reversal of characteristic exponents "on the contrary," namely, the solution of the first approximation system has a positive LCE while the solution of the original system with the same initial data has a negative LCE [52].

Consider the following continuous system [53]:

$$\dot{x}_1 = -ax_1$$
$$\dot{x}_2 = -2ax_2 \tag{7.31}$$
$$\dot{x}_3 = \left(\sin(\ln(t+1)) + \cos(\ln(t+1)) - 2a\right)x_3 + x_2 - x_1^2$$

and its discrete analog

$$x_1(t+1) = e^{-a}x_1(t)$$
$$x_2(t+1) = e^{-2a}x_2(t) \tag{7.32}$$
$$x_3(t+1) = \frac{\exp\left((t+2)\sin\ln(t+2) - 2a(t+1)\right)}{\exp\left((t+1)\sin\ln(t+1) - 2at\right)}x_3(t) + x_2(t) - x_1(t)^2$$

on the invariant manifold

$$M = \{x_3 \in \mathbb{R}^1, \ x_2 = x_1^2\}.$$

Here, the value a satisfies condition (7.24).

The solutions of Equations 7.31 and 7.32 on the manifold M take the form

$$x_1(t) = \exp\left(-at\right)x_1(0)$$
$$x_2(t) = \exp\left(-2at\right)x_2(0)$$
$$x_3(t) = \exp\left((t+1)\sin(\ln(t+1)) - 2at\right)x_3(0), \tag{7.33}$$
$$x_1(0)^2 = x_2(0).$$

Obviously, these solutions have negative characteristic exponents.

For system (7.31) in the neighborhood of its zero solution, consider the first approximation system

$$\dot{x}_1 = -ax_1$$
$$\dot{x}_2 = -2ax_2 \tag{7.34}$$
$$\dot{x}_3 = \left(\sin(\ln(t+1)) + \cos(\ln(t+1)) - 2a\right)x_3 + x_2.$$

The solutions of this system are the following:

$$x_1(t) = \exp\left(-at\right)x_1(0)$$
$$x_2(t) = \exp\left(-2at\right)x_2(0)$$
$$x_3(t) = \exp\left((t+1)\sin(\ln(t+1)) - 2at\right) \tag{7.35}$$
$$\times \left(x_3(0) + x_2(0)\int_0^t \exp\left(-(\tau+1)\sin(\ln(\tau+1))\right)d\tau\right).$$

For system (7.32) in the neighborhood of its zero solution, the first approximation system is as follows:

$$x_1(t+1) = \exp(-a)x_1(t)$$
$$x_2(t+1) = \exp(-2a)x_2(t) \tag{7.36}$$
$$x_3(t+1) = \frac{\exp\left((t+2)\sin\ln(t+2) - 2a(t+1)\right)}{\exp\left((t+1)\sin\ln(t+1) - 2at\right)}x_3(t) + x_2(t).$$

Then the solutions of system (7.36) take the form

$$x_1(t) = \exp(-at)x_1(0)$$
$$x_2(t) = \exp(-2at)x_2(0)$$
$$x_3(t) = \exp((t+1)\sin\ln(t+1) - 2at) \tag{7.37}$$
$$\times \left(x_3(0) + x_2(0)^2 \sum_{k=0}^{t-1} \exp(-(k+2)\sin\ln(k+2) + 2a) \right).$$

By estimates (7.26) and (7.30) for solutions (7.35) and (7.37) for $x_2(0) \neq 0$, one obtains

$$\limsup_{t\to+\infty} \mathcal{X}[x_3(t)] > 0.$$

It is easily shown that for solutions of systems (7.31) and (7.34), the following relations

$$\left(x_1(t)^2 - x_2(t)\right)^\bullet = -2a\left(x_1(t)^2 - x_2(t)\right)$$

are valid. Similarly, for system (7.36), one has

$$x_1(t+1)^2 - x_2(t+1) = \exp(-2a)\left(x_1(t)^2 - x_2(t)\right).$$

Then

$$x_1(t)^2 - x_2(t) = \exp(-2at)\left(x_1(0)^2 - x_2(0)\right).$$

It follows that the manifold M is an invariant exponentially attractive manifold for solutions of continuous systems (7.31) and (7.34), and for solutions of discrete systems (7.32) and (7.36).

This means that the relation $x_1(0)^2 = x_2(0)$ yields the relation $x_1(t)^2 = x_2(t)$ for all $t \in \mathbb{R}^1$, and for any initial data, one has

$$\left|x_1(t)^2 - x_2(t)\right| \leq \exp(-2at)\left|x_1(0)^2 - x_2(0)\right|.$$

Thus, systems (7.31) and (7.34) have the same invariant exponentially attractive manifold M on which almost all solutions of the first approximation system (7.34) have a positive LCE and all solutions of the original system (7.31) have negative characteristic exponents. The same result can be obtained for discrete systems (7.32) and (7.36).

The Perron effect occurs here on the whole manifold

$$\{x_3 \in \mathbb{R}^1, \ x_2 = x_1^2 \neq 0\}.$$

To construct exponentially stable system, the first approximation of which has a positive LCE, we represent system (7.31) in the following way:

$$\dot{x}_1 = F(x_1, x_2)$$
$$\dot{x}_2 = G(x_1, x_2) \tag{7.38}$$
$$\dot{x}_3 = \left(\sin\ln(t+1) + \cos\ln(t+1) - 2a\right)x_3 + x_2 - x_1^2.$$

Here, the functions $F(x_1, x_2)$ and $G(x_1, x_2)$ have the form

$$F(x_1, x_2) = \pm 2x_2 - ax_1, \quad G(x_1, x_2) = \mp x_1 - \varphi(x_1, x_2),$$

in which case the upper sign is taken for $x_1 > 0$, $x_2 > x_1^2$ and for $x_1 < 0$, $x_2 < x_1^2$, the lower one for $x_1 > 0$, $x_2 < x_1^2$ and for $x_1 < 0$, $x_2 > x_1^2$.

The function $\varphi(x_1, x_2)$ is defined as

$$\varphi(x_1, x_2) = \begin{cases} 4ax_2 & \text{for } |x_2| > 2x_1^2 \\ 2ax_2 & \text{for } |x_2| < 2x_1^2. \end{cases}$$

The solutions of system (7.38) are regarded in the sense of A.F. Filippov [22]. By definition of $\varphi(x_1, x_2)$, the following system

$$\dot{x}_1 = F(x_1, x_2)$$
$$\dot{x}_2 = G(x_1, x_2) \tag{7.39}$$

on the lines of discontinuity $\{x_1 = 0\}$ and $\{x_2 = x_1^2\}$ has sliding solutions, which are given by the equations

$$x_1(t) \equiv 0, \quad \dot{x}_2(t) = -4ax_2(t)$$

and

$$\dot{x}_1(t) = -ax_1(t), \quad \dot{x}_2(t) = -2ax_2(t), \quad x_2(t) \equiv x_1(t)^2.$$

In this case, the solutions of system (7.39) with the initial data $x_1(0) \neq 0$, $x_2(0) \in \mathbb{R}^1$ attain the curve $\{x_2 = x_1^2\}$ in a finite time, which is less than or equal to 2π.

This implies that for the solutions of system (7.38) with the initial data $x_1(0) \neq 0$, $x_2(0) \in \mathbb{R}^1$, $x_3(0) \in \mathbb{R}^1$, for $t \geq 2\pi$, one obtains the relations $F(x_1(t), x_2(t)) = -ax_1(t)$, $G(x_1(t), x_2(t)) = -2ax_2(t)$. Therefore, based on these solutions for $t \geq 2\pi$, system (7.34) is a system of the first approximation for system (7.38).

System (7.34), as was shown above, has a positive LCE. At the same time, all solutions of system (7.38) tend exponentially to zero. ∎

The considered technique permits us to construct the different classes of nonlinear continuous and discrete systems for which the Perron effects occur.

7.5 Stability Criteria by the First Approximation

Let us describe the most famous stability criteria by the first approximation.

Consider the continuous case. Assume that there exists a number $C > 0$ and a piecewise continuous function $p(t)$ such that for the Cauchy matrix $X(t)X(\tau)^{-1}$, the estimate

$$|X(t)X(\tau)^{-1}| \leq C \exp \int_\tau^t p(s)\, ds, \quad \forall t \geq \tau \geq 0 \tag{7.40}$$

is valid.

Theorem 7.1: [56]

If condition (7.9) with $\nu = 1$ and the inequality

$$\lim_{t \to +\infty} \sup \frac{1}{t} \int_0^t p(s)\, ds + C\kappa < 0$$

are satisfied, then the solution $x(t) \equiv 0$ of system (7.7) is asymptotically Lyapunov stable.

Consider a discrete analog of this theorem. In the discrete case, it is assumed that in place of inequality (7.40), one has

$$|X(t)X(\tau)^{-1}| \leq C \prod_{s=\tau}^{t-1} p(s), \quad \forall t > \tau \geq 0, \tag{7.41}$$

where $p(s)$ is a positive function. ∎

In the discrete case, one has a similar theorem

Theorem 7.2: [45,60]

If condition (7.9) with $v = 1$ and the inequality

$$\lim_{t \to +\infty} \sup \frac{1}{t} \ln \prod_{s=0}^{t-1} (p(s) + C\kappa) < 0 \qquad (7.42)$$

are satisfied, then the solution $x(t) \equiv 0$ of system (7.8) is asymptotically Lyapunov stable.   ∎

Corollary 7.1

For the first-order system (7.7) or (7.8), the negativeness of LCE (or LE) implies the asymptotic stability of its zero solution.

Assume that for the Cauchy matrix $X(t)X(\tau)^{-1}$ the following estimate

$$|X(t)X(\tau)^{-1}| \le C \exp\left(-\alpha(t - \tau) + \gamma\tau\right), \quad \forall t \ge \tau \ge 0, \qquad (7.43)$$

where $\alpha > 0$, $\gamma \ge 0$, is satisfied.

Theorem 7.3: [13,64,65]

Let condition (7.9) with sufficiently small κ and condition (7.43) be valid. Then, if the inequality

$$(v - 1)\alpha - \gamma > 0 \qquad (7.44)$$

holds, then the solution $x(t) \equiv 0$ of corresponding nonlinear system (7.7) or (7.8) is asymptotically Lyapunov stable.   ∎

Theorem 7.3 strengthens the well-known Lyapunov theorem [63] on stability by the first approximation for regular systems.

7.5.1 Stability Criteria for the Flow and Cascade of Solutions

Consider system (7.4) or (7.5), where $F(\cdot, \cdot)$ is a twice continuously differentiable vector-function. Consider corresponding linearizations (7.10) and (7.11) of these systems along solutions with the initial data $y = x(0, y)$ from the open set Ω, which is bounded in $\mathbb{R}^n$. Thus,

$$A(t) = \frac{\partial F(x, t)}{\partial x}\bigg|_{x=x(t,y)}$$

is a Jacobian matrix of the vector-function $F(x, t)$ on the solution $x(t, y)$. Let $X(t, y)$ be a fundamental matrix of a linear system and $X(0, y) = I_n$.

Assume that for the largest singular value $\sigma_1(t, y)$ of systems (7.10) and (7.11) for all t, the following estimate

$$\sigma_1(t, y) < \sigma(t), \quad \forall y \in \Omega, \qquad (7.45)$$

where $\sigma(t)$ is a scalar function, is valid.

Theorem 7.4: [44,50]

Suppose, the function $\sigma(t)$ is bounded on the interval $(0, +\infty)$. Then the flow (cascade) of solutions $x(t, y)$, $y \in \Omega$, of systems (7.4) and (7.5) are Lyapunov stable. If, in addition,

$$\lim_{t \to +\infty} \sigma(t) = 0,$$

then the flow (cascade) of solutions $x(t, y)$, $y \in \Omega$, is asymptotically Lyapunov stable. ∎

Corollary 7.2

We consider flow (cascade) of solutions, all these solutions are stable by the first approximation, Perron effects are possible only on the boundary of such flow (cascade).

Consider the flow of solutions of system (7.22) with the initial data in a neighborhood of the point $x_1 = x_2 = 0$: $x_1(0, x_{10}, x_{20}) = x_{10}$, $x_2(0, x_{10}, x_{20}) = x_{20}$.

Hence, it follows easily that

$$x_1(t, x_{10}, x_{20}) = \exp\left(-at\right)x_{10}.$$

Therefore, for a continuous system, the matrix $A(t)$ of linear system takes the form

$$A(t) = \begin{pmatrix} -a & 0 \\ 2\exp(-at)x_{10} & r(t) \end{pmatrix}, \tag{7.46}$$

where

$$r(t) = \sin(\ln(t+1)) + \cos(\ln(t+1)) - 2a.$$

For the discrete system, one has

$$A(t) = \begin{pmatrix} e^{-a} & 0 \\ 2\exp(-at)x_{10} & r(t) \end{pmatrix}, \tag{7.47}$$

$$r(t) = \frac{\exp\left((t+2)\sin\ln(t+2) - 2a(t+1)\right)}{\exp\left((t+1)\sin\ln(t+1) - 2at\right)}.$$

The solutions of system (7.10) and (7.11) with matrices (7.46) and (7.47), respectively, are the following:

$$\begin{aligned} z_1(t) &= \exp\left(-at\right)z_1(0), \\ z_2(t) &= p(t)(z_2(0) + 2x_{10}z_1(0))q(t). \end{aligned} \tag{7.48}$$

Here

$$p(t) = \exp\left((t+1)\sin(\ln(t+1)) - 2at\right),$$

$$q(t) = \int_0^t \exp\left(-(\tau+1)\sin(\ln(\tau+1))\right) d\tau$$

in the continuous case and

$$p(t) = \exp\left((t+1)\sin(\ln(t+1)) - 2at\right),$$

$$q(t) = \sum_{k=0}^{t-1} \exp\left(-(k+2)\sin\ln(k+2) + 2a\right)$$

in the discrete case.

As was shown above, if relations (7.24) are satisfied and

$$z_1(0)x_{10} \neq 0,$$

then the LCE of $z_2(t)$ is positive (see Equation 7.27).

Hence, in an arbitrary small neighborhood of the trivial solution $x_1(t) \equiv x_2(t) \equiv 0$, there exist the initial data x_{10}, x_{20} such that for $x_1(t, x_{10}, x_{20})$, $x_2(t, x_{10}, x_{20})$, the first approximation system has the positive largest LCE (and the largest LE).

Therefore, in this case, there does not exist a neighborhood Ω of the point $x_1 = x_2 = 0$ such that uniform estimates (7.45) are satisfied. Thus, for systems (7.22) and (7.23), the Perron effect occurs.

7.6 Instability Criteria by the First Approximation

Consider instability in the sense of Krasovsky for the solution $x(t) \equiv 0$ of continuous system (7.7) and of discrete system (7.8).

Theorem 7.5: [43,51,53]

If the relation

$$\sup_{1 \leq k \leq n} \lim_{t \to +\infty} \inf \left[\frac{1}{t} \left(\ln \left| \det X(t) \right| - \sum_{j \neq k} \ln \left| x_j(t) \right| \right) \right] > 1 \tag{7.49}$$

is satisfied, then the solution $x(t) \equiv 0$ of corresponding nonlinear system (7.7) or (7.8) is unstable in the sense of Krasovsky.   ∎

Corollary 7.3

Condition (7.49) of Theorem 7.5 is satisfied if the following inequality

$$\Lambda - \Gamma > 0 \tag{7.50}$$

is valid, where Λ is the largest LCE and Γ is the irregularity coefficient.

Consider now Lyapunov instability of the solution $x(t) \equiv 0$ of multidimensional continuous system (7.7) and of discrete system (7.8).

Theorem 7.6: [43,53,60]

Let for certain values $C > 0$, $\beta > 0$, $\alpha_1, ..., \alpha_{n-1}$ ($\alpha_j < \beta$ for $j = 1, ..., n-1$), the following conditions hold:
1.

$$|x_j(t)| \leq C \exp(\alpha_j(t - \tau))|x_j(\tau)|,$$
$$\forall t \geq \tau \geq 0, \quad j = 1, ..., n-1, \tag{7.51}$$

2.

$$\frac{1}{(t - \tau)} \ln |\det X(t)| > \beta + \sum_{j=1}^{n-1} \alpha_j, \quad \forall t \geq \tau \geq 0 \tag{7.52}$$

and, if $n > 2$,

3.

$$\prod_{j=1}^{n} |x_j(t)| \leq C |\det X(t)|, \quad \forall t \geq 0. \tag{7.53}$$

Then the zero solution of the considered nonlinear system is Lyapunov unstable. ∎

Corollary 7.4

For the first-order system (7.7) or (7.8) with bounded coefficients, the positiveness of lower LCE of the first approximation system results in exponential instability of zero solution of the original system.

7.6.1 Instability Criterion for the Flow and Cascade Solutions

The problem arises naturally as to the weakening of instability conditions, which are due to Theorems 7.5 and 7.6. However, the Perron effects impose restrictions on such weakening.

Consider continuous and discrete systems (7.4) and (7.5), respectively.

Suppose, for a certain vector-function $\xi(t)$, the following relations hold

$$|\xi(t)| = 1, \quad \inf_{y \in \Omega} |X(t, y)\xi(t)| \geq \sigma(t), \quad \forall t \geq t_0. \tag{7.54}$$

Theorem 7.7: [43,50]

Let for the function $\alpha(t)$ the following condition

$$\lim_{t \to +\infty} \sup \sigma(t) = +\infty \tag{7.55}$$

be satisfied.

Then the flow (cascade) of solutions $x(t, y)$, $y \in \Omega$ is Lyapunov unstable.

Proof. Holding a certain pair $x_0 \in \Omega$ and $t \geq t_0$ fixed, choose the vector y_0 in any δ-neighborhood of the point x_0 in such a way that

$$x_0 - y_0 = \delta\xi(t). \tag{7.56}$$

Let δ be so small that the ball of radius δ centered at x_0 is entirely placed in Ω.

For any fixed values t, j and for the vectors x_0, y_0, there exists a vector $w_j \in \mathbb{R}^n$ such that

$$|x_0 - w_j| \leq |x_0 - y_0|,$$

$$x_j(t, x_0) - x_j(t, y_0) = X_j(t, w_j)(x_0 - y_0). \tag{7.57}$$

Here, $x_j(t, x_0)$ is the jth component of the vector-function $x(t, x_0)$ and $X_j(t, w)$ is the jth row of the matrix $X(t, w)$.

By Equation 7.57 one has

$$|x(t, x_0) - x(t, y_0)| = \sqrt{\sum_j |X_j(t, w_j)(x_0 - y_0)|^2}$$

$$\geq \delta \max\{|X_1(t, w_1)\xi(t)|, \dots, |X_n(t, w_n)\xi(t)|\}$$

$$\geq \delta \max_j \inf_\Omega |X_j(t, x_0)\xi(t)| = \delta \inf_\Omega \max_j |X_j(t, x_0)\xi(t)|$$

$$\geq \frac{\delta}{\sqrt{n}} \inf_\Omega |X(t, x_0)\xi(t)| \geq \frac{\sigma(t)\delta}{\sqrt{n}}.$$

This estimate and conditions (7.55) imply that for any positive numbers ε and δ, there exist a number $t \geq t_0$ and a vector y_0 such that

$$|x_0 - y_0| = \delta, \quad |x(t, x_0) - x(t, y_0)| > \varepsilon.$$

The latter means that the solution $x(t, x_0)$ is Lyapunov unstable.   ■

Consider the hypotheses of Theorem 7.7.

The hypotheses of Theorem 7.7 is, in essence, the requirement that, at least, one LE of the linearizations of the flow of solutions with the initial data from Ω is positive under the condition that the "unstable directions $\xi(t)$" (or unstable manifolds) of these solutions depend continuously on the initial data x_0. Actually, if this property holds, then, regarding (if necessary) the domain Ω as the union of the domains Ω_i, of arbitrary small diameter, on which conditions (7.54) and (7.55) are valid, one obtains Lyapunov instability of the whole flow of solutions with the initial data from Ω.

Apply Theorem 7.7 to systems (7.31) and (7.32).

For the solutions $x(t, t_0, x_0)$ with the initial data $t_0 = 0$,

$$x_1(0, x_{10}, x_{20}, x_{30}) = x_{10},$$
$$x_2(0, x_{10}, x_{20}, x_{30}) = x_{20},$$
$$x_3(0, x_{10}, x_{20}, x_{30}) = x_{30}$$

in the continuous case, one has the following relations:

$$x_1(t, x_{10}, x_{20}, x_{30}) = \exp(-at)x_{10},$$

$$\frac{\partial F(x, t)}{\partial x}\Big|_{x=x(t,0,x_0)} = \begin{pmatrix} -a & 0 & 0 \\ 0 & -2a & 0 \\ -2\exp(-at)x_{10} & 1 & r(t) \end{pmatrix}, \tag{7.58}$$

where

$$r(t) = \sin(\ln(t+1)) + \cos(\ln(t+1)) - 2a.$$

For discrete system, one obtains

$$\frac{\partial F(x, t)}{\partial x}\Big|_{x=x(t,0,x_0)} = \begin{pmatrix} \exp(-a) & 0 & 0 \\ 0 & \exp(-2a) & 0 \\ -2\exp(-at)x_{10} & 1 & r(t) \end{pmatrix}, \tag{7.59}$$

where

$$r(t) = \frac{\exp\big((t+2)\sin\ln(t+2) - 2a(t+1)\big)}{\exp\big((t+1)\sin\ln(t+1) - 2at\big)}.$$

Solutions of Equations 7.10 and 7.11 with matrices (7.58) and (7.59), respectively, have the form

$$z_1(t) = \exp(-at)z_1(0),$$
$$z_2(t) = \exp(-2at)z_2(0), \tag{7.60}$$
$$z_3(t) = p(t)\big(z_3(0) + (z_2(0) - 2x_{10}z_1(0))\, q(t)\big).$$

Here, in the continuous case, one has

$$p(t) = \exp\big((t+1)\sin(\ln(t+1)) - 2at\big),$$

$$q(t) = \int_0^t \exp\big(-(\tau+1)\sin(\ln(\tau+1))\big)\, d\tau.$$

and in the discrete case

$$p(t) = \exp\left((t+1)\sin(\ln(t+1)) - 2at\right),$$

$$q(t) = \sum_{k=0}^{t-1} \exp(-(k+2)\sin\ln(k+2) + 2a).$$

Relations (7.60) give

$$X(t, 0, x_0) = \begin{pmatrix} \exp(-at) & 0 & 0 \\ 0 & \exp(-2at) & 0 \\ -2x_{10}p(t)q(t) & p(t)q(t) & p(t) \end{pmatrix}.$$

If it is assumed that

$$\xi(t) = \begin{pmatrix} 0 \\ 1 \\ 0 \end{pmatrix},$$

then for $\Omega = \mathbb{R}^n$ and

$$\alpha(t) = \sqrt{\exp(-4at) + (p(t)q(t))^2}$$

relations (7.54) and (7.55) are satisfied (see estimate (7.26)).

Thus, by Theorem 7.7, any solution of system (7.31) is Lyapunov unstable.

We restrict ourselves to the consideration of the manifold

$$M = \{x_3 \in \mathbb{R}^1, \quad x_2 = x_1^2\}.$$

In this case, the initial data of the unperturbed solution x_0 and the perturbed solution y_0 belong to the manifold M:

$$x_0 \in M, \quad y_0 \in M. \tag{7.61}$$

The analysis of the proof of Theorem 7.7 (see (7.56)) implies that the vector-function $\xi(t)$ satisfies the following additional condition: if Equations 7.56 and 7.61 hold, then the inequality $\xi_2(t) \neq 0$ yields the relation $\xi_1(t) \neq 0$.

In this case, Equations 7.54 and 7.55 are not valid since for either $2x_{10}\xi_1(t) = \xi_2(t) \neq 0$ or $\xi_2(t) = 0$, the value $|X(t, x_0)|$ is bounded on $[0, +\infty)$.

Thus, since in conditions (7.54) and (7.55) the uniformity with respect to x_0 is violated, for system (7.31) on the set M, the Perron effects are possible under certain additional restrictions on the vector-function $\xi(t)$.

7.7 Zhukovsky Stability

Zhukovsky stability is simply the Lyapunov stability of reparametrized trajectories. To study it, we may apply the arsenal of methods and tools that were developed for the study of Lyapunov stability. The reparametrization of trajectories permits us to introduce additional tool for investigation, the *moving Poincaré section*. The classical Poincaré section is the transversal $(n-1)$-dimensional surface S in the phase space $\mathbb{R}^n$, which possesses a recurring property. The latter means that for the trajectory of a dynamical system $x(t, x_0)$ with the initial data $x_0 \in S$, there exists a time instant $t = T > 0$ such that $x(T, x_0) \in S$. The transversal property means that

$$n(x)^* f(x) \neq 0, \quad \forall x \in S.$$

Here, $n(x)$ is a normal vector of the surface S at the point x, and $f(x)$ is the right-hand side of the differential equation generating a dynamical system.

$$\frac{dx}{dt} = f(x), \quad t \in \mathbb{R}^1, \quad x \in \mathbb{R}^n, \tag{7.62}$$

We now "force" the Poincaré section to move along the trajectory $x(t, x_0)$. We assume further that the vector-function $f(x)$ is twice continuously differentiable and that the trajectory $x(t, x_0)$, whose the Zhukovsky stability (or instability) will be considered, is wholly situated in a certain bounded domain $\Omega \subset \mathbb{R}^n$ for $t \geq 0$. Suppose also that $f(x) \neq 0, \forall x \in \overline{\Omega}$. Here, $\overline{\Omega}$ is a closure of the domain Ω. Under these assumptions, there exist positive numbers δ and ε such that

$$f(y)^* f(x) \geq \delta, \quad \forall y \in S(x, \varepsilon), \quad \forall x \in \overline{\Omega}.$$

Here

$$S(x, \varepsilon) = \{ y \mid (y - x)^* f(x) = 0, \quad |x - y| < \varepsilon \}.$$

Definition 7.15

The set $S(x(t, x_0), \varepsilon)$ is called a *moving Poincaré section*.

Note that for small ε it is natural to restrict oneself to the family of segments of the surfaces $S(x(t, x_0), \varepsilon)$ rather than arbitrary surfaces. From this point of view, a more general consideration does not give new results. It is possible to consider the moving Poincaré section more generally by introducing the set

$$S(x, q(x), \varepsilon) = \{ y \mid (y - x)^* q(x) = 0, \quad |x - y| < \varepsilon \},$$

where the vector-function $q(x)$ satisfies the condition $q(x)^* f(x) \neq 0$. We treat the most interesting and descriptive case $q(x) \equiv f(x)$.

The classical Poincaré section allows us to clarify the behavior of trajectories using the information at their discrete times of crossing the section. Reparametrization makes it possible to organize the motion of trajectories so that at time t all trajectories are situated on the same moving Poincaré section $S(x(t, x_0), \varepsilon)$:

$$x(\varphi(t), y_0) \in S(x(t, x_0), \varepsilon). \tag{7.63}$$

Here, $\varphi(t)$ is a reparametrization of the trajectory $x(t, y_0)$, $y_0 \in S(x_0, \varepsilon)$. This consideration has, of course, a local property and is only possible for t satisfying

$$|x(\varphi(t), y_0) - x(t, x_0)| < \varepsilon. \tag{7.64}$$

Let us consider the system of the first approximation

$$\frac{dw}{dt} = \frac{\partial f}{\partial x}(x(t, x_0))w \tag{7.65}$$

System (7.65) has one null LCE: let $LCE_1 = 0$. Denote by $\mathrm{LCE}_2 \geq \cdots \geq \mathrm{LCE}_n$ the remaining LCEs, and Γ is the coefficient of irregularity.

Theorem 7.8: [54]

If for system (7.65), the inequality

$$\mathrm{LCE}_2 + \Gamma < 0$$

is satisfied, then the trajectory $x(t, x_0)$ is asymptotically Zhukovsky stable.   ∎

This result generalizes the well-known Andronov–Witt theorem.

Theorem 7.9: [17]

If the trajectory $x(t, x_0)$ is periodic and differs from equilibria, and for system (7.65) the inequality

$$LCE_2 < 0$$

is satisfied, then the trajectory $x(t, x_0)$ is asymptotically orbitally stable (asymptotically Poincaré stable).   ∎

Theorem 7.9 is a corollary of Theorem 7.8 since system (7.65) with the periodic matrix

$$\frac{\partial f}{\partial x}(x(t, x_0))$$

is regular.

Recall that for periodic trajectories, asymptotic stability in the senses of Zhukovsky and Poincaré are equivalent.

The theorem of Demidovich is also a corollary of Theorem 7.10.

Theorem 7.10: [17]

If system (7.65) is regular (i.e., $\Gamma = 0$) and $LCE_2 < 0$, then the trajectory $x(t, x_0)$ is asymptotically orbitally stable.   ∎

7.8 Attractor Dimensions

Harmonic oscillations are characterized by an amplitude, period, and frequency, and periodic oscillations by a period. Numerous investigations have shown that more complex oscillations also have numerical characteristics. These are the dimensions of attractors, corresponding to ensembles of such oscillations.

The theory of topological dimension [29,39], developed in the first half of the twentieth century, is of little use in giving the scale of dimensional characteristics of attractors. The point is that the topological dimension can take integer values only. Hence, the scale of dimensional characteristics compiled in this manner turns out to be quite poor.

For investigating attractors, the Hausdorff dimension of a set is much better. This dimensional characteristic can take any nonnegative value, and on such customary objects in Euclidean space as a smooth curve, a surface, or a countable set of points, it coincides with the topological dimension. Let us proceed to the definition of Hausdorff dimension.

Consider an invariant set K of dynamical system, and numbers $d \geq 0$, $\varepsilon > 0$. We cover K by balls of radius $r_j < \varepsilon$ and denote

$$\mu_H(K, d, \varepsilon) = \inf \sum_j r_j^d,$$

where the infimum is taken over all such ε-coverings K. It is obvious that $\mu_H(K, d, \varepsilon)$ does not decrease with decreasing ε. Therefore, there exists the limit (perhaps infinite), namely

$$\mu_H(K, d) = \lim_{\varepsilon \to 0} \mu_H(K, d, \varepsilon).$$

Definition 7.16

The function $\mu_H(\cdot, d)$ is called the *Hausdorff d-measure*.

For a fixed set K, the function $\mu_H(K, \cdot)$ has the following property. It is possible to find $d_{kp} \in [0, \infty]$ such that

$$\mu_H(K, d) = \infty, \quad \forall d < d_{kp},$$
$$\mu_H(K, d) = 0, \quad \forall d > d_{kp}.$$

If a dynamical system is defined in $\mathbb{R}^n$, then here $d_{kp} \leq n$.

We put

$$\dim_H K = d_{kp} = \inf\{d \mid \mu_H(K, d) = 0\}.$$

Definition 7.17

We call $\dim_H E$ the *Hausdorff dimension* of the set K.

While the topological dimension is invariant with respect to homeomorphisms, Hausdorff dimension is invariant with respect to diffeomorphisms, and the noninteger Hausdorff dimension is not invariant with respect to homeomorphisms [29].

In studying the attractors of dynamical systems in phase space, the smooth change of coordinates is often used. Therefore, in such considerations, it is sufficient to assume invariance with respect to diffeomorphisms.

We now give two equivalent definitions of fractal dimension. Denote by $\mathcal{N}_\varepsilon(K)$ the minimal number of balls of radius ε needed to cover the set K. Consider the numbers $d \geq 0, \varepsilon > 0$ and put

$$\mu_F(K, d, \varepsilon) = \mathcal{N}_\varepsilon(K)\varepsilon^d,$$
$$\mu_F(K, d) = \limsup_{\varepsilon \to 0} \mu_F(K, d, \varepsilon).$$

Definition 7.18

The *fractal dimension* of the set E is the value

$$\dim_F K = \inf\{d \mid \mu_F(K, d) = 0\}.$$

Note that this definition is patterned after that for the Hausdorff dimension. However, in this case, the covering is by the balls of the same radius ε only.

Definition 7.19

The fractal dimension of K is the value

$$\dim_F K = \limsup_{\varepsilon \to 0} \frac{\log \mathcal{N}_\varepsilon(K)}{\log(1/\varepsilon)}.$$

It is easy to see that

$$\dim_H K \leq \dim_F K.$$

It turns out [8,9,21,28,58] that the upper estimate of the Hausdorff and fractal dimension of invariant sets is the Lyapunov dimension [21,32,48], which is based on ordered Lyapunov exponents (see Equation 7.15).

Consider $j(t, x_0) \in [1, n]$, which is equal to the largest natural number m such that

$$\mathrm{LE}_1^o(t, x_0) + \ldots + \mathrm{LE}_m^o(t, x_0) \geq 0, \quad \mathrm{LE}_{m+1}^o(t, x_0) < 0,$$
$$\frac{\mathrm{LE}_1^o(t, x_0) + \ldots + \mathrm{LE}_m^o(t, x_0)}{|\mathrm{LE}_{m+1}^o(t, x_0)|} < 1.$$

Define the function $LD(t, x_0) = 0$, if $LE_1^o(t, x_0) \leq 0$, and $LD(t, x_0) = n$, if $\sum_{i=1}^{n} LE_i^o(t, x_0) \geq 0$, otherwise

$$LD(t, x_0) = j(t, x_0) + \frac{LE_1^o(t, x_0) + \ldots + LE_j^o(t, x_0)}{|LE_{j+1}^o(t, x_0)|}. \tag{7.66}$$

Definition 7.20

The Lyapunov dimension of invariant compact set K of dynamical system is defined as

$$\dim_L K = \sup_{x_0 \in K} \limsup_{t \to +\infty} LD(t, x_0).$$

Similar definition can be constructed based on LCEs, but, in general, the values may be different.

The properties of the Lyapunov dimension are considered in detail in References 9, 70, and 72 (see also the recent surveys [5,57]). Note that the Lyapunov dimension of a bounded invariant set K is invariant under the diffeomorphism of the phase space (see, e.g., the study of so-called Lyapunov transformations in Reference 63 and the work [41]).

The Lyapunov dimension is not a dimensional characteristic in the classical sense (see, e.g., [24]). However, it does permit us to estimate from above the topological, Hausdorff, and fractal dimensions. It is also the characteristic of instability of dynamical systems. Finally, it is well "adapted" for investigations by the methods of classical stability theory, since the Lyapunov functions can be used for the estimate of the Lyapunov dimension. The idea of introducing Lyapunov functions in the estimate of dimensional characteristics first appeared in Reference 49 (see also [9,58]).

Acknowledgment

Authors were supported by Saint-Petersburg State University (6.39.416.2014).

References

1. H.D.I. Abarbanel, R. Brown, J.J. Sidorowich, and L.Sh. Tsimring. The analysis of observed chaotic data in physical systems. *Reviews of Modern Physics*, 65(4):1331–1392, 1993.
2. L.Ya. Andrianova. *Introduction to Linear Systems of Differential Equations*. American Mathematical Society, Providence, Rhode Island, 1998.
3. P. Augustova, Z. Beran, and S. Celikovsky. On some false chaos indicators when analyzing sampled data. In *ISCS 2014: Interdisciplinary Symposium on Complex Systems, Emergence, Complexity and Computation* (Eds. A. Sanayei et al.), pages 249–258. Springer International Publishing, Switzerland, 2015.
4. F. Balibrea and M. Caballero. Stability of orbits via Lyapunov exponents in autonomous and non-autonomous systems. *International Journal of Bifurcation and Chaos*, 23, art. no. 1350127, 2013.
5. L. Barreira and K. Gelfert. Dimension estimates in smooth dynamics: A survey of recent results. *Ergodic Theory and Dynamical Systems*, 31:641–671, 2011.
6. L. Barreira and Y. Pesin. Lectures on Lyapunov exponents and smooth ergodic theory. In *Proceedings of Symposia in Pure Mathematics*, volume 69, pages 3–89. AMS, 2001.
7. G. Benettin, L. Galgani, A. Giorgilli, and J.-M. Strelcyn. Lyapunov characteristic exponents for smooth dynamical systems and for hamiltonian systems. A method for computing all of them. Part 2: Numerical application. *Meccanica*, 15(1):21–30, 1980.
8. V.A. Boichenko, G.A. Leonov, A. Franz, and V. Reitmann. Hausdorff and fractal dimension estimates for invariant sets of non-injective maps. *Zeitschrift für Analysis und ihre Anwendung*, 17(1):207–223, 1998.

9. V.A. Boichenko, G.A. Leonov, and V. Reitmann. *Dimension Theory for Ordinary Differential Equations.* Teubner, Stuttgart, 2005.

10. H.W. Broer, F. Dumortier, S.J. Strien, and F. Takens. *Structures in Dynamics.* Elsevier Science Publishers, North-Holland, 1991.

11. B.E. Bylov, R.E. Vinograd, D.M. Grobman, and V.V. Nemytskii. *Theory of Characteristic Exponents and Its Applications to Problems of Stability.* Nauka, Moscow [in Russian], 1966.

12. D.N. Cheban, P.E. Kloeden, and B. Schmalfuss. The relationship between pullback, forwards and global attractors of nonautonomous dynamical systems. *Nonlinear Dynamics and Systems Theory,* 2:9–28, 2002.

13. N.G. Chetaev. *Stability of Motion.* Gostekhizdat, Moscow [in Russian], 1955.

14. I. Chueshov. *Introduction to the Theory of Infinite-Dimensional Dissipative Systems.* Electronic Library of Mathematics. ACTA, 2002.

15. P. Cvitanović, R. Artuso, R. Mainieri, G. Tanner, and G. Vattay. *Chaos: Classical and Quantum.* Niels Bohr Institute, Copenhagen, 2012. http://ChaosBook.org.

16. A. Czornik, A. Nawrat, and M. Niezabitowski. Lyapunov exponents for discrete time-varying systems. *Studies in Computational Intelligence,* 440:29–44, 2013.

17. B.P. Demidovich. *Lectures on Mathematical Theory of Stability.* Nauka, Moscow, 1967.

18. V.B. Demidovich. Stability criterion for difference equations. *Differentsial'nye uravneniya,* 5(7):1247–1255, 1969.

19. B. Demir and S. Koak. A note on positive Lyapunov exponent and sensitive dependence on initial conditions. *Chaos, Solitons and Fractals,* 12:2119–2121, 2013.

20. L. Dieci, M.S. Jolly, and E.S. Van Vleck. Numerical techniques for approximating Lyapunov exponents and their implementation. *Journal of Computational and Nonlinear Dynamics,* 6(1), 2011. art. num. 011003.

21. A. Douady and J. Oesterle. Dimension de Hausdorff des attracteurs. *Comptes Rendus de l'Académie des Sciences, Série,* 290(24):1135–1138, 1980.

22. A.F. Filippov. *Differential Equations with Discontinuous Right-Hand Sides.* Kluwer, Dordrecht, 1988.

23. I.V. Gayschun. *Systems with Discrete Time.* Inst. Matemat. RAN Belarusi, Minsk [in Russian], 2001.

24. K. Gelfert and A.E. Motter. (Non)invariance of dynamical quantities for orbit equivalent flows. *Communications in Mathematical Physics,* 300(2):411–433, 2010.

25. P. Hartman. *Ordinary Differential Equation.* John Wiley, New York, 1964.

26. R. Hegger, H. Kantz, and T. Schreiber. Practical implementation of nonlinear time series methods: The TISEAN package. *Chaos,* 9:413–435, 1999.

27. R.A. Horn and C.R. Johnson. *Topics in Matrix Analysis.* Cambridge University Press, UK, 1994.

28. B.R. Hunt. Maximum local Lyapunov dimension bounds the box dimension of chaotic attractors. *Nonlinearity,* 9(4):845–853, 1996.

29. W. Hurewicz and H. Wallman. *Dimension Theory.* Princeton University Press, Princeton, 1941.

30. N.A. Izobov. *Lyapunov Exponents and Stability.* Cambridge Scientific Publishers, Cambridge, 2012.

31. N.A. Izobov and S.K. Korovin. Multidimensional analog of the two-dimensional Perron effect of sign change of characteristic exponents for infinitely differentiable differential systems. *Differential Equations,* 48(11):1444–1460, 2011.

32. J.L. Kaplan and J.A. Yorke. Chaotic behavior of multidimensional difference equations. In *Functional Differential Equations and Approximations of Fixed Points* (Eds. H.-O. Peitgen and H.-O. Walther), pages 204–227. Springer, Berlin, 1979.

33. P.E. Kloeden and M. Rasmussen. *Nonautonomous Dynamical Systems.* American Mathematical Society, Rhode Island, 2011.

34. S.K. Korovin and N.A. Izobov. On the Perron sign change effect for Lyapunov characteristic exponents of solutions of differential systems. *Differential Equations,* 46(10):1395–1408, 2010.

35. S.K. Korovin and N.A. Izobov. The Perron effect of change of values of characteristic exponents for solutions of differential systems. *Doklady Mathematics,* 82(2):798–800, 2010.

36. S.K. Korovin and N.A. Izobov. Realization of the Perron effect whereby the characteristic exponents of solutions of differential systems change their values. *Differential Equations*, 46(11):1537–1551, 2010.

37. S.K. Korovin and N.A. Izobov. Generalization of the Perron effect whereby the characteristic exponents of all solutions of two differential systems change their sign from negative to positive. *Differential Equations*, 47(7):942–954, 2011.

38. S.K. Korovin and N.A. Izobov. Multidimensional Perron sign change effect for all characteristic exponents of differential systems. *Doklady Mathematics*, 85(3):315–320, 2012.

39. K. Kuratowski. *Topology*. Academic Press, New York, 1966.

40. N.V. Kuznetsov. *Stability and Oscillations of Dynamical Systems: Theory and Applications*. Jyvaskyla University Printing House, Jyvaskyla, Finland, 2008.

41. N.V. Kuznetsov, T.A. Alexeeva, and G.A. Leonov. Invariance of Lyapunov characteristic exponents, Lyapunov exponents, and Lyapunov dimension for regular and non-regular linearizations. *arXiv:1410.2016v2*, 2014.

42. N.V. Kuznetsov and G.A. Leonov. Counterexample of Perron in the discrete case. *Izv. RAEN, Diff. Uravn.*, 5:71, 2001.

43. N.V. Kuznetsov and G.A. Leonov. Criteria of stability by the first approximation for discrete nonlinear systems. *Vestnik Sankt-Peterburgskogo Universiteta. Ser 1. Matematika Mekhanika Astronomiya*, (3):30–42, 2005.

44. N.V. Kuznetsov and G.A. Leonov. Criterion of stability to first approximation of nonlinear discrete systems. *Vestnik Sankt-Peterburgskogo Universiteta. Ser 1. Matematika Mekhanika Astronomiya*, (2):55–63, 2005.

45. N.V. Kuznetsov and G.A. Leonov. On stability by the first approximation for discrete systems. In *2005 International Conference on Physics and Control, PhysCon 2005*, volume 2005, pages 596–599. IEEE, 2005.

46. N.V. Kuznetsov, T.N. Mokaev, and P.A. Vasilyev. Numerical justification of Leonov conjecture on Lyapunov dimension of Rossler attractor. *Communications in Nonlinear Science and Numerical Simulation*, 19:1027–1034, 2014.

47. O.A. Ladyzhenskaya. Determination of minimal global attractors for the navier-stokes equations and other partial differential equations. *Russian Mathematical Surveys*, 42(6):25–60, 1987.

48. F. Ledrappier. Some relations between dimension and lyapounov exponents. *Communications in Mathematical Physics*, 81(2):229–238, 1981.

49. G.A. Leonov. On estimations of the Hausdorff dimension of attractors. *Vestnik St. Petersburg University, Mathematics*, 24(3):41–44, 1991.

50. G.A. Leonov. On stability in the first approximation. *Journal of Applied Mathematics and Mechanics*, 62(4):511–517, 1998.

51. G.A. Leonov. Instability in the first approximation for time–dependent linearization. *Journal of Applied Mathematics and Mechanics*, 66(2):323–325, 2002.

52. G.A. Leonov. A modification of Perron's counterexample. *Differential Equations*, 39(11):1651–1652, 2003.

53. G.A. Leonov. Criteria of instability in the first approximation for time–dependent linearizations. *Journal of Applied Mathematics and Mechanics*, 68(6):858–871, 2004.

54. G.A. Leonov. Generalization of the Andronov–Vitt theorem. *Regular and Chaotic Dynamics*, 11(2):281–289, 2006.

55. G.A. Leonov. Strange attractors and classical stability theory. *Nonlinear Dynamics and System Theory*, 1(8):49–96, 2008.

56. G.A. Leonov. *Strange Attractors and Classical Stability Theory*. St. Petersburg University Press, St. Petersburg, 2008.

57. G.A. Leonov. Lyapunov functions in the attractors dimension theory. *Journal of Applied Mathematics and Mechanics*, 76(2):129–141, 2012.

58. G.A. Leonov and V.A. Boichenko. Lyapunov's direct method in the estimation of the Hausdorff dimension of attractors. *Acta Applicandae Mathematicae*, 26(1):1–60, 1992.

59. G.A. Leonov, I.M. Burkin, and A.I. Shepelyavy. *Frequency Methods in Oscillation Theory*. Kluwer, Dordretch, 1996.

60. G.A. Leonov and N.V. Kuznetsov. Time-varying linearization and the Perron effects. *International Journal of Bifurcation and Chaos*, 17(4):1079–1107, 2007.

61. G.A. Leonov and N.V. Kuznetsov. Lyapunov exponent sign reversal: Stability and instability by the first approximation. In *Nonlinear Dynamics and Complexity* (Eds V. Afraimovich, A.C.J. Luo, and X. Fu), volume 8, Springer International Publishing, Switzerland, 2014.

62. G.A. Leonov, N.V. Kuznetsov, and T.N. Mokaev. Homoclinic orbits, and self-excited and hidden attractors in a Lorenz-like system describing convective fluid motion, *European Physical Journal Special Topics*, 224:1421–1458, 2015.

63. A.M. Lyapunov. *The General Problem of the Stability of Motion*. Kharkov Mathematical Society, Kharkov, 1892.

64. I.G. Malkin. *Stability of Motion*. Nauka, Moscow [in Russian], 1966.

65. J.L. Massera. Contributions to stability theory. *Annals of Mathematics*, 64:182–206, 1956.

66. J.W. Milnor. Attractor. *Scholarpedia*, 1(11), 2006. 1815.

67. V.I. Oseledec. Multiplicative ergodic theorem: Characteristic Lyapunov exponents of dynamical systems. In *Transactions of the Moscow Mathematical Society*, volume 19, pages 179–210. 1968.

68. O. Perron. Die stabilitatsfrage bei differentialgleichungen. *Mathematische Zeitschrift*, 32(5):702–728, 1930.

69. Ya. B. Pesin. Characteristic Lyapunov exponents and smooth ergodic theory. *Russian Mathematical Surveys*, 32(4):55–114, 1977.

70. Ya. B. Pesin. Dimension type characteristics for invariant sets of dynamical systems. In *Russian Mathematical Surveys, 43:4*, pages 111–151, 1988.

71. M.T. Rosenstein, J.J. Collins, and C.J. De Luca. A practical method for calculating largest Lyapunov exponents from small data sets. *Physica D: Nonlinear Phenomena*, 65(1-2):117–134, 1993.

72. R. Temam. *Infinite-Dimensional Dynamical Systems*. Springer-Verlag, New York, 1993.

73. A. Wolf, J.B. Swift, H.L. Swinney, and J.A. Vastano. Determining Lyapunov exponents from a time series. *Physica D: Nonlinear Phenomena*, 16(D):285–317, 1985.

74. N.E. Zhukovsky. *On Stability of Motion*. Uchenye Zapiski Moskovskogo Universiteta, 4:10–21, 1882.

75. N.E. Zhukovsky. *On Stability of Motion. Collected Works, V.1*. Gostehkizdat, Moscow [in Russian], 1948.

8

Numerical Visualization of Attractors: Self-Exciting and Hidden Attractors

8.1 Self-Exciting and Hidden Attractors

An oscillation in a dynamical system can be easily localized numerically if initial conditions from its open neighborhood in the phase space (with the exception of a minor set of points of measure zero) lead to long-time behavior that approaches the oscillation. From a computational point of view, such an oscillation (or a set of oscillations) is called an attractor and its attracting set is called a basin of attraction (i.e., a set of initial data for which the trajectories numerically tend to the attractor).

Usually, the study of a dynamical system begins with the analysis of equilibria, which can easily be done numerically or analytically. Therefore, from a computational point of view, it is natural to suggest [31,33,39,44,45,47] the following classification of attractors, based on the simplicity of finding their basin of attractions in the phase space.

Definition 8.1

An attractor is called a *self-excited attractor* if its basin of attraction intersects with any open neighborhood of an equilibrium, otherwise it is called a *hidden attractor* [31,39,40,45].

Similar to the autonomous systems, for perturbed dynamical systems, depending on the physical problem statement, the notion of self-excited and hidden attractors can be considered, for example, with respect either to the stationary states of the considered nonautonomous system or to the equilibria of the corresponding dynamical system without time-varying excitation.

8.1.1 Self-Exciting Attractor Localization

For a *self-excited attractor*, its basin of attraction is connected with an unstable equilibrium and, therefore, self-excited attractors can be localized numerically by the *standard computational procedure*, in which after a transient process a trajectory, started from a point of unstable manifold in a neighborhood of unstable equilibrium, is attracted to the state of oscillation and traces it. Thus, self-excited attractors can be easily visualized.

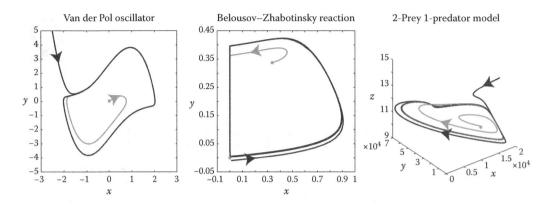

FIGURE 8.1 Standard computation of self-excited periodic oscillations.

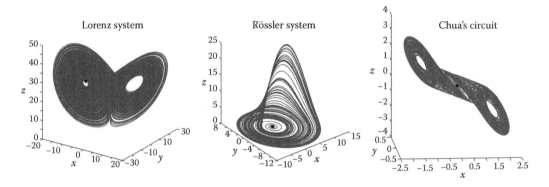

FIGURE 8.2 Standard computation of self-excited chaotic attractors.

In the first half of the last century, during the initial period of the development of the theory of nonlinear oscillations (see, e.g., [5,63,64]), much attention was given to analysis and synthesis of oscillating systems, for which the problem of the existence of oscillations can be studied with relative ease since the oscillations were excited from unstable equilibria.

In Figure 8.1 is shown the computation of classical self-excited oscillations: van der Pol oscillator [66], one of the modifications of Belousov–Zhabotinsky reaction [8], and two preys and one predator model [19].

In Figure 8.2 is shown the computation of classical self-excited chaotic attractors: Lorenz system [50], Rössler system [61], and "double-scroll" attractor in Chua's circuit [11].

The following classical example of a self-excited chaotic attractor (Figure 8.3) in the forced Duffing system $\ddot{x} + 0.05\dot{x} + x^3 = 7.5\sin(t)$ was numerically constructed by Y. Ueda in 1961 [65]. For the construction of a self-excited chaotic attractor in this system, we use a transient process from zero equilibrium of a unperturbed dynamical system (i.e., without $\sin(t)$) to the attractor (see Figure 8.3) in the perturbed dynamical system.

Nowadays, there is enormous number of publications devoted to the computation and analysis of various self-excited attractors.

8.1.2 Hidden Attractor Localization

For a hidden attractor, its basin of attraction is not connected with any equilibria. For example, the hidden attractors are attractors in systems with no equilibria or with only one equilibrium, which is stable (a special

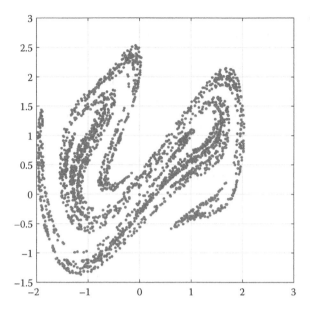

FIGURE 8.3 The $(x, \dot{x})$ plane is mapped into itself by following the trajectory for time $0 \leq t \leq 2\pi$. A trajectory from the vicinity of zero stationary point of unperturbed Duffing oscillator (without $7.5 \sin(t)$) visualizes after transition processes a self-excited chaotic attractor in the forced oscillator.

case of multistability: coexistence of attractors in multistable systems). Remark that multistability may cause inconvenience in various practical applications (see, e.g., discussion on the problems related to the synchronization of coupled multistable systems in References 23, 24, 33, and 57). Coexisting self-excited attractors can be found by the standard computational procedure while there is no effective regular way to predict the existence or coexistence of hidden attractors in a system.

For localization of a hidden attractor, it is necessary to use a special numerical method since the basin of the attractor does not intersect small neighborhoods of unstable manifolds of equilibria. One of the effective methods for the numerical localization of hidden attractors is the method based on *homotopy* and *numerical continuation*: it is necessary to construct a sequence of similar systems such that for the first (starting) system the initial data for numerical computation of the oscillating solution (starting oscillation) can be obtained analytically, for example, it is often possible to consider a starting system with self-excited starting oscillation. Then, the transformation of this starting oscillation is tracked numerically while passing from one system to another. In a scenario of transition to chaos in dynamical system, there is usually a parameter $\lambda \in [a_1, a_2]$, whose variation gives this scenario. One can also introduce the parameter λ artificially, and then consider its variation on interval $[a_1, a_2]$ such that $\lambda = a_2$ corresponds to the initial system, and choose a parameter a_1 in such a way that for $\lambda = a_1$ it is possible to find analytically or compute a certain attractor (often such an attractor is of simple form). In other words, here instead of the analysis of the scenario of transition to chaos, a scenario of transition to chaos can be synthesized. Further, a sequence λ_j, $\lambda_1 = a_1$, $\lambda_m = a_2$, $\lambda_j \in [a_1, a_2]$ such that the distance between λ_j and λ_{j+1} is sufficiently small is considered. Then the change of the structure of the attractor, obtained for $\lambda_1 = a_1$, is considered numerically. If during the change of the parameter λ (from λ_j to λ_{j+1}) there is no loss of stability bifurcation of the considered attractor, then for $\lambda_m = a_2$ at the end of the procedure an attractor of the initial system is localized.

In general, for the construction of a transient process leading to the discovery of a chaotic set, it can be used another object, which can be constructed for the considered system or its modifications (i.e., instead of the analysis of scenario of transition to chaos in system, a new scenario of transition to chaos may be synthesized). For example, as such object for autonomous systems, in Reference 59 it is suggested to use so-called perpetual points and in Reference 53 the equilibria of complexified system is used. In a similar way, a periodic solution or a homoclinic trajectory can be used.

It should be noted that if the attracting domain is a whole state space, then the attractor can be visualized by any trajectory and the only difference between computations is the time of transient process.

The study of hidden attractors arises in connection with various fundamental problems and applied models. At first, the problem of analyzing hidden periodic oscillations arose in the second part of Hilbert's 16th problem (1900) on the number and mutual disposition of limit cycles in two-dimensional polynomial systems [21]. The first nontrivial results obtained in Bautin's works [7] were devoted to theoretical construction of three nested limit cycles around one equilibrium in quadratic systems. While Bautin's method allows one to construct only nested small-amplitude limit cycles (see, e.g., [30,37,39,41,42]), which can hardly be visualized, recently an analytical approach, which allows one to visualize effectively nested normal amplitude nested cycles in quadratic systems, was developed [28,39].

Later in the 1950–1960s twentieth, the investigations of widely known Markus–Yamabe's [52], Aizerman's [1], and Kalman's [22] conjectures on absolute stability led to the discovery of possible coexistence of a hidden periodic oscillation and the unique stable stationary point in automatic control systems with nonlinearities, which belong to the sector of linear stability (see, e.g., [2,6,9,12,18,38,58]).

One of the first known systems in which a hidden chaotic attractor (see Figures 8.4 and 8.5) can be found is a Lorenz-like model [46,47]

$$\begin{cases} \dot{x} = -\sigma(x - y) - ayz \\ \dot{y} = rx - y - xz \\ \dot{z} = -bz + xy \end{cases} \tag{8.1}$$

describing the process of interaction between waves in plasma (the Rabinovich system [56,60]) and convective fluid motion (the Glukhovsky–Dolghansky system [20]).

After the notion "*hidden attractor*" was introduced and the first hidden Chua attractor was discovered [27,31,44,45], the study of hidden attractors has received much attention.

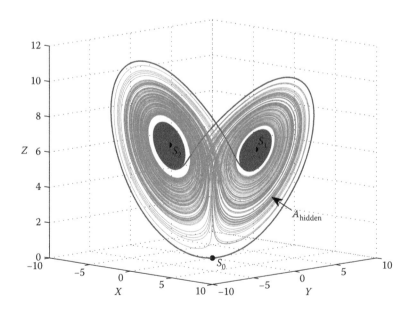

FIGURE 8.4 Hidden attractor in Lorenz-like system (8.1) with $r = 6.8$, $a = -0.5$, $b = 1$, and $\sigma = -ar$ (the Rabinovich system). Separatrices of the zero saddle are attracted to stable equilibria.

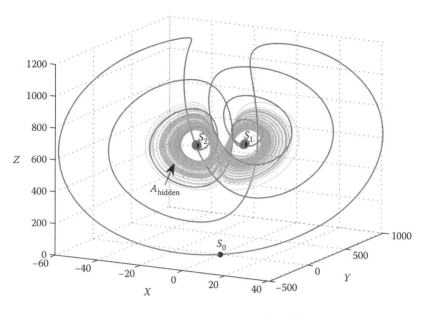

FIGURE 8.5 Hidden attractor in Lorenz-like system (8.1) with $r = 700$, $a = 0.0052$, $\sigma = 4$, and $b = 1$ (the Glukhovsky–Dolghansky system). Separatrices of the zero saddle are attracted to stable equilibria.

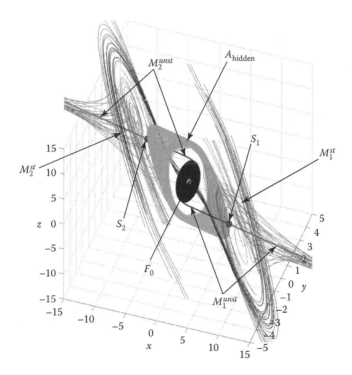

FIGURE 8.6 Hidden chaotic attractor (A_{hidden}) in a classical Chua circuit: locally stable zero equilibrium F_0 attracts trajectories from stable manifolds $M_{1,2}^{st}$ of two saddle points $S_{1,2}$; trajectories from unstable manifolds $M_{1,2}^{unst}$ tend to infinity; $\alpha = 8.4562$, $\beta = 12.0732$, $\gamma = 0.0052$, $m_0 = -0.1768$, $m_1 = -1.1468$.

Until recently, only self-excited attractors have been found in Chua circuits (see, e.g., Figure 8.2). Note that L. Chua himself, in analyzing various cases of attractors existence in Chua's circuit [17], does not admit the existence of a hidden attractor (being discovered later) in his circuits. Now, it is shown that Chua's circuit and its various modifications [27,29,31,33,35,43–45], can exhibit hidden chaotic attractors (see Figure 8.6[*]).

Currently, there are a number of papers of various authors devoted to the study of hidden attractors. Recently, hidden periodic oscillations and hidden chaotic attractors have been studied in such models as phase-locked loop [32,34], Costas loop [10], drilling systems [25,40], DC–DC converters [73], aircraft control systems [3], launcher stabilization systems [4], and many other models (see, e.g., [13–16,26,36,48, 53–55,62,67–72]).

References

1. M. A. Aizerman. On a problem concerning the stability in the large of dynamical systems. *Uspekhi Matematicheskikh Nauk* [in Russian], 4:187–188, 1949.
2. R. Alli-Oke, J. Carrasco, W. P. Heath, and A. Lanzon. A robust Kalman conjecture for first-order plants. *IFAC Proceedings Volumes (IFAC-PapersOnline)* 7 (PART 1), pp. 27–32, 2012.
3. B. R. Andrievsky, N. V. Kuznetsov, G. A. Leonov, and A. Yu. Pogromsky. Hidden oscillations in aircraft flight control system with input saturation. *IFAC Proceedings Volumes (IFAC-PapersOnline)*, 5(1):75–79, 2013.
4. B. R. Andrievsky, N. V. Kuznetsov, G. A. Leonov, and S. M. Seledzhi. Hidden oscillations in stabilization system of flexible launcher with saturating actuators. *IFAC Proceedings Volumes (IFAC-PapersOnline)*, 19(1):37–41, 2013.
5. A. A. Andronov, E. A. Vitt, and S. E. Khaikin. *Theory of Oscillators*. Pergamon Press, Oxford, 1966.
6. N. E. Barabanov. On the Kalman problem. *Siberian Mathematical Journal*, 29(3):333–341, 1988.
7. N. N. Bautin. On the number of limit cycles appearing on varying the coefficients from a focus or centre type of equilibrium state. *Matematicheskii Sbornik (N.S.)* [in Russian], 30(72):181–196, 1952.
8. B. P. Belousov. A periodic reaction and its mechanism. *Collection of Short Papers on Radiation Medicine for 1958*. Med. Publ., Moscow [in Russian], 1959.
9. J. Bernat and J. Llibre. Counterexample to Kalman and Markus-Yamabe conjectures in dimension larger than 3. *Dynamics of Continuous, Discrete and Impulsive Systems*, 2(3):337–379, 1996.
10. R. E. Best, N.V. Kuznetsov, O.A. Kuznetsova, G.A. Leonov, M.V. Yuldashev, and R.V. Yuldashev. A short survey on nonlinear models of the classic costas loop: Rigorous derivation and limitations of the classic analysis. In *2015 American Control Conference*, pp. 1296–1302, 2015.
11. E. Bilotta and P. Pantano. *A Gallery of Chua Attractors*, volume Series A. 61. World Scientific, Singapore, 2008.
12. V. O. Bragin, V. I. Vagaitsev, N. V. Kuznetsov, and G. A. Leonov. Algorithms for finding hidden oscillations in nonlinear systems. The Aizerman and Kalman conjectures and Chua's circuits. *Journal of Computer and Systems Sciences International*, 50(4):511–543, 2011.
13. I. M. Burkin and N.N. Khien. Analytical–numerical methods of finding hidden oscillations in multi-dimensional dynamical systems. *Differential Equations*, 50(13):1695–1717, 2014.
14. U. Chaudhuri and A. Prasad. Complicated basins and the phenomenon of amplitude death in coupled hidden attractors. *Physics Letters, Section A: General, Atomic and Solid State Physics*, 378(9):713–718, 2014.
15. M. Chen, M. Li, Q. Yu, B. Bao, Q. Xu, and J. Wang. Dynamics of self-excited attractors and hidden attractors in generalized memristor-based Chua's circuit. *Nonlinear Dynamics*, doi: 10.1007/s11071-015-1983-7, 2015.

---

[*] Using symmetry and applying more accurate analytical–numerical methods [51], one might distinguish a few close coexisting attractors (see e.g., [49]).

16. M. Chen, J. Yu, and B.-C. Bao. Finding hidden attractors in improved memristor-based Chua's circuit. *Electronics Letters*, 51:462–464, 2015.

17. L. O. Chua. A zoo of strange attractors from the canonical Chua's circuits. *Proceedings of the IEEE 35th Midwest Symposium on Circuits and Systems (Cat. No. 92CH3099-9)*, 2:916–926, 1992.

18. R. E. Fitts. Two counterexamples to Aizerman's conjecture. *IEEE Transactions on Automatic Control*, 11(3):553–556, 1966.

19. K. Fujii. Complexity–stability relationship of two-prey-one-predator species system model; local and global stability. *Journal of Theoretical Biology*, 69(4):613–623, 1977.

20. A. B. Glukhovskii and F. V. Dolzhanskii. Three-component geostrophic model of convection in a rotating fluid. *Academy of Sciences, USSR, Izvestiya, Atmospheric and Oceanic Physics*, 16:311–318, 1980.

21. D. Hilbert. Mathematical problems. *Bulletin of the American Mathematical Society*, (8):437–479, 1901–1902.

22. R. E. Kalman. Physical and mathematical mechanisms of instability in nonlinear automatic control systems. *Transactions of ASME*, 79(3):553–566, 1957.

23. T. Kapitaniak. *Chaotic Oscillators: Theory and Applications*. World Scientific, Singapore, 1992.

24. T. Kapitaniak. Uncertainty in coupled chaotic systems: Locally intermingled basins of attraction. *Physical Review E*, 53:6555–6557, 1996.

25. M. A. Kiseleva, N. V. Kuznetsov, G. A. Leonov, and P. Neittaanmäki. Drilling systems failures and hidden oscillations. In *IEEE 4th International Conference on Nonlinear Science and Complexity, NSC 2012—Proceedings*, pages 109–112, 2012.

26. A. P. Kuznetsov, S. P. Kuznetsov, E. Mosekilde, and N. V. Stankevich. Co-existing hidden attractors in a radio-physical oscillator system. *Journal of Physics A: Mathematical and Theoretical*, 48:125101, 2015.

27. N. Kuznetsov, O. Kuznetsova, G. Leonov, and V. Vagaitsev. Analytical–numerical localization of hidden attractor in electrical Chua's circuit. *Informatics in Control, Automation and Robotics, Lecture Notes in Electrical Engineering*, Volume 174, Part 4, pages 149–158. Springer-Verlag, Berlin Heidelberg, 2013.

28. N. V. Kuznetsov, O. A. Kuznetsova, and G. A. Leonov. Visualization of four normal size limit cycles in two-dimensional polynomial quadratic system. *Differential Equations and Dynamical Systems*, 21 (1–2):29–34, 2013.

29. N. V. Kuznetsov, O. A. Kuznetsova, G. A. Leonov, and V. I. Vagaytsev. Hidden attractor in Chua's circuits. *ICINCO 2011—Proceedings of the 8th International Conference on Informatics in Control, Automation and Robotics*, 1:279–283, 2011.

30. N. V. Kuznetsov and G. A. Leonov. Lyapunov quantities, limit cycles and strange behavior of trajectories in two-dimensional quadratic systems. *Journal of Vibroengineering*, 10(4):460–467, 2008.

31. N. V. Kuznetsov, G. A. Leonov, and V. I. Vagaitsev. Analytical–numerical method for attractor localization of generalized Chua's system. *IFAC Proceedings Volumes (IFAC-PapersOnline)*, 4(1):29–33, 2010.

32. N. V. Kuznetsov, O. A. Kuznetsova, G. A. Leonov, P. Neittaanmaki, M. V. Yuldashev, and R. V. Yuldashev. Limitations of the classical phase-locked loop analysis. *Proceedings–IEEE International Symposium on Circuits and Systems*, pp. 533–536, art. no. 7168688, 2015. (doi: 10.1109/ISCAS.2015.7168688).

33. N. V. Kuznetsov and G. A. Leonov. Hidden attractors in dynamical systems: Systems with no equilibria, multistability and coexisting attractors. *IFAC World Congress*, 19(1):5445–5454, 2014.

34. N. V. Kuznetsov, G. A. Leonov, M. V. Yuldashev, and R. V. Yuldashev. Phase-detector characteristic of classical PLL for general case of linear filter. *IFAC Proceedings Volumes (IFAC-PapersOnline)*, 19(1):8253–8258, 2013.

35. N. V. Kuznetsov, V. I. Vagaitsev, G. A. Leonov, and S. M. Seledzhi. Localization of hidden attractors in smooth Chua's systems. *Proceedings of the 2011 International Conference on Applied and Computational Mathematics*, pp. 26–33, 2011.

36. S.-K. Lao, Y. Shekofteh, S. Jafari, and J. C. Sprott. Cost function based on Gaussian mixture model for parameter estimation of a chaotic circuit with a hidden attractor. *International Journal of Bifurcation and Chaos*, 24(1):art. no. 1450010, 2014.

37. G. A. Leonov and N. V. Kuznetsov. Computation of the first Lyapunov quantity for the second-order dynamical system. *IFAC Proceedings Volumes (IFAC-PapersOnline)*, 3:87–89, 2007.

38. G. A. Leonov and N. V. Kuznetsov. Algorithms for searching for hidden oscillations in the Aizerman and Kalman problems. *Doklady Mathematics*, 84(1):475–481, 2011.

39. G. A. Leonov and N. V. Kuznetsov. Hidden attractors in dynamical systems. From hidden oscillations in Hilbert-Kolmogorov, Aizerman, and Kalman problems to hidden chaotic attractors in Chua circuits. *International Journal of Bifurcation and Chaos*, 23(1): art. no. 1330002, 2013.

40. G. A. Leonov, N. V. Kuznetsov, M. A. Kiseleva, E. P. Solovyeva, and A. M. Zaretskiy. Hidden oscillations in mathematical model of drilling system actuated by induction motor with a wound rotor. *Nonlinear Dynamics*, 77(1–2):277–288, 2014.

41. G. A. Leonov, N. V. Kuznetsov, and E. V. Kudryashova. Cycles of two-dimensional systems: Computer calculations, proofs, and experiments. *Vestnik St. Petersburg University. Mathematics*, 41(3):216–250, 2008.

42. G. A. Leonov, N. V. Kuznetsov, and E. V. Kudryashova. A direct method for calculating Lyapunov quantities of two-dimensional dynamical systems. *Proceedings of the Steklov Institute of Mathematics*, 272 (Suppl. 1):S119–S127, 2011.

43. G. A. Leonov, N. V. Kuznetsov, O. A. Kuznetsova, S. M. Seledzhi, and V. I. Vagaitsev. Hidden oscillations in dynamical systems. *Transaction on Systems and Control*, 6(2):54–67, 2011.

44. G. A. Leonov, N. V. Kuznetsov, and V. I. Vagaitsev. Localization of hidden Chua's attractors. *Physics Letters A*, 375(23):2230–2233, 2011.

45. G. A. Leonov, N. V. Kuznetsov, and V. I. Vagaitsev. Hidden attractor in smooth Chua systems. *Physica D: Nonlinear Phenomena*, 241(18):1482–1486, 2012.

46. G. A. Leonov, N. V. Kuznetsov, and T. N. Mokaev. Homoclinic orbit and hidden attractor in the Lorenz-like system describing the fluid convection motion in the rotating cavity. *Communications in Nonlinear Science and Numerical Simulation*, doi:10.1016/j.cnsns.2015.04.007, 2015.

47. G. A. Leonov, N. V. Kuznetsov, and T. N. Mokaev. Homoclinic orbits, and self-excited and hidden attractors in a Lorenz-like system describing convective fluid motion, *European Physical Journal Special Topics*, 224:1421–1458, 2015.

48. C. Li and J. C. Sprott. Coexisting hidden attractors in a 4-D simplified Lorenz system. *International Journal of Bifurcation and Chaos*, 24(03):art. no. 1450034, 2014.

49. Q. Li, H. Zeng, and X.-S. Yang. On hidden twin attractors and bifurcation in the Chua's circuit. *Nonlinear Dynamics*, 77(1–2):255–266, 2014.

50. E. N. Lorenz. Deterministic nonperiodic flow. *Journal of the Atmospheric Sciences*, 20(2):130–141, 1963.

51. R. Lozi and S. Ushiki. The theory of confinors in Chua's circuit: Accurate anatysis of bifurcations and attractors. *International Journal of Bifurcation and Chaos*, 3(2):333–361, 1993.

52. L. Markus and H. Yamabe. Global stability criteria for differential systems. *Osaka Journal of Mathematics*, 12:305–317, 1960.

53. V.-T. Pham, S. Jafari, C. Volos, X. Wang, and S. M. R. H. Golpayegani. Is that really hidden? The presence of complex fixed-points in chaotic flows with no equilibria. *International Journal of Bifurcation and Chaos*, 24(11):art. no. 1450146, 2014.

54. V.-T. Pham, F. Rahma, M. Frasca, and L. Fortuna. Dynamics and synchronization of a novel hyper-chaotic system without equilibrium. *International Journal of Bifurcation and Chaos*, 24(06):art. no. 1450087, 2014.

55. V.-T. Pham, C. Volos, S. Jafari, X. Wang, and S. Vaidyanathan. Hidden hyperchaotic attractor in a novel simple memristive neural network. *Optoelectronics and Advanced Materials – Rapid Communications*, 8(11–12):1157–1163, 2014.

56. A. S. Pikovski, M. I. Rabinovich, and V. Yu. Trakhtengerts. Onset of stochasticity in decay confinement of parametric instability. *Soviet Physics JETP*, 47:715–719, 1978.

57. A.N. Pisarchik and U. Feudel. Control of multistability. *Physics Reports*, 2014.

58. V. A. Pliss. *Some Problems in the Theory of the Stability of Motion*. Izd LGU, Leningrad [in Russian], 1958.
59. A. Prasad. Existence of perpetual points in nonlinear dynamical systems and its applications. arXiv:1409.4921v1, 2014.
60. M.I. Rabinovich. Stochastic autooscillations and turbulence. *Uspehi Physicheskih Nauk* [in Russian], 125(1):123–168, 1978.
61. O. E. Rossler. An equation for continuous chaos. *Physics Letters A*, 57(5):397–398, 1976.
62. J.C. Sprott, X. Wang, and G. Chen. Coexistence of point, periodic and strange attractors. *International Journal of Bifurcation and Chaos*, 23(5):art. no. 1350093, 2013.
63. J. J. Stoker. *Nonlinear Vibrations in Mechanical and Electrical Systems*. Interscience, New York, 1950.
64. S. Timoshenko. *Vibration Problems in Engineering*. Van Nostrand, New York, 1928.
65. Y. Ueda, N. Akamatsu, and C. Hayashi. Computer simulations and non-periodic oscillations. *IEICE Transactions Japan*, 56A(4):218–255, 1973.
66. B. van der Pol. On relaxation-oscillations. *Philosophical Magazine and Journal of Science*, 7(2):978–992, 1926.
67. X. Wang and G. Chen. Constructing a chaotic system with any number of equilibria. *Nonlinear Dynamics*, 71:429–436, 2013.
68. Z. Wei, I. Moroz, and A. Liu. Degenerate Hopf bifurcations, hidden attractors and control in the extended Sprott E system with only one stable equilibrium. *Turkish Journal of Mathematics*, 38(4):672–687, 2014.
69. Z. Wei, R. Wang, and A. Liu. A new finding of the existence of hidden hyperchaotic attractors with no equilibria. *Mathematics and Computers in Simulation*, 100:13–23, 2014.
70. Z. Wei and W. Zhang. Hidden hyperchaotic attractors in a modified Lorenz-Stenflo system with only one stable equilibrium. *International Journal of Bifurcation and Chaos*, 24(10):art. no. 1450127, 2014.
71. Z. Wei, W. Zhang, Z. Wang, and M. Yao. Hidden attractors and dynamical behaviors in an extended Rikitake system. *International Journal of Bifurcation and Chaos*, 25(2):art. no. 1550028, 2015.
72. H. Zhao, Y. Lin, and Y. Dai. Hidden attractors and dynamics of a general autonomous van der Pol-Duffing oscillator. *International Journal of Bifurcation and Chaos*, 24(6):art. no. 1450080, 2014.
73. Z. T. Zhusubaliyev and E. Mosekilde. Multistability and hidden attractors in a multilevel DC/DC converter. *Mathematics and Computers in Simulation*, 109:32–45, 2015.

9

Bifurcation Analysis of a Simple 3D BVP Oscillator and Chaos Synchronization of Its Coupled Systems

Yusuke Nishiuchi and
Tetsushi Ueta

9.1 Introduction

The Bonhoffer–van der Pol (BVP) oscillator is a two-dimensional autonomous system, which can exhibit many nonlinear phenomena for various inputs [3,4]. The extended BVP oscillator forms a simple circuit by addition of a capacitor to the BVP oscillator, further exhibiting a three-dimensional autonomous system. Oscillations, non-scintillation, and many nonlinear phenomena were observed in case of the extended BVP oscillator [5]. Even torus was observed by using the specific nonlinear resistor [1]. In this chapter, we investigate in detail, the bifurcation phenomena in the extended BVP oscillator. First, we reveal bifurcations of the parameter region of chaos attractors with two-parameter bifurcation diagrams computed by adopting the shooting method [6,7], based on numerical integration of variational equations. Second, we discussed an application with relation to the oscillator–chaos synchronization. When we coupled these two oscillators by a resistor, inphase and antiphase chaos synchronization was obtained. We illustrated a bifurcation diagram for these coupled oscillator.

9.2 The Extended BVP Oscillator

Figure 9.1 shows the circuit of extended BVP oscillator. The circuit equation is as follows:

$$\begin{cases} C\dfrac{dv_1}{dt} = -i - g(v_1) \\[2mm] C\dfrac{dv_2}{dt} = i - \dfrac{v_2}{r} \\[2mm] L\dfrac{di}{dt} = v_1 - v_2 \end{cases} \tag{9.1}$$

The circuit (Figure 9.2a) consists of an FET and an op-amp as a nonlinear resistor. The v–i characteristics of the nonlinear resistor is shown in Figure 9.2b. The characteristics $g(v)$ is approximated by the following formula:

$$g(v) = -a \tanh bv$$

Equation 9.2 is the normalized differential equations of Equation 9.1.

$$\begin{cases} \dot{x} = -z + \tanh \gamma x \\ \dot{y} = z - \delta y \\ \dot{z} = x - y \end{cases} \tag{9.2}$$

where,

$$\cdot = d/d\tau, \quad \tau = \frac{1}{\sqrt{LC}}t, \quad \gamma = ab\sqrt{\frac{L}{C}}, \quad \delta = \frac{1}{r}\sqrt{\frac{L}{C}}, \quad x = \frac{v_1}{a}\sqrt{\frac{C}{L}}, \quad y = \frac{v_2}{a}\sqrt{\frac{C}{L}}, \quad z = \frac{i}{a}.$$

Figure 9.3 shows a chaos attractor that can be observed in the extended BVP oscillator.

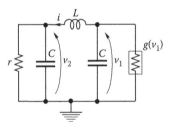

FIGURE 9.1   The extended BVP oscillator.

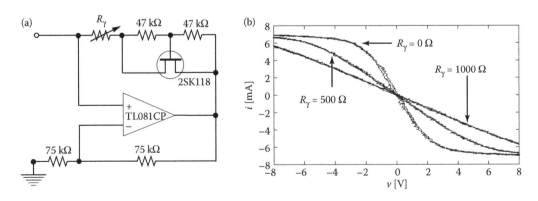

FIGURE 9.2   The nonlinear resistor and its characteristics. (a) The nonlinear resistor circuit. (b) The v–i characteristics of the nonlinear resistor.

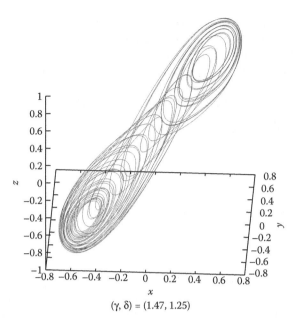

$(\gamma, \delta) = (1.47, 1.25)$

FIGURE 9.3 A chaos attractor of the extended BVP oscillator.

9.3 Bifurcations of the Extended BVP Oscillator

Figure 9.4a shows the bifurcation diagram of the extended BVP oscillator. The origin $(x, y, z) = (0, 0, 0)$ is an equilibrium in all the parameter regions. For the studied circuit, if the Hopf bifurcation of the origin led to generation of an amplitude of limit cycle, then the circuit was considered to be at oscillation state. Hopf h_0 and pitchfork bifurcation d_0 are bifurcation phenomena of the origin, and two equilibria generated by d_0 are symmetric about the origin. All bifurcations of these symmetric equilibria occur in the same parameters.

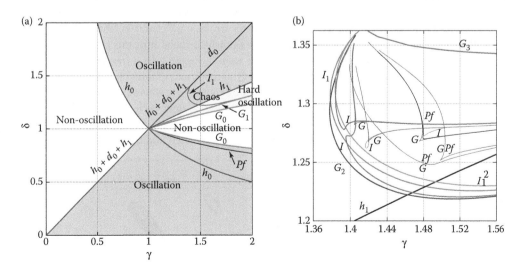

FIGURE 9.4 Bifurcation diagrams of the extended BVP oscillator.

To track bifurcation sets of periodic solutions, the Poincaré mapping is applied. Suppose the system is described by $\dot{x} = f(x)$, where $x = (x, y, z)$, and $x(t) = \varphi(t, x_0)$ be a solution with the initial condition $x(0) = x_0 = \varphi(0, x_0)$. A periodic solution is expressed as $x(t) = \varphi(t + \tau, x_0)$, where τ is the period. We define the Poincaré section for Equation 9.2 as $\prod = \{x \in R^3 \mid x = 0\}$, thus the Poincaré mapping T is written as follows:

$$T : R^3 \to R^3; \quad x \mapsto T(x), x \in R^n \tag{9.3}$$

Thereby, the condition of the fixed point is written as $T(x_0) = x_0$. The Jacobian matrix $\partial\varphi/\partial x_0$ is a principal matrix solution obtained by numerical integration of the following variational equation from $t = 0$ to $t = \tau(x_0)$:

$$\frac{d}{dt}\frac{\partial\varphi}{\partial x_0} = \frac{\partial f}{\partial x}\frac{\partial\varphi}{\partial x_0}, \left.\frac{\partial\varphi}{\partial x_0}\right|_{t=0} = I_3 \tag{9.4}$$

where I_3 is the 3×3 identical matrix. Stability of the fixed point of the Poincaré mapping depends on the roots of the characteristic equation:

$$\chi_\mu = \left|\frac{\partial\varphi}{\partial x_0} - \mu I_3\right| = 0. \tag{9.5}$$

Suppose that $u \in \Sigma \subset R^2$ is a location on the local coordinate, then there is a projection satisfying $p(x_0) = u_0$. Let u_1 be a point on Σ, and $\varphi(t, x_1)$ be the solution starting in $h^{-1}(u_1) = x_1 \in \prod$. Let also the $x_2 \in \prod$ be a point at which $\varphi(t, x_1)$ intersects with the return time $\tau(x_1)$, thus we have $x_2 = \varphi(\tau(x_1), x_1)$. Then, we define the Poincaré mapping on the local coordinate system:

$$\begin{aligned} T_l &: \Sigma \to \Sigma \\ u_1 &\mapsto u_2 = p(\varphi(\tau(h^{-1}(u_1)), p^{-1}(u_1))) = p \circ T \circ p^{-1}(u_1) \end{aligned} \tag{9.6}$$

The fixed point of the mapping T_l is given by

$$T_l(u_0) = u_0 \tag{9.7}$$

A bifurcation set is tracked by a shooting method, that is, a bifurcation parameter value is obtained by solving simultaneous Equations 9.5 and 9.7 with Newton's method [7].

G_0 is the tangent bifurcation of the limit cycle generated by the origin as shown in Figure 9.4a. G_1 and I_1 are the tangent bifurcation and the period-doubling bifurcation, respectively, of limit cycles generated by the symmetric equilibria. All bifurcations of symmetric limit cycles occur in the same parameters. Figure 9.3 shows the chaos attractor observed in a parameter area near the character "chaos" of Figure 9.4a. Figure 9.4b shows a magnified view of the parameter area in which chaos attractors are observed. When parameters are changed from the bottom left to the top right in Figure 9.4b, a transition was observed from the limit cycles to chaos attractors. Figure 9.5 shows the conditions of limit cycles due to the changing parameters. If parameters are fixed in the bottom left of Figure 9.4b, where $(\gamma, \delta) = (1.36, 1.25)$, two symmetric stable limit cycles were found to exist inside the stable limit cycle generated by the origin (Figure 9.5b). When parameters were increased, new symmetric stable limit cycles were generated by the tangent bifurcation G_2 (Figure 9.5a). When parameters were increased additionally, stable limit cycles

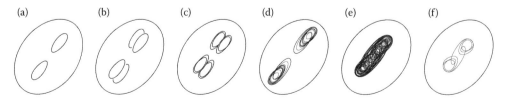

FIGURE 9.5 A transition from the limit cycles to chaos attractors.

changed to period-two limit cycles through period-doubling I_1, I (Figure 9.5c). Then limit cycles were changed to chaos attractors by period-doubling cascades (Figure 9.5d). Additionally, these chaos attractors were fused into one of the chaos attractor by parameter changing (Figure 9.5e). Tangent bifurcation G and pitchfork bifurcation Pf or period-doubling bifurcation I sets were observed in the parameter region of chaos. Period-locking phenomena could also be observed in parameter region near these bifurcation curves (Figure 9.5f). If parameters were changed over G_3, the chaos attractor could not be observed. All limit cycles inside the stable limit cycle generated by the origin disappeared.

9.4 Chaos Synchronization in Coupled Oscillators

Our next interest was finding out various applications of the chaos circuit. We studied chaos synchronization of the coupled circuits. Chaos synchronization is considered to be the most successful application of secure communication methods. Bifurcation phenomena of coupled and simple chaos generator circuits assisted in the study of chaos synchronization. In this section, we investigated the existence of complete chaos synchronization in diffusively coupled two extended BVP oscillators.

Figure 9.6 shows the coupled extended BVP oscillators. Two identical extended BVP oscillators were selected. The terminal of v_1 was chosen as an interface port. The circuit equation is written as follows:

$$\begin{cases} C\frac{dv_1}{dt} = -i_1 - g(v_1) - \frac{v_1 - v_3}{R} \\ C\frac{dv_2}{dt} = i_1 - \frac{v_2}{r} \\ L\frac{di_1}{dt} = v_1 - v_2 \\ C\frac{dv_3}{dt} = -i_2 - g(v_3) + \frac{v_1 - v_3}{R} \\ C\frac{dv_4}{dt} = i_2 - \frac{v_4}{r} \\ L\frac{di_2}{dt} = v_3 - v_4 \end{cases} \tag{9.8}$$

The variable transformations are stated as follows:

$$\cdot = d/d\tau, \quad \tau = \frac{1}{\sqrt{LC}}t, \quad \gamma = ab\sqrt{\frac{L}{C}}, \quad \delta = \frac{1}{r}\sqrt{\frac{L}{C}}, \quad \alpha = \frac{1}{R}\sqrt{\frac{L}{C}},$$

$$x_1 = \frac{v_1}{a}\sqrt{\frac{C}{L}}, \quad y_1 = \frac{v_2}{a}\sqrt{\frac{C}{L}}, \quad z_1 = \frac{i_1}{a}, \quad x_2 = \frac{v_3}{a}\sqrt{\frac{C}{L}}, \quad y_2 = \frac{v_4}{a}\sqrt{\frac{C}{L}}, \quad z_2 = \frac{i_2}{a}$$

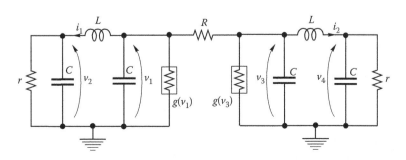

FIGURE 9.6   Coupled extended BVP oscillators.

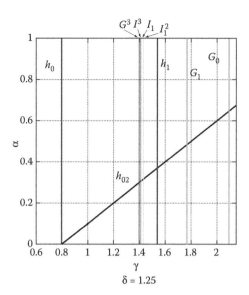

FIGURE 9.7   Bifurcation diagram of coupled extended BVP oscillators.

Thus, we have

$$\begin{cases} \dot{x}_1 = -z_1 + \tanh \gamma x_1 - \alpha(x_1 - x_2) \\ \dot{y}_1 = z_1 - \delta y_1 \\ \dot{z}_1 = x_1 - y_1 \\ \dot{x}_2 = -z_2 + \tanh \gamma x_2 + \alpha(x_1 - x_2) \\ \dot{y}_2 = z_2 - \delta y_2 \\ \dot{z}_2 = x_2 - y_2 \end{cases} \qquad (9.9)$$

Figure 9.7 shows the bifurcation diagram for Equation 9.9. Subsequently, the parameter δ was fixed to 1.25. α is a parameter rerated to the coupling strength. If α was large, oscillators were considered to be strongly influenced by each other. Bifurcations of Figure 9.7 unaffected by α are correspondence bifurcations of Figure 9.4 in $\delta = 1.25$. All attractors generated through bifurcations, unaffected by α, were complete inphase synchronization, and the parameter γ changing showed a transition similar to Figure 9.5. Figure 9.8 shows the limit cycles and chaos attractors generated by these bifurcations. The complete inphase chaos synchronization generated by these bifurcations remained unaffected due to the coupling strength.

The h_{02} of Figure 9.7 shows the Hopf bifurcation of the origin, and a complete antiphase synchronized unstable limit cycle is generated by this bifurcation. There are some bifurcations of complete antiphase synchronized limit cycles in parameter region under h_{02}. Figure 9.9 shows the complete antiphase synchronized limit cycles and chaos synchronization. Complete antiphase synchronized attractors were observed to be coexisting with the complete inphase synchronized attractors, but complete antiphase synchronized attractors were affected by the coupling strength. By parameter α changing, complete antiphase synchronized attractors showed a change in state and size. Though, these synchronized attractors only have a weak attraction force of phase space.

Figure 9.10 represents non-synchronous chaos and torus. These are observed in the parameter area of small α. The single BVP oscillator did not have a chaos attractor, but complete antiphase and inphase synchronization and its bifurcations, chaos, and torus in the weakly coupling strength could be observed in the coupled BVP oscillators. Coupled extended BVP oscillators have similar characteristics to coupled BVP oscillators [2].

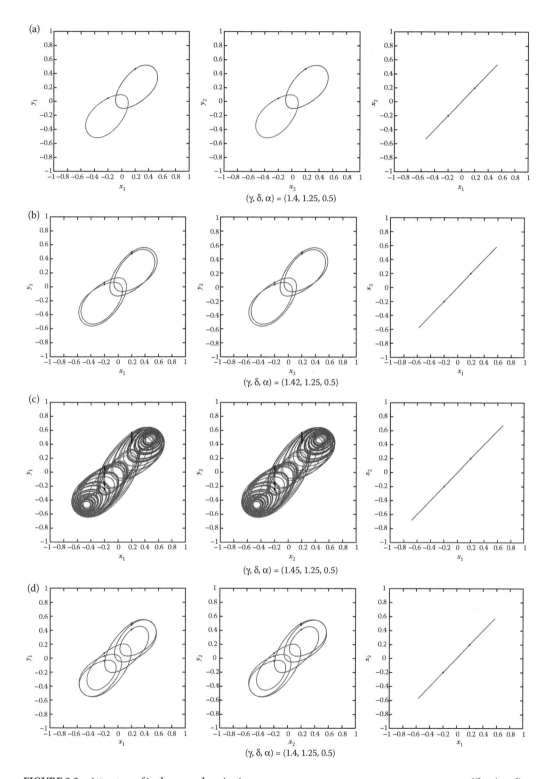

FIGURE 9.8 Attractors of inphase synchronization. *(Continued)*

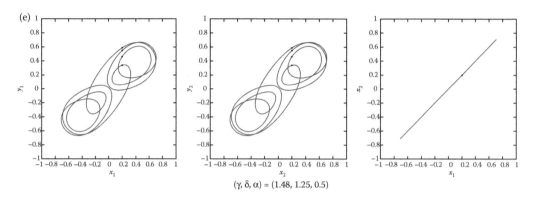

(γ, δ, α) = (1.48, 1.25, 0.5)

FIGURE 9.8 (*Continued*)   Attractors of inphase synchronization.

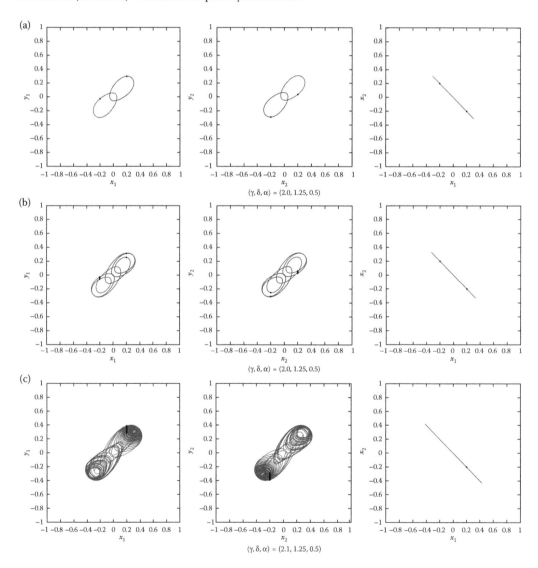

FIGURE 9.9   Attractors of antiphase synchronization.

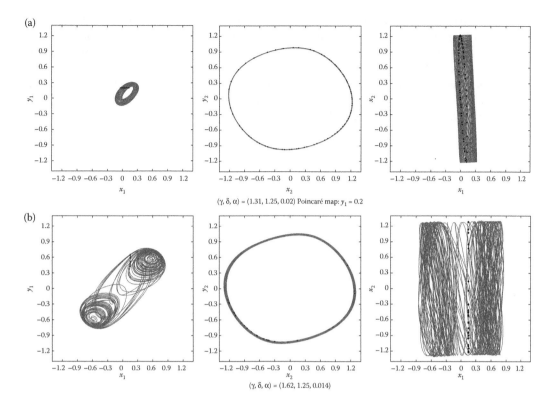

FIGURE 9.10   A chaos attractor and a torus.

9.5 Conclusion

We conducted a detailed investigation on the bifurcation structure of the extended BVP oscillator. By implementing the shooting method featuring variational equations, accurate bifurcation sets have been traced. Some bifurcation sets of periodic window have been shown in chaotic parameter regions.

Next, we studied the coupled extended BVP oscillators which were a six-dimensional autonomous system. In the studied system, complete antiphase and inphase chaos synchronization, chaos, and torus in the weakly coupling strength have been confirmed. These attractors and their transitions were similar to the coupled BVP oscillators. We envisaged the coupled extended BVP oscillators also having the characteristics of connection type.

References

1. Y. Nishiuchi, T. Ueta, and H. Kawakami. Stable torus and its bifurcation phenomena in a simple three-dimensional autonomous circuit. *Chaos, Solitons & Fractals*, 27:941–951, 2006.
2. Y. Nishiuchi, T. Ueta, and H. Kawakami. Classification of bifurcations and oscillations in coupled BVP oscillators. *Journal of Signal Processing*, 18:39–48, 2014.
3. T. Nomura, S. Sato, S. Doi, JP. Segundo, and MD. Stiber. A Bonhoeffer–van der pol oscillator model of locked and non-locked behaviors of living pacemaker neurons. *Biological Cybernetics*, 69:429–437, 1993.
4. T. Nomura, S. Sato, S. Doi, JP. Segundo, and MD. Stiber. Global bifurcation structure of a Bonhoeffer–van der pol oscillator driven by periodic pulse trains. *Biological Cybernetics*, 72:55–67, 1994.

5. T. Ueta and H. Kawakami. Bifurcation and chaos in the extended BVP oscillator. In *Proceedings of International Conference on Nonlinear Dynamics of Electronic Systems (NDES2002)*, Izmir, Turkey, 2002.

6. T. Ueta and A. Tamura. Bifurcation analysis of a simple 3D oscillator and chaos synchronization of its coupled systems. *Chaos, Solitons & Fractals*, 45:1460–1468, 2012.

7. T. Ueta, M. Tsueike, H. Kawakami, T. Yoshinaga, and Y. Katsuta. A computation of bifurcation parameter values for limit cycles. *IEICE Transactions on Fundamentals of Electronics, Communications and Computer Sciences*, E80-A(9):1725–1728, 1997.

10

Capture of a Particle into Resonance

Oleg Mikhailovich
Kiselev

Typically, the dynamics of a particle in a potential well is defined by the following equation:

$$u'' + g(u) = \epsilon f(t) - \mu u'. \tag{10.1}$$

Here, $g(u)$ is a potential force; ϵ and μ are the small positive parameters, s; and $f(t)$ is an outer oscillating force. The potential well is defined by potential energy $G(u)$, so that $G'(u) = g(u)$.

This model seems one of the simplest to study the chaotic and deterministic approaches in mechanics. The full energy of the particle falls due to the dissipation, and the outer force is not influenced when the period is far from resonance. One can obtain this drift of the energy using averaging in perturbation theory [5,7,17]. The period of the oscillations changes due to the changing of the energy. Therefore, the particle goes through close to the resonance with the outer force.

In small distance of the resonance, the system is determined by the equation of nonlinear resonance (see Reference 6). But the dissipation plays a main role in this case as well. The equation for nonlinear resonance with dissipation has two typical solutions. Some of them captured into resonance stay in the resonance for a long time, and some of them pass through the resonance with small changes of parameters. The capture into resonance was studied in the quality theory of ordinary differential equations (see Reference 14).

The capture into resonance is involved by passing through a separatrix for the equation of nonlinear resonance. This process was intensively studied from different points of view. First, the passing is considered from symbolic dynamics [1] and separatrix maps (see review [20]). Second, is calculating the change of an action variable [3,15,21] and phase variable [9]. Another approach that gives uniform asymptotic using the matching method [2,8] is developed in a series of work [12,18,19].

The problem is to determine the trajectories that will be captured into the resonance and the trajectories that will pass through the resonance layer. Numerical and mechanical models in such a case demonstrate an unstable behavior and are traditionally classified as deterministic models with chaotic behavior [4,16].

The capture is allowed due to a gap between separatrices for dissipative equation for nonlinear resonance. The width of the gap is defined by Melnikov's integral [13], which was found to be very thin. Therefore, to obtain an analytical approach for the capture, the initial data for the trajectory should be given with small error. The asymptotic approach for the capture was given in References 10 and 11.

10.1 Resonant Levels of Energy

The nonperturbed dynamics of the particle is defined by the equation in the form

$$U'' + g(U) = 0.$$

The nonperturbed equation has a periodic solution $U(t - t_0, E)$, which depends on time shift t_0 and full energy:

$$E = \frac{(U')^2}{2} + G(U).$$

The period of motion is defined by the following formula:

$$T = \int_{\mathcal{L}} \frac{dU}{U'}, \quad \mathcal{L} = \{(U, U') : E = \text{const}\}.$$

We will study a typical case, when the period depends on the energy and $dT/dE > 0$.

Let us consider the perturbed force that is a conditionally periodic function:

$$f(t) = \sum_{k=0}^{N} f_k(t), \quad f_k(t + T_k) = f_k(t), \quad T_k = \text{const} > 0, \quad \int_t^{t+T_k} f_k(t)dt = 0.$$

The levels of energy $E = E_{n,m}^k$ are resonant if

$$nT(E_{n,m}^k) = mT_k, \quad n, m \in \mathbb{N}.$$

the rational digits are everywhere dense, then the resonant levels are everywhere dense as well. But the capture into resonance depends on a balance of the energy:

The energy of solutions for perturbed equations is

$$\frac{dE}{dt} = \epsilon f(t)u' - \mu(u')^2.$$

The dissipative term $\mu(u')^2$ leads to decreasing of the energy. Therefore, the capture into resonance at the resonant level of energy $E_{n,m}^k$ can be realized if the dissipation of the energy does not exceed the increasing of the energy due to the force f_k. The energy balance for the observed resonant level of energy $E_{n,m}^k$ is

$$\epsilon \int_0^{mT_k} f_k(t)U_t(t, E_{n,m}^k)dt \geq \mu \int_0^{mT_k} (U_t(t, E_{n,m}^k))^2 dt.$$

10.2 Equation for Neighborhood of the Resonant Level

To study the solution near the resonant level of energy, we use the two-scaling method and averaging over fast variable. Let us define the slow-time variable as $\tau = \sqrt{\epsilon}t$.

One should consider the asymptotic expansion for the solution as follows:

$$u = u_0(t + \phi(\tau), E_{n,m}^k).$$

Here, $\phi(\tau)$ is a new function, which defines the phase shift for the asymptotic. The indexes m, n, and k will be omitted for simplicity.

Using the two-scale method, one can obtain a new form of the equation for the particle

$$(1 + 2\sqrt{\epsilon}\phi' + \epsilon(\phi')^2)u'' + \epsilon\phi''u' + g(u) = \epsilon f(t) - \mu(1 + \sqrt{\epsilon}\phi')u'.$$

Then, the equation for the energy

$$\frac{dE}{dt} = -2\sqrt{\epsilon}\phi'u''u' + \epsilon(f(t)u' - \phi''u'^2 - \frac{\mu}{\epsilon}(1 + \sqrt{\epsilon}\phi')(u')^2).$$

The change of energy is small on captured trajectories. Then, an averaging of the equation for the energy over fast variable t leads to an equation for ϕ. Define

$$I = \lim_{S \to \infty} \frac{1}{S} \int_{t_0}^{T} (u')^2 dt$$

and

$$A(\phi) = -\lim_{S \to \infty} \frac{1}{S} \int_{t_0}^{S} f(t)u'(t + \phi)dt.$$

The parameter I is averaging of the action for the particle, and $A(\phi)$ is a projection of $f(t)$ on $u'(t + \phi)$. Function $A(s)$ is periodic: $A(s + T_k) = A(s)$.

As a result, one can obtain the equations for the function ϕ:

$$I\phi'' + A(\phi) + \frac{\mu}{\epsilon}I + \sqrt{\epsilon}I\phi' = 0.$$

It is convenient to use new definitions: $\alpha = A/I$, $\gamma = \mu/\epsilon$. It yields

$$\phi'' + \alpha(\phi) + \gamma = -\sqrt{\epsilon}\phi'. \tag{10.2}$$

10.3 Capture of Trajectories

To study trajectories of Equation 10.2, we consider a phase plane (ϕ, ϕ'). The trajectories are solutions of the first-order equation:

$$\frac{d\phi'}{d\phi} = \frac{-\alpha(\phi) - \gamma}{\phi'} - \sqrt{\epsilon}.$$

In case $\gamma > |\alpha(\phi)|$ for $\forall \phi \in \mathbb{R}$, the capture does not exist and the resonant layer is not observed for such a level of E. If γ such that $\exists \phi \in \mathbb{R}$, $|\alpha(\phi)| > \gamma$, then there exist singular points of the equation $\alpha(\phi_j) = -\gamma$ and $\phi'_j = 0$. The singular points are stable and unstable focuses that alternate on line $\phi' = 0$.

Nonsingular trajectories of the equation are close to trajectories of integrable equation for $\epsilon = 0$. They are close to parabolas $\phi'^2 = -2\gamma\phi$ as $\phi' \to \pm\infty$ or close to spirals in neighbors of stable focuses. Each stable focus is concluded into two separatrices. The separatrices strive to unstable focus but one of the separatrices has almost a closed loop around the stable focus. The trajectories concluded between these two separatices strive to achieve a stable focus. The size of a gap between the separatrix is defined by Melnikov's integral:

$$M = \sqrt{\epsilon} \int_{\mathcal{L}} \phi'd\phi, \quad \mathcal{L} : (\phi, \phi').$$

The trajectories passed through the gap are captured into the resonance. The gap in the phase manifold has coordinates $\phi = \phi_j$, $0 < \phi' < \sqrt{2M}$. Therefore, the trajectories with data $\phi = 0$ and $0 < \phi' < \sqrt{2M}$ at any moment of time τ are captured into resonance. To find such trajectories, one should solve the family of Cauchy problems with such initial data.

As a result, one can see that the capture into resonance is fully a deterministic process. This process is defined by the Cauchy problem for a second-order differential equation (see Figure 10.1).

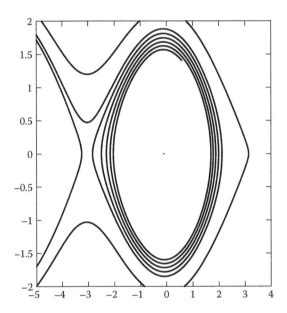

FIGURE 10.1 Captured and passed trajectories of Equation 10.2 on phase plane, as $\alpha(\phi) \equiv \sin(\phi)$, $\gamma = 0.1$, and $\epsilon = 0.0001$.

10.4 Structure of Captured Manifold

To determine the structure of the set of captured trajectories, we need to invert asymptotic formulas for the inner asymptotics near the unstable focus.

Far from the resonant level of the energy, the solution has the form

$$u \sim U(S,E), \quad S \sim \frac{\Omega(\theta)}{\mu} + \Phi(\theta), \quad \Omega' \int_{\mathcal{L}} \frac{dU}{U'} = 2\pi.$$

Evolution of E is described by averaging equation for an action of unperturbed equation:

$$\frac{dI}{dt} = -\mu I.$$

The beam of the captured trajectories looks like a thin spiral with width of order $O(\sqrt{\epsilon})$. Such beams of the trajectories start from small distance of saddles ϕ_j of each resonance level of energy $E^k_{n,m}$.

This implies the accuracy of initial data for the Cauchy problem, which give an opportunity to solve the problem of capture by using asymptotic methods. To predict the capture, the initial data should be done with accuracy $o(\mu\sqrt{\epsilon})$.

References

1. V.M. Alekseev. Quasirandom dynamical systems. i, ii, iii. *Matematicheskii Sbornik (N.S.)*, 76,77,78(1,4,1):72–134,545–601,3–50, 1968,1968,1969.
2. A.M. Il'in. *Matching of Asymptotic Expansions of Solutions of Boundary Value Problem*. AMS, Providence, RI, 1992.
3. A. Neishtadt and A. Vasiliev. Phase change between separatrix crossings in slow-fast hamiltonian systems. *Nonlinearity*, 18:1393–1406, 2005.
4. Vladimir I. Arnold. Applicability conditions and an error bound for the averaging method for systems in the process of evolution through a resonance. *Doklady Akademii nauk SSSR*, 161:9–12, 1965.

5. F.J. Bourland and R. Haberman. The modulated phase shift for strongly nonlinear, slowly varying and weakly damped oscillators. *SIAM Journal on Applied Mathematics*, 48(4):737–748, 1988.
6. Boris V. Chirikov. Resonance processes in magnetic traps. *Atomnaya Energiya*, 6(1):630–638, 1959.
7. G.E. Kuzmak. Asymptotic solutions of nonlinear second order differential equations with variable coefficients. *Journal of Applied Mathematics and Mechanics*, 23(3):730–744, 1959.
8. J. Kevorkian and J.D. Cole. Multiple scale and singular perturbation methods, *Applied Mathematical Science*, Vol. 114. Springer, New York, 1996.
9. J.R. Cary and R.T. Skodje. Phase change between separatrix crossing. *Physica D*, 36, 1989.
10. Oleg Kiselev and Nikolai Tarkhanov. The capture of a particle into resonance at potential hole with dissipative perturbation. *Chaos, Solitons & Fractals*, 58:27–39, 2014.
11. Oleg Kiselev and Nikolai Tarkhanov. Scattering of trajectories at a separatrix under autoresonance. *Journal of Mathematical Physics*, 55:063502, 2014.
12. O.M. Kiselev. Hard loss of stability in painleve-2 equation. *Journal of Nonlinear Mathematical Physics*, 8(1):65–95, 2001.
13. V.K. Mel'nikov. On the stability of a center for time-periodic perturbations. *Trudy Moskovskogo Matematicheskogo Obshchestva*, 12:3–52, 1963.
14. A.D. Morozov. *Quasi-Conservative Systems. Cycles, Resonances and Chaos.* World Scientific Publishing Co., Inc., River Edge, 1998.
15. A.I. Neishtadt. Passage through a separatrix in a resonance problem with a slowly-varying parameter. *Journal of Applied Mathematics and Mechanics*, 39:594–605, 1975.
16. A.I. Neishtadt. Passing through resonances in a two-frequency problem. *Doklady Akademii Nauk SSSR*, 221(2):301–304, 1975.
17. N.N. Bogolyubov and Yu.A. Mitropolskii. *Asymptotic Methods in the Theory of Non-linear Oscillations.* Gordon and Breach Science Publishers, New York, 1961.
18. O.M. Kiselev. Oscillations near a separatrix in the Duffing equation. *Proceedings of the Steklov Institute of Mathematics*, 281:82–94, 2013.
19. O.M. Kiselev and S.G. Glebov. An asymptotic solution slowly crossing the separatrix near a saddle-center bifurcation point. *Nonlinearity*, 16:327–362, 2003.
20. G.N. Piftankin and D.V. Treshchev. Separatrix maps in hamiltonian systems. *Russian Mathematical Surveys*, 62:219–322, 2007.
21. R. Haberman. Slowly varying jump and transition phenomena associated with algebraic bifurcation problem. *SIAM Journal of Applied Mathematics*, 37:69–109, 1979.

III

Chaotic Data Analysis, Equations, and Applications

11

Integral Equations and Applications

Alexander G. Ramm

The goal of this chapter is to formulate some of the basic results on the theory of integral equations and mention some of its applications. The literature of this subject is very large. Proofs are not given due to space restriction. The results are taken from the works mentioned in the references.

11.1 Fredholm Equations

11.1.1 Fredholm Alternative

One of the most important results of the theory of integral equations is the Fredholm alternative. The results in this subsection are taken from References 5, 6, 20, and 21.

Consider the equation

$$u = Ku + f, \quad Ku = \int_D K(x,y)u(y)dy. \tag{11.1}$$

Here, f and $K(x,y)$ are given functions, $K(x,y)$ is called the kernel of the operator K. Equation 11.1 is usually considered in $H = L^2(D)$ or $C(D)$. The first is a Hilbert space with the norm $||u|| := \left(\int_D |u|^2 dx\right)^{1/2}$ and the second is a Banach space with the norm $||u|| = \sup_{x \in D} |u(x)|$. If the domain $D \subset \mathbb{R}^n$ is assumed bounded, then, K is compact in H if, for example, $\int_D \int_D |K(x,y)|^2 dx dy < \infty$, and K is compact in $C(D)$ if, for example, $K(x,y)$ is a continuous function, or, more generally, $\sup_{x \in D} \int_D |K(x,y)| dy < \infty$.

For Equation 11.1 with compact operators, the basic result is the Fredholm alternative. To formulate it one needs some notations. Consider Equation 11.1 with a parameter $\mu \in \mathbb{C}$:

$$u = \mu K u + f \tag{11.2}$$

and its adjoint equation

$$v = \bar{\mu} K^* v + g, \quad K^*(x, y) = \overline{K(y, x)}, \tag{11.3}$$

where the overbar stands for a complex conjugate. By (11.2_0) and (11.3_0), the corresponding equations with $f = 0$, respectively, $g = 0$ are denoted. If K is compact so is K^*. In H one has $(Ku, v) = (u, K^*v)$, where $(u, v) = \int_D u\bar{v} dx$. If equation (11.2_0) has a nontrivial solution u_0, $u_0 \neq 0$, then the corresponding parameter μ is called the characteristic value of K and the nontrivial solution u_0 is called an eigenfunction of K. The value $\lambda = \dfrac{1}{\mu}$ is then called an eigenvalue of K. Similar terminology is used for equation (11.3_0). Let us denote by $N(I - \mu K) = \{u : u = \mu K u\}$, $R(I - \mu K) = \{f : f = (I - \mu K)u\}$. Let us formulate the Fredholm alternative.

Theorem 11.1

If $N(I - \mu K) = \{0\}$, then $N(I - \bar{\mu} K^*) = \{0\}$, $R(I - \mu K) = H$, $R(I - \bar{\mu} K^*) = H$.

If $N(I - \mu K) \neq \{0\}$, then $\dim N(I - \mu K) = n < \infty$, $\dim N(I - \bar{\mu} K^*) = n$, $f \in R(I - \mu K)$ if and only if $f \perp N(I - \bar{\mu} K^*)$, and $g \in R(I - \bar{\mu} K^*)$ if and only if $g \perp N(I - \mu K)$.

Remark 11.1

Conditions $f \perp N(I - \bar{\mu} K^*)$ and $g \perp N(I - \mu K)$ are often written as $(f, v_{0j}) = 0, 1 \le j \le n$, and, respectively, $(g, u_{0j}) = 0, 1 \le j \le n$, where $\{v_{0j}\}_{j=1}^n$ is a basis of the subspace $N(I - \bar{\mu} K^*)$, and $\{u_{0j}\}_{j=1}^n$ is a basis of the subspace $N(I - \mu K)$.

Remark 11.2

Theorem 11.1 is very useful in applications. It says that Equation 11.1 is solvable for any f as long as equation (2_0) has only the trivial solution $v_0 = 0$. A simple proof of Theorem 11.1 can be found in References 20 and 21.

Remark 11.3

Theorem 11.1 can be stated similarly for Equations 11.2 and 11.3 in $C(D)$.

Remark 11.4

If $n > 0$ and $f \in R(I - \mu K)$, then the general solution to Equation 11.2 is $u = \tilde{u} + \displaystyle\sum_{j=1}^{n} c_j u_{0j}$, where c_j, $1 \le j \le n$, are arbitrary constants, $u_{0j} \in N(I - \mu K)$ and $\tilde{u}$ is a particular solution to Equation 11.2. A similar result holds for v.

11.1.2 Degenerate Kernels

Suppose that

$$K(x, y) = \sum_{m=1}^{M} a_m(x) \overline{b_m(y)}, \tag{11.4}$$

where the functions a_m and b_m are linearly independent, $a_m, b_m \in H$, and the overline in this subsection denotes the complex conjugate. The operator K with the degenerate kernel 11.4 is called a finite-rank operator. Equation 11.2 with degenerate kernel 11.4 can be reduced to a linear algebraic system (LAS). Denote

$$u_m = \int_D u\overline{b_m}dx. \tag{11.5}$$

Multiply Equation 11.2 by $\overline{b_p}$, integrate over D, and get

$$u_p = \mu \sum_{m=1}^{M} a_{pm}u_m + f_p, \quad 1 \le p \le M, \tag{11.6}$$

where $a_{pm} := (a_m, b_p), f_p := (f, b_p)$. There is a one-to-one correspondence between solutions to Equations 11.2 and 11.6:

$$u(x) = \mu \sum_{m=1}^{M} a_m(x)u_m + f(x). \tag{11.7}$$

Exercise 11.1

Solve the equation $u(x) = \mu \int_0^{\pi} \sin(x - t)u(t)dt + 1$.

Hint: Use the formula $\sin(x - t) = \sin x \cos t - \cos x \sin t$. Find characteristic values and eigenfunctions of the operator with the kernel $\sin(x - t)$ on the interval $(0, \pi)$.

Remark 11.5

If K is a compact integral operator in $L^2(D)$ then there is a sequence $K_n(x, y)$ of degenerate kernels such that $\lim_{n\to\infty} ||K - K_n|| = 0$, where $||K||$ is the norm of the operator K, $||K|| := \sup_{u \ne 0} \dfrac{||Ku||}{||u||}$.

11.1.3 Volterra Equations

If $K(x, y) = 0$ for $y > x$ and $\dim D = 1$ then Equation 11.2 is called a Volterra equation. It is of the form

$$u(x) = \mu \int_a^x K(x, y)u(y)dy + f(x), \quad a \le x \le b. \tag{11.8}$$

This is a particular case of the Fredholm equations. But the Volterra equation (11.8) has a special property.

Theorem 11.2

Equation 11.8 is uniquely solvable for any μ and any f in $C(D)$, if the function $K(x, y)$ is continuous in the region $a \le y \le x, a \le x \le b$. The solution to Equation 11.8 can be obtained by iterations:

$$u_{n+1}(x) = \mu Vu_n + f, \quad u_1 = f; \quad Vu := \int_a^x K(x, y)u(y)dy. \tag{11.9}$$

11.1.4 Self-Adjoint Operators

If $K^* = K$ then K is called self-adjoint. In this case, Equation 11.2 has at least one characteristic value, all characteristic values are real numbers, and the corresponding eigenfunctions are orthogonal. One can solve Equation 11.2 with the self-adjoint operator K by the formula

$$u = \sum_{j=1}^{\infty} \frac{\mu_j f_j}{\mu_j - \lambda} u_{0j}(x), \quad f_j = (f, u_{0j}), \tag{11.10}$$

where $u_{0j} = \mu_j K u_{0j}, j = 1, 2, \dots$.

For the eigenvalues of a self-adjoint compact operator K the following result (minimax representation) holds. We write $K \ge 0$ if $(Ku, u) \ge 0, \forall u \in H$, and $A \le B$ if $(Au, u) \le (Bu, u), \forall u \in H$.

Theorem 11.3

If $K = K^* \geq 0$ is compact in H, $Ku_j = \lambda_j u_j$, $u_j \neq 0$ then

$$\lambda_{j+1} = \max_{u \perp L_j} \frac{(Ku, u)}{||u||^2}, \quad j = 0, 1, \ldots \tag{11.11}$$

Here $L_0 = \{0\}$, L_j is the subspace with the basis $\{u_1, \ldots, u_j\}$ and $(u_i, u_m) = \delta_{im}$. Also

$$\lambda_{j+1} = \min_M \max_{u \perp M} \frac{(Ku, u)}{||u||^2}, \quad \dim M = j, \tag{11.12}$$

where M runs through the set of j-dimensional subspaces of H.

One has $\lambda_1 \geq \lambda_2 \geq \cdots \geq 0$. The set of all eigenfunctions of K, including the ones corresponding to the zero eigenvalue, if such an eigenvalue exists, forms an orthonormal basis of H.

Remark 11.6

If $0 \leq A \leq B$ are compact operators in H, then $\lambda_j(A) \leq \lambda_j(B)$.

Remark 11.7

If A is a linear compact operator in H, then $A = U|A|$, where $|A| = (A^*A)^{1/2} \geq 0$ and U is a partial isometry

$$U : R(|A|) \to R(A), \quad ||U|A|f|| = ||Af|| \; \forall f \in H, \; N(U) = N(A)$$

and $\overline{R(|A|)} = \overline{R(A^*)}$, where the overline denotes the closure in H. One has

$$Au = \sum_{j=1}^{r(A)} s_j(A)(u, u_j)v_j. \tag{11.13}$$

Here $r(A) = \dim R(|A|)$, $s_j \geq 0$ are the eigenvalues of $|A|$. They are called the s-values of A, $\{u_j\}$ and $\{v_j\}$ are orthonormal systems, $|A|u_j = s_j u_j$, $s_1 \geq s_2 \geq \ldots$, and $v_j = Uu_j$.

Remark 11.8

If A is a linear compact operator in H then

$$s_{j+1}(A) = \min_{K \in K_j} ||A - K||, \tag{11.14}$$

where K_j is the set of finite-rank operators of dimension $\leq j$, that is, $\dim R(K_j) = j$.

11.1.5 Equations of the First Kind

If $Ku = \lambda u + f$ and $\lambda = 0$, then one has

$$Ku = f. \tag{11.15}$$

If K is compact in H, then Equation 11.15 is called Fredholm equation of the first kind. The linear set $R(K)$ is not closed unless K is a finite-rank operator. Therefore, a small perturbation of $f \in R(K)$ may lead to an equation that has no solutions, or, if it has a solution, this solution may differ very much from the solution to Equation 11.15. Such problems are called ill-posed. They are important in many applications, especially in inverse problems [4,15]. The usual statement of the problem of solving an ill-posed equation

(11.15) consists of the following. It is assumed that Equation 11.15 is solvable for the exact data f, possibly nonuniquely, that the exact data f are not known but the "noisy" data f_δ are known, and $||f_\delta - f|| \leq \delta$, where $\delta > 0$ is known. Given f_δ, one wants to find u_δ such that $\lim\limits_{\delta \to 0} ||u - u_\delta|| = 0$.

There are many methods to solve the above problem: the DSM (dynamical systems method) [17,18], iterative methods, variational regularization, and their variations [4,15].

Examples of ill-posed problems of practical interest include stable numerical differentiation, stable summation of the Fourier series and integrals with perturbed coefficients, stable solution to LAS with large condition numbers, solving Fredholm and Volterra integral equations of the first kind, various deconvolution problems, various inverse problems, the Cauchy problem for the Laplace equation and other elliptic equations, the backward heat problem, tomography, and many other problems [15].

11.1.6 Numerical Solution of Fredholm Integral Equations

Let us describe the projection method for solving Equation 11.2 with a compact operator K. Let $L_n \subset H$ be a linear subspace of H, $\dim L_n = n$. Assume throughout that the sequence $L_n \subset L_{n+1}$ is limit dense in H, that is, $\lim\limits_{a \to \infty} \rho(f, L_n) = 0$ for any $f \in H$, where $\rho(f, L_n) = \inf\limits_{u \in L_n} ||f - u||$. Let P_n be an orthogonal projection operator on L_n, that is, $P_n^2 = P_n$, $P_n f \in L_n$, $||f - P_n f|| \leq ||f - u||$, $\forall u$ in L_n.

Assume that $N(I - K) = \{0\}$. Then $||(I - K)^{-1}|| \leq c$. Consider the projection method for solving Equation 11.1:

$$P_n(I - K)P_n u_n = P_n f. \tag{11.16}$$

Theorem 11.4

If $N(I - K) = \{0\}$, K is compact, and $\{L_n\}$ is limit dense, then Equation 11.16 is uniquely solvable for all n sufficiently large, and $\lim\limits_{n \to \infty} ||u_n - u|| = 0$, where u is the solution to Equation 11.1.

Remark 11.9

The projection method is valid also for solving equations $Au = f$, where $A : X \to Y$ is a linear operator, X and Y are Hilbert spaces. Let $Y_n \subset Y$ and $X_n \subset X$ be n-dimensional subspaces, P_n and Q_n are projection operators on X_n and Y_n respectively. Assume that the sequences $\{X_n\}$ and $\{Y_n\}$ are limit dense and consider the projection equation

$$Q_n A P_n u_n = Q_n f, \quad u_n \in X_n. \tag{11.17}$$

Theorem 11.5

Assume that A is a bounded linear operator, $||A^{-1}|| \leq c$, and

$$||Q_n A P_n u|| \geq c||P_n u||, \quad \forall u \in X, \tag{11.18}$$

where $c > 0$ stands for various positive constants independent of n and u. Then Equation 11.17 is uniquely solvable for any $f \in Y$ for all sufficiently large n, and

$$\lim\limits_{n \to \infty} ||u - u_n|| = 0. \tag{11.19}$$

Conversely, if Equation 11.17 is uniquely solvable for any $f \in Y$ for all sufficiently large n, and Equation 11.19 holds, then Equation 11.18 holds.

Remark 11.10

If Equation 11.18 holds for A, then it will hold for $A + B$ if $||B|| > 0$ is sufficiently small. If Equation 11.18 holds for A, then it holds for $A + B$ if B is compact and $||(A + B)^{-1}|| \leq c$.

Iterative methods for solving integral equation (11.1) are popular. If $||K|| < 1$, then Equation 11.1 has a unique solution for every $f \in H$, and this solution can be obtained by the iterative method

$$u_{n+1} = Ku_n + f, \quad \lim_{n \to \infty} ||u_n - u|| = 0, \tag{11.20}$$

where $u_1 \in H$ is arbitrary.

Define the spectral radius of a linear-bounded operator A in a Banach space by the formula $\rho(A) = \lim_{n \to \infty} ||A^n||^{1/n}$. It is known that this limit exists [5,12]. If a is a number, then $\rho(aA) = |a|\rho(A)$. The basic fact is: If $|\lambda| > \rho(A)$, then the equation $\lambda u = Au + f$ is uniquely solvable for any f, and the iterative process $\lambda u_{n+1} = Au_n + f$ converges to $u = (\lambda I - A)^{-1}f$ for any $f \in H$ and any initial approximation u_1.

Given an equation $Au = f$, can one transform it into an equation $u = Bu + g$ with $\rho(B) < 1$, so that it can be solved by iterations. To do so, one may look for an operator T such that $(A + T)u = Tu + f$, $(A + T)^{-1}$ is bounded; so, $u = (A + T)^{-1}Tu + (A + T)^{-1}f$, and $\rho((A + T)^{-1}T) < 1$. Finding such an operator T is, in general, not simple. If $A = A^*$, $0 \leq m \leq A \leq M$, where m and M are constants, one may transform equation $Au = f$ to the equivalent form $u = u - \frac{2Au}{m+M} + \frac{2}{m+M}f$, and get $\rho(I - \frac{2A}{m+M}) \leq \frac{M-m}{M+m} < 1$. If $A \neq A^*$ then equation $A^*Au = A^*f$ is useful since A^*A is a self-adjoint operator. The above results can be found, for example, in References 6 and 30.

Equation $Au = f$ is often convenient to consider in a cone of a Banach space X [30]. A cone is a closed convex set $K \subset X$ that contains with a point u all the points $tu, t \geq 0$, and such that $u \in K$ and $-u \in K$ imply $u = 0$. One writes $u \leq v$ if $v - u \in K$. This defines a semiorder in X. If K is invariant with respect to A, that is, $u \in K$ implies $Au \in K$, then one can give conditions for the solvability of the equation $Au = f$ in K. For example, let $X = C(D), D \subset \mathbb{R}^m$ is a bounded domain, $K = \{u : u \geq 0\}$,

$$Au = \int_D A(x,y)u(y)dy, \quad A(x,y) \geq 0.$$

Remark 11.11

The following result is useful:

Suppose that there exist functions v and w in K such that $Av \geq v$, $Aw \leq w$, and $v \leq w$. Let us assume that every monotone-bounded sequence in K, $u_1 \leq u_2 \leq \cdots \leq U$ has a limit: $\exists u \in K$ such that $\lim_{n \to \infty} ||u_n - u|| = 0$. Then there exists $u \in K$ such that $u = Au$, where $v \leq u \leq w$.

A cone is called normal if $0 \leq u \leq v$ implies $||u|| \leq c||v||$, where $c > 0$ is constant independent of u and v.

11.2 Integral Equations with Special Kernels

11.2.1 Equations with Displacement Kernel

The results in this subsection are taken from References 2 and 30.

Let us mention some equations with displacement kernels. We start with the equation

$$\lambda u(t) - \int_0^t K(t-s)u(s)ds = f(t), \quad t \geq 0.$$

This equation is solved by taking the Laplace transform: $\lambda \bar{u} - \bar{K}\bar{u} = \bar{f}$, $\bar{u} = (\lambda - \bar{K})^{-1}\bar{f}$, where $\bar{u} := Lu := \int_0^\infty e^{-pt}u(t)dt$, and

$$u(t) = L^{-1}\bar{u} = \frac{1}{2\pi i} \int_{\sigma-i\infty}^{\sigma+i\infty} e^{pt}\bar{u}(p)dp.$$

Equation

$$\lambda u(t) - \int_{-\infty}^{\infty} K(t-s)u(s)ds = f(t), \quad -\infty < t < \infty,$$

is solved by taking the Fourier transform: $\tilde{u}(\xi) := \int_{-\infty}^{\infty} u(t)e^{-i\xi t}dt$. One has $\tilde{u}(\xi) = (\lambda - \tilde{K}(\xi))^{-1}\tilde{f}$, and

$$u(t) = \frac{1}{2\pi} \int_{-\infty}^{\infty} \tilde{u}(\xi)e^{i\xi t}d\xi.$$

In these formal calculations one assumes that $1 - \tilde{K}(\xi) \neq 0$ [2,11].

More complicated is the Wiener–Hopf equation

$$u(t) - \int_0^{\infty} K(t-s)u(s)ds = f(t), \quad t \geq 0. \tag{11.21}$$

Assume that $K(t) \in L(-\infty, \infty)$ and $1 - \tilde{K}(\xi) \neq 0$, $\forall \xi \in (-\infty, \infty)$. Define the index:

$$\kappa := -\frac{1}{2\pi} \arg[1 - \tilde{K}(\xi)]|_{-\infty}^{\infty}.$$

The simplest result is

If and only if $\kappa = 0$ Equation 11.21 is uniquely solvable in $L^p(-\infty, \infty)$, $p > 1$.

The solution can be obtained analytically by the following scheme. Denote $u_+(t) := \begin{cases} u(t), & t \geq 0 \\ 0, & t < 0 \end{cases}$.

Write Equation 11.21 as

$$u_+(t) = \int_{-\infty}^{\infty} K(t-s)u_+(s)ds + u_-(t) + f_+, \quad -\infty < t < \infty,$$

where $u_-(t) = 0$ for $t \geq 0$ is an unknown function. Take the Fourier transform of this equation to get $[1 - \tilde{K}(\xi)]\tilde{u}_+ = \tilde{f}_+ + \tilde{u}_-$. Write $1 - \tilde{K}(\xi) = \tilde{K}_+(\xi)\tilde{K}_-(\xi)$, where $\tilde{K}_+$ and $\tilde{K}_-$ are analytic in the upper, respectively, lower half-plane functions. Then

$$\tilde{K}_+(\xi)\tilde{u}_+(\xi) = \tilde{f}_+(\xi)\tilde{K}_-^{-1}(\xi) + \tilde{u}_-(\xi)\tilde{K}_-(\xi).$$

Let P_+ be the projection operator on the space of functions analytic in the upper half-plane. Then

$$\tilde{u}_+ = \tilde{K}_+^{-1}(\xi)P_+(\tilde{f}_+(\xi)\tilde{K}_-^{-1}(\xi)).$$

If $\tilde{u}_+$ is known, then $u_+(t)$ is known, and the Wiener–Hopf equation is solved. The nontrivial part of this solution is the factorization of the function $1 - \tilde{K}(\xi)$.

11.2.2 Equations Basic in Random Fields Estimation Theory

In this subsection the results from Reference 16 are given.

Let $u(x) = s(x) + n(x)$, $x \in \mathbb{R}^r$, r is a positive integer, $s(x)$ and $n(x)$ are random fields, $s(x)$ is a "useful signal," and $n(x)$ is "noise." Assume that the mean values $\overline{s(x)} = \overline{n(x)} = 0$, where the overbar denotes the mean value. Assume that covariance functions

$$\overline{u^*(x)u(y)} := R(x,y), \quad \overline{u^*(x)s(y)} := f(x,y)$$

are known. The star denotes a complex conjugate. Suppose that a linear estimate $\mathcal{L}u$ of the observations in a domain D is

$$\mathcal{L}u = \int_D h(x,y)u(y)dy,$$

where $h(x,y)$ is, in general, a distributional kernel. Then the optimal estimate of As, where A is a linear operator, is found by solving the optimization problem

$$\epsilon := \overline{(\mathcal{L}u - As)^2} = \min. \tag{11.22}$$

A necessary condition for the kernel (filter) $h(x,y)$ to solve this problem is the integral equation:

$$Rh := \int_D R(x,y)h(z,y)dy = f(x,z), \quad x,z \in \bar{D} := D \cup \Gamma, \tag{11.23}$$

where $\Gamma := \partial D$ is the boundary of D. Since z enters into Equation 11.23 as a parameter, one has to study the following integral equation:

$$\int_D R(x,y)h(y)dy = f(x), \quad x \in \bar{D}. \tag{11.24}$$

The questions of interest are:

1. Under what assumptions does Equation 11.24 have a unique solution in some space of distributions?
2. What is the order of singularity of this solution?
3. Is this solution stable under small (in some sense) perturbations of f? Does it solve estimation problem (11.22)?
4. How does one calculate this solution analytically and numerically?

If a distribution $h = D^l h_1$, where $h_1 \in L^2_{\text{loc}}$, then the order of singularity of h, $\text{ords}ing h$, is $|l|$ if h_1 is not smoother than L^2_{loc}. By $|l|$ one denotes $l_1 + l_2 + \cdots + l_r$, $D^l = \dfrac{\partial^{|l|}}{\partial x_1^{l_1} \dots \partial x_r^{l_r}}$. For the distribution theory one can consult, for example, Reference 3.

Let us define a class of kernels $R(x,y)$, or, which is the same, a class $\mathcal{R}$ of integral equations (11.24), for which questions 1 through 4 can be answered.

Let L be a self-adjoint elliptic operator of order s in $L^2(\mathbb{R}^r)$, Λ, $d\rho(\lambda)$ and $\Phi(x,y,\lambda)$ are, respectively, its spectrum, spectral measure, and spectral kernel. This means that the spectral function E_λ of the operator L has the kernel $E_\lambda(x,y) = \int_\Lambda \Phi(x,y,\lambda)d\rho(\lambda)$ and a function $\phi(L)$ has the kernel

$$\phi(L)(x,y) = \int_\Lambda \phi(\lambda)\Phi(x,y,\lambda)d\rho(\lambda),$$

where $\phi(\lambda)$ is an arbitrary function such that $\int_\Lambda |\phi(\lambda)|^2 d(E_\lambda f, f) < \infty$ for any $f \in D(\phi(L))$.

Define class $\mathcal{R}$ of the kernels $R(x,y)$ by the formula

$$R(x,y) = \int_\Lambda \frac{P(\lambda)}{Q(\lambda)}\Phi(x,y,\lambda)d\rho(\lambda), \tag{11.25}$$

where $P(\lambda) > 0$ and $Q(\lambda) > 0$ are polynomials of degree p and, respectively, q, and $p \le q$. Let

$$\alpha := 0.5(q - p)s.$$

The number $\alpha \ge 0$ is an integer because p and q are even integers if the polynomials $P > 0$ and $Q > 0$ are positive for all values of λ. Denote by $H^\alpha = H^\alpha(D)$ the Sobolev space and by $\dot{H}^{-\alpha}$ its dual space with respect to the $L^2(D)$ inner product. Our basic results are formulated in the following theorems, obtained in Reference 16.

Theorem 11.6

If $R(x, y) \in \mathcal{R}$, then the operator R in Equation 11.24 is an isomorphism between the spaces $\dot{H}^{-\alpha}$ and H^{α}. The solution to Equation 11.24 of minimal order of singularity, ord sing$h \leq \alpha$, exists, is unique, and can be calculated by the formula

$$h(x) = Q(L)G, \; G(x) = \begin{cases} g(x) + v(x) & \text{in } D \\ u(x) & \text{in } D' := \mathbb{R}^r \setminus D, \end{cases} \quad (11.26)$$

where $g(x) \in H^{0.5s(p+q)}$ is an arbitrary fixed solution to the equation

$$P(L)g = f \text{ in } D, \quad f \in H^{\alpha} \quad (11.27)$$

and the functions $u(x)$ and $v(x)$ are the unique solution to the problem

$$Q(L)u = 0 \text{ in } D', \quad u(\infty) = 0, \quad (11.28)$$

$$P(L)v = 0 \text{ in } D, \quad (11.29)$$

$$\partial_N^j u = \partial_N^j (v + g) \text{ on } \partial D, \; 0 \leq j \leq 0.5s(p+q) - 1. \quad (11.30)$$

Here $\partial_N u$ is the derivative of u along the outer unit normal N.

Theorem 11.6 gives answers to questions 1 through 4, except the question of a numerical solution of Equation 11.24. This question is discussed in Reference 16.

Remark 11.12

If one considers the operator $R \in \mathcal{R}$ as an operator in $L^2(D)$, $R : L^2(D) \to L^2(D)$ and denotes by $\lambda_j = \lambda_j(D)$ its eigenvalues, $\lambda_1 \geq \lambda_2 \geq \cdots \geq 0$, then

$$\lambda_j = cj^{-(q-p)s/r}[1 + o(1)] \quad \text{as } j \to \infty, \quad (11.31)$$

where $c = \gamma^{(q-p)s/r}$, $\gamma := (2\pi)^{-r} \int_D \eta(x)dx$, and the $\eta(x)$ is defined in Reference 16.

Example 11.1

Consider Equation 11.24 with $D = [-1, 1]$, $R(x, y) = \exp(-|x - y|)$, $L = -i\frac{d}{dx}$, $r = 1$, $s = 1$, $P(\lambda) = 1$, $Q(\lambda) = (\lambda^2 + 1)/2$, $\Phi(x, y, \lambda) = (2\pi)^{-1} \exp\{i\lambda(x - y)\}$, and $d\rho(\lambda = d\lambda)$. Then

$$\int_{-1}^{1} \exp(-|x - y|)h(y)dy = f, \quad -1 \leq x \leq 1,$$

$\alpha = 1$, and Theorem 11.6 yields the following formula for the solution h:

$$h(x) = \frac{-f'' + f}{2} + \frac{f'(1) + f(1)}{2}\delta(x - 1) + \frac{-f'(-1) + f(-1)}{2}\delta(x + 1),$$

where δ is the delta function.

11.3 Singular Integral Equations

11.3.1 One-Dimensional Singular Integral Equations

The results in this section are taken from References 10 and 30.

Consider the equation

$$Au := a(t)u(t) + \frac{1}{i\pi} \int_L \frac{M(t,s)}{s-t} u(s)ds = f(t). \tag{11.32}$$

Here the functions $a(t)$, $M(t,s)$, and $f(t)$ are known. Assume that they satisfy the Hölder condition. The contour L is a closed smooth curve on the complex plane, L is the boundary of a connected smooth domain D, and D' is its complement on the complex plane. Since $M(t,s)$ is assumed Hölder-continuous, one can transform Equation 11.32 to the form

$$Au := a(t)u + \frac{b(t)}{i\pi} \int_L \frac{u(s)ds}{s-t} + \int_L K(t,s)u(s)ds = f(t), \tag{11.33}$$

where the operator with kernel $K(t,s)$ is of the Fredholm type, $b(t) = M(t,t)$, and $K(t,s) = \frac{1}{i\pi} \frac{M(t,s)-M(t,t)}{s-t}$. The operator A in Equation 11.33 is of the form $A = A_0 + K$, where $A_0 u = a(t)u + Su$, and $Su = \frac{1}{i\pi} \int_L \frac{u(s)ds}{s-t}$. The singular operator S is defined as the limit $\lim_{\epsilon \to 0} \frac{1}{i\pi} \int_{|s-t|>\epsilon} \frac{u(s)ds}{s-t}$.

The operator S maps $L^p(L)$ into itself and is bounded, $1 < p < \infty$. Let us denote the space of Hölder-continuous functions on L with exponent γ by $\mathrm{Lip}_\gamma(L)$. The operator S maps the space $\mathrm{Lip}_\gamma(L)$ into itself and is bounded in this space. If $\Phi(z) = \frac{1}{2\pi i} \int_L \frac{u(s)ds}{s-z}$ and $u \in \mathrm{Lip}_\alpha(L)$, then

$$\Phi^\pm(t) = \pm \frac{u(t)}{2} + \frac{1}{2\pi i} \int_L \frac{u(s)ds}{s-t}, \tag{11.34}$$

where $+(-)$ denotes the limit $z \to t$, $z \in D\,(D')$.

If $u \in L^p(L)$, $1 < p < \infty$, then

$$\frac{d}{dt} \int_L u(s) \ln \frac{1}{|s-t|} ds = i\pi u(t) + \int_L \frac{u(s)ds}{s-t}. \tag{11.35}$$

One has $S^2 = I$, that is, for $1 < p < \infty$,

$$\frac{1}{(i\pi)^2} \int_L \frac{1}{\tau - t} \left(\int_L \frac{u(s)ds}{s-\tau} ds \right) d\tau = u(t), \quad \forall u \in L^p(L). \tag{11.36}$$

Let $n = \dim N(A)$ and $n^* := \dim N(A^*)$. Then the index of the operator A is defined by the formula $\mathrm{ind}A := n - n^*$ if at least one of the numbers n or n^* is finite.

If $A = B + K$, where B is an isomorphism and K is compact, then A is the Fredholm operator and $\mathrm{ind}A = 0$. If A is a singular integral operator (11.33), then its index may not be zero

$$\kappa := \mathrm{ind}A = \frac{1}{2\pi} \int_L d \arg \frac{a(t)-b(t)}{a(t)+b(t)}, \tag{11.37}$$

where $\arg f$ denotes the argument of the function f. The operator A with nonzero index is called a Noether operator. Formula (11.37) was derived by F. Noether in 1921. Equation 11.33 with $K(x,y) = 0$ can be reduced to the Riemann problem. Namely, if $\Phi(z) = \frac{1}{2\pi i} \int_L \frac{u(s)}{s-z} ds$, then $u(t) = \Phi^+(t) - \Phi^-(t)$ and $\Phi^+(t) + \Phi^-(t) = Su$ by formula (11.34).

Thus, Equation 11.33 with $K(x, y) = 0$ can be written as

$$a(t)(\Phi^+ - \Phi^-) + b(t)(\Phi^+ + \Phi^-) = f$$

or as

$$\Phi^+(t) = \frac{a(t) - b(t)}{a(t) + b(t)} \Phi^- + \frac{f}{a(t) + b(t)}, \quad t \in L. \tag{11.38}$$

This problem is usually written as

$$\Phi^+(t) = G(t)\Phi^-(t) + g(t), \quad t \in L \tag{11.39}$$

and is a Riemann problem for finding a piecewise analytic function $\Phi(z)$ from the boundary condition (11.39).

Assume that $a(t) \pm b(t) \neq 0, \forall t \in L$, and that L is a closed smooth curve, the boundary of a simply connected bounded domain D.

If one solves problem (11.39) then the solution to Equation 11.33 (with $K(x, y) = 0$) is given by the formula $u = \Phi^+ - \Phi^-$. Let $\kappa = \mathrm{ind}\,G$.

Theorem 11.7

If $\kappa \geq 0$ then problem (11.39) is solvable for any g and its general solution is

$$\Phi(z) = \frac{X(z)}{2\pi i} \int_L \frac{g(s)ds}{X^+(s)(s - z)} + X(z)P_\kappa(z), \tag{11.40}$$

where

$$X^+(z) = \exp(\Gamma_+(z)), \quad X^-(z) = z^{-\kappa} \exp(\Gamma_-(z)), \tag{11.41}$$

$$\Gamma(z) = \frac{1}{2\pi i} \int_L \frac{\ln[s^{-\kappa}G(s)]}{s - z} ds, \tag{11.42}$$

where $P_\kappa(z)$ is a polynomial of degree κ with arbitrary coefficients. If $\kappa = -1$, then problem (11.39) is uniquely solvable and $X^-(\infty) = 0$. If $\kappa < -1$ then problem (11.39), in general, does not have a solution. For its solvability, it is necessary and sufficient that

$$\int_L \frac{g(s)}{X^+(s)} s^{k-1} ds = 0, \quad k = 1, 2, \ldots, -\kappa - 1. \tag{11.43}$$

If these conditions hold, then the solution to problem (11.39) is given by formula (11.40) with $P_\kappa(z) \equiv 0$.

This theorem is proved, for example, in References 2 and 10. It also gives an analytic formula for solving integral equation (11.33) with $k(x, y) = 0$.

If one considers Equation 11.33 with $k(x, y) \neq 0$, then this equation can be transformed to a Fredholm-type equation by inverting the operator A_0. A detailed theory can be found in References 2 and 11.

11.3.2 Multidimensional Singular Integral Equations

Consider the singular integral

$$Au := \int_{\mathbf{R}^m} r^{-m} f(x, \theta) u(y) dy, \quad x, y \in \mathbf{R}^m, \quad r = |x - y|, \tag{11.44}$$

Assume that $u(x) = O(\frac{1}{|x|^k})$ for large $|x|$, that $k > 0$, and f is a bounded function continuous, with respect to θ. Let $\int_S f(x, \theta) dS = 0$, where S stands in this section for the unit sphere in $\mathbf{R}^m$.

Suppose that $|u(x) - u(y)| \leq c_1 r^\alpha (1 + |x|^2)^{-k/2}$ for $r \leq 1$, where $c_1 = const$, and $|u(x)| \leq c_2(1 + |x|^2)^{-k/2}$, where $c_2 = const$. Denote this set of functions u by $A_{\alpha,k}$. If the first inequality holds, but the second is replaced by the inequality $|u(x)| \leq c_3(1 + |x|^2)^{-k/2} \ln(1 + |x|^2)$, then this set of u is denoted $A'_{\alpha,k}$.

Theorem 11.8

The operator (11.44) maps $A_{\alpha,k}$ with $k \leq m$ into $A'_{\alpha,k}$, and $A_{\alpha,k}$ with $k > m$ into $A_{\alpha,m}$.

Let $D \subset \mathbf{R}^m$ be a domain, possibly $D = \mathbf{R}^m$, the function $f(x,\theta)$ be continuously differentiable with respect to both variables, and u satisfies the Dini condition, that is $\sup_{||x-y||<\delta;x,y\in D} |u(x) - u(y)| = w(u,t)$, where $\int_0^\delta \frac{w(u,t)}{t} dt < \infty$, $\delta = const > 0$.
    Then

$$\frac{\partial}{\partial x_j} \int_D \frac{f(x,\theta)}{r^{m-1}} u(y) dy = \int_D u(y) \frac{\partial}{\partial x_j} \left[\frac{f(x,\theta)}{r^{m-1}} \right] dy - u(x) \int_S f(x,\theta) \cos(r, x_j) ds. \qquad (11.45)$$

Assume that

$$\int_S |f(x,\theta)|^{p'} dS \leq C_0 = const, \quad \frac{1}{p} + \frac{1}{p'} = 1 \qquad (11.46)$$

and let

$$A_0 u = \int_{\mathbf{R}^m} \frac{f(x,\theta)}{r^m} u(y) dy. \qquad (11.47)$$

Theorem 11.9

If Equation 11.46 holds then

$$||A_0||_{L^p} \leq c \sup_x ||f(x,\theta)||_{L^{p'}(S)}.$$

The operator (11.47) is a particular case of a pseudodifferential operator defined by the formula

$$Au = \int_{\mathbf{R}^m} \int_{\mathbf{R}^m} e^{i(x-y)\cdot\xi} \sigma_A(x,\xi) u(y) dy d\xi, \qquad (11.48)$$

where the function $\sigma_A(x,\xi)$ is called the symbol of A. For the operator $Au = a(x)u(x) + A_0 u$, where A_0 is defined in formula (11.47), the symbol of A is defined as $\sigma(x,\xi) = a(x) + \widetilde{K}(x,\xi)$, where $K(x, x-y) = \frac{f(x,\theta)}{r^m}$, and $\widetilde{K}(x,\xi)$ is the Fourier transform of $K(x,z)$ with respect to z. One has

$$\sigma_{A+B}(x,\xi) = \sigma_A(x,\xi) + \sigma_B(x,\xi), \quad \sigma_{AB}(x,\xi) = \sigma_A(x,\xi)\sigma_B(x,\xi).$$

It is possible to estimate the norm of the operator (11.47) in terms of its symbol,

$$\sigma_{A_0}(x,\xi) = \int_{\mathbf{R}^m} \frac{f(x,y^0)}{|y|^m} e^{-iy\cdot z} dy, \quad y^0 = \frac{y}{|y|}, \quad \xi = \frac{z}{|z|}.$$

If T is a compact operator in $L^2(\mathbf{R}^m)$ then its symbol is equal to zero. Let $Au = a(x)u + A_0 u$, and $\sigma_A(x,\xi) = a(x) + \sigma_{A_0}(x,\xi)$. If $\sigma_A(x,\xi)$ is sufficiently smooth and does not vanish, then the singular integral equation $Au + Tu = f$ is of Fredholm type and its index is zero. In general, the index of a system of singular integral equations (and of more general systems) is calculated in Reference 1. Additional material about singular integral equations and pseudodifferential operators is given in References 10, 28, and 29.

11.4 Nonlinear Integral Equations

The results in this section are taken from References 5, 7, 8, and 30.
    Let us call the operators

$$Uu = \int_D K(x,t,u(t))dt, \quad Hu = \int_D K(x,t)f(t,u(t))dt,$$

and $Fu := f(t,u(t))$, respectively, Urysohn, Hammerstein, and Nemytskii operators.

The operator F acts from $L^p := L^p(D)$ into $L^q := L^q(D)$ if and only if

$$|f(t, u)| \leq a(t) + b|u|^{p/q}, \quad p < \infty, \quad a(t) \in L^q, \quad b = const.$$

If $p = \infty$, then $F : L^\infty \to L^q$ if and only if $|f(t, u)| \leq a_h(t)$, $a_h(t) \in L^q$, and $|u(t)| \leq h$, $0 \leq h < \infty$. The function $f(t, u)$, $t \in D$, and $u \in (-\infty, \infty)$, is assumed to satisfy the Caratheodory conditions, that is, this function is continuous with respect to u for almost all $t \in D$ and is measurable with respect to $t \in D$ for all $u \in (-\infty, \infty)$.

If $F : L^p \to L^q$ and $f(t, u)$ satisfies the Caratheodory conditions, then F is compact provided that one of the following conditions holds: (1) $q < \infty$, (2) $q = \infty$, $p < \infty$, $f(t, u) \equiv a(t)$, and (3) $q = p = \infty$, $|f(t, u) - f(t, v)| \leq \phi_h(u - v)$ with respect to z for any $h > 0$.

If f satisfies the Caratheodory conditions, and $K(x, t)$ is a measurable function on $D \times D$, and $F : L^p \to L^r$, $K : L^r \to L^q$, then the operator $H = KF : L^p \to L^q$, H is continuous if $r < \infty$, and H is compact if $r < \infty$ and K is compact.

One can find these results, for example, in Reference 30.

A nonlinear operator $A : X \to Y$ is Fréchet differentiable if

$$A(u + h) - A(u) = Bh + w(u, h),$$

where $B : X \to Y$ is a linear bounded operator and $\lim_{\|h\|_X \to 0} \frac{\|w(u,h)\|_Y}{\|h\|_X} = 0$.

Consider the equation

$$u(x) = \mu \int_D K(x, t, u(t)) dt + f(x) := Au, \tag{11.49}$$

where $D \subset \mathbf{R}^m$ is a bounded closed set, meas $D > 0$, the functions $K(x, t, u)$ and $f(x)$ are given, the function u is unknown, and μ is a number.

Theorem 11.10

Let us assume that A is a contraction on a set M, that is, $\|A(u) - A(v) \leq q\|u - v\|$, where $u, v \in M$, M is a subset of a Banach space X, $0 < q < 1$ is a number, and $A : M \to M$. Then Equation 11.49 has a unique solution u in M, $u = \lim_{n \to \infty} u_n$, where $u_{n+1} = A(u_n)$, $u_0 \in M$, and $\|u_n - u\| \leq \frac{q^n}{1-q} \|u_1 - u_0\|$, $n = 1, 2, 3 \dots$.

This result is known as the contraction-mapping principle.

Theorem 11.11

Assume that A is a compact operator, $M \subset X$ is a bounded closed convex set, and $A : M \to M$. Then Equation 11.49 has a solution.

This result is called the Schauder principle. The solution is a fixed point of the mapping A, that is, $u = A(u)$. Uniqueness of the solution in Theorem 11.11 is not claimed: in general, there can be more than one solution.

Remark 11.13

A version of Theorem 11.11 can be formulated as follows:

Theorem 11.12

Assume that A is a continuous operator that maps a convex and closed set M into its compact subset. Then A has a fixed point in M.

Let us formulate the Leray–Schauder principle.

Theorem 11.13

Let A be a compact operator, $A : X \to X$, X be a Banach space. Let all the solutions $u(\lambda)$ of the equation $u = A(u, \lambda)$, $0 \leq \lambda \leq 1$, satisfy the estimate $\sup_{0 \leq \lambda \leq 1} ||u(\lambda)|| \leq a < \infty$. Let $||A(u, 0)|| \leq b$ for $b > a$ and $||u|| = b$. Then the equation $u = A(u, 1)$ has a fixed point in the ball $||u|| \leq a$.

Let $D \subset X$ be a bounded convex domain, $A^n u \subset \overline{D}$ if $u \in \overline{D}$, $n = 1, 2, \ldots, \overline{D}$ is the closure of D, $A^n u \neq u$ for $n > n_0$, and $u \in \partial D$, where ∂D is the boundary of D. Finally, let us assume that A is compact.

Theorem 11.14

Under the above assumptions A has a fixed point in D.

This theorem is proved in Reference 7.

Consider the equation $u = H(u), H : L^2(D) \to L^2(D)$, and assume that $K(x, t)$ is a positive-definite, continuous kernel, $f(t, u)$ is a continuous function of $t \in D$ and $u \in (-\infty, \infty)$ such that $\int_0^u f(t, s)ds \leq 0.5 \, au^2 + b$, where $a < \Lambda^{-1}$, and Λ is the maximal eigenvalue of the linear operator K in $L^2(D)$. Then there exists a fixed point of the operator H. This result can be found in Reference 30.

11.5 Applications

11.5.1 General Remarks

The results in this subsection are taken from References 13, 22, and 24.

There are many applications of integral equations. In this section, there is no space to describe in detail applications to solving boundary value problems by integral equations involving potentials of single and double layers, applications of integral equations in the elasticity theory [9,11], applications to acoustics electrodynamics, etc.

We will restrict this subsection to some questions not considered by other authors. The first question deals with the possibility to express potentials of the single layer by potentials of the double layer and vice versa. The results are taken from Reference 13. The second application deals with the wave scattering by small bodies of an arbitrary shape. These results are taken from References 22 and 24.

11.5.2 Potentials of Single and Double Layers

$$V(x) = \int_S g(x, t)\sigma(t)dt, \quad g(x, t) = \frac{e^{ik|x-t|}}{4\pi|x-t|}, \tag{11.50}$$

where S is a boundary of a smooth-bounded domain $D \subset \mathbf{R}^3$ with the boundary S, $\sigma(t) \in \mathrm{Lip}_\gamma(S)$, and $0 < \gamma \leq 1$.

It is known that $V(x) \in C(\mathbf{R}^3)$, $V(\infty) = 0$

$$V_N^{\pm}(s) = \frac{A\sigma \pm \sigma(s)}{2}, \quad A\sigma := 2\int_S \frac{\partial g(s, t)}{\partial N_s}\sigma(t)dt, \tag{11.51}$$

where $N = N_s$ is the unit normal to S at the point s pointing out of D, $+(-)$ denotes the limiting value of the normal derivative of V when $x \to s \in S$, and $x \in D$ (D'), where $D' := \mathbf{R}^3 \backslash D$.

Potential of the double layer is defined as follows:

$$W(x) := \int_S \frac{\partial g(x,t)}{\partial N_t} \mu(t)dt.$$

One has the following properties of W:

$$W^{\pm}(s) = W(s) \mp \frac{\mu(t)}{2}, \quad W(s) := \int_S \frac{\partial g(s,t)}{\partial N_t} \mu(t)dt \qquad (11.52)$$

$$W^+_{N_s} = W^-_{N_s}. \qquad (11.53)$$

Theorem 11.15

For any $V(W)$ there exists a $W(V)$ such that $W = V$ in D. The $V(W)$ is uniquely defined.

Theorem 11.16

A necessary and sufficient condition for $V(\sigma) = W(\mu)$ in D' is

$$\int_S V h_j dt = 0, \quad 1 \leq j \leq r', \quad (I+A)h_j = 0, \qquad (11.54)$$

where the set $\{h_j\}$ forms a basis of $N(I+A)$, and A is defined in Equation 11.51.

A necessary and sufficient condition for $W(\mu) = V(\sigma)$ in D' is

$$\int_S W \sigma_j dt = 0, \quad 1 \leq j \leq r, \quad A\sigma_j - \sigma_j = 0, \qquad (11.55)$$

where the set $\{\sigma_j\}$ forms a basis of $N(I-A)$.

Theorems 11.15 and 11.16 are proved in Reference 13.

11.5.3 Wave Scattering by Small Bodies of an Arbitrary Shape

The results in this subsection are taken from References 14, 22–27. Consider the wave-scattering problem:

$$(\nabla^2 + k^2)u = 0 \text{ in } D' := \mathbf{R}^3 \backslash D, \quad u_N = \zeta u \text{ on } S = \partial D, \qquad (11.56)$$

$$u = u_0 + v, \quad u_0 = e^{ik\alpha \cdot x}, \quad \alpha \in S^2, \qquad (11.57)$$

$$\frac{\partial v}{\partial r} - ikv = o\left(\frac{1}{r}\right), \quad r = |x| \to \infty. \qquad (11.58)$$

Here $k = const > 0$, S^2 is the unit sphere in $\mathbf{R}^3$, $\zeta = const$ is a given parameter, the boundary impedance, D is a small body, a particle, S is its boundary, which we assume Hölder continuous, and N is the unit normal to S pointing out of D. If $Im\zeta \leq 0$ then problems (11.56) through (11.58) have exactly one solution, and v is the scattered field

$$v = A(\beta, \alpha, \kappa)\frac{e^{ikr}}{r} + o\left(\frac{1}{r}\right), \quad |x| = r \to \infty, \quad \beta = \frac{x}{r}, \qquad (11.59)$$

$A(\beta, \alpha, \kappa)$ is called the scattering amplitude.

Our basic assumption is the smallness of the body D: this body is small if $ka \ll 1$, where $a = 0.5$ diameter D.

In applications a is called the characteristic size of D.

Let us look for the solution to problems (11.56) through (11.58) of the form

$$u(x) = u_0(x) + \int_S g(x,t)\sigma(t)dt, \quad g(x,t) = \frac{e^{ik|x-t|}}{4\pi|x-t|}, \tag{11.60}$$

and write

$$\int_S g(x,t)\sigma(t)dt = g(x,x_1)Q + \int_S [g(x,t) - g(x,x_1)]\sigma(t)dt, \tag{11.61}$$

where $x_1 \in D$ is an arbitrary fixed point inside D, and

$$Q := \int_S \sigma(t)dt. \tag{11.62}$$

If $ka \ll 1$ and $|x - x_1| := d \gg a$, then

$$\left| \int_S [g(x,t) - g(x,x_1)]\sigma(t)dt \right| \ll |g(x,x_1)|Q. \tag{11.63}$$

Therefore, the scattering problem (11.56) through (11.58) has an approximate solution of the form

$$u(x) = u_0(x) + g(x,x_1)Q, \quad ka \ll 1, \quad |x - x_1| \gg a. \tag{11.64}$$

The solution is reduced to finding just one number Q in contrast with the usual methods, based on the boundary integral equation for the unknown function $\sigma(t)$.

To find the main term of the asymptotic of Q as $a \to 0$, one uses the exact boundary integral equation

$$u_{0N}(s) - \zeta u_0(s) + \frac{A\sigma - \sigma}{2} - \zeta \int_S g(s,t)\sigma(t)dt = 0, \quad s \in S, \tag{11.65}$$

where formula (11.51) was used.

We do not want to solve Equation 11.65 which is only numerically possible, but want to derive an asymptotically exact analytic formula for Q as $a \to 0$.

Integrate both sides of formula (11.65) over S. The first term is equal to $\int_S u_{0N}ds = \int_D \nabla^2 u_0 dx \simeq O(a^3)$. The sign $\simeq$ stands for the equality up to the terms of higher order of smallness as $a \to 0$. The second term is equal to $-\zeta \int_S u_0 ds \simeq -\zeta u_0(x_1)|S|$, where $|S| = O(a^2)$ is the surface area of S. The third term is equal to $\frac{1}{2} \int_S A\sigma ds - \frac{1}{2}Q$. When $a \to 0$ one checks that $A\sigma \simeq A_0\sigma$, where $A_0 = A|_{k=0}$, and $\int_S A_0\sigma dt = -\int_S \sigma dt$. Therefore the third term is equal to $-Q$ up to the term of higher order of smallness. The fourth term is equal to $-\zeta \int_S dt\sigma(t) \int_S g(s,t)ds = o(Q)$, as $a \to 0$. Thus

$$Q \simeq -\zeta|S|u_0(x_1), \quad a \to 0 \tag{11.66}$$

and $u_0(x_1) \simeq 1$

$$A(\beta,\alpha,\kappa) = -\frac{\zeta|S|u_0(x_1)}{4\pi}. \tag{11.67}$$

For the scattering by a single small body one can choose $x_1 \in D$ to be the origin, and take $u_0(x_1) = 1$. But in the many-body scattering problem the role of u_0 is played by the effective field which depends on x. Formulas (11.64) through (11.66) solve the scattering problem (11.56) through (11.58) for one small body D of an arbitrary shape if the impedance boundary condition (11.56) is imposed. The scattering in this case is isotropic and $A(\beta,\alpha,\kappa) = O(a^2|\zeta|)$.

If the boundary condition is the Dirichlet one, $u|_S = 0$, then one derives that

$$Q \simeq -Cu_0, \quad a \to 0; \quad u_0 \simeq 1 \tag{11.68}$$

and

$$A(\beta,\alpha,\kappa) = -\frac{C}{4\pi}, \quad a \to 0, \quad u_0 \simeq 1, \quad C = O(a). \tag{11.69}$$

Here C is the electrical capacitance of the perfect conductor with the shape D.

For the Dirichlet boundary condition the scattering is isotropic and $|A| = O(a), a \to 0$. Thus, the scattered field is much larger than for the impedance boundary condition.

For the Neumann boundary condition $u_N = 0$ on S, the scattering, is anisotropic, and $A(\beta, \alpha, \kappa) = O(a^3)$, which is much smaller than for the impedance boundary condition.

One has for the Neumann boundary condition the following formula:

$$A(\beta, \alpha, \kappa) = \frac{|D|}{4\pi}\left(ik\beta_{pq}\frac{\partial u_0}{\partial x_q}\beta_p + \nabla^2 u_0\right), \tag{11.70}$$

where β_{pq} is some tensor, $|D|$ is the volume of D, $\beta_p = \lim_{|x|\to\infty}\frac{x_p}{|x|}$, $x_p := x \cdot e_p$ is the pth coordinate of a vector $x \in \mathbf{R}^3$, and u_0 and its derivatives are calculated at an arbitrary point inside D. This point can be chosen as the origin of the coordinate system. Since D is small, the choice of this point does not influence the results. The tensor β_{pq} is defined for a body with volume V and boundary S as follows:

$$\beta_{pq} = \frac{1}{V}\int_S t_p\sigma_q dt,$$

where σ_q solves the equation $\sigma_q = A\sigma_q - 2N_q$, $N_q := N \cdot e_q$, and $\{e_q\}_{q=1}^3$ is a Cartesian basis of $\mathbf{R}^3$.

Finally, consider the many-body scattering problem. Its statement can also be written as Equations 11.56 through 11.58 but now $D = \cup_{m=1}^M D_m$ is the union of many small bodies, $\zeta = \zeta_m$ on $S_m = \partial D_m$.

One looks for the solution of the form

$$u(x) = u_0(x) + \sum_{m=1}^M \int_{S_m} g(x, t)\sigma_m(t)dt. \tag{11.71}$$

Let us define the effective field in the medium by the formula

$$u_e(x) = u_0(x) + \sum_{m\neq j} g(x, x_m)Q_m, \quad |x - x_j| \leq a. \tag{11.72}$$

Here $x_m \in D_m$ are arbitrary fixed points. If the point x is not too close to any of the points x_m, $1 \leq m \leq M$, then formula (11.72) gives the field $u_e(x)$ that is asymptotically, as $a \to 0$, equal to the field $u_0(x) + \sum_{m=1}^M g(x, x_m)Q_m$.

Assume that the points x_m are distributed by the formula

$$\mathcal{N}(\Delta) = \frac{1}{a^{2-\kappa}}\int_\Delta N(x)dx[1 + o(1)], \quad a \to 0. \tag{11.73}$$

Here $\mathcal{N}(\Delta) = \sum_{x_m \in \Delta} 1$ is the number of points in an arbitrary open set Δ, $N(x) \geq 0$ is a given function, $0 \leq \kappa < 1$ is a parameter, and an experimentalist can choose $N(x)$ and κ as he wishes. Let us assume that $\zeta_m = \frac{h(x_m)}{a^\kappa}$, where $h(x)$ is an arbitrary continuous in D function such that $Imh(x) \leq 0$. This function can be chosen by an experimenter as he wishes.

Under these assumptions in Reference 24 the following results are proved.

Assume that $|S_m| = c_m a^2$, where $c_m > 0$ are constants depending on the shape of D_m. Then

$$Q_m \simeq -c_m h(x_m)u_e(x_m)a^{2-\kappa}.$$

Denote $u_e(x_m) := u_m, h(x_m) := h_m$. Then, assuming for simplicity that $c_m = c$ for all m, one can find the unknown numbers u_m from the following LAS:

$$u_j = u_{0j} - c\sum_{m\neq j} g_{jm}h_m u_m a^{2-\kappa}, \quad 1 \leq j \leq M, \quad g_{jm} := \frac{e^{ik|x_j-x_m|}}{4\pi|x_j - x_m|}. \tag{11.74}$$

LAS (11.74) can be reduced to an LAS of a much smaller order. Namely, denote by Ω a finite domain in which all the small bodies are located. Partition Ω into a union of P cubes Δ_p with the side $b = b(a)$, the cubes are nonintersecting in the sense that they do not have common interior points, and they can have only pieces of a common boundary. Denote by $d = d(a)$ the smallest distance between neighboring bodies D_m. Assume that

$$a \ll d \ll b, \quad ka \ll 1. \tag{11.75}$$

Then system (11.74) can be reduced to the following LAS:

$$u_q = u_{0q} - c \sum_{p \neq q}^{P} g_{qp} h_p u_p N_p |\Delta_p|, \quad 1 \leq q \leq P, \tag{11.76}$$

where $N_p := N(x_p)$, $x_p \in \Delta_p$, is an arbitrary point, $N(x)$ is the function from formula (11.73), and $|\Delta_p|$ is the volume of the cube Δ_p.

The effective field has a limit as $a \to 0$, and this limit solves the equation

$$u(x) = u_0(x) - c \int_\Omega g(x,y)h(y)N(y)u(y)dy, \tag{11.77}$$

where the constant c is the constant in the definition $|S_m| = ca^2$. One may consider by the same method the case of small bodies of various sizes. In this case $c = c_m$, but we do not go into details. If the small bodies are spheres of radius a then $c = 4\pi$.

Applying the operator $\nabla^2 + k^2$ to Equation 11.77 one gets

$$(\nabla^2 + k^2 n^2(x))u := (\nabla^2 + k^2 - cN(y)h(y))u = 0 \text{ in } \mathbf{R}^3. \tag{11.78}$$

Therefore, embedding of many small impedance particles, distributed according to formula (11.73) leads to a medium with a new refraction coefficient

$$n^2(x) = 1 - k^{-2}cN(x)h(x). \tag{11.79}$$

Since $N(x)$ and $h(x)$ can be chosen as one wishes, with the only restrictions $N(x) \geq 0$, $Imh(x) \leq 0$, one can create a medium with a *desired refraction coefficient* by formula (11.79) if one chooses a suitable $N(x)$ and $h(x)$ [27].

A similar theory is developed for electromagnetic wave scattering by small impedance bodies in References 19, 23, and 27.

References

1. M. Atiyah, I. Singer, The index of elliptic operators on compact manifolds, *Bull. Am. Math. Soc.*, 69, 1963, 422–433.
2. F. Gahov, *Boundary Value Problems,* Pergamon Press, New York, 1966.
3. I. Gel'fand, G. Shilov, *Generalized Functions*, vol. 1, Academic Press, New York, 1964.
4. V. Ivanov et al., *Theory of Linear Ill-Posed Problems and Its Applications*, VSP, Utrecht, 2002.
5. L. Kantorovich, G. Akilov, *Functional Analysis*, Pergamon Press, New York, 1982.
6. M. Krasnoselskii et al., *Approximate Solution of Operator Equations,* Walters-Noordhoff, Groningen, 1972.
7. M. Krasnoselskii, P. Zabreiko, *Geometric Methods of Nonlinear Analysis*, Springer-Verlag, Berlin, 1984.
8. M. Krasnoselskii et al., *Integral Operators in the Spaces of Summable Functions,* Noordhoff International, Leiden, 1976.
9. V.D. Kupradze et al., *Three-Dimensional Problems of Mathematical Theory of Elasticity and Thermo-Elasticity*, Nauka, Moscow, 1976 (in Russian).
10. S. Mikhlin, S. Prössdorf, *Singular Integral Operators,* Springer-Verlag, Berlin, 1980.

11. N. Muskhelishvili, *Singular Integral Equations. Boundary Problems of Functions Theory and Their Applications to Mathematical Physics*, Wolters-Noordhoff Publishing, Groningen, 1972.

12. G. Polya, G. Szegö, *Isoperimetric Inequalities in Mathematical Physics*, Princeton University Press, Princeton, 1951.

13. A.G. Ramm, *Scattering by Obstacles*, D. Reidel, Dordrecht, 1986.

14. A.G. Ramm, *Wave Scattering by Small Bodies of Arbitrary Shapes*, World Science Publishers, Singapore, 2005.

15. A.G. Ramm, *Inverse Problems*, Springer, New York, 2005.

16. A.G. Ramm, *Random Fields Estimation,* World Science Publishers, Singapore, 2005.

17. A.G. Ramm, *Dynamical Systems Method for Solving Operator Equations*, Elsevier, Amsterdam, 2007.

18. A.G. Ramm, N.S. Hoang, *Dynamical Systems Method and Applications. Theoretical Developments and Numerical Examples,* Wiley, Hoboken, 2012.

19. A.G. Ramm, Scattering of electromagnetic waves by many small perfectly conducting or impedance bodies, *J. Math. Phys.*, 56, N9, 2015, 091901.

20. A.G. Ramm, A simple proof of the Fredholm alternative and a characterization of the Fredholm operators, *Am. Math. Mon.*, 108, N9, 2001, 855–860.

21. A.G. Ramm, A characterization of unbounded Fredholm operators, *Cubo Math. J.*, 5, N3, 2003, 91–95.

22. A.G. Ramm, Wave scattering by many small bodies and creating materials with a desired refraction coefficient, *Afr. Mathematika*, 22, N1, 2011, 33–55.

23. A.G. Ramm, Electromagnetic wave scattering by small impedance particles of an arbitrary shape, *J. Appl. Math. Comput.*, 43, N1, 2013, 427–444.

24. A.G. Ramm, Many-body wave scattering problems in the case of small scatterers, *J. Appl. Math. Comput.*, 41, N1, 2013, 473–500.

25. A.G. Ramm, Wave scattering by many small bodies: Transmission boundary conditions, *Rep. Math. Phys.*, 71, N3, 2013, 279–290.

26. A.G. Ramm, Scattering of electromagnetic waves by many nano-wires, *Mathematics*, 1, 2013, 89–99. Open access Journal: http://www.mdpi.com/journal/mathematics.

27. A.G. Ramm, *Scattering of Acoustic and Electromagnetic Waves by Small Bodies of Arbitrary Shapes. Applications to Creating New Engineered Materials*, Momentum Press, New York, 2013.

28. M. Shubin, *Pseudodifferential Operators and Spectral Theory*, Springer-Verlag, New York, 1987.

29. E. Stein, *Harmonic Analysis: Real-Variable Methods, Orthogonality and Oscillatory Integrals*, Princeton University Press, Princeton, 1993.

30. P. Zabreiko et al., *Integral Equations, Reference Book*, Nauka, Moscow, 1968.

12

Large-Time Behavior of Solutions to Evolution Equations

Alexander G. Ramm

12.1 Introduction

Large-time behavior of solutions to abstract differential equations is studied. The results give sufficient condition for the solution to an abstract dynamical system (evolution problem) not to exhibit chaotic behavior. The corresponding evolution problem is

$$\dot{u} = A(t)u + F(t, u) + b(t), \quad t \geq 0; \quad u(0) = u_0. \tag{*}$$

Here, $\dot{u} := \frac{du}{dt}$, $u = u(t) \in H$, H is a Hilbert space, $t \in \mathbb{R}_+ := [0, \infty)$, $A(t)$ is a linear dissipative operator: $Re(A(t)u, u) \leq -\gamma(t)(u, u)$, and $F(t, u)$ is a nonlinear operator, $\|F(t, u)\| \leq c_0\|u\|^p, p > 1$, c_0 and p are positive constants, $\|b(t)\| \leq \beta(t)$, and $\beta(t) \geq 0$ is a continuous function.

Sufficient conditions are given for the solution $u(t)$ to problem (*) to exist for all $t \geq 0$, to be bounded uniformly on $\mathbb{R}_+$, and a bound on $\|u(t)\|$ is given. This bound implies the relation $\lim_{t \to \infty} \|u(t)\| = 0$ under suitable conditions on $\gamma(t)$ and $\beta(t)$.

The basic technical tool in this work is the following *new nonlinear inequality*:

$$\dot{g}(t) \leq -\gamma(t)g(t) + \alpha(t, g(t)) + \beta(t), \ t \geq 0; \quad g(0) = g_0,$$

which holds on any interval $[0, T)$ on which $g(t) \geq 0$ exists and has bounded derivative from the right, $\dot{g}(t) := \lim_{s \to +0} \frac{g(t+s) - g(t)}{s}$. It is assumed that $\gamma(t)$, and $\beta(t)$ are real-valued, continuous functions of t, defined on $\mathbb{R}_+ := [0, \infty)$, the function $\alpha(t, g)$ is defined for all $t \in \mathbb{R}_+$, locally Lipschitz with respect to g uniformly with respect to t on any compact subsets $[0, T]$, $T < \infty$.

If there exists a function $\mu(t) > 0$, $\mu(t) \in C^1(\mathbb{R}_+)$, such that

$$\alpha\left(t, \frac{1}{\mu(t)}\right) + \beta(t) \leq \frac{1}{\mu(t)}\left(\gamma(t) - \frac{\dot{\mu}(t)}{\mu(t)}\right), \quad \forall t \geq 0; \quad \mu(0)g(0) \leq 1,$$

then $g(t)$ exists on all of $\mathbb{R}_+$, that is, $T = \infty$, and the following estimate holds:

$$0 \leq g(t) \leq \frac{1}{\mu(t)}, \quad \forall t \geq 0.$$

If $\mu(0)g(0) < 1$, then $0 \leq g(t) < \frac{1}{\mu(t)}, \forall t \geq 0$.

A classical area of study is stability of solutions to evolution equations. We identify an evolution problem with an abstract dynamical system. An evolution problem is described by an equation

$$\dot{u}(t) = F_1(t, u), \quad u(0) = u_0. \tag{12.1}$$

Here, $F_1 : X \to X$ is a nonlinear operator in a Banach space X, $\dot{u} = \dot{u}(t) = \frac{du}{dt}$. Quite often it is convenient to assume X to be a Hilbert space H, because the energy is often interpreted as a quantity (u, u) in a suitable Hilbert space. Suppose that $F_1(t, 0) = 0$ and $u_0 = 0$. Then $u = 0$ is a solution to Equation 12.1. A. M. Lyapunov in 1892 published a classical work on stability of motion [1], where he studied Equation 12.1 in the case $X = \mathbb{R}^n$ and F_1 analytic function of u. If $F_1(t, 0) = 0$, and F_1 is twice Fréchet differentiable, then one can write $F_1(t, u) = A(t)u + F(t, u)$, where $A(t)$ is a linear operator in X and $\|F(t, u)\| = O(\|u\|^2)$, $\|u\| \to 0$. This representation is a linearization of F_1 around the point $u = 0$. Lyapunov defined the notion of stability (Lyapunov stability) of the equilibrium solution $u = 0$ toward small perturbations of the data u_0. He calls this solution stable (Lyapunov stable), if for any $\epsilon > 0$ there is a $\delta = \delta(\epsilon) > 0$ such that if inequality $\|u_0\| < \delta$ holds then $\sup_{t \geq 0} \|u(t)\| < \epsilon$. Note that this definition implies the global existence of the solution to problem (12.1) for all u_0 in the ball $\|u_0\| < \delta$. There is a very large literature on the stability of solutions to ordinary differential equations, see References 2–8, where hundreds of references can be found. It is not possible in this section to discuss most of the methods dealing with the large-time behavior of solutions to evolution equations.

In this section, the emphasis is on the new methods in a study of the large-time behavior of solutions to nonlinear evolution equations based on applications of the differential inequality introduced and discussed in Section 12.2. These applications are illustrated by several examples, but many more applications could be given. The results presented are obtained by the author.

The equilibrium solution $u = 0$ is called unstable if it is not Lyapunov stable. This means that there is an $\epsilon > 0$ such that for any $\delta > 0$ there is a u_0, $\|u_0\| < \delta$, and a $t_\delta > 0$ such that $\|u(t_\delta)\| \geq \epsilon$.

One can give similar definitions for stability and instability of a solution to problem (12.1) with $u_0 \neq 0$. In this case, one calls the solution $u = u(t; u_0)$ stable if all the solutions $u(t; w_0)$ to problem (12.1), with w_0 in place of u_0, exist for all $t \geq 0$ and satisfy the inequality $\sup_{t \geq 0} \|u(t; u_0) - u(t; w_0)\| < \epsilon$ provided that $\|u_0 - w_0\| < \delta$.

A solution $u(t; u_0)$ is called asymptotically stable if it is stable and there is a $\delta > 0$ such that all the solutions $u(t; w_0)$ with $\|u_0 - w_0\| < \delta$ satisfy the relation $\lim_{t \to \infty} \|u(t; u_0) - u(t; w_0)\| = 0$.

The equilibrium solution $u = 0$ is asymptotically stable if it is stable and there is a $\delta > 0$ such that all the solutions $u(t; u_0)$ with $\|u_0\| < \delta$ satisfy the relation $\lim_{t \to \infty} \|u(t; u_0)\| = 0$.

Consider problem (12.1) with $F_1(t, u) + \phi(t, u)$ in place of $F_1(t, u)$. The term $\phi(t, u)$ is called persistently acting perturbations. The equilibrium solution $u = 0$ is called stable with respect to persistently acting perturbations if for any $\epsilon > 0$ there exists a $\delta = \delta(\epsilon) > 0$ such that if $\|\phi(t, u)\| < \delta$ and $\|u_0\| < \delta$, then $\sup_{t \geq 0} \|u(t; u_0)\| < \epsilon$.

Stability of the solutions and their behavior as $t \to \infty$ is of interest in a study of dynamical systems. For example, if the equilibrium solution is asymptotically stable, then it does not have chaotic behavior.

If $A(t) = A$ is independent of time and $X = \mathbb{R}^n$, then Lyapunov obtained classical results on the stability of the equilibrium solution to problem (12.1). He assumed that F is analytic with respect to $u \in \mathbb{R}^n$, that $|F(t, u)| \leq c|u|^2$ in a neighborhood of the origin, and $c > 0$ is a constant. Lyapunov has proved that *if the spectrum $\sigma(A)$ of A lies in the half-plane $\mathrm{Re}\,z < 0$, then the equilibrium solution $u = 0$ is asymptotically stable, and if at least one eigenvalue of A lies in the half-plane $\mathrm{Re}\,z > 0$, then the equilibrium solution is unstable.*

If some of the eigenvalues of A lie on the imaginary axis and $F = 0$, so that problem (12.1) is linear, and if all the Jordan cells of the Jordan canonical form of the matrix A, corresponding to the operator A in $\mathbb{R}^n$, consist of just one element, then the equilibrium solution is stable. Otherwise it is unstable.

Thus, a necessary and sufficient condition for Lyapunov stability of the equilibrium solution of the linear equation $\dot{u} = Au$ in $\mathbb{R}^n$ is known: the spectrum of A has to lie in the left complex half-plane:

$\sigma \subset \{Z : Rez \leq 0\}$, and the Jordan cells corresponding to purely imaginary eigenvalues of A have to consist of just one element.

If $F \not\equiv 0$, then, in general, when the spectrum of A lies in the left half-plane of the complex plane, and some eigenvalues of A lie on the imaginary axis, the stability cannot be decided by the linearized part A of F_1 only. One can give examples of A such that the nonlinear part F can be chosen so that the equilibrium solution $u = 0$ is stable, and F can also be chosen so that this solution is unstable. For instance, consider $\dot{u} = cu^3$, where $c = const$. This equation can be solved analytically by separation of variables. The result is $u(t) = [u^{-2}(0) - 2ct]^{-0.5}$. Therefore, if $c < 0$ and $|u(0)| \leq \delta$, $\delta > 0$, then the solution exists for all $t \geq 0$, and is asymptotically stable. But if $c > 0$, then the solution blows up at a finite time t_b, the blow-up time, and $t_b = [2cu^2(0)]^{-1}$. In this case, the zero solution is unstable.

If $A = A(t)$ the stability theory is more complicated. The case of periodic $A(t)$ was studied much due to its importance in many applications [9].

The stability theory in infinite-dimensional spaces, for example, in Hilbert and Banach spaces, was developed in the second half of the twentieth century (see Reference 10 and references therein). Again, the location of the spectrum of $A(t)$ plays an important role in this theory.

The basic *novel points* of the theory presented below include sufficient conditions for the stability and asymptotic stability of the equilibrium solution to abstract evolution problem (12.1) in a Hilbert space when $\sigma(A(t))$ may lie in the right half-plane for some or all moments of time $t > 0$, but $\sup \sigma(ReA(t)) \to 0$ as $t \to \infty$. Therefore, our results are *new* even in the finite-dimensional spaces.

The technical tool, on which our study is based, is a new nonlinear differential inequality. The results are stated in several theorems and illustrated by several examples. These results are taken from References 11–22, and, especially, from Reference 19. In the joint papers by the author's student N. S. Hoang and the author one can find various additional results on nonlinear inequalities [23–28]. Some versions of this inequality have been used in the monographs [11,12], where the dynamical systems method (DSM) for solving operator equations was developed. In Reference 29, the relation between spectral properties of the Schrödinger-type operators and large-time behavior of the solution to the related Cauchy problem is established.

The literature on the stability of solutions to evolution problems and their behavior at large times is enormous, and we refer the reader to the author's papers and books directly related to the novel points mentioned above.

Consider an abstract nonlinear evolution problem

$$\dot{u} = A(t)u + F(t, u) + b(t), \quad \dot{u} := \frac{du}{dt}, \tag{12.2}$$

$$u(0) = u_0, \tag{12.3}$$

where $u(t)$ is a function with values in a Hilbert space H and $A(t)$ is a linear bounded dissipative operator in H, which satisfies inequality

$$Re(A(t)u, u) \leq -\gamma(t)\|u\|^2, \quad t \geq 0; \quad \forall u \in H, \tag{12.4}$$

where $F(t, u)$ is a nonlinear map in H,

$$\|F(t, u)\| \leq c_0 \|u(t)\|^p, \quad p > 1, \tag{12.5}$$

$$\|b(t)\| \leq \beta(t), \tag{12.6}$$

$\gamma(t)$ and $\beta(t) \geq 0$ are continuous real-valued functions, defined on all of $\mathbb{R}_+ := [0, \infty)$, $c_0 > 0$ and $p > 1$ are constants. We allow $\gamma(t)$ to be not only positive but also negative, see examples below.

Recall that a linear operator A in a Hilbert space is called dissipative if $Re(Au, u) \leq 0$ for all $u \in D(A)$, where $D(A)$ is the domain of definition of A. Dissipative operators are important because they describe systems in which energy is dissipating, for example, due to friction or other physical reasons. Passive nonlinear networks can be described by Equation 12.2 with a dissipative linear operator $A(t)$ (see References 17 and 18, Chapter 3, and Reference 30).

Let $\sigma := \sigma(A(t))$ denote the spectrum of the linear operator $A(t)$, $\Pi := \{z : Rez < 0\}$, $\ell := \{z : Rez = 0\}$, and $\rho(\sigma, \ell)$ denote the distance between sets σ and ℓ. We assume that

$$\sigma \subset \Pi, \tag{12.7}$$

but we allow $\lim_{t\to\infty} \rho(\sigma, \ell) = 0$. This is the basic *novel* point in our theory. The usual assumption in stability theory (see, e.g., Reference 10) is $\sup_{z\in\sigma} Rez \leq -\gamma_0$, where $\gamma_0 = const > 0$. For example, if $A(t) = A^*(t)$, where A^* is the adjoint operator, and if the spectrum of $A(t)$ consists of eigenvalues $\lambda_j(t)$, $0 \geq \lambda_j(t) \geq \lambda_{j+1}(t)$, then, we allow $\lim_{t\to\infty} \lambda_1(t) = 0$. This is in contrast with the usual theory, where the assumption is $\lambda_1(t) \leq -\gamma_0$, $\gamma_0 > 0$ is a constant, is used.

Moreover, our results cover the case, *apparently not considered earlier in the literature*, when $Re(A(t)u, u) \leq \gamma(t)\|u\|^2$ with $\gamma(t) > 0$, $\lim_{t\to\infty} \gamma(t) = 0$. This means that the spectrum of $A(t)$ may be located in the half-plane $Rez \leq \gamma(t)$, where $\gamma(t) > 0$, but $\lim_{t\to\infty} \gamma(t) = 0$.

Our goal is to give sufficient conditions for the existence and uniqueness of the solution to problems (12.2) and (12.3) for all $t \geq 0$, that is, for global existence of $u(t)$, for boundedness of $\sup_{t\geq 0} \|u(t)\| < \infty$, or to the relation $\lim_{t\to\infty} \|u(t)\| = 0$.

If $b(t) = 0$ in Equation 12.2, then $u(t) = 0$ solves Equation 12.2 and $u(0) = 0$. This equation is called zero solution to Equation 12.2 with $b(t) = 0$.

If $b(t) \not\equiv 0$, then one says that problems (12.2) and (12.3) are the problems with persistently acting perturbations. The zero solution is called Lyapunov stable for problems (12.2) and (12.3) with persistently acting perturbations if for any $\epsilon > 0$, however small, one can find a $\delta = \delta(\epsilon) > 0$, such that if $\|u_0\| \leq \delta$, and $\sup_{t\geq 0} \|b(t)\| \leq \delta$, then the solution to Cauchy problems (12.2) and (12.3) satisfies the estimate $\sup_{t\geq 0} \|u(t)\| \leq \epsilon$.

We do not discuss here the method of Lyapunov functions for a study of stability [8,13].

The approach, developed here, consists of reducing the stability problems to some nonlinear differential inequality and estimating the solutions to this inequality.

In Section 12.2, the formulation and a proof of two theorems, containing the result concerning this inequality and its discrete analog, are given. In Section 12.3, some results concerning Lyapunov stability of zero solution to Equation 12.2 are obtained with the help of the nonlinear differential inequality mentioned above. In Section 12.4, we derive stability results in the case when $\gamma(t) < 0$ in formula (12.4). This means that the linear operator $A(t)$ in Equation 12.2 may have spectrum in the half-plane $Rez > 0$.

Our results are closely related to the DSM [11–14]. Recently, these results were applied to biological problems [22] and to evolution equations with delay [21].

In the theory of chaos, one of the reasons for the chaotic behavior of a solution to an evolution problem to appear is the lack of stability of solutions to this problem [32,33]. The results presented in Section 12.3 can be considered as sufficient conditions for chaotic behavior not to appear in the evolution system described by problems (12.2) and (12.3).

Our results are formulated in Theorems 12.1–12.6 and illustrated by several examples.

12.2 A Nonlinear Differential Inequality

In this section, an essentially self-contained proof is given of an estimate for nonnegative solutions of a nonlinear inequality

$$\dot{g}(t) \leq -\gamma(t)g(t) + \alpha(t, g(t)) + \beta(t), \ t \geq 0; \ g(0) = g_0; \quad \dot{g} := \frac{dg}{dt}. \tag{12.8}$$

In Section 12.3, some of the many possible applications of this estimate (see, e.g., estimate (12.12)) are demonstrated.

It is not assumed *a priori* that solutions $g(t) \geq 0$ to inequality (12.8) are defined on all of $\mathbb{R}_+$, that is, that these solutions exist globally. In Theorem 12.1, we give sufficient conditions for the global existence of $g(t)$. Moreover, under these conditions, a bound on $g(t)$ is given, see estimate (12.12) in Theorem 12.1. This bound yields the relation $\lim_{t\to\infty} g(t) = 0$ if $\lim_{t\to\infty} \mu(t) = \infty$.

Let us formulate our assumptions. We assume that $g(t) \geq 0$. We *do not assume that the functions* γ, α, *and* β *are nonnegative*. However, in many applications, the functions α and β are bounds on some norms, and then these functions are nonnegative. The function $\gamma(t)$ is often (but not always) nonnegative. For example, this happens if $\gamma(t)$ comes from an estimate of the type $(Au, u) \geq \gamma(u, u)$. If the functions α and β are bounds from above on some norms, then one may assume without loss of generality that these functions are smooth, because one can approximate a nonsmooth function with an arbitrary accuracy by an infinitely smooth function, and choose this smooth function to be greater than the function it approximates.

Assumption A_1

We assume that the function $g(t) \geq 0$ is defined on some interval $[0, T)$, has a bounded derivative $\dot{g}(t) := \lim_{s\to+0} \frac{g(t+s)-g(t)}{s}$ from the right at any point of this interval, and $g(t)$ satisfies inequality (12.8) at all t at which $g(t)$ is defined. The functions $\gamma(t)$, and $\beta(t)$, are real-valued, defined on all of $\mathbb{R}_+$ and continuous there. The function $\alpha(t, g)$ is continuous on $\mathbb{R}_+ \times \mathbb{R}_+$ and locally Lipschitz with respect to g. This means that

$$|\alpha(t, g) - \alpha(t, h)| \leq L(T, M)|g - h|, \tag{12.9}$$

if $t \in [0, T]$, $|g| \leq M$ and $|h| \leq M$. Here $M = const > 0$ and $L(T, M) > 0$ is a constant independent of g, h, and t.

Assumption A_2

There exists a $C^1(\mathbb{R}_+)$ function $\mu(t) > 0$, such that

$$\alpha\left(t, \frac{1}{\mu(t)}\right) + \beta(t) \leq \frac{1}{\mu(t)}\left(\gamma(t) - \frac{\dot{\mu}(t)}{\mu(t)}\right), \quad \forall t \geq 0, \tag{12.10}$$

and

$$\mu(0)g(0) \leq 1. \tag{12.11}$$

One can replace the initial point $t = 0$ by some point $t_0 \in \mathbb{R}$, and assume that the interval of time is $[t_0, t_0 + T)$, and that inequalities hold for $t \geq t_0$, rather than for $t \geq 0$. The proofs and the conclusions remain unchanged.

Theorem 12.1

If Assumptions A_1 and A_2 hold, then any solution $g(t) \geq 0$ to inequality (12.8) exists on all of $\mathbb{R}_+$, i.e. $T = \infty$, and satisfies the following estimate:

$$0 \leq g(t) \leq \frac{1}{\mu(t)} \quad \forall t \in \mathbb{R}_+. \tag{12.12}$$

If $\mu(0)g(0) < 1$, then $0 \leq g(t) < \frac{1}{\mu(t)}$ $\quad \forall t \in \mathbb{R}_+$.

Remark 12.1

If $\lim_{t\to\infty} \mu(t) = \infty$, then $\lim_{t\to\infty} g(t) = 0$.

Proof of Theorem 12.1. Let us rewrite inequality for μ as follows:

$$-\gamma(t)\mu^{-1}(t) + \alpha(t, \mu^{-1}(t)) + \beta(t) \leq \frac{d\mu^{-1}(t)}{dt}. \tag{12.13}$$

Let $\phi(t)$ solve the following Cauchy problem:

$$\dot{\phi}(t) = -\gamma(t)\phi(t) + \alpha(t, \phi(t)) + \beta(t), \quad t \geq 0; \quad \phi(0) = \phi_0. \tag{12.14}$$

The assumption that $\alpha(t, g)$ is locally Lipschitz with respect to g guarantees local existence and uniqueness of the solution $\phi(t)$ to problem (12.14). From the comparison result (see *A Comparison Lemma* proved below), it follows that

$$\phi(t) \leq \mu^{-1}(t) \quad \forall t \geq 0, \tag{12.15}$$

provided that $\phi(0) \leq \mu^{-1}(0)$, where $\phi(t)$ is the unique solution to problem (12.15). Let us take $\phi(0) = g(0)$. Then $\phi(0) \leq \mu^{-1}(0)$ by the assumption (12.11), and an inequality, similar to Equation 12.15, implies that

$$g(t) \leq \phi(t) \quad t \in [0, T). \tag{12.16}$$

Inequalities $\phi(0) \leq \mu^{-1}(0)$, Equations 12.15 and 12.16 imply

$$g(t) \leq \phi(t) \leq \mu^{-1}(t), \quad t \in [0, T). \tag{12.17}$$

By the assumption, the function $\mu(t)$ is defined for all $t \geq 0$ and is bounded on any compact subinterval of the set $[0, \infty)$. Consequently, the functions $\phi(t)$ and $g(t) \geq 0$ are defined for all $t \geq 0$, and estimate (12.12) is established.

If $g(0) < \mu^{-1}(0)$, then one obtains by a similar argument the strict inequality $g(t) < \mu^{-1}(t)$, $t \geq 0$. Theorem 12.1 is proved. ∎

Let us now prove the comparison result that was used above, see, for example, Reference 34, Theorem III.4.1.

A Comparison Lemma.
    Let

$$\dot{\phi}(t) = f(t, \phi), \quad \phi(0) = \phi_0, \tag{*}$$

and

$$\dot{\psi}(t) = g(t, \psi), \quad \psi(0) = \psi_0. \tag{**}$$

Assume $\psi_0 \geq \phi_0$, and

$$g(t, x) \geq f(t, x) \tag{***}$$

for any t and x for which both f and g are defined. Assume that f and g are continuous functions in a set $[0, s) \times (a, b)$, $\phi_0 \in (a, b)$, ψ is the maximal solution to (**) and ϕ is any solution to (*). Then $\phi(t) \leq \psi(t)$ on the maximal interval $[0, T)$ of the existence of both ϕ and ψ.

Proof of the Comparison Lemma. First, let us assume for simplicity that problems (*) and (**) have a unique solution. Later we will discard this simplifying assumption. If f and g satisfy a local Lipschitz condition with respect to ϕ, respectively, ψ, then our simplifying assumption holds. Second, also for simplicity assume that $g(t, x) > f(t, x)$. Under this simplifying assumption, it is easy to prove the conclusion of the Lemma, because the graph of ψ must lie above the graph of ψ for $t > 0$. Indeed, in a small neighborhood $[0, \delta)$, where $\delta > 0$ is sufficiently small, the graph of ψ lies above the graph of ϕ. This is obviously true if $\phi_0 < \psi_0$, because of the continuity of ϕ and ψ. If $\phi_0 = \psi_0$, then the graph of ψ lies above the graph of ϕ because $\dot{\phi}(0) < \dot{\psi}(0)$ due to the assumption $f(0, \phi_0) < g(0, \phi_0) = g(0, \psi_0)$. To check the last claim assume that there is a point $t_1 \in [0, T)$ such that $\phi(t_1) = \psi(t_1)$, and $\phi(t) < \psi(t)$ for $t \in (0, t_1)$. Then $\phi(t) - \phi(t_1) < \psi(t) - \psi(t_1)$. Divide this inequality by $t - t_1 < 0$ and get

$$\frac{\phi(t) - \phi(t_1)}{t - t_1} > \frac{\psi(t) - \psi(t_1)}{t - t_1}.$$

Pass to the limit $t \to t_1$, $t < t_1$, in the above inequality, use the differential equations for ϕ and ψ and the equality $\phi(t_1) = \psi(t_1)$, and obtain the following relation:

$$f(t_1, \phi(t_1)) = \dot\phi(t_1) \geq \dot\psi(t_1) = g(t_1, \psi(t_1)) = g(t_1, \phi(t_1)).$$

This relation contradicts the assumption $f(t, x) < g(t, x)$. This contradiction proves the conclusion of the Comparison Lemma under the additional assumption $f(t, x) < g(t, x)$.

To prove the Comparison Lemma under the original assumption $f(t, x) \leq g(t, x)$, let us consider problem (*) with f replaced by $f_n := f - \frac{1}{n} < f$. Let ϕ_n solve problem (*) with f replaced by f_n, and with the same initial condition as in (*). Since $f_n(t, x) < g(t, x)$, then, by what we have just proved, it follows that $\phi_n(t) \leq \psi(t)$ on the common interval $[0, T_n)$ of the existence of ϕ_n and ψ. By the standard result about continuous dependence of the solution to (*) on a parameter, one concludes that $\lim_{n\to\infty} T_n = T$ and $\lim_{n\to\infty} \phi_n(t) = \phi(t)$ for any $t \in [0, T)$. Therefore, passing to the limit $n \to \infty$ in the inequality $\phi_n(t) \leq \psi(t)$ one gets the conclusion of the Comparison Lemma under the original assumption $f(t, x) \leq g(t, x)$.

If the simplifying assumption concerning uniqueness of the solutions to (*) and (**) is dropped, then (*) and (**) may have many solutions. The limit of the solution ϕ_n is the minimal solution to (*). If one considers problem (**) with g replaced by $g_n := g + \frac{1}{n} > g$, and denotes by ψ_n the corresponding solution, then the limit $\lim_{n\to\infty} \psi_n(t) = \psi(t)$ is the maximal solution to (**). In this case, the above argument yields the conclusion of the Lemma with $\psi(t)$ being the maximal solution to (**), and $\phi(t)$ being any solution to (*). The Comparison Lemma is proved. ∎

Remark 12.2

If $\phi(t)$ is bounded from below for all $t \geq 0$, so that $c \leq \phi(t)$ for all $t \geq 0$, and $\psi(t)$ exists globally, that is, for all $t \geq 0$, then the inequality $c \leq \phi(t) \leq \psi(t)$ and the continuity of $f(t, x)$ on the set $[0, \infty) \times \mathbb{R}$ imply that any solution ϕ to (*) exists globally. Indeed, if it would exist only on a finite interval $[0, T)$ then it has to tend to infinity as $t \to T$, but this is impossible because the bound $c \leq \phi(t) \leq \psi(t)$ and the global existence and continuity of ψ do not allow $\phi(t)$ to grow to infinity as $t \to T$.

Let us formulate and prove a *discrete version* of Theorem 12.1.

Theorem 12.2

Assume that $g_n \geq 0$, $\alpha(n, g_n) \geq 0$,

$$g_{n+1} \leq (1 - h_n\gamma_n)g_n + h_n\alpha(n, g_n) + h_n\beta_n; \quad h_n > 0, \quad 0 < h_n\gamma_n < 1, \tag{12.18}$$

and $\alpha(n, g_n) \geq \alpha(n, p_n)$ if $g_n \geq p_n$. If there exists a sequence $\mu_n > 0$ such that

$$\alpha\left(n, \frac{1}{\mu_n}\right) + \beta_n \leq \frac{1}{\mu_n}\left(\gamma_n - \frac{\mu_{n+1} - \mu_n}{h_n\mu_n}\right), \tag{12.19}$$

and

$$g_0 \leq \frac{1}{\mu_0}, \tag{12.20}$$

then

$$0 \leq g_n \leq \frac{1}{\mu_n}, \quad \forall n \geq 0. \tag{12.21}$$

Proof of Theorem 12.2. For $n = 0$, inequality (12.21) holds because of Equation 12.20. Assume that it holds for all $n \leq m$ and let us check that then it holds for $n = m + 1$. If this is done, Theorem 12.2 is proved.

Using the inductive assumption, one gets

$$g_{m+1} \leq (1 - h_m \gamma_m) \frac{1}{\mu_m} + h_m \alpha \left(m, \frac{1}{\mu_m} \right) + h_m \beta_m.$$

This and inequality (12.19) imply

$$g_{m+1} \leq (1 - h_m \gamma_m) \frac{1}{\mu_m} + h_m \frac{1}{\mu_m} \left(\gamma_m - \frac{\mu_{m+1} - \mu_m}{h_m \mu_m} \right)$$

$$= \mu_m^{-1} - \frac{\mu_{m+1} - \mu_m}{\mu_m^2} \leq \mu_{m+1}^{-1}.$$

The last inequality is obvious since it can be written as

$$-(\mu_m - \mu_{m+1})^2 \leq 0.$$

Theorem 12.2 is proved.   ∎

Theorem 12.2 was formulated in Reference 23 and proved in Reference 24. We included a proof, which is slightly shorter than the one in Reference 24.

Let us give a few simple examples of applications of Theorem 12.1.

Example 12.1

Consider the inequality

$$\dot{g}(t) \leq tg - (t+1)^2 g^2 - 2(t+1)^{-2}. \tag{12.22}$$

Assume $g \geq 0$. Choose $\mu(t) = t + 1$. Then inequality (12.10) holds if

$$(t+1)[-(t+1)^2 - 2(t+1)^{-2}] \leq -t - (t+1)^{-1},$$

and $g(0) \leq 1$. Thus, inequality (12.10) holds if

$$-(t+1)^3 - 2(t+1)^{-1} \leq -t - (t+1)^{-1}.$$

This inequality holds obviously. Therefore, any $g \geq 0$, that satisfies inequalities (12.22) and $g(0) \leq 1$, exists for all $t \geq 0$ and satisfies the estimate

$$0 \leq g(t) \leq \frac{1}{t+1}.$$

In this example, the linearized problem

$$\dot{g}(t) = tg - 2(t+1)^{-2}, \quad g(0) = g_0,$$

has a unique solution

$$g(t) = e^{t^2/2} \left[g(0) - 2 \int_0^t e^{-\frac{s^2}{2}} (s+1)^{-2} ds \right].$$

This solution tends to infinity as $t \to \infty$.

Example 12.2

Consider a classical problem

$$\dot{u}(t) = A(t)u + F(t, u), \quad u(0) = u_0, \tag{12.23}$$

where $A(t)$ is a linear operator in $\mathbb{R}^n$ and F is a nonlinear operator. Assume that $Re(A(t)u, u) \leq -\gamma(u, u)$, where $\gamma = const > 0$, and $\|F(t, u)\| \leq c\|u\|^p$, $p = const > 1$, $c = const > 0$, and $\|\cdot\|$ is the norm of a vector in $\mathbb{R}^n$. We also assume that Equation 12.23 has the following property:

Property P: If a solution to Equation 12.23 is defined on the maximal interval of its existence $[0, T)$ and $T < \infty$, then $\lim_{t \to T-0} \|u(t)\| = \infty$.

It is known (see, e.g., Reference 34) that Property P holds if $F(t, u)$ is a continuous function on $[0, T] \times \mathbb{R}^n$.

By Peano's theorem the Cauchy problem

$$\dot{u}(t) = f(t, u), \quad u(0) = u_0, \quad u \in \mathbb{R}^n, \tag{12.24}$$

has a local solution on an interval $[0, a)$, provided that f is a continuous function on $[0, T] \times D(u_0)$, where $a \in (0, T)$ and $D(u_0)$ is a neighborhood of u_0. This solution is nonunique, in general. One can give an explicit estimate of the length a of the interval on which the solution does exist. Namely, $a = min(T, \frac{b}{M})$, where $M := max_{|u-u_0| \leq b, t \in [0,T]} |f(t, u)|$, and the neighborhood $D(u_0)$ is taken to be the set $\{u : |u - u_0| \leq b\}$.

It is known that *in every infinite-dimensional Banach space the Peano theorem fails*. Therefore, in an infinite-dimensional Banach space, we assume that problems (12.23) and (12.24) have a solution, and if $[0, T)$ is the maximal interval of the existence of the solution, then property P holds. This happens, for example, if $f(t, u)$ satisfies a local Lipschitz condition with respect to u and is continuous with respect to $t \in [0, T]$. Indeed, if a local Lipschitz condition holds, then the local interval of the existence of the solution to the Cauchy problem (12.24) is of the length $b = min(RM^{-1}, L)$, provided that f is continuous with respect to t and satisfies the estimates $\|f(t, u)\| \leq M$, $\|f(t, u) - f(t, v)\| \leq L\|u - v\|$, in the region $[0, T] \times B(u_0, R)$, $B(u_0, R) := \{u : \|u - u_0\| \leq R\}$. Under these assumptions the solution to problem (12.24) is unique and stays in the ball $B(u_0, R)$ for all $t \in [0, b]$.

To see that property P holds for problem (12.24) if f satisfies a local Lipschitz condition with respect to u, assume that the solution to Equation 12.24 does not exist for $t > T$. Under our assumptions, if the solution u of problem (12.24) satisfies the inequality $\sup_{0 \leq t < T} \|u(t)\| < \infty$, then the constants M, L, and R are finite. Therefore, $b > 0$. Take the initial point $t_0 = T - 0.5b$. By the local existence theorem, the solution $u(t)$ exists on the interval $[T - 0.5b, T + 0.5b]$. This is a contradiction, since we have assumed that this solution does not exist for $t > T$. This contradiction proves that Property P holds for problem (12.24) if f satisfies a local Lipschitz condition.

Let us use Theorem 12.1 to prove asymptotic stability of the zero solution to Equation 12.23 and to illustrate the application of our general method for a study of stability of solutions to abstract evolution problems, the method that we develop below.

Let $g(t) := \|u(t)\|$, where the norm is taken in $\mathbb{R}^n$. Take a dot product of Equation 12.23 with u, then take the real part of both sides of the resulting equation and get

$$Re(\dot{u}, u) = g\dot{g} = Re(Au, u) + Re(F(t, u), u) \leq -\gamma g^2 + cg^{p+1}.$$

Since $g \geq 0$, one obtains from the above inequality an inequality of the type (12.8), namely,

$$\dot{g}(t) \leq -\gamma g(t) + cg^p(t), \quad p = const > 1,$$

where γ and c are positive constants. Choose

$$\mu(t) = \lambda e^{at},$$

where $\lambda = const > 0$, $a = const \in (0, \gamma)$, $\alpha(t, g) = cg^p$, $\beta(t) = 0$, and $\gamma > 0$ is a constant. Note that a can be chosen arbitrarily close to γ. We choose λ later. Denote $b := \gamma - a > 0$. Then inequality (12.11) holds for any $g(0)$ if $\lambda > 0$ is sufficiently small. Inequality (12.10) holds if

$$c\lambda^{-(p-1)}e^{-(p-1)at} \leq \gamma - a = b.$$

Since $p > 1$ this inequality holds if

$$c\lambda^{-(p-1)} \leq b.$$

The last inequality holds for an arbitrary fixed $c > 0$ and an arbitrary small fixed $b > 0$ provided that $\lambda > 0$ is sufficiently large.

One concludes that *for any initial data u_0 the solution to Equation 12.23 exists globally and admits an estimate $\|u(t)\| \leq \lambda^{-1}e^{-at}$, where the positive constant $a < \gamma$ can be chosen arbitrarily close to γ if the positive constant c is sufficiently small.*

The above argument remains valid also for unbounded, closed, densely defined linear operators $A(t)$, provided that property P holds.

If $A(t)$ is a generator of a C_0 semigroup $T(t)$, and F satisfies a local Lipschitz condition, then problem (12.23) is equivalent to the equation $u = T(t)F(t, u)$, and this equation may be useful for a study of the global existence of the solution to problem (12.23) [35].

Example 12.3

Consider an example in which *the solution blows up in a finite time*, so it does not exist globally. Consider the problem

$$\dot{u} - \Delta u = u^2 \quad in \quad [0, \infty) \times D \subset \mathbb{R}^n; \quad u_N = 0; \quad u(0, x) = u_0(x). \tag{12.25}$$

Here D is a bounded domain with a smooth boundary S, N is an outer unit normal to S, $u_0 > 0$ is a smooth function. Let

$$g_0 := \int_D u_0(x)dx, \quad g(t) := \int_D u(t, x)dx.$$

Integrate Equation 12.25 over D and get $\dot{g}(t) = \int_D u^2 dx$. Use the inequality

$$\left(\int_D u \, dt \right)^2 \leq c \int_D u^2 dx,$$

where $c = c(D) = const > 0$, and get $\dot{g} \geq g^2/c$. Integrating this inequality, one obtains $g(t) \geq [\frac{1}{g_0} - ct]^{-1}$. Since $c > 0$ and $g_0 > 0$ it follows that:

$$\lim_{t \to t_b} g(t) = \infty,$$

where $t_b := \frac{1}{cg_0}$ is the blow-up time, and $t < t_b$. Consequently, *for any initial data with $g_0 > 0$ the solution to Equation 12.25 does not exist globally.*

Example 12.4

Consider the following equation:

$$\dot{u} + A(t)u + \phi(u) - \psi(t, u) = f(t, u), \quad u(0, x) = u_0(x), \tag{12.26}$$

where $u = u(t, x)$, ϕ and $\psi(t, u)$ are smooth functions growing to infinity as $|u| \to \infty$. Let us assume that

$$u\phi(u) \geq 0, \quad \forall t \geq 0,$$

and

$$(u, \psi(t, u)) \leq \alpha ||u||^3, (u, f(t, u)) \leq \beta(t)||u||^2,$$

where $\alpha(t) > 0$ and $\beta(t) > 0$ are continuous functions, $x \in D \subset \mathbb{R}^n$, D is a bounded domain,

$$Re(Au, u) \geq \gamma(u, u) \quad \forall u \in D(A), \quad \gamma = const > 0,$$

A is an operator in a Hilbert space $H = L^2(D)$, the domain of definition of A, $D(A)$, is a dense in H linear set, (u, v) is an inner product in H, $||u||^2 = (u, u)$. An example of A is $A = -\Delta$, the Laplacean with the Dirichlet boundary condition on S, the boundary of D. Denote $g(t) := ||u(t)||$. We want to estimate the large-time behavior of the solution u to Equation 12.26.

Take the inner product in H of Equation 12.26 and u, then take real part of both sides of the resulting equation and get

$$g\dot{g} \leq -\gamma g^2 + \alpha(t)g^3 + \beta(t)g.$$

Since $g \geq 0$ one obtains an inequality of the type (12.8), namely

$$\dot{g} \leq -\gamma g + \alpha(t)g^2 + \beta(t).$$

Now it is possible to use Theorem 12.1.

Choose $\mu(t) = \lambda e^{kt}$, where λ and k are positive constants, $k < \gamma$. Assume that $\lambda g_0 \leq 1$, where $g_0 := ||u_0(x)||$. Then inequality (12.11) holds for any initial data $u_0(x)$, that is, for any g_0, if λ is sufficiently small. Inequality (12.10) holds if

$$\frac{\alpha(t)e^{-kt}}{\lambda} + \lambda e^{kt}\beta(t) \leq \gamma - k.$$

One can easily impose various conditions on $\alpha(t)$ and $\beta(t)$ so that the above inequality holds. For example, assume that α decays monotonically as t grows, $\frac{\alpha(0)}{\lambda} < (\gamma - k)/2$, and $\beta(t) \leq \nu e^{-k't}$, where $k' > k$, $k' = const$, $\nu > 0$ is a constant, $\lambda\nu \leq (\gamma - k)/2$. Then inequality (12.10) holds, and it implies that

$$||u(t)|| \leq \frac{e^{-kt}}{\lambda},$$

so that the exponential decay of $||u(t)||$ as $t \to \infty$ is established.

In Sections 12.3 and 12.4, some stability results for abstract evolution problems are presented in detail. These results are formulated in four theorems. The basic ideas are similar to the ones discussed in examples in this section, but new assumptions and new technical tools are used.

12.3 Stability Results

In this section, we develop a method for a study of stability of solutions to the evolution problems described by the Cauchy problems (12.2) and (12.3) for abstract differential equations with a dissipative bounded linear operator $A(t)$ and a nonlinearity $F(t, u)$ satisfying inequality (12.5). Condition (12.5) means that for sufficiently small $\|u(t)\|$ the nonlinearity is of the higher order of smallness than $\|u(t)\|$. We also study the large-time behavior of the solution to problems (12.2) and (12.3) with persistently acting perturbations $b(t)$.

In this section, we assume that $A(t)$ is a bounded linear dissipative operator, but our methods are valid also for unbounded linear dissipative operators $A(t)$, for which one can prove global existence of the solution to problems (12.2) and (12.3). We do not go into further detail.

Let us formulate the first stability result.

Theorem 12.3

Assume that $Re(Au, u) \leq -k\|u\|^2 \; \forall u \in H$, $k = const > 0$, and inequality (12.4) holds with $\gamma(t) = k$. Then the solution to problems (12.2) and (12.3) with $b(t) = 0$ satisfies an estimate $\|u(t)\| = O(e^{-(k-\epsilon)t})$ as $t \to \infty$. Here $0 < \epsilon < k$ can be chosen arbitrarily small if $\|u_0\|$ is sufficiently small.

This theorem implies asymptotic stability in the sense of Lyapunov of the zero solution to Equation 12.2 with $b(t) = 0$. Our proof of Theorem 12.3 is new and very short.

Proof of Theorem. Multiply Equation 12.2 (in which $b(t) = 0$ is assumed) by u, denote $g = g(t) := \|u(t)\|$, take the real part, and use assumption (12.4) with $\gamma(t) = k > 0$, to get

$$g\dot{g} \leq -kg^2 + c_0 g^{p+1}, \quad p > 1. \tag{12.27}$$

If $g(t) > 0$ then the derivative $\dot{g}$ does exist, and

$$\dot{g}(t) = Re\left(\dot{u}(t), \frac{u(t)}{\|u(t)\|} \right),$$

as one can check. If $g(t) = 0$ on an open subset of $\mathbb{R}_+$, then the derivative $\dot{g}$ does exist on this subset and $\dot{g}(t) = 0$ on this subset. If $g(t) = 0$ but in any neighborhood $(t - \delta, t + \delta)$ there are points at which g does not vanish, then by $\dot{g}$ we understand the derivative from the right, that is,

$$\dot{g}(t) := \lim_{s \to +0} \frac{g(t+s) - g(t)}{s} = \lim_{s \to +0} \frac{g(t+s)}{s}.$$

This limit does exist and is equal to $\|\dot{u}(t)\|$. Indeed, the function $u(t)$ is continuously differentiable, so

$$\lim_{s \to +0} \frac{\|u(t+s)\|}{s} = \lim_{s \to +0} \frac{\|s\dot{u}(t) + o(s)\|}{s} = \|\dot{u}(t)\|.$$

The assumption about the existence of the bounded derivative $\dot{g}(t)$ from the right in Theorem 12.3 was made because the function $\|u(t)\|$ does not have, in general, the derivative in the usual sense at the points t at which $\|u(t)\| = 0$, no matter how smooth the function $u(t)$ is. Indeed,

$$\lim_{s \to -0} \frac{\|u(t+s)\|}{s} = \lim_{s \to -0} \frac{\|s\dot{u}(t) + o(s)\|}{s} = -\|\dot{u}(t)\|,$$

because $\lim_{s \to -0} \frac{|s|}{s} = -1$. Consequently, the right and left derivatives of $\|u(t)\|$ at the point t at which $\|u(t)\| = 0$ do exist, but are different. Therefore, the derivative of $\|u(t)\|$ at the point t at which $\|u(t)\| = 0$ does not exist in the usual sense.

However, as we have proved above, the derivative $\dot{g}(t)$ from the right does exist always, provided that $u(t)$ is continuously differentiable at the point t.

Since $g \geq 0$, inequality (12.27) yields inequality (12.8) with $\gamma(t) = k > 0$, $\beta(t) = 0$, and $\alpha(t, g) = c_0 g^p$, $p > 1$. Inequality (12.10) takes the form

$$\frac{c_0}{\mu^p(t)} \leq \frac{1}{\mu(t)} \left(k - \frac{\dot{\mu}(t)}{\mu(t)} \right), \quad \forall t \geq 0. \tag{12.28}$$

Let

$$\mu(t) = \lambda e^{bt}, \quad \lambda, b = const > 0. \tag{12.29}$$

We choose the constants λ and b later. Inequality (12.10), with μ defined in Equation 12.29, takes the form

$$\frac{c_0}{\lambda^{p-1} e^{(p-1)bt}} + b \leq k, \quad \forall t \geq 0. \tag{12.30}$$

This inequality holds if it holds at $t = 0$, that is, if

$$\frac{c_0}{\lambda^{p-1}} + b \leq k. \tag{12.31}$$

Let $\epsilon > 0$ be an arbitrary small number. Choose $b = k - \epsilon > 0$. Then Equation 12.31 holds if

$$\lambda \geq \left(\frac{c_0}{\epsilon} \right)^{\frac{1}{p-1}}. \tag{12.32}$$

Condition (12.11) holds if

$$\|u_0\| = g(0) \leq \frac{1}{\lambda}. \tag{12.33}$$

We choose λ and b so that inequalities (12.32) and (12.33) hold. This is always possible if $b < k$ and $\|u_0\|$ is sufficiently small.

By Theorem 12.1, if inequalities (12.31) through (12.33) hold, then one gets estimate (12.12):

$$0 \leq g(t) = \|u(t)\| \leq \frac{e^{-(k-\epsilon)t}}{\lambda}, \quad \forall t \geq 0. \tag{12.34}$$

Theorem 12.3 is proved. ∎

Remark 12.3

One can formulate the result differently. Namely, choose $\lambda = \|u_0\|^{-1}$. Then inequality (12.33) holds and becomes an equality. Substitute this λ into Equation 12.31 and get

$$c_0 \|u_0\|^{p-1} + b \leq k.$$

Since the choice of the constant $b > 0$ is at our disposal, this inequality can always be satisfied if $c_0 \|u_0\|^{p-1} < k$. Therefore, condition

$$c_0 \|u_0\|^{p-1} < k$$

is a sufficient condition for the estimate

$$\|u(t)\| \leq \|u_0\| e^{-(k-c_0\|u_0\|^{p-1})t},$$

to hold.

Let us formulate the second stability result.

Theorem 12.4

Assume that inequalities (12.4) through (12.6) hold and

$$\gamma(t) = \frac{c_1}{(1+t)^{q_1}}, \quad q_1 \leq 1; \quad c_1, q_1 = const > 0. \tag{12.35}$$

Suppose that $\epsilon \in (0, c_1)$ is an arbitrary small fixed number,

$$\lambda \geq \left(\frac{c_0}{\epsilon}\right)^{1/(p-1)} \quad \text{and} \quad \|u(0)\| \leq \frac{1}{\lambda}.$$

Then the unique solution to Equations 12.2 and 12.3 with $b(t) = 0$ exists on all of $\mathbb{R}_+$ and

$$0 \leq \|u(t)\| \leq \frac{1}{\lambda(1+t)^{c_1-\epsilon}}, \quad \forall t \geq 0. \tag{12.36}$$

Theorem 12.4 gives the size of the initial data, namely, $\|u(0)\| \leq \frac{1}{\lambda}$, for which estimate (12.36) holds. For a fixed nonlinearity $F(t, u)$, that is, for a fixed constant c_0 from assumption (12.5), the maximal size of $\|u(0)\|$ is determined by the minimal size of λ.

The minimal size of λ is determined by the inequality $\lambda \geq \left(\frac{c_0}{\epsilon}\right)^{1/(p-1)}$, that is, by the maximal size of $\epsilon \in (0, c_1)$. If $\epsilon < c_1$ and $c_1 - \epsilon$ is very small, then $\lambda > \lambda_{min} := \left(\frac{c_0}{c_1}\right)^{1/(p-1)}$ and λ can be chosen very close to λ_{min}.

Proof of Theorem 12.4. Let

$$\mu(t) = \lambda(1+t)^\nu, \quad \lambda, \nu = const > 0. \tag{12.37}$$

We will choose the constants λ and ν later. Inequality (12.10) (with $\beta(t) = 0$) holds if

$$\frac{c_0}{\lambda^{p-1}(1+t)^{(p-1)\nu}} + \frac{\nu}{1+t} \leq \frac{c_1}{(1+t)^{q_1}}, \quad \forall t \geq 0. \tag{12.38}$$

If

$$q_1 \leq 1 \quad \text{and} \quad (p-1)\nu \geq q_1, \tag{12.39}$$

then inequality (12.38) holds if

$$\frac{c_0}{\lambda^{p-1}} + \nu \leq c_1. \tag{12.40}$$

Let $\epsilon > 0$ be an arbitrary small number. Choose

$$\nu = c_1 - \epsilon. \tag{12.41}$$

Then inequality (12.40) holds if inequality (12.32) holds. Inequality (12.11) holds because we have assumed in Theorem 12.4 that $\|u(0)\| \leq \frac{1}{\lambda}$. Combining inequalities (12.32), (12.33), and (12.12), one obtains the desired estimate:

$$0 \leq \|u(t)\| = g(t) \leq \frac{1}{\lambda(1+t)^{c_1-\epsilon}}, \quad \forall t \geq 0. \tag{12.42}$$

Condition (12.32) holds for any fixed small $\epsilon > 0$ if λ is sufficiently large. Condition (12.33) holds for any fixed large λ if $\|u_0\|$ is sufficiently small.

Theorem 12.4 is proved. ∎

Let us formulate a stability result in which we assume that $b(t) \not\equiv 0$. The function $b(t)$ has physical meaning of persistently acting perturbations.

Theorem 12.5

Let $b(t) \not\equiv 0$, conditions (12.4) through (12.6) and (12.35) hold, and

$$\beta(t) \leq \frac{c_2}{(1+t)^{q_2}}, \tag{12.43}$$

where $c_2 > 0$ and $q_2 > 0$ are constants. Assume that

$$q_1 \leq \min\{1, q_2 - v, v(p-1)\}, \quad \|u(0)\| \leq \lambda_0^{-1}, \tag{12.44}$$

where $\lambda_0 > 0$ is a constant defined in Equation 12.51, see below, and

$$c_2^{1-\frac{1}{p}} c_0^{\frac{1}{p}} (p-1)^{\frac{1}{p}} \frac{p}{p-1} + v \leq c_1. \tag{12.45}$$

Then problem (12.2)–(12.3) has a unique global solution $u(t)$, and the following estimate holds:

$$\|u(t)\| \leq \frac{1}{\lambda_0(1+t)^v}, \quad \forall t \geq 0. \tag{12.46}$$

Proof of Theorem 12.5. Let $g(t) := \|u(t)\|$. As in the proof of Theorem 12.4, multiply Equation 12.2 by u, take the real part, use the assumptions of Theorem 12.5, and get the inequality:

$$\dot{g} \leq -\frac{c_1}{(1+t)^{q_1}} g + c_0 g^p + \frac{c_2}{(1+t)^{q_2}}. \tag{12.47}$$

Choose $\mu(t)$ by formula (12.37). Apply Theorem 12.1 to inequality (12.47). Condition (12.10) takes now the form

$$\frac{c_0}{\lambda^{p-1}(1+t)^{(p-1)v}} + \frac{\lambda c_2}{(1+t)^{q_2-v}} + \frac{v}{1+t} \leq \frac{c_1}{(1+t)^{q_1}} \quad \forall t \geq 0. \tag{12.48}$$

If assumption (12.44) holds, then inequality (12.48) holds for all $t \geq 0$ provided that it holds for $t = 0$, that is, provided that

$$\frac{c_0}{\lambda^{p-1}} + \lambda c_2 + v \leq c_1. \tag{12.49}$$

Condition (12.11) holds if

$$g(0) \leq \frac{1}{\lambda}. \tag{12.50}$$

The function $h(\lambda) := \frac{c_0}{\lambda^{p-1}} + \lambda c_2$ attains its global minimum in the interval $[0, \infty)$ at the value

$$\lambda = \lambda_0 := \left(\frac{(p-1)c_0}{c_2} \right)^{1/p}, \tag{12.51}$$

and this minimum is equal to

$$h_{min} = c_0^{\frac{1}{p}} c_2^{1-\frac{1}{p}} (p-1)^{\frac{1}{p}} \frac{p}{p-1}.$$

Thus, substituting $\lambda = \lambda_0$ in formula (12.49), one concludes that inequality (12.49) holds if the following inequality holds:

$$c_0^{\frac{1}{p}} c_2^{1-\frac{1}{p}} (p-1)^{\frac{1}{p}} \frac{p}{p-1} + v \leq c_1. \tag{12.52}$$

Inequality (12.50) holds if

$$\|u(0)\| \leq \frac{1}{\lambda_0}. \tag{12.53}$$

Therefore, by Theorem 12.1, if conditions (12.52) and (12.53) hold, then estimate (12.12) yields

$$\|u(t)\| \leq \frac{1}{\lambda_0(1+t)^v}, \quad \forall t \geq 0, \tag{12.54}$$

where λ_0 is defined in Equation 12.51.

Theorem 12.5 is proved. ∎

12.4 Stability Results Under Non-classical Assumptions

Let us assume that $Re(A(t)u, u) \leq \gamma(t)\|u\|^2$, where $\gamma(t) > 0$. This corresponds to the case when the linear operator $A(t)$ may have spectrum in the right half-plane $Re z > 0$. Our goal is to derive under this assumption sufficient conditions on $\gamma(t)$, $\alpha(t, g)$, and $\beta(t)$, under which the solution to problem (12.2) is bounded as $t \to \infty$, and stable. We want to demonstrate new methodology, based on Theorem 12.1. By this reason, we restrict ourselves to a derivation of the simplest results under simplifying assumptions. However, our derivation illustrates the method applicable to many other problems.

Our assumptions in this section are:

$$\beta(t) = 0, \quad \gamma(t) = c_1(1+t)^{-m_1}, \quad \alpha(t, g) = c_2(1+t)^{-m_2}g^p, \ p > 1.$$

Let us choose

$$\mu(t) = d + \lambda(1+t)^{-n}.$$

The constants c_j, m_j, λ, d, and n are assumed positive.

We want to show that a suitable choice of these parameters allows one to check that basic inequality (12.10) for μ is satisfied. If this is verified, then one obtains by Theorem 12.1 inequality (12.12) for $g(t)$. This inequality allows one to derive global boundedness of the solution to Equation 12.2, and the Lyapunov stability of the zero solution to Equation 12.2 (with $u_0 = 0$). Note that under our assumptions $\dot{\mu} < 0, \lim_{t \to \infty} \mu(t) = d$. We choose $\lambda = d$. Then $(2d)^{-1} \leq \mu^{-1}(t) \leq d^{-1}$ for all $t \geq 0$. The basic inequality (12.10) takes the form

$$c_1(1+t)^{-m_1} + c_2(1+t)^{-m_2}[d + \lambda(1+t)^{-n}]^{-p+1} \leq n\lambda(1+t)^{-n-1}[d + \lambda(1+t)^{-n}]^{-1}, \quad (12.55)$$

and

$$g_0(d + \lambda) \leq 1. \quad (12.56)$$

Since we have chosen $\lambda = d$, condition (12.56) is satisfied if

$$d = (2g_0)^{-1}. \quad (12.57)$$

Choose n so that

$$n + 1 \leq \min\{m_1, m_2\}. \quad (12.58)$$

Then Equation 12.55 holds if

$$c_1 + c_2 d^{-p+1} \leq n\lambda d^{-1}. \quad (12.59)$$

Inequality (12.59) is satisfied if c_1 and c_2 are sufficiently small. Let us formulate our result, which follows from Theorem 12.1.

Theorem 12.6

If inequalities (12.59) and (12.58) hold, then

$$0 \leq g(t) \leq [d + \lambda(1+t)^{-n}]^{-1} \leq d^{-1}, \quad \forall t \geq 0. \quad (12.60)$$

Estimate (12.60) proves global boundedness of the solution $u(t)$, and implies Lyapunov stability of the zero solution to problem (12.2) with $b(t) = 0$ and $u_0 = 0$.

Indeed, by the definition of Lyapunov stability of the zero solution, one should check that for an arbitrary small fixed $\epsilon > 0$ estimate $\sup_{t \geq 0} \|u(t)\| \leq \epsilon$ holds provided that $\|u(0)\|$ is sufficiently small. Let $\|u(0)\| = g_0 = \delta$. Then estimate (12.60) yields $\sup_{t \geq 0} \|u(t)\| \leq d^{-1}$, and Equation 12.57 implies $\sup_{t \geq 0} \|u(t)\| \leq 2\delta$. So, $\epsilon = 2\delta$, and the Lyapunov stability is proved. ∎

References

1. A. M. Lyapunov, *Probléme général de la stabilité du mouvement*, Annals of Mathematics Studies, Vol. 17, Princeton University Press, Princeton, New Jersey, 1947.
2. L. Cesari, *Asymptotic Behavior and Stability Problems in Ordinary Differential Equations*, Springer-Verlag, New York, 1971.
3. S. Chow and J. Hale, *Methods of Bifurcation Theory*, Springer-Verlag, New York, 1982.
4. J. Guckenheimer and P. Holmes, *Nonlinear Oscillations, Dynamical Systems, and Bifurcations of Vector Fields*, Springer-Verlag, New York, 1983.
5. R. Z. Has'minskii, *Stochastic Stability of Differential Equations*, Sijthoff and Noordhoff, Alphen aan den Rijn, Germantown, 1980.
6. M. A. Krasnosel'skii, *The Operator of Translation along the Trajectories of Differential Equations*, Translations of Mathematical Monographs, Vol. 19, American Mathematical Society, Providence, Rhode Island, 1968.
7. I. G. Malkin, *Theory of Stability of Motion*, Nauka, Moskow, 1966 (in Russian).
8. N. Rouche, P. Habets, and M. Laloy, *Stability Theory by Lyapunov's Direct Method*, Springer-Verlag, New York, 1977.
9. V. Yakubovich, V. Starzhinskii, *Linear Differential Operators with Periodic Coefficients and Their Applications*, Nauka, Moscow, 1972 (in Russian). Translated from Russian to English in 1975, Halsted Press, New York-Toronto, Israel Program for Scientific Translations, Jerusalem-London, Vols. 1, 2.
10. Y. L. Daleckii and M. G. Krein, *Stability of Solutions of Differential Equations in Banach Spaces*, American Mathematical Society, Providence, Rhode Island, 1974.
11. A. G. Ramm, *Dynamical Systems Method for Solving Operator Equations*, Elsevier, Amsterdam, 2007.
12. A. G. Ramm, N. S.Hoang, *Dynamical Systems Method and Applications: Theoretical Developments and Numerical Examples*, Wiley, Hoboken, New Jersey, 2012.
13. A. G. Ramm, Dynamical systems method (DSM) and nonlinear problems, in J. Lopez-Gomez (ed.) *Spectral Theory and Nonlinear Analysis*, World Scientific Publishers, Singapore, 2005, 201–228.
14. A. G. Ramm, How large is the class of operator equations solvable by a DSM Newton-type method? *Appl. Math. Lett.*, 24, N6, 2011, 860–865.
15. A. G. Ramm, A nonlinear inequality and evolution problems, *J. Inequal. Spec. Funct.*, 1, N1, 2010, 1–9.
16. A. G. Ramm, Asymptotic stability of solutions to abstract differential equations, *J. Abs. Differ. Equ., (JADEA)*, 1, N1, 2010, 27–34.
17. A. G. Ramm, Stationary regimes in passive nonlinear networks, in P. Uslenghi (ed.), *Nonlinear Electromagnetics*, Academic Press, New York, 1980, 263–302.
18. A. G. Ramm, *Theory and Applications of Some New Classes of Integral Equations*, Springer-Verlag, New York, 1980.
19. A. G. Ramm, Stability of solutions to some evolution problems, *Chaotic Model. Simul. (CMSIM)*, 1, 2011, 17–27.
20. A. G. Ramm, Stability result for abstract evolution problems, *Math. Meth. Appl. Sci.*, 36 (4), 2013, 422–426.
21. A. G. Ramm, Stability of solutions to abstract evolution equations with delay, *J. Math. Anal. Appl. (JMAA)*, 396, 2012, 523–527.
22. A. G. Ramm and V. Volpert, Convergence of time-dependent Turing structures to a stationary solution, *Acta Appl. Math.*, 123, N1, 2013, 31–42.
23. N. S. Hoang and A. G. Ramm, DSM of Newton-type for solving operator equations $F(u) = f$ with minimal smoothness assumptions on F, *Int. J. Comput. Sci. Math.* (IJCSM), 3, N1/2, 2010, 3–55.
24. N. S. Hoang and A. G. Ramm, A nonlinear inequality and applications, *Nonlinear Anal. Theory Methods Appl.*, 71, 2009, 2744–2752.

25. N. S. Hoang and A. G. Ramm, The dynamical systems method for solving nonlinear equations with monotone operators, *Asian Eur. Math. J.*, 3, N1, 2010, 57–105.
26. N. S. Hoang and A. G. Ramm, DSM of Newton-type for solving operator equations $F(u) = f$ with minimal smoothness assumptions on F, *Int. J. Comput. Sci. Math. (IJCSM)*, 3, N1/2, 2010, 3–55.
27. N. S. Hoang and A. G. Ramm, Nonlinear differential inequality, *Math. Inequal. Appl.* (MIA), 14, N4, 2011, 967–976.
28. N. S. Hoang and A. G. Ramm, Some nonlinear inequalities and applications, *J. Abs. Differ. Equ. Appl.*, 2, N1, 2011, 84–101.
29. A. G. Ramm, Spectral properties of Schrödinger-type operators and large-time behavior of the solutions to the corresponding wave equation, *Math. Model. Nat. Phenom.*, 8, N1, 2013, 207–214.
30. R. Temam, *Infinite-Dimensional Dynamical Systems in Mechanics and Physics*, Springer-Verlag, New York, 1997.
31. N. Krasovski, *Problems of the Theory of Stability of Motion*, Stanford University Press, Stanford, California, 1963.
32. B. Davies, *Exploring Chaos*, Perseus Books, Reading, Massachusetts, 1999.
33. R. L. Devaney, *An Introduction to Chaotic Dynamical Systems*, Addison-Wesley, Reading, Massachusetts, 1989.
34. P. Hartman, *Ordinary Differential Equations*, Wiley, New York, 1964.
35. A. Pazy, *Semigroups of Linear Operators and Applications to Partial Differential Equations*, Springer-Verlag, New York, 1983.

13

Empirical Wavelet Coefficients and Denoising of Chaotic Data in the Phase Space

Matthieu Garcin

13.1 Introduction

Among the techniques used to eliminate a measurement noise disrupting a chaotic dataset, wavelet-based techniques are extremely efficient. However, a proper use of wavelets in the framework of a chaotic signal needs some adaptations from the classical linear case. Indeed, in the phase space, the impact of the noise on a chaotic system is nonlinear and the observations are non-equally spaced. These two specific features impose particular wavelet-based denoising rules, which are strongly related on how the wavelet coefficients are empirically computed. We present these denoising rules in the present chapter.

We consider a chaos, or more generally a dynamical system, defined in discrete time, (X_t). Two successive states of that dynamical system are linked by an evolution function z [1]:

$$\forall t \in \{1, \ldots, T\}, \; X_{t+1} = z(X_t), \tag{13.1}$$

where X_t is a vector of dimension in the embedding dimension of the system. For simplicity, we limit our analysis to one-dimensional (1D) systems, even though the adaptation to multidimensional frameworks is possible [2].

Besides, we assume that we do not observe directly the dynamical system but, instead, a noisy version of it. The noise considered is an additive measurement noise. Therefore, the observations are u_t:

$$\forall t \in \{1, \ldots, T\}, \; u_t = X_t + \varepsilon_t,$$

where $\varepsilon_1, \ldots, \varepsilon_T$ are independent identically distributed (i.i.d.) random variables. Concomitantly, the observed evolution function is not z but a noisy function z^ε defined by

$$\forall t \in \{1, \ldots, T\}, \ u_{t+1} = z^\varepsilon(u_t) = z(u_t - \varepsilon_t) + \varepsilon_{t+1}.$$

We observe a sample of N states of the noisy dynamical system, which is supposed to evolve in the interval $[0, 1]$. As we assume that the system is unidimensional, we can rank all the observations and we write $u_1 \leq \cdots \leq u_N$ instead of $u_{1:N} \leq \cdots \leq u_{N:N}$, for conciseness. As a consequence, we observe N values of the noisy evolution function, $z^\varepsilon(u_1), \ldots, z^\varepsilon(u_N)$, such that

$$\forall n \in \{1, \ldots, N\}, \ z^\varepsilon(u_n) = z(u_n - \varepsilon_n^\star) + \varepsilon_n, \tag{13.2}$$

where $\varepsilon_1, \ldots, \varepsilon_N, \varepsilon_1^\star, \ldots, \varepsilon_N^\star$ are $2N$ i.i.d random variables [2].

From these observations of z^ε, we want to estimate z in order to make predictions for any time trajectory of the dynamical system. To achieve this goal, several methods are possible, such as singular value decomposition, kernel or maximum likelihood techniques [3,4]. Alternatively, wavelet filtering techniques have shown a great efficiency when the impact of the noise on the observations is linear and it is particularly well adapted to spatially inhomogeneous signals [5]. However, we are facing a nonlinear impact of the noise on the observations of the chaos. Hence, the estimation of the chaos in the phase space, that is the estimation of z, requires a few adaptations of the standard wavelet method. Furthermore, one of the main specificities of the chaos-adapted wavelet denoising method is the irregularity of the observation grid. As it is non-equispaced, the calculation of the wavelet coefficients as classically defined is not appropriate. We will then present how to cope with this peculiarity.

In this chapter, we present the practical steps of a denoising of the evolution function of a chaos. First, we briefly explain how the wavelet method works in the classical case, why it must be transformed in the chaotic framework and why it should then be rather used in the phase space than in the time domain (Section 13.2). Then, since the observation sample is non-equispaced, various methods are possible to calculate empirical wavelet coefficients. Besides generic formulas, we will present some of these methods (distortion of the observation grid, signal interpolation, and Voronoi cells) as well as the consequences in the denoising technique. More precisely, the question we want to answer is: Which calculation method should be used depending on the nature of the observation sample? (See Section 13.3.)

13.2 Denoising Chaotic Data in the Phase Space

The decomposition of a pure and quite smooth function in a wavelet basis enables us to describe it accurately and with few nonzero coefficients. When this function is disrupted by some noise, we can use its wavelet decomposition to denoise the noisy signal, that is to estimate the initial function. Indeed, the erratic nature of the noise adds many nonzero noisy wavelet coefficients that can be shrunk toward zero. We detail this technique in the next subsection. Then, we explain how it should be applied to chaos.

13.2.1 Theoretical Wavelet Denoising

Let z be a squared-integrable function: $z \in \mathcal{L}^2(\mathbb{R})$. Thanks to the Fourier analysis, we can decompose this function as a sum of functions of various frequencies. However, if the function z is inhomogeneous, we would prefer a decomposition in localized functions: instead of a unique function corresponding to a given frequency (e.g., a sinusoid), we prefer a set of functions of this same resolution but, for each function, only a limited number of oscillations. This can be obtained by means of a windowed Fourier transform. But, for better accuracy, the size of the window should depend on the resolution level. Therefore, we prefer wavelets, which are obtained from a real mother wavelet, $\Psi \in \mathcal{L}^2(\mathbb{R})$, by dilatation and translation:

$$\psi_{j,k} : t \in \mathbb{R} \mapsto 2^{j/2} \Psi \left(2^j t - k \right),$$

where $j \in \mathbb{Z}$ is the resolution parameter and $k \in \mathbb{Z}$ is the translation parameter. Then, the theoretical wavelet coefficient of parameters j and k of the function z is defined by

$$\hat{z}_{j,k} = \int_{\mathbb{R}} z(x)\psi_{j,k}(x)dx. \tag{13.3}$$

The function z can be written as an infinite sum of wavelets weighted by the wavelet coefficients:

$$z = \sum_{j \in \mathbb{Z}} \sum_{k \in \mathbb{Z}} \hat{z}_{j,k} \psi_{j,k}, \tag{13.4}$$

or, equivalently, whatever $j_0 \in \mathbb{Z}$:

$$z = \sum_{k \in \mathbb{Z}} \beta_{j_0,k} \phi_{j_0,k} + \sum_{j=j_0}^{+\infty} \sum_{k \in \mathbb{Z}} \hat{z}_{j,k} \psi_{j,k}, \tag{13.5}$$

where $\phi_{j_0,k}$ is the scaling function corresponding to the chosen wavelet basis and

$$\beta_{j_0,k} = \int_{\mathbb{R}} z(x)\phi_{j,k}(x)dx.$$

This decomposition in gross structure ($\beta_{j_0,k}$) and details ($\hat{z}_{j,k}$) is inherent to the multiresolution analysis and allows to calculate fewer coefficients [5]. Moreover, in Equations 13.4 and 13.5, the sums may contain even fewer terms if one uses wavelets defined on a compact support: indeed, only a limited number of translation parameters k would then lead to nonzero coefficients. As a consequence, the Daubechies wavelets, which have a compact support, form an interesting class of mother wavelets [6]. Concerning the other infinite sum, over the resolution parameter j, a good approximation of z can be obtained by truncating this sum. This is due to the decrease of the amplitude of the wavelet coefficients when j grows [5]. Incidentally, a finite dataset is also a reason for truncating the sum. Thus, Equation 13.5 yields the following estimate of z:

$$\sum_{k \in \mathbb{Z}} \beta_{j_0,k} \phi_{j_0,k} + \sum_{j=j_0}^{J} \sum_{k \in \mathbb{Z}} \hat{z}_{j,k} \psi_{j,k},$$

for a given maximum resolution level J.

If the function z is linearly disrupted by some i.i.d. noise, then the wavelet coefficients $\hat{z}^{\varepsilon}$ of the noisy signal z^{ε} are also disrupted by the noise. We can then filter them in order to diminish the impact of the noise. Thus, the estimate of z from the noisy signal z^{ε} is

$$\sum_{k \in \mathbb{Z}} \beta^{\varepsilon}_{j_0,k} \phi_{j_0,k} + \sum_{j=j_0}^{J} \sum_{k \in \mathbb{Z}} F(\hat{z}^{\varepsilon}_{j,k}) \psi_{j,k}, \tag{13.6}$$

where F is the filtering function. The scaling coefficients $\beta^{\varepsilon}_{j_0,k}$ are not filtered since they correspond to the gross structure of the signal. As we will explain in the next section, the wavelet coefficients used in practice for a discrete observation grid are not the theoretical ones but empirical ones.

The filter F may be chosen in several ways. In particular, thresholding filters are popular filters. The hard threshold filter consists in eliminating noisy wavelet coefficients close to zero, that is to say coefficients below a certain threshold $\lambda \geq 0$:

$$F_{\lambda}^{hard} : x \in \mathbb{R} \mapsto x\mathbf{1}_{|x|\geq\lambda}.$$

The soft threshold filter, in addition to eliminating small wavelet coefficients, also shrinks towards zero all the coefficients so as to get a continuous filtering function:

$$F_{\lambda}^{soft} : x \in \mathbb{R} \mapsto (x-\lambda)\mathbf{1}_{x\geq\lambda} + (x+\lambda)\mathbf{1}_{x\leq\lambda}.$$

The underlying idea of the thresholding is that, for smooth enough functions, most of the wavelet coefficients are equal to zero. Therefore, when some noise disrupts the smooth function z, the signal z^{ε} is not as smooth as z and many of its wavelet coefficients slightly diverge from zero. These small wavelet coefficients account for more noise than pure signal. If we use a thresholding filter, our estimate of z, defined as in Equation 13.6, will have a parsimonious decomposition in the wavelet basis, like z.

What about the value of λ? In this framework of a linear impact of the noise, the *universal threshold*, also known as *VisuShrink*, is such that $\lambda = \sigma\sqrt{2\log N}$, where σ^2 is the noise variance. It is often used in the scientific literature. This threshold value has the advantage of not depending on the signal. That is why it is called *universal*. In fact, in the case of a Gaussian noise, filtering with this threshold leads to an expected reconstruction error of the same *magnitude* than oracular linear filters, which are the best possible linear filters [7]. However, even though these expected errors are of the same magnitude, it is often possible to do better than the universal threshold in practice and one would rather use an unbiased estimate of the error in order to numerically choose the quasi-optimal threshold which minimizes the estimated error. This is known as SURE threshold, or *SureShrink* [5,8,9]. The estimate of the error is suitable for a soft thresholding and a Gaussian noise. By definition, it is equal to

$$\sum_{j}\sum_{k}\mathcal{S}(\hat{z}_{j,k}^{\varepsilon}),$$

where

$$\mathcal{S} : x \mapsto \begin{cases} \lambda^2 + \sigma^2 & \text{if } |x| \geq \lambda \\ x^2 - \sigma^2 & \text{else.} \end{cases}$$

13.2.2 The Case of Chaos

As we explained in the previous paragraph, a performing denoising can be achieved when filtering wavelet coefficients, provided that the pure function to recover is smooth enough. Therefore, if one wants to apply this method to a chaos, one should first choose a quite smooth function to estimate. For example, a time trajectory of a chaos is often erratic and thus not much adapted to a wavelet denoising. It would indeed be complicated to distinguish the pure trajectory of the chaos from the noise component: neither the wavelet decomposition of the pure time trajectory nor the decomposition of the noisy one will be parsimonious, so that a thresholding would either be useless, or severely alter the pure trajectory itself.

On the contrary, the phase space representation of the chaos, by means of the evolution function like in Equation 13.1, is often much smoother. As a consequence, we will apply the wavelet method to this evolution function rather than to the time trajectory.

We can check this assertion about the smoothness with an example. We consider a logistic chaos, defined by the evolution function:

$$z : x \in [0, 1] \mapsto 4x(1 - x). \tag{13.7}$$

The corresponding time series has more small wavelet coefficients than the observations in the phase space. Therefore, the same filter will eliminate more pertinent wavelet coefficients in the case of the time series than in the case of the observations in the phase space, as seen in Figure 13.1. In this graph, the logistic chaos is considered in the absence of noise. When some noise is added, its elimination implies a certain

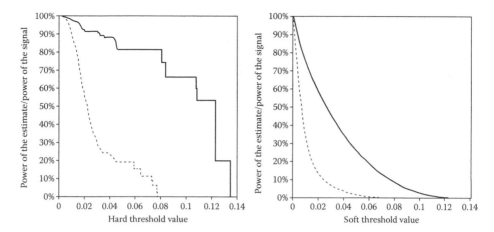

FIGURE 13.1   Ratio of the power of the signal filtered using a hard (on the left) or a soft (on the right) threshold, on the power of the signal, which is defined in Equation 13.7. The sample consists of 1025 observations. The ratio for the phase space is the thick line and the ratio for the time series is the thin dotted line. As the second one is always below the first one, it is more vulnerable to the thresholding. For example, for a hard threshold equal to 0.08, all the wavelet coefficients of the time series are eliminated whereas 81% of the power of the observations of z in the phase space is still intact.

value for the threshold and therefore the concomitant elimination of a bigger part of the pure signal in the case of the time series than in the case of the observations in the phase space.

Then, as we have decided to observe the dynamical system in the phase space and more precisely the noisy evolution function z^{ε}, two special features make the denoising using wavelets different from the standard case:

- The impact of the noise is nonlinear
- The observation grid is non-equispaced

Therefore, the denoising technique has to be adapted to the framework of dynamical systems in order to take into account both the nonlinear impact of the noise and the irregularity of the sample. This will be detailed in the next sections.

13.3 Empirical Wavelet Coefficients

The integral expression in the formula 13.3 of the theoretical wavelet coefficients $\hat{z}_{j,k}$ must be approximated if the function z is only observed on a discrete grid. Therefore, in practice, theoretical wavelet coefficients are replaced by empirical wavelet coefficients, calculated to be as close as possible to the theoretical ones.

If the observation sample is equispaced, then the empirical wavelet coefficients have a simple expression. For the parameters j and k, the empirical wavelet coefficient of the function z, noted $\langle z, \psi_{j,k} \rangle$, is defined by

$$\langle z, \psi_{j,k} \rangle = \frac{1}{N} \sum_{n=1}^{N} z(n/N) \psi_{j,k}(n/N), \tag{13.8}$$

where we assumed equispaced observations of the form n/N, for $n \in \{1, 2, \ldots, N\}$.

However, in the case of a chaos observed in the phase space, the observations are not equispaced. The calculation of the empirical wavelet coefficients must then be adapted to that particular grid design. Several ways of doing this are possible. Three of them are detailed hereafter. If the observations are equispaced, all these three methods lead to the standard Equation 13.8.

13.3.1 Wavelet Tools for Non-Equispaced Samples

Besides the various calculation methods we will present later, we can define a general framework for empirical wavelet coefficients. We assume that the (non-equispaced) observations are ranked: $u_1 \leq u_2 \leq \cdots \leq u_N$. Then, for parameters j and k, the empirical wavelet coefficient of z is defined by

$$\langle z, \bar{\psi}_{j,k} \rangle = \sum_{n=1}^{N} z(u_n) \bar{\psi}_{j,k}(u_n), \tag{13.9}$$

where $\bar{\psi}_{j,k}$ is a function consisting in a transformation of $\psi_{j,k}$ related to the calculation method employed. This expression is also suitable for equispaced samples, with $u_i = i/N$ for $i \in \{1, 2, \ldots, N\}$ and $\bar{\psi}_{j,k} = \psi_{j,k}/N$. The estimate of z using a filtering method on the empirical wavelet coefficients is then

$$\tilde{z} = \sum_j \sum_k F\left(\langle z, \bar{\psi}_{j,k} \rangle\right) \psi_{j,k} \circ h, \tag{13.10}$$

where F is the filtering function and h is a function allowing to take into account a distortion of the observation grid. Again, this formula may be used in the standard case of an equispaced sample, with h equal to the identity $Id_{[0,1]}$. We could also write Equation 13.10 as the sum of a gross structure and filtered details, like in Equation 13.6.

13.3.2 Various Methods for Calculating Empirical Wavelet Coefficients

We present thereafter three methods for calculating empirical wavelet coefficients using non-equispaced samples, with their pros and cons. They are all particular cases of the general framework of empirical wavelet coefficients defined in the previous paragraph.

13.3.2.1 Using a Distortion of the Observation Grid

Fast algorithms calculating wavelet coefficients are based on the dyadic structure of the sample. In these algorithms, the sample must contain $2^\nu + 1$ equispaced observations, for $\nu \in \mathbb{N}$. Indeed, the cascade algorithm allows a fast calculation of the wavelet functions at these dyadic points [5], whereas the interpolation of these functions takes more time. However, in our chaotic framework, the observations are irregularly spaced and fast algorithms cannot be applied directly. Then, a possible solution consists in the distortion of the non-equispaced sample in order to match each observation point with a dyadic abscissa of the wavelet functions. More precisely, the functions $\bar{\psi}_{j,k}$ and h, in the general formulas for non-equispaced samples, are defined in this case by

$$\begin{cases} \bar{\psi}_{j,k} : u_n, n \in \mathbb{N} \mapsto N^{-1} \psi_{j,k}(n/N) \\ h : x \in [0,1] \mapsto N^{-1} \min\{n \in \mathbb{N} | x \geq u_n\}. \end{cases}$$

This method is equivalent to the standard calculation of the wavelet coefficients of $z \circ g$ where $g : n/N, n \in \mathbb{N} \mapsto u_n$.

Cai and Brown proposed an improvement of this method, based on multiresolution analysis and the projection method, leading to estimators with better convergence rates when g is smoother than z [10].

The advantage of this method, in addition to its simplicity and accuracy, is the fast calculation due to the dyadic structure of the distorted grid. However, there is an important drawback as this method is not much adapted to multidimensional samples. Indeed, there is no natural definition of the order statistics in dimension higher than 1, even though some methods may be tested to overcome this difficulty. One of them is the easy path wavelet transform (EPWT), in which a 1D wavelet transform is used along path vectors in a 2D grid of equispaced observations [11]. Nevertheless, we are not aware of any published use of the EPWT on non-equispaced samples of dimension higher than 2.

13.3.2.2 Using a Signal Interpolation

Another way of using the value of wavelet functions at dyadic abscissas and therefore to use fast algorithms consists in the interpolation of the function z at these dyadic points [12]. This enables us to overcome the difficulty of multidimensional samples mentioned in the previous paragraph. Indeed, the interpolation can be implemented in dimension 1 as well as in multidimensional frameworks. We may define several interpolation rules such as a local averaging or a local linear interpolation [12]. We propose the following function Z, whose restriction to $\mathbb{R}\backslash\{u_1, u_2, \ldots, u_N\}$ interpolates the observed part of z, that is to say $z_{|\{u_1,u_2,\ldots,u_N\}}$:

$$Z(x) = \frac{\sum_{n=1}^N z(u_n)w(|x - u_n|)}{\sum_{i=1}^N w(|x - u_i|)},$$

where w is a weighting function. When $x = u_n$, we aim to have $Z(x)$ very close to the observed value $z(u_n)$. Ideally, w is thus a decreasing function, with a very high limit at 0, so that the value of the function Z at any u_n is close to $z(u_n)$. For example, we may be interested in the following class of weighting functions:

$$w_\alpha : \mathbb{R}^+ \to \mathbb{R}^+, x \mapsto \exp\left(\frac{1}{x+\alpha}\right), \tag{13.11}$$

with $\alpha > 0$. When $\alpha \to 0$, then w_α has a high value in 0 and $Z(u_n)$ is very close to $z(u_n)$.

As a consequence, the functions $\bar{\psi}_{j,k}$ and h, in the general formulas for non-equispaced samples, are now defined by

$$\begin{cases} \bar{\psi}_{j,k} : u_n, n \in \mathbb{N} \mapsto N^{-1} \sum_{m=1}^N \psi_{j,k}(m/N)\dfrac{w(|m/N - u_n|)}{\sum_{i=1}^N w(|m/N - u_i|)} \\ h : x \in [0,1] \mapsto x. \end{cases}$$

The interpolation method smooths the observations, particularly when w decreases at a slow rate. Therefore, inhomogeneities of the function z may be mitigated and some errors be generated. This fact is illustrated in Figure 13.2.

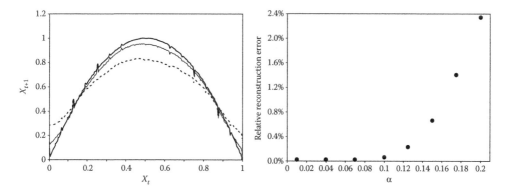

FIGURE 13.2 On the left: Reconstructed evolution function of the logistic chaos defined in Equation 13.7 with the interpolation method for a weighting function like in Equation 13.11, with α equal to 0.2 (dotted line), 0.125 (solid thin line), or 0.01 (solid thick line). When α grows, the method underestimates the value of z around $X_t = 0.5$ because of the smoothing effect of the interpolation. This impairs the chaotic features of the estimate. On the right: Reconstruction error relatively to the integral of the pure signal over $[0, 1]$, as a function of the weighting parameter α. All the wavelet decompositions are truncated at resolution level 3.

13.3.2.3 Using Voronoi Cells

In this last method, we focus on the theoretical definition of wavelet coefficients, Equation 13.3, in which an integral appears. Therefore, we would like to use an approximation of this integral. We achieve this goal thanks to Voronoi cells [2,13]. More precisely, we approach the integral by a discrete weighted sum of the observed values of the function z, with a weight corresponding to the size of the region *dominated* by each observation point. In dimension d, for $u_n \in \mathbb{R}^d$, this region is called the Voronoi cell of u_n. It is equal to the set of points of $\mathbb{R}^d$ closer to u_n than to any other observation point. We note V_n the size of the Voronoi cell of u_n. In dimension $d = 1$, we simply have

$$V_n = \frac{1}{2}u_{n+1} - u_n + \frac{1}{2}u_{n-1}.$$

The formula is not as simple in dimension higher than 1, but a few algorithms help to determine the Voronoi diagram, from which the size of the Voronoi cells can be calculated, like Fortune's algorithm in dimension 2 [14] or other fast algorithms proposing an approximation of the Voronoi diagram in higher dimension [15].

This discrete approximation of the integral leads to the following formulas for the generic functions $\bar{\psi}_{j,k}$ and h of the non-equispaced sample framework:

$$\begin{cases} \bar{\psi}_{j,k} : u_n, n \in \mathbb{N} \mapsto \psi_{j,k}(u_n)V_n \\ h : x \in [0,1] \mapsto x. \end{cases}$$

The main drawback of this method is its need of oversampling. Indeed, the formula will poorly approximate the integral if only few observation points are available, as one can see in Figure 13.3. This may be more patent than in both the previous calculation methods, because irregular sampling implies irregular spacings between observations and potentially, among all these spacings, there may be some particularly big spacings which generate a substantial error. In the chaotic framework, the knowledge of the maximal spacing between neighbor observations, depending on the invariant measure of an observed ergodic chaos, is therefore a good indication of the error made using this method [16]. In addition, the computation of the wavelet functions at the observation points is not as fast as in both the previous methods. However, it gives accurate approximations when there are many observations, even in dimension higher than 1.

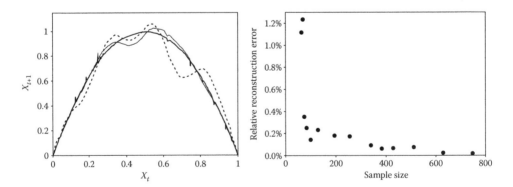

FIGURE 13.3   On the left: Reconstructed evolution function of the logistic chaos defined in Equation 13.7 with the Voronoi method, for 65 observations (dotted line), 257 observations (solid thin line), and 750 observations (solid thick line). The case of 257 observations is particularly eloquent since the error seems concentrated around $X_t = 0.5$, where the spacings between observations are at the highest [16]. On the right: Positive impact of the oversampling on the accuracy with the reconstruction error relatively to the integral of the pure signal over [0,1], as a function of the number of observations. All the wavelet decompositions are truncated at resolution level 3.

13.3.3 Filtering Empirical Wavelet Coefficients of a Noisy Chaos

We now assume that the observed dynamical system is disrupted by a measurement noise. Therefore, we observe z^ε as defined in Equation 13.2. We have already highlighted the fact that the noise has a nonlinear impact on these observations. This particularity implies that the standard filtering of wavelet coefficients must be adjusted.

All the standard thresholds such as *VisuShrink* or *SureShrink*, are defined for a Gaussian noise. In fact, these thresholds are directly related to the probability distribution of the wavelet coefficients. It is quite easy to obtain this distribution in the standard case, when the noise has a linear impact on the signal. However, in the framework of dynamical systems, the noise has a nonlinear impact on the signal and the probability density of wavelet coefficients are thus different from the standard case. The calculation of this density can even be very challenging and we would rather use an approximation. This approximation consists in the local linearization of the noise impact, when the evolution function z is smooth enough. It is indeed much easier to determine the probability density of the wavelet coefficients $\hat{z}^{\mathcal{L}}_{j,k}$ of the linearized noisy evolution function $z^{\mathcal{L}}$, which is defined by

$$\forall n \in \{1, \ldots, N\}, \ z^{\mathcal{L}}(u_n) = z(u_n) - \varepsilon_n^\star z'(u_n) + \varepsilon_n.$$

We can indeed write the probability distribution of the corresponding wavelet coefficients for an α-stable noise [2]. We limit here our analysis to a Gaussian noise of mean 0 and variance σ^2 and we consider the generic case of empirical wavelet coefficients in non-equispaced samples defined by Equation 13.9. We can then assert that the empirical wavelet coefficient $\hat{z}^{\mathcal{L}}_{j,k}$, of resolution parameter $j \in \mathbb{Z}$ and translation parameter $k \in \mathbb{Z}$, is a Gaussian random variable of mean zero and of variance $\bar{\sigma}^2_{j,k}$, where

$$\bar{\sigma}_{j,k} = \sigma\sqrt{\langle 1 + |z'|^2, |\bar{\psi}_{j,k}|^2 \rangle}.$$

Furthermore, we can gauge the error made by linearizing the noise impact thanks to the following inequality:

$$\mathbb{E}[|\hat{z}^\varepsilon_{j,k} - \hat{z}^{\mathcal{L}}_{j,k}|] \leq \frac{\sigma^2}{2} \max_{x \in Supp(z)} \left(|z''(x)|\right) \langle 1, |\bar{\psi}_{j,k}(u_n)| \rangle.$$

Concerning the denoising of the signal using wavelets, we now have all the tools. Indeed, we can perform a standard filtering of the wavelet coefficients by replacing the noise variance σ^2 by $\bar{\sigma}^2_{j,k}$ for the appropriate parameters j and k related to each wavelet coefficient. This can be done either in *VisuShrink* or in *SureShrink*. However, $\bar{\sigma}^2_{j,k}$ is an approximation of the true variance of the noisy wavelet coefficients and its use is mostly relevant for denoising the linearized version $z^{\mathcal{L}}$ of the noisy signal z^ε. Therefore, it is a decent tool for practical use, but theoretically, the filtering of the wavelet coefficients of z^ε may lead to a slightly different reconstruction error. Fortunately, we are able to bound the difference between both errors [13]. The bound is related to the smoothness of z as well as to the noise variance. Therefore, when these features make the bound small enough, the use of $\bar{\sigma}^2_{j,k}$ in the standard threshold expressions is well justified.

13.4 Conclusion

In this chapter, we have presented how the wavelet denoising method should be adapted in the context of a chaotic dataset. First, we have highlighted that we should consider observations in the phase space rather than time series. Then, we have given generic formulas adapting the wavelet method to the framework of non-equispaced samples, which are inherent to dynamical systems observed in the phase space, whatever the embedding dimension. In particular, the definition of empirical wavelet coefficients is preponderant:

- If the embedding dimension is 1, then we may prefer to distort the grid
- If the signal is particularly homogeneous, we can use the interpolation method
- For all the other cases, provided that the dataset is oversampled, we should use the Voronoi approach

For equispaced samples, all these methods are equivalent.

Classical filtering methods are thereby clearly adaptable to chaos and allows us to estimate evolution functions in the presence of an additive measurement noise. This estimation in the phase space can then be useful for making predictions of chaotic time series.

References

1. D. Guégan. *Les Chaos en Finance : Approche Statistique*. Economica, Paris, France, 2003.

2. M. Garcin and D. Guégan. Probability density of the empirical wavelet coefficients of a noisy chaos. *Physica D: Nonlinear Phenomena*, 276(1):28–47, 2014.

3. H. Abarbanel, R. Brown, J. Sidorowich, and L. Tsimring. The analysis of observed chaotic data in dynamical systems. *Reviews of Modern Physics*, 65(4):1331–1392, 1993.

4. E.J. Kostelich and J.A. Yorke. Noise reduction: Finding the simplest dynamical system consistent with the data. *Physica D: Nonlinear Phenomena*, 41(2):183–196, 1990.

5. S. Mallat. *Une exploration des signaux en ondelettes*. Ellipses, Éditions de l'École Polytechnique, Paris, France, 2000.

6. I. Daubechies. *Ten Lectures on Wavelets*. SIAM, Philadelphia, Pennsylvania, 1992.

7. D. Donoho and I. Johnstone. Ideal spatial adaptation by wavelet shrinkage. *Biometrika*, 81(3):425–455, 1994.

8. D. Donoho and I. Johnstone. Adapting to unknown smoothness via wavelet shrinkage. *Journal of the American Statistical Association*, 90(432):1200–1244, 1995.

9. C. Stein. Estimation of the mean of a multivariate normal distribution. *The Annals of Statistics*, 9(6):1135–1151, 1981.

10. T. Cai and L. Brown. Wavelet shrinkage for nonequispaced samples. *The Annals of Statistics*, 26(5):1783–1799, 1998.

11. G. Plonka. The easy path wavelet transform: A new adaptive wavelet transform for sparse representation of two-dimensional data. *Multiscale Modeling and Simulation*, 7(3):1474–1496, 2009.

12. P. Hall and B. Turlach. Interpolation methods for nonlinear wavelet regression with irregularly spaced design. *The Annals of Statistics*, 25(5):1912–1925, 1997.

13. M. Garcin and D. Guégan. Optimal wavelet shrinkage of a noisy dynamical system with non-linear noise impact. Working Paper, 2015.

14. S. Fortune. A sweepline algorithm for Voronoi diagram. *Algorithmica*, 2(1–4):153–174, 1987.

15. J. Vleugels and M. Overmars. Approximating generalized Voronoi diagrams in any dimension. Technical Report 14, Department of Computer Science, Utrecht University, Utrecht, The Netherlands 1995.

16. M. Garcin and D. Guégan. Extreme values of random or chaotic discretization steps and connected networks. *Applied Mathematical Sciences*, 119(6):5901–5926, 2012.

Characterization of Time Series Data

Christopher W. Kulp
and Brandon J. Niskala

14.1 Introduction: What Is Characterization?

A time series is a list of successive data points measured from a system. The system can be physical, biological, economic, or of any other type. For complicated systems, such as a stock market, it is often the case that the equations that govern a system are unknown. Hence, one can think of the system as a "black box" that produces values (measurements). The goal of time series analysis (TISEAN) [7,17] is to use the measurements to obtain information about the data and their statistics in the hope of developing an understanding of what is going on inside the "black box." In other words, TISEAN consists of a set of tools that, when applied to time series data, provides information about the data and the system itself.

The two basic pieces of information about a time series are what is being measured and how often is it measured. In this chapter, we will focus on real, single-variable time series; in other words, each element of the time series is either a real number or an integer. Much of the field of TISEAN is dedicated to what are called *regularly sampled time series*. Regularly sampled time series are time series of the form

$$X = \{x_1, x_2, \ldots, x_N\}, \tag{14.1}$$

where N is the number of elements of the time series, and x_i is the value measured at time t_i. For regularly sampled time series, the sampling time, $\Delta t = t_{i+1} - t_i$, is a fixed quantity for all $i \in [1, N-1]$. An example of a regularly sampled time series would be the daily closing price of a commodity. In this example, x_i would be the value of the commodity and Δt is 1 day. The vast majority of the tools in TISEAN are designed for long, regularly sampled time series because the assumptions of infinite length and regular sampling make the theoretical development of TISEAN algorithms easier.

Real-world data, however, are often *irregularly sampled*, meaning that there is not a fixed sampling time between all measurements in the time series. There are many practical reasons as to why this might happen. For example, in astrophysical measurements, observations may be impaired for some days due to weather conditions; hence, measuring a star's brightness at a regular interval may not be possible. Irregularly sampled time series often take the form

$$X = \{\{t_1, x_1\}, \{t_2, x_2\}, \ldots, \{t_N, x_N\}\}, \tag{14.2}$$

where the value x_i is measured at time t_i. Irregularly sampled time series are difficult to work with. The sampling irregularities make the theoretical development of analysis algorithms challenging and, therefore, the number of algorithms specifically for irregularly sampled time series is less than those available for regularly sampled series. It is important to note that algorithms designed for regularly sampled time series are not always reliable when applied to irregularly sampled time series [22,25].

One of the most important problems in TISEAN is that of *characterization*. Time series characterization is the process of determining the dynamics of the system from time series data. In other words, does the system follow a set of deterministic rules or is the system random? Furthermore, if the system is deterministic, can we tell if the system is periodic, quasiperiodic, or chaotic? Understanding the nature of a system's dynamics is important for several reasons. First is the matter of basic science; understanding the system's dynamics is part of what a scientist does to learn more about the behavior of the system. Once a basic understanding of a system is obtained, the next typical step is to build predictive models of that system. It is difficult to develop reliable predictive models of a system without knowledge of the underlying dynamics of a system. If chaos is present in a system, then a model's ability to make long-term predictions is limited. Such limitations must be known before models can be applied in real-world settings. Finally, one may already have a model of a system. Time series characterization of data measured from the system can help to determine the validity of the model. For example, if chaos is detected in its time series data, then the model needs to also display chaotic behavior. In fact, one can use time series characterization techniques on data generated by the model to test for consistency with the results of those same tests on real-world data from the system.

This chapter contains several time series algorithms. To demonstrate the algorithms, we use model data generated from the Lorenz equations [28]

$$\dot{x} = \sigma(y - x),$$
$$\dot{y} = rx - y - xz, \tag{14.3}$$
$$\dot{z} = xy - bz.$$

Two sets of parameters are used to generate data from Equation 14.3. For a nonchaotic series, the parameters used are $r = 200$, $b = 8/3$, and $\sigma = 100$. For a chaotic series, the parameters used are $r = 45.92$, $b = 4$, and $\sigma = 16$. In each case, the equations were solved using TISEAN's [15] *lorenz* command. After a transient time of 50 time units, the solution was sampled with a sampling rate of $\Delta t = 0.05$ for the chaotic series and $\Delta t = 0.02$ for the periodic series. In each case, the time series has a length of $N = 10,000$ elements. Figure 14.1 shows a short sample of each series. The lines in Figure 14.1 are the solution to the equation and the dots represent the points sampled in the time series.

The chapter is organized as follows. Section 14.2 discusses two tests for determinism and presents typical results for those tests. Section 14.3 contains a brief review of some common methods for detecting chaos in regularly sampled time series. Section 14.3 also contains typical results for each method. In Section 14.4, a method for detecting chaos in irregularly sampled time series is presented, as is a discussion of the method's robustness to noise. Concluding remarks and a brief discussion of future work is presented in Section 14.5.

14.2 Tests for Determinism

As mentioned previously, characterization of time series data typically begins with a test for determinism [18,26,39]. A test for determinism attempts to determine whether the data are stochastic or deterministic. It can be difficult to distinguish chaotic behavior from randomness [9,10,26,31]. As we will see, tests

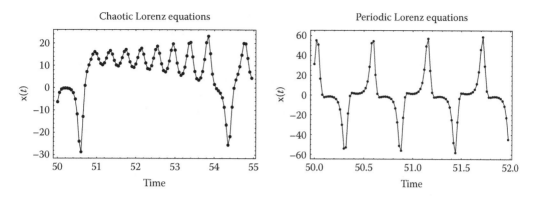

FIGURE 14.1   Short sample of the model time series generated from Equation 14.3. The chaotic series is on the left and the periodic series is on the right.

for determinism are an important element to being able to interpret tests for chaos. In some cases, tests for chaos can falsely identify randomness as being chaotic [24].

It would be impossible to provide an exhaustive study of all tests for determinism in this chapter. Instead, we will focus on two tests, the symbol tree test presented by Yang and Zhou [39] and the permutation spectrum test (PST) introduced by one of the authors (Kulp) and Zunino [26]. For other tests, the interested reader is directed to References 10, 18, 21, 31, and 41, and also Reference 26.

Both the symbol tree test and the PST work by detecting patterns in time series data. The basic idea behind the tests is that a deterministic time series will have frequently recurring patterns as well as patterns that never occur, called forbidden patterns [31,40], while stochastic time series will not have consistently forbidden patterns. The difference between the two tests is in how patterns are detected.

14.2.1 The Symbol Tree Test

The symbol tree test was first proposed by Yang and Zhou [39] and more recently used by Kulp and Smith [24]. The symbol tree test identifies patterns in a symbolic time series $\{s_i\}$ generated from the original series $\{x_i\}$. A symbolic time series is one with a small set of distinct symbols called an alphabet. The symbolic series is created by partitioning the range of the series and assigning each partition its own symbol. Elements in the original series are then replaced by the symbol representing the partition to which the element belongs. For example, a binary symbolization is one where the new symbolic series consists of 0 and 1. Such a symbolization is done by choosing a threshold, t, which could be the mean of the series. Then the symbolization could be carried out using $s_i = 1$ if $x_i \geq t$ and $s_i = 0$ if $x_i < t$. For more discussion about choosing a threshold to partition a series, the interested reader is directed to References 5 and 7 and the references therein. The symbol tree test uses a binary symbolization of the time series being tested.

The symbol tree test begins by partitioning a binary series of length, N, into subsets of length ℓ. The partitioning can be done in two ways. First, the series can be broken up into ℓ disjoint subsets. Or, one can randomly choose ℓ subsets essentially by randomly choosing ℓ starting positions for each subset from the list $\{1, 2, \ldots, N - \ell\}$. Note that the list of starting positions ends at $N - \ell$ so that a subset of length ℓ can be formed from the last possible starting position. The choice of ℓ will be discussed later and is dependent on the next step.

Next, the level, L, of the symbol tree is chosen. The choice of L groups the elements of each partition into "words" of length L. For example, in the case $L = 2$, there will be four possible words in the partition 00, 01, 10, and 11. For a binary series, there will be 2^L words at the Lth level of the symbol tree. For each partition of length ℓ, there will be $\ell - (L - 1)$ words. This second partition has an overlap of one element. Hence, the second element of the first partition is the same as the first element in the second partition, and so on. This next step is optional, but it helps to speed up computation if each word is converted into base-10. Hence, for $L = 2$, each word is equivalent to 0, 1, 2, and 3, respectively.

As an example of the steps for the symbol tree test, suppose our time series has been partitioned such that $\ell = 6$ and that one of those partitions is $\{0, 1, 0, 0, 1, 0\}$. Then, if we choose $L = 2$, the partition becomes $\{01, 10, 00, 01, 10\}$. After conversion to base-10, the partition is $\{1, 2, 0, 1, 2\}$.

The next step is to count the number of times each word appears in the partition. That is called the symbol spectrum. We then plot the symbol spectrum for each length-ℓ partition on the same graph. Deterministic series will have spectra with significant overlap, whereas random series will have little overlap from one spectrum to the next.

The results of the symbol tree test applied to chaotic and periodic series generated from Equation 14.3 and for a series of random numbers with a normal distribution with mean zero and standard deviation of $1/3$ are shown in Figure 14.2. Each symbol was symbolized by taking the range of the series and dividing by two. Elements in the upper half of the range were assigned the symbol 1 and those in the lower half are assigned the symbol 0. In Figure 14.2, $\ell = 500$ and $L = 5$ were used for each series; hence, we are using 20 partitions in each case. Notice that the chaotic series (top left) and periodic series (top right) have symbol spectra that have similar shapes, i.e., they have peaks and zeroes in similar places. Also notice that the periodic symbol spectra lay almost exactly on top of one another, a result of the series being periodic and hence often returning to the same or similar values. The chaotic spectra are not as similar to each other as the periodic ones because of the aperiodic nature of chaotic dynamics. But the chaotic series does have certain words that appear frequently and other words that do not appear at all; hence, illustrating the repeating patterns of a deterministic series. For both chaotic and periodic series, the words 0 and 32 appear very often. The word 0 corresponds to $\{0, 0, 0, 0, 0\}$ and 32 corresponds to $\{1, 1, 1, 1, 1\}$. The frequency of

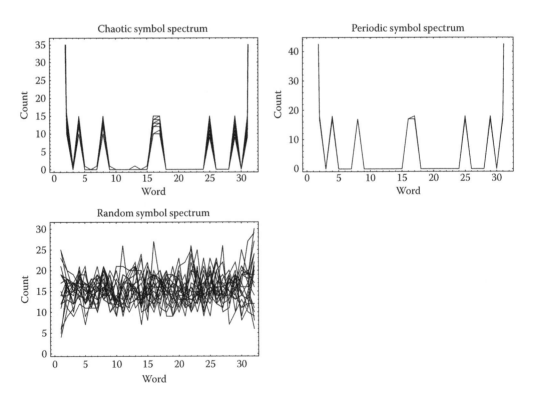

FIGURE 14.2 Results of the symbol spectrum test for chaotic (top left) and periodic (top right) series generated from Equation 14.3 and random series (bottom).

these words is due to the binary symbolization. Note from Figure 14.1 that it will be expected to have many elements in a row that are above or below the threshold. Because of the high frequency of 0 and 32 for the chaotic and periodic dynamics, we scaled the graph so that the spread of the spectra for the other words is more visible. Unlike the periodic and chaotic spectra, the random spectra in Figure 14.2 have no common patterns and all words occur; this is a typical result for a random series.

As mentioned previously, the choice of word length, L, depends on the choice of partition size ℓ. While there is no formula for choosing these parameters, our experience with the test has helped us develop a few guidelines [24]. When it comes to short data sets, and often real-world time series are short, it is important to choose a reasonable partition. One needs to be able to generate a large number of partitions to create a reasonable comparison of symbol spectra. Hence, we focus on generating a specific number of symbol spectra and let that choice dictate ℓ and L. In our experience, 20 spectra tend to be enough for the symbol tree test to determine whether the spectra are similar or not. With too few spectra, it is difficult to distinguish between a deterministic or a stochastic series because it is difficult to establish a pattern (or lack thereof) between the spectra. Choosing too many partitions leads to the possibility that the partition length might be too short to produce a reliable spectrum. Once ℓ is determined, then one should choose an L such that $2^L \ll \ell$ and there are enough elements in the partition to provide a reasonable sample of the number of possible words.

14.2.2 The PST

The PST was developed in Reference 26 as another means of testing for determinism. Like the symbol tree test, the PST detects patterns in time series data using a symbol spectrum. The difference between the PST and the symbol tree test is how the series is symbolized.

In the PST, the time series is symbolized using ordinal patterns [3], which are the set of permutations of the list $\{1, 2, \ldots, D\}$, where D is an integer known as the embedding dimension. Similar to the symbol tree test, the PST begins with partitioning the time series either into disjoint subsets or by randomly choosing initial elements (as described above). However, unlike the symbol tree test, the PST does not symbolize the series before partitioning. After choosing the partition length ℓ, and the embedding dimension D, the partitions of length ℓ, are then further partitioned into D-tuples with an overlap of one element (similar to the symbol tree test's word partitions). Each element in each D-tuple is then replaced with an integer that denotes its rank in the partition resulting in replacing that D-tuple with an ordinal pattern. For example, suppose we have a series where $D = 4$ was chosen and one of the D-tuples is $\{5.1, 2.3, 7.0, 3.5\}$. The resulting ordinal pattern would then be $\{3, 1, 4, 2\}$. For computational convenience, we can then assign each ordinal pattern an integer value called an *ordinal pattern index*. There will be $D!$ such indexes. The choice of D and ℓ for a given series can be done by keeping in mind the comments made in Section 14.2.1 for choosing ℓ and L.

Once each D-tuple in each ℓ-length partition has been assigned an ordinal pattern index, then the permutation spectrum for each partition can be found. The permutation spectrum is a count of the number of times each ordinal pattern occurs in the ℓ-length partition. In that regard, it is similar to the symbol spectrum of Section 14.2.1. A time series is considered deterministic if its permutation spectra have significant overlap. However, in addition to the permutation spectrum, a standard deviation of the spectra as a function of ordinal pattern index is also computed. The standard deviation gives a means of comparing the variability of permutation spectra and allows us to identify forbidden patterns. Forbidden patterns are ordinal patterns that consistently never appear in the time series. In other words, the spectral value of the forbidden ordinal pattern is always zero, and hence, has a standard deviation of zero. The presence of forbidden patterns is a known indicator of deterministic dynamics [2].

Figure 14.3 shows the result of the PST applied to chaotic (top) and periodic (middle) series generated from Equation 14.3 and the random series from Section 14.2.1. The results of the PST are interpreted in a similar manner to those of the symbol tree test. Notice that the deterministic series (top two) have significant overlap in their spectra. The chaotic series has a greater standard deviation than the periodic

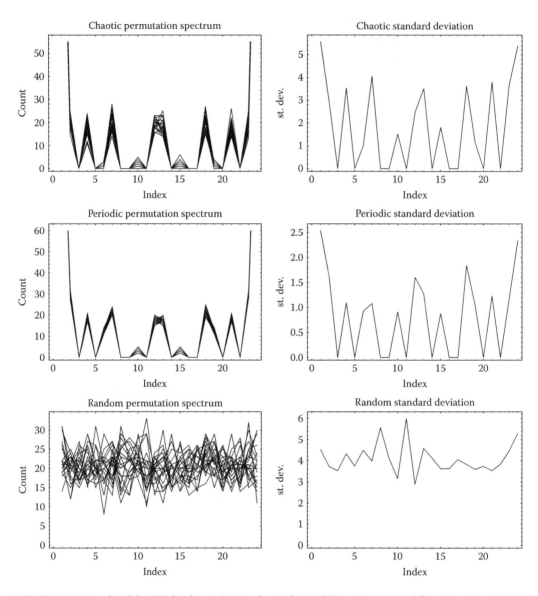

FIGURE 14.3 Results of the PST for chaotic (top) and periodic (middle) series generated from Equation 14.3 and random series (bottom).

series. The lower standard deviation for the periodic series is to be expected due to the series periodicity. Similar to the symbol spectrum test, there are two patterns with a higher count than the others, 1 and 24. This is due to the nature of the time series. The ordinal pattern index, 0, corresponds to the ordinal pattern $\{1, 2, 3, 4\}$ where elements in the series are increasing. The ordinal pattern index, 24, corresponds to the ordinal pattern $\{4, 3, 2, 1\}$, where elements in the series are decreasing. From Figure 14.1, we can see that the time series will have many segments where the elements are increasing or decreasing. Hence, the frequency of ordinal patterns 0 and 24 is not surprising. Just like the symbol spectrum test, the random series has significant variation in its permutation spectra as is demonstrated both visually by the spectra and the higher standard deviation. Note that no indices have a zero standard deviation for the random series. This suggests that there are no consistently forbidden patterns in the random series unlike in the deterministic series.

The ordinal pattern symbolization has many advantages [6]. First, it avoids using the amplitude thresholds for symbolization that can be problematic [4]. Instead, the symbolization is done by a difference-based method that takes into account causal relationships within the time series. For a more detailed discussion of the advantages of the ordinal pattern symbolization, the interested reader is referred to Reference 26 and the references therein.

Finally, it should be noted that in this section, and Section 14.2.1, the D- or L-tuples were created using consecutive elements in the ℓ-length partitions. This need not be the case. One can use every other or every-third (and so on) element to create the tuples. This is known as choosing an embedding delay, τ, different from unity. This may lead to additional relevant information in the time series. For more information, the interested reader is referred to Reference 26.

14.3 Methods of Detecting Chaos in Regularly Sampled Data

In this section, a few of the methods of detecting chaos for regularly sampled time series are presented. What follows is not meant to be an exhaustive study of tests for chaos. We begin by presenting one of the first methods of detecting chaos that uses phase space reconstruction. Then, we present a more recent test called the 0–1 test for chaos. Finally, another test called the heuristic method is presented because of its ability to be modified for irregularly sampled series.

14.3.1 Phase Space Reconstruction

One method of detecting chaos is by measuring the maximal Lyapunov exponent, λ, of a system from its time series data. The Lyapunov exponent measures the exponential rate of divergence of initially close trajectories in phase space. If the largest Lyapunov exponent is positive, then those trajectories diverge exponentially. Hence, the system displays a sensitive dependence on initial conditions, a hallmark of chaos. The measurement of λ can be done by reconstructing the phase space using time series data.

The Takens Embedding Theorem [36] shows that it is possible to take a time series of scalar measurements and recreate the system's phase space. For the work in this chapter, the details of the theorem are not necessary; however, the interested reader is referred to page 17 of Reference 1 for more information and references. The basic idea is that the original scalar time series $\{x_1, x_2, \ldots, x_N\}$ can be used to create a set of vectors $\{\mathbf{y}(i)\}$ that define motion in a finite-dimensional Euclidean space that we call the *reconstructed space*. As stated in Reference 1, the reconstructed space, $\{\mathbf{y}(i)\}$, is related to the system's true phase space by smooth, differentiable transformations. Hence, all the invariants of the motion observed in the reconstructed space will be the same as those in the original phase space. For our purposes, this means that the maximal Lyapunov exponent measured in the reconstructed space will be the same as the system's maximal Lyapunov exponent.

The set of vectors $\{\mathbf{y}(i)\}$ is created by grouping elements of the original time series, x_i, together

$$\mathbf{y}(i) = \{x_i, x_{i+\tau}, x_{i+2\tau}, \ldots, x_{i+(d-1)\tau}\}. \tag{14.4}$$

In effect, the elements that are grouped together are separated in the original series by a time τ. The question here is how to choose the delay τ and the embedding dimension d. Note that this is not the same as the embedding dimension, D, used in the PST. The embedding dimension used here will be the dimension of the reconstructed space and need not be the dimension of the system's phase space. Choosing τ and d can be difficult and there are many different techniques for each. It is not possible to give a comprehensive review in this chapter of all the possible techniques of choosing τ and d. In this chapter, we will use the mutual information to obtain τ and the method of false-nearest neighbors (FNNs) to obtain d. For more information on other techniques, the interested reader is directed to References 1, 7, and 17.

To choose τ, we will use the time-delayed mutual information [8] that is based on the Shannon entropy [33]. The time-delayed mutual information tells us how much information we know about the

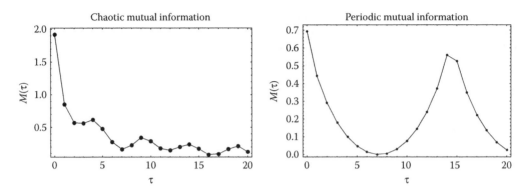

FIGURE 14.4 Time-delayed mutual information for the chaotic (left) and periodic (right) Lorenz series.

measurement $x_{i+\tau}$ once we have measured x_i. In other words, with each measurement, how much do we learn about measurements that will be made at time τ later. Before the time-delayed mutual information can be computed, the original time series, $\{x_i\}$ must be symbolized into a new series $\{s_i\}$ as described in Section 14.2.1. The time-delayed mutual information is defined as

$$
M(\tau) = \sum p\left(s_i, s_{i+\tau}\right) log_2 \left(\frac{p\left(s_i, s_{i+\tau}\right)}{p\left(s_i\right)p\left(s_{i+\tau}\right)} \right),
\tag{14.5}
$$

where the sum is done over all possible elements in the symbolic series (e.g., in the case of a binary series, the sum is done over 0 and 1). The function $p(s_i)$ in Equation 14.5 is the probability that s_i appears in the series and $p(s_i, s_{i+\tau})$ is a joint probability. A common choice of τ obtained from $M(\tau)$ is the first local minimum. The idea is that at the first local minimum, s_i and $s_{i+\tau}$ will be relatively independent of each other, however, not so independent that they would not be useful as coordinates. It should be noted that the choice of the first minimum is a guideline and not a hard-and-fast rule. Figure 14.4 illustrates typical results for a time-delayed mutual information analysis. The analysis was done with the TISEAN [15] program, *mutual*. The results of the time-delayed mutual information suggest choosing a time delay of $\tau = 3$ for the chaotic series and $\tau = 7$ for the periodic series.

After a delay is chosen, then the embedding dimension can be found. One method for finding the embedding dimension is known as the method of FNNs [20]. The basic idea behind FNN is to use τ to create the delay vectors $\{\mathbf{y}(i)\}$ in a dimension d, known as the embedding dimension, and then find the nearest neighbor for each vector, $\mathbf{y}^{NN}(i)$. The distance between each $\mathbf{y}(i)$ and $\mathbf{y}^{NN}(i)$ can be computed. Next, recreate the delay vectors using an embedding dimension $d+1$ and compute the new distance between each vector and its nearest neighbor. If the distance between the vector and its nearest neighbor does not significantly change, then $\mathbf{y}(i)$ and $\mathbf{y}^{NN}(i)$ were truly neighbors in the phase space. However, if the distance between the two vectors changes significantly, then the two vectors proximity was a result of projecting the attractor to a dimension that is too small and the vectors are considered FNNs. As an example, consider drawing a cube on a chalkboard. Some of the points on opposite faces are close to one another because the three-dimensional cube is projected down onto a two-dimensional surface. For more details on the FNN algorithm and how it is applied, the interested reader is directed to Reference 1. In addition to FNN, there is another technique called *false-nearest strands* [19] that eliminate problems that oversampling can introduce with FNN.

Figure 14.5 shows the results of an FNN analysis of the Lorenz time series generated from Equation 14.3 with $\tau = 3$ for the chaotic series and $\tau = 7$ for the periodic series. The TISEAN [15] algorithm, *false-nearest*, was used to produce Figure 14.5. Note that in the case of the chaotic series, the percentage of FNN drops to about 6% at $d = 3$ and progresses toward zero at $d = 5$. In general, it is not necessary to choose an

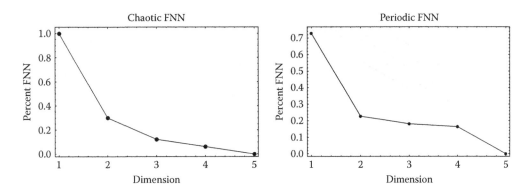

FIGURE 14.5 False nearest neighbor (FNN) analysis of the chaotic (left) and periodic (right) Lorenz series.

embedding dimension that has zero FNNs. However, it is typically wise to choose an embedding dimension with a small number of FNNs. In the case of the chaotic series, the number of FNNs drops significantly from $d = 1$ to $d = 3$ and there is a bit of a slower decrease with higher dimension. That observation coupled with the knowledge of the Lorenz attractor's dimension of approximately 2.05 [35], suggests that an embedding dimension of $d = 3$ is an appropriate choice for the chaotic series. With the periodic series, we see that the percentage of FNNs flattens out from $d = 2$ to $d = 4$ with a very large drop from $d = 1$ to $d = 2$. Again, using the prior knowledge of the dimension of the Lorenz system, we chose to use $d = 3$ for an embedding dimension for the periodic series. Had we not known the equations of motion generating the time series (as is often the case in the real world), then we could try embedding dimensions of $d = 3$, 4, and 5 for future analyses and look for consistency of results. If the results of future analysis using the two dimensions are similar, then we can probably rely on using the small embedding dimension.

Now that a time delay, τ, and embedding dimension, d, has been chosen for each series, the final step is to compute the Lypaunov exponent from the time series data. The first algorithm for computing the Lyapunov exponent was developed by Wolf in 1985 [38]. The Wolf algorithm assumes exponential divergence of trajectories in the phase space and, hence, does not allow one to test for the presence of such a divergence [17]. Instead, we will use an algorithm developed independently by Kantz [16] and Rosenstein et al. [30] that tests directly for exponential divergence. The basic idea behind the Kantz algorithm is to compute the average distance of all neighbors to a reference point on a trajectory in the reconstructed phase space as a function of time Δi. The distance calculation is done for a large number of reference points. The average is sometimes referred to as the stretching factor, $S(\Delta i)$ [17], and found using the formula

$$S(\Delta i) = \frac{1}{N} \sum_{i_0=1}^{N} \ln \left(\frac{1}{|U(\mathbf{y}(i_0))|} \sum_{\mathbf{y}_i \in U(\mathbf{y}(i_0))} |y_{i_0+\Delta i} - y_{i+\Delta i}| \right),$$ (14.6)

where $\mathbf{y}(i_0)$ is the reference point in the reconstructed phase space, $U(\mathbf{y}(i_0))$ is the neighborhood of $\mathbf{y}(i_0)$ with diameter ϵ, and y_{i_0} is the last element of $\mathbf{y}(i_0)$. Hence, an exponential divergence of initially close trajectories will be represented by a region of linear growth in the graph of $S(\Delta i)$. The slope of the linear region of $S(\Delta i)$ gives the Lyapunov exponent, λ. For more information on the application of the Kantz algorithm to time series data, the interested reader is referred to Reference 17.

Figure 14.6 shows the results of computing $S(\Delta i)$ for the chaotic (left) and periodic (right) Lorenz series. In each case, the TISEAN [15] algorithm, *lyap_k* was used. For the chaotic series, we used $\epsilon = 0.060, 0.11$, 0.19, 0.34, and 0.60. Each line on the graph represents one value of ϵ. Notice that for each line, there is a region of linearity approximately between $\Delta i = 20$ and $\Delta i = 50$. From the slope of the best-fit line to each of these regions, we estimate a Lyapunov exponent of 1.33 ± 0.0606 that is in agreement with the Lyapunov exponent of 1.38 for the parameters used in Equation 14.3. It should be noted that the TISEAN command,

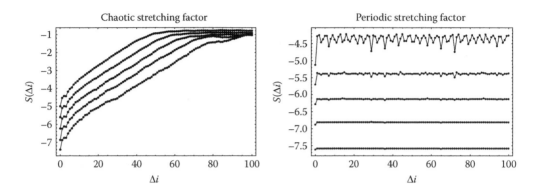

FIGURE 14.6 $S(\Delta i)$ for the chaotic (left) and periodic (right) Lorenz series.

lorenz gives the Lyapunov exponent for the Lorenz equations with the given parameters r, σ, and b. For the periodic Lorenz results in Figure 14.6, we see that there is no region of linear growth, and hence the Lyapunov exponent is zero, as expected for a periodic system.

As with any algorithm, there are advantages and disadvantages to phase space reconstruction. The advantage to this algorithm is that it gives the Lyapunov exponent. Knowledge of the Lyapunov exponent is important when developing a predictive model from time series data. However, the method described in this section requires a lot of data, often more data than are available from real-world systems. Furthermore, the algorithms that are involved in phase space reconstruction require regular sampling. The algorithms are also susceptible to noise. For example, see pages 43 and 44 of Reference 1 for a discussion of noise effects on the FNN algorithm and pages 69–72 of Reference 17 for a discussion of noise effects on $S(\Delta i)$. With these limitations aside, phase space reconstruction is arguably one of the most important tools available for detecting chaos in time series data.

14.3.2 The 0–1 Test for Chaos

Another algorithm that detects chaos in time series data is called the 0–1 test for chaos and was developed for deterministic series by Gottwald and Melbourne in 2004 [11]. The name is from the fact that if the time series is chaotic, the test outputs the value 1. If the series is periodic or quasiperiodic, then the test produces the result of 0. The 0–1 test was given a theoretical justification in Reference 13 and an implementation guide was developed in Reference 14.

The 0–1 test is performed as follows [14]. Consider a time series $\{x_1, x_2, \ldots, x_N\}$. The 0–1 test begins with the computation of "translation variables,"

$$p_c(n) = \sum_{j=1}^{n} x_j \cos(jc),$$

$$q_c(n) = \sum_{j=1}^{n} x_j \sin(jc), \tag{14.7}$$

where c is randomly chosen from the interval $\in (0, \pi)$. It is common to restrict the choice of c to be $c \in (\pi/5, 4\pi/5)$ to avoid resonances in the test [14].

The next step is to compute the mean square displacement of the translation variables

$$M_c(n) = \lim_{N \to \infty} \frac{1}{N} \sum_{j=1}^{N} \left[p_c(j+n) - p_c(j) \right]^2 + \left[q_c(j+n) - q_c(j) \right]^2. \tag{14.8}$$

In Reference 14, it was mentioned that Equation 14.8 requires $n \ll N$; hence, we only compute $M_c(n)$ for $n < n_{cut} = N/10$. The behavior of $M_c(n)$ is central to the test's ability to detect chaos. For a given value of c, $M_c(n)$ takes the form [14]

$$M_c(n) = V(c)n + V_{osc}(c, n) + e(c, n), \qquad (14.9)$$

where $e(c, n)$ is an error term ($e(c, n) \to 0$ as $n \to \infty$) and

$$V_{osc}(c, n) = \langle x \rangle^2 (1 - \cos(nc))/(1 - \cos(c)). \qquad (14.10)$$

The term $\langle x \rangle$ is the mean of the time series. Hence, without the error term, $M_c(n)$ has the form of a cosine wave with a slope given by $V(c)$. Note that $V(c)$ is constant for a given value of c. It is this slope that characterizes the dynamics. To find this slope, we subtract off the V_{osc} term from $M_c(n)$ to create a modified mean square displacement

$$D_c(n) = M_c(n) - V_{osc}(c, n). \qquad (14.11)$$

Finally, we find the asymptotic growth rate K_c of the modified mean square displacement $D_c(n)$. The asymptotic growth rate turns out to be the correlation coefficient of the vectors

$$\xi = (1, 2, \ldots, n_{cut}), \qquad (14.12)$$
$$\Delta = (D_c(1), D_c(2), \ldots, D_c(n_{cut})). \qquad (14.13)$$

It can be shown (see Reference 14 and references therein) that $K_c = 1$ for chaotic dynamics and $K_c = 0$ for nonchaotic dynamics.

Note the randomly chosen variable, c, in the equations above. The test requires that we compute K_c for many values of c and take the median of the result. In Reference 12, it was found that 100 different values of c chosen at random are sufficient for computing the median of K_c. The median value is used to suppress the effects of resonance (high values of K_c generated by isolated values of c when studying periodic systems). We have also found it useful to compute the "spectrum" of K_c.

Figure 14.7 contains the results of the 0–1 test when applied to model data generated from Equation 14.3. Notice that K_c takes on a value close to 1 for all values of c. The median value of the growth rate for the chaotic series is $K = 0.99$. For the periodic series, we see a few examples of the resonance that can occur near $c = 0$, 1.1, and 2.4, hence the need for multiple K_c calculations. However, the vast majority of the K_c values is near or slightly below 0 and the median value of the growth rate for the periodic series is $K = -0.083$.

The 0–1 test for chaos is useful because it can detect chaos without reconstructing the phase space. It can work on shorter time series as compared to estimating the maximal Lyapunov exponent from the reconstructed space. However, the test is designed for deterministic data. If applied to random data, the

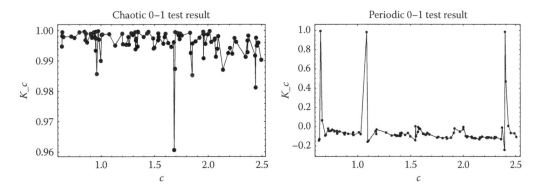

FIGURE 14.7 0–1 test result for the chaotic (left) and periodic (right) Lorenz series.

0–1 test returns $K = 1$ [24], giving a false positive for chaos. In the real world, the level of noise may not be known. This can be remedied, in part, by pairing the 0–1 test with a test for determinism [24]. However, like estimating the Lyapunov exponent from phase space reconstruction, the 0–1 test does not work if the series is irregularly sampled.

14.3.3 The Heuristic Method

The final test for detecting chaos in regularly sampled time series that we will discuss is known as the heuristic method [37]. The heuristic method uses the discrete Fourier transform (DFT) to detect chaos. The DFT of a time series $\{x_j\}$ is defined as

$$X_k = \sum_{j=1}^{N} x_j \exp\left(\frac{-2\pi i k j}{N}\right), \tag{14.14}$$

where $i = \sqrt{-1}$ and $k \in \mathbb{Z}$. The basic idea behind the heuristic method is a visual inspection of the power spectrum, i.e., the square magnitude of the DFT. The power spectrum of a periodic or quasiperiodic time series will have distinct separate peaks. The power spectrum of a chaotic series, however, will consist of a larger number of peaks that are much closer together than the spectra of a periodic or quasiperiodic series. The power spectrum of a chaotic series will decay with increasing frequency. There have been studies to suggest that decay is related to the positive Lyapunov exponents of the series [34].

Figure 14.8 shows the normalized power spectrum for the chaotic (left) and periodic (right) Lorenz series. Notice that each power spectrum takes the shapes mentioned in the paragraph above. The normalization is done such that the maximum of the power spectrum is set to 1. In Reference 37, it was mentioned that to address the issue of noise, one can set an amplitude threshold. In other words, if a peak amplitude is below the threshold it is not counted. The justification for the threshold is that low noise will, in general, introduce low power at high frequencies. Hence, by ignoring small peaks below some threshold, say 0.05, one can study the distribution of the peaks with an amplitude above the threshold and characterize the time series.

One of the weaknesses of the heuristic method is that it does not provide a quantifiable output such as phase space reconstruction or the 0–1 test. However, the DFT is well understood and easily computed. Hence, the heuristic method remains a useful tool in identifying chaos especially in experimental data. The heuristic method cannot be applied to irregularly sampled time series because the frequencies used in the DFT are not well defined for those series. In the next section, we will see how we can modify the heuristic method so that it can be used on irregularly sampled time series.

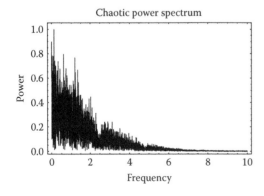 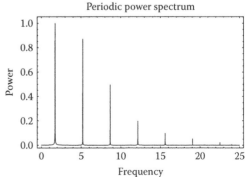

FIGURE 14.8 Normalized power spectrum for the chaotic (left) and periodic (right) Lorenz series.

14.4 Detecting Chaos in Irregularly Sampled Data

It is common for real-world time series to be irregularly sampled. The presence of irregular sampling poses a challenge to the aforementioned methods of detecting chaos. The methods discussed in Section 14.3 can be thought of as being composed of algorithms that, when combined, allow one to determine whether or not chaos is present in a time series. For example, phase space reconstruction can be thought of as the combination of the time-delayed mutual information, FNN, and Kantz algorithms. Each one of the algorithms require regularly spaced time series to work. One method of addressing the issue of irregular sampling is to replace one or more of those algorithms with an algorithm that computes the same or similar quantity but is designed to be used for irregularly sampled time series. The modified heuristic method (MHM) [23] does exactly that with the heuristic method. The MHM replaced the DFT with the Lomb–Scargle periodogram (LSP) [27,32].

The LSP computes the power spectrum for irregularly sampled data. The LSP is

$$P_N(f) = \frac{1}{2\sigma^2} \left(\frac{\left[\sum_{i=1}^{N} (x_i - \bar{x}) \cos \left(2\pi f \left(t_i - \tau \right) \right) \right]^2}{\sum_{i=1}^{N} \cos^2 \left(2\pi \left(t_i - \tau \right) \right)} + \frac{\left[\sum_{i=1}^{N} (x_i - \bar{x}) \sin \left(2\pi f \left(t_i - \tau \right) \right) \right]^2}{\sum_{i=1}^{N} \sin^2 \left(2\pi \left(t_i - \tau \right) \right)} \right), \quad (14.15)$$

where τ is defined as

$$\tan \left(4\pi f \tau \right) = \frac{\sum_{i=1}^{N} \sin \left(4\pi f t_i \right)}{\sum_{i=1}^{N} \cos \left(4\pi f t_i \right)}. \quad (14.16)$$

The properties of the LSP are discussed in Reference 23 and the references therein. However, it is important to note that as commented in Reference 29, τ makes Equation 14.15 "identical to the result one would obtain if one estimated the harmonic content at a frequency, f, by a linear least-squares fitting to the model: $x(t) = A \cos(2\pi f t) + B \sin(2\pi f t)$."

For convenience, the LSP is normalized in this chapter such that the maximum value of the LSP is unity. The normalization of the power spectrum is done simply by dividing each element of the LSP by the maximum value of the LSP after the LSP has been calculated for the series. However, it should be noted that when using the LSP on irregularly sampled data, the highest frequency for which the power spectrum can be computed is $f_{max} = N/(2T)$, where N is the length of the series and $T = t_N - t_1$ is the time range of the series [29]. The choice of f_{max} is analogous to the Nyquist frequency for the irregularly sampled series.

Before applying the LSP to irregularly sampled time series in an attempt to detect chaos, it is important to see if the LSP gives similar results to the DFT for both chaotic and periodic series. Figure 14.9 shows the results of the LSP when applied to model data created from Equation 14.3. Notice that the power spectrum found by the LSP and shown in Figure 14.9 is similar to those of the DFT shown in Figure 14.8. Note that the

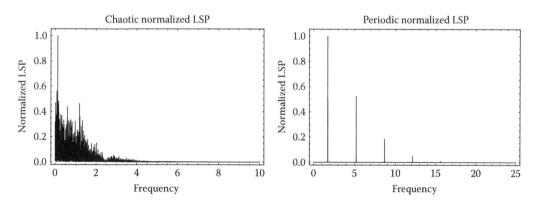

FIGURE 14.9 Normalized LSP for the chaotic (left) and periodic (right) Lorenz series.

peaks in the periodic DFT near $f = 15$ and $f = 19$ are present in the LSP but do not appear in Figure 14.9 due to their amplitude being too small. The important thing is that the chaotic LSP and periodic LSP have similar qualities to their DFT counterparts. In particular, the chaotic LSP contains many closely spaced peaks whose amplitude decays with increasing frequency and that the periodic LSP contains distinct peaks.

Because of the similarity between the power spectra computed by the LSP and DFT, we can begin to test the ability of the LSP to detect chaos in irregularly sampled time series. The test will involve computing the LSP and looking for the characteristics of a chaotic power series. We will remove the DFT from the heuristic method and, in its place, use the LSP. We refer to this new method as the MHM and it was first presented by one of the authors (Kulp) in Reference 23. Originally in Reference 23, the LSP was paired with something called a point count plot (PCP) that attempted to help distinguish a chaotic LSP from a periodic one. The PCP graphs the number of points in the LSP above a threshold as a function of the threshold. The threshold would range from 0.05 to 1 for a normalized LSP. The nonzero-starting value for the threshold is done to avoid a large number of small-amplitude peaks that can be present in both noisy and chaotic time series. The PCP associated with a chaotic LSP would take the form of a decaying exponential, whereas the PCP associated with a periodic LSP would look like a descending staircase. Since the publication of Reference 23, the authors have found that LSPs are often easy to characterize without creating a PCP. Hence, in this chapter, we will forego the PCP and focus only on the LSP. The MHM, would then refer to characterizing the time series using only the qualitative properties of the LSP.

First, we will verify that the DFT does not work well with irregularly sampled time series. Figure 14.10 shows a comparison of the LSP (top row) and DFT (bottom row) for irregularly sampled chaotic (left column) and periodic (right column) data generated from Equation 14.3. To generate irregularly sampled data, we took the data generated from Equation 14.3 and randomly removed 50% of each series. Note that it is difficult to assign frequencies to the DFT for irregularly sampled data; hence, the subscript k is used from Equation 14.4 on the bottom row of Figure 14.10. When it comes to replicating the power spectrum from the irregularly sampled series, we see that the LSP performs much better than the DFT. Comparing Figure 14.9 to Figure 14.10, we see that the LSP does a good job of extracting the correct frequency content from the irregularly sampled data. However, comparing the DFT of the original series in Figure 14.8 to the DFT of the irregularly sampled time series in Figure 14.10, we see that the DFT struggles to reproduce the frequency content from the irregularly sampled series. The deficiency of the DFT is most evident in the periodic series (left row of Figure 14.10). In fact, the DFT of both the periodic series and the chaotic series appear to have a decaying amplitude with increasing frequency. This suggests that it is possible to get false positives for chaos when using the DFT to characterize an irregularly sampled series. The LSP, however, produces distinctly different power spectra from the irregularly sampled time series. The chaotic power spectrum generated by the LSP from irregularly sampled data in Figure 14.10 has the decaying amplitude expected of chaotic power spectra. Furthermore, the periodic power spectrum in Figure 14.10 looks largely like the one from the regularly sampled series shown in Figure 14.9. Note that the frequencies of the LSP in Figure 14.10 do not have the same range as those in Figure 14.9. This is due to the fact f_{max}, that the highest frequency for which we can compute the LSP, decreases as we remove elements from the original regularly sampled time series.

Now that we have seen that the MHM can be used to successfully detect chaos in irregularly sampled time series, the next step is to understand the MHM's modes of failure. There are two avenues through which the MHM can fail, excessive sampling irregularity and noise. Understanding these modes of failure is critical to being able to use the MHM for analyzing real-world data. Real-world data are often irregularly sampled and have noise present.

We begin our analysis of the modes of failure by applying the LSP to time series with successively higher levels of depletion. The idea here is that by randomly removing more data, we are using our model to simulate higher and higher degrees of sampling irregularity. Figure 14.11 shows the result of the LSP performed on data generated from Equation 14.3 with 25%, 75%, and 95% of the data removed. Recall that the case of 50% of the data removed is shown in Figure 14.10. We can see that the chaos can be detected in series up to about 50% of the data removed. In both the 25% and 50% cases, there is a fairly clear decay in the amplitude

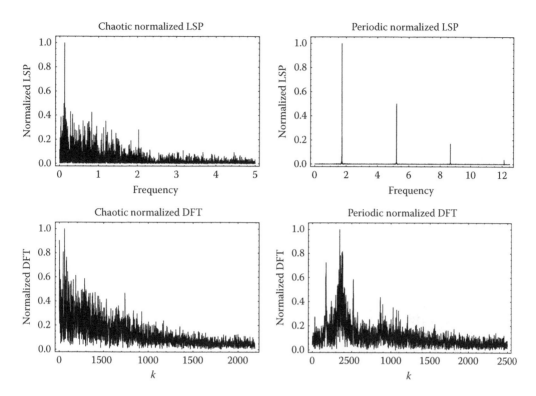

FIGURE 14.10   Comparison of the LSP (top) and DFT (bottom) for the chaotic (left) and periodic (right) Lorenz series.

of the power spectrum as the frequencies increase. In the 75% and 95% cases, the results are more like we would expect from noise (or a random series) where we have high power at all frequencies. The reader may be wondering why the LSP was not computed at higher frequencies. Recall that $f_{max} = N/(2T)$, is determined by the series' length, N, and time range, T and is in analogy to the Nyquist sampling frequency [29]. There is simply not enough data in the irregularly sampled time series to reliably compute the power spectrum to frequencies greater than f_{max}. If we were to compute the power spectrum out to those higher frequencies, would we see the decay? Maybe. However, we cannot be sure if that decay is due to the dynamics of the system or the result of aliasing.

In Figure 14.11, the periodic series is much more robust to sampling. The periodicity of the signal appears to be clear even up to the case where 75% of the data is removed. There is a clear failure to identify periodicity at 95%. The robustness of the periodic signal when compared to the chaotic signal is not surprising. As we have seen, chaotic signals have many more frequencies than periodic signals in their power spectrum. The isolated low-frequency peaks in the power spectrum of a periodic series will be better preserved as f_{max} decreases with an increasing level of data depletion. Figure 14.11 suggests that the MHM can tolerate significant levels of sampling irregularity. However, the MHM is much more robust to sampling irregularities when it comes to detecting regular behavior as compared to chaotic. Hence, the MHM may not be able to distinguish chaos from noise at high levels of sampling irregularity.

Next, we study the robustness of the MHM to noise. The noise robustness test is done by adding random numbers chosen from a normal distribution to the series generated from Equation 14.3. First, the mean is subtracted from each series. This allows us to define the signal-to-noise ratio as

$$SNR = \frac{\sigma_{signal}^2}{\sigma_{noise}^2}, \tag{14.17}$$

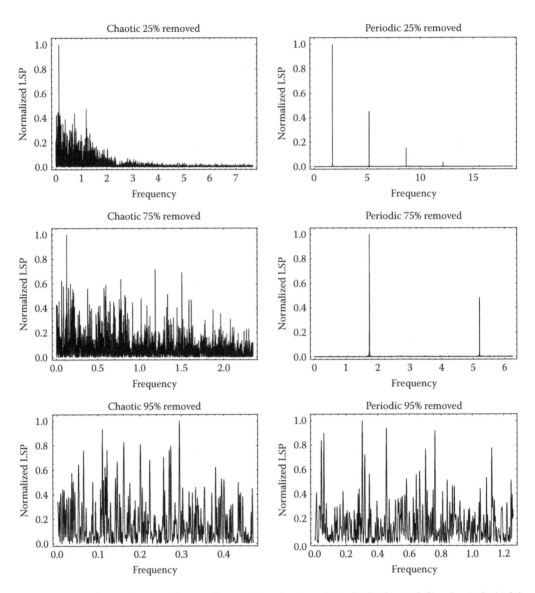

FIGURE 14.11   The LSP for series of varying degrees of sampling irregularity for the chaotic (left) and periodic (right) Lorenz series.

where σ^2_{signal} is the variance of the signal, in this case the time series, and σ^2_{noise} is the variance of the noise. The standard deviation of the normally distributed random numbers added to each series is found using Equation 14.17 for various values of *SNR*. Note that a low value of *SNR* means that the series has a high level of noise. Figure 14.12 shows the results of the LSP applied to series with various values of *SNR*. First note that, unsurprisingly, the MHM fails to distinguish chaos from noise at higher levels of *SNR* (and hence lower levels of noise) as compared to its ability to distinguish periodic signals from noise. Both series fail in similar ways, by adding power across all frequencies. Hence, the LSP can be used as a tool to identify the presence of noise in time series data by looking for the consistent presence of small-amplitude peaks across the frequency spectrum. Notice that the characteristics of each type of dynamics persists until failure. The chaotic power spectrum has many peaks at low frequency that do decay, although not

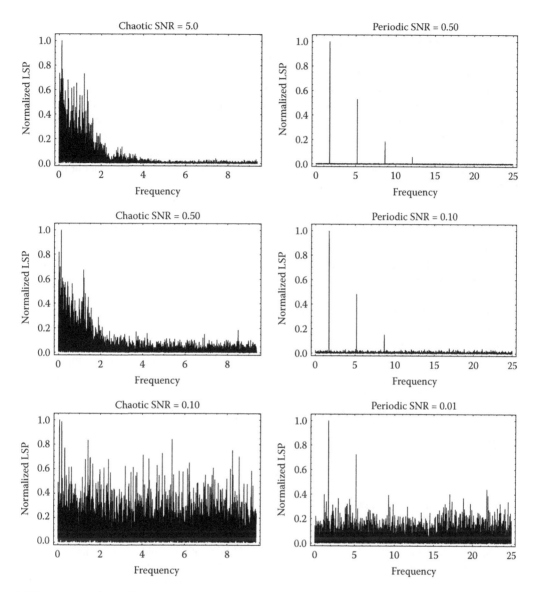

FIGURE 14.12 The LSP for series of varying degrees of noise for the chaotic (left) and periodic (right) Lorenz series.

to zero due to the noise. The periodic series continues to contain distinct peaks; however, these peaks are surrounded by many lower-amplitude peaks.

When comparing Figure 14.11 to Figure 14.12, it is difficult to determine the particular mode of failure. While the graphs on the bottom row of each figure appear similar, the horizontal scales are very different. Hence, in practice, it will be difficult to determine whether the MHM test fails because of sampling or noise. However, the test does display clear signs of failure even if the reason may be unknown. Determining the cause of failure when the MHM is applied to real-world time series data is a future avenue of research.

14.5 Conclusion

Characterization is an important problem in nonlinear dynamics and TISEAN. In effect, it is at the heart of what scientists do in terms of developing models and making predictions of systems. Often, the systems in question are complex and the governing equations are unknown. The methods of time series characterization provide the tools with which scientists can develop reliable, predictive mathematical models of complex systems. Without effective tools for, say, distinguishing between randomness and chaos, it would be impossible to develop such models.

While it may seem that those of us who study TISEAN have it all "figured out," the truth is that there are still many open questions in the field of time series characterization. This is especially true when it comes to irregularly sampled time series. Currently, the authors are studying the effects of irregular sampling on tests for determinism such as the PST. Furthermore, a more quantitative means of detecting chaos in irregularly sampled time series data is also needed. There is evidence to link the Lyapunov exponent to the power spectrum, as computed by the DFT, of a time series [34]. It is unknown whether or not such a connection exists for the power spectrum as computed by the LSP.

Presented in this chapter is a brief list of tools that can be used to characterize the dynamics of a system from time series data. The list of algorithms presented here is not intended to be comprehensive. There is a large body of literature discussing the application of these, and other TISEAN algorithms to experimental data. Experimental data provide a significant challenge to TISEAN algorithms. The algorithms are often developed and tested using model data, i.e., the best-case scenario. However, real-world experimental data typically are short and have noise. The reader interested in the application of time series algorithms to experimental data is encouraged to begin with Reference 17.

We close the chapter with a warning that applies not only to characterization algorithms, but also to all TISEAN algorithms. Beware of treating the algorithm as a black box. It is often tempting to throw data into an algorithm and trust the results without thinking about the algorithm itself. For example, one cannot necessarily use an algorithm designed for regularly sampled series on irregularly sampled series, even if the sampling irregularities are low, i.e., they have few missing elements. Always keep in mind what the algorithm was designed for and what assumptions are made about the data being inputted into the algorithm. In the end, these algorithms are tools and all tools have design limitations.

References

1. H. D. I. Abarbanel. *Analysis of Observed Chaotic Data.* Springer, New York, 1996.
2. J. M. Amigó, S. Zambrano, and M. A. F. Sanjuán. True and false forbidden patterns in deterministic and random dynamics. *Europhys. Lett.*, 79(5):50001, 2007.
3. C. Bandt and B. Pompe. Permutation entropy: A natural complexity measure for time series. *Phys. Rev. Lett.*, 88(17):174102, 2002.
4. E. M. Bollt, T. Stanford, Y.-C. Lai, and K. Życzkowski. Validity of threshold-crossing analysis of symbolic dynamics from chaotic time series. *Phys. Rev. Lett.*, 85(16):3524–3527, 2000.
5. E. M. Bollt, T. Stanford, Y.-C. Lai, and K. Życzkowski. What symbolic dynamics do we get with a misplaced partition? On the validity of threshold crossings analysis of chaotic time-series. *Phys. D*, 154(3–4):259–286, 2001.
6. Y. Cao, W. Tung, J. B. Gao, V. A. Protopopescu, and L. M. Hively. Detecting dynamical changes in time series using the permutation entropy. *Phys. Rev. E*, 70(4):046217, 2004.
7. C. S. Daw, C. E. A. Finney, and E. R. Tracy. A review of symbolic analysis of experimental data. *Rev. Sci. Instrum.*, 74(2):915–930, 2003.
8. A. M. Fraser and H. L. Swinney. Independent coordinates for strange attractors from mutual information. *Phys. Rev. A*, 33:1134, 1986.
9. U. S. Freitas, C. Letellier, and L. A. Aguirre. Failure in distinguishing colored noise from chaos using the "noise titration" technique. *Phys. Rev. E*, 79(3):035201(R), 2009.
10. J. Gao, J. Hu, X. Mao, and W. W. Tung. Detecting low-dimensional chaos by the "noise-titration" technique: Possible problems and remedies. *Chaos, Solitons & Fractals*, 45(3):213–223, 2012.
11. G. A. Gottwald and I. Melbourne. A new test for chaos in deterministic systems. *Proc. R. Soc. London Ser. A*, 460(2042):603–611, 2004.

12. G. A. Gottwald and I. Melbourne. Testing for chaos in deterministic systems with noise. *Phys. D*, 212:100–110, 2005.

13. G. A. Gottwald and I. Melbourne. On the validity of the 0–1 test for chaos. *Nonlinearity*, 22:1367–1382, 2009.

14. G. A. Gottwald and I. Melbourne. Reliability of the 0–1 test for chaos. *SIAM J. Appl. Dyn. Syst.*, 8(1):129–145, 2009.

15. R. Hegger, H. Kantz, and T. Schreiber. Practical implementation of nonlinear time series methods: The TISEAN package. *Chaos*, 9:413, 1999.

16. H. Kantz. A robust method to estimate the maximal Lyapunov exponent of a time series. *Phys. Lett. A*, 185(1):77–87, 1994.

17. H. Kantz and T. Schreiber. *Nonlinear Time Series Analysis*. Cambridge University Press, London, second edition, 2004.

18. D. T. Kaplan and L. Glass. Direct test for determinism in a time series. *Phys. Rev. Lett.*, 68(4):427–430, 1992.

19. M. B. Kennel and H. D. I. Abarbanel. False neighbors and false strands: A reliable minimum embedding dimension algorithm. *Phys. Rev. E*, 66:026209, 2002.

20. M. B. Kennel, R. Brown, and H. D. I. Abarbanel. Determining embedding dimension for phase space reconstruction using a geometrical construction. *Phys. Rev. A*, 45(6):3403–3411, 1992.

21. M. B. Kennel and S. Isabelle. Method to distinguish possible chaos from colored noise and to determine embedding parameters. *Phys. Rev. A*, 46(6):3111–3118, 1992.

22. D. M. Kreindler and C. J. Lumsden. The effects of the irregular sample and missing data in time series analysis. *Nonlinear Dyn. Psychol. Life Sci.*, 10(2):187–214, 2006.

23. C. W. Kulp. Detecting chaos in irregularly sampled time series. *Chaos*, 23:033110, 2013.

24. C. W. Kulp and S. Smith. Characterization of noisy symbolic time series. *Phys. Rev. E*, 83(2):026201, 2011.

25. C. W. Kulp and E. R. Tracy. The application of the transfer entropy to gappy time series. *Phys. Lett. A*, 373(14):1261–1267, 2009.

26. C. W. Kulp and L. Zunino. Discriminating chaotic and stochastic dynamics through the permutation spectrum test. *Chaos*, 24(3):033116, 2014.

27. N. R. Lomb. Least squares frequency analysis of unequally spaced data. *Astrophys. Space Sci.*, 39:447–162, 1976.

28. E. N. Lorenz. Deterministic nonperiodic flow. *J. Atmos. Sci.*, 20:130, 1963.

29. W. H. Press, S. A. Teukolsky, W. T. Vetterling, and B. P. Flannery. *Numerical Recipes in C: The Art of Scientific Computing*. Cambridge University Press, London, second edition, 1997.

30. M. G. Rosenstein, J. J. Collins, and C. J. De Luca. A practical method for calculating largest Lyapunov exponents from small data sets. *Phys. D*, 65(15):117–134, 1993.

31. O. A. Rosso, H. A. Larrondo, M. T. Martín, A. Plastino, and M. A. Fuentes. Distinguishing noise from chaos. *Phys. Rev. Lett.*, 99(15):154102, 2007.

32. J. D. Scargle. Studies in astronomical time series analysis ii. Statistical aspects of spectral analysis of unevenly spaced data. *Astrophys. J.*, 263:835–853, 1982.

33. C. E. Shannon. A mathematical theory of communication. *Bell Syst. Tech. J.*, 27(3):379–423, 1948.

34. D. E. Sigeti. Exponential decay of power spectra at high frequency and positive Lyapunov exponents. *Phys. D*, 82:136–153, 1995.

35. S. H. Strogatz. *Nonlinear Dynamics and Chaos*. Perseus, Cambridge, MA, first edition, 1994.

36. F. Takens. Detecting strange attractors in turbulence. In D. Rand and L. S. Young, ed. *Dynamical Systems and Turbulence, Warwick 1980*, p. 366. Springer, Berlin, 1981.

37. R. Wiebe and L. N. Virgin. A heuristic method for identifying chaos from frequency content. *Chaos*, 22:013136, 2012.

38. A. Wolf, J. B. Swift, H. L. Swinney, and J. A. Vastano. Determining Lyapunov exponents from a time series. *Phys. D*, 16(3):285–317, 1985.

39. Z. Yang and G. Zhao. Application of symbolic techniques in detecting determinism in time series. *Proc. 20th Annu. Int. Conf. IEEE Eng. Med. Biol. Soc.*, 20(5):2670–2673, 1998.

40. M. Zanin. Forbidden patterns in financial time series. *Chaos*, 18(1):013119, 2008.

41. L. Zunino, M. C. Soriano, and O. A. Rosso. Distinguishing chaotic and stochastic dynamics from time series by using a multiscale symbolic approach. *Phys. Rev. E*, 86(4):046210, 2012.

15

Geometry of Local Instability in Hamiltonian Dynamics

M. Lewkowicz,
J. Levitan, Y. Ben Zion,
and L. Horwitz

15.1 Introduction

In this chapter, we discuss a geometrical embedding method for the analysis of the stability of time-dependent Hamiltonian systems using geometrical techniques familiar from general relativity. This method has proven to be very effective in numerous examples, predicting correctly the stability/instability of motions, sometimes contrary to indications of the Lyapunov method. For example, although the application of *local* Lyapunov analysis predicts the completely integrable Kepler motion to be unstable, this geometrical analysis predicts the observed stability. The general theory of the structure and application of this method for time-independent and time-dependent potential problems is given in this chapter, with criteria for its applicability in these cases. As time-independent examples, the perturbed 2D coupled harmonic oscillator and a fifth-order expansion Toda lattice are studied in more detail and the controlling effect is demonstrated. Two time-dependent examples are presented: the restricted three-body problem and the two-dimensional Duffing oscillator with a time-dependent coefficient. The first represents an important class of geophysical and astrophysical problems, and the second, the result of a perturbed bistable system with analogs in electric circuit theory. In both cases the local Lyapunov analysis fails to predict the correct limiting behavior.

Hadamard seems to have been the first to use a geometrical approach to describe chaos. In 1898, when he studied geodesics on surfaces of negative curvature, he wrote a seminal paper "On the Billard on a Surface with Negative Curvature"[1]. Hadamard was keen on differential geometry and the embedding of two-dimensional surfaces in the three-dimensional Euclidian space. He was able to show the strong connection between chaotic motion and negative curvature, not necessarily constant. He studied the so-called Hadamard Billiard, in which a free particle moves on a frictionless surface of constant negative curvature. He was able to show that nearby trajectories diverge strongly. Hadamard made extensive use of the celebrated Gauss–Bonnet theorem, connecting geometry, topology, and the Euler characteristic; among other results it yields that for S being a smooth, closed surface, there are points on S where the Gaussian curvature is positive, zero, or negative. If one bends and deforms S, its Euler characteristic, being a topological invariant, will not change, while the curvatures at some points will. The theorem states, somewhat surprisingly, that the total integral of all curvatures will remain constant, no matter how the deforming is done. We shall show that instabilities *are expected* to be found in areas with negative curvature.

For surfaces with positive Gaussian curvature everywhere the Euler coefficient is also positive (due to the Gauss–Bonnet Theorem), which implies that the surfaces are homeomorphic and hence also diffeomorphic and, moreover, if a surface is isometrically embedded in R^3, the Gauss map provides a diffeomorphism. Hadamard proved that the imbedded surface then is convex and that this example establishes a generalization of the second derivative criterion for convexity for planar curves. Hilbert extended the proof to contain the important result that every isometrically embedded closed surface must have a point of positive curvature and found that a closed Riemannian 2-manifold of non-positive curvature cannot be embedded in R^3, but is only possible for quasi-conformal mappings with use of some conformal equivalent metric.

There are many examples of chaotic behavior which cannot be analyzed by classical differential geometry, but only within the framework of Riemannian geometry like the restricted three-body problem or (higher-order expansions of) the two-body Toda lattice Hamiltonian. In the traditional geometrical approach describing Hamiltonian chaos with tools from Riemannian geometry [2–4] the natural motions of Hamiltonian systems are viewed as geodesics of the configuration space manifold equipped with a metric g (generally either the Jacobi or Eisenhart metric). The stability properties of the geodesics are investigated by use of the Jacobi–Levi-Civita (JLC) equation for the geodesic deviation. Stable and unstable motions are thus determined by the curvature properties, positive, or negative, of the manifold.

Maffione et al. [5] and Cincotta et al. [6] have provided an excellent discussion of recent developments in the theory of dynamical stability and its applications. They discuss methods of deriving criteria for global stability based on local stability tests. Maffione et al. compare the reliability of different indicators of chaos in mappings and show that the mean exponential growth factor of nearby orbits and the relative Lyapunov indicator, which is based on the evolution of the difference between two close orbits, are reliable chaos indicators to analyze a general mapping. Cincotta et al. also studied the mean exponential growth factor of nearby orbits and showed that it is an efficient tool for the investigation of both regular and stochastic components of phase space. It provides information about location of stable and unstable periodic orbits and is a measure of hyperbolicity in chaotic domains, which coincides with that given by the Lyapunov characteristic number. They applied the mean exponential growth factor of nearby orbits in a simple model, the 3D perturbed quartic oscillator and were able to visualize the structure of its phase space, obtaining a clear picture of its resonance structure.

This chapter concerns itself with the construction of tests of *local* stability, which enter as a basic ingredient of the procedures for determining *global* stability. The methods worked out here to determine local stability have been shown to be very sensitive and in many cases are well correlated with the outcome of the global analysis, as emerges from the study of the Poincaré plots in examples cited herein. These results do not necessarily preclude the application of the global treatments discussed by Maffione et al. and Cincotta et al., but lend support to their effectiveness in this framework.

The traditional approaches for determining local stability have some serious problems in many cases. Casetti et al. [3] derived from the JLC equation an effective stability equation which formally describes a parametrically unstable oscillator. They conjectured that some "average" global geometric property should provide information about the degree of chaos and applied the geometric method to the Fermi–Pasta–Ulam beta-model and to a chain of oscillators. They made an analytic calculation of the largest Lyapunov exponent and found in both cases agreement between the theoretical predictions from the geometrical approach and the numerical values for the largest Lyapunov exponent. However, for models undergoing stochastic transitions, as, for example, the Hénon–Heiles model, the scalar curvature of the configuration space manifold provided with a suitable metric is positive and does not depend on the energy value and hence not on whether the model exhibits ordered or chaotic behavior [7].

It has been shown [8] by a numerical example that in contrast to the Newtonian two-center problem, where the dynamics is completely integrable, relativistic null-geodesic motion on the two black-hole space-time exhibits chaotic behavior. Yurtsever identified the geometric sources of this chaotic dynamics by reducing the problem to that of geodesic motion on a negatively curved (Riemannian) surface.

Yurtsever further considered the geodesic flow on a Riemannian surface with only one extremal black hole. The geodesic flow on this surface is completely integrable but a calculation of such a scenario yielded strictly positive Lyapunov exponents and revealed that one cannot identify chaos with merely the presence of positive Lyapunov exponents.

Safaai and Saadat [9] have used the geometric method developed in References 2–4 in an attempt to predict chaos in the restricted three-body problem, where the third body is restricted to move on a circle of large radius inducing an adiabatic time-dependent potential on the second body. This causes the second body to move in a periodic trajectory.

The authors analyzed their results both by use of an indicator of chaos that measures the divergence of the nearby geodesics and by calculation of the largest Lyapunov exponent. Both methods point to chaotic behavior and the authors concluded that the fluctuations of the curvature of the manifold along the geodesics yield parametric instability of the trajectories and hence chaos. In fact, the motion is stable.

It was recently shown [10,11] that there is a possibility to characterize chaos in Hamiltonian systems by a geometrical approach which takes its point of origin in the curvature associated with a conformal Riemann metric tensor (different from the Jacobi or Eisenhart metric). The method takes its starting point in the equivalence between motions generated by a standard Hamiltonian with a quadratic, kinetic term, and an additional potential term, and a Hamiltonian described by a metric type function of the coordinates multiplied with the momenta in a bilinear form. The Hamilton equations of the original potential model are in this approach contained in the geodesics equations through an inverse map in the tangent space in terms of a geometric embedding. The curvature is obtained from the second covariant derivative of the geodesic deviation and leads to a local criterion for unstable behavior different from the criterion rendered by the Lyapunov criterion. The method can be adapted to include a large class of potential models. Our approach represents a new direction in the use of Riemannian geometry by associating instabilities with an energy-dependent negative curvature appropriate for the geodesic motion and different from that implied by the Jacobi metric. It appears to be more sensitive than calculating the largest Lyapunov exponents or using the Jacobi metric, and it establishes a natural connection between Hamiltonian flows and Anosov flows [12].

Calderon et al. [14] showed that under the proper assignment of a metric and a connection, the (classical) dynamical trajectories can be identified as geodesics of the underlying manifold. They demonstrated how the correspondence between geometry and dynamics can be applied to study the conserved quantities of a dynamical system and determined how the mean curvature of the energy level-sets in phase-space correlate with strongly chaotic behavior.

Levitan et al. [13] extended the method to include time-dependent Hamiltonians, as, for example, the restricted three-body problem.

Strauss et al. [15] extended the method to include a quantized treatment of the geodesic deviation. They showed that the geodesic deviation equation, constructed with a second covariant derivative, is unitarily equivalent to that of a parametric harmonic oscillator, and studied the second quantization of this oscillator. The excitations of the Fock space modes correspond to the emission and absorption of quanta into the dynamical medium, thus associating unstable behavior of the dynamical system with calculable fluctuations in an ensemble with possible thermodynamic consequences.

15.2 The Geometrical Method

15.2.1 Time-Independent Potentials

A Hamiltonian system of the form (the summation convention is used)

$$H = \frac{p^i p^j}{2M}\delta_{ij} + V(x) \tag{15.1}$$

where V is a function of space variables alone, can be put into the equivalent form considered by Gutzwiller [16]

$$H_G = \frac{1}{2M}g_{ij}p^i p^j \tag{15.2}$$

where g_{ij} is a function of the coordinates alone [17,18]. Hamilton's equations applied to Equation 15.2 result in the geodesic form

$$\ddot{x}_l = -\Gamma_l^{mn}\dot{x}_m\dot{x}_n; \tag{15.3}$$

where

$$\Gamma_l^{mn} = \frac{1}{2}g_{lk}\left\{\frac{\partial g^{km}}{\partial x_n} + \frac{\partial g^{kn}}{\partial x_m} - \frac{\partial g^{nm}}{\partial x_k}\right\}, \tag{15.4}$$

and g^{ij} is the inverse of g_{ij}.

The formulation of Hamiltonian dynamics of the type of Equation 15.1 in the form of Equation 15.2 is carried out by requiring that Equation 15.1 be equivalent to Equation 15.2. For a metric of conformal form

$$g_{ij} = \varphi\delta_{ij} \tag{15.5}$$

on the hypersurface defined by $H = E = constant$, the requirement of equivalence implies that

$$\varphi = \frac{E}{E - V(x)}. \tag{15.6}$$

Substituting this result in the geodesic Equation 15.3, one obtains an equation that does not coincide in form with the Hamilton equations obtained from Equation 15.1. However, the correspondence can be obtained by introducing a *velocity field* defined by

$$u^j \equiv g^{ji}\dot{x}_i = \frac{p^j}{M}, \tag{15.7}$$

coinciding formally with one of the Hamilton equations implied by (15.1). From this definition, it can be recognized that one is dealing with two manifolds, each characterized by a different connection form, but related by

$$dx^i = g^{ij}dx_j \tag{15.8}$$

on a common tangent space at each point (for which g^{ij} is non-singular).

The correspondence is completed by considering the Hamilton equation for $\dot{p}^l$:

$$\dot{p}^l = -\frac{\partial H}{\partial x_l} = -\frac{1}{2M}\frac{\partial g_{ij}}{\partial x_l}p^i p^j. \tag{15.9}$$

With g_{ij} given in Equations 15.5 and 15.6, one obtains in the particular coordinate system in which Equation 15.1 is defined

$$\dot{p}^l = -\frac{E}{E - V}\frac{\partial V}{\partial x_l}. \tag{15.10}$$

The partial derivative of V occurs here with a factor that belongs to the conformal metric. The small quantity dx^i defined in Equation 15.8 cannot be considered as a differential on some coordinate space

$\{x^i\}$, since dx^i is not an exact differential and it is not uniquely integrable. In the Appendix, it is shown in what sense one can, however, consider (15.10) as corresponding effectively to the (non-covariant since it is derived in the particular choice of coordinates for which (15.5) is valid) relation

$$\dot{p}^l = -g_{lm}\frac{\partial V}{\partial x_m} = -\frac{\partial V}{\partial x^l}, \tag{15.11}$$

the second Hamilton equation in the usual form, where V may be considered a function of a coordinate set $\{x^i\}$, defined in a neighborhood of the orbit, see Appendix. Moreover, our dynamical model for the Hamiltonian motion is constructed with V as considered as a function of the $\{x^i\}$.

The coordinate set $\{x^i\}$ can be considered as the *Hamilton manifold*, for which local variations are defined in terms of the manifold $\{x_l\}$, called the *Gutzwiller manifold*. As indicated by the expression 15.11, it is in terms of this Hamilton manifold that, according to the construction here, the formulations of mechanics of Hamilton and Lagrange are expressed. The local relation (15.8) induces, from the geometry of the Gutzwiller manifold, a corresponding geometry on the Hamilton manifold (i.e., subject to well-defined diffeomorphisms).

The geodesic equation (15.3) can be transformed directly from an equation for $\ddot{x}_j$ to an equation for $\ddot{x}^j$, the motion defined in the Hamilton manifold (the notation $\ddot{x}^j$, $\dot{x}^j$ corresponds to the quantities $\dot{u}^j$, u^j in accordance with the interpretation of the Hamilton manifold). From Equation 15.8 it follows that

$$\ddot{x}_l = g_{lj}\ddot{x}^j + \frac{\partial g_{lj}}{\partial x_n}\dot{x}_n\dot{x}^j$$

$$= -\frac{1}{2}g_{lk}\left\{\frac{\partial g^{km}}{\partial x_n} + \frac{\partial g^{kn}}{\partial x_m} - \frac{\partial g^{nm}}{\partial x_k}\right\}\dot{x}_n\dot{x}_m. \tag{15.12}$$

Now, since

$$\frac{\partial g_{lj}}{\partial x_n} = -g_{lk}\frac{\partial g^{km}}{\partial x_n}g_{mj}, \tag{15.13}$$

it follows that, with the symmetry of $\dot{x}_n\dot{x}_m$,

$$\frac{\partial g_{lj}}{\partial x_n}\dot{x}_n\dot{x}^j = -\frac{1}{2}g_{lk}\left(\frac{\partial g^{km}}{\partial x_n} + \frac{\partial g^{kn}}{\partial x_m}\right)\dot{x}_n\dot{x}_m. \tag{15.14}$$

Thus, the term on the left side of Equation 15.14 containing the derivative of g_{lj} cancels the first two terms of the connection form; multiplying the result by the inverse of g_{lj} and applying the identity (15.13) to lower the indices of g^{nm} in the remaining term on the right side of Equation 15.12, one obtains

$$\ddot{x}^l = -\mathcal{M}^l_{mn}\dot{x}^m\dot{x}^n \tag{15.15}$$

where

$$\mathcal{M}^l_{mn} \equiv \frac{1}{2}g^{lk}\frac{\partial g_{nm}}{\partial x^k}. \tag{15.16}$$

Equation 15.15 has the form of a geodesic equation, with a truncated connection form. (Note that performing parallel transport on the local flat tangent space of the Gutzwiller manifold [for which Γ^{mn}_l and g_{ij} are compatible], the resulting connection, after raising the tensor index [as in Equation 15.8] to reach the Hamilton manifold, is exactly the "truncated" connection (15.16)). Substituting Equations 15.5 and 15.6 into Equations 15.15 and 15.16, the Kronecker deltas identify the indices of $\dot{x}^m$ and $\dot{x}^n$; the resulting square of the velocity cancels a factor of $(E - V)^{-1}$, leaving the Hamilton–Newton law as in Equation 15.11. Equation 15.15 is, therefore, a geometrically covariant form of the Hamilton–Newton law, exhibiting what can be considered an underlying geometry for the standard Hamiltonian motion.

Since the coefficients $\mathcal{M}^l_{mn}$ constitute a connection form, they can be used to construct a covariant derivative. It is this covariant derivative that must be used to compute the rate of transport of the geodesic deviation $\xi^l = x'^l - x^l \equiv \Delta x^l$ along the (approximately common) motion of neighboring orbits in the Hamilton manifold, since it follows the geometrical structure of the geodesic curves in $\{x^l\}$, as defined above.

The second-order geodesic deviation equations [19]

$$\ddot{\xi}^l = -2\mathcal{M}^l_{mn}\dot{x}^m\dot{\xi}^n - \frac{\partial \mathcal{M}^l_{mn}}{\partial x^q}\dot{x}^m\dot{x}^n\xi^q, \tag{15.17}$$

obtained from Equation 15.15, can be factorized in terms of the covariant derivative

$$\xi^l_{;n} = \frac{\partial \xi^l}{\partial x^n} + \mathcal{M}^l_{nm}\xi^m. \tag{15.18}$$

One obtains

$$\frac{D_\mathcal{M}^2}{D_\mathcal{M}t^2}\xi^l = R_\mathcal{M}{}^l_{qmn}\dot{x}^q\dot{x}^n\xi^m, \tag{15.19}$$

where the index $\mathcal{M}$ refers to the connection (15.16), and the *dynamical curvature* is given by

$$R^l_{\mathcal{M}qmn} = \frac{\partial \mathcal{M}^l_{qm}}{\partial x^n} - \frac{\partial \mathcal{M}^l_{qn}}{\partial x^m} + \mathcal{M}^k_{qm}\mathcal{M}^l_{nk} - \mathcal{M}^k_{qn}\mathcal{M}^l_{mk}. \tag{15.20}$$

This expression does not coincide with the curvature of the Gutzwiller manifold (given by this formula with Γ^{qm}_l in place of $\mathcal{M}^l_{qm}$), but is a *dynamical curvature* which is appropriate for geodesic motion in $\{x^l\}$.

With the conformal metric in noncovariant form of (15.5) and (15.6) (in the coordinate system in which (15.6) is defined), the dynamical curvature (15.20) can be written in terms of derivatives of the potential V, and the geodesic deviation equation (15.19) becomes

$$\frac{D_\mathcal{M}^2\xi}{D_\mathcal{M}t^2} = -\mathcal{V}P\xi, \tag{15.21}$$

where the matrix $\mathcal{V}$ is given by

$$\mathcal{V}_{li} = \frac{3}{M^2v^2}\frac{\partial V}{\partial x^l}\frac{\partial V}{\partial x^i} + \frac{1}{M}\frac{\partial^2 V}{\partial x^l\partial x^i}. \tag{15.22}$$

and

$$P^{ij} = \delta^{ij} - \frac{v^iv^j}{v^2}, \tag{15.23}$$

with $v^i \equiv \dot{x}^i$, defining the projection into a direction orthogonal to v^i.

Instability should occur if at least one of the eigenvalues of $P\mathcal{V}P$ is negative, in terms of the second covariant derivatives of the transverse component of the geodesic deviation. This condition is easily seen to be equivalent to the same condition imposed on the spectrum of $\mathcal{V}$, and is thus independent of the direction of the motion on the orbit.

The condition implied by the geodesic deviation equation (15.21), in which the orbits are viewed geometrically as geodesic motion [10], provides new insight into the behavior of Hamiltonian dynamical systems. The method has been successfully applied to many potential models [20,21]. Two examples are outlined in the following:

15.2.1.1 The Perturbed 2D Coupled Harmonic Oscillator

The first example is the simple and important case of two coupled harmonic oscillators with perturbation for which the potential is given by

$$V(x,y) = \frac{1}{2}(x^2 + y^2) + 6x^2y^2 \qquad (15.24)$$

Figure 15.1 shows the surface of section for several energies. Figure 15.1a and b corresponding to $E = 1/10, 1/8$, show stable orbits, while as the energy increases, as in Figure 15.1c–e corresponding to $E = 1/7, 1/6, 1/5, 1$, the orbits have become more and more unstable and show a chaotic signature. The calculation using geometrical methods as shown in Figure 15.2 is parallel to Figure 15.1. Figure 15.2a and b, corresponding to $E = 1/10, 1/8$, show completely positive eigenvalues in the physical region. In comparison, at higher energy, Figure 15.2c–e, corresponding to $E = 1/7, 1/6, 1/5$, show regions of negative eigenvalues appearing at the boundary of the physical area. These regions increase as the energy increases, and in Figure 15.2f, at $E = 1$, the unstable region becomes significant. Both show $E_c \simeq 1/7$ as the critical energy of transition.

15.2.1.2 Two-Dimensional Toda Lattice

The next illustration is a slight modification of the fifth-order expansion of a two-body Toda lattice Hamiltonian:

$$V(x,y) = \frac{1}{2}(x^2 + y^2) + x^2y - \frac{1}{3}y^3 + \frac{3}{2}x^4 + \frac{1}{2}y^4 \qquad (15.25)$$

For sufficiently low energy this system shows stabilization, while in the case of higher energies the system becomes less stable and a chaotic signature appears. Figure 15.3 shows the surface of section for several energies. For $E = 1/5$, Figure 15.3a shows stable orbits; on the other hand, at higher energies $E = 1/2, 3/2, 5$, Figure 15.3b–d show clearly unstable regions corresponding to chaotic motion.

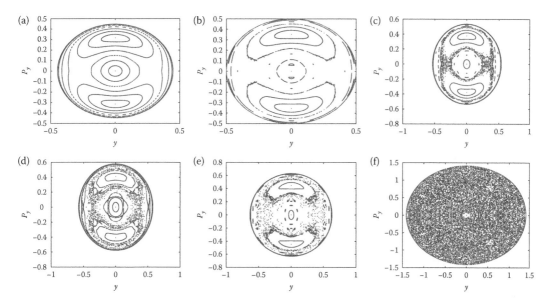

FIGURE 15.1 Poincaré plots for the perturbed 2D oscillator (15.24) in the y, p_y plane. (a,b) For $E = 1/10, 1/8$, indicating regular motion. (c–e) For $E = 1/7, 1/6, 1/5$ indicating chaotic behavior. (f) For $E = 1$, indicating a strongly chaotic behavior.

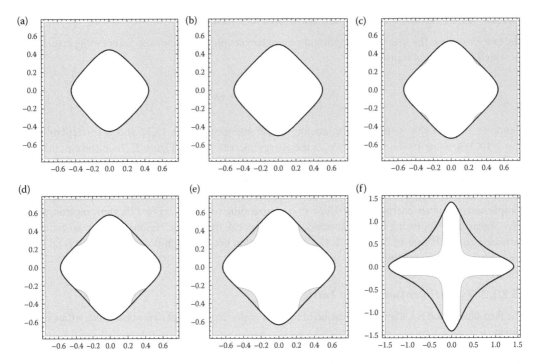

FIGURE 15.2   The dark area shows the region of negative eigenvalues for the matrix $\mathcal{V}$. The light area corresponds to positive eigenvalues where the boundaries are the limits of the physical region. (a,b) Correspond to $E = 1/10, 1/8$. The region of negative eigenvalues does not penetrate the physically accessible region in this case. (c) Corresponds to $E = 1/7$. The dark areas appear on the boundary of the physically accessible region. (d,e) Correspond to $1/6, 1/5$. The dark areas correspond to the existence of at least one negative eigenvalue in the physical region. (f) The dark area of negative eigenvalues for the matrix $\mathcal{V}$ is seen to penetrate deeply into the light region of physically allowable motions for $E = 1$.

Examining the criterion (15.22) for this potential, one finds the eigenvalues as a function of space x, y. Figure 15.4 shows the result in comparison to Figure 15.3. One observes that for $E = 1/5$, Figure 15.4a shows two positive eigenvalues in the physical area, while for energies $E = 1/2, 3/2, 5$, Figure 15.4b–d show that there are one or two negative eigenvalues in the physical area and hence chaotic motion. Both show $E_c \simeq 1/5$ as the critical energy of transition.

15.2.2 Controlling Effect

Computer investigations of this local criterion show that local modifications of the Hamiltonian in regions where negative eigenvalues occur can be used to control the stability of the system [21]. Removal or modification of the nonlinear and symmetry breaking terms in just those local regions have dramatic effects on the Poincaré plots, completely stabilizing the global motion in the examples studied there.

Figure 15.5 shows the effects on the dynamical behavior of the change of the coupling to a stable value in the regions of negative eigenvalues for the perturbed oscillator potential (15.24). The results for different values of energy are shown; these values correspond to different sizes of the physically accessible region. Note that the radius of the region of positive eigenvalues does not change in this example.

The first row shows the physical region and the location of negative eigenvalues. The interior of the circle corresponds to a region in which no negative eigenvalues occur, chosen in this way to simplify the computation. The second row shows the surface of section Poincaré plot (surface of section) of the original

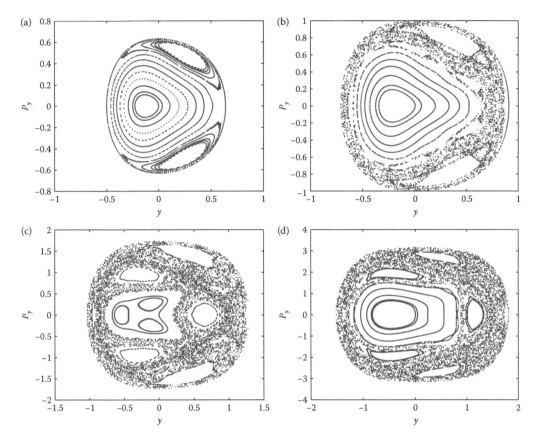

FIGURE 15.3 Poincaré plots for the Toda model (15.25) in the y, p_y plane. (a) For $E = 1/5$ indicating regular motion. (b–d) For $E = 1/2, 3/2, 5$, indicating chaotic behavior.

uncontrolled Hamiltonian indicating chaotic dynamics. The third row shows the Poincaré plot generated by the controlled potential:

$$V(x, y) = \begin{cases} \dfrac{1}{2}(x^2 + y^2) + 6x^2y^2 & x^2 + y^2 < r^2 \\[2mm] \dfrac{1}{2}(x^2 + y^2) & x^2 + y^2 \geq r^2 \end{cases} \qquad (15.26)$$

where r stands for the radius of the region of positive eigenvalues.

Figure 15.6 shows the effect of control on the second system (15.25), using the controlled potential:

$$V(x, y) = \begin{cases} \dfrac{1}{2}(x^2 + y^2) + x^2 y - \dfrac{1}{3}y^3 + \dfrac{3}{2}x^4 + \dfrac{1}{2}y^4 & y > -\alpha \\[2mm] \dfrac{1}{2}(x^2 + y^2) & y \leq -\alpha \end{cases} \qquad (15.27)$$

where α stands for the limit (a horizontal dashed line) of the region of positive eigenvalues.

In both cases, for sufficiently high energies the uncontrolled systems become less stable and a chaotic signature appears. Yet, one can easily see that the Poincaré plots in the controlled systems present almost completely regular motion.

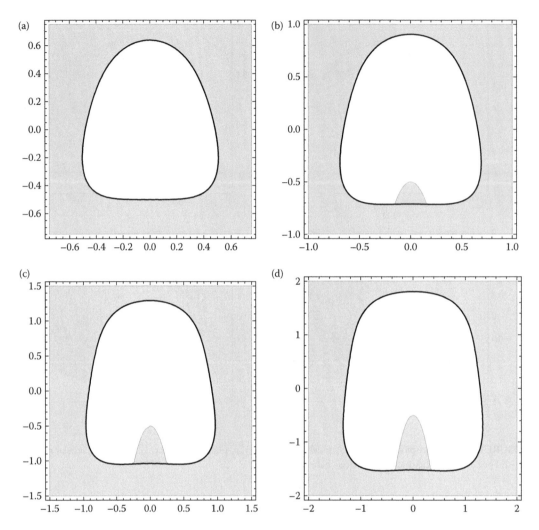

FIGURE 15.4 The dark area shows the region of negative eigenvalues for the matrix $\mathcal{V}$. The lighter area corresponds to positive eigenvalues where the boundaries are the limits of the physical region. (a) Corresponds to $E = 1/5$. The region of negative eigenvalues will not penetrate the physically accessible region in this case. (b,c) Correspond to $E = 1/2, 3/2$. The dark areas correspond to the existence of at least one negative eigenvalue in the physical region. (d) The dark area of negative eigenvalues for the matrix $\mathcal{V}$ is seen to penetrate deeply into the lighter region of physically allowable motion for $E = 5$.

15.2.3 Generalization to Weakly Time-Dependent Potentials

In this case E is not precisely conserved, but one can make an adiabatic approximation in which E can be considered to be time independent. If instability occurs, it should be evident over time intervals small enough to be able to neglect the time dependence of E. The previous analysis then applies with the only exception that the explicit time dependence in V must be taken into account. This time dependence appears in the form of the Gutzwiller space geodesics as a partial time derivative of the metric, but in the geometrical imbedding of the Hamiltonian motion, this term cancels out in the geodesic evolution of the velocity vector field of the Hamilton motion. It then has the same form as for the time-independent potential problem. The formulation of geodesic deviation in terms of the second covariant derivative, however, introduces another time derivative, introducing an additional term in the stability matrix. Arguing

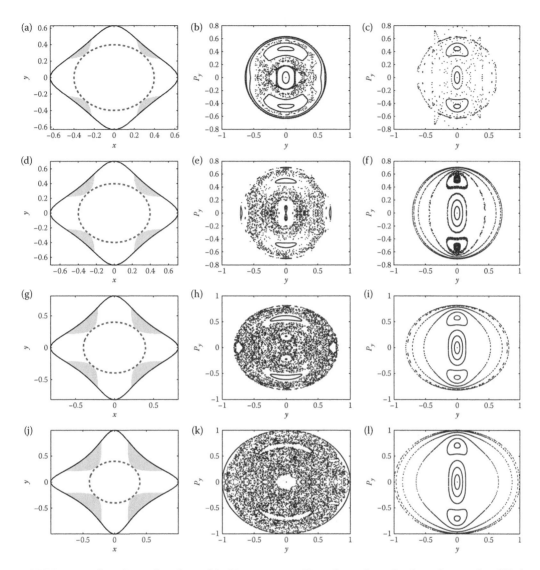

FIGURE 15.5 Effect of control on the model of Equation 15.24. First column shows the physical region closed black curves corresponding to the sequence of energies $E = 1/5, 1/4, 1/3, 1/2$ corresponding to Figures (a,d,g,j). The regions of negative eigenvalues instability are shown in gray. The second column shows the Poincaré plots for the uncontrolled system. The third column shows the Poincaré plots for the controlled system (Equation 15.26 with $r = 0.4$ for all cases). The dashed line in first column shows the boundary of the control modification.

that this term is small, one can apply the stability condition for the time-independent problem directly, modified to take into account the additional time-dependent term, as will be done below for the time-dependent Duffing oscillator and the restricted three-body problem.

In the time-dependent case, the metric g_{ij} is a function of x and explicitly also of t. It then follows from the Hamilton equations as above,

$$\dot{x}_r = \frac{\partial H_G}{\partial p^r} = \frac{1}{m} g_{rs} p^s \tag{15.28}$$

so that

$$p^s = m g^{sr} \dot{x}_r \tag{15.29}$$

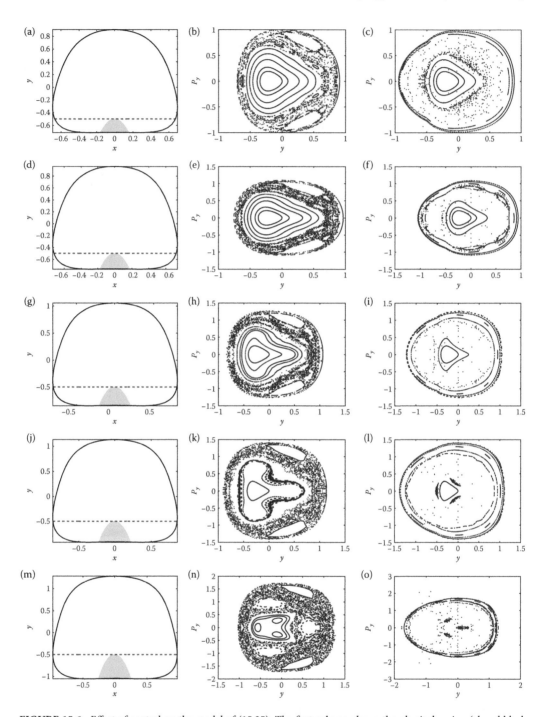

FIGURE 15.6 Effect of control on the model of (15.25). The first column shows the physical region (closed black curves) corresponding to the sequence of energies, 0.5, 0.6, 0.8, 1.0, 1.5 corresponding to (a,d,g,j,m). The regions of negative eigenvalues instability are shown in gray. Second column shows the Poincaré plots for the uncontrolled system. Third column shows the Poincaré plots for the controlled system (Equation 15.27 with $\alpha = 0.5$ for all cases). The dashed line in first column shows the boundary of the control modification.

Then,

$$
\begin{aligned}
\dot{p}^k &= -\frac{\partial H_G}{\partial x_k} = -\frac{1}{2m} p^i p^j \frac{\partial g_{ij}}{\partial x_k} \\
&= -\frac{m}{2} g^{il} g^{jm} \dot{x}_\ell \dot{x}_m \frac{\partial g_{ij}}{\partial x_k} \\
&= \frac{m}{2} \dot{x}_l \dot{x}_m \frac{\partial g^{lm}}{\partial x_k}.
\end{aligned}
\tag{15.30}
$$

Now, compute from Equation 15.29 the time derivative of p^k, which has an additional term due to time dependence in V:

$$
\dot{p}^k = m \frac{\partial g^{km}}{\partial x_l} \dot{x}_l \dot{x}_m + m g^{kr} \ddot{x}_r + m \frac{\partial g^{kr}}{\partial t} \dot{x}_r
\tag{15.31}
$$

Comparing Equations 15.30 and 15.31 gives

$$
\ddot{x}_s = -\Gamma_s^{lm} \dot{x}_l \dot{x}_m - g_{sk} \frac{\partial g^{kr}}{\partial t} \dot{x}_r.
\tag{15.32}
$$

The geodesic in this Gutzwiller space, therefore, has an extra term due to the time dependence of g^{kr}. However, defining from Equation 15.29 a velocity field as in Equation 15.7 the time derivative will cancel the extra term, and one obtains (as in Equation 15.7)

$$
\dot{u}^q = -\frac{1}{2} \frac{\partial g_{rt}}{\partial x_q} u^r u^t
\tag{15.33}
$$

Equation 15.33 provides a relation controlling the velocity field in the tangent space in the neighborhood of some point x along the geodesic. It is remarkable that this result is valid for any arbitrary t dependence in the potential. The geodesic deviation of the flow on this coordinate surface, with

$$
\xi^q = x^q - x'^q,
\tag{15.34}
$$

for which $dx^q = g^{qp} dx^p$, so that

$$
\dot{\xi}^q = u^q - u'^q
\tag{15.35}
$$

corresponds to the rate of local separation of the neighboring orbits generated by the $\{x^q\}$.

Now take the derivative of Equation 15.35 and write the difference on the right-hand side with the geodesic equations (15.33), resulting in the same form as Equation 15.17.

Define a covariant derivative using the M-connection as in Equation 15.18. One obtains

$$
\begin{aligned}
\frac{D_M^2 \xi^q}{D_M t^2} &= \frac{d}{dt} \left\{ \frac{d\xi^q}{dt} + M_{rt}^q \xi^r \dot{x}^t \right\} + M_{rt}^q \dot{x}^t \left\{ \frac{d\xi^r}{dt} + M_{sl}^r \xi^s \dot{x}^l \right\} \\
&= R_M{}^q{}_{lrt} \dot{x}^l \dot{x}^t \xi^r + \frac{\partial M_{rt}^q}{\partial t} \dot{x}^t \xi^r,
\end{aligned}
\tag{15.36}
$$

which is just the result obtained in Equation 15.19, with the additional term

$$
\frac{\partial M_{rt}^q}{\partial t} \dot{x}^t \xi^r = \frac{1}{2} \frac{\partial^2 g_{rt}}{\partial t \partial x_q} \dot{x}^t \xi^r.
\tag{15.37}
$$

Using

$$
g_{rt} = \frac{E}{E - V(x,t)} \delta_{rt},
\tag{15.38}
$$

and assuming adiabatically that E does not depend on t, we have

$$\frac{\partial^2 g_{rt}}{\partial x_q \partial t} = \frac{2E}{(E - V(x,t))^3} \frac{\partial V}{\partial t} \frac{\partial V}{\partial x_q} + \frac{E}{(E - V(x,t))^2} \frac{\partial^2 V}{\partial x_q \partial t}. \tag{15.39}$$

Near the boundary of the physical region, these terms can become large, but they are generally small since they involve the time derivative of V, in the same approximation in which E is considered as constant. One may therefore write Equation 15.36 as

$$\frac{D_M^2 \xi^q}{D_M t^2} = -\mathcal{V}_{qi} P^{ij} \xi^j + L_q \dot{x}^r \xi^r, \tag{15.40}$$

where

$$L_q = \left[\frac{1}{(E - V)^2} \frac{\partial V}{\partial t} \frac{\partial V}{\partial x^q} + \frac{1}{2} \frac{1}{E - V} \frac{\partial^2 V}{\partial t \partial x^q} \right]. \tag{15.41}$$

Now, multiply by P^{sq}

$$\frac{D_M^2 \xi_\perp^s}{D_M t^2} = -(PVP)_{sr} \xi_\perp^r + (P^{sq} L_q) \xi_\parallel, \tag{15.42}$$

where

$$\xi_\parallel = \dot{x}^r \xi^r$$

and

$$\xi_\perp^r = P^{rs} \xi^s.$$

Multiplying Equation 15.40 by $\dot{x}^q$ one gets (these operations all commute with the covariant derivatives since the contracted M terms are diagonal):

$$\frac{D_M^2 \xi_\parallel}{D_M t^2} = -(\dot{x}VP)_j \xi_\perp^j + (\dot{x}^q L_q) \xi_\parallel \tag{15.43}$$

These equations are coupled. Call

$$T_r = -(\dot{x}VP)_r \tag{15.44}$$

and

$$Q_s = P_{sq} L^q \tag{15.45}$$

and

$$N = \dot{x}^q L_q \tag{15.46}$$

More:

$$M_{rs} = -(PVP)_{rs}. \tag{15.47}$$

One can then write these coupled equations in matrix form as

$$\frac{D_M^2}{D_M t^2} \begin{pmatrix} \xi_\perp \\ \xi_\parallel \end{pmatrix} = \begin{pmatrix} M & Q \\ T & N \end{pmatrix} \begin{pmatrix} \xi_\perp \\ \xi_\parallel \end{pmatrix} \tag{15.48}$$

Here, M is a matrix (symmetric, corresponding to PVP in the time-independent case), Q is a vector, T is a vector, and N is a scalar, as defined above; $\xi_\perp$ is a vector, and $\xi_\parallel$ is a scalar. Studying the eigenvalue problem shows that the eigenvalues for the full motion are dominated by the eigenvalues of M as in the t-independent case.

Denoting the eigenvalues of the full matrix $\hat{\lambda}$, the eigenvalues of M alone λ (as for the t-independent case), and write the eigenvalue equation by studying the determinant of the right-hand matrix minus λ.

In the two-dimensional case, of interest here, let $\hat{\lambda} = \lambda + \mu$, and suppose that μ is small; then check for consistency. Working out the determinant, one finds the exact eigenvalue condition

$$0 = \mu^2 - \mu(M_{11} + M_{12} - 2\lambda) + \frac{1}{N - \lambda - \mu}[(Q_1 T_2 + Q_2 T_1)(M_{12} + \lambda + \mu) - Q_1 T_2 M_{22} - Q_2 T_2 M_{11}] \tag{15.49}$$

Assuming μ is small, and working to first order in μ, one obtains

$$\mu \cong -\frac{Q_1 T_1 M_{22} + Q_2 T_2 M_{11} - (Q_1 T_2 + Q_2 T_1)(M_{12} + \lambda)}{(N - \lambda)(M_{11} + M_{22} - 2\lambda) + Q_1 T_2 + Q_2 T_1(M_{12} + \lambda)} \tag{15.50}$$

Thus,

$$\mu = O(Q)\frac{1}{N - \lambda}(TrM - 2\lambda) \tag{15.51}$$

This implies that μ is consistently small, provided that N is not close to λ. Both Q and N contain L_q, which is singular near $E = V$.

Another, weaker, bound can be found for which $\xi_\perp$ passes through a value on the orbit for which it is an exact eigenvector for the matrix M, with eigenvalues λ. The eigenvalues then must satisfy the equation

$$\det \begin{pmatrix} \lambda - \hat{\lambda} & 0 & Q_1 \\ 0 & \lambda - \hat{\lambda} & Q_2 \\ T_1 & T_2 & N - \hat{\lambda} \end{pmatrix} \tag{15.52}$$

The determinant condition is

$$(\lambda - \hat{\lambda})[(\lambda - \hat{\lambda})(N - \hat{\lambda}) - T_2 Q_2] - Q_1 T_1(\lambda - \hat{\lambda}) = 0 \tag{15.53}$$

Clearly, 15.49 has the solution $(\lambda - \hat{\lambda}) = 0$, but that would imply the $\xi_\parallel$ is zero (if $Q \neq 0$). That would be unlikely. The second solution is

$$(\lambda - \hat{\lambda}) = \frac{T_2 Q_2 - Q_1 T_1}{N - \hat{\lambda}}, \tag{15.54}$$

showing that $(\lambda - \hat{\lambda})$ is small, provided that, as above, the small quantity $N \neq \hat{\lambda}$.

The estimates given here show, except for the boundaries of the physical region, that the eigenvalues of the full motion should be essentially those of the static case. The eigenvalue for the full motion can, of course, be computed exactly to compare.

15.3 Examples

The method will be illustrated with two examples: a time-dependent two-dimensional Duffing oscillator and the restricted three-body problem.

15.3.1 The Time-Dependent Two-Dimensional Duffing Oscillator

Consider a time-dependent coefficient of the quadratic term in the Hamiltonian:

$$H = \frac{p_x^2 + p_y^2}{2} + \frac{(x^2 + y^2)^2}{4} - \beta \cos(t)\frac{x^2 + y^2}{2}. \tag{15.55}$$

It turns out that in this case two of the three eigenvalues of the matrix in Equation 15.48 are zero or negative. The third eigenvalue is negative for $\beta = 0$, indicating stable trajectories, albeit positive for $\beta > 0$. The magnitude of these positive eigenvalues increases with β, as seen in Figure 15.7, suggesting that the trajectories become more and more unstable.

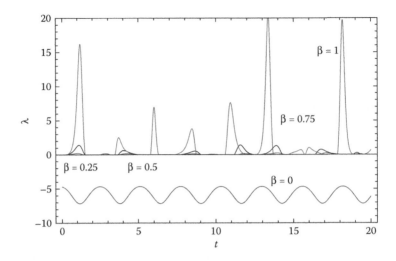

FIGURE 15.7 The "third" eigenvalue for various values of $\beta = 0, 0.25, 0.5, 0.75,$ and 1.

Figure 15.7 shows the "third" eigenvalue as a function of time for $\beta = 0, 0.25, 0.5, 0.75,$ and 1, demonstrating the change in sign. For small β the eigenvalue's magnitude is rather small, but growing fast for $\beta > 0.5$. The instability can be acknowledged in Figure 15.8a, showing the diverging trajectories with two nearby initial conditions for $\beta = 1$, as well as in the phase-space plot in Figure 15.8b bearing clearly a chaotic signature.

In Figure 15.9, we correlate the eigenvalue for $\beta = 1$ with the scalar curvature R_M contracted from the dynamical curvature tensor R^ℓ_{Mqmn} defined in Equation 15.21:

$$R_M = -\frac{1}{2E} \left\{ \left(\frac{\partial^2 V}{\partial x^2} + \frac{\partial^2 V}{\partial y^2} \right) + \frac{3}{2} \frac{1}{E - V(x,y)} \left[\left(\frac{\partial V}{\partial x} \right)^2 + \left(\frac{\partial V}{\partial y} \right)^2 \right] \right\} \tag{15.56}$$

The moments in time when the positive eigenvalue reaches its temporal maximal magnitude coincide with the times of maximal negative curvature, reassuring that the instability of the motion is indeed associated with negative curvature.

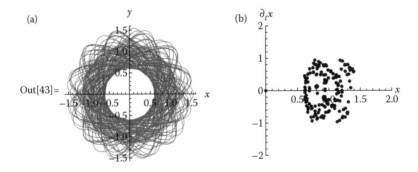

FIGURE 15.8 (a) Trajectories for two nearby initial conditions for $\beta = 1$ (solid: $x_0 = 0.94, \dot{x}_0 = 0.04$, dashed: $x_0 = 0.93, \dot{x}_0 = 0.05$; in both cases $y_0 = 0, \dot{y}_0 = 1$). (b) The Poincaré map for the time-dependent two-dimensional Duffing oscillator with $\beta = 1$.

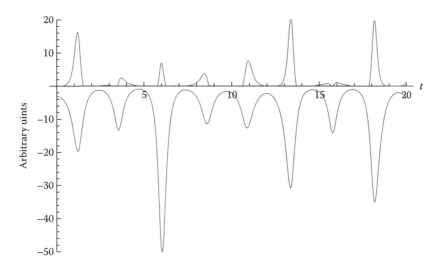

FIGURE 15.9 The eigenvalue for $\beta = 1$ (upper curve) correlated with the scalar curvature R_M (lower curve).

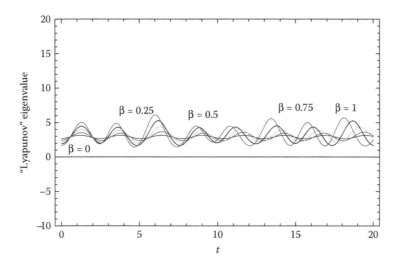

FIGURE 15.10 The local Lyapunov eigenvalue as a function of time for various values of β

Contrary to the results of the geometrical analysis which predicts the stability or instability of the orbit, the local Lyapunov eigenvalue [22], calculated from the deviation of two nearby tajectories (see Equation 15.21 and Appendix A in Reference 11), does not vary significantly with the strength of the time dependence β, see Figure 15.10, and indicates stability also for $0 < \beta \leq 1$, which are obviously unstable.

15.3.2 The Restricted Three-Body Problem

Euler's three-body problem is to solve for the motion of a particle (the "Earth") that is acted upon by the gravitational field of two other point masses, one fixed in space (the "Sun") and the other one ("Jupiter") moving in circular coplanar orbit around the former. This system was considered previously in Reference 11 neglecting the time dependence in the potential. Thus the eigenvalues analyzed were those of the matrix (15.22). Three cases with different "Jupiter" masses were studied: zero (the Kepler problem), a realistic Jupiter mass, and a thousand-fold Jupiter mass (this value is close to the mass of the Sun, and therefore the dynamics is close to that of a binary solar system). The analysis there showed stability for the Kepler problem as well as for the three-body problem (unlike the above-mentioned local Lyapunov eigenvalue), yet for the "large" Jupiter short unstable excursions in one of the eigenvalues of (15.22) were noticeable (see Figure 15.11 in Reference 11).

To include the time dependence, the eigenvalues of the matrix in Equation 15.48 are examined. Once more two out of the three eigenvalues are zero or negative, however, the third one can change sign. For the Kepler problem and for the realistic Jupiter mass two of the eigenvalues are zero while one is negative, whereas for the thousand-fold Jupiter mass two eigenvalues are negative and the third one is prominently positive suggesting instability of the system. The onset of a noticeable instability is found around 500 Jupiter masses. Figure 15.11 shows the positive eigenvalue as a function of time for various "Jupiter" masses; their average value over 100 time units are depicted in Figure 15.12 together with the negative

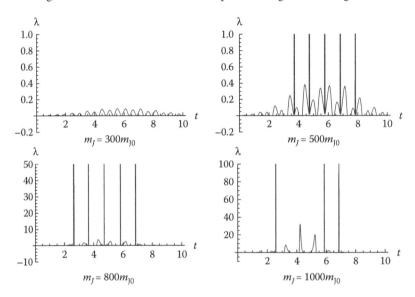

FIGURE 15.11   The positive eigenvalue as a function of time for various "Jupiter" masses.

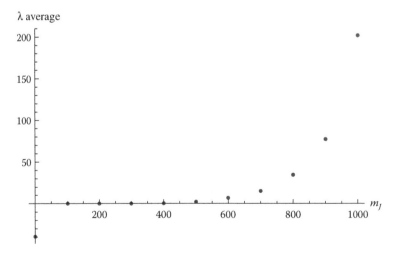

FIGURE 15.12   The average value of the eigenvalues for various "Jupiter" masses. The negative value for $m_J = 0$ (the Kepler limit) is noted.

average value for the stable Kepler problem. There is a dramatic change in magnitude between 300 and 500 Jupiter masses.

The inception of instability is also manifested from Figure 15.13, showing the diverging trajectories with two nearby initial conditions for various "Jupiter" masses as well as in the phase-space plots in Figure 15.14.

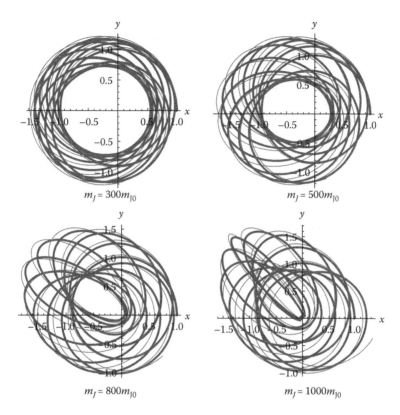

FIGURE 15.13 Trajectories for two nearby initial conditions, various "Jupiter" masses (thick: $x_0 = 0.983, \dot{x}_0 = 0$, thin: $x_0 = 1, \dot{x}_0 = 0.0167$; in both cases $y_0 = 0, \dot{y}_0 = 1.01684 \cdot 2\pi$).

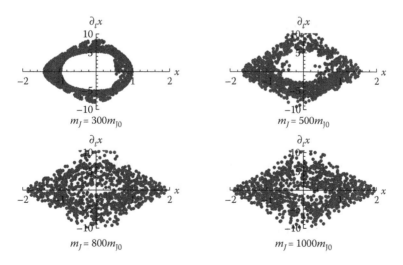

FIGURE 15.14 Phase space plots for various "Jupiter" masses.

15.4 Conclusions

Two important examples of Hamiltonian systems with weak explicit time dependence were treated. It was shown that the geometrical criterion for stability recently developed for time-independent systems provides an effective criterion for these systems as well. Taking into account the time dependence, within an adiabatic approximation for the conformal factor involved in the geometric embedding, one finds a coupling between the longitudinal and transverse deviations of the geometrically imbedded orbit. Diagonalizing this coupled system, the resulting eigenvalues provide, from the point of view of the behavior of the orbit as well as Poincaré plots, a reliable criterion of stability. The method of geometrical embedding, therefore, seems to provide a measure for stability that is robust in the case of weak explicit time dependence as well, significantly widening its range of application. In application to relativistic problems, the singular regions do not necessarily correspond to boundaries of the physical region, but are associated with transitions from timelike to spacelike evolution, involved in pair production and annihilation processes, and are of independent interest [23].

Let us remark on the adaptation of our approach to systems with many degrees of freedom as in nuclear physics, or large molecules, where the number of particles is large, and the center of momentum can be separated canonically, leaving total momentum squared plus the sum of relative momenta squared, for which the conformal map can be defined. The very comprehensive analysis of Casetti et al. discusses in detail the application of geometrical methods to such systems. They used the Jacobi metric and the related Eisenhart metric to apply these methods; our procedure for the case of many degrees of freedom would be precisely the same, but using the conformal metric 15.5. The resulting connection form will, therefore, be quite different. Such systems will be investigated in subsequent work.

Appendix

Here, some aspects of the identification of an effective Hamiltonian manifold are discussed. By considering u^s to be the derivative of a coordinate on the local tangent space, one has

$$u^s = \dot{x}^s \tag{A15.1}$$

or

$$\dot{x}_r = g_{rs}\dot{x}^s$$

which implies the nonintegrable relation (15.8),

$$dx^s = g^{sr} dx_r. \tag{A15.2}$$

However, define a first-order variation with some of the properties of a derivative in the tangent space by, for some function $F(x)$,

$$F(x + \Delta x) - F(x) = \frac{\partial F(x)}{\partial x_s} \Delta x_s = g_{sr} \Delta x^r \frac{\partial F(x)}{\partial x_s} \tag{A15.3}$$

so that one can define an effective derivative (note that as as a derivative, the $\partial/\partial x^r$ do not commute)

$$\frac{\partial F(x)}{\partial x^r} = g_{rs} \frac{\partial F(x)}{\partial x_s}. \tag{A15.4}$$

With this, the geodesic on the tangent space as in Equations 15.15 and 15.16 is defined, and, using Equation A15.3, the geodesic deviation is calculated. Initially, labelling the orbits by points in $\{x_r\}$, what has been done is to induce an effective *Hamilton manifold* in terms of which the conventional Hamilton dynamics is constructed.

References

1. J. Hadamard. Les surfaces à courbures opposées et leurs lignes géodésiques. *J. Math. Pures Appl.* **4**, 27: 1898.

2. L. Casetti and M. Pettini. Analytic computation of the strong stochasticity threshold in Hamiltonian dynamics using Riemannian geometry. *Phys. Rev. E* **48**, 4320: 1993.

3. L. Caiani, L. Casetti, C. Clementi, and M. Pettini. Geometry of dynamics, Lyapunov exponents, and phase transitions. *Phys. Rev. Lett.* **79**, 4361: 1997.

4. L. Casetti, C. Clementi, and M. Pettini. Riemannian theory of Hamiltonian chaos and Lyapunov exponents. *Phys. Rev. E* **54**, 5969: 1996.

5. N. P. Maffione, L. A. Darriba, P. M. Cincotta, and C. M. Giordano. A comparison of different indicators of chaos based on the deviation vectors: Application to symplectic mappings. *Celest. Mech. Dyn. Astron.* **111**, 285: 2011.

6. P. M. Cincotta, C. M. Giordano, and C. Simó. Phase space structure of multi-dimensional systems by means of the mean exponential growth factor of nearby orbits. *Physica D* **182**, 151: 2003.

7. M. Pettini. *Geometry and Topology in Hamiltonian Dynamics and Statistical Mechanics.* Springer, Berlin, 2007.

8. U. Yourtsever. Geometry of chaos in the two-center problem in general relativity. *Phys. Rev. D* **52**, 3176: 1995.

9. H. Safaai, M. Hasan and G. Saadat. *Understanding Complex Systems.* Springer, Berlin, 2006.

10. L. Horwitz, Y. Ben Zion, M. Lewkowicz, M. Schiffer, and J. Levitan. Geometry of Hamiltonian chaos. *Phys. Rev. Lett.* **98**, 234301: 2007.

11. A. Yahalom, J. Levitan, M. Lewkowicz and L. Horwitz. Lyapunov vs. geometrical stability analysis of the Kepler and the restricted three body problems. *Phys. Lett. A* **375**, 2111: 2011.

12. D. Anosov. *Geodesic Flows on Closed Riemannian Manifolds with Negative Curvature.* Proc. Steklov Instit. Math. **90**, Amer. Math. Soc. Providence, Rhode Island, 1969.

13. J. Levitan, A. Yahalom, L. Horwitz, and M. Lewkowicz. On the stability of Hamiltonian systems with weakly time dependent potentials. *Chaos* **23**, 023122: 2013.

14. E. Calderon, L. Horwitz, R. Kupferman, and S. Shnider. On the geometric formulation of Hamiltonian dynamics. *Chaos* **23**, 013120: 2013.

15. Y. Strauss, L. P. Horwitz, J. Levitan, and A. Yahalom. Quantum field theory of classically unstable hamiltonian dynamics. *J. Math. Phys.* **56**(7), id.072701: 2015.

16. M. C. Gutzwiller. *Chaos in Classical and Quantum Mechanics*, Springer-Verlag, New York, 1990. See also W. D. Curtis and F. R. Miller. *Differentiable Manifolds and Theoretical Physics.* Academic Press, New York. 1985; J. Moser and E. J. Zehnder. *Notes on Dynamical Systems.* Amer. Math. Soc., Providence, 2005; L. P. Eisenhardt. *A Treatise on the Differential Geometry of Curves and Surfaces.* Ginn, Boston, 1909 [Dover, N.Y., 2004]; V. I. Arnold. *Mathematical Methods of Classical Mechanics.* Springer-Verlag, New York, 1978; A. Oloumi and D. Teychenne. Controlling Hamiltonian chaos via Gaussian curvature. *Phys. Rev. E* **60**, R6279: 1999.

17. M. Szydlowski and J. Szczesny. Invariant chaos in mixmaster cosmology. *Phys. Rev. D* **50**, 819: 1994. See also M. Szydlowski and A. Krawiec. Description of chaos in simple relativistic systems. *Phys. Rev. D* **53**, 6893: 1996, who have studied a somewhat generalized Gutzwiller form (with the addition of a scalar potential) which accommodates the Jacobi metric.

18. P. Appell. *Dynamique des Systemes Mécanique Analytique*, Gauthier-Villars, Paris, 1953; H. Cartan. *Calcul Différential et Formes Différentielle*, Herman, Paris, 1967; L. D. Landau, *Mechanics*, Mir, Moscow, 1969. The method was utilized for the relativistic case by D. Zerzion, L. P. Horwitz, and R. Arshansky. Classical mechanics of special relativity in a Riemannian space-time. *J. Math. Phys.* **32**, 1788: 1991.

19. Substituting the conformal metric 15.5 into 15.17, and taking into account the constraint that both trajectories x'^l and x^l have the same energy E, one sees that 15.17 becomes the orbit deviation equation

based on 15.5, the Hamilton equations generated by H. Thus, the introduction of the second covariant derivative 15.19 in the framework of the embedding general geometric structure carries more information on stability than the stability analysis applied to the Hamilton equations derived directly from 15.1 (which, as it is easy to see, corresponds to a Lyapunov analysis).

20. Y. Ben Zion and L. P. Horwitz. Detecting order and chaos in three-dimensional Hamiltonian systems by geometrical methods. *Phys. Rev. E* **76,** 046220: 2007; Y. Ben Zion and L. P. Horwitz. Applications of geometrical criteria for transition to Hamiltonian chaos. *Phys. Rev. E* **78**, 036209: 2008.

21. Y. Ben Zion and L. P. Horwitz. Controlling effect of geometrically defined local structural changes on chaotic Hamiltonian systems. *Phys. Rev. E* **81**, 046217: 2010.

22. W. Siegert, *Local Lyapunov Exponents*. Lecture Notes in Mathematics Vol. 1963. Springer, 2009.

23. A. Gershon and L. P. Horwitz. Kaluza-Klein theory as a dynamics in a dual geometry. *J. Math. Phys.* **50**, 102704: 2009; L. P. Horwitz, A. Gershon, and M. Schiffer. Hamiltonian map to conformal modication of spacetime metric: Kaluza-Klein and TeVeS. *Found. Phys.* **41**, 141: 2011.

24. L. Casetti, M. Pettini, and E. G. D. Cohen. Geometric approach to Hamiltonian dynamics and statistical mechanics. *Phys. Rep.* **337**, 237: 2000.

16

Chaos Analysis of ER-Network Dynamics in Microscopy Imaging

Tuan D. Pham and
Ikuo Wada

16.1 Introduction

Metazoa are equipped with extensively developed intramembranous structures that create a microenvironment, enabling various fundamental reactions for living activities. Each organelle, which is a structure unit delineated by the membranes, has a distinct dynamic shape to fulfill the highly interlinked reactions. Among them, the endoplasmic reticulum (ER) is the compartment having the largest surface that creates a single-large lumen by interconnecting tubules and sheets. The ER is continuous to the nuclear envelop and contains several microdomains required for the export of secretory cargo or physical coupling to mitochondria, endosomes, or plasma membranes. The sheet structure provides a suitable surface for attachment of polysomes-synthesizing secretory proteins. The nascent proteins are sequestered into the lumen of the ER and extensively edited to obtain the stable conformation in the oxidative environment, which are then transported to the Golgi apparatus through the COPII-coated vesicles budded from one of the ER microdomain, ER exit sites. Extensive studies have identified the essential ER factors for the folding, maturation, and transport of secretory cargo [1,2]. Recent studies also elucidated the physiological role of the attachment sites to the other organelles [3,4]. Still, it remains elusive why such chains of reactions can be efficiently carried out in the microenvironment of nanometer scale.

While the structure of the ER is totally dependent on cell types, the ER except the nuclear envelop is morphologically classified into two structures, tubules and sheets. A series of elaborate studies have revealed that the ER contains sets of membrane proteins creating a different membrane curvature determining the shape and stability of the ER. For example, the structure with the least curvature of the sheet region is thought to be maintained by several coiled-coil proteins including Climp-63 while reticulons and DP1 are involved in the high-curvature edge of the sheet [5]. As ribosomes are abundantly found in the sheets, translocation of nascent secretory cargo proteins is particularly active in this region. Consistently, tubule structures are dominant in such cells that have low-secretion capability.

The ER is also characterized by its complicated motion. The most prominent dynamics is extension, shrinkage, and fusion of the tubules [6]. The first two processes are reminiscent of microtubules and, indeed, these types of motion are regulated by their direct association with tubulin. Formation of the extensive network throughout the cells requires fusion of the tubules [7,8]. The homotypic fusion of the ER tubules is mediated by a family of GTPases, atlastin. Also, myosin1c-mediated, actin-based motion of the sheet has been recently described [9]. Importantly, these morphologically distinct structures interconvert between each other, which are highly affected by various pathological events including folding or oxidative stress. Also, direct etiological links have been found in mutations of the shape-forming proteins. For example, mutations in atlastin-1 are assigned as responsible for early-onset hereditary spastic paraplegia [10]. Thus, the ER network dynamics, which occurs in large time and space scale, could provide a variety of crucial information to understand not only the ER responses but also organization of the reactions; however, the quantitative measure of the ER network is difficult as conventional tracking methods are not applicable. It was the motivation of the contribution presented in this chapter.

Here, we are interested in applying the theory of chaos to identify the nature of the dynamic behavior of the ER network as well as to quantify its measure. Such a quantification can be helpful for gaining a deeper understanding of the spatial and temporal organization of the ER. A study, which is closely related to the work reported in this chapter, is the chaos analysis of ER dynamics and ER–mitochondrial contacts [11]. The previous research examined the ER dynamics using 10 time-lapse microcopy image datasets of COS-7 cell experiments reported in Reference 12. The research presented herein investigated richer time-lapse microscopy image datasets of the ER network. We applied the largest Lyapunov exponent [13] to determine a predictability for the ER-network dynamics in time. Being motivated by an earlier work [14], we further studied the dynamics of the ER network in space using the concept of the fuzzy Kolmogorov–Sinai entropy [15]. The following sections present the concepts and numerical calculations of two measures of chaos employed in this chapter as well as discussions on experimental results carried out on ER-network image data.

16.2 Time-Lapse Microscopy Imaging of the ER Network

The ER was visualized in COS-7 cells by the transient expression of SGFP2-fused Lnp1p, a membrane protein abundantly found in the junctions of the ER tubules. The details of the live cell recording were described in Reference 16 except that the superresolution microscopic images were captured by a structured illumination microscope (SIM) as in Reference 17. Dataset #1: sequence of 300 images at 0.5 frame per second, each image is of size 821×883 (pixels). Dataset #2: sequence of 350 images, each image is of size 821×883. Dataset #3: sequence of 350 images, each image is of size 821×883. The width and height of a pixel is 0.031 μm.

16.3 Chaos Analysis of ER-Network Dynamics in Time

16.3.1 Image Motion Detection

The two-dimensional (2-D) motion detection of the time-lapse microscopy videos of the ER was carried out by converting all color-image frames into grayscale image sequences, and then a differential motion analysis method was applied for extracting motion magnitude from the image sequences. The conversion of a sequence of 2-D images into a single time series is carried out by extracting the intensity difference between pairs of the image frames, which works as follows.

A difference image $\delta(i, j)$ is a binary image, where nonzero values represent image areas with motion, and are defined as [18]

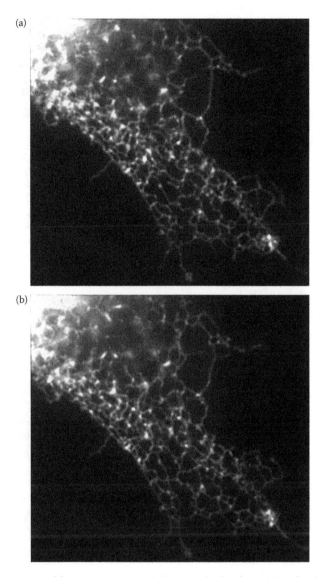

FIGURE 16.1 Typical images of dataset #1: ER network shown in the first frame (a), and 10th frame (b) of the time-lapse microscopy video. The top-left corner of each image frame is the region of the ER network being adjacent to the nucleus.

$$\delta(i,j) = 0 \quad \text{if } |f_1(i,j) - f_2(i,j)| \leq \epsilon$$
$$= 1 \quad \text{otherwise,} \tag{16.1}$$

where ϵ is a small positive number; and $f_1(i,j)$ and $f_2(i,j)$ are the intensity values of image frames 1 and 2 at location (i,j), respectively. In this chapter, the value of ϵ is set to be zero due to the constant illumination of the microscopy image sequence.

The total motion magnitude ϕ aggregates the difference over the whole pair of image frames

$$\phi = \sum_{i=1}^{R} \sum_{j=1}^{C} \delta(i,j), \tag{16.2}$$

where $R \times C$ is the image size.

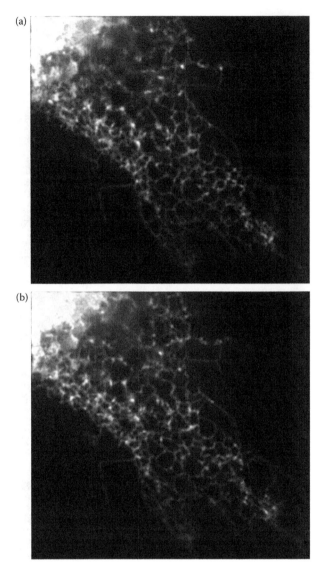

FIGURE 16.2   Typical images of dataset #2: ER network shown in the first frame (a), and 10th frame (b) of the time-lapse microscopy video. The top-left corner of each image frame is the region of the ER network being adjacent to the nucleus.

16.3.2 The Largest Lyapunov Exponent

One of the most well-known methods for quantitative measures of chaos is the largest Lyapunov exponent (LLE) [19,20]. The value of the LLE can be negative, zero, or positive. A negative LLE indicates the convergence of two trajectories. The LLE of zero means on the average rate, the two trajectories keep at the same distance from each other. A positive LLE implies the divergence of the trajectories, which is an indicator of chaos. In other words, a positive LLE expresses sensitive dependence on initial conditions by presenting the average rate over the whole attractor, at which two nearby trajectories become exponentially separate with evolution. A practical numerical technique for calculating the LLE is the method developed by Rosenstein et al. [13], which works well with small datasets; and is robust to changes in the embedding dimension, reconstruction delay, and noise level. Therefore, this method was adopted in this chapter to calculate the LLE of ER dynamics and ER–mitochondrial contacts, and is briefly described as follows.

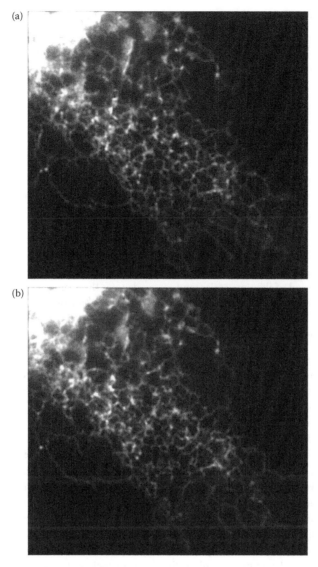

FIGURE 16.3 Typical images of dataset #3: ER network shown in the first frame (a), and 10th frame (b) of the time-lapse microscopy video. The top-left corner of each image frame is the region of the ER network being adjacent to the nucleus.

Given a sequence of motion magnitudes of length N, the first step is to reconstruct the phase space of the dynamical system. Let m and L be the embedding dimension and time delay (lag). The reconstructed phase space can be expressed in matrix form as

$$\mathbf{X} = (\mathbf{X}_1, \mathbf{X}_2, \ldots, \mathbf{X}_M)^T, \tag{16.3}$$

where $\mathbf{X}$ is matrix of size $M \times m$, $M = N - (m-1)L$, and $\mathbf{X}_i = (x_i, \ldots, x_{i+(m-1)L})$ that is the state of the system at discrete time i

$$d_j(0) = \min_{\mathbf{X}_{j*}} ||\mathbf{X}_j - \mathbf{X}_{j*}||, \tag{16.4}$$

where $|j - j^*| > MP$ where MP is the mean period that is the reciprocal of the mean frequency of the power spectrum.

TABLE 16.1 Values of the LLE of ER-Network Dynamics in Time Obtained from Three Datasets

Dataset #1	Dataset #2	Dataset #3
0.484	0.359	0.363

The basic idea is that the LLE (λ_1) for a dynamical system can be defined as [13]

$$d(t) = ce^{\lambda_1 t}, \tag{16.5}$$

where $d(t)$ is the average divergence of two randomly chosen initial conditions at time t, and c is a constant that normalizes the initial separation between neighboring points.

After some mathematical arrangements, the LLE can be computed as the slope of a straight-line fit to the logarithmic average divergence curve defined by [13]

$$s(i) = \frac{1}{i\Delta t} < \ln[d_j(i)] >_j, \tag{16.6}$$

where $d_j(i)$ is the distance between the j pair of nearest neighbors after i discrete-time steps, which is $i\Delta t$, Δt is the sampling period of the time series, and $< \cdot >_j$ denotes the average over all values of j, which is essential to accurately estimate λ_1 using small, noisy data (Figures 16.1 through 16.3).

16.3.3 Experimental Results and Discussion

The embedding dimension of 3, and time delay of 1 were used in all three time-lapse microscopy imaging datasets for the reconstruction of the phase space of the time series, and the calculation of the LLE. Figures 16.4 through 16.6 show the sequences of motion magnitudes of the ER-network, three-dimensional (3-D) reconstructed phase spaces, and divergence curves used to estimate the LLE values of the ER dynamics using datasets #1–#3, respectively.

The LLE values of the three datasets are presented in Table 16.1, which shows that dataset #1 exhibits the largest degree of chaos in comparison with datasets #2 (the least) and #3. The Lyapunov exponent of a dynamical system is a quantitative measure that characterizes the rate of separation of infinitesimally close trajectories, where two trajectories in a phase space with an initial separation diverge at a rate expressed by Equation 16.5. The mathematical interpretation of this property is known as sensitive dependence. In addition to exhibiting sensitive dependence, chaotic systems have two other implications that chaotic systems are deterministic and nonlinear [21]. These three properties of chaotic systems may relate to the potential role of dynamics. One of the important functions of the structure motion must be able to find optimal coordinates or make a physical link to the other organelles. It is conceivable that the ER has a machinery to test and examine all the possibilities. The observed sensitivity to the initial condition and the nonlinear properties may reflect the property of the ER to explore different spatial conditions. LLE analysis could be helpful for gaining insights into the hidden machineries as the complexity developed along time can be quantified.

16.4 Chaos Analysis of ER-Network Dynamics in Space

16.4.1 Fuzzy K–S Entropy

Metric entropy or K–S entropy, denoted as H_{KS}, measures the entropy difference of a dynamical system as time approaches infinity and bin size ϵ that subdivides the phase space of the system reduces to zero [19,22]:

$$H_{KS} = \lim_{\epsilon \to 0} \lim_{t \to \infty} (H_t - H_{t-1}). \tag{16.7}$$

where H_t is the entropy at time t, and $t-1$ refers to the observation prior to time t.

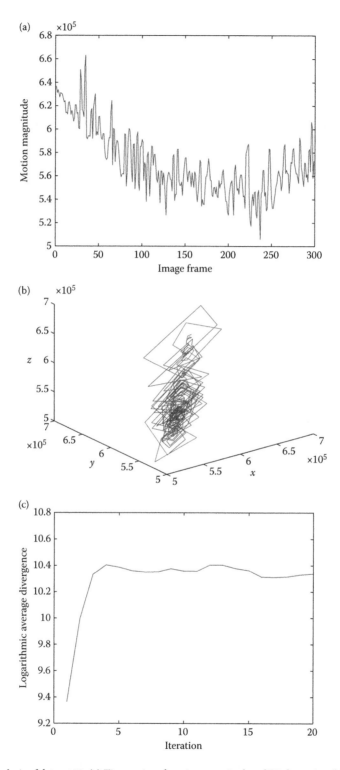

FIGURE 16.4 Analysis of dataset #1: (a) Time series of motion magnitudes of ER dynamics; (b) 3-D reconstructed phase space, where *x*, *y*, and *z* are motion-magnitude coordinates; and (c) LLE estimated as a slope of a straight-line fit to a linear segment at the beginning of the curve.

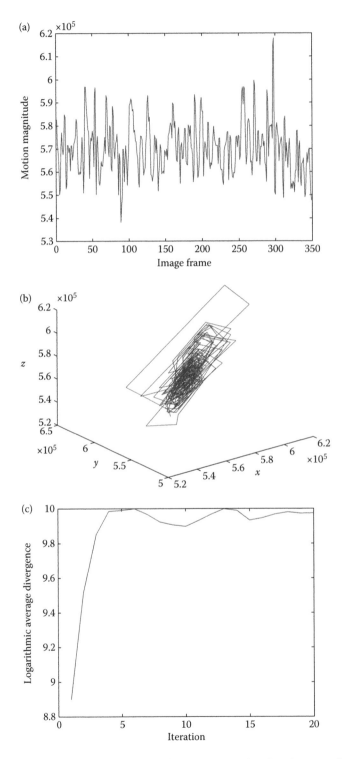

FIGURE 16.5 Analysis of dataset #2: (a) Time series of motion magnitudes of ER dynamics; (b) 3-D reconstructed phase space, where x, y, and z are motion-magnitude coordinates; and (c) LLE estimated as a slope of a straight-line fit to a linear segment at the beginning of the curve.

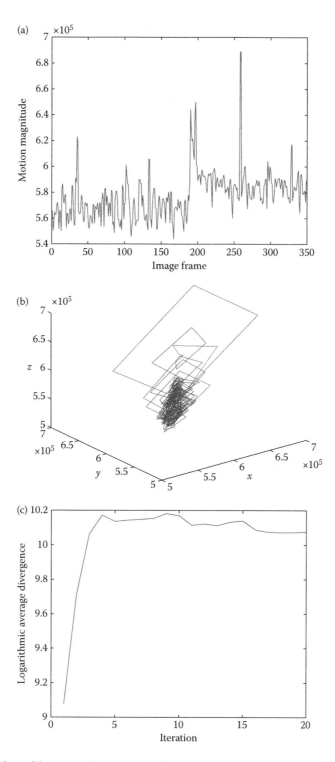

FIGURE 16.6 Analysis of dataset #3: (a) Time series of motion magnitudes of ER dynamics; (b) 3-D reconstructed phase space, where *x*, *y*, and *z* are motion-magnitude coordinates; and (c) LLE estimated as a slope of a straight-line fit to a linear segment at the beginning of the curve.

Let $X = \{x\}$ be a collection of points. A fuzzy set $A \in X$ is characterized by a membership function $\mu_A(x)$, which maps $x \in X$ to a real number in the interval $[0, 1]$ [23]. The entropy of the fuzzy set A, denoted by $D(A)$, is a measure of the degree of its fuzziness, which has the following three properties [24]:

1. $D(A) = 0$ if $\mu_A(x) = 0$ or 1 (nonfuzzy).
2. $D(A)$ is maximum if and only if $\mu_A(x) = 0.5$ (most fuzzy).
3. $D(A) \geq D(A^*)$ where $\mu_A^*(x)$ is any "sharpened" (less fuzzy) version of $\mu_A(x)$, such that $\mu_A^*(x) \geq \mu_A(x)$ if $\mu_A(x) \geq 0.5$ and $\mu_A^*(x) \leq \mu_A(x)$ if $\mu_A(x) \leq 0.5$.

On the basis of the definition of the entropy of a fuzzy set, the uncertainty of a fuzzy system in the context of the K–S entropy measured with a sequence of observations after m units of time can be defined as

$$D_m = \sum_{i=1}^{N_m} F_i, \tag{16.8}$$

where F_i is the Shannon entropy

$$F_i = -\mu_i \log(\mu_i) - (1 - \mu_i) \log(1 - \mu_i) \tag{16.9}$$

in which $\mu_i \in [0, 1]$ is the fuzzy membership grade of a trajectory in the ith cell (it is not necessary that $\sum_i \mu_i = 1$ because the notion of a fuzzy set is nonstatistical in nature).

On the basis of Equation 16.7, the K–S entropy of fuzzy sets is expressed as [15]

$$G_{KS} = \lim_{\epsilon \to 0} \lim_{t \to \infty} (G_t - G_{t-1}). \tag{16.10}$$

On the sequence probabilities for calculating H_{KS}, which can be of any type of probability and are the likelihoods that the system will follow each of the various possible routes, for successive time intervals; these probabilities are computed using the product rule. For the sequence fuzzy membership values associating with Equation 16.9, which are the degrees expressing possible routes the system will pass through, the product rule for combining conditional probabilistic events is replaced with the intersection of fuzzy sets. Let A and B be two fuzzy sets, each of which associates with each x in a Euclidean n-space R^n. The intersection of A and B is defined as [23]

$$A \cap B \Leftrightarrow \mu_{A \cap B}(x) = \min[\mu_A(x), \mu_B(x)] \forall x. \tag{16.11}$$

Two alternatives have been suggested to get the inner limit of Equation 16.7 [19], or similarly to get the inner limit of Equation 16.10, that is, $\lim_{t \to \infty}$. As for the K–S entropy of fuzzy sets, the first alternative is to divide each entropy by the associated time to get an entropy rate, and the plot of the entropy rates versus time allows the estimate of the asymptotic entropy rate as time increases. However, this procedure can be impractical because the data must be large enough for the time events to reach the asymptotic limit or convergence. The other alternative resorts to an entropy difference [25], denoted by Δ_m, as follows:

$$\Delta_m = D_{m+1} - D_m. \tag{16.12}$$

It has been shown, when plotted against time, that both the average entropy rate and the entropy difference reach the same asymptotic value of the average entropy rate; but the plot of Δ_m versus time is the better approximation and converges sooner [25]. Therefore, the use of the entropy difference is more favorable because it can alleviate the demand of a large dataset and consequently save computational effort. The mathematical expression for the K–S entropy of fuzzy sets estimated by the method of the entropy difference becomes

$$G_{KS} = \lim_{\epsilon \to 0} \lim_{m \to \infty} \Delta_m. \tag{16.13}$$

The procedure for calculating the K–S entropy of fuzzy sets using Equation 16.13 is outlined as follows:

Estimating Fuzzy K–S Entropy

1. Select a bin size for the partition of the phase space.
2. For a given time step, estimate sequence fuzzy membership grades for all possible routes, and compute D_m.
3. Repeat Step 2 for the next time step to obtain D_{m+1}.
4. Compute the entropy difference Δ_m.
5. Repeat Steps 2–4 for higher successive times.
6. Plot the entropy differences versus the time/space steps and observe the graphical behavior, particularly if the curve converges to some constant.

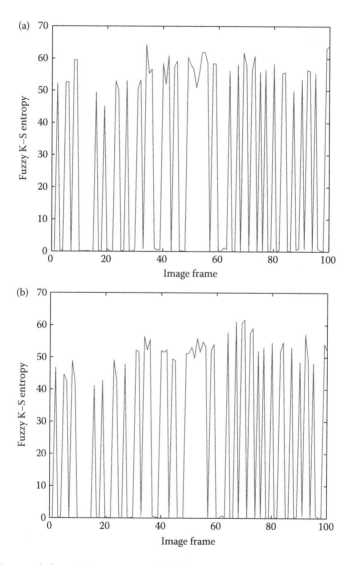

FIGURE 16.7 ER-network dynamics in space using the first 100-image frames of dataset #1: (a) fuzzy K–S entropy in the row-wise direction, and (b) fuzzy K–S entropy in the row-wise direction.

16.4.2 Measure of Chaos in the ER Network by Fuzzy K–S Entropy for Images

Sequence fuzzy membership grades required for the calculation of the fuzzy K–S entropy to quantify uncertainty in an image, which is inherently due to the imprecise description of the image content, can be obtained by the partition of the image intensities using the fuzzy c-means (FCM) algorithm [26]. Thus, given fuzzy c-partitions, the FCM assigns the intensity value of each pixel to the c clusters with its respective membership grades.

In the context of cluster analysis of a grayscale image, let $\mathbf{x} = \{x_1, x_2, \ldots, x_n\}$ be a set of pixel intensity values. The FCM algorithm seeks to minimize the following fuzzy objective function J_F [26]:

$$J_F(\mathbf{U}, \mathbf{v}) = \sum_{i=1}^{n} \sum_{j=1}^{c} (\mu_{ij})^q \, [d(x_i, v_j)]^2, \tag{16.14}$$

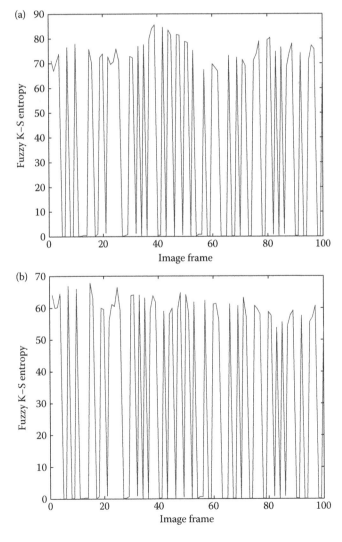

FIGURE 16.8 ER-network dynamics in space using the first 100-image frames of dataset #2: (a) fuzzy K–S entropy in the row-wise direction, and (b) fuzzy K–S entropy in the row-wise direction.

where c is the number of clusters, $1 < c < n$, $q \in [1, \infty)$ is the fuzzy-weighting exponent, $\mathbf{U}$ is the matrix representation of the fuzzy c-partition of $\mathbf{x}$, $\mathbf{v} = (v_1, v_2, \dots, v_c)$ is the vector of cluster centers, v_j is the center of cluster j, and $d(x_i, v_j)$ is any inner-product-induced norm metric.

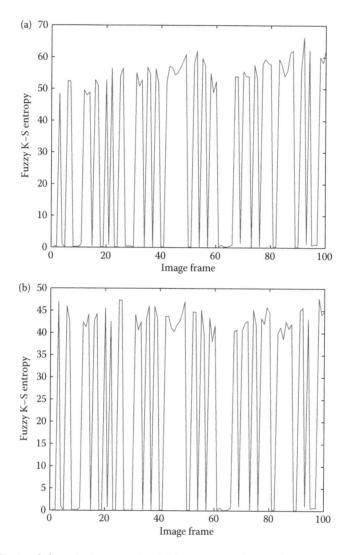

FIGURE 16.9 ER-network dynamics in space using the first 100-image frames of dataset #3: (a) fuzzy K–S entropy in the row-wise direction, and (b) fuzzy K–S entropy in the row-wise direction.

TABLE 16.2 Values of Fuzzy K–S Entropy (Mean, Standard Deviation) of ER-Network Dynamics in Space Obtained from First 100 Image Frames of Three Datasets

Orientation	Dataset #1	Dataset #2	Dataset #3
Row wise	(27.204, 28.301)	(39.829, 37.527)	(31.913, 27.666)
Column wise	(24.919, 26.103)	(32.143, 30.301)	(24.756, 21.458)

The fuzzy objective function expressed in Equation 16.14 is subject to the following constraints:

$$\sum_{j=1}^{c} \mu_{ij} = 1, i = 1, \dots, n,$$

(16.15)

where

$$\mu_{ij} \in [0,1], i = 1, \dots, n, j = 1, \dots, c.$$

The objective function $J_F(\mathbf{U}, \mathbf{v})$ is a squared error-clustering criterion and is to be minimized to optimally determine $\mathbf{U}$, and $\mathbf{v}$. A solution to the minimization of the objective function is by a process of iteratively updating $\mathbf{U}$ and $\mathbf{v}$ until some convergence is reached [26]. The fuzzy membership grades and

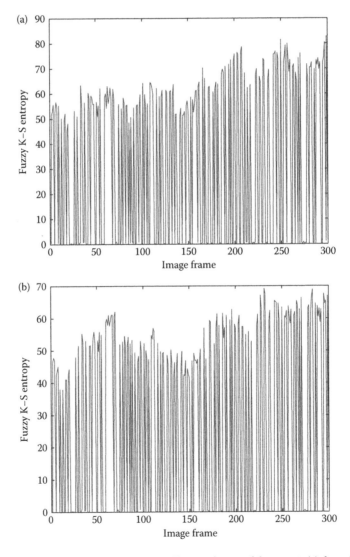

FIGURE 16.10 ER-network dynamics in space using all image frames of dataset #1: (a) fuzzy K–S entropy in the row-wise direction, and (b) fuzzy K–S entropy in the row-wise direction.

cluster centers of the FCM are updated as

$$\mu_{ij} = \frac{1}{\sum_{j=1}^{c} \left[\frac{d(x_i, v_u)}{d(x_i, v_j)} \right]^{2/(q-1)}}, \; 1 \le u \le c; \tag{16.16}$$

$$v_j = \frac{\sum_{i=1}^{n} (\mu_{ij})^q \, x_i}{\sum_{i=1}^{n} (\mu_{ij})^q}, \; \forall j. \tag{16.17}$$

The selection of the size of c used in the FCM is equivalent to Step 1 of the procedure for estimating the K–S entropy of fuzzy sets. To construct the sequence membership grades of an image, the FCM partition is performed with a given c. The sequence fuzzy membership grades at each route at $m + 1$ are computed by taking the minimum values of the fuzzy membership grades of the corresponding gray levels determined by the FCM at m and $m + 1$ (Step 2), based on which the entropy difference Δ_m can be calculated (Step 3). A practical way is to calculate the entropy difference defined in Equation 16.12 using the fuzzy membership

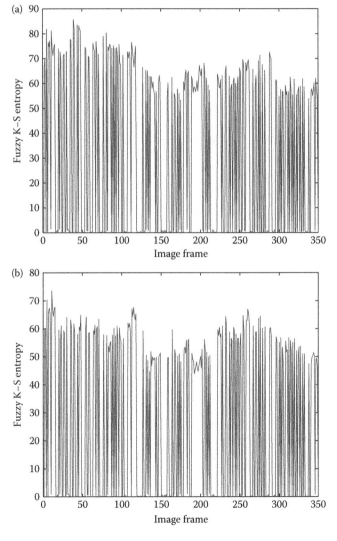

FIGURE 16.11 ER-network dynamics in space using all image frames of dataset #2: (a) fuzzy K–S entropy in the row-wise direction, and (b) fuzzy K–S entropy in the row-wise direction.

grades for each cluster obtained from the FCM in any forward orientation of the image. Typical orientations are the vertical and horizontal directions of the image, which refer to the entropy differences in the image rows and columns, respectively. The plot of the sum of all the entropy differences in the image rows or image columns against the corresponding image rows or image columns, which are the iterative steps, allows the estimate of G_{KS} when a plateau (indicating chaos) is visible on the curve.

16.4.3 Experimental Results and Discussion

Being analogous to the fuzzy K–S analysis of biological aging images [15], it can be seen that each frame of the three time-lapse microscopy image datasets appears to have two fuzzy clusters of bright and dark pixels, referring to the corresponding intensities of object and background of the image; therefore, for the FCM, two clusters, $c=2$, was chosen to partition the foreground (object) and background pixels, and the selection for the weighting exponent $q=2$ was based on the suggestion of the study in Reference 27.

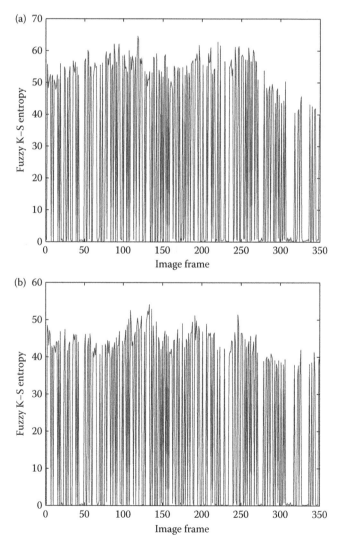

FIGURE 16.12 ER-network dynamics in space using all frames of dataset #3: (a) fuzzy K–S entropy in the row-wise direction, and (b) fuzzy K–S entropy in the row-wise direction.

TABLE 16.3 Values of Fuzzy K–S Entropy (Mean, Standard Deviation) of ER-Network Dynamics in Space Obtained from All Image Frames of Three Datasets

Orientation	Dataset #1	Dataset #2	Dataset #3
Row wise	(31.820, 32.002)	(32.345, 33.217)	(26.861, 27.029)
Column wise	(27.673, 27.893)	(27.670, 28.430)	(21.930, 22.115)

The selection of two fuzzy clusters ($c=2$) results in the fuzzy K–S entropy vector of four elements, two for each cluster in the row and column orientations of the image.

To gain an insight into the morphological dynamics of the ER network over time, we applied the fuzzy K–S entropy to measure the potential chaos using the first 100 frames, and then all frames of the three datasets. Figures 16.7 through 16.9 show the values of the fuzzy K–S entropy of the ER dynamics in row-wise and column-wise directions for the first 100 image frames of datasets #1–#3, respectively. Datasets #1 and #3 have similar average fluctuation patterns of the fuzzy K–S entropy. The means and standard deviations of the fuzzy K–S entropy of the ER network in space using the partial datasets are shown in Table 16.2. These fuzzy K–S entropy means suggest that the dynamic morphological patterns of the ER network in datasets #1 and #3 are similar in both row-wise and column-wise directions, whereas the spatial structure of the ER network in dataset #2 is most unpredictable.

Figures 16.10 through 16.12 show the values of the fuzzy K–S entropy of the ER dynamics in row-wise and column-wise directions for all the image frames of datasets #1 through #3, respectively. In this all-frame analysis, datasets #1 and #2 have similar higher average fluctuation patterns of the fuzzy K–S entropy in both row-wise and column-wise directions. The means and standard deviations of the fuzzy K–S entropy of the ER network in space using full datasets are shown in Table 16.3. These fuzzy K–S entropy means now suggest that the dynamic morphological pattern of the ER network in dataset #3 is more predictable than datasets #1 and #2. The test result provides a way to obtain information about anisotropy of the structural dynamics, and the current analysis implicates that the reactions causing the motion of the structure were not polarized and occurred throughout the ER in a way that is difficult to predict.

16.5 Conclusion

The results reported in this chapter further confirm that the motion of the ER network is deterministic and nonlinear. The predictability of the ER dynamics is subject to chaos. Although we did not apply this method with the segmentation of the ER network, further chaos analysis using segmentation can provide more accurate results, particularly in case of a noisy background. This chaos analysis of the ER has a unique implication as a computational tool for studying dynamics in time-lapse images, while other methods are not ready to handle. Furthermore, the presented approach is applicable to a variety of microscopic image analysis including drug effects. This is particularly useful for the large collection of chemical compounds to identify potential drug candidates [28]. The methods for chaos analysis discussed in this chapter can also be further applied for studying motion in molecular imaging to advance its applications in several fields of medicine and biotechnology.

References

1. Buchberger A, Bukau B, Sommer T. Protein quality control in the cytosol and the endoplasmic reticulum: Brothers in arms. *Mol Cell* 2010; 40: 238–252.
2. Ruggiano A, Foresti O, Carvalho P. ER-associated degradation: Protein quality control and beyond. *J Cell Biol* 2014; 204: 869–879.
3. Rowland AA, Chitwood PJ, Phillips MJ, Voeltz GK. ER contact sites define the position and timing of endosome fission. *Cell* 2014; 159: 1027–1041.

4. Rowland AA, Voeltz GK. Endoplasmic reticulum–mitochondria contacts: Function of the junction. *Nat Rev Mol Cell Biol* 2012; 13: 607–625.

5. Shibata Y, Shemesh T, Prinz WA, Palazzo AF, Kozlov MM, Rapoport TA. Mechanisms determining the morphology of the peripheral ER. *Cell* 2010; 143: 774–788.

6. Friedman JR, Voeltz GK. The ER in 3D: A multifunctional dynamic membrane network. *Trends Cell Biol* 2011; 21: 709–717.

7. Hu J, Shibata Y, Zhu PP, Voss C, Rismanchi N, Prinz WA, Rapoport TA, Blackstone C. A class of dynamin-like GTPases involved in the generation of the tubular ER network. *Cell* 2009; 138: 549–561.

8. Orso G, Pendin D, Liu S, Tosetto J, Moss TJ, Faust JE, Micaroni M et al. Homotypic fusion of ER membranes requires the dynamin-like GTPase atlastin. *Nature* 2009; 460: 978–983.

9. Joensuu M, Belevich I, Ramo O, Nevzorov I, Vihinen H, Puhka M, Witkos TM, Lowe M, Vartiainen MK, Jokitalo E. ER sheet persistence is coupled to myosin 1c-regulated dynamic actin filament arrays. *Mol Biol Cell* 2014; 25: 1111–1126.

10. Salinas S, Proukakis C, Crosby A, Warner TT. Hereditary spastic paraplegia: Clinical features and pathogenetic mechanisms. *Lancet Neurol* 2008; 7: 1127–1138.

11. Pham TD. The butterfly effect in ER dynamics and ER–mitochondrial contacts. *Chaos Solitons Fractals* 2014; 65: 5–19.

12. Friedman JR, Webster BM, Mastronarde DN, Verhey KJ, Voeltz GK. ER sliding dynamics and ER–mitochondrial contacts occur on acetylated microtubules. *J Cell Biol* 2010; 190: 363–375.

13. Rosenstein MT, Collins JJ, DeLuca CJ. A practical method for calculating largest Lyapunov exponents from small data sets. *Physica D: Nonlinear Phen* 1993; 65: 117–134.

14. Pham TD, Ichikawa K. Spatial chaos and complexity in the intracellular space of cancer and normal cells. *Theor Biol Med Model* 2013; 10:62. Doi:10.1186/1742-4682-10-62.

15. Pham TD. Classification of complex biological aging images using fuzzy Kolmogorov–Sinai entropy. *J Phys D: Appl Phys* 2014; 47: 485402 (12pp).

16. Suzuki T, Arai S, Takeuchi M, Sakurai C, Ebana H, Higashi T, Hashimoto H, Hatsuzawa K, Wada I. Development of cysteine-free fluorescent proteins for the oxidative environment. *PLoS ONE* 2012; 7(5): e37551. Doi: 10.1371/journal.pone.0037551.

17. Sakurai C, Hashimoto H, Nakanishi H, Arai S, Wada Y, Sun-Wada GH, Wada I, Hatsuzawa K. SNAP-23 regulates phagosome formation and maturation in macrophages. *Mol Biol Cell* 2012; 23: 4849–4863.

18. Sonka M, Hlavac V, Boyle R. *Image Processing, Analysis, and Machine Vision*. Pacific Grove: PWS Publishing; 1999.

19. Williams GP. *Chaos Theory Tamed*. Washington DC: Joseph Henry Press; 1997.

20. Dingwell JB. Lyapunov exponents, in: Metin A, e.d., *Wiley Encyclopedia of Biomedical Engineering*. New York: John Wiley & Sons, 12pp.; 2006.

21. Smith LA. *Chaos: A Very Short Introduction*. Oxford: Oxford University Press; 2007.

22. Hilborn RC. *Chaos and Nonlinear Dynamics*. New York: Oxford University Press; 2000.

23. Zadeh LA. Fuzzy sets. *Inf Control* 1965; 8: 338–353.

24. De Luca A, Termini S. A definition of a nonprobabilistic entropy in the setting of fuzzy sets theory. *Inf Control* 1972; 20: 301–312.

25. Shannon CE, Weaver W. *The Mathematical Theory of Communication*. Urbana: University of Illinois Press; 1949.

26. Bezdek JC. *Pattern Recognition with Fuzzy Objective Function Algorithms*. New York: Plenum Press; 1981.

27. Pal NR, Bezdek JC. On cluster validity for the fuzzy c-means model. *IEEE Trans Fuzzy Syst* 1995; 3: 370–379.

28. Gong Z, Zhao H, Zhang T, Nie F, Pathak P, Cui K, Wang Z, Wong S, Que L. Drug effects analysis on cells using a high throughput microfluidic chip. *Biomed Microdevices* 2011; 13: 215–219.

Supersymmetric Theory of Stochastics: Demystification of Self-Organized Criticality

Igor V. Ovchinnikov

17.1 Introduction

Among all the classes of mathematical models, stochastic (partial) (integro-) differential equations (SDEs) have the widest applicability in modern science. For example, SDEs describe all natural dynamical systems (DSs) above the scale of quantum degeneracy/coherence because (i) time is continuous in nature (above the Planck scale) and (ii) no natural DS can be completely isolated from the stochastic influence of its environment. In quantum models, on the other hand, SDEs are often used as additional tools.

The theory of SDEs [26] and random DSs [1] has a long history. Many important insights on stochastic dynamics have been provided so far [5,16], including those on the interplay between chaos and noise [18]. Nevertheless, some of the most fundamental questions remained unanswered. For example, there was no theoretical explanation for the ubiquitous chaotic or dynamical long-range order (DLRO) that reveals itself through such well-established phenomena as 1/f or pink noise [19] and power-law statistics of avalanche-type processes in astrophysics [2], geophysics [15], neurodynamics [6,7,9,21], econodynamics [34], and many other branches of modern science.

This conspicuous gap in our theoretical understanding of natural dynamics was partly covered by phenomenology: the DLRO itself had quite a few names, including dynamical complexity [36] and self-organization [3], whereas the associated emergence of the scale-free behavior was understood via the concept of self-organized criticality (SOC)—the proclaimed *mysterious* tendency of some stochastic DSs to fine-tune themselves into the phase transition into chaos [4].

The picture of chaotic dynamics was significantly more clear in the domain of ordinary (or partial) differential equations (ODEs) [25]. There, the deterministic chaos had a rigorous mathematical definition of nonintegrability in the sense of DS theory.[*] This definition, however, did not explain the most celebrated chaotic property of the highest sensitivity to perturbations and/or initial conditions known as the butterfly effect [25]. As a result, the butterfly effect was often viewed as a necessary part of the definition of chaos or even as its alternative definition. The necessity to introduce yet another definition of chaos in order to explain its most celebrated property reveals that our previous understanding of deterministic chaos was incomplete, let alone that of stochastic chaos.

A breakthrough understanding of the theoretical essence of stochastic (and deterministic) chaos came from the theory of supersymmetric field theories. Soon after the proposition of the Parisi–Sourlas [32, 33] stochastic quantization procedure for Langevin SDEs, it was realized [13] that general form SDEs possess at least one supersymmetry, unlike Langevin SDEs and classical mechanics [10] that have two supersymmetries. This type of supersymmetry was later identified as a definitive feature of Witten-type topological or cohomological field theories [8,12,20,37–40]. Yet another important piece of the puzzle was the development of the theory of pseudo-Hermitian evolution operators [24], because the stochastic evolution operator of a general form SDE is pseudo-Hermitian. At last, in Reference 27, it was hypothesized that stochastic chaos, or rather SOC, could as well be the result of the spontaneous breakdown of this topological supersymmetry that all SDEs possess. Further work in this direction resulted [28–31] in the formulation of the approximation-free and coordinate-free supersymmetric theory of SDEs or stochastics (STS). The first part of this chapter (Sections 17.3, 17.4, and 17.5) contains a brief and self-consistent exposition of the current state of the STS.

From the mathematical point of view, the most interesting results that will be discussed is the demonstration that the sharp-trace of the generalized transfer operator (GTO) of the DS theory, the Witten index, and the stochastic Lefschetz index are essentially the same object representing up to a topological factor[†] the partition function of the stochastic noise.

From the physical point of view, the most valuable results addressed in this chapter are (iii) identification of stochastic/deterministic chaos and thermodynamic equilibrium (TE) as complementary situations with respectively spontaneously broken and unbroken topological supersymmetry and (iv) theoretical explanation of "DLRO" and its signatures such as the butterfly effect.

In the last section of this chapter (Section 17.6), the STS will be used to propose the general phase diagram of stochastic DSs. It will be discussed that one of the types of stochastic dynamics is the low-temperature noise-induced (or intermittent) chaos. This is the famous stochastic dynamical behavior on the "border of ordinary chaos" featured by the power-law or algebraic statistics of instantonic (e.g., avalanche-type) processes. For nonzero temperatures, the noise-induced chaos has a finite "width" so that there is no need for phenomenonlogical approaches such as SOC to explain the fact that the algebraic statistics of avalanches is observed for a whole range of model's parameters.

---

[*] Within this definition, global unstable manifolds of the integrable flow vector fields provide well defined foliations of the phase space, whereas those of nonintegrable or chaotic flows do not.

[†] This topological factor equals the Euler characteristic of the phase space for a wide range of models, including models with closed phase spaces.

In concluding Section 17.7, we briefly discuss the potentially fruitful directions of further work such as the methodology of the low-energy effective theories for chaotic stochastic DSs.

17.2 Models of Interest

The following general class of SDEs covers most of the models in the literature:[*]

$$\dot{x}(t) = F(x(t)) + (2T)^{1/2} e_a(x(t)) \xi^a(t) \equiv \tilde{F}(t). \tag{17.1}$$

Here, $x \in X$ is a point from a D-dimensional topological differentiable manifold called phase space, X; $F(x) \in TX_x$ is the flow vector field from the tangent space of X at point x; $\xi^a \in \mathbb{R}^1, a = 1, 2 \ldots$ are variables of the noise, T is the noise intensity or temperature; $e_a(x) \in TX_x, a = 1, 2 \ldots$ is a set of vector fields. In case when the number of e's equals the dimensionality of X and $g^{ij}(x) = e_a^i(x)e_a^j(x)$ is nondegenerate, g^{ij} and e's have the meanings of the noise-induced metric and veilbeins, respectively. The position-dependent/independent e's are often called multiplicative/additive noise.

In most cases, the noise will be assumed Gaussian white. The probability of realization is

$$P(\xi) = Ce^{-\int dt(\xi^a(t)\xi^a(t))/2}, \tag{17.2}$$

with C being the normalization constant such that

$$\langle 1 \rangle_{\mathrm{Ns}} \equiv \iint D\xi \cdot 1 \cdot P(\xi) = 1, \tag{17.3}$$

where the functional integration is over all the noise configurations. Stochastic expectation value of some functional $f(\xi)$ is defined as

$$\langle f(\xi) \rangle_{\mathrm{Ns}} \equiv \iint D\xi f(\xi) P(\xi). \tag{17.4}$$

For example,

$$\langle \xi^a(t) \xi^b(t') \rangle_{\mathrm{Ns}} = \delta^{ab} \delta(t - t'). \tag{17.5}$$

The case of white noise is important from the mathematical point of view as it is the domain of the Ito–Stratonovich dilemma. We will address this problem in Section 17.4.1. In Section 17.5, turning to the pathintegral representation of the theory will allow for the generalization from the Gaussian white noise to noises of any form. Furthermore, in Section 17.5.2, the generalization to higher-dimensional continuous-space models will be briefly discussed.

Stochastic dynamics is intrinsically probabilistic and thus must be described in terms of temporal evolution of generalized probability distributions (GPDs). Our approach to the derivation of the law of the temporal evolution of the GPDs is as follows. First, in the next section, we will derive the law of the deterministic temporal evolution of the GPDs defined by the unambiguous resolution of ODE obtained from Equation 17.1 by fixing the configuration of the noise. The stochastic evolution will be addressed in Section 17.4.

---

[*] Summation over the repeated indices is assumed throughout this chapter.

17.3 Deterministic Dynamics

17.3.1 Temporal Evolution as Maps

For a fixed configuration of noise, Equation 17.1 is an ODE with time-dependent right-hand side (r.h.s.). This ODE defines a family of maps of a phase space onto itself, $M_{tt'} : X \to X$,

$$M_{tt'} : x' \mapsto x = M_{tt'}(x'). \tag{17.6}$$

These maps have straightforward interpretation: $M_{tt'}(x')$ is the solution of the ODE with x' being the initial condition at time moment t'. Obviously,

$$M_{tt} = \mathrm{Id}_X, M_{tt'} \circ M_{t't''} = M_{tt''}, \text{ and } M_{t't} = M_{tt'}^{-1}. \tag{17.7}$$

We consider only physical models so that maps are invertible (for finite-time propagation time). On the mathematical level, this means that the r.h.s. of the ODE is sufficiently smooth and, in particular, the Picard–Lindelöf theorem about the existence and uniqueness of the solution of the ODE for any initial condition is applicable.

Let us assume now that at time t', the DS is described by a total probability function $P(x)$, so that the expectation value of some function $f(x)$ is

$$\bar{f}(t') = \int_X f(x)P(x)dx^1 \ldots dx^D. \tag{17.8}$$

According to Equation 17.6, the same expectation value at later time moment, $t > t'$, becomes

$$\bar{f}(t) = \int_X f(M_{tt'}(x))P(x)dx^1 \ldots dx^D = \int_X f(x)M_{t't}^*(P(x)dx^1 \ldots dx^D). \tag{17.9}$$

Here, we made the transformation of the dummy variable of integration, $x \to M_{t't}(x)$, and $M_{t't}^*$ is the operation of the variable transformation applied to the coordinate-free object comprising of $P(x)$ and the collection of all the differentials $dx^1 \ldots dx^D$:

$$M_{t't}^*(P(x)dx^1 \ldots dx^D) = P\left(M_{t't}(x)\right) J(TM_{t't}(x))dx^1 \ldots dx^D, \tag{17.10}$$

where J is the Jacobian of the tangent map, $TM_{t't}(x) : TX_x \to TX_{M_{t't}(x)}$,

$$TM_{t't}(x) : dx^i \mapsto d(M_{t't}(x))^i = TM_{t't}(x)_k^i dx^k, \tag{17.11}$$

with

$$TM_{t't}(x)_k^i = \partial(M_{t't}(x))^i / \partial x^k, \tag{17.12}$$

being the coordinate representation of the tangent map.

Equation 17.9 tells us that forward temporal propagation of variables of the DS is equivalent to backward temporal propagation of the "coordinate-free" total probability distribution (TPD), $P(x)dx^1 \ldots dx^D$. In terms of algebraic topology, this coordinate-free object representing the TPD is a top differential form or D-form, whereas the operation $M_{t't}^*$ in Equation 17.10 is the pullback or action induced on D-form by map $M_{t't}$.

17.3.2 Hilbert Space

Conventional description of stochastic DSs in terms of TPDs only as in previous section is insufficient in general case. This can be seen from the following qualitative example. Consider the simplest Langevin SDE with $X = \mathbb{R}^1$ (see Section 17.6.2 for details). Consider also the case of a "stable" Langevin potential (see Figure 17.1a). It is clear that after long-enough temporal evolution, this DS will forget its initial condition

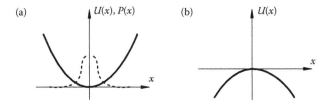

FIGURE 17.1 One-variable Langevin SDE with stable (a) and unstable (b) potential. For the stable case, the DS eventually forgets its initial condition and the ground state is the steady-state probability distribution of x, $P(x)$ (dotted bell-shaped curve). In case of unstable potential, the position is highly sensitive to its initial condition that the DS never forgets. No meaningful steady-state probability distribution can be associated with x in this case.

and its (only) variable will be distributed according to some steady-state TPD, which is the ground state of the model in this case. On the contrary, when the Langevin potential is unstable (Figure 17.1b), the DS will never forget its initial condition[*] and no meaningful steady-state TPD can be prescribed to its unstable variable. This example suggests that meaningful steady-state probability distribution that forgets initial condition can be associated only with a stable variable.

Previous example of unstable dynamics is not quite physical as for all initial conditions the DS escapes to infinity and does not come back. A better example for our purposes is the (deterministic) chaotic DSs, in which unstable variables revealed by positive (global) Lyapunov exponents exist even after infinitely long temporal evolution, that is, in the ground state of the DS. Such chaotic ground states must not be (probability) distributions in unstable variables (see Section 17.6.2 below). The DSs theory predecessors of such ground states are the Sinai–Ruelle–Bowen conditional probability functions on unstable manifolds [11].

In Section 17.4.5, it will be seen on a rigorous level that the extension of the Hilbert space[†] from the conventional description in terms of only TPDs to the description in terms of GPD is actually a necessity. This by no means contradicts the intuitive understanding that it must be possible to associate a TPD with any wavefunction. Just like in quantum theory, the TPD associated with a wavefunction is not the wavefunction itself but rather the bra–ket combination. It will be clear below that the bra–ket combination of any eigenstate of a stochastic DS is a D-form representing the TPD associated with it (see, e.g., Section 17.5.4 and the discussion following Equations 17.52 and 17.53).

In the coordinate-free setting, the GPDs, including total, conditional, and marginal distributions, are differential forms ($0 \leq k \leq D$):

$$\psi^{(k)} = (1/k!)\psi^{(k)}_{i_1...i_k}dx^{i_1} \wedge ... \wedge dx^{i_k} \in \Omega^k(X). \tag{17.13}$$

Here, $\psi^{(k)}_{i_1...i_k} \equiv \psi^{(k)}_{i_1...i_k}(x)$ is an antisymmetric tensor, $\wedge$ is the wedge or antisymmetrized product of differentials, e.g., $dx^1 \wedge dx^2 = -dx^2 \wedge dx^1 = dx^1 \otimes dx^2 - dx^2 \otimes dx^1$, and $\Omega^k(X)$ is the space of all differential forms of degree k, or k-forms. The space of all differential forms is called the exterior algebra of X, $\Omega(X) = \bigoplus_{k=0}^{D} \Omega^k(X)$.

The following example demonstrates how conditional probability distributions in conventional notations can be represented as a differential form (the dimensionality of X is $D = 3$),

$$\psi^{(2)} = (1/2!)\psi^{(2)}_{i_1 i_2}dx^{i_1} \wedge dx^{i_2}$$
$$= P(x^2 x^3 | x^1)dx^2 \wedge dx^3 + P(x^1 x^3 | x^2)dx^3 \wedge dx^1 + P(x^1 x^2 | x^3)dx^1 \wedge dx^2 \in \Omega^2(X),$$

---

[*] In deterministic case, the small difference in two close initial conditions will grow exponentially.

[†] In the DS theory, the concept of Banach space is often used instead of Hilbert space. In this chapter, we will use the terminology of the quantum theory.

where $\psi_{12}^{(2)} = -\psi_{21}^{(2)} = P(x^1 x^2 | x^3)$, $\psi_{31}^{(2)} = -\psi_{13}^{(2)} = P(x^1 x^3 | x^2)$, and $\psi_{23}^{(2)}(x) = -\psi_{32}^{(2)} = P(x^2 x^3 | x^1)$. The D-form of the TPD introduced in the previous section is

$$\psi^{(D)} = (1/D!)\psi_{i_1\ldots i_D}^{(D)} dx^{i_1} \wedge \ldots \wedge dx^{i_D} = P dx^1 \wedge \ldots \wedge dx^D \in \Omega^D(X),$$

where $\psi_{i_1\ldots i_D}^{(D)} = P\epsilon_{i_1\ldots i_D}$ with $\epsilon_{i_1\ldots i_D} = (-1)^{p(i_1\ldots i_D)}$ being the Levi–Civita antisymmetric tensor and with $p(i_1 \ldots i_D)$ being the parity of the permutation of indices.

The geometrical meaning of a k-form is a differential of a k-dimensional "oriented" volume. An integration of a k-form over a k-dimensional submanifold of X, often called a k-chain, c_k, gives a number:

$$\int_{c_k} \psi^{(k)} = p_{c_k} \in \mathbb{R}^1. \tag{17.14}$$

This number can be interpreted as follows. If we introduce local coordinates such that the k-chain belongs to the k-dimensional submanifold cut out by, $(x^{k+1}, \ldots, dx^D) = (\mathrm{Const}^{(k+1)}, \ldots, \mathrm{Const}^{(D)})$, then Equation 17.14 is the probability of finding variables $(x^1, \ldots, x^k)$, within this k-chain given that all the other variables are known with certainty to be equal $(\mathrm{Const}^{(k+1)}, \ldots, \mathrm{Const}^{(D)})$.

To establish the law of the temporal evolution of k-forms, we assume that at time moment, t', the DS is described by $\psi^{(k)}(x)$. By analogy with Equation 17.9, the quantity in Equation 17.14 at later time moment $t > t'$ is

$$p_{c_k}(t) = \int_{M_{tt'}(c_k)} \psi^{(k)}(x) = \int_{c_k} M_{t't}^* \psi^{(k)}(x). \tag{17.15}$$

Here, $M_{t't}^* : \Omega^k(X) \to \Omega^k(X)$ is the generalization of the pullback from Equation 17.10 to pullbacks acting on $\Omega^{(k)}(X)$. Explicitly,

$$M_{t't}^* \psi^{(k)}(x) = (1/k!)\psi_{i_1\ldots i_k}^{(k)}(M_{t't}(x)) d(M_{t't}(x))^{i_1} \wedge \ldots \wedge d(M_{t't}(x))^{i_1}, \tag{17.16}$$

where the k-form is from Equation 17.13 and $d(M_{t't}(x))^i$ is from Equation 17.11.

Pullbacks act in the "opposite" direction as compared to the maps inducing them.[*] This statement can be clarified via the following diagram ($t > t' > t''$):

$$
\begin{array}{ccccc}
t & \xleftarrow{\text{flow of time}} & t' & \xleftarrow{\text{flow of time}} & t'', \\
X & \xrightarrow{M_{t't}} & X' & \xrightarrow{M_{t''t'}} & X'', \\
\Omega(X) & \xleftarrow{M_{t't}^*} & \Omega(X') & \xleftarrow{M_{t''t'}^*} & \Omega(X'').
\end{array}
\tag{17.17}
$$

Accordingly, the composition law for pullbacks is

$$M_{t''t}^* = M_{t't}^* M_{t''t'}^*. \tag{17.18}$$

17.3.3 Operator Algebra

Infinitesimal pullback is known as physical or Lie derivative:

$$\hat{\mathcal{L}}_{\tilde{F}} \psi^{(k)}(x) = \lim_{\Delta t \to 0} \frac{1}{\Delta t}\left(M_{t+\Delta t,t}^* - \hat{1}_{\Omega(X)}\right)\psi^{(k)}(x). \tag{17.19}$$

Using the definition in Equation 17.16, that the infinitesimal map is

$$M_{t+\Delta t,t}(x) \approx x + \Delta t \tilde{F}(t), \tag{17.20}$$

---

[*] This is actually the reason for the term "pullback."

with $\tilde{F}$ being the r.h.s. of Equation 17.1, and that the infinitesimal tangent map in Equation 17.12 is

$$TM_{t+\Delta t,t}(x)^i_k \approx \delta^i_k + \Delta t T\tilde{F}^i_k(t), \tag{17.21}$$

with $T\tilde{F}^i_k = \partial \tilde{F}^i/\partial x^k$, one arrives at the following expression for the Lie derivative:

$$\hat{\mathcal{L}}_{\tilde{F}}\psi^{(k)}(x) = \frac{1}{k!}\left(\tilde{F}^i\left(\frac{\partial}{\partial x^i}\right)\psi^{(k)}_{i_1...i_k} + \sum_{j=1}^{k} T\tilde{F}^{\tilde{i}_j}_{i_j}\psi^{(k)}_{i_1...\tilde{i}_j...i_k}\right)(x)dx^{i_1}\wedge\ldots\wedge dx^{i_k}. \tag{17.22}$$

with $\psi^{(k)}$ from Equation 17.13.

Let us introduce now two other fundamental operators. The first operator is the exterior multiplication, $dx^i\wedge : \Omega^k(X) \to \Omega^{k+1}(X)$, defined via its action on a k-form from Equation 17.13 as

$$dx^i \wedge \psi^{(k)} = (1/k!)\psi_{i_1...i_k}dx^i \wedge dx^{i_1} \wedge \ldots \wedge dx^{i_k} \in \Omega^{k+1}(X). \tag{17.23}$$

Viewing differentials in the definition of a k-form in Equation 17.13 as operators, one can also define exterior multiplication or exterior product of differential forms such that

$$\psi^{(k)} \wedge \psi^{(n)} \in \Omega^{k+n}(X). \tag{17.24}$$

The second operation is the interior multiplication, $\hat{\imath}_i : \Omega^k(X) \to \Omega^{k-1}(X)$, defined as

$$\hat{\imath}_i\psi^{(k)} = (1/k!)\sum_{j=1}^{k}(-1)^{j+1}\psi_{i_1...i_{j-1}ii_{j+1}...i_k}dx^{i_1} \wedge \ldots \widehat{dx^{i_j}} \ldots \wedge dx^{i_k} \in \Omega^{k-1}(X), \tag{17.25}$$

where $\widehat{dx^{i_j}}$ denotes a missing differential. As can be readily checked, the (anti)commutation relations for these operators are

$$\left[dx^{i_1}\wedge, dx^{i_2}\wedge\right] = 0, \left[\hat{\imath}_{j_1}, \hat{\imath}_{j_2}\right] = 0, \left[dx^i\wedge, \hat{\imath}_j\right] = \delta^i_j. \tag{17.26}$$

Here and in the following square bracket's denote bi-graded commutator defined as

$$[\hat{X}, \hat{Y}] = \hat{X}\hat{Y} - (-1)^{deg(\hat{X})F(\hat{Y})}\hat{Y}\hat{X}, \tag{17.27}$$

with $deg(\hat{X}) = \#(dx\wedge) - \#(\hat{\imath})$ being the degree of operator $\hat{X}$, that is, the difference between the numbers of exterior and interior multiplication operators in $\hat{X}$. For example, $deg(dx\wedge) = 1$ and $deg(\hat{\imath}) = -1$ so that the bi-graded commutators in Equation 17.26 are anticommutators.

Let us also introduce the exterior derivative or De Rahm operator

$$\hat{d} = dx^i \wedge \frac{\partial}{\partial x^i}. \tag{17.28}$$

The exterior derivative is a bi-graded differentiation, that is, for any operators $\hat{X}$ and $\hat{Y}$,

$$[\hat{d}, \hat{X}\hat{Y}] = [\hat{d}, \hat{X}]\hat{Y} + (-1)^{deg(\hat{X})}\hat{X}[\hat{d}, \hat{Y}]. \tag{17.29}$$

In newly introduced notations, the Lie derivative is given via the Cartan formula

$$\hat{\mathcal{L}}_{\tilde{F}} = \tilde{F}^i\partial/\partial x^i + T\tilde{F}^i_j dx^j \wedge \hat{\imath}_i = [\hat{d}, \hat{\imath}_{\tilde{F}}], \tag{17.30}$$

where $\hat{\imath}_{\tilde{F}} = \tilde{F}^i\hat{\imath}_i$ is the interior multiplication by $\tilde{F}$, and we used that $[\hat{d}, \hat{\imath}_i] = \partial/\partial x^i$ and $[\hat{d}, \tilde{F}^i] = T\tilde{F}^i_j dx^j \wedge$.

With the help of Equations 17.18 and 17.19, it can be shown that the SDE-defined pullbacks satisfy the following equation:

$$\partial_t M^*_{t't} = -\hat{\mathcal{L}}_{\tilde{F}(t)} M^*_{t't}. \tag{17.31}$$

This equation can be formally integrated with the condition $M^*_{tt} = \hat{1}_{\Omega(X)}$:

$$M^*_{t't} = \mathcal{T} e^{-\int_{t'}^t d\tau \hat{\mathcal{L}}_{\tilde{F}(\tau)}} = \hat{1}_{\Omega(X)} - \int_{t'}^t d\tau \hat{\mathcal{L}}_{\tilde{F}(\tau)} + \int_{t'}^t d\tau_1 \hat{\mathcal{L}}_{\tilde{F}(\tau_1)} \int_{t'}^{\tau_1} d\tau_2 \hat{\mathcal{L}}_{\tilde{F}(\tau_2)} - \ldots, \tag{17.32}$$

where $\mathcal{T}$ denotes chronological ordering, which is necessary because $\hat{\mathcal{L}}_{\tilde{F}(\tau)}$ at different τ's do not commute since $\tilde{F}$ is time-dependent.

17.4 Stochastic Dynamics

17.4.1 Generalized Fokker–Planck Evolution

Let us now turn to stochastic evolution. The stochastic generalization of Equation 17.9 is

$$\bar{f}(t) = \left\langle \int_X f(x) M^*_{t't} P(x) dx^1 \ldots dx^D \right\rangle_{\text{Ns}} = \int_X f(x) \hat{\mathcal{M}}_{tt'}(P(x) dx^1 \ldots dx^D), \tag{17.33}$$

and that of Equation 17.15 is

$$p_{c_k}(t) = \left\langle \int_{c_k} M^*_{t't} \psi^{(k)}(x) \right\rangle_{\text{Ns}} = \int_{c_k} \hat{\mathcal{M}}_{tt'} \psi^{(k)}(x), \tag{17.34}$$

where the notation for stochastic averaging is defined in Equation 17.4, and the new operator, $\hat{\mathcal{M}}_{t't}$: $\Omega(X) \to \Omega(X)$, is defined as

$$\hat{\mathcal{M}}_{tt'} = \langle M^*_{t't} \rangle_{\text{Ns}}. \tag{17.35}$$

The operation of stochastic averaging is legitimate here because pullbacks are linear operators on the linear infinite-dimensional space, $\Omega(X)$.

In the DS theory, operator Equation 17.35 is known as the GTO [35]. The only new element in Equation 17.35 as compared with its standard definition is that pullbacks in Equation 17.35 are those of *inverse* maps.

In case of white noise, the noise variables at different times do not correlate, and the GTOs satisfy the conventional composition law of evolution operators:

$$\hat{\mathcal{M}}_{tt''} = \langle M^*_{t''t} \rangle_{\text{Ns}} = \langle M^*_{t't} M^*_{t''t'} \rangle_{\text{Ns}} = \langle M^*_{t't} \rangle_{\text{Ns}} \langle M^*_{t''t'} \rangle_{\text{Ns}} = \hat{\mathcal{M}}_{tt'} \hat{\mathcal{M}}_{t't''}, \tag{17.36}$$

where we used the composition law for pullbacks in Equation 17.18.

The quantum mechanical analog of the GTO is the finite-time evolution operator denoted typically as $\hat{U} = e^{-it\hat{H}_q}$, with $\hat{H}_q$ being some Hermitian Hamiltonian. Capitalized U signifies that quantum evolution is unitary. The current notation for the GTO borrowed from Reference 35 seems more appropriate for the nonunitary stochastic evolution.

In order to establish the explicit form of the GTO, we proceed in the spirit of pathintegral representation of evolution operators. We split the time domain of the temporal evolution, (t, t'), into the large number, $N \gg 1$, of intervals: $(t, t') = \bigcup_{n=0}^{N-1} (t_{n+1}, t_n)$, where $t_n = t' + n\Delta t$ and $\Delta t = (t - t')/N$. We assume that the noise is piece-wise constant on each interval, that is, the value of the noise variable, $\xi^a(\tau) = \xi^a_n$, for $\tau \in (t_n, t_{n-1})$. Accordingly, within each interval, the r.h.s. of Equation 17.1 is also a constant vector field,

$$\tilde{F}(xt)\big|_{t_n > t > t_{n-1}} = \tilde{F}_n(x) \equiv F(x) + (2T)^{1/2} e_a(x) \xi^a_n. \tag{17.37}$$

The GTO of the temporal evolution during this short but finite-time interval is

$$\hat{\mathcal{M}}_{t_{n+1} t_n} = \langle M^*_{t_n t_{n+1}} \rangle_{\text{Ns}} = \langle e^{-\Delta t \hat{\mathcal{L}}_{\tilde{F}_n}} \rangle_{\text{Ns}}, \tag{17.38}$$

as follows from the definition of the Lie derivative in Equation 17.19.

The discrete-time version of the probability distribution of the Gaussian white noise in Equation 17.2 is

$$P(\xi) \propto e^{-\Delta t \sum_n \xi_n^a \xi_n^a / 2}, \tag{17.39}$$

and that of the expectation value in Equation 17.5 is

$$\langle \xi_n^a \xi_{n'}^b \rangle_{Ns} = \Delta t^{-1} \delta^{ab} \delta_{nn'}, \tag{17.40}$$

whereas all the other (even) order averages

$$\langle \xi_{n_1}^{a_1} \dots \xi_{n_{2k}}^{a_{2k}} \rangle_{Ns} \propto \Delta t^{-k}. \tag{17.41}$$

Now, in the continuous-time limit, $N \to \infty$, $\Delta t \to 0$, the infinitesimal temporal evolution of a wavefunction takes the form of the (generalized) Fokker–Planck (FP) equation,

$$\partial_t \psi(t) = -\hat{H}\psi(t), \tag{17.42}$$

where the (generalized) FP operator is obtained by Taylor expansion of Equation 17.38:

$$\hat{H} = -\lim_{\Delta t \to 0} \frac{1}{\Delta t}(\hat{\mathcal{M}}_{t_{n+1} t_n} - \hat{1}_{\Omega(X)}) = -\lim_{\Delta t \to 0} \frac{1}{\Delta t}\langle -\Delta t \hat{\mathcal{L}}_{\tilde{F}_n} + \Delta t^2 \frac{1}{2}\hat{\mathcal{L}}_{\tilde{F}_n}^2 + \dots \rangle_{Ns}. \tag{17.43}$$

Using the linearity of the Lie derivative in its vector field

$$\hat{\mathcal{L}}_{\tilde{F}_n} = \hat{\mathcal{L}}_F + (2T)^{1/2}\xi_n^a \hat{\mathcal{L}}_{e_a}, \tag{17.44}$$

and Equations 17.40 and 17.41, one arrives at

$$\hat{H} = \hat{\mathcal{L}}_F - T\hat{\mathcal{L}}_{e_a}\hat{\mathcal{L}}_{e_a}. \tag{17.45}$$

Accordingly, the GTO is the finite-time (generalized) FP evolution operator:

$$\hat{\mathcal{M}}_{tt'} = e^{-\hat{H}(t-t')}. \tag{17.46}$$

The model under consideration is time-translation invariant. This is the reason why the GTO depends only on the duration of the temporal evolution, $t - t'$.

The physical meaning of the FP operator (17.45) is clear. The first term is the deterministic flow along F and the second term represents the noise-induced diffusion. Operator

$$\hat{\mathcal{L}}_{e_a}\hat{\mathcal{L}}_{e_a}\psi(x) = \left(g^{ij}(x)\frac{\partial}{\partial x^i}\frac{\partial}{\partial x^j} + \dots\right)\psi(x), \tag{17.47}$$

can be called diffusion Laplacian. It is a member of the family of Laplace operators. In general case, this operator is neither Hodge (or De Rahm) nor Bochner (or Beltrami) Laplacian. Nevertheless, just as the Hodge Laplacian, the diffusion Laplacian has the important property of being $\hat{d}$-exact, that is, $\hat{\mathcal{L}}_{e_a}\hat{\mathcal{L}}_{e_a} = [\hat{d}, \hat{i}_i e_a^i \hat{\mathcal{L}}_{e_a}].^*$

---

* One of the early-stage mistakes in the development of the STS was the identification of the diffusion Laplacian as of the Hodge Laplacian [29].

17.4.1.1 Ito–Stratonovich Dilemma

The time-interval picture used above is a trick or rather a tool but it is not an approximation. Therefore, the so-obtained FP operator is exact. This, in particular, resolves the long-standing Ito–Stratonovich dilemma (see, e.g., Reference 23). Indeed, using the Cartan formula (17.30) and noting that $\hat{d}\psi^{(D)} = 0$ for any $\psi^{(D)} \in \Omega(X)$ so that $\hat{\mathcal{L}}_G\psi^{(D)} = \partial/\partial x^i G^i \psi^{(D)}$ for any vector field, $G \in TX$, we find that the FP Equation 17.42 for D-forms and/or TPD is

$$\partial_t \psi^{(D)}(xt) = -\left(\frac{\partial}{\partial x^i} F^i(x) - T\frac{\partial}{\partial x^i} e_a^i(x)\frac{\partial}{\partial x^j} e_a^j(x)\right)\psi^{(D)}(xt). \qquad (17.48)$$

This is the well-known FP equation for the TPD in the Stratonovich interpretation of the white Gaussian noise. What resolves the issue here is the understanding that the temporal evolution is a stochastically averaged pullback induced by the SDE-defined maps and that the TPD in the coordinate-free setting is a top differential form.

In Section 17.5.3, we will see that the Ito–Stratonovich dilemma is the well-known problem of operator ordering in turning from the pathintegral to the operator representation of the model and that the correct Stratonovich approach is equivalent to the Weyl quantization rule of the (bi-graded in our case) symmetrization of operators. A more detailed discussion of the Ito-Stratonovich dilemma in the context of STS can be found in Reference 31.

17.4.1.2 De Rahm Supersymmetry

The FP operator is $\hat{d}$-exact, that is, it has the form of the bi-graded commutator:

$$\hat{H} = [\hat{d}, \hat{\bar{d}}], \qquad (17.49)$$

where

$$\hat{\bar{d}} = F^i\hat{\imath}_i - Te_a^i\hat{\imath}_i\hat{\mathcal{L}}_{e_a}. \qquad (17.50)$$

The exterior derivative is commutative with the FP operator,

$$[\hat{d}, \hat{H}] = 0. \qquad (17.51)$$

This can be seen from the nilpotency property of exterior derivative, $\hat{d}^2 = 0$, leading to the conclusion that $\hat{d}$ commutes with any $\hat{d}$-exact operator, $[\hat{d}, [\hat{d}, \hat{X}]] = 0, \forall \hat{X}$.

In Section 17.4.4, we will see that the exterior derivative can be represented in such a way that its action substitutes commuting variables by anticommuting variables. Therefore, it can be identified as a supersymmetry of the model. The easiest way to understand the unconditional presence of this supersymmetry is to recall that temporal evolution is the stochastically averaged pullback and that any pullback commutes with the exterior derivative.

The commutativity of an operator with the evolution operator says that this operator is a symmetry of the model. In particular, the expectation value of this operator must not evolve in time. Note, however, that not all the possible evolution operators that commute with $\hat{d}$ are necessarily $\hat{d}$-exact as Equation 17.49. In fact, a $\hat{d}$-exact evolution operator implies more than just the commutativity with $\hat{d}$. As we will discuss below, the additional implication of the $\hat{d}$-exact evolution operator is that all the d-symmetric eigenstates have zero eigenvalues.

17.4.2 Spectrum

In this section, we discuss some of the most important properties of the eigensystem of the FP operator. For simplicity, we assume that the phase space is closed, that g^{ij} is nondegenerate, and T > 0 so that the FP operator is elliptic and the eigenstates are countable. We believe, however, that most of the claims below hold true or at least transformative to more general classes of models.

First of all, the FP operator is real and consequently its spectrum consists of real eigenvalues and pairs of complex conjugate eigenvalues that in DS theory are called Ruelle–Pollicott resonances. Such evolution operators are pseudo-Hermitian [24]. This means, in particular, that the eigensystem constitutes a complete biorthogonal basis in the Hilbert space,

$$\hat{H}|\psi_n\rangle = \mathcal{E}_n|\psi_n\rangle, \quad \langle\psi_n|\hat{H} = \langle\psi_n|\mathcal{E}_n, \tag{17.52}$$

$$\sum_n |\psi_n\rangle\langle\psi_n| = \hat{1}_{\Omega(X)}, \quad \langle\psi_n|\psi_m\rangle = \int_X \psi_m \wedge \bar\psi_n = \delta_{nm}. \tag{17.53}$$

Here we introduced the bra–ket notation: $|\psi_n\rangle \equiv \psi_n$ and $\langle\psi_n| \equiv \bar\psi_n$. The integration in the last formula over X is nonzero only if $\psi_k \wedge \bar\psi_n \in \Omega^D(X)$. This suggests that the bra–ket combination of any eigenstate is a top differential form, $\psi_n \wedge \bar\psi_n \in \Omega^D(X)$, having the meaning of the TPD associated with this eigenstate. In particular, if $\psi_n \in \Omega^k(X)$, then $\bar\psi_n \in \Omega^{D-k}(X)$. Because the FP operator is non-Hermitian, the relation between bra and ket is not trivial: $\bar\psi_n = \sum_m \star(\psi_m^*)\eta_{mn}$, where η_{mn} is the metric on the Hilbert space and $\star$ denotes the Hodge conjugation.

From group-theoretic point of view, the topological supersymmetry is a continuous one-parameter group of transformations,

$$\hat{G}_\alpha = (\hat{G}_{-\alpha})^{-1} = e^{\alpha\hat{d}} = 1 + \alpha\hat{d}, \alpha \in \mathbb{R}^1, \tag{17.54}$$

that the FP operator is invariant of

$$\hat{G}_\alpha\hat{H}\hat{G}_{-\alpha} = \hat{H}. \tag{17.55}$$

Just as in case of any other symmetry, the eigenstates must be irreducible representations of this group. There are only two types of irreducible representations of this symmetry: most of the eigenstates are non-$\hat{d}$-symmetric "bosonic-fermionic" doublets or pairs of eigenstates, whereas some of the eigenstates are $\hat{d}$-symmetric singlets.

Each pair of non-$\hat{d}$-symmetric eigenstates, which we denote as $|\vartheta\rangle$'s and $|\vartheta'\rangle$'s, can be defined via a single bra–ket pair, $|\tilde\vartheta_n\rangle$ and $\langle\tilde\vartheta_n|$, such that $\langle\tilde\vartheta_n|\hat{d}|\tilde\vartheta_n\rangle = 1$, so that the bra's and ket's of the corresponding non-$\hat{d}$-symmetric pair of eigenstates are given as

$$|\vartheta_n\rangle = |\tilde\vartheta_n\rangle, \langle\vartheta_n| = \langle\tilde\vartheta_n|\hat{d}, \tag{17.56}$$

$$|\vartheta'_n\rangle = \hat{d}|\tilde\vartheta_n\rangle, \langle\vartheta'_n| = \langle\tilde\vartheta_n|. \tag{17.57}$$

The $\hat{d}$-symmetric singlets, which we call $|\theta\rangle$'s, are such that

$$\hat{d}|\theta_k\rangle = 0, \text{ but } |\theta_k\rangle \neq \hat{d}|\text{something}\rangle, \tag{17.58}$$

and the same for the bra, $\langle\theta_k|\hat{d} = 0$, but $\langle\theta_k| \neq \langle\text{something}|\hat{d}$. Equation 17.58 is nothing else but the condition for a state to be nontrivial in De Rahm cohomology.

All expectation values of $\hat{d}$-exact operators vanish on $\hat{d}$-symmetric states:

$$\langle\theta_k|[\hat{d}, \hat{X}]|\theta_l\rangle = 0. \tag{17.59}$$

The FP operator is also $\hat{d}$-exact so that all $\hat{d}$-symmetric eigenstates have zero-eigenvalues, $0 = \langle\theta_k|\hat{H}|\theta_k\rangle = \mathcal{E}_{\theta_k}\langle\theta_k|\theta_k\rangle = \mathcal{E}_{\theta_k}$.

As we show next, it reasonable (but not necessary for our purposes) to believe that each De Rahm cohomology class must provide one $\hat{d}$-symmetric eigenstate of the form

$$|\theta_k\rangle = |h_k\rangle + \hat{d}|\tilde\theta_n\rangle, \tag{17.60}$$

where $|h_k\rangle$ is the harmonic form (see below) from this class of De Rahm cohomology and $|\tilde\theta_n\rangle$ will be defined later. That this may indeed be true can be demonstrated using perturbation theory in the following manner. Let us introduce the eigensystem of the Hodge Laplacian

$$\hat{\triangle}_H = [\hat{d}, \hat{d}^\dagger], \tag{17.61}$$

where $\hat{d}^\dagger$ is the Hodge conjugate of $\hat{d}$. The Hodge Laplacian has a real and nonnegative spectrum. Each De Rahm cohomology class provides one $\hat{d}$-symmetric harmonic eigenstate from the kernel of the Hodge Laplacian already used in Equation 17.60,

$$\hat{\triangle}_H|h_k\rangle = 0, \langle h_k|\hat{\triangle}_H = 0. \tag{17.62}$$

All the other eigenstates of the Hodge Laplacian are non-$\hat{d}$-symmetric and they have real and positive eigenvalues. By analogy with Equations 17.56 and 17.57, let us denote these non-$\hat{d}$-symmetric pairs of eigenstates of $\triangle_H$ as

$$|\zeta_n\rangle = |\tilde{\zeta}_n\rangle, \langle \zeta_n| = \langle \tilde{\zeta}_n|\hat{d}, \tag{17.63}$$

$$|\zeta'_n\rangle = \hat{d}|\tilde{\zeta}_n\rangle, \langle \zeta'_n| = \langle \tilde{\zeta}_n|, \tag{17.64}$$

and their eigenvalues as $\triangle_n > 0$.

Let as also represent the FP operator as

$$\hat{H} = \hat{H}_0 + \hat{V}, \ \hat{H}_0 = T\hat{\triangle}_H, \ \hat{V} = [\hat{d}, \hat{v}], \hat{v} = \hat{\bar{d}} - T\hat{d}^\dagger, \tag{17.65}$$

and view $\hat{V}$ as a perturbation. The zeroth-order FP operator is elliptic, whereas the perturbation operator is only linear in derivative, $\hat{V} = \hat{f}^i \frac{\partial}{\partial x^i} + \hat{g}$. This implies that the perturbation series must be well defined at least for some class of models.

Because $\hat{V}$ is $\hat{d}$-exact, the following is true:

$$\langle \zeta_n|\hat{V}|h_k\rangle = \langle \tilde{\zeta}_n|\hat{d}[\hat{d}, \hat{v}]|h_k\rangle = 0,$$

$$\langle \zeta_n|\hat{V}|\zeta'_m\rangle = \langle \tilde{\zeta}_n|\hat{d}[\hat{d}, \hat{v}]\hat{d}|\tilde{\zeta}_m\rangle = 0,$$

$$\langle h_k|\hat{V}|h_i\rangle = \langle h_k|[\hat{d}, \hat{v}]|h_i\rangle = 0,$$

$$\langle h_k|\hat{V}|\zeta'_i\rangle = \langle h_k|[\hat{d}, \hat{v}]\hat{d}|\tilde{\zeta}_i\rangle = 0,$$

where we used Equations 17.63 and 17.64. It is easy to see now that to all orders of the perturbation series, each harmonic form remains a $\hat{d}$-symmetric eigenstate:

$$|\theta_n\rangle = |h_n\rangle + \hat{d}|\tilde{\theta}_n\rangle, \tag{17.66}$$

where

$$|\tilde{\theta}_n\rangle = \sum_{n_1} |\tilde{\zeta}_{n_1}\rangle \frac{1}{-T\triangle_{n_1}} \left(\langle \tilde{\zeta}_{n_1}|\hat{V}|h_n\rangle + \sum_{n_2} \frac{1}{-T\triangle_{n_2}} \langle \tilde{\zeta}_{n_1}|\hat{V}\hat{d}|\tilde{\zeta}_{n_2}\rangle \langle \tilde{\zeta}_{m_2}|\hat{V}|h_n\rangle + \cdots \right). \tag{17.67}$$

Similarly, the bra of this $\hat{d}$-symmetric state is

$$\langle \theta_n| = \langle h_n| + \langle \tilde{\theta}_n|\hat{d}. \tag{17.68}$$

The statement that each De Rahm cohomology class must provide one $\hat{d}$-symmetric eigenstate must be correct even outside the domain of the applicability of the perturbation theory. One reason is the completeness argument, that is, if this is not so, the space of eigenstates is not complete in $\Omega(x)$. Another indication on that this statement must be true will be provided later by establishing that the Witten index equals the Euler characteristic of X.

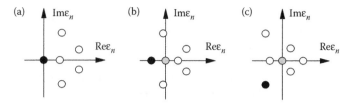

FIGURE 17.2 Possible spectra of the FP operator. $\hat{d}$-symmetry is spontaneously broken for (b) and (c), because the ground state (black dots) has nonzero eigenvalue and thus is non-$\hat{d}$-symmetric. The pseudo-time reversal symmetry is spontaneously broken in (c), because the ground state is one of a pair of Rulle–Pollicott resonances. Dots with gray filling in (b) and (c) represent $\hat{d}$-symmetric states that are not ground states of the model anymore.

Whether or not each De Rahm cohomology class provides one $\hat{d}$-symmetric eigenstate is actually not that important for our further discussion. What is important is the existence of at least one $\hat{d}$-symmetric eigenstate. This eigenstate is a D-form. Its existence can be established through the physical version of the completeness argument. Indeed, all the non-$\hat{d}$-symmetric eigenstates from $\Omega^D(X)$ are $\hat{d}$-exact, that is, of type (17.57). This means that the integral of all such eigenstates over X is zero, $\int_X \hat{d}\tilde{\vartheta} = 0$. On the other hand, a wavefunction from $\Omega^D(X)$ has the meaning of the TPD. The integral of a physically meaningful TPD over X must not vanish. This suggests that at least one $\hat{d}$-symmetric eigenstate from $\Omega^D(X)$ must exist if the model is physically meaningful. This eigenstate is the steady-state (zero eigenvalue) TPD that must be associated with the *TE* of the DS.

The ground states of the model are those with the least attenuation rate, $\mathrm{Re}\mathcal{E}_n$. Only the ground states contribute into the dynamical partition function (see below) in the long time limit. If there are two eigenvalues with the same least attenuation rate as in Figure 17.2c, then in order to identify the unique ground state (up to $\hat{d}$-degeneracy), one can employ the argumentation used in quantum mechanics. There, the ground state is the one with the least "energy" which in our terms is the imaginary part of the eigenvalue. Note also that in the situation given in Figure 17.2c, the so-called pseudo-time-reversal symmetry (or ηT-symmetry) of the pseudo-Hermitian FP operator [24] must also be spontaneously broken because the Ruelle–Pollicott resonances are ηT-companions and we chose only one of them as the ground state of the model.[*]

The procedure of the identification of the ground state from a spectrum is exemplified in Figure 17.2. There, spectra *b* and *c* correspond to the spontaneously broken $\hat{d}$-symmetry because the ground states have nonzero eigenvalues and thus are non-$\hat{d}$-symmetric.

In the DS theory, there are theorems [35] stating that in a wide class of models the ground state of the GTO has real eigenvalue. In light of these theorems, Figure 17.2b seems to be the most likely spectrum of DSs with spontaneously broken $\hat{d}$-symmetry.

17.4.3 Witten Index and Dynamical Partition Function

The sharp and counting traces of the GTO that in the supersymmetric quantum theory are called the Witten index and the dynamical partition function are defined as

$$W_{tt'} = \mathrm{Tr}(-1)^{\hat{k}}\hat{\mathcal{M}}_{tt'} = \mathrm{Tr}(-1)^{\hat{k}}e^{-(t-t')\hat{H}} = \sum_n (-1)^{k_n}e^{-(t-t')\mathcal{E}_n}, \qquad (17.69)$$

$$Z_{tt'} = \mathrm{Tr}\hat{\mathcal{M}}_{tt'} = \mathrm{Tr}e^{-(t-t')\hat{H}} = \sum_n e^{-(t-t')\mathcal{E}_n}. \qquad (17.70)$$

---

[*] The other Ruelle–Pollicott resonance can be identified as a physical state because just like the ground state it survives infinitely long temporal evolution.

In Equation 17.69, $\hat{k} : \Omega^k(X) \to \Omega^k(X)$, $\hat{k} = dx^i \wedge \hat{i}_i$, is the operator of the degree of a differential form, $\hat{k}\psi^{(k)} = k\psi^{(k)}$. This operator is commutative with the FP evolution operator, $[\hat{k}, \hat{H}] = 0$, so that its eigenvalue, $\hat{k}\psi_n = k_n\psi_n$, is a good quantum number.

From the previous discussion, we know that all the eigenstates with nonzero eigenvalue are non-$\hat{d}$-symmetric, they come in pairs of even and odd degrees, and thus cancel each other in the expression for the Witten index. Therefore, only the $\hat{d}$-symmetric eigenstates that all have zero eigenvalues contribute into the Witten index, which is thus independent of the duration of time evolution

$$W_{tt'} \equiv W = \sum_{k=0}^{D} (-1)^k b_k,\qquad(17.71)$$

where b_k is the number of $\hat{d}$-symmetric states of degree k. If we believe that each De Rahm cohomology class provides only $\hat{d}$-symmetric eigenstate, then b_k are the Betti numbers and W equals the Euler characteristic of X, $Eu(X)$. In the next subsection, we will discuss the physical meaning of Equations 17.69 and 17.70 and provide an alternative proof of $W = Eu(X)$ by identifying it with the stochastic Lefschetz index.

17.4.4 The Meaning of W

Let us now introduce fermionic variables. As was pointed out in Reference 38, (anti-)commutation relations in Equation 17.26 are equivalent to those of Grassmann or anticommuting numbers, χ^i, and derivatives over them, $\partial/\partial\chi^j$:

$$\left[\chi^{i_1}, \chi^{i_2}\right]_+ = 0, \left[\frac{\partial}{\partial\chi^{j_1}}, \frac{\partial}{\partial\chi^{j_2}}\right]_+ = 0, \left[\chi^i, \frac{\partial}{\partial\chi^j}\right]_+ = \delta^i_j,\qquad(17.72)$$

Therefore, we can make a formal substitution

$$dx^i \wedge \to \chi^i, \text{ and } \hat{i}_j \to \frac{\partial}{\partial\chi^j},\qquad(17.73)$$

and a wavefunction in new notations becomes

$$(1/k!)\psi^{(k)}_{i_1\ldots i_k}(x)dx^{i_1} \wedge \ldots \wedge dx^{i_k} \to (1/k!)\psi^{(k)}_{i_1\ldots i_k}(x)\chi^{i_1}\ldots\chi^{i_k} \equiv \psi^{(k)}(x\chi),\qquad(17.74)$$

whereas the expression for the exterior derivative, $\hat{d} = \chi^i\partial/\partial x^i$, reveals why $\hat{d}$ is a supersymmetry—it destroys a bosonic or commuting variable, x, and creates a fermionic or anticommuting variable, χ.

Equation 17.74 can be viewed as a kth term in the Taylor expansion in χ's of a general wavefunction,

$$\psi(x\chi) = \sum_{k=0}^{D} \psi^{(k)}(x\chi),\qquad(17.75)$$

which is a function of a pair of variables that are supersymmetric partners, x and χ.

Some properties of fermionic variables are similar to those of bosonic variables. For example, one can introduce the fermionic δ-function

$$\int d^D\chi\, \delta^D(\chi - \chi')f(\chi) = f(\chi').\qquad(17.76)$$

Here, $f(\chi)$ is an arbitrary function of a fermionic variable, χ' is yet another arbitrary fermionic variable, the differential is $d^D\chi = d\chi^D \ldots d\chi^1$, and*

$$\delta^D(\chi - \chi') = (-1)^D(\chi - \chi')^1 \ldots (\chi - \chi')^D.\qquad(17.77)$$

The above property of fermionic variables and their δ-function can be established using the Berezin rules of integration over fermionic variables that include identities such as $\int d\chi^1 = 0$, $\int \chi^1 d\chi^1 = -\int d\chi^1\chi^1 = 1$.

---

* Note that the definition of fermionic δ-function depends on the relative position of the differentials because $\int d^D\chi\,\delta^D$ $(\chi - \chi') = (-1)^D \int \delta^D(\chi - \chi')d^D\chi$, and we used $(-1)^{D^2} = (-1)^D$.

Other properties of fermionic variables are in a sense opposite to those of bosonic variables, e.g.,

$$\int d^D\chi\, \delta^D(\hat{A}\chi) = \det\hat{A}, \tag{17.78}$$

whereas for bosonic variables we would have $\int d^D x\, \delta^D(\hat{A}x) = |\det\hat{A}|^{-1}$, with $x \in \mathbb{R}^D$.

With the use of fermionic variables, Equation 17.16 can be given as

$$M^*_{t't}\psi(x\chi) = \int d^D x'\, d^D\chi'\, M^*_{t't}(x\chi, x'\chi')\psi(x'\chi'), \tag{17.79}$$

$$M^*_{t't}(x\chi, x'\chi') = \delta^D(x' - M_{t't}(x))\delta^D(\chi' - TM_{t't}(x)\chi), \tag{17.80}$$

where the bosonic and fermionic δ-functions substitute x' and χ' by $M_{t't}(x)$ and $TM_{t't}(x)\chi$, with $TM_{t't}(x)$ being the tangent map from Equation 17.12. The GTO now is an operator with the "in-" and "out-" variables

$$\hat{\mathcal{M}}_{tt'}(x\chi, x'\chi') = \langle M^*_{t't}(x\chi, x'\chi')\rangle_{\text{Ns}}. \tag{17.81}$$

The Witten index in Equation 17.69 takes the form of the trace of the GTO with the periodic boundary conditions (PBC) for the bosonic and fermionic fields,

$$W = \int d^D x\, d^D\chi\, \hat{\mathcal{M}}_{tt'}(x\chi, x\chi) = \left\langle \sum_{x = M_{t't}(x)} \frac{\det(\hat{1}_{TX} - TM_{t't}(x))}{|\det(\hat{1}_{TX} - TM_{t't}(x))|}\right\rangle_{\text{Ns}}$$

$$= \sum_{k=0}^{D}(-1)^k\left\langle \sum_{x = M_{t't}(x)} \frac{m_k(x)}{|\det(\hat{1}_{TX} - TM_{t't}(x))|}\right\rangle_{\text{Ns}}, \tag{17.82}$$

where we used the characteristic polynomial formula, $\det(\hat{1}_{TX} + \lambda TM_{t't}(x)) = \sum_{k=0}^{D} \lambda^k m_k(x)$,

$$m_k(x) = \sum_{i_1 < i_2 < \ldots < i_k} \det \begin{pmatrix} TM_{t't}(x)^{i_1}_{i_1} & \cdots & TM_{t't}(x)^{i_1}_{i_k} \\ \vdots & \ddots & \vdots \\ TM_{t't}(x)^{i_k}_{i_1} & \cdots & TM_{t't}(x)^{i_k}_{i_k} \end{pmatrix}. \tag{17.83}$$

The denominator in Equation 17.82 comes from the trace over bosonic variables, whereas m_k can be viewed as a fermionic trace over $\Omega^k(X)$. Indeed, the basis of the differentials in $\Omega^k(X)$, that the fermionic variables represent, is given by the C_D^k ordered combinations of the differentials: $dx^{i_1} \wedge \ldots \wedge dx^{i_k}$, $i_1 < \ldots < i_k$. Thus, the trace of fermionic variables over $\Omega^k(X)$ is

$$\text{Tr}^{ferm}_{\Omega^k(X)} M^*_{t't} = \sum_{i_1 < \ldots < i_k} \hat{\imath}_{i_k}\ldots\hat{\imath}_{i_1} d(M_{t't}(x))^{i_1} \wedge \ldots \wedge d(M_{t't}(x))^{i_k} \tag{17.84}$$

$$= \sum_{i_1 < \ldots < i_k} \frac{\partial}{\partial\chi^{i_k}}\cdots\frac{\partial}{\partial\chi^{i_1}} TM_{t't}(x)^{i_1}_{i_1}\chi^{\bar{i}_1}\ldots \wedge TM_{t't}(x)^{i_k}_{i_k}\chi^{\bar{i}_k} = m_k(x), \tag{17.85}$$

where $d(M_{t't}(x))$ is defined in Equation 17.11. Now one has

$$W = \sum_{k=0}^{D}(-1)^k \text{Tr}_{\Omega^k(X)}\langle M^*_{t't}\rangle_{\text{Ns}}, \tag{17.86}$$

which is equivalent to Equation 17.69.

Accordingly, the dynamical partition function from Equation 17.70 is the trace of the GTO with periodic/antiperiodic boundary conditions (APBC) for the bosonic/fermionic variables:

$$Z_{tt'} = \int d^D x d^D \chi \hat{\mathcal{M}}_{tt'}(x(-\chi), x\chi) = \left\langle \sum_{x=M_{t't}(x)} \frac{\det(\hat{1}_{TX} + TM_{t't}(x))}{|\det(\hat{1}_{TX} - TM_{t't}(x))|} \right\rangle_{Ns}$$

$$= \sum_{k=0}^{D} \left\langle \sum_{x=M_{t't}(x)} \frac{m_k(x)}{|\det(\hat{1}_{TX} - TM_{t't}(x))|} \right\rangle_{Ns} = \mathrm{Tr}\langle M_{t't}^* \rangle_{Ns}. \tag{17.87}$$

The Witten index in Equation 17.82 can be rewritten as

$$W = \langle I_L \rangle_{Ns}, \tag{17.88}$$

where

$$I_L = \sum_{x=M_{tt'}(x)} \mathrm{sign} \det(\hat{1}_{TX} - TM_{t't}(x)) \tag{17.89}$$

is known as the Lefschetz index of map $M_{t't}$. According to the Lefschetz–Hopf theorem, under some general conditions,

$$I_L = \sum_{k=0}^{D} (-1)^k \mathrm{Tr}_{H^k(X)} M_{t't}^*, \tag{17.90}$$

where the trace is over the De Rahm cohomology, $H^k(X)$. In the limit $t' \to t$, when $M_{t't} \to \mathrm{Id}_X$, Lefschetz index evaluates to the sum of signed Betti numbers, that is, to the Euler characteristics of X. On the other hand, we know that W is independent of t, which leads to the conclusion that Witten index (under some general conditions) equals the Euler characteristic of X for any duration of temporal evolution.

The topological origin of W can be qualitatively interpreted as follows (see Figure 17.3). For each configuration of noise, there may exist many periodic solutions of the SDE, that is, the fixed points of $M_{t't}$. As one gradually changes the noise configuration, periodic solutions appear and disappear in pairs with positive and negative determinants of matrix $\hat{1}_{TX} - TM_{t't}$. Therefore, constant, I_L, remains the same. Up to this constant, W represents the (normalized) partition function of the stochastic noise.

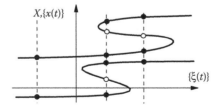

FIGURE 17.3 Qualitative explanation of the topological character of Witten index. The x-axis is that of the noise configurations. The y-axis is that of either all the points in X for the Lefschetz index interpretation of the Witten index in Section 17.4.4, or all the possible closed paths in X for the Mathai–Quillen interpretation in Section 17.5.1. For each noise configuration, there are multiple closed solutions of the SDE. As one changes the noise, these solutions appear and disappear in pairs of positive (filled dots) and negative (hollow dots) determinants. Therefore, the sum of the signs of the determinants is independent of the noise configuration.

17.4.5 The Meaning of Z

As to the dynamical partition function, its meaning is seen in the opposite limit of the infinitely long temporal evolution. Consider models, in which the absolute values of the eigenvalues of the tangent map, $Spec(TM_{t't}) = (\mu_1(t, t'), ..., \mu_D(t, t'))$, in the long time limit, $t - t' \to \infty$, are such that, $|\mu_i(t, t')| \approx e^{\lambda_i(t'-t)}$, with λ's being the stochastic version of the (global) Lyapunov exponents. In the assumption that none of λ's vanish, one has for $t - t' \to \infty$,

$$
\begin{aligned}
Z_{tt'} &= \left\langle \sum_{x=M_{t't}(x)} \frac{\det(\hat{1}_{TX} + TM_{t't}(x))}{|\det(\hat{1}_{TX} - TM_{t't}(x))|} \right\rangle_{\mathrm{Ns}} \\
&= \left\langle \sum_{x=M_{t't}(x)} \frac{\prod_{i=1}^{D}(1 + \mu_i(t, t'))}{|\prod_{i=1}^{D}(1 - \mu_i(t, t'))|} \right\rangle_{\mathrm{Ns}} \\
&\approx \left\langle \sum_{x=M_{t't}(x)} \mathrm{sign}\left(\prod_{i,\lambda_i<0} \mu_i(t, t')\right) \right\rangle_{\mathrm{Ns}} \\
&\leq \left\langle \sum_{x=M_{t't}(x)} 1 \right\rangle_{\mathrm{Ns}} = \left\langle \# \text{ of fixed points of } M_{t't} \right\rangle_{\mathrm{Ns}}.
\end{aligned}
\tag{17.91}
$$

In other words, in this class of models, the DPF grows slower than the stochastically averaged number of the fixed points of the SDE-induced diffeomorphisms, or, equivalently, of the number of periodic solutions of the SDE.

In the DS theory, there exist the so-called Shub conjecture (see, e.g., Reference 22 and references therein) stating that for a smooth enough map, $M : X \to X$, the spectral radius of $M_* : H_*(X) \to H_*(X)$, where H_* denotes the homology group, provides a lower bound for the topological entropy, the central measure of chaos. The spectral radius of the finite-time stochastic evolution operator, $\langle M^*_{0t} \rangle_{\mathrm{Ns}} = e^{-t\hat{H}}$, is (up to the sign) the real part of its ground state eigenvalue, which can therefore be recognized as the stochastic generalization of the lower bound for the topological entropy in the Shub conjecture. Therefore, for the models in Figure 17.2b and 17.2c that have the negative real part of the ground state eigenvalue, the topological entropy is positive and the models must be identified as chaotic. The ground states eigenvalues for these models are nonzero and thus are non-$\hat{d}$-symmetric. Thus we came to the conclusion that the stochastic generalization of dynamical chaos is the phenomenon of the spontaneous breakdown of topological supersymmetry.

Besides being a proof of that the spontaneous breakdown of topological supersymmetry is the stochastic generalization of the concept of deterministic chaos, Equation 17.90 has yet another important implication. It says that in order for the dynamical partition function to count all the periodic solution in models where inequality (17.90) saturates, one must take trace of the evolution operator over the entire exterior algebra. In other words, the conventional description of SDEs in terms of TPD only is actually incomplete. The lower-degree differential forms must also be viewed as rightful wavefunctions. This is not an artificial invention within the STS, but rather a necessity. Without it, the dynamical partition function will fail to represent (in the large time limit) the stochastically averaged number of periodic solutions for all DSs.

The same conclusion follows from the physical interpretation of the Witten index as of the partition function of the noise (up to the topological factor, see Equation 17.88). Indeed, the partition function of the noise is one of the most fundamental objects in the theory that appears at the very level of the formulation of the SDE. Such object must certainly have its representative in a consistent theory of the SDE. This representative would not exist at all unless one viewed the entire exterior algebra as the Hilbert space of the model.

We would also like to mention that in this section, we sloppily used the notation for the summation over fixed points of the maps (see, e.g., Equation 17.82) as though these fixed points are isolated in X. This is not

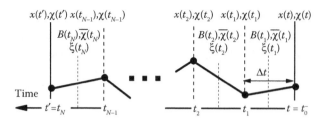

FIGURE 17.4 Structure of pathintegral representation of the finite-time FP evolution operator. At each t_n, there is a bosonic–fermionic pair of variables $x(t_n) \in X$ and $\chi(t_n) \in TX_x(t_n)$. In between the time slices, there are pairs of Langrange multiplier and yet another fermionic variable both from the cotangent space $B(t_n), \bar{\chi}(t_n) \in TX^*_{x(t_n)}$ (or $TX^*_{x(t_{n-1})}$ as this choice does not make difference in the continuous time limit. This choice does matter on the way out of the pathintegral representation (see Section 17.5.3). The noise variables also belong in between the time slices. Just like in quantum theory, the time flows from right to left because the FP operator acts on the wavefunction from the left. Integration of all the variables except $x(t'), \chi(t')$, and $x(t), \chi(t)$ results in the GTO. The additional integration over $x(t), \chi(t)$ with the PBC $x(t') = x(t), \chi(t') = \chi(t)$ results in the Witten index, whereas with the APBC for the fermionic variables $x(t') = x(t), \chi(t') = -\chi(t)$ results in the dynamical partition function.

true in general. Fixed points of maps may come in submanifolds of X.[*] How to count all the fixed points in general situation within the methodology of this section is not clear. This problem, however, does not exist in the operator representation of the theory we considered previously as well as in its pathintegral representation that we turn to next.

17.5 Pathintegral Representation

In this section, we would like to discuss the pathintegral representation of the STS. Pathintegrals is a well-developed methodology that greatly extends the set of available analytical tools as compared to, say, operator representation. In our case, the use of pathintegrals will allow for the generalization of the discussion to noises of any form, not only the Gaussian white noises.

17.5.1 Evolution Operator

Let us now turn back to the time-interval picture already used in Section 17.4.1 and visualized now in Figure 17.4. That is, the domain of the temporal evolution is split into $N \gg 1$ intervals with boundaries at $t_n = t_0 + n\Delta t$, $\Delta t = (t - t')/N$, $t \equiv t_N$ and $t' \equiv t_0$. The GTO can be now given as

$$\hat{\mathcal{M}}_{tt'} = \langle M^*_{t_0 t_N} \rangle_{\text{Ns}} = \langle M^*_{t_{N-1} t_N} M^*_{t_{N-2} t_{N-1}} \cdots M^*_{tt_1} \rangle_{\text{Ns}}, \tag{17.92}$$

where we made multiple use of the composition law for the pullbacks from Equation 17.18. Using fermionic variables introduced in the previous section, Equation 17.92 can be also given as

$$\hat{\mathcal{M}}_{tt'}(x(t)\chi(t), x(t')\chi(t')) = \Big\langle M^*_{t_{N-1}t}(x(t)\chi(t), x(t_{N-1})\chi(t_{N-1})) \times$$
$$\times \prod_{n=1}^{N-1} d^D x(t_n) d^D \chi(t_n) M^*_{t_{n-1}t_n}(x(t_n)\chi(t_n), x(t_{n-1})\chi(t_{n-1})) \Big\rangle_{\text{Ns}}. \tag{17.93}$$

where the pullbacks on the r.h.s. are the combinations of the bosonic and fermionic δ-functions as in Equation 17.80, and at each t_n, we introduce a pair of bosonic and fermionic variables, $x(t_n), \chi(t_n)$.

---

[*] One example is the maps defined by the Morse–Bott flow vector fields.

In order to exponentiate bosonic δ-functions, we introduce an additional bosonic variable called the Lagrange multiplier from the cotangent space of X, $B(t_n) \in TX^*_{x(t_n)}$:

$$\delta^D(x(t_{n-1}) - M_{t_{n-1}t_n}(x(t_n))) = \int \frac{d^D B(t_n)}{(2\pi)^D} e^{-iB_i(t_n)(x(t_{n-1}) - M_{t_{n-1}t_n}(x(t_n)))^i}. \tag{17.94}$$

In order to exponentiate the fermionic δ-functions, we use the following identity:

$$\delta^D(\chi) = \int d^D \bar{\chi} e^{\bar{\chi}_i \chi^i},$$

valid (up to a sign) for any fermionic variable, $\chi \in TX_x$, and the dummy fermionic variable of the additional integration from the cotangent space, $\bar{\chi} \in TX^*_x$. This gives

$$\delta^D(\chi(t_{n-1}) - TM_{t_{n-1}t_n}\chi(t_n)) = \int d^D(i\bar{\chi}(t_n)) e^{i\bar{\chi}_i(t_n)(\chi(t_{n-1}) - TM_{t_{n-1}t_n}(x(t_n))\chi(t_n))^i}. \tag{17.95}$$

Here, the imaginary unity is needed for the purpose of bringing the action of the model into the form conventional for the literature on cohomological field theories.

In the continuous-time limit, $N \to \infty$, we have

$$M_{t_{n-1}t_n}(x(t_n))^i \approx x^i(t_n) - \Delta t \tilde{F}^i(x(t_n)), \tag{17.96}$$

$$(TM_{t_{n-1}t_n}(x(t_n))\chi(t_n))^i \approx \chi^i(t_n) - \Delta t T\tilde{F}^i_j(x(t_n))\chi^j(t_n), \tag{17.97}$$

where $T\tilde{F}(x)$ is introduced in Equation 17.21. Combining Equations 17.94 through 17.97, we get

$$\hat{\mathcal{M}}_{tt'}(x(t)\chi(t), x(t')\chi(t')) = \left\langle \iint D'\Phi e^{\tilde{S}(x(t)\chi(t)\dots x(t')\chi(t'))} \right\rangle_{Ns}. \tag{17.98}$$

Here,

$$\begin{aligned}
\tilde{S}(\Phi) &= \lim_{N\to\infty} i \sum_{n=1}^{N} \Delta t \left(B_i(t_n) \left(\frac{x(t_n) - x(t_{n-1})}{\Delta t} - \tilde{F}(x(t_n)) \right)^i \right. \\
&\left. - \bar{\chi}_i(t_n) \left(\frac{\chi(t_n) - \chi(t_{n-1})}{\Delta t} - T\tilde{F}(x(t_n))\chi(t_n) \right)^i \right), \\
&= i \int_{t'}^{t} d\tau \left(B_i(\tau)(\dot{x}(\tau) - \tilde{F}(x(\tau)))^i - \bar{\chi}_i(\tau)(\dot{\chi}(\tau) - T\tilde{F}(x(\tau))\chi(\tau))^i \right) \tag{17.99}
\end{aligned}$$

is the action of the model, $\Phi = (x, \chi, B, \bar{\chi})$ denotes the collection of all the fields introduced so far, and the dots in Equation 17.98 denote all the intermediate variables over which the path integration takes place with the differential

$$D'\Phi = \lim_{N\to\infty} \frac{d^D B(t_N)}{(2\pi)^D} d(i\bar{\chi}^D(t_N)) \prod_{n=1}^{N-1} d^{4D}\Phi(t_n), \tag{17.100}$$

where

$$d^{4D}\Phi(t_n) = d^D x(t_n) d^D \chi(t_n) \frac{d^D B(t_n)}{(2\pi)^D} d^D(i\bar{\chi}(t_n)). \tag{17.101}$$

The action can be expressed in the so-called Q-exact form:

$$\tilde{S}(\Phi) = \{Q, \tilde{\Psi}(\Phi)\}, \tag{17.102}$$

where

$$\tilde{\Psi}(\Phi) = \lim_{N\to\infty} \sum_{n=1}^{N} \Delta t \left(i\bar{\chi}_i(t_n) \left(\frac{x(t_n) - x(t_{n-1})}{\Delta t} - \tilde{F}(x(t_n)) \right)^i \right),$$

$$= i \int_{t'}^{t} d\tau \bar{\chi}_i(\tau)(\dot{x}(\tau) - \tilde{F}(x(\tau)))^i, \tag{17.103}$$

is known as gauge fermion and the curly brackets denote the operator of the topological supersymmetry

$$\{\mathcal{Q}, \tilde{\Psi}\} = \lim_{N\to\infty} \left(\sum_{n=0}^{N} \chi^i(t_n) \frac{\partial}{\partial x^i(t_n)} + \sum_{n=1}^{N} B_i(t_n) \frac{\partial}{\partial \bar{\chi}_i(t_n)} \right) \Psi,$$

$$= \int_{t'}^{t} d\tau \left(\chi^i(\tau) \frac{\delta}{\delta x^i(\tau)} + B_i(\tau) \frac{\delta}{\delta \bar{\chi}_i(\tau)} \right) \Psi. \tag{17.104}$$

This operator is the pathintegral version of the exterior derivative. In particular, it has similar properties: it is nilpotent, $\{\mathcal{Q}, \{\mathcal{Q}, X(\Phi)\}\} = 0$, for any $X(\Phi)$, and it is a bi-graded differentiation:

$$\{\mathcal{Q}, XY\} = \{\mathcal{Q}, X\}Y + (-1)^{deg(X)} X\{\mathcal{Q}, Y\}, \tag{17.105}$$

where X and Y are some functionals of Φ, and $deg(X)$ is the degree of X defined as the number of χ's minus the number of $\bar{\chi}$'s. Equation 17.105 is the pathintegral version of Equation 17.29.

In order to perform stochastic averaging, we separate the noise term as

$$\tilde{S}(\Phi) = S_0(\Phi) + \int_{t'}^{t} d\tau y_a(\tau)\xi^a(\tau), \quad S_0(\Phi) = \{\mathcal{Q}, \Psi_0(\Phi)\}, \tag{17.106}$$

with

$$\Psi_0(\Phi) = i \int_{t'}^{t} d\tau \bar{\chi}_i(\tau)(\dot{x}(\tau) - F(x(\tau)))^i, \tag{17.107}$$

and

$$y_a(\tau) = \{\mathcal{Q}, -i(2T)^{1/2}\bar{\chi}_i(\tau)e_a^i(x(\tau))\}. \tag{17.108}$$

We can now integrate out the noise field that we do not consider Gaussian white anymore.[*] This leads from Equation 17.98 to

$$\hat{\mathcal{M}}_{tt'} = \iint D'\Phi e^{S(\Phi)}, \tag{17.109}$$

---

[*] Though we believe that the noise is physical and pathintegral over the noise variables is well defined.

where the arguments of the GTO are dropped for brevity and the new action is

$$S(\Phi) = \log\langle e^{\tilde{S}(\Phi)}\rangle_{\mathrm{Ns}} = S_0(\Phi) + \log\langle e^{\int_{t'}^{t} d\tau y_a(\tau)\xi^a(\tau)}\rangle_{\mathrm{Ns}}$$

$$= S_0(\Phi) + \sum_{k=1}^{\infty} \frac{1}{k!} \int_{t'}^{t} \left(\prod_{i=1}^{k} d\tau_i y_{a_i}(\tau_i)\right) c_{(k)}^{a_1\ldots a_k}(\tau_1\ldots\tau_k),$$

with $c_{(k)}$ being irreducible correlators of the noise. The zeroth-order term in the Taylor series vanishes because the partition function of the noise is normalized, that is, for $y=0$, we have $\log\langle e^0\rangle_{\mathrm{Ns}} = \log 1 = 0$.

Using now the property of the nilpotency of $\mathcal{Q}$, the differentiation rule in Equation 17.105, and the fact that all y's in Equation 17.108 are $\mathcal{Q}$-exact, we arrive at

$$S(\Phi) = \{\mathcal{Q}, \Psi(\Phi)\}, \tag{17.110}$$

where the new gauge fermion is

$$\Psi(\Phi) = \sum_{k=0}^{\infty} \Psi_k(\Phi), \tag{17.111}$$

with Ψ_0 defined in Equation 17.107, and for $k \geq 1$,

$$\Psi_k(\Phi) = -i\sum_{k=1}^{\infty} \frac{(2\mathrm{T})^{\frac{k}{2}}}{k!} \int_{t'}^{t} \left(\prod_{i=1}^{k} d\tau_i\right) \bar{\chi}_i(\tau_1) e_{a_1}^i(x(\tau_1)) y_{a_2}(\tau_2) \ldots y_{a_k}(\tau_k) c_{(k)}^{a_1\ldots a_k}(\tau_1\ldots\tau_k). \tag{17.112}$$

In other words, even after stochastic averaging, the action is still $\mathcal{Q}$-exact.

A $\mathcal{Q}$-exact action as in Equations 17.109 through 17.112 is a definitive feature of cohomological field theories [8,12,20,39,40]. Their standard pathintegral representations are with the PBC for fermionic variables. This is related to the fact that the PBC are consistent with the $\mathcal{Q}$-operator,

$$\{\mathcal{Q}, \Psi(\Phi)\}_{PBC} \equiv \{\mathcal{Q}, \Psi(\Phi)\}|_{x(t)=x(t'),\chi(t)=\chi(t')} = \{\mathcal{Q}, \Psi(\Phi)_{PBC}\}. \tag{17.113}$$

This equality and Equations 17.80 and 17.109 through 17.112 lead to

$$W = \oiint_{PBC} D\Phi e^{\{\mathcal{Q},\Psi(\Phi)\}} = \oiint D\Phi e^{\{\mathcal{Q},\Psi(\Phi)_{PBC}\}}, \tag{17.114}$$

where the pathintegration is over

$$D\Phi = d^D x(t) d^D \chi(t) D'\Phi, \tag{17.115}$$

with $D'\Phi$ defined in Equation 17.100.

The integrand in Equation 17.114 belongs to the class of mathematical objects known as Mathai–Quillen forms, and its integral is of topological character. This can be clarified by representing Equation 17.88 as

$$W = \left\langle \oiint D\Phi e^{\tilde{S}(\Phi)} \right\rangle_{\mathrm{Ns}}$$

$$= \left\langle \oiint_{\text{closed paths}} Dx \left(\prod_{\tau} \delta(\dot{x}(\tau) - \tilde{F}(x(\tau)))\right) \mathrm{Det}(\partial_\tau - T\tilde{F}(x(\tau))) \right\rangle_{\mathrm{Ns}}$$

$$= \left\langle \sum_{\text{closed solutions of SDE}} \mathrm{sign}\, \mathrm{Det}(\partial_\tau - T\tilde{F}(x(\tau))) \right\rangle_{\mathrm{Ns}}. \tag{17.116}$$

Here, in the first line, the functional δ-function that limits the path integration only to closed solutions of SDE follows after integrating out the Lagrange multiplier, B, whereas the functional determinant of the infinite-dimensional matrix of the functional derivatives of the SDE is provided by out-integration of the fermionic fields, $\chi, \bar{\chi}$. The second line is the pathintegral analog of the Lefschetz index in Equation 17.88.

Equation 17.116 has the meaning of the infinite-dimensional generalization of the Poincaré–Hopf theorem. The latter states that (under certain and general conditions) the sum of the signs of the determinants of the matrix of derivatives of a vector field over the isolated critical points of this vector field equals the Euler characteristic of the manifold. In our case, the manifold is the space of all the closed paths, $x(t') = x(t)$, the vector field is the SDE, $\dot{x} - \tilde{F}$, and the critical points are the closed solutions, that is, the closed paths that satisfy the SDE, $\dot{x} - \tilde{F} = 0$. The resulting constant is independent of the configuration of the noise. In other words, as we change the noise configuration, the closed solutions of the SDE appear/disappear in pairs with opposite signs of determinants. This situation can be illustrated graphically similarly to the Lefschetz index interpretation of W in the previous section (see Figure 17.3).

Equation 17.116 is the generating functional of the Parisi–Sourlas stochastic quantization procedure [32]. It is a typical mistake to treat it as the "generating functional" of the model, within which various expectation values and correlators can be obtained. The real generating functional, within which various correlators must be calculated, corresponds to the APBC for the fermionic fields. These boundary conditions are not consistent with the $\mathcal{Q}$-operator,

$$\{\mathcal{Q}, \Psi(\Phi)\}_{APBC} \equiv \{\mathcal{Q}, \Psi(\Phi)\}|_{x(t')=x(t), \chi(t')=-\chi(t)} \neq \{\mathcal{Q}, \Psi(\Phi)_{APBC}\}. \tag{17.117}$$

As a result, the topological nature is lost for the dynamical partition function,

$$Z_{tt'} = \iint_{APBC} D\Phi e^{\{\mathcal{Q}, \Psi(\Phi)\}} \neq \iint D\Phi e^{\{\mathcal{Q}, \Psi(\Phi)_{APBC}\}}. \tag{17.118}$$

One important difference between W and Z is as follows: only $\hat{d}$-symmetric states contribute to W, whereas the contribution to Z (in the long time limit) is only from the ground states, which in case of spontaneously broken topological supersymmetry are not $\hat{d}$-symmetric.

17.5.2 Continuous Space Generalization

The pathintegral formulation allows for further generalization of the theory to the class of spatio-temporal models. These models are defined by the following SDEs:

$$\dot{x}(rt) = F(x, rt) + (2T)^{1/2} e_a(x, rt) \xi^a(rt) \equiv \tilde{F}(x, \xi, rt), \tag{17.119}$$

where r is the spatial coordinate of the "base-space." In the general case, the flow vector field and the veilbeins are some temporarily and spatially nonlocal functionals of $x(rt)$ that can also have explicit dependence on the "base-space" coordinates, rt.

The relation between models 17.119 with the previously discussed models with time being the only "base-space" coordinate is the same as the relation between quantum nonlinear sigma models (or field theories) with quantum mechanics. In other words, Equation 17.119 is the infinite-dimensional version of Equation 17.1. The phase space now is the infinite-dimensional space of all the possible configurations $x(r)$.

The procedure of the stochastic quantization of Equation 17.119 goes along the same lines, so that the action is $\mathcal{Q}$-exact with the topological supersymmetry operator being

$$\mathcal{Q} = \int dr d\tau \left(\chi^i(r\tau) \frac{\delta}{\delta x^i(r\tau)} + B_i(r\tau) \frac{\delta}{\delta \bar{\chi}_i(r\tau)} \right), \tag{17.120}$$

and the gauge fermion before integrating out the noise variables being

$$\tilde{\Psi}(\Phi) = i \int dr d\tau \bar{\chi}_i(r\tau) \left(\dot{x}(r\tau) - \tilde{F}(x, \xi, r\tau) \right)^i. \tag{17.121}$$

After integrating out the noise, one arrives at a model with a Q-exact action. If one restricts their attention only to expectation values of Q-closed operators evaluated on supersymmetric states, they get a full-fledged cohomological (pseudo-Hermitian) nonlinear sigma model.

Equation 17.119 is the most general class of models covered by the STS. Having said that, in the rest of this chapter, we consider only models from Equation 17.1 with the Gaussian white noise.

17.5.3 Weyl–Stratonovich Quantization

Let us now see how the pathintegral representation of the theory with white (Gaussian) noise can be turned back into the operator representation. This exercise will reveal the close relation between the Ito–Stratonovich dilemma (see Section 17.4.1) and the Weyl quantization rules of quantum theory.

The gauge fermion can be obtained from Equation 17.112 and the fact that the only irreducible correlator of the Gaussian white noise is the one given in Equation 17.5:

$$\Psi = \int d\tau \left(i\bar{\chi}_i(\tau)\dot{x}^i(\tau) - \bar{d}(\Phi(\tau)) \right), \tag{17.122}$$

where

$$\bar{d}(\Phi) = i\bar{\chi}_i F^i(x) - \mathrm{T}i\bar{\chi}_i e_a^i(x)\{Q, i\bar{\chi}_j e_a^j(x)\}, \tag{17.123}$$

is the pathintegral version of operator 17.50.

Accordingly, the action is

$$S = \{Q, \Psi\} = \int d\tau \left(iB_i(\tau)\dot{x}^i(\tau) - i\bar{\chi}_i(\tau)\dot{\chi}^i(\tau) - H(\Phi(\tau)) \right), \tag{17.124}$$

where

$$H(\Phi) = \{Q, \bar{d}(\Phi)\}. \tag{17.125}$$

Function $H(\Phi)$ is the pathintegral version of the FP operator (17.45) and it is a stochastic analog of the Hamilton function in the pathintegral representation of quantum mechanics. For these reasons, it can be called the FP Hamilton function. It also reveals that Q is the pathintegral version of the commutator with the exterior derivative as is seen from comparison of Equations 17.49 and 17.125.

Consider now one slice of temporal evolution, say, between $t = t_0$ and $t_1 = t + \Delta t$. In the $\Delta t \to 0$ limit, the pathintegral representation of the evolution operator takes the following approximate form:

$$\psi(x\chi t_1) = \int \frac{d^D B}{(2\pi)^D} d^D i\bar{\chi} \, d^D y \, d^D \varphi \, e^{iB_i(x-y)^i - i\bar{\chi}_i(\chi-\varphi)^i - \Delta t H(B\bar{\chi}x_\alpha\chi_\alpha)} \psi(y\varphi t). \tag{17.126}$$

Here, for the sake of brevity, we used variables different from those defined in Figure 17.4. The relation between these variables and those introduced in Figure 17.4 is as follows: $(x, \chi) = (x(t_1), \chi(t_1))$, $(B, \bar{\chi}) \equiv (B(t_1), \bar{\chi}(t_1))$ and $(y, \varphi) \equiv (x(t), \chi(t))$.

In Equation 17.126, we also introduced $x_\alpha = (1-\alpha)y + \alpha x$ and $\chi_\alpha = (1-\alpha)\varphi + \alpha\chi$. Parameter α controls at which point of the time interval, $(t + \Delta t, t)$, the arguments of function $H(\Phi)$ are evaluated: $\alpha = 0$ and $\alpha = 1/2$ correspond respectively to the Ito and Stratonovich choices of the very beginning of the time interval and of the middle point. The reason why we did not speak about these different possible choices before is that on the way into the pathintegral representation, the choice of α does not matter. It is only on the way back from the pathintegral prepresentation to the operator representation that different choices of α give different expressions for the evolution operators as we will see below.

Using Equation 17.126, the infinitesimal evolution of the wavefunction is

$$\partial_t \psi(x\chi t) = \lim_{\Delta t \to 0} (\Delta t)^{-1}(\psi(x\chi t_1) - \psi(x\chi t)). \tag{17.127}$$

Taylor expanding the exponent in Equation 17.126 in $(-\Delta t H)$, we get

$$\partial_t \psi(x\chi t) = -\int \frac{d^D B}{(2\pi)^D} d^D(i\bar\chi) d^D y\, d^D \varphi\, e^{iB_i(x-y)^i - i\bar\chi_i(\chi-\varphi)^i} H(B\bar\chi x_\alpha \chi_\alpha) \psi(y\varphi t). \tag{17.128}$$

This version of the FP equation clarifies the roles of variables involved: the first integrations over y and φ brings the wavefunction into the Fourier space, where B and $\bar\chi$ are diagonal, whereas the second integration over B and $\bar\chi$ brings the wavefunction back into the real space, where x and χ are diagonal. The first conclusion from this observation is that in the real space B and $\bar\chi$ are the operators of the bosonic and fermionic momenta:

$$i\hat B_i = \partial/\partial x^i, i\hat{\bar\chi}_i = \partial/\partial \chi^i. \tag{17.129}$$

The sign in front of the bosonic momentum is unambiguous because this operator acts on ordinary commuting variables. On the contrary, the sign in front of the fermionic momentum operator is ambiguous or rather can be established unambiguously only by a tedious exercise of carefully tracking all the signs associated with relative positions of anticommuting variables and their differentials in the pathintegral representation of the theory. There is a simpler way of performing this task. One can demand that the bi-graded commutator of the exterior derivative in the operator representation must act just as the operator of the Q-differentiation in the pathintegral representation. That is, as long as

$$\{Q, (x^i, \chi^i, B_i, \hat{\bar\chi}_i)\} = (\chi^i, 0, 0, B_i) \tag{17.130}$$

in the pathintegral representation, in the operator representation, we must have

$$[\hat d, (\hat x^i, \hat\chi^i, \hat B_i, \hat{\bar\chi}_i)] = (\hat\chi^i, 0, 0, \hat B_i). \tag{17.131}$$

It is readily seen that the choice of sign in Equation 17.129 satisfies this requirement.

Another important observation from Equation 17.128 is about the order of the operators. Both variables x and y (and χ and φ) correspond to operator $\hat x$ (and $\hat\chi$). The difference between, say, x's and y's is that y's act on the wavefunction before the momentum operators, B's, whereas x's act after B's. In particular, if the FP Hamilton function H had a term $B_i x_\alpha^i$; this term would transform in the operator representation into

$$B_i x_\alpha^i \xrightarrow{B_i \to \hat B_i} \alpha \hat x^i \hat B_i + (1-\alpha)\hat B_i \hat x^i. \tag{17.132}$$

The same can be said about fermionic operators with the only correction that the symmetrization must be "fermionic", that is,

$$\bar\chi_i \chi_\alpha^j \xrightarrow{\bar\chi_i \to \hat{\bar\chi}_i} (1-\alpha)\hat{\bar\chi}_i \hat\chi^i - \alpha\hat\chi^i \hat{\bar\chi}_i. \tag{17.133}$$

This reveals that if in the Ito case ($\alpha = 0$) all the momentum operators must act after all the position operators, in the Stratonovich case ($\alpha = 1/2$) the operators must be symmetrized.

This symmetrization is the well-known Weyl quantization rule of quantum theory. It guarantees that any real Hamilton function in the pathintegral representation results in a Hermitian Hamiltonian in operator representation of the theory. In our case, the Weyl–Stratonovich bi-graded symmetrization of FP Hamilton function of the pathintegral representation will result (see Reference 30 for details) in the FP operator in Equation 17.45, which is correct as we already established outside the pathintegral formulation of the theory (see Section 17.4.1). In other words, the Stratonovich interpretation of SDEs is correct and it corresponds to the bi-graded Weyl symmetrization on passing from the pathintegral to operator representation of the model.

17.5.4 Ergodicity, Response, and the Butterfly Effect

Let us consider now models with spectra given in Figure 17.2a and b. Physical expectation value of an operator, $\hat{O}(t)$, at time moment, t, can be defined as

$$\langle \hat{O}(t) \rangle = \lim_{t_\pm \to \pm\infty} Z_{t_+ t_-}^{-1} \text{Tr} \hat{\mathcal{M}}_{t_+ t} \hat{O}(t) \hat{\mathcal{M}}_{tt_-} = N_g^{-1} \sum_g \langle g | \hat{O}(t) | g \rangle. \tag{17.134}$$

Here, N_g is the number of the ground states and we used that in the limit of long temporal evolution only the ground states, g's, contribute into the nominator and that the exponential factors that come from the temporal evolution of bra's and ket's of the ground states, $\hat{\mathcal{M}}_{tt_-} | g \rangle = e^{-(t-t_-)\mathcal{E}_g} | g \rangle$ and $\langle g | \hat{\mathcal{M}}_{t_+ t} = \langle g | e^{-(t_+ - t)\mathcal{E}_g}$, are compensated by the same exponential factor in the denominator, $Z_{t_+ t_-} = N_g e^{-(t_+ - t_-)\mathcal{E}_g}$.

In case when $\hat{O}$ does not have explicit dependence on time, $\hat{O}(t) \to \hat{O}(\hat{\Phi}(t))$, its expectation value is independent of time argument:

$$\langle \hat{O}(\hat{\Phi}(t)) \rangle = \langle \hat{O}(\hat{\Phi}) \rangle, \tag{17.135}$$

because in the Schrödinger picture, that we used so far, operators are independent of time, $\hat{\Phi}(t) \equiv \hat{\Phi}$. The same is true for any other representation of the theory. For example, in the Heisenberg representation, where the temporal evolution is passed onto the operators,

$$\hat{O}(\Phi) \to \hat{O}_H(t) = \lim_{t_- \to -\infty} e^{(t-t_-)\hat{H}} \hat{O}(\Phi) e^{-(t-t_-)\hat{H}}, \tag{17.136}$$

while wavefunctions are viewed independent of time, the expectation value is

$$\langle \hat{O}_H(t) \rangle = N_g^{-1} \sum_g \langle g | \hat{O}_H(t) | g \rangle = N_g^{-1} \sum_g \langle g | e^{(t-t_-)\mathcal{E}_g} \hat{O}(t) e^{-(t-t_-)\mathcal{E}_g} | g \rangle, \tag{17.137}$$

which is exactly the r.h.s. of Equation 17.134.

Furthermore, for operators that are functions on $X, \hat{f} = f(x) \in \Omega^0(X)$, the physical expectation value is,

$$\langle f(x) \rangle = N_g^{-1} \sum_g \int_X f(x) \psi_g(x) \wedge \bar{\psi}_g(x) = N_g^{-1} \sum_g \int_X f(x) P_g^{(D)}(x), \tag{17.138}$$

where $P_g^{(D)}$ are the TPD associated with the ground states. Time-independence of expectation values such as Equation 17.138 is the ergodicity property, which can be defined as the property of reduction in the long time limit of (stochastic) expectation values to averages over the ground state(s) only. Thus, ergodicity is equivalent to the existence of the ground states of unique eigenvalue as in Figure 17.2b.

As to DSs with the spectrum of type given in Figure 17.2c, the ergodicity here is more subtle. The point is that there are at least two different eigenvalues that contribute into the dynamical partition function in the limit of the infinitely long temporal evolution. One of the ways to deal with this situation as we already mentioned in Section 17.4.5 can be borrowed from quantum theory. There, the trace of evolution operator has the following form, $Z = \sum_n e^{-itE_n}$, and all the eigenvalues of the Hermitian Hamiltonian, E_n, are real. None of the eigenstates will vanish from Z on its own. However, after Wick rotating time "a little," $t \to (1 + i0^+)t$, only the state with the lowest E_n will contribute into the dynamical partition function in the infinitely long time limit, and the expectation values will become vacuum expectation values. Having said that, from now on we only consider spectra of type Figure 17.2b.

Beware that in the literature, ergodicity is often misinterpreted as being equivalent to TE. This is not correct. The probability distribution in Equation 17.138 are not the wavefunctions themselves but rather the bra–ket combinations of the ground states' wavefunctions. This situation is similar with quantum theory, in which it is the bra–ket combination of eigenstates that are the TPD. At this, the fact that this combination (the diagonal element of the density matrix) is stationary in time by no means imply that the eigenstate has zero eigenvalue.

The fundamental difference between DSs at TE and with spontaneously broken topological supersymmetry can be revealed by studying their response to external perturbations. The physical way to couple a DS to external influence is only on the level on the SDE. This can be done by the introduction of the probing fields, $\phi^c(t), c = 1, 2 \ldots$, into the flow vector field in Equation 17.1:

$$F^i(x(t)) \rightarrow F^i(x(t)) - \phi^c(t)f_c^i(x(t)), \tag{17.139}$$

where f's is a set of some vector fields on X. The action of the model transforms accordingly,

$$S \rightarrow S + \int dt\phi^c(t) \left\{ Q, i\bar{\psi}_i(t)f_c^i(x(t)) \right\}. \tag{17.140}$$

The response of the DS is characterized by the set of the following expectation values:

$$\lim_{t_\pm \rightarrow \pm\infty} Z_{t_+ t_-}^{-1}(\phi) \frac{\delta^l Z_{t_+ t_-}(\phi)}{\delta\phi^{c_1}(t_1) \ldots \delta\phi^{c_l}(t_l)} \bigg|_{\phi=0}$$

$$= N_g^{-1} \sum_g \langle g | \mathcal{T} \prod_{k=1}^l \left[\hat{d}, i\hat{\bar{\psi}}_{i_k}(t_k)\hat{f}_{c_k}^{i_k}(x(t_k)) \right]_H | g \rangle. \tag{17.141}$$

Here, we used that Q-exact pieces in Equation 17.140 become $\hat{d}$-exact in the operator representation, operators in Equation 17.141 are in the Heisenberg representation, $\hat{d}(t) = \hat{d}$ is independent of time since $\hat{d}$ is commutative with $\hat{H}$, and $\mathcal{T}$ denotes chronological ordering.

When topological supersymmetry is not broken and all the ground state are $\hat{d}$-symmetric, all the perturbation correlators in Equation 17.141 vanish by the definition of the $\hat{d}$-exact states in Equation 17.59. In other words, in the long time limit, the DS does not respond to the perturbations or rather forgets them.[*] On the contrary, if the topological supersymmetry is spontaneously broken, some of the expectation values of the perturbation operators do not vanish. This means that the DS remembers perturbations even in the limit of infinite-time evolution. This is how the STS reveals the famous butterfly effect. Noteworthy, the butterfly effect is often viewed as a part of the definition of deterministic chaos, whereas within the STS it is a derivable consequence of the phenomenon of the spontaneous breakdown of topological supersymmetry.

17.6 Types of Ergodic Stochastic Dynamics

In this section, we apply the STS for the purpose of classification of ergodic stochastic DSs on the most general level pertinent to the topological supersymmetry that all (continuous-time) stochastic DSs possess. This analysis will reveal the theoretical essence of stochastic dynamical behavior that is known under such names as intermittency, complex dynamics, and SOC.

17.6.1 Transient versus Ergodic Dynamics

Before we proceed with the discussion of ergodic dynamics, let us address the following issue. One important type of dynamics is called transient dynamics. Roughly speaking, transient dynamics begins at one point of the phase space and ends at another. Physical examples of transient dynamics include quenches of various types such as Barkhausen effect, crumpling paper, etc. Another example is glasses: it is often said that (at nonzero temperature) a glass will eventually crystalize, that is, it will reach the state of TE corresponding to the crystal lattice structure. This process, however, may take a very long time and at this

---

[*] At this, on short times, the DS will respond to the perturbation but not in the long time limit, where only the ground states contribute.

very moment, an external observer sees transient dynamics from some initial point in the phase space to the crystal structure state.

In transient processes, boundary conditions for the bosonic fields are open as opposed to the PBC implied by the operation of taking trace of the evolution operator. The building blocks of transient dynamics are instantons. The latter are classical solutions that lead from less stable invariant manifolds to more stable ones. In terms of DS theory, instantons are the pieces of unstable manifolds connecting invariant manifolds of different stability. In the pathintegral representation, invariant manifolds correspond to perturbative (not global) ground states that are Poincaré-duals of unstable manifolds of these invariant manifolds (see below). Instantons in their turn are the matrix elements between such perturbative ground states. We will have more to say about instantons below.

It is well known that quenches and other transient processes often exhibit signatures of DLRO. One example of this behavior is the power-law statistics of Barkhausen jumps in ferromagnets. The mathematical origin of this long-range behavior in quenches has never been understood in the general case.[*] Within the STS, this long-range behavior can be attributed to the intrinsic breakdown of the topological supersymmetry within instantons. On a more technical level, it has been shown [12] that topological field theories are log-conformal when instantons condense. In case of the STS, this condensation of only instantons (and not anti-instatons) corresponds to dynamics which is a composite instanton or rather a sequence of "fundamental" instantons.

Transient dynamics is often referred to as out-of-equilibrium dynamics. The same term is often used for the characterization of chaotic behavior. The term "out-of-equilibrium" has two different meanings here. In the former case, it means nonergodic dynamics out of the global ground state of the DS, whereas in the latter case it denotes dynamics out of the $\hat{d}$-symmetric state of TE but within the global non-$\hat{d}$-symmetric ground state.

In this section, we are interested only in the properties of ergodic dynamics, which is also called sometimes "self-sustained" dynamics that last infinitely long. Transient dynamics is out of the scope of this chapter. In this context, it is also worth mentioning that the very concept of spontaneous symmetry breakdown formally applies only to ground states.

17.6.2 Unstable Manifolds and Ground States

17.6.2.1 Langevin SDE

In this section, we would like to discuss the relation between $\hat{d}$-symmetric ground states in the weak noise limit and unstable manifolds of flow vector fields. Let us first address Langevin SDEs—the most studied class of SDEs that is closely related to the N = 2 supersymmetric quantum mechanics.

For simplicity, we consider the case of flat phase space and additive "Euclidian" noise, $e_a^i = \delta_a^i$. The flow vector field, $F^i(x) = -\delta^{ij}U_{,j}(x)$, is defined via the Langevin potential, $U(x)$, and $U_{,j} = \partial U / \partial x^j$. The FP operator is given by Equation 17.49 with

$$\hat{\hat{d}} = \frac{\partial}{\partial \chi^i} \delta^{ij} \left(-U_{,j} - T \frac{\partial}{\partial x^j} \right). \tag{17.142}$$

Let us now apply the similarity transformation defined as, $\hat{A} \to \hat{A}_H = e^{U/(2T)} \hat{A} e^{-U/(2T)}$, to the FP evolution operator,

$$\hat{H} \to \hat{H}_U = T[\hat{d}_U, \hat{d}_U^{\dagger}], \tag{17.143}$$

---

[*] In quenches across phase transitions, this long-range behavior is often attributed to the "criticality" of the DS, that is, to the proximity of the phase transition. This may be a misleading explanation because quenches that are not across phase transitions also exhibit signatures of the long-range behavior and the criticality arguments are not applicable here. At this, it is natural to believe that the origin of this long-range behavior must be the same for all quenches.

where

$$\hat{d}_U = e^{U/(2\mathrm{T})}\hat{d}e^{-U/(2\mathrm{T})} = \chi^i\left(\frac{\partial}{\partial x^i} - U_{,i}/2\mathrm{T}\right), \tag{17.144}$$

$$\hat{\bar{d}}_U = \frac{\partial}{\partial \chi^i}\delta^{ij}\left(-U_{,j}/2 - \mathrm{T}\frac{\partial}{\partial x^j}\right) = \mathrm{T}\hat{d}_U^{\dagger}, \tag{17.145}$$

and we used $(\chi^i)^{\dagger} = \delta^{ij}\partial/\partial\chi^j$ and $(\partial/\partial x^i)^{\dagger} = -\partial/\partial x^i$.

Since $\hat{H}$ and $\hat{H}_U$ are related via the similarity transformation, their spectra are identical, whereas the eigenstates relate as

$$|\psi\rangle = e^{-U/2\mathrm{T}}|\psi_U\rangle, \text{ and } \langle\psi| = \langle\psi_U|e^{U/2\mathrm{T}}. \tag{17.146}$$

Up to the overall factor, T, operator $\hat{H}_U$ is the Hamiltonian of the N=2 supersymmetric quantum mechanics. It is Hermitian and its spectrum is real and nonnegative. This implies that topological supersymmetry is never broken in this class of models as long as we believe that there must always exist the $\hat{d}$-symmetric ground state of the TE (see discussion at the end of Section 17.4.2).[*]

In the single variable case with the harmonic potential, $U = \omega x^2$, the zero-eigenvalue ground state of $\hat{H}_U$ is (see, e.g., Section 10.2.4 of Reference 17):

$$|\psi_{g,U}\rangle = \langle\psi_{g,U}|^* \propto \begin{cases} \chi e^{-|\omega|x^2/2\mathrm{T}}, & \omega > 0, \\ e^{-|\omega|x^2/2\mathrm{T}}, & \omega < 0. \end{cases} \tag{17.147}$$

Here, the relation between bra's and ket's is trivial because $\hat{H}_U$ is Hermitian. In terms of the eigensystem of the non-Hermitian $\hat{H}$, the bra and ket look different. With the help of Equation 17.146, we have,

$$|\psi_g\rangle \propto \begin{cases} \chi e^{-|\omega|x^2/\mathrm{T}}, & \omega > 0, \\ 1, & \omega < 0, \end{cases} \text{ and } \langle\psi_g| \propto \begin{cases} 1, & \omega > 0, \\ \chi e^{-|\omega|x^2/\mathrm{T}}. & \omega < 0. \end{cases} \tag{17.148}$$

These are the ground state wavefunctions of the two models graphically presented in Figure 17.1. For the stable variable case ($\omega > 0$), the ket of the ground state is the narrow distribution around stationary position, $x = 0$, and the bra is not a distribution but rather is a constant function. In the unstable case ($\omega < 0$), the bra and ket swap places.

This analysis can be extended now to multiple-variable Langevin SDEs. Consider a vicinity of a nondegenerate critical point where the Langevin potential can be approximated as a quadratic form. With the appropriate coordinate rotation, this quadratic form can be diagonalized, $U = \sum_i \omega_i(x^i)^2/2, \omega_i \neq 0$, $i = 1 \ldots D$. The wavefunction of the (local) $\hat{d}$-symmetric ground state factorizes in all coordinates and each coordinate provides a factor of the form (17.148). In result, the wavefunction is a narrow distribution in stable variables and it is a constant function in unstable variables of the unstable manifold of this critical point as illustrated in Figure 17.5a.

The wavefunctions we just obtained are known as Poincaré duals. They appear in one of the versions of Poincaré duality stating that for each k-dimensional submanifold, c_k, there exist a differential form $\bar{\psi}_{c_k} \in \Omega^{D-k}$ such that $\int_{c_k} \varphi^{(k)} = \int_X \varphi^{(k)} \wedge \bar{\psi}_{c_k}$ for all $\varphi^{(k)} \in \Omega^k$. Using this terminology, the bra/ket of the local $\hat{d}$-symmetric ground state on a nondegenerate critical point of a Langevin SDE and in the weak noise limit are the Poincaré duals of the local stable/unstable manifolds of this critical point.

Local unstable manifolds have boundaries on the lower-dimensional unstable manifolds of more stable critical points. For example, in Figure 17.5b, local unstable manifold of the saddle point in the middle of

---

[*] In the literature, there are examples of N = 2 supersymmetric quantum models that do not have a single $\hat{d}$-symmetric global ground state and this situation is posed as the spontaneous supersymmetry breaking. One example is the Langevin SDE on $\mathbb{R}^1$ with the Langevin potential going to negative infinity at one of the phase space infinities. Clearly, this is an unphysical model from the point of view of stochastic dynamics as the system will eventually escape to the negative infinity and never come back. We do not consider unphysical models here.

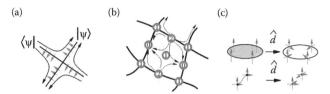

FIGURE 17.5 (a) In the deterministic limit, the bra, $\langle g'|$, and ket, $|g'\rangle$, of pertubative or local $\hat{d}$-symmetric ground state on a saddle are Poincaré duals of stable and unstable manifolds of the flow vector field represented here by four long curved arrows. Short arrows represent fermions or differentials of the Poincaré duals in transverse directions. (b) For integrable flow vector field, as the one represented here by curved dashed arrows, local (un)stable manifolds can be glued into global boundaryless (un)stable manifolds as the ones represented here by thick curves. Circled numbers are the indices of the corresponding critical points. (c) The exterior derivative acts on Poincaré duals of submanifolds as the boundary operator would have acted on the manifolds themselves. As a result, exterior derivative annihilates Poincaré duals of boundaryless global (un)stable manifolds of integrable flow vector field. These Poincaré duals are the bra's and ket's ($\langle g|$ and $|g\rangle$) of global supersymmetric ground states of deterministic integrable models.

the figure terminates at two nearest stable critical points. In the field-theoretic terms, this local unstable manifold consists of two instantons connecting this saddle and the two stable critical points.

The collection of all the local unstable manifolds of different dimensionality is known as the Morse complex, whereas the collection of the corresponding local $\hat{d}$-symmetric states as the Morse–Witten complex. Operator $\hat{d}$ acts on local $\hat{d}$-symmetric states as the boundary operator would have acted on the local unstable manifolds themselves (see, Figure 17.5c). Since local unstable manifolds have boundaries, the corresponding local $\hat{d}$-symmetric ground states are not $\hat{d}$-symmetric in the global sense for operator $\hat{d}$ does not annihilate them.

In order to get the global $\hat{d}$-symmetric ground states, one must glue local unstable manifolds into the global boundaryless unstable manifolds. The Poincaré duals of these global unstable manifolds are the global $\hat{d}$-symmetric ground states of this deterministic DS.

The discussion can be generalized to Morse–Bott situations then the critical points of the gradient flow vector field are not isolated but rather form closed submanifolds of X. In this case, the local $\hat{d}$-symmetric states must be complemented by the factors from the De Rahm cohomology of the critical submanifolds.

17.6.2.2 Deterministic Integrable Models

The existence of well-defined global (un)stable manifolds that are said to provide foliations of the phase space is the mathematical definition of integrability of a flow vector field in the sense of the DS theory. From the point of view of the STS, this means that Poincaré duals of these unstable manifolds are the ket's of the $\hat{d}$-symmetric global ground states.

Global $\hat{d}$-symmetric ground states are invariant with respect to the flow. Indeed, by definition, an (un)stable manifold consists of points that remain on them at all times. Therefore, the Poincaré dual of an unstable manifold, being a constant function on it, is unchanged by the flow. Another effect of the flow on a Poincaré dual of (un)stable manifold is squeezing in transverse directions. This will provide the corresponding Jacobian from the delta-functional dependence on transverse directions. Nevertheless, this Jacobian will be compensated in the supersymmetric manner by the Jacobian provided by the corresponding transformation of the differentials/ghosts in the transverse directions. As a result, the Poincaré duals of the global (un)stable manifolds are invariant under the flow, that is, they have zero eigenvalues.[*] In other words, the integrability of the flow vector field in the sense of the DS theory is equivalent to the unbroken topological supersymmetry in the corresponding STS.

---

[*] That these states are $\hat{d}$-symmetric follows trivially from the fact that the global (un)stable manifolds have no boundaries.

Each De Rahm cohomology class may contain more than one global $\hat{d}$-symmetric ground states. Each of these ground states is a superposition of one $\hat{d}$-symmetric ground state and a $\hat{d}$-exact piece. This means that pairs of non-$\hat{d}$-symmetric states accidentally have zero eigenvalues. This is true only in the strict deterministic limit, as any weak noise will introduce exponentially weak tunneling effects (instanton–anti-instanton configurations) that must lift this accidental degeneracy and leave only one $\hat{d}$-symmetric ground state in each De Rahm cohomology class.

17.6.2.3 Deterministic Chaotic Models

Let us now turn to the analysis of the ground states of chaotic or nonintegrable deterministic models. We believe that the model is ergodic, that is, the ground state(s) with the least attenuation rate exist. The ground state(s) must represent dynamics on strange attractors. Just like in the integrable models, the strange attractors are formed by the intersection of corresponding stable and unstable manifolds and the bra/ket of the ground state's wavefunction must represent (or rather be) the Poincaré duals of these submanifolds of X.

In chaotic models, however, (un)stable manifolds are not well-defined topological manifolds. They can fold on themselves in a recursive manner as qualitatively illustrated for "homoclinic tangle" in Figure 17.6a. As a result, the straightforward attempt to come up with a Poincaré dual for such an unstable manifold leads to the ambiguity in the orientation of the manifold at the point of the accumulation of the self-folding.

This situation has its analogs in quantum theory. For example, nonrotationally symmetric electron wavefunctions on a rotationally symmetric atom (p,d,f... orbitals) will have an ambiguity at the origin unless it vanishes there. Another example is from the theory of superfluids, where the superfluidic order parameter of the Bose condensate at the core of a vortex must vanish because otherwise its value will be ambiguous.

In our case, the ambiguity of the Poincaré dual of the unstable manifold of a homoclinic tangle can be remedied by modifying it with a continuous function that goes to zero at the point of the accumulation of the self-folding. At this, the coordinate dependence along the manifold would automatically suggest that the wavefunction is not annihilated by $\hat{d}$. In other words, the wavefunction of the ground state is not $\hat{d}$-symmetric.

Another way to understand why in chaotic deterministic models the ground state is not $\hat{d}$-symmetric is provided by the topological theory of chaos [14]. In this theory, the unstable manifold of a chaotic flow can be qualitatively represented as a branched manifold that has self-intersections as illustrated for the case of Rössler model in Figure 17.6b. Acting by $\hat{d}$ on the Poincaré dual of this branched manifold gives Poincaré dual of its self-intersection. That is, such wavefunction is also not $\hat{d}$-symmetric.

(a)   (b)

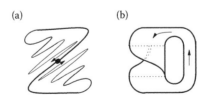

FIGURE 17.6 (a) The qualitative representation of (the Poincaré section) of unstable manifold of the deterministic chaotic behavior known as homoclinic tangle. The unstable manifold folds on itself in a recursive manner and this folding accumulates at the origin. There, the orientation of the unstable manifold is not well defined and the wavefunction must vanish as is indicated by the fading width of the curve. This suggests that $\hat{d}|\psi\rangle \neq 0$ and that the topological supersymmetry is broken. (b) The effective representation of the unstable manifold of the Rössler model in the topological theory of chaos (see, e.g., Reference 14). The unstable manifold is a branching manifold that has self-intersection. An attempt to construct a Poincaré dual will lead to either the requirement that the wavefunction vanishes at self-intersections or to the effective existence of a boundary at the self-intersection. In either case, the corresponding wavefunction is non-$\hat{d}$-symmetric.

The above qualitative analysis of the ground states of deterministic chaotic models is only an indication that the topological supersymmetry breaking is the field-theoretic essence of deterministic chaos. The rigorous proof of this statement is given by Equation 17.90 that establishes the exponential growth of periodic solutions, which is definitive for chaos.

17.6.3 Phase Diagram

The picture of the spontaneous breakdown of $\hat{d}$-symmetry in deterministic models discussed above is relatively simple: the $\hat{d}$-symmetry is spontaneously broken or not depending of whether its flow vector field is nonintegrable (chaotic) or integrable in the sense of the DS theory. In this section, we discuss the stochastic generalization of this picture.

Let us first note that in the very high temperature limit, the FP operator (17.45) is dominated by the diffusion Laplacian. In a wide class of models such as models with torsion-free vielbeins 28, the diffusion Laplacian equals the Hodge Laplacian (17.61). The latter has real and nonnegative spectrum, which corresponds to unbroken $\hat{d}$-symmetry. In such models, the $\hat{d}$-symmetry is not broken at sufficiently high temperatures irrespective of whether it was broken at lower temperatures or not. Here, we only consider models of this type, that is, the models in which the diffusion Laplacian alone does not break the $\hat{d}$-symmetry. It can be said that in these models the noise will eventually destroy the DLRO as one rises the temperature.

Two qualitatively different types of the phase diagram exist for this class of models (see Figure 17.7). For the first type, the $\hat{d}$-broken phase gradually narrows as one rises the temperature, until it completely disappears. This is the situation discussed, for example, in Reference 18.

The second type of the phase diagram Figure 17.7b is more involved. There, the $\hat{d}$-broken phase first widens with the temperature before it starts shrinking and then disappears at even higher temperatures. For this type of models, there exists a phase with integrable flow vector fields but with the spontaneously broken $\hat{d}$-symmetry. This phase can be called the noise-induced chaos because in the deterministic limit, it collapses onto the boundary of the deterministic chaos.

17.6.3.1 Low-Temperature Regime

Let us first address the low-temperature regime. The $\hat{d}$-broken phase consists of the two major subphases: the ordinary chaotic phase (C-phase) where the $\hat{d}$-symmetry is broken by nonintegrability of the flow vector field, and the noise-induced or intermittent chaotic phase (I-phase) where the flow vector field is integrable so that the $\hat{d}$-symmetry is broken by some other mechanism.

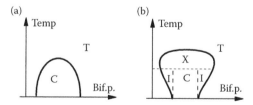

FIGURE 17.7 Two qualitatively different types of general phase diagram. As compared to the first type (a), consisting of chaotic phase (C) and the phase of the thermodynamics equilibrium (T), the second type (b) has the phase of noise-induced chaos. In this phase, the flow vector field is integrable but the topological supersymmetry is spontaneously broken by the condensation of (anti-)instantons, that is, the noise-induced tunneling processes between, for examples, different attractors. This phase can be called intermittent (I) because an external observer will see sporadic jumps between different patterns of "regular" behavior, that is, the perturbative supersymmetric ground states. At higher temperatures, the sharp boundary between the I- and C-phases must smear out because perturbative supersymmetric ground states will overlap significantly and it will be hard to tell one (anti-)instanton from another. Under these condition, the I- and C-phases will merge into one complicated phase (X) with spontaneously broken supersymmetry.

There are two known additional mechanisms that may lead to the spontaneous breakdown of a symmetry. The first one is anomaly, that is, the possibility to break a symmetry by perturbative corrections. Supersymmetries, however, are very hard (if at all possible) to break pertubratively because of what is generally known as non-renormalization theorems. Therefore, the $\hat{d}$-symmetry in the I-phase must be broken by the other remaining mechanism. This mechanism is known as the condensation of (anti-)instanton configurations [37].[*] In terms of stochastic dynamics, these (anti-)instanton configurations are the tunneling processes that appear due to the exponentially weak overlap between perturbative ground states localized on global invariant manifolds.

The first effect that the noise-induced tunneling processes will provide is the lifting of the degeneracy of the deterministic zero-eigenvalue eigenstates corresponding to Poincaré duals of "parallel" global unstable manifolds within the same De Rahm cohomology class.[†] As a result, each De Rahm cohomology class will have only one $\hat{d}$-symmetric eigenstate, whereas other eigenstates will acquire (exponentially small) nonzero eigenvalues.

A more subtle question is whether or not the condensation will lead to the spontaneous breakdown of the $\hat{d}$-symmetry, that is, to the emergence of eigenstate(s) with negative (real part of its) eigenvalue. Clearly, the very existence of the noise-induced tunneling processes is not enough for this to happen. For example, the class of Langevin SDEs discussed in Section 17.6.2 and with Langevin functions that have multiple local minima, the tunneling processes between these local minima certainly exist. At this, the $\hat{d}$-symmetry is never broken for this class of models (the eigenvalues of the evolution operator are real and nonnegative).

In other words, the weak-noise tunneling effects can only help the spontaneous $\hat{d}$-symmetry breaking in models with "nearly chaotic" flow vector fields. This is the reason why the I-phase exists on the "border" of the T- and C-phases.

The physical picture of the weak-noise stochastic ergodic dynamics in the I-phase is as follows. The dynamics is mostly alone the global unstable manifolds (e.g., attractors) of the integrable flow vector field. This dynamics is interrupted sporadically by the noise-induced tunneling processes between "parallel" global unstable manifolds. Moreover, since it is the noise-induced tunneling processes that break the $\hat{d}$-symmetry, they must exhibit signatures of the DLRO. This signature might as well be an algebraic statistics of the tunneling processes.

If in the discussion of the I-phase dynamics in the previous paragraph we substitute the term "weak-noise-induced tunneling processes" by the term "avalanches," we will arrive at the classical physical description of the self-organized critical (SOC) behavior [2,4].

17.6.3.2 High-Temperature Regime

In the previous discussion of the weak-noise regime, the concept of the noise-induced tunneling processes is well defined because the overlap between the perturbative ground states is exponentially weak. As a result, an external observer will be able to tell one tunneling process from another. At higher temperatures, the overlap is not weak anymore and it may become hard for an external observer to tell one tunneling event from another. This suggests that the sharp boundary between the C- and I-phases must smear out into a crossover. Note that the boundary between the I- and C-phases is not a $\hat{d}$-symmetry breaking phase transition in the first place so that its disappearance does not contradict any symmetry-based argument.

It can be said that above a certain temperature, the C- and I-phases must merge into a complicated phase with the spontaneously broken $\hat{d}$-symmetry. In Figure 17.7b, this phase is indicated as an X-phase. Borrowing from the high-energy physics terminology, this phase can probably be identified as the phase of stochastic chaos in the "strongly-coupled" regime.

---

[*] Owing to the renormalization theorems, the dynamical supersymmetry breaking by (anti-)instantons is considered one of the most reliable mechanism of supersymmetry breaking in the high-energy physics models.

[†] See the discussion in the last paragraph of Section 17.6.2.2.

17.7 Conclusion

In this chapter, the current state of the recently proposed approximation-free and coordinate-free supersymmetric theory of SDEs is presented. It was discussed that the STS has provided a few novel mathematical insights on stochastic dynamics. In particular, it revealed that the field-theoretic essence of stochastic (and deterministic) chaos (SC) is the phenomenon of the spontaneous breakdown of the topological supersymmetry ($\hat{d}$-symmetry) pertinent to all SDEs, and that the SC is complementary to the concept of the TE corresponding, in its turn, to the unbroken $\hat{d}$-symmetry. It is worth noting that this approximation-free picture of the SC is in a certain sense opposite to the semantics of word "chaos": if in the daily language "chaos" denotes the absence of order, in the STS, it is the SC that is the ordered (or low-symmetry) phase and not the TE phase. In other words, the SC phase has a DLRO, whereas the TE does not.

The presence of the DLRO is the reason why many natural nonlinear DSs exhibit emergent scale-free behavior such as 1/f noise, butterfly effect, and algebraic (or scale-free) statistics of instantonic or avalanche-type processes. This understanding and a few other qualitative findings such as the clarification of the concept of egrodicity are the main outcomes from the STS so far. It is expected, however, that further development of the STS will lead to more specific and valuable results and one of the most fruitful directions of further investigation is the work on the methodology for the (i) identification of the DLRO parameter and (ii) construction of the low-energy effective theory (LEET) for it. We do not have this methodology beyond a few thoughts related to this important issue. Allow us to discuss those here briefly. To be more concrete, we will speak about continuous-space models such as hydrodynamical models.

The most important or qualitative part of the dynamics in models with spontaneously broken global symmetry occupies a reduced phase space and/or in a reduced number of "low-energy" variables and the LEETs are the theories describing this dynamics. For example, in ferromagnets, these variables are the local magnetization of electron spins and the LEET or rather the equation of motion of the LEET is the Landau–Lifshitz–Gilbert equation. In superconductors, the order parameter is the local wavefunction of the condensate of the Cooper pairs and the LEET is the corresponding Ginzburg–Landau functional. In solids, in which the global translational symmetry is broken by the lattice structure of the ground state, the order parameter is the local displacement of atoms from their average positions in the lattice and the LEET is the low-energy theory describing, say, the propagation of transverse sound, which is the Goldstone mode in this case.

As to the DLRO, the most important variables are the unstable and/or unthermalized variables of the non-d-symmetric ground state, in which a chaotic DS has an infinite memory of initial conditions and/or perturbations. The order parameter must be a gapless fermionic field (because the $\hat{d}$-symmetry is a fermionic symmetry) consisting of supersymmetric partners of these unstable variables. For a spontaneously broken supersymmetry, this fermionic field is often called goldstinos to emphasize that their gaplessness is the consequence of the Goldstone theorem. In many interesting continuous-space models, unstable variables are the modulii of solitonic configurations comprising of fundamental solitons such as kinks, domain walls, vortices, etc. The processes of the creation/annihilation of (pairs of) these fundamental solitons are the (anti-)instatonic processes, the condensation of which is the essence of the noise-induced (or intermittent) chaotic behavior.

Generally speaking, there are actually two major types of orders in physics: the local order and the global or topological order. In the case of a global continuous symmetry breaking as in the case of the d-symmetry, a local order parameter picture discussed above typically suffices. On the other hand, it is intuitively appealing to believe that the DLRO of the topological supersymmetry breaking must also have a topological or global character (at least partially). On the level of the LEET, this global character may probably appear through the nonlocal interactions between goldstinos, such as the interaction via chiral gauge fields. This points to the possibility that there may exist a class of continuous-space DSs with the LEETs being goldstinos interacting via a chiral gauge field. One of the candidates for this class of model is a two-dimensional topological-defects-induced turbulence where the goldstinos must represent the spatial positions of (anti-)vortices and the chiral gauge field must represent the long-term memory of braining between them. This picture of dynamics looks qualitatively similar to the Schwartz-type topological field theories used in models related to the concept of topological quantum computing.

Apart from the methodology of LEET, there are a few other open questions even on the interpretational side of the STS. For example, what happens with the wavefunction when one variable is observed/measured? The interpretation of the wavefunction as of the GPD suggests that the fermionic content of the wavefunction may as well change suddenly on observation. If this is indeed true, does this sudden change has anything to do with the wavefunction collapse on observation in quantum theory or the Bayesian update of the probability distribution in the probability theory?

Acknowledgment

We thank Kang L. Wang for encouragement and discussions.

References

1. L. Arnold. *Random Dynamical Systems*. Springer-Verlag, Berlin, Heidelberg, New York, 2003.
2. M. Aschwanden. *Self-Organized Criticallity in Astrophysics: Statistics of Nonlinear Processes in the Universe*. Springer-Verlag, Berlin, Heidelberg, 2011.
3. W. R. Ashby. Principles of the self-organizing system, In *Principles of Self-Organization: Transactions of the University of Illinois Symposium*, H. Von Foerster and G. W. Zopf, Jr. (eds.), Pergamon Press: London, UK, pp. 255–278, 1962.
4. P. Bak, C. Tang, and K. Wiesenfeld. Self-organized criticality: An explanation of the 1/f noise. *Physical Review Letters*, 59:381, 1987.
5. P. Baxendale and S. V. Lototsky. *Stochastic Differential Equations: Theory and Applications*. World Scientific Publishing, Singapore, 2007.
6. J. M. Beggs and D. Plenz. Neuronal avalanches in neocortical circuits. *The Journal of Neuroscience*, 23(35):11167–11177, 2003.
7. J. M. Beggs and D. Plenz. Neuronal avalanches are diverse and precise activity patterns that are stable for many hours in cortical slice cultures. *The Journal of Neuroscience*, 24:5216, 2004.
8. D. Birmingham, M. Blau, M. Rakowski, and G. Thompson. Topological field theory. *Physics Reports*, 209:129–340, 1991.
9. D. R. Chialvo. Emergent complex neural dynamics. *Nature Physics*, 6:744–750, 2010.
10. E. Deotto and E. Gozzi. On the "Universal" $N = 2$ supersymmetry of classical mechanics. *International Journal of Modern Physics A*, 16:2709–2746, 2001.
11. J. P. Eckmann and D. Ruelle. Ergodic theory of chaos and strange attractors. *Reviews of Modern Physics*, 57:617–656, Jul 1985.
12. E. Frenkel, A. Losev, and N. Nekrasov. Notes on instantons in topological field theory and beyond. *Nuclear Physics B–Proceedings Supplements*, 171:215, 2007.
13. K. Gawedzki and A. Kupiainen. Critical behaviour in a model of stationary flow and supersymmetry breaking. *Nuclear Physics B*, 269:45, 1986.
14. R. Gilmore. Topological analysis of chaotic dynamical systems. *Reviews of Modern Physics*, 70:1455–1529, 1998.
15. B. Gutenberg and C. F. Richter. Magnitude and energy of earthquakes. *Nature*, 176:795, 1955.
16. T. Hida, R. L. Karandikar, H. Kunita, B. S. Rajput, S. Watanabe, and J. Xiong. *Stochastics in Finite and Infinite Dimensions*. Springer Science+Business Media, New York, 2001.
17. K. Hori, S. Katz, A. Klemm, R. Pandharipande, R. Thomas, C. Vafa, R. Vakil, and E. Zaslow. *Mirror Symmetry*. American Mathematical Society, Clay Mathematics Institute, Cambridge, Massachusetts, 2003.

18. T. Kapitaniak. *Chaos in Systems with Noise.* World Scientific, Singapore, New Jersey, London, Hong Kong, 1990.

19. S. Kogan. *Electronic Noise and Fluctuations in Solids.* Cambridge University Press, Cambridge, 1996.

20. J. M. F. Labastida. Morse theory interpretation of topological quantum field theories. *Communications in Mathematical Physics,* 123:641, 1989.

21. A. Levina, J. M. Herrmann, and T. Geisel. Dynamical synapses causing self-organized criticality in neural networks. *Nature Physics,* 3:857–860, 2007.

22. A. Manning. Topological entropy and the first homology group, in *Dynamical Systems,* Lecture Notes in Mathematics, Vol. 486, pp. 185–190, 2006.

23. W. Moon and J. S. Wettlaufer. On the interpretation of Stratonovich calculus. *New Journal of Physics,* 16(5):055017, 2014.

24. A. Mostafazadeh. Pseudo-supersymmetric quantum mechanics and isospectral pseudo-hermitian hamiltonians. *Nuclear Physics B,* 640(3):419–434, 2002.

25. A. E. Motter and D. K. Campbell. Early chaos theory. *Physics Today,* 66:27, 2014.

26. B. Øksendal. *Stochastic Differential Equations: An Introduction with Applications.* Springer-Verlag, Berlin, Heidelberg, 2010.

27. I. V. Ovchinnikov. Self-organized criticality as Witten-type topological field theory with spontaneously broken Becchi–Rouet–Stora–Tyutin symmetry. *Physical Review E,* 83:051129, 2011.

28. I. V. Ovchinnikov. Topological field theory of dynamical systems. *Chaos,* 22:033134, 2012.

29. I. V. Ovchinnikov. Topological field theory of dynamical systems. II. *Chaos,* 23:013108, 2013.

30. I. V. Ovchinnikov. Transfer operators and topological field theory. *arXiv:1308.4222,* 2014.

31. I. V. Ovchinnikov. Introduction to supersymmetric theory of stochastics, *Entropy,* 18:108, 2016.

32. G. Parisi and N. Sourlas. Random magnetic fields, supersymmetry, and negative dimensions. *Physical Review Letters,* 43:744–745, 1979.

33. G. Parisi and N. Sourlas. Supersymmetric field theories and stochastic differential equations. *Nuclear Physics B,* 206:321–332, 1982.

34. T. Preis, J. J. Schneider, and H. E. Stanley. Switching processes in financial markets. *Proceedings of the National Academy of Sciences,* 108(19):7674–7678, 2011.

35. D. Ruelle. Dynamical zeta functions and transfer operators. *Notices of AMS,* 49:887–895, 2002.

36. P. Sibani and H. J. Jensen. *Stochastic Dynamics of Complex Systems.* World Scientific Publishing, Singapore, 2013.

37. E. Witten. Dynamical breaking of supersymmetry. *Nuclear Physics B,* 188(3):513–554, 1981.

38. E. Witten. Supersymmetry and morse theory. *Journal of Differential Geometry,* 17(4):661–692, 1982.

39. E. Witten. Topological quantum field theory. *Communications in Mathematical Physics,* 117(3):353, 1988.

40. E. Witten. Topological sigma models. *Communications in Mathematical Physics,* 118(3):411–449, 1988.

18

New Robust Stability of Discrete-Time Hybrid Systems

Grienggrai Rajchakit

18.1 Introduction

Stochastic modeling has come to play an important role in many branches of science and industry. An area of particular interest has been the automatic control of stochastic systems, with consequent emphasis being placed on the analysis of stability in stochastic models. One of the most useful stochastic models that appear frequently in applications is the stochastic differential delay equations. In practice, we need to estimate the parameters of systems. If the parameters are estimated using point estimations, the systems are described precisely and hence the study of the systems becomes relatively easier. On the other hand, if the parameters are estimated using confidence intervals, the systems become stochastic interval equations and the study of such systems is much more complicated.

Switched systems constitute an important class of hybrid systems. Such systems can be described by a family of continuous-time subsystems (or discrete-time subsystems) and a rule that orchestrates the switching between them. It is well known that a wide class of physical systems in power systems, chemical process control systems, navigation systems, automobile speed change system, and so forth may be appropriately described by the switched model [1–8]. In the study of switched systems, most works have been centralized on the problem of stability. In the last two decades, there has been increasing interest in the stability analysis for such switched systems; see, for example, References 9 through 16 and the references cited therein. Two important methods are used to construct the switching law for the stability analysis of the switched systems. One is the state-driven switching strategy [17–22]; the other is the time-driven switching strategy [23–25]. A switched system is a hybrid dynamical system consisting of a finite number of subsystems and a logical rule that manages switching between these subsystems (see, e.g., References 22 through 26 and the references therein).

The main approach for stability analysis relies on the use of Lyapunov–Krasovskii functional and linear matrix inequality (LMI) approach for constructing a common Lyapunov function [21–30]. Although many important results have been obtained for switched linear continuous-time systems, there are few results concerning the stability of switched linear discrete systems with time-varying delays. In References 20 through 29, a class of switching signals has been identified for the considered switched discrete-time delay systems to be stable under the average dwell time scheme.

This chapter studies a robust mean square stability problem for an uncertain stochastic-switched linear discrete-time delay with interval time-varying delays. Specifically, our goal is to develop a constructive way to design the switching rule to robustly mean square stable the uncertain stochastic linear discrete-time delay systems. By using an improved Lyapunov–Krasovskii functional combined with LMI techniques, we propose new criteria for the robust mean square stability of the uncertain stochastic linear discrete-time delay system. Compared with the existing results, our result has its own advantages. First, the time delay is assumed to be a time-varying function belonging to a given interval, which means that the lower and upper bounds for the time-varying delay are available, and the delay function is bounded but not restricted to zero. Second, the approach allows us to design the switching rule for a robust mean square stability in terms of LMIs. Finally, some examples are exploited to illustrate the effectiveness of the proposed schemes.

The chapter is organized as follows: Section 18.2 presents definitions and some well-known technical propositions needed for the proof of the main results. Switching rule for the robust mean square stability is presented in Section 18.3. Numerical examples are provided to illustrate the theoretical results in Section 18.4, and the conclusions are drawn in Section 18.5.

18.2 Preliminaries

The following notations will be used throughout this chapter. R^+ denotes the set of all real nonnegative numbers; R^n denotes the n-dimensional space with the scalar product of two vectors $\langle x, y \rangle$ or $x^T y$; and $R^{n \times r}$ denotes the space of all matrices of $(n \times r)$—dimension. N^+ denotes the set of all nonnegative integers; A^T denotes the transpose of A; and matrix A is symmetric if $A = A^T$.

Matrix A is semipositive definite ($A \geq 0$) if $\langle Ax, x \rangle \geq 0$, for all $x \in R^n$; A is positive definite ($A > 0$) if $\langle Ax, x \rangle > 0$ for all $x \neq 0$; and $A \geq B$ means $A - B \geq 0$. $\lambda(A)$ denotes the set of all eigenvalues of A; $\lambda_{\min}(A) = \min\{Re\lambda : \lambda \in \lambda(A)\}$.

Consider an uncertain stochastic discrete systems with interval time-varying delay of the form

$$x(k+1) = (A_\gamma + \Delta A_\gamma(k))x(k) + (B_\gamma + \Delta B_\gamma(k))x(k - d(k)) + \sigma_\gamma(x(k), x(k - d(k)), k)\omega(k), \quad (18.1)$$
$$k \in N^+, \quad x(k) = v_k, \quad k = -d_2, -d_2 + 1, \ldots, 0,$$

where $x(k) \in R^n$ is the state, $\gamma(.) : R^n \to \mathcal{N} := \{1, 2, \ldots, N\}$ is the switching rule, which is a function depending on the state at each time, and will be designed. A switching function is a rule that determines a switching sequence for a given switching system. Moreover, $\gamma(x(k)) = i$ implies that the system realization is chosen as the ith system, $i = 1, 2, \ldots, N$. It is seen that system (18.1) can be viewed as an autonomous switched system in which the effective subsystem changes when the state $x(k)$ hits predefined boundaries. $A_i, B_i, i = 1, 2, \ldots, N$ are the given constant matrices and the time-varying uncertain matrices $\Delta A_i(k)$ and $\Delta B_i(k)$ are defined by $\Delta A_i(k) = E_{ia}F_{ia}(k)H_{ia}, \Delta B_i(k) = E_{ib}F_{ib}(k)H_{ib}$, where E_{ia}, E_{ib}, H_{ia}, and H_{ib} are known constant real matrices with appropriate dimensions. $F_{ia}(k), F_{ib}(k)$ are unknown uncertain matrices satisfying

$$F_{ia}^T(k)F_{ia}(k) \leq I, \quad F_{ib}^T(k)F_{ib}(k) \leq I, \quad k = 0, 1, 2, \ldots, \quad (18.2)$$

where I is the identity matrix of appropriate dimension, $\omega(k)$ is a scalar Wiener process (Brownian motion) on $(\Omega, \mathcal{F}, \mathcal{P})$ with

$$E[\omega(k)] = 0, \quad E[\omega^2(k)] = 1, \quad E[\omega(i)\omega(j)] = 0(i \neq j), \quad (18.3)$$

and $\sigma_i : R^n \times R^n \times R \to R^n, i = 1, 2, \ldots, N$ is the continuous function, and is assumed to satisfy that

$$\sigma_i^T(x(k), x(k - d(k)), k)\sigma_i(x(k), x(k - d(k)), k) \leq \rho_{i1}x^T(k)x(k) + \rho_{i2}x^T(k - d(k))x(k - d(k)),$$
$$x(k), x(k - d(k) \in R^n, \quad (18.4)$$

where $\rho_{i1} > 0$ and $\rho_{i2} > 0, i = 1, 2, \ldots, N$ are known constant scalars. The time-varying function $d(k)$: $N^+ \rightarrow N^+$ satisfies the following condition:

$$0 < d_1 \leq d(k) \leq d_2, \quad \forall k \in N^+.$$

Remark 18.1

It is worth noting that the time delay is a time-varying function belonging to a given interval, in which the lower bound of delay is not restricted to zero.

Definition 18.1

The uncertain stochastic-switched system (18.1) is robustly stable if there exists a switching function $\gamma(.)$ such that the zero solution of the uncertain stochastic-switched system is robustly stable.

Definition 18.2

The system of matrices $\{J_i\}, i = 1, 2, \ldots, N$ is said to be strictly complete if for every $x \in R^n \backslash \{0\}$ there is $i \in \{1, 2, \ldots, N\}$ such that $x^T J_i x < 0$.

It is easy to see that system $\{J_i\}$ is strictly complete if and only if

$$\bigcup_{i=1}^{N} \alpha_i = R^n \backslash \{0\},$$

where

$$\alpha_i = \{x \in R^n : \quad x^T J_i x < 0\}, i = 1, 2, \ldots, N.$$

Definition 18.3

The discrete-time system (18.1) is robustly stable in the mean square if there exists a positive definite scalar function $V(k, x(k)) : R^n \times R^n \rightarrow R$ such that

$$E[\Delta V(k, x(k))] = E[V(k+1, x(k+1)) - V(k, x(k))] < 0,$$

along any trajectory of solution of system (18.1).

Proposition 18.1: [31]

System $\{J_i\}, i = 1, 2, \ldots, N$ is strictly complete if there exist $\delta_i \geq 0, i = 1, 2, \ldots, N, \sum_{i=1}^{N} \delta_i > 0$ such that

$$\sum_{i=1}^{N} \delta_i J_i < 0.$$

If $N = 2$ then the above condition is also necessary for strict completeness.

Proposition 18.2: (Cauchy inequality)

For any symmetric positive definite matrix $N \in M^{n \times n}$ and $a, b \in R^n$ we have

$$\pm a^T b \leq a^T N a + b^T N^{-1} b.$$

Proposition 18.3: [31]

Let E, H, and F be any constant matrices of appropriate dimensions and $F^T F \leq I$. For any $\epsilon > 0$, we have

$$EFH + H^T F^T E^T \leq \epsilon EE^T + \epsilon^{-1} H^T H.$$

18.3 Main Results

Let us set

$$W_i = \begin{bmatrix} W_{i11} & W_{i12} & W_{i13} \\ * & W_{i22} & W_{i23} \\ * & * & W_{i33} \end{bmatrix},$$

where

$$
\begin{aligned}
W_{i11} &= Q - P, \\
W_{i12} &= S_1 - S_1 A_i, \\
W_{i13} &= -S_1 B_i, \\
W_{i22} &= P + S_1 + S_1^T + H_{ia}^T H_{ia} + S_1 E_{ib} E_{ib}^T S_1^T, \\
W_{i23} &= -S_1 B_i, \\
W_{i33} &= -Q + 2 H_{ib}^T H_{ib} + 2 \rho_{i2} I, \\
J_i &= (d_2 - d_1) Q - S_1 A_i - A_i^T S_1^T + 2 S_1 E_{ia} E_{ia}^T S_1^T + S_1 E_{ib} E_{ib}^T S_1^T + H_{ia}^T H_{ia} + 2 \rho_{i1} I, \\
\alpha_i &= \{x \in R^n : \quad x^T J_i x < 0\}, \ i = 1, 2, \ldots, N, \\
\bar{\alpha}_1 &= \alpha_1, \quad \bar{\alpha}_i = \alpha_i \setminus \bigcup_{j=1}^{i-1} \bar{\alpha}_j, \quad i = 2, 3, \ldots, N.
\end{aligned}
$$

(18.5)

The main result of this chapter is summarized in the following theorem.

Theorem 18.1

The uncertain stochastic-switched system (18.1) is robustly stable in the mean square if there exist symmetric positive definite matrices $P > 0$, $Q > 0$, and matrix S_1 satisfying the following conditions:

1. $\exists \delta_i \geq 0, i = 1, 2, \ldots, N, \sum_{i=1}^{N} \delta_i > 0 : \sum_{i=1}^{N} \delta_i J_i < 0$.
2. $W_i < 0, \quad i = 1, 2, \ldots, N$.

The switching rule is chosen as $\gamma(x(k)) = i$, whenever $x(k) \in \bar{\alpha}_i$.

Proof. Consider the following Lyapunov–Krasovskii functional for any ith system (18.1):

$$V(k) = V_1(k) + V_2(k) + V_3(k),$$

where

$$V_1(k) = x^T(k) P x(k), \quad V_2(k) = \sum_{i=k-d(k)}^{k-1} x^T(i) Q x(i),$$

$$V_3(k) = \sum_{j=-d_2+2}^{-d_1+1} \sum_{l=k+j+1}^{k-1} x^T(l) Q x(l).$$

We can verify that

$$\lambda_1 \|x(k)\|^2 \leq V(k). \tag{18.6}$$

Let us set $\xi(k) = [x(k)\,x(k+1)\,x(k-d(k))\,\omega(k)]^T$ and

$$H = \begin{pmatrix} 0 & 0 & 0 & 0 \\ 0 & P & 0 & 0 \\ 0 & 0 & 0 & 0 \\ 0 & 0 & 0 & 0 \end{pmatrix}, \quad G = \begin{pmatrix} P & 0 & 0 & 0 \\ I & I & 0 & 0 \\ 0 & 0 & I & 0 \\ 0 & 0 & 0 & I \end{pmatrix}.$$

Then, the difference of $V_1(k)$ along the solution of system (18.1) and taking the mathematical expectation, we obtained

$$E[\Delta V_1(k)] = E[x^T(k+1)Px(k+1) - x^T(k)Px(k)]$$

$$= E\left[\xi^T(k)H\xi(k) - 2\xi^T(k)G^T \begin{pmatrix} 0.5x(k) \\ 0 \\ 0 \\ 0 \end{pmatrix} \right]. \tag{18.7}$$

because of

$$\{\xi^T(k)H\xi(k)\} = x(k+1)Px(k+1),$$

$$2\xi^T(k)G^T \begin{pmatrix} 0.5x(k) \\ 0 \\ 0 \\ 0 \end{pmatrix} = x^T(k)Px(k).$$

Using the expression of system (18.1)

$$0 = -S_1 x(k+1) + S_1(A_i + E_{ia}F_{ia}(k)H_{ia})x(k) + S_1(B_i + E_{ib}F_{ib}(k)H_{ib})x(k-d(k)) + S_1\sigma_i\omega(k),$$

$$0 = -\sigma_i^T x(k+1) + \sigma_i^T(A_i + E_{ia}F_{ia}(k)H_{ia})x(k) + \sigma_i^T(B_i + E_{ib}F_{ib}(k)H_{ib})x(k-d(k)) + \sigma_i^T\sigma_i\omega(k),$$

we have

$$E\left[-2\xi^T(k)G^T \begin{pmatrix} 0.5x(k) \\ [-S_1 x(k+1) + S_1(A_i + E_{ia}F_{ia}(k)H_{ia})x(k) \\ +S_1(B_i + E_{ib}F_{ib}(k)H_{ib})x(k-d(k)) + S_1\sigma_i\omega(k)] \\ 0 \\ [-\sigma_i^T x(k+1) + \sigma_i^T(A_i + E_{ia}F_{ia}(k)H_{ia})x(k) \\ +\sigma_i^T(B_i + E_{ib}F_{ib}(k)H_{ib})x(k-d(k)) + \sigma_i^T\sigma_i\omega(k)] \end{pmatrix} \right].$$

Therefore, from Equation 18.7 it follows that

$$\begin{aligned}
E[\Delta V_1(k)] = E[&x^T(k)[-P - S_1 A_i - S_1 E_{ia}F_{ia}(k)H_{ia} - A_i^T S_1^T - H_{ia}^T F_{ia}^T(k)E_{ia}S_1^T]x(k) \\
&+ 2x^T(k)[S_1 - S_1 A_i - S_1 E_{ia}F_{ia}(k)H_{ia}]x(k+1) + 2x^T(k)[-S_1 B_i \\
&- S_1 E_{ib}F_{ib}(k)H_{ib}]x(k-d(k)) + 2x^T(k)[-S_1\sigma_i - \sigma_i^T A_i - \sigma_i^T E_{ia}F_{ia}(k)H_{ia}]\omega(k) \\
&+ x(k+1)[S_1 + S_1^T]x(k+1) + 2x(k+1)[-S_1 B_i - S_1(E_{ib}F_{ib}(k)H_{ib})]x(k-d(k)) \\
&+ 2x(k+1)[\sigma_i^T - S_1\sigma_i]\omega(k) + x^T(k-d(k))[-\sigma_i^T B_i - \sigma_i^T E_{ib}F_{ib}(k)H_{ib}]\omega(k) \\
&+ \omega^T(k)[-2\sigma_i^T \sigma_i]\omega(k)],
\end{aligned}$$

By assumption (18.3), we have

$$
\begin{aligned}
E[\Delta V_1(k)] = E[x^T(k)[-P - S_1 A_i - S_1 E_{ia} F_{ia}(k) H_{ia} - A_i^T S_1^T - H_{ia}^T F_{ia}^T(k) E_{ia} S_1^T] x(k) \\
+ 2x^T(k)[S_1 - S_1 A_i - S_1 E_{ia} F_{ia}(k) H_{ia}] x(k+1) + 2x^T(k)[-S_1 B_i \\
- S_1 E_{ib} F_{ib}(k) H_{ib}] x(k - d(k)) + x(k+1)[S_1 + S_1^T] x(k+1) + 2x(k+1)[-S_1 B_i \\
- S_1 E_{ib} F_{ib}(k) H_{ib}] x(k - d(k)) - 2\sigma_i^T \sigma_i],
\end{aligned}
$$

Applying Propositions 18.2, 18.3, condition (18.2), and assumption (18.4), the following estimations hold:

$$
\begin{aligned}
&- S_1 E_{ia} F_{ia}(k) H_{ia} - H_{ia}^T F_{ia}^T(k) E_{ia}^T S_1^T \leq S_1 E_{ia} E_{ia}^T S_1^T + H_{ia}^T H_{ia}, \\
&- 2x^T(k) S_1 E_{ia} F_{ia}(k) H_{ia} x(k+1) \leq x^T(k) S_1 E_{ia} E_{ia}^T S_1^T x(k) + x(k+1)^T H_{ia}^T H_{ia} x(k+1), \\
&- 2x^T(k) S_1 E_{ib} F_{ib}(k) H_{ib} x(k - d(k)) \leq x^T(k) S_1 E_{ib} E_{ib}^T S_1^T x(k) + x(k - d(k))^T H_{ib}^T H_{ib} x(k - d(k)), \\
&- 2x^T(k+1) S_1 E_{ib} F_{ib}(k) H_{ib} x(k - d(k)) \leq x^T(k+1) S_1 E_{ib} E_{ib}^T S_1^T x(k+1) + x(k - d(k))^T H_{ib}^T H_{ib} x(k - d(k)), \\
&- \sigma_i^T(x(k), x(k - d(k)), k) \sigma_i(x(k), x(k - d(k)), k) \leq \rho_{i1} x^T(k) x(k) + \rho_{i2} x^T(k - d(k)) x(k - d(k).
\end{aligned}
$$

Therefore, we have

$$
\begin{aligned}
E[\Delta V_1(k)] = E[x^T(k)[-P - S_1 A_i - A_i^T S_1^T + 2 S_1 E_{ia} E_{ia}^T S_1^T \\
+ S_1 E_{ib} E_{ib}^T S_1^T + S_2 E_{ia} E_{ia}^T S_2^T + H_{ia}^T H_{ia} + 2\rho_{i1} I] x(k) \\
+ 2x^T(k)[S_1 - S_1 A_i] x(k+1) + 2x^T(k)[-S_1 B_i - S_2 A_i] x(k - d(k)) \\
+ x(k+1)[S_1 + S_1^T + H_{ia}^T H_{ia} + S_1 E_{ib} E_{ib}^T S_1^T] x(k+1) \\
+ 2x(k+1)[S_2 - S_1 B_i] x(k - d(k)) + x^T(k - d(k))[2 H_{ib}^T H_{ib} + 2\rho_{i2} I] x(k - d(k))],
\end{aligned}
\tag{18.8}
$$

The difference of $V_2(k)$ is given by

$$
\begin{aligned}
E[\Delta V_2(k)] &= E\left[\sum_{i=k+1-d(k+1)}^{k} x^T(i) Q x(i) - \sum_{i=k-d(k)}^{k-1} x^T(i) Q x(i) \right] \\
&= E\left[\sum_{i=k+1-d(k+1)}^{k-d_1} x^T(i) Q x(i) + x^T(k) Q x(k) - x^T(k - d(k)) Q x(k - d(k)) \right. \\
&\qquad \left. + \sum_{i=k+1-d_1}^{k-1} x^T(i) Q x(i) - \sum_{i=k+1-d(k)}^{k-1} x^T(i) Q x(i) \right].
\end{aligned}
\tag{18.9}
$$

Since $d(k) \geq d_1$ we have

$$
\sum_{i=k+1-d_1}^{k-1} x^T(i) Q x(i) - \sum_{i=k+1-d(k)}^{k-1} x^T(i) Q x(i) \leq 0,
$$

and hence from Equation 18.9 we have

$$
E[\Delta V_2(k)] \leq E\left[\sum_{i=k+1-d(k+1)}^{k-d_1} x^T(i) Q x(i) + x^T(k) Q x(k) - x^T(k - d(k)) Q x(k - d(k)) \right].
\tag{18.10}
$$

The difference of $V_3(k)$ is given by

$$
\begin{aligned}
E[\Delta V_3(k)] &= E\left[\sum_{j=-d_2+2}^{-d_1+1}\sum_{l=k+j}^{k} x^T(l)Qx(l) - \sum_{j=-d_2+2}^{-d_1+1}\sum_{l=k+j+1}^{k-1} x^T(l)Qx(l)\right] \\
&= E\left[\sum_{j=-d_2+2}^{-d_1+1}\left[\sum_{l=k+j}^{k-1} x^T(l)Qx(l) + x^T(k)Q(\xi)x(k)\right.\right. \\
&\qquad\qquad\left.\left. - \sum_{l=k+j}^{k-1} x^T(l)Qx(l) - x^T(k+j-1)Qx(k+j-1)\right]\right] \\
&= E\left[\sum_{j=-d_2+2}^{-d_1+1}[x^T(k)Qx(k) - x^T(k+j-1)Qx(k+j-1)]\right] \\
&= E\left[(d_2-d_1)x^T(k)Qx(k) - \sum_{j=k+1-d_2}^{k-d_1} x^T(j)Qx(j)\right].
\end{aligned}
$$

(18.11)

Since $d(k) \le d_2$, and

$$
\sum_{i=k=1-d(k+1)}^{k-d_1} x^T(i)Qx(i) - \sum_{i=k+1-d_2}^{k-d_1} x^T(i)Qx(i) \le 0,
$$

we obtain from Equations 18.10 and 18.11 that

$$
E[\Delta V_2(k) + \Delta V_3(k)] \le E[(d_2-d_1+1)x^T(k)Qx(k) - x^T(k-d(k))Qx(k-d(k))].
$$

(18.12)

Therefore, combining inequalities (18.8) and (18.12) gives

$$
E[\Delta V(k)] \le E[x^T(k)J_i x(k) + \psi^T(k)W_i\psi(k)],
$$

(18.13)

where

$$
\psi(k) = [x(k)\,x(k+1)\,x(k-d(k))]^T,
$$

$$
W_i = \begin{bmatrix} W_{i11} & W_{i12} & W_{i13} \\ * & W_{i22} & W_{i23} \\ * & * & W_{i33} \end{bmatrix},
$$

$$
W_{i11} = Q - P,
$$

$$
W_{i12} = S_1 - S_1 A_i,
$$

$$
W_{i13} = -S_1 B_i,
$$

$$
W_{i22} = P + S_1 + S_1^T + H_{ia}^T H_{ia} + S_1 E_{ib} E_{ib}^T S_1^T,
$$

$$
W_{i23} = -S_1 B_i,
$$

$$
W_{i33} = -Q + 2H_{ib}^T H_{ib} + 2\rho_{i2}I,
$$

$$
J_i = (d_2-d_1)Q - S_1 A_i - A_i^T S_1^T + 2S_1 E_{ia} E_{ia}^T S_1^T + S_1 E_{ib} E_{ib}^T S_1^T + H_{ia}^T H_{ia} + 2\rho_{i1}I.
$$

Therefore, we finally obtain from Equation 18.13 and condition (ii) that

$$
E[\Delta V(k)] < E[x^T(k)J_i x(k)], \quad \forall i = 1,2,\ldots,N, k = 0,1,2,\ldots.
$$

We now apply condition (i) and Proposition 18.1, the system J_i is strictly complete, and the sets α_i and $\bar{\alpha}_i$ by Equation 18.5 are well defined such that

$$\bigcup_{i=1}^{N} \alpha_i = R^n \backslash \{0\},$$

$$\bigcup_{i=1}^{N} \bar{\alpha}_i = R^n \backslash \{0\}, \quad \bar{\alpha}_i \cap \bar{\alpha}_j = \emptyset, i \neq j.$$

Therefore, for any $x(k) \in R^n, k = 1, 2, \ldots$, there exists $i \in \{1, 2, \ldots, N\}$ such that $x(k) \in \bar{\alpha}_i$. By choosing the switching rule as $\gamma(x(k)) = i$ whenever $x(k) \in \bar{\alpha}_i$, from condition (18.13) we have

$$E[\Delta V(k)] \leq E[x^T(k)J_i x(k)] < 0, \quad k = 1, 2, \ldots,$$

which, combining condition (18.6), Definition 18.3, and the Lyapunov stability theorem [31], concludes the proof of the theorem in the mean square. ∎

Remark 18.2

Note that the result is proposed in References 7 through 14 and 18 through 23 for switching systems to be asymptotically stable under an arbitrary-switching rule. The asymptotic stability for switching linear discrete time-delay systems studied in References 12 through 19 was limited to constant delays. In References 23 through 29, some examples are exploited to illustrate the effectiveness of the proposed schemes.

18.4 Numerical Examples

Example 18.1 (stability)

Consider the uncertain switched discrete-time system (18.1), where the delay function $d(k)$ is given by

$$d(k) = 1 + 4sin^2 \frac{k\pi}{2}, \quad k = 0, 1, 2, \ldots.$$

and

$$(A_1, B_1) = \left(\begin{bmatrix} -0.1 & 0.01 \\ 0.02 & -0.2 \end{bmatrix}, \begin{bmatrix} -0.7 & 0.01 \\ 0.02 & 0.3 \end{bmatrix} \right), (A_2, B_2) = \left(\begin{bmatrix} -0.2 & 0.02 \\ 0.03 & -0.3 \end{bmatrix}, \begin{bmatrix} -0.5 & 0.02 \\ 0.04 & 0.12 \end{bmatrix} \right),$$

$$(H_{1a}, H_{1b}) = \left(\begin{bmatrix} 0.1 & 0 \\ 0 & 0.2 \end{bmatrix}, \begin{bmatrix} 0.2 & 0 \\ 0 & 0.3 \end{bmatrix} \right), (H_{2a}, H_{2b}) = \left(\begin{bmatrix} 0.4 & 0 \\ 0 & 0.5 \end{bmatrix}, \begin{bmatrix} 0.1 & 0 \\ 0 & 0.2 \end{bmatrix} \right),$$

$$(E_{1a}, E_{1b}) = \left(\begin{bmatrix} 5.3 & 0 \\ 0 & 3.4 \end{bmatrix}, \begin{bmatrix} 3.2 & 0 \\ 0 & 5.5 \end{bmatrix} \right), (E_{2a}, E_{2b}) = \left(\begin{bmatrix} 3.5 & 0 \\ 0 & 3.3 \end{bmatrix}, \begin{bmatrix} 2.2 & 0 \\ 0 & 4.3 \end{bmatrix} \right),$$

$$(F_{1a}, F_{1b}) = \left(\begin{bmatrix} 0.1 & 0 \\ 0 & 0.2 \end{bmatrix}, \begin{bmatrix} 0.2 & 0 \\ 0 & 0.3 \end{bmatrix} \right), (F_{2a}, F_{2b}) = \left(\begin{bmatrix} 0.2 & 0 \\ 0 & 0.5 \end{bmatrix}, \begin{bmatrix} 0.1 & 0 \\ 0 & 0.2 \end{bmatrix} \right).$$

By LMI toolbox of MATLAB, we find that conditions (i) and (ii) of Theorem 18.1 are satisfied with $d_1 = 1, d_2 = 5, \delta_1 = 1, \delta_2 = 1$, and

$$P = \begin{bmatrix} 1.1329 & -0.0010 \\ -0.0010 & 1.7289 \end{bmatrix}, Q = \begin{bmatrix} 0.0506 & -0.0011 \\ -0.0011 & 0.4454 \end{bmatrix}, S_1 = \begin{bmatrix} -0.0169 & 0.0002 \\ 0 & -0.0798 \end{bmatrix}.$$

In this case, we have

$$(J_1(S_1, Q), J_2(S_1, Q)) = \left(\begin{bmatrix} -0.2170 & -0.0026 \\ -0.0026 & -1.8633 \end{bmatrix}, \begin{bmatrix} -0.3591 & -0.0016 \\ -0.0016 & -2.0531 \end{bmatrix} \right).$$

Moreover, the sum

$$\delta_1 J_1(R,Q) + \delta_2 J_2(R,Q) = \begin{bmatrix} -0.5761 & -0.0042 \\ -0.0042 & -3.9164 \end{bmatrix},$$

is negative definite; i.e. the first entry in the first row and the first column $-0.5761 < 0$ is negative and the determinant of the matrix is positive. Sets α_1 and α_2 are given as

$$\alpha_1 = \{(x_1,x_2) : -0.2170x_1^2 - 0.0052x_1x_2 - 1.8633x_2^2 < 0\},$$
$$\alpha_2 = \{(x_1,x_2) : 0.3591x_1^2 + 0.0032x_1x_2 + 2.0531x_2^2 > 0\}.$$

Obviously, the union of these sets is equal to $R^2 \setminus \{0\}$. The switching regions are defined as

$$\overline{\alpha}_1 = \{(x_1,x_2) : -0.2170x_1^2 - 0.0052x_1x_2 - 1.8633x_2^2 < 0\},$$
$$\overline{\alpha}_2 = \alpha_2 \setminus \overline{\alpha}_1.$$

By Theorem 18.1, the uncertain system is robustly stable and the switching rule is chosen as $\gamma(x(k)) = i$ whenever $x(k) \in \overline{\alpha}_i$.

18.5 Conclusion

This chapter has proposed a switching design for the robust stability of uncertain stochastic-switched discrete time-delay systems with interval time-varying delays. On the basis of the discrete Lyapunov functional, a switching rule for the robust stability for the uncertain stochastic-switched discrete-time-delay system is designed via linear matrix inequalities. Numerical examples are provided to illustrate the theoretical results.

Acknowledgment

This chapter was supported by the Thailand Research Fund Grant, the Higher Education Commission, and by the Faculty of Science, Maejo University, Thailand.

References

1. K. Ratchagit, Stability criteria of LPD system with time-varying delay, *International Journal of Pure and Applied Mathematics*, 78(6), 2012, 857–866.
2. K. Ratchagit, Stability analysis of linear systems with time delays, *International Journal of Pure and Applied Mathematics*, 76(1), 2012, 21–28.
3. K. Ratchagit, Stability of linear time-varying systems, *International Journal of Pure and Applied Mathematics*, 63(4), 2010, 411–417.
4. K. Ratchagit, Exponential stability of switched linear systems, *International Journal of Pure and Applied Mathematics*, 58(3), 2010, 361–371.
5. K. Ratchagit, The sufficient conditions for stability of linear time-varying systems with state delays, *International Journal of Pure and Applied Mathematics*, 65(1), 2010, 65–72.

6. K. Ratchagit and V.N. Phat, Stability criterion for discrete-time systems, *Journal of Inequalities and Applications*, 2010, 2010, 201459. DOI: 10.1155/2010/201459.

7. M. De la Sen and A. Ibeas, Stability results of a class of hybrid systems under switched continuous-time and discretet-time control, *Discrete Dynamics in Nature and Society*, 2009, 2009, 28. DOI: 10.1155/2009/315713.

8. K. Ratchagit, A switching rule for the asymptotic stability of discrete-time systems with convex polytopic uncertainties, *Asian–European Journal of Mathematics*, 5, 2012. DOI: 10.1142/S1793557112500258.

9. G. Rajchakit, Stabilization of switched discrete-time systems with convex polytopic uncertainties, *Journal of Computational Analysis and Applications*, 16, 2014, 20–29.

10. M. Rajchakit, P. Niamsup, and G. Rajchakit, A constructive way to design a switching rule and switching regions to mean square exponential stability of switched stochastic systems with non-differentiable and interval time-varying delay, *Journal of Inequalities and Applications*, 2013, 2013, 499. DOI: 10.1186/1029-242X-2013-499.

11. G. Rajchakit, Delay-dependent asymptotical stabilization criterion of recurrent neural networks, *Applied Mechanics and Materials*, 330, 2013, 1045–1048. DOI: 10.4028/www.scientific.net/AMM.330.1045.

12. K. Ratchagit, Asymptotic stability of nonlinear delay-difference system via matrix inequalities and application, *International Journal of Computational Methods*, 6, 2009, 389–397. DOI: 10.1142/S0219876209001899.

13. M. de la Sen, Global stability of polytopic linear time-varying dynamic systems under time-varying point delays and impulsive controls, *Mathematical Problems in Engineering*, 2010, 2010, 33. DOI: 10.1155/2010/693958.

14. G. Rajchakit, Exponential stability of switched linear systems with interval time-varying delays, *Proceedings of the 2012 IEEE International Conference on Robotics and Biomimetics*, Guangzhou, China, December 11–14, 2012, 1502–1506. DOI: 10.1109/ROBIO.2012.6491181.

15. K. Ratchagit and V.N. Phat, Stability and stabilization of switched linear discrete-time systems with interval time-varying delay, *Nonlinear Analysis Hybrid Systems*, 5, 2011, 605–612. DOI: 10.1016/j.nahs.2011.05.006.

16. M. Rajchakit, P. Niamsup, and G. Rajchakit, A constructive way to design a switching rule and switching regions to mean square exponential stability of switched stochastic systems with non-differentiable and interval time-varying delay, *Journal of Inequalities and Applications*, 3, 2013, 34–45. DOI: 10.1186/1029-242X-2013-499.

17. V.N. Phat, Y. Kongtham, and K. Ratchagit, LMI approach to exponential stability of linear systems with interval time-varying delays, *Linear Algebra Applications*, 436, 2012, 243–251. DOI: 10.1016/j.laa.2011.07.016.

18. P. Niamsup, M. Rajchakit, and G. Rajchakit, Guaranteed cost control for switched recurrent neural networks with interval time-varying delay, *Journal of Inequalities and Applications*, 43, 2013, 147–158. DOI: 10.1186/1029-242X-2013-292.

19. P. Niamsup and G. Rajchakit, New results on robust stability and stabilization of linear discrete-time stochastic systems with convex polytopic uncertainties, *Journal of Applied Mathematics*, 23, 2013, 78–89. DOI: 10.1155/2013/368259.

20. D. Liberzon and A.S. Morse, Basic problems in stability and design of switched systems, *IEEE Control Systems Magazine*, 19, 1999, 57–70.

21. A.V. Savkin and R.J. Evans, *Hybrid Dynamical Systems: Controller and Sensor Switching Problems*, Springer, New York, 2001.

22. Z. Sun and S.S. Ge, *Switched Linear Systems: Control and Design*, Springer, London, 2005.

23. F. Gao, S. Zhong, and X. Gao, Delay-dependent stability of a type of linear switching systems with discrete and distributed time delays, *Applied Mathematics Computation*, 196, 2008, 24–39.

24. C.H. Lien, K.W. Yu, Y.J. Chung, Y.F. Lin, L.Y. Chung, and J.D. Chen, Exponential stability analysis for uncertain switched neutral systems with interval-time-varying state delay, *Nonlinear Analysis: Hybrid Systems*, 3, 2009, 334–342.

25. G. Xie and L. Wang, Quadratic stability and stabilization of discrete-time switched systems with state delay, In: *Proceedings of the IEEE Conference on Decision and Control*, Atlantics, December 2004, 3235–3240.

26. S. Boyd, L.E. Ghaoui, E. Feron, and V. Balakrishnan, *Linear Matrix Inequalities in System and Control Theory*, SIAM, Philadelphia, 1994.

27. D.H. Ji, J.H. Park, W.J. Yoo, and S.C. Won, Robust memory state feedback model predictive control for discrete-time uncertain state delayed systems, *Applied Mathematics Computation*, 215, 2009, 2035–2044.

28. G.S. Zhai, B. Hu, K. Yasuda, and A. Michel, Qualitative analysis of discrete-time switched systems, In: *Proceedings of the American Control Conference*, Guangdong, China, May 18–20, 2002, 1880–1885.

29. W.A. Zhang and L. Yu, Stability analysis for discrete-time switched time-delay systems, *Automatica*, 45, 2009, 2265–2271.

30. F. Uhlig, A recurring theorem about pairs of quadratic forms and extensions, *Linear Algebra Applications*, 25, 1979, 219–237.

31. R.P. Agarwal, *Difference Equations and Inequalities*, second edition, Marcel Dekker, New York, 2000.

IV

Chaos in Plasma

19

Chaos in Plasma Physics

Dan-Gheorghe Dimitriu
and Maricel Agop

19.1 Dynamics of Plasmas

19.1.1 Introduction

Plasma is one of the four fundamental states of matter, being a conductive assembly of charged particles (electrons and ions), neutrals, and fields that exhibit collective effects. Plasmas carry electrical currents and generate magnetic fields. Plasmas are radically multiscale in at least two senses: (1) most plasma systems involve electrodynamic coupling across micro-, meso-, and macroscale and (2) plasma systems occur over most of the physically possible ranges in space, energy, and density scales.

The study of plasma dynamics focuses on the properties of classical, collective, and many-body systems and finds applications in plasma processing, fusion, intense particle beams, and fluid dynamics. Currently, active areas of research include advanced accelerators, space plasmas, basic and theoretical plasma physics, heavy-ions fusion, high-power laser plasma interaction, ion beam generation, plasma processing, plasma simulations, as well as dissipative and Hamiltonian dynamical systems, bifurcation theory and chaos, linear and nonlinear waves and pattern formation, particularly in fluid dynamics, and continuum mechanics.

In complex systems, and thus in plasma, deterministic chaos arises in association with the emergence of spatiotemporal structures. For temporal scales that are large with respect to the inverse of the highest Lyapunov, the deterministic trajectories can be replaced by a collection of potential trajectories and the concept of definite positions can be replaced by that of probability density. This concept was introduced in the framework of the scale relativity theory (SRT), which states that the particle movement takes place on continuous but nondifferentiable curves, i.e., on fractal curves [87,90]. Subsequently, all physical phenomena become dependent not only on the spatiotemporal coordinates but also on the spatiotemporal scales (i.e., they are invariant to scale transformations). From such a perspective, the physical quantities that describe a system dynamics may be considered as fractal functions.

In this theoretical framework, it is not necessary to endow a point-like particle with mass, energy, momentum, or velocity. The particle may be reduced to and identified with its own trajectory, i.e., its geodesics (for details, see Reference 90). Moreover, the physical systems behave as a special interactionless "fluid" by means of geodesics in a nondifferentiable/fractal space. To better understand this affirmation, let us consider the trajectory associated with the fluid particle undergoing different types of collisions (elastic and inelastic). This trajectory is described by a fractal curve. Since the fractality implies a certain statistic [87,90], it results that a certain type of collision has to be described by a certain random process. For example, in the case of elastic collisions, the dynamics of the fluid particles can be described by Brownian-type movements: between two successive elastic collisions, the particle trajectory is a straight line, the trajectory becoming nondifferentiable in the impact point (there are left and right derivatives in this point). Considering that all the elastic collisions impact points from an innumerable set of points, it results that the trajectories become fractals. The random process able to describe the Brownian motion could be, for example, the Wiener process.

19.1.2 Fractal Hydrodynamics

19.1.2.1 Consequences of Nondifferentiability

Let us suppose that the plasma particles movements (electrons, ions, and neutrals) take place on continuous but nondifferentiable curves (fractal curves). The nondifferentiability implies the following [86,87,90]:

1. A continuous and a nondifferentiable curve (or almost nowhere differentiable) is explicitly scale dependent, and its length tends to infinity, when the scale interval tends to zero. In other words, a continuous and nondifferentiable space is fractal, in the general meaning given by Mandelbrot to this concept [72].
2. There is an infinity of fractal curves (geodesics) relating any couple of its points (or starting from any point), and this is valid for all scales.
3. The breaking of local differential time reflection invariance. The time derivative of an arbitrary field Q (speed, concentration, etc.) can be written as twofold

$$\frac{dQ}{dt} = \lim_{\Delta t \to 0_+} \frac{Q(t + \Delta t) - Q(t)}{\Delta t}, \tag{19.1}$$

$$\frac{dQ}{dt} = \lim_{\Delta t \to 0_-} \frac{Q(t + \Delta t) - Q(t)}{\Delta t}. \tag{19.2}$$

Both definitions are equivalent in the differentiable case. In the nondifferentiable situation, these definitions fail, since the limits are no longer defined. In the framework of fractal theory, the physics is related to the behavior of the function during the "zoom" operation on the time resolution δt, here identified with the differential element dt ("substitution principle"), which is considered as an independent variable. The standard arbitrary field $Q(t)$ is therefore replaced by a fractal arbitrary field $Q(t, dt)$, explicitly dependent on the time resolution interval, whose derivative is undefined only at the unobservable limit $dt \to 0$ (from a mathematical point of view, these fields are described by fractal functions, for details [definition, properties, etc.] see Reference 86). As a consequence, this leads us to define the two derivatives of the fractal arbitrary field as explicit functions of the two variables t and dt

$$\frac{d_+ Q}{dt} = \lim_{\Delta t \to 0_+} \frac{Q(t + \Delta t, \Delta t) - Q(t, \Delta t)}{\Delta t}, \tag{19.3}$$

$$\frac{d_-Q}{dt} = \lim_{\Delta t \to 0_-} \frac{Q(t, \Delta t) - Q(t - \Delta t, \Delta t)}{\Delta t}. \tag{19.4}$$

The sign "+" corresponds to the forward process and to the backward process, respectively.

4. The differential of the coordinates, $d_\pm X(t, \Delta t)$, can be decomposed as follows:

$$d_+X(t, \Delta t) = d_+x(t) + d_+\xi(t, dt), \tag{19.5}$$

$$d_-X(t, \Delta t) = d_-x(t) + d_-\xi(t, dt), \tag{19.6}$$

where $d_\pm x(t)$ is the "classical part" and $d_\pm\xi(t, dt)$ is the "fractal part."

5. The differential of the "fractal part" components $\xi^i(t, dt)$, $i = \overline{1, 3}$, satisfies the relation (the fractal equation)

$$d_+\xi^i = \lambda_+^i (dt)^{1/D_F}, \tag{19.7}$$

$$d_-\xi^i = \lambda_-^i (dt)^{1/D_F}, \tag{19.8}$$

where $\lambda_\pm^i$ are some constant coefficients, and D_F is a constant fractal dimension. We note that for the fractal dimension, we can use any definition (Kolmogorov, Hausdorff, etc.) [48,55,72,90,106].

6. The local differential time reflection invariance is recovered by combining the two derivatives, d_+/dt and d_-/dt, in the complex operator:

$$\frac{\hat{d}}{dt} = \frac{1}{2}\left(\frac{d_+ + d_-}{dt}\right) - \frac{i}{2}\left(\frac{d_+ - d_-}{dt}\right). \tag{19.9}$$

By applying this operator to the "position vector," a complex speed yields

$$\hat{V} = \frac{\hat{d}X}{dt} = \frac{1}{2}\left(\frac{d_+X + d_-X}{dt}\right) - \frac{i}{2}\left(\frac{d_+X - d_-X}{dt}\right) = \frac{V_+ + V_-}{2} - i\frac{V_+ - V_-}{2} = V - iU, \tag{19.10}$$

with

$$V = \frac{V_+ + V_-}{2}, \tag{19.11}$$

$$U = \frac{V_+ - V_-}{2}. \tag{19.12}$$

The real part, V, of the complex speed, $\hat{V}$, represents the standard classical speed, which is independent of resolution, while the imaginary part, U, is a new quantity arising from fractality, which is resolution dependent.

7. The average values of the quantities must be considered in the sense of a generalized statistical fluid-like description. Particularly, the average of $d_\pm X$ is

$$\langle d_+X \rangle = d_+x, \tag{19.13}$$

$$\langle d_-X \rangle = d_-x, \tag{19.14}$$

with

$$\langle d_+\xi \rangle = 0, \tag{19.15}$$

$$\langle d_-\xi \rangle = 0. \tag{19.16}$$

In such an interpretation, the "particles" are identified with the geodesics themselves. As a consequence, any measurement is interpreted as a sorting out (or selection) of the geodesics by the measuring devices.

19.1.2.2 Covariant Total Derivative

Let us now assume that the curves describing the particle movements (continuous but nondifferentiable) are immersed in a three-dimensional (3D) space, and that X of components $X^i(i = \overline{1,3})$ is the position vector of a point on the curve. Let us also consider the fractal arbitrary field $Q(X, t)$ and expand its total differential up to the second order:

$$d_+Q = \frac{\partial Q}{\partial t}dt + \nabla Q \cdot d_+X + \frac{1}{2}\frac{\partial^2 Q}{\partial X^i \partial X^j}d_+X^i d_+X^j, \qquad (19.17)$$

$$d_-Q = \frac{\partial Q}{\partial t}dt + \nabla Q \cdot d_-X + \frac{1}{2}\frac{\partial^2 Q}{\partial X^i \partial X^j}d_-X^i d_-X^j. \qquad (19.18)$$

Equations 19.17 and 19.18 valid in any point of the space manifold and also for the point X on the fractal curve which we have selected in Equations 19.17 and 19.18. From here, the forward and backward average values of this relation take the form

$$\langle d_+Q \rangle = \left\langle \frac{\partial Q}{\partial t}dt \right\rangle + \langle \nabla Q \cdot d_+X \rangle + \frac{1}{2}\left\langle \frac{\partial^2 Q}{\partial X^i \partial X^j}d_+X^i d_+X^j \right\rangle, \qquad (19.19)$$

$$\langle d_-Q \rangle = \left\langle \frac{\partial Q}{\partial t}dt \right\rangle + \langle \nabla Q \cdot d_-X \rangle + \frac{1}{2}\left\langle \frac{\partial^2 Q}{\partial X^i \partial X^j}d_-X^i d_-X^j \right\rangle. \qquad (19.20)$$

We make the following stipulation: the mean value of Q and its derivatives coincide with themselves, and the differentials $d_\pm X^i$ and dt are independent; therefore, the average of their products coincides with the product of averages. Thus, Equations 19.19 and 19.20 becomes

$$d_+Q = \frac{\partial Q}{\partial t}dt + \nabla Q \cdot \langle d_+X \rangle + \frac{1}{2}\frac{\partial^2 Q}{\partial X^i \partial X^j}\langle d_+X^i d_+X^j \rangle, \qquad (19.21)$$

$$d_-Q = \frac{\partial Q}{\partial t}dt + \nabla Q \cdot \langle d_-X \rangle + \frac{1}{2}\frac{\partial^2 Q}{\partial X^i \partial X^j}\langle d_-X^i d_-X^j \rangle, \qquad (19.22)$$

or more, by using Equations 19.5 and 19.6 with properties (19.15) and (19.16)

$$d_+Q = \frac{\partial Q}{\partial t}dt + \nabla Q \cdot d_+x + \frac{1}{2}\frac{\partial^2 Q}{\partial X^i \partial X^j}\left(d_+x^i d_+x^j + \langle d_+\xi^i d_+\xi^j \rangle\right), \qquad (19.23)$$

$$d_-Q = \frac{\partial Q}{\partial t}dt + \nabla Q \cdot d_-x + \frac{1}{2}\frac{\partial^2 Q}{\partial X^i \partial X^j}\left(d_-x^i d_-x^j + \langle d_-\xi^i d_-\xi^j \rangle\right). \qquad (19.24)$$

Even the average value of $d_\pm\xi^i$ is null (see Equations 19.15 and 19.16); for the higher order of these average coordinates, the situation can be different. First, let us focus on the mean $\langle d_\pm\xi^i d_\pm\xi^j \rangle$. If $i \neq j$, this average is zero because of the independence of $d_\pm\xi^i$ and $d_\pm\xi^j$. So, by using Equations 19.7 and 19.8, we can write

$$\langle d_+\xi^i d_+\xi^j \rangle = \lambda_+^i \lambda_+^j \left(dt\right)^{(2/D_F)-1}dt, \qquad (19.25)$$

$$\langle d_-\xi^i d_-\xi^j \rangle = \lambda_-^i \lambda_-^j \left(dt\right)^{(2/D_F)-1}dt. \qquad (19.26)$$

Then, Equations 19.23 and 19.24 may be written under the form

$$d_+Q = \frac{\partial Q}{\partial t}dt + d_+x \cdot \nabla Q + \frac{1}{2}\frac{\partial^2 Q}{\partial X^i \partial X^j}d_+x^i d_+x^j + \frac{1}{2}\frac{\partial^2 Q}{\partial X^i \partial X^j}\lambda_+^i \lambda_+^j \left(dt\right)^{(2/D_F)-1}dt, \qquad (19.27)$$

$$d_-Q = \frac{\partial Q}{\partial t}dt + d_-x \cdot \nabla Q + \frac{1}{2}\frac{\partial^2 Q}{\partial X^i \partial X^j}d_-x^i d_-x^j + \frac{1}{2}\frac{\partial^2 Q}{\partial X^i \partial X^j}\lambda_-^i \lambda_-^j \left(dt\right)^{(2/D_F)-1}dt. \qquad (19.28)$$

If we divide by dt and neglect the terms which contain differential factors (see the method from Reference 3), Equations 19.27 and 19.28 are reduced to

$$\frac{d_+Q}{dt} = \frac{\partial Q}{\partial t} + V_+ \cdot \nabla Q + \frac{1}{2}\frac{\partial^2 Q}{\partial X^i \partial X^j}\lambda_+^i \lambda_+^j \left(dt\right)^{(2/D_F)-1}, \qquad (19.29)$$

$$\frac{d_- Q}{dt} = \frac{\partial Q}{\partial t} + \mathbf{V}_- \cdot \nabla Q + \frac{1}{2} \frac{\partial^2 Q}{\partial X^i \partial X^j} \lambda_-^i \lambda_-^j \, (dt)^{(2/D_F)-1}. \tag{19.30}$$

These relations also allow us to define the operator

$$\frac{d_+}{dt} = \frac{\partial}{\partial t} + \mathbf{V}_+ \cdot \nabla + \frac{1}{2} \frac{\partial^2}{\partial X^i \partial X^j} \lambda_+^i \lambda_+^j \, (dt)^{(2/D_F)-1}, \tag{19.31}$$

$$\frac{d_-}{dt} = \frac{\partial}{\partial t} + \mathbf{V}_- \cdot \nabla + \frac{1}{2} \frac{\partial^2}{\partial X^i \partial X^j} \lambda_-^i \lambda_-^j \, (dt)^{(2/D_F)-1}. \tag{19.32}$$

Under these circumstances, let us calculate $\hat{d}/dt$. By taking into account Equations 19.9, 19.10, 19.31, and 19.32, we obtain

$$\frac{\hat{d}Q}{dt} = \frac{1}{2} \left[\frac{d_+ Q}{dt} + \frac{d_- Q}{dt} - i \left(\frac{d_+ Q}{dt} - \frac{d_- Q}{dt} \right) \right]$$

$$= \frac{1}{2} \frac{\partial Q}{\partial t} + \frac{1}{2} \mathbf{V}_+ \cdot \nabla Q + \frac{1}{4} \frac{\partial^2 Q}{\partial X^i \partial X^j} \lambda_+^i \lambda_+^j \, (dt)^{(2/D_F)-1}$$

$$+ \frac{1}{2} \frac{\partial Q}{\partial t} + \frac{1}{2} \mathbf{V}_- \cdot \nabla Q + \frac{1}{4} \frac{\partial^2 Q}{\partial X^i \partial X^j} \lambda_-^i \lambda_-^j \, (dt)^{(2/D_F)-1}$$

$$- \frac{i}{2} \frac{\partial Q}{\partial t} - \frac{i}{2} \mathbf{V}_+ \cdot \nabla Q - \frac{i}{4} \frac{\partial^2 Q}{\partial X^i \partial X^j} \lambda_+^i \lambda_+^j \, (dt)^{(2/D_F)-1}$$

$$+ \frac{i}{2} \frac{\partial Q}{\partial t} + \frac{i}{2} \mathbf{V}_- \cdot \nabla Q + \frac{i}{4} \frac{\partial^2 Q}{\partial X^i \partial X^j} \lambda_-^i \lambda_-^j \, (dt)^{(2/D_F)-1}$$

$$= \frac{\partial Q}{\partial t} + \left(\frac{\mathbf{V}_+ + \mathbf{V}_-}{2} - i \frac{\mathbf{V}_+ - \mathbf{V}_-}{2} \right) \cdot \nabla Q$$

$$+ \frac{1}{4} \frac{\partial^2 Q}{\partial X^i \partial X^j} \left[\left(\lambda_+^i \lambda_+^j + \lambda_-^i \lambda_-^j \right) - i \left(\lambda_+^i \lambda_+^j - \lambda_-^i \lambda_-^j \right) \right] (dt)^{(2/D_F)-1}$$

$$= \frac{\partial Q}{\partial t} + \hat{\mathbf{V}} \cdot \nabla Q + \frac{1}{4} \frac{\partial^2 Q}{\partial X^i \partial X^j} \left[\left(\lambda_+^i \lambda_+^j + \lambda_-^i \lambda_-^j \right) - i \left(\lambda_+^i \lambda_+^j - \lambda_-^i \lambda_-^j \right) \right] (dt)^{(2/D_F)-1}. \tag{19.33}$$

This relation also allows us to define the fractal operator

$$\frac{\hat{d}}{dt} = \frac{\partial}{\partial t} + \hat{\mathbf{V}} \cdot \nabla + \frac{1}{4} \frac{\partial^2}{\partial X^i \partial X^j} \left[\left(\lambda_+^i \lambda_+^j + \lambda_-^i \lambda_-^j \right) - i \left(\lambda_+^i \lambda_+^j - \lambda_-^i \lambda_-^j \right) \right] (dt)^{(2/D_F)-1}. \tag{19.34}$$

Particularly, by choosing

$$\lambda_+^i \lambda_+^j = -\lambda_-^i \lambda_-^j = 2D\delta^{ij}, \tag{19.35}$$

relations (19.25) and (19.26) become

$$\langle d_+ \xi^i d_+ \xi^j \rangle = 2D \, (dt)^{(2/D_F)-1} \, dt, \tag{19.36}$$

$$\langle d_- \xi^i d_- \xi^j \rangle = 2D \, (dt)^{(2/D_F)-1} \, dt. \tag{19.37}$$

We note the following:

1. The fractal processes [3,48,55,72,87,106] given by Equations 19.36 and 19.37 with $D_F \neq 2$ are known as anomalous diffusion (subdiffusion for $D_F < 2$ and superdiffusion for $D_F > 2$). Usually, the Fokker–Planck equations for anomalous diffusion do not have the form of the ordinary diffusion equation. Indeed, it is well known that the "Fokker–Planck equations" for anomalous diffusion have the form of the fractional derivative equations, and the equations are called fractional Fokker–Planck equations [48,55,106].

2. The Nottale's theory is formulated in the fractal dimension $D_F = 2$, i.e., for movements on Peano curves, and for Wiener's stochastic processes [3,48,55,72,87,106].

In these conditions, the fractal operator (19.34) takes the simple form

$$\frac{\hat{d}}{dt} = \frac{\partial}{\partial t} + \hat{V} \cdot \nabla - iD \left(dt \right)^{(2/D_F)-1} \Delta. \tag{19.38}$$

We now apply the principle of scale covariance, and postulate that the passage from classical (differentiable) physics to the fractal physics can be implemented by replacing the standard time derivative operator, d/dt, by the complex operator $\hat{d}/dt$. As a consequence, we are now able to write the equation of the field flow in its covariant form

$$\frac{\hat{d}Q}{dt} = \frac{\partial Q}{\partial t} + \left(\hat{V} \cdot \nabla \right) Q - iD \left(dt \right)^{(2/D_F)-1} \Delta Q. \tag{19.39}$$

This means that at any point of a fractal path, the local temporal term, $\partial_t Q$, the nonlinearly (convective) term, $\left(\hat{V} \cdot \nabla \right) Q$, and the dissipative one, ΔQ, make their balance. Moreover, the behavior of a fractal fluid is of a viscoelastic or of hysteretic type, i.e., the fractal fluid has memory. Such a result is in agreement with the opinion given in References 3, 35, 51, 57, and 84: the fractal fluid can be described by Kelvin–Voight or Maxwell rheological model with the aid of complex quantities, e.g., the complex field, Q, the complex structure coefficients, $iD \left(dt \right)^{(2/D_F)-1}$, etc.

19.1.2.3 Geodesics. Fractal Hydrodynamics

We are now able to write the equation of geodesics (a generalization of the first Newton's principle) in the form

$$\frac{\hat{d}\hat{V}}{dt} = \frac{\partial \hat{V}}{\partial t} + \left(\hat{V} \cdot \nabla \right) \hat{V} - \overline{\eta} \Delta \hat{V}. \tag{19.40}$$

Formally, at the global scale (with its differentiable and fractal components), Equation 19.40 is a Navier–Stokes-type equation with the imaginary viscosity coefficient

$$\overline{\eta} = iD \left(dt \right)^{(2/D_F)-1}. \tag{19.41}$$

This result evidences the rheological properties of the fractal fluid. If motions of the fractal fluid are irrotational, i.e., $\nabla \times \hat{V} = 0$, we can choose $\hat{V}$ of the form

$$\hat{V} = -2iD \left(dt \right)^{(2/D_F)-1} \nabla \ln \psi, \tag{19.42}$$

with ψ being the scalar potential of the complex speed. Then, by substituting Equation 19.42 in Equation 19.40 and using the method from References 57 and 84, it results in

$$\frac{d\hat{V}}{dt} = -2iD \left(dt \right)^{(2/D_F)-1} \nabla \left(\frac{\partial \ln \psi}{\partial t} - 2iD \left(dt \right)^{(2/D_F)-1} \frac{\Delta \psi}{\psi} \right) = 0. \tag{19.43}$$

Equation 19.43 can be integrated in a universal way, which yields

$$\hat{L}\psi = 0, \tag{19.44}$$

$$\hat{L} = 4D^2 \left(dt \right)^{(4/D_F)-2} \Delta + 2iD \left(dt \right)^{(2/D_F)-1} \frac{\partial}{\partial t}, \tag{19.45}$$

up to an arbitrary phase factor which may be set to zero by a suitable choice of the phase of ψ. Equation 19.44, where $\hat{L}$ is the differential operator (19.45), is of the Schrödinger type.

For $\psi = \sqrt{\rho}e^{iS}$, with $\sqrt{\rho}$ the amplitude and S the phase of ψ, and by using Equation 19.42, the complex speed field (19.10) takes the explicit form

$$\hat{V} = 2D\left(dt\right)^{(2/D_F)-1}\nabla S - iD\left(dt\right)^{(2/D_F)-1}\nabla\ln\rho, \tag{19.46}$$

$$V = 2D\left(dt\right)^{(2/D_F)-1}\nabla S, \tag{19.47}$$

$$U = D\left(dt\right)^{(2/D_F)-1}\nabla\ln\rho. \tag{19.48}$$

By substituting Equations 19.46 through 19.48 in Equation 19.40 and separating the real and imaginary parts, up to an arbitrary phase factor which may be set at zero by a suitable choice of the phase of ψ, we obtain

$$m_0\left(\frac{\partial V}{\partial t} + (V\cdot\nabla)\,V\right) = -\nabla\overline{Q}, \tag{19.49}$$

$$\frac{\partial\rho}{\partial t} + \nabla\cdot\left(\rho V\right) = 0, \tag{19.50}$$

with $\overline{Q}$ the fractal potential

$$\overline{Q} = -2m_0 D^2\left(dt\right)^{(4/D_F)-2}\frac{\Delta\sqrt{\rho}}{\sqrt{\rho}} = -\frac{m_0 U^2}{2} - m_0 D\left(dt\right)^{(2/D_F)-1}\nabla\cdot U. \tag{19.51}$$

Equation 19.49 is the law of momentum conservation, Equation 19.50 is the law of density conservation, and m_0 is the rest mass of the fractal fluid particle. Together, these two equations define the fractal hydrodynamics.

19.1.3 Predictability through Fractality. Fractal Fluid Self-Structuring

In the one-dimensional case, Equations 19.49 through 19.51 with the initial conditions

$$V(x, t = 0) = c, \tag{19.52}$$

$$\rho(x, t = 0) = \frac{1}{\sqrt{\pi}\alpha}e^{-\left(\frac{x}{\alpha}\right)^2} = \rho_0(x), \tag{19.53}$$

and the boundary ones

$$V(x = ct, t) = c, \tag{19.54}$$

$$\rho(x = -\infty, t) = \rho(x = +\infty, t) = 0, \tag{19.55}$$

imply the solutions (for details, see the method described in Reference 108)

$$\rho(x, t) = \frac{1}{\sqrt{\pi\left[\alpha^2 + \left(\frac{2\tilde{D}}{\alpha}\right)^2 t^2\right]}}\exp\left[-\frac{(x - ct)^2}{\alpha^2 + \left(\frac{2\tilde{D}}{\alpha}\right)^2 t^2}\right], \tag{19.56}$$

$$V = \frac{c\alpha^2 + \left(\frac{2\tilde{D}}{\alpha}\right)^2 xt}{\alpha^2 + \left(\frac{2\tilde{D}}{\alpha}\right)^2 t^2}, \tag{19.57}$$

where c is a constant speed, α is the distribution parameter, and

$$\widetilde{D} = D(dt)^{(2/D_F)-1}. \tag{19.58}$$

Then, it results in the complex speed

$$\hat{V} = V - iU = \frac{c\alpha^2 + \left(\frac{2\widetilde{D}}{\alpha}\right)^2 xt}{\alpha^2 + \left(\frac{2\widetilde{D}}{\alpha}\right)^2 t^2} + 2\widetilde{D}i \frac{x - ct}{\alpha^2 + \left(\frac{2\widetilde{D}}{\alpha}\right)^2 t^2}, \tag{19.59}$$

the fractal potential, and force

$$\overline{Q} = -2m_0\widetilde{D}^2 \frac{x - ct}{\left[\alpha^2 + \left(\frac{2\widetilde{D}}{\alpha}\right)^2 t^2\right]^2} + \frac{2m_0\widetilde{D}^2}{\alpha^2 + \left(\frac{2\widetilde{D}}{\alpha}\right)^2 t^2}, \tag{19.60}$$

$$\overline{F} = -\frac{\partial \overline{Q}}{\partial x} = 4m_0\widetilde{D}^2 \frac{x - ct}{\left[\alpha^2 + \left(\frac{2\widetilde{D}}{\alpha}\right)^2 t^2\right]^2}. \tag{19.61}$$

In nondimensional coordinates

$$\tau = \omega t, \tag{19.62}$$

$$\xi = \frac{x}{\lambda}, \tag{19.63}$$

where ω is a specific frequency of the fractal fluid and λ is a characteristic length of the fractal fluid, and with the substitutions

$$\mu = \left(\frac{\alpha}{\lambda}\right)^2, \tag{19.64}$$

$$\nu = \frac{2\widetilde{D}}{\alpha\omega\lambda}, \tag{19.65}$$

relations (19.59) through (19.61) become

$$\hat{\overline{V}} = \frac{\mu^2 + \nu^2\tau\xi}{\mu^2 + \nu^2\tau^2} - i\frac{\xi - \tau}{\mu^2 + \nu^2\tau^2}, \tag{19.66}$$

$$\overline{Q} = -\frac{(\xi - \tau)^2}{\left(\mu^2 + \nu^2\tau^2\right)^2} + \frac{1}{\mu^2 + \nu^2\tau^2}, \tag{19.67}$$

$$\overline{F} = \frac{\xi - \tau}{\left(\mu^2 + \nu^2\tau^2\right)^2}. \tag{19.68}$$

The complex current density field is also obtained

$$\hat{\overline{J}} = \frac{\mu^2 + \nu^2\tau\xi}{\left(\mu^2 + \nu^2\tau^2\right)^{3/2}} \exp\left[-\frac{(\xi - \tau)^2}{\mu^2 + \nu^2\tau^2}\right] - i\frac{\xi - \tau}{\left(\mu^2 + \nu^2\tau^2\right)^{3/2}} \exp\left[-\frac{(\xi - \tau)^2}{\mu^2 + \nu^2\tau^2}\right]. \tag{19.69}$$

Figure 19.1 shows the dependences: (a) $\overline{\rho}\left(\xi, \tau, \mu = \nu = 1\right)$, (b) $\text{Re}\hat{\overline{V}}\left(\xi, \tau, \mu = \nu = 1\right)$, (c) $\text{Im}\hat{\overline{V}}$ $\left(\xi, \tau, \mu = \nu = 1\right)$, (d) $\overline{Q}\left(\xi, \tau, \mu = \nu = 1\right)$, (e) $\overline{F}\left(\xi, \tau, \mu = \nu = 1\right)$, (f) $\text{Re}\hat{\overline{J}}\left(\xi, \tau, \mu = \nu = 1\right)$, and (g) $\text{Im}\hat{\overline{J}}$ $\left(\xi, \tau, \mu = \nu = 1\right)$. It results:

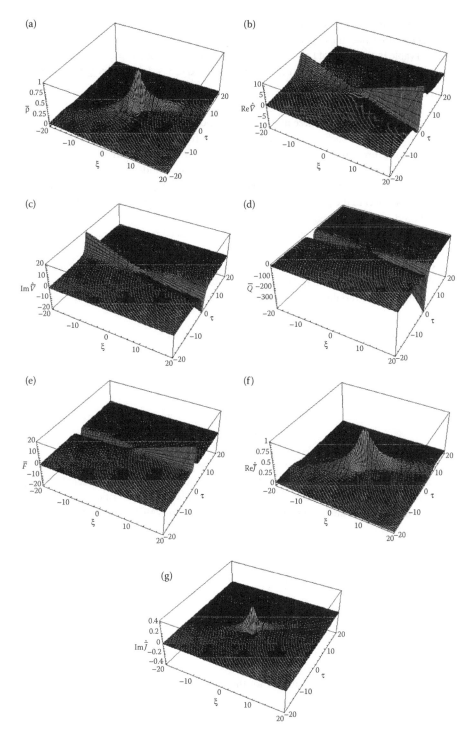

FIGURE 19.1 The dependence on the normalized spatial coordinate and time of (a) normalized density $\overline{\rho}(\xi, \tau, \mu = \nu = 1)$, (b) normalized differential speed $\mathrm{Re}\widehat{\overline{V}}(\xi, \tau, \mu = \nu = 1)$, (c) normalized fractal speed $\mathrm{Im}\widehat{\overline{V}}(\xi, \tau, \mu = \nu = 1)$, (d) normalized fractal potential $\overline{Q}(\xi, \tau, \mu = \nu = 1)$, (e) normalized fractal force $\overline{F}(\xi, \tau, \mu = \nu = 1)$, (f) normalized differential current density $\mathrm{Re}\widehat{\overline{J}}(\xi, \tau, \mu = \nu = 1)$, and (g) normalized fractal current density $\mathrm{Im}\widehat{\overline{J}}(\xi, \tau, \mu = \nu = 1)$, respectively.

1. The force field induces fractal characteristics to the quantities which define the system dynamics. Consequently, they become dependent on the spatiotemporal coordinates.

2. The observable in the form of the rectilinear and uniform motion, $V = c$, is obtained by annulling the force field. The fractal forces on the semispaces $-\infty < x \le \langle x \rangle$ and $\langle x \rangle \le x < \infty$, with $\langle x \rangle$ the mean position, compensate each other $\left\langle m_0 \frac{d\overline{V}}{dt} \right\rangle_{x=-\infty}^{\langle x \rangle} = \left\langle m_0 \frac{d\overline{V}}{dt} \right\rangle_{\langle x \rangle}^{x=+\infty}$.

This means that a fluid particle on free motion locally "polarizes" the fractal fluid behind itself, $x \le ct$, and ahead of itself, $x \ge ct$, in such a way that the resulting fractal forces are symmetrically distributed with respect to a plane through the observable particle position $\langle x \rangle = ct$ at any time t (see the symmetry of the curves from Figure 19.1). In this case, the quantities become independent on the spatiotemporal coordinates. The presence of an external perturbation induces an asymmetry in the distribution of the fractal force field in respect to the plane where the particle is, having as a result the "excitation" of a specific mode of fractal fluid self-structuring.

Therefore, in the particular case of plasma, the collisions induced by the interactions of their particles can be "substituted" by the fractal field (we will return later in the chapter to this subject), while the presence of an external constraint (as, e.g., a voltage) "excites" a specific mode of plasma self-structuring, that could lead to the generation of an electric double layer (DL) or multiple double layer (MDL).

By using the normalized variables (19.62) and (19.63) and

$$\widetilde{N} = \frac{\rho}{\rho_0}, \tag{19.70}$$

$$\widetilde{V} = \frac{V}{u}, \tag{19.71}$$

$$u^2 = \frac{k_B T}{m_0}, \tag{19.72}$$

$$\overline{v_0}^2 = \frac{2\widetilde{D}^2}{u\omega} \frac{1}{\lambda^3}, \tag{19.73}$$

where ρ_0 is the equilibrium density, u is a specific propagation speed of a perturbation in the fractal fluid, k_B is the Boltzmann constant, and T is the "temperature" of the fluid particle (in this model, the fluid particles are identified with the geodesics of the fractal space and their distribution satisfies a certain statistic. We associate the temperature T to such a statistical ensemble), Equations 19.49 and 19.50 become

$$\frac{\partial \widetilde{N}}{\partial \tau} + \frac{\partial \left(\widetilde{N} \widetilde{V} \right)}{\partial \xi} = 0, \tag{19.74}$$

$$\frac{\partial \widetilde{V}}{\partial \tau} + \widetilde{V} \frac{\partial \widetilde{V}}{\partial \xi} = \overline{v_0}^2 \frac{\partial}{\partial \xi} \left[\frac{1}{\sqrt{\widetilde{N}}} \frac{\partial^2}{\partial \xi^2} \left(\sqrt{\widetilde{N}} \right) \right]. \tag{19.75}$$

For a localized stationary solution, let us choose a transformed coordinate q in the moving frame, such that

$$q = \xi - M\tau, \tag{19.76}$$

where M is the equivalent of the Mach number for the fractal fluid

$$M = \frac{\widetilde{V}_0}{u}, \tag{19.77}$$

and $\widetilde{V}_0$ is the speed of a perturbation moving together with the frame. After integration, from the continuity equation, it results

$$\widetilde{V} = M \left(1 - \frac{1}{\widetilde{N}} \right), \tag{19.78}$$

and from the momentum equation it results

$$\frac{\widetilde{V}^2}{2} - \widetilde{V}M = \bar{v}_0^2 \frac{1}{\sqrt{\widetilde{N}}} \frac{d^2}{dq^2} \left(\sqrt{\widetilde{N}} \right), \tag{19.79}$$

where we have used the restrictions

$$q \to \pm\infty, \tag{19.80}$$

$$\widetilde{V} \to 0, \tag{19.81}$$

$$\widetilde{N} \to 1, \tag{19.82}$$

$$\frac{d\sqrt{\widetilde{N}}}{dq} \to 0, \tag{19.83}$$

$$\frac{d^2\sqrt{\widetilde{N}}}{dq^2} \to 0. \tag{19.84}$$

Now, by substituting Equation 19.78 in Equation 19.79 and taking into account restrictions (19.80) through (19.84), we successively obtain

$$\frac{M^2}{2\bar{v}_0^2} \left(\frac{1}{Z^4} - 1 \right) = \frac{1}{Z} \frac{d^2 Z}{dq^2}, \tag{19.85}$$

$$Z = \sqrt{\widetilde{N}}, \tag{19.86}$$

$$-\frac{M^2}{4\bar{v}_0^2} \left(\frac{1}{Z^2} + Z^2 \right) = \frac{1}{2} \left(\frac{dZ}{dq} \right)^2, \tag{19.87}$$

$$\int \frac{d(Z)^2}{\sqrt{1 + \left(Z^2\right)^2}} = -\frac{M}{\bar{v}_0} \left(q - q_0\right), \tag{19.88}$$

$$q_0 = \text{const.}, \tag{19.89}$$

which implies the solution

$$\widetilde{N} = \sinh\left[-\frac{M}{\bar{v}_0} \left(q - q_0\right) \right], \tag{19.90}$$

with q_0 an integration constant. In these conditions, the current density is

$$\bar{J} = \widetilde{N}\widetilde{V} = M\left(\widetilde{N} - 1\right) = M\sinh\left[-\frac{M}{\bar{v}_0} \left(q - q_0\right) \right] - M, \tag{19.91}$$

the fractal potential is

$$\overline{Q} = \frac{M^2}{2} \left(\frac{1}{\widetilde{N}^2} - 1 \right) = \frac{M^2}{2} \coth\left[-\frac{M}{\bar{v}_0} \left(q - q_0\right) \right], \tag{19.92}$$

and the "voltage–current characteristic" is

$$\overline{Q} = \frac{M^2}{2} \left[\left(1 + \frac{\bar{J}}{M}\right)^{-2} - 1 \right]. \tag{19.93}$$

Now, if the structured fractal fluid is equivalent with a circuit element, for example, as it happens with the electric DL in plasma, since for $\bar{J}/M \ll 1$, $\overline{Q} \approx -\bar{J}M$ and for $\bar{J}/M \gg 1$, $\overline{Q} \approx -M^2/2$, it results that it behaves as a "nonlinear element of the circuit." Moreover, the restriction $d\overline{Q}/d\bar{J} = -M\left(1 + \bar{J}/M\right)^{-3} < 0$ marks the beginning of the self-structuring mechanism, for example, the generation of a DL in the case of plasma.

We note that if ω is the ion plasma frequency, λ is the Debye length, and u is the ion-acoustic speed, the previous results can describe the dynamics of plasma.

19.1.4 Synchronous Movements in the Fractal Systems. Types of Dynamics in Plasma at Differential Scale (Macroscopic Scale)

Let us allow the next assumptions:

1. The movements at the two scales, differential (through V) and fractal (through U), are synchronous, which implies

$$V = -U = -D\left(dt\right)^{(2/D_F)-1} \nabla \ln n, \tag{19.94}$$

$$\rho \equiv n. \tag{19.95}$$

2. The movements take place on Peano curves, i.e., with the fractal dimension $D_F = 2$ [55,72]. In this situation, Equation 19.58 takes the form

$$\widetilde{D} = D. \tag{19.96}$$

For an ionized gas with only one type of charge carriers, the total current density is null [32,94]

$$j = \rho V = en\mu E - e\overline{D}\nabla n = 0, \tag{19.97}$$

where μ is the mobility of the charge carriers, e is the elementary electrical charge, $\overline{D}$ is the diffusion coefficient, and E is the electric field. In this case, the field current, $en\mu E$ is totally compensated by the diffusion current $e\overline{D}\nabla n$, and the vectors E and ∇n are parallel. By limiting to the case of a nondegenerate gas, the partial pressure of the charge carriers is related to their density and temperature through the relation

$$p = nk_B T, \tag{19.98}$$

so that, from Equation 19.97, it results

$$\frac{\mu}{\overline{D}} = \frac{|\nabla n|}{nE} = \frac{|\nabla p|}{pE} = \frac{enE}{pE} = \frac{en}{p} = \frac{e}{k_B T}, \tag{19.99}$$

where we used the relation $dp = eEndl$ between the two expressions of the force on a layer of dl thickness and unit surface, normal on the pressure gradient. From Equations 19.96 and 19.99, it results first

$$\widetilde{D} = \overline{D} = \frac{\mu k_B T}{e}, \tag{19.100}$$

and then, through Equations 19.50, 19.94, 19.95, and 19.100, the diffusion equation

$$\frac{\partial n}{\partial t} = \widetilde{D}\Delta n. \tag{19.101}$$

Let us consider an ionized gas with the electron density n_e and ion density n_p. Usually, the gradients of the electron and ion densities, ∇n_e and ∇n_p, as well as the electric field E, are different from zero, so that the total current density

$$\rho V = j_e + j_p, \tag{19.102}$$

has a diffusion component, as well as a field component [32,94]

$$j_e = e\left(\mu_e n_e E + D_e \nabla n_e\right), \tag{19.103}$$

$$j_p = e\left(\mu_p n_p E - D_p \nabla n_p\right). \tag{19.104}$$

The field current appears even if no external electric field is applied. Indeed, because $D_e \gg D_p$, the electrons radially diffuse outside plasma, leaving behind an excess of positive charge. This leads to the appearance of a radial electric field, E_r, which retards the electrons and accelerates the positive ions, the total radial current being null in the stationary regime. By taking into account the quasineutrality of plasma

$$n_e \cong n_p = n, \tag{19.105}$$

from Equations 19.102 through 19.105, it results

$$E_r = -\frac{D_e - D_p}{\mu_e + \mu_p}\frac{1}{n}\frac{\partial n}{\partial r}, \tag{19.106}$$

as well as the density of the particles current

$$G_e = \frac{j_e}{e} = G_p = \frac{j_p}{e} = -D_a\frac{\partial n}{\partial r}, \tag{19.107}$$

with

$$D_a = \frac{\mu_e D_p + \mu_p D_e}{\mu_e + \mu_p}, \tag{19.108}$$

as an ambipolar diffusion coefficient. It results first

$$\tilde{D} = \overline{D} = D_a, \tag{19.109}$$

and then, from Equations 19.50 and 19.107, the ambipolar diffusion equation

$$\frac{\partial n}{\partial t} = D_a \Delta n. \tag{19.110}$$

The presence of collisions dramatically changes the expression of the diffusion coefficient. For example, for weak-ionized plasma and in the approximation of small density gradients, the electron-free diffusion coefficient is [32,94]

$$\overline{D} = D_e = \frac{k_B T_e}{m_e \nu_{en}}, \tag{19.111}$$

where ν_{en} is the frequency of the elastic electron-neutral collisions, m_e is the electron mass, and T_e is the electrons temperature.

In the case of elastic collisions, the dynamics of the plasma particles can be described by Brownian-type movements. The random process that can describe the Brownian motion could be, for example, the Wiener process [55,72,87]. In this case, the mean square distance covered by a particle in the mean time τ, can be assimilated to a diffusion coefficient (up to a numeric factor)

$$\tilde{D} = \overline{D} \approx \frac{\langle x \rangle^2}{\tau}. \tag{19.112}$$

Moreover, by taking into account the statistic meaning of the collision cross section, σ, a correspondence with the diffusion coefficient can be established in the form [32,94]

$$\tilde{D} = \overline{D} = \frac{\overline{c}}{n_0 \sigma}, \tag{19.113}$$

where n_0 is an equilibrium density and $\overline{c}$ is a specific propagation speed of a perturbation in plasma. Because in the general case σ is a function of the charge carrier energy, the scale dependence (19.58) can be replaced by the normalized energy ε dependence

$$\tilde{D} = D(\varepsilon)^{(2/D_F)-1}, \tag{19.114}$$

having in mind the fractal characteristics of a relation of type $\sigma = \sigma(E)$ [55,72].

19.1.5 Conclusion

By considering that the particle movements in plasma take place on fractal curves, a fractal hydrodynamic model was developed in order to describe its dynamics. Thus

1. The scale relativity model was presented in more detail compared with those presented by Nottale in References 48, 86, 87, 90, and 106 (consequences of nondifferentiability, covariant total derivative, geodesics via Schrödinger-type equation, or fractal hydrodynamic model).
2. Through a fractal hydrodynamic model, we show that the predictability is imposed by fractality and the conditions in which a fractal fluid can self-structure were specified.
3. In the frame of the fractal hydrodynamic model, the synchronous fractal movements were analyzed and some types of plasma dynamics were presented, which satisfies such a condition (ionized gas with only one type of charge carriers, ionized gas with two types of charge carriers, and ambipolar diffusion).
4. The correspondence between the collisions and fractality was established, as well as the way in which this correspondence functions in plasma.

19.2 Complex Space Charge Structures and Chaos in Plasma

19.2.1 Introduction

Plasma is a nonlinear system very favorable for developing of spatial, temporal, or spatio-temporal structures such as DL, MDL, solitons, etc.

DLs are narrow localized nonlinear potential structures consisting of two adjacent layers of positive and negative charges which sustain large potential jumps, i.e., electric fields [30]. When a positively biased electrode is immersed into plasma, for a sufficiently bias voltage, a luminous nearly spherical fireball (see Figure 19.2a) is generated in contact with the electrode [99,103]. Experimental investigation revealed that such a complex structure consists of a positive core (an ion-enriched plasma) confined by a DL. The potential drop across the DL is almost equal to the ionization potential of the gas atoms. At high values of the potential applied on the electrode, the structure passes into a dynamic state (see Figure 19.2b), consisting of periodic disruptions and reaggregations of the DL at its border [84,99]. When the DL disrupts, the initially trapped particles (electrons and positive ions) are released into plasma, triggering different types of instabilities such as ion-acoustic instability (in nonmagnetized plasmas) [34] or potential relaxation instability and electrostatic ion–cyclotron instability (in magnetized plasma) [44].

Under certain experimental conditions (gas nature and pressure, plasma density), a more complex structure in the form of two or more subsequent DLs, called MDLs, was observed [29,37,42,61,62,79,104].

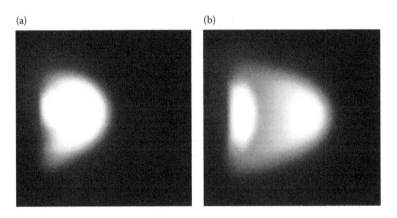

(a)   (b)

FIGURE 19.2   Photos of the fireball in stationary (a) and dynamic (b) states, respectively.

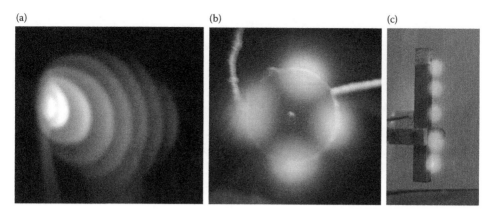

FIGURE 19.3 Concentric (a) and nonconcentric (b and c) MDLs in plasma.

It appears as several bright and concentric plasma shells attached to the anode of a glow discharge or to a positively biased electrode immersed into plasma (see Figure 19.3a). The successive DLs are located precisely at the abrupt changes of luminosity between two adjacent plasma shells. Langmuir probe measurements indicated that the axial profile of the plasma potential has a stair step shape, with potential drops close to the ionization potential of the used gas [37,62].

If the electrode is large with respect to the characteristic length of plasma, or if it is strongly asymmetric (e.g., with almost one-dimensional geometry), the MDL structure appears nonconcentrically, as a network of intense luminous plasma spots, located near each other, almost equally distributed on the electrode surface (see Figure 19.3b and c) [11,17,18,42,96,109]. This kind of structure was called nonconcentric MDLs [63].

Experimental investigations have proven the important role of elementary processes such as electron-neutral impact excitations and ionizations in the formation and dynamics of the DLs and MDLs [62].

In certain experimental conditions, at high values of the potential applied on the electrode, the dynamics of such complex space charge structures (CSCSs) pass into chaotic states through different scenarios [2, 34,43]. These routes to chaos can be theoretically modeled in the frame of SRT, by a fractal hydrodynamic model as the one described in Section 19.1 of this chapter.

19.2.2 Experimental Transition to Chaos of the Fireball Dynamics through Type I Intermittency

The experiment was conducted in a hot-cathode discharge plasma diode, schematically shown in Figure 19.4. Plasma is created by volume ionization processes between energetic electrons from the hot filament (marked by "F" in Figure 19.4) and gas atoms. The anode (marked by "A" in Figure 19.4 and made from nonmagnetic stainless steel) is grounded and the discharge current was $I_d = 40$ mA. The plasma diffuses into the chamber, where the supplementary electrode E (3 cm in diameter) is positively biased with respect to the plasma potential (and also to the ground). The plasma parameters, measured by emissive and cold probes were plasma density $n_{pl} \cong 5 \times 10^8$ cm^{-3}; electron temperature $T_e \cong 2$–3 eV for an argon pressure, $p = 5 \times 10^{-3}$ mbar.

Figure 19.5 shows the static current–voltage characteristic of the electrode, obtained by gradually increasing and subsequently decreasing the potential on the electrode, V_E. The sudden jumps of the current collected by E, marked as I_E, are related to the generation and dynamics of the DLs [99] and show hysteresis. This phenomenon proves that DLs can maintain their static or dynamic states for conditions weaker than those required for their creation. After the first sudden jump, marked (c to d) in the static current–voltage characteristic of the electrode in Figure 19.5, a quasi-spherical fireball appears in front

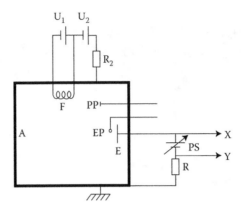

FIGURE 19.4 Hot-cathode plasma diode (F filament; E additional electrode; A anode; U_1 power supply for heating the filament, U_2 power supply for discharge, PS power supply for the electrode bias, R, R_2 load resistors, EP emissive probe, PP plane probe, and X, Y to the oscilloscope).

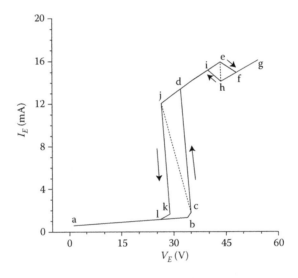

FIGURE 19.5 Static current–voltage characteristic of the additional electrode E (see Figure 19.4), where the small letters mark the positions on the characteristic where the behavior of the plasma changes.

of E (see the photo in Figure 19.6). Emissive probe measurements prove that this fireball consists of a positive core (positive ion-rich plasma) confined by a DL (see Figure 19.7). Its appearance implies a process by which thermal energy of the electrons extracted from a population with quasi-Maxwellian distribution is converted into the electrical field energy of the DL [69].

The stability of the DL is ensured by the balance between processes of charge production (by electron-neutral impact excitations and ionizations) and charge loss (by recombination and diffusion). When this equilibrium is affected, the DL shifts away from the electrode assuring its autonomy by continuing to extract particles and thermal energy from plasma, in order to keep its potential drop constant. In this way, a new DL starts to develop in the region between the first one and the electrode. The new negative space charge acts as a current barrier, related to the development of excitation processes in that region, reducing the electron flux through the DL. Thus, the stability is affected and the DL disrupts. During this process, ions are released and injected into the surrounding plasma. At the same time, electrons, initially

FIGURE 19.6 Photo of the luminous fireball obtained in front of the additional electrode E (see Figure 19.4).

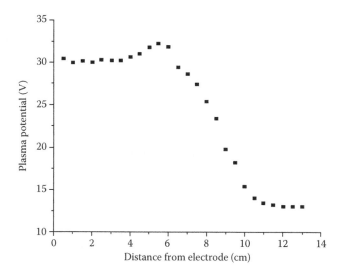

FIGURE 19.7 Axial profile of the plasma potential in front of the additional electrode E (see Figure 19.4), revealing the presence of the DL structure at the border of the fireball.

trapped in the DL structure, also become free, being accelerated toward E. Reaching the region where a new DL is emerging, these electrons enhance the ionization rate to the value for which the new DL starts its detachment process. In this way, an internal feedback mechanism ensures the periodicity of the phenomenon. The periodic injection of bunches of ions into the surrounding plasma stimulates the onset of the ion-acoustic instability there.

The decrease of the current (e to f) in Figure 19.5 is not a sudden jump like the one associated with the transition from c to d, but occurs gradually. The periodical detachment and reaggregation of DLs cause a modulation of the current collected by the electrode, with a frequency marked by f_1 in Figure 19.8a. Figure 19.8a shows the dynamic current–voltage characteristic of the electrode E, by recording the current to E versus the increase of the voltage on it with a rate of 2 V/ms, from $V_E = 45$ V to $V_E = 55$ V, in order to show the transition from an oscillating DL to the ion-acoustic instability. The oscillating DL, with the fundamental frequency marked by f_1 in Figure 19.8a, appears in the region (e to f) of the current–voltage characteristic (with a negative differential resistance [NDR]) in Figure 19.5, while the ion-acoustic

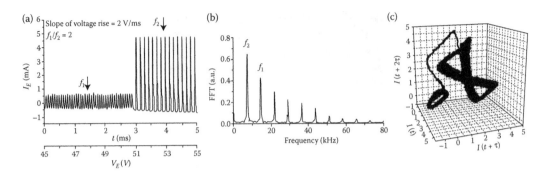

FIGURE 19.8 (a) Dynamic current–voltage characteristic obtained by fast varying the potential on the additional electrode E (see Figure 19.4) between 45 and 55 V, with a rate of 2 V/ms; (b) fast Fourier transform (FFT) amplitude spectra of the ac component of the current from (a); and (c) 3-D-reconstructed states space of the plasma system dynamics.

instability, with the fundamental frequency marked by f_2 in Figure 19.8a, appears in the region (f to g) of the current–voltage characteristic (with a positive resistance slope). The ion-acoustic instability appears as a nonlinear periodic variation of the current with almost 100% modulation.

The ac components of the current are recorded with a sampling rate of 500 kHz, in order to also show the upper harmonics of the fundamental frequency. By using the time delay method (extensively described in Reference 44), we reconstructed the 3D state space attractors of the plasma system dynamics (see Figure 19.8c).

The transition indicated in Figure 19.5, from the frequency f_1 to f_2, proves that the moving DL dissolves and sustains ion-acoustic-like oscillations in the background plasma. The fundamental frequency of these oscillations (~14 kHz) is small compared to the ion plasma frequency; so, the dispersion relation of the ion-acoustic wave is linear. The ion-acoustic waves are compressive spherical ones [93], which are usually proved by measuring the electronic saturation current of the probe during one period of the ion-acoustic instability and also by localized measurements of the ion density where the DL dissolves [9].

The ion-acoustic instability is very sensitive to the background plasma parameters. Any small variation of these parameters modifies the frequency and amplitude of the instability and can also cause its suppression (see Figure 19.9). The onset of chaotic states is due to intermittencies, as one can see in Figure 19.9. The fast Fourier transform (FFT) amplitude graphs indicate the evolution to chaotic states by embedding the fundamental instability frequency in broadband noise, associated with the onset of the intermittencies. The largest Lyapunov exponent calculated from the time series corresponding to the signal in Figure 19.9f has the value 0.015492. Its positive value proves that this signal is a chaotic one. The reconstructed 3D state spaces indicate the loss of stability of a periodic attractor through a succession of bursts. During their appearance, the ion-acoustic instability is suppressed. The mechanism of reinsertion of trajectories in the closed loop of the attractor is relevant for proving the intermittency route to chaos. By performing log–log plots of the spectral power density versus frequency, as indicated in Figure 19.10, we obtained a slope which confirms an f^{-1} power law associated with the chaotic development state of the instability. Another typical fingerprint of type I intermittency is the presence of a tangent bifurcation (saddle-node point bifurcation), represented in the return map shown in Figure 19.11. We reconstructed this map by plotting the maxima and minima of the time series. We can define a control parameter for the intermittent state, $\varepsilon = V_E/V_{EC}$, where V_{EC} is the critical voltage for which the intermittencies set in, and $V_E > V_{EC}$ is the voltage for which the intermittencies develop more and more. Thus, when $\varepsilon = 0$, the first intermittency appears. This is related to the onset of a new unstable direction in the phase space. The orbits generated by the map are trapped in the narrow channel formed between the fitting curve and the identity map. When $\varepsilon \to 0$, the channel becomes narrower and finally disappears, revealing the tangent bifurcation of

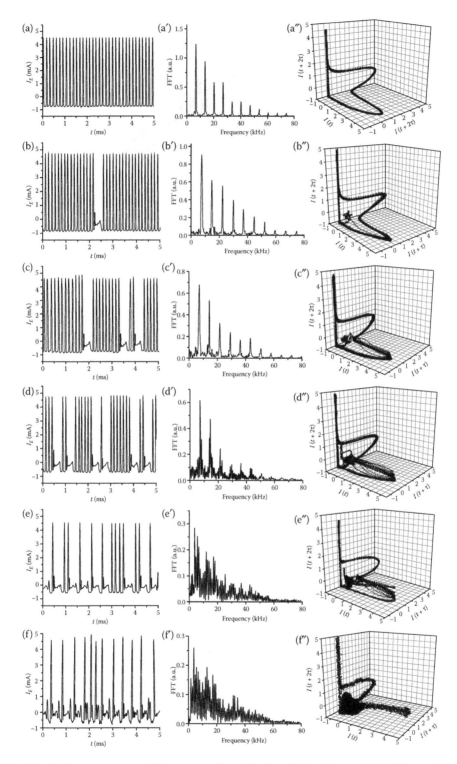

FIGURE 19.9 (a–f) ac components of the current collected by the additional electrode, for different values of the voltage applied on it: (a) 55, (b) 57, (c) 58, (d) 60, (e) 62, and (f) 64 V; (a′–f′) FFT amplitude spectra of the corresponding signals, for the same values of the voltage applied on the additional electrode; and (a″–f″) 3-D-reconstructed states space of the plasma system dynamics, for the same values of the voltage applied on the additional electrode.

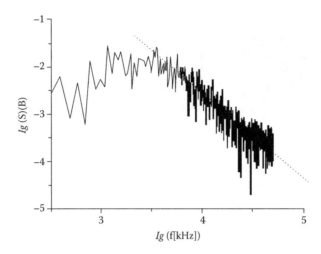

FIGURE 19.10   Log–log plot of the power spectral density versus frequency, for $V_E = 62$ V.

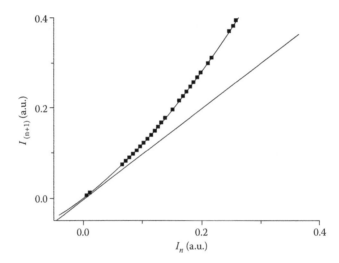

FIGURE 19.11   The Return map obtained from maxima and minima of the time series with intermittencies.

this intermittent chaotic state. The probability distribution of a laminar phase versus laminar length in regions of type I intermittency is also found to obey scaling laws with exp(1/2).

19.2.3 Experimental Transition to Chaos of the MDL Dynamics through Cascade of Sub-Harmonic Bifurcations

The experiment was conducted in the same hot-cathode discharge plasma diode, schematically shown in Figure 19.4. In this case, a supplementary electrode E with 1 cm diameter was used, under the following experimental conditions: argon pressure $p = 10^{-2}$ mbar, plasma density $n_{pl} \cong 10^9$ cm^{-3}, and electron temperature $kT_e \cong 2$ eV.

When the voltage on the electrode reaches $V_E \cong 55$ V, a DL structure appears in front of the electrode (see the photo in Figure 19.12a). Owing to the experimental conditions, this structure directly appears in a dynamic state, as explained in Reference 41. The oscillations of the current collected by the electrode, their FFT, and the reconstructed attractors of the system dynamics in the states space (by the time delay

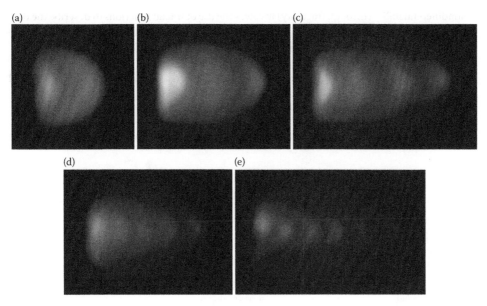

FIGURE 19.12 Photos of the MDL structure in different stages (at different increasing values of the potential applied on the electrode) of its formation.

method) are shown in Figure 19.13a–c, respectively. By a further increase in the voltage on the electrode, new DLs develop in front of the electrode, giving rise to a dynamic MDL structure (see photos in Figure 19.12b–e). Simultaneously with every new DL formation, a new sub-harmonic appears in the FFT spectrum of the current oscillations collected by the electrode, the corresponding attractor becoming more and more complex (see Figure 19.13, second and third columns). Thus, we recorded, in fact, spatiotemporal bifurcations in the plasma system (sudden changes in the spatial symmetry and in the temporal dynamics of the plasma system). At high values of the applied potential, the plasma system passes into a chaotic state, characterized by uncorrelated and intermittent oscillations (see Figure 19.13s–u, respectively).

Figure 19.14a and b shows the axial profiles of the electric field and electric charge density in front of the electrode (the electrode is placed at the position $x = 0$ on the graphs), respectively, obtained as the first and second derivatives of the axial profile of the plasma potential, which was experimentally recorded with the help of an axially movable emissive probe. The two graphs correspond to the situation in which an MDL structure consisting of two DLs exists in front of the electrode in a static state (no oscillations of the current collected by the electrode being recorded). The DLs are placed in the positions where the electric field has the maximum values, marked by arrows in Figure 19.14a. From Figure 19.14b we observe that the space charge density decreases as the distance to the electrode increases.

19.2.4 Order to Chaos Transitions in Plasma via Nondifferentiability

The theoretical analysis of the above-described experimental results is made in the frame of the fractal hydrodynamic model, extensively described in Section 19.1 of this chapter. Once accepted that the plasma particles move on continuous but nondifferentiable curves (fractal curves), some specific consequences through SRT are stated (adapted to the dynamics of a discharge plasma) [52,87,90]:

1. A continuous and nondifferentiable curve is explicitly scale dependent. Its length tends to infinity when the scale interval tends to zero. Since the generalization of this theorem to three dimensions is straightforward, it follows that a continuous and nondifferentiable space is fractal, under the general meaning of scale divergence [72].

2. Nondifferentiability (through fractality) involves the use of fractal functions to describe the dynamics of a discharge plasma. This means that the quantities describing the dynamics depend on both coordinate (x), time (t), and the scale resolution (dt), in the general form [87,90]

$$F(x, t, dt) = P(x, t) + Q(x, t, dt), \qquad (19.115)$$

where $P(x, t)$ is a differentiable function, while

$$Q(x, t, dt) = \overline{Q}(x, t)(dt)^{\tau}, \qquad (19.116)$$

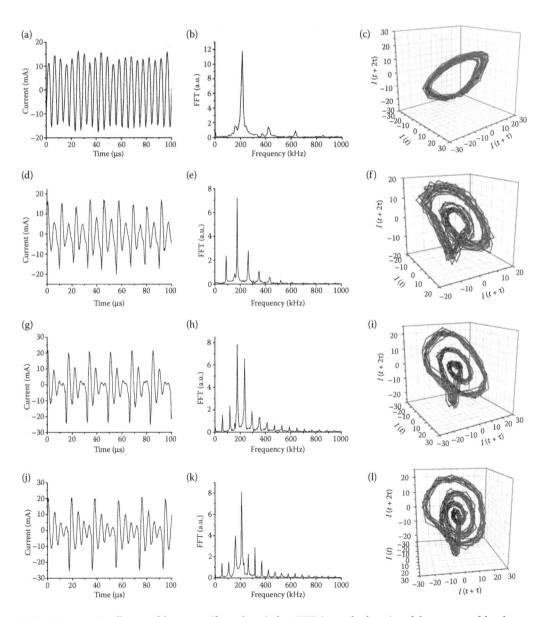

FIGURE 19.13 Oscillations of the current (first column), their FFTs (second column), and the attractor of the plasma system dynamics in the reconstructed state space (third column), at different increasing values of the voltage applied on the electrode. *(Continued)*

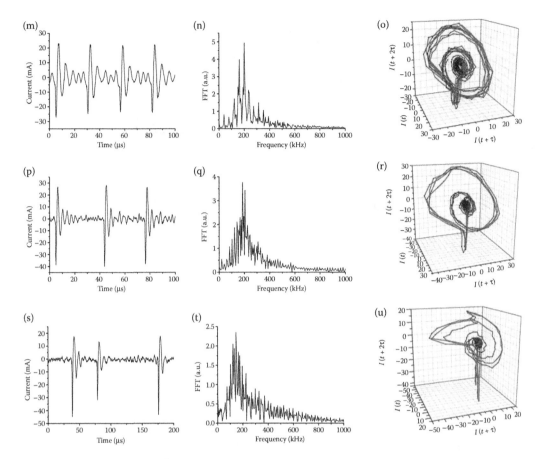

FIGURE 19.13 (Continued) Oscillations of the current (first column), their FFTs (second column), and the attractor of the plasma system dynamics in the reconstructed state space (third column), at different increasing values of the voltage applied on the electrode.

is a nondifferentiable function and τ is a parameter depending on the fractal dimension D_F of the movement curve. We can distinguish the following limit dynamic regime:

a. Asymptotic regime for small-scale resolutions $dt \ll 1$ and constant τ, case in which relation (19.115) reduces to the power-type law

$$F(x, t, dt) \to \overline{Q}(x, t)(dt)^{\tau}.$$ (19.117)

In particular, the Child–Langmuir conduction, $j \sim V^{3/2}$, is a power-type law.

b. Asymptotic regime for large-scale resolutions $dt \gg 1$ and constant τ, case in which relation (19.115) involves

$$F(x, t, dt) \to P(x, t).$$ (19.118)

For example, the ohmic conduction is such a regime. According to Reference 6 and through relations (19.115) and (19.116), we can choose the fractal-normalized current–voltage characteristic in the form

$$\overline{\phi} = \overline{I}\left(1 + \frac{\overline{a}}{1 + \overline{I}^2}\right),$$ (19.119)

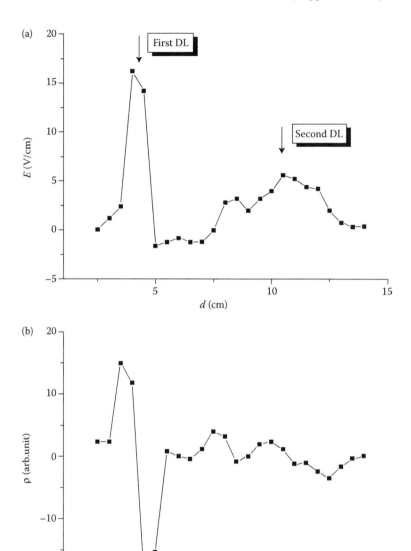

FIGURE 19.14 Axial profiles of the electric field (a) (the position of the individual double layers is marked by arrows) and charge density (b), respectively, in front of the electrode.

where $\overline{\phi}$ is the fractal-normalized voltage, $\overline{I}$ is the fractal-normalized current, and $\overline{a}$ is a parameter depending on the scale resolution. This relation induces a conduction bistability (see Figure 19.15) as follows:

i. According to Reference 6, the restriction $\overline{a} \geq 8$ implies the bistability.
ii. The value of $\overline{a}$ set the scale resolution through the ionization and recombination rates.
iii. Once $\overline{a}$ is fixed (with $\overline{a} \geq 8$), for values of the fractal-normalized current in the interval AB on the characteristic (see Figure 19.15), the fractal-normalized voltage can have two distinct stable values.

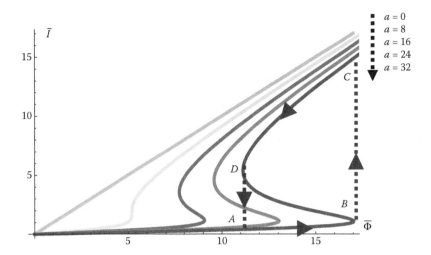

FIGURE 19.15 Theoretical dependence of the fractal-normalized current on the fractal-normalized potential.

    iv. The conduction bistability is associated with the NDR (or hysteresis).

    v. Since in relation (19.119) $\bar{\phi}$ and $\bar{I}$ are fractal functions, they show the propriety of self-similarity. Consequently, the conduction bistability from Figure 19.15 can be found at any scale resolution (i.e., for different ionization and recombination rates). Thus, we state the possible correspondence between the multiplicity order of the DL and the one of the conduction bistability.

3. There is an infinity of fractal trajectories (geodesics) relating to any couple of points at all scales. In order to account for the infinity of geodesics, for their fractality, and for two valuedness of the derivative which all come from the nondifferentiable geometry of the space (for details, see References 87 and 90), one therefore adopts a generalized statistical fluid-like description (fractal fluid).

Indeed, according to SRT, the dynamics of the plasma discharge are described by the complex operator (19.38) (see also References 4, 8, 57, and 84). In the standard models [76,92], the movements of the plasma particles take place on continuous and differentiable curves, so that its dynamics are described by the classic operator d/dt. The presence of an external field with the specific scalar potential U modifies the law of momentum conservation (19.49) in the form

$$\frac{\partial V}{\partial t} + (V \cdot \nabla) V = -\nabla (\overline{Q} + U).$$ (19.120)

The fractal potential $\overline{Q}$ can generate a viscosity stress-type tensor. Indeed, written in the form

$$\overline{Q} = -m_0 D^2 (dt)^{(4/D_F)-2} \left[\frac{\nabla^2 \rho}{\rho} - \frac{1}{2} \left(\frac{\nabla \rho}{\rho} \right)^2 \right],$$ (19.121)

the fractal potential induces the symmetric tensor

$$\sigma_{il} = m_0 D^2 (dt)^{(4/D_F)-2} \rho \nabla_i \nabla_l \ln \rho = m_0 D^2 (dt)^{(4/D_F)-2} \left[\nabla_i \nabla_l \rho - \frac{(\nabla_i \rho)(\nabla_l \rho)}{\rho} \right].$$ (19.122)

The divergence of this tensor is equal to the force density associated with $\overline{Q}$:

$$\nabla \cdot \overline{\sigma} = -\rho \nabla \overline{Q}.$$ (19.123)

The quantity $\bar{\sigma}$ can be identified with the tensor of viscosity stress in a Navier–Stokes-type equation:

$$m_0 \rho \frac{d\mathbf{V_D}}{dt} = \nabla \cdot \bar{\sigma}. \tag{19.124}$$

The momentum flux density-type tensor is

$$\pi_{il} = m_0 \rho V_{Di} V_{Dl} - \sigma_{il}, \tag{19.125}$$

and it satisfies the momentum flow-type equation

$$m_0 \frac{\partial}{\partial t} \left(\rho \mathbf{V_D} \right) = -\nabla \cdot \bar{\pi}. \tag{19.126}$$

We formally introduce the kinematical and dynamical-type viscosities

$$\nu = \frac{1}{2} D(dt)^{(2/D_F)-1}, \tag{19.127}$$

$$\bar{\nu} = \frac{1}{2} m_0 \rho D(dt)^{(2/D_F)-1}. \tag{19.128}$$

The quantities ν and $\bar{\nu}$ are formal viscosities, both of them being induced by the fractal scale. Then, the tensor σ_{il} takes the usual form

$$\sigma_{il} = \bar{\nu} \left(\frac{\partial V_{Fi}}{\partial x_l} + \frac{\partial V_{Fl}}{\partial x_i} \right). \tag{19.129}$$

In particular, if σ_{il} is diagonal

$$\sigma_{il} = \sigma \delta_{il}, \tag{19.130}$$

Equations 19.49 and 19.50 take the form

$$m_0 \left[\frac{\partial \mathbf{V_D}}{\partial t} + \left(\mathbf{V_D} \cdot \nabla \mathbf{V_D} \right) \right] = -\frac{\nabla \sigma}{\rho}, \tag{19.131}$$

$$\frac{\partial \rho}{\partial t} + \nabla \cdot \left(\rho \mathbf{V_D} \right) = 0. \tag{19.132}$$

Let us consider now that the movements of the plasma particles take place on Peano-type curves (for details, see References 6, 71, 72, 87, and 90). Then, if we assimilate tensor (19.129) with the gas pressure, i.e., $\sigma_{il} = p\delta_{il}$, Equations 19.131 and 19.132 are reduced to the classical plasma hydrodynamics.

Further, using Equations 19.131 and 19.132 for an axial symmetry, we will analyze the dynamics of a discharge plasma by considering that the movements of the plasma particles take place on fractal curves with $D_F = 2$. The external constraint on the discharge plasma, in the form of the potential applied on the electrode, is specified by adequate initial and boundary conditions. Let us introduce the normalized quantities

$$\omega t = \tau, \tag{19.133}$$

$$kr = \xi, \tag{19.134}$$

$$kz = \eta, \tag{19.135}$$

$$\frac{V_{Dr}}{c} = V_\xi, \tag{19.136}$$

$$\frac{V_{Dz}}{c} = V_\eta, \tag{19.137}$$

$$\frac{\rho}{\rho_0} = N, \tag{19.138}$$

$$\frac{kc}{\omega} = 1, \tag{19.139}$$

where ω is the plasma pulsation, k is the inverse of the Debye length, c is the ion-acoustic speed, and ρ_0 is the equilibrium plasma density. Then, Equations 19.131 and 19.132 become

$$\frac{\partial}{\partial \tau}\left(NV_\xi\right) + \frac{1}{\xi}\frac{\partial}{\partial \xi}\left(\xi NV_\xi^2\right) + \frac{\partial}{\partial \eta}\left(NV_\xi V_\eta\right) = -\frac{\partial N}{\partial \xi},\tag{19.140}$$

$$\frac{\partial}{\partial \tau}\left(NV_\eta\right) + \frac{1}{\xi}\frac{\partial}{\partial \xi}\left(\xi NV_\xi V_\eta\right) + \frac{\partial}{\partial \eta}\left(NV_\eta^2\right) = -\frac{\partial N}{\partial \eta},\tag{19.141}$$

$$\frac{\partial N}{\partial \tau} + \frac{1}{\xi}\frac{\partial}{\partial \xi}\left(\xi NV_\xi\right) + \frac{\partial}{\partial \eta}\left(NV_\eta\right) = 0.\tag{19.142}$$

For the numerical integration, we shall impose the initial conditions

$$V_\xi(0,\xi,\eta) = 0,\tag{19.143}$$

$$V_\eta(0,\xi,\eta) = 0,\tag{19.144}$$

$$N(0,\xi,\eta) = 1/5,\tag{19.145}$$

$$1 \le \xi \le 2,\tag{19.146}$$

$$0 \le \eta \le 1,\tag{19.147}$$

as well as the boundary conditions

$$V_\xi(\tau,1,\eta) = V_\xi(\tau,2,\eta) = 0,\tag{19.148}$$

$$V_\eta(\tau,1,\eta) = V_\eta(\tau,2,\eta) = 0,\tag{19.149}$$

$$V_\xi(\tau,\xi,0) = V_\xi(\tau,\xi,1) = 0,\tag{19.150}$$

$$V_\eta(\tau,\xi,0) = V_\eta(\tau,\xi,1) = 0,\tag{19.151}$$

$$N(\tau,1,\eta) = N(\tau,2,\eta) = 1/5,\tag{19.152}$$

$$N(\tau,\xi,0) = \frac{1}{10}\exp\left[-\left(\frac{\tau-1/5}{1/5}\right)^2\right]\exp\left[-\left(\frac{\xi-3/2}{1/5}\right)^2\right],\tag{19.153}$$

$$N(\tau,\xi,1) = 1/5.\tag{19.154}$$

The equation systems (19.140) through (19.142) with the initial conditions (19.143) through (19.147) and the boundary conditions (19.148) through (19.154) was numerically resolved by using the finite differences [110]. Figures 19.16 through 19.18 show the 3D dependences (Figures 19.16a through 19.18a) and two-dimensional (2D) contours (Figures 19.16b through 19.18b) of the normalized density N and the normalized speeds V_ξ and V_η, respectively, on the normalized coordinates ξ and η at the normalized time value $\tau = 0.74$. Using these numerical solutions, the theoretical profiles of the electrical field and density, respectively, are given in Figure 19.19a and b.

It results in

1. Generation of multiple structures in plasma (Figure 19.16a and b), corresponding to the MDL like those shown in Figure 19.12
2. Symmetry of the normalized speed field V_ξ with respect to the symmetry axis of the spatiotemporal Gaussian (Figure 19.17a and b)
3. Shock waves and vortices at the structure's periphery for the normalized speed field V_η (Figure 19.18a and b)
4. A good agreement between the experimental curves in Figure 19.14a and b and the theoretical ones in Figure 19.19a and b

If the discharge plasma is placed into the external field U, then Equations 19.44 and 19.45 become

$$\hat{L}\psi = 0,\tag{19.155}$$

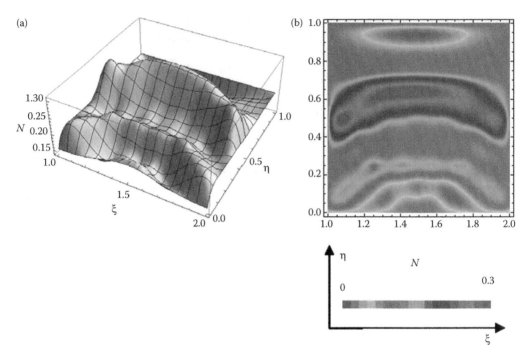

FIGURE 19.16 3D dependence of the normalized density field N on the normalized spatial coordinates (ξ, η) (a) and 2D contour of the same normalized density field (b) at normalized time $\tau = 0.74$.

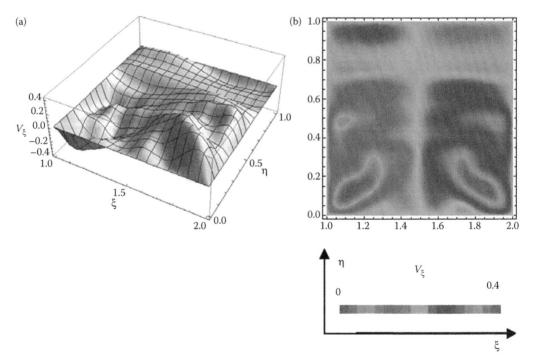

FIGURE 19.17 3D dependence of the normalized speed field V_ξ on the normalized spatial coordinates (ξ, η) (a) and 2D contour of the same normalized speed field (b) at normalized time $\tau = 0.74$.

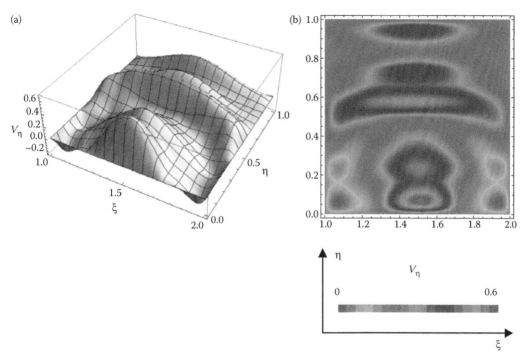

FIGURE 19.18 3D dependence of the normalized speed field V_η on the normalized spatial coordinates (ξ, η) (a) and 2D contour of the same normalized speed field (b) at normalized time $\tau = 0.74$.

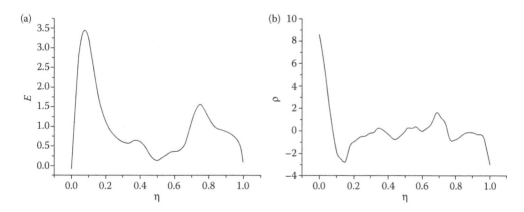

FIGURE 19.19 Profile of the normalized electric field (a) and charge density (b), respectively, for $\xi = 0.3$.

$$\hat{L} = 4D^2 \left(dt\right)^{(4/D_F)-2} \Delta + 2iD \left(dt\right)^{(2/D_F)-1} \frac{\partial}{\partial t} - U. \tag{19.156}$$

Since the position vector of the particle is assimilated with a stochastic process of Wiener type (for details, see References 87 and 90), it is not only the scalar potential of a complex speed (through ψ), but also the density of probability (through $|\psi|^2$) in the frame of a Schrödinger-type theory. It results in the equivalence between the fractal hydrodynamics—see Equations 19.49 through 19.51—and the Schrödinger-type equation—see Equations 19.44 and 19.45, as well as the chaoticity through stochasticization via fractalization. Now, by considering that the potential applied on the electrode immersed into plasma simulates, in our opinion, an infinite one-dimensional potential well, according to the method described in References 4 and 43, we obtain the discrete energy eigenvalues after solving either the time-independent fractal hydro-dynamic Equations 19.49 through 19.51, or the time-independent Schrödinger-type Equation 19.155:

$$E_n \equiv Q_n = 2m_0 D^2 (dt)^{(4/D_F)-2} \left(\frac{n\pi x}{a}\right)^2, \quad n = 1, 2, \ldots,$$ (19.157)

and the eigenfunctions

$$\phi_n = \begin{cases} \left(\frac{2}{a}\right)^{1/2} \sin\left(\frac{n\pi x}{a}\right), & n \text{ even}, \ |x| \le \frac{a}{2}, \\ \left(\frac{2}{a}\right)^{1/2} \cos\left(\frac{n\pi x}{a}\right), & n \text{ odd}, \ |x| \le \frac{a}{2}, \end{cases}$$ (19.158)

where a is the well's width and m_0 is the mass of the fluid particle. Some timescales of a speed potential evolution are contained in the coefficients of the Taylor series of the quantized energy levels E_n around the main energy $E_{\bar{n}}$ (by generalization of the results from Reference 16)

$$E_n = E_{\bar{n}} + 4\pi m_0 D \left[\frac{n - \bar{n}}{T_\alpha} + \frac{(n - \bar{n})^2}{T_\beta} + \cdots \right],$$ (19.159)

where often the zero of energy is shifted to remove the $E_{\bar{n}}$ term. Regrouping the infinite square-well energies (19.157) in this form gives

$$E_n = E_1 n^2 = E_1 \bar{n}^2 + 2E_1 \bar{n} (n - \bar{n}) + E_1 (n - \bar{n})^2,$$ (19.160)

and comparing Equations 19.159 and 19.160 we relate

$$T_\alpha = \frac{2\pi m_0 D(dt)^{(2/D_F)-1}}{\bar{n} E_1},$$ (19.161)

$$T_\beta = \frac{4\pi m_0 D(dt)^{(2/D_F)-1}}{E_1}.$$ (19.162)

We note that the timescale T_β does not depend on the mean energy level $\bar{n}$. This will provide us with a universal timescale for describing speed potential evolution that does not depend on the particle average energy.

Now, the full and fractional revivals formalism may be applied. A full and fractional revivals of a speed scalar potential in the infinite square well occurs when a speed scalar potential evolves in time to a state that can be described as a collection of spatially distributed subspeed scalar potentials that each closely reproduces the initial speed scalar potential shape—see for details Reference 16. Therefore, the full and fractional revivals of a speed scalar potential in the infinite square well implies either

$$\psi\left(x, t = t_0 + 2^k T_\beta\right) = \psi(x, t = t_0),$$ (19.163)

or

$$\psi\left(x, t = t_0 + \frac{p}{q} T_\beta\right) = \psi(x, t = t_0),$$ (19.164)

for any time t_0 and k, p, and q integers. In any of the situations above, either for $t = T_F = 2^k T_\beta$ or for $t = T_{SH} = \left(\frac{p}{q}\right) T_\beta$, we can introduce Reynolds-type criterions

$$Re_F = \frac{V_F L_F}{\nu_F} = 2^k,$$ (19.165)

$$Re_{SH} = \frac{V_{SH} L_{SH}}{\nu_{SH}} = \frac{p}{q},$$ (19.166)

where

$$E_1 \equiv E_{F/SH} = \frac{1}{2}m_0 V_{F/SH}^2, \tag{19.167}$$

$$L_{F/SH} = V_{F/SH} T_{F/SH}, \tag{19.168}$$

$$v_{F/SH} = 8\pi D(dt)^{(2/D_F)-1}. \tag{19.169}$$

have the usual significations from fluid mechanics [66]. Up to the critical values $Re_{F/SH}^c$ the plasma discharge becomes turbulent. Then through $\frac{T_F}{T_\beta} = 2^k$ and $Re_F = \frac{V_F L_F}{v_F} = 2^k$ it formally simulated the criterion of evolution to chaos via Feigenbaum scenario (cascade of period-doubling bifurcations), while through $\frac{T_\beta}{T_{SH}} = \frac{\omega_{SH}}{\omega_\beta} = \frac{q}{p}$ with $p > q$ and $Re_{SH} = \frac{V_{SH} L_{SH}}{v_{SH}} = \frac{p}{q}$ the criterion of evolution to chaos via a cascade of subharmonic bifurcations (similar to the experimental scenario described above—see Figure 19.13).

The fractional revival of a speed scalar potential can be extended even for an infinite square-well potential using a fractional Schrödinger-type equation (for details, see References 21, 53, 59, and 67). Then, for $t = T_{RT} = \left(\frac{p}{q}\right)^\alpha T_\beta$ in the form $\frac{\omega_{RT}}{\omega_\beta} = \left(\frac{p}{q}\right)^\alpha$ with $1 < \alpha < 2, p > q$, and p, q integers, we can introduce the Reynolds number

$$Re_{RT} = \frac{V_{RT} L_{RT}}{v_{RT}} = \left(\frac{p}{q}\right)^\alpha, \tag{19.170}$$

where

$$E_1 \equiv E_{RT} = \frac{1}{2}m_0 V_{RT}^2, \tag{19.171}$$

$$L_{RT} = V_{RT} T_{RT}, \tag{19.172}$$

$$v_{RT} = 8\pi D(dt)^{(2/D_F)-1}. \tag{19.173}$$

If the above values are greater than a critical value, the system becomes turbulent simulating the scenario of evolution to chaos via Ruelle–Takens mechanism.

We admit that in any of the three situations mentioned above, the fractal velocity (19.48) is null, since $\rho = |\psi|^2 = $ const., meanwhile the differential velocity (19.47) is not zero, since the phase S is not constant, the increase of the systems phase incoherence being associated with the increase in turbulence of a fractal fluid.

We note that in the standard model (Landau's scenario [40]), the Fourier spectrum is always discrete and cannot approximate a continuum spectrum that in case of a large number of frequencies will generate an unlimited number of spectral components as a result of their beats which appear thanks to the presence of nonlinearities in the system. Yet, considering the standard model, the flow can never be truly chaotic because, in case of multiple periodic functions, correlations tend to be not null, but have an oscillating character. Therefore, the Landau's scenario can describe the transition toward chaotic behavior only in a system with an infinite number of degrees of freedom, such as a fluid. In our case, because the Reynolds numbers (19.165), (19.166), and (19.170) present scale dependencies (19.167) through (19.169) and (19.171) through (19.173), when $dt \to 0$ for $D_F \neq 2$ the fractals physical values that describe the dynamics of the system are no longer defined. So, in this approximation, a simulation of a system with an infinite number of degrees of freedom is used. Moreover, dynamic states could be generated, characterized by windows of regular oscillations interrupted by chaotic bursts, the transition between the two states being spontaneous, unpredictable, and independent of any of the control parameters variation (criterion of evolution to chaos through intermittency—for details, see References 87 and 90).

19.2.5 Conclusions

Two scenarios of transition to chaos were experimentally evidenced: by type I intermittency, related to the nonlinear dynamics of a fireball in plasma and to the ion-acoustic instability, and by a cascade of subharmonic bifurcations in connection to the generation and dynamics of MDLs in plasma.

By considering that the particle movements in plasma take place on fractal curves, a mathematical model according to the SRT was developed in order to describe the dynamics of plasma. The potential applied on the electrode is modeled as a one-dimensional square well. By using the full and fractional revivals formalism, different criteria of evolution from order to chaos are obtained. A good agreement is observed between the experimental results and the theoretical ones.

19.3 Oscillations and Instabilities in Plasma

19.3.1 NDR in Discharge Plasma

Many systems in physical science and technology, and also in chemical or biological domains, in which transport of electrical charges occurs, show transitions between distinctly stable states. These usually appear as bistabilities, experimentally manifested by the presence of an S- or N-shaped NDR. The physical origin of these effects is still a disputed subject in nonlinear physics. In plasma physics, it is well known that the S-type NDR is related to the appearance and disappearance of a CSCS (e.g., fireball or anode DL, MDL, etc.) [26,33,68], whereas the N-type NDR is related to the spatiotemporal dynamics of the CSCS [95,100], or to the onset of low-frequency instabilities [19,20,34,41,44].

For its appearance, a negative resistance requires an active component in the electrical circuit able to act as a source of energy. In the case of plasma, this component is the self-consistent DL existing at the border of a fireball. The potential drop across the self-consistent DL is almost equal with the ionization potential of the working gas. The initially thermal electrons are accelerated when passing through the DL, gaining enough energy to produce electron-neutral excitations and ionization impacts. In this way, an enhancement of the charged particles (electrons and ions) production takes place, leading to a sudden increase of the current collected by the electrode. The DL works as a nonlinear circuit element able to convert the thermal energy into electrical energy, i.e., it is a source of energy in the electrical circuit, creating all the conditions necessary for the appearance of the S-type NDR effect in the current–voltage characteristic of a plasma conductor.

19.3.1.1 Experimental Results

Figure 19.20 shows a family of current–voltage (I–U) characteristics of an electrode immersed into plasma, obtained for different values of the discharge current, under the following experimental conditions: argon pressure $p = 10^{-2}$ mbar, plasma density $n_{pl} \cong 10^7 - 10^8$ cm^{-3} and the electron temperature $kT_e = 2$ eV. Such an argon plasma presents the following characteristics: the inverse of the plasma electron pulsation is $\omega_{pe}^{-1} = \left(\varepsilon_0 m_e / n_0 e^2\right)^{1/2} \approx 1.8 \cdot 10^{-9}$ s, the inverse of the ion pulsation is $\omega_{pi}^{-1} = \left(\varepsilon_0 M / n_0 e^2\right)^{1/2} \approx 4.6 \cdot 10^{-7}$ s, the Debye length is $\lambda_D = \left(\varepsilon_0 k T_e / n_0 e^2\right)^{1/2} \approx 1.05 \cdot 10^{-3}$ m, the electron-acoustic speed is $c_e = \lambda_D \omega_{pe} \approx 1.89 \cdot 10^6$ m/s, and the ion-acoustic speed is $c_i = \lambda_D \omega_{pi} \approx 4.83 \cdot 10^4$ m/s. The I–U characteristics were recorded by using an X–Y plotter, which averages the small amplitude fluctuations of the current collected by the electrode. They start with an ohmic branch (linear dependence between I and U), ending at a critical value of the voltage applied on the electrode, when a sudden increase of the current appears, associated with an NDR effect. In other experimental conditions (higher plasma densities and lower values of the load resistor), a sudden jump of the current was recorded [34]. Figure 19.21 shows such a characteristic recorded by increasing and subsequently decreasing the voltage applied on the electrode. The hysteresis phenomenon is also present in this case. Simultaneously with the current increase, a fireball develops in front of the electrode. The development of the fireball corresponds to the NDR region of the current–voltage characteristic. Similar characteristics were experimentally obtained in microhollow cathode discharges [102], as well as theoretically, by using a chemical reaction model for the gas discharge plasma [73], and by simulation [23]. The critical value of the voltage at which the fireball aggregates depends on the discharge current (proportional to the equilibrium plasma density) as shown in Figure 19.22.

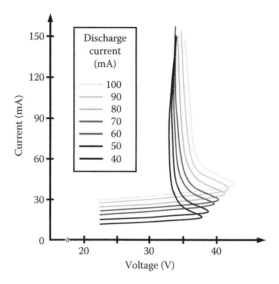

FIGURE 19.20 Set of current–voltage characteristics of the electrode, for different values of the discharge current (proportional to plasma density).

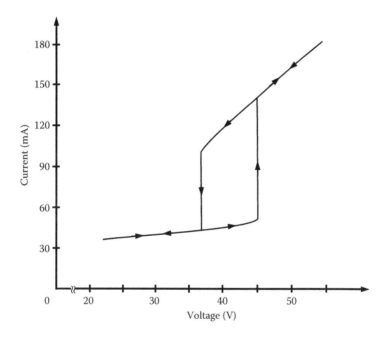

FIGURE 19.21 Current–voltage characteristic of the electrode E at the value of the discharge current $I_d = 150$ mA.

19.3.1.2 Theoretical Model

Since the instability analyzed here is generated at spatiotemporal scales at which the ionic component of plasma is inertial, while the electronic one is not inertial, we will analyze in the following just the electronic component of plasma. Moreover, all the physical quantities that describe the dynamics will be normalized with respect to the quantities characteristic of the electronic component of plasma. In this sense, we can bring some argumentations: (1) the NDR implies the plasma response to an external field. As the electron

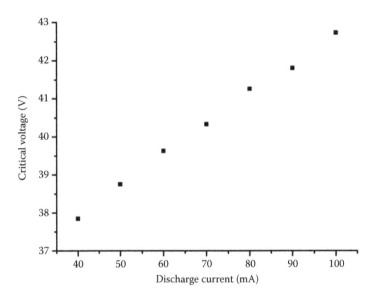

FIGURE 19.22 Dependence of the critical voltage at which the fireball appears on the discharge current (proportional to the plasma density).

mobility is much higher than that of the ions, it results that the electronic component reacts faster to the external field action; (2) the dynamics leading to the NDR phenomenon occurrence involve electron-neutral collisions, the ions being only a result of this process (see the elementary processes responsible for the NDR phenomenon [68,95]).

Now, by employing the one-dimensional hydrodynamic model, we can write the dynamic equations for particles as follows [31,74]:

1. The continuity equation

$$\frac{\partial n}{\partial t} + \frac{\partial (nv)}{\partial x} = 0. \tag{19.174}$$

2. The momentum equation

$$\frac{\partial v}{\partial t} + v\frac{\partial v}{\partial x} = \frac{q}{m}\frac{\partial \varphi}{\partial x} - \frac{1}{nm}\frac{\partial p}{\partial x} + v_0 v, \tag{19.175}$$

where n, m, q, and v are the density, mass, charge, and speed of electrons, respectively, while p is the pressure, φ is the electrostatic potential, and v_0 is the electron-neutral collision frequency.

We mention that, in the general case, by considering the electron-neutral collisions as dominant, the "force" corresponding to the momentum transfer from electrons to neutrals has the expression [31,49,74, 101]

$$f = v_{en} m_e n_e (v_e - v_n), \tag{19.176}$$

where v_{en} is the frequency of the electron-neutral collisions, v_e is the speed of the electrons, v_n is the speed of the neutrals, m_e is the electron mass, and n_e is the electron density. Because in most cases the speed of the neutrals can be neglected, the neutrals being considered as inertial, with the notations

$$v_{en} = v_0, \tag{19.177}$$

$$m_e = m, \tag{19.178}$$

$$n_e = n, \tag{19.179}$$

$$v_e = v, \tag{19.180}$$

the collision term becomes

$$f = \nu_0 mn\mathbf{v}. \tag{19.181}$$

In the one-dimensional case, by multiplying Equation 19.181 with $1/nm$, we found the collision term from Equation 19.175 in the form $f = \nu_0 v$. Using the normalized variables

$$\tau = \omega_p t, \tag{19.182}$$

$$\xi = x/\lambda_D, \tag{19.183}$$

$$N = n/n_0, \tag{19.184}$$

$$V = v/c, \tag{19.185}$$

$$\phi = q\varphi/(k_B T), \tag{19.186}$$

$$c^2 = \gamma k_B T/m, \tag{19.187}$$

$$\bar{\nu} = \nu/\omega_p, \tag{19.188}$$

where ω_p is the plasma frequency, λ_D is the Debye length, n_0 is the equilibrium density of particles, c is the acoustic speed, k_B is the Boltzmann constant, T is the particle temperature, and γ is the adiabatic index, the equation systems (19.174) and (19.175) become

$$\frac{\partial N}{\partial \tau} + \frac{\partial (NV)}{\partial \xi} = 0, \tag{19.189}$$

$$\frac{\partial V}{\partial \tau} + V\frac{\partial V}{\partial \xi} = \frac{\partial \phi}{\partial \xi} - \frac{1}{N}\frac{\partial N}{\partial \xi} + \bar{\nu}V. \tag{19.190}$$

In the following, we will build an analytical solution for the equations system (19.189) and (19.190), in the stationary case, by considering that the quasi-neutrality condition for plasma is fulfilled at any time (which involves the abandon of the Poisson equation), while the electron pressure adiabatically varies (for details, see References 31 and 74). For this, in the equations system (19.189) and (19.190), we make the following change of variable:

$$\theta = \xi - W\tau, \tag{19.191}$$

where W is a quantity that will be later specified. With this transformation, we find

$$-W\frac{dN}{d\theta} + \frac{d(NV)}{d\theta} = 0, \tag{19.192}$$

$$-W\frac{dV}{d\theta} + V\frac{dV}{d\theta} = \frac{d\phi}{d\theta} - \frac{1}{N}\frac{dN}{d\theta} + \bar{\nu}V. \tag{19.193}$$

After integration, from the continuity equation, it results

$$N(V - W) = -c_1, \tag{19.194}$$

$$c_1 = \text{const.}, \tag{19.195}$$

while from the momentum equation, it results

$$\frac{V^2}{2} - WV = \phi - \ln N - c_2(\theta), \tag{19.196}$$

$$c_2(\theta) = -\int \bar{\nu}V(\theta)d\theta. \tag{19.197}$$

By reducing V between Equations 19.194 and 19.196, we first obtain the dependence

$$\phi(N, \theta) = -\frac{W^2}{2} + \frac{1}{2}\frac{c_1^2}{N^2} + \ln N - c_2(\theta). \tag{19.198}$$

Moreover, from Equation 19.198, through the current density

$$J = NV = -c_1 + NW, \tag{19.199}$$

we obtain the voltage–current characteristic $\phi = \phi(J, \theta)$ in the form

$$\phi(J, \theta) = -\frac{W^2}{2} - \ln\frac{W}{c_1} + \frac{W^2}{2\left(\frac{J}{c_1}+1\right)^2} + \ln\left(\frac{J}{c_1}+1\right) - c_2(\theta). \tag{19.200}$$

Particularly, by imposing the theoretical restrictions

$$\theta \to \pm\infty, \tag{19.201}$$

$$\phi \to 0, \tag{19.202}$$

$$V \to 0, \tag{19.203}$$

$$N \to 1, \tag{19.204}$$

which imply $c_1 \to W$ and $c_2 \to 0$, relations (19.194), (19.196), (19.198), (19.199), and (19.200) become

$$N(V - W) = -W, \tag{19.205}$$

$$\frac{V^2}{2} - WV = \phi - \ln N, \tag{19.206}$$

$$\phi = -\frac{W^2}{2} + \frac{1}{2}\frac{W^2}{N^2} + \ln N, \tag{19.207}$$

$$J = NV = NW - W, \tag{19.208}$$

$$\phi = -\frac{W^2}{2} + \frac{W^2}{2\left(\frac{J}{W}+1\right)^2} + \ln\left(\frac{J}{W}+1\right). \tag{19.209}$$

These results generalize the ones from Reference 70.

From Equation 19.209, for $J/W \ll 1$, a linear dependence between J and ϕ is obtained, $J \approx \phi W/(1 - W^2)$, corresponding to the ohmic behavior of the plasma conductor (first part of the current–voltage characteristic in Figure 19.20). For $J/W \gg 1$, an exponential dependence of the current density on the applied potential is obtained, $J \approx W \exp \phi$, i.e., a nonlinear behavior of the plasma conductor (described by the final part of the current–voltage characteristic in Figure 19.20). We mention that the theoretical restrictions (19.201) through (19.204) correspond to the experimental conditions

$$x - ct \gg \lambda_D, \tag{19.210}$$

$$\varphi \to 0, \tag{19.211}$$

$$v \to 0, \tag{19.212}$$

$$n \to n_0. \tag{19.213}$$

Let us now return to the current–voltage characteristic given by Equation 19.200. In this equation, it is very difficult to find the explicit form of the collision term $c_2(\theta)$ (for details, see References 31, 49, 74, and 101). We can simplify the problem by supposing that the dynamics of the electronic component of the diode plasma display chaotic behaviors (self-similarity and strong fluctuations at all possible scales [72, 86,87,90]), so that the electrons move on continuous but nondifferentiable curves, i.e., fractal curves. Once such a hypothesis is accepted, specific mechanisms started [1,84,87,90] leading to the following results:

1. The explicit form of the collision term is assimilated by the fractality of the electron trajectories, so that the electrons motion becomes "free" [1,84,87–90]. Mathematically, this means that c_2 is independent of the variable θ, i.e., $c_2(\theta) \to c_2 = \text{const.}$, case in which relation (19.200) becomes

$$\phi(J) = -\frac{W^2}{2} - \ln\frac{W}{c_1} + \frac{W^2}{2\left(\frac{J}{c_1}+1\right)^2} + \ln\left(\frac{J}{c_1}+1\right) - c_2. \tag{19.214}$$

2. The dynamics of the electronic component of the diode plasma can be described through fractal functions [72,86,87,90], i.e., functions that depend on both standard coordinates (time and space coordinates) and resolution scale. The description of these dynamics involves the mathematical extension of the movement parameters in the complex space. In such a conjecture, let us consider the approximation $J/c_1 \ll 1$ in relation (19.214) with the choice $W = c_1 \exp(-c_2)$. We obtain

$$\phi + \frac{W^2}{2} \approx \frac{W^2}{2\left(1+2\frac{J}{c_1}\right)} + \frac{J}{c_1}. \tag{19.215}$$

By extending in the complex space and by recalibrating the coordinates origin

$$2\frac{J}{c_1} \to i\bar{J}, \tag{19.216}$$

$$W \to ia, \tag{19.217}$$

$$2\phi \to i\bar{\phi}, \tag{19.218}$$

relation (19.215) becomes

$$\bar{\phi} \approx \bar{J}\left(1+\frac{a}{1+\bar{J}^2}\right). \tag{19.219}$$

We mention that quantities $\bar{\phi}$ and $\bar{J}$ are measurable, for example, $\bar{\phi}$ is proportional to the normalized potential, while $\bar{J}$ is proportional to the normalized current (for details, see References 1 and 84). Moreover, the normalized quantities from Equations 19.182 through 19.188 keep their standard signification.

Equation 19.219 is of the third order in $\bar{J}$; so, it could have three real roots, i.e., to a single value of the field $\bar{\phi}$ that corresponds three different values of the field $\bar{J}$. The curves in Figure 19.15 show a maximum and a minimum when the parameter a surpasses a certain value. These maximum and minimum values can be found by canceling the first derivative of the function (19.219)

$$\frac{d\bar{\phi}}{d\bar{J}} = 0 \quad \Rightarrow \quad 1+\frac{a\left(1-\bar{J}^2\right)}{\left(1+\bar{J}^2\right)^2} = 0. \tag{19.220}$$

This is a biquadratic equation with the solutions

$$\bar{J}_{1,2}^2 = \frac{1}{2}\left[(a-2)\pm\sqrt{a(a-8)}\right]. \tag{19.221}$$

Equation 19.221 will have two different real (positive) solutions only if $a > 8$. For such values of the parameter a, the curve $\bar{J} = F(\bar{\phi})$ has two extremes, so that the system shows bistability. This situation is easy to be understood if we look at the curve corresponding to $a = 32$ in Figure 19.15. If $\bar{\phi}$ slowly increases from $\bar{\phi} = 0$, $\bar{J}$ increases until it reaches the point marked by B. Further increasing $\bar{\phi}$ will determine a sudden jump of $\bar{J}$ until the point marked by C is attained, because the region BD of the curve contains unstable states, which cannot be experimentally accomplished. When $\bar{\phi}$ decreases from values greater than the critical value corresponding to the point marked by C, $\bar{J}$ decreases until the point marked by D is reached.

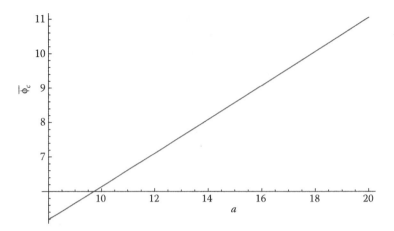

FIGURE 19.23 Theoretical dependence of the normalized critical potential on the normalized discharge current.

If we further decrease the value of $\bar{\phi}$, $\bar{J}$ will jump to the point marked by A, following the curve to the origin. Thus, for values of the field $\bar{\phi}$ in the interval AB, the field $\bar{J}$ can have two different stable values. This bistable behavior determines the NDR.

A good qualitative agreement can be observed between the experimental current–voltage characteristics from Figures 19.20 and 19.21 and those obtained from the theoretical model in Figure 19.15. In such a context, the quantity W is proportional (through the quantity a) to the normalized discharge current density. Moreover, also a good qualitative agreement can be observed between the dependence of the critical voltage at which the fireball appears $\bar{\phi}_c$, on the discharge current from Figure 19.22 and that obtained from the theoretical model in Figure 19.23. This theoretical dependence was obtained by substituting the value $\bar{J}_2$ given by Equation 19.221 in Equation 19.219:

$$\bar{\phi}_c = \bar{J}_2 \left(1 + \frac{a}{1 + \bar{J}_2^2} \right).$$
(19.222)

19.3.2 Oscillations and Ion-Acoustic Instability in Nonmagnetized Plasma

As we mentioned in the preceding section, a fireball can perform a proper dynamics, consisting of periodic aggregations and disruptions of the DL at its border. During this process, bunches of ions are periodically injected into the surrounding plasma, stimulating here the onset of the ion-acoustic instability. This instability was experimentally evidenced as bursts appearing in the current–voltage characteristic of a probe immersed into plasma, having a frequency equal with the frequency of the fireball disruptions (see Figure 19.24).

In order to model these phenomena, we will apply the fractal hydrodynamics, described in Section 19.1 of this chapter. We start the analysis from the equation of geodesics (19.40). In the case of irrotational motion $\left(\nabla \times \hat{V} = 0 \right)$, the speed field can be expressed through the gradient of the scalar potential of a complex speed field Φ

$$\hat{V} = \nabla \Phi.$$
(19.223)

Substituting Equation 19.223 in Equation 19.40 and using the operational relationship $\partial_t \nabla = \nabla \partial_t$, it results

$$\nabla \left[\frac{\partial \Phi}{\partial t} + \frac{1}{2} (\nabla \Phi)^2 - 2iD\Delta\Phi \right] = 0,$$
(19.224)

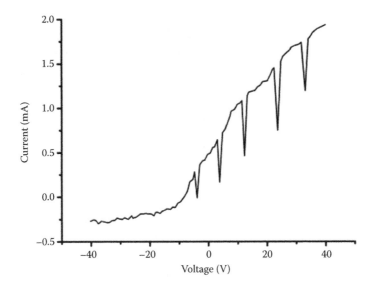

FIGURE 19.24 Dynamic current–voltage characteristic of the cylindrical probe recorded in the presence of a dynamic fireball in plasma.

and by integration, a Bernoulli-type equation

$$\frac{\partial \Phi}{\partial t} + \frac{1}{2}(\nabla \Phi)^2 - 2iD\Delta\Phi = F(t), \tag{19.225}$$

with $F(t)$ a function which depends only on time. Particularly, for Φ of the form

$$\Phi = -2iD \ln \psi, \tag{19.226}$$

where ψ is a new complex scalar function, Equation 19.225 with the operatorial identity $\Delta\psi/\psi = \Delta \ln \psi + (\nabla \ln \psi)^2$ takes the form

$$D^2 \Delta\psi + 2iD\frac{\partial \psi}{\partial t} + \frac{F(t)}{2}\psi = 0. \tag{19.227}$$

From here, Schrödinger-type geodesics result for $F(t) \equiv 0$, i.e.,

$$D\Delta\psi + 2i\frac{\partial \psi}{\partial t} = 0. \tag{19.228}$$

Moreover, in an external field of scalar potential U, this equation becomes

$$D\Delta\psi + 2i\frac{\partial \psi}{\partial t} = \frac{U}{2m_0}\psi, \tag{19.229}$$

where m_0 is the rest mass of the fractal fluid particle. Particularly, for movements at Compton's scale on fractal curves of Peano's type [87] and fractal dimension $D_F = 2$, Equation 19.228 takes the form of the standard Schrödinger equation (see also Reference 84):

$$\frac{\hbar^2}{2m_0}\Delta\psi + i\hbar\frac{\partial \psi}{\partial t} = 0, \tag{19.230}$$

with $\hbar$ the reduced Planck's constant.

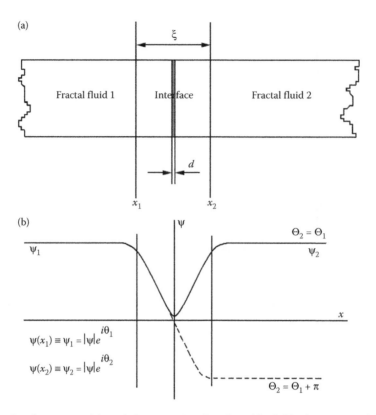

FIGURE 19.25 Interface generated through the interaction of two fractal fluids (d is the geometrical thickness, before the self-structuring of the interface and ξ is the physical thickness, after the self-structuring of the interface) (a) and the variation of the speed field with the fractal coordinates (b), respectively.

Let us now consider the DL at the border of the fireball in plasma as an interface between two fractal fluids. Consider two fractal fluids, 1 and 2, separated by an interface as shown in Figure 19.25. If the interface is thick enough so that the fractal fluids are "isolated" from each other, the time-dependent Schrödinger-type equation for each side is

$$im_0 2D \frac{d\psi_1}{dt} = H_1 \psi_1, \tag{19.231}$$

$$im_0 2D \frac{d\psi_2}{dt} = H_2 \psi_2, \tag{19.232}$$

where ψ_i and H_i, $i = 1, 2$ are the scalar potentials of the complex speed fields and respectively the "Hamiltonians" on either side of the interface. We assume that a voltage $2V$ is applied between the two fractal fluids. If the zero point of the potential is assumed to occur in the middle of the interface, fractal fluid 1 will be at the potential $-V$, while fractal fluid 2 will be at the potential $+V$.

The presence of the interface couples together the two previous equations (19.231) and (19.232) in the form

$$im_0 2D \frac{d\psi_1}{dt} = eV\psi_1 + \Gamma\psi_2, \tag{19.233}$$

$$im_0 2D \frac{d\psi_2}{dt} = eV\psi_2 + \Gamma\psi_1, \tag{19.234}$$

where Γ is the coupling constant for the scalar potentials of the complex speed fields across the interface. Since the square of each scalar potential of the complex speed fields is also a probability density [38,39,72,87,88], the two scalar potentials of the complex speed fields can be written in the form

$$\psi_1 = \sqrt{\rho_1}e^{i\theta_1}, \tag{19.235}$$

$$\psi_2 = \sqrt{\rho_2}e^{i\theta_2}, \tag{19.236}$$

$$\Theta = \theta_2 - \theta_1, \tag{19.237}$$

where ρ_1 and ρ_2 are the densities of particles in the two fractal fluids and Θ is the phase difference across the interface. If the two scalar potentials of the complex speed fields (19.235) and (19.236) are substituted in the coupled Equations 19.233 and 19.234 and the results separated into real and imaginary parts, we obtain equations for the time dependence of the particle densities and the phase difference:

$$\frac{d\rho_1}{dt} = \frac{\Gamma}{m_0 D}\sqrt{\rho_1\rho_2}\sin\Theta, \tag{19.238}$$

$$\frac{d\rho_2}{dt} = -\frac{\Gamma}{m_0 D}\sqrt{\rho_1\rho_2}\sin\Theta, \tag{19.239}$$

$$\frac{d\Theta}{dt} = \frac{eV}{m_0 D} = \overline{\Omega}. \tag{19.240}$$

We can specify the current density in terms of the difference between Equations 19.238 and 19.239 times e

$$j = e\frac{d}{dt}(\rho_1 - \rho_2), \tag{19.241}$$

which has the value

$$j = j_c\sin\Theta, \tag{19.242}$$

where

$$j_c = \frac{2e\Gamma\sqrt{\rho_1\rho_2}}{m_0 D}. \tag{19.243}$$

Equations 19.240 and 19.243 define the carriers transport inside the interface.

If the voltage from Equation 19.240 is zero, a dc current density of any value between $-j_c$ and j_c may flow through the junction according to Equation 19.243. If we apply a constant voltage V_0 to the junction, from Equations 19.240 and 19.243, it results

$$\Theta(t) = \frac{e}{m_0 D}V_0 t + \Theta_0, \quad \Theta_0 = \text{const.}, \tag{19.244}$$

$$j(t) = j_c\sin(\Omega_0 t + \Theta_0), \tag{19.245}$$

$$\Omega_0 = \frac{e}{m_0 D}V_0. \tag{19.246}$$

Thus, an ac current density results, although a constant voltage is applied. Ω_0 given in Equation 19.246 corresponds to the disruption pulsation of the DL.

Now, if one overlays an alternative voltage over the constant voltage

$$V(t) = V_0 + \overline{V_0}\cos(\Omega t), \tag{19.247}$$

one obtains a frequency modulation of the ac density current

$$j(t) = j_c\sin\left(\Omega_0 t + \frac{e\overline{V_0}}{m_0\Omega D}\sin(\Omega t) + \overline{\Theta_0}\right) =$$

$$= j_c\sum_{n=-\infty}^{+\infty}(-1)^n J_n\left(\frac{e\overline{V_0}}{m_0\Omega D}\right)\sin\left[(\Omega_0 - n\Omega)t + \overline{\Theta_0}\right], \tag{19.248}$$

$$\overline{\Theta_0} = \text{const.} \tag{19.249}$$

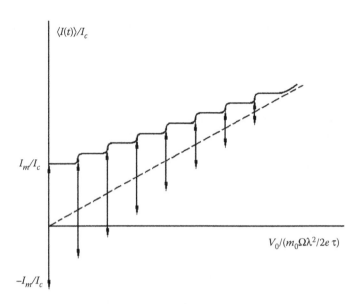

FIGURE 19.26 Theoretical current–voltage characteristic (the dashed line is for $\langle I(t) \rangle = V_0 G$, with G a "conductance").

J_n is the Bessel function of the integer index [85]. We note that, in the first approximation, for any "arbitrary" signal, and so for the external signal applied on the cylindrical probe as in our experiment, we can always perform a Fourier's decomposition [64]. Since I versus V characteristic, where $I = jS$ is the current and S is the area of the junction, is drawn for the average current $I \approx \langle I(t) \rangle$, and since the sine term averages to zero unless $\Omega_0 = n\Omega$, there are spikes appearing on this characteristic for voltages equal to

$$V_n = n\frac{m_0 D}{e}\Omega, \tag{19.250}$$

with the maximum amplitude

$$I_{max} = I_c J_n \left(\frac{e\overline{V_0}}{m_0 \Omega D} \right), \tag{19.251}$$

$$I_c = j_c S, \tag{19.252}$$

occurring for the phase $\overline{\Theta} = \pi/2$. Figure 19.26 shows these spikes at intervals proportional to the source frequency and indicates their maximum amplitude range. The value of the current can be anywhere along a particular current spike, depending on the initial phase.

19.3.3 Instabilities in Magnetized Plasma

19.3.3.1 The Inertial 1-Form and Cartan's Motion Principle

According to the SRT [87,90], we will show that a limit speed is specific to every resolution scale. To demonstrate this, the mathematical formalism of Cartan is used, described in Reference 91. The differential of the action S written as

$$dS = p_l dq^l - \left(p_l dq^l - Ldt \right), \quad l = \overline{1, n}, \tag{19.253}$$

where L is the Lagrangian, $p_l = \left(\partial L/\partial \dot{q}^l \right)$ is the generalized momentum, q^l is the generalized coordinate, and $\dot{q}^l$ is the generalized speed, is introduced by the Hamiltonian H

$$H = p_l \dot{q}^l - L, \tag{19.254}$$

an inertial 1-form ω^i. For "displacements"d, the 1-form ω^i is the inertial 1-form ω^i_d (Cartan 1-form) [91]

$$\omega^i_d = p_l dq^l - H dt, \tag{19.255}$$

and for a variation δ, different from d, the 1-form ω^i has the expression

$$\omega^i_\delta = p_l \delta q^l - H \delta t. \tag{19.256}$$

Now the motion principle of Cartan (see Reference 91) forces the external differential $D(\omega^i_\delta)$

$$D\left(\omega^i_\delta\right) = d\left(\omega^i_\delta\right) - \delta\left(\omega^i_\delta\right), \tag{19.257}$$

to vanish, hence $D\omega^i = 0$. More explicitly this becomes

$$D\omega^i = dp_l \delta q^l - \delta p_l dq^l - dH\delta t + \delta H dt, \tag{19.258}$$

where we took into account that d and δ commute, $d\delta = \delta d$. Knowing that H admits the functional dependence

$$H = H\left(q^l, p_l, t\right), \tag{19.259}$$

the variation of the Hamiltonian

$$\delta H = \frac{\partial H}{\partial q^l}\delta q^l + \frac{\partial H}{\partial p^l}\delta p^l + \frac{\partial H}{\partial t}\delta t, \tag{19.260}$$

replaced in Equation 19.258, grouping the terms and multiplying with $(dt)^{-1}$, gives

$$\left(\dot{p}_l + \frac{\partial H}{\partial q^l}\right)\delta q^l + \left(-\dot{q}_l + \frac{\partial H}{\partial p^l}\right)\delta p^l + \left(-\frac{dH}{dt} + \frac{\partial H}{\partial t}\right)\delta t = 0. \tag{19.261}$$

Since the variation δq^l, δp^l, and δt are arbitrarily chosen, by equating the coefficients of the variations to zero one gets

1. The canonical equations, $\dot{q}^l = \partial H/\partial p_l$, $\dot{p}^l = \partial H/\partial q_l$
2. The integral of motion, $dH/dt = \partial H/\partial t$

The motion principle of Cartan is equivalent to the closure of the inertial 1-form ω^i, $D\omega^i = dp_l \wedge dq_l - dH \wedge dt = 0$. This is easy to verify directly.

19.3.3.2 The Inertial Motion of a Material Point and the Fractal Structure of the Space–Time Manifold

Postulate. Local, the space and time of the motion are structured as an $E_3 \times T$ manifold (E_3 corresponds to the Euclidean space and T to time). Therefore

1. The speed $\boldsymbol{v}$ is defined by $\boldsymbol{v} = d\boldsymbol{r}/dt$, where t is the time
2. The inertial 1-form (19.255) has the expression

$$\omega^i = \boldsymbol{p}d\boldsymbol{r} - H dt. \tag{19.262}$$

3. The space isotropy and homogeneity impose the functional dependence

$$H = H\left[\alpha = (1/2)\boldsymbol{p}^2\right]. \tag{19.263}$$

Postulate. The motion is described by equating to zero the external differential of the inertial 1-form (19.262)

$$D\left(\omega^i\right) = dp\delta r - \delta p dr - dH\delta t + \delta H dt. \tag{19.264}$$

Knowing that $\delta H = (\partial H/\partial\alpha)p\delta p$, relation (19.264) given more explicitly becomes

$$D\left(\omega^i\right) = dp\delta r - dH\delta t + \left(-dr + \frac{\partial H}{\partial\alpha}pdt\right)\delta p = 0. \tag{19.265}$$

Since the variations δr, δt, and δp are arbitrary, from relation (19.265) by multiplication with $(dt)^{-1}$ and equating the coefficients of this variation to zero, one gets

1. The momentum conservation law

$$\frac{dp}{dt} = 0. \tag{19.266}$$

2. The energy conservation law

$$\frac{dH}{dt} = 0. \tag{19.267}$$

3. Momentum definition relation

$$p = mv, \tag{19.268}$$

where $m = (\partial H/\partial\alpha)^{-1}$ is named inertial mass. If the inertial mass is constant, from the principle of motion (19.264) by means of Equation 19.268 and multiplication with m one finds $-p\delta p + m\delta H = 0$. From here, integrating this relation with the condition $H = 0$ for $p = 0$ one gets the classical mechanics result $H = p^2/2m$.

The material point being a physical object of irreducible simplicity, its inertial motion on $E_3 \times T$ manifold is to be differentiated from the one of another point, if making abstraction of position and speed, by mass. Therefore, the kinematical 1-form $\omega^c = vdr - (H/m)dt$, obtained from Equation 19.262, $\omega^i = m\omega^c$, must be the same for any material point. This observation is concretized as follows:

Postulate. The ratio H/m is a universal constant. This ratio having the dimension of a squared speed, $H/m = \omega^2$, it results that the Hamiltonian of the free particle has the expression

$$H = m\omega^2. \tag{19.269}$$

In such a context, from the principle of motion (19.265) by multiplication with m and by integration one obtains $p^2 - \omega^2 m^2 = \text{const}$. The value of the integration constant is determined by the condition $m = m_0$ for $p = 0$, where m_0 is the rest mass. One finds $p^2 - \omega^2 m^2 = -\omega^2 m_0^2$, from which, taking into account Equation 19.268, it results in the dependence on speed of the inertial mass of the free material point

$$m = m_0\left(1 - \frac{v^2}{\omega^2}\right)^{-1/2}. \tag{19.270}$$

With Equation 19.270, relation (19.268) gives the expression of the momentum

$$p = m_0 v\left(1 - \frac{v^2}{\omega^2}\right)^{-1/2}, \tag{19.271}$$

and Equation 19.269 gives the expression of the energy

$$H = m_0\omega^2\left(1 - \frac{v^2}{\omega^2}\right)^{-1/2}. \tag{19.272}$$

The mass m has a physical meaning for $v < \omega$. It results that ω is a limiting speed. In other words, ω is any limiting speed, being an attribute of the light only by extrapolation of the electromagnetic phenomena to the other phenomena of the physics world. The value of ω is not subjected *a priori* to any restriction; it can be practically both inferior and superior to the speed of light. The constant ω may be in principle finite and may have any value, this situation remaining to be settled in the last instance only by the experiment. In particular for $\omega \equiv c$, where c is the speed of light in vacuum, the model leads to the expressions from special relativity.

The alternative $H/m = -\omega^2$ must be also taken into account. Then, $m = m_0 \left[1 + \left(v^2/\omega^2\right)\right]^{-1/2}$, $p = m_0 v \left[1 - \left(v^2/\omega^2\right)\right]^{-1/2}$, and $H = -\omega^2 \left[m_0^2 + \left(p^2/\omega^2\right)\right]^{-1/2}$, situation corresponding to "noninertia"; the mass of the body will decrease to zero with the increasing speed, $m = \left[m_0^2 - \left(p^2/\omega^2\right)\right]^{-1/2}$, m_0 being the heaviest mass, $m \leq m_0$. In this case, m can be named an antimass of ω-type.

The fact that in this alternative the inertial 1-form has the expression $\omega^i = p dr + H dt$ is equivalent to consider that the time fluctuations are negative. It results by means of the motion principle (19.264) with $m = \text{const.}$, $H = -p^2/2m$, i.e., the inertial mass is negative. The result is in agreement with Nottale's assumption from References 87 and 90 according to which the negative fluctuations of time induce the negative mass. Moreover, the reversibility of time achieved previously indicates that, by means of inertial 1-form (19.262) and the principle of motion (19.264), the space–time manifold has a fractal structure.

19.3.3.3 Nonlinear Equations of a Beam

Let us follow the results from References 5, 13 through 15, and 36. We consider a relativistic-type monoenergetic beam of charged particles that move in an external uniform magnetic field B_0 with components given by

$$B_0 = \{B_{0x}, B_{0y}, B_{0z}\} = \{0, 0, B_0\}. \tag{19.273}$$

However, a charged particle moving at an angle to an external magnetic field behaves as a nonlinear oscillator which can generate an electromagnetic field. Thus, we observe a relativistic-type beam of phased oscillators in a self-consistent electromagnetic field described by the following system of equations:

- Maxwell equations for the electromagnetic field

$$\nabla \times B_b = \mu j_b + \mu \varepsilon \frac{\partial E_b}{\partial t}, \tag{19.274}$$

$$\nabla \times E_b = -\frac{\partial B_b}{\partial t}, \tag{19.275}$$

$$\nabla \cdot E_b = \frac{n_b q}{\varepsilon}, \tag{19.276}$$

$$\nabla \cdot B_b = 0. \tag{19.277}$$

- Relativistic-type equation of motion for a single particle of the beam

$$\frac{dr}{dt} = \frac{p}{m_0 \gamma} = \frac{p_c - qA}{m_0 \gamma} = \frac{q}{m_0 \gamma}(R_c - A) \equiv v, \tag{19.278}$$

$$\frac{dp}{dt} = qE_b + q^2 \frac{R}{m_0 \gamma} \times (B_b + B_0), \tag{19.279}$$

where q and m_0 are respectively the charge and rest mass of a beam particle, n_b is the particle density of the beam, $R = p/q$ is the rigidity (momentum per unit charge), μ and ε are respectively the permeability and permittivity of a "medium" associated to the beam, and $p = m_0 \gamma v$ is the (kinetic) relativistic-type particle momentum. The quantity

$$p_c = \frac{\partial L}{\partial \dot{r}} = m_0 \gamma v + qA = q(R + A), \tag{19.280}$$

provides the canonical momentum of a particle or momentum conjugate to r through the Lagrangian L. A is the vector potential of the electromagnetic field and E_b and B_b are the electric and magnetic field intensities generated by the beam. Furthermore

$$\gamma = \left(1 + \frac{p_\perp^2}{m_0^2 c_0^2}\right)^{1/2} \equiv \left(1 - \frac{v_\perp^2}{c_0^2}\right)^{-1/2}, \tag{19.281}$$

is the relativistic Lorentz-type factor (or the total particle energy normalized by the rest mass energy), $p_\perp$ and $v_\perp$ being respectively the momentum and velocity components perpendicular to the stationary uniform magnetic field $\boldsymbol{B_0}$. The relativistic-type expression of a cyclotron frequency (gyrofrequency) is

$$\omega_c = \frac{1}{\gamma}\omega_B = \frac{qB_0}{m_0\gamma}. \tag{19.282}$$

We consider that the beam possesses a symmetry such that it generates a perturbation of an electromagnetic field with a harmonic space dependence and, generally, nonharmonic time dependence of the form

$$\boldsymbol{E_b}(x, t) = \{E_{bx}, E_{by}, E_{Bz}\} = \left\{\Re\left[E_x(t)e^{ikx}\right], \Re\left[E_y(t)e^{ikx}\right], 0\right\}, \tag{19.283}$$

$$\boldsymbol{B_b}(x, t) = \{B_{bx}, B_{by}, B_{Bz}\} = \left\{0, 0, \Re\left[B_b(t)e^{ikx}\right]\right\}, \tag{19.284}$$

where

$$\boldsymbol{k} = \{k_x, k_y, k_z\} = \{k_x, 0, 0\}, \tag{19.285}$$

is the corresponding wave vector and $\Re$ applied to a quantity refers to the real part of that quantity. Using the dimensionless variables

$$T = kc_0 t, \tag{19.286}$$

$$X = kx, \tag{19.287}$$

$$\boldsymbol{P} = \frac{\boldsymbol{p}}{m_0 c_0}, \tag{19.288}$$

$$\gamma = \left(1 + P_x^2 + P_y^2\right)^{1/2} \equiv \left(1 + P_\perp^2\right)^{1/2}, \tag{19.289}$$

and

$$H_b(T) = \frac{1}{kc}\omega_b(T) = \frac{qB_b(t)}{m_0 c_0 k} \equiv \Omega_b, \tag{19.290}$$

$$\varepsilon_{x,y} = \frac{qE_{x,y}(t)}{m_0 c_0^2 k}, \tag{19.291}$$

the equations of motion (19.278) and (19.279) become

$$\frac{dX}{dT} = \frac{1}{\gamma}P_x = \frac{1}{\left(1 + P_x^2 + P_y^2\right)^{1/2}}P_x \equiv \frac{1}{\left(1 + P_\perp^2\right)^{1/2}}P_x, \tag{19.292}$$

$$\frac{dP_x}{dt} = \frac{1}{\gamma}\left\{\Re\left[H_b(T)e^{iX}\right] + \Omega_B\right\}P_y + \Re\left[\varepsilon_x(T)e^{iX}\right], \tag{19.293}$$

$$\frac{dP_y}{dt} = -\frac{1}{\gamma}\left\{\Re\left[H_b(T)e^{iX}\right] + \Omega_B\right\}P_x + \Re\left[\varepsilon_y(T)e^{iX}\right], \tag{19.294}$$

where

$$\Omega_B = \frac{1}{kc_0}\omega_B = \frac{\gamma}{kc_0}\omega_c \equiv \gamma\Omega_c. \tag{19.295}$$

We parenthetically note that Ω_B represents the dimensionless gyrofrequency of the magnetic field generated by the electromagnetic wave. Applying the same scheme to the Maxwell equations (19.274) and (19.275) we find

$$\frac{dH_b}{dT} = -i\varepsilon_y, \tag{19.296}$$

$$\frac{d\varepsilon_y}{dT} = -iH_b - \frac{1}{\gamma}\Omega_p^2 P_y e^{-iX}, \tag{19.297}$$

$$\frac{d\varepsilon_x}{dT} = -\frac{1}{\gamma}\Omega_p^2 P_x e^{-iX}, \tag{19.298}$$

where

$$\Omega_p = \frac{1}{kc_0}\omega_p = \frac{1}{kc_0}\left(\frac{4\pi q^2 n_b}{m_0}\right)^{1/2}, \tag{19.299}$$

is the dimensionless plasma-beam frequency and ω_p is the plasma frequency. Equations 19.292 through 19.294 can be combined to yield

$$\Omega_B \frac{dX}{dT} + \Im\left[\frac{d}{dT}\left(H_b e^{iX}\right)\right] + \frac{dP_y}{dT} = 0. \tag{19.300}$$

Consequently, we can write the following integral of motion:

$$\Omega_B X(T) + \Im\left[H_b(T)e^{iX}\right] + P_y = \text{const.}, \tag{19.301}$$

where the symbol $\Im$ applied to a quantity defines the imaginary part of this quantity.

We note that the system of Equations 19.292 through 19.294 and 19.296 through 19.298 describes a nonlinear interaction between three coupled (mechanical and electromagnetic) oscillators. Following numerical experiments, we will demonstrate that there exists a continuous exchange of energy between these oscillators which leads to a self-organization process with a spontaneous formation of special ordered structures (e.g., gun effects and phase Larmor spirals). There three oscillators emerge, which are

1. A nonlinear-forced electrodynamic oscillator described by Equations 19.293 and 19.294 which are collectively contained in

$$\frac{d^2 P_x}{dT^2} + \frac{1}{\gamma}\left\{\Omega_p^2 + \frac{1}{\gamma}\left[\Re\left(H_b e^{iX}\right) + \Omega_B\right]^2\right\} P_x$$
$$= \frac{1}{\gamma}\left\{\Re\left[H_b(T)e^{iX}\right] + \Omega_B\right\}\Re\left[\varepsilon_y(T)e^{iX}\right] + \frac{1}{\gamma}P_y\frac{d}{dT}\left\{\Re\left[H_b(T)e^{iX}\right]\right\}. \tag{19.302}$$

2. A transverse electromagnetic oscillator generated by the system and described by Equations 19.296 and 19.297 which lead to the expression

$$\frac{d^2 H_b}{dT^2} + H_b = \frac{i}{\gamma}\Omega_p^2 P_y e^{-iX}. \tag{19.303}$$

3. A longitudinal electromagnetic oscillator described by Equations 19.293 and 19.294 which are combined into

$$\frac{d^2}{dt^2}\Re\left[\varepsilon_x(T)e^{iX}\right] + \frac{1}{\gamma}\Omega_p^2\Re\left[\varepsilon_x(T)e^{iX}\right] = -\frac{1}{\gamma}\Omega_p^2 P_y\left\{\Re\left[H_b(T)e^{iX}\right] + \Omega_B\right\}. \tag{19.304}$$

It is clear that a beam of charged particles moving in a magnetic field represents a complex system of radiating oscillators.

In order to obtain the values of parameters which define an efficient mechanism of acceleration, we consider in what follows the motion of a single particle of the beam in the field as generated by the remaining particles. Thus, the initial system with a self-consistent field is transformed in to a system with external fields. This situation can be simulated in a laboratory since the present-day light-wave technology permits the generation of almost any type of electromagnetic field. Thus, we will first refer to the motion of a charged particle in a constant external magnetic field and in the presence of a transverse electromagnetic wave. If we apply Equations 19.292 through 19.294 and 19.296 through 19.298 to this particular case for which

$$H_b(T) = \varepsilon_y(T) = He^{-iT}, \tag{19.305}$$

$$H = \text{const.}, \tag{19.306}$$

$$\Omega_p = 0, \tag{19.307}$$

$$\varepsilon_x = 0, \tag{19.308}$$

we obtain the following equations which describe the nonlinear dynamics of our model:

$$\dot{X} = \frac{dX}{dT} = \frac{1}{\gamma} P_x = \frac{1}{\left(1 + P_x^2 + P_y^2\right)^{1/2}} P_x, \tag{19.309}$$

$$\dot{P}_x = \frac{dP_x}{dT} = \Omega_c \left[\beta \cos(X - T) + 1\right] P_y, \tag{19.310}$$

$$\dot{P}_y = \frac{dP_y}{dT} = -\Omega_c \left[\beta \cos(X - T) + 1\right] P_x + H \cos(X - T), \tag{19.311}$$

or in vectorial (matrix) form

$$\begin{bmatrix} \dot{X} \\ \dot{P}_x \\ \dot{P}_y \end{bmatrix} = V \equiv \begin{bmatrix} \frac{1}{\gamma} P_x \\ \frac{1}{\gamma} \left[H \cos(X - T) + \Omega_B\right] P_y \\ -\frac{1}{\gamma} \left[H \cos(X - T) + \Omega_B\right] P_x + H \cos(X - T) \end{bmatrix}, \tag{19.312}$$

where

$$\beta = \frac{H}{\Omega_B} \equiv \frac{|\Omega_b|}{\Omega_B}, \tag{19.313}$$

is in fact the ratio of two gyrofrequencies associated with the magnetic field of the wave and, respectively, to the external uniform magnetic field. First, we determine the divergence div$V$ or, equivalently, the trace of the Jacobi matrix M_J of the nonlinear system (19.309) through (19.311) in order to deduce a global concept of the configuration of flow in the phase space. We find

$$M_J = \begin{bmatrix} 0 & \frac{1}{\gamma} & 0 \\ \frac{1}{\gamma} H \sin(X - T) P_y & 0 & \frac{1}{\gamma} \left[H \cos(X - T) + \Omega_B\right] \\ \left(1 - \frac{P_x}{\gamma}\right) H \sin(X - T) & -\frac{1}{\gamma} \left[H \cos(X - T) + \Omega_B\right] & 0 \end{bmatrix}, \tag{19.314}$$

and

$$\text{div}V = \text{trace}\, M_J = 0. \tag{19.315}$$

Thus, during the evolution of the system, the phase volume is conserved and no dissipation takes place. In order to find the conditions of resonance, we consider in the sequel a map F which, if expressed in vector notation reads

$$
\begin{bmatrix} X \\ P_x \\ P_y \end{bmatrix} = F \begin{bmatrix} \xi \\ P_\perp \\ \Phi \end{bmatrix} = \begin{bmatrix} \xi - \alpha \sin \Phi \\ P_\perp \cos \Phi \\ P_\perp \sin \Phi \end{bmatrix}, \tag{19.316}
$$

where

$$
\alpha = \frac{P_\perp}{\Omega_B}. \tag{19.317}
$$

Expressed in the ("action-angle") variables $P_\perp$, Φ, and ξ, Equations 19.309 through 19.311 become

$$
\dot{P_\perp} = H \sum_{n=-\infty}^{\infty} \frac{dJ_n(\alpha)}{d\alpha} \sin(n\Phi + T - \xi), \tag{19.318}
$$

$$
\dot{\Phi} = -\frac{\Omega_B}{\gamma} + H \sum_{n=-\infty}^{\infty} J_n(\alpha) \left(\frac{n\Omega_B}{P_\perp^2} - \frac{1}{\gamma} \right) \cos(n\Phi + T - \xi), \tag{19.319}
$$

$$
\xi + \beta \sin(\xi - \alpha \sin \Phi - T) = \text{const.}, \tag{19.320}
$$

where $J_n(\alpha)$ is a Bessel function of order n with

$$
n = -\infty, \ldots, -2, -1, 0, 1, 2, \ldots, +\infty. \tag{19.321}
$$

We deduce from Equations 19.318 and 19.319 that in the presence of a phase velocity of the wave equal to the velocity of light (i.e., if $\omega = kc$) the wave–particle interaction is most effective in the vicinity of a single point on the Larmor circle on which the particle moves in the direction of the propagation of the wave. This also assumes that the (relativistic) cyclotron resonance is satisfied. There follows:

$$
\frac{\omega}{kc} = n\Omega_c = n\frac{\Omega_B}{\gamma_n} = 1 \quad \text{or} \quad \omega = n\omega_c. \tag{19.322}
$$

As a result, we have to consider the case of an "on-resonance" wave, for which the wave frequency is an integer multiple of the cyclotron frequency and the acceleration may take place through a "web structure" with a low limit of the wave amplitude H. However, by raising the wave amplitude, we shall demonstrate that a new type of an effective individual acceleration becomes feasible. In such a case, the term stochastic or chaotic is understood to refer to the chaotic motion of a single particle.

As an immediate consequence of Equations 19.289 and 19.322, the variations of transverse momentum between neighboring resonances, for instance at a resonance of the fourth order ($n = 4$) with $\Omega_B = 0.5$, are given by the following expressions:

$$
P_{\perp 5} - P_{\perp 4} = \sqrt{\gamma_5^2 - 1} - \sqrt{\gamma_4^2 - 1} = \frac{\sqrt{21}}{2} - \sqrt{3} = 0.55923704, \tag{19.323}
$$

$$
P_{\perp 4} - P_{\perp 3} = \sqrt{\gamma_4^2 - 1} - \sqrt{\gamma_3^2 - 1} = \sqrt{3} - \frac{\sqrt{5}}{2} = 0.61401682. \tag{19.324}
$$

Thus, if a particle possesses an initial transverse momentum of $P_\perp \approx \sqrt{3}$ and as a result of a nonlinear wave–particle interaction the variation of its momentum is of the order $\delta P_\perp \approx 0.5$. We conclude that the phase oscillations of the particle correspond to a motion at an isolated resonance of the fourth order.

19.3.3.4 Isolated and Overlap Resonances

When the separation Ω_B between adjacent resonances is much larger than the change $\Delta\gamma$ in the particle energy, we may conclude that the conditions of an *isolated resonance* exist and we find

$$
\Omega_B \gg \Delta\gamma. \tag{19.325}
$$

Then, Equations 19.318 through 19.322 yield

$$
\dot{P_\perp} = H\frac{dJ_n(\alpha)}{d\alpha} \sin(n\Phi + T), \tag{19.326}
$$

$$\dot{\Phi} = -\frac{\Omega_B}{\gamma}. \tag{19.327}$$

Considering the *resonance phase*

$$\Phi_n = n\Phi + T, \tag{19.328}$$

Equations 19.326 and 19.327 become

$$\dot{P}_\perp = H\frac{dJ_n(\alpha)}{d\alpha}\sin\Phi_n, \tag{19.329}$$

$$\dot{\Phi}_n = -\frac{n\Omega_B}{\gamma} + 1. \tag{19.330}$$

We remind that in Equation 19.327, we neglected terms of order H since we considered only the isolated resonances which arise when the wave amplitudes are small.

If we expand $1/\gamma$ in a Taylor series at a resonance value

$$P_{\perp n} = P_\perp - \delta P_\perp = \left(\gamma_n^2 - 1\right)^{1/2} \equiv \left(n^2\Omega_B^2 - 1\right)^{1/2}, \tag{19.331}$$

Equations 19.329 and 19.330 become

$$\delta\dot{P}_\perp = H\frac{dJ_n(\alpha_n)}{d\alpha}\sin\Phi_n, \tag{19.332}$$

$$\dot{\Phi}_n = -\frac{P_{\perp n}}{\gamma_n^2}\delta P_\perp. \tag{19.333}$$

These are the equations of a mathematical pendulum. The width of the isolated nonlinear cyclotron resonance (NCR) is then obtained as

$$\Delta_{NCR} = 4\gamma_n\left[\frac{H\frac{dJ_n(\alpha_n)}{d\alpha}}{P_{\perp n}}\right]^{1/2}. \tag{19.334}$$

We emphasize that a resonance is indeed an isolated one if

$$\Delta_{NCR} \ll \Omega_B. \tag{19.335}$$

Thus, for relativistic particles, we obtain

$$4\left[HP_{\perp n}\frac{dJ_n(\alpha_n)}{d\alpha}\right] \ll \Omega_B. \tag{19.336}$$

If

$$4\left[HP_{\perp n}\frac{dJ_n(\alpha_n)}{d\alpha}\right] \gtrsim \Omega_B, \tag{19.337}$$

the resonances overlap. In our case of a resonant nonlinear wave–particle interaction, Equation 19.337 represents the (necessary) condition for the onset of a stochastic instability of motion due to an overlap of resonances.

We remark that the definitions and conditions evolved in this section possess merely an approximate character since any approximation of the mathematical model of a system with a possible chaotic behavior may lead to false physical conclusions. In other words, however much justifiable is an approximation, it may be equivalent to a change of initial conditions which may lead to a "butterfly effect" in a chaotic system. In this context, an isolated resonance appears to be an ideal concept since the background of the slow motions within the domain of an "isolated resonance" is perturbed by fast motions of a small amplitude as generated by the effect of other resonances. In order to construct a real image of the nonlinear interaction, we must not apply to the fast motion any averaging procedure in the numerical integration. As numerical experiments demonstrate, the motion of particles within the conditions of overlap resonances becomes very complicated and possibly chaotic.

19.3.3.5 Numerical Experiments

Since there are no efficient analytic methods for describing stochastic components of a motion, it is indicated to apply a numerical integration of Equations 19.309 through 19.311. In this section, we propose to numerically analyze the process of the stochastization (chaotization or randomization) of the wave–particle interaction. We initiate the calculus for the following values of parameters:

Cyclotron frequency	$\Omega_B = 0.5$
Order of resonance	$n = 4$
Initial energy	$\gamma_0 \equiv \gamma_4 = 2$
Initial momentum	$P_{x0} = P_{y0} = \sqrt{1.5}$

The dynamic system of Equations 19.309 through 19.311 generates a flow in a 3D phase space (P_x, P_y, X). Numerical solutions have been obtained by applying a fifth-order Runge–Kutta algorithm with an adaptive stepsize control. The initial values for $[T, P_x, P_y, X]$ are $[0.001, \sqrt{1.5}, \sqrt{1.5}, 0.001]$.

This analysis was performed by Argyris et al. [14,15], but in an incomplete manner. The complete time series, Poincaré sections, complete phase space, Lyapunov exponents, and bifurcation diagrams, i.e., the entire nonlinear dynamics "arsenal," is missing. This is the reason why, in the following, we will perform a complete and detailed nonlinear dynamics analysis of the system evolution described in Equations 19.309 through 19.311.

For small amplitudes of the electromagnetic field, e.g., $H = 0.05$, the particle motion is complex but retains a regular character (see Figure 19.27a–g). Then, one can associate spiral-type "trajectories" with a periodic movement having a large amplitude (see Figure 19.27a) and harmonic time increasing momentum (see Figure 19.27b and c), either in a 2D phase space (see Figure 19.27d–f) or in a 3D phase space (see Figure 19.27g).

The onset of fractalization (by means of stochastization) is observed once H exceeds 0.5 (see Figure 19.28a–g). The debut of chaos through period doubling (see Figure 19.28b and c), as a natural result of the field–particle nonlinear interaction, initiates the particle rotation-phase stochastization process in the considered magnetic field (for details, see References 14 and 15). As a consequence, one can associate modified spiral-type "trajectories" with a periodic movement having a small amplitude (see Figure 19.28a) and harmonic time increasing momentum, modified through period doubling (see Figure 19.28b and c), either in a 2D phase space (see Figure 19.28d–f) or in a 3D phase space (see Figure 19.28g).

When $H = 1.5$, a gun effect is initiated (see Figure 19.29a–g). The stochastization expansion process (Figure 19.29a–c) induces the "stochastic medium" (for details, see References 14 and 15), which is responsible for particle acceleration (the curve or curves connecting the "trajectories," either in a 2D phase space [Figure 19.29d–f] or in a 3D phase space [Figure 19.29g], corresponds to the gun effect). The gun effect is generated through the resonant field–particle interaction.

An extensive chaotic regime is obtained for $H = 2.5$ (see Figure 19.30a–g). Spontaneously and unpredictably, regular behaviors intercalate with chaotic ones (emphasized either through time series [see Figure 19.30a–c] or through "trajectories" in a 2D phase space [see Figure 19.30d–f] or through "trajectories" in a 3D phase space [see Figure 19.30g]). The system passes into an extended chaotic regime.

A chaotic gun effect erupts for $H = 3.5$ (see Figure 19.31a–g). In the chaotic gun effect dynamics, we distinguish the following sequences: first, a localized chaotic regime emerges, subsequently a high-frequency oscillation with chaotic modulation of the amplitude appears, and finally, we observe the sharp, rectilinear part of the trajectory, which in fact displays the gun effect. These sequences are emphasized through time series (see Figure 19.31a–c) or through "trajectories" in a 2D phase space (see Figure 19.31d–f) or through "trajectories" in a 3D phase space (see Figure 19.31g).

A multigun effect results for $H = 4.5$ (see Figure 19.32a–g). This is clearly illustrated by the time series (see Figure 19.32a–c) or by "trajectories" in a 2D phase space (see Figure 19.32d–f) or by "trajectories" in a 3D phase space (see Figure 19.32g). The multigun effect is correlated with the jumps between different

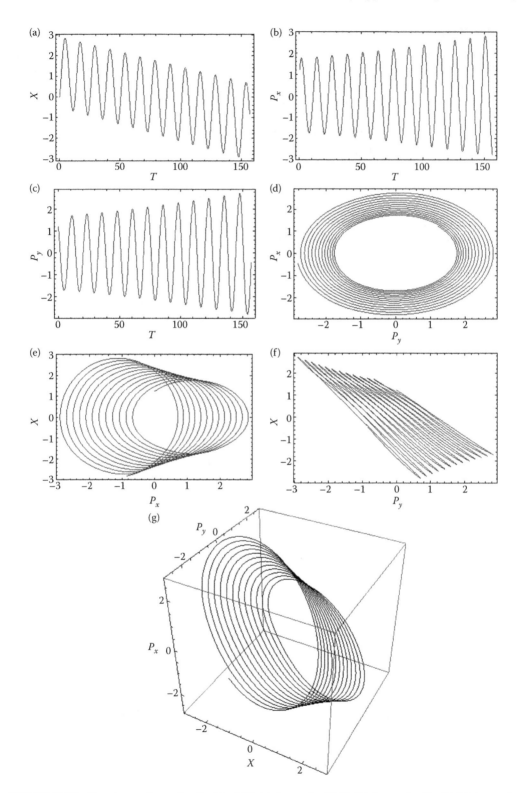

FIGURE 19.27 Complex regular motion for $H = 0.05$: (a) time series (X, T), (b) time series (P_x, T), (c) time series (P_y, T), (d) Poincaré sections (P_x, P_y), (e) phase space (X, P_x), (f) phase space (X, P_y), and (g) 3D phase space (X, P_x, P_y).

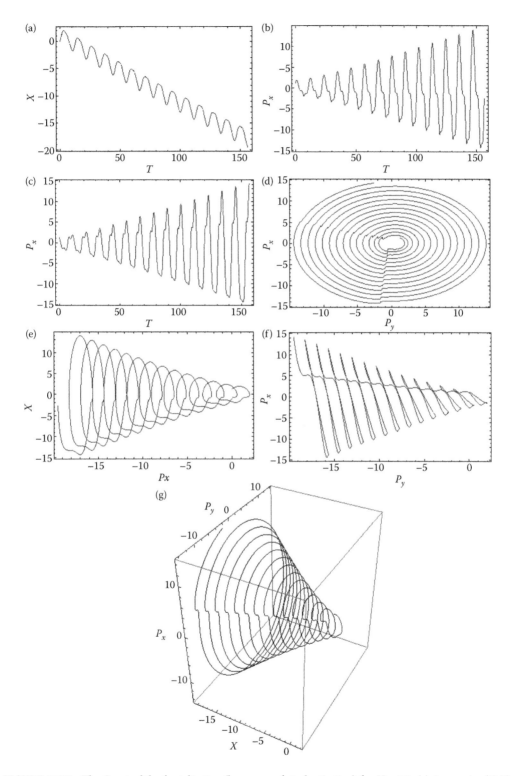

FIGURE 19.28 The Onset of the fractalization (by means of stochastization) for $H = 0.5$: (a) time series (X, T), (b) time series (P_x, T), (c) time series (P_y, T), (d) Poincaré sections (P_x, P_y), (e) phase space (X, P_x), (f) phase space (X, P_y), and (g) 3D phase space (X, P_x, P_y).

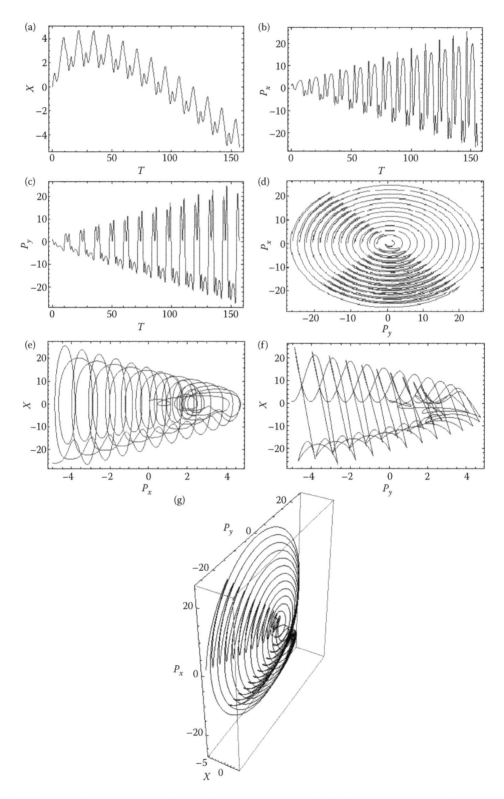

FIGURE 19.29   The Onset of a gun effect for $H = 1.5$: (a) time series (X, T), (b) time series (P_x, T), (c) time series (P_y, T), (d) Poincaré sections (P_x, P_y), (e) phase space (X, P_x), (f) phase space (X, P_y), and (g) 3D phase space (X, P_x, P_y).

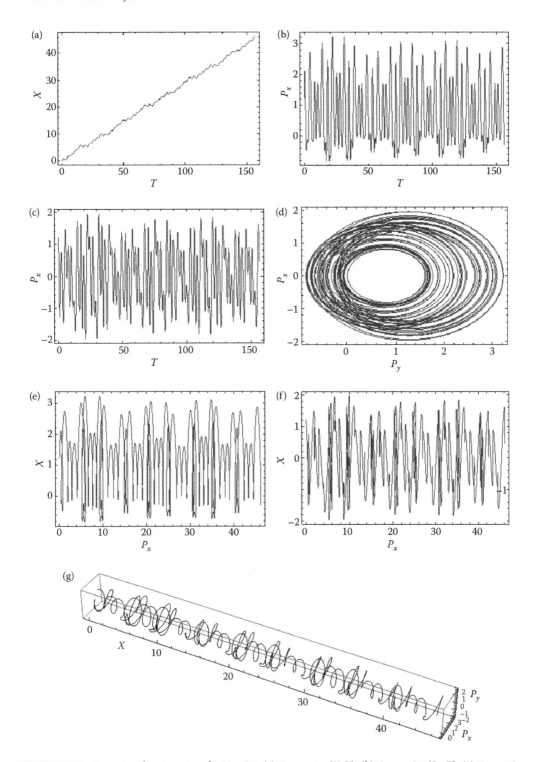

FIGURE 19.30 Extensive chaotic regime for $H = 2.5$: (a) time series (X, T), (b) time series (P_x, T), (c) time series (P_y, T), (d) Poincaré sections (P_x, P_y), (e) phase space (X, P_x), (f) phase space (X, P_y), and (g) 3D phase space (X, P_x, P_y).

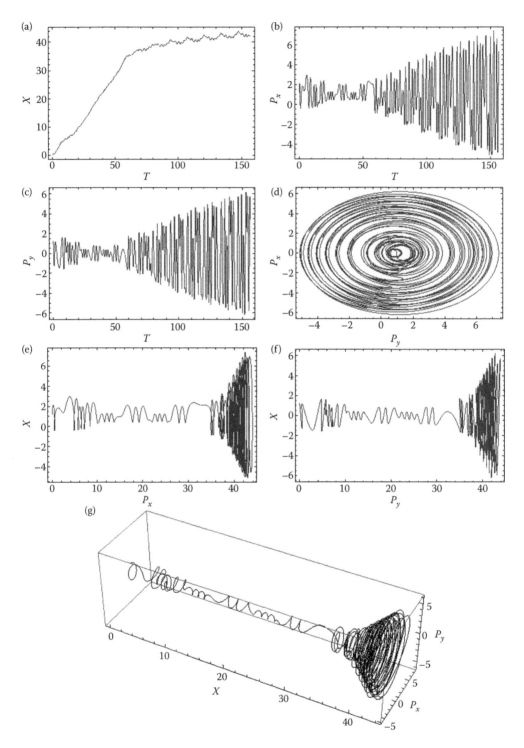

FIGURE 19.31 Chaotic gun effect for $H = 3.5$: (a) time series (X, T), (b) time series (P_x, T), (c) time series (P_y, T), (d) Poincaré sections (P_x, P_y), (e) phase space (X, P_x), (f) phase space (X, P_y), and (g) 3D phase space (X, P_x, P_y).

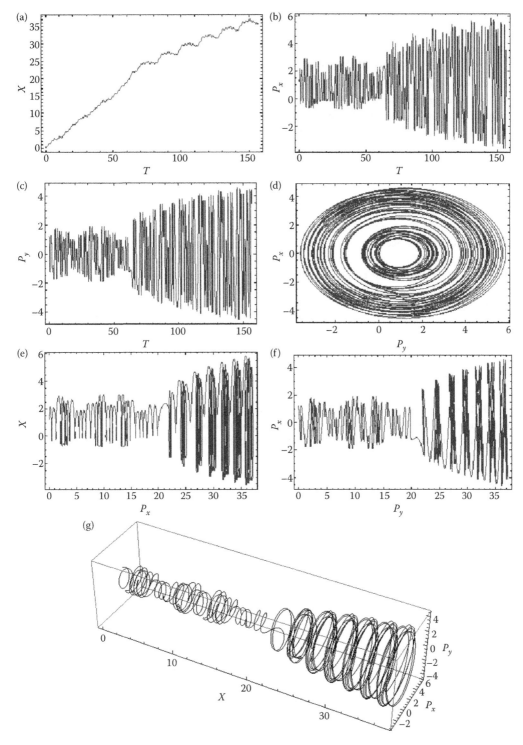

FIGURE 19.32 Multigun effect for $H = 4.5$: (a) time series (X, T), (b) time series (P_x, T), (c) time series (P_y, T), (d) Poincaré sections (P_x, P_y), (e) phase space (X, P_x), (f) phase space (X, P_y), and (g) 3D phase space (X, P_x, P_y).

Larmor orbits. Vertical and large-amplitude oscillations correspond to different Larmor orbits, while the curves connecting them correspond to the chaotic gun effect.

19.3.3.6 Lyapunov Exponent Analysis

After analyzing the trajectories in the phase space and the laws of movement, the value $H = 2.5$ was suggested for an extensive chaotic regime. By decreasing the intervals of focus (in our case $H = 4.75$), we searched for zones in which the dynamic of the proposed system enters a chaotic regime. It is difficult to directly observe an evolution toward chaos because the form of circular trajectories does not change significantly (only in certain specific cases the particle returns). Nevertheless, in the zone $H = 4.75$, one can observe successive dedoubling of the frequencies as precursors of the routes to chaos. These analyses were performed by studying the Fourier transform of the signal for different values of the parameters.

In order to have a more complete image of the chaotic zones, the Lyapunov exponents are further discussed. We note that the Lyapunov exponents are a measure of chaos. More precisely, they show the rate at which the trajectories converge or diverge in the phase space. The number of Lyapunov exponents is equal to the phase-space dimension. Usually, the one with the highest value is the one that is most important, because, if it is positive, then the system will be characterized by a high sensitivity to initial conditions in the vicinity of the initially considered point. There are several ways of defining the Lyapunov exponents as can be seen from References 50, 55, and 72. In this chapter, we used the following definition of the Lyapunov exponent:

$$\lambda = \lim_{t \to \infty} \frac{1}{t} \ln \left| \frac{\delta x(t)}{\delta x_0} \right|, \tag{19.338}$$

where $\delta x_0 = x_2(0) - x_1(0)$ is a small variation of the initial position and $\delta x(t) = x_2(t) - x_1(t)$ is the distance between the positions resulting from the system evolution at the moment t. In the framework of numerical analysis, we considered in relative units $t/T \cong 10^3$ (T is the specific period of movement for $4 < H < 5$ and $\Omega_B = 0.5$) and $\delta x_0 / x_2(0) \cong 10^{-3}$. Under such conditions, t is big enough and δx_0 is sufficiently small. Analyses for increasing precision were performed, but no significant modifications of the results were found.

In such circumstances, we analyzed the Lyapunov exponent in three cases, according to the modifications of initial conditions for the three parameters X, P_x, and P_y (see Figure 19.33a–c). These graphs illustrate 2D mappings of the Lyapunov exponent as a function of the two important parameters: H (along the Ox-axis) and Ω_B (along the Oy-axis); as expected, a strong similarity between these three graphs was observed.

Figure 19.34 illustrates an analysis of the chaotic zones. The darker areas represent the zones with a higher Lyapunov exponent, i.e., the chaotic zones. Thus, we can confirm the fact that in order to obtain extensive chaotic regimes, we must have $H > 2.5$. Thus, three important regions are outlined (see Figure 19.34) and isles 2 and 3 are between $4 < H < 5$, i.e., in the same zone resulting from the previous analyses. Through 3D representation, one can observe a finer variation of the Lyapunov exponent (see Figure 19.35). The stable zones ($\lambda < 0$) were eliminated from the graph; thus, only the positive values of the exponent are represented.

19.3.3.7 Bifurcation Diagrams

The scenario of evolution toward chaos through resonances overlapping is accomplished: stochastic layers (inside which the particle executes random walks) surrounding the cyclotron resonances are separated from one another by invariant curves. Some of these invariant curves may vanish (intersecting resonances) and a stochastic (fractal) web-like network arises in the phase portrait, where the layers do overlap each other. The stochastic web may disappear quickly because of a strong nonlinearity of the physical system. The instability is illustrated by the fact that an elliptical point transforms into a hyperbolic one. Two new elliptical points of the doubled period are then formed and this represents an island-doubling bifurcation. If the energy of the particle grows continuously, there occurs a cascade of successive island-doubling bifurcations (a characteristic of the Hamiltonian systems) in the vicinity of the elliptical points, which form

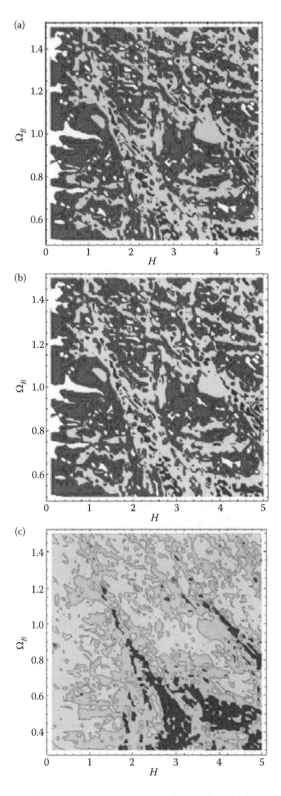

FIGURE 19.33 2D mappings of the Lyapunov exponent, according to the modification of initial conditions for the parameters X (a), P_x (b), and P_y (c), respectively.

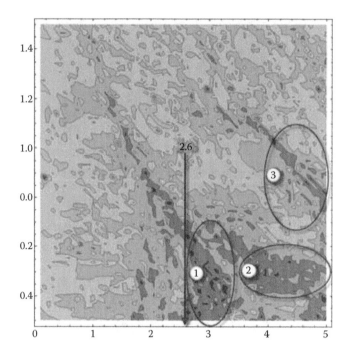

FIGURE 19.34　Analysis of the chaotic zones by investigating the Lyapunov exponents.

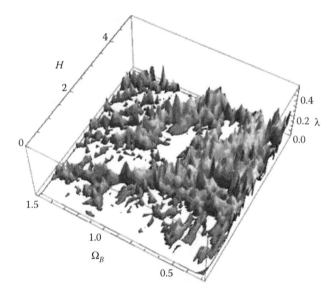

FIGURE 19.35　3D representation of the Lyapunov exponents as a function of H and Ω_B.

necklaces of new (smaller) stability islands corresponding to higher-order resonances in the interaction of the particle with the field. When the amplitude of the field increases, the stochastic net covers the whole phase plane and even the particles with small energies can diffuse into the region of very high energies [5].

　　The bifurcation diagrams confirm this scenario of transition to chaos (see Figure 19.36a–c). Let us note that, in order to numerically obtain bifurcation diagrams, we must find the frequencies for each evolution of X, P_x, or P_y as functions of H. For this purpose, several methods can be used [50,55,72]. In our case,

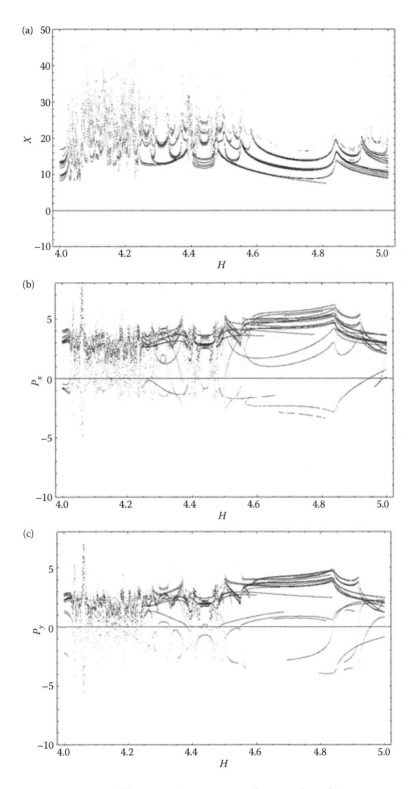

FIGURE 19.36   3D representation of the Lyapunov exponents as a function of H and Ω_B.

we did the mapping of X, P_x, or P_y values versus time in a certain time interval. This time interval is swept along the time axis and the maxima are searched in this interval. If we find only one frequency of the same oscillation amplitude, we find only one point. If we find two different frequencies, we find two different maxima and hence the doubling of the frequencies results. In essence, the method is based on the intermediate value theorem [50,55].

19.3.4 Conclusion

A theoretical fractal hydrodynamic model was built in order to explain the appearance of the NDR phenomenon in plasma (associated with bistability), a condition necessary for the onset of plasma instabilities.

The DL oscillations and ion-acoustic instability in nonmagnetized plasma were described by a set of time-dependent Schrödinger equations and the self-structuring of plasma is given by means of the NDR.

We described a new localized chaotic effect which is generated when a charged particle moves into the field of a plane electromagnetic wave with a sufficiently large amplitude. This entails an electromagnetic gun effect and refers to the chaotic motion of a single particle but it may also represent a chaotisizing element on a charged beam subject to a stochastic acceleration. This gun effect may be important for a novel laboratory scheme of accelerated particles. It may also be important in the study of the stochastic acceleration of particles in astrophysical systems such as solar flares, blazars, etc. It can become a major element in the study of the stability of dense beams of charged particles.

Any experimenter should not be afraid of the large amplitude ($H = 0.9$) of electromagnetic waves or equivalently of the high beam densities necessary to generate gun effects. Applying numerical experiments, we demonstrated that our model is very flexible in the sense that the essential parameter in obtaining the gun effect is the gyrofrequency ratio β. We obtained the gun effect even for $H = 0.2$ and $\Omega_B = 0.05$ and over extended domains of space–time. Such situations with $\beta > 1$ may arise frequently in astrophysical systems.

Acceleration of charged particles in electromagnetic fields leads to the extraction of energy from these fields. The release of electromagnetic energy by stochastic processes is accompanied by stochastic topological changes in the electromagnetic field configuration which probably occur through some form of magnetic reconnection but the details of this process are still not well understood. The importance of this problem becomes more evident if we remember that the magnetic reconnection may generate turbulent waves. The chaotic gun effect may be one of the sources for magnetic reconnections.

We have reproduced this effect by applying a direct numerical integration of the nonlinear dynamic system of equations without any approximation and without any averaging process. The importance of the parameter β does not appear in any analytic approximation and we discovered it only through numerical experiments. As a further development, it would be interesting to study the dynamics of charged particles in the case when the gyrofrequency ratio β derives from the magnetic fields generated by two electromagnetic waves. Since the gun effect is a pure result of numerical experiments, our final conclusion is that any approximation and any ignored physical parameters could be, if incorporated in to the analysis, the source of a chaotic behavior. Little physical insight is gained by trying to impose the analytic method through an artificial effort without any secure assumptions.

Such a theoretical model qualitatively describes the dynamics of instabilities in magnetized plasma, such as potential relaxation instability [19,44,45,60], electrostatic ion–cyclotron instability [20,44,45,98], or Kelvin–Helmholtz instability [45], as well as the nonlinear effects related to these instabilities.

19.4 Laser Ablation Plasma

19.4.1 Introduction

Laser-produced plasma (LPP) is a topic of growing interest in different fields such as material processing [22,54], diagnostic techniques [97], and space applications. Pulsed laser deposition has been successfully employed for the deposition of thin films of classical and novel materials [10,54]. The possibility of producing species in LPP with electronic states far from chemical equilibrium enlarges the potential of making novel materials that would be unattainable under thermal conditions.

For the experiments involving the interaction of intense laser pulses with matter, it is desirable to know directionality, velocity, and other parameters of an ejected plasma plume far from the target. The plume evolution includes two important parts: in the initial stage, i.e., during the laser energy deposition, the one-dimensional expansion takes place (large laser spot size compared with the skin depth); after some time, the plasma cloud becomes truly 3D [12,24,65,75,77,78]. According to Reference 57, both the elementary physical processes which require different timescales, and the patterns evolution [28] that requires different degrees of freedom (e.g., from 1, at the initial stages, to 3, at the final stages of the patterns induced by LPP), imply a nondifferentiable space–time, i.e., fractals [48,87,90]. Moreover, the dynamics of the plasma transition from "disorder" to "order" also needs a fractal description. In this section, a mathematical model to describe some characteristics of the LPP expansion is established in the frame of SRT for an arbitrary topological dimension D_F. In the following section, we will extensively present the results from References 80 and 83.

19.4.2 Experimental Results

Experimentally (see Figure 19.37), the plasma was generated by laser ablation of an aluminum target using 532 nm radiation pulses from a Nd:YAG laser. The laser pulse energy ranges from 10 to 80 mJ, in a pulse time width of 10 ns, and it has been focused by an $f = 25$ cm lens at normal incidence on an aluminum target placed in a vacuum chamber (pressure $< 10^{-6}$ Torr), to obtain the spot diameter at the impact point of 300 μm. The formation and dynamics of the plasma plume have been studied by means of an intensified charge-coupled device (ICCD) camera (PI MAX 576 × 384, gating time 20 ns) placed orthogonal to the plasma expansion direction. In Figure 19.38, the ICCD images of the plasma plume at different times after the laser pulse are given. Successive pulses of energy 40 mJ were used. The images of these structures reveal a splitting process of the plasma blobs. Analyzing Figure 19.38, we conclude the following: (1) In the range time 10–50 ns after the laser pulse, the visible emitting regions of plasma are almost stationary and they form two structures, each having a maximum emissive region and (2) by measuring the position of the maximum emissivity at different times, a linear dependence resulted. Then, the velocities of the two plasma formations have been calculated as being $v_1 = 4.66 \cdot 10^4$ m/s for the first plasma formation and $v_2 = 6.9 \cdot 10^3$ m/s for the second structure [56].

To validate the previous experimental results, additional electrical measurements were made using a cylindrical Langmuir probe (stainless steel 0.8 mm diameter, 5 mm length), biased at -30 V and placed at various distances relatively to the plasma plume. The transient ionic current plotted in Figure 19.39 for various probe-target distances shows that the previous two plasma structures are evidenced as two arrival times. Thus, the first part of the transient signal has a maximum that corresponds to the first structure, while the second structure gives a hump (see the marks in Figure 19.39). Moreover, using these time values and taking into account the probe positions, a good agreement with the previous velocities resulted.

The plasma splitting in fast and slower parts was previously reported in many papers (e.g., see Reference 58) as an effect of the high ambient pressure. Moreover, it was observed for each ion species [46]. Thus, this effect is reasonably described using the phenomenology of the supersonic underexpanded gaseous jet [27], while a possible explanation is that the two well-localized plasma patterns have an electric DLs structure [28]. Indeed, as it is shown in the next paragraph, the previous analogy is valid for any interaction scale if we consider the fractal structure of space–time. Such treatment is achieved by means of the hydrodynamic model of SRT [48,57,87,90]. We note that the experiment shows that the plume splitting occurs even at low ambient pressure. In our opinion, this takes place because the second structure of "mushroom" shape is expanding in the atmosphere of an evaporated material at the early stages of laser target interaction.

Using an electrical highly stabilized dc power source, the aluminum target (isolated from the vacuum chamber) has been biased at a certain voltage V_T in the range between -30 and $+30$ V. The laser beam

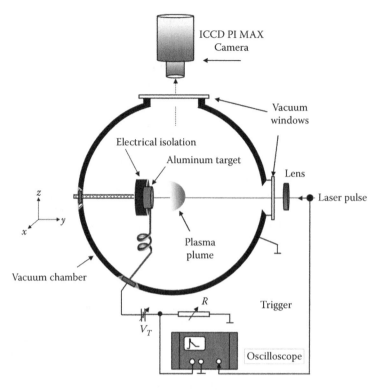

FIGURE 19.37 Experimental setup used for studying laser-produced plasma expansion.

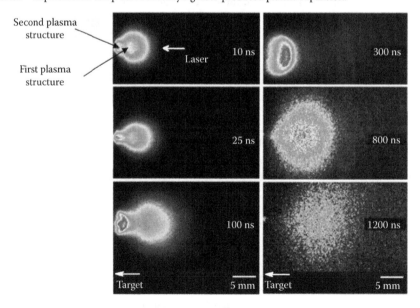

FIGURE 19.38 Evolution of visible emission from the aluminium laser-plasma plume recorded using an ICCD PI MAX camera (gating time 20 ns). The laser beam energy was 40 mJ/pulse and background pressure $p = 10^{-6}$ mbar.

energy (1–60 mJ/pulse) has been continuously monitored by a Joulemeter (OphirNovaII). Plasma formation, dynamics, and influence of the target biasing upon the plume expansion were investigated by measuring the target current. The transient target currents recorded for various laser beam energies (E_L) are plotted in Figure 19.40. The positive charging can be postulated to arise through the electrons escaping

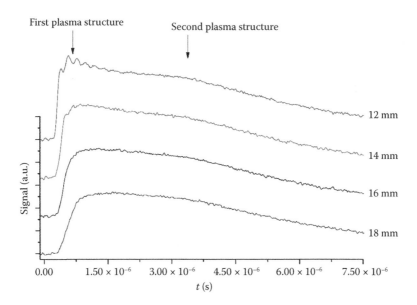

FIGURE 19.39 Transient ion current recorded by Langmuir probe at various distances from the target. The two plasma structures are evidenced as two arrival times.

from the expanding plasma to the grounded chamber [47], i.e., having sufficient energies to overcome the plasma work function, while the negative charging is given by the ion contribution. From these experimental results, we conclude the following: (1) at laser beam energies below 11 mJ, a double-peak structure of the positive part is observed and it corresponds to the hot and cold electrons of the expanding plasma; (2) the amplitude of the fast peak increases rapidly with the laser energy, together with the time required to reach this value, and the fast electron contribution becomes dominant above $E_L = 16$ mJ; (3) the amplitudes of the slow electronic peak and of the ionic (negative) peak are also increasing with the laser energy, but times of these values are decreasing; (4) at low laser fluences, the positive part extends over a longer time due to the lower expansion velocity; (5) integrating over time, the positive part of the current is equal with the negative one.

Since in general the plasma temperature, and consequently the thermal expansion velocity, are increasing with the laser energy, the behaviors of the slow electrons and of the ions can thus be justified. However, having in view that the time corresponding to the maximum current of fast electrons is increasing with the laser energy (Figure 19.40), it experimentally resulted that their velocity is decreasing, even though their temperature (as will be shown below) is higher. In our opinion, this can be explained by different mechanisms of ejection. The fast part is dominated by Coulomb interactions; the hot electrons which have absorbed a large part of the laser energy during the interaction create a charge separation field, which pulls ions toward vacuum, and progressively transfers the energy from the electrons to the ions. Diverse models have been proposed to explain these processes [75,77]. The slow part can be related to thermal processes. Since a higher laser energy gives a higher ion average charge state, we can conclude that the electrostatic force may reduce the fast electrons expansion velocity.

For comparison, in Reference 81, using a cylindrical Langmuir probe biased at −30 V and placed at various distances from the target, the typical transient ion current for various probe aluminum target distances also showed the splitting of plasma in to two structures of different velocities, which were evidenced as two arrival times. Thus, the first part had an oscillatory character, while the second structure gave a hump. Such behavior was in agreement with the plasma images recorded by an ICCD camera (for details, see Reference 56). Moreover, space- and time-resolved optical emission spectroscopy revealed different charge states: the first part is dominated by the Al^{++} and Al^{+} ions, while the second part is dominated

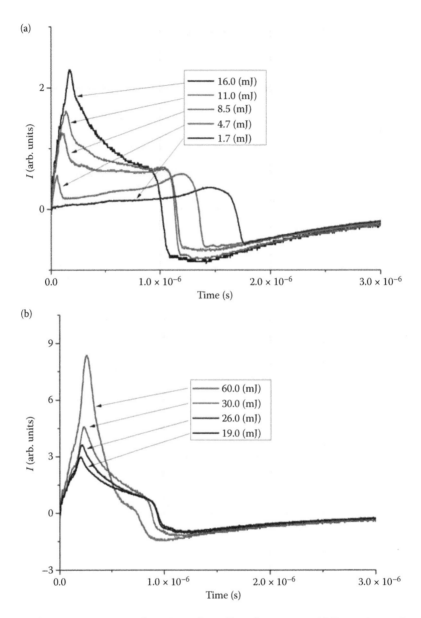

FIGURE 19.40 Transient target current for various values of laser beam energy: (a) $E_L = 1.7-16$ mJ, and (b) $E_L = 19-60$ mJ.

by Al^+ ions and neutrals [105]. These point two distinct mechanisms for the ejection of these packets: the highly charged ions would be ejected on a very short timescale through a Coulomb process in the very intense field left by laser-excited electron oscillation and detachment, while the neutrals and low-charged ions would come from a subsequent thermal process [105]. Therefore, we conclude that the results in Figure 19.40 are in a good agreement with previous measurements.

Simple evaluations of the plasma temperature can be made by changing the target voltage, V_T. For a Maxwellian electron energy distribution, the electron current is given by [47]

$$i_T = i_0 \exp\left(-\frac{eV_T}{k_B T_e}\right), \tag{19.339}$$

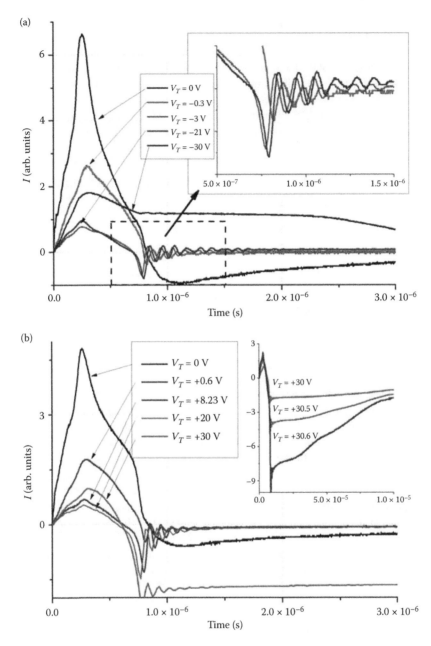

FIGURE 19.41 Transient target current for (a) positive and (b) negative external target polarizations, respectively, and $E_L = 47$ mJ.

where T_e is the electron temperature and i_0 is a constant. A plot $\ln i_T = f(V_T)$ should thus be linear. In Figure 19.41a and b, the transient target current for positive and negative polarizations is given for $E_L = 47$ mJ. According to Figure 19.41a, the time of the electronic peak is not changed by the target negative or positive biasing, but the maximum value of the current is decreasing in both cases since the charge separation, responsible for plasma acceleration, is obstructed by repelling or attracting the electrons. For the target voltage of $V_T = -30$ V, a negative (ionic) part of the transient current is completely removed, and the positive (electronic) part is extending over a longer time (see Figure 19.41a). For $V_T = +30$ V an opposite behavior was observed: the positive part is almost removed and the negative part becomes dominant (see Figure 19.41b).

An interesting feature appears in the ionic part: the applied voltages (either positive or negative) induce a damped oscillatory behavior of the ionic (negative) part, of the same averaged frequency, $f = (9.1 \pm 0.5)$ MHz (see the inset in Figure 19.41a). The initial amplitude is slightly increasing with the target voltage. Similar oscillations were reported for the current recorded by a cylindrical Langmuir probe [56]. They were studied for various laser energies and radial distances with respect to the expansion axis, and a weak variation of the frequency with these parameters was evidenced. Threshold energy and maximum radial distance from the target for which such oscillations appear were observed at $E_L = 2.5$ mJ and $r = 1.4$ cm, respectively. For these measurements, this threshold laser energy corresponds exactly to the value for which the fast electronic (positive) peak appears. Since the oscillation frequency was found to depend on the atomic mass of the target, a possible explanation was the oscillatory motion of the ions in the applied electric field, with the plasma ion frequency [83]

$$f_{pi} = 210z\sqrt{\frac{n_i}{A}}, \tag{19.340}$$

where z is the ionic charge state, n_i (cm^{-3}) is the ion density, and A (amu) is the atomic mass. For this experimental data, if we consider the aluminum target ($A = 27$) and doubly ionized species ($z = 2$), Equation 19.340 gives the ion density, $n_i = 1.27 \cdot 10^{10}$ cm^{-3}. Moreover, because the oscillations are observed only for the ions, the influence of the target polarization circuit is excluded, and possible oscillations given by electrons might be evidenced by a faster detection system.

We also note that in Reference 25 oscillations of kilovolt magnitude in the potential of laser-irradiated targets have been measured. The highest frequency of oscillation was around 10 GHz, but for laser intensities of 10^{14} W/cm^2, i.e., for higher values of n_i and z. The oscillations appeared to be associated with the plasma instabilities but, since they were too fast to be measured or detected by the experimental system, the authors suggested that the existence of DLs and turbulence in the plasma give rise to a virtual charged capacitor in the plasma on the target, oscillating with frequencies of the same order of magnitude as plasma ion frequency.

Now, returning to the dependency given by Equation 19.339, the dependencies of time-integrated current versus target voltage for the laser energies $E_L = 25$ and 47 mJ (Figure 19.42) confirm the existence of non-Maxwellian electron distribution through the hot and cold electrons, as also predicted by Figure 19.40a. Their temperatures were evaluated as being $T_h = 24.3$ eV and $T_c = 4.98$ eV for $E_L = 47$ mJ, and $T_h = 19.94$ eV and $T_c = 4.86$ eV for $E_L = 25$ mJ, i.e., almost the same temperature for the cold electrons, but a significantly higher temperature of the hot electrons for an increased laser energy. These values are of the same order of magnitude as previously reported in Reference 105.

19.4.3 Theoretical Modeling

Let us study the vacuum plasma expansion in the frame of a one-dimensional fractal hydrodynamic model, extensively described in Section 19.1 of this chapter. Then, using the substitutions

$$\omega t = \tau, \tag{19.341}$$

$$kx = \xi, \tag{19.342}$$

$$v/v_0 = V, \tag{19.343}$$

$$\rho/\rho_0 = N, \tag{19.344}$$

$$\mu^2 = \frac{k^3 D^2}{\omega v_0}(dt)^{(4/D_F)-1}. \tag{19.345}$$

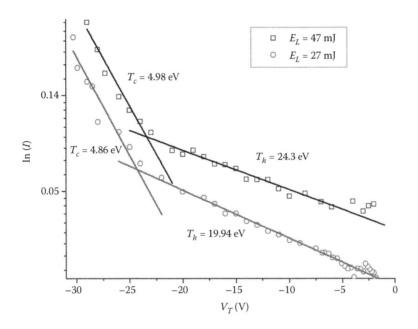

FIGURE 19.42 Dependencies of time-integrated target current versus target voltage and laser energies $E_L = 25$ and 47 mJ.

Equations 19.49 and 19.50 take the form

$$\frac{\partial V}{\partial t} + V\frac{\partial V}{\partial \xi} = \frac{\partial}{\partial \xi}\left[\frac{2\mu^2}{\sqrt{N}}\frac{\partial^2}{\partial \xi^2}\sqrt{N}\right], \tag{19.346}$$

$$\frac{\partial N}{\partial t} + \frac{\partial(NV)}{\partial \xi} = 0. \tag{19.347}$$

In Equations 19.341 through 19.345, ω is the plasma pulsation, k is the inverse of Debye length, v_0 is the ion-acoustic speed, and ρ_0 is the density corresponding to thermodynamic equilibrium. The parameter μ defines a normalized diffusion coefficient.

Through the method from Reference 7, with the initial conditions

$$V(\xi - \xi_0, \tau = 0) = c, \tag{19.348}$$

$$N(\xi - \xi_0, \tau = 0) = \frac{1}{\sqrt{\pi}\alpha}\exp\left[-\left(\frac{\xi - \xi_0}{\alpha}\right)^2\right] = N_0(\xi - \xi_0), \tag{19.349}$$

and the boundary ones

$$V(\xi - \xi_0 = c\tau, \tau) = c, \tag{19.350}$$

$$N(\xi - \xi_0 = -\infty, \tau) = N(\xi - \xi_0 = +\infty, \tau) = 0, \tag{19.351}$$

the solutions of Equations 19.346 and 19.347 have the explicit form

$$V(\xi - \xi_0, \tau) = \frac{c\alpha^2 + \left(\frac{2\mu}{\alpha}\right)^2 \tau(\xi - \xi_0)}{\alpha^2 + \left(\frac{2\mu}{\alpha}\right)^2 \tau^2}, \tag{19.352}$$

$$N\left(\xi - \xi_0, \tau\right) = \frac{1}{\sqrt{\pi\left[\alpha^2 + \left(\frac{2\mu}{\alpha}\right)^2 \tau^2\right]}} \exp\left[-\frac{(\xi - \xi_0 - c\tau)^2}{\alpha^2 + \left(\frac{2\mu}{\alpha}\right)^2 \tau^2}\right]. \tag{19.353}$$

In the previous relations c, N_0, and ξ_0 are the initial normalized speed, density and position, respectively; α is the normalized distribution parameter. These solutions induce the complex speed field

$$\hat{V}\left(\xi - \xi_0, \tau\right) = \frac{c\alpha^2 + \left(\frac{2\mu}{\alpha}\right)^2 \tau(\xi - \xi_0)}{\alpha^2 + \left(\frac{2\mu}{\alpha}\right)^2 \tau^2} - 2i\mu\frac{(\xi - \xi_0 - c\tau)^2}{\alpha^2 + \left(\frac{2\mu}{\alpha}\right)^2 \tau^2}, \tag{19.354}$$

the complex current density speed field

$$\hat{j}\left(\xi - \xi_0, \tau\right) = \frac{c\alpha^2 + \left(\frac{2\mu}{\alpha}\right)^2 \tau(\xi - \xi_0)}{\sqrt{\pi}\left[\alpha^2 + \left(\frac{2\mu}{\alpha}\right)^2 \tau^2\right]^{3/2}} \exp\left[-\frac{(\xi - \xi_0 - c\tau)^2}{\alpha^2 + \left(\frac{2\mu}{\alpha}\right)^2 \tau^2}\right]$$

$$- 2i\mu\frac{\xi - \xi_0 - c\tau}{\sqrt{\pi}\left[\alpha^2 + \left(\frac{2\mu}{\alpha}\right)^2 \tau^2\right]^{3/2}} \exp\left[-\frac{(\xi - \xi_0 - c\tau)^2}{\alpha^2 + \left(\frac{2\mu}{\alpha}\right)^2 \tau^2}\right], \tag{19.355}$$

the complex specific potential

$$\hat{Q}\left(\xi - \xi_0, \tau\right) = 0 - 2i\mu^2\frac{(\xi - \xi_0 - c\tau)^2}{\left[\alpha^2 + \left(\frac{2\mu}{\alpha}\right)^2 \tau^2\right]^2} + 2i\mu^2\frac{1}{\alpha^2 + \left(\frac{2\mu}{\alpha}\right)^2 \tau^2}, \tag{19.356}$$

and the complex force field

$$\hat{F}\left(\xi - \xi_0, \tau\right) = 0 + 4i\mu^2\frac{\xi - \xi_0 - c\tau}{\left[\alpha^2 + \left(\frac{2\mu}{\alpha}\right)^2 \tau^2\right]^2}. \tag{19.357}$$

The force field $\hat{F}\left(\xi - \xi_0, \tau\right)$ imposes fractal characteristics to the global dynamics of the system. Consequently, the complex speed and the complex current density fields are nonhomogenous in the coordinates $\xi - \xi_0$ and τ. The predictable (observable) global dynamics, for example, of linear uniform motion form, are obtained by making the force field zero, $\hat{F}\left(\xi - \xi_0, \tau\right)$. In this case, the complex speed field takes the simple form

$$\hat{V}\left(\xi - \xi_0 = c\tau, \tau\right) \to c + i0. \tag{19.358}$$

This means that the particle of fluid in free motion polarizes the "fractal medium" behind itself, $\xi - \xi_0 \leq c\tau$, and ahead of itself, $\xi - \xi_0 \geq c\tau$, in such a form that the resulting forces are symmetrically distributed with respect to the plane through the observable particle position, $\langle\xi - \xi_0\rangle = c\tau$ at any time τ. This behavior is also induced to the other physical quantities. In such a context, the presence of a "perturbation," for example, given by the laser beam, induces an asymmetry of the force field, having the excitation of a specific mode of matter structuring as an effect.

Equation 19.353 corresponds to the speed scalar potential associated to a free particle

$$\psi\left(\xi - \xi_0, \tau\right) = \frac{1}{\pi^{1/4}\sqrt{\alpha + i\frac{2\mu\tau}{\alpha}}} \exp\left[-\frac{[(\xi - \xi_0) - c\tau]^2}{2\alpha^2\left(1 + i\frac{2\mu\tau}{\alpha^2}\right)}\right] \exp\left[-i\frac{c^2\tau}{4\mu} + i\frac{c(\xi - \xi_0)}{2\mu}\right]. \tag{19.359}$$

Through Equation 19.47, the phase S of this speed scalar potential is connected with the real part (observable) of a complex velocity field.

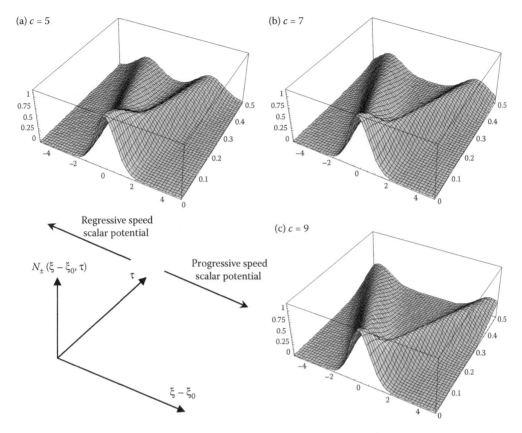

(a) $c = 5$

(b) $c = 7$

(c) $c = 9$

Regressive speed scalar potential

Progressive speed scalar potential

$N_\pm(\xi - \xi_0, \tau)$

τ

$\xi - \xi_0$

FIGURE 19.43 Progressive speed scalar potential ($\tau > 0$) and the regressive one ($\tau < 0$) for different values of the normalized speed c.

Now, let us assume the particle interaction with an external potential (which can be approximately modeled, e.g., by an infinite wall potential), resulting in a discontinuous change in momentum. Localized time-dependent solutions for this problem, i.e., the bouncing speed scalar potential, can be constructed in a very straightforward way from solutions of the free-particle problem [7]. With $\xi - \xi_0 = 0$ the wall position, the simple difference solutions of the form $\psi(\xi - \xi_0, \tau) - \psi(-(\xi - \xi_0), \tau)$ not only satisfy the free-particle Schrödinger equation, or the equivalent form (given by Equations 19.49 and 19.50 of the fractal hydrodynamics) for all $(\xi - \xi_0)$ values (if $\psi(\xi - \xi_0, \tau)$ does), but also accommodate the new boundary condition at the wall, namely that $\psi(0, \tau) = 0$. Then, the density is

$$N_R(\xi - \xi_0, \tau) = |\psi(\xi - \xi_0, \tau) - \psi(-(\xi - \xi_0), \tau)|^2 = N_+(\xi - \xi_0, \tau) + N_-(\xi - \xi_0, \tau)$$

$$- 2\sqrt{N_+(\xi - \xi_0, \tau)N_-(\xi - \xi_0, \tau)} \cos\left\{\frac{\xi}{\mu} \frac{\left[\xi_0\tau(2\mu/\alpha)^2 - c\alpha^2\right]}{\alpha^2 + (2\mu\tau/\alpha)^2}\right\}, \quad (19.360)$$

where $N_+(\xi - \xi_0, \tau) = N(\xi - \xi_0, \tau), N_-(\xi - \xi_0, \tau) = N(-(\xi - \xi_0), \tau)$, and $N(\xi - \xi_0, \tau)$ are given by Equation 19.353. The result (19.360) is equivalent with the interference of a progressive speed scalar potential (for $\tau > 0$) with a regressive one (for $\tau < 0$) (see Figure 19.43a–c).

The term of interaction from Equation 19.360 gives local maxima and minima that can be associated with the multipeak structure of the normalized plasma charge density $N_R(\xi - \xi_0, \tau)$ (see Figures 19.44a–c and 19.45a–c) and implicitly of the normalized current density. Their space–time positions depend on the initial velocity c. From Figures 19.44 and 19.45, it results that the multipeak structure can be observed

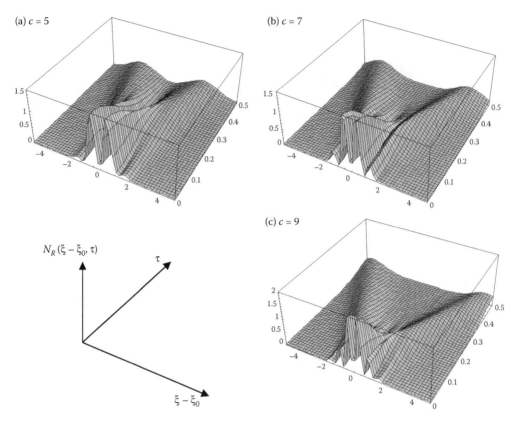

FIGURE 19.44 Spatial "interference" of a progressive speed scalar potential with a regressive one for different values of the normalized speed c.

both for the spatial component and for the temporal one, but in the last case only for $\alpha \ll 2\mu\tau$ (see Figure 19.45). We note the similarity between the oscillatory (ionic) part of the experimental curves shown in Figure 19.41a and b and those generated by the theoretical model shown in Figure 19.45.

Let us rewrite the equations of fractal hydrodynamics (19.131) and (19.132) for the 2D case (x, y) in the form

$$\frac{\partial}{\partial t}(\rho V_x) + \frac{\partial}{\partial x}(\rho V_x^2) + \frac{\partial}{\partial y}(\rho V_x V_y) = -\frac{\partial \sigma}{\partial x}, \tag{19.361}$$

$$\frac{\partial}{\partial t}(\rho V_y) + \frac{\partial}{\partial x}(\rho V_x V_y) + \frac{\partial}{\partial y}(\rho V_y^2) = -\frac{\partial \sigma}{\partial y}, \tag{19.362}$$

$$\frac{\partial \rho}{\partial t} + \frac{\partial}{\partial x}(\rho V_x) + \frac{\partial}{\partial y}(\rho V_y) = 0, \tag{19.363}$$

$$\frac{\partial(\rho e)}{\partial t} + \frac{\partial}{\partial x}(\rho e V_x) + \frac{\partial}{\partial y}(\rho e V_y) = -\rho e \left(\frac{\partial V_x}{\partial x} + \frac{\partial V_y}{\partial y}\right), \tag{19.364}$$

where the energy density (ρe) continuity equation was added. With the nondimensional variables

$$\omega t = \tau, \tag{19.365}$$

$$kx = \xi, \tag{19.366}$$

$$ky = \eta, \tag{19.367}$$

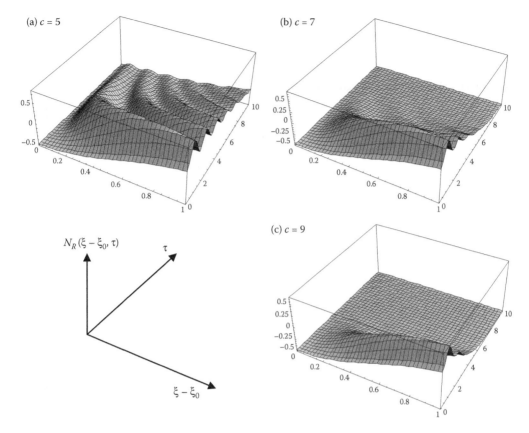

(a) $c = 5$

(b) $c = 7$

(c) $c = 9$

$N_R(\xi - \xi_0, \tau)$

τ

$\xi - \xi_0$

FIGURE 19.45 Temporal interference of a progressive speed scalar potential with a regressive one for different values of the normalized speed c.

$$\frac{V_x k}{\omega} = V_\xi, \tag{19.368}$$

$$\frac{V_y k}{\omega} = V_\eta, \tag{19.369}$$

$$\frac{\rho}{\rho_0} = N, \tag{19.370}$$

assuming the ideal gas case, and that the variation of σ is induced by the variation of density and temperature

$$\nabla\sigma = \alpha\nabla(NT), \tag{19.371}$$

$$\alpha = \text{const.} \tag{19.372}$$

Equations 19.361 through 19.364 become

$$\frac{\partial}{\partial\tau}(NV_\xi) + \frac{\partial}{\partial\xi}(NV_\xi^2) + \frac{\partial}{\partial\eta}(NV_\xi V_\eta) = -\frac{\partial(NT)}{\partial\xi}, \tag{19.373}$$

$$\frac{\partial}{\partial\tau}(NV_\eta) + \frac{\partial}{\partial\xi}(NV_\xi V_\eta) + \frac{\partial}{\partial\eta}(NV_\eta^2) = -\frac{\partial(NT)}{\partial\eta}, \tag{19.374}$$

$$\frac{\partial N}{\partial\tau} + \frac{\partial}{\partial\xi}(NV_\xi) + \frac{\partial}{\partial\eta}(NV_\eta) = 0, \tag{19.375}$$

$$\frac{\partial(NT)}{\partial\tau} + \frac{\partial}{\partial\xi}(NTV_\xi) + \frac{\partial}{\partial\eta}(NTV_\eta) = -NT\left(\frac{\partial V_\xi}{\partial\xi} + \frac{\partial V_\eta}{\partial\eta}\right), \tag{19.376}$$

where in Equations 19.373 through 19.376 the functional scaling relation

$$\alpha \frac{k^2}{\omega^2} = 1, \tag{19.377}$$

was considered. For numerical integration, the initial conditions

$$V_\xi(0, \xi, \eta) = 0, \tag{19.378}$$

$$V_\eta(0, \xi, \eta) = 0, \tag{19.379}$$

$$N(0, \xi, \eta) = 1/4, \tag{19.380}$$

$$T(0, \xi, \eta) = 1/4, \tag{19.381}$$

$$0 \times 0 \le \xi \times \eta \le 1 \times 1, \tag{19.382}$$

and boundary ones

$$V_\xi(\tau, 0, \eta) = 0, \tag{19.383}$$

$$V_\xi(\tau, 1, \eta) = 0, \tag{19.384}$$

$$V_\eta(\tau, 0, \eta) = 0, \tag{19.385}$$

$$V_\eta(\tau, 1, \eta) = 0, \tag{19.386}$$

$$N(\tau, 0, \eta) = 1/4, \tag{19.387}$$

$$N(\tau, 1, \eta) = 1/4, \tag{19.388}$$

$$T(\tau, 0, \eta) = 1/4, \tag{19.389}$$

$$T(\tau, 1, \eta) = 1/4, \tag{19.390}$$

$$V_\xi(\tau, \xi, 0) = 0, \tag{19.391}$$

$$V_\xi(\tau, \xi, 1) = 0, \tag{19.392}$$

$$V_\eta(\tau, \xi, 0) = 0, \tag{19.393}$$

$$V_\eta(\tau, \xi, 1) = 0, \tag{19.394}$$

$$N(\tau, \xi, 0) = N_0 \exp\left[-\frac{(\tau - 1/4)^2}{(1/4)^2}\right] \exp\left[-\frac{(\xi - 1/2)^2}{(1/4)^2}\right], \tag{19.395}$$

$$N(\tau, \xi, 1) = 1/4, \tag{19.396}$$

$$T(\tau, \xi, 0) = T_0 \exp\left[-\frac{(\tau - 1/4)^2}{(1/4)^2}\right] \exp\left[-\frac{(\xi - 1/2)^2}{(1/4)^2}\right], \tag{19.397}$$

$$T(\tau, \xi, 1) = 1/4, \tag{19.398}$$

are considered. In the boundary conditions (19.395) and (19.397), we assumed that the laser pulse which "hits" the target induces a plasma source that has a spatial–temporal Gaussian profile, similarly with the laser beam. The maximum normalized atom density is N_0, while T_0 is the maximum normalized temperature which is assumed to be proportional with the laser beam energy, by preserving the number of total released atoms.

Equations 19.373 through 19.376 with the initial conditions (19.378) through (19.382) and the boundary ones (19.383) through (19.398) were numerically integrated using finite differences [110]. In Figure 19.46, the 2D contour curves of the normalized density N, normalized temperature T, and normalized velocities V_ξ and V_η, are given for the normalized time $\tau = 1/2$, $N_0 = 1$, and various maximum normalized temperatures: Figure 19.46a $T_0 = 1$, Figure 19.46b $T_0 = 0.7$, and Figure 19.46c $T_0 = 0.4$. The following results: (1) the generation of two plasma structures is more obvious for high maximum normalized temperatures (see Figure 19.46a); (2) the symmetry of the normalized speed V_ξ, with respect to symmetry axis of the spatial–temporal Gaussian; and (3) vertices at the plume periphery for the normalized speed

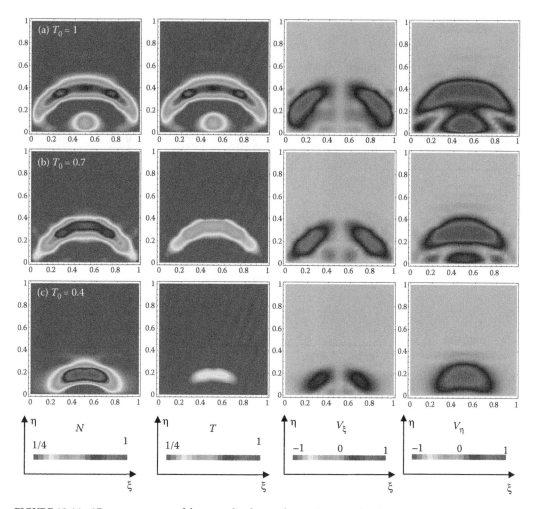

FIGURE 19.46 2D contour curves of the normalized atom density, N, normalized temperature, T, and normalized velocities, V_ξ and V_η, as resulted from the numerical simulations of Equations 19.173 through 19.376 for the normalized time, $\tau = 1/2$, $N_0 = 1$, and various maximum normalized temperatures: (a) $T_0 = 1$, (b) $T_0 = 0.7$, and (c) $T_0 = 0.4$.

field V_η. These observations were in agreement with the experimental images at various stages of evolution [56]. Moreover, on the symmetry axis ($\xi = 1/2$), plotting the particle current density (see Figure 19.47) for various maximum reduced temperatures, a multipeak structure as in Figure 19.41a and b can be observed. Increasing the value of the control parameter T_0, we conclude the following: (a) the arrival time of the first peak is decreasing; and (b) the ratio between the first and the second maximum is increasing. Moreover, we conclude that the plume splitting is a hydrodynamic process, being similar with the propagation of a Gaussian perturbation. We note that similar numerical results were obtained in Reference 82 for a cylindrical symmetry, but assuming the adiabatic expansion.

We note that, while the simple planar self-similar solutions (e.g., see Reference 75) predict a constant value of the current density for $\tau \to \infty$; in our model, the theoretical dependence is in a good qualitative agreement with the experimental data given in Figure 19.39. Similar theoretical results were obtained by Murakami et al. [77], where also a Gaussian dependence on position of the plasma density is considered; for comparison, in Figure 19.48, the dependencies on time of the plasma particle current density show the same behavior for an arbitrary position ξ. We note that in Reference 77, the self-similar solutions of Euler

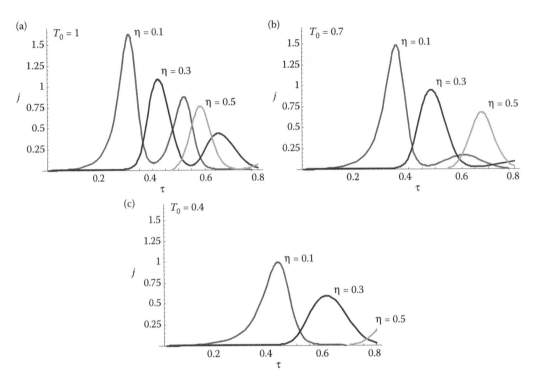

FIGURE 19.47 The Particle current density on the symmetry axis ($\xi = 1/2$), as resulted from the numerical simulations for various maximum reduced temperatures: (a) $T_0 = 1$, (b) $T_0 = 0.7$, and (c) $T_0 = 0.4$.

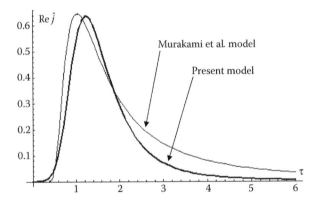

FIGURE 19.48 Comparison between the fractal hydrodynamic model and the one proposed by M. Murakami et al. *Phys. Plasmas*, 12:062706, 2005.

equation were derived for the isothermal and adiabatic plasma expansion, where the ion acceleration is given by the electrostatic potential and by pressure, respectively. In our case, it is not required to specify the type of the expansion, since the fractal force (associated with the fractal potential gradient) is responsible for the acceleration.

The two plasma structures that were experimentally evidenced can be mathematically described by using two different initial velocities which may correspond to two electron species, belonging to hot and cold Maxwellian velocity distributions (having the temperatures T_h and T_c) [107]. Under these circumstances, in Figure 19.49, the time dependence of the current density of plasma particle is given as the

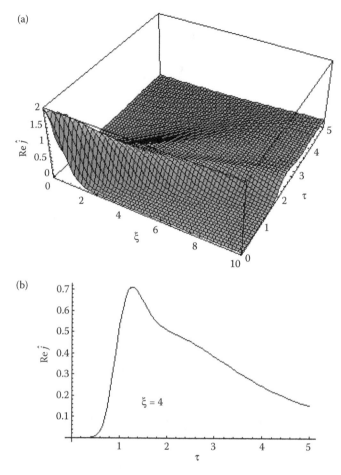

FIGURE 19.49 Time dependence of the current density as a sum of two current densities of normalized speeds $c_h = 3$ and $c_c = 1$.

superposition of two current densities of normalized velocities $c_h = 3$ and $c_c = 1$. Such behavior is in a good agreement with the previous experimental data. For example, in Figure 19.50, the current recorded at 14 mm from the sample (see Figure 19.39) and the theoretical dependence (see Figure 19.49) are plotted for comparison. We note here that, for a better fitting with experimental data, the maximum values of the current densities were arbitrarily chosen having the ratio 1:5.

Consequently, the analytical solutions (19.352) and (19.353) qualitatively describe the plasma expansion into vacuum for a microseconds timescale. The fact that the same equations system describes the plasma expansion process for two different timescales (tens of nanoseconds and microseconds, respectively) shows the universality of a fractal hydrodynamic model. By comparison with the usual treatment of the microparticle motion (Euclidean, Riemannian, stochastic, and quantum approximations), the microparticle motion on fractal curves is a more consistent new approximation.

19.4.4 Conclusions

The main conclusions of this section are the following:

1. The expansion of LPP was experimentally studied by measuring the target current. Positive charging arises through the electrons escaping to the grounded chamber, while the negative one is given by

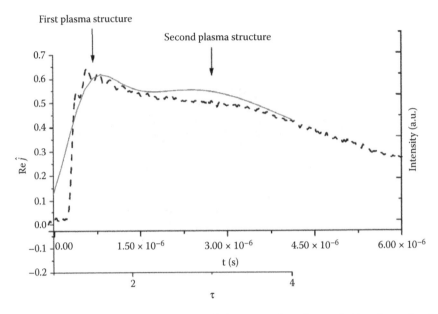

FIGURE 19.50 Comparison between the current recorded at 14 mm from the sample (dotted curve) and the theoretical dependence (continuous curve).

   the ion contribution. A double-peak structure of the positive part is observed, being connected with
   the laser beam energy.

2. By applying an external potential on the target, the time of the electronic peak is not changed, but
   its value is decreasing both for positive and negative polarizations. In this case, the ionic part of the
   transient target current shows an oscillatory behavior that can be explained by oscillatory motion
   of the ions in the applied electric field with the plasma ion frequency.

3. The dependencies of time-integrated target current versus the external applied voltage confirm the
   existence of cold and hot electrons.

4. A model of plasma expansion was theoretically built by assuming that particle motions take place
   on continuous and nondifferentiable curves. Thus (a) for the one-dimensional case, the analytical
   solutions of a fractal hydrodynamic model are obtained; they impose complex velocity, current density, potential, and force fields. The global dynamics are induced by fractality, while the deterministic
   dynamics, for example, the uniform motions, are induced by the absence of fractal force. It results in
   a polarization mechanism through which the particle is preserving the motion characteristics (symmetrical distribution of the force field with respect to a plane that contains the particle, at any time).
   The presence of an external perturbation "breaks" the symmetrical distribution of the force field and
   excites a specific mode of matter structuring; (b) the interaction of a Gaussian speed scalar potential
   with an infinite wall potential gives local maxima and minima that can be a reasonable description
   of the multipeak structure of the transient target current; and (c) assuming that the fractal potential is connected with the ideal gas pressure, the ablation plasma expansion is studied by numerical
   simulations. The theoretical results are in reasonable agreement with the experimental ones.

References

1. M. Agop, D. Alexandroaei, A. Cerepaniuc, and S. Bacaita. El Naschie's $\varepsilon^{(\infty)}$ space–time and patterns
   in plasma discharge. *Chaos Solitons Fractals*, 30:470–489, 2006.

2. M. Agop, D. G. Dimitriu, O. Niculescu, E. Poll, and V. Radu. Experimental and theoretical evidence for the chaotic dynamics of complex structures. *Phys. Scr.*, 87:045501, 2013.

3. M. Agop, N. Forna, I. Casian-Botez, and I. C. Bejenariu. New theoretical approach of the physical processes in nanostructures. *J. Comput. Theor. Nanosci.*, 5:483–489, 2008.

4. M. Agop, G. V. Munceleanu, O. Niculescu, and T. Dandu-Bibire. Static and free time-dependent fractal system through an extended hydrodynamic model of the scale relativity theory. *Phys. Scr.*, 82:015010, 2010.

5. M. Agop, P. Nica, S. Gurlui, C. Focsa, D. Magop, and Z. Borsos. The chaotic atom model via a fractal approximation of motion. *Phys. Scr.*, 84:045017, 2011.

6. M. Agop, P. Nica, O. Niculescu, and D. G. Dimitriu. Experimental and theoretical investigations of the negative differential resistance in a discharge plasma. *J. Phys. Soc. Jpn.*, 81:064502, 2012.

7. M. Agop, P. E. Nica, S. Gurlui, C. Focsa, V. P. Paun, and M. Colotin. Implications of an extended fractal hydrodynamic model. *Eur. Phys. J. D*, 56:405–419, 2010.

8. M. Agop, O. Niculescu, A. Timofte, L. Bibire, A. S. Ghenadi, A. Nicuta, C. Nejneru, and G. V. Munceleanu. Non-differentiable mechanical model and its implications. *Int. J. Theor. Phys.*, 49:1489–1506, 2010.

9. D. Alexandroaei and M. Sanduloviciu. Dynamics of a double layer formed in the transition region between two negative plasma glows. *Phys. Lett. A*, 122:173–177, 1987.

10. S. Ameer-Beg, W. Perrie, S. Rathbone, J. Wright, W. Weaver, and H. Champoux. Femtosecond laser microstructuring of materials. *Appl. Surf. Sci.*, 127–129:875–880, 1998.

11. E. Ammelt, D. Schweng, and H. G. Purwins. Spatio-temporal pattern formation in a lateral high-frequency glow discharge system. *Phys. Lett. A*, 179:348–354, 1993.

12. S. I. Anisimov, B. S. Luk'yanchuk, and A. Luches. An analytical model for three-dimensional laser plume expansion into vacuum in hydrodynamic regime. *Appl. Surf. Sci.*, 96–98:96–98, 1996.

13. A. Antici, C. Marin, and M. Agop. *Chaos through Stochastization*. Ars Longa, Iasi, 2009.

14. J. Argyris and C. Ciubotariu. A chaotic gun effect for relativistic charged particles. *Chaos Solitons Fractals*, 11:1001–1014, 2000.

15. J. Argyris, C. Marin, and C. Ciubotariu. *Physics of Gravitation and the Universe*. Tehnica-Info and Spiru Haret, Iasi, 2006.

16. D. L. Aronstein and C. R. Stroud. Fractional wave-function revivals in the infinite square well. *Phys. Rev. A*, 55:4526–4537, 1997.

17. Y. Astrov, E. Ammelt, S. Teperick, and H. G. Purwins. Hexagon and stripe turing structures in a gas discharge system. *Phys. Lett. A*, 283:184–190, 1996.

18. Y. A. Astrov and H. G. Purwins. Plasma spots in a gas discharge system: Birth, scattering and formation of molecules. *Phys. Lett. A*, 283:349–354, 2001.

19. C. Avram, R. Schrittwieser, and M. Sanduloviciu. Nonlinear effects in the current–voltage characteristic of a low-density Q-machine plasma: I. Related to the potential relaxation instability. *J. Phys. D: Appl. Phys.*, 32:2750–2757, 1999a.

20. C. Avram, R. Schrittwieser, and M. Sanduloviciu. Nonlinear effects in the current–voltage characteristic of a low-density Q-machine plasma: II. Related to the electrostatic ion–cyclotron instability. *J. Phys. D: Appl. Phys.*, 32:2758–2762, 1999b.

21. S. Ş. Bayın. On the consistency of the solutions of the space fractional Schrödinger equation. *J. Math. Phys.*, 53:042105, 2012.

22. M. F. Becker, J. R. Brock, H. Cai, D. E. Henneke, J. W. Keto, J. Lee, W. T. Nichols, and H. D. Glicksman. Metal nanoparticles generated by laser ablation. *Nanostruct. Mater.*, 10:853–863, 1998.

23. J. L. Boeuf, L. C. Pitchford, and K. H. Schoenbach. Predicted properties of microhollow cathode discharge in xenon. *Appl. Phys. Lett.*, 86:071501, 2005.

24. A. Bogaerts, Z. Chen, R. Gijbels, and A. Vertes. Laser ablation for analytical sampling: What can we learn from modeling? *Spectrochim. Acta B*, 58:1867–1893, 2003.

25. J. L. Borowitz, S. Eliezer, Y. Gazit, M. Givon, S. Jackel, A. Ludmirsky, D. Salzmann, E. Yarkoni, A. Zigler, and B. Arad. Temporally resolved target potential measurements in laser-target interactions. *J. Phys. D: Appl. Phys.*, 20:210–214, 1987.

26. R. A. Bosch and R. L. Merlino. Sudden jumps, hysteresis and negative resistance in an argon plasma discharge. Part I: Discharges with no magnetic field. *Contrib. Plasma Phys.*, 26:1–12, 1986.

27. A. V. Bulgakov and N. M. Bulgakova. Gas-dynamic effects of the interaction between a pulsed laser-ablation plume and the ambient gas: Analogy with an underexpanded jet. *J. Phys. D: Appl. Phys.*, 31:693–703, 1998.

28. N. M. Bulgakova, A. V. Bulgakov, and O. F. Bobrenok. Double layer effects in laser-ablation plasma plumes. *Phys. Rev. E*, 62:5624–5635, 2000.

29. C. Chan and N. Hershkowitz. Transition from single to multiple double layers. *Phys. Fluids*, 25:2135–2137, 1982.

30. C. Charles. A review of recent laboratory double layer experiments. *Plasma Sources Sci. Technol.*, 16:R1–R25, 2007.

31. F. F. Chen. *Introduction to Plasma Physics*. Plenum, New York, 1974.

32. F. F. Chen. *Introduction to Plasma Physics,* 2nd edition. Plenum Press, New York, 1984.

33. S. Chiriac, M. Aflori, and D. G. Dimitriu. Investigation of the bistable behaviour of multiple anodic structures in dc discharge plasma. *J. Optoelectron. Adv. Mater.*, 8:135–138, 2006.

34. S. Chiriac, D. G. Dimitriu, and M. Sanduloviciu. Type I intermittency related to the spatiotemporal dynamics of double layers and ion-acoustic instabilities in plasma. *Phys. Plasmas*, 14:072309, 2007.

35. V. Chiroiu, P. Stiuca, L. Munteanu, and S. Danescu. *Introduction in Nanomechanics*. Romanian Academy Publishing House, Bucharest, 2005.

36. C. Ciubotariu, V. Stancu, and C. Ciubotariu. A chaotic–stochastic model of an atom. In R. L. Amoroso, G. Hunter, M. Kafatos, and J. P. Vigier, eds. *Gravitation and Cosmology: From the Hubble Radius to the Planck Scale (Fundamentals Theories of Physics)*, vol. 126, p. 357–366. Kluwer, Dordrecht, 2003.

37. L. Conde and L. Leon. Multiple double layers in a glow discharge. *Phys. Plasmas*, 1:2441–2447, 1994.

38. J. Cresson. Non-differentiable variational principles. *J. Math. Anal. Appl.*, 307:48–64, 2005.

39. J. Cresson. Non-differentiable deformations of R^n. *Int. J. Geom. Methods Mod. Phys.*, 3:1395–1415, 2006.

40. C. P. Cristescu. *Nonlinear Dynamics and Chaos: Theoretical Fundaments and Applications*. Romanian Academy Publishing House, Bucharest, 2008.

41. D. G. Dimitriu. Physical processes related to the onset of low-frequency instabilities in magnetized plasma. *Czech. J. Phys.*, 54:C468–C474, 2004.

42. D. G. Dimitriu, M. Aflori, L. M. Ivan, C. Ionita, and R. W. Schrittwieser. Common physical mechanism for concentric and non-concentric multiple double layers in plasma. *Plasma Phys. Control Fusion*, 49:237–248, 2007.

43. D. G. Dimitriu, M. Aflori, L. M. Ivan, V. Radu, E. Poll, and M. Agop. Experimental and theoretical investigations of plasma multiple double layers and their evolution to chaos. *Plasma Sources Sci. Technol.*, 22:035007, 2013.

44. D. G. Dimitriu, V. Ignatescu, C. Ionita, E. Lozneanu, M. Sanduloviciu, and R. W. Schrittwieser. The influence of electron impact ionizations on low frequency instabilities in magnetized plasma. *Int. J. Mass Spectrom.*, 223–224:141–158, 2003.

45. D. G. Dimitriu, C. Ionita, and R. Schrittwieser. Nonlinear effects related to the simultaneous excitation of three instabilities in magnetized plasma. *Contrib. Plasma Phys.*, 51:554–559, 2011.

46. D. Doria, A. Lorusso, F. Belloni, V. Nassisi, L. Torrisi, and S. Gammino. A study of the parameters of particles ejected from a laser plasma. *Laser Part. Beams*, 22:461–467, 2004.

47. P. E. Dyer. Electrical characterization of plasma generation in KrF laser Cu ablation. *Appl. Phys. Lett.*, 55:1630–1632, 1989.

48. M. S. El Naschie, O. E. Roessler, and I. Prigogine, eds. *Quantum Mechanics, Diffusion and Chaotic Fractals*. Elsevier, Oxford, 1995.

49. Y. Elskens and D. Escande. *Microscopic Dynamics of Plasma and Chaos*. IOP Publishing, Bristol, 2002.

50. J. Feder and A. Aharony, eds. *Fractals in Physics*. North Holland, Amsterdam, 1990.

51. D. K. Ferry and S. M. Goodnick. *Transport in Nanostructures*. Cambridge University Press, Cambridge, 1997.

52. R. P. Feynman and A. R. Hibbs. *Quantum Mechanics and Path Integrals*. McGraw-Hill, New York, 1965.

53. X. Gao and M. Xu. Some physical applications of fractional Schrödinger equation. *J. Math. Phys.*, 47:082104, 2006.

54. D. B. Geohegan. Diagnostics and characteristics of pulsed laser deposition laser plasmas. In D. B. Chrisey and G. K. Hubler, eds. *Pulsed Laser Deposition of Thin Films*, p. 115–165. Wiley, New York, 1994.

55. J. F. Gouyet. *Physique et Structures Fractals*. Masson, Paris, 1992.

56. S. Gurlui, M. Agop, P. Nica, M. Ziskind, and C. Focsa. Experimental and theoretical investigations of a laser-produced aluminium plasma. *Phys. Rev. E*, 78:026405, 2008.

57. S. Gurlui, M. Agop, M. Strat, G. Strat, S. Bacaita, and A. Cerepaniuc. Some experimental and theoretical results on the anodic patterns in plasma discharge. *Phys. Plasmas*, 13:063503, 2006.

58. S. S. Harilal, C. V. Bindhu, M. S. Tillack, F. Najmabadi, and A. C. Gaeris. Internal structure and expansion dynamics of laser ablation plumes into ambient gases. *J. Appl. Phys.*, 93:2380–2388, 2003.

59. R. Herrmann. *Fractional Calculus: An Introduction for Physicist*. World Scientific, Singapore, 2011.

60. S. Iizuka, P. Michelsen, J. Rasmussen, R. Schrittwieser, R. Hatakeyama, K. Saeki, and N. Sato. Dynamics of a potential barrier formed on the tail of a moving double layer in a collisionless plasma. *Phys. Rev. Lett.*, 48:145–148, 1982.

61. T. Intrator, J. Menard, and N. Hershkowitz. Multiple magnetized double layers in the laboratory. *Phys. Fluids B*, 5:806–811, 1993.

62. C. Ionita, D. G. Dimitriu, and R. W. Schrittwieser. Elementary processes at the origin of the generation and dynamics of multiple double layers in DP machine plasma. *Int. J. Mass Spectrom.*, 233:343–354, 2004.

63. L. M. Ivan, G. Amarandei, M. Aflori, M. Mihai-Plugaru, C. Gaman, D. G. Dimitriu, and M. Sanduloviciu. Experimental observation of multiple double layers structures in plasma: II. Non-concentric multiple double layers. *IEEE Trans. Plasma Sci.*, 33:544–545, 2005.

64. E. A. Jackson. *Perspectives in Nonlinear Dynamics*. Cambridge University Press, Cambridge, 1991.

65. A. M. Komashko, M. D. Feit, and A. M. Rubenchik. Modeling of long-term behavior of ablation plumes produced with ultrashort laser pulses. *Proc. SPIE*, 3935:97–103, 2000.

66. L. Landau and E. M. Lifshitz. *Fluid Mechanics*, 2nd edition. Butterworth-Heinemann, Oxford, 1987.

67. N. Laskin. Fractional Schrödinger equation. *Phys. Rev. E*, 66:056108, 2002.

68. E. Lozneanu, V. Popescu, and M. Sanduloviciu. Negative differential resistance related to the self-organization phenomena in dc gas discharge. *J. Appl. Phys.*, 92:1195–1199, 2002.

69. E. Lozneanu and M. Sanduloviciu. Self-organization scenario acting as physical basis of intelligent complex systems created in laboratory. *Chaos Solitons Fractals*, 30:125–132, 2006.

70. S. Mahmood and A. Mushtaq. Quantum ion acoustic solitary waves in electron–ion plasmas: A Sagdeev potential approach. *Phys. Lett. A*, 372:3467–3470, 2008.

71. B. B. Mandelbrot. *Multifractals and 1/f Noise*. Springer, New York, 1999.

72. B. B. Mandelbrot. *The Fractal Geometry of Nature*. Freeman, San Francisco, 1983.

73. Y. Matsunaga and T. Kato. Simple model analysis of hysteresis phenomenon of gas discharge plasma. *J. Phys. Soc. Jpn.*, 66:115–119, 1997.

74. S. S. Moiseev, V. Oraevsky, and V. Pungin. *Non-Linear Instabilities in Plasmas and Hydrodynamics*. IOP Publishing, Bristol, 1999.

75. P. Mora. Plasma expansion into a vacuum. *Phys. Rev. Lett.*, 90:185002, 2003.

76. A. I. Morozov. *Introduction to Plasma Dynamics*. CRC Press, Boca Raton, FL, 2012.

77. M. Murakami, Y.-G. Kang, K. Nishikara, S. Fujioka, and H. Nishimura. Ion energy spectrum of expanding laser-plasma with limited mass. *Phys. Plasmas*, 12:062706, 2005.

78. I. V. Nemchinov. Expansion of a tri-axial gas ellipsoid in a regular behavior. *J. Appl. Math. Mech.*, 29:134–140, 1965.

79. O. A. Nerushev, S. A. Novopashin, V. V. Radchenko, and G. I. Sukhinin. Spherical stratification of a glow discharge. *Phys. Rev. E*, 58:4897–4902, 1998.

80. P. Nica, M. Agop, S. Gurlui, C. Bejinariu, and C. Focsa. Characterization of aluminium laser produced plasma by target current measurements. *Jpn. J. Appl. Phys.*, 51:106102, 2012.

81. P. Nica, M. Agop, S. Gurlui, and C. Focsa. Oscillatory Langmuir probe ion current in laser-produced plasma expansion. *Europhys. Lett.*, 89:65001, 2010.

82. P. Nica, M. Agop, S. Miyamoto, S. Amano, A. Nagano, T. Inoue, E. Poll, and T. Mochizuki. Multi-peak structure of the ion current in laser produced plasma. *Eur. Phys. J. D*, 60:317–323, 2010.

83. P. Nica, P. Vizureanu, M. Agop, S. Gurlui, C. Focsa, N. Forna, P. D. Ioannou, and Z. Borsos. Experimental and theoretical aspects of aluminium expanding laser plasma. *Jpn. J. Appl. Phys.*, 48:066001, 2009.

84. O. Niculescu, D. G. Dimitriu, V. P. Paun, P. D. Matasaru, D. Scurtu, and M. Agop. Experimental and theoretical investigations of a plasma fireball dynamics. *Phys. Plasmas*, 17:042305, 2010.

85. A. Nikitov and V. Ouvanov. *Élémentes de la Théorie des Functions Spéciales*. Mir, Moscow, 1974.

86. L. Nottale. Fractals and the quantum theory of spacetime. *Int. J. Mod. Phys. A*, 4:5047–5117, 1989.

87. L. Nottale. *Fractal Space–Time and Microphysics: Towards a Theory of Scale Relativity*. World Scientific, Singapore, 1993.

88. L. Nottale. Scale relativity, fractal space–time and quantum mechanics. *Chaos Solitons Fractals*, 4:361–388, 1994.

89. L. Nottale. On the transition from the classical to the quantum regime in fractal space–time theory. *Chaos Solitons Fractals*, 25:797–803, 2005.

90. L. Nottale. *Scale Relativity and Fractal Space–Time: A New Approach to Unifying Relativity and Quantum Mechanics*. Imperial College Press, London, 2011.

91. O. Onicescu. *Invariantive Mechanics and Cosmology*. Romanian Academy Publishing House, Bucharest, 1974.

92. A. Piel. *Plasma Physics: An Introduction to Laboratory, Space, and Fusion Plasma*. Springer, Berlin, 2010.

93. V. Pohoata, G. Popa, R. Schrittwieser, C. Ionita, and M. Čerček. Properties and control of anode double layer oscillations and related phenomena. *Phys. Rev. E*, 68:016405, 2003.

94. G. Popa and L. Sirghi. *Fundamentals of Plasma Physics*. Alexandru Ioan Cuza University Publishing House, Iasi, 2000.

95. S. Popescu. Turing structures in dc gas discharge. *Europhys. Lett.*, 73:190–196, 2006.

96. C. Radehaus, T. Dirksmeyer, H. Willebrand, and H. G. Purwins. Pattern formation in gas discharge systems with high impedance electrodes. *Phys. Lett. A*, 125:92–94, 1987.

97. L. J. Radziemski. From laser to LIBS, the path of technology development. *Spectrochim. Acta B*, 57:1109–1113, 2002.

98. J. J. Rasmussen and R. W. Schrittwieser. On the current-driven electrostatic ion–cyclotron instability: A review. *IEEE Trans. Plasma Sci.*, 19:457–501, 1991.

99. M. Sanduloviciu and E. Lozneanu. On the generation mechanism and the instability properties of anode double layers. *Plasma Phys. Control Fusion*, 28:585–595, 1986.

100. M. Sanduloviciu, E. Lozneanu, and S. Popescu. On the physical basis of pattern formation in nonlinear systems. *Chaos Solitons Fractals*, 17:183–188, 2003.

101. A. S. Sharma and P. Kaw, eds. *Nonequilibrium Phenomena in Plasmas*. Springer, Berlin, 2005.

102. W. Shi, R. H. Stark, and K. H. Schoenbach. Parallel operation of microhollow cathode discharge. *IEEE Trans. Plasma Sci.*, 27:16–17, 1999.

103. B. Song, N. D'Angelo, and R. L. Merlino. On anode spots, double layers and plasma contactors. *J. Phys. D: Appl. Phys.*, 24:1789–1795, 1991.

104. M. Strat, G. Strat, and S. Gurlui. Ordered plasma structures in the interspace of two independently working discharges. *Phys. Plasmas*, 10:3592–3600, 2003.

105. C. Ursu, S. Gurlui, C. Focsa, and G. Popa. Space- and time-resolved optical diagnosis for the study of laser ablation plasma dynamics. *Nucl. Instrum. Methods Phys. Res. B*, 267:446–450, 2009.

106. P. Weibel, G. Ord, and O. E. Roessler, eds. *Space Time Physics and Fractality*. Springer, Wien, New York, 2005.

107. L. M. Wickens, J. E. Allen, and P. T. Rumsby. Ion emission from laser-produced plasma with two electron temperatures. *Phys. Rev. Lett.*, 41:243–246, 1978.

108. H. E. Wilhelm. Hydrodynamic model of quantum mechanics. *Phys. Rev. D*, 1:2278–2285, 1970.

109. A. L. Zanin, E. L. Gurevich, A. S. Moskalenko, H. U. Bodeker, and H. G. Purwins. Rotating hexagonal pattern in a dielectric barrier discharge system. *Phys. Rev. E*, 70:036202, 2004.

110. O. C. Zienkiewicz and R. L. Taylor. *The Finite Element Method*, 4th edition, vol. 2. McGraw-Hill, London, 1991.

20

Plasma Harmonic and Overtone Coupling

Victor J. Law

20.1 Introduction

This chapter considers the complementary phenomena of overtones and partial signals that are present in technological low-pressure and atmospheric-pressure plasma jets. The term "technological" is used to define plasmas that are designed for the manufacturing of semiconductor devices [1], modification of engineering surfaces [2], and the treatment of biological material [3]. Equivalent electrical models that describe the generation of integer harmonics of the fundamental drive frequency (f_o) have been considered for many years and are now a means of plasma process control [4–9]. The physical origin of these plasma harmonic arises from the rapid oscillatory momentum of electrons and ions within the imposed electric field of the plasma ion sheath(s) that separates the plasma bulk from the reactor walls [10–13]. The plasma may also be considered to be a nonlinear power device and therefore standard radio frequency (RF) terminology is used throughout this chapter to describe the observations and measurements. Indeed, the harmonic distortion of a pure tone sinusoidal signal may be compared with the behavior of a thermionic valve or diode valve* and the solid-state diode mixer. More recently, a number of papers have proposed that low-frequency (kHz) generated harmonics may also be used to impart energy into the plasma [14,15].

The aim of this chapter is to provide an insight into the interaction (coupling) of electrically generated harmonics and sound-pressure generated acoustic overtones that have been recently presented for linear-field atmospheric-pressure plasma jet that is electrically driven in the kilohertz frequency range. This is achieved by organizing the chapter as follows: Section 20.2 reviews the Taylor† and Maclaurin‡ expansion

---

* Sir John Ambrose Fleming (1845–1945) patented the thermionic valve in 1904.
† Originally discovered by the mathematician James Gregory (1638–1675).
‡ Also attributed to Colin Maclaurin who extensively used the expansion around the point $x = 0$ (1698–1746).

theorems, followed by Fourier[*] analysis of electrical waveforms to explain many of the observed nonlinear characteristic signals emanating from the plasma. Implicit within the Taylor, Maclaurin, and Fourier analyses, the phase noise that is present within real-world oscillator's circuits and plasma is absent. Section 20.3 looks at three basic equations for the speed of sound (c_{sound}), Strouhal number (St), and the emission frequencies (f_n) from closed and open air columns that govern the audible acoustic sound production and emission (1–20 kHz) from an atmospheric-pressure plasma jet nozzle.

Using the knowledge outlined in Sections 20.2 and 20.3, Section 20.4 examines the electro-acoustic emission emanating from a helium linear-field atmospheric-pressure plasma jet that is operated within a fundamental drive frequency range of 3–16 kHz. This drive frequency range and the generated harmonics of the fundamental overlay the acoustic emission of the plasma jet nozzle. The outcome of the study provides an electro-acoustic coupling model that may be used for other plasma jets operating in the audible frequency range.

Throughout this chapter, many of the presented harmonics and overtone measurements may be found in the author's published peer-reviewed papers and PhD thesis [16].

20.2 Mathematical Representation of Harmonic Series

The production of integer numbers may be mathematically expressed using the Tayler power series expansion that represents a function as a sum of multiple polynomials that are calculated from the function derivatives (1st, 2nd, 3rd, and 4th) about a single point ($x = a$). This is shown in Equation 20.1, where a single point represents the fundamental drive frequency (f_o) and the coefficients $\alpha_{2,3,4}$ and the exponents characterize the nonlinear term at multiples frequencies of f_o:

$$f(x) = a_0 + a_1 x + a_2 x^2 + a_3 x^3 + a_4 x^4 + \cdots \tag{20.1}$$

For the specific case of expansion evaluated at a single point of $x = 0$ (known as the Maclaurin series), the polynomials are calculated from the function derivatives. For the purpose, we are only interested in a harmonic series where the function is $\sin(x)$ and when $f(-x) = f(x)$. In this case, the first four derivatives are $\sin x = 0$, $\cos x = 1$, $-\sin x = 0$, and $-\cos x = -1$. We need only to compute these four derivatives as they have repeat periodic sequence of 4. The polynomial representation of this series is given in Equation 20.2.

$$p(x) = x - \frac{x^3}{3!} + \frac{x^5}{5!} - \frac{x^7}{7!} + \frac{x^9}{9!} \cdots \tag{20.2}$$

The polynomial representation of the $\cos(x)$ function when $f(x) = f(-x)$, using also the first four derivatives ($-\sin x = 0$, $-\cos x = -1$, $\sin x = 0$, and $\cos x = -1$), is given in Equation 20.3:

$$p(x) = 1 - \frac{x^2}{2!} + \frac{x^4}{4!} - \frac{x^6}{6!} + \frac{x^8}{8!} \cdots \tag{20.3}$$

Comparing Equations 20.2 and 20.3, it can be seen that there is a complementary relationship as the sign switching is the same in both polynomial representations and they each fill in the gaps of other. Moreover, the $\cos(x)$ function produces even numbers and the $\sin(x)$ function produces odd numbers.

From a harmonic point of view, the interesting information here is the function symmetry imposes odd and even polynomials. To exemplify this symmetry (and asymmetry), the functions of $f(x) = x^2$ and $f(x) = x^3$ are plotted in Figure 20.1. It is seen here that the geometry argument states that even functions are symmetrical around the y-axis and remains unchanged after reflection, whereas an odd function is rotated by 180° through the origin. The analytical convergence of the expansion series to values along a curve is of little importance as there is no physical meaning relating to harmonic production.

---

[*] The mathematical transformation of time-domain signals into frequency-domain was first attributed to Joseph Fourier (1768–1830).

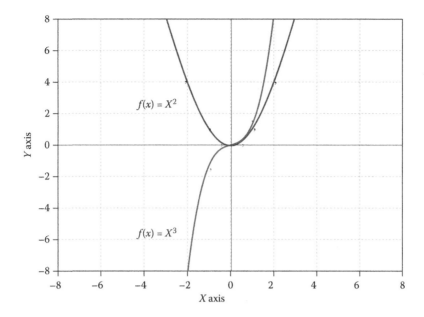

FIGURE 20.1 Odd and even functions plotted in the orthogonal *XY* plane.

20.2.1 Electrical Circuit Harmonic Generation

In the RF and microwave spectrum, a diode does more than simply rectify an AC signal. The diode can also oscillate, mix, detect, and attenuate when used with the appropriate external circuit.

First, let us consider the difference between linear and nonlinear circuits. In the linear amplifier, for example, the power output versus power input is a straight line with no curve or distortion in the current–voltage relationship: the voltage output signal is proportional and predictably linear. For example, a voltage amplifier may be described using Equation 20.4, where A is the gain of the amplifier and ω is the angular frequency ($2\pi f_0$) of the voltage output waveform, which is the same as the voltage input waveform:

$$V_{\text{input}}(\omega t) \cdot A = V_{\text{out}}(\omega t) \tag{20.4}$$

Under these conditions, no harmonics are produced (see Figure 20.2a). Now consider a nonlinear device such as a diode valve or a solid-state diode mixer. In these devices, the input power versus the output power relationship breaks down at a region where the expected increase in output power either increases due to avalanche effects or falls due to saturation of charge carriers. This does not mean that the input power is lost, but is frequency translated as the system is driven into a nonlinear region. This frequency translation may therefore be considered to be the production source of the integer harmonic series.

20.2.1.1 Counting the Harmonic Number

At this point, we need to establish the method of counting integer harmonics as the terminology sometimes gets confused with the method of counting overtones and partials of atmospheric-pressure plasma jets [17,18]. The term "harmonic" has a precise meaning: that of an integer (whole number) multiple of the fundamental frequency of a vibrating object, whereas the terms "overtones" and "partials" (that is generally preferred by musicians) do not include the fundamental frequency. Thus, the correct counting procedure (from an RF engineering view point) is as follows: the fundamental frequency is f_0 and corresponds to a harmonic number of $n = 1$, the harmonics above the fundamental frequency are thus denoted $n = 2, 3, 4 \ldots$. For example, when the input power is increased in the nonlinear device, more and more of

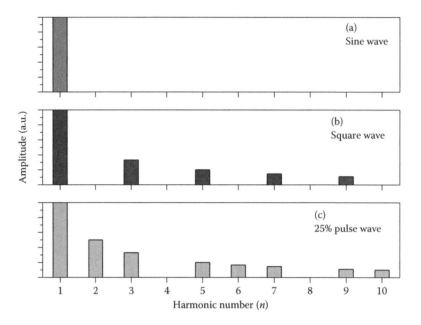

FIGURE 20.2 Simple predication of harmonic number amplitude as a function of voltage waveform. Note the fundamental drive frequency (f_o) is represented here as $n = 1$.

the power is frequency translated into the $2f_o$, $3f_o$ (or, $n = 2$ and 3), and so on. Given this numeric terminology, we are now in a position to examine the production of even and odd harmonics and intermediate products (IP) and two-tone intermodulation.

20.2.1.2 Odd Harmonic Production

It is well known that when a square waveform, or a time pulse-modulated waveform, is generated, the time-dependent voltage level alternates (within a finite discontinuity) between two voltage levels around an average voltage (V/2). This repeating waveform has symmetry both in time and around the average voltage level. Under these conditions, every other even harmonic is suppressed. The even harmonics are missing because of the wave duty cycle (D), which has 50%, or 1/2, of the waveform displaced above the zero value of the time axis. Mathematically, the duty cycle may be expressed as in Equation 20.5, where T is the pulse-on time and P is the total time of the wave period:

$$D = \frac{T}{P} \tag{20.5}$$

In addition, each harmonic will be reduced in amplitude as energy is lost to the frequency translation process as the harmonic series progress to infinity. A simple nonlinear electrical circuit conversion loss mode is given here as the ratio of the harmonic number to f_o. For example, the harmonic $n = 3$ will be 1/3 the amplitude of f_o and the even harmonics ($n = 2, 4 \ldots$) are suppressed to a value $= 0$.

Figure 20.2 graphically compares the harmonic production from a pure sine wave (plot A) with the two concepts of square waveform odd harmonic production and conversion loss (plot B). It can be seen from this figure that $f_o = n = 1$ and has an amplitude $= 1$. The features worth noting here is that only the fundamental is present for the pure sine waveform and the fundamental and odd harmonics are present for the square waveform.

20.2.1.3 Non-50% Duty Cycle Wave-Related Harmonic

When the duty cycle is any percentage other than 50%, the result is a rectangular waveform, with asymmetric on and off time periods within one period of the repeating waveform. Again under these conditions, the wave's harmonic spectrum is related to its duty cycle. For example, if a rectangle wave has a duty cycle of 25%, or 1/4, every fourth harmonic is missing. If the duty cycle is 20%, or 1/5, every fifth harmonic would be missing. Given a duty cycle of 12.5%, or 1/8, then every eighth harmonic would be missing. The results of the 25% pulse waveform are shown in Figure 20.2c. Note again in this figure $f_o = n = 1$ and has an amplitude $= 1$.

20.2.2 Harmonic Dropout

In this section, two electromagnetic mechanisms are presented to explain the phenomena of a single harmonic dropout. The experimental observations were made in the year 2000 on the University College London circular parallel-plate ($\lambda_D = 0.3$ m diameter) OIPT PD80 plasma reactor which is driven at 13.56 MHz [19]. In this reactor, the electrode has four lead-throughs that produce nonstationary radial surface waves (regions of bright intensity) and an associated harmonic dropout of $f_n = 2$ (27.1 MHz) was observed. For a single-electrode lead-through, Lieberman et al. [20] made extensive studies of similar surface waves, both at low and high frequencies (13.56 MHz), and found that surface waves are excited outwards from the vacuum gap (electrode electrical lead-through) and back along through the plasma volume.

20.2.2.1 Transmission-Line Short Circuit Theory

For an electromagnetic surface wave transmission-line model to hold true, it is assumed that the standing wave flows along the electrode surface and the plasma bulk has uniform slab geometry and therefore does not interact with the standing waves. Thus, the geometry of the circular parallel-plate reactor is modeled rather than the plasma. Under these boundary conditions, the inductive (parallel to the electrode surface) electric field needs to be invoked and the electrical length calculated at one-half wavelength, where a short circuit (minimum voltage) is found. For a dropout at 27.1 MHz, the $\lambda/2$ wavelength transmission-line representation is given in Equation 20.6, where c is the velocity of light (2.998×10^8). In this case, the plasma effective velocity factor (v_p) is calculated to be 0.054:

$$v_p = \frac{2\lambda_D \, 27.1 \, \text{MHz}}{c} = 0.054 \tag{20.6}$$

20.2.2.2 Harmonic Phase Cancellation

An alternative electromagnetic theory to the transmission-line model would be that second-order mixing is responsible for the harmonic dropout. The theory is based on two facts: First, the plasma is driven by a single frequency and therefore all the products occupy the same frequency lines and add together to produce the vector sum of their parts. Second, one must consider the swinging phase of the two main sidebands, as the harmonic at any position will view other harmonics as sidebands of itself and incorporate them in its own remixing process. In addition to these two facts, the mixing process is most likely to occur on the even harmonic numbers.

Now consider that the plasma produces a small voltage amplitude at $n = 2$ and that the harmonic $n = 3$ is $\gg n = 2$. Then, on remixing, the third harmonic will view both the second and the fourth as sidebands. This is now going to grossly overmodulate the second harmonic as it is much lower in amplitude, the phase of the sum and difference signals will be inverted, and as the difference signal occupies the same spectral line as the true second harmonic, then when the two signal levels are equal, they will cancel as the vector sum $= 0$. Figure 20.3 shows this phenomenon in the three spectrum analyzer screen shots as a function of argon pressure.

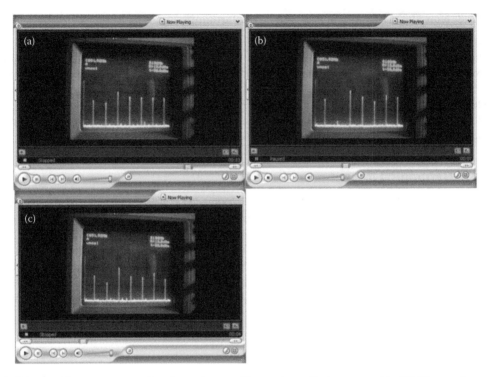

FIGURE 20.3 Three spectrum analyzer screen images of harmonic amplitudes generated by 100 W argon plasma as a function of chamber pressure from a, to b, to c.

20.2.3 RF Mixing

RF mixing is one of the principal processes within the discipline of RF engineering that enables a signal to be processed efficiently. Consequently, the production of IP within a harmonic series and two of their conversion processes (reciprocal mixing and down-conversion) are examined. It is important to note that RF mixing is not like audio mixing where several signals are added together in a linear way to produce an audible sound. RF mixing is a nonlinear process that involves the instantaneous level of one signal affecting another signal. The process involves the two signal levels multiplying together at any instant in time to process a complex waveform consisting of products of the original signals.

20.2.3.1 IP within a Harmonic Series

In Section 20.2.1, we have seen that a simple model of harmonic number amplitude may be expressed by its ratio to f_0. However, given the sufficient frequency power transfer, the generated harmonics are able to alter the neighboring harmonics amplitude by the production of IP. This may be visualized by considering the following argument. Suppose the nonlinear system produces a small level of $n = 2$ harmonic and the $n = 3$ harmonic is $\gg n = 2$ harmonic. Then, on remixing, the third harmonic will view the fundamental as a modulating frequency. This may overmodulate the third harmonic as it is much lower in level, the phase of the sum and difference signals will then alter the $n = 2$ and $n = 4$ harmonic amplitudes as the sum and difference occupies the same spectral line as the true $n = 2$ and $n = 4$ harmonic. This remixing process may be mathematically expressed according to the second-order mixing as $f_0 \pm f_3$. Under plasma conditions, the IP can lead to odd-numbered harmonics being greater in amplitude than the even-numbered harmonics [21].

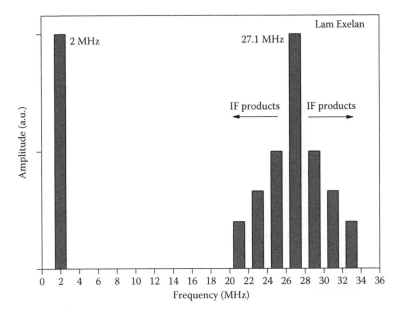

FIGURE 20.4 Reciprocal mixing of two frequencies (2 and 27.1 MHz) to produce sum and difference IF products around the 27.1 MHz drive frequency.

20.2.3.2 Reciprocal Mixing within the Dual-Frequency Plasma Reactor

The second important mixing mechanism that has been observed is reciprocal mixing within a dual-frequency plasma reactor where the two drive frequencies are used to individually control the plasma density and ion bombardment energy in the processing of materials. A good example of this has been observed in the Exelan® parallel-plate plasma etcher (manufactured by Lam Research Corporation), where the two capacitive-coupled power sources (2 and 27.1 MHz) act upon the wafer platen electrode [22]. Under these conditions, a 100% depth modulation of the 27.1 MHz by the 2 MHz bias is imposed. In this case, reciprocal mixing occurs as the sum and difference intermediate frequency (IF) triplets of $\pm\Delta f = 2$ MHz sidebands around the 27.1 MHz frequency and its harmonics. This process of reciprocal mixing is graphically shown in Figure 20.4.

20.2.3.3 Mixer Down-Conversion

The deliberate process of RF mixer down-conversion that allows out-of-band frequency information to be acquired is well established, the purpose of which is to allow greater ease of signal processing at a reduced cost. Note that the opposite of down-conversion (up-conversion) is equally established. This down-conversion process may be employed in the monitoring of the drive frequency or its plasma harmonic, where each of the frequencies is out of band to existing test equipment. In this case, modulation around the carrier of interest (e.g., 13.56 MHz or 2.54 GHz) to another low-frequency band that is in the passband of the test equipment may be achieved. This is performed by injecting the frequency of interest into the input port of the RF mixer and an offset frequency ($\Delta f \sim \pm 120$ kHz) into the local oscillator (LO) port of the mixer. The IF output port then produces the sum (RF + LO) and the difference (LO − RF) between the two signals at 120 kHz. [N.B. Experimentally, 120 kHz may be considered to be the minimum practicable IF due to the −3 dB bandwidth of the RF and LO and the requirement for non-overlap of the two signals. A workable IF criterion is 6 times the −3 dB bandwidth of the LO frequency].

Mathematically, the down-conversion process can be represented by Equations 20.7 through 20.9:

$$V_{RF} = A(t)\cos(w_0 t + \varnothing(t)) \tag{20.7}$$

$$V_{LO} = A_{LO}\cos(\omega_{LO} t) \tag{20.8}$$

Applying the trigonometric identity, we have

$$V_{out} = V_{RF} \times V_{LO} = \frac{A(t)A_{LO}}{2}$$
$$\times \left\{ \cos\varnothing \left(\cos(\omega_{LO}\breve{a} + \omega_0)t + \cos(\omega_{LO} - \omega_0)t\right) - \sin\varnothing \left(\sin(\omega_{LO} + \omega_0)t + \sin(\omega_{LO} - \omega_0)t\right)\right\} \tag{20.9}$$

At this point, a band-pass filter is used to select either the sum or difference so that the down-converted signal is positioned in the passband of the test equipment.

20.3 Overtones and Partials

Traditionally, the terms "overtone" and "partials" have been the preserve of string instruments (violin, guitar, etc.), wind instruments (flute, clarinet, etc.), and percussion instruments (drum, bell, etc.). With the increase in the development of nonthermal atmospheric-pressure plasma jets for processing of materials in the last decade, the terms "overtone" and "partials" have found a new convention outside the world of music. It is now worth clarifying what is meant by the terms "overtone" and "partial."

Overtones and partials are acoustic resonances (or sound vibrations) that are perceived by the human ear as a whole. The frequency and bandwidth of these acoustic signals are directly altered by temperature and the medium which the sound vibration is traveling through to reach the ear. Within these physical constraints, overtones are perceived to form a quasi-harmonic series. In addition, historically, the counting of the overtone number differs from the harmonic number as the first overtone is considered to be twice the frequency of the fundamental or 1 octave higher (where the fundamental is 1 times itself). Partials have an inharmonic ratio of the fundamental. Moreover, the fundamental and overtones are called partials.

Finally, it is worth noting that as the overtones and partials share similar acoustic properties with harmonics, software programs that have been specifically developed for plasma monitoring have found uses in the realm of dance [23] and particle deposition systems [24] and sheep herding [25].

20.3.1 Atmospheric-Pressure Plasma Jet Acoustic Overtone

It has been known for over a century now that atmospheric-pressure plasma (and plasma pressures down to $\sim 1 \times 10^{-1}$ Pa [26]) readily supports acoustic pressure waves (50 Hz to ~ 20 kHz [27]) and ultrasound pressure waves (~ 20 to ~ 600 kHz [28,29]). Within the cold limit of ions, it has been shown that the speed of an ion-acoustic wave is related to the kinetic gas temperature (see Equation 20.10 and Reference 28):

$$c_{sound} = \sqrt{\frac{\gamma R T_{gas}}{M}} \tag{20.10}$$

where C_{sound} is the speed of sound in the gas medium, R is the gas constant (8.314 J K^{-1} mol^{-1}), T_{gas} is the gas temperature in kelvin, M is the molar mass in kilograms per mole of the gas (argon $= 39.94 \times 10^{-3}$ kg mol^{-1} and helium $= 4.00 \times 10^{-3}$ kg mol^{-1}), and γ is the adiabatic constant of the gas (argon and helium $= 1.6$).

With regard to plasma jets, it is reasonable to assume that the role of ions is broadly the same as that of neutral atoms in sound production. Ions, however, can additionally absorb electrical energy from the alternating electric field of the imposed drive frequency. In so doing, the plasma gas gains energy as it expands in the pulse-on period and loses electrical energy as the plasma gas contracts in the pulse-off period of the repetitive drive frequency. Note, in this simple view of expanding and contracting plasma gas, the electrons are not considered due to their low mass. This periodic fluctuation in ion speed generates a longitudinal electronic wind along the direction of gas flow within the reactor tube. Owing to the high collisionality of charged particles and neutrals at atmospheric pressure, the charge particles quickly transfer all momentum energy to the neutrals thus enabling a quasi-synchronized condition to occur. Under these conditions, the electronic wind enhances the velocity of the neutral gas molecules traveling through the reactor to the vibrational antinode (nozzle exit) where the air is free to undergo increasing alternating compression and rarefaction, which is perceived as an increased loudness in the radiated sound energy without affecting the position of the nozzle's fundamental resonant frequency. An estimation of the acoustic emission boundary limit of a plasma jet nozzle can be found by computing the dimensionless Strouhal number (*St*) as defined in Equation 20.11 and Reference 30:

$$St = \frac{f_d D}{v} \tag{20.11}$$

In this equation, f_d is the drive frequency, D is the length scale of the nozzle diameter, and v is the gas velocity. Thus, for $St \sim 1$, the drive frequency is synchronized through the nozzle orifice to the velocity of the gas exiting the nozzle. For low St, the quasi-steady state of the gas dominates the oscillation within the nozzle and thus the frequency of the acoustic emission, whereas at high values of St, the viscosity of the gas dominates the fluid flow.

Table 20.1 provides a limited matrix of published atmospheric-pressure plasma jets of different electrode constructions, along with their Strouhal number information for f_d, v, and D. A reference for each plasma jet is also given. From this limited data, it may be assumed that the acoustic boundary is between $St = 6.8$ and 27. Note, for the kINPen med plasma jet [30], the acoustic response is solely dependent on the 2.5 kHz modulation on the 1.1 MHz drive frequency.

For constant value of C_{sound} and the plasma nozzle acoustic emission range as defined by the St boundary, the zero bandwidth frequency of the plasma nozzle may be computed using Equation 20.12:

$$f_n \approx \frac{nc}{4(L + 0.6)r} \tag{20.12}$$

In this equation, the modulo character (*n*) enables the overtone frequencies to be identified, otherwise L is the physical length of the nozzle, $0.6r$ is the end correction [33], where r is the internal radius of the nozzle. The denominator value 4 describes the quarter-wavelength longitudinal mode of a clarinet. For the quarter-wave model, the nozzle exit aperture defines the maximum pressure vibration (or acoustic antinode), whereas the internal nozzle aperture is the acoustic node due to the flow of gas is being compressed with respect to the nozzle exit. In the case of a flute, or half-wavelength model, the denominator is 2.

TABLE 20.1 Strouhal Data for Six Plasma Jet Nozzle Designs

F_d (kHz)	V (m s^{-1})	D (m)	St	Audio	Electrode Configuration	Reference
0.1	25	0.002	0.008	Yes	Arc	[31]
2.5	36.78	0.0017	0.1	Yes	Cross-field	[30]
3.6	20	0.002	0.36	Yes	Linear-field	[32]
18	5	0.002	1.35	Yes	Linear-field	[13]
19–23	37.5–76.6	0.005	5–6.8	Yes	Arc	[18]
1100	36.78	0.0017	27	No	Cross-field	[30]

The acoustic overtone frequency however has a real bandwidth; this is because gas molecules at any given temperature do not all have the same translational speed (or kinetic energy) but have a distribution of translation speeds. The result is that the gas molecules are moving haphazardly and bouncing off each other and the reactor walls. The probability distribution of these collisions is classically defined according to the Maxwell–Boltzmann distribution function as shown in Equation 20.13:

$$\frac{dN}{Nv} = 4\pi N \left(\frac{M}{2\pi RT}\right)^{3/2} v^2 e^{-(Mv^2/2RT)} \tag{20.13}$$

In Equation 20.13, N is the number of molecules in the sample; dN/Nv is the fraction of molecules with a speed between $v = $ minimum and $v = $ maximum; M is the molecular weight of the gas (kg mol^{-1}); R is the gas constant (8.314 J K^{-1} mol^{-1}); and T is the gas temperature in kelvin. In Figure 20.5, the computed Maxwell–Boltzmann distribution curves for helium, one at 300 K (narrow peak) and the other at 600 K (broad peak), is plotted. In this figure, the molecule translational speed (energy) is plotted on the x-axis and on the y-axis is the probability or the fraction of molecules having that speed or energy. It can be seen that each of the two curves has a prominent peak, which corresponds to the most probable speed of the helium gas molecules (1117 and 1579 m s^{-1}, respectively). The area under each curve is the same and the area equals the total number of molecules. Owing to the skewness of the curves, the translational average speed is located to the right of the peak (1260 and 1782 m s^{-1}, respectively).

Two additional terms that have not been computed here and may be reasonably assumed to affect the distributions of translational speeds and hence energy is the electric field modulation of the ions and the geometry (nozzle diameter and electrode spacing) of the plasma reactor tube. This speed distribution and hence energy distribution is very different from the situation of electromagnetic generated harmonics that travel at the speed of light and thus are not affected by the material and temperature that the harmonics are traveling through.

In Section 20.4, f_o and its harmonics are used to probe the electro-acoustic coupling within a linear-field plasma jet.

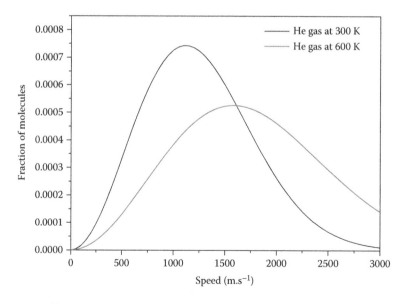

FIGURE 20.5 Maxwell–Boltzmann distribution for helium at 300 and 600 K.

20.4 Electro-Acoustic Coupling within a Linear-Field Atmospheric-Pressure Plasma Jet

20.4.1 Plasma Jet

The linear-field atmospheric-pressure plasma comprises a glass tube of 60 mm in length with a 4.1 mm outside diameter and a 2.7 mm inside diameter (nozzle area = 5.72 mm^2). Here, three copper foil electrode pair geometries are investigated. Each electrode pair is wrapped around the glass tube, each with a width of 10 mm and a linear spacing of 7.5, 10, and 12.5 mm, making a total linear electrode spacing of 27, 30, and 32.5 mm. The linear configurations are such that each electrode pair is separated from the jet nozzle by 5 mm. Helium is used at the carrier gas and is controlled using an electronic mass flow controller at 1 bar pressure. The gas is supplied to the plasma reactor at a rate of 1 and 3 SLM. For these gas flow conditions, the gas velocities are approximately 5 and 0.9 m s^{-1} and the *St* varies from 1.2 to 2, respectively. A schematic of the plasma jet is shown in Figure 20.6.

An impulse frequency power source (Haiden Laboratory Inc.) is used to maintain the plasma over an impulse frequency range of 3, 4, 6, 12, and 16.5 kHz at 5–6.5 kV. As the electric field and gas flow are parallel, this type of plasma jet is referred to as a linear-field device [34]. This electrode configuration is in contrast to the cross-field jet design where the electrodes are coaxially arranged to produce an electric field that is perpendicular to the gas flow. References for both types of electric field can be found in Table 20.1.

A condenser mini-microphone is placed within 2 cm of the plasma plume that is exiting the chamber. In this configuration, the microphone acts as both a near-field emission E-probe and a sound energy sensor, where both measured quantities are distance dependent. The microphone is plugged directly in the sound-card of a Dell laptop computer where a purpose-designed and built National Instruments LabVIEW 5.8 software program is used to capture the frequency spectrum from the electrical and sound energy. The signals are sampled at a rate of 4400 kS s^{-1} over a 1 s period. Table 20.2 provides a matrix of the design of experiments (DoE) described here and in Section 20.4.2.

20.4.2 Results

Experimental measurement results for a 7.5 mm electrode spacing, 6.5 kV supply voltage, and 1 SLM of helium gas are shown in Figure 20.7. In this figure, the 3 and 6 kHz drive frequencies are depicted. Under these conditions and using the knowledge of Equation 20.12, the linear-field plasma jet electrodes are expected to resonate in the region of 15 kHz for the half-wavelength model and 7.5 kHz for the quarter-wavelength model.

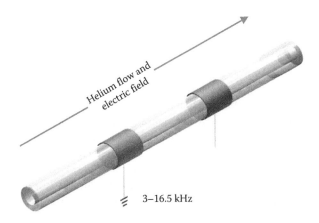

Helium flow and electric field

3–16.5 kHz

FIGURE 20.6 Schematic of the linear-field plasma jet.

TABLE 20.2   Table 20.2 Matrix of DoE (X)

f_0 (kHz)	7.5 mm @ 1 SLM	10 mm @ 1 SLM	12.5 mm @ 3 SLM
3	X		
4			X
6	X		
12	X		X
16.5	X	X	X

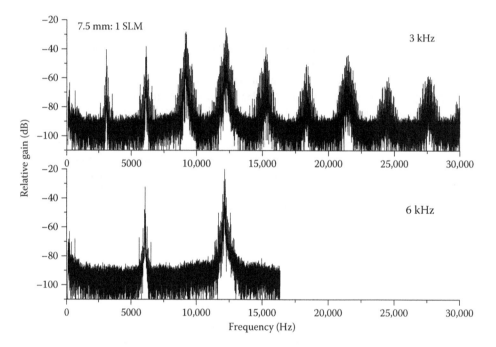

FIGURE 20.7   Frequency spectrum of $f_0 = 3$ and 6 kHz at an electrode spacing of 7.5 mm and 1 SLM of helium gas.

For the 3 kHz drive frequency, the spectral density data reveals that between 6 kHz to 15 kHz the harmonics have a greater amplitude and bandwidth than the fundamental. For $n = 3$ and 4, the harmonic magnitudes are typically enhanced by some 12 dB with respect to f_0 and their -10 dB bandwidths have typically increased by 2 kHz. Above the $n = 4$ harmonic, there is a subsequent fall in magnitude of the harmonics.

For $f_0 = 6$ kHz, there is a similar response of the $n = 2$ harmonic (12 kHz) as seen in the previous experiment. That is, the second harmonic is enhanced both in magnitude and bandwidth.

These electro-acoustic measurements show that there is coupling in the frequency range between 9 and 16 kHz. Moreover, when we consider the two resonance models, the half-wavelength model where the denominator $= 2$ is preferred over the quarter-wavelength model (denominator $= 4$).

To explore this coupling range further, electro-acoustic measurements for 7.5 mm electrode spacing and 1 SLM of helium gas were performed at impulse drive frequencies of $f_0 = 12.5$ and 16.5 kHz. The results are shown in Figure 20.8. In the case of $f_0 = 12.5$ kHz, the magnitude of f_0 is enhanced once again by 12 dB to reach a maximum magnitude of -20 dB with evidence of ± 2.5 kHz sidebands. The signal is also sufficiently strong to raise the noise floor in this region and which partially obscures the sidebands. The $n = 2$ harmonic has a decreased magnitude (-40 dB) with respect to f_0 with clear ± 2.5 kHz sidebands.

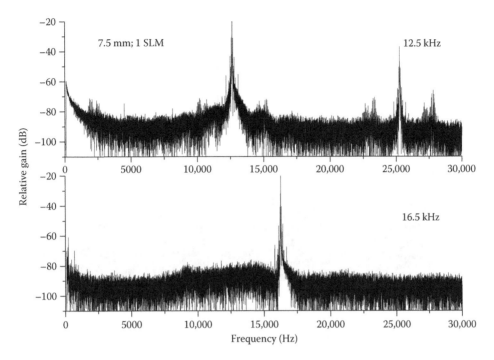

FIGURE 20.8 Frequency spectrum for $f_o = 12.5$ and $16.5\,\text{kHz}$ at a total linear electrode spacing of $27.5\,\text{mm}$ and 1 SLM of helium.

For the $f_o = 16.5\,\text{kHz}$ case (lower spectrum), the drive signal is positioned above a broadband (10–15 kHz) noise signal, which is again raised by some 8–10 dB above the surrounding noise floor. It is also noticed that f_o does not have any sideband noise.

With f_o shifted to a higher frequency, the broadband noise signal that is partly obscured in Figure 20.6 is now identified easily. Moreover, in the absence of $\pm$ sideband noise around f_o, there appears to be also intermodulation processes associated with the broadband noise region. From an acoustic response point of view, the coupling in the 9–16 kHz range strongly suggests the half-wavelength model where a denominator $= 2$ is preferred.

The next series of experimental measurements were designed to investigate the linear increase in electrode spacing from 27 to 30 mm while positioning f_o above the broadband noise structure. Under these conditions, the half-wavelength longitudinal model of the electrode configuration is expected to be of the order of 14 kHz. For this purpose, Figure 20.9 depicts the plasma jet frequency response of the $f_o = 16.5\,\text{kHz}$ at 10 mm electrode spacing and 1 SLM of helium gas.

The electro-acoustic measurements reveal that the broadband noise structure has a spectral range similar to the 27 mm linear electrode spacing experiments as depicted in Figures 20.6 and 20.7. However, within the noise floor, there is a sharp rise in noise at 12.5 kHz, which approximates to the half-wavelength model frequency response. The two spikes around the 25 kHz region are from random signals and take no part in the experiment analysis.

Finally, the varying f_o (4, 6, and 16.5 kHz) is investigated at fixed total electrode spacing (32.5 mm) and fixed helium gas flow of 3 SLM. The results of the electro-acoustic measurements are shown in Figure 20.10 as a triplet of frequency response plots. The first spectra (top plot) of the triplet clearly show the enhanced magnitude of harmonics of $n = 3, 4$, and 5 of the drive frequency ($f_o = 4\,\text{kHz}$). The middle plot shows the $n = 2$ of the drive frequency ($f_o = 6\,\text{kHz}$) is enhanced in magnitude and a broadening of harmonic on top of the raised noise floor is noticed. Lastly, the bottom plot shows the raised noise floor in the 10–15 kHz range with the drive frequency ($f_o = 16\,\text{kHz}$) at a position that is not affected by the noise.

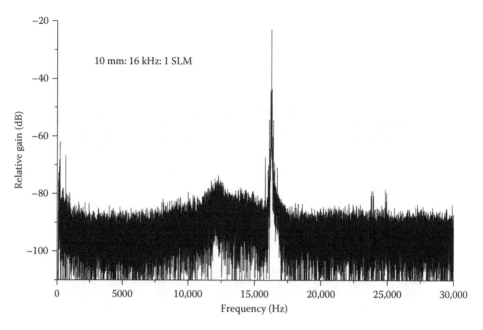

FIGURE 20.9 Frequency spectrum for $f_o = 16.5$ kHz drive frequency at a total linear electrode spacing of 30 mm and a helium flow rate of 1 SLM.

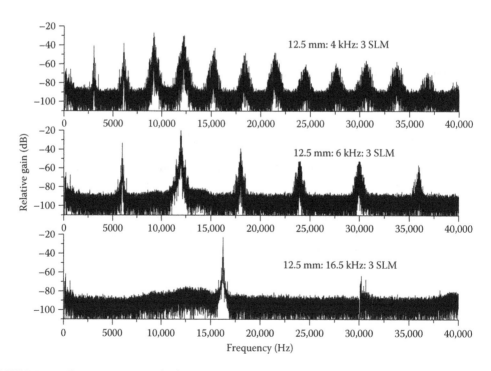

FIGURE 20.10 Frequency spectrum for $f_o = 4, 6,$ and 16.5 kHz at a total linear electrode spacing of 32.5 mm and at a helium flow rate of 3 SLM.

20.4.3 Discussion of Experimental Results

The eight DoE measurements have revealed that the 1–3 SLM of helium gas, at an injection pressure of 1 bar, has little effect on the acoustic response of the linear-field plasma jet.

More informative is the alteration of the total linear dual-electrode configuration from 27 to 32.5 mm.

First, the zero-band acoustic response of the dual electrodes can be predicted using a half-wavelength longitudinal model, suggesting that the helium gas is insufficiently compressed at the gas input to the dual electrodes; thus the gas dynamic behavior action is like a "flute." Comparing this outcome with previously published plasma jets that have a "clarinet-"like acoustics response due to their cross-field or end-field design [17,18,30,31], no additional overtones are observed. It is reasonable to assume that due to the open-ended nature of the electrode pair design of the linear-field plasma jet, there is no physical or gas boundary within the reactor tube; consequently, the overtones and partial merge into one single broadband spectrum.

Second, for this linear-field plasma jet, the movement of the electrode spacing does induce frequency structures within the bandwidth acoustic spectrum without shifting the bandwidth or the frequency registration (9–16 kHz).

In 2011, Law et al. [14] first reported harmonic coupling with the electric circuit resonance of plasma jet, followed by Zaplotink et al. in 2014 [15], who also observed this interaction. In these works, the coupling between plasma jet acoustics and the nonlinear harmonic may be observed when the harmonic frequencies, including f_o, are in the same frequency range as that of the acoustic response of the linear-field plasma jet. It can be put forward that the linear acoustic production by the vibration of neutral gas atoms and molecules is modified by the contribution of ions that make up the electric wind along the direction of gas flow within the reactor tube at each repetitive impulse of the drive frequency and its harmonics.

This nonlinear coupling between the electric drive power circuit and plasma jet acoustics (see Figures 20.7 through 20.10) occurs when the drive frequency (f_o), or a harmonic of f_o, falls within the plasma jet acoustic band. The coupling manifests itself as an increase in both magnitude and bandwidth of the selected harmonic(s). The coupling also allows intermodulation with the selected harmonic. Figure 20.8 shows this clearly where the drive frequency ($f_o = 12.5$ kHz) and at $n = 2$ have ± sidebands. These sidebands mathematically correspond to the periodic time-varying signals at 2 and 2.6 kHz through the second-order relationship $f_{o-n} \pm X$, where X is -2 kHz and $+2.6$ kHz, respectively.

20.5 Summary

Section 20.2 presented both a mathematical construct (Tayler and Maclaurin expansion series) and a graphical representation to define an integer series. Fourier analysis of a time-based pattern in the form of electrical circuit waveforms has also been performed. Implicit within these examples, the fundamental and its integer series have zero bandwidth and so are positioned on a finite frequency line without a phase noise component. It is shown that the three mathematical approaches exhibit symmetry and asymmetry in their representations, which account for the production of odd- and even-number integers.

In addition, it has been shown that electromagnetic transmission-line theory relates to chamber geometry, and phase cancellation due to RF remixing of products may be used for the singular event of harmonic dropout. Intermodulation within a harmonic series, reciprocal mixing between two signals that have no harmonic relation, and mixer down-conversion have also been presented.

Section 20.3 has explored the production of overtones and partials with respect to plasma jets through the calculation of the speed of sound in a gas medium, the dimensionless Strouhal number of the plasma nozzle, and the concept of zero-band quarter-wavelength and half-wavelength longitudinal mode models of plasma jets have been presented.

To elucidate harmonic and overtone coupling, Section 20.4 describes the use of the knowledge presented in Sections 20.2 and 20.3 and has been applied to inform the DoE and measurement for a linear-field plasma jet operating in the 3–16 kHz range. It is found that either the fundamental drive frequency (f_o) or multiple harmonics couple the broadband acoustic noise emanating from the plasma jet reactor tube. In the case of a linear-field plasma jet, the zero-band half-wavelength longitudinal mode model is found to best fit the frequency response of the total linear length of the electrode pair. The nonlinear harmonic and acoustic coupling manifests itself as an increase in both harmonic magnitude and bandwidth. In addition, RF mixing in the form of intermodulation is present.

Acknowledgment

Dr. Law acknowledges the financial support by Enterprise Ireland under grant number CFTD/7/IT/304 extended to visiting Professor Bill W. Graham at Queen's University Belfast in the course performing electro-acoustic measurements.

References

1. H Abe, M Yoneda, and N Fujiwara. Developments of plasma etching technology for fabricating semiconductor devices. *Jpn. J. Appl. Phys.*, 47(3R), 1435–1455, 2008.

2. V J Law, J Mohan, F T O'Neill, A Ivankovic, and D P Dowling. Air-based atmospheric pressure plasma jet removal of FreKote 710-NC prior to composite-to-composite bonding. *Int. J. Adhesion Adhesives*, 54(C), 72–81, 2014.

3. A V Nastuta, I Topala, C Grigoras, V Pohoata, and G Popa. Stimulation of wound healing by helium atmospheric pressure plasma treatment. *J. Phys. D, Appl. Phys.*, 44, 105204, 2012.

4. S Bushman, F Edgar, and I Trachtenberg. Radio frequency diagnostics for plasma etch systems. *J. Electrochem. Soc.*, 144(2), 721–732, 1997.

5. A T-C Koh, N F Thornhill, and V J Law. Principal components analysis of plasma harmonics in endpoint detection of photoresist stripping. *Electron. Lett.*, 35(16), 1383–1385, 1999.

6. A Pagliarani, A J Kenyon, N F Thornhill, E Sirisena, K Lee, and V J Law. Process harmonic pulling in a RIE plasma-tool. *Electron. Lett.*, 42(2), 120–121, 2006.

7. Y Yamazawa, M Nakaya, M Iwata, and A Shimizu. Control of the harmonics generation in a capacitively coupled plasma reactor. *Jpn. J. Appl. Phys.*, 46(11R), 7453–7459, 2007.

8. J V Ringwood, S Lynn, G Bacelli, M Beibei, E Ragnoli, and S Mcloone. Estimation and control in semiconductor etch practice and possibilities. *IEEE Trans. Plasma Sci.*, 23(1), 87–98, 2010.

9. I Batty, M Cooke, and V J Law. Harmonic characterisation of a plasma-tool using a diplexer. *Vacuum*, 52(4), 509–514, 1999.

10. P Linardakis and G Borg. Harmonic and intermodulation distortion output of a radio-frequency plasma capacitor. *IEEE Microw. Compon. Lett.*, 18(3), 164–166, 2008.

11. H Conrads and M Schmidt. Plasma generation and sources. *Plasma Sources Sci. Technol.*, 9, 441–454, 2000.

12. A Schütze, J Y Jeong, S E Babayan, J Park, G S Selwyn, and R F Hicks. The atmospheric-pressure plasma jet: A review and comparison to other plasma sources. *IEEE Trans. Plasma Sci.*, 26(6), 1685–1694, 1998.

13. J L Walsh, F Iza, N B Janson, V J Law, and M G Kong. Three distinct modes in a cold atmospheric pressure plasma jet. *J. Phys. D, Appl. Phys.* 43(7), 075201 (14pp), 2010.

14. V J Law and S D Anghel. Compact atmospheric pressure plasma self-resonant drive circuits. *J. Phys. D, Appl. Phys.*, 45(7), 075202 (14pp), 2012.

15. R Zaplotink, Z Kregar, M Bišćan, A Vesel, U Cvelbar, M Mozetič, and S Milošević. Multiple vs. single harmonic AC-driven atmospheric plasma jet. *Eur. Electron Lett.*, 106(2), 25001 (6pp), 2014.

16. V J Law. Radio frequency plasma power spectroscopy for semiconductor device processing. PhD thesis, University of Ulster, Jordantown, May, 2005.

17. V J Law, C E Nwankire, D P Dowling, and S Daniels. Acoustic emission within an atmospheric helium discharge jet. In C H Skiadas, I. Dimotikalis, and C Skiadas (Eds), *Chaos Theory: Modeling, Simulation and Applications*, pp. 255–264 (World Scientific, Singapore), 2011. ISBN: 9814350338.

18. V J Law, F T O'Neill, and D P Dowling. Evaluation of the sensitivity of electro-acoustic measurements for process monitoring and control of an atmospheric pressure plasma jet system. *Plasma Sources Sci. Technol.*, 20(3), 035024, 2011.

19. V J Law, A Kenyon, D C Clary, and I Batty. A non-invasive RF probe for the study of ionization and dissociation processes in technological plasmas. *J. Appl. Phys.*, 86(8), 4100–4106, 1999.

20. M A Lieberman, J Booth, P Chabert, J M Rax, and M M Turner. Standing wave and skin effects in large-area, high frequency capacitive discharges. *Plasma Sources Sci. Technol.*, 11(3), 283–293, 2002.

21. V J Law, N F Thornhill, A J Kenyon, A Pagliarani, K Lee, M Watkins, and L Lea. Harmonic monitoring of the switched silicon etch process. *J. Phys. D, Appl. Phys.*, 36(7), 2146–2151, 2003.

22. V Milosavljevic, A R Elingboe, C Garman, and J V Ringwood. Real-time plasma control in a dual-frequency, confined plasma etcher. *J. Appl. Phys.*, 103, 083302, (10pp), 2008.

23. V J Law, M Donegan, and B Creaven. Acoustic metrology: From atmospheric plasma to solo percussive Irish dance. *CMSIM J.*, 4, 663–670, 2012.

24. K A McDonnell, V J Law, N J English, P Dobbyn, and D P Dowling. Process control of particle deposition systems using acoustic and electrical response signals. *Adv. Powder Technol.*, 25(5), 1560–1570, 2014.

25. V J Law. Acoustic decoding of a sheep bell and trotters within a herd of sheep. *7th International Chaos Conference, Lisbon, Portugal, June 7–10. Proceedings of Chaos*, 2014, pp. Kaw-L 277–284.

26. M Yasaka, M Takeshita, and R Miyagawa. Detection of supersonic wave emitted from anomalous arc discharge in plasma processing equipment. *Jpn. J. Appl. Phys.*, 39, L1286–L1288, 2000.

27. M Fitaire and T D Mantei. Some experimental results on acoustic wave propagation in a plasma. *Phys. Fluids*, 15(3), 464–469, 1972.

28. T Nakane. Discharge phenomenon in a high-intensity acoustic standing wave field. *IEEE Trans. Plasma Sci.*, 33(2), 356–357, 2005.

29. V S Soukhomlinov, V Y Kolosov, V A Sheverev, and M V Ötügen. Acoustic dispersion in glow discharge plasma: A phenomenological analysis. *Phys. Fluids*, 14(1), 427–429, 2002.

30. V J Law, A Chebbi, F T O'Neill, and D P Dowling. Resonances and patterns within the atmospheric pressure plasma jet kINPen-med. *CMSIM J.*, 1, 3–10, 2014.

31. V J Law and D P Dowling. Active control metrology for preventing induced thermal damage during atmospheric pressure plasma processing of thermal sensitive materials. In A Sanyei, I Zelinka, and O Rossier (Eds), *ISCS 2013: International Symposium on Complex Systems. Emergence of Complexity and Computation*, Vol. 8, Chapter 32, pp. 321–332 (Springer-Verlag, Berlin Heidelberg), 2014. ISBN: 978-3-642-45437-0.

32. N O'Connor and S Daniels. Passive acoustic diagnostics of an atmospheric pressure linear field jet including analysis in the time–frequency domain. *J. Appl. Phys.*, 110(1) 013308, 2011.

33. H Levine and J Schwinger. On the radiation of sound from an unflanged circular pipe. *Phys. Rev.*, 73(4), 383–406, 1948.

34. J L Walsh and M G Kong. Contrasting characteristics of linear-field and cross-field atmospheric plasma jets. *Appl. Phys. Lett.*, 93, 111501, 2008.

V

Chaos in Flows and Turbulence

21

Wave Turbulence in Vibrating Plates

Olivier Cadot,
Michele Ducceschi,
Thomas Humbert,
Benjamin Miquel,
Nicolas Mordant,
Christophe Josserand,
and Cyril Touzé

21.1 Introduction

Turbulence is a general term used for describing the erratic motions displayed by nonlinear systems that are driven far from their equilibrium position and thus display complicated motions involving different time and length scales. Without other precision, the term generally refers to hydrodynamic turbulence, as the main field of research has been directed toward irregular motions of fluids and the solutions of Navier–Stokes equations. During the twentieth century, theoretical developments showed important breakthroughs thanks to the qualitative ideas of Richardson and the quantitative arguments of Kolmogorov that culminated in the so-called K41 theory [25,30,31]. This statistical approach, although giving successful predictions, still faces an irreducible obstacle due to the lack of closure in the infinite hierarchy of moment equations.

Wave turbulence (WT) shares many common ideas with turbulence, in particular as being a statistical theory for out-of-equilibrium systems. A main difference resides in the fact that the persistence of waves is assumed. By considering a sea of weakly interacting dispersive wave trains, it has been shown that a natural asymptotic closure may be derived from the hierarchy of moment equations [2,47,63]. Furthermore, this closure equation, referred to as the kinetic equation, has been shown to admit two sets of stationary solutions [61–63]. The first one is the classical equipartition of energy. Most importantly, the second one describes an energy flux through the scales and thus recovers Richardson's picture of turbulence with a cascade of energy from the injection to the dissipative scales. In between, an inertial range with a conservative Hamiltonian dynamics is assumed.

A salient feature of the WT theory is that solutions of the kinetic equation are analytic, hence yielding accurate predictions for the stationary repartition of energy through scales in a given out-of-equilibrium system. The solutions of the kinetic equation have been derived starting from the mid-1960s and correspond to energy spectra with power-law dependence on the wavenumber. The theory has been successfully applied to capillary [50,62] or gravity [49,61] waves on the surface of liquids, to plasmas [43], to nonlinear optics [22], to magnetohydrodynamics [46], or even to Bose–Einstein condensates [34]. The theoretical bases are now firmly established and the reader is referred to the existing books [45,63] or the review paper [48] for a complete picture of the existing literature.

The application of WT to vibrating plates started with the theoretical derivation of the kinetic equation from the dynamical von Kármán equations [12,32,57] that describe large-amplitude motions of thin plates [20]. Since then, numerous papers have been published covering experimental, theoretical, and numerical materials. In fact, it appears that the vibrating plate is a perfect candidate for a thorough comparison of experiments with theoretical predictions. As compared to other physical systems such as capillary or gravity waves, for example, an experimental setup with a fine control of energy injection and a comfortable range of wavelength is not too difficult to put in place. Second, the available measurement techniques allow one to get a complete and precise picture of the dynamics through the scales, in both the space and frequency domains. Finally, numerical codes with good accuracy have been developed so that all the underlying assumptions of the theory as well as its predictions have been tested, on both the experimental and the numerical levels.

The first experimental papers reported a discrepancy between the theoretical predictions and the measurements [6,41]. An important research effort has then been undertaken in order to understand the origin of these differences. The aim of this chapter is to sum up the most important results obtained so as to give an overview of the solved problems and open issues. The chapter is organized as follows:

- Section 21.2 is devoted to theoretical results. It starts with a description of the von Kármán equations for geometrically nonlinear vibrations of plates, and recalls the main assumptions underlying the mechanical model. The application of the WT theory to the von Kármán model is then overviewed. Stationary solutions of the kinetic equation are given, as well as self-similar laws for nonstationary turbulence.
- Section 21.3 sums up the numerical results obtained for stationary and nonstationary turbulence, including the effect of a simple imperfection.
- Section 21.4 gathers the experimental confrontations to the theory, and reviews all the underlying assumptions of WT and their experimental verifications, in order to explain the origin of the discrepancies first reported.
- A general discussion is given and the conclusions are drawn.

21.2 Theoretical Results

21.2.1 Nonlinear Vibrations of Plates: Von Kármán Model

Mechanical models for large-amplitude, geometrically nonlinear vibrations of thin plates are numerous and rely on a set of assumptions and approximations that give the level of needed accuracy, depending on the vibratory state, the thickness, and the frequency range. The von Kármán model is one of the simplest and relies on strong assumptions that may be violated, especially when the plate is not too thin, or when the frequency range of the vibrations includes high-frequency components. It is largely used as it writes as a simple set of coupled nonlinear partial differential equations that are amenable to analytical approaches [1,32,44,51,56]. Moreover, it is known to produce very accurate results for very thin plates and for vibration amplitudes up to one to five times the thickness [53–55,60].

As a description of geometrically nonlinear vibrations, the material is assumed to be linear elastic. Isotropy is also assumed for simplicity here, so that the material is fully described thanks to its Young modulus E, Poisson ratio ν, and density ρ. The main assumptions for deriving the von Kármán model are the following [12,32,44,52,57]:

- The Kirchhoff–Love kinematical assumptions are fulfilled. This implies in particular that the transverse shear stresses are neglected. Rotation angles are assumed to be small so that the displacement of any point of the mid-surface of the plate is parameterized with the three displacements (u, v, w) only.
- The normal stress along the transverse direction is neglected.
- A particular truncation in the longitudinal part of the Green–Lagrange strain tensor, due to von Kármán [13,36,57], is used.
- Rotatory inertia is neglected.

With these assumptions, a von Kármán model with the three displacements (u, v, w) is obtained, see e.g. [5,29,60]. A last assumption consists of neglecting the in-plane inertia. In this case, an Airy stress function F can be introduced and the two longitudinal displacements (u, v) can be condensated, as proposed by Föppl [1,24]. This model, generally referred to as Föppl–von Kármán, depends only on the two unknowns $w(\mathbf{x}, t)$ and $F(\mathbf{x}, t)$ (where $\mathbf{x}$ is the two-dimensional space position and t the time), and reads, for an undamped perfect plate without external forcing:

$$\rho h \ddot{w} + D \Delta \Delta w = L(w, F), \tag{21.1}$$

$$\Delta \Delta F = -\frac{Eh}{2} L(w, w). \tag{21.2}$$

In these equations, h is the thickness, $D = Eh^3/12(1 - \nu^2)$ the bending rigidity, Δ the Laplacian, and L a bilinear differential operator that reads, in Cartesian coordinates: $L(f, g) = f_{,xx}g_{,yy} + f_{,yy}g_{,xx} - 2f_{,xy}g_{,xy}$.

21.2.2 WT Theory for Vibrating Plates

This section is devoted to the application of the WT theory to the von Kármán equations for large-amplitude vibrations of thin plates. The underlying assumptions needed for deriving the theory are the following:

1. The linearized system is composed of dispersive waves.
2. Weak nonlinearity is assumed, so that the nonlinear terms can be ordered and considered small as compared to the linear ones.
3. A clear separation of time scales exists between the linear oscillations and the nonlinear time upon which energy is nonlinearly exchanged between wavetrains.
4. An inertial range with a conservative dynamics exists.
5. The system is of infinite size.

Assumptions 2 and 3 are correlated and state that the dynamics is composed of a sea of weakly interacting, persistent waves. Formally speaking, this framework should thus be referred to as a "weak, WT theory." However, for the sake of simplicity, we will name it under the general term "wave turbulence" in the remainder of the chapter. Thanks to the second assumption, one can derive a hierarchy of equations for the different moments of the field. An equation for the wave spectrum is eventually deduced using long time asymptotics. This kinetic equation can be obtained equivalently using slightly different arguments: the so-called random phase approximation [45,63] or an asymptotic expansion for the cumulants [48].

More precisely, the von Kármán (VK) equations are first written in Fourier space by introducing the transforms $W_k(t) = \frac{1}{2\pi} \int w(x, t) e^{ikx} \mathrm{d}^2 x$ and $F_k(t) = \frac{1}{2\pi} \int F(x, t) e^{ikx} \mathrm{d}^2 x$ for the two unknown fields, yielding

$$\rho h \ddot{W}_k = -Dk^4 W_k + L_k(w, F), \tag{21.3}$$

$$k^4 F_k = -\frac{Eh}{2} L_k(w, w), \tag{21.4}$$

where L_k denotes the Fourier transform of the bilinear operator L. From the linear terms, the dispersion relation between the wavenumber k and the radian frequency of the wave ω_k is retrieved as

$$\omega_k = \sqrt{\frac{Eh^2}{12\rho(1-\nu^2)}k^2}. \tag{21.5}$$

In order to write the VK equations under its canonical Hamiltonian form, one introduces the momentum $Y_k = \rho\partial_t W_k$, as well as the canonical variables A_k defined by $W_k = \frac{X_k}{\sqrt{2}}\left(A_k + A_{-k}^\star\right)$ and $Y_k = \frac{-i}{\sqrt{2X_k}}\left(A_k - A_{-k}^\star\right)$ with $X_k = 1/\sqrt{\rho\omega_k}$. This results in a formulation with a diagonalized linear part where the nonlinear term appears clearly as a perturbation to the linear wave equation for small wave amplitudes. Finally, separating the dynamics within the linear time scale of the waves and the long time nonlinear interactions allows us to write $A_k = a_k\,e^{i\omega_k t}$, where the variation of the amplitude a_k is slow as compared to the oscillation time $1/\omega_k$. Formally, the linear dynamics of the modes can be suppressed in this slow time scale analysis so that we end up with a kinetic equation for the second-order moment defined by $\langle a_{k_1}a_{k_2}^\star\rangle = n_{k_1}\delta^{(2)}(k_1 + k_2)$, which reads for $n_k \equiv n(k,t)$:

$$\frac{\partial n_k}{\partial t} = I(k). \tag{21.6}$$

The expression of the collision integral $I(k)$ comes from the nonlinear interaction between four waves (see Reference 20 for more details), and reads:

$$I(k) = 12\pi \int |J_{k123}|^2 f_{k123}\delta(k + s_1 k_1 + s_2 k_2 + s_3 k_3)$$
$$\times \delta(\omega_k + s_1\omega_1 + s_2\omega_2 + s_3\omega_3)dk_1\,dk_2\,dk_3, \tag{21.7}$$

where J_{k123} stands for the interaction term itself and f_{k123} is such that

$$f_{k123} = \sum_{s_1,s_2,s_3} n_k n_{k_1} n_{k_2} n_{k_3}\left(\frac{1}{n_k} + \frac{s_1}{n_{k_1}} + \frac{s_2}{n_{k_2}} + \frac{s_3}{n_{k_3}}\right). \tag{21.8}$$

Here, the notation s_i stands for $s_i = \pm 1$ so that the collision integral corresponds to four waves interaction, with both $2 \leftrightarrow 2$ and $3 \leftrightarrow 1$ waves mechanisms [20,45]. As it has been shown for water waves, two distinct types of stationary solutions exist in general for such a kinetic equation. First, the equipartition of energy between the modes, since $n_k \propto 1/\omega_k$ is a trivial root of the integral term (21.7). This is the so-called Rayleigh–Jeans (RJ) spectrum, and it writes in Fourier space:

$$n_k \propto \frac{1}{k^2} \quad \text{giving for the Fourier spectrum of the displacement} \quad \langle|W_k|^2\rangle \propto \frac{1}{k^4}. \tag{21.9}$$

In addition, another solution can be exhibited that exactly removes the full integral as first shown for water waves by Zakharov [61,62] and therefore called the Kolmogorov–Zakharov (KZ) spectrum. It involves a constant flux of energy ϵ that is transferred from the large scales (formally $k = 0$ in the mathematical solution) to the small scales (mathematically $k \to \infty$ and practically toward a scale where dissipation becomes dominant). This solution including a cascade of energy to the small scales reads

$$n_k^{KZ} = C\frac{h\epsilon^{1/3}\rho^{2/3}}{(12(1-\nu^2))^{2/3}}\frac{\ln^{1/3}(k_\star/k)}{k^2} \quad \text{or} \quad \langle|W_k|^2\rangle \propto \frac{\ln^{1/3}(k_\star/k)}{k^4}. \tag{21.10}$$

This solution has a particular structure compared to the usual power laws: indeed, this solution is the same as the RJ spectrum but a logarithm correction. This peculiar structure, similar to the one observed for the nonlinear Schrödinger equation [22], comes from the degeneracy of the KZ solution so that the next order term in the collision integral has to be considered (see References 19 through 21 for more details). This logarithmic correction involves a cut-off wave number $k_\star$ above which the mathematical solution is no more valid. The mathematical function vanishes at $k = k_\star$ so that everything works as if the constant flux ϵ would be absorbed at $k = k_\star$. In fact, since the logarithmic correction is obtained as a second-order expansion, it is valid only for $k \ll k_\star$, so that one expects practically that the spectrum simply decreases rapidly around and above $k_\star$. Such KZ solution can be observed when energy is injected in the system at large scale (small k) and dissipated at small scale (large k) so that one observes a transition between the inertial regime (where the constant flux holds) and the dissipative range.

This KZ spectrum can be written in terms of the energy spectrum, as a function of k, or as a function of ω (or f), following,

$$P_v(k) = \frac{\bar{C}h}{(1 - v^2)^{2/3}} \varepsilon^{1/3} \log^{1/3}\left(\frac{k_\star}{k}\right) \text{ and } P_v(f) = \frac{\bar{C}h}{(1 - v^2)^{2/3}} \varepsilon^{1/3} \log^{1/3}\left(\frac{f^\star}{f}\right), \qquad (21.11)$$

where

$$f^\star = \frac{1}{2\pi} \sqrt{\frac{Eh^2}{12\rho(1 - v^2)}} k_\star^2.$$

We notice that the energy spectrum is almost flat since it varies only through the logarithmic correction.

21.2.3 Nonstationary WT

Other properties of the solutions of the kinetic equation can be derived by considering nonstationary evolutions. In this particular case, one is able to exhibit self-similarity laws that must be fulfilled by the solutions [16,17,23,63]. Let us consider a self-similar solution for the wave spectrum, which depends only on the wavevector modulus, $n(k, t) = t^{-q} f(kt^{-p}) = t^{-q} f(\eta)$. Introducing this expression in the kinetic equation (21.6), and taking into account the expression of $|J_{k123}|^2$ [17,19–21], one obtains the following relationship:

$$-t^{-q-1}\left[qf(\eta) + p\eta f'(\eta)\right] = I(\eta)t^{-3q+2p}, \qquad (21.12)$$

so that a self-similar solution for the wave spectrum exists if the condition $-q - 1 = -3q + 2p$ is satisfied.

Let us now consider two different cases of nonstationary evolution of the WT spectrum for the system, in order to derive another condition that must be fulfilled by the unknowns p and q:

Case 1, forced turbulence: We assume that the plate is forced by a sinusoidal pointwise forcing of constant amplitude and excitation frequency. In this case, the total energy $\xi = \int \omega n_k \, dk$ increases linearly with time so that $\xi \sim t$.

Case 2, free turbulence: The plate is left free to vibrate, given an amount of energy as initial condition. In this case, the total energy is constant so that $\xi \sim t^0$.

Using the self-similar assumption for the wave vector in the energy equation obtained for the two cases considered yields the following relationships:

$$4p - q = \begin{cases} 1 & \text{for the forced turbulence (case 1)} \\ 0 & \text{for the free turbulence (case 2)} \end{cases} \qquad (21.13)$$

Solving for (p, q) in both cases gives

$$\text{Forced turbulence:} \quad p = 1/2, \quad q = 1, \qquad (21.14)$$

$$\text{Free turbulence:} \quad p = 1/6, \quad q = 2/3. \qquad (21.15)$$

The self-similar laws obtained for the wave spectrum can be translated for the power spectrum of the transverse velocity $P_v(\omega)$, where $v = \dot{w}$. One obtains for forced vibrations [17]:

$$P_v(\omega, t) \sim f_1\left(\sqrt{\frac{\omega}{t}}\right) = g_1\left(\frac{\omega}{t}\right), \qquad (21.16)$$

and for free vibrations (case 2):

$$P_v(\omega, t) \sim t^{-1/3} f_2\left(\sqrt{\frac{\omega}{t^{1/3}}}\right) = t^{-1/3} g_2\left(\frac{\omega}{t^{1/3}}\right),   \tag{21.17}$$

where $g_{1,2}$ (or $f_{1,2}$) have been indexed with respect to cases 1 and 2, and are functions to be defined. The first relationship underlines the fact that, when the plate is excited by an external harmonic forcing of constant amplitude and frequency, the cascade should show a front propagating linearly in frequency with respect to time. More specifically, denoting ω_c the frequency of the front, one should observe that $\omega_c \propto t$, and that the amplitude of the power spectrum at the front $P_v(\omega_c)$ should stay constant. In the second case where the plate is left free to vibrate, given an initial amount of energy, one should observe the front of the cascade propagating as $t^{1/3}$ while the amplitude of the spectrum at the front should decrease as $t^{-1/3}$. These two predictions will be confronted to numerical simulations in Section 21.3.2.

21.3 Numerical Results

21.3.1 KZ Spectrum

Numerical simulations of the VK equation can be performed using a pseudo-spectral method that takes advantage of the linear dynamics of the vibrating plate. For that purpose, we use in Fourier space the reduced variable $Z_k = W_k e^{i\omega_k t}$ so that the dynamical equation for Z_k involves only the nonlinear terms. This terms are then computed in real space first and then transformed in the Fourier space in order to integrate the dynamics using an Adams–Bashford scheme.

In order to test the KZ spectrum, we need to impose the system conditions that are consistent with the energy cascading from large to small scales. We thus add two terms in the VK equation (21.1) in the Fourier space:

$$\rho h \ddot{W}_k = -D k^4 W_k + L_k(w, F) + I_k + d_k \dot{W}_k,   \tag{21.18}$$

$$k^4 F_k = -\frac{Eh}{2} L_k(w, w).   \tag{21.19}$$

In these equations, the injection term I_k is modeled by a random forcing valid at large scale (for $|k| < k_i$ only where $2\pi/k_i$ is the injection scale):

$$I_k \propto \Theta \text{ for } |k| < k_i \text{ and } I_k = 0 \text{ elsewhere}$$

where Θ is a classical random process. On the other hand, an idealized linear dissipation is used, acting only for $|k| > k_d$, where $2\pi/k_d$ is the dissipative or Kolmogorov scale, yielding

$$d_k \propto (k^2 - k_d^2) \text{ for } |k| > k_d \text{ and } d_k = 0 \text{ elsewhere.}$$

We investigate numerically the dynamics of this system by changing only the forcing amplitude, all the other terms remaining the same. Moreover, we use the dimensionless version of the VK equations where the lengthes are expressed in unit of $\ell = h/\sqrt{3(1-\nu^2)}$, the times in unit of $\sqrt{\rho/E\ell}$, and the Airy function F in units of $Eh\ell^2$. Figure 21.1a shows the evolution of the energy in the system for different values of the injection amplitude. After a transient, a stationary regime is reached as illustrated in Figure 21.1b where the dissipated energy rate ϵ is plotted as a function of time. In this steady state, this constant dissipated energy corresponds to the flux of energy that cascades from the large scale toward the small scales.

Finally, Figure 21.2 shows the energy spectra in the stationary regime for these different forcing amplitudes as functions of frequency. The energy spectra have been rescaled by $\epsilon^{1/3}$, following the theoretical prediction (21.11). We observe that the different spectra collapse in a single curve that is well fitted by the WT solution $\log^{1/3}(f_c^\star/f)$. Amazingly, $f_c^\star$ corresponds exactly to the frequency of the wavenumber k_d.

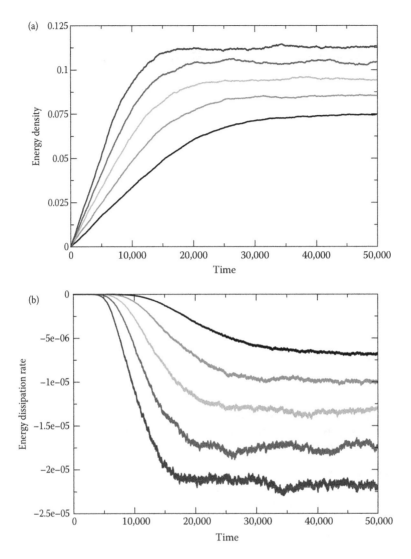

FIGURE 21.1 (a) Evolution of the energy density with time for different forcing amplitudes obtained by solving numerically the dimensionless VK equations. The mesh size is $dx = 0.5$ on a 1024×1024 grid so that the size of the plate is 512×512 in dimensionless unit ℓ. The random forcing holds for $|k| < k_i = 0.1$ while the dissipation acts for $|k| > k_d = 5.5$. (b) The energy dissipation rate as a function of time for the same forcing amplitudes.

21.3.2 Nonstationary WT

This section offers numerical illustrations of the theoretical results for nonstationary turbulence introduced in Section 21.2.3. The numerical method chosen here is a finite-difference, energy-conserving scheme. As opposed to the pseudo-spectral method used in section 21.3.1, finite-difference simulations take place entirely in physical space. A family of such algorithms has been provided in Reference 3, where discrete energy conservation properties give a strict stability condition and a bound on the solution growth. Such schemes, originally thought for use in sound synthesis [4], have found interesting applications in WT simulations [17,53].

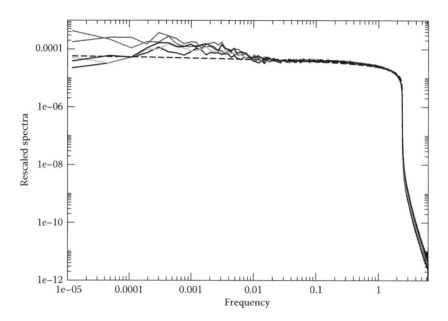

FIGURE 21.2   Rescaled spectra for different forcing amplitudes (same parameters as in Figure 21.1). The spectra are rescaled by the dissipation rate $\epsilon^{1/3}$ and collapse well into a single curve for $f_i < f < f_c^\star$, where f_i is the injection frequency corresponding to the wave number k_i. The black dotted line shows the fit of this curve by the theoretical law (21.11), exhibiting a good agreement.

The two cases where self-similar laws have been derived from the kinetic equation are numerically considered with the following assumptions: in both cases, a rectangular plate is selected, transversely simply supported with in-plane movable edges. More precisely, we have for:

Forced vibrations (case 1): The plate is excited with a pointwise harmonic forcing for all the duration of the simulation. Both the excitation frequency and the forcing amplitude are kept constant. The excitation frequency is close to the 4th eigenfrequency of the plate. Damping is not taken into account.

Free vibrations (case 2): The plate is forced pointwise impulsively at the start on a very short time interval, and then left free to vibrate. By doing so, a free turbulence regime settles down with a given, controllable amount of initial energy. Damping is not considered so that the dynamics is conservative.

Forced vibrations. A first case is considered with a 0.4×0.6 m^2 rectangular plate of thickness $h = 1$ mm. The forcing, located at an arbitrary point, has excitation frequency at 75 Hz, at a steady amplitude of 10 N (after an initial transient where the amplitude is increased from zero to the steady value). 102×153 grid points and a sampling rate of 400 kHz are used. Figure 21.3a shows the spectrogram of the displacement at an arbitrary output point: the absence of damping in addition to the steady forcing creates a nonstationary turbulent cascade, with a front propagating to the high frequencies. A characteristic frequency defined as $f_c = \frac{\int_0^\infty P_v(f)f\,df}{\int_0^\infty P_v(f)\,df}$ is introduced so as to quantitatively investigate the front, where $P_v(f)$ represents the velocity power spectrum in the frequency domain.

Figure 21.3c and d shows that $f_c \propto t$ and that the power spectrum at this characteristic frequency, $P_v(f_c)$, is constant. Plotting successive velocity spectra normalized by f_c and $P_v(f_c)$ gives rise to Figure 21.3b, from which a self-similar dynamics is deduced. Therefore

$$P_v(f) = P_v(f_c)\phi_P\left(\frac{f}{f_c}\right) = g_1\left(\frac{f}{t}\right). \tag{21.20}$$

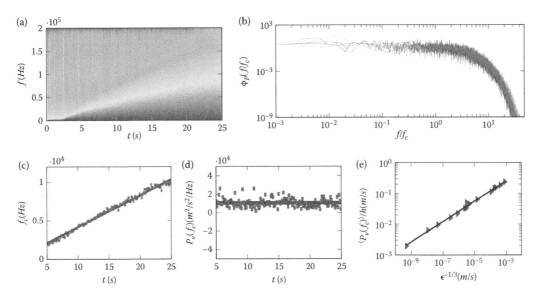

FIGURE 21.3 Nonstationary turbulence with steady forcing (case 1). (a) Spectrogram. (b) Normalized velocity power spectra. (c) Evolution of f_c versus t. (d) Evolution of $P_v(f_c)$ versus t. (e) Scaling of spectral amplitude with injected power.

This last equality, along with the fact that the energy in the system grows linearly over time, is in accordance with the theoretical prediction of nonstationary turbulence, Equation 21.16. Other than agreeing formally with the theory, these simulations offer a visualization of the shape of the self-similar function g_1, which is not given by the theory. Note that g_1 has the shape of ϕ_P used in Equation 21.20 and shown in Figure 21.3b, where the subscript P refers to the periodic, constant-amplitude forcing (case 1, forced vibrations). The nonstationary turbulence is in this case associated with a constant mean injected power $\bar{\epsilon}$ with increasing dispersion ϵ_{rms} (see Reference 17). The cascade possesses various scaling laws. An important one, depicted in Figure 21.3e, relates the value of the mean injection $\bar{\epsilon}$ with the spectral amplitude $P_v(f_c)$. The figure, obtained from 15 simulations of plates with varying geometrical and forcing parameters, leads to the following expression [17]:

$$P_v(f_c) \propto h\bar{\epsilon}^{1/3}. \tag{21.21}$$

This relationship shows that the power spectrum is proportional to the injected power at the power one-third, in accordance with the fact that a four-waves interaction process is at hand [17,20,45].

Free turbulence. For this case, where the plate is left free to vibrate, given an amount of energy as initial condition, a plate of sides 0.4×0.6 m^2 and thickness $h = 0.1$ mm is considered. The spectrogram of the displacement of an arbitrary output point is shown in Figure 21.4a. The behavior of the front of the cascade $f_c(t)$ and the spectral amplitude $P_v(f_c)$ are shown in Figure 21.4c and d with logarithmic scales, clearly exhibiting that $f_c \propto t^{1/3}$, and $P_v(f_c) \propto t^{-1/3}$. The normalized power spectra, in Figure 21.4b, show again a self-similar dynamics, so that $P_v(f) = P_v(f_c)\phi_F\left(\frac{f}{f_c}\right) = t^{-1/3}g_2\left(\frac{f}{t^{1/3}}\right)$, which is in accordance with the theoretical predictions, Equation 21.17. Again, thanks to the simulations, one can appreciate the shape of the self-similar function g_2, represented through the function ϕ_F in Figure 21.4b, where the subscript F stands now for free turbulence (case 2). Note that the shape of ϕ_P and ϕ_F are not exactly the same; when represented on the same figure, one observes clearly that whereas the slope is almost perfectly flat in the free vibration case (function ϕ_F, Figure 21.4b), a small slope behaving as $f^{-0.2}$ is at hand for forced vibrations (function ϕ_P, Figure 21.3b). This highlights the fact that the pointwise forcing has a small effect on the slope of the spectra in the low-frequency range, as observed numerically as well as experimentally, see References 17 and 40.

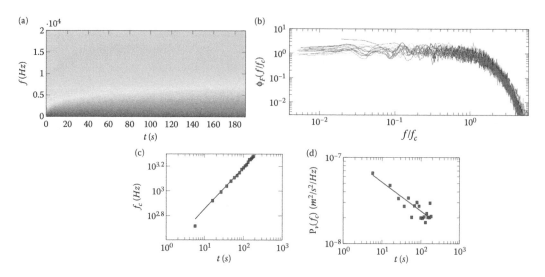

FIGURE 21.4 Nonstationary free turbulence (case 2). (a) Spectrogram. (b) Normalized velocity power spectra. (c) Evolution of f_c versus t. (d) Evolution of $P_v(f_c)$ versus t.

21.3.3 Effect of an Imperfection

Section 21.2.2 shows that the von Kármán equations for plates yield a 4-waves interaction term in the collision integral. Such equations apply exclusively to perfectly flat plates. In practice, real plates present local deformations that may alter their dynamical response [10,53]. When deformations are static, the von Kármán equations can be modified to account for them. Let w_0 denote a static deformation, then

$$\rho h \ddot{w} + D \Delta \Delta w = \mathcal{L}(w + w_0, F), \tag{21.22}$$

$$\Delta \Delta F = -\frac{Eh}{2} \mathcal{L}(w, w + 2w_0). \tag{21.23}$$

The quadratic nonlinearity of the equations for imperfect plates translates in a 3-waves process as a correction to the 4-waves interactions of the perfectly flat plate. The question of whether or not such correction modifies the statistics of the turbulence has been investigated in Reference 17, with a selected deformation of the form of a raised cosine at the center of the plate. The amplitude of the imperfection has been chosen in a range from 1 to 10 times the thickness, and is shown in Figure 21.5a (axis not in scale). Such a choice is based on the consideration that real deformations in plates have the same order of magnitude, in general, at large wavelengths. Remarkably, the statistics of the turbulent regime is not affected by the presence of the imperfection. For obtaining this result, numerous simulations with different setups have been computed in the framework of case 1, i.e., when the plate is constantly forced with a harmonic external excitation, and the different scaling laws have been found to be analogous to those obtained for the perfect plate: the front propagates to high frequencies proportionally to time and the amplitude of the power spectrum at the front is constant. Figure 21.5b shows again a self-similar dynamics whose self-similar function does not differ from that of Figure 21.3b. The scaling law for the injected power is also unaffected: Figure 21.5c shows that $P_v(f_c) = 2.30 \, h\bar{e}^{1/3}$, statistically the same as Equation 21.21 [17].

These numerical results show that in the WT regime, the cubic nonlinearities dominate the quadratic ones, so that the effect of a small imperfection should not affect the statistical properties of the dynamical solutions.

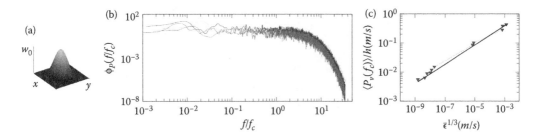

FIGURE 21.5 Nonstationary turbulence with steady forcing for imperfect plates. (a) Example of raised cosine imperfection w_0. (b) Normalized velocity power spectra. (c) Scaling of spectral amplitude with injected power.

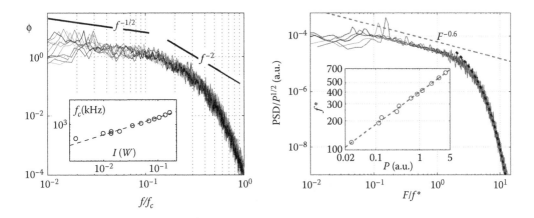

FIGURE 21.6 Rescaled power spectra of the transverse velocity obtained on two different experimental setups and reported in References 6 (left) and 41 (right). The power spectrum is rescaled by the square root of the injected power (denoted as I on the left and P on the right), while the frequency is rescaled by the cut-off frequency (denoted as f_c on the left and $f^\star$ on the right). Both inserts show the dependence of the cut-off frequency with respect to the injected power.

21.4 Real Plates

The two previous sections present the theoretical results and their numerical verification in two different cases, showing that when all the WT assumptions are fulfilled, the analytical predictions are numerically retrieved. We now turn to the case of real plates in order to see how the theoretical predictions confront with experiments.

21.4.1 First Experimental Results

The first experimental investigations on the WT in plates have been reported in 2008 on two different setups [6,41]. In each case, a large (lateral dimensions 1×2 m) and thin (thickness around $0.4 - 0.5$ mm) plate, made of a homogeneous metallic material, has been selected. For exciting the plate in the turbulent regime, a shaker is used and creates a pointwise forcing. At this point, thanks to the measurement of the velocity, one is able to retrieve the experimental injected power. For the first measurements reported, the velocity of the transverse displacement was recorded with a laser vibrometer.

The two papers evidenced a discrepancy between the predicted power spectrum and the measurements. The turbulent behavior is nonetheless confirmed, with the appearance of a large bandwidth and continuous spectrum having a cut-off frequency that gets larger when one increases the injected power. Figure 21.6

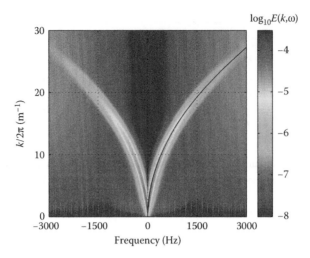

FIGURE 21.7 The angle-integrated space–time spectrum $E(k, \omega)$. The continuous line is the linear dispersion relation for the considered plate $\omega \propto k^2$. (Adapted from Mordant, N., *Eur. Phys. J. B*, 76, 537–545, 2010.)

summarizes the obtained results by showing the rescaled power spectra of the transverse velocity. A clear self-similar behavior is obtained in experiments and is highlighted by dividing the power spectrum by the square root of the injected power, while the frequency axis is made nondimensional by using the cut-off frequency. Both experiments show a scaling for the power spectrum as $P_v(\omega) \propto \varepsilon^{0.66}\omega^{-0.6}$, while the theory predicts $P_v(\omega) \propto \varepsilon^{0.33}\omega^0$. Considering the statistics of velocity increments, no intermittency has been measured [6]. Furthermore, using the correlations in a two-points measurement shows that the waves seem to be persistent [41]. Hence, the general framework of the WT was not called into question, but further investigation was needed to understand the origin of these discrepancies.

21.4.2 Weak Nonlinearity and Separation of Timescales

Considering the previous observations that the spectrum of the vibration of the plate does not fulfill the predictions, the experimental investigations logically turn toward the verification of all the assumptions underlying the WT theory. In order to get a definitive answer to the very first assumption, stating that a wide spectrum of weakly nonlinear waves is at hand, a dedicated experimental technique was developed in order to check whether a dispersion relation exists that is characteristic of propagating waves. The experimental difficulty of such a task is that a 2D space, and time-resolved measurement is required. This can be achieved by using a high-speed profilometry technique developed by Maurel et al. [14,35]. The basic principle is to record with a high-speed camera a pattern projected on the plate. When the plate is deformed, the observed pattern is altered as well. For some patterns and for an adequate optical configuration, the images of the deformed pattern can be inverted to obtain the deformation of the plate. Thanks to this technique, movies of the deformation of the plate over a large area (typically over 1 m^2 for a 1 × 2 m^2 steel plate) are obtained [15,42]. The Fourier spectrum $E(\mathbf{k}, \omega)$ of the deformation can then be easily computed; it is found to be isotropic in $\mathbf{k}$. Hence, we display in Figure 21.7 a spectrum that has been summed over the directions of the wave vector $\mathbf{k}$. Two important conclusions can be drawn from this measurement: (i) the energy of the spectrum is concentrated on a dispersion relation, hence confirming that the motion is composed of waves, and (ii) the energy concentration lies in the very vicinity of the linear dispersion relation, evidencing that the waves are weakly nonlinear. The vibration of the plate is thus truly a turbulent state of weakly nonlinear waves.

One of the main hypothesis of the WT theory is the scale separation between the linear dynamics (oscillation of the waves) and the nonlinear dynamics (energy exchanges between waves). In fact, in the course

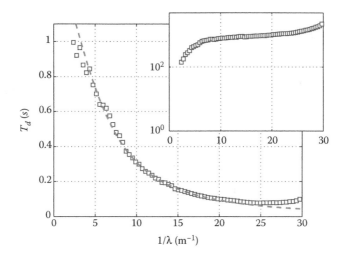

FIGURE 21.8 Squares: measured dissipative time T_d versus $k/2\pi = 1/\lambda$. Dashed line: Lorentzian fit $T_d = (0.73 + 0.025(k/2\pi)^2)^{-1}$ used in the following. Insert: $T_d\omega$ versus $k/2\pi$ (semilog scale). (Adapted from Miquel, B. and N. Mordant., *Phys. Rev. Lett.*, 107(3), 034501, 2011.)

of the derivation of the kinetic equation, the decomposition of the amplitudes as $A_k = a_k(t)\,e^{i\omega_k t}$ is valid only if the time scale T_{NL} of the wave modulation $a_k(t)$ is large compared to the wave period $T = 2\pi/\omega_k$. For real plates, an additional condition arises naturally: the dissipation time scale T_d must be even longer so that nonlinear dynamics can develop. The final condition thus reads

$$T \ll T_{NL} \ll T_d. \tag{21.24}$$

The dissipation time scale can be measured in various ways [26,40]. One possibility is to run decay experiments from a fully turbulent state [40]. The final decay state is exponential and corresponds to a linear dissipative decay of the wave amplitude. The dissipation rate can be extracted from this curve. Figure 21.8 shows the dissipative time scales obtained with this method. The insert shows the product $T_d\omega$ that compares the dissipative time scale and the linear period of the wave. From this measurement, two important conclusions can be drawn. First, the inequality $T \ll T_d$ is fulfilled as two orders of magnitude separates the two time scales for the wavelengths of interest. Second, the product $T_d\omega$ is almost constant with the wavenumber, showing that dissipation is weak but present at all frequencies.

In order to compare the nonlinear time scale with the linear ones, one has to extract the nonlinear dynamics from the experimental data, a challenging task since the measurement is operated only on a part of the full plate. Furthermore, the real plate is finite so that reflections occur on the boundaries. The goal is to compute the time correlations of $a_k(t)$ in order to extract the characteristic time scale of the nonlinear dynamics. The correlation coefficient can be reconstructed by performing a wave packet analysis [40]. The deformation field is projected over Gaussian wavelets at various positions and for a given wavevector. This projection allows one to follow the propagation of the wave packets by correlating the wavelets coefficients at a given wavevector over various positions and times. Reflections on the boundaries can be used to follow the wave packets over longer times. When tracking the wave packets along their trajectory at the group velocity, the magnitude of the correlation of the wave packet magnitude is seen to decay with the propagation time. Two reasons account for this decay: (i) the dispersion of the wave packet, this effect can be easily corrected for Gaussian wave packets, and (ii) the decay of the correlation due to the nonlinear energy exchanges among waves that destroy the coherence of the wave packet. Figure 21.10 shows the dispersion-corrected decay of the temporal correlation of the wavelet coefficients for various positions. The decoherence of the wave packet is exponential so that the extraction of T_{NL} is straightforward. Using this

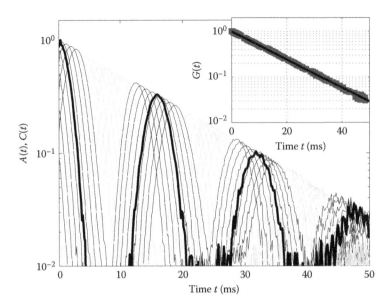

FIGURE 21.9 Decay of the coherence of a wave packet computed by correlation of wavelet coefficient at a given wavevector **k** for various positions. The abscissa axis is the propagation time from an initial position at the center of the plate. The black curves are the correlations of wave packets with the same **k** (propagating in the same direction) for a set of positions on the trajectory of the wave packet. Hence, these curves correspond to wavepackets that have undergone an even number of reflections (or no reflection at all). The light gray curves are the correlation between counterpropagating wave packets that have an opposite wavevector −**k**, as a result of an odd number of reflections. As time flows, the wavepacket propagates nearly ballistically and bounces on the edges so that a succession of dark blue and cyan curves groupings is observed. The envelop of the correlations curves is seen in the insert that shows that the overall decay of the coherence of the wave packet is exponential so that the extraction of the nonlinear time scale T_{NL} is simply the characteristic time of the exponential decay. (Adapted from Miquel, B. and N. Mordant., *Phys. Rev. Lett.*, 107(3), 034501, 2011.)

procedure, a time–space diagram of the decoherence of the wave packet is reconstructed on its trajectory (Figure 21.10). When the forcing strength is increased, the decoherence of the wave packet is seen to be faster consistently with the expectation that the nonlinear dynamics is more developed when the forcing is stronger.

The extracted nonlinear timescale T_{NL} is shown in Figure 21.11 and compared to the dissipation time scale and the period of the wave. Thanks to the large separation between the two latter timescales, an intermediate range for the nonlinear timescale is possible. Indeed, T_{NL} lies in between these two scales. T_{NL} is seen to vary very little with the wavenumber. It decays when the forcing strength is increased as mentioned above. Thus, the double scale separation $T \ll T_{NL} \ll T_d$ is observed so that a true weak turbulence can develop in this system. Note a major difference with the canonical situation of the weak turbulence theory: dissipation is weak but present at all scales and this leads naturally to a steepening of the wave spectrum as the energy flux is progressively decreased by dissipation. This point will be further addressed in Section 21.4.4.

21.4.3 Finite-Size Effects

Another major hypothesis of the weak turbulence theory is that the system is asymptotically large, so that the discreteness of the modes, which is typical of finite-size systems, is avoided. For small systems, a chaotic dynamics may develop in which the nonlinear time scales shows a distinct scaling [28,33]. Indeed, for a small system, the discreteness of the mode may restrict significantly the number of solutions of the

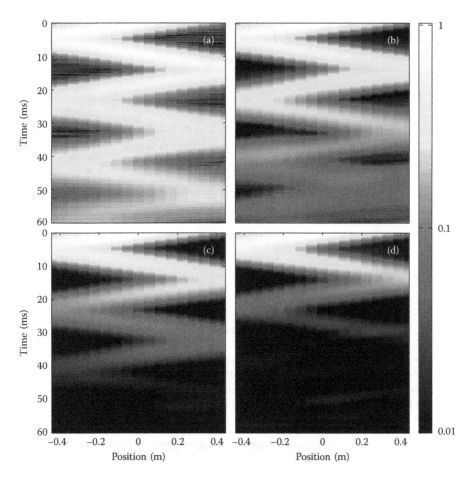

FIGURE 21.10 Space–time representation of the decoherence of the wave packets over their trajectory in the plate, including the bounces on the boundary. The initial position is at the center of the plate. Wavenumber $k/2\pi = 13 \text{ m}^{-1}$ for this specific example. Gray levels are log-coded and represent the correlation of the wave packets with its initial position. The forcing power is P_0, $7P_0$, $27P_0$, and $59P_0$ (from (a) to (d)). (Adapted from Miquel, B. and N. Mordant., *Phys. Rev. Lett.*, 107(3), 034501, 2011.)

resonance equations. Thus, energy injected at the forcing scales must find a path to small scales through connected resonant quadruplets (or triplets for a three-waves interaction process). When the size of the system decreases, the topology of this clusters can become sparse, hence rendering the energy cascade difficult.

When the size of the system is increased, the transition from a chaotic regime to a weak turbulence regime is actually an open question [48]. A qualitative description is the following: nonlinear effects induce a spectral widening of the modes. If the system is large enough, the spectral separation of the modes can become of the same order as this nonlinear widening. Consequently, the discreteness of the modes is destroyed by nonlinearity, and all frequency and wavenumber values are then possible, so that eventually a weak turbulence regime can develop.

Discrete modes can be observed for the experimental plate as can be seen in Figure 21.12. The dispersion relation is continuous at low frequency but shows discrete peaks at the highest frequencies. This observation is related to the fact that the magnitude of the nonlinearity decreases when the frequency is increased. This can also be observed in Figure 21.11: the ratio T_{NL}/T increases with the frequency.

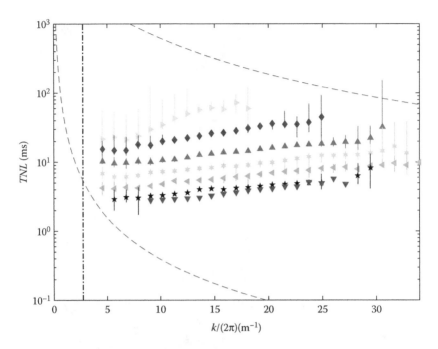

FIGURE 21.11 Comparison of the nonlinear time scale with other relevant time scales. Symbols are experimental data for T_{NL} at different values of the forcing intensity. Upper dashed line: twice the dissipative time T_d measured by the decay of energy in the unforced case (see Figure 21.8). Lower dashed line: $1/\omega$. The forcing is operated at 30 Hz and is shown with the vertical dash-dotted line. (Adapted from Miquel, B. and N. Mordant., *Phys. Rev. Lett.*, 107(3), 034501, 2011.)

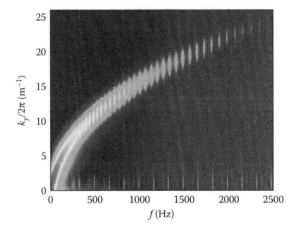

FIGURE 21.12 Cut of the space–time spectrum $E(\mathbf{k}, \omega)$ at $k_x = 0$. Discrete frequencies are clearly visible. (Adapted from Mordant, N., *Eur. Phys. J. B*, 76, 537–545, 2010.)

For a given frequency interval, Figure 21.13 shows the evolution of the crest line of energy along the dispersion relation when the forcing intensity is increased. At very low forcing, the nonlinearity is very weak and the discrete peaks of the modes are clearly visible. When the forcing is increased, the peaks enlarge (and their position shifts slightly as well) due to the effect of nonlinearity. At the highest forcing intensity, the spectrum is continuous. Note that the discrete character of the mode is visible for all forcing intensities but it appears at higher frequencies for larger forcing intensities. This feature can also be observed in

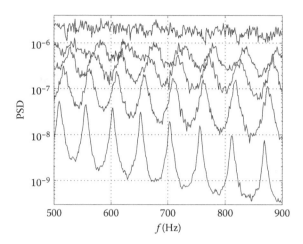

FIGURE 21.13 Evolution of the spectral energy on the dispersion relation in the plane $k_x = 0$ for $P = 1, 4, 9, 16,$ and 36 (from bottom to top). The curves have been shifted vertically by a factor 1.5 for clarity. (Adapted from Mordant, N., *Eur. Phys. J. B*, 76, 537–545, 2010.)

numerical simulations of the plate. Thus, a regime of weak turbulence develops at the lowest frequencies but always evolve into a regime of discrete turbulence at the highest frequencies that are observed in the spectrum. As a conclusion, the finite-size effect in the considered experimental set-up, though observable, should not be responsible for the important discrepancies found between theory and measurements.

21.4.4 Effect of Damping on the Slope of the Power Spectra

The last assumption to be tested is the transparency window or inertial range, i.e., the system shall display a large frequency range between the forcing and dissipative scales, where a conservative Hamiltonian dynamics can be assumed, so that the energy flux shall be conserved when cascading through scales. In solid plates, the dissipation has different physical origins, and in the set up considered, the main causes of losses are thermoelasticity, viscoelasticity, and dissipation at the boundaries. As already noted in the previous section, the damping, though small, is present at all scales, see also, e.g., [11].

In Reference 26, the effect of damping on the WT regime has been experimentally and numerically studied. The experimental plate is made of steel, the lateral dimensions are 2×1 m^2, and the thickness is $h = 0.5$ mm. The plate is set into a turbulent regime with a sinusoidal forcing at frequency $f_0 = 30$ Hz. The natural damping of the plate is increased using different techniques based on paint and edge dampers. Four different configurations are studied and characterized by the measurements of the damping factors in the linear regime, denoted γ and displayed as functions of the frequency in Figure 21.14. Interestingly, despite the different attenuation sources, the damping factors exhibit always the same qualitative behavior, which can be characterized by a power law normalized so that

$$\gamma(f) = \gamma^* f^{0.6}, \tag{21.25}$$

where γ^* varies between 1 and 5.

Figure 21.15a displays the experimental power spectral densities of the normal velocity measured at similar injected powers for the four configurations. They all behave roughly as power laws in the cascade regime with frequency exponents that become clearly smaller as γ^* increases. For the natural plate the exponent (-0.5) is consistent with the previous results [6,41], while for the most damped plate, the exponent is almost twice this value. In Reference 26, numerical simulations of the von Kármán plate equations are also performed, using the same pseudo-spectral method than in Reference 20. Within this framework,

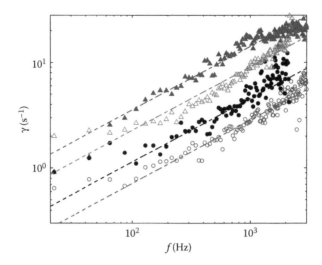

FIGURE 21.14 Evolution of the damping factor γ as a function of the frequency. Empty gray circles: natural plate. Filled black circles: one side painted. Empty light gray triangles: two sides painted. Filled dark gray triangles: two sides painted + edge dampers. Dashed lines: fitted power laws $\gamma \propto f^{0.6}$. (Adapted from Humbert, T., O. Cadot, G. Düring, C. Josserand, S. Rica, and C. Touzé, *EPL*, 102, 30002, 2013.)

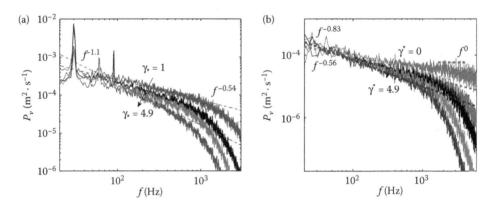

FIGURE 21.15 Power spectral density of the transverse velocity for the four configurations. Dashed lines + equations: smallest and largest slopes. (a) Experiments. From top to bottom: gray: $\gamma^* = 1$, $\varepsilon_I = 0.56 \times 10^{-3}\,\mathrm{m}^3 \cdot \mathrm{s}^{-3}$. Black: $\gamma^* = 1.6$, $\varepsilon_I = 0.54 \times 10^{-3}\,\mathrm{m}^3 \cdot \mathrm{s}^{-3}$. Light gray: $\gamma^* = 3.1$, $\varepsilon_I = 0.52 \times 10^{-3}\,\mathrm{m}^3 \cdot \mathrm{s}^{-3}$. Dark gray: $\gamma^* = 4.9$, $\varepsilon_I = 0.48 \times 10^{-3}\,\mathrm{m}^3 \cdot \mathrm{s}^{-3}$. (b) Numerics. From top to bottom: light gray: $\gamma^* = 0$, $\varepsilon_I = 0.057\,\mathrm{m}^3 \cdot \mathrm{s}^{-3}$. Other cases: $\varepsilon_I = 0.024\,\mathrm{m}^3 \cdot \mathrm{s}^{-3}$. (b) Numerics. From top to bottom: light gray: $\gamma^* = 0$, $\varepsilon_I = 0.057\,\mathrm{m}^3 \cdot \mathrm{s}^{-3}$. (Adapted from Humbert, T., O. Cadot, G. Düring, C. Josserand, S. Rica, and C. Touzé, *EPL*, 102, 30002, 2013.)

it is straightforward to inject energy at controlled scales and to mimic the measured experimental dissipation by using the fitted power laws displayed in Figure 21.14. Figure 21.15b shows the power spectral densities of the normal velocity obtained by numerical simulations for similar injected powers. The same behavior as in the experiments is observed. The ideal case $\gamma^* = 0$ with the almost flat KZ spectrum is drawn for comparison, arguing that the difference between the experimental and the theoretical values of the slope is mainly due to the existence of damping at all scales. Finally, in Reference 38, numerical simulations of the von Kármán equations are performed using three damping laws (displayed in Figure 21.16a) going from the experimental dissipation to the one used in the theory of WT. Figure 21.16b draws the

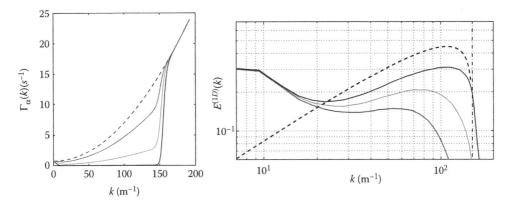

FIGURE 21.16 (Left) Dissipation $\Gamma_\alpha(k)$ (solid lines) compared to experimental dissipation γ_k^{EXP} (black dashed line). (Right) Power spectral density $E^{(1D)}(k)$. The black dashed line indicates the Kolmogorov–Zakharov spectrum. (Adapted from Miquel, B., A. Alexakis, and N. Mordant, *Phys. Rev. E*, 89(6), 062925, 2014.)

numerical spectra associated to these three damping laws, showing that a transition from the experimental spectrum to the predicted KZ spectrum is observed when dissipation is decreased at large and medium scales.

These results clearly highlight that the slope of the turbulent power spectra in vibrating plates depends strongly on the damping, indicating that it must be retained as a pertinent feature to explain the difference between theory and experiments. Moreover, one can argue that because the dissipation is relevant at each scale, no inertial range (or transparency window) exists in these turbulent regimes so that the flux of energy is not constant over the cascade.

21.4.5 Dependence on Injected Power

A fundamental result of the WTT is the relationship between the number of resonant waves and the power exponent λ of the energy flux in the power spectral density expression [63]. For three-waves interactions, the theory predicts $\lambda = \frac{1}{2}$ while four-waves interactions imply $\lambda = \frac{1}{3}$. For plates, the cubic nonlinearity of the restoring force implies four-waves interactions, such that one would expect a $\frac{1}{3}$ power law on the energy flux. In the first reported experiments, see Section 21.4.1 and [6,41], the dependence of the power spectra on injected power was found to be near 0.6, i.e., very far from the expected value of $\frac{1}{3}$.

In order to obtain more quantitative results from the experiments, the energy budget of the cascade has been considered in References 6 and 26. The budget assumes that the cascade stops when the magnitude of the injected power has been completely dissipated by all the excited modes of the plate. Introducing ε_D as the dissipated power, one can write

$$\varepsilon_D = h \int_0^\infty \gamma(f) P_v(f) df \simeq h \int_0^{f_c} \gamma(f) P_v(f) df \propto \varepsilon_I. \tag{21.26}$$

Considering a dissipation given by Equation 21.25 and a spectrum of the form $P_v(f) \propto \epsilon_I^\lambda (f/f_c)^{-\beta}$ with the exponent λ and the slope β unknown, the following expression for the cut-off frequency is obtained:

$$\gamma_*^{1/1.6} f_c \propto \epsilon_I^{\frac{1-\lambda}{1.6}}. \tag{21.27}$$

From the dependence of the cut-off frequency with respect to the injected power, one is able to retrieve the coefficient λ. It is found to vary from 0.36 to 0.57 in experiments and from 0.33 to 0.39 in numerics, depending on the damping coefficient. This result again shows that the presence of damping at all scales precludes for a direct comparison with the analytical predictions of the theory. A systematic deviation is observed, and the amount of discrepancy directly depends on the level of the damping in the system.

21.5 Discussion

Turbulence in a solid. Whereas the term "turbulence" generally refers to hydrodynamic turbulence and is somehow strongly related to the irregular motions of a fluid, the nonlinear interactions of a sea of coupled waves may also produce a regime of WT. In this context, the nonlinear vibrations of large and thin plates can be interpreted as an example of turbulence in a solid. The case of geometrically nonlinear vibrations of thin plates and shells, where the material's behavior is considered as linear elastic so that the nonlinearity comes from the large-amplitude motions of the transverse deflection $w(x, t)$, offers a correct framework to derive theoretical results thanks to the von Kármán model, as well as experimental results. Within this framework, the nonlinearity of the vibratory state is assessed by the ratio w/h of the transverse displacement w with respect to the thickness h of the plate. When $w \ll h$, linear vibrations occurs. The von Kármán model has first been derived for vibration amplitudes that are of the order of the thickness. Though it may be valid for a larger range of amplitudes, numerical studies clearly assessing its validity limits in terms of amplitude are still missing. From the experimental viewpoint, the range of amplitudes where the turbulence has been observed is $h < w < 50h$.

Cascade of energy and turbulence. For the range of large-amplitude motions where $w > h$, the behavior of the plate can be said as being truly turbulent. This fact is assessed by the broadband Fourier spectrum of the vibrations, the upper frequency of which depends on the energy level, evidencing a cascade of energy from the injection to a dissipative scale. This particular regime has musical applications in the understanding of the sound of gongs, for example, where the higher frequency is reached a few milliseconds after the strike (the time the cascade takes to progress to higher frequencies) and the larger frequency is defined by the amplitude of the strike [18]. It has also been used in the past in theaters where large metallic plates were vigorously shaken to simulate the sound of thunder.

The turbulent behavior is also testified by the relation between the injected power and the injected velocity. Figure 21.17 shows the injected power ε_I as a function of the root mean square value of the velocity at the injection point V_{RMS}, both in experiment and in numerical simulation, for the set up described in Section 21.4.4. For both cases, and for a damping coefficient γ^* varying between 1 and 5 as in Section

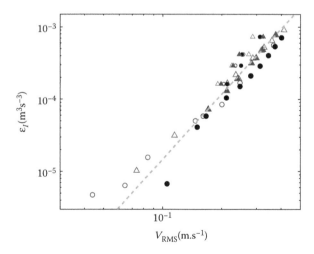

FIGURE 21.17 Injected power ε_I versus root mean square value of the velocity at the injection point V_{RMS}. Large markers are used for experiments whereas small markers refers to numerical simulations. Variations of the damping as in Section 21.4.4: Empty light gray circles: $\gamma^* = 1$, full black circles: $\gamma^* = 1.6$, empty light gray triangles: $\gamma_* = 3.1$, full dark gray triangles: $\gamma_* = 4.9$, light gray: $\gamma^* = 0$. Dashed line : $\epsilon_I \propto V_{RMS}^3$. (Adapted from T. Humbert et al. *EPL*, 102:30002, 2013.)

21.4.4, the dependence clearly exhibits the power law: $\varepsilon_I \propto V_{RMS}^3$. This shows that the mechanism of power injection is inertial as observed in hydrodynamic turbulence when varying the viscosity [8].

Theory and experiments. The WT theory provides analytical results predicting the statistical repartition of energy through scales. Moreover, the self-similar solutions of the kinetic equation can be derived for predicting behavior laws in the non stationary case, though the theory does not yield the exact functional shape of the self-similar functions.

The discrepancy between the first experimental results reported in 2008 in References 6 and 41 and the analytical prediction derived in 2006 [20] shall thus be attributed to the violation of an assumption used in the theoretical framework.

Checking the assumptions. Thanks to refined measurements, all hypothesis have been checked in a series of papers [6,7,9,15,17,26,37–42]. The persistence of waves and the correct separation of timescales has been fully documented in References 15, 41, and 39. As shown in Section 21.4.3, finite-size effects are noticeable on the measurements but only for large frequencies where the vibration amplitude is getting exponentially small so that the nonlinearity is not sufficiently excited. In the frequency range of interest where the cascade develops, the energy spectrum is dense and sufficiently flat to assume that the finite-size effects are negligible in the current experimental set up.

Experimental set up. Finally, typical effects induced by the experimental set up could be invoked for explaining the observed discrepancies. Among these, the presence of imperfections in the static position are unavoidable with a real plate. However, numerical simulations shown in Reference 17 clearly highlight that simple imperfections with large wavelength have no incidence on the dynamics of the power spectrum. As a matter of fact, at such vibration amplitudes, cubic nonlinearities dominate the quadratic ones so that the effect of the imperfection does not profoundly modify the results. Another experimental bias is brought by the presence of a shaker, a device used for creating a pointwise excitation of the plate. The presence of this pointwise forcing has been shown to modify a little the slope of the spectrum around the forcing frequency. In experiments, this has been emphasized in Reference 40 in a non stationary framework where the forcing was suddenly stopped once the plate was in the WT regime. In the decay of turbulence, the slope of the spectrum was observed to flatten a little. Similar observations are shown in numerical nonstationary WT in Reference 17, where the effect of the forcing is noticeable. However, this effect, though observable, remains small and is not the most salient feature for explaining the discrepancy between theory and measurements.

Effect of damping. The last assumption to validate is the transparency window. Thanks to experiments with varying damping ratios, and numerical simulations, recent studies undoubtedly shows that in real plates, all loss terms, though small, can not be neglected [26,38]. As a consequence, the energy flux through scales is not constant, but slowly decreasing so that the amount of damping as a function of the frequency defines both the slope of the power spectrum and the cut-off frequency.

21.6 Conclusion

WT in nonlinearly vibrating plates have been thoroughly studied in this review chapter, which summarizes the most important results found since the publication of the theoretical results in 2006 [20]. The picture of turbulence given in the theoretical framework is an idealization that is difficult or even to reproduce in a real experiment. A peculiar feature brought by the logarithmic correction in the KZ theoretical spectrum is that energy is assumed to completely disappear at the small scale introduced, $k_\star$, and which has no experimental counterpart. Thanks to measurements combined with numerical simulations, it has been clearly demonstrated that the presence of damping at all scales is the major effect for explaining the observed discrepancies. More precisely, the energy flux in the cascade is not constant, and this phenomenon precludes for a direct comparison with the theoretical results, as its effects is of prime importance on both the slope of the power spectrum and the dependence with injected power.

Further research in the direction of a better agreement between WT theory and experiments must then mandatorily consider the effect of the damping in the WT calculations, as a first-order effect that needs to be taken into account in the kinetic equation. For that purpose, a phenomenological model may be used, see for example, Reference 27.

Other aspects may need further developments, in either numerical simulation or experimental verification. Recently, it has been shown numerically that a strongly nonlinear regime can be obtained for amplitudes of vibration w that are larger than $100h$, with h being the thickness [37,58,59]. This nonlinear regime is dominated by ridges connecting developable cones (D-cones), and scales linearly with the injected power. This observation supports the idea that the mechanisms responsible for spectral energy transfers differ strongly from the collisions between resonant waves stated by weak WT. Emergence of intermittency has also been reported in this context. Experimental measurements show that evidences of this regime may be observed when the gradient of the displacement is large enough [37]. However, for this range of amplitudes, the validity of the von Kármán model is questionable, so that further work is needed, both experimentally and numerically, in order to get insight into this strongly nonlinear regime.

Finally, the existence of an inverse cascade appears as a question witnessing new recent developments [21]. On the theoretical viewpoint, no stationary inverse cascade should exist in the framework of WT, in particular due to the non conservation of the wave action. However, when artificially neglecting the $3 \leftrightarrow 1$ interactions in the collision integral and thus conserving the number of waves in the interactions, an inverse cascade of wave action with constant flux can be derived [21]. Numerical simulations clearly exhibit this inverse cascade, thus questioning the nature of the involved interactions, as well the emergence of pattern and the possibility of self-organization.

References

1. Z. P. Bazant and L. Cedolin. *Stability of Structures*. World Scientific, New York, 1991.

2. D. J. Benney and A. C. Newell. Sequential time closures of interacting random waves. *J. Math. Phys.*, 363–393, 1967.

3. S. Bilbao. A family of conservative finite difference schemes for the dynamical von Kármán plate equations. *Num. Meth. Partial Diff. Eq.*, 24(1):193–216, 2008.

4. S. Bilbao. *Numerical Sound Synthesis*. Wiley, Chichester, 2008.

5. S. Bilbao, O. Thomas, C. Touzé, and M. Ducceschi. Conservative numerical methods for the full von Kármán plate equations. *Num. Meth. Partial Diff. Eq.*, 31(6):1948–1970, 2015.

6. A. Boudaoud, O. Cadot, B. Odille, and C. Touzé. Observation of wave turbulence in vibrating plates. *Phys. Rev. Lett.*, 100:234504, 2008.

7. O. Cadot, A. Boudaoud, and C. Touzé. Statistics of power injection in a plate set into chaotic vibration. *Eur. Phys. J. B*, 66:399–407, 2008.

8. O. Cadot, Y. Couder, A. Daerr, S. Douady, and A. Tsinober. Energy injection in closed turbulent flows: Stirring through boundary layers versus inertial stirring. *Phys. Rev. E*, 56(1):427–433, 1997.

9. O. Cadot, C. Touzé, and A. Boudaoud. Linear versus nonlinear response of a forced wave turbulence system. *Phys. Rev. E*, 82:046211, 2010.

10. C. Camier, C. Touzé, and O. Thomas. Non-linear vibrations of imperfect free-edge circular plates and shells. *Eur. J. Mech - A/Solids*, 28:500–515, 2009.

11. A. Chaigne and C. Lambourg. Time-domain simulation of damped impacted plates. I: Theory and experiments. *J. Acoustic. Soc. Am.*, 109:1422, 2001.

12. H. N. Chu and G. Herrmann. Influence of large amplitudes on free flexural vibrations of rectangular elastic plates. *J. Appl. Mech.*, 23:532–540, 1956.

13. P. G. Ciarlet. A justification of the von Kármán equations. *Arch. Rat. Mech. Anal.*, 73:349–389, 1980.

14. P. J. Cobelli, A. Maurel, V. Pagneux, and P. Petitjeans. Global measurement of water waves by Fourier transform profilometry. *Exp. Fluids*, 6(6):1037–1047, 2009.

15. P. Cobelli, P. Petitjeans, A. Maurel, V. Pagneux, and N. Mordant. Space-time resolved wave turbulence in a vibrating plate. *Phys. Rev. Lett.*, 103(20):204301, 2009.
16. C. Connaughton, A. Newell, and Y. Pomeau. Non-stationary spectra of local wave turbulence. *Physica D*, 184:64–85, 2003.
17. M. Ducceschi, O. Cadot, C. Touzé, and S. Bilbao. Dynamics of the wave turbulence spectrum in vibrating plates: A numerical investigation using a conservative finite difference scheme. *Physica D*, 280-281:73–85, 2014.
18. M. Ducceschi and C. Touzé. Modal approach for nonlinear vibrations of damped impacted plates: Application to sound synthesis of gongs and cymbals. *J. Sound Vibration*, 344:313–331, 2015.
19. G. Düring. *Non-Equilibrium Dynamics of Nonlinear Wave Systems: Turbulent Regime, Breakdown and Wave Condensation*. PhD thesis, Université Pierre et Marie Curie, Paris 6, 2010.
20. G. Düring, C. Josserand, and S. Rica. Weak turbulence for a vibrating plate: Can one hear a Kolmogorov spectrum? *Phys. Rev. Lett.*, 97:025503, 2006.
21. G. Düring, C. Josserand, and S. Rica. Self-similar formation of an inverse cascade in vibrating elastic plates. *Physical Review E*, 91:052916, 2015.
22. S. Dyachenko, A. C. Newell, A. Pushkarev, and V. E. Zakharov. Optical turbulence: weak turbulence, condensates and collapsing filaments in the nonlinear Schrödinger equation. *Physica D*, 57:96–160, 1992.
23. G. E. Falkovich and A. V. Shafarenko. Nonstationary wave turbulence. *J. Nonlinear Sci.*, 1:457–480, 1991.
24. A. Föppl. *Vorlesungen über technische Mechanik*, volume 5. Druck und Verlag von B. G. Teubner, München, 1907.
25. U. Frisch. *Turbulence. The Legacy of A.N. Kolmogorov.* Cambridge University Press, Cambridge, 1995.
26. T. Humbert, O. Cadot, G. Düring, C. Josserand, S. Rica, and C. Touzé. Wave turbulence in vibrating plates: The effect of damping. *EPL*, 102:30002, 2013.
27. T. Humbert, C. Josserand, C. Touzé, and O. Cadot. Phenomenological model for predicting stationary and non-stationary spectra of wave turbulence in vibrating plates. *Physica D: Nonlinear Phenomena*, 316:34–42, 2016.
28. E. Kartashova. Weakly nonlinear theory of finite-size effects in resonators. *Phys. Rev. Lett.*, 72(13):2013–2016, Jan 1994.
29. R. M. Kirby and Z. Yosibash. Solution of von Kármán dynamic non-linear plate equations using a pseudo-spectral method. *Comput. Methods Appl. Mech. Eng.*, 193:575–599, 2004.
30. A. N. Kolmogorov. On degeneration (decay) of isotropic turbulence in an incompressible viscous liquid. *Dokl. Akad. Nauk. SSSR*, 31:538–540, 1941.
31. A. N. Kolmogorov. The local structure of turbulence in incompressible viscous fluid for very large Reynolds number. *Dokl. Akad. Nauk. SSSR*, 30:9–13, 1941.
32. L. D. Landau and E. M. Lifshitz. *Theory of Elasticity*. Pergamon Press, 1959.
33. V. S. L'vov and S. Nazarenko. Discrete and mesoscopic regimes of finite-size wave turbulence. *Phys. Rev. E*, 82(5):056322, 2010.
34. Y. L'vov, S. V. Nazarenko, and R. West. Wave turbulence in Bose-Einstein condensates. *Physica D*, 184:333–351, 2003.
35. A. Maurel, P. Cobelli, V. Pagneux, and P. Petitjeans. Experimental and theoretical inspection of the phase-to-height relation in Fourier transform profilometry. *Appl. Optics*, 48:380–392, 2009.
36. O. Millet, A. Hamdouni, and A. Cimetière. Justification du modèle bidimensionnel non-linéaire de plaque par développement asymptotique des équations d'équilibre. *C. R. Académie des Sciences IIb*, 324:349–354, 1997.
37. B. Miquel, A. Alexakis, C. Josserand, and N. Mordant. Transition from wave turbulence to dynamical crumpling in vibrated elastic plates. *Phys. Rev. Lett.*, 111(5):054302, 2013.
38. B. Miquel, A. Alexakis, and N. Mordant. The role of dissipation in flexural wave turbulence: from experimental spectrum to Kolmogorov-Zakharov spectrum. *Phys. Rev. E*, 89(6):062925, 2014.

39. B. Miquel and N. Mordant. Nonlinear dynamics of flexural wave turbulence. *Phys. Rev. E*, 84(6):066607, 2011.

40. B. Miquel and N. Mordant. Non stationary wave turbulence in an elastic plate. *Phys. Rev. Lett.*, 107(3):034501, 2011.

41. N. Mordant. Are there waves in elastic wave turbulence? *Phys. Rev. Lett.*, 100(23):234505, 2008.

42. N. Mordant. Fourier analysis of wave turbulence in a thin elastic plate. *Eur. Phys. J. B*, 76:537–545, 2010.

43. S. L. Musher, A. M. Rubenchik, and V. E. Zhakarov. Weak Langmuir turbulence. *Phys. Rep.*, 252:177–274, 1995.

44. A. H. Nayfeh and D. T. Mook. *Nonlinear Oscillations*. John Wiley & Sons, New York, 1979.

45. S. Nazarenko. *Wave Turbulence*. Springer-Verlag, Berlin, Heidelberg, 2011.

46. S. V. Nazarenko, A. C. Newell, and S. Galtier. Non-local MHD turbulence. *Physica D*, 152–153:646–652, 2001.

47. A. C. Newell, S. Nazarenko, and L. Biven. Wave turbulence and intermittency. *Physica D*, 152–153, 2001.

48. A. C. Newell and B. Rumpf. Wave turbulence. *Annu. Rev. Fluid Mech.*, 43:59–78, 2011.

49. M. Onorato, A. R. Osborne, M. Serio, D. Resio, A. Pushkarev, V. E. Zakharov, and C. Brandini. Freely decaying weak turbulence for sea surface gravity waves. *Phys. Rev. Lett.*, 89:144501, 2002.

50. A. N. Pushkarev and V. E. Zakharov. Turbulence of capillary waves. *Phys. Rev. Lett.*, 76:3320–3323, 1996.

51. S. Sridhar, D. T. Mook, and A. H. Nayfeh. Non-linear resonances in the forced responses of plates, Part I: Symmetric responses of circular plates. *J. Sound Vibration*, 41(3):359–373, 1975.

52. O. Thomas and S. Bilbao. Geometrically nonlinear flexural vibrations of plates: In-plane boundary conditions and some symmetry properties. *J. Sound Vibration*, 315(3):569–590, 2008.

53. C. Touzé, S. Bilbao, and O. Cadot. Transition scenario to turbulence in thin vibrating plates. *J. Sound Vibration*, 331(2):412–433, 2012.

54. C. Touzé, M. Vidrascu, and D. Chapelle. Calcul direct de la raideur non linéaire géométrique pour la réduction de modèles de coques en éléments finis. In *Proceedings of CSMA 2013, Colloque national en calcul de structures*, Giens, May 2013.

55. C. Touzé, M. Vidrascu, and D. Chapelle. Direct finite element computation of non-linear modal coupling coefficients for reduced-order shell models. *Comput. Mech.*, 54(2):567–580, 2014.

56. C. Touzé, O. Thomas, and A. Chaigne. Asymmetric non-linear forced vibrations of free-edge circular plates, Part I: theory. *J Sound Vibration*, 258(4):649–676, 2002.

57. T. von Kármán. Festigkeitsprobleme im maschinenbau. *Encyklopädie der Mathematischen Wissenschaften*, 4(4):311–385, 1910.

58. N. Yokoyama and M. Takaoka. Weak and strong wave turbulence spectra for elastic thin plate. *Phys. Rev. Lett.*, 110(10):105501, 2013.

59. N. Yokoyama and M. Takaoka. Identification of a separation wave number between weak and strong turbulence spectra for a vibrating plate. *Phys. Rev. E*, 89:012909, 2014.

60. Z. Yosibash and R. M. Kirby. Dynamic response of various von Kármán non-linear plate models and their 3-d counterparts. *Int. J. Solids Structures*, 42:2517–2531, 2005.

61. V. E. Zakharov and N. N. Filonenko. Energy spectrum for stochastic oscillations of surface of a liquid. *Sov. Phys. Dokl.*, 11:881–884, 1967.

62. V. E. Zakharov and N. N. Filonenko. Weak turbulence of capillary waves. *J. Appl. Mech. Tech. Phys.*, 8:37–42, 1967.

63. V. E. Zakharov, V. S. L'vov, and G. Falkovich. *Kolmogorov Spectra of Turbulence 1: Wave Turbulence*. Series in Nonlinear Dynamics, Springer, 1992.

22

Nonlinear Dynamics of the Oceanic Flow

S.V. Prants,
M.V. Budyansky, and
M.Yu. Uleysky

22.1 Introduction

Despite the fascinatingly complex behavior of ocean flows, mesoscale (a few hundreds of km) and submesoscale (a few decades of km) coherent structures are clearly visible on satellite images of the sea surface temperature (SST) and ocean color. These are major currents like the Gulf Stream in the Atlantic and the Kuroshio in the Pacific and many other jet currents in different seas and oceans. The other examples are long-lived rings, the mesoscale eddies that detach regularly from the Gulf Stream, the Kuroshio, and the other major currents. These coherent structures impact the waters through which they pass because they exchange their waters with specific temperature, salinity, chlorophyll concentration, etc., with ambient waters. Such a fluid exchange may have a great impact on climate, biota, and pollution propagation. Thus, transport and mixing of water masses are important processes in the ocean.

By transport, we mean Lagrangian transport, i.e., advection of water mass with its conserved properties due to the fluid's bulk motion. Traditionally, fluid flows are characterized by studying their velocity fields, the approach known as an Eulerian one. The more convenient approach to study transport and mixing in the ocean, the fate, and origin of water masses is the Lagrangian one (which, however, belongs to L. Euler as well), when one integrates trajectories for a large number of passive particles advected by an Eulerian velocity field

$$\frac{d\mathbf{r}}{dt} = \mathbf{v}(\mathbf{r}, t). \tag{22.1}$$

The velocity field, $\mathbf{v}(\mathbf{r}, t)$, is supposed to be known analytically, numerically, estimated from satellite altimetry or radar measurements. In what follows, we will deal only with two-dimensional (2D) surface ocean flows on the sphere, i.e., $\mathbf{v} = (u, v)$ and $\mathbf{r} = (x, y)$. While in the Eulerian approach we get frozen snapshots of data, Lagrangian diagnostics enable to quantify spatio-time variability of the velocity field.

The advection equations (22.1) are, in general, nonlinear because the velocity, $\mathbf{v}$ is a nonlinear function of the particle's position $\mathbf{r}$. Thus, we deal with a nonlinear dynamical system with one and half degrees of freedom. It is well known from dynamical systems theory that solutions of the system (22.1) with a non-steady deterministic velocity field can be chaotic in the sense of exponential sensitivity to small variations in initial conditions and/or control parameters. It means that even a simple time-periodic deterministic velocity field may cause practically unpredictable particle trajectories, the phenomenon known as chaotic advection (for a review on chaotic advection in the ocean see References 14, 17, and 30). The real oceanic flows are not, of course, deterministic and regular, but if the Eulerian correlation time is large as compared to the Lagrangian one, the problem may be treated in the framework of the chaotic advection concept.

The general method we used is based on the Lagrangian approach to study mixing and transport at the sea surface [3,8,11,12,15–17,19,20,22,25] when one follows fluid particle trajectories in a velocity field generated by analytical kinematic or dynamically consistent models, calculated from satellite altimetric measurements, or obtained as an output of one of the numerical circulation model. The important notion in that approach is the so-called Lagrangian coherent structures (LCS) [11]. The well-known coherent structures in the ocean are eddies and jet currents that can be visible in Eulerian velocity fields and at satellite images of the SST and/or chlorophyll concentration. The LCSs, in general, are not visible at snapshots but can be computed with a given velocity field by special methods. LCS are operationally defined as local extrema of the finite-time Lyapunov exponent (FTLE) field [11]. They are the most influential attracting and repelling hyperbolic material surfaces which are curves in 2D velocity fields. The LCS are Lagrangian because they are invariant material curves consisting of the same fluid particles. They are coherent because they are comparatively long lived and more robust than the other adjacent structures.

In order to study transport and mixing in the ocean, it is reasonable to begin with simple analytical models of the velocity field because simplified models enable to elucidate the underlying mechanisms by which the waters are mixed. A number of simple kinematic and dynamically consistent model stream functions have been proposed to study large-scale chaotic mixing and transport in geophysical meandering jet flows and in eddy systems (for a review see References 13, 14, 17, and 30). Analytical models do not pretend to quantify transport fluxes in the real ocean but they are useful in revealing space–time structures that qualitatively and quantitatively specify the mixing and transport of water masses.

This chapter is devoted to a brief review of the application of dynamical systems theory methods in physical oceanography. We start with a toy model of meandering strong jet currents in Section 22.2. It is a simple 2D kinematic model with a streamfunction, playing the role of a Hamiltonian, that mimics water motion in such currents. In the framework of this model, we study the origin and bifurcations of typical classes of the unstable periodic orbits (UPOs) in 2D incompressible flows. In Section 22.3, we analyze a phenomenological connection between dynamical, topological, and statistical properties of chaotic mixing and zonal transport in the same kinematic model and explain zonal transport properties by the phenomenon of the so-called dynamical traps. A more adequate picture of chaotic mixing and transport in the ocean can be reached with more realistic models, namely, with dynamically consistent models conserving the potential vorticity. In Section 22.4, we study an analytical and numerical cross-jet chaotic transport in one of these models with two Rossby waves.

In the past 20 years, it became possible to measure the velocity field at the ocean surface (more exactly, at the layer just below the upper mixed layer) with the help of satellite altimeters. They are able to measure the sea surface heights on a regular basis in the World Ocean except for a band around the equator with a width of $\pm 5°-7°$. After solving equations of the geostrophic balance, one gets the velocity field in the geostrophic approximation that can be used, in turn, to solve the advection equations for passive particles.

In Section 22.5, we have used this product, freely distributed by AVISO (http://www.aviso. oceanobs.com), to simulate radionuclide propagation in the ocean from the Fukushima–Daiichi nuclear plant accident in 2011. We develop an approach, based on searching for specific Lagrangian features in the altimetric geostrophic velocity fields, which indicate the presence of convergence of waters of different properties. We call them Lagrangian fronts (LFs), which are boundaries between surface waters with different Lagrangian properties. In this section, we report on the connection of the LFs with fishing grounds with maximal catches. To be concrete, we focus on fishing grounds of Pacific saury (*Cololabis saira*), one of the most commercial pelagic fishes in the Northwest Pacific where the cold Oyashio Current flows out of the Arctic along the Kamchatka Peninsula and the Kuril Islands and converges with the warmer Kuroshio Current off the eastern shore of Japan. This frontal zone is known to be one of the richest fisheries in the world due to the large nutrient content in the Oyashio waters and high tides.

22.2 Chaos in a Kinematic Model of a Meandering Jet Current

22.2.1 The Model Flow and the Return Map

It is well known that a dynamical system is chaotic if it displays sensitivity to initial conditions, has a dense orbit, and a dense set of periodic orbits. Periodic orbits play an important role in organizing dynamical chaos both in Hamiltonian and dissipative systems. Stable periodic orbits (SPOs) organize a regular motion inside islands of stability in the phase space. UPOs form a skeleton around which chaotic dynamics is organized. The motion nearby an UPO is governed by its stable and unstable manifolds. Owing to the density property, the UPOs influence even the asymptotic dynamics. Order and disorder in a chaotic regime are produced eventually by an interplay between sensitivity to initial conditions and regularity of the periodic motion.

The equations of motion of passive particles advected by any 2D-incompressible planar flow are known to have a Hamiltonian form

$$\frac{dx}{dt} = u(x, y, t) = -\frac{\partial \Psi}{\partial y}, \qquad \frac{dy}{dt} = v(x, y, t) = \frac{\partial \Psi}{\partial x}, \tag{22.2}$$

where the streamfunction Ψ plays the role of a Hamiltonian, and the particle's coordinates x and y are canonically conjugated variables. The phase space of Equation 22.2 is a physical space for advected particles. If the velocity field, $u = u(x, y)$ and $v = v(x, y)$, is stationary, then fluid particles move along streamlines, and the motion is completely regular with any Eulerian stationary field whatever its complexity. A time-periodic velocity field, $u(x, y, t) = u(x, y, t + T)$ and $v(x, y, t) = v(x, y, t + T)$, can produce chaotic particle's trajectories, the phenomenon known as "chaotic advection" [2].

In this section, we study the origin and bifurcations of typical classes of the UPOs in a 2D-incompressible flow that has been introduced and analyzed in References 18, 24, and 27 as a toy kinematic model of transport and mixing of passive particles in meandering jet currents in the ocean, like the Gulf Stream, Kuroshio, and other main oceanic currents. Among the variety of models of shear flows, one of the simplest ones is a Bickley jet with the velocity profile $\sim \text{sech}^2 y$ and a running wave imposed. The phase portrait of such a flow in the frame, moving with the phase velocity of the running wave, is shown in Figure 22.1. The flow consists of three distinct regions, the eastward jet (J), the circulations (C), and the westward peripheral currents (P) to the north and south from the jet, separated from each other by the northern and southern ∞-like separatrices. A simple periodic modulation of the wave's amplitude breaks up these separatrices, produces stochastic layers in place of them, and chaotic mixing and transport of passive particles may occur.

The aim of this section is to study in detail the origin, properties, and bifurcations of typical UPOs in the flow considered in Reference 18 with the following streamfunction in the laboratory frame of reference:

$$\Psi'(x', y', t') = -\Psi'_0 \tanh \left(\frac{y' - a \cos k(x' - ct')}{\lambda \sqrt{1 + k^2 a^2 \sin^2 k(x' - ct')}} \right), \tag{22.3}$$

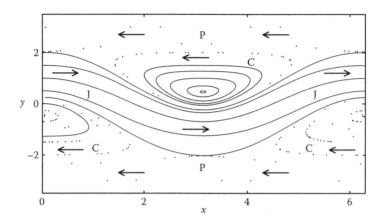

FIGURE 22.1 The phase portrait of the model flow (22.4) in the comoving frame of reference. The first frame with streamlines in the circulation (C), jet (J), and peripheral currents (P) zones is shown. The parameters are: $A = 0.785$, $C = 0.1168$, and $L = 0.628$.

where the hyperbolic tangent produces the Bickley-jet profile, the square root provides a constant width of the jet λ, and a, k, and c are amplitude, wave number, and phase velocity of the running wave, respectively. The normalized streamfunction in the frame moving with c is

$$\Psi = -\tanh\left(\frac{y - A\cos x}{L\sqrt{1 + A^2 \sin^2 x}}\right) + Cy, \tag{22.4}$$

where $x = k(x' - ct')$ and $y = ky'$ are new scaled coordinates. The normalized jet's width $L = \lambda k$, wave's amplitude $A = ak$, and phase velocity $C = c/\Psi_0' k$ are the control parameters. The advection equations (22.2) with the streamfunction (22.4) have the following form in the comoving frame:

$$\dot{x} = \frac{1}{L\sqrt{1 + A^2 \sin^2 x}\cosh^2\theta} - C, \qquad \dot{y} = -\frac{A\sin x(1 + A^2 - Ay\cos x)}{L\left(1 + A^2 \sin^2 x\right)^{3/2}\cosh^2\theta},$$

$$\theta = \frac{y - A\cos x}{L\sqrt{1 + A^2 \sin^2 x}}, \tag{22.5}$$

where dot denotes differentiation with respect to the scaled time $t = \Psi_0' k^2 t'$.

The flow with the streamfunction (22.3) is steady in the comoving frame and its phase portrait is shown in Figure 22.1. There are southern and northern sets of elliptic fixed points: $x_e^{(s)} = 2\pi n$, $y_e^{(s)} = -L\operatorname{arcosh}\sqrt{1/LC} + A$ and $x_e^{(n)} = (2n+1)\pi$, $y_e^{(n)} = L\operatorname{arcosh}\sqrt{1/LC} - A$, respectively, and the southern and northern sets of hyperbolic (saddle) fixed points: $x_s^{(s)} = (2n+1)\pi$, $y_s^{(s)} = -L\operatorname{arcosh}\sqrt{1/LC} - A$ and $x_s^{(n)} = 2\pi n$, $y_s^{(n)} = L\operatorname{arcosh}\sqrt{1/LC} + A$, respectively, where $n = 0, \pm1, \ldots$.

A perturbation is provided by a periodic modulation of the wave's amplitude

$$A(t) = A_0 + \varepsilon\cos(\omega t + \varphi). \tag{22.6}$$

The equations of motion (22.5) are symmetric under the following transformations: $t \to t$, $x \to \pi + x$, $y \to -y$, and $t \to -t$, $x \to -x$, and $y \to y$. Owing to these symmetries, the motion can be considered in the northern chain of the circulation cells on the cylinder with $0 \le x \le 2\pi$. The part of the phase space with $2\pi n \le x \le 2\pi(n+1)$, $n = 0, \pm1, \ldots$, is called *a frame*. The first frame is shown in Figure 22.1. The values of the following control parameters are fixed in our simulation: $L = 0.628$, $A_0 = 0.785$, $C = 0.1168$, $T_0 = 2\pi/\omega = 24.7752$, and $\varphi = \pi/2$. The only varying parameter is the perturbation amplitude ε.

In fluid mechanics, an infinite number of initial conditions comes into play simultaneously and a number of fluid elements, launched in different places, may follow the same orbit on the flow plane. Essentially, a number of particles with different initial positions may move along the same orbit. More precisely, *an orbit is a set of points* $\{x_i, y_i\}$ ($i = 1, 2, \ldots$) on the phase plane (on the flow plane) with the following two properties: (i) there exists for $\forall i, j$ an integer k (positive or negative) such that

$$\begin{pmatrix} x_i \\ y_i \end{pmatrix} = \hat{U}(kT_0) \begin{pmatrix} x_j \\ y_j \end{pmatrix}, \tag{22.7}$$

where $\hat{U}$ is an evolution operator; and (ii) there exists for $\forall i, k$ an integer j such that

$$\hat{U}(kT_0) \begin{pmatrix} x_i \\ y_i \end{pmatrix} = \begin{pmatrix} x_j \\ y_j \end{pmatrix}. \tag{22.8}$$

A *period-m orbit* is a finite set of points on the phase plane with the properties (22.7) and (22.8) consisting of m elements. Thus, any period-m orbit contains m points whose trajectories belong to this orbit.

To locate the UPOs in the phase space, we fix values of the control parameters and compute with a large number of particles the Euclidean distance $d = \sqrt{[x(t_0 + mT_0) - x(t_0)]^2 + [y(t_0 + mT_0) - y(t_0)]^2}$ between the particle's position at an initial moment of time t_0 and at the moments of time $T = mT_0$, where $m = 1, 2, \ldots$. The data are plotted as *a period-m return map* (RM) that shows by color the values of d for particles with initial positions $[x(t_0), y(t_0)]$. At the first stage, we select a large number of points where the function $d(x(t_0), y(t_0))$ may have local minima. Then we apply the method of a downhill simplex to localize the minima in the neighborhoods of those points. There are such minima among them for which $d = 0$ with a given value of m. The procedure allows us to detect both UPOs and SPOs not only in periodically perturbed Hamiltonian systems but also in any chaotic system.

RMs with $m = 1, 2, 3, 4$, and 12 have been computed. In the next section, we consider the saddle orbit (SO) which is a period-1 UPO. The main efforts are devoted to analysis of period-4 UPOs because it is not a trivial task to detect the UPOs when the corresponding resonances cannot be identified on Poincaré sections. It is the case with $m = 4$. The period-4 RM is shown in Figure 22.2a with the cross-marking location of the period-1 saddle trajectory and black dots marking initial positions of 26 trajectories of the period-4 UPOs with a relative accuracy $10^{-13} \cdot 10^{-14}$. For comparison, we demonstrate in Figure 22.2b locations of those dots on the Poincaré section. It is evident that they cannot be prescribed to any structures in the phase space and cannot be identified by inspection of the Poincaré section.

22.2.2 Saddle Orbit

The saddle points of the unperturbed equations of motion (22.5), $(x_s^{(n)}, y_s^{(n)})$, and $(x_s^{(s)}, y_s^{(s)})$ become period-1 SOs under the periodic perturbation (22.6). To analyze such a SO, we linearize the perturbed equations

$$\dot{x} = X(x, y), \quad \dot{y} = Y(x, y), \tag{22.9}$$

in a neighborhood of the saddle point

$$x_s = 0, \quad y_s = L \operatorname{arcosh} \sqrt{\frac{1}{LC}} + A_0, \tag{22.10}$$

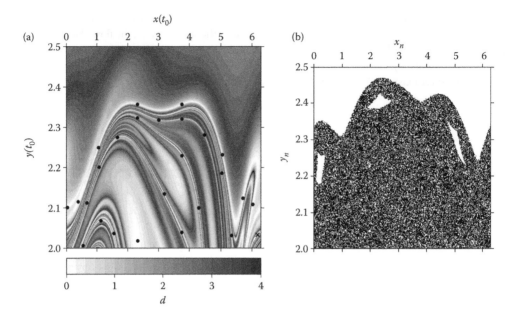

FIGURE 22.2 (a) Period-4 return map representing the distance d between particle's position $x(t_0)$ and $y(t_0)$ at t_0 and its position at $t_0 + 4T_0$. Cross marks the location of the saddle trajectory and dots mark initial positions of trajectories of the period-4 UPOs. (b) Poincaré section of the northern separatrix layer with positions of those dots.

where $X(x, y)$ and $Y(x, y)$ are the right-hand sides of the corresponding equations in the set (22.5) with A being the time-periodic amplitude (22.6). After linearizing, we get the following equations for small deviations $\eta \equiv x - x_s$ and $\xi \equiv y - y_s$:

$$\dot{\eta} = \frac{1}{L \cosh^2 \Theta} - C - \frac{2 \tanh \Theta}{L^2 \cosh^2 \Theta} \xi, \qquad \dot{\xi} = \frac{A(AL\Theta - 1)}{L \cosh^2 \Theta} \eta, \qquad (22.11)$$

where

$$\Theta \equiv \operatorname{arcosh} \sqrt{\frac{1}{LC} - \frac{\varepsilon \cos \Phi}{L}}, \qquad \Phi \equiv \omega t + \varphi.$$

For small values of the perturbation amplitude ε, it is possible to simplify the set (22.11) expanding the right-hand sides in a series in powers of $\varepsilon \cos \Phi$ and neglecting terms above the first-order. In terms of the variables

$$X_s = \lim_{\varepsilon \to 0} X(x_s, y_s), \quad X_{ys} = \lim_{\varepsilon \to 0} X_y(x_s, y_s), \quad Y_{xs} = \lim_{\varepsilon \to 0} Y_x(x_s, y_s),$$

$$X_\varepsilon = \lim_{\varepsilon \to 0} \frac{\partial X(x_s, y_s)}{\partial(\varepsilon \cos \Phi)}, \quad X_{y\varepsilon} = \lim_{\varepsilon \to 0} \frac{\partial X_y(x_s, y_s)}{\partial(\varepsilon \cos \Phi)}, \quad Y_{x\varepsilon} = \lim_{\varepsilon \to 0} \frac{\partial Y_x(x_s, y_s)}{\partial(\varepsilon \cos \Phi)},$$

$$X_x(x, y) = \frac{\partial X(x, y)}{\partial x}, \qquad X_y(x, y) = \frac{\partial X(x, y)}{\partial y},$$

$$Y_x(x, y) = \frac{\partial Y(x, y)}{\partial x}, \qquad Y_y(x, y) = \frac{\partial Y(x, y)}{\partial y}$$

$$(22.12)$$

we get the following equations of the first-order in $\varepsilon \cos \Phi$:

$$\dot{\eta} = X_s + X_{ys}\xi + \varepsilon \cos \Phi(X_\varepsilon + X_{y\varepsilon}\xi), \quad \dot{\xi} = (Y_{xs} + \varepsilon \cos \Phi Y_{x\varepsilon})\eta, \qquad (22.13)$$

where

$$X_s = 0, \quad X_{ys} = -\frac{2C\sqrt{1-LC}}{L}, \quad Y_{xs} = A_0 C \left(A_0 L \operatorname{arsech} \sqrt{LC} - 1 \right),$$

$$X_\varepsilon = \frac{2C\sqrt{1-LC}}{L}, \quad X_{y\varepsilon} = \frac{2C(3LC-2)}{L^2}, \tag{22.14}$$

$$Y_{x\varepsilon} = \frac{C \left(2A_0 L \left(L + A_0\sqrt{1-LC} \right) \operatorname{arsech} \sqrt{LC} - 2A_0\sqrt{1-LC} - L(1+A_0^2) \right)}{L}.$$

Coming back to set (22.11) and expanding the equations up to the second-order in $\varepsilon \cos \Phi$, we obtain the equations

$$\dot{\eta} = X_s + X_{ys}\xi + \varepsilon \cos \Phi (X_\varepsilon + X_{y\varepsilon}\xi) + \frac{\varepsilon^2}{2} \cos^2 \Phi (X_{\varepsilon^2} + X_{y\varepsilon^2}\xi),$$

$$\dot{\xi} = (Y_{xs} + \varepsilon \cos \Phi Y_{x\varepsilon} + \frac{\varepsilon^2}{2} \cos^2 \Phi Y_{x\varepsilon^2})\eta, \tag{22.15}$$

with the following notations:

$$Y_{x\varepsilon^2} = \lim_{\varepsilon \to 0} \frac{\partial^2 Y_x(x_s, y_s)}{\partial (\varepsilon \cos \Phi)^2} = -\frac{2C}{L^2} \left[A_0(2L^2 - 3LC + 2) + 2L(1+A_0^2)\sqrt{1-LC} - \right.$$

$$\left. - L \left(L^2 + A_0^2(2 - 3LC) + 4A_0 L\sqrt{1-LC} \right) \operatorname{arsech} \sqrt{LC} \right],$$

$$X_{\varepsilon^2} = \lim_{\varepsilon \to 0} \frac{\partial^2 X(x_s, y_s)}{\partial (\varepsilon \cos \Phi)^2} = \frac{2C(2 - 3LC)}{L^2}, \tag{22.16}$$

$$X_{y\varepsilon^2} = \lim_{\varepsilon \to 0} \frac{\partial^2 X_y(x_s, y_s)}{\partial (\varepsilon \cos \Phi)^2} = \frac{8C\sqrt{1-LC}(3LC-1)}{L^3}.$$

In the following, we numerically compare some properties of the SOs, generated from the same fixed point (x_s, y_s), with the main equations of motion (22.9), with linearized equations (22.11), with linearized Equations 22.13 with $O(\varepsilon)$, and Equations 22.15 with $O(\varepsilon^2)$. Simulation shows that dependence of x-coordinate of the initial position of the SO on the perturbation amplitude ε is practically the same with all the versions of the advection equations listed above. As ε increases, the x-coordinate shifts to the west from the point $x_s = 0$. As to the y-coordinate, it moves to the north with increasing ε in a similar way for Equations 22.9, 22.11, and 22.15, but its behavior differs strongly for the first-order equations in ε (22.13). We may conclude that even with small values of the perturbation amplitude, the effect of the second harmonic 2ω is not small.

22.2.3 Origin and Bifurcations of UPOs

In this section, we analyze the origin of the period-4 UPOs and their bifurcations that occur with changing the perturbation amplitude ε. Using the period-4 RM (Figure 22.2a), we located all the period-4 UPOs and the initial positions of four trajectories for each of them. They differ by the type of motion of the passive particles and their length l which is a length of the corresponding curve between two positions on the orbit separated by the time interval $4T_0$. On the cylinder $0 \le x \le 2\pi$, all the UPOs are closed curves whose topology may be very complicated. In the physical space, the UPOs can be classified as rotational ones (particles rotate in the same frame), ballistic ones (particles move ballistically from frame to frame), and rotation-ballistic ones (particles may rotate for a while in a frame, then move ballistically through a few frames, and change their direction of motion).

22.2.3.1 $C_{WB}^{4:1}$ Class: Western Ballistic UPOs Associated with the 4:1 Western Ballistic Resonance

The shortest ones among all the period-4 UPOs are western ballistic UPOs. The particles belonging to the $C_{WB}^{4:1}$ class move in a periodic way to the west along such an orbit in the northern separatrix layer which

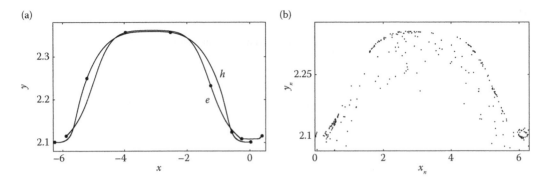

FIGURE 22.3 (a) Two $C_{WB}^{4:1}$ western ballistic period-4 UPOs at $\varepsilon = 0.0785$ with initial positions of four unstable periodic trajectories belonging to each of them. The e and h orbits were born from the elliptic and hyperbolic points of the northern ballistic resonance 4:1, respectively. (b) The Poincaré section at $\varepsilon = 0.005$ with the northern ballistic resonance 4:1 consisting of four ballistic islands along the northern border of the separatrix layer.

appears between the northern (C) and (P) regions in Figure 22.1 as a result of the perturbation. With the help of the RM in Figure 22.2a, we located two $C_{WB}^{4:1}$ orbits with initial positions of four periodic trajectories belonging to each of them. In order to track out the origin of the e and h orbits, shown in Figure 22.3a at $\varepsilon = 0.0785$, we decrease the value of the perturbation amplitude ε, compute the corresponding period-4 RMs, locate the $C_{WB}^{4:1}$ orbits, and measure their length.

The result may be resumed as follows. The western ballistic resonance 4:1 with four elliptic and four hyperbolic points appears under a perturbation with a very small value of ε. On the Poincaré section in Figure 22.3b at $\varepsilon = 0.005$, it is manifested as four ballistic islands along the northern border of the separatrix layer. Two of the islands are so thin that they are hardly visible in the figure. With increasing ε, the size of the resonance decreases and at the critical value $\varepsilon \approx 0.016$ it vanishes (see Figure 22.2b where there are no signs of that resonance at $\varepsilon = 0.0785$). The orbit, associated with the elliptic fixed points of the western resonance 4:1, loses its stability and bifurcates into the $C_{WB}^{4:1}$ UPO of period-4 which we denote by the symbol e. Its length practically does not change with increasing ε. The length l of the h orbit, associated with the hyperbolic points of that resonance, changes dramatically at $\varepsilon \approx 0.0715$ increasing rapid after this point because of the appearance of a meander and a loop on the h orbit nearby the SO.

22.2.3.2 $C_{EB}^{4:1}$ Class: Eastern Ballistic UPOs Associated with the 4:1 Eastern Ballistic Resonance

The eastern ballistic UPOs of period-4 lie in the southern separatrix layer between the (J) and (C) regions (see Figure 22.1) to the north from the jet. Particles move along such orbits to the east in a periodic way. The origin and bifurcations of the $C_{EB}^{4:1}$ orbits are similar to the western ballistic ones. They appear from elliptic and hyperbolic points of the eastern ballistic resonance 4:1 as ε increases. The elliptic orbit of this resonance loses it stability and bifurcates into the UPO of the e-type. The hyperbolic orbit changes its topology, transforming from the bell-like curve at $\varepsilon = 0.01$ to the curve with a meander and a loop nearby the SO at $\varepsilon = 0.0785$ (see Figure 22.4). The length l of the h orbit decreases with increasing ε up to the bifurcation point $\varepsilon \approx 0.0275$, after which it increases rapid due to the complexification of the orbit's form (see Figure 22.4).

22.2.3.3 $C_{R}^{4:1}$ Class: Orbits Associated with the 4:1 Rotational Resonance

Four different orbits in the $C_{R}^{4:1}$ class, located with the help of the period-4 RM, are shown in Figure 22.5 at $\varepsilon = 0.0785$ along with initial positions of four unstable periodic trajectories on each of them. The corresponding fluid particles rotate in the same frame along closed curves. The genesis of the $C_{R}^{4:1}$ orbits is

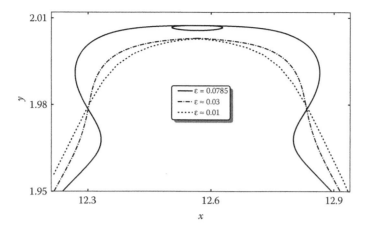

FIGURE 22.4   Metamorphoses of the h $C_{EB}^{4:1}$ eastern ballistic UPO as the perturbation amplitude ε changes: solid line, $\varepsilon = 0.0785$; dashed line, $\varepsilon = 0.03$; and dotted line, $\varepsilon = 0.01$.

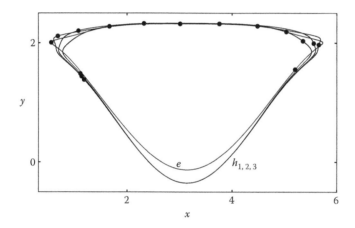

FIGURE 22.5   Four 4:1 rotational UPOs of the $C_R^{4:1}$ class with initial positions of four unstable periodic trajectories belonging to each of them. The e and $h_{1,2,3}$ orbits were born from the elliptic and hyperbolic points of the 4:1 rotational resonance, respectively ($\varepsilon = 0.0785$).

the following. A 4:1 rotational resonance appears under a small perturbation (22.6). As the amplitude ε increases, its elliptic orbit loses stability and bifurcates at $\varepsilon \approx 0.003$ into the UPO, which we denote by the symbol e. The hyperbolic orbit of the 4:1 resonance h bifurcates at $\varepsilon \approx 0.040945$ into two hyperbolic orbits, h_1, h_2, and an elliptic orbit h_3 in the centers of four stability islands. It is a pitchfork bifurcation. In Figure 22.6a, we show by arrows a movement of initial positions of the $C_R^{4:1}$ orbits as ε decreases. The pitchfork bifurcation point is shown as a black circle. As ε increases further, the elliptic orbit h_3 loses its stability and becomes a hyperbolic $C_R^{4:1}$ orbit.

22.2.3.4 $C_R^{2:1}$ Class: Orbits Associated with the 2:1 Rotational Resonance

The class $C_R^{2:1}$ consists of two rotational UPOs, associated with the 2:1 rotational resonance, with elliptic and hyperbolic orbits, and two periodic trajectories on each of them. The resonance appears under a small perturbation. At $\varepsilon \approx 0.0665$, the elliptic orbit of this resonance undergoes a period-doubling bifurcation into a period-4 elliptic orbit with four trajectories and a period-2 hyperbolic orbit with two trajectories.

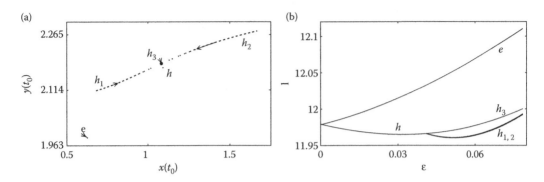

FIGURE 22.6 (a) Movement of initial positions $(x(t_0), y(t_0))$ of the rotational 4:1 UPOs with decreasing the perturbation amplitude is shown by arrows. The black circle is a pitchfork bifurcation point at $\varepsilon \approx 0.040945$ where the hyperbolic orbit of the resonance h bifurcates into three different orbits h_1, h_2, and h_3. (b) Bifurcation diagram $l(\varepsilon)$.

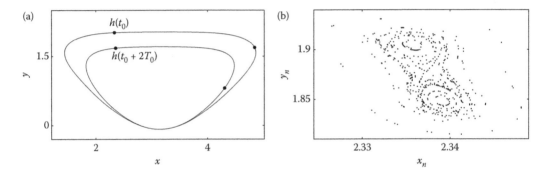

FIGURE 22.7 (a) 2:1 rotational UPOs with initial positions of four unstable periodic trajectories ($\varepsilon = 0.0785$). (b) Poincaré section of the 2:1 rotational resonance at $\varepsilon = 0.0668$ (just after the period-doubling bifurcation).

The latter is shown in Figure 22.7a at $\varepsilon = 0.0785$. The Poincaré section of the 2:1 rotational resonance at $\varepsilon = 0.0668$ (just after the bifurcation) is shown in Figure 22.7b. By further increasing ε, we see that the resonance 2:1 vanishes, and the period-4 elliptic orbit loses its stability and bifurcates into a period-4 UPO. To clarify this bifurcation, we fix a point at the moment t_0 on the upper branch of the orbit in Fig. 22.7a and another point on the lower branch at $t = t_0 + 2T_0$ and scan their positions when decreasing the perturbation amplitude ε. Figure 22.8 demonstrates clearly that the points move to the position of the elliptic orbit of the 2:1 resonance and merge with it at the bifurcation point.

22.2.3.5 $C_{RB}^{4:1}$ Class: Rotation-Ballistic UPOs Associated with the Rotational-Ballistic Resonance

The class $C_{RB}^{4:1}$ consists of four period-4 UPOs shown in Figure 22.9a. We call those orbits rotational-ballistic (RB) ones because the corresponding particles begin to move to the west in one frame, then turn to the east, and travel in the southern separatrix layer to the next frame, fulfill one turnover in this frame, and repeat their motion to the east. The genesis of the $C_{RB}^{4:1}$ orbits differs from the genesis of the other classes of period-4 UPOs. Each of the resonances, associated with $C_{RW}^{4:1}$, $C_{RE}^{4:1}$, $C_{R}^{4:1}$, and $C_{R}^{2:1}$, appears under an infinitely small perturbation amplitude ε. Rotation-ballistic motion of period-m and corresponding orbits cannot in principle appear in the flow below some critical value of ε (which depends on m) because the width of the stochastic layers (which increases with increasing ε) should be large enough in order that particles would have enough time to travel the corresponding distance to the east and west. The genesis

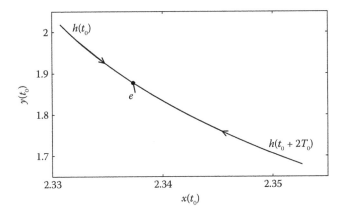

FIGURE 22.8 Emergency of the doubly branched UPO, shown in Figure 22.7a, from the elliptic orbit of the 2:1 rotational resonance at the bifurcation point (black circle). Movement of fixed initial points on the upper $h(t_0)$ and lower $h(t_0 + 2T_0)$ branches to the bifurcation point with decreasing ε is shown.

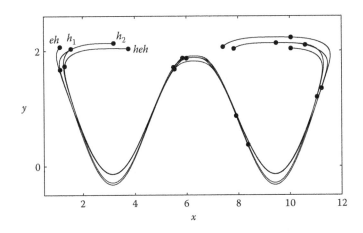

FIGURE 22.9 Four rotational-ballistic period-4 UPOs with their initial positions in the phase space eh, h_1, h_2, and heh. The other points are the initial positions of the corresponding trajectories.

and evolution of the $C_{RB}^{4:1}$ UPOs are shown schematically on the bifurcation diagrams in Figure 22.10. In Figure 22.10a, we demonstrate movement in the phase space of initial positions of the $C_{RB}^{4:1}$ orbits with decreasing the perturbation amplitude ε. The dependence of the lengths of those orbits on ε is shown in Figure 22.10b.

The RB 4:1 resonance appears at a critical value of the perturbation amplitude $\varepsilon = \varepsilon_1 \approx 0.040715$ as the result of the saddle-center bifurcation and manifests itself as four small islands of stability on the corresponding Poincaré section. One of these islands is shown in Figure 22.11a. If ε increases further, the elliptic orbit eh in the centers of the RB 4:1 resonance loses its stability and becomes a hyperbolic RB period-4 orbit. The hyperbolic orbit h of the RB resonance at $\varepsilon = \varepsilon_2 \approx 0.0433$ undergoes a pitchfork bifurcation into two hyperbolic RB period-4 orbits h_1 and h_2 and the elliptic orbit heh in the centers of the new period-4 RB resonance one of whose stability islands is shown in Figure 22.11b. Under further increasing ε, the elliptic orbit heh loses its stability and becomes a hyperbolic RB period-4 orbit.

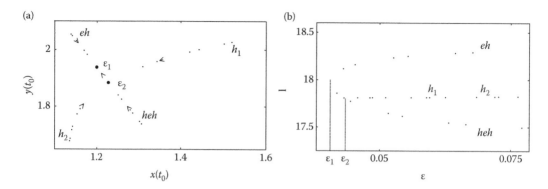

FIGURE 22.10   (a) At the saddle-center bifurcation point $\varepsilon = \varepsilon_1 \approx 0.040715$, there appears an RB 4:1 resonance with the elliptic orbit *eh* which loses its stability with increasing ε. The hyperbolic orbit of the RB resonance bifurcates at $\varepsilon = \varepsilon_2 \approx 0.0433$ into two hyperbolic period-4 UPOs h_1 and h_2 and the elliptic orbit *heh* which loses its stability with further increasing ε. (b) Bifurcation diagram $l(\varepsilon)$.

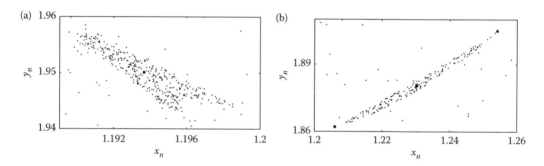

FIGURE 22.11   (a) One of the stability islands of the RB 4:1 resonance appearing after the saddle-center bifurcation at $\varepsilon = 0.04072 \gtrsim \varepsilon_1$ with the elliptic point *eh* in its center. (b) One of the stability islands of the RB 4:1 resonance appearing after a pitchfork bifurcation at $\varepsilon = 0.0435 \gtrsim \varepsilon_2$ with the elliptic point *heh* in its center.

The results of this section may be concluded as follows.

We have developed a method to track UPOs based on computing and analyzing local minima of the distance function $d(x(t_0), y(t_0))$. The method is not restricted to our simple Hamiltonian system with a periodic perturbation and is applicable to a variety of chaotic systems.

We detected and located the UPOs of periods 1 and 4 with our specific model of a meandering jet. Varying the perturbation amplitude, we have found bifurcations of the period-4 orbits, the UPOs of the lowest period with nontrivial origin. Those orbits have been grouped into five classes by their origin and bifurcations. $C_{\text{WB}}^{4:1}$ ($C_{\text{EB}}^{4:1}$) class consists of the western (eastern) ballistic UPOs associated with the 4:1 western (eastern) ballistic resonance. $C_{\text{R}}^{4:1}$ ($C_{\text{R}}^{2:1}$) class consists of the rotational UPOs associated with the 4:1 (2:1) rotational resonance. $C_{\text{RB}}^{4:1}$ class consists of specific RB UPOs associated with 4:1 RB resonance.

We would like to stress the following. Rotational and ballistic resonant islands are well-known objects in the phase space of Hamiltonian systems. As far as we know, rotation-ballistic resonant islands of stability have not been found before. They are expected to appear as well in other jet-flow models (kinematic and dynamic ones) under specific values of control parameters. Like ballistic islands, the rotation-ballistic islands should affect transport and its statistical characteristics will be demonstrated in the next section. Owing to the presence of rotation phase of motion, the mean drift velocity of RB particles is smaller than that for ballistic particles. It is expected that boundaries of the RB islands are specific dynamical traps that should affect transport and statistical properties of passive particles (see the next section). Properties of RB islands traps may differ from properties of rotational-islands traps (RITs) and ballistic-islands traps (BITs) because of different locations of the corresponding hyperbolic points around the islands. In the chains of rotational and ballistic islands, they appear to be between the islands, whereas in the case of RB islands they are situated either on one (a saddle-center bifurcation) or both sides (a pitchfork bifurcation) of a given island.

In addition, by linearizing the advection equations, we have studied the properties of the period-1 SOs playing a crucial role in chaotic transport and mixing of passive particles. Stable and unstable manifolds of UPOs are material curves of complicated forms which cannot be crossed by the particles' trajectories. Their role in transport and mixing of passive particles in the meandering jet flow is considered significant because they separate trajectories with different dynamical and topological properties. The UPOs found in this section act as dynamical traps where particles and their trajectories may spend a rather long time before escaping. It was checked by computing distributions of the escape times for the period-4 UPOs and the SO. The rotational UPOs should contribute to statistics of comparatively short flights (see Figure 5 in Reference 27). The UPOs with larger values of the period can be classified into three big groups: ballistic, rotational, and rotation-ballistic ones. The origin and bifurcations of the orbits in each class can be studied in a similar way, but may require larger computational efforts.

The results obtained do not depend critically on the exact form of the streamfunction and chosen values of the control parameters. The UPOs in other kinematic and dynamical models of geophysical jets, known in the literature (for a review see References 14 and 17), can be detected, located, and classified in a similar way. The UPOs form the skeleton for chaotic advective mixing and transport in fluid flows, and the knowledge of them (at least, lower-period ones) allows us to analyze complex, albeit rather regular, stretching and folding structures in the flows.

22.3 Effect of Dynamical Traps on Chaotic Zonal Transport in a Meandering Jet Current

In this section, we study a phenomenological connection between dynamical, topological, and statistical properties of chaotic mixing and transport in the meandering jet flow considered in the preceding section and explain transport properties by a phenomenon of the so-called dynamical traps. Following Zaslavsky, the dynamical trap is a domain in the phase space of a Hamiltonian system where a particle (or, its trajectory) can spend an arbitrary-long finite time, performing almost regular motion, despite the fact that the full trajectory is chaotic in any appropriate sense. In fact, it is the definition of a quasi-trap. Absolute traps, where particles could spend an infinite time, are possible in Hamiltonian systems only with a zero measure set. The dynamical traps are due to a stickiness of trajectories to some singular domains in the phase space, largely, to the boundaries of resonant islands, saddle trajectories, and cantori. There are no classification and description of the dynamical traps. Zaslavsky described two types of dynamical traps in Hamiltonian systems: hierarchical-islands traps around chains of resonant islands and stochastic-layer traps which are stochastic jets inside a stochastic sea where trajectories can spend a very long time. It is expected that classification and description of the most typical dynamical traps would help us to construct kinetic equations which will be able to describe transport properties of chaotic systems including anomalous ones.

The model of a meandering jet current has been introduced in the preceding section. It is convenient to characterize chaotic mixing and transport in terms of zonal flights. A zonal flight is a motion of a particle between two successive changes of signs of its zonal velocity, i.e., the motion between two successive events $\dot{x} = u = 0$. Particles (and corresponding trajectories) in chaotic jet flows can be classified in terms of the lengths of flights x_f as follows. The trajectories with $|x_f| < 2\pi$ correspond to the particles moving in the same frame or in neighbor frames. In the global stochastic layer, there are particles moving chaotically forever in the same frame but they are of a zero measure. Among the particles with interframe motion, there are regular and chaotic ballistic ones. Regular ballistic trajectories can be defined as those which cannot have two flights with $|x_f| > 2\pi$ in succession. They correspond to particles moving in regular regions of the phase space persisting under the perturbation, (eastward motion in the jet and western motion in the peripheral current) and those moving in the stochastic layer (trajectories belonging to ballistic islands). Typical chaotic trajectories have complicated distributions over the lengths and durations of flights.

In the laboratory frame of reference, all the fluid particles move to the east together with the jet flow and a flight is a motion between two successive events when the particle's zonal velocity U is equal to the meander's phase velocity c. If $U < c$, the corresponding particle is left behind the meander (it is a western flight in the comoving frame), if $U > c$, it passes the meander (an eastern flight in the comoving frame). Short flights with $|x_f| < 2\pi$ (motion in the same spatial frame in the comoving frame of reference) correspond to the motion in the laboratory frame when two successive events $U = c$ occur on the space interval less than the meander's spatial period $2\pi/k$. Ballistic flights between the spatial frames in the comoving frame with $|x_f| > 2\pi$ correspond to the motion in the laboratory frame when the particles move through more than one meander's crest between two successive events $U = c$.

22.3.1 Turning Points

As in Reference 18, we will characterize statistical properties of chaotic transport by probability density functions (PDFs) of lengths of flights $P(x_f)$ and durations of flights $P(T_f)$ for a number of very long chaotic trajectories. Both regular and chaotic particles may change many times the sign of their zonal velocity $\dot{x} = u$. From the condition $\dot{x} = 0$ in Equation 22.5, it is easy to find the equations for the curves which are loci of turning points

$$Y_{\pm}(x, A) = \pm L\sqrt{1 + A^2 \sin^2 x} \times \operatorname{arsech} \sqrt{LC\sqrt{1 + A^2 \sin^2 x} + A \cos x}. \tag{22.17}$$

We consider the northern curve, i.e., Equation 22.17 with the positive sign. Taking into account that the perturbation has the form (22.6), we realize that all the northern turning points are inside a strip confined by the two curves of the form (22.17) with $A = A_0 \pm \varepsilon$. Let us analyze the derivative over the varying parameter A

$$\frac{\partial Y}{\partial A} = \cos x + \frac{ACL^2 \sin^2 x}{2D}\left(2\operatorname{arsech}\sqrt{D} - \frac{1}{\sqrt{1-D}}\right), \tag{22.18}$$

where $D = LC\sqrt{1 + A^2 \sin^2 x}$. If the derivative at a fixed value of x does not change its sign on the interval $A_0 - \varepsilon \le A \le A_0 + \varepsilon$, then Y varies from $Y(x, A_0 - \varepsilon)$ to $Y(x, A_0 + \varepsilon)$, and for each value of y, we have a single value of the perturbation parameter A. However, there may exist such values of x for which the equation $\partial Y/\partial A = 0$ has a solution on the interval mentioned above. In this case, one may have more than 1 value of A for a single value of y. Thus, the width of the strip, containing turning points, is defined by the values of Y at the extremum points and at the end points of the interval of the values of A. In Figure 22.12b, we show the turning points of a single chaotic trajectory on the cylinder $0 \le x \le 2\pi$ confined between the two corresponding curves.

In the numerical simulation, we use the Runge–Kutta integration scheme of the fourth-order with the constant time step $\Delta t = 0.0247$. To study chaotic transport, we have carried out numerical experiments with tracers initially placed in the stochastic layer. It was found that statistical properties of chaotic transport practically do not depend on the number of tracers provided that the corresponding trajectories are sufficiently long ($t \simeq 10^8$). The PDFs for the lengths x_f and durations T_f of flights for five tracers with the computation time $t = 5 \cdot 10^8$ for each tracer are shown in Figure 22.13a and b, respectively, for both the eastward (e) and westward (w) motion. Both $P(x_f)$ and $P(T_f)$ are complicated functions with local extrema decaying in a different manner for different ranges of x_f and T_f. The main aim of our study of chaotic transport is to figure out the basic peculiarities of the statistics and attribute them to specific zones in the phase space, namely, to dynamical traps strongly influencing the transport.

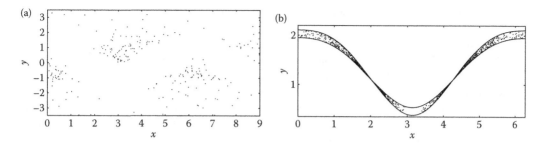

FIGURE 22.12 (a) Poincaré section of the perturbed meandering jet in the co-moving frame. The parameters of the steady flow are: the jet's width $L = 0.628$, the meander's amplitude $A_0 = 0.785$ and its phase velocity $C = 0.1168$. The perturbation amplitude and frequency are: $\varepsilon = 0.0785$ and $\omega = 0.2536$. (b) Turning points of a single chaotic trajectory on the cylinder $0 \leq x \leq 2\pi$ are in a strip confined by two curves (22.17) with $A = A_0 \pm \varepsilon$.

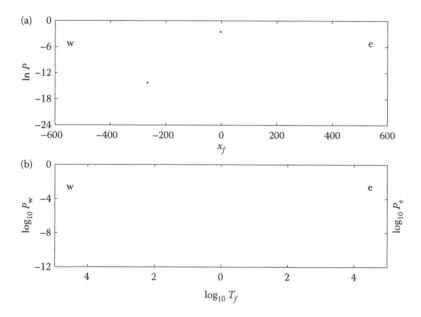

FIGURE 22.13 PDFs of (a) lengths x_f and (b) durations T_f of the w and e flights. The PDFs $P_w(T_f)$ and $P_e(T_f)$ are normalized to the number of w $(4.23 \cdot 10^7)$ and e $(4 \cdot 10^7)$ flights, respectively. Statistics for five tracers with the computation time $t = 5 \cdot 10^8$ for each one.

22.3.2 Rotational-Islands Traps

It is well known that in nonlinear Hamiltonian systems, a complicated structure of the phase space with islands, stochastic layers, and chains of islands, immersed in a stochastic sea, arises under a perturbation due to a variety of nonlinear resonances and their overlapping. The motion is quasi-periodic and stable in the islands. The boundaries of the islands are absolute barriers to transport: particles cannot go through them either from inside or from outside. Invariant curves of the unperturbed system (see Figure 22.1) are destroyed under the perturbation (22.6) (see Figure 22.12a). As the perturbation strength ε increases, a closed invariant curve with frequency f is destroyed at some critical value of ε. If the f/ω is a rational number, the corresponding curve is replaced by an island chain, whereas the curves with irrational frequencies are replaced by cantori. There are uncountably many cantori forming a complicated hierarchy. Numerical experiments with a variety of Hamiltonian systems with a different number of degrees of freedom provide an evidence for the presence of strong partial barriers to transport around the island's boundaries (for a review, see Reference 31) which manifest themselves on Poincaré sections as domains with increased density of points.

In Reference 18, we have found that chosen values of the control parameters exist in each frame of a vortex core (which is an island of the primary resonance $\omega = f$) immersed into a stochastic sea, where there are six islands of a secondary resonance emerged from three islands of the primary resonance $3f = 2\omega$ (see Figure 13 in Reference 18). Chains of smaller-size islands are present around the vortex core and the secondary-resonance islands. Particles belonging to all these islands (including the vortex core) rotate in the same frame performing short flights with the lengths $|x_f| < 2\pi$. So, we will call them *rotational islands* and distinguish from the so-called *ballistic islands* to be considered below.

Stickiness of particles to boundaries of the rotational islands has been demonstrated in Reference 18. It means that real fluid particles can be trapped for a long time in a singular zone near the borders of the rotational islands which we will call RITs. To illustrate the effect of the RITs, we demonstrate in Figures 22.14 and 22.15 the Poincaré sections of a chaotic trajectory in the frame $0 \le x \le 2\pi$ sticking to the vortex core and to the secondary-resonance islands, respectively. The contour of the vortex core is shown in Figure 22.15 by a thick line. The small points are tracks of the particle's position at the moments of time $t_n = 2\pi n/\omega$ (where $n = 1, 2, \dots$) and the thin curves are fragments of the corresponding trajectory on the phase plane. Increased density of points indicates the presence of dynamical traps near the boundaries of the rotational islands. Contribution of the vortex-core RIT (Figure 22.14) to chaotic transport is expected to be much more significant than the one of the RITs of the other islands (Figure 22.15).

It is reasonable to suppose that RITs contribute to the statistics of short flights. By short flights, we mean the flights with the length shorter than 2π. In Figure 22.16, we show the part of the full PDF $P(T_f)$ (Figure 22.13b) for the e and w short flights separately. There are a comparatively small number of the e flights with $T_f < 11$. Let us note the prominent peak of the corresponding PDF at $T_f \simeq 11$ followed by an exponential decay. As to the w short flights, there are two small local peaks around $T_f \simeq 17$ and 21.

To estimate the contribution of the vortex-core RIT to the statistics of short flights, we compute and compare the statistics of the durations of flights T_f for two trajectories: a regular quasi-periodic one with the initial position close to the inner border of the vortex core (Figure 22.17a) and a chaotic one with the initial position close to the vortex-core border from outside (Figure 22.17b). Each full rotation of a particle in a frame consists of two flights, e and w, with different values of T_f because of the zonal asymmetry of the flow. The statistics for the chaotic trajectory, sticking to the vortex core (Figure 22.17b), may be considered

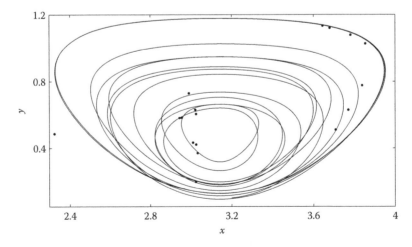

FIGURE 22.14 The vortex-core trap. Poincaré section of a chaotic trajectory in the frame $0 \le x \le 2\pi$ with a fragment of a trajectory.

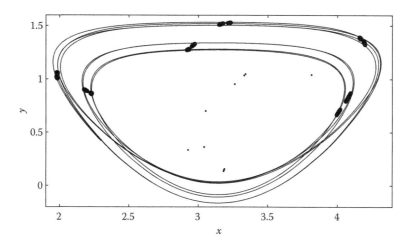

FIGURE 22.15 The secondary-resonance islands trap. A fragment of a chaotic trajectory sticking to the islands is shown.

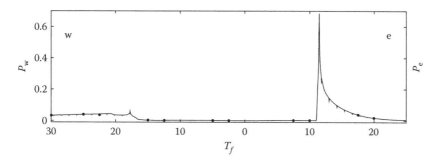

FIGURE 22.16 The PDFs for the e and w flights with the length shorter than 2π. The PDFs $P_w(T_f)$ and $P_e(T_f)$ are normalized to the number of w $(4.19 \cdot 10^7)$ and e $(3.7 \cdot 10^7)$ flights, respectively. Statistics for five tracers with the computation time $t = 5 \cdot 10^8$ for each one.

as a distribution of the durations of flights in the vortex-core RIT. The minimal flight duration in this RIT is $T_f \simeq 11$ (the flights with smaller values of T_f are rare and they occur outside the trap). Positions of the local maxima of the PDF for the sticking trajectory in Figure 22.17b correlate approximately with the corresponding local maxima of the PDF for the regular trajectory inside the core in Figure 22.17a. The similar correlations have been found (but not shown here) between the local maxima of the PDFs for the lengths of flights $P(x_f)$ for the interior regular and sticking chaotic trajectories. These correlations and positions of the peaks prove numerically that short flights with $|x_f| < 2\pi$ and $11 \lesssim T_f \lesssim 21$ may be caused by the effect of vortex-core RIT. We conclude from Figure 22.17b that the vortex-core RIT contributes to the statistics of the short flights in the range $11 \lesssim T_f \lesssim 20$ for the e flights with the prominent peak at $T_f \simeq 11$ and in the range $15 \lesssim T_f \lesssim 21$ for the w flights with small peaks at $T_f \simeq 17$ and 21.

The effect of the RIT of the secondary-resonance islands is illustrated in Figure 22.15. To find the characteristic times of this RIT, we compute two trajectories: a regular quasi-periodic one with the initial position inside one of these islands and a chaotic one with the initial position close to the outer border of the island. The respective PDFs $P(T_f)$, shown in Figure 22.18a and b, demonstrate strong correlations between the corresponding peaks at $T_f \simeq 12, 23$, and 27. Computed (but not shown here) PDFs $P(x_f)$ for these trajectories confirm the effect of the islands RIT on the statistics of short flights.

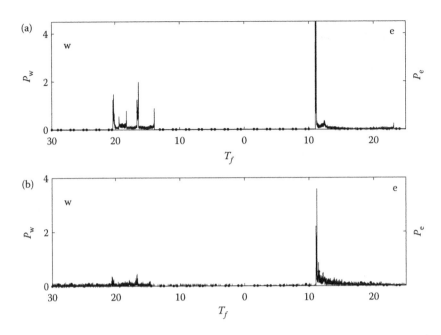

FIGURE 22.17   The vortex-core trap PDFs of durations T_f of the e and w flights. (a) Regular quasi-periodic trajectory with the duration $t = 2 \cdot 10^5$ inside the vortex core close to its boundary. Both the PDFs are normalized to the number $8 \cdot 10^3$ of corresponding flights. (b) Chaotic trajectory with the duration $t = 2 \cdot 10^5$ sticking to the boundary of the vortex core from outside. Both the PDFs are normalized to the number $4 \cdot 10^3$ of corresponding flights.

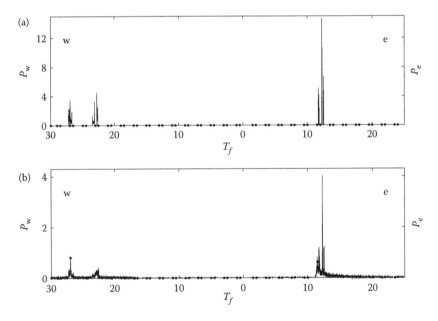

FIGURE 22.18   The secondary-resonance islands trap. The PDFs of durations T_f of the e and w flights. (a) Regular quasi-periodic trajectory inside the islands with the duration $t = 5 \cdot 10^5$. Both the PDFs are normalized to the number $1.5 \cdot 10^4$ of corresponding flights. (b) Chaotic trajectory sticking to the island's boundary from outside with the duration $t = 5 \cdot 10^5$. $P_w(T_f)$ and $P_e(T_f)$ are normalized to the number of w ($1.1 \cdot 10^4$) and e ($9 \cdot 10^3$) flights, respectively.

22.3.3 Saddle Traps

As a result of the periodic perturbation (22.6), the saddle points of the unperturbed system at $x_s^{(n)} = 2\pi n$, $y_s^{(n)} = L \operatorname{arcosh} \sqrt{1/LC} + A$ and at $x_s^{(s)} = (2n+1)\pi$, $y_s^{(s)} = -L \operatorname{arcosh} \sqrt{1/LC} - A$ $(n = 0, \pm 1, \dots)$ become periodic saddle trajectories. These hyperbolic trajectories have their own stable and unstable manifolds and play a role of specific dynamical traps which we call *saddle traps* (STs). In this section, we demonstrate that the STs strongly influence chaotic mixing and transport of passive particles and contribute, mainly, in the short-time statistics of flights.

Tracers with initial positions close to a stable manifold of a saddle trajectory are trapped for a while performing a large number of revolutions along it. To illustrate the effect of the STs, Figure 22.19a and b shows fragments of two chaotic trajectories sticking to the saddle trajectory and performing about 20 full revolutions before escaping to the east (Figure 22.19a) and to the west (Figure 22.19b). We have managed to detect and locate the corresponding periodic unstable saddle trajectory which is situated in Figure 22.19a and b in the domain where a few fragments of the chaotic trajectory imposed on each other. Because of the flow asymmetry, the duration of eastern flights of a particle along the saddle trajectory $T_e \simeq 11.9$ is shorter than the duration of western flights $T_w \simeq 12.9$. The black points are the tracks of the particle's positions on the flow plane at the moments of time $t_n = 2\pi n/\omega \simeq 24.8\,n$ (where $n = 1, 2, \dots$). They belong to smooth

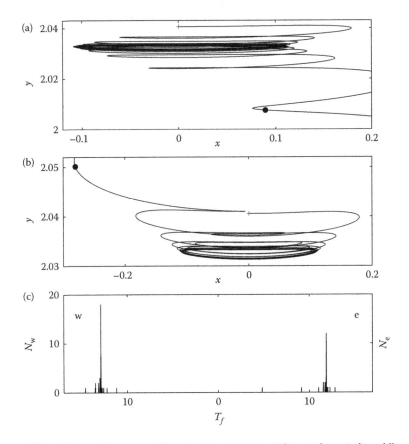

FIGURE 22.19 The saddle trap. Fragments of two chaotic trajectories sticking to the periodic saddle trajectory, one of which escapes to the east (a) and another one to the west (b). (c) The number of the eastward (N_e) and westward (N_w) short flights with duration T_f for those two trajectories. Statistics with two trajectories with the duration $t = 10^3$ and the total number of western $N_w = 55$ and eastern $N_e = 51$ flights.

curves which are fragments of the stable and unstable manifolds of the saddle trajectory at the chosen initial phase $\phi = \pi/2$.

To estimate the contribution of the STs to the statistics of short flights shown in Figure 22.16, we compute and plot in Figure 22.19c the number of the eastward (N_e) and westward (N_w) short flights with a given duration T_f for those two chaotic trajectories sticking to the saddle trajectory arising from the saddle point with the position $x_s = 0$, $y_s \simeq 2.02878$. Each full rotation of the particles consists of an e flight with the duration $T_e \simeq 11.9$ and w flight with the duration $T_w \simeq 12.9$. The flights with $T_e \simeq 11.9$ contribute to the main peak in Figure 22.16 and the flights with $T_w \simeq 12.9$ contribute to "the westward" plateau in that figure.

The mechanism of operation of the STs can be describe as follows. Each saddle trajectory $\gamma(t)$ possesses time-dependent stable $W_s(\gamma(t))$ and unstable $W_u(\gamma(t))$ material manifolds composed of a continuous sets of points through which pass at time t trajectories of fluid particles that are asymptotic to $\gamma(t)$ as $t \to \infty$ and $t \to -\infty$, respectively. Under a periodic perturbation, the stable and unstable manifolds oscillate with the period of the perturbation. It was firstly proved by Poincaré that W_s and W_u may intersect each other transversally at an infinite number of homoclinic points through which pass doubly asymptotic trajectories. To give an image of a fragment of the stable manifold of the periodic saddle trajectory, we distribute homogeneously $2.5 \cdot 10^5$ particles in the rectangular $[-0.4 \le x \le 0.45; 2 \le y \le 2.1]$ and compute the time the particles need to escape the rectangular. The color in Figure 22.20 modulates the time T when particles with given initial positions (x_0, y_0) reach the western line at $x = -1$ or the eastern line at $x = 1$. The particles with initial positions marked by the black and white colors move close to the stable manifold of the saddle trajectory and spend a maximal time near it before escaping. The black and white diagonal curve in Figure 22.20 is an image of a fragment of the corresponding stable manifold. The particles with initial positions to the north from the curve escape to the west along the unstable manifold of the saddle trajectory whereas those with initial positions to the south from the curve escape to the east along its another unstable manifold.

We have found that particles quit the ST along the unstable manifolds in accordance with specific laws. We distribute a large number of particles along the segment with $x_0 = 0$ and $y_0 = [2.02; 2.06]$, crossing the stable manifold W_s, and compute the time T particles with given initial latitude positions y_0 that need to

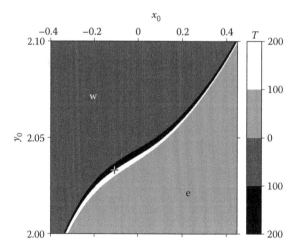

FIGURE 22.20 The saddle-trap map. Color modulates the time T which $2.5 \cdot 10^5$ particles with given initial positions (x_0, y_0) need to reach the lines at $x = -1$ or $x = 1$ escaping to the west (w) and to the east (e), respectively. The black and white diagonal curve is an image of a fragment of the stable manifold of the saddle trajectory. The cross is a position of a particle on that trajectory at the initial time moment. The integration time is $t = 500$.

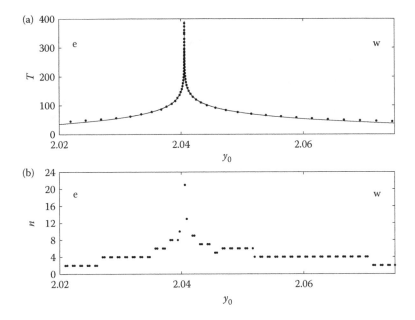

FIGURE 22.21 (a) Time T of particle with an initial latitude position y_0 needs to quit the saddle trap. (b) The number of short flights n such a particle performs before quitting the saddle trap. The ranges of y_0 from which particles quit the trap moving to the west and east are denoted by "w" and "e," respectively.

quit the ST. More precisely, $T(y_0)$ is a time moment when a particle with the initial position y_0 reaches the lines with $x = -1$ or $x = 1$. The "experimental" points in Figure 22.21a fit the law $T_e = (-85.81 \pm 0.04) - (31.216 \pm 0.007) \ln(y_{0s} - y_0)$ for the particles which quit the trap moving to the east and the law $T_w = (-60.61 \pm 0.03) - (28.933 \pm 0.006) \ln(y_0 - y_{0s})$ for those particles which move to the west when quitting the trap, where $y_{0s} = 2.0405755472$ is a crossing point of W_s with the segment of initial positions.

The ST attracts particles and force them to rotate in its zone of influence performing short flights, the number of which n depends on particle's initial positions y_0. The $n(y_0)$ is a steplike function (see Figure 22.21b) with the lengths of the steps decreasing in a geometric progression in the direction to the singular point, $l_j = l_0 \, q^{-j}$, where l_j is the length of the jth step, $q \simeq 2.27$ for the western exits, and $q \simeq 2.20$ for the eastern ones. The seeming deviation from this law in the range $y_0 = [2.045; 2.046]$ (see a small western segment between two larger ones in Figure 22.21b) is explained by crossing the initial line $y_0 = [2.02; 2.06]$ by the curve of zero zonal velocity u. To have the correct law for the western exits, it is necessary to add the two segments of that cut step. The asymmetry of the functions $T(y_0)$ and $n(y_0)$ is caused by the asymmetry of the flow.

22.3.4 Ballistic-Islands Traps

Besides the rotational islands with particles moving around the corresponding elliptic points in the same frame, we have found in Reference 18 ballistic islands situated both in the stochastic layer and in the peripheral currents. Regular ballistic modes correspond to a stable quasi-periodic interframe motion of particles. Only the ballistic islands in the stochastic layer are important for chaotic transport. Mapping positions of the regular ballistic trajectories at the moments of time $t_n = 2\pi n/\omega$ ($n = 1, 2, \dots$) onto the first frame, we obtain chains of ballistic islands both in the northern and southern stochastic layers, i.e., between the borders of the northern (southern) peripheral currents and of the corresponding vortex cores. A chain with three large ballistic islands is situated in those stochastic layers. The particles, belonging to these islands, move to the west, and their mean zonal velocity can be easily calculated to be

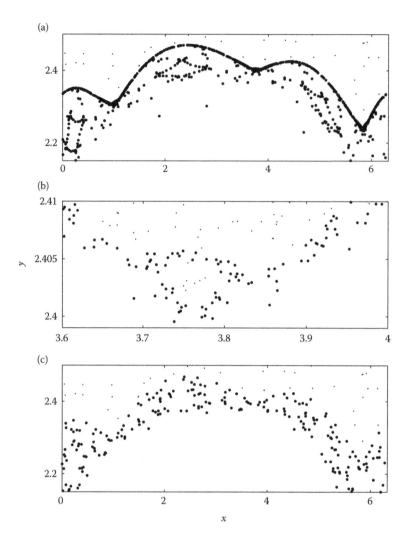

FIGURE 22.22 (a) Poincaré section of the northern stochastic layer where stickiness to the very border with the regular w current and to three large ballistic islands are shown. Increased density of points along the border with the peripheral current is caused by the traps of the border ballistic islands, one of which is shown in (b). (c) The trap of the large ballistic islands.

$\langle u_f \rangle = -2\pi/3T = -\omega/3 \simeq -0.0845$. There are also chains of smaller-size ballistic islands along the very border with the peripheral currents.

We have demonstrated in Reference 18 a stickiness of chaotic trajectories to the borders of those three large ballistic islands (see Figures 6 and 7 in Reference 18). The Poincaré section with fragments of two chaotic trajectories in the northern stochastic layer is shown in Figure 22.22a. One particle performs a long flight sticking to the very border with the regular w current, and another one moves to the west sticking to the very boundaries of three large ballistic islands. A magnification of a fragment of the border and tracks of a sticking trajectory around a smaller-size ballistic island are demonstrated in Figure 22.22b. Figure 22.22c demonstrates the effective size of the trap of the large ballistic islands with tracks of a sticking trajectory around them.

It is reasonable to suppose that the BITs contribute, largely, to the statistics of long flights with $|x_f| \gg 2\pi$. All the ballistic particles, moving both to the west and to the east, can finish a flight and make a turn

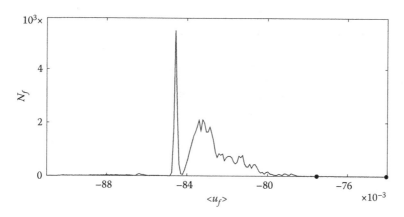

FIGURE 22.23 The distribution of a number of long w flights with $T_f \geq 10^3$ over their mean zonal velocities $\langle u_f \rangle$. The sharp peak corresponds to the trap connected with the very boundaries of the large ballistic islands, the left wing—to a number of traps of families of the border ballistic islands, and the right wing—to the trap situated around the large ballistic islands. Statistics for five tracers with the total number of long w flights $N_f = 5 \cdot 10^4$ and the computation time $t = 5 \cdot 10^8$ for each tracer.

only in the strip shown in Figure 22.12b. The loci of the corresponding turning points have a complicated fractal-like structure. We further consider only long w flights, taking place in the northern stochastic layer, because it is much wider than the stochastic layer between the regular central jet and the southern parts of the vortex cores where e flights take place.

To distinguish between contributions of the traps of different ballistic islands (and, maybe, other zones in the phase space) to the statistics of long flights, we compute for five long chaotic trajectories (up to $t = 5 \cdot 10^8$) the distribution of a number of w flights with $T_f \geq 10^3$ over the mean zonal velocities $\langle u_f \rangle = x_f / T_f$ of the particles performing such flights. The distribution in Figure 22.23 has a prominent peak centered at the mean zonal velocity $\langle u_f \rangle \simeq -0.0845$ which corresponds to a large number of long flights of those particles (and their trajectories) which stick to the very boundaries of the large ballistic islands (see Figure 22.22a) moving with the mean velocity $\langle u_f \rangle \simeq -0.0845$. The flat left wing of the distribution $N(\langle u_f \rangle)$ corresponds to the traps of smaller-size ballistic islands nearby the border with the peripheral current. There are different families of these islands (see one of them in Figure 22.22b) with their own values of the mean zonal velocity which are in the range $-0.092 \lesssim \langle u_f \rangle \lesssim -0.0845$. Stickiness to the boundaries of the border islands is weaker because they are smaller than the large islands and their contribution to the statistics of long flights is comparatively small.

The right wing of the distribution $N(\langle u_f \rangle)$ with $-0.084 \lesssim \langle u_f \rangle \lesssim -0.075$ deserves further investigation. The value $\langle u_f \rangle \simeq -0.075$ is a minimal value of the zonal velocity for long w flights possible in the northern stochastic layer. Increasing the minimal duration of a flight from $T_f = 10^3$ to $T_f = (2 \cdot 5) \cdot 10^3$, we have found splitting of the broad distribution with $-0.084 \lesssim \langle u_f \rangle \lesssim -0.08$ into a number of small distinct peaks. Comparing trajectories with the values of $\langle u_f \rangle$ corresponding to these peaks, we have found that all they move around the large ballistic islands. The particles with smaller values of $\langle u_f \rangle$ used to penetrate further to the south from the islands more frequently than those with larger values of $\langle u_f \rangle$ which prefer to spend more time in the northern part of the dynamical trap connected with those islands. Thus, we attribute the right wing of the distribution $N(\langle u_f \rangle)$ to an effect of the trap situated around the large ballistic islands.

To estimate the contribution of different BITs to the statistics of long w flights in Figure 22.13, we have computed the PDFs $P(x_f)$ and $P(T_f)$ for particles performing w flights with $x_f \geq 100$ and $T_f \geq 1000$ and with the mean zonal velocity $\langle u_f \rangle$ to be chosen in three different ranges shown in Figure 22.23: $-0.092 \lesssim \langle u_f \rangle \lesssim -0.085$ (particles sticking to the border islands) $-0.085 < \langle u_f \rangle \lesssim -0.084$ (particles sticking to the very boundary of three large islands), and $-0.084 < \langle u_f \rangle \lesssim -0.075$ (the trap of the three large islands).

All the PDFs $P(x_f)$ decay exponentially but with different values of the exponents equal to $\nu \simeq -0.005$ and $\nu \simeq -0.0018$. The -0.0014 for the traps of border and the large ballistic islands, respectively. The tail of the PDF $P(x_f)$ for w flights, shown in Figure 22.13, decays exponentially with $\nu \simeq -0.0014$. Thus, the contribution of the large island's BIT to the statistics of long w flights is dominant. As to temporal PDFs $P(T_f)$ for w long flights, they are neither exponential nor power-law like with strong oscillations at the very tails. The slope for the border BITs is again smaller than the large BIT.

A meandering jet is a fundamental structure in oceanic and atmospheric flows. We describe in this section statistical properties of chaotic mixing and transport of passive particles in a kinematic model of a meandering jet flow in terms of dynamical traps in the phase (physical) space. The boundaries of rotational islands (including the vortex cores) in circulation zones are dynamical traps (RITs) contributing, mainly, to the statistics of short flights with $|x_f| < 2\pi$. Characteristic times and spatial scales of the RITs have been shown to correlate with the PDFs for the lengths x_f and durations T_f of short flights. The stable manifolds of periodic saddle trajectories play a role of STs with the specific values of the lengths and durations of short flights of the particles sticking to the saddle trajectories. The boundaries of ballistic islands in the stochastic layers (including those situated along the border with the peripheral current) are dynamical traps (BITs) contributing, mainly, to the statistics of very long flights with $|x_f| \gg 2\pi$.

Dynamical traps are robust structures in the phase space of dynamical systems in the sense that they present at practically all values of the corresponding control parameters. We never know the exact values of the parameters in real flows, especially, in geophysical ones. We do not know exactly the structure of the corresponding phase space; however, we know that typical features, such as islands of regular motion, vortices, and jets, exist in real flows. In this section, we chose specific values of the control parameters for which specific PDFs have been computed and explained by the effect of those dynamical traps that exist under the chosen parameters. We have carried out computer experiments with different values of the control parameters and found that the phase space structure has been changed, of course, with changing the values of the parameters, but the corresponding RITs, STs, and BITs with specific temporal and spatial characteristics have been found to contribute to the corresponding statistics.

22.3.5 Fractal Geometry of Mixing

Poincaré sections provide a good impression about the structure of the phase space but not about the geometry of mixing. In this section, we consider the evolution of a material line consisting of a large number of particles distributed initially on a straight line that transverses the stochastic layer at $x = 0$. A typical stochastic layer consists of an infinite number of unstable periodic and chaotic orbits with islands of regular motion to be embedded. All the unstable invariant sets are known to possess stable and unstable manifolds. When time progresses a particle's trajectories nearby a stable manifold of an invariant set tend to approach the set whereas the trajectories close to an unstable manifold go away from the set. Because of such a very complicated heteroclinic structure, we expect a diversity of the particle's trajectories. Some of them are trapped forever in the first eastern frame $0 \leq x \leq 2\pi$ rotating around the elliptic point along heteroclinic orbits. Other ones quit the frame through the lines $x = 0$ or $x = 2\pi$, and then either are trapped there or move to the neighbor frames (including the first one), and so on to infinity.

To get a more deep insight into the geometry of chaotic mixing, we follow the methodology of our works [4,6] and compute the time T, particles spend in the neighbor circulation zones $-2\pi \leq x \leq 2\pi$ before reaching the critical lines $x = 0, x = \pm 2\pi$, and the number of times $n/2$ they wind crossing the lines $x = \pm \pi$. In the upper panel in Figure 22.24, the functions $n(y_0)$ and $T(y_0)$ are shown. The upper parts of each function (with $n > 0$ and $T > 0$) represent the results for the particles with initial positive zonal velocities which they have simply due to their locations on the material line at $x = 0$. These particles enter the eastern frame and may change the direction of their motion many times before leaving the frame through the lines $x = 0$ or $x = 2\pi$. The time moments of those events we fix for all the particles with $1.9 \leq y_0 \leq 2.045$. The lower "negative" parts of the functions $n(y_0)$ and $T(y_0)$ represent the results for the particles with initial negative zonal velocities ($y_0 \geq 2.045$) which move initially to the first western frame

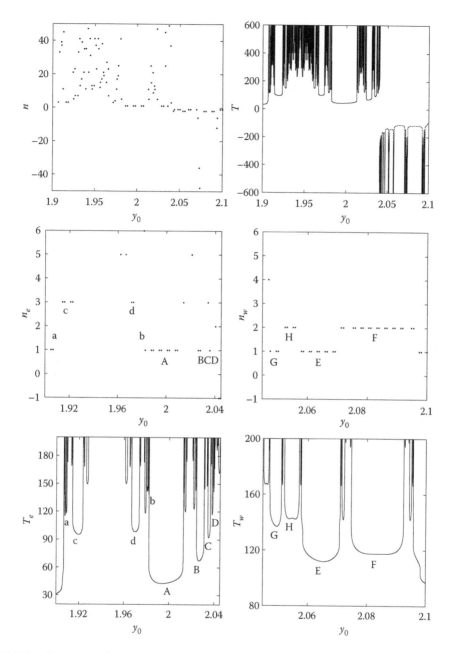

FIGURE 22.24 Fractal set of initial positions y_0 of particles that reach the lines $x = 0, \pm 2\pi$ after $n/2$ turns around the elliptic points. T is a time particles need to reach the lines $x = 0, \pm 2\pi$. Indices e and w mean particles moving in the e and w directions, respectively.

$(-2\pi \le x \le 0)$. In fact, $T_e(y_0)$ and $T_w(y_0)$ are the time moments when a particle with the initial position y_0 quits the eastern or western frames, respectively. Both the functions consist of a number of smooth U-like segments intermittent with poorly resolved ones. Border points of each U-like segments separate particles belonging to stable and unstable manifolds of the heteroclinic structure. The corresponding initial y-positions is a set (of zero measure) of particles to be trapped forever in the respective frame. A fractal-like

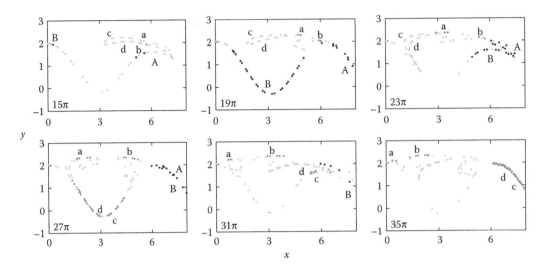

FIGURE 22.25 Fragments of the evolution of a material line in the first eastern frame. The fragments of the fractal in Figure 22.24 with $n_e = 1, 2, 3$ are marked by the respective letters.

structure of chaotic advection in both the frames is shown in the upper panel in Figure 22.24, and its fragments for the first levels are shown in the middle panel for the eastern and the western fragments separately. Particles with even values of n quit one of the frames through the border $x = 0$, those with odd n—quit through the border $x = 2\pi$ for the eastern frame and $x = -2\pi$ for the western one.

Let us consider in detail the fractal-like structure in the eastern frame, keeping in mind that the results are similar with any other frame. The $n_e(y_0)$-dependence is a complicated hierarchy of sequences of segments of the material line. Following Reference 6, we call an epistrophe a sequence of segments of the $(n + 1)$-level, converging to the ends of a segment of a sequence of the nth level, whose length decreases in accordance with a law. At $n_e = 1$ we see in Figure 22.24 an epistrophe with segment's length A, B, C, D, and so on decreasing as $l_m = l_0 q^m$ with $q \approx 0.46$. Letters a and b in Figure 22.24 denote the first segments of the epistrophes at the level $n_e = 2$, whereas d and c—denote the first segments of the epistrophes at the level $n_e = 3$. The respective laws for all those epistrophes are not exponential.

In Figure 22.25, we demonstrate fragments of the evolution of the material line in the first eastern frame at the moments indicated in the figure. Letters on the line mark the corresponding segments of the $n_e(y_0)$ and $T_e(y_0)$ functions in Figure 22.24. As an example, let us explain formation of the epistrophe ABCD at the level $n_e = 1$. With the period of perturbation $T_0 = 2\pi/\omega \simeq 8\pi$, a portion from the north end of the material line leaves the frame through its eastern border. Look at the segments A and B at $t = 15\pi$ and $t = 23\pi$. They quit the first frame as a fold through the period $T_0 \simeq 8\pi$. The other segments—C and D (not shown in Figure 22.25) do the same job. The epistrophe's segments at the odd levels ($n = 2k - 1 > 1$) quit the frame with the period of perturbation T_0 one by one being folded (c and d segments). The folds of the segments of the $(2k - 1)$-level are exterior with respects to the folds of the segments of the $(2k + 1)$-level. The following empirical law is valid: $T_{2k-1} - T_{2k+1} \simeq 2T_0$, where T_{2k-1} is a time when the first segments of the epistrophes at the level $(2k - 1)$ (A with $n_e = 1$) reach the line $x = 2\pi$, and T_{2k+1} the respective time for the first segments of the epistrophes at the level $2k + 1$ (c and d segments with $n_e = 3$).

Segments of the epistrophes of the even levels ($n = 2k$) leave the frame with the period T_0 as well but through the border $x = 0$ moving to the west. We show the evolution of some of them at the moments $t = 31\pi$ and $t = 35\pi$ in Figure 22.25. Thus, the material line evolves by stretching and folding, and folds quit the frame in both directions with the period of perturbation.

22.4 Chaotic Cross-Jet Transport in a Dynamical Model of a Meandering Jet Current

Strong oceanic and atmospheric currents, the Gulf Stream in the Atlantic, the Kuroshio in the Pacific, and the polar night Antarctic jet in the atmosphere, are meandering jets separating water and air masses with distinct physical properties. For example, the Gulf Stream separates the colder and fresher slope ocean waters from the salty and warmer Sargasso sea ones. Transport of particles across a geophysical jet is of crucial importance and may cause depletion of ozone in the atmosphere and heating and freshing of waters in the ocean.

Zonal chaotic mixing and transport in a jet flow (i.e., chaos along the jet) have been studied with the kinematic model to be considered in the preceding section. In this section, we pay attention to meridional chaos, i.e., a chaotic cross-jet transport, not in a kinematic model but in the dynamical one [9,23]. It has been found both numerically and experimentally that fluid is effectively mixed along the jet, but in common opinion [9,23] a large gradient of the potential vorticity in the central part prevents transport across the jet. The transport barrier has been numerically shown to be broken only with large values of wave amplitudes that are beyond the validity of linear models and can be hardly observed in real flows.

However, in References 28 and 29 it has been proven analytically and numerically that chaotic cross-jet transport under appropriate conditions is possible at comparatively small values of the wave amplitudes and, therefore, may occur in geophysical jets. A general method has been elaborated in those papers to detect a core of the transport barrier and find a mechanism of its destruction using the dynamical model of a zonal jet flow with two propagating Rossby waves. The method comprises the identification of a central invariant curve (CIC), which is an indicator of existence of the barrier, finding certain resonant conditions for its destruction at given values of the wave numbers, and detection of cross-jet transport.

Motion of 2D-incompressible fluid on the rotating Earth is governed by the equation for conserving potential vorticity $(\partial/\partial t + \vec{v} \cdot \vec{\nabla})\Pi = 0$. In the quasi-geostrophic approximation, one gets $\Pi = \nabla^2\Psi + \beta y$, where β is the Coriolis parameter. The x-axis is chosen along the zonal flow, from the west to the east and y—along the gradient from the south to the north. Barotropic perturbations of zonal flows produce planetary Rossby waves propagating to the west and producing an essential impact on transport and mixing in the ocean and the atmosphere. It is possible to find in a linear approximation an exact solution for the stream function obeying the equation for conserving potential vorticity [9] and consisting of steady zonal flow with the velocity profile of a Bickley jet and two propagating Rossby waves with the amplitudes A_1 and A_2 and wave numbers $n_1 = mN_1$ and $n_2 = mN_2$, where $m \neq 1$ is the greatest common divisor and N_1/N_2 is an irreducible fraction. The normalization, introduced in References 28 and 29, is very convenient because it reduces the number of control parameters to the main ones N_1 and N_2. The stream function in the frame moving with the phase velocity of the first wave is

$$\Psi(x, y, t) = -\tanh y + A_1 \operatorname{sech}^2 y \cos(N_1 x) + A_2 \operatorname{sech}^2 y \cos(N_2 x + \omega_2 t) + C_2 y. \tag{22.19}$$

At odd values of N_1 and N_2, the advection equations have two symmetries $\hat{I}_0$ and $\hat{S}$. Solving the equation $\hat{I}_0(x_j, y_j) = \hat{S}(x_j, y_j)$, $j = 1, 2$, one gets indicator points [26]: $(x_1 = \pi/2, y_1 = 0)$ and $(x_2 = 3\pi/2, y_2 = 0)$. Iterating them, we construct a CIC [5] in the central part of the jet, which is the last transport barrier in the sense that the CIC breaks down and is replaced by a stochastic layer with variation of the wave amplitudes. The algorithm for calculating the CIC has been developed in References 28 and 29.

We illustrate this in Figure 22.26. Figure 22.26a shows the phase portrait of the steady flow ($A_2 = 0$) in a frame moving with the phase velocity of the first wave. There is an e jet sandwiched between two chains of five vortices, while the adjacent flows to its south and north are w in the moving frame. In a steady flow, tracers move along streamlines. With the second-wave amplitude, A_2, increasing, the onset of chaos follows a scenario typical for nonlinear Hamiltonian systems: as the separatrices of an integrable

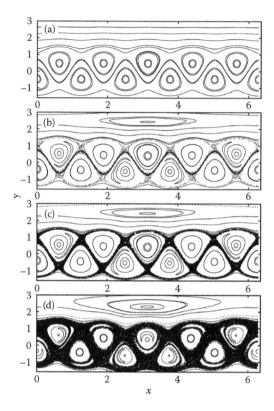

FIGURE 22.26   Poincaré sections of a zonal current perturbed by two Rossby waves with $N_1 = 5$ and $N_2 = 1$ in the moving frame at $A_1 = 0.2416$. (a) $A_2 = 0$: steady flow with streamlines. (b) $A_2 = 0.09$: chaotic advection in the zonal direction, but the CIC (bold curve) acts as a barrier to cross-jet transport. (c) $A_2 = 0.095$: destruction of CIC and onset of local cross-jet transport. (d) $A_2 = 0.2$: global cross-jet transport.

system break down, they are replaced by stochastic layers. As the wave amplitudes exceed certain threshold values, the CIC in the central jet region breaks up and is replaced by a stochastic layer. The layer is confined between invariant curves and its average width increases with the wave amplitudes. In Figure 22.26b, the CIC is represented by a bold curve with stochastic layers to its north and south, where zonal chaotic transport takes place. This Poincaré section is obtained by setting $A_2 = 0.09$, in which case the region occupied by the CIC and adjacent invariant curves acts as a barrier to meridional or cross-jet transport. With increasing A_2, the CIC breaks up and is replaced by a stochastic layer where local transport is observed (Figure 22.26c). The layer width increases with disturbance amplitude, and the cross-jet transport becomes global (Figure 22.26d).

There are two mechanisms of the onset of chaotic cross-jet transport. As the Rossby wave amplitudes increase, the southern and northern stochastic layers around the CIC grow wider. As certain threshold values of the amplitudes are exceeded, the layers overlap and the CIC breaks up. It is an amplitude mechanism of CIC breakup and onset of chaotic cross-jet transport, which is in general a global one. To illustrate the amplitude mechanism of destruction of CIC, the topology of the phase space near the islands of the resonance 7:3 has been studied. In the range $0 < A_2 < 0.088$ the smooth CIC and neighboring invariant curves form a transport barrier (Figure 22.27a). At $A_2 \simeq 0.088$, invariant manifolds of hyperbolic orbits of the resonance 7:3 cross each other, the CIC breaks down, and there appears at its place a narrow stochastic layer locked between remained invariant curves (Figure 22.27b). When A_2 increases further islands of the resonance 7:3 diverge, and a meandering CIC appears again between them (see Figure 22.27c at

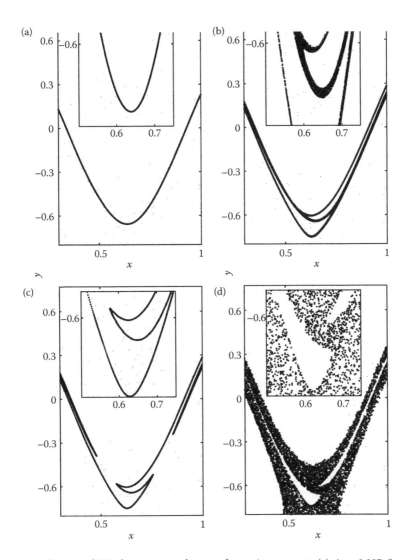

FIGURE 22.27 Mechanism of CIC destruction and onset of cross-jet transport. (a) $A_2 = 0.087$. Smooth CIC and neighboring invariant curves form a transport barrier. (b) $A_2 = 0.088$. A narrow stochastic layer (shadowed strip) appears at the place of the broken CIC. (c) $A_2 = 0.09$. CIC appears again as a meandering curve. (d) $A_2 = 0.1$. Breakdown of CIC and onset of cross-jet transport. Insets show magnification of the phase space region nearby the resonance 7:3.

$A_2 = 0.09$). At $A_2 > 0.095$, CIC and surrounding invariant curves are destroyed, and cross-jet transport becomes possible in a wide range of the y coordinate (Figure 22.27d). Thus, existence of a CIC is a sufficient but not necessary condition for existence of a transport barrier. The other, resonant mechanism of CIC breakup and onset of local cross-jet transport based on the processes of resonance overlapping and separatrix reconnection, has been studied in detail in Reference 28.

If both wavenumbers are odd, the advection equations have the symmetry that makes it possible to develop a regular and efficient method for finding a CIC, whose breakup signifies the onset of chaotic cross-jet transport. In the case of an even–odd wavenumber pair, this method is inapplicable because of the lack of flow symmetry, and an alternative algorithm for detecting cross-jet transport, based on an overlap of northern and southern stochastic layers, has been developed in Reference 28.

22.5 Mixing and Transport in the Real Ocean with Altimetric Velocity Fields

The ocean seems as a highly turbulent medium and presents a variety of dynamical phenomena with different space and time scales ranging from millimeters to a few thousands of kilometers and from milliseconds to thousands of years. The ocean is subjected to a variety of small- and large-scale random perturbations. Some of them generate well-ordered and long-lived coherent structures by means of the following factors: the Earth's rotation, density stratification, wind stress, and bottom topography. Large-scale surface coherent structures are visible at satellite images of the sea color, surface temperature, chlorophyll concentration, and can be identified along with some deep coherent structures by means of *in situ* measurements.

The major western boundary oceanic currents, such as the Gulf Stream in the Atlantic and the Kuroshio in the Pacific, separate waters with different physical (temperature, salinity, density, etc.), chemical (nitrates, silicates, phosphates, etc.), and biological characteristics (the content of nutrients, phyto- and zooplankton, etc.). They are warm deep "rivers" in the ocean with the width on the order of 100–200 km and the maximal speed of current at the surface of 1–2 m/s. These currents are not straight channels with warm water but strongly meandering features with eddies (the so-called rings) to be pinched off from the main jet. Some of the rings (with the size of a few hundreds of km) live for a comparatively short time, a month or 2 months, and then merge with the jet. The others can transport water preferably to the west over thousands of kilometers and can survive up to a few years before breaking down. Being coherent features, they do not contain the same waters but exchange them with the surrounding ocean, the process known as mixing.

In the preceding sections, we have studied mixing and transport in analytical models of the jet currents with the streamfunctions to be constructed "by hand" (kinematic models) or to be satisfied to some linearized equations of fluid motion (dynamic models). It enabled us to get a deep insight into the mechanisms of onset of chaos and its properties in such over simplified versions of oceanic flows. There exist global and regional numerical circulation models (POMs, ROMs, and many others) which provide more realistic description of oceanic circulation, including main currents and eddies. The velocity fields with a given resolution in a given region are obtained as an output of the corresponding circulation model. The advantage of those models is a possibility to get three-dimensional (3D) velocity fields along with 3D temperature, pressure, and salinity scalar fields. Some recent examples of Lagrangian study of mixing and transport with numerical regional prognostic models can be found in References 19 and 22.

Now, we can get global surface velocity fields from direct measurements with the help of satellite altimeters. The new branch of oceanography, operational oceanography, has been developed in the last two decades. It can be defined as the activity of systematic and routine measurements of the World Ocean and atmosphere, their rapid interpretation with the help of numerical forecasting models, and providing final products on the present state of the sea and continuous forecasts of the future condition of the sea. At different websites (http://www.aviso.oceanobs.com, http://oceancolor.gsfc.nasa.gov, http://www.nodc.noaa.gov, and others) one can monitor day by day how quiet the ocean may be in some regions and how complicated and unpredictable in others. In the following sections, we address the problem of surface large-scale mixing and transport of ocean water masses in the regions with strong activity.

The real oceanic flows are not, of course, strictly time periodic as the model flows, we have studied in the preceding sections. However, in aperiodic flows there exist under some mild conditions hyperbolic points and trajectories of a transient nature. In aperiodic flows, it is possible to identify aperiodically moving hyperbolic points with stable and "effective" unstable manifolds [10]. Unlike the manifolds in steady and periodic flows, defined in the infinite time limit, the "effective" manifolds of aperiodic hyperbolic trajectories have a finite lifetime. The point is that they play the same role in organizing oceanic flows as do invariant manifolds in simpler flows. The effective manifolds in course of their life undergo stretching and folding at progressively small scales and intersect each other in the homoclinic points in the vicinity of which fluid particles move chaotically. Trajectories of initially close fluid particles diverge rapidly in these regions, and particles from other regions appear there. It is the mechanism for effective transport and mixing of water masses in the ocean. Moreover, stable and unstable effective manifolds constitute Lagrangian transport barriers between different regions because they are material invariant curves that cannot be crossed by purely advective processes.

To introduce the reader to the problem, we briefly mention those oceanographic notions and terms we will be using. A geostrophic current is an oceanic flow in which the pressure gradient force is balanced by the Coriolis effect. Its direction is parallel to the isobars, with the high pressure to the right of the flow in the northern hemisphere, and the high pressure to the left in the southern hemisphere. Geostrophy allows to infer ocean currents from measurements of the sea-surface height by satellite altimeters. The major currents, the Gulf Stream, the Kuroshio Current, the Agulhas Current, and the Antarctic Circumpolar Current, are examples of geostrophic currents.

Altimetry is a technique for measuring height. Satellite altimetry measures the time taken by a radar pulse to travel from the satellite antenna to the surface and back to the satellite receiver. Combined with precise satellite location data, altimetry measurements yield sea-surface heights. Solving the equation for a geostrophical balance, one gets a surface geostrophic velocity field.

Dynamic topography refers to the topography of the sea surface related to the dynamics of its own flow. In hydrostatic equilibrium, the surface of the ocean would have no topography, but due to the ocean currents, its maximum dynamic topography is on the order of 2 m and are influenced by ocean circulation, temperature, and salinity. A clockwise rotation (anticyclone) is found around "hills" in the northern hemisphere and "valleys" in the southern hemisphere. Conversely, a counterclockwise rotation (cyclone) is found around "valleys" in the northern hemisphere and "hills" in the southern hemisphere.

Geostrophic velocities to be used in the following sections were obtained from the AVISO database (http://www.aviso.oceanobs.com). The data are gridded on a $1/3° \times 1/3°$ Mercator grid. Bicubical spatial interpolation and third-order Lagrangian polynomials in time have been used to provide accurate numerical results. Lagrangian trajectories have been computed by integrating the advection equations with a fourth-order Runge–Kutta scheme with a fixed time step of 0.001th part of a day. The SST data (http://oceancolor.gsfc.nasa.gov) were used to illustrate oceanographic conditions in the region studied.

In References 17, 19, 20, 22, and 32–40 we have introduced a number of Lagrangian indicators such as a distance passed by fluid particles for a given time, absolute, meridional, and zonal displacements of particles from their initial positions, the time of residence of fluid particles in a given region, etc. Those quantities can be computed by solving advection equations forward and backward in time in order to know the fate and origin of the corresponding water masses, respectively. Of particular importance in detecting Lagrangian structures is the absolute displacement, D, that is simply the distance between the final, (x_f, y_f), and initial, (x_0, y_0), positions of advected particles on the Earth sphere with the radius R

$$D \equiv R\,\mathrm{arcosh}[\sin y_0 \sin y_f + \cos y_0 \cos y_f \cos(x_f - x_0)]. \tag{22.20}$$

This quantity and zonal, D_x, and meridional, D_y, drifts have been shown to be useful in quantifying transport of radionuclides in the Northern Pacific after the accident at the Fukushima atomic plant station [20].

The FTLE field quantitatively characterizes mixing along with directions of maximal stretching and contracting, and it is applied to identify LCSs in irregular velocity fields. The FTLE is computed here by the method of the singular-value decomposition of an evolution matrix for the linearized advection equations to be introduced in Reference 22 and is given by the formula

$$\lambda(t, t_0) = \frac{\ln \sigma(t, t_0)}{t - t_0}, \tag{22.21}$$

which is the ratio of the logarithm of the maximal possible stretching in a given direction to a time interval $t - t_0$. Here $\sigma(t, t_0)$ is the maximal singular value of the evolution matrix. The method proposed enables to compute accurately the FTLE in altimetric velocity fields.

22.5.1 Lagrangian Look at Radionuclide Propagation from the Fukushima–Daiichi Nuclear Plant Accident

A great earthquake on March 11, 2011 off Miyagi prefecture (Japan) followed by a tsunami inflicted heavy damage on the Fukushima–Daiichi nuclear plant. A large amount of water contaminated with radionuclides leaked directly into the ocean. Moreover, the radioactive pollution of the sea was caused by conveyance via rivers of the radioactive pollutants deposited on the ground following atmospheric release and subsequent rainwater runoff [20].

In this section, we report on direct numerical simulation of propagation of contaminated water in the ocean from the Lagrangian point of view. Based on the AVISO geostrophic velocity field, we compute the absolute drift map (22.20) by integrating the advection equations for $2.25 \cdot 10^6$ synthetic particles backward in time for 15 days. In Figure 22.28a, shadows of the gray color code the value of D in kilometers on June 17, 2012. This map with 15-days history clearly reveals the Kuroshio Current (the black band in the figure up to $x_0 \simeq 142°E, y_0 \simeq 36°N$), its extension with large meanders and the filaments of different size, and a number of mesoscale eddies pinched off from the meanders. The synoptic Lyapunov map on the same day (Fig. 22.28b) gives the image of unstable manifolds in the whole region which delineate the same eddies as in Fig. 22.28a.

Because of lack of exact knowledge of initial area with contaminated water, we model the initial distribution of radionuclides to be a patch with concentration decreasing logarithmically in distance from the Fukushima plant. To simulate propagation of radionuclides we compute the advection equation starting on April 1, 2011 with the AVISO velocity field. The "instantaneous" distribution of the radioactive concentration, χ, on June 17, 2012 is shown in Figure 22.29a. As expected, the radionuclides in the surface layer were transported mainly along the Kuroshio extension to the east. The concentration is larger on the north flank of the jet because the larger part of the initial radioactive patch was situated to the north from the Kuroshio extension jet. Transport of the radionuclides to the southern flank may be explained partly by tracer advection from the southern part of the initial patch and mainly by meridional cross-jet transport due to pinching off eddies from the northern flank of the Kuroshio jet and their subsequent merging with

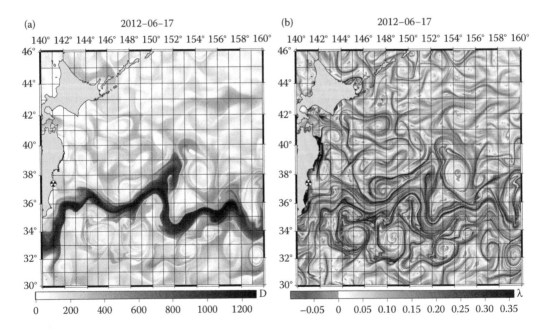

FIGURE 22.28   (a) Drift and (b) Lyapunov maps on June 17, 2012 after the accident at the Fukushima nuclear plant marked by the radioactive sign.

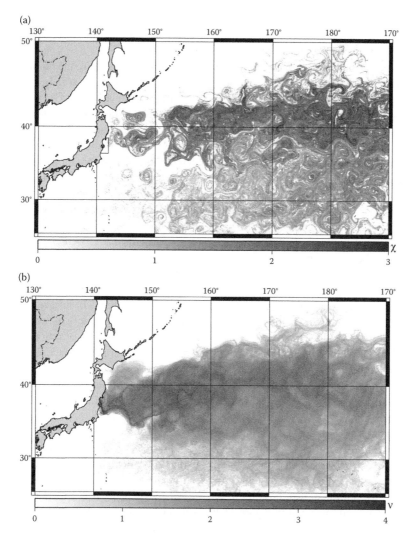

FIGURE 22.29 (a) Distribution of radionuclide concentration χ and (b) visiting map on June 17 2012 in the logarithmic scale. Concentration χ and the number of visits ν are in relative units.

the jet. We have not found a significant impact of the initial patch size on the distribution of the radioactive concentration which is formed over a sufficiently long time by the system of stable manifolds in the region.

The D, λ, and χ maps show that there is an eddy structure to the north from the Fukushima plant which might have transported the contaminated water to the coasts of Honshu and Hokkaido Islands (Japan) and even to the southern Kuril Islands (Russia). On the day of the accident, there was a complicated eddy system to the north from the nuclear plant with a large anticyclonic warm-core Kuroshio ring approximately at the traverse of the plant, a small anticyclonic eddy to the north from it, and a medium cyclonic eddy at the traverse of the Tsugaru strait. This eddy system with a complicated structure of invariant manifolds has defined mixing and transport of radioactive water, part of which has been trapped by the stable manifold of the Kuroshio ring and rounded about it as a streamer. The radionuclides have advected to the north along the unstable manifold of the Kuroshio ring that is, in turn, a stable manifold of the adjacent eddy. The concentration of radionuclides along the invariant manifolds may be significantly greater than in other places because they are a kind of attractors. It has been shown in the preceding section, unstable manifolds are the LFs attractive for plankton, pelagic fishes and other marine organisms, and the larger radioactive concentration at the fronts may be dangerous for them.

A new information provides the map of the integrated quantity ν [32,33,39], the number of times the radioactive particles visit different places in the region for the time of integration. We partition the region in a large number of small cells and calculate how many times the particles visit each cell for the period from April 1, 2011 to June 17, 2012. The visiting map allows us to reveal the transport corridors for radioactive tracers. It is clear from Figure 22.29b that the Kuroshio extension is the main corridor. However, there are another transport corridors seen as black filaments. Moreover, this map allows to resolve black loops with increased number of visits which delineate eddies of different size trapping contaminated water.

We conclude this section by a brief review on in situ measurements of concentration of Fukushima-derived ^{134}Cs ($t_{1/2} = 2.07$ y) and ^{137}Cs ($t_{1/2} = 30.07$ y) isotopes in the Northwest Pacific Ocean. With the short half-lives, any ^{134}Cs observed in the sea could only be derived from the 2011 Fukushima releases. Before the accident, ^{137}Cs levels off Japan were $\approx$0.001–0.002 Bq/kg while ^{134}Cs was not detectable. Concentrations at the nuclear power plant discharge channels in early April 2011 were more than 50 million times greater than the preexisting ocean level of ^{137}Cs.

During the cruise with R/V "Mirai" (April 14 to May 5, 2011) about 1 month after the accident [1], seawater, suspended solids, and zooplankton examples were collected from the surface-mixed layer and subsurface layers in a number of stations 200–2000 km offshore from Fukushima. In surface water, ^{137}Cs concentrations were from several times to two orders of magnitude higher than before the accident. ^{134}Cs isotope was also detected with the ratio ^{134}Cs/^{137}Cs to be about 1. The highest concentrations, from $\approx$0.15 to $\approx$0.35 Bq/kg, have been found off Fukushima ($\sim$200 km from the nuclear power) and Miyagi (earthquake source). ^{137}Cs concentrations to the east of this region, at the stations at $x_0 \simeq$ 146–147°E, $y_0 \simeq$ 37–38°N, were also high ($\approx$0.05–0.06 Bq/kg). The ^{137}Cs concentrations in the Kuroshio extension ($<$0.01 Bq/kg) were unexpectedly low because it was considered to be the main potential pathway for contaminated water to the open ocean.

The R/V "Ka'imikai-o-Kanaloa" cruise has been conducted on June 4–18, 2011 [7] investigating horizontal and vertical distribution of gamma-emitting radionuclides in the seawater, zooplankton, and micronectonic fishes 30–600 km offshore from Fukushima. In June, the highest surface-water concentrations for both isotopes, 3.9 Bq/kg, have been detected in an eddy, 150 km offshore, centered at $x_0 \simeq$ 142.5°E, $y_0 \simeq$ 37°N, not the nearest location to the nuclear power plant. It is remarkable that we see a closed black loop around this point in Figure 22.29b with an increased number of radioactive particles' visits. Activities up to 0.325 Bq/kg were found more than 600 km offshore. As to ^{137}Cs, the highest level (except for the discharge channels), 0.6–0.8 Bq/kg, has been detected 30 km offshore. In June, Fukushima-derived Cs did not generally penetrate below 100–200 m. Over time, it is expected to find a deeper penetration proving a means to study the rates of vertical mixing processes in the Pacific. Fukushima-derived isotopes have also been detected in zooplankton (with the maximal level about 50 Bq/kg dry weight comparable with the recommended value of 40) and jellyfish but not in micronectonic fishes.

The recent R/V "Professor Gagarinsky" cruise, headed by Dr. V. Lobanov (Pacific Oceanological Institute, Vladivostok, Russia), has been conducted from June 11 to July 10 2012, more than a year after the accident. Only a few preliminary results were known at the time of writing this section. The ratio ^{134}Cs/^{137}Cs has been measured to be about 1. The maximal ^{137}Cs concentrations in surface waters at the stations along the 155°E longitude track from 43°N to 37°N have been detected to be around 0.02 Bq/kg, 50 times smaller than around the same places in the cruise with R/V "Mirai" [1] about 1 month after the accident. The penetration of Cs isotopes up to 500 m has been detected at two stations.

The concentrations of Cs isotopes soon after the accident have been detected in some places to be 10–1000 times higher than preexisting levels in waters off Japan. However, radiation risks due to these radionuclides were below those generally considered harmful to marine animals and human consumers, and even below those from naturally occurring radionuclides, such as ^{40}K with the background level of $\simeq$10 Bq/kg.

22.5.2 LFs in the Ocean

In this section, we introduce the notion of LFs and discuss their connection with potential fishing grounds [35–37,39]. Our approach is based on searching for specific Lagrangian features in the altimetric geostrophic velocity fields which indicate the presence of convergence of waters of different properties. We call them LFs, which are boundaries between surface waters with different Lagrangian properties. It may be, for example, a physical property, such as SST, salinity, density, etc., or concentration of chlorophyll-a. Lateral maximal gradients of those properties would indicate on specific oceanic fronts, thermal, salinity, density, and chlorophyll ones, which are often connected with each other. However, one may consider more specific Lagrangian indicators such as absolute, meridional, and zonal displacements of particles from their initial positions, etc. Even in the situation where the water itself is indistinguishable, say, in temperature, and the corresponding SST image does not show a thermal front there may exist a LF separating waters with the other distinct properties. In particular, strong gradients of SST provide specific LFs.

Atmospheric fronts are well-known features studied and used in weather forecasts. Besides the currents and eddies, there are other coherent Eulerian structures in the deep and coastal ocean analogous to atmospheric fronts. Oceanic (hydrological) fronts are comparatively narrow regions ranged from a kilometer to a hundred of kilometer size, where a horizontal gradient of a given property goes through a maximum. Changes in temperature, salinity, density, or other properties across the front can be an order of magnitude larger than those on both sides of the front. Fronts are associated with convergent flows which bring waters with different Lagrangian properties into the frontal region. The waters, flowing advectively into the region with strong horizontal gradients, are forced to move diffusively across isolines of the respective characteristics providing intense turbulent mixing of momentum and other properties.

A hydrological front is formed by two dissimilar water masses, usually light (warm and fresh) and heavy (cold and salty). The pycnocline there (a layer with the maximal vertical density gradient) becomes sloped leading to a strong cross-frontal horizontal density gradient, which, in turn, causes strong along-front flows. By continuity reasons, there should be a vertical flow to compensate for the horizontal convergence, the process known as downwelling. Owing to the surface convergence and the downwelling, there must be a divergence at a depth, which is accompanied by a vertical water rise (upwelling) to compensate for the waters which sink along the front. Thus, a cell-like circulation pattern arises at fronts .

The hydrological fronts in the ocean are regions of enhanced physical energy with strong large-scale advective convergence flows, lower-scale vertical motion, and turbulent mixing. The convergence flows increase the concentration of nutrients in frontal zones. Moreover, floating and surface-active materials accumulate at convergent fronts where they can attract and affect marine organisms. However, the smaller vertical displacement has a greater effect on biota at all trophic levels because of changes in the light seen by phytoplankton and its mixing with nutrients along the front. Phytoplankton is a food for zooplankton. Small fish aggregate at the front for zooplankton and is, in turn, a food for larger fishes and animals. In this sense, we can say that the enhanced physical energy at oceanic fronts is converted into the biological one.

In the satellite era, it becomes possible to see thermal and chlorophyll fronts on the images of the SST and ocean color, respectively. Recently, a general method to identify LFs [21] using a given velocity field has been proposed. The definition of LF given above describes any frontal feature, even the one where similar waters from different places converge. For example, one may mark in a given region those synthetic particles that enter the region through its northern border by one color and those which enter it through a southern border by another color. The border between the colors on the synoptic map will be an LF and it does not matter how different the properties of convergent waters are. However, in practice any LF is a convergence of dissimilar waters.

LFs can be accurately detected in a given velocity field by computing synoptic maps of the displacement of synthetic tracers and other Lagrangian indicators. The questionas as to how they correlate with the

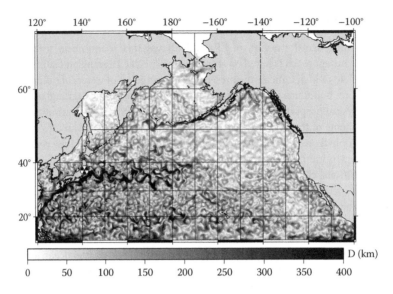

FIGURE 22.30 Lagrangian drift map of the absolute displacements of particles D (in km) in the North Pacific on May 15 2011 to be computed backward in time for 2 weeks before that date.

LCS is open till now. It is well known that the curves of local maxima ("ridges") on the FTLE map, computed backward in time, delineate the unstable manifolds, which are by definition the curves of maximal stretching. The adjacent particles, belonging to them, have a different origin because the larger the FTLE value the larger is an initial distance between the corresponding particles (integration backward in time). The particles with maximal FTLE came to their locations from very different places. Thus, the maximal gradients of the FTLE field at the "ridges" demarcate the corresponding LFs. We would like to stress the important role of the LFs because, differently from rather-abstract geometric objects such as invariant manifolds, the LFs are fronts of real physical quantities that can be measured directly.

In Figure 22.30 we show the Lagrangian drift map of the absolute displacements D of 2.25 millions of particles in the North Pacific on May 15, 2011 to be computed backward in time for 2 weeks before that date. The curves of the maximal (local) gradients of D visualize LFs in the region. The two powerful currents are manifested on the map as black-meandering belts: the Kuroshio and its extension to the east of the Japan coast and the Aleut Current to the north. Practically, all the region is covered by mesoscale eddies of different sizes along with dipole and mushroom structures. A few currents (the Kamchatka, the Oyashio, and the Californian ones) look like a street with moving mesoscale eddies, each of which is surrounded by a black collar which demarcates the LF separating the eddy's core from the surrounding waters passed for 2 weeks a much longer distance than the core's waters. The map in Figure 22.30 demonstrates the LFs of the planetary and synoptic scales, including the subarctic frontal zone in the Japan Sea and the regions with very weak fronts such as the Okhotsk Sea.

The methods developed allow us to resolve a much more finer structure of LFs. We restrict our analysis by the region to the east off Hokkaido (Japan) and southern Kuril Islands (Russia) coasts where the daily saury catch data from the Russian fishery were collected for a few fishery seasons. With the aim to detect the LFs separating waters of different origin and histories, we distribute a large number of synthetic particles over that region and integrate advection equations (22.2) backward in time for 2 weeks to compute the absolute, D, particle's displacements from their positions on the initial day. Coding the particle's displacements by color, one gets the synoptic maps that provide the evidence of the origin of water masses present in the region on a given day. The main LFs are visualized in Figure 22.31a including the SOLF (Soya–Oyashio Lagrangian Front), separating waters of the Soya Current between the straits and the Oyashio waters, and the SARLF (Soya–Anticycloni Ring Lagrangian Front), separating the Soya waters from the

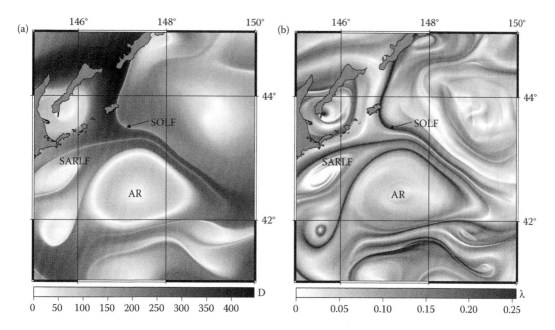

FIGURE 22.31 Backward-in-time (a) drift map D (in km) and (b) Lyapunov map λ (in days) on October 1 2004.

anticyclonic Kuroshio ring (AR). Each of the LFs can be identified as well by a narrow white band along the corresponding curve of the maximal gradients of D. White color means that the corresponding particles have experienced very small displacements over 2 weeks. In order to understand why it is so, we have computed the FTLE in the region. The map in Figure 22.31b clearly demonstrates the Kuroshio ring surrounded by black ridges which are known to approximate unstable manifolds of the hyperbolic trajectories in the region. The black ridges in Figure 22.31b are situated along the corresponding white curves of the maximal gradients of D because the motion of particles near unstable manifolds slows down due to the presence of the so-called saddle dynamical traps (see Section 22.3 and Reference 27).

22.5.3 Identifying LFs with Favorable Fishery Conditions

We restrict our analysis here to the region to the east off Hokkaido (Japan) and southern Kuril Islands (Russia) coasts where the daily saury catch data from the Russian fishery were collected for a few fishery seasons. In some seasons, there exists the First Oyashio Intrusion along the eastern coast of Hokkaido. The data of saury catch for September 23–25, 2002 overlaid allow to conclude that the fishery grounds have been concentrated in the waters of the First Oyashio Intrusion (Figure 22.32a). It is seen that productive cold waters of the Oyashio Current and warmer waters of the southern branch of the Soya Current flowing through the straits between the southern Kuril Islands converge at the LFs with the maximal catches. This LF demarcates the boundary between "dark" and "gray" waters.

The oceanographic situation in the region cardinally differs in the years with the Second Oyashio Intrusion when a large anticyclonic eddy, formed as a warm-core Kuroshio ring, approaches the Hokkaido eastern coast and forces the Oyashio Current to shift to the east rounding the eddy. The map in Figure 22.31a averaged for October 16–18, 2004 demonstrates that the fishery grounds have been concentrated mainly at the convergence front between the Oyashio and Soya waters and the periphery of the AR with the center around $x_0 = 147°30'E$, $y_0 = 42°30'N$. The absolute drift map in Figure 22.32b with the circles of saury catch locations overlaid clearly show that the fishery grounds with maximal catches are located along the LFs where productive cold waters of the Oyashio Current, warmer waters of the southern branch of the Soya Current, and warm and salty waters of the Kuroshio ring converge. Animation of

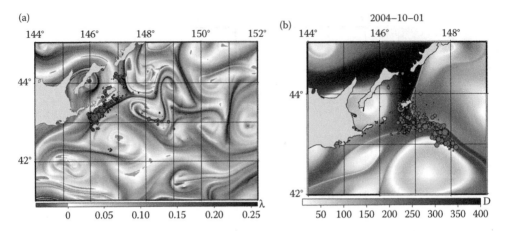

FIGURE 22.32 Backward-in-time drift maps with the season with (a) the First Oyashio Intrusion (September, 2002) and (b) the Second Oyashio Intrusion (October, 2004). Catch locations are marked by circles with the radius to be proportional to the catch in tons per a given ship.

the daily Lagrangian maps for August–December 2004 with the fishery grounds overlaid is available at http://dynalab.poi.dvo.ru/data/GRL12/2004.

We demonstrated above a relationship between LFs and locations of fishing boats with saury catches in different oceanographic situations. In order to quantitatively determine whether saury is actively associating with the LFs in the region studied or not, we compute the frequency distribution of the distances, r, between locations of fishing boats with saury catches and the strong LFs for available fishery seasons. The maximal gradient of the absolute displacement, $\nabla D = \sqrt{(\partial D/\partial x)^2 + (\partial D/\partial y)^2}$, is supposed to be an identificator of the presence of an LF. Such gradients delineate boundaries between waters that passed distances that may differ in two orders of magnitude. It has been shown above that the contrast boundaries are the strongest LFs separating productive cold waters of the Oyashio Current, warmer waters of the southern branch of the Soya Current, and waters of warm-core Kuroshio rings. In order to get rid of ephemeral LFs, we choose a threshold, $\nabla D_{th} = 60$, and only the LFs with $\nabla D_{th} \geq 60$ are supposed to be "strong." We have found that such gradient values correspond to permanent LFs in the Kuroshio–Oyashio frontal area. Then, we compute on each day the distance r between the location of each boat with saury catch and the nearest geographical point where $\nabla D_{th} \geq 60$. The corresponding PDF for each season is compared with the random PDF which is computed by the same way but with 10,000 points randomly distributed over the same region. The fishery strategy depends on the oceanographic situation in the region, the captain's experience, other subjective factors, and differs during the fishery period. As a rule, the Russian captains begin to catch saury in August in the northern part of the region studied where saury schools start their migration to the south. It is profitable to begin the fishery near the ports, off the coasts of the southern Kuril Islands. The fishing grounds there are connected with the strong upwelling due to high tides rather than with LFs. Then, in September, the fishing boats moved to the south following the saury schools. For safety and other reasons, the boats prefer to stick together. After the reporting of a good catch by one of them, the others may move to that place if they are around. The combination of different factors in fishery strategy, including subjective ones, makes it difficult to find statistically significant correlations between locations of fishing boats with saury catches and any oceanic features. After all, even if we were able to find precisely potential fishing grounds with favorable hydrological conditions, it would not necessarily imply good real catches if the saury or fishing boats do not reach the place.

The results of our statistical analysis are shown in Figure 22.33 for all available fishery seasons. The number of events, i.e., the number of locations of the boats with saury catches, varies from season to season and in average was about 1000 per season. As expected, the random PDFs (thin curves) are rather smooth

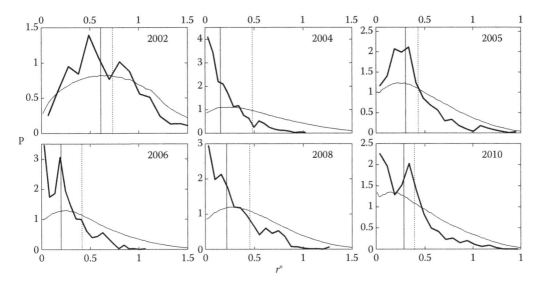

FIGURE 22.33 PDFs for six fishery seasons with real locations of the fishing boats with saury catches (solid curves) and the fishing boats randomly distributed over the same region (dashed curves).

curves with long tails. The real PDFs (bold curves) have a tooth-like structure that can be explained partly by congregation of boats near strong LFs with large D gradients and partly by the fishery strategy to stick together. The vertical solid and dashed lines represent the medians for real and random PDFs, respectively. By definition, the median of a finite list of numbers can be found by arranging all the observations from lowest value to highest one and picking the middle one. It is such a location value that there exists in the list the same number of locations smaller and larger than the median value. The median is a more robust statistical indicator than the mean value and can be used as a measure of location when a distribution is skewed having, for example, a heavy tail. In all the seasons, the medians were closer to the LFs for the real PDFs than for the corresponding random ones. Moreover, the random PDFs have longer tails than the corresponding real ones proving that the fishing boats really tend to be closer to the LFs than be randomly distributed over the region. The plot in Figure 22.34 demonstrates the relations between medians (stars) for the real and random PDFs and between mean values (crosses) for the real and random PDFs. In order to take into account the effect of a choice of the threshold value for the displacement gradients, ∇D_{th}, we compute the median and mean values for all the seasons at $\nabla D_{th} = 60, 100$, and 130. The points below the slope line in Figure 22.34 give evidence that the corresponding median or mean value is closer to LFs for the real fishing boats, r_b, than for the randomly distributed ones, r_r. It is evident that both the medians and mean values are closer to LFs in the first case.

The oceanographic situation in 2002 was not typical being the single fishery season, among the studied ones, with the First Oyashio Intrusion. Moreover, the oceanographic situation in the region had changed significantly during that season. Animation of daily Lagrangian maps in 2002 with the fishery grounds overlaid (http://dynalab.poi.dvo.ru/data/GRL12/2002) clearly demonstrates that in addition to the intrusion of Soya Current waters along the north-eastern coast of Hokkaido Island there appeared in October–November the intrusion of cold Oyashio waters from the north into the fishery region which cardinally changed the oceanographic situation. The fishery grounds moved to the east and south in that time. The most stable oceanographic situation among the years with available fishery data was in the year 2004 with the Second Oyashio Intrusion, when the quasi-stationary warm Kuroshio anticyclonic ring was situated in the region for the whole fishery period. The strong LFs have been found to be the permanent ones in that year. The real PDF in Figure 22.33 (2004) significantly exceeds the random one and decays rapidly. The tail of the random PDF extends significantly over the distance r as compared with the real one. All

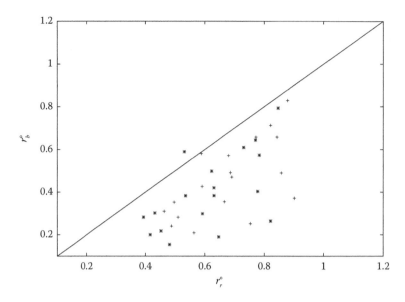

FIGURE 22.34 Median (stars) and mean (crosses) values for locations of real boats, r_b, versus randomly distributed ones, r_r.

these facts prove that saury fishing grounds really were located mainly along the strong LFs in that year. In the other years with available fishery data, the oceanographic situations resemble the 2004 case with the Second Oyashio Intrusion. PDFs for those fishery seasons in Figure 22.33 provide the statistical evidence that saury fishing locations are not randomly distributed over the region but are concentrated near the strongest LFs in the region studied. Based on statistical results, we may conclude that the more stable the oceanographic situation in the fishery region, the closer fishing boats tend to be to the LFs.

Oceanic fronts are areas with strong horizontal and vertical mixing. Highly turbid waters, however, are unsuitable for saury because it is a visual predator hunting in comparatively clear waters outside the exact locations of fronts. They avoid highly turbid waters and waters with large phytoplankton concentration, more than 5 g/m^3, which are turbid due to organic matter. On the other hand, extremely oligotrophic waters contain little food. Food abundance and water clarity are known to be two factors affecting the rate of food encounter. As to physical and biological reasons that may cause saury aggregation near LFs, we suggest the following ones. LFs in the Kuroshio–Oyashio frontal area demarcate the convergence of water masses with different productivity. They are zones with increased lateral and vertical mixing and often with increased primary and secondary production. Mixture of the nutrient-rich Oyashio waters with more oligotrophic Kuroshio waters can locally simulate phytoplankton photosynthesis and thus sustains higher phyto- and zooplankton concentrations with a net effect of aggregation of saury to forage on the lower trophic-level organisms. Stretching of material lines in the vicinity of hyperbolic objects in the ocean, a hallmark of chaotic advection (for a review see Reference 14), is one of the possible mechanisms providing effective intrusions of the nutrient-rich Oyashio cold waters into the more oligotrophic Kuroshio warm waters. Those filament-like intrusions may expand over hundreds of kilometers and are easily captured by the Lagrangian diagnostics but may be not visible on SST or chlorophyll-a images.

We illustrate that the mechanism of transport of nutrients in Figure 22.35 where evolution of the patches with synthetic particles selected on September 15, 2004 at five hyperbolic trajectories in the region is shown. It is the fishery season with the Second Oyashio Intrusion and a prominent quasi-stationary Kuroshio warm anticyclonic ring in the region (Figure 22.31). Let us suppose that some of the patches are rich in food and trace their evolution. Comparing Figure 22.35 with the backward-time FTLE map in Figure 22.31b, it is clear that all the patches in the course of time delineate the corresponding ridges on

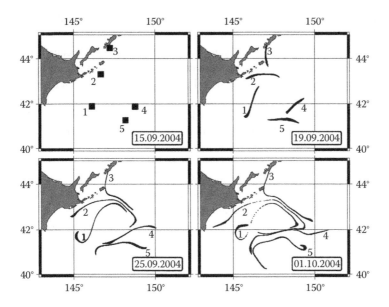

FIGURE 22.35 Evolution of the synthetic food patches selected at 5 hyperbolic trajectories in the 2004 fishery season with the Second Oyashio Intrusion. For 2 weeks, the patches rapidly delineate the corresponding unstable manifolds (the black ridges on the FTLE map in Figure 22.31) around the Kuroshio warm ring.

the map which approximate the unstable manifolds of selected five hyperbolic trajectories. The passive marine organisms in those fluid patches have been advected attracting saury for feeding along with them. The perimeter of some patches increased more than 100 times for only 2 weeks significantly increasing the chance for saury to find food. Such a mechanism of export of nutrient-rich waters into more poor ones is supposed to be typical because of a large number of hyperbolic trajectories in the frontal oceanic zones with increased mixing activity, rich in eddies.

At the end of this section, we would like to discuss the connection between potential fishing grounds, LFs, and thermal fronts visible on SST images. SST fronts have long been the main indicators used to find places in the ocean rich in marine resources. So, in order to find potential fishing grounds, it is instructive to use as the first guess the strong thermal gradients visible on satellite SST images if they are available. Unfortunately, SST images are not available on cloudy and rainy days which often occur in the fishery period in the region studied. In some fishery seasons in the Kuroshio–Oyashio frontal area, up to half the days have been found to be cloudy or rainy. Typically, the saury fishing grounds with maximal catches have been found not exactly at the SST fronts but in a comparatively large area around that front where a few LFs may be detected. Computation of LFs is a simple way to visualize a fine structure of the frontal zone which is a problem when using SST and/or chlorophyll-a images. Any strong large-scale LFs can be accurately detected in a given altimetric geostrophic velocity field by computing synoptic maps of the drift of synthetic tracers, their Lyapunov exponents, and the other Lagrangian indicators. Moreover, by computing meridional and zonal drift maps, one gets an information on the origin and history of convergent waters that may be useful to determine by which waters of a particular LF have been formed. Thus, the LFs, that can always be computed with AVISO altimetric geostrophic velocity fields, may serve as additional indicators of potential fishing grounds along with satellite SST images.

The results of this section can be summarized as follows. It is well known that oceanic fronts are sites of enhanced biomass that affect populations of organisms at all trophic levels. What is new we proposed in this section was a method to identify and analyze those oceanic fronts in altimetric (or given by another way) velocity fields. We introduced the notion of an LF, the region where surface waters of a different origin and history converge, and showed how to find them computing the zonal, meridional, and absolute drift maps and the FTLE maps for a large number of synthetic particles in a given region.

Based on satellite-derived surface velocities, we have integrated advection equations for a large number of synthetic tracers backward in time and computed the vorticity and FTLE maps, zonal, meridional, and absolute drift maps in the region to the east off Hokkaido and the southern Kuril Islands coasts, one of the richest fishery spots in the world. The data on fishing locations and daily catches of the Russian ships were imposed on the SST, vorticity, FTLE, and drift daily maps. To quantitatively determine whether saury was actively associating or not with the LFs in the region studied, we computed the frequency distribution of the distances between locations of fishing boats with saury catches and the strong LFs for available fishery seasons. It has been statistically shown that the saury fishing grounds with catches were not randomly distributed over the region but located mainly along those LFs where the productive cold waters of the Oyashio Current, warmer waters of the southern branch of the Soya Current, and waters of warm-core Kuroshio rings converged. We proposed a mechanism of effective export of nutrient-rich waters to more poor ones based on stretching of material lines in the vicinity of hyperbolic objects in the ocean. Those filament-like intrusions may expand over hundreds of kilometers and are easily captured by the Lagrangian diagnostics but not visible on SST or chlorophyll-a images. Thus, it has been shown that the strong LF locations may serve as good indicators of potential fishing grounds in rather different oceanographic conditions.

The method proposed seems to be quite general and may be applied to forecast potential fishing grounds for the other pelagic fishes in different regions of the World Ocean. On the other hand, our ability to recognize areas where pelagic fishes and marine animals prefer to congregate may help to create protectable marine reservations there. The Lagrangian tools can be useful toward this aim.

References

1. M. C. Honda, T. Aono, M. Aoyama, Y. Hamajima, H. Kawakami, M. Kitamura, Y. Masumoto, Y. Miyazawa, M. Takigawa, and T. Saino. Dispersion of artificial caesium-134 and -137 in the western North Pacific one month after the Fukushima accident. *Geochemical Journal*, 46:e1–e9, 2012.
2. H. Aref. Stirring by chaotic advection. *Journal of Fluid Mechanics*, 143(1):1, 1984.
3. F. J. Beron-Vera, M. J. Olascoaga, and G. J. Goni. Oceanic mesoscale eddies as revealed by Lagrangian coherent structures. *Geophysical Research Letters*, 35(12):L12603, 2008.
4. M. Budyansky, M. Uleysky, and S. Prants. Hamiltonian fractals and chaotic scattering of passive particles by a topographical vortex and an alternating current. *Physica D: Nonlinear Phenomena*, 195(3–4):369–378, 2004.
5. M. Budyansky, M. Uleysky, and S. Prants. Detection of barriers to cross-jet Lagrangian transport and its destruction in a meandering flow. *Physical Review E*, 79(5):056215, 2009.
6. M. V. Budyansky, M. Y. Uleysky, and S. V. Prants. Chaotic scattering, transport, and fractals in a simple hydrodynamic flow. *Journal of Experimental and Theoretical Physics*, 99(5):1018–1027, 2004.
7. K. O. Buesseler, S. R. Jayne, N. S. Fisher, I. I. Rypina, H. Baumann, Z. Baumann, C. F. Breier, E. M. Douglass, J. George, and A. M. Macdonald. Fukushima-derived radionuclides in the ocean and biota off Japan. *Proceedings of the National Academy of Sciences*, 109(16):5984–5988, 2012.
8. F. d Ovidio, V. Fernández, E. Hernández-García, and C. López. Mixing structures in the Mediterranean Sea from finite-size Lyapunov exponents. *Geophysical Research Letters*, 31(17), 2004.
9. D. del Castillo-Negrete and P. J. Morrison. Chaotic transport by Rossby waves in shear flow. *Physics of Fluids A: Fluid Dynamics*, 5(4):948, 1993.
10. G. Haller. Lagrangian coherent structures from approximate velocity data. *Physics of Fluids*, 14(6):1851, 2002.
11. G. Haller and A.C. Poje. Finite time transport in aperiodic flows. *Physica D: Nonlinear Phenomena*, 119(3–4):352–380, 1998.
12. A. D. Kirwan. Dynamics of "critical" trajectories. *Progress in Oceanography*, 70(2–4):448–465, 2006.

13. K. V. Koshel, M. A. Sokolovskiy, and P. A. Davies. Chaotic advection and nonlinear resonances in an oceanic flow above submerged obstacle. *Fluid Dynamics Research*, 40(10):695–736, 2008.

14. K. V. Koshel and S. V. Prants. Chaotic advection in the ocean. *Physics-Uspekhi*, 49(11):1151–1178, 2006.

15. Y. Lehahn, F. d Ovidio, M. Lévy, and E. Heifetz. Stirring of the northeast Atlantic spring bloom: A Lagrangian analysis based on multisatellite data. *Journal of Geophysical Research*, 112(C8):C08005, 2007.

16. A. M. Mancho, D. Small, and S. Wiggins. Computation of hyperbolic trajectories and their stable and unstable manifolds for oceanographic flows represented as data sets. *Nonlinear Processes in Geophysics*, 11(1):17–33, 2004.

17. S. V. Prants. Dynamical systems theory methods to study mixing and transport in the ocean. *Physica Scripta*, 87(3):038115, 2013.

18. S. V. Prants, M. V. Budyansky, M. Y. Uleysky, and G. M. Zaslavsky. Chaotic mixing and transport in a meandering jet flow. *Chaos: An Interdisciplinary Journal of Nonlinear Science*, 16(3):033117, 2006.

19. S. V. Prants, V. I. Ponomarev, M. V. Budyansky, M. Y. Uleysky, and P. A. Fayman. Lagrangian analysis of mixing and transport of water masses in the marine bays. *Izvestiya, Atmospheric and Oceanic Physics*, 49(1):82–96, 2013.

20. S. V. Prants, M. Y. Uleysky, and M. V. Budyansky. Numerical simulation of propagation of radioactive pollution in the ocean from the Fukushima Daiichi nuclear power plant. *Doklady Earth Sciences*, 439(2):1179–1182, 2011.

21. S. V. Prants, M. Y. Uleysky, and M. V. Budyansky. Numerical simulation of propagation of radioactive pollution in the ocean from the Fukushima Daiichi nuclear power plant. *Doklady Earth Sciences*, 439(2):1179–1182, 2011.

22. S. V. Prants, M. V. Budyansky, V. I. Ponomarev, and M. Y. Uleysky. Lagrangian study of transport and mixing in a mesoscale eddy street. *Ocean Modelling*, 38(1–2):114–125, 2011.

23. I. I. Rypina, M. G. Brown, F. J. Beron-Vera, H. Koçak, M. J. Olascoaga, and I. A. Udovydchenkov. On the Lagrangian dynamics of atmospheric zonal jets and the permeability of the stratospheric polar vortex. *Journal of the Atmospheric Sciences*, 64(10):3595–3610, 2007.

24. R. M. Samelson. Fluid exchange across a meandering jet. *Journal of Physical Oceanography*, 22(4): 431–444, 1992.

25. S. C. Shadden, F. Lekien, and J. E. Marsden. Definition and properties of Lagrangian coherent structures from finite-time Lyapunov exponents in two-dimensional aperiodic flows. *Physica D: Nonlinear Phenomena*, 212(3–4):271–304, 2005.

26. S. Shinohara and Y. Aizawa. The breakup condition of shearless KAM curves in the quadratic map. *Progress of Theoretical Physics*, 97(3):379–385, 1997.

27. M. Y. Uleysky, M. V. Budyansky, and S. V. Prants. Effect of dynamical traps on chaotic transport in a meandering jet flow. *Chaos: An Interdisciplinary Journal of Nonlinear Science*, 17(4):043105, 2007.

28. M. Y. Uleysky, M. V. Budyansky, and S. V. Prants. Chaotic transport across two-dimensional jet streams. *Journal of Experimental and Theoretical Physics*, 111(6):1039–1049, 2010.

29. M. Y. Uleysky, M. V. Budyansky, and S. V. Prants. Mechanism of destruction of transport barriers in geophysical jets with Rossby waves. *Physical Review E*, 81(1):017202, 2010.

30. S. Wiggins. The dynamical systems approach to Lagrangian transport in oceanic flows. *Annual Review of Fluid Mechanics*, 37(1):295–328, 2005.

31. G. M. Zaslavsky. Dynamical traps. *Physica D: Nonlinear Phenomena*, 168–169:292–304, 2002.

32. S. V. Prants, M. V. Budyansky, and M. Yu. Uleysky. Lagrangian study of surface transport in the Kuroshio Extension area based on simulation of propagation of Fukushima-derived radionuclides. *Nonlinear Processes in Geophysics*, 21:279–289, 2014.

33. M. V. Budyansky, V. A. Goryachev, D. D. Kaplunenko, V. B. Lobanov, S. V. Prants, A. F. Sergeev, N. V. Shlyk, and M. Yu. Uleysky. Role of mesoscale eddies in transport of Fukushima-derived cesium isotopes in the ocean. *Deep Sea Research I*, 98:15–27, 2015.

34. S. V. Prants, A. G. Andreev, M. Yu. Uleysky, and M. V. Budyansky. Lagrangian study of temporal changes of a surface flow through the Kamchatka Strait. *Ocean Dynamics*, 64(6):771–780, 2014.

35. S. V. Prants, M. V. Budyansky, and M. Yu. Uleysky. Lagrangian fronts in the ocean. Izvestiya. *Atmospheric and Oceanic Physics*, 50(3):284–291, 2014.

36. S. V. Prants, M. V. Budyansky, and M. Yu. Uleysky. Identifying Lagrangian fronts with favourable fishery conditions. *Deep Sea Research I*, 90:27–35, 2014.

37. S. V. Prants. Chaotic Lagrangian transport and mixing in the ocean. *The European Physical Journal Special Topics*, 223(13):2723–2743, 2014.

38. S. V. Prants, M. V. Budyansky, V. I. Ponomarev, M. Yu. Uleysky, and P. A. Fayman. Lagrangian analysis of the vertical structure of eddies simulated in the Japan Basin of the Japan/East Sea. *Ocean Modelling*, 86:128–140, 2015.

39. S. V. Prants. Backward-in-time methods to simulate chaotic transport and mixing in the ocean. *Physica Scripta*, 90:074054, 2015.

40. S. V. Prants, A. G. Andreev, M. V. Budyansky, and M. Yu. Uleysky. Impact of the Alaskan Stream flow on surface water dynamics, temperature, ice extent, plankton biomass and walleye pollock stocks in the eastern Okhotsk Sea. *Journal of Marine Systems*, 151:47–58, 2015.

23

The Suspensions of Maps to Flows

John Starrett

23.1 Introduction

Roughly speaking, a suspension of a map f to a flow ϕ is a smooth interpolation with parameter t between the domain and the range of f. We will make this notion precise in the following section. More specifically, we are interested in ordinary differential equations whose solutions at integer values of t are solutions to difference equations.

Flows and diffeomorphisms are related by surface of section maps, and while not every flow has a surface of section map, it is the case that every diffeomorphism has a suspension flow. If M is a manifold and f a diffeomorphism $f : M \to M$, we know a suspension exists, so why bother finding specific flows? We may want to illustrate a concrete version of a theoretical result, as when we find templates associated with certain braids, or we may want an example of a flow with precisely the set of periodic orbits that the Hénon map has. To a limited degree, we can construct differential equations with specified periodic solution sets; we can custom-build attractors, strange, and otherwise.

There are many different ways of going about this task, and we will highlight the ones with which we are most familiar. We will certainly miss some interesting methods, fail to cite important references, overlook applications or techniques out of our field of expertise, and otherwise slight those who have made significant contributions. We hope to hear from researchers who can educate us on the things we missed, the topics of which we failed to give fair expositions, and enlighten us with new ideas and improvements on the methods we lay out.

23.2 Basic Definitions

Definition 23.1

A *map* $f : S \to T$ is a function from a set S to a set T. In the context of suspensions of maps, the sets are *manifolds*, sets that are locally homeomorphic to $n-$dimensional Euclidean space.

Definition 23.2

An *orbit* of a point $\mathbf{x}_0$ under a mapping $f : M \to M$ is the infinite sequence $\{\mathbf{x}_i\}$, $i = 0 \ldots \infty$ of iterates $\lim_{n \to \infty} f^n(\mathbf{x}_0)$.

Definition 23.3

A *flow* is a continuous one parameter group of diffeomorphisms $\phi(\mathbf{x}, t)$ of a manifold M with identity ϕ_0, composition $\phi_t \circ \phi_s = \phi_{t+s}$ and inverse $\phi_{-t}\phi_t = \phi_0$.

Flows are nonsingular: $\phi_t(\mathbf{x}_1) \neq \phi_t(\mathbf{x}_2)$ for $\mathbf{x}_1 \neq \mathbf{x}_2$.

Definition 23.4

A *semiflow* is a continuous one parameter group of diffeomorphisms $\phi(\mathbf{x}, t)$ of a manifold M with identity ϕ_0 and composition $\phi_t \circ \phi_s = \phi_{t+s}$.

In other words, a semiflow is an irreversible flow: it has singular points.

Definition 23.5

An *orbit* of a point $\mathbf{x}(0) \in M$ under a flow or a semiflow $\phi(\mathbf{x}(0), t): M \to M$ is the codimension 1 submanifold of M given by $\lim_{t \to \infty} \phi(\mathbf{x}(0), t)$.

Definition 23.6

Let M be a smooth manifold, and $g : M \to V$ be a smooth invertible map. The quotient manifold $Q = M \times [0, 1]/\sim$ with $M \times 0$ and $M \times 1$ identified, is called a *suspension manifold*, and the flow $\phi(\mathbf{x}, t)$ on Q with $\phi(\mathbf{x}, 1) = \phi(g(\mathbf{x}), 0) \, \forall \mathbf{x} \in M$ is called a *suspension flow*.

Consider a system of ordinary differential equations (ODEs) $\dot{\mathbf{x}} = F(t, \mathbf{x}(t), \theta(t))$, where $\theta(t)$ is a periodic function of t. The ODEs define a vector field $\mathbf{F}$ that generates the *flow* $\phi(\mathbf{x}, t)$, the group structure on the set of all solution curves $\mathbf{x}(t)$ of the system. If F is nonvanishing, these curves foliate a solution manifold M.

Definition 23.7

A *Poincaré section* of the flow is the restriction of ϕ to a codimension-1 submanifold P, which intersects each solution curve $\mathbf{x}(t)$ transversely.

Definition 23.8

A *Poincaré map* $g_\phi : P \to P$ takes $\mathbf{x} \in P$ to $g(\mathbf{x}) \in P$ with the condition $g(\mathbf{x}(t)) = \mathbf{x}(t + \tau)$ with $\mathbf{x}(t + \gamma) \notin P$ for any $0 < \gamma < \tau$.

For a periodically forced system with period τ, a family of distinguished Poincaré maps, with parameter δ, can always be extracted by sampling the solution $\mathbf{x}(t)$ at equally spaced time intervals to get $\mathbf{x}(n\tau + \delta)$, $n = 0, 1, 2, ...$, where δ offsets the phase. Similarly, a Poincaré map can often be extracted from an experimental system as a sequence of vector measurements $\mathbf{x}(t_k)$, or a vector of time-delayed scalar measurements $\langle x(t), x(t - \tau), x(t - 2\tau), ... \rangle$ if there is some fundamental time scale τ.

Let $g : M \to M$ be continuous, one-to-one, and orientation preserving. Let Φ be a flow on $Q = M \times [0, 1]/ \sim$ with $g(\mathbf{x}) = \phi(\mathbf{x}, 1)$. Then g is the Poincaré map of Φ.

23.3 Simplest Examples

We begin in one dimension, with maps $f : \mathbb{R} \to \mathbb{R}$. Many simple maps of the real line can be "suspended," after a fashion, to a flow if there is a solution $x_n = g(n, x_0, x_1, \ldots, x_k)$ to the recursion $x_{n+1} = f(n, x_n, x_{n-1}, \ldots, x_{n-p})$. We simply replace the discrete index n with the continuous time t. However, orbits may cross in the two-dimensional suspension flow. Individual orbits, beginning from some initial condition *can* be values at integer times of their "suspended" functions.

The simplest example is the constant map $x_{n+1} = x_n$, which has constant solution $g(n) = x_0$. Replacing n with t, we obtain $g(t) = x_0$, which is the solution to the differential equation $\dot{x} = 0$ with initial condition $x(0) = x_0$. Of course, the same suspension can be a solution to a number of different differential equations. For instance, the system

$$\dot{r} = r(C - r) \tag{23.1}$$

$$\dot{\theta} = 1 \tag{23.2}$$

has a solution constant in r for $r = 0$ or $r = C$, as does $\dot{r} = 0$ with initial condition $r(0) = C$, and the differential equation $\dot{x} = -2\pi y$, $\dot{y} = 2\pi x$ has solutions that are constant at integer times. The doubling map $x_{n+1} = 2x_n$ has solution $x_n = x_0 2^n$, so a suspension is $x(t) = x_0 2^t$, which is the general solution of the ODE $\dot{x} = x \ln 2$ with initial condition $x(0) = x_0$.

As a more complicated example whose derivation is not as obvious, consider the map

$$r_{n+1} = \frac{r_n}{r_n - e^{-1}(r_n - 1)}, \quad r_n \geq 0. \tag{23.3}$$

This map suspends to a flow, i.e., it is a time one map of the solution

$$r(t) = \frac{1}{1 + ke^{-t}}, \quad r \geq 0 \tag{23.4}$$

to the differential equation

$$\dot{r} = r(1 - r).$$

The solutions can be viewed as a suspension in the cylinder $\mathbb{R}^+ \times \mathbb{I}$, with $t \in \mathbb{I} = [0, 1]$ or as a suspension in the infinite cylinder $\mathbb{R}^+ \times \mathbb{R}$, see Figure 23.1. However, because of the natural time scale of the time one map, we may interpret the flow as a solution to the polar system

$$\dot{r} = r(1 - r)$$

$$\dot{\theta} = 2\pi.$$

Writing Equation 23.4 in Cartesian coordinates

$$x = \cos 2\pi t (1 + ke^{-t})^{-1} \tag{23.5}$$

$$y = \sin 2\pi t (1 + ke^{-t})^{-1}, \tag{23.6}$$

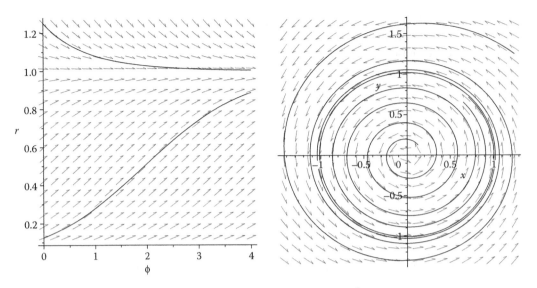

FIGURE 23.1 Suspension to a cylinder $\mathbb{R}^+ \times \mathbb{R}$ and suspension to a plane $\mathbb{R}^2$ of Equation 23.3 via and embedding of $\mathbb{R}^+ \times \mathbb{S}^1$ into $\mathbb{R}^2$.

we see that Equations 23.5 and 23.6 satisfy the autonomous planar system

$$\dot{x} = -2\pi y + x\left(1 - \sqrt{x^2 + y^2}\right)$$

$$\dot{y} = 2\pi x + y\left(1 - \sqrt{x^2 + y^2}\right).$$

Unfortunately, there are very few recursions with closed-form solutions. However, linear iterated maps $x_{n+1} = c_n x_n + c_{n-1} + \cdots + c_{n-p} x_{n-p}$ have solutions that can be computed explicitly, so a simple suspension flow can be constructed by replacing n by t. Solutions can be computed by hand using generating functions [12] or other methods, but in practice a computer algebra system such as Maple or Mathematica may be more practical.

To briefly illustrate the method of solution by generating functions, consider the recursion $F_n = F_{n-1} + F_{n-2}$ for the Fibonacci sequence $0, 1, 1, 2, 3, 5, 8, 13, \ldots$. To find the general solution F_n, we form the function $F(x) = \sum_{n=0}^{\infty} F_n x^n$, that is, the power series with the Fibonacci numbers as coefficients. Because $F_0 = 0, F_1 = 1$, we can express this as

$$F(x) = 0 + x + \sum_{n=2}^{\infty} F_n x^n = x + \sum_{n=2}^{\infty} (F_{n-1} + F_{n-2}) x^n$$

$$= x + \sum_{n=2}^{\infty} F_{n-1} x^n + \sum_{n=2}^{\infty} F_{n-2} x^n.$$

We aim to express the right-hand side entirely in terms of F_n, so we write

$$\sum_{n=2}^{\infty} F_{n-1} x^n = x \sum_{n=2}^{\infty} F_{n-1} x^{n-1} = x \sum_{n=0}^{\infty} F_n x^n$$

and

$$\sum_{n=2}^{\infty} F_{n-2} x^n = x^2 \sum_{n=2}^{\infty} F_{n-2} x^{n-2} = x^2 \sum_{n=0}^{\infty} F_n x^n.$$

Then $F(x) = x + xF(x) + x^2 F(x)$ and $F(x) = \frac{x}{1-x-x^2}$. An explicit power series for $F(x)$ remains to be written, and we will recover the general term, the coefficient of x^n. Writing

$$F(x) = \frac{x}{(1 - xr_1)(1 - xr_2)} = \frac{A}{(1 - xr_1)} + \frac{B}{(1 - xr_2)},$$

we determine that

$$F(x) = A \sum_{n=0}^{\infty} r_1^n x^n + B \sum_{n=0}^{\infty} r_2^n x^n = \sum_{n=0}^{\infty} (A r_1^n + B r_2^n) x^n,$$

so the general recursion is

$$F_n = A \left(\frac{1 + \sqrt{5}}{2}\right)^n + B \left(\frac{1 - \sqrt{5}}{2}\right)^n,$$

with A and B determined by F_0 and F_1. Specifically,

$$F_n = \left(\frac{\sqrt{5}}{5}\right) \left(\left(\frac{1 + \sqrt{5}}{2}\right)^n - \left(\frac{1 - \sqrt{5}}{2}\right)^n\right).$$

Replacing the discrete index n by the continuous variable t, we have a suspension of the Fibonacci sequence, whose real part has the graph shown below.

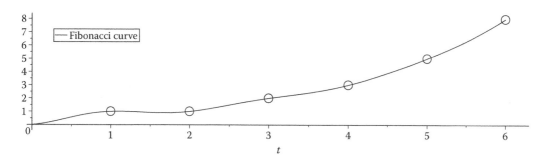

As a short additional example, we have solved $h(n) = h(n - 1) - 2h(n - 2) + 2n$, with $h(0) = 1$, $h(1) = 1$ using Maple, and plotted the (rather complicated) real part of the suspension in below figure.

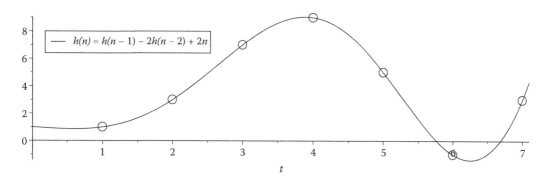

23.4 Suspensions of the Logistic Map

Definition 23.9

Two maps f and g are topologically conjugate if there exists an h so that $f = h^{-1} \circ g \circ h$.

Iterates of conjugate maps are conjugate: $f(f(x)) = h^{-1} \circ g \circ h \circ h^{-1} \circ g \circ h = h^{-1} \circ g^2 \circ h$, and also $f^n = h^{-1} \circ g^n \circ h$.

Given a map f, we may be able to find a conjugacy that simplifies calculations. For instance, the logistic map $f(x) = 4x(1 - x)$ is topologically conjugate to the tent map

$$g(x) = \begin{cases} 2x & \text{for } 0 \le x \le 1/2 \\ 2(1-x) & \text{for } 1/2 \le x \le 1 \end{cases} \tag{23.7}$$

through the conjugacy $h(x) = 2/\pi sin^{-1}(\sqrt{x})$. Using this conjugacy, we find that for the logistic map $x_{n+1} = 4x_n(1 - x_n)$, we obtain $x_{n+1} = \sin^2(2 \arcsin \sqrt{x_n})$, $x_{n+2} = \sin^2(2^2 \arcsin \sqrt{x_n})$, and in general $x_n = \sin^2(2^n \arcsin \sqrt{x_0})$. Replacing n by the continuous variable t, we have the suspension $x(t) = \sin^2(2^t \arcsin \sqrt{x_0})$ (see below figure).

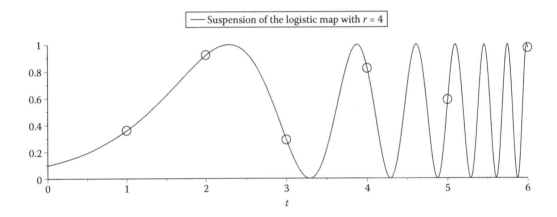

Suspension of the logistic map with $r = 4$

Definition 23.10

A *half iterate* of a map $f^{1/2}(\mathbf{x})$ is a function g with the property $g(g(\mathbf{x})) = f(\mathbf{x})$. A *fractional iterate of order* n $f^{1/n}(\mathbf{x})$ is a function h with the property $h^n(\mathbf{x}) = \underbrace{h(h(...(h(\mathbf{x})))...)}_{n \text{ times}} = f(\mathbf{x})$.

If we can find a general form for the fractional iterate of order n of a map f, we may replace n with t to obtain the suspension of f.

Schröder's equation [9] $\Phi(f(x)) = s\Phi(x)$ expresses the conjugacy $f(x) = \Phi^{-1} \circ (s\Phi(x))$, a simple dilation by s. Similarly, $f^{1/n}(x) = \Phi^{-1} \circ (s^{1/n}\Phi(x))$, and in the same way, we can pass to the continuous case $f(t, x) = \Phi^{-1} \circ (s^t\Phi(x))$. Therefore, if we can find a Φ for our function $f(x) = 4x(1 - x)$, we can transform the logistic map to a continuous function of t, whose argument changes as a continuous dilation. Begin with the half iterate $f^{1/2}(x) = \Phi^{-1} \circ (s^{1/2}\Phi(x))$, and let $\Phi = \arcsin^2(\sqrt{x})$, $s = 4$, and $\Phi^{-1} = \sin^2(\sqrt{y})$. Then $f^{1/2}(x) = \sin^2(2^{1/2} \arcsin(\sqrt{x}))$, and passing to the continuous limit, $f(t, x) =$

$\sin^2(2^t \arcsin(\sqrt{x}))$, as above. Madea writes the suspension of the logistic map $1/2(1 - \cos((2^t) \arccos (1 - 2x_0)))$.

Schröder also determined a continuous solution for the nonchaotic logistic map $f(x) = 2x(1 - x)$, obtaining $x(t) = -\frac{1}{2}((1 - 2x_0)^{2^t} - 1)$ (see below figure).

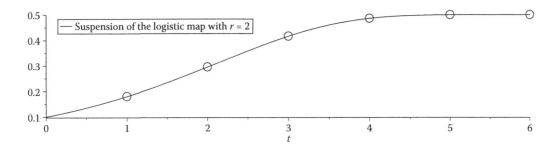

Again, there are few recursions that have explicit suspensions through these methods [11], and for the logistic map, solutions for multiplier 2 and 4 are the only ones known.

23.5 Other Exact Methods

23.5.1 Suspension of Two-Dimensional Maps

Channell [2] gave the first specific prescription of the conditions that must hold for a practical suspension of an n-dimensional map to a flow.

Hénon [4] studied the orientation reversing map

$$x_{n+1} = a + by_n - cx_n^2$$
$$y_{n+1} = dx_n,$$

$a = 1$, $b < 1$, c, $d > 0$, for which there is an attracting set, either periodic or chaotic depending on the parameters. For $d = -1$, the map is orientation preserving and can be suspended to a flow $\phi(x, y, t)$ in the cylinder $\mathbb{R}^2 \times [0, 1]$ of one dimension greater.

Mayer-Kress and Haken [8] constructed a differentiable suspension of an orientation preserving Hénon map $g : \mathbb{R}^2 \to \mathbb{R}^2$ to a flow $\phi(t)$ on the cylinder $\mathbb{R}^2 \times [0, 1]$, $t \in [0, 1]$ by interpolating the initial and final conditions of the Hénon map $\Phi(\mathbf{x}_0, t) = \xi\mathbf{x}_0 + \eta g(\mathbf{x}_0)$ using sigmoidal polynomial interpolating diffeomorphisms ξ and η with continuity conditions $\xi(0) = \eta(1) = 1$, $\xi(1) = \eta(0) = 0$ differentiability conditions $D_t\xi(0) = D_t\eta(1)$, $D_t\xi(1) = D_t\eta(0) = 0$, and with the identification $\Phi(\mathbf{x}, 1) = \Phi(g(\mathbf{x}), 0)$. The initial conditions $\mathbf{x}_0$ were explicitly solved for in terms of functions of t, then substituted into the derivative $\dot{\Phi} = \frac{d\phi}{dt}$ of the interpolation, giving the differential equation $\dot{\Phi} = F(\mathbf{x}, t)$.

The criteria were chosen to yield a flow coming as close as possible to the flow of a driven anharmonic oscillator with position independent dissipation rate $\ln(j)$. Specifically, they chose $\xi(t) = (3 - 2t)t^2$ and $\eta(t) = 1 + \ln(j)t - (2\ln(j) + 3)t^2 + (2 + \ln(j))t^3$, where j is set to $j = 0.3$. Solutions can be found by numerical integration of the (now specific) differential equations

$$\dot{x} = hx + gy + \xi\dot{\xi}f(x_0)$$
$$\dot{y} = gx + fy + \eta\dot{\xi}f(x_0)$$
$$\dot{z} = 1,$$

where

$$g(\xi, \eta) = \frac{1}{\xi^2 + \eta^2} \left(\dot{\xi}\eta - \xi\dot{\eta} - \xi\eta \left(\ln(j) - 2\frac{\dot{\xi}\eta - \xi\dot{\eta}}{\xi^2 + \eta^2} \right) \right)$$

$$h(\xi, \eta) = \frac{1}{\xi^2 + \eta^2} \left((\dot{\xi}\xi - \eta\dot{\eta})\frac{\xi^2 - \eta^2}{\xi^2 + \eta^2} + \eta^2 \ln(j) \right).$$

Time t does appear in the interpolating diffeomorphisms, however, and they stated that it would be interesting to find a suspension where time did not appear explicitly.

23.5.2 An Autonomous Suspension of the Hénon and Duffing Maps

We will demonstrate a different approach, using geometric methods, suspending the Hénon map and the Duffing map. In these suspensions, time does not appear explicitly, and we obtain a system of autonomous differential equations whose attractors lie in a torus surrounding the z axis in $\mathbb{R}^3$.

We suspend the orientation preserving maps

$$x_{n+1} = \frac{4}{3}(1 - x_n^2) + \frac{1}{9}z_n \tag{23.8}$$

$$z_{n+1} = -x_n \tag{23.9}$$

and

$$x_{n+1} = 2x_n - x_n^3 - \frac{2}{5}z_n \tag{23.10}$$

$$z_{n+1} = -x_n \tag{23.11}$$

of the x, z plane to $\mathbb{R}^3$.

Let $x_{n+1} = F_1(x_n, z_n), z_{n+1} = F_2(x_n, z_n) = -x_n$ be the components of our two dimensional mappings of the plane. Then the action of the maps may be broken down into two geometric transformations. First the (x, z) plane is deformed by

$$\begin{bmatrix} \hat{x}_{n+1} \\ \hat{z}_{n+1} \end{bmatrix} = \begin{bmatrix} -F_2 \\ F_1 \end{bmatrix},$$

then the result is rotated by $\pi/2$ clockwise

$$\begin{bmatrix} x_{n+1} \\ z_{n+1} \end{bmatrix} = \begin{bmatrix} \cos\left(\frac{\pi}{2}\right) & \sin\left(\frac{\pi}{2}\right) \\ -\sin\left(\frac{\pi}{2}\right) & \cos\left(\frac{\pi}{2}\right) \end{bmatrix} \begin{bmatrix} -F_2 \\ F_1 \end{bmatrix} = \begin{bmatrix} F_1 \\ F_2 \end{bmatrix}.$$

The map is suspended to a flow by differentiable interpolations between the initial and the final conditions. We break the process into these two steps to avoid the self-intersections that could occur is we used a direct interpolation.

The interpolation is accomplished via the functions $C = \cos(\frac{\pi}{2}t)$ and $S = \sin(\frac{\pi}{2}t)$, which also appear in a rotation matrix $R = \begin{bmatrix} C & S \\ -S & C \end{bmatrix}$. The functions $C^2 = \cos^2\left(\frac{\pi}{2}t\right), S^2 = \sin^2\left(\frac{\pi}{2}t\right)$ are smooth, and evolve initial to final conditions by $C^2 x_0 + S^2 x_1$ as t varies from 0 to 1. Because of the scaling of the argument, a quarter clockwise rotation and a smooth interpolation between initial and final conditions takes place as t goes from 0 to 1, suspending the map to a cylinder $(0, \infty) \times \mathbb{R} \times [0, 1]$.

Because we will be wrapping the suspension around the z axis, we write it $\mathbb{R}^3$ in a torus surrounding the z axis. Embed the suspension cylinder in $\mathbb{R}^3$ by $(0, \infty) \times \mathbb{R} \times [0, 1]$ as

$$\begin{bmatrix} x \\ y \\ z \end{bmatrix} = \begin{bmatrix} C & 0 & S \\ 0 & 1 & 0 \\ -S & 0 & C \end{bmatrix} \begin{bmatrix} C^2 x_0 - S^2 F_2 \\ 0 \\ C^2 z_0 + S^2 F_1 \end{bmatrix} = \begin{bmatrix} Cx_0 + S(C^2 z_0 + S^2 F_1) \\ 0 \\ -Sx_0 + C(C^2 z_0 + S^2 F_1) \end{bmatrix},$$

where we have used the fact that $F_2 = -x$ and $C^2 + S^2 = 1$.

We shift the function in the x direction by an amount sufficient to include the attractor in the left half plane and to avoid self-intersection (3, in these examples), and rotate counterclockwise about the z axis:

$$
\begin{bmatrix} x \\ y \\ z \end{bmatrix} = \begin{bmatrix} \cos(2\pi t) & -\sin(2\pi t) & 0 \\ \sin(2\pi t) & \cos(2\pi t) & 0 \\ 0 & 0 & 1 \end{bmatrix} \begin{bmatrix} 1 + Cx_0 + S(C^2 z_0 + S^2 F_1) \\ 0 \\ -Sx_0 + C(C^2 z_0 + S^2 F_1) \end{bmatrix}
$$
$$
= \begin{bmatrix} \cos(2\pi t)(1 + Cx_0 + S(C^2 z_0 + S^2 F_1)) \\ \sin(2\pi t)(1 + Cx_0 + S(C^2 z_0 + S^2 F_1)) \\ -Sx_0 + C(C^2 z_0 + S^2 F_1) \end{bmatrix}. \tag{23.12}
$$

To obtain a differential equation we solve Equation 23.12 for the initial conditions in terms of x, z, and t, take the derivative of Equation 23.12 with respect to t, and substitute the expressions for the initial conditions into the resulting differential equation. For example, the initial conditions for the Hénon suspension are given by

$$
x_0 = (Cx - 3cC - cSz)/c
$$
$$
z_0 = 3(4c^2 S^4 z^2 + 24c^2 CS^3 z - 8cCS^3 xz - 36c^2 S^4 + 24cS^4 x - 4S^4 x^2 + 3c^2 Cz
$$
$$
+ 32c^2 S^2 - 24cS^2 x + 4S^2 x^2 - 9c^2 S + 3cSx)/c^2(9 - 8S^2),
$$

where we have used the abbreviations $c = \cos(2\pi t)$ and $s = \sin(2\pi t)$.

The differential equation at this point is nonautonomous, so we use the half angle formulas twice to write the interpolation and rotation operators with the same arguments:

$$
C = \begin{cases} \sqrt{\frac{1}{2} + \frac{1}{4}\sqrt{2 + 2\cos(2\pi t)}} & 0 \le t \le \frac{1}{2} \\ \sqrt{\frac{1}{2} - \frac{1}{4}\sqrt{2 + 2\cos(2\pi t)}} & \frac{1}{2} \le t \le 1 \end{cases} \quad S = \begin{cases} \sqrt{\frac{1}{2} - \frac{1}{4}\sqrt{2 + 2\cos(2\pi t)}} & 0 \le t \le \frac{1}{2} \\ \sqrt{\frac{1}{2} + \frac{1}{4}\sqrt{2 + 2\cos(2\pi t)}} & \frac{1}{2} \le t \le 1 \end{cases}.
$$

Using the identity $\cos(2\pi t) = \frac{x}{\sqrt{x^2 + y^2}}$, and the equivalence $0 \le t \le 1/2 \Rightarrow y \ge 0$, we obtain

$$
C = \begin{cases} \sqrt{\frac{1}{2} + \frac{1}{4}\sqrt{2 + \frac{2x}{\sqrt{x^2 + y^2}}}} & y \ge 0 \\ \sqrt{\frac{1}{2} - \frac{1}{4}\sqrt{2 + \frac{2x}{\sqrt{x^2 + y^2}}}} & y < 0 \end{cases} \quad S = \begin{cases} \sqrt{\frac{1}{2} - \frac{1}{4}\sqrt{2 + \frac{2x}{\sqrt{x^2 + y^2}}}} & y \ge 0 \\ \sqrt{\frac{1}{2} + \frac{1}{4}\sqrt{2 + \frac{2x}{\sqrt{x^2 + y^2}}}} & y < 0 \end{cases}. \tag{23.13}
$$

Substitution of Equation 23.13 into the nonautonomous differential equation transforms the nonautonomous differential equations to autonomous ones.

At the plane $y = 0$, the interpolating diffeomorphisms are discontinuous, but the tangent vector to the solution curves are

$$
\lim_{t \to 0^+} \begin{bmatrix} \dot{x} \\ \dot{y} \\ \dot{z} \end{bmatrix} = \lim_{t \to 0^-} \begin{bmatrix} \dot{x} \\ \dot{y} \\ \dot{z} \end{bmatrix} = \lim_{t \to 1^+} \begin{bmatrix} \dot{x} \\ \dot{y} \\ \dot{z} \end{bmatrix} = \lim_{t \to 1^-} \begin{bmatrix} \dot{x} \\ \dot{y} \\ \dot{z} \end{bmatrix} = \left\langle \frac{\pi}{2}z, 2\pi x, \frac{\pi}{2}(3 - x) \right\rangle,
$$

so the derivative is continuous there.

Integrating initial conditions in the attractor for the Hénon map and the Duffing map forward results in strange attractors whose Poincaré maps are identical with the original Hénon and Duffing maps (see Figure 23.2).

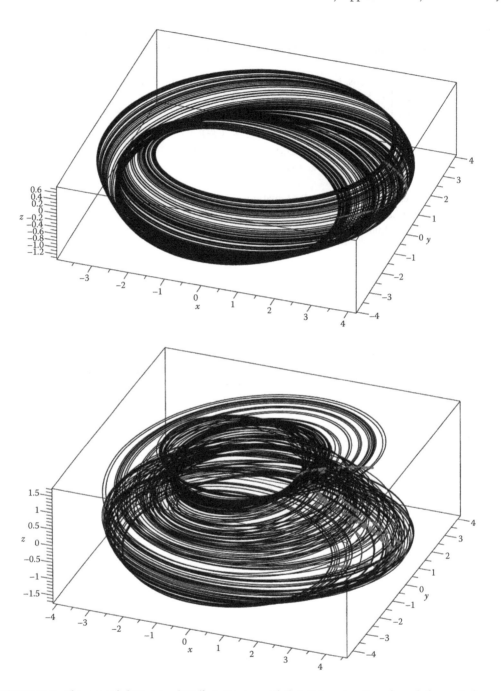

FIGURE 23.2 The suspended Hénon and Duffing attractors with their Poincaré sections through the $y = 0$ plane.

23.5.3 Suspension of One-Dimensional Maps of the Unit Interval to Semiflows

It is well known that a true suspension of two-to-one maps, such as the tent map and logistic map, to a flow, cannot exist in any dimension less than four, so here we will suspend to a semiflow instead.

Consider the difference equations

$$x_{n+1} = 4x_n(1 - x_n) + \alpha z_n \quad z_{n+1} = -x_n \tag{23.14}$$

$$x_{n+1} = x_n^3 - 13/5\, x_n + \alpha z_n \quad z_{n+1} = -x_n, \tag{23.15}$$

where the first, a modification of the fully chaotic logistic map, is an analog of the Hénon map, and the second, a modification of the cubic map, is an analog of the Duffing map. For a range of parameter α, these systems are chaotic with fractal attractors. For $\alpha = 0$, these maps have one-dimensional attractors in the x, z plane; they are not fractal. The iterates are of the form $(x_{n+1}, -x_n)$, a clockwise rotation by $\frac{\pi}{2}$ of the functional form of the map (x_n, x_{n+1}). The projection of the dynamics onto the x axis give the dynamics of the original one-dimensional maps (see Figures 23.3 and 23.4).

Suspending these one-dimensional attracting sets to semiflows results in a two-dimensional attracting set; the unstable manifold of the semiflow. The attracting sets are (almost) topologically equivalent to a *template*, a branched two manifold that is a standard tool of knot and braid theory: the original map attractor is two to one, and the surfaces to either side of the extrema of the functions join smoothly at the $y = 0$ plane in a one-dimensional set. At the critical points of the time one iterate of the maps (the xz plane), there are nondifferentiable folds. To form a true template from this stable manifold, we "unzip" it by flowing the critical point(s) at the Poincaré section backward in time to the previous Poincaré section

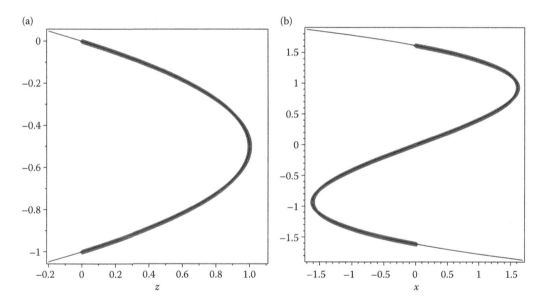

FIGURE 23.3 (a) The logistic map and (b) the cubic map. The attractors, shown in thick gray, are subsets of the rotated functions along which the attractors lie, thin black lines.

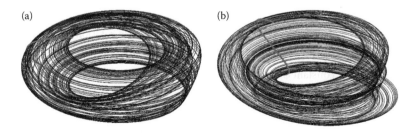

FIGURE 23.4 (a) The suspended logistic map and (b) the cubic map.

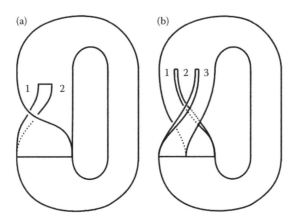

FIGURE 23.5   (a) The template for the logistic map and (b) the template for the cubic map.

and remove this piece of orbit. What remains is a branched two-manifold, a *template* in the usual sense of the word, except for the open boundaries where the orbit of the critical point(s) were removed. If we perturb the surface near the critical point to separate the closures of the sets to each side of the critical orbit, we can close this set to obtain a classic template. In the case of the logistic suspension, the template is (topologically) exactly the template of the suspended Smale horseshoe (see Figure 23.5).

The suspension of these maps is done in exactly the same way as we suspended the Hénon and Duffing maps, by a displacement, and a simultaneous $\pi/2$ clockwise rotation, smooth interpolation between initial and final conditions, and a rotation about the z axis.

23.5.4 Suspension of the Plykin Attractor

The Plykin map is an important example of a map with a hyperbolic attractor, where all the orbits, periodic and chaotic, are of saddle type. It is a member of the small family of uniformly hyperbolic mappings (including the Smale horseshoe), whose significance is that they are *structurally stable*, that is, for small enough variations in parameters of the maps, the dynamics are essentially unchanged. There is quite a bit known about the dynamics of hyperbolic maps, and the theorems about hyperbolic maps can be easily extended to hyperbolic flows.

Kuznetsov [5] constructed a diffeomorphic mapping of the unit sphere to itself with an attractor that is a Plykin attractor. Kuznetsov built a system of ODEs based on a sequence of geometric deformations of a unit sphere centered at the origin, as follows: Let the sphere be parameterized by ϕ and θ with Cartesian coordinates

$$x = \cos\theta\sin\phi$$
$$y = \sin\theta\sin\phi$$
$$z = \cos\phi,$$

where θ increases from 0 at the positive x axis with counterclockwise rotation about the z axis and ϕ increases from zero at $z = -1$ axis, and make this sequence of geometric transformations:

1. Flow along circles of constant ϕ by

$$\dot{x} = -\epsilon xy^2$$
$$\dot{y} = \epsilon x^2 y$$
$$\dot{z} = 0.$$

The great circle through $\theta = 0$ from $\phi = -\pi$ to $\phi = \pi$ is a repeller and the great circle through $\theta = \pi/2$ from $\phi = -\pi$ to $\phi = \pi$ is an attractor.

2. Differential rotation about the z axis by

$$
\begin{aligned}
\dot{x} &= \pi(z/\sqrt{2} + 1/2)y \\
\dot{y} &= -\pi(z/\sqrt{2} + 1/2)x \\
\dot{z} &= 0.
\end{aligned}
$$

Rotation is greatest at $z = 1$ and decreases to zero at $z = -1$.

3. Flow as in (1), but with constant values of an angle measured from the x axis in the xz plane by

$$
\begin{aligned}
\dot{x} &= 0 \\
\dot{y} &= \epsilon y z^2 \\
\dot{z} &= -\epsilon y^2 z.
\end{aligned}
$$

Here, the circle $\phi = \pi/2$ is an attractor, and the great circle in the xy plane is a repeller.

4. Differential rotation about the z axis by

$$
\begin{aligned}
\dot{x} &= 0 \\
\dot{y} &= -\pi(x/\sqrt{2} + 1/2)z \\
\dot{z} &= \pi(x/\sqrt{2} + 1/2)y.
\end{aligned}
$$

Rotation is greatest at $x = 1$ and decreases to zero at $x = -1$.

23.6 Geometric Suspension by Globalization of Local Vector Fields

23.6.1 Vector Field Reconstruction

The method of vector field reconstruction has been used successfully to model dynamical systems, especially those with chaotic dynamics, from data [1,3,6,10]. The most basic approach is to take a vector valued time series $\langle x_i, y_i, z_i \rangle$ in $\mathbb{R}^3$, or $\langle x_i, x_{i-n}, x_{i-2n} \rangle$ if the vector is a time delay reconstruction, and fit each component of the tangent vector $\langle \dot{x}, \dot{y}, \dot{z} \rangle$ to a curve through or near these points with some set of basis functions. The functions may be polynomials, radial basis functions, Fourier series, etc., and their globalizing the local (tangent vector) field. The resulting vector field is of the form $\dot{x}_k(t) = F_k(x_i)$ and depends sensitively on the type and number of the basis functions, as well as on the number of data points, the length of the time series, and many other factors. If the original system can be well expressed with the same basis with which the fit is made, the reconstruction can be quite good, as in Letellier's reconstruction of the Rössler equations from a single period one orbit [7].

23.6.2 Suspension by Periodic Orbits

Because the structure of attractors in chaotic systems is organized by their periodic orbits, we may use a set of periodic orbits as data from which to derive an approximate suspension of the whole system. Note that even if we used the entire infinite set of periodic orbits in a chaotic map, we most likely will not have enough data to give a perfect reconstruction of the Poincaré map: the method by which we suspend the map to the flow will most likely have terms that cannot be well approximated by a small set of basis functions. However, given a complete basis, in theory we could exactly reproduce attractor of the Poincaré map, assuming the suspension equations were analytic.

We suspended periodic orbits of the Hénon map by the same method as in the previous two sections, and determined a minimal subset that generated autonomous polynomial differential equations that resulted in a good approximation of the original map in their Poincaré section. A set of periodic orbits of periods one, two, and five from the Hénon map were computed numerically and used as initial conditions in equations for the flow (see Figures 23.6 and 23.7). We computed tangents to the flow, which together with position data in $\mathbb{R}^3$ was used to find global polynomial functions giving the x, y, and z time derivatives $\dot{x}$, $\dot{y}$, and $\dot{z}$ by a least squares fit.

Comparing the Poincaré section of this attractor and the attractor of the original Hénon map, we see that the dynamics of the original map were well reproduced. We determined the success of fitting on-attractor dynamics by comparing the orbits used to build the differential equations with the "same" orbits generated by the equations, and by comparing a set of sections of chaotic orbits from the exact suspension with those generated by the differential equations. We also compared sections of true orbits beginning off the attractor with the "same" orbits generated by the differential equations.

In the limit of a large number of periodic orbits, basis functions, and data points, the suspension should become exact *on* the attractor.

Because the data is from periodic orbits in the attractor, we cannot expect the off-attractor map dynamics to be reproduced exactly. A fit to off-attractor data, such as orbits in the stable manifolds of periodic or chaotic orbits, would better approximate the dynamics in the map's basin of attraction.

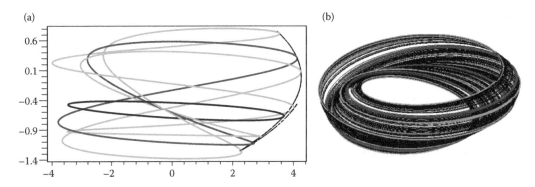

FIGURE 23.6   (a) The Hénon map $x_{n+1} = \frac{4}{3} + \frac{1}{12}y_n - \frac{4}{3}x_n^2$, $y_{n+1} = -x_n$ along with suspended period one, two, and five orbits. (b) The flow generated by the differential equations in the appendix, along with the period one, two, and five orbits that generated the vector field.

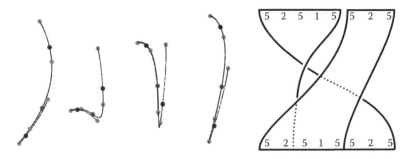

FIGURE 23.7   Stretching and folding of Poincaré section of the Hénon suspension at four equally spaced cuts. On the right, the template extracted from the Poincaré sections of the attractor. The period one orbit is indicated by the dark gray dot, period two by black, and period five by light gray.

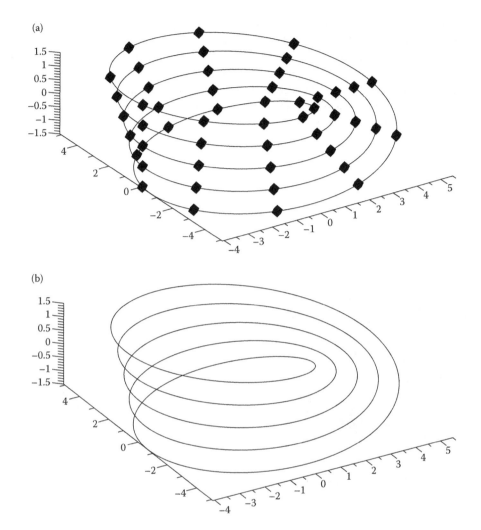

FIGURE 23.8 A suspension of a set of five points on a line to curves on a Möbius band. (a) The curves and data points and (b) the solution curves to the differential equation.

23.6.3 Other Inexact Suspensions Using Globalization of Local Tangent Data

Beginning with an odd number of evenly spaced points to the right of the origin on the x axis, we can rotate them about their center point by $\pi/2$ while rotating the set about the z axis. The result is a set of equally spaced orbits parallel to the core of a Möbius band. Computing tangents to the curves and globalizing the vector field via a least squares fit to a third degree polynomial gives us a differential equation that reproduces the original curves quite well (see Figure 23.8).

23.6.4 Other Approximate Suspensions

23.6.4.1 Approximate Suspension of the Tent Map

In theory, we should be able to suspend the tent map exactly in the same way that we suspended the logistic and cubic one-dimensional maps. However, we need to use a computer algebra system (we used Maple) to keep track of the extremely long equations generated by our methods, and when the map is piecewise defined, the governing equations are much more complicated. However, we can suspend the tent map equations approximately by generating suspensions of individual periodic orbits, possibly with additional guide orbits that are not periodic, and globalizing their local tangent vectors to vector fields.

We begin by generating the surfaces along which we wish the solutions to flow (see Figure 23.9). Then we choose a small set of periodic orbits that evenly covers the template. We supplement this set with some additional "guide curves" along the edges of the template. This works moderately well, producing an attractor that, while not exact, roughly mimics what we expect of a suspension of the tent map: it is similar to a Rössler attractor (see Figure 23.10).

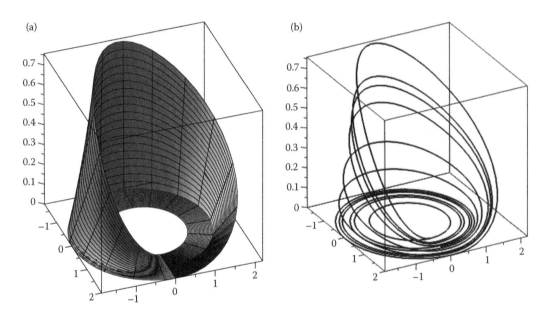

FIGURE 23.9 (a) The template for the tent map. (b) The periodic orbit and guide orbit data.

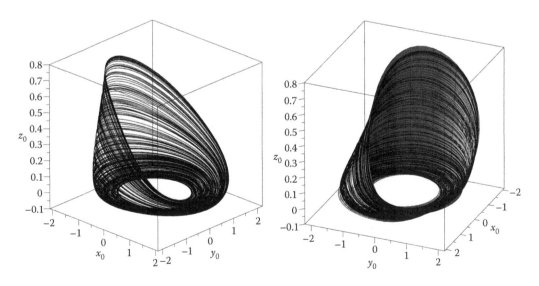

FIGURE 23.10 Two views of a suspension of the tent map by periodic orbits and guide curves.

23.7 Summary

We have tried to give a relatively complete accounting of practical methods of suspending maps to flows. Much work remains to be done. Refinements of approximate methods will likely result in better fits at the Poincaré section, so that the Poincaré map will more closely approximate the original map. There are likely many simpler exact autonomous suspension methods to be discovered. While the method of Mayer-Kress and Hacken is relatively simple, it is not quite autonomous, due to the presence of time t in the interpolating diffeomorphisms. The exact suspension method we presented, while autonomous, is overly complex, so complex that it would take several pages just to write down the system of three differential equations. The results of approximately suspending permutations of point sets deserves to be studied more extensively. While we did not mention it above, the differential equations produced, when integrated forward, frequently give rise to new and interesting strange attractors. The method of geometric templates, as used in the approximate suspension of the tent map, also give rise to new strange attractors. It would be interesting to suspend the Smale horseshoe, by both exact and approximate methods. The result of an exact suspension would be the simplest kind of uniformly hyperbolic flow.

References

1. B. P. Bezruchko and D. A. Smirnov. Constructing nonautonomous differential equations from experimental time series. *Physical Review E*, 63:1–7, 2000.
2. P. J. Channell. Explicit suspensions of diffeomorphisms—An inverse problem in classical dynamics. *Journal of Mathematical Physics*, 24:823, 1983.
3. G. Gousebet and C. Letellier. Global vector field reconstruction by using a multivariate polynomial l2 approximation on nets. *Physical Review E*, 49(6):4955–4972, 1994.
4. M. Hénon. A two-dimensional mapping with a strange attractor. *Communications in Mathematical Physics*, 50:69–77, 1976.
5. S. P. Kuznetsov. A non-autonomous flow system with plykin type attractor. *Communications in Nonlinear Science and Numerical Simulation*, 14:3487–3491, 2009.
6. C. Letellier, L. Le Sceller, E. Marchal, P. Dutertre, B. Maheu, G. Gouesbet, Z. Fei, and J. L. Hudson. Global vector field reconstruction from a chaotic experimental signal in copper electrodissolution. *Physical Review E*, 51(5):4262–4266, 1995.
7. C. Letellier, E. Ringuet, B. Maheu, J. Maquet, and G. Gouesbet. Global vector field reconstruction of chaotic attractors from one unstable periodic orbit. *Entropie*, 202:147–153, 1997.
8. G. Mayer-Kress and H. Haken. An explicit construction of a class of suspensions and autonomous differential equations for diffeomorphisms in the plane. *Communications in Mathematical Physics*, 111:63–74, 1987.
9. E. Schrder. Ueber iterirte functionen. *Mathematische Annalen*, 3(2):296–322, 1870.
10. L. Wei-Dong, K. F. Ren, S. Meunier-Guttin-Cluzel, and G. Gouesbet. Global vector-field reconstruction of nonlinear dynamical system from a time series with SVD method and validation with Lyapunov exponent. *Chinese Physics*, 12(12):1366–1373, 2003.
11. Wikipedia. Iterated function, 2015.
12. H. S. Wilf. *Generatingfunctionology*. Academic Press, Inc., San Diego, 2005.

24

Lagrangian Coherent Structures at the Onset of Hyperchaos in Two-Dimensional Flows

Rodrigo A. Miranda,
Erico L. Rempel,
Abraham C.-L. Chian,
and Adriane B. Schelin

24.1 Introduction

A remarkable feature of two-dimensional (2D) flows is the spontaneous emergence of large-scale coherent structures. If energy is injected at a constant rate at some intermediate scale, the energy transfer in 2D flows is characterized by an inverse cascade toward large scales, leading to the development of coherent structures [18,25]. These structures have a direct impact on the dynamics of particles advected by the flow, which covers a vast range of applications. For instance, in plankton blooms in the ocean [9,29], filamentary flow patterns are known to trap the advected particles for long periods of time, increasing the access to nutrients for different plankton species, and inducing changes to their traditional population dynamics. Likewise, in References 52–54, it was shown that particles carried by the blood flow in 2D models of blood vessels can behave chaotically. Such process also leads to particle trapping, a mechanism that boosts blood coagulation and accelerates the development of diseases as stenosis and aneurysms. The movement of oil spill ashore is also influenced by such structures. Mezić et al. [37] developed a diagnostic tool, based on the theory of 2D chaotic advection in fluids, which is able to estimate the location of oil spreading. Another important application is in the process of ozone depletion [28], where it is known that the filamentary structure of the polar vortex affects the dispersion of chemically active particles in the stratosphere, with similar behavior also observed in the atmospheres of Venus [42] and Saturn [12]. Several other examples can also be found in microfluids [57], combustion [62], and even in the theory of the evolution of life [55] (for a review, see Reference 60).

The 2D flows are modeled by 2D Navier–Stokes equations (2D NSE), which can be solved numerically with the spectral method by transforming the equations to Fourier space, and truncating the Fourier series to a finite number of modes for numerical implementation. This method has been used in a series of works to study the transition from a laminar state to a chaotic regime. It was Boldrighini and Franceschini [3], who first showed that the numerical solutions of a 5-mode truncation of the 2D NSE can exhibit chaotic behavior arising from a series of bifurcations observed with increasing values of the Reynolds number. Lee [27] presented a detailed work on the complex route from laminar to chaotic behavior by varying the Reynolds number and strength of the forcing term, and compared the effect of an external force acting on a single mode and on multiple modes in Fourier space. A systematic study of the transition to chaos in the 2D NSE was presented by Feudel and Seehafer [15,16]. They characterized a wealth of regimes with increasing amplitude of the forcing term, including steady states with broken symmetries, traveling waves, modulated traveling waves, torus, and chaos, and compared the influence of these regimes on the trajectory of a test particle in real space. The transition to chaos and the mixing properties of the 2D NSE were also examined by Braun et al. [7]. They analyzed the evolution of a set of tracer particles in the presence of an external forcing consisting of an array of vortices, and demonstrated that the rate of stretching of the initial patch of tracers is stronger in the chaotic and quasiperiodic regimes and is weaker in the periodic and laminar regimes. Braun et al. [6] compared numerical simulations of the 2D NSE using stress-free and no-slip boundary conditions in the vertical direction, and periodic boundary conditions in the horizontal direction. They found that the bifurcation scenario is relatively robust to changes of the boundary conditions, and that the final route to chaos in the 2D NSE with stress-free boundary conditions occurs by a period-doubling cascade, whereas the destruction of a two-frequency torus may be responsible for the onset of chaos in the no-slip case. The transition to chaos of a 2D flow confined on a square domain with no-slip walls was explored by Molenaar et al. [38]. They found that the route to chaos corresponds to the Ruelle–Takens–Newhouse scenario [39,50] and the flow is dominated by a large circulation cell in real space.

Numerical studies of nonlinear systems modeled by partial differential equations show that they may exhibit a crisis-like transition involving random switches between periods of transient temporal chaos (TC) and spatiotemporal chaos (STC). TC refers to the regime in which the patterns of numerical solution are chaotic in time and regular in space, and is characterized by a low fractal dimension [11,44], narrowband power spectrum [49], and one positive Lyapunov exponent [11]. STC refers to the regime in which the patterns are chaotic in time and disordered in space, and is characterized by a high fractal dimension [11,44], broadband power spectrum [49], and the presence of hyperchaos (i.e., two or more positive Lyapunov exponents) [11]. The coupling between two chaotic saddles was shown to be responsible for the TC–STC intermittency [44,48]. Chaotic saddles are nonattracting chaotic sets responsible for transient chaos [5,22,24,47]. A thorough overview on the subject of transient chaos and its applications can be found in Reference 26.

The Lagrangian description of fluids has attracted much attention during the past decade because it can provide a consistent definition of coherent structures [33], and as a result, several concepts and new techniques have been developed to detect them. In this chapter, we study a crisis-like transition to hyperchaos (HC) in the 2D NSE with periodic boundary conditions and an external force. We construct bifurcation diagrams and detect chaotic saddles at the onset of hyperchaos based on an Eulerian approach. Next, we characterize the spatiotemporal patterns of the fluid using a heuristic Lagrangian diagnostic proposed by Madrid and Mancho [30] and Mendoza and Mancho [34] to detect distinguished trajectories (DTs), the maximum finite-time Lyapunov exponent (FTLE), which is a popular tool for detecting Lagrangian coherent structures (LCSs) [21,56], and a mathematical theory to detect hyperbolic LCS as smooth, parametrized curves derived by Haller [19].

This chapter is organized as follows. The derivation of the spectral form of the 2D NSE and the details of the numerical implementation are presented in Section 24.2. The Lagrangian techniques used to characterize the transport of particles are briefly reviewed in Section 24.3. Our numerical analysis begins in

Section 24.4, where we identify a crisis-like transition to hyperchaos and detect chaotic saddles in a fixed frame of reference in Fourier space (i.e., the Eulerian approach). In Section 24.5, we combine the Eulerian bifurcation scenario with the phenomenology of chaotic advection using the Lagrangian approach. Finally, we present our conclusion in Section 24.6.

24.2 2D Navier–Stokes Equations

The dynamics of 2D incompressible fluids are governed by the NSE, which in nondimensional form are [16]

$$\partial_t \mathbf{u} + (\mathbf{u} \cdot \nabla)\,\mathbf{u} = -\nabla p + \frac{1}{\text{Re}} \nabla^2 \mathbf{u} + \mathbf{f}, \tag{24.1}$$

$$\nabla \cdot \mathbf{u} = 0, \tag{24.2}$$

where $\mathbf{u} = \mathbf{u}(\mathbf{x}, t) = (u_x(\mathbf{x}, t), u_y(\mathbf{x}, t))$ denotes the fluid velocity; $\mathbf{x} = (x, y)$ is the position vector in the fixed frame of reference, $p = p(\mathbf{x}, t)$ is the pressure; Re represents the Reynolds number; and $\mathbf{f} = (f_x(\mathbf{x}, t), f_y(\mathbf{x}, t))$ represents an external force. In the Fourier representation, we can write $\mathbf{u}$, p, and $\mathbf{f}$ as

$$\mathbf{u}(\mathbf{x}, t) = \sum_{\substack{\mathbf{k} \in \mathbb{Z}^2, \\ \mathbf{k} \neq 0}} \hat{\mathbf{u}}_{\mathbf{k}}(t) e^{i\mathbf{k} \cdot \mathbf{x}}, \tag{24.3}$$

$$p(\mathbf{x}, t) = \sum_{\substack{\mathbf{k} \in \mathbb{Z}^2, \\ \mathbf{k} \neq 0}} \hat{p}_{\mathbf{k}}(t) e^{i\mathbf{k} \cdot \mathbf{x}}, \tag{24.4}$$

$$\mathbf{f}(\mathbf{x}, t) = \sum_{\substack{\mathbf{k} \in \mathbb{Z}^2, \\ \mathbf{k} \neq 0}} \hat{\mathbf{f}}_{\mathbf{k}}(t) e^{i\mathbf{k} \cdot \mathbf{x}}, \tag{24.5}$$

where $\hat{\ }$ indicates the complex Fourier coefficient of the corresponding quantity; $\mathbf{k} = (k_x, k_y)$ is the wavevector; $k_x = 2\pi n_x / L_x$; $k_y = 2\pi n_y / L_y$, $n_x, n_y \in \mathbb{Z}$; and $i = \sqrt{-1}$. In this chapter, we assume $L_x = L_y = 2\pi$ and the periodic boundary conditions in x and y directions; hence $\mathbf{k} \in \mathbb{Z}^2$. Substituting Equations 24.3 through 24.5 into Equation 24.1, we have

$$\sum_{\substack{\mathbf{k} \in \mathbb{Z}^2, \\ \mathbf{k} \neq 0}} \left(\frac{d\hat{\mathbf{u}}_{\mathbf{k}}}{dt} \right) e^{i\mathbf{k} \cdot \mathbf{x}} = -i \sum_{\substack{\mathbf{k} \in \mathbb{Z}^2, \\ \mathbf{k} \neq 0}} \mathbf{k} \hat{p}_{\mathbf{k}} e^{i\mathbf{k} \cdot \mathbf{x}} - \frac{1}{\text{Re}} \sum_{\substack{\mathbf{k} \in \mathbb{Z}^2, \\ \mathbf{k} \neq 0}} k^2 \hat{\mathbf{u}}_{\mathbf{k}} e^{i\mathbf{k} \cdot \mathbf{x}} + \sum_{\substack{\mathbf{k} \in \mathbb{Z}^2, \\ \mathbf{k} \neq 0}} \hat{\mathbf{f}}_{\mathbf{k}} e^{i\mathbf{k} \cdot \mathbf{x}}$$

$$- i \sum_{\substack{\mathbf{p} \in \mathbb{Z}^2, \\ \mathbf{p} \neq 0}} \sum_{\substack{\mathbf{q} \in \mathbb{Z}^2, \\ \mathbf{q} \neq 0}} (\hat{\mathbf{u}}_{\mathbf{p}} \cdot \mathbf{k})\, \hat{\mathbf{u}}_{\mathbf{q}} e^{i(\mathbf{p}+\mathbf{q}) \cdot \mathbf{x}}. \tag{24.6}$$

If $\mathbf{k} = 0$, the nonlinear term vanishes, and as a result, the evolution equation for $\hat{\mathbf{u}}_0$ is decoupled from other Fourier modes. From Equation 24.6, we obtain the evolution equation for $\mathbf{k} \neq 0$:

$$\frac{d\hat{\mathbf{u}}_{\mathbf{k}}}{dt} = -i\mathbf{k}\hat{p}_{\mathbf{k}} - \frac{1}{\text{Re}} k^2 \hat{\mathbf{u}}_{\mathbf{k}} + \hat{\mathbf{f}}_{\mathbf{k}} - i \sum_{\substack{\mathbf{p} \in \mathbb{Z}^2, \\ \mathbf{p} \neq 0, \mathbf{k}}} (\hat{\mathbf{u}}_{\mathbf{p}} \cdot \mathbf{k})\, \hat{\mathbf{u}}_{\mathbf{k}-\mathbf{p}}. \tag{24.7}$$

Inserting Equation 24.3 into Equation 24.2, the incompressibility requirement takes the form

$$\hat{\mathbf{u}}_{\mathbf{k}}(t) \cdot \mathbf{k} = 0. \tag{24.8}$$

Note from Equation 24.8, that the incompressibility in Fourier space restricts the dynamics of the Fourier modes $\hat{\mathbf{u}}_{\mathbf{k}}$ to the direction perpendicular to the wavevector $\mathbf{k}$. To impose this restriction, we define a real unit vector $\mathbf{e_k}$ perpendicular to $\mathbf{k}$ [15,16,27]:

$$\mathbf{e_k} \cdot \mathbf{k} = 0, \qquad e_\mathbf{k}^2 = \mathbf{e_k} \cdot \mathbf{e_k} = 1, \qquad \mathbf{e_{-k}} = \mathbf{e_k}, \tag{24.9}$$

and project the velocity in Fourier space along the vector $\mathbf{e_k}$:

$$\hat{\mathbf{u}}_\mathbf{k} = \hat{u}_\mathbf{k} \mathbf{e_k}, \tag{24.10}$$

where $\hat{u}_\mathbf{k} = u_\mathbf{k}^R + i u_\mathbf{k}^I$ is a complex scalar quantity. As a result, the complex 2D vector $\hat{\mathbf{u}}_\mathbf{k}$ is reduced to a complex one-dimensional (i.e., scalar) quantity $\hat{u}_\mathbf{k}$ because $\hat{\mathbf{u}}_\mathbf{k}$ vanishes in the direction parallel to $\mathbf{k}$ [15,16]. Substituting Equation 24.10 into Equation 24.7 and projecting the resulting equation into $\mathbf{e_k}$, we obtain

$$\frac{d\hat{u}_\mathbf{k}}{dt} = -\frac{1}{\mathrm{Re}} k^2 \hat{u}_\mathbf{k} + \hat{f}_\mathbf{k} - i \sum_{\substack{\mathbf{p} \in \mathbb{Z}^2, \\ \mathbf{p} \neq 0, \mathbf{k}}} (\mathbf{e_p} \cdot \mathbf{k})(\mathbf{e_{k-p}} \cdot \mathbf{e_k}) \hat{u}_\mathbf{p} \hat{u}_{\mathbf{k-p}}, \tag{24.11}$$

where $\hat{f}_\mathbf{k} = \hat{\mathbf{f}}_\mathbf{k} \cdot \mathbf{e_k}$. Note that the pressure term vanishes. Since $\mathbf{u}(\mathbf{x}, t)$ in Equation 24.1 is a real variable,

$$\hat{u}_\mathbf{k} = \hat{u}_{-\mathbf{k}}^*,$$

where the asterisk indicates the complex conjugate. We restrict the wavevectors to a subset of $\mathbb{Z}^2$ defined by

$$\mathbb{K}^2 \equiv \{\mathbf{k} \in \mathbb{Z}^2 : k_x > 0\} \cup \{\mathbf{k} \in \mathbb{Z}^2 : k_x = 0 \wedge k_y > 0\}.$$

In addition, we define

$$\bar{\mathbf{k}} \equiv \begin{cases} \mathbf{k} & : \mathbf{k} \in \mathbb{K}^2, \\ -\mathbf{k} & : \mathbf{k} \notin \mathbb{K}^2, \end{cases} \tag{24.12}$$

and

$$\mathrm{sgn}(\mathbf{k}) \equiv \frac{\mathbf{k} \cdot \bar{\mathbf{k}}}{k^2}. \tag{24.13}$$

Introducing Equations 24.12 and 24.13 into Equation 24.11, and separating into real and imaginary parts, we obtain the following set of ordinary differential equations (ODEs) [16]:

$$\frac{du_\mathbf{k}^R}{dt} = -\frac{1}{\mathrm{Re}} k^2 u_\mathbf{k}^R + f_\mathbf{k}^R + \sum_{\substack{\mathbf{p} \in \mathbb{K}^2, \\ \mathbf{p} \neq \mathbf{k}}} (\mathbf{e_p} \cdot \mathbf{k}) \left\{ \left(\mathbf{e_{\overline{k-p}}} \cdot \mathbf{e_k} \right) \left[\mathrm{sgn}(\mathbf{k-p}) u_\mathbf{p}^R u_{\overline{\mathbf{k-p}}}^I + u_\mathbf{p}^I u_{\overline{\mathbf{k-p}}}^R \right] \right.$$

$$\left. + (\mathbf{e_{k+p}} \cdot \mathbf{e_k}) \left[u_\mathbf{p}^R u_{\mathbf{k+p}}^I - u_\mathbf{p}^I u_{\mathbf{k+p}}^R \right] \right\}, \tag{24.14}$$

$$\frac{du_\mathbf{k}^I}{dt} = -\frac{1}{\mathrm{Re}} k^2 u_\mathbf{k}^R + f_\mathbf{k}^I - \sum_{\substack{\mathbf{p} \in \mathbb{K}^2, \\ \mathbf{p} \neq \mathbf{k}}} (\mathbf{e_p} \cdot \mathbf{k}) \left\{ \left(\mathbf{e_{\overline{k-p}}} \cdot \mathbf{e_k} \right) \left[u_\mathbf{p}^R u_{\overline{\mathbf{k-p}}}^R - \mathrm{sgn}(\mathbf{k-p}) u_\mathbf{p}^I u_{\overline{\mathbf{k-p}}}^I \right] \right.$$

$$\left. + (\mathbf{e_{k+p}} \cdot \mathbf{e_k}) \left[u_\mathbf{p}^R u_{\mathbf{k+p}}^R + u_\mathbf{p}^I u_{\mathbf{k+p}}^I \right] \right\}, \tag{24.15}$$

where $\hat{f}_\mathbf{k} = f_\mathbf{k}^R + i f_\mathbf{k}^I$. Following References 15 and 16, we apply the external forcing to the wavevector $\mathbf{k} = (4, 1)$:

$$\hat{f}_\mathbf{k} = \begin{cases} f_{(4,1)}^R + i f_{(4,1)}^I & : \mathbf{k} = (4, 1), \\ 0 & : \mathbf{k} \neq (4, 1). \end{cases} \tag{24.16}$$

and set $f_{(4,1)}^R = f_{(4,1)}^I = 0.13666$. For this type of external forcing Equations 24.1 and 24.2 remain invariant with respect to the lines parallel to the external force, and can lead to the coexistence of symmetrical attractors in phase space. Note that our choice of the forcing term implies that the Fourier coefficient $\hat{\mathbf{u}}_0$ is constant in time; therefore, we set $\hat{\mathbf{u}}_0 = (0, 0)$, which is equivalent to assume a vanishing mean flow [51].

We solve Equations 24.14 and 24.15 using an isotropic truncation of wavenumbers in Fourier space, which means that the Fourier space is segmented into successive rings $n^2 - n < \mathbf{k}^2 \leq n^2 + n, n = 1, 2, \ldots$. The rings up to $n = 8$ are considered, which gives a set of 112 complex Fourier coefficients, and 224 ODEs after separating into real and imaginary parts. Numerical integration is performed using a fourth-order Runge–Kutta method.[*]

24.3 Lagrangian Techniques

The transport and mixing properties of a fluid can be studied following the trajectories of tracer particles advected by the fluid velocity field [40]. Neglecting molecular diffusion, and assuming passive tracers (i.e., the effect of particles on the flow is negligible), the trajectory $\mathbf{x}_i$ of a particle labeled i, starting at $\mathbf{x}_{i0}$ at time t_0, is given by

$$\dot{\mathbf{x}}_i(\mathbf{x}_{i0}, t_0, t) = \mathbf{u}(\mathbf{x}_i(\mathbf{x}_{i0}, t_0, t), t), \tag{24.17}$$

where

$$\mathbf{x}_{i0} = \mathbf{x}_i(\mathbf{x}_{i0}, t_0, t_0). \tag{24.18}$$

In the special case of steady flows (i.e., flows with time-independent velocity fields), the transport of particles can be characterized by identifying the fixed points in space where

$$\mathbf{u}(\mathbf{x}) = 0. \tag{24.19}$$

The local stability of fixed points, also called stagnation points (SPs) [32,40] can be characterized by evaluating the eigenvalues of the Jacobian matrix of the velocity field at position $\mathbf{x}$. For 2D flows, the Jacobian matrix is given by

$$\frac{\partial \mathbf{u}(\mathbf{x})}{\partial \mathbf{x}} = \begin{bmatrix} \dfrac{\partial u_x(\mathbf{x})}{\partial x} & \dfrac{\partial u_x(\mathbf{x})}{\partial y} \\ \dfrac{\partial u_y(\mathbf{x})}{\partial x} & \dfrac{\partial u_y(\mathbf{x})}{\partial y} \end{bmatrix}. \tag{24.20}$$

If the two eigenvalues are real, and have opposite signs, then the stability of the SP is hyperbolic. If the two eigenvalues are purely imaginary and complex conjugates of each other, then the stability of the SP is elliptic [32]. Hyperbolic SPs and elliptic SPs are the two most common types of SPs in incompressible fluids [32]. Hyperbolic SPs are related to regions of enhanced particle diffusion and mixing, whereas elliptic SPs are related to particle-trapping vortices [40]. Note that this is an Eulerian description of the flow, since SPs of steady flows remain fixed in space, as indicated by Equation 24.19. In unsteady flows, one can define the instantaneous stagnation points (ISPs) as fixed points in space where, for a given time t,

$$\mathbf{u}(\mathbf{x}, t) = 0. \tag{24.21}$$

The local stability of ISPs can be determined by evaluating Equation 24.20 at position $\mathbf{x}$ and time t. Note that this corresponds to an Eulerian description of the flow. Since the ISPs are identified by inspecting the spatial pattern of the velocity field "frozen" in time, their local stability properties will depend on the selected value of t. For example, hyperbolic ISPs can lose hyperbolicity and even disappear as time increases [31]. For this reason, ISPs are not solutions of Equation 24.17 and do not follow particle trajectories. As a consequence, the characterization of particle dynamics of unsteady flows through the identification of ISPs can give misleading information [31,32].

---

[*] Our numerical code is freely available at https://github.com/rmiracer/Jade.

24.3.1 Distinguished Trajectories

A Lagrangian generalization of fixed points for a time-dependent velocity field is given by the concept of DTs, which can be regarded as fixed points that "move" following particle trajectories that depend on the spatiotemporal dynamics of the flow [23,32]. Hence, DTs must also be solutions of Equation 24.17. A definition of a DT introduced by Madrid and Mancho [30] is briefly stated as follows. The Euclidian arc length of a particle trajectory $\mathbf{x}_i$ passing through the point $\mathbf{x}_{i0}$ at time t_0 can be measured by

$$M(\mathbf{x}_{i0}, t_0, \tau) = \int_{t_0-\tau}^{t_0+\tau} \left[\left(\frac{dx_i(\mathbf{x}_{i0}, t_0, t')}{dt} \right)^2 + \left(\frac{dy_i(\mathbf{x}_{i0}, t_0, t')}{dt} \right)^2 \right]^{1/2} dt'. \qquad (24.22)$$

A particle trajectory $\mathbf{x}_j$ is τ-distinguished at time t_0 if there exists an open set B around $\mathbf{x}_j$ on which, for any initial condition $\mathbf{x}_{i0} \in B$, $M(\mathbf{x}_{i0}, t_0, \tau)$ has a minimum and [30]

$$\min(M(\mathbf{x}_{i0}, t_0, \tau)) = M(\mathbf{x}_j, t, \tau), \qquad \forall \mathbf{x}_{i0} \in B. \qquad (24.23)$$

In principle, a τ-distinguished trajectory qualifies as a DT if the minimum of M converges to a point, or a "limit coordinate," with increasing τ, and the trajectory exists for all times t. However, numerical errors can accumulate if $\tau \to \infty$, imposing constraints on the convergence of the minimum of M [30]. Moreover, in practical situations (e.g., laboratory and numerical experiments), we deal with data represented by discrete datasets of finite time duration. In our numerical simulations, a value of $\tau = 10$ time units allows us to compare the results obtained using this technique with those obtained from the computation of the maximum FTLE. In the following, we will refer to a τ-distinguished trajectory as a DT. Note that DTs can be classified as hyperbolic DTs ("moving" saddle point) and elliptic DTs ("moving" elliptic point).

The detection of DTs has some advantages in comparison to other Lagrangian techniques, for instance, Equation 24.22 can be easier to be numerically implemented, and its computation is relatively faster [35,36,46]. However, the results of this technique are not preserved under arbitrary Euclidian coordinate transforms of the velocity field [20,36]. For example, by changing from an inertial frame to a frame that follows an arbitrary particle trajectory will make the arc length of that trajectory equal to zero. Therefore, any particle trajectory can lead to a minimum of M in a given frame, and thus can be incorrectly identified as a DT [20].

24.3.2 Lagrangian Coherent Structures

The term "Lagrangian coherent structure" describes fluid structures that act as barriers of particle transport moving with the flow [19–21, 56]. A common approach to detect LCSs is through the computation of the maximum FTLE [19,21,56]. A brief review of this technique follows. The solution of Equation 24.17 can be viewed as a mapping process that takes points from their position $\mathbf{x}_{i0}$ at time t_0 to a new position $\mathbf{x}_i$ at time $t = t_0 + \tau$, where $\tau > 0$. This process is referred to as the flow map $\phi_{t_0}^t$, and satisfies [19,56]

$$\phi_{t_0}^t : D \to D$$
$$\mathbf{x}_{i0} \to \phi_{t_0}^t(\mathbf{x}_{i0}) = \mathbf{x}_i(\mathbf{x}_{i0}, t_0, t), \qquad (24.24)$$

where $D \subset \mathbb{R}^2$ is the domain of the fluid in real space. Consider the evolution of a perturbed particle trajectory

$$\mathbf{x}_i' = \mathbf{x}_i + \delta\mathbf{x}_0, \qquad (24.25)$$

where $\delta\mathbf{x}_0$ is infinitesimal and arbitrarily oriented. At t, the perturbation becomes

$$\delta\mathbf{x}_i^t = \phi_{t_0}^t(\mathbf{x}_i') - \phi_{t_0}^t(\mathbf{x}_i). \qquad (24.26)$$

Expanding $\phi_{t_0}^t$ into a Taylor series in the neighborhood of $\mathbf{x}_i$, and neglecting the higher-order terms, we obtain

$$\delta\mathbf{x}_i^t = \frac{d\phi_{t_0}^t(\mathbf{x}_i)}{d\mathbf{x}} \delta\mathbf{x}_0. \qquad (24.27)$$

The growth of the infinitesimal perturbation will be given by

$$||\delta \mathbf{x}_i^t|| = \sqrt{\delta \mathbf{x}_0 \cdot \left(\left[\frac{d\phi_{t_0}^t(\mathbf{x}_i)}{d\mathbf{x}} \right]^T \left[\frac{d\phi_{t_0}^t(\mathbf{x}_i)}{d\mathbf{x}} \right] \delta \mathbf{x}_0 \right)}. \tag{24.28}$$

where the superscript T denotes the matrix transpose. Let us define a finite-time version of the right Cauchy–Green deformation tensor as

$$C_{t_0}^t = \left[\frac{d\phi_{t_0}^t(\mathbf{x}_i)}{d\mathbf{x}} \right]^T \left[\frac{d\phi_{t_0}^t(\mathbf{x}_i)}{d\mathbf{x}} \right]. \tag{24.29}$$

The tensor $C_{t_0}^t$ is symmetric and positive definite, hence it admits two real positive eigenvalues and orthogonal eigenvectors. Denote the eigenvectors of $C_{t_0}^t$ as $\xi_1(\mathbf{x}_{i0}, t_0, t)$ and $\xi_2(\mathbf{x}_{i0}, t_0, t)$ with the corresponding eigenvalues $\lambda_1(\mathbf{x}_{i0}, t_0, t) > \lambda_2(\mathbf{x}_{i0}, t_0, t)$, satisfying

$$C_{t_0}^t \xi_i = \lambda_i \xi_i, \qquad i = 1, 2. \tag{24.30}$$

If we assume that the perturbation is aligned with the direction of maximum stretching, then the growth of the infinitesimal perturbation is given by

$$||\delta \mathbf{x}_i^t|| = e^{\sigma_{t_0}^t(\mathbf{x}_i)\tau} ||\delta \mathbf{x}_0||, \tag{24.31}$$

where

$$\sigma_{t_0}^t(\mathbf{x}_i) = \frac{1}{2\tau} \ln \lambda_1 \tag{24.32}$$

is the maximum FTLE. In the following, we will refer to the maximum FTLE as the FTLE. The computation of the FTLE is a common technique to detect hyperbolic LCSs, which are defined as the locally strongest repelling or attracting material surfaces over a finite-time interval, acting as barriers of particle transport. Taking $t = t_0 + \tau$ in the flow map (Equation 24.24), repelling LCSs are identified as ridges in the forward-time FTLE, associated with finite-time stable manifolds [56]. One can also take $t = t_0 - \tau$ in Equation 24.24, in which case attracting LCSs are identified as ridges in the backward-time FTLE, associated with finite-time unstable manifolds.

We compute $\sigma_{t_0}^t$ by first covering the $[0, 2\pi] \times [0, 2\pi]$ domain with an equally spaced grid of size 512×512. At each point of the grid, we place four particles perturbed in x and y directions following Equation 24.25 where the perturbation $\delta \mathbf{x}_0$ is given by

$$\delta \mathbf{x}_0 = \{(\Delta, 0), (-\Delta, 0), (0, \Delta), (0, -\Delta)\}. \tag{24.33}$$

In our case, $\Delta = 2\pi/1024$. The trajectory of each perturbed particle is followed by solving Equation 24.17. After that, we use the resulting trajectories to solve Equation 24.29 using the finite-difference method.

The FTLE is a frame-independent technique frequently used to identify hyperbolic LCSs as ridges of the forward and backward FTLE. However, simple counterexamples demonstrate that not all ridges of the FTLE field are hyperbolic LCSs, and LCSs need not be ridges of the FTLE field [19]. As a result, FTLE ridges can yield both false negatives and false positives in the detection of LCS [19,41].

A mathematical theory of hyperbolic LCSs was developed by Haller [19] and Farazmand and Haller [14]. In this theory, the hyperbolic LCSs are defined as the locally strongest repelling or attracting material lines (in the 2D case) over a finite-time interval τ. A numerical implementation of this theory was presented by Farazmand and Haller [13], which is briefly described as follows. Hyperbolic LCSs can be extracted from trajectories of the ODE defined by the first eigenvector of the Cauchy–Green deformation tensor

$$\mathbf{r}' = \xi_1(\mathbf{r}), \tag{24.34}$$

Trajectories of Equation 24.34 are referred as strainlines [13]. A compact segment γ_0 of a strainline qualifies as an LCS if the following is satisfied for all $\mathbf{x}_0 \in \gamma_0$:

1. $\lambda_1(\mathbf{x}_0) \neq \lambda_2(\mathbf{x}_0) > 1$.
2. $\langle \xi_2(\mathbf{x}_0), \nabla^2 \lambda_2 \xi_2(\mathbf{x}_0) \rangle \leq 0$.
3. $\xi_1(\mathbf{x}_0) \parallel \gamma_0$.
4. The averaged repulsion rate $\bar{\lambda}_2(\gamma)$, which is the average of λ_2 over a strainline segment γ_0, must be maximal among the averaged repulsion rates of all nearby curves γ satisfying $\gamma \parallel \xi_1(\mathbf{x}_0)$.

These conditions can be summarized as follows. Condition 1 ensures that the segment γ_0 of a strainline has a repulsion rate normal to γ_0 (i.e., in the direction of $\xi_2(\mathbf{x}_0)$) larger than the repulsion rate in the direction tangent to γ_0 (i.e., in the direction of $\xi_1(\mathbf{x}_0)$). This condition guarantees that the detected LCSs are hyperbolic (repelling material lines) and not elliptic or parabolic (e.g., due to shear). Conditions 3 and 4 ensure that the repulsion rate normal to γ_0 has a local extremum along the LCS relative to all nearby material lines, and Condition 2 ensures that this extremum is a strict local maximum [13].

We follow the numerical implementation presented by Farazmand and Haller [13] to detect hyperbolic LCS. We solve Equation 24.34 using a fourth-order Runge–Kutta method with a step of 0.01. Since small-scale LCSs are expected to have a negligible effect on the resulting pattern of the flow, we discard LCSs with length smaller than a suitable threshold following Reference 13.

24.4 Eulerian Chaos

24.4.1 Bifurcation Diagram of Attractors

We begin our nonlinear analysis by constructing a bifurcation diagram for the solution of Equations 24.14 and 24.15 as a function of the Reynolds number Re. This diagram was constructed by defining a Poincaré hyperplane as $u^R_{(8,2)} = -0.0002$, and plotting the intersections between the trajectory and the Poincaré hyperplane, where $du^R_{(8,2)}/dt > 0$. In the following, we will refer to these intersections as the "Poincaré points." Figure 24.1 shows the bifurcation diagram using two different projections, the real part of $\hat{u}_{(0,1)}$ (upper panel) and the real part of the forced mode $\hat{u}_{(4,1)}$ (lower panel). Starting from Re = 51.40, there is a coexistence of two attractors, denoted A_1 and A_2. Depending on the initial condition, the dynamics of Equations 24.14 and 24.15 can converge to A_1 or A_2. From the upper panel of Figure 24.1, it seems that

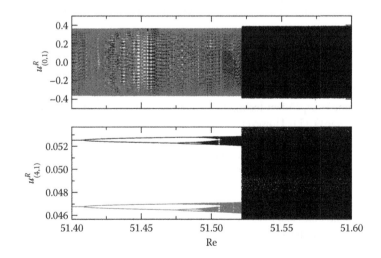

FIGURE 24.1 Bifurcation diagram of $u^R_{(0,1)}$ (upper panel) and $u^R_{(4,1)}$ (lower panel) as a function of the Reynolds number Re, for two precrisis attractors A_1 (black) and A_2 (dark gray). After crisis, A_1 and A_2 merge to form an enlarged attractor (black).

the two attractors occupy the same region, and that the two attractors maintain the same dynamics with increasing Re, up to the critical value Re = $Re_c \sim 51.5216$, where they merge and form an enlarged chaotic attractor via a global bifurcation know as attractor merging crisis [10,43]. However, several hidden features of the bifurcation diagram are unveiled by the projection onto the forced mode, shown in the lower panel of Figure 24.1, in which the two coexisting attractors can be clearly distinguished. At Re = 51.40, the computation of the two largest Lyapunov exponents λ_1 and λ_2 using the method described by Bennetin et al. [1] gives $\lambda_1 = \lambda_2 = 0$ for A_1 and A_2, which indicates that the dynamics of the two coexisting attractors is quasiperiodic. At Re $\sim$ 51.41, a quasiperiodic doubling bifurcation occurs for A_1 and A_2. This bifurcation occurs when a quasiperiodic attractor loses stability, and a new quasiperiodic attractor is created with one fundamental frequency, which is half the fundamental frequency of the previous quasiperiodic attractor [4,8,61]. Quasiperiodic doubling bifurcations have been observed in numerical studies of low-dimensional models of the 2D NSE [17] and in simulations of three-dimensional highly symmetric flows [61]. As the Reynolds number is increased, the two attractors undergo a cascade of quasiperiodic doubling bifurcations until Re $\sim$ 51.49, where A_1 and A_2 become chaotic. At Re $\sim$ 51.5216, the two attractors merge and form an enlarged chaotic attractor in a crisis-like transition. Note that the scale of the vertical axis in the bottom panel of Figure 24.1 is smaller than the size of the enlarged chaotic attractor. This scale was chosen to clearly show the coexistence of A_1 and A_2 before Re_c, and the quasiperiodic doubling bifurcations.

We compute the two largest Lyapunov exponents λ_1 and λ_2 to characterize the different flow regimes shown in the bifurcation diagram as a function of Re. Since for Re < Re_c the evolution of the bifurcation diagram of the two attractors is identical, we choose to display the Lyapunov exponents of A_1. The upper panel of Figure 24.2 shows the projection of the bifurcation diagram of A_1 onto $u^R_{(4,1)}$, and the lower panel shows the computed values of λ_1 and λ_2. From this figure, it is clear that $\lambda_1 = \lambda_2 = 0$ for Re < 51.48, demonstrating that the dynamics in this regime are in fact quasiperiodic, and that the doubling bifurcations of A_1 and A_2 observed in the lower panel of Figure 24.1 must be quasiperiodic doubling bifurcations. For 5.48 < Re < 51.525, we observe that $\lambda_1 > 0$ and $\lambda_2 = 0$, which is indicative of chaotic dynamics. For Re > 51.525, we have $\lambda_1 > 0$ and $\lambda_2 > 0$, which indicates that the dynamics is hyperchaotic.

Next, we inspect the projections of the phase space on the Poincaré section for different values of Re. Figure 24.3 shows the Poincaré points of A_1 and A_2 projected onto the real and imaginary parts of the complex Fourier modes $\hat{u}_{(0,1)}$ and $\hat{u}_{(4,1)}$. For Re = 51.4 (left-hand side panels), the Poincaré

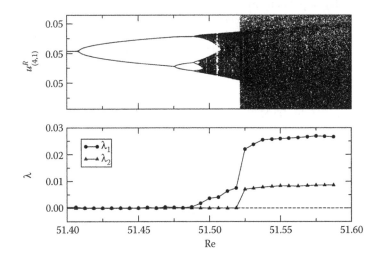

FIGURE 24.2 Upper panel: Bifurcation diagram of $u^R_{(4,1)}$ as a function of the Reynolds number Re, for A_1. Lower panel: The two largest Lyapunov exponents as a function of Re.

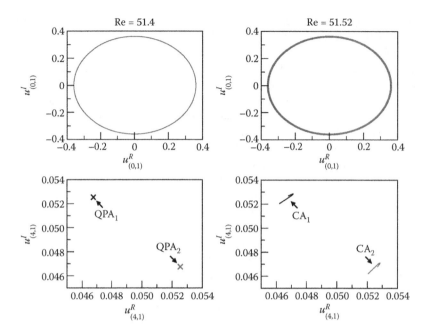

FIGURE 24.3 Poincaré points of the quasiperiodic attractors QPA$_1$ and QPA$_2$ for Re = 51.4 (left-side panels) and chaotic attractors CA$_1$ and CA$_2$ for Re = 51.52 (right-side panels) projected using Fourier modes $\hat{u}_{(0,1)}$ and $\hat{u}_{(4,1)}$.

points of A_1 and A_2 projected onto $\hat{u}_{(0,1)}$ form a closed curve, which is a typical feature of a quasiperiodic attractor. Note that, in this projection, it is apparent that A_1 and A_2 occupy the same region of phase space, and as a result, only the Poincaré points of A_2 become visible. The projection onto the Fourier mode $\hat{u}_{(4,1)}$ clearly shows that A_1 and A_2 occupy different regions of phase space, and that each quasiperiodic attractor is represented by a single point. This indicates that the quasiperiodic attractor can be classified as a nongeneric torus in which the trajectory of the attractor projected on the forcing mode displays a periodic orbit, whereas the other modes display quasiperiodic orbits with two frequencies, namely, one lower frequency related to a traveling wave, and one higher frequency related to the frequency of the forcing mode [15,16]. The nongeneric torus in Fourier space generates a modulated traveling wave in real space [15,16]. The right-hand side panels of Figure 24.3 shows the Poincaré points of A_1 and A_2 for Re = 51.52, in the chaotic regime, hereafter denoted by CA$_1$ and CA$_2$, respectively. The two coexisting chaotic attractors can be clearly distinguished in the lower panel. Note that the two attractors are symmetric with respect to the line $u^I_{(4,1)} = u^R_{(4,1)}$. This is due to a reflection symmetry of the 2D NSE with respect to the lines parallel to the external force (Equation 24.16), where the modulus of the force is maximal [15]. Symmetric solutions have been previously reported in the numerical studies of the 2D NSE with periodic boundary conditions and constant single-mode forcing [7,15,16] and in numerical simulations of the three-dimensional Rayleigh–Bénard convection [51].

   The bifurcation diagrams shown in Figures 24.1 and 24.2 indicate that at Re = Re$_c$ ∼ 51.5216, the attractors CA$_1$ and CA$_2$ merge to form a hyperchaotic attractor (HCA). The Poincaré points of the enlarged attractor at Re = 51.525 > Re$_c$ projected onto the Fourier modes $\hat{u}_{(0,1)}$ and $\hat{u}_{(4,1)}$ are shown in the upper and middle panels of Figure 24.4, respectively. Evidently, the HCA occupies a larger region of phase space. A detailed view of the projection on mode $\hat{u}_{(4,1)}$ is shown in the lower panel of Figure 24.4. Compared with the lower, right-hand side panel of Figure 24.3, it becomes clear that A_1 and A_2 are merged into the HCA.

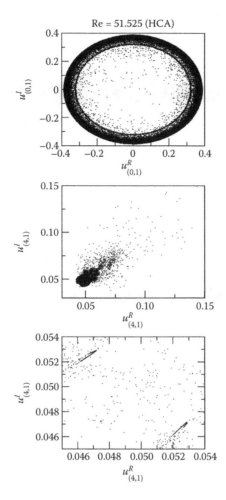

FIGURE 24.4 Poincaré points of the hyperchaotic attractor (HCA) in the hyperchaotic regime projected using Fourier modes $\hat{u}_{(0,1)}$ (upper panel) and $\hat{u}_{(4,1)}$ (middle panel). The lower panel shows an enlargement of the HCA projected using the Fourier mode $\hat{u}_{(4,1)}$.

24.4.2 Chaotic Saddles

The left-hand side panels of Figure 24.5 show that, prior to the transition to hyperchaos, the time series of the kinetic energy

$$E(t) = \frac{1}{(2\pi)^2} \int_0^{2\pi} \int_0^{2\pi} u^2(\mathbf{x},t)\,dx\,dy, \tag{24.35}$$

and the enstrophy

$$\Omega(t) = \frac{1}{(2\pi)^2} \int_0^{2\pi} \int_0^{2\pi} \omega^2(\mathbf{x},t)\,dx\,dy, \tag{24.36}$$

where $\omega = \nabla \times \mathbf{u}$ is the vorticity, display chaotic transients before the trajectory converges to either CA_1 or CA_2. The dynamics of the chaotic transient shows a higher variability than the regular dynamics of the asymptotic state, and sudden decreases of kinetic energy associated with strong "bursts" of enstrophy occur intermittently. This behavior strongly resembles the dynamics of the HCA (right-hand side panels

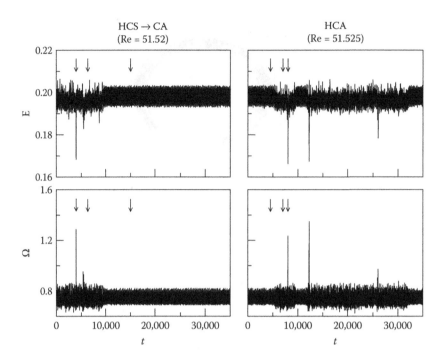

FIGURE 24.5 Time series of the kinetic energy E (upper panels) and the enstrophy Ω (lower panels) in the chaotic regime (Re = 51.52) and the hyperchaotic regime (Re = 51.525). The arrows indicate the selected values of t used to detect Lagrangian coherent structures (Figures 24.7 and 24.8).

of Figure 24.5), in which the time series randomly alternates between dynamics resembling the "laminar" behavior prior to Re_c, and dynamics with higher variability and intermittent "bursts."

The chaotic transients shown in the left-hand side panels of Figure 24.5 are due to the presence of a chaotic saddle in phase space. We use the sprinkler method to find chaotic saddles [22,24]. This method works by defining a restraining region in the Poincaré section, which contains a chaotic saddle and no attractor. The trajectory of any initial condition arbitrarily close to the chaotic saddle will eventually leave the restraining region, except for initial conditions located exactly in the chaotic saddle or its stable manifold. Define the escape time as the time it takes for a trajectory to leave this region. The restraining region is covered by a grid of initial conditions, and the trajectory of each initial condition is followed until some time t_c, which should be larger than the average escape time from the restraining region, and must be adjusted after some trial and error. We define the restraining region by covering the Poincaré points of CA_1, projected using the real and imaginary parts of the forced mode $\hat{u}_{(4,1)}$, with two narrow boxes superposing the upper and the lower part of the attractor. A similar region was defined for the CA_2 attractor. We select those trajectories that remain in the restraining region after $t_c = 16585$ time units, and plot the Poincaré points obtained at $t = 0.5t_c = 8292.5$, which approximate the chaotic saddle [22,24]. The left-hand side panels of Figure 24.6 show the Poincaré points of a chaotic saddle superposed by the Poincaré points of the chaotic attractors CA_1 and CA_2, for Re = 51.52 < Re_c, and projected onto Fourier modes $\hat{u}_{(0,1)}$ (upper panel) and $\hat{u}_{(4,1)}$ (middle panel). The bottom panel shows an enlarged view of the $\hat{u}_{(4,1)}$ projection, clearly showing the coexistence of CA_1, CA_2, and the chaotic saddle. We used the stagger-and-step algorithm [58] to compute the two largest Lyapunov exponents of this chaotic saddle, and obtained the values $\lambda_1 = 0.026$ and $\lambda_2 = 0.008$. Hence, we call this chaotic saddle a hyperchaotic saddle (HCS). The HCS governs the dynamics of chaotic transients in the time series of the kinetic energy and enstrophy before converging to either CA_1 or CA_2, as exemplified in the left-hand side panels of Figure 24.5. A

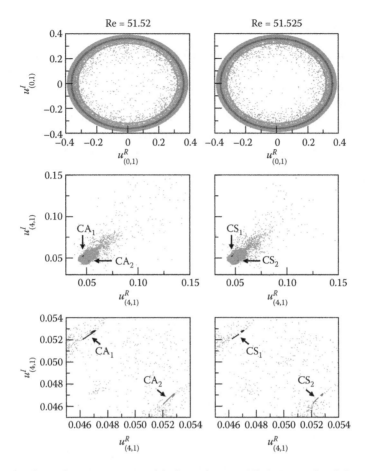

FIGURE 24.6 Left-side panels: Poincaré points of the hyperchaotic saddle (HCS, gray) and chaotic attractors CA_1 and CA_2 in the chaotic regime (Re = 51.52) projected using Fourier modes $\hat{u}_{(0,1)}$ (upper panels) and $\hat{u}_{(4,1)}$ (middle panels). The lower panel shows an enlargement of the middle panel. Right-side panels: Poincaré points of the HCS (gray), and chaotic saddles CS_1 and CS_2, in the hyperchaotic regime (Re = 51.525).

comparison between the HCS and the HCA shown in Figure 24.4 indicates that, after the crisis, the two chaotic attractors merge with the HCS, which explains the sudden creation of an enlarged attractor.

The right-hand side panels of Figure 24.6 show the Poincaré points of the HCS and two chaotic saddles found after applying the sprinkler method for Re = 51.525 > Re_c. These two chaotic saddles are located within the same region in phase space previously occupied by CA_1 and CA_2 prior to the transition. For this reason, the chaotic saddles are properly referenced as CS_1 and CS_2. They are the continuation of CA_1 and CA_2 in the hyperchaotic regime, after losing stability and becoming nonattracting chaotic sets. This change of stability (i.e., the transition from an attracting chaotic set with a corresponding basin of attraction to a nonattracting chaotic set with an associated fractal stable manifold) explains the small "gaps" or empty spaces that appear in the Poincaré projections of both CS_1 and CS_2. From the right-hand side panels of Figure 24.6, it becomes clear that the HCA shown in Figure 24.4 is composed by the HCS, CS_1, and CS_2. In this regime, the CS_1 and the CS_2 are responsible for the laminar intervals shown in the time series of the right-hand side panels of Figure 24.5, while the HCS governs the dynamics with higher variability and occasional "bursts," which appear intermittently in the time series of the enstrophy of Figure 24.5 for Re = 51.525 > Re_c. This scenario is the same described by Szabó and Tél [59] for temporal chaos and by Rempel and Chian [44] for one-dimensional STC.

24.5 Lagrangian Chaos

The Lagrangian techniques discussed in Section 24.3 require the integration of particle trajectories backwards in time when $t = t_0 - \tau$. To avoid numerical instabilities, we choose to interpolate recorded datasets of our numerical results. The datasets consist of snapshots of the velocity field in real space recorded every 0.1 time units. Following Reference 34, we use third-order Lagrange polynomials, for interpolation in time. We interpolate in space using Hermite polynomials, which is a third-order interpolation scheme commonly used for the integration of particle trajectories [21,45,56]. We plot the resulting values of Equations 24.22 and 24.32 against the initial condition of the respective particle, resulting in a scalar field. In the following, we refer to the "M field" as the field obtained from Equation 24.22, and represented using a gray scale. The "FTLE field" is obtained by representing the numerical values of the backward-time FTLE using a color scale varying from black (smaller values) to dark gray (larger values), and the forward-time FTLE is represented using a color scale from black (smaller values) to light gray (large values). After that, the forward- and backward-time FTLE are merged to form an image for visualization. We set $\tau = 10$, which allows a better comparison between the M field and the FTLE field.[*]

An important drawback of the M field is its lack of objectivity, i.e., it may yield different results in a given frame, as mentioned in Section 24.3. Therefore, the application of this technique in complex flows must be treated with caution. For this reason, we check the consistency of the M field with the FTLE field through a visual comparison.

Figure 24.7 shows the M and the FTLE fields computed from the HCA after transition, corresponding to the time series of kinetic energy and enstrophy shown on the right-hand side of Figure 24.5. The upper panel corresponds to a laminar interval at $t_0 = 4500$, the middle panel corresponds to a period of higher variability than the laminar period at $t_0 = 7000$, and the lower panel corresponds to a burst at $t_0 = 8005$. The selected values of t_0 are indicated by arrows in Figure 24.5. The M field during the laminar period is characterized by smooth regions. Several vortices can be easily identified as elliptic "islands" in which the M field has a local minimum (i.e., elliptical DTs) and are related to regions of particle confinement. Localized patches of minima resembling "crosses," which correspond to regions of particle dispersion (i.e., hyperbolic DTs) can be distinguished as well. The FTLE field is superposed by hyperbolic LCSs obtained by solving Equation 24.34 and checking Conditions 1–4 in Section 24.3. The backward-time LCSs are represented by white lines, whereas the forward-time LCSs are indicated by light gray lines. The spatiotemporal patterns of the FTLE field in the laminar period are organized by several large-scale vortices surrounded by hyperbolic LCS. The middle panels of Figure 24.7 display similar smooth patterns of the M field and the FTLE field, indicating that many features of the fluid are robust, e.g., the vortices in the middle of the spatial domain. The difference between the upper and the middle panels of Figure 24.7 becomes clear in the statistics of the FTLE field, which will be discussed later. The lower panels show the M field and the FTLE field during a burst, showing a sudden increase in fluid complexity. The M field indicates that the vortices have been disrupted, and the FTLE field displays a complex entanglement of material lines highlighted by the hyperbolic LCSs that accounts for a higher degree of disorder in space and time. From Figure 24.7 we can conclude that the intermittent bursts observed in the time series of Figure 24.5 after the transition to hyperchaos are responsible for episodic enhancements of fluid complexity.

Recall that, before the transition to hyperchaos, the chaotic transients are governed by an HCs. Figure 24.8 shows the M field and the FTLE field for $\text{Re} = 51.52 < \text{Re}_c$, at three different values of t_0 indicated by arrows in the time series shown on the left-hand side panels of Figure 24.5. The upper panels correspond to the middle of a burst at $t = 3960$, the middle panels correspond to $t = 6316$, and the lower panels represent the dynamics after converging to the CA_1 at $t = 15{,}000$. The upper panels indicate that the bursty behavior of the HCS is characterized by irregular patterns of the M and the FTLE fields, similar to the burst of the

---

[*] The Lagrangian techniques applied in this chapter are implemented in a numerical code freely available at https://github.com/rmiracer.

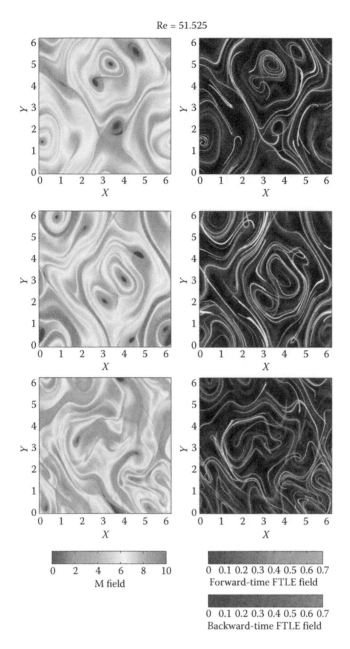

FIGURE 24.7 The M field (left-hand side panels) and the FTLE field (right-hand side panels) of the hyperchaotic attractor (Re = 51.525) at $t = 4500$ (upper panels), $t = 7000$ (middle panels), and $t = 8500$ (bottom panels). The FTLE field is superposed by forward-time hyperbolic LCSs (light gray lines) and backward-time hyperbolic LCSs (white lines).

HCA shown in the lower panels of Figure 24.7. The spatiotemporal patterns of the middle and bottom panels resemble the patterns of the middle and upper panels of Figure 24.7, respectively.

The spatiotemporal patterns of Figures 24.7 and 24.8 can be compared quantitatively by constructing the probability distribution function (PDF) of the FTLE field [2]. The left-side panel of Figure 24.9 shows the PDF of the backward-time FTLE field $\sigma_{t_0}^t$ computed from chaotic saddles for Re = $51.52 < $ Re$_c$.

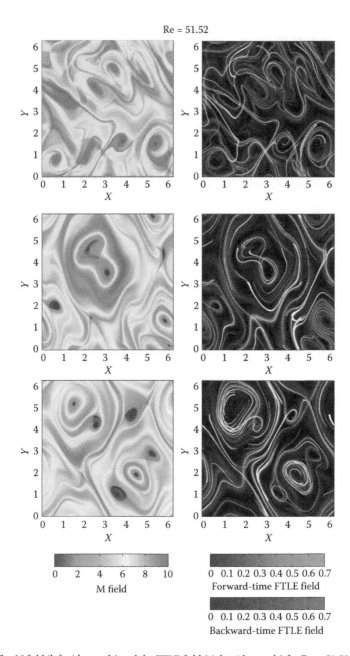

FIGURE 24.8 The M field (left-side panels) and the FTLE field (right-side panels) for Re = 51.52 at $t = 3960$ (upper panels), $t = 6316$ (middle panels), and $t = 15,000$ (bottom panels). The FTLE field is superposed by forward-time hyperbolic LCSs (light gray lines) and backward-time hyperbolic LCSs (white lines).

The dashed line represents the strong burst during the chaotic transient around $t_0 = 3960$ (see Figure 24.6), the dotted line corresponds to $t_0 = 6316$, and the continuous line corresponds to the laminar period after convergence to CA_1 ($t_0 = 15,000$). All PDFs exhibit an asymmetric shape with a fat tail toward large values of $\sigma_{t_0}^t$. A similar shape was observed by Beron-Vera et al. [2] for the backward-time FTLE computed from velocity fields of surface ocean currents inferred using satellite data. Clearly, the PDF is narrower during the laminar period, and becomes broader during the burst period. This figure allows for a clear

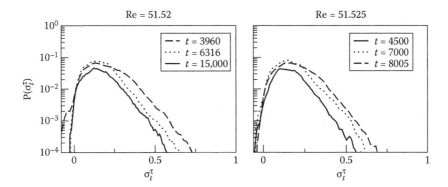

FIGURE 24.9   Probability distribution function of the FTLE field for Re $= 51.52 <$ Re$_c$ (left-side panels) and Re $= 51.525 >$ Re$_c$ (right-side panels). The continuous line represent laminar periods, the dotted line corresponds to the high-variability periods, and the dashed line represents strong bursts.

differentiation between the laminar state at $t_0 = 15,000$ and the high-variability state at $t_0 = 6316$ with a slightly broader PDF compared with the laminar PDF.

The right-hand side panels of Figure 24.9 show the PDFs of the backward-time $\sigma^t_{t_0}$ for Re $= 51.525 >$ Re$_c$. The laminar state at $t = 4500$ is represented by a continuous line, the higher variability state at $t = 7000$ is represented by a dotted line, and the strong burst observed at $t = 8005$ is represented by a dashed line. The shape and width of the PDF of each state are very similar to the corresponding PDF prior to the transition to HCA. From this figure, we conclude that the spatiotemporal patterns of the flow after transition to hyperchaos can be predicted by the hyperchaotic transient observed before the transition, in agreement with the conclusion of Rempel and Chian [44] for a one-dimensional regularized long-wave equation.

24.6 Conclusion

In this chapter, we performed numerical simulations of the 2D incompressible NSE with external forcing and periodic boundary conditions. By constructing bifurcation diagrams, we showed the transition from a quasiperiodic regime to a chaotic regime, and then to a hyperchaotic regime with increasing Reynolds number. Prior to the transition to hyperchaos, we show that the chaotic transient observed in the time series of the kinetic energy and enstrophy is due to the presence of an HCS. After transition, there is chaos–hyperchaos intermittency due to the coupling between the HCS and two chaotic saddles, which are the continuation of the two symmetric chaotic attractors CA$_1$ and CA$_2$. The Lagrangian mixing properties of the fluid were characterized using the FTLE field and a mathematical theory recently developed to detect LCSs as parametrized curves in 2D flows. The laminar periods associated with the chaotic saddles CS$_1$ and CS$_2$ display smoother patterns of the FTLE field, and are organized by several large-scale vortices surrounded by hyperbolic LCSs. In contrast, during the bursty periods, there is a sudden increase on the fluid complexity associated with the HCS. The PDFs of the backward-time FTLE field demonstrate that the enhanced complexity of the spatiotemporal patterns during the intermittent bursts after transition can be predicted by the HCS prior to transition, in agreement with the conclusion of Rempel and Chian [44].

The transition to hyperchaos described in this chapter has a strong similarity with the onset of STC reported in a number of numerical simulations of nonlinear partial differential equations [5,44,48,49]. However, we were unable to observe a transition from narrowband to broadband power spectrum in the wavenumber domain, which is a characteristic of STC, due to the regime of low Reynolds number studied in this chapter. Nonetheless, the Lagrangian techniques applied here are applicable to high Reynolds number fluid turbulence where STC is expected.

Acknowledgments

This work was supported by CAPES, CNPq, FAPESP, FAPDF, and DPP/UnB in Brazil.

References

1. G. Benettin, L. Galgani, A. Giorgilli, and J.-M. Strelcyn. Lyapunov characteristic exponents for smooth dynamical systems and for Hamiltonian systems; A method for computing all of them. Part 1: Theory. *Meccanica*, 15:9–20, 1980.
2. F. J. Beron-Vera, M. J., Olascoaga, and G. J. Goni. Surface ocean mixing inferred from different multi-satellite altimetry measurements. *J. Phys. Ocean.*, 40:2466–2480, 2010.
3. C. Boldrighini and V. Franceschini. A five-dimensional truncation of the plane incompressible Navier-Stokes Equations. *Commun. Math. Phys.*, 64:159–170, 1979.
4. B. L. J. Braaksma, H. W. Broer, and G. B. Huitema. Toward a quasi-periodic bifurcation theory. *Mem. AMS*, 83:83, 1990.
5. R. Braun and F. Feudel. Supertransient chaos in the two-dimensional complex Ginzburg-Landau equation. *Phys. Rev. E*, 53:6562, 1996.
6. R. Braun, F. Feudel, and P. Guzdar. Route to chaos for a two-dimensional externally driven flow. *Phys. Rev. E*, 58:1927–1932, 1998.
7. R. Braun, F. Feudel, and N. Seehafer. Bifurcations and chaos in an array of forced vortices. *Phys. Rev. E*, 55:6979–6984, 1997.
8. H. W. Broer, G. B. Huitema, F. Takens, and B. L. J. Braaksma. Unfoldings of quasi-periodic tori. *Mem. AMS*, 83:1, 1990.
9. P. H. R. Calil and K. J. Richards. Transient upwelling hot spots in the oligotrophic North Pacific. *J. Geophys. Res.*, 115:C02003, 2010.
10. A. C.-L. Chian, F. A. Borotto, E. L. Rempel, and C. Rogers. Attractor merging crisis in chaotic business cycles. *Chaos Solitons Fractals*, 24:869–875, 2005.
11. A. C.-L. Chian, R. A. Miranda, E. L. Rempel, Y. Saiki, and M. Yamada. Amplitude-phase synchronization at the onset of permanent spatiotemporal chaos. *Phys. Rev. Lett.*, 104:254102, 2010.
12. U. A. Dyudina, A. P. Ingersoll, S. P. Ewald, A. R. Vasavada, R. A. West, A. D. Del Genio, J. M. Barbara, C. C. Porco, R. K. Achterberg, F. M. Flasar et al. Dynamics of Saturn's south polar vortex. *Science*, 319:1801–1801, 2008.
13. M. Farazmand and G. Haller. Computing Lagrangian coherent structures from their variational theory. *Chaos*, 22, 2012.
14. M. Farazmand and G. Haller. Erratum and addendum to: A variational theory of hyperbolic Lagrangian coherent structures. *Physica D*, 241:439–441, 2012.
15. F. Feudel and N. Seehafer. Bifurcations and pattern formation in a two-dimensional Navier-Stokes fluid. *Phys. Rev. E*, 52:3506–3511, 1995.
16. F. Feudel and N. Seehafer. On the bifurcation phenomena in truncations of the 2D Navier-Stokes equations. *Chaos Solitons Fractals*, 5:1805–1816, 1995.
17. V. Franceschini. Bifurcations of tori and phase locking in a dissipative system of differential equations. *Physica D*, 6:285–304, 1983.
18. U. Frisch. *Turbulence: The Legacy of A. N. Kolmogorov*. Cambridge University Press, Cambridge, 1995.
19. G. Haller. A variational theory of hyperbolic Lagrangian coherent structures. *Physica D*, 240:574–598, 2011.
20. G. Haller. Lagrangian coherent structures. *Annu. Rev. Fluid Mech.*, 47:137–161, 2015.
21. G. Haller and G. Yuan. Lagrangian coherent structures and mixing in two-dimensional turbulence. *Physica D*, 147:352–370, 2000.

22. G.-H. Hsu, E. Ott, and C. Grebogi. Strange saddles and the dimensions of their invariant manifolds. *Phys. Lett. A*, 127:199–204, 1988.

23. K. Ide, D. Small, and S. Wiggins. Distinguished hyperbolic trajectories in time-dependent fluid flows: Analytical and computational approach for velocity fields defined as data sets. *Nonlin. Proc. Geophys.*, 9:237–263, 2002.

24. H. Kantz and P. Grassberger. Repellers, semi-attractors and long-lived chaotic transients. *Physica D*, 17:75–86, 1985.

25. R. H. Kraichnan. Inertial ranges in 2-dimensional turbulence. *Phys. Fluids*, 10:1417, 1967.

26. Y.-C. Lai and T. Tél. *Transient Chaos: Complex Dynamics on Finite-Time Scales*. Springer, New York, 2011.

27. J. Lee. Topology of trajectories of the 2D Navier-Stokes equations. *Chaos*, 2:537–563, 1992.

28. F. Lekien and S. D. Ross. The computation of finite-time Lyapunov exponents on unstructured meshes and for non-Euclidian manifolds. *Chaos*, 20:017505, 2010.

29. S. C. Leterme and R. D. Pingree. The Gulf Stream, rings and North Atlantic eddy structures from remote sensing (Altimeter and SeaWiFS). *J. Mar. Syst.*, 69:177–190, 2008.

30. J. A. J. Madrid and A. M. Mancho. Distinguished trajectories in time dependent vector fields. *Chaos*, 19:013111, 2009.

31. A. M. Mancho, D. Small, and S. Wiggins. Computation of hyperbolic trajectories and their stable and unstable manifolds for oceanographic flows represented as data sets. *Nonlin. Proc. Geophys.*, 11:17–33, 2004.

32. A. M. Mancho, D. Small, and S. Wiggins. A tutorial on dynamical systems concepts applied to Lagrangian transport in oceanic flows defined as finite time data sets: Theoretical and computational issues. *Phys. Rep.*, 437:55–124, 2006.

33. M. Mathur, G. Haller, T. Peacock, J. E. Ruppert-Felsot, and H. L. Swinney. Uncovering the Lagrangian skeleton of turbulence. *Phys. Rev. Lett.*, 98:144502, 2007.

34. C. Mendoza and A. M. Mancho. Hidden geometry of ocean flows. *Phys. Rev. Lett.*, 105:038501, 2010.

35. C. Mendoza, A. M. Mancho, and S. Wiggins. Lagrangian descriptors and the assessment of the predictive capacity of oceanic data sets. *Nonlin. Proc. Geophys.*, 21:677–689, 2014.

36. C. Mendoza and M. Mancho. The Lagrangian description of aperiodic flows: A case study of the kuroshio current. *Nonlin. Proc. Geophys.*, 19:449–472, 2012.

37. I. Mezić, S. Loire, V. A. Fonoberov, and P. Hogan. A new mixing diagnostic and Gulf oil spill movement. *Science*, 330:486–489, 2010.

38. D. Molenaar, H. J. H. Clercx, and G. J. F. van Heijst. Transition to chaos in a confined two-dimensional flow. *Phys. Rev. Lett.*, 95:104503, 2005.

39. S. Newhouse, D. Ruelle, and F. Takens. Occurrence of strange axiom A attractors near quasi periodic flows on T^m, m $\geq$ 3. *Commun. Math. Phys.*, 64:35–40, 1978.

40. J. M. Ottino. *The Theory of Mixing: Stretching, Chaos and Transport*. Cambridge University Press, Cambridge, England, 1989.

41. T. Peacock and G. Haller. Lagrangian coherent structures: The hidden skeleton of fluid flows. *Physics Today*, 41–47, 2013.

42. G. Piccioni, P. Drossart, A. Sanchez-Lavega, R. Hueso, F.W. Taylor, C.F. Wilson, D. Grassi, L. Zasova, M. Moriconi, A. Adriani et al. South-polar features on Venus similar to those near the north pole. *Nature*, 450:637–640, 2007.

43. E. L. Rempel and A. C.-L. Chian. Intermittency induced by attractor-merging crisis in the Kuramoto-Sivashinsky equation. *Phys. Rev. E*, 71:016203, 2005.

44. E. L. Rempel and A. C.-L. Chian. Origin of transient and intermittent dynamics in spatiotemporal chaotic systems. *Phys. Rev. Lett.*, 98:014101, 2007.

45. E. L. Rempel, A. C.-L. Chian, and A. Brandenburg. Lagrangian coherent structures in nonlinear dynamos. *Astrophys. J.*, 735:L9, 2011.

46. E. L. Rempel, A. C.-L. Chian, A. Brandenburg, P. R. Muñoz, and S. C. Shadden. Coherent structures and the saturation of a nonlinear dynamo. *J. Fluid Mech.*, 729:309–329, 2013.

47. E. L. Rempel, A. C.-L. Chian, E. E. N. Macau, and R. R. Rosa. Analysis of chaotic saddles in high-dimensional dynamical systems: The Kuramoto-Sivashinsky equation. *Chaos*, 14:545–556, 2004.

48. E. L. Rempel, A. C.-L. Chian, and R. A. Miranda. Chaotic saddles at the onset of intermittent spatiotemporal chaos. *Phys. Rev. E*, 76:056217, 2007.

49. E. L. Rempel, R. A. Miranda, and A. C.-L. Chian. Spatiotemporal intermittency and chaotic saddles in the regularized long-wave equation. *Phys. Fluids*, 21:074105, 2009.

50. D. Ruelle and F. Takens. On the nature of turbulence. *Commun. Math. Phys.*, 20:167–192, 1971.

51. S. Scheel and N. Seehafer. Bifurcation to oscillations in three-dimensional Rayleigh-Bènard convection. *Phys. Rev. E*, 56:5511–5516, 1997.

52. A. B. Schelin, G. Károlyi, A. P. S. de Moura, N. A. Booth, and C. Grebogi. Chaotic advection in blood flow. *Phys. Rev. E*, 80:016213, 2009.

53. A. B. Schelin, G. Károlyi, A. P. S. de Moura, N. A. Booth, and C. Grebogi. Fractal structures in stenoses and aneurysms in blood vessels. *Philos. Trans. R. Soc. A Math. Phys. Eng. Sci.*, 368:5605–5617, 2010.

54. A. B. Schelin, G. Károlyi, A. P. S. de Moura, N. A. Booth, and C. Grebogi. Are the fractal skeletons the explanation for the narrowing of arteries due to cell trapping in a disturbed blood flow? *Comput. Biol. Med.*, 42:276–281, 2012.

55. I. Scheuring, T. Czárán, P. Szabó, G. Károlyi, and Z. Torocz-kai. Spatial models of prebiotic evolution: Soup before pizza? *Origins Life Evol. Biosphere*, 33:319–355, 2003.

56. S. C. Shadden, F. Lekien, and J. E. Marsden. Definition and properties of Lagrangian coherent structures from finite-time Lyapunov exponents in two-dimensional aperiodic flows. *Physica D*, 212:271–304, 2005.

57. A. D. Stroock, S. K. W. Dertinger, A. Ajdari, I. Mezić, H. A. Stone, and G. M. Whitesides. Chaotic mixer for microchannels. *Science*, 295:647–651, 2002.

58. D. Sweet, H. E. Nusse, and J. A. Yorke. Stagger-and-step method: Detecting and computing chaotic saddles in higher dimensions. *Phys. Rev. Lett.*, 86:2261, 2001.

59. K. G. Szabó and T. Tél. Transient chaos as the backbone of dynamics on strange attractors beyond crisis. *Phys. Lett. A*, 196:173–180, 1994.

60. T. Tél, A. P. S. de Moura, C. Grebogi, and G. Károlyi. Chemical and biological activity in open flows: A dynamical system approach. *Phys. Rep.*, 413:91–196, 2005.

61. L. van Veen. The quasi-periodic doubling cascade in the transition to weak turbulence. *Physica D*, 210:249–261, 2005.

62. F. A. Williams. *Combustion Theory: The Fundamental Theory of Chemically Reacting Flow Systems*. Benjamin-Cummings, Menlo Park, CA, 1985.

VI

Chaos and Quantum Theory

IV

Chaotic Interference versus Decoherence: External Noise, State Mixing, and Quantum–Classical Correspondence

Valentin V. Sokolov
and Oleg V. Zhirov

25.1 Outline

The famous Niels Bohr's quantum–classical correspondence principle states that classical mechanics is a limiting case of the more general quantum mechanics. This implies that "under certain conditions" quantum laws of motion become equivalent to classical ones. One of the conditions is fairly obvious: the corresponding classical action should be very large as compared with Planck's constant $\hbar$. But is this the *sufficient* condition? In fact, *it is not!*

The quantum laws show up in two different, although not entirely independent ways:

1. Discrete spectrum of finite motion
2. Interference phenomena

Even if the energy spectrum of a finite closed quantum system becomes continuous in the formal limit of vanishing $\hbar$, the interference effects cannot disappear in similar manner. Indeed, a quantum wave functions has no definite classical counterpart. Meanwhile, suppression of effects of quantum interference ("decoherence") is a key requirement for the classical laws to appear. Being, in essence, of a quite general nature, this problem takes on special significance in the nontrivial case of nonlinear classically chaotic quantum systems.

A number of typical manifestations of the quantum coherence in the time evolution as well as eigenstates' properties are widely discussed in the scientific literature:

- Wave packet dynamics and decay of quantum fidelity.
- Universal local spectral fluctuations.
- Scars in the stationary eigenfunctions.
- Elastic enhancement in chaotic resonance scattering.
- Weak localization in transport phenomena.

The specific features of quantum dynamics of classically chaotic systems seem to be in striking contrast with those of genuine classical chaos. Since these features can even question the validity of the quantum–classical correspondence principle by itself, a more profound analysis is needed for understanding the bridge between the classical and quantum chaotic worlds.

25.1.1 Chaotic Time Evolution

In the case of regular classical dynamics, the system's response to a weak external perturbation is proportional to its strength and the system may be still treated as a *closed* one for a sufficiently long time. In contrast, chaotic classical dynamics is exponentially unstable and therefore it is extremely sensitive to any uncontrollable external influence. We can never neglect the influence of the environment. This therefore stipulates the *self-mixing* property of classical dynamical chaos and, as a consequence, a very fast decay of the phase correlations (here and throughout the chapter we use the language of action-angle variables).

While the exponential decay of the phase correlations is an underlying feature of the classical dynamical chaos [14], the so-called "quantum chaos," i.e., quantum dynamics of classically chaotic systems is not by itself capable of destroying the *quantum phase* coherence. Strictly speaking, any initially pure quantum state remains pure during arbitrary long evolution. Quantization of the phase space removes exponential instability and makes the quantum dynamics substantially more stable than the classical one. There exists a threshold $\sigma_c(t)$ of the external noise strength σ below which the coherence survives up to the time t [43–45]. Only appreciably strong noise or finite measurement accuracy can produce a mixture of quantum states sufficient for noticeable suppression of quantum coherence. A number of appropriate characteristics: Peres fidelity $F(t)$, purity $\mathbb{P}(t)$, Shennon (information) $\mathcal{I}(t)$, and von Neumann (correlational) $\mathcal{S}(t)$ entropies are used to demonstrate the gradual loss of quantum coherence during the system's evolution in the presence of noisy background. Being sensitive to the quantum coherence, the von Neumann entropy remains smaller than the Shennon entropy but runs up monotonically to the latter when the evolution time approaches some moment $t_{(dec)}$ when decoherence becomes complete. After this time, the system occupies the whole phase space volume accessible at the running degree of excitation energy, thus reaching a sort of equilibrium. Henceforth, the phase volume expands "adiabatically": both entropies remain almost constant. The evolution after the time $t_{(dec)}$ is Markovian [45].

25.1.2 2D Billiards

The plain two-dimensional (2D) areas with closed irregular borders called billiards are pet systems often used to illustrate characteristic features of the classical dynamical chaos. With the advent of the ability to fabricate mesoscopic analogs of the classical billiards, the opportunity appeared to experimentally observe the signatures of chaos in classically chaotic quantum systems. An excellent possibility thus has arisen to verify the theoretical concepts developed in numerous theoretical investigations of "quantum chaos" phenomena. Extensive study of the electron transport through ballistic 2D meso-structures [27,31] (see also [2,10] and references therein) has fully confirmed correctness of the basic ideas [21,22,26] of the theory of the chaotic quantum interference and relevance [2,10,11] of the so-called random-matrix approach

[29,30,46] to the problem of the universal spectral fluctuations as well as conductance fluctuations in open mesoscopic set ups (see Section 25.1.5).

The energy E is the only integral of motion in a closed billiard. Repeating reflections from the border produces a countless manifold of unclosed exponentially unstable trajectories and, as a consequence, chaotic dynamics. At the same time, there exists a countable set of specific, but also exponentially unstable, periodic orbits. The number $N(T)$ of the latter trajectories grows exponentially with the period T as $N(T) \sim e^{\frac{T}{\tau_c}}$ with the characteristic time τ_c directly connected, $\tau_c \sim 1/\Lambda$, to the Lyapunov exponent Λ that describes the exponential instability [47]. Meanwhile, in contravention with the classical exponential instability, stable interference maxima (scars) along the periodic classical orbits (alongside with irregularly scattered point-like interference maxima [specks]) are discovered numerically [23] in the stationary billiard's eigenstates even in the very deep semiclassical region. Does this fact compromise the quantum–classical correspondence principle?

To answer the afore-posed question, let us first consider an elementary example of one-dimensional (1D) finite motion. A particle with a mass m and energy E moves in a potential well. The classical position probability density is proportional to the fraction of the period of oscillations the particle spends near a point x and is easily found to be

$$w_c(x) = \begin{cases} 0 & x \notin [a,b], \\ \dfrac{m\,\omega_c(E)}{\pi\, p_c(x)}, & x \in [a,b] \end{cases} \tag{25.1}$$

where $p_c(x) = \sqrt{2m[E - U(x)]}$ is the particle's classical momentum at the point x and $\omega_c(E)$ is the frequency of classical oscillations between the turning points a and b.

On the other hand, the semiclassical solution of the corresponding quantum problem yields

$$w_n(x) = |\psi_n(x)|^2$$
$$= \begin{cases} e^{\left\{ -\frac{2}{\hbar}\int_x^a dx'|p_c(x')|\right\}},\, e^{\left\{ -\frac{2}{\hbar}\int_b^x dx'|p_c(x')|\right\}} & \to 0, \\ \dfrac{m\,\omega_c(E_n)}{\pi\, p_c(x)}\, 2\cos^2\left\{ \frac{1}{\hbar}\int_a^x dx'p_c(x') - \frac{\pi}{4}\right\} & \to ? \end{cases} \tag{25.2}$$

As expected (see the simplest of a linear oscillator, Figure 25.1, bottom frame), the probability density (25.2) vanishes in the classically forbidden region when the semiclassical parameter goes to infinity but there exists no reasonable result in the *classically allowed* interval. The wildly fluctuating without

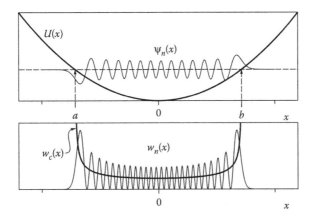

FIGURE 25.1 Top: wave function of a harmonic oscillator for the energy level $n = 25$; bottom: corresponding probability distributions are shown by thick and thin lines for the classical and quantum case, respectively.

approaching a certain limit cosine-square factor appears from the interference of two waves running in the opposite directions. To attain the classical result (25.1), an additional *averaging* over some finite either position Δx or energy Δn intervals around fixed $|x\rangle$ or $|n\rangle$ states is necessary.

$$\overline{w_n(x)}^{(x)} = \langle n| \left(\int_{x' \in \Delta x} dx' \, \mathbf{p}_x(x')|x'\rangle\langle x'| \right) |n\rangle = \langle n|\hat{\rho}^{(x)}|n\rangle, \tag{25.3}$$

$$\overline{w_n(x)}^{(n)} = \langle x| \left(\sum_{n' \in \Delta n} \mathbf{p}_n(n')|n'\rangle\langle n'| \right) |x\rangle = \langle x|\hat{\rho}^{(n)}|x\rangle. \tag{25.4}$$

The density matrices $\hat{\rho}^{(q)}$ ($q = n, x$) describe *incoherent mixtures* of the quantum states within the indicated intervals. The real and normalized to unity quantities $\mathbf{p}$ characterize the weights of these states. Obviously, the range of averaging Δx should satisfy the condition $\Delta x \geqslant \pi\hbar/p_c(x)$ to meet the classical behavior. Similar reasoning leads to the condition $\Delta E \approx \Delta n\,\hbar\omega_c \geqslant 2\pi\hbar/T$ where $T = 2\int_a^b dx/v_c(x)$ is the period of classical oscillations. Similarly, averaging over the energy interval ΔE wipes off all scars of eigenstates of a 2D billiard with the periods $T > \hbar/\Delta E$, whereas those with smaller periods still survive [13].

Strictly speaking, decoherence is not perfect as long as off-diagonal matrix elements of the density matrix still exist. They are complex and their phases carry some information on the more subtle interference effects. Complete decoherence is achieved only when the number of mixing states is so large that the density matrix becomes proportional to the unit matrix and ceases to depend on the basis in the Hilbert space of states. As a matter of fact, decoherence can originate only from: (i) the process of preparation of initially mixed state and (ii) mixing induced during the time evolution by a persisting external noise.

25.1.3 The Basics of Quantum Mixed States

The concept of mixed states plays a paramount role in the problem of decoherence. A pure quantum state is specified at some moment of time t by its wave vector $|\psi(t)\rangle$ in the Hilbert space. This allows, in particular, calculation of the mean value $O(t) = \langle\psi(t)|\hat{O}|\psi(t)\rangle$ (in what follows we suggest this vector to be normalized to unity, $\langle\psi(t)|\psi(t)\rangle = 1$) of any dynamical quantity represented by a Hermitian operator $\hat{O}$. Equivalently, the same state can be described by the density matrix $\hat{\rho}(t) = |\psi(t)\rangle\langle\psi(t)|$ that satisfies two obvious conditions:

$$Tr\hat{\rho}(t) = 1, \quad \hat{\rho}^2(t) = \hat{\rho}(t). \tag{25.5}$$

The second relation is the necessary and sufficient condition of purity of a quantum state. With these definitions at hand, the mean values can be presented in one of the following forms:

$$O(t) = \langle\psi(t)|\hat{O}|\psi(t)\rangle = Tr\left[\hat{O}\,\hat{\rho}(t)\right] = \sum_{m,n} O_{nm}(t)\rho_{mn}(t). \tag{25.6}$$

The expression on the right-hand side(rhs) represents the same mean value on the basis of a complete set of motionless vectors $|n\rangle$ in the Hilbert space. Since the density matrix is Hermitian, all its diagonal elements $\rho_{nn}(t)$ are real.

Like any Hermitian matrix, the density matrix $\rho_{mn}(t)$ can be diagonalized with the help of some unitary transformation. On account of the conditions (25.5), the density matrix of a pure quantum state has only one nonzero eigenvalue that equals one. Obviously, the diagonalization returns us to the original form $\hat{\rho}(t) = |\psi(t)\rangle\langle\psi(t)|$. The transformation matrix depends on time if the system undergoes time evolution. In practice, a fixed basis of states $|n\rangle$ is, as a rule, a more relevant choice from the physical point of view.

A mixed quantum state is described at any moment of time t by an incoherent sum of binary contributions

$$\hat{\rho}(t) = \sum_k \mathbf{p}_k(t)|v_k(t)\rangle\langle v_k(t)|, \tag{25.7}$$

where the states $|v_k(t)\rangle$ are the eigenvectors of the density matrix when $0 \leq \mathbf{p}_k(t) \leq 1$ stand for the weights of the corresponding pure fragments. These weights do not actually depend on time during dynamical evolution described by some unitary evolution matrix $\hat{U}(t)$. It is not the case, however, if the system interacts with a noisy background.

In the fixed basis $|n\rangle$, an initially diagonal incoherent mixture develops off-diagonal elements even during unitary dynamical evolution. There exists, however, a convenient invariant measure of state purity called *Purity*

$$\mathbb{P}(t) = \text{Tr}\left[\hat{\rho}^2(t)\right] = \sum_k \mathbf{p}_k^2 = \mathbb{P}(0), \tag{25.8}$$

that remains constant as long as the noise is absent. The Purity is restricted to the interval $0 \leq \mathbb{P} \leq 1$ and is mounting to one in the limit of a perfectly pure state. In many respects, similar properties are inherent in the invariant von Neumann entropy

$$\mathcal{S}(t) = -\text{Tr}\left[\hat{\rho}(t) \ln \hat{\rho}(t)\right] = \mathcal{S}(0), \tag{25.9}$$

which vanishes only in the case of a perfectly pure state.

It is very useful to transfer the story on the, generally mixed, quantum states evolution to the language of the phase space [3,4,44]. This elucidates analogy and distinctions between classical and quantum dynamics. A double Fourier transformation of the density matrix

$$\begin{aligned} W(\alpha^*, \alpha; t) &= \frac{1}{\pi^2 \hbar} \int d^2\eta \, \exp\left(\eta \frac{\alpha^*}{\sqrt{\hbar}} - \eta^* \frac{\alpha}{\sqrt{\hbar}}\right) \tilde{\rho}(\eta^*, \eta; t) \\ &= \frac{1}{\pi^2 \hbar} \int d^2\eta \, \exp\left(\eta^* \frac{\alpha}{\sqrt{\hbar}} - \eta \frac{\alpha^*}{\sqrt{\hbar}}\right) \text{Tr}\left[\hat{\rho}(t) \hat{D}(\eta)\right], \end{aligned} \tag{25.10}$$

with the operator $\hat{D}(\eta) = \exp(\eta \hat{a}^\dagger - \eta^* \hat{a})$ being the coherent states displacement defines the *Wigner function* $W(\alpha^*, \alpha; t)$ that is a direct quantum counterpart of the classical phase space distribution function $W^{(c)}(\alpha^*, \alpha; t)$. The complex variables α^*, α are connected with the standard action-angle variables I, θ by the canonical transformations $\alpha = \sqrt{I}e^{-i\theta}$, $\alpha^* = \sqrt{I}e^{i\theta}$.

25.1.4 Peres Fidelity

Response of an evolving in time quantum system to a weak external perturbation is of prime interest in the context of the problem of stability and reversibility of quantum motion. The customary quantitative characteristic of sensitivity of classically chaotic quantum dynamics to such perturbations is the Peres fidelity [32]:

$$F(t) = \frac{\text{Tr}\left[\hat{\rho}_H(t) \, \hat{\rho}_{H'}(t)\right]}{\mathbb{P}(t)} = \frac{\sum_{kl} \mathbf{p}_k \mathbf{p}_l' \left|\langle v_k(t) | v_l'(t) \rangle\right|^2}{\mathbb{P}(t)}. \tag{25.11}$$

Fidelity measures the weighted mean distance between two, generally mixed, quantum states evolving according to slightly different Hamiltonians $\hat{H}$ and $\hat{H}'$ bounded in the interval $[0, 1]$ and equals one when $\hat{H}' = \hat{H}$. Its time decay due to diminishing of the overlap of v-eigenvectors elucidates the sensitivity of the motion to an external influence. In particular, this quantity enables one to directly connect stability and reversibility of quantum dynamics with complexity of the quantum Wigner function. More than that, the notion of the Peres fidelity directly extends to classical mechanics (see Section II in [8,9]). The number of θ-harmonics can serve (see below) as a natural quantitative measure of complexity of both, classical and quantum phase space (quasi-)distributions [44] (see, however, Reference 7).

25.1.5 Open Mesoscopic Billiards and Electron Quantum Transport

Scattering of quantum particles by 2D billiard-like mesoscopic structures connected to the continuum by a long lead is an intriguing issue that attracted a lot of attention of theorists as well as experimentalists for at least the last two decades. If the mean free path of such a particle exceeds the typical size of the structure, the particle's dynamics inside it strongly depends on the shape of its border. The classical motion in this case becomes stochastic and quantum scattering is expected to be described well within the framework of the random-matrix approach to the resonance-scattering theory. Intensive experimental studies confirmed in many respects these expectations. Nevertheless, electron transport experiments with ballistic quantum dots reveal noticeable and persisting up to zero temperature loss of the quantum-mechanical coherence in contravention of predictions of the random matrix as well as semiclassical scattering theories.

There exists a number of different methods of accounting for the decoherence effect in the ballistic quantum transport processes. By the highest standards, all of them originate from the pioneering Büttiker's ideas [5]. The dot is supposed to be connected by one way or another with a bath of electrons so that the dot and the bath can exchange electrons in such a manner that the mean exchange electric current vanishes. Since the incoming electron carries no information of the phase of the wave function in the dot, the coherence turns out to be suppressed. The cost pai is that the number of particles is conserved during the scattering only in average. An alternative approach has been therefore proposed in Reference 12 with a closed long stub instead of an opening lead. Thereby unitarity of the scattering matrix is guaranteed and none of the electrons is lost at any individual measurement. Decoherence takes place in this case because of a spatially random time-dependent external electric field that acts in the stub. As a result, an electron that once penetrated the stub returns in the dot without whatever phase memory.

In spite of the advantages of the stub model, the necessity of introducing ad hoc an external time-dependent potential seems to be somewhat artificial. Still another possibility arises if a time-independent weak interaction is taken into account with relatively rare irregular impurities in the semiconductor heterostructure to whose interface region the electrons are confined. At that, unavoidable experimental averaging over energy on account of finite experimental accuracy in measuring the cross sections is a point of primary importance.

Owing to interaction with the environment, each "doorway" resonance state excited in the structure via external channels gets fragmented onto a large number $\sim \Gamma_s/d$ (the spreading width Γ_s characterizes the strength of the coupling to the environment when d is the single-quasi-particle mean level spacing) of very narrow resonances [40]. Only the cross sections averaged over the fine ($\sim d$) structure scale are observable. Owing to such an averaging, the doorway resonance states are damped not only because of escape through such channels but also due to the ulterior population of the long-lived environmental states. As a result, transmission of an electron with a given incoming energy E_{in} through the structure turns out to be an incoherent sum of the flow formed by the interfering damped doorway resonances and the retarded flow of particles reemitted into the structure by the environment. Being delayed, the returning electrons do not interfere with those that escape directly through the external leads.

We suppose the temperature of the environment to be zero whereas the energy of the incoming particle E_{in} can be close to or somewhat above the Fermi surface of electrons in the environment. Therefore, though the number of the particles is definitely conserved in each individual event of transmission, there exists a probability that some part of the electron's energy can be absorbed due to environmental many-body effects. Both decoherence and absorption phenomena are treated below within the framework of a unit microscopic model based on the general theory of resonance scattering. Both these effects are controlled by the only parameter: the spreading width of the doorway resonances.

If the energy E_{in} noticeably exceeds the environment's Fermi surface and the doorway resonances overlap, the random-matrix approach becomes relevant and ensemble averaging in the doorway sector is appropriate. Such an averaging, being equivalent to the energy averaging over the doorway scale D, suppresses all interference effects save the elastic enhancement phenomenon. The latter is a direct consequence of the time reversal symmetry and manifests itself in the so-called *weak localization effect*. However, the energy absorption in the environment violates this symmetry and suppresses the weak localization.

25.2 Particulars

25.2.1 An Example of Chaotic Classical Dynamics

As an instructive and typical example of a chaotic classical system, we consider below a periodically kicked quartic nonlinear oscillator

$$H(\alpha^*, \alpha; t) = H^{(0)} + H^{(k)}, \tag{25.12}$$

where the unperturbed Hamiltonian function reads

$$H^{(0)}(\alpha^*, \alpha) = \frac{p^2}{2m} + \frac{\omega_0^2}{4} \tan^2(\sqrt{2m}\, x) = \omega_0 |\alpha|^2 + |\alpha|^4, \tag{25.13}$$

and the time-dependent perturbation

$$H^{(k)}(\alpha^*, \alpha; t) = g(t) = g_0 \sum_\tau \delta(t - \tau)(\alpha^* + \alpha). \tag{25.14}$$

constitutes a sequence of periodic kicks with strength g_0 that are acting at the instances $\tau = \pm 1, \pm 2, \dots$.

There are two convenient choices of the canonical variables in this case: I, θ or $\alpha, i\alpha^*$ related by the canonical transformation $\alpha = \sqrt{I}\, e^{-i\theta}$, $\alpha^* = \sqrt{I}\, e^{i\theta}$. The action-angle variables I, θ are ordinarily used in the classical considerations. On the other hand, the advantage of α-variables is that they are directly related to the quantum creation–annihilation operators $\hat{a}^\dagger, \hat{a}$. The action-angle variables satisfy, along a given phase trajectory, two coupled nonlinear equations [38]:

$$I(t) = \left| \sqrt{\overset{\circ}{I}} + i \int_0^t d\tau g(\tau) e^{i[\theta(\tau) - \overset{\circ}{\theta}]} \right|^2, \tag{25.15}$$

$$\theta(t) = \int_0^t d\tau [\omega_0 + 2I(\tau)], \tag{25.16}$$

(here and in what follows we mark the initial values by the overset circle). A great majority of trajectories becomes exponentially unstable and, the corresponding motion is globally chaotic when the strength of the kicks exceeds 1: $|g_0| > 1$. Under this condition, the phase correlations decay exponentially fast with time [35],

$$\left| \int d\overset{\circ}{I} \, d\overset{\circ}{\theta} \, W^{(c)}(\overset{\circ}{I}, \overset{\circ}{\theta}; t = 0) \, e^{i\left(\theta(t) - \overset{\circ}{\theta}\right)} \right|^2 = \exp(-t/\tau_c), \tag{25.17}$$

where the characteristic time $\tau_c \sim 1/\Lambda$ is directly connected to the Lyapunov exponent Λ. Here, $W^{(c)}(I, \theta; t = 0)$ is the initial probability distribution in the phase space. This exponential decay of phase correlations is an universal fingerprint of the classical dynamical chaos.

In fact, almost all trajectories are alike when the motion is globally chaotic and, actually, only the behavior of manifolds of them is of real interest. Therefore, the phase space methods appear to be most relevant in the chaotic regime. The classical phase space distribution function $W^{(c)}(\alpha^*, \alpha; t)$ satisfies the linear Liouville equation

$$i\frac{\partial}{\partial t} W^{(c)}(\alpha^*, \alpha; t) = \hat{\mathcal{L}}_c(t) \, W^{(c)}(\alpha^*, \alpha; t), \tag{25.18}$$

with the unitary Liouville operator $\hat{\mathcal{L}}_c(t)$ that is a sum of two, stationary and time-dependent, parts: $\hat{\mathcal{L}}_c(t) = \hat{\mathcal{L}}_c^{(0)} + \hat{\mathcal{L}}^{(k)}(t)$. The first unperturbed part reads [44]

$$\hat{\mathcal{L}}_c^{(0)} = \left(\omega_0 + 2|\alpha|^2\right) \left(\alpha^* \frac{\partial}{\partial \alpha^*} - \alpha \frac{\partial}{\partial \alpha}\right), \tag{25.19}$$

where the operator $\left(\alpha^* \frac{\partial}{\partial \alpha^*} - \alpha \frac{\partial}{\partial \alpha}\right) = -i\frac{\partial}{\partial \theta}$ formally coincides with the quantum-mechanical angular momentum operator $\frac{1}{\hbar}\hat{L}_z$. In fact, this operator describes rotation in the phase space around the origin with a local angular velocity $\left(\omega_0 + 2|\alpha|^2\right)$.

The time-dependent perturbation (kick) operator

$$\hat{L}^{(k)}(t) = g_0 \sum_{\tau} \delta(t - \tau) \left(\frac{\partial}{\partial \alpha^*} - \frac{\partial}{\partial \alpha} \right), \qquad (25.20)$$

describes the sequence of instant shifts by the distance g_0 in the α-plane. The alternating twists and shifts develop an unpredictably complicated pattern of the density distribution $W^{(c)}(\alpha^*, \alpha; t)$ when the perturbation strength constant $|g_0| > 1$ [14]. It is of primary importance here that the unperturbed part $\hat{L}_c^{(0)}$ of the Liouville operator has a *continuous* spectrum of eigenvalues. As a consequence, the classical phase distribution is structuring exponentially fast on a finer and finer scale during chaotic evolution. Such a behavior is the paramount property of the classical dynamical chaos.

Fourier analysis in the phase plane provides a natural tool for elucidating the process of this structuring. Taking into account periodicity of the distribution function $W^{(c)}(\alpha^*, \alpha; t)$ with respect to the angle θ

$$W^{(c)}(I, \theta; t) = \frac{1}{2\pi} \sum_{m=-\infty}^{\infty} W_m^{(c)}(I; t)\, e^{im\theta}, \qquad (25.21)$$

the most simple and efficient idea is just to follow the upgrowth of the number of its θ-harmonics. It can be easily quantified with the help of notion of the Peres fidelity. We suppose for simplicity that the initial distribution has been isotropic, $W_m^{(c)}(I; 0) = 0$ if $m \neq 0$, and let the system evolve autonomously for some time t_r. At this moment, we probe the system with the help of an instant weak perturbation $\xi I \delta(t - t_r)$ that produces rotation in the phase plane by some angle ξ. To slightly simplify the subsequent formulae, we suggest the parameter ξ to be a Gaussian random variable. The probe then results in an instant change of the distribution $W^{(c)}(I, \theta; t_r - 0) \rightarrow W^{(c)}(I, \theta + \sigma; t_r + 0)$, where the parameter σ sets the mean-squared strength of the perturbation. The Peres fidelity defined in this case as

$$F_{(sen)}(\sigma; t_r) = \frac{\int d^2\alpha\, W^{(c)}\left(\alpha^*, \alpha; t_r\right)\, W^{(c)}\left(\sigma | \alpha^*, \alpha; t_r + 0\right)}{\int d^2\alpha\, [W^{(c)}\left(\alpha^*, \alpha; t_r\right)]^2} = \sum_{m=0}^{\infty} e^{-\frac{1}{2}\sigma^2 m^2} \mathcal{W}_m^{(c)}(t_r), \qquad (25.22)$$

characterizes sensitivity of the motion to an external influence.

The quantities $\mathcal{W}_m^{(c)}(t)$

$$\mathcal{W}_m^{(c)}(t) = (2 - \delta_{m0}) \frac{\int_0^\infty dI\, \left| W_m^{(c)}(I; t) \right|^2}{\sum_{m=-\infty}^{+\infty} \int_0^\infty dI |W_m^{(c)}(I; t)|^2}, \qquad (25.23)$$

that are expressed in terms of the Fourier harmonics and satisfy the normalization condition

$$\sum_{m=0}^{\infty} \mathcal{W}_m^{(c)}(t) = 1, \qquad (25.24)$$

can naturally be interpreted in the probabilistic manner as the weights of the corresponding θ-harmonics. Therefore, we can define in the spirit of the linear response approach the mean number of them $\langle |m| \rangle_t = \sqrt{\langle m^2 \rangle_t}$ at a moment of time t with the aid of the relation

$$\langle m^2 \rangle_t = \sum_{m=0}^{\infty} m^2\, \mathcal{W}_m^{(c)}(t) = -\frac{d^2 F_{(sen)}(\sigma; t)}{d\sigma^2}\Big|_{\sigma=0}. \qquad (25.25)$$

The number $\langle |m| \rangle_t$ can serve as a convenient quantitative measure of complexity of a phase space distribution at a given moment of time t. Owing to exponential instability of classical dynamics, the number of harmonics proliferates exponentially $\langle |m| \rangle_t \propto e^{t/\tau_c}$ [44].

Let us suppose now that the motion has been reversed at the moment t_r, immediately following after the time of probing. In view of the fact that the Liouville evolution operator is a unitary one, we can transform the expression (25.22) into the form

$$F_{(sen)}(\sigma; t_r) = \frac{\int d^2\alpha \, W^{(c)}\left(\alpha^*, \alpha; 0\right) W^{(c)}\left(\sigma | \alpha^*, \alpha; t_r + 0, 0\right)}{\int d^2\alpha \, [W^{(c)}\left(\alpha^*, \alpha; 0\right)]^2} = F_{(rev)}(\sigma; t_r) \equiv F(\sigma; t_r). \tag{25.26}$$

Transformed in this form, fidelity measures the overlap of the isotropic initial distribution function and the distribution $W^{(c)}\left(\sigma | \alpha^*, \alpha; t_r + 0, 0\right)$ that, after being changed by the probing perturbation at the moment t_r, has reverted to the initial moment $t = 0$. This formula connects the reversibility of the motion with sensitivity to external perturbations or, by other words, with complexity of the distribution function at the reversal moment t_r.

25.2.1.1 Deterministic Diffusion and Onset of Irreversibility

The phenomenon of the so-called deterministic diffusion is one of the simplest manifestation of the classical dynamical chaos. The mean value of the action $I(t)$ at any given time t is calculated as follows:

$$\langle I \rangle_t = \int_0^\infty dI \, I \, W_0^{(c)}(I; t) = \int_0^\infty dI \, I \, \overline{W_0^{(c)}(I; t)} = \langle I \rangle_0 + g_0^2 \, t, \tag{25.27}$$

see, e.g., References 1, 14, 37. The averaging is performed in two steps here. We have smoothed first the very irregular function $W_0^{(c)}(I; t)$ over a small interval surrounding a fixed value I within which the factor I does not appreciably change. Successive I-integration with the "coarse grained" distribution $\overline{W_0^{(c)}(I; t)}$ results in the diffusive increase of the mean action. Similarly, we can also calculate the moments $\langle I^k \rangle_t$, $k = 2, 3, ..., k_t$. The longer the duration t of the evolution, the larger the power k_t at which the two step integration procedure is still valid.

It is obvious that, formally, the deterministic diffusion (25.27) is perfectly time-reversible. However, the exponential instability makes this statement impractical. Even a very weak external noise destroys reversibility and turns the motion into irreversible process.

To illustrate this statement, we add to our Hamiltonian function a new term

$$H^{(noise)}(\alpha^*, \alpha; t) = |\alpha|^2 \sum_\tau \xi_\tau \, \delta(t - \tau), \quad \langle \xi_\tau \rangle = 0, \langle \xi_\tau \xi_{\tau'} \rangle = \sigma^2 \delta_{\tau\tau'}. \tag{25.28}$$

that describes a stationary Gaussian noise with the strength parameter σ. Each period of unperturbed evolution is immediately followed by a phase plane rotation by a random angle ξ. Averaging over the noise realizations sets up the coarse grained distribution function

$$W^{(c)}(\sigma | I, \theta; t) = \overline{W^{(c)}(\{\xi\} | I, \theta; t)}^{(noise)}. \tag{25.29}$$

The averaging suppresses Fourier harmonics with respect to both canonical variables thus extenuating irregular oscillations of the distribution. They are perfectly smoothed away in the limit of very strong noise $\sigma \to \infty: W_{|m| \geqslant 1}^{(c)}(\infty | I, ; t) = 0$ and

$$W_0^{(c)}(\infty | I; t) = \frac{1}{\langle I \rangle_0 + g_0^2 \, t} \exp\left(\frac{I}{\langle I \rangle_0 + g_0^2 \, t}\right). \tag{25.30}$$

In this case, the fidelity (25.26) equals $F_{(rev)}(\infty | t_r) = e^{-\frac{t_r}{\tau_c}}$. Indeed, the backward evolution starts with the isotropic distribution (25.30) at the moment t_r and develops $m(t_r) \propto e^{t_r/\tau_c}$ θ-harmonics by the time $t = 0$. Thus, the overlap with the initial isotropic distribution $W^{(c)}\left(\alpha^*, \alpha; 0\right)$ is exponentially small. The standard diffusion takes place in the backward evolution as well. Under an influence of the noise of a moderate level σ, the backward diffusion is delayed for some time $\tau_d \propto \ln \sigma$ [24] during which the system partly recovers its preceding states. But after that the diffusion recommences.

25.2.2 Quantum Evolutions of Classically Chaotic System

The Hamiltonian of the quantum counterpart of the classical oscillator considered above is immediately obtained by the substitutions: $\alpha \Rightarrow \sqrt{\hbar}\,\hat{a}$, $\alpha^* \Rightarrow \sqrt{\hbar}\,\hat{a}^\dagger$.

$$\hat{H}(\hat{a}^\dagger, \hat{a}; t) = \hbar\omega_0 \hat{n} + \hbar^2 \hat{n}^2 + g_0 \sum_\tau \delta(t - \tau)\sqrt{\hbar}\,(\hat{a}^\dagger + \hat{a}). \tag{25.31}$$

(In the chosen units all parameters: $\hbar$, ω_0, g_0 are dimensionless.)

The quantum evolution $\hat{\rho}(t) = \hat{\mathcal{U}}(t)\hat{\rho}(0)\hat{\mathcal{U}}^\dagger(t)$ of the density matrix is described by the unitary operator $\hat{\mathcal{U}}(t)$ that is t successive repetitions of the one period Floquet operator $\hat{U}$: $\hat{\mathcal{U}}(t) = \hat{U}^t$, the latter being a sequence of instant change of the excitation number (coherent state displacement operator) and subsequent free rotation in the phase plane induced by the time-independent Hamiltonian $\hat{H}^{(0)}(\hat{a}^\dagger, \hat{a}) = \hbar\omega_0 \hat{n} + \hbar^2 \hat{n}^2$:

$$\hat{U} = e^{-\frac{i}{\hbar}\hat{H}^{(0)}} \hat{D}\left(i\frac{g_0}{\sqrt{\hbar}}\right). \tag{25.32}$$

The Wigner function (25.10) satisfies the quantum Liouville equation:

$$i\frac{\partial}{\partial t} W(\alpha^*, \alpha; t) = \hat{\mathcal{L}}_q(t) W(\alpha^*, \alpha; t), \tag{25.33}$$

where the quantum Liouville operator is, similar to the classical one, the sum $\hat{\mathcal{L}}_q(t) = \hat{\mathcal{L}}_q^{(0)} + \hat{\mathcal{L}}^{(k)}(t)$. The only, but of primary importance, distinction from the classical case is appearance of a new second-derivative term in the rotation operator

$$\hat{\mathcal{L}}_q^{(0)} = \left(\omega_0 - \hbar - \frac{1}{2}\hbar^2 \frac{\partial^2}{\partial\alpha^*\partial\alpha} + 2|\alpha|^2\right)\left(\alpha^*\frac{\partial}{\partial\alpha^*} - \alpha\frac{\partial}{\partial\alpha}\right). \tag{25.34}$$

The combination $-\frac{1}{2}\hbar^2 \frac{\partial^2}{\partial\alpha^*\partial\alpha} + 2|\alpha|^2$ formally coincides with the quantum Hamiltonian of a 2D isotropic linear oscillator. As a result, the spectrum of the operator (25.34) becomes discrete, in contrast to its classical counterpart. This important fact is a manifestation of *the quantization of the phase space*.

Below, we will consider evolution of an initially isotropic and, generally, mixed state. If we choose the density matrix in the form

$$\hat{\rho}(0) = \frac{\hbar}{\Delta + \hbar} \sum_{n=0}^\infty \left(\frac{\Delta}{\Delta + \hbar}\right)^n |n\rangle\langle n|, \tag{25.35}$$

the corresponding Wigner function

$$W(\alpha^*, \alpha; 0) = \frac{1}{\Delta + \hbar/2} e^{-\frac{|\alpha|^2}{\Delta + \hbar/2}}, \tag{25.36}$$

is a Poissonian distribution with respect to the action variable $I = |\alpha|^2$. In particular, in the case $\Delta = 0$, this state turns into the pure ground state $\hat{\rho}(0) = |0\rangle\langle 0|$ whose Wigner function $W(\alpha^*, \alpha; 0) = \frac{2}{\hbar} e^{-\frac{2|\alpha|^2}{\hbar}}$ occupies the minimal quantum cell $\hbar/2$.

As compared with the classical Liouville equation, the quantum one

$$i\frac{\partial}{\partial t} W(\alpha^*, \alpha; t) = \hat{\mathcal{L}}_c W(\alpha^*, \alpha; t) - \frac{1}{2}\hbar^2 \frac{\partial^2}{\partial\alpha^*\partial\alpha} W(\alpha^*, \alpha; t); \tag{25.37}$$

contains an additional proportional to $\hbar^2$ term with higher (second) order derivative. Even being initially very small, this term can be neglected only because of the classical exponential instability, during a short time interval, which is called the Ehrenfest time $t_E = \tau_c \ln \frac{2\langle I\rangle t_E}{\hbar}$. After this interval, the quantum effects dominate and the difference between the two dynamics becomes crucial. Indeed, meshing of the Wigner function stops quite soon on account of the quantization of the phase space.

25.2.3 Stability and Reversibility versus Complexity of Quantum States

Relying upon the analogy mentioned above between the classical phase space distribution function on the one hand and the quantum Wigner function on the other, below, we will confront the typical features of the two corresponding "chaotic" dynamics. Quantization of the phase space implies a much simpler structure of the Wigner function than that of the corresponding classical distribution. The Liouvillian phase space approach that we use allows us to measure complexity of this function using the same tools that have been utilized in the classical case considered in Section 25.2.1.

25.2.3.1 Response to a Single Probe

In much the same manner as in Equations 25.22 through 25.26, the response of a quantum system to an instant probe at a moment of time t is measured with the help of the quantum Peres fidelity (25.11) defined by the relation

$$F(\sigma|t) = \frac{\text{Tr}\left[\hat{\rho}(t)\,\hat{\rho}(\sigma|t)\right]}{\mathbb{P}(t)} = \sum_{m=0}^{\infty} e^{-\frac{1}{2}\sigma^2 m^2}\mathcal{W}_m(t), \tag{25.38}$$

where the quantities

$$\mathcal{W}_m(t) = \frac{(2-\delta_{m0})}{\mathbb{P}(t)} \sum_{n=0}^{\infty}\left|\langle n+m|\hat{\rho}(t)|n\rangle\right|^2, \quad \sum_{m=0}^{\infty}\mathcal{W}_m(t) = 1, \tag{25.39}$$

are direct quantum analogs of the classical weights (25.23) of the θ-harmonics. Indeed, they can be expressed [43,44] in the terms of Fourier amplitudes of the Wigner function by the formulae:

$$\mathcal{W}_m(t) = \frac{(2-\delta_{m0})}{\mathbb{P}(t)}\,\hbar \int_0^{\infty} dI\left|W_m(I;t)\right|^2, \tag{25.40}$$

(see Figure 25.2).

$$\mathbb{P}(t) = \text{Tr}\left[\hat{\rho}^2(t)\right] = \sum_{m=-\infty}^{+\infty}\int_0^{\infty} dI|W_m(I;t)|^2 = \mathbb{P}(0). \tag{25.41}$$

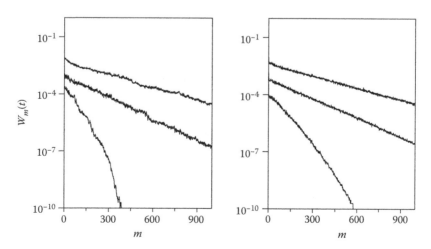

FIGURE 25.2 Distribution of harmonics $\mathcal{W}_m(t)$ as a function of m, at different times $t = 10, 30$, and 50 from bottom to top (these curves are scaled by a factor 0.01, 0.1, and 1, respectively). Left and right panels correspond to pure ($\Delta = 0$) and mixed ($\Delta = 25$) initial states, see Equation 25.35. Other parameters of simulations are $g_0 = 2$, $\hbar = 1$.

These relations differ from Equations 25.23 and 25.24 only by replacement $W_m^{(c)}(I;t) \Rightarrow W_m(I;t)$ of the Fourier amplitudes of the classical distribution function by those of the quantum Wigner function. It is worth noting that the phases of the off-diagonal matrix elements of the density matrix at the moment t fall out of the weights (25.39). Nevertheless such phases do play a certain, though not significant (see below), role during the preceding evolution.

25.2.3.2 Complexity of a Quantum State

Just as it has been in the case of a classical phase space distribution, the complexity of a quantum state can be characterized by the number $\langle |m| \rangle_t$ of θ-harmonics of the Wigner function. The corresponding weights are now given now by Equations 25.40 and 25.41 and again

$$\langle m^2 \rangle_t = \sum_{m=0}^{\infty} m^2 \, \mathcal{W}_m(t) = -\left.\frac{d^2 F(\sigma;t)}{d\sigma^2}\right|_{\sigma=0}. \tag{25.42}$$

It should be emphasized that the chosen measure is *equally valid* in the quantum as well as classical cases. This allows direct comparison of the main features of both dynamics. In particular, whereas the number of harmonics of the classical distribution function increases, due to the exponential instability, exponentially during the *whole* time of evolution, in the quantum case the exponential regime is, generally speaking, restricted to the Ehrenfest time interval. These statements are illustrated in the Figure 25.3 where the dependence of the quantity $\langle m^2 \rangle_t$ on the time is shown for a set of different values of the effective Plank's constant.

Plank's constant plays a twofold role here: on one part, it fixes the phase volume of the elementary quantum cell that is occupied in the phase space by the pure ground state $\hat{\rho}(0) = |0\rangle\langle0|$, and, on the other, it governs the dynamics via the evolution equation. In the Figure 25.3 we keep the size $1/2$ of the initial Wigner distribution (25.36) constant by choosing $\Delta = 1/2 - \hbar/2$. The initial state is pure ($\Delta = 0$) when $\hbar = 1$ (triangles) but becomes more and more mixed ($\Delta \approx 1/2 \gg \hbar$) in the cases $\hbar = 0.1$ (diamonds) and $\hbar = 0.01$ (squares) correspondingly. It is clearly seen that the smaller is the value of *dynamical* Plank's constant the longer the classical exponential regime lasts. In other words, the mixing of the initial state suppresses the quantum interference effect and restores the classical behavior.

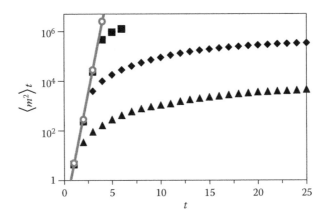

FIGURE 25.3 $\langle m^2 \rangle_t$ before and after the Ehrenfest time. Triangles, diamonds, and squares: $\hbar = 1$, 0.1, and 0.01, respectively. Classical dynamics is shown by empty gray circles, and the gray line presents an exponential fit. The initial phase area *holds constant* $1/2$ in all three cases. The kick strength used here is $g_0 = 1.5$.

25.2.3.3 Information Entropy versus von Neumann Entropy

The probabilistic meaning of the quantities $\mathcal{W}_m(t)$ allows us to introduce the Shennon or *information entropy*:

$$\mathcal{I}(t) = -\sum_{m=0}^{\infty} \mathcal{W}_m(t) \ln \mathcal{W}_m(t) \tag{25.43}$$

$$= \begin{cases} 0, & t = 0, \\ \ln\langle|m|\rangle_t + 1 - \dfrac{\ln 2}{2} + O(1/\langle|m|\rangle_t), & \langle|m|\rangle_t \gg 1. \end{cases} \tag{25.44}$$

This is another, though equivalent, possible way to characterize the complexity of a quantum state. Such a choice turns out to be even more convenient for our further purposes. This entropy starts from zero (the initial state has no harmonics but zero), increases linearly (with a slope defined by the classical characteristic time τ_c) during the Ehrefest time and then slows down to the quantum logarithmic regime. Since the number of harmonics at sufficiently large time weakly depends on the peculiar properties of the initial state, being practically the same for pure and mixed ones, this entropy is practically insensitive to quantum correlations.

On the contrary, the invariant von Neumann entropy

$$S(t) = -\mathrm{Tr}\left[\hat{\rho}(t) \ln \hat{\rho}(t)\right] = S(0), \tag{25.45}$$

is *perfectly* sensitive to quantum correlations (hence the name "correlational" [36]). This entropy does not depend on time as long as the evolution remains unitary and equals zero $S(t) = 0$ when the state is pure. The coherence and quantum correlations can be destroyed only in the presence of a persistent external noise or during the process of preparation of the initial state (see below).

25.2.3.4 Persistent Noise

The stationary noise is described in our model by the Hamiltonian operator

$$\hat{H}^{(noise)}(\hat{a}^\dagger, \hat{a}; t) = \hbar \hat{n} \sum_\tau \xi_\tau \delta(t - \tau), \quad \langle\xi_\tau\rangle = 0, \ \langle\xi_\tau\xi_{\tau'}\rangle = \sigma^2\delta_{\tau\tau'}. \tag{25.46}$$

As a result, the evolution operator takes the form

$$\hat{\mathcal{U}}(\{\xi\}; t) = \prod_{\tau=1}^{\tau=t} \left[e^{-i\xi_\tau\hat{n}} \hat{U}\right]. \tag{25.47}$$

This operator remains unitary for any fixed noise realization $\{\xi\}$ (history). Accordingly, a pure initial state remains pure during the whole time of evolution. At a running moment t, the excitation of the oscillator and the degree of anisotropy of the Wigner function are characterized by the probability distributions [45]

$$w_n(\xi; t) = \langle n|\hat{\rho}(\xi; t)|n\rangle, \tag{25.48}$$

and

$$\mathcal{W}_m(t) = \frac{(2 - \delta_{m0})}{\mathbb{P}(t)} \sum_{n=0}^{\infty} \left|\langle n + m|\hat{\rho}(\xi t)|n\rangle\right|^2, \quad \sum_{m=0}^{\infty} \mathcal{W}_m(t) = 1, \tag{25.49}$$

where the density matrix

$$\hat{\rho}(\xi; t) = \hat{\mathcal{U}}(\{\xi\}; t)\hat{\rho}(t = 0)\hat{\mathcal{U}}^\dagger(\{\xi\}; t), \tag{25.50}$$

is defined for some fixed noise realization $\{\xi\}$. The initial state can be chosen to be the ground one $\hat{\rho}(t = 0) = |0\rangle\langle 0|$, because after a few first kicks the distributions (25.48 and 25.49) acquire a practically general

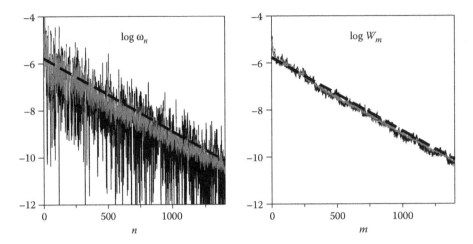

FIGURE 25.4 Probability distributions (25.48) and (25.49) at the moment $t = 80$ with no noise ($\sigma = 0$, thin black lines), weak noise ($\sigma = 0.001$, thin gray lines), and strong noise ($\sigma = 1$, thick dashed line). In the two latter cases, the distributions are averaged over 10^3 realizations. The initial state is pure ($\Delta = 0$, see Equation 25.35). Other parameters of simulations are $g_0 = 2$, $\hbar = 1$.

exponential form (see Figure 25.4). Averaging over noise realizations keeps the slopes of these distributions unchanged but kills the wild fluctuations around their regular exponential decay. In the limit of strong noise $\sigma \gg 1$ (see thick dashed lines in Figure 25.4) the noise acts as a "coarse graining" that reproduces the behavior of the corresponding classical distributions [45]. There exist "self-averaging" quantities such as mean excitation number $\langle n \rangle_t$ or the mean number $\langle |m| \rangle_t$ of θ-harmonics that do not depend, in fact, on the noise realization [45]

$$\langle n(\{\xi\}; t) \rangle = n(\sigma; t), \quad \langle m^2(\{\xi\}; t) \rangle = m^2(\sigma; t) = n(\sigma; t)\left(n(\sigma; t) + 1\right). \tag{25.51}$$

On the contrary, the quantum Peres fidelity defined as

$$F(\{\xi\}; t) = \frac{\text{Tr}\left[\hat{\rho}(t)\,\hat{\rho}(\{\xi\}; t)\right]}{\mathbb{P}(t)}, \tag{25.52}$$

is not a self-averaging quantity and wildly fluctuates from one noise history to another. Therefore, averaging over all possible noise realizations is necessary to obtain a reasonably simple and adequate measure:

$$F(\sigma; t) = \overline{F(\{\xi\}; t)}^{\{\xi\}} = \text{Tr}\left[\hat{\rho}(t)\,\hat{\rho}^{(av)}(\sigma; t)\right]. \tag{25.53}$$

This procedure brings into consideration the average density matrix $\rho^{(av)}(t)$ whose one step evolution is described by the transformation

$$\langle n' | \hat{\rho}^{(av)}(\sigma; \tau) | n \rangle = e^{-\frac{1}{2}\sigma^2(n'-n)^2}\,\langle n' | \hat{U}\hat{\rho}^{(av)}(\sigma; \tau - 1)\hat{U}^\dagger | n \rangle. \tag{25.54}$$

The noise suppresses off-diagonal matrix elements of the density matrix thus gradually cutting down the number of harmonics of the corresponding Wigner function. The evolution is not unitary anymore. The latter entails state mixing, loss of memory on the initial state, and suppression of the quantum interference. These effects show up in the behavior of the von Neumann entropy defined in terms of the averaged density matrix as

$$S(\sigma; t) = -\text{Tr}\left[\hat{\rho}^{(av)}(\sigma; t)\ln\hat{\rho}^{(av)}(\sigma; t)\right]. \tag{25.55}$$

Figure 25.5 illustrates this behavior in comparison with that of the information Shennon entropy $\mathcal{I}(t)$ (see Equation 25.43). The faster the entropy $S(\sigma; t)$ increases, the larger is the level of noise (the full lines from

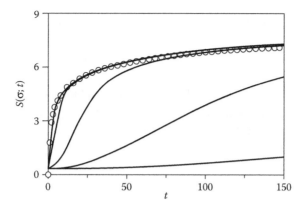

FIGURE 25.5 von Neumann entropy: $\sigma = (0.125, 1, 8, 64, 512) \cdot 10^{-3}$, solid lines from bottom to top. Circles: information entropy $\mathcal{I}(t)$; $S(\sigma; t > t_{(dec)}) \rightarrow \mathcal{I}(t)$. In these simulations, we use $g_0 = 2$, $\hbar = 1$.

bottom to top) and approaches the information Shennon $\mathcal{I}(\sigma; t)$ entropy (circles) from below. At some time $t_{(dec)}$, both entropies coincide and continue together. All coherent effects are washed away by this time. Therefore, the entropy (25.55) is suitable *for tracing the gradual loss of the quantum coherence.*

The decoherence time $t_{(dec)}$ can be estimated as [45]

$$t_{(dec)}(\sigma) \sim \sqrt{\frac{\hbar}{\sigma^2 D}}, \tag{25.56}$$

where D is the classical diffusion coefficient. Henceforth, the system occupies the maximal phase area accessible at the running value of excitation thus reaching a sort of equilibrium. Finally, the system expands "adiabatically," both entropies being almost constant. The further evolution turns out to be *Markovian* [45].

25.2.3.5 Time-Independent Perturbation: Mixed Initial State

As it has already been noted above, the quantum interference turns out to be somewhat suppressed if the evolution started from a mixed initial state. Below, we consider another kind of decoherence that takes place in the case of a time-independent perturbation

$$\hat{H}_V = \hat{H}^{(0)} + \varepsilon \hat{V}, \tag{25.57}$$

where the unperturbed Hamiltonian describes, as before, dynamics of the classically chaotic nonlinear oscillator. The perturbed and unperturbed motions are juxtaposed by means of the unitary Loschmidt echo operator $\hat{f}(t)$ [35]

$$\hat{f}(t) = \hat{U}^\dagger(t)\hat{U}_V(t), \tag{25.58}$$

$$\hat{f}(t) = T \exp\left[-i\frac{\varepsilon}{\hbar}\int_0^t d\tau \hat{\mathcal{V}}(\tau)\right], \tag{25.59}$$

$$\hat{\mathcal{V}}(\tau) = \hat{U}^\dagger(t)\,\hat{V}\hat{U}(t). \tag{25.60}$$

In spite of the fact that the perturbation $\hat{\mathcal{V}}(\tau)$ evolves chaotically, the quantum coherence *is in no way spoiled* as long as the initial state is pure. More than that, one might think that even if the initial state is an incoherent mixture, quantum coherence can be rapidly generated by producing complex off-diagonal matrix elements during dynamical evolution. Nevertheless, we will show below that if the system is classically chaotic and the evolution *starts from a wide incoherent mixed state*, then the initial incoherence persists due to the intrinsic classical chaos so that the quantum phases remain irrelevant [35].

In the case of a pure initial state $\hat{\rho}(0) = |\overset{\circ}{\psi}\rangle\langle\overset{\circ}{\psi}|$, the Peres fidelity (25.11) is simply the probability

$$F_{\overset{\circ}{\psi}}(t) = |\langle\overset{\circ}{\psi}|\hat{f}(t)|\overset{\circ}{\psi}\rangle|^2 = \left|\mathrm{Tr}\left[\hat{f}(t)\overset{\circ}{\rho}\right]\right|^2, \tag{25.61}$$

to survive in this state under influence of a chaotically evolving perturbation $\hat{V}(\tau)$ till the time t. When the evolution starts from a mixed state $\hat{\rho}(0) = \sum_k p_k|\overset{\circ}{\psi}_k\rangle\langle\overset{\circ}{\psi}_k|$, $\sum_k p_k = 1$, the expression (25.61) can be generalized in two different ways. The first of them leads to the standard definition (25.11) that can be rewritten identically in the form

$$F(t) = \frac{1}{\mathrm{Tr}\hat{\rho}^2(0)} \sum_{k,k'} p_k p_{k'} W_{kk'}(t), \tag{25.62}$$

where the quantities $W_{kk'}(t) = |\langle\overset{\circ}{\psi}_k|\hat{f}(t)|\overset{\circ}{\psi}_{k'}\rangle|^2$ are probabilities of transitions induced by the unitary transformation $\hat{f}(t)$. The influence of coherent effects is hidden in the dynamics of the complex matrix elements $f_{kk'}(t) = \langle\overset{\circ}{\psi}_k|\hat{f}(t)|\overset{\circ}{\psi}_{k'}\rangle$.

Another way of generalization is suggested by the experimental configuration with periodically kicked ion traps proposed in Reference 18. In such Ramsey-type interferometry experiments, one directly accesses the fidelity amplitudes (see Reference 18 rather than their square moduli). Motivated by this consideration, we will further consider the quantity

$$\mathcal{F}(t) = \left|\mathrm{Tr}\left[\hat{f}(t)\hat{\rho}(0)\right]\right|^2 = \left|\sum_k p_k f_k(t)\right|^2 = \sum_k p_k^2 F_k(t) + \sum_{k,k'}(1 - \delta_{kk'})p_k p_{k'} f_k(t) f_{k'}^*(t), \tag{25.63}$$

that is obtained by directly extending formula (25.61). Below we refer to this new quantity as *allegiance*. The first term on the rhs is the sum of fidelities $F_k = |f_k|^2 = |\langle\overset{\circ}{\psi}_k|\hat{f}|\overset{\circ}{\psi}_k\rangle|^2$ of the individual pure initial states with weights p_k^2, while the second, interference term, depends on the relative phases of fidelity amplitudes. If the number K of pure states $|\overset{\circ}{\psi}_k\rangle$ that form the initial mixed state is large, $K \gg 1$, so that $p_k \simeq 1/K$ for $k \leqslant K$ and zero otherwise, the first term is $\backsim 1/K$ at the initial moment $t = 0$ while the second term $\backsim 1$. Therefore, in the case of a wide mixture, the decay of the function $\mathcal{F}(t)$ is determined by the second sum of interfering contributions. Therefore, in contrast to the standard Peres fidelity (25.11 and 25.62), the allegiance $\mathcal{F}$ directly *accounts for the quantum interference* and can be expected to retain quantal features even in the deep semiclassical region. This is not, however, the case as will be shown below.

Analytical calculation of the pure-state fidelity $F_{\overset{\circ}{\alpha}}(t)$ for a pure coherent quantum state $|\overset{\circ}{\alpha}\rangle$ as well as the allegiance $\mathcal{F}(t)$ for an incoherent mixed state can be performed with the help of expressing both quantities in terms of Feynman's path integral in the oscillator's phase plane. A method of semiclassical evaluation of this integral has been worked out in Reference 38. Referring the reader to this chapter for all technical details, we present below the main results of the calculations [35].

These results are quite different from the two cases of our interest. If the initial state is a pure coherent one $|\overset{\circ}{\alpha}\rangle$, the fidelity amplitude is found to be

$$f_{\overset{\circ}{\alpha}}(t) = \frac{2}{\pi\hbar} \int d^2\delta\, e^{-\frac{2}{\hbar}|\delta|^2} \exp\left\{i\frac{\sigma}{2}\left[\tilde{\theta}_c(t) - \overset{\circ}{\theta}_c\right]\right\}, \tag{25.64}$$

where the phase $\tilde{\theta}_c(t) = \theta_c(\omega_0 - 2|\delta|^2; \overset{*}{\alpha} + \delta^*, \overset{\circ}{\alpha} + \delta; t) = \int_0^t d\tau[\omega_0 - 2|\delta|^2 + 2\tilde{I}_c(\tau)]$. It should be stressed that the fidelity $F_{\overset{\circ}{\alpha}}(t) = |f_{\overset{\circ}{\alpha}}(t)|^2$ does not decay in time if the quantum fluctuations described by the integral over δ in Equation 25.64 are neglected.

On the initial stage of the evolution $t \ll \frac{1}{\Lambda} \ln \frac{2}{\varepsilon}$, while the phases $\tilde{\theta}_c(t)$ are not yet perfectly randomized the fidelity $F_{\overset{\circ}{\alpha}}$ decays, because of classical exponential instability, super exponentially [25,34]:

$$F_{\overset{\circ}{\alpha}}(t) \approx \exp\left(-\frac{\varepsilon^2}{4\hbar}e^{\Lambda t}\right).$$

During this time, the contribution of the averaging over the initial Gaussian distribution in the classical $\overset{\circ}{\alpha}$ phase plane dominates while the influence of the quantum fluctuations of the linear frequency remains negligible. Such a decay has, basically, a classical nature [16] and Planck's constant appears only as the size of the initial distribution. For larger times, the quantum fluctuations reduce the fidelity decay to exponential law $F_{\overset{\circ}{\alpha}}(t) = \exp(-2\Lambda t)$.

The situation changes dramatically if the initial state is an incoherent mixture. More precisely, we consider a mixed initial state represented by a Glauber's diagonal expansion [19,20] $\hat{\rho}(0) = \int d^2\overset{\circ}{\alpha} \mathcal{P}(|\overset{\circ}{\alpha} - \overset{\circ}{\alpha}_c|^2)|\overset{\circ}{\alpha}\rangle\langle\overset{\circ}{\alpha}|$ with a wide positive definite weight function $\mathcal{P}$ which covers a large number of quantum cells. Note that here and in the following, we assume that the initial mixture is isotropically distributed in the phase plane around a fixed point $\overset{\circ}{\alpha}_c$, with the density $\mathcal{P}_{\overset{\circ}{\alpha}_c}(\overset{\circ}{\alpha}^*, \overset{\circ}{\alpha}) = \mathcal{P}(|\overset{\circ}{\alpha} - \overset{\circ}{\alpha}_c|^2)$. Then, allegiance equals [35] $\mathcal{F}(t; \overset{\circ}{\alpha}_c) = |f(t; \overset{\circ}{\alpha}_c)|^2$, where

$$\begin{aligned}
f(t; \overset{\circ}{\alpha}_c) &= \int d^2\overset{\circ}{\alpha} \, \mathcal{P}(|\overset{\circ}{\alpha} - \overset{\circ}{\alpha}_c|^2) f_{\overset{\circ}{\alpha}}(t) \\
&\approx \int d^2\overset{\circ}{\alpha} \, \mathcal{P}(|\overset{\circ}{\alpha} - \overset{\circ}{\alpha}_c|^2) \exp\left\{i\frac{\varepsilon}{2\hbar}\left[\theta_c(t) - \overset{\circ}{\theta}_c(0)\right]\right\}.
\end{aligned} \tag{25.65}$$

This formula directly relates the decay of a *quantum* quantity, the allegiance, to that of a correlation function of the *classical* phases. No quantum features are present on the rhs of Equation 25.65.

Summarizing, the decay pattern of the allegiance $\mathcal{F}(t)$ depends on the value of the parameter $\sigma = \varepsilon/\hbar$. In particular, for $\sigma \ll 1$, we recover the well-known Fermi Golden Rule (FGR) regime. Indeed, in this case, the cumulant expansion can be used, $\ln f(t; \overset{\circ}{\alpha}_c) = \sum_{\kappa=1}^{\infty} \frac{(i\sigma)^\kappa}{\kappa!}\chi_\kappa(t)$. All the cumulants are real, hence, only the even ones are significant. The lowest of them

$$\chi_2(t) = \int_0^t d\tau_1 \int_0^t d\tau_2 \langle[I_c(\tau_1) - \langle I_c(\tau_1)\rangle][I_c(\tau_2) - \langle I_c(\tau_2)\rangle]\rangle \equiv \int_0^t d\tau_1 \int_0^t d\tau_2 K_I(\tau_1, \tau_2), \tag{25.66}$$

is positive. Assuming that the classical autocorrelation function decays exponentially, $K_I(\tau_1, \tau_2) = \langle(\Delta I_c)^2\rangle \exp(-|\tau_1 - \tau_2|/\tau_I)$ with some characteristic time τ_I, we obtain $\chi_2(t) = 2\langle(\Delta I_c)^2\rangle\tau_I t = 2Kt$ for the times $t > \tau_I$ and arrive, finally, at the FGR decay law $\mathcal{F}(t; \overset{\circ}{\alpha}_c) = \exp(-2\sigma^2 Kt)$ [15,28,33]. Here, $K = \int_0^\infty d\tau K_I(\tau, 0) = \langle(\Delta I_c)^2\rangle\tau_I$.

The significance of the higher connected correlators $\chi_{\kappa \geq 4}(t)$ grows with the increase of the parameter σ. When this parameter roughly exceeds one, the cumulant expansion fails and the FGR approximation is no longer valid. In the regime $\sigma \gtrsim 1$, the decay rate of the function $\mathcal{F}(t; \overset{\circ}{\alpha}_c) = |f(t; \overset{\circ}{\alpha}_c)|^2$ ceases to depend on σ [41] and coincides with the decay rate $1/\tau_c$ of the classical correlation function (25.17)

$$\mathcal{F}(t; \overset{\circ}{\alpha}_c) = \exp(-t/\tau_c). \tag{25.67}$$

This rate is intimately related to the local instability of a chaotic classical motion though it is not necessarily given by the Lyapunov exponent Λ. Quantum interference *does not show up* at all.

25.2.4 Ballistic Electron Quantum Transport in the Presence of Weakly Disordered Background

Finally, we will discuss the decoherence phenomenon in the electron transport through an open ballistic quantum dot as this problem is seen from the point of view of the general resonance-scattering theory [40].

Peculiarities of this transport reflect the properties of eigenstates of the quantum billiards, whose spectra are highly nontrivial in the classically chaotic regimes.

The loss of coherence that is the main problem of our concern is attributed below to interaction with a weakly disordered many-body environment ("walls"). So, the whole system consists of an open cavity and walls and is described by the following non-Hermitian effective Hamiltonian:

$$\hat{\mathcal{H}} = \begin{pmatrix} \mathcal{H}^{(s)} & V^{\dagger} \\ V & H^{(e)} \end{pmatrix}. \tag{25.68}$$

The upper-left block stands for the non-Hermitian effective Hamiltonian of the irregularly shaped cavity (dot) with two similar leads supporting each $M/2$ equivalent channels

$$\mathcal{H}^{(s)} = H^{(s)} - \frac{i}{2} A A^{\dagger}. \tag{25.69}$$

This non-Hermitian effective Hamiltonian describes a set of $N^{(s)}$ electron doorway resonance states with complex eigenenergies $\mathcal{E}_n = E_n - \frac{i}{2}\Gamma_n$ separated by mean level spacing D. The Hermitian matrix $H^{(e)}$ represents the environment with a very dense discrete spectrum of $N^{(e)}$ ($>>> N^{(s)}$) real energy levels ϵ_e (mean level spacing $\delta <<< D$). These states get excess to the continuum only due to the coupling V to the doorway states in the cavity.

We further exploit the single-particle approximation in the environmental sector: $H^{(e)} \Rightarrow H_{sp}^{(e)}$. The mean level spacing of a quasi-electron $d \propto 1/N_{sp}^{(e)}$ is much greater than the many-body spacing δ but still much smaller than the doorway spacing, $\delta \ll d \ll D$. The interaction V with irregular impurities is described by a rectangular $N_{sp}^{(e)} \times N^{(s)}$ matrix with random-matrix elements

$$\langle V_{\nu n} \rangle = 0, \quad \langle V_{\mu m}^* V_{\nu n} \rangle = \frac{1}{2} \Gamma_s \frac{d}{\pi} \delta_{\mu\nu} \delta_{mn}. \tag{25.70}$$

The second relation defines the spreading width

$$\Gamma_s = 2\pi \frac{\langle |V|^2 \rangle}{d}, \tag{25.71}$$

that satisfies the condition $\Gamma_s \gg d$, so that the influence of the disorder lies beyond validity of the standard perturbation theory.

The unitary $M \times M$ scattering matrix has the form

$$S(E) = I - iT(E) = I - iA^{\dagger} \mathcal{G}_D(E) A. \tag{25.72}$$

Therefore, the evolution of a scattered electron inside the cavity is described by the ($N^{(s)} \times N^{(s)}$) doorway resolvent (doorway propagator)

$$\mathcal{G}_D(E) = \frac{I}{E - \mathcal{H}^{(s)} - \Sigma(E)}. \tag{25.73}$$

Here, the Hermitian $N^{(s)} \times N^{(s)}$-matrix

$$\Sigma(E) = V^{\dagger} \frac{I}{E - H_{sp}^{(e)}} V, \tag{25.74}$$

accounts for transitions cavity $\leftrightarrow$ environment. Being averaged over the random coupling amplitudes V, this matrix is, with accuracy $1/N_{sp}^{(e)}$, diagonal and is proportional to the trace in the single-quasi-particle space,

$$\Sigma(E) \Rightarrow \frac{1}{2} \Gamma_s g(E); \quad g(E) = \frac{d}{\pi} Tr \frac{1}{E - H_{sp}^{(e)}}. \tag{25.75}$$

As a result, a given doorway resonance is fragmented in a large number of narrow resonances whose complex energies are found by solving the equation

$$\mathcal{E}_\nu^n - \mathcal{E}_n - \frac{1}{2}\Gamma_s g(\mathcal{E}_\nu^n) = 0, \tag{25.76}$$

so that Γ_s/d fine-structure resonances originate from any given doorway state.

The transition matrix now transforms into

$$T^{ab}(E) = \sum_n \frac{\mathcal{A}_n^a \mathcal{A}_n^b}{E - \mathcal{E}_n - \frac{1}{2}\Gamma_s g(E)} = \sum_\nu \frac{\tilde{\mathcal{A}}_\nu^a \tilde{\mathcal{A}}_\nu^b}{E - \mathcal{E}_\nu}. \tag{25.77}$$

The resulting transition amplitudes are now sums of interfering contributions of all narrow fine-structure resonances. The new pole residues are complex and, therefore, interfere! *No loss of coherence on this stage!*

In fact, however, the spectrum of the fine-structure resonances are extremely dense so that this structure cannot be experimentally resolved. In fact, only cross sections averaged over some energy interval $d \ll \Delta E \ll D$ are observed

$$\overline{\sigma^{ab}(E)} = \frac{1}{\Delta E} \int_{E-\frac{1}{2}\Delta E}^{E+\frac{1}{2}\Delta E} dE' \left|T^{ab}(E')\right|^2. \tag{25.78}$$

To carry out the energy averaging explicitly, we neglect the level fluctuations on the fine-structure scale and assume the uniform spectrum, $\varepsilon_\mu = \mu d$ (*the picket fence approximation*). This yields immediately $g(E) = \cot\left(\frac{\pi E}{d}\right)$.

25.2.4.1 Isolated Doorway Resonance Near the Fermi Energy

If the incoming electron with energy $E \approx E_{res}$ excites an isolated ($\Gamma = \sum_c \Gamma^c \ll D$) resonance state with energy $E_{res} \approx 0$ very close to the Fermi surface in the environment, the transition cross section equals

$$\sigma^{ab}(E) = \left|T^{ab}(E)\right|^2 = \frac{\Gamma^a \Gamma^b}{\left[E - \frac{1}{2}\Gamma_s \cot\left(\frac{\pi E}{d}\right)\right]^2 + \frac{1}{4}\Gamma^2}, \tag{25.79}$$

and the fine-scale energy averaging yields

$$\overline{\sigma^{ab}(E)} = \frac{\Gamma^a \Gamma^b}{E^2 + \frac{1}{4}\left(\Gamma + \Gamma_s\right)^2} + \frac{\Gamma^a \Gamma^b}{\Gamma} \frac{\Gamma_s}{E^2 + \frac{1}{4}\left(\Gamma + \Gamma_s\right)^2}. \tag{25.80}$$

The averaging has destroyed the coherence and decomposed the cross section on sum of two incoherent contributions. The first of them corresponds to excitation and subsequent decay via one of the outer channels of the doorway resonance widened because of leaking into the environment. The latter effect is described by additional shift in the upper part of the complex energy plane by the distance $\frac{1}{2}\Gamma_s$. The second term accounts for the particle reinjected from the background. There is no net loss of the electrons. The environment looks from outside like a black box which swallows a particle and spits it back into the cavity after some time. This time is characterized by the mean Wigner time delay that also consists of two contributions

$$\overline{\tau_W}(E) = \frac{\Gamma + \Gamma_s}{E^2 + \frac{1}{4}(\Gamma + \Gamma_s)^2} + \frac{2\pi}{d}. \tag{25.81}$$

The first term describes the delay on the resonance level inside the dot damped by the internal "friction" due to the environment when the second one accounts for the electron delayed in the environment by the time $\tau_d = \frac{2\pi}{d}$ proportional to the quasi-particle's level density.

The conductivity of the dot is proportional to the transport cross section

$$G(E) = \sum_{a \in 1, b \in 2} \overline{\sigma^{ab}(E)} = \frac{\Gamma_1 \Gamma_2}{\Lambda(E)} + \frac{\Gamma_1 \Gamma_2}{\Gamma_1 + \Gamma_2} \frac{\Gamma_s}{\Lambda(E)} \qquad (25.82)$$

$$= T_{12} + \frac{T_{1s} T_{s2}}{T_{1s} + T_{s2}}, \qquad (25.83)$$

where $\Gamma_k = \sum_{c \in k} \Gamma^c$ $k = 1, 2$, $\Gamma_1 + \Gamma_2 = \Gamma$; and

$$T_{sk}(E) = \frac{\Gamma_s \Gamma_k}{\Lambda(E)}, \quad \Lambda(E) = E^2 + \frac{1}{4}(\Gamma + \Gamma_s)^2. \qquad (25.84)$$

The term T_{12} describes transition from the first to the second lead via the broadened intermediate doorway resonance when the additional contribution incorporates the interchanges with the environment. The latter can be naturally interpreted by introducing an additional fictitious $(M+1)$th channel that connects the resonance state with the environment. The corresponding extended scattering matrix remains unitary.

The found expression is formally identical to that obtained within the framework of Büttiker's voltage-probe model [5,6] of the decoherence phenomenon. The corresponding dimensionless decoherence rate turns out to be equal in our case to $\gamma_s = \frac{2\pi}{D}\Gamma_s = \Gamma_s \tau_D$.

The single-particle approximation used up to now is well justified only when the scattering energy E is very close to the Fermi surface in the environment. For higher scattering energies, many-body effects should be taken into account. They show up, in particular, in a finite lifetime of the quasi-electron with the energy $E > E_F = 0$. The simplest way to account for this effect is to attribute some imaginary part to the quasi-particle's energy, $\varepsilon_\mu = \mu d - \frac{i}{2}\Gamma_e$. The resonant denominator then looks as [42]

$$\mathcal{D}_{res}(E) = E - E_{res} - \frac{1}{2}\Gamma_s(1 - \xi^2)\frac{\eta}{1 + \xi^2\eta^2} + \frac{i}{2}\left(\Gamma + \Gamma_s \xi \frac{1 + \eta^2}{1 + \xi^2\eta^2}\right), \qquad (25.85)$$

where E_{res} is the position of the doorway resonance and the following notations have been used:

$$\xi = \tanh\left(\frac{\pi\Gamma_e}{2d}\right), \quad \eta = \cot\left(\frac{\pi E}{d}\right).$$

The transport cross section $G(E)$ still retains its form (25.82) but the subsidiary transition probabilities look as

$$T_{sk}(E) \Rightarrow T_{sk}(E; \kappa) = \frac{\Gamma_s \Gamma_k}{\Lambda(E; \kappa)}, \qquad (25.86)$$

instead of Equation 25.84. The factor

$$\frac{1}{\Lambda(E; \kappa)} = \frac{1}{\Lambda(E)} \frac{1}{1 + \kappa \frac{\Lambda(E)}{\Gamma\Gamma_s}}, \qquad (25.87)$$

depends on the new parameter κ which accounts for inelastic effects in the background

$$\kappa = \frac{4\xi}{(1 - \xi)^2} = e^{\gamma_e} - 1$$
$$\approx \begin{cases} \gamma_e \ll 1, & \text{if } \tau_e \gg \tau_d, \\ e^{\gamma_e} \gg 1, & \text{if } \tau_e < \tau_d, \end{cases} \quad (\gamma_e = \tau_d \Gamma_e) \qquad (25.88)$$

where $\tau_e = 1/\Gamma_e$ is the lifetime of the quasi-electron in the environment.

Strictly speaking, the assumed quasi-electron decay, that implies infinite density of the final states in the background, seems to destroy the unitarity of the scattering matrix in contradiction with what has been stated before. In fact, a single-particle state once excited in the environment with a very dense but, nevertheless, discrete spectrum evolves after that quite similar to a quasi-stationary state till the time $2\pi/\delta \gg \tau_e = 1/\Gamma_e$. After this time, recovery of the initial nonstationary state begins. An electron preserves to a certain extent its individuality in the environment. It can only lose, because of the many-body effects, a part of its energy but inevitably returns sooner or later in the cavity and escapes finally via one of the outer channels. A good probability there exists for the electron to be reemitted in the cavity with some intermediate energy $E_{out} < E_{in} \approx E_{res}$ within the much shorter time interval $\tau_d = \frac{2\pi}{d}$. The portion of energy lost by such a retarded electron dissipates inside the environment. As a result, the background temperature jumps slightly up during each act of the scattering. However, supposing that the environment system is bulky enough, we can disregard this very slow increase of the environment temperature. Alternatively, we can suppose that a special cooling technique is in use.

Near the doorway resonance energy E_{res}, the influence of the finite lifetime effects is negligible within the range $0 \leqslant \kappa \leqslant \kappa_c = \frac{4\Gamma\Gamma_s}{(\Gamma+\Gamma_s)^2}$. The critical value κ_c reaches its maximum possible, $\kappa_c = 1$, when $\Gamma = \Gamma_s$ and becomes small if one out of the two widths noticeably exceeds another. In these cases, the interval of weak absorption is very restricted and the absorption begins to play an important role. If the resonance is so narrow that $\Gamma \ll \Gamma_s$, then $\kappa_c \approx 4\frac{\Gamma}{\Gamma_s} \ll 1$ and the subsidiary probabilities (25.86) at the resonance energy $E = E_{res}$ and $\kappa \gtrsim \kappa_c$ are small, $T_{sk}(E = E_{res}; \kappa) \approx \frac{16}{\kappa}\frac{\Gamma\Gamma_k}{\Gamma_s^2} \lesssim \frac{16}{\kappa_c}\frac{\Gamma\Gamma_k}{\Gamma_s^2} \approx 4\frac{\Gamma_k}{\Gamma_s} \ll 1$. On the other hand, the very quasi-particle concept is self-consistent only if $\gamma_e = \tau_d\Gamma_e \lesssim 1$ so that the physically feasible interval of the strong absorption regime is $\kappa_c \approx 4\frac{\Gamma}{\Gamma_s} \lesssim \kappa \lesssim 1$. In this interval, only the contribution $T_{12}(E)$ remains in Equation 25.82 and our approach reproduces the result of Efetov's homogeneous imaginary potential model [17] with the strength of this potential $-\frac{i}{2}\gamma_s$ and, correspondingly, the decoherence rate γ_s.

25.2.4.2 Overlapping Doorway Resonances

A number of overlapping doorway states can be excited if the incoming electron energy E_{in} appreciably exceeds the Fermi energy. As before, the cross sections averaged over the fine-structure scale consist of incoherent contributions that directly scattered and penetrated into the environment and then reemitted particles

$$\overline{\sigma^{ab}(E)} = \sigma_d^{ab}(E) + \sigma_r^{ab}(E), \tag{25.89}$$

where the direct and reemitted contributions are

$$\sigma_d^{ab}(E) = \left| \sum_n \frac{A_n^a A_n^b}{\mathcal{D}_n(E)} \right|^2, \tag{25.90}$$

$$\sigma_r^{ab}(E) = \Gamma_s \int_0^\infty dt_r \, \sigma_r^{ab}(E; t_r), \tag{25.91}$$

$$\sigma_r^{ab}(E; t_r) = \left| \sum_n \frac{A_n^a A_n^b}{\mathcal{D}_n(E)} e^{-i\mathcal{E}_n t_r} \right|^2, \tag{25.92}$$

$$\mathcal{D}_n(E) = E - E_n + \frac{i}{2}\left(\Gamma_n + \Gamma_s\right). \tag{25.93}$$

The particles delayed within the environment for different times contribute incoherently.

Since the electron motion in the cavity is supposed to be classically chaotic, the ensemble averaging $\langle ... \rangle$ in the doorway sector is appropriate. It is easy to see that, as long as the inelastic effects in the background are fully neglected, such an averaging perfectly eliminates dependence of all mean cross sections on the spreading width. Indeed, the ensemble-averaged cross section is expressed in terms of the S-matrix two-point correlation function $C_V^{ab}(\varepsilon) = C_0^{ab}(\varepsilon - i\Gamma_s)$ as

$$\langle \sigma_d^{ab}(E) \rangle = C_V^{ab}(0) = C_0^{ab}(-i\Gamma_s) = \int_0^\infty dt \, e^{-\Gamma_s t} K_0^{ab}(t). \tag{25.94}$$

The subscript V indicates the coupling to the background and the function $K_0^{ab}(t)$ is the Fourier transform of the correlation function $C_0^{ab}(\varepsilon)$. On the other hand, it is easy to show that

$$\langle \sigma_r^{ab}(E; t_r)\rangle = \int_0^\infty dt\, e^{-\Gamma_s t} K_0^{ab}(t + t_r). \tag{25.95}$$

Therefore, finally,

$$\begin{aligned}
\overline{\langle \sigma^{ab}(E)\rangle} &= \int_0^\infty dt\, e^{-\Gamma_s t} K_0^{ab}(t) + \Gamma_s \int_0^\infty dt_r \int_0^\infty dt\, e^{-\Gamma_s t} K_0^{ab}(t + t_r) \\
&= \int_0^\infty dt\, K_0^{ab}(t) = \langle \sigma_0^{ab}(E)\rangle.
\end{aligned} \tag{25.96}$$

The ensemble averaging, being in fact equivalent to the energy averaging over the doorway scale D, suppresses all interference effects that save the elastic enhancement because of the time reversal symmetry. The latter effect manifests itself in the weak localization phenomenon. The time reversal symmetry is violated owing to the energy absorption in the environment.

25.2.4.3 Energy Absorption and Suppression of the Weak Localization

Taking into account Equation 25.96, we rewrite the ensemble-averaged cross sections as $\overline{\langle \sigma^{ab}(E)\rangle} = \langle \sigma_0^{ab}(E)\rangle + \Delta\sigma^{ab}(E; \kappa)$, where the second contribution can be reduced [40] to the following compact expression:

$$\Delta\sigma^{ab}(E; \kappa) = -\sqrt{\frac{\kappa\Gamma_s}{4}} \int_0^\infty \frac{dt}{\sqrt{-\frac{d}{dt} + \frac{\kappa}{4\Gamma_s}\left(\frac{d}{dt} - \Gamma_s\right)^2}} K_0^{ab}(t). \tag{25.97}$$

Being presented in such a form, this result is equally valid for both the orthogonal (GOE, time reversal symmetry) and the unitary (GUE, no time reversal symmetry) cases.

To simplify further calculation, we will consider the case of an appreciably large number $M \gg 1$ of statistically equivalent scattering channels, all of them with the maximal transmission coefficient $T = 1$. Then, the channel indices a, b can be dropped. Independently of time reversal symmetry, the function $K_0(t)$ is real, positive definite, monotonously decreases with time t, and satisfies the conditions $K_0(t < 0) = 0$, $K_0(0) = 1$. This allows us to represent this function in the form of the mean-weighted decay exponent [39]:

$$K_0(t) = \int_0^\infty d\Gamma\, e^{-\Gamma t} w(\Gamma), \quad \int_0^\infty d\Gamma\, w(\Gamma) = K_0(0) = 1. \tag{25.98}$$

Rigorously speaking, the weight functions $w(\Gamma)$ have different forms before $(t < \tau_D)$ and after $(t > \tau_D)$ the Heisenberg time τ_D. However, contribution of the latter interval is as small as e^{-M} [39]. Neglecting such a contribution, we obtain in any inelastic channel

$$\Delta\sigma(E; \kappa) = -\sqrt{\frac{\kappa\Gamma_s}{4}} \int_0^\infty d\Gamma\, \frac{w(\Gamma)}{\Gamma} \frac{1}{\sqrt{\Gamma + \frac{\kappa}{4\Gamma_s}(\Gamma + \Gamma_s)^2}}. \tag{25.99}$$

In the strong absorption limit $\kappa \gg \frac{4\Gamma_s\Gamma_W}{(\Gamma_s + \Gamma_W)^2}$, the parameter κ disappears from the found expression and the latter reduces to

$$\Delta\sigma(E; \kappa) \Rightarrow -\Gamma_s \int_0^\infty d\Gamma\, \frac{w(\Gamma)}{\Gamma(\Gamma + \Gamma_s)} = -\Gamma_s \int_0^\infty dt_r \int_0^\infty dt\, e^{-\Gamma_s t} K_0^{ab}(t + t_r). \tag{25.100}$$

According to Equations 25.94 and 25.96, this brings us to the result

$$\overline{\langle \sigma(E)\rangle} = \int_0^\infty dt\, e^{-\Gamma_s t} K_0(t) = \int_0^\infty d\Gamma\, \frac{w(\Gamma)}{\Gamma + \Gamma_s} = \langle \sigma_d(E)\rangle. \tag{25.101}$$

The averaged cross section approaches in this limit the value $1/\Gamma_s$ *independently of the symmetry class* when the spreading width Γ_s noticeably exceeds the typical widths contributing to the integral over Γ.

In the opposite limit of weak absorption $\kappa \ll \frac{4\Gamma_s\Gamma_W}{(\Gamma_s+\Gamma_W)^2}$, the expression (25.99) reads

$$\Delta\sigma(E;\kappa) \Rightarrow -\sqrt{\frac{\kappa\Gamma_s}{4}} \int_0^\infty d\Gamma\, \frac{w(\Gamma)}{\Gamma^{\frac{3}{2}}}, \tag{25.102}$$

so that

$$\overline{\langle\sigma(E)\rangle} = \int_0^\infty d\Gamma\, \frac{w(\Gamma)}{\Gamma}\left(1-\sqrt{\frac{\kappa\Gamma_s}{4\Gamma}}\right). \tag{25.103}$$

For the case of time reversal symmetry (GOE), the asymptotic expansion [46] of the two-point correlation function gives [35]

$$w^{(GOE)}(\Gamma) = \delta(\Gamma-\Gamma_W) - \frac{2}{t_H}\delta'(\Gamma-\Gamma_W) + \frac{M}{2t_H^2}\delta''(\Gamma-\Gamma_W) + \cdots \tag{25.104}$$

whereas for the case of absence of such a symmetry (GUE), a similar expansion results in

$$w^{(GUE)}(\Gamma) = \delta(\Gamma-\Gamma_W) + \cdots. \tag{25.105}$$

In both cases, contributions of the omitted terms are estimated as $O(1/(M^{-7/2}))$. With such an accuracy, the formula (25.99) yields for the weak localization the expression

$$\Delta G \equiv G^{(GUE)} - G^{(GOE)}$$
$$= M_1 M_2 \left(2\frac{d}{d\mu} + \frac{\mu}{2}\frac{d^2}{d\mu^2}\right)\left\{\frac{1}{\mu}\left[1-\frac{\sqrt{\frac{\kappa\gamma_s}{4}}}{\sqrt{\mu+\frac{\kappa}{4\gamma_s}(\mu+\gamma_s)^2}}\right]\right\}\Bigg|_{\mu=M}, \tag{25.106}$$

which is valid for arbitrary values of the parameters κ, γ_s, and M. The unfolded explicit expression is a bit too lengthy. Visualization of this result is presented in Figure 25.6 for two different values of the (dimensionless) spreading width γ_s and $M_1 = M_2 = 2$; $M = 4$. The reduction of the difference ΔG displays suppression of the quantum coherence. Note that the effect becomes more pronounced as the number of channels decreases.

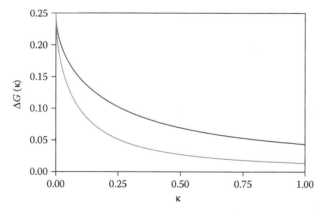

FIGURE 25.6 Weak localization versus absorption parameter κ. Lines correspond to $\gamma_s = 25$ (top) and $\gamma_s = 64$ (bottom); $M = 4$.

Acknowledgments

We greatly appreciate financial support from the federal program "Personnel of the Innovative Russia" (Grant 14.740.11.0082) as well as countenance by the RAS Joint Scientific Program "Nonlinear Dynamics and Solitons."

References

1. D.I.H. Abarbanel and J.D. Crawford. Strong coupling expansions for nonintegrable Hamiltonian systems. *Physica D: Nonlinear Phenomena*, 5(2–3):307–321, 1982.
2. Y. Alhassid. The statistical theory of quantum dots. *Reviews of Modern Physics*, 72(4):895–968, 2000.
3. G. Agarwal and E. Wolf. Calculus for functions of noncommuting operators and general phase-space methods in quantum mechanics. I. Mapping theorems and ordering of functions of noncommuting operators. *Physical Review D*, 2(10):2161–2186, 1970.
4. G. Agarwal and E. Wolf. Calculus for functions of noncommuting operators and general phase-space methods in quantum mechanics. II. Quantum mechanics in phase space. *Physical Review D*, 2(10):2187–2205, 1970.
5. M. Büttiker. Role of quantum coherence in series resistors. *Physical Review B*, 33(5):3020–3026, 1986.
6. M. Büttiker. Coherent and sequential tunneling in series barriers. *IBM Journal of Research and Development*, 32:63–75, 1988.
7. G. Benenti, G.G. Carlo, and T. Prosen. Wigner separability entropy and complexity of quantum dynamics. *Physical Review E*, 85(5):051129, 2012.
8. G. Benenti, G. Casati, and G. Veble. Decay of the classical Loschmidt echo in integrable systems. *Physical Review E*, 68(3):036212, 2003.
9. G. Benenti, G. Casati, and G. Veble. Stability of classical chaotic motion under a systems perturbations. *Physical Review E*, 67(5):055202, 2003.
10. C. Beenakker. Random-matrix theory of quantum transport. *Reviews of Modern Physics*, 69(3):731–808, 1997.
11. H. Baranger and P. Mello. Mesoscopic transport through chaotic cavities: A random S-matrix theory approach. *Physical Review Letters*, 73(1):142–145, 1994.
12. C.W.J. Beenakker and B. Michaelis. Stub model for dephasing in a quantum dot. *Journal of Physics A: Mathematical and General*, 38(49):10639–10646, 2005.
13. E.B. Bogomolny. Smoothed wave functions of chaotic quantum systems. *Physica D: Nonlinear Phenomena*, 31(2):169–189, 1988.
14. B.V. Chirikov. A universal instability of many-dimensional oscillator systems. *Physics Reports*, 52:263–379, 1979.
15. N. Cerruti and S. Tomsovic. Sensitivity of wave field evolution and manifold stability in chaotic systems. *Physical Review Letters*, 88(5):054103, 2002.
16. B. Eckhardt. Echoes in classical dynamical systems. *Journal of Physics A: Mathematical and General*, 36(2):371–380, 2003.
17. K. Efetov. Temperature effects in quantum dots in the regime of chaotic dynamics. *Physical Review Letters*, 74(12):2299–2302, 1995.
18. S. Gardiner, J. Cirac, and P. Zoller. Quantum chaos in an ion trap: The delta-kicked harmonic oscillator. *Physical Review Letters*, 79(24):4790–4793, 1997.
19. R. Glauber. Coherent and incoherent states of the radiation field. *Physical Review*, 131(6):2766–2788, 1963.
20. R. Glauber. Photon correlations. *Physical Review Letters*, 10(3):84–86, 1963.

21. M.C. Gutzwiller. *Chaos in Classical and Quantum Mechanics (Interdisciplinary Applied Mathematics, vol. 1)*. Springer-Verlag, New York, 1990.

22. F. Haake. *Quantum Signatures of Chaos (Springer Series in Synergetics, vol 54)*. Springer-Verlag, Berlin, 1991.

23. E. Heller. Bound-state eigenfunctions of classically chaotic Hamiltonian systems: Scars of periodic orbits. *Physical Review Letters*, 53(16):1515–1518, 1984.

24. K.S. Ikeda. Time irreversibility of classically chaotic quantum dynamics. In G. Casati and B. Chirikov, eds, *Quantum Chaos: Between Order and Disorder*, pp. 147–155. Cambridge University Press, Cambridge, 1995.

25. A. Iomin. Loschmidt echo for a chaotic oscillator. *Physical Review E*, 70(2):026206, 2004.

26. F.M. Izrailev. Simple models of quantum chaos: Spectrum and eigenfunctions. *Physics Reports*, 196(5–6):299–392, 1990.

27. R. Jalabert, H. Baranger, and A. Stone. Conductance fluctuations in the ballistic regime: A probe of quantum chaos? *Physical Review Letters*, 65(19):2442–2445, 1990.

28. Ph. Jacquod, P.G. Silvestrov, and C.W.J. Beenakker. Golden rule decay versus Lyapunov decay of the quantum Loschmidt echo. *Physical Review E*, 64(5):055203(R), 2001.

29. L. Mehta. *Random Matrices*. Academic Press, New York, 1991.

30. P.A. Mello, P. Pereyra, and T.H. Seligman. Information theory and statistical nuclear reactions. I. General theory and applications to few-channel problems. *Annals of Physics*, 161(2):254–275, 1985.

31. C. Marcus, A. Rimberg, R. Westervelt, P. Hopkins, and A. Gossard. Conductance fluctuations and chaotic scattering in ballistic microstructures. *Physical Review Letters*, 69(3):506–509, 1992.

32. A. Peres. Stability of quantum motion in chaotic and regular systems. *Physical Review A*, 30(4):1610–1615, 1984.

33. T. Prosen and M. Znidaric. Stability of quantum motion and correlation decay. *Journal of Physics A: Mathematical and General*, 35(6):1455–1481, 2002.

34. P.G. Silvestrov and C.W.J. Beenakker. Ehrenfest times for classically chaotic systems. *Physics Review E*, 65:035208(R), 2002.

35. V. Sokolov, G. Benenti, and G. Casati. Quantum dephasing and decay of classical correlation functions in chaotic systems. *Physical Review E*, 75(2):026213, 2007.

36. V. Sokolov, B. Brown, and V. Zelevinsky. Invariant correlational entropy and complexity of quantum states. *Physical Review E*, 58(1):56–68, 1998.

37. V.V. Sokolov. Moments of the distribution function and kinetic equation for stochastic motion of a nonlinear oscillator. *Theoretical and Mathematical Physics*, 59(1):396–403, 1984.

38. V.V. Sokolov. On the nature of the quantum corrections in the case of stochastic motion of a nolinear oscillator. *Theoretical and Mathematical Physics*, 61(1):1041–1048, 1984.

39. V.V. Sokolov. Decay rates statistics of unstable classically chaotic systems. In *Nuclei and Mesoscopic Physics: Proceedings International Workshop on Nuclei and Mesoscopic Physics WNMP 2007*, vol. 995 of *AIP Conference Proceedings*, pp. 85–91. American Institute of Physics, 2008. East Lansing, Michigan, October 20–22, 2007.

40. V.V. Sokolov. Ballistic electron quantum transport in the presence of a disordered background. *Journal of Physics A: Mathematical and Theoretical*, 43(26):265102, 2010.

41. R.Z. Sagdeev, D.A. Usikov, and G.M. Zaslavsky. *Nonlinear Physics: From Pendulum to Turbulence and Chaos*, vol. 4. Harwood Academic, Chur, Switzerland, 1988.

42. V. Sokolov and V. Zelevinsky. Simple mode on a highly excited background: Collective strength and damping in the continuum. *Physical Review C*, 56(1):311–323, 1997.

43. V.V. Sokolov and O.V. Zhirov. How well a chaotic quantum system can retain memory of its initial state? *Europhysics Letters*, 84(3):30001, 2008.

44. V. Sokolov, O. Zhirov, G. Benenti, and G. Casati. Complexity of quantum states and reversibility of quantum motion. *Physical Review E*, 78(4):046212, 2008.

45. V.V. Sokolov, O.V. Zhirov, and Y.A. Kharkov. Quantum dynamics against a noisy background. *Europhysics Letters*, 88(6):60002, 2009.

46. J.J.M. Verbaarschot, H.A. Weidenmller, and M.R. Zirnbauer. Grassmann integration in stochastic quantum physics: The case of compound-nucleus scattering. *Physics Reports*, 129(6):367–438, 1985.

47. G. Zaslavsky. *Chaos in Dynamic Systems*. Harwood Academic Publishers, New York, 1985.

26

Application of Microwave Networks to Simulation of Quantum Graphs

Michał Ławniczak,
Szymon Bauch, and
Leszek Sirko

26.1 Introduction

Quantum graphs of connected one-dimensional wires were introduced almost 80 years ago by Pauling [1]. The idea of quantum graphs was used later by Kuhn [2] in order to describe organic molecules with free electron models. Quantum graphs are often used as idealizations of physical networks in the limit where the widths of the wires are much smaller than their lengths, assuming that the propagating waves remain in a single transversal mode. Quantum graphs were successfully applied to model a variety of physical problems, e.g., electromagnetic optical waveguides [3,4], quantum wires [5,6], mesoscopic systems [7,8], excitation of fractons in fractal structures [9,10], and isoscattering systems [11] (see also Reference 12 and references cited therein). Quantum graphs can also be realized experimentally. Recent developments in various epitaxy techniques also allowed for the fabrication and design of quantum nanowire networks [13,14]. The statistical properties of spectra of quantum graphs were studied in the series of theoretical papers by Kottos and Smilansky [15–17]. They have shown that quantum graphs are excellent paradigms of quantum chaos. Their findings have been confirmed in numerous theoretical investigations of this topic [18–27] and in the experiments with microwave networks simulating quantum graphs [28–32].

    The aim of this chapter is to demonstrate that using a simple experimental setup consisting of microwave networks, one may successfully simulate quantum graphs. The bidirectional microwave networks (circuits) are constructed of connected coaxial cables (annular waveguides). In order to construct directed microwave networks, it is sufficient to add the microwave circulators or Faraday isolators into the circuits.

We discuss the application of microwave networks to the investigation of periodic orbits on the graphs, spectral statistics of the graphs [28], and the distributions of the reflection coefficient R and Wigner reaction matrix K for systems with absorption [29,33]. We present the results of the experimental study of the elastic enhancement factor $W_{S,\beta}$ for microwave networks simulating quantum graphs with preserved and broken time reversal symmetry (TRS) in the presence of moderate and strong absorption. Finally, we address an important mathematical problem whether scattering properties of wave systems are uniquely connected to their shapes [11].

26.2 Telegraph Equation on a Microwave Network

The analogy between quantum graphs and microwave networks is based on the equivalency of the Schrödinger equation describing a quantum system and the telegraph equation describing an ideal microwave circuit.

A general microwave network consists of N vertices connected by bonds, e.g., coaxial cables. The $N \times N$ connectivity matrix C_{ij} of a network takes the value 1 if the vertices i and j are connected and 0 otherwise [16]. Each vertex i of a network is connected to the other vertices by v_i bonds; v_i is called the valency of the vertex i.

A coaxial cable consists of an inner conductor of radius r_1 surrounded by a concentric conductor of inner radius r_2. The space between the inner and outer conductors is filled with a homogeneous material having the dielectric constant ε.

Below the onset of the next TE_{11} mode [34], inside a coaxial cable can propagate only a wave in the fundamental transverse electromagnetic (TEM) mode, in the literature often called a Lecher wave.

The continuity equation for the charge and the current is used to find the propagation of a Lecher wave inside the coaxial cable joining the ith and the jth vertex of the microwave network [28,35]

$$\frac{de_{ij}(x,t)}{dt} = -\frac{dJ_{ij}(x,t)}{dx},$$

$$(26.1)$$

where $e_{ij}(x,t)$ and $J_{ij}(x,t)$ are the charge and the current per unit length on the surface of the inner conductor of a coaxial cable.

The potential difference $U_{ij}(x,t)$ between the conductors is given by

$$U_{ij}(x,t) = V_2^{ij}(x,t) - V_1^{ij}(x,t) = \frac{e_{ij}(x,t)}{\mathcal{C}},$$

$$(26.2)$$

where $V_1^{ij}(x,t)$ and $V_2^{ij}(x,t)$ are the potentials of the inner and the outer conductors of a coaxial cable and $\mathcal{C}$ is the capacitance per unit length of a cable.

The spatial derivative of Equation 26.2 gives [35]

$$\frac{d}{dx}U_{ij}(x,t) = -\mathcal{Z}J_{ij}(x,t),$$

$$(26.3)$$

where $\mathcal{Z}$ is the impedance per unit length. Calculation of the second spatial derivative of $U_{ij}(x,t)$ leads to the equation

$$\frac{d^2}{dx^2}U_{ij}(x,t) + \mathcal{Z}\frac{d}{dx}J_{ij}(x,t) = 0.$$

$$(26.4)$$

Using Equations 26.1 and 26.2, Equation 26.4 can be transformed to

$$\frac{d^2}{dx^2}U_{ij}(x,t) - \mathcal{Z}\mathcal{C}\frac{d}{dt}U_{ij}(x,t) = 0.$$

$$(26.5)$$

For a monochromatic wave propagating along the cable, the time dependence of $e_{ij}(x,t)$ and $U_{ij}(x,t)$ is given by $e_{ij}(x,t) = e^{-i\omega t}e_{ij}(x)$ and $U_{ij}(x,t) = e^{-i\omega t}U_{ij}(x)$, where the angular frequency $\omega = 2\pi\nu$ and ν is the microwave frequency. The impedance per unit length is given by $\mathcal{Z} = \mathcal{R} - \frac{i\omega\mathcal{L}}{c^2}$ [35], where $\mathcal{R}$ and $\mathcal{L}$ denote the resistance and the inductance per unit length, respectively, and c stands for the speed of light in a vacuum.

For an ideal lossless coaxial cable with the resistance $\mathcal{R} = 0$, Equation 26.5 leads to the telegraph equation on the microwave network

$$\frac{d^2}{dx^2}U_{ij}(x) + \frac{\omega^2\varepsilon}{c^2}U_{ij}(x) = 0, \tag{26.6}$$

where $\varepsilon = \mathcal{L}C$ [36].

The continuity equation for the potential difference requires that for every $i = 1, \ldots, N$

$$U_{ij}(x)|_{x=0} = \varphi_i, \quad U_{ij}(x)|_{x=L_{ij}} = \varphi_j, \quad i < j, \quad C_{ij} \neq 0. \tag{26.7}$$

The current conservation condition may be written in the form

$$-\sum_{j<i} C_{ij}\frac{d}{dx}U_{ji}(x)|_{x=L_{ij}} + \sum_{j>i} C_{ij}\frac{d}{dx}U_{ij}(x)|_{x=0} = 0, \tag{26.8}$$

where

$$\frac{dU_{ij}(x)}{dx} = -\mathcal{Z}J_{ij}(x). \tag{26.9}$$

L_{ij} represents the length of the bond joining the ith and jth vertices of the network.

Assuming the following correspondence: $\Psi_{ij}(x) \Leftrightarrow U_{ij}(x)$ and $k^2 \Leftrightarrow \frac{\omega^2\varepsilon}{c^2}$, Equation 26.6 is formally equivalent to the one-dimensional Schrödinger equation (with $\hbar = 2m = 1$) on the graph possessing TRS [16]

$$\frac{d^2}{dx^2}\Psi_{ij}(x) + k^2\Psi_{ij}(x) = 0. \tag{26.10}$$

Moreover, Equations 26.7 and 26.8 are equivalent to the equations derived in Reference 16 for quantum graphs with Neumann boundary conditions and vanishing magnetic vector potential $A_{ij} = A_{ji} = 0$.

One should point out that the introduction of one-dimensional microwave networks simulating quantum graphs extends substantially the number of systems that are used to verify wave effects predicted on the basis of quantum physics [37–43]. The other systems used for this purpose include, e.g., two-dimensional and three-dimensional microwave chaotic billiards and experiments with highly excited hydrogen and helium atoms. Experiments for two-dimensional microwave systems were pioneered by Reference 44 and further developed by References 45–55. In the case of two dimensions, the Schrödinger equation for quantum billiards is equivalent to the Helmholtz equation for microwave cavities of corresponding shape. The three-dimensional chaotic billiards have also been studied experimentally in the microwave frequency domain [56–59] but for these systems there is no direct analogy between the vectorial Helmholtz equation and the Schrödinger equation.

26.3 Applications of Microwave Networks

The equivalence of microwave networks and quantum graphs was carefully checked in Reference 28. For this purpose, the spectra of 10 tetrahedral microwave networks were measured in the frequency range 0.0001–16 GHz. In this frequency range, only a Lecher wave may propagate in the network because the cutoff frequency ν_c of the TE_{11} mode is much above the specified range: $\nu_c \simeq \frac{c}{\pi(r_1+r_2)\sqrt{\varepsilon}} = 32.9$ GHz [34], where $r_1 = 0.05$ cm is the inner wire radius of the coaxial cable (SMA-RG402), $r_2 = 0.15$ cm is the inner radius of the surrounding conductor, and $\varepsilon = 2.08$ is the Teflon dielectric constant.

The experimental setup for measurements of spectra of the microwave networks is shown in Figure 26.1a. To simplify the notation, the lengths of the six bonds of the four-vertex tetrahedral network shown in Figure 26.1a are labeled by letters $(a, \ldots, f)$. The total "optical" lengths of the microwave

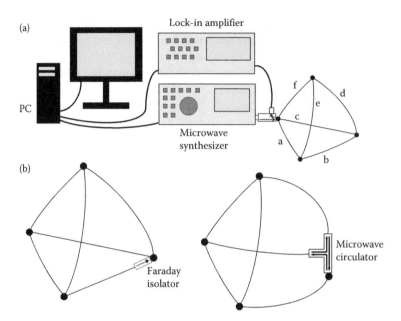

FIGURE 26.1 (a) Experimental setup for measurement of the spectra of microwave networks. (b) Two examples of the directed microwave networks containing a Faraday isolator and a microwave circulator.

networks, including T-joints (vertices), varied from 171.7 to 262.2 cm, which allowed for the observation of 156–264 eigenfrequencies in the frequency range 0.0001–16 GHz. To avoid the degeneracy of eigenvalues, the lengths $L_{i,j}$ of the bonds (cables) were chosen not to be commensurable. Figure 26.1b shows the examples of the directed networks containing a Faraday isolator and a microwave circulator. The directed microwave networks will be discussed in detail in Section 26.6.1.

In Figure 26.2a, a fragment of a typical reflection spectrum of the graph is presented in the frequency range 4–5 GHz. Figure 26.2b shows frequency differences between experimentally measured eigenfrequencies of the microwave network and numerically calculated eigenvalues of the graph having the same bond lengths as the experimental one. The agreement between experimental and theoretical results is quite good (the relative errors are of the order of 10^{-3}), which justifies the assumption that microwave networks can be described by quantum graphs with Neumann boundary conditions [16,60,61].

The presented results show that relatively short microwave networks consisting of coaxial cables, at least as it concerns the eigenvalues positions, can be approximately treated as ideal lossless graphs with the resistance $\mathcal{R} = 0$. This finding is not very surprising. A similar situation can be found in the experiments with microwave cavities [44,62].

26.4 Level Spacing Statistics for Networks with TRS

In Reference 28, the statistical properties of spectra of the microwave networks such as the integrated nearest-neighbor spacing (INNS) distribution $I(s)$ and the spectral rigidity $\Delta_3(L)$ [63,64] were investigated.

If the system is classically chaotic, it is expected that the nearest-neighbor spacing distribution $P(s)$ should follow the prediction of random matrix theory (RMT). The nearest-neighbor spacing distribution $P(s)$ is well suited to study the short-range spectral correlations [63,64]. The unfolded variable s is defined as follows: $s \equiv (k_n - k_{n-1})\bar{\rho}((k_n + k_{n-1})/2)$, where $k_n - k_{n-1}$ is the nearest-neighbor separation between a pair of consecutive eigenvalues of the system and $\bar{\rho}(k) = d\bar{N}(k)/dk$ is the mean density of states. The

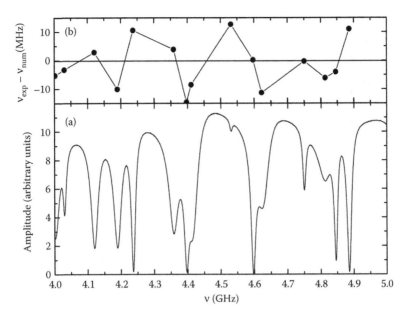

FIGURE 26.2 (a) A fragment of the reflection spectrum of the microwave network of the "optical" length 223.6 cm in the frequency range 4–5 GHz. (b) Frequency differences between experimental eigenvalues of the above microwave network and numerically calculated eigenvalues of the graph having the same bond lengths as the experimental one. Solid lines joining the points are plotted to guide the eyes.

mean $\bar{N}(k)$ of the staircase function, i.e., the number of resonances up to the wave number k, in the case of microwave networks can be obtained from a least squares fit $\bar{N}(k) = \alpha k + \beta$ of the measured staircase $N(k)$, where α and β were the fit parameters.

The INNS distribution $I(s)$ that measures the cumulative probability of finding a pair in $[0, s]$ is therefore given by

$$I(s) = \int_0^s P(s')ds'. \tag{26.11}$$

The $\Delta_3(L)$ statistic is used to investigate the long-range correlations. The function $\Delta_3(L)$, averaged over intervals, measures the deviations of the spectrum from a true equidistant spectrum

$$\Delta_3(L) = < \frac{1}{L} \min_{A,B} \int_s^{s+L} [N(s') - As' - B]^2 ds' > . \tag{26.12}$$

Figure 26.3 presents the INNS distributions. The solid line represents predictions of random matrix theory obtained for Gaussian orthogonal ensemble (GOE), applicable for systems with a TRS. The dashed line denotes theoretical results characteristic of Gaussian unitary ensemble (GUE), valid if the TRS is broken [63]. Experimental curve (empty triangles) was obtained by averaging over the set of 10 microwave graphs that differed in the length of one bond. Numerical curve (empty circles) shows results averaged for 10 quantum graphs having the same bond lengths as the experimental ones. Eigenfrequencies were calculated by numerically solving the secular equations for quantum graphs Equations 26.6 through 26.8 in Reference 16). Figure 26.3 shows that for both experimental and numerical results, the INNS distributions are in a very good agreement with the GOE predictions.

Figure 26.4 demonstrates the spectral rigidity $\Delta_3(L)$ obtained for the microwave network of the "optical" length 223.6 cm. Experimental curve (empty triangles) was based on 229 identified eigenfrequencies, while the numerical data (empty circles) were computed out of 237 eigenfrequencies. In both cases, the frequency range was 0.0001–16 GHz. Deviations of the experimental and numerical rigidity from the GOE

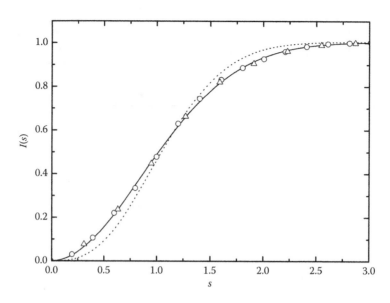

FIGURE 26.3 INNS distribution $I(s)$ averaged for 10 microwave networks. Results of the experiment (empty triangles) are compared with the numerical results (empty circles) and theoretical prediction for GOE (solid line) and GUE (dashed line).

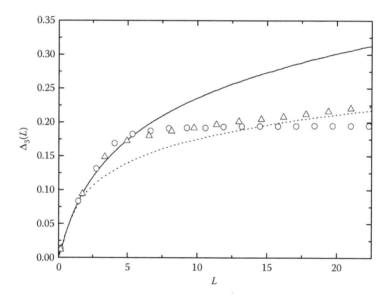

FIGURE 26.4 Spectral rigidity $\Delta_3(L)$ for the microwave network of the "optical" length 223.6 cm. Results of the experiment (empty triangles) are compared with the numerical results (empty circles) and theoretical prediction for GOE (solid line) and GUE (dashed line).

predictions (solid line) are visible. The dashed line in Figure 26.4 shows the RMT prediction for GUE. Experimental and numerical results lay above the GOE prediction for L between 2.5 and 5. For a higher value of L, a saturation of the numerical value of the spectral rigidity is observed in agreement with the predictions of Berry [65]. The experimental rigidity for $L > 10$ is located below the GOE curve and above the numerical results. The departure of the experimental results from the numerical one can be probably attributed to the loss of about 3% of eigenfrequencies in the experiment.

26.5 Lengths of Periodic Orbits in the Network

The measurement of the spectrum of the microwave network allows the calculation of the lengths of its periodic orbits [28]. They are computed from the Fourier transform

$$F(l) = \int_0^{k_{max}} \tilde{\rho}(k)\omega(k)e^{-ikl}dk, \tag{26.13}$$

where $\tilde{\rho}$ is the oscillating part of the level density and $\omega(k) = \sin^2(\pi \frac{k}{k_{max}})$ is a window function that suppresses the Gibbs overshoot phenomenon [50,51]. Here, k_{max} is the maximal value of the wave number within the interval where the eigenvalues of the network were evaluated. The oscillating part of the level density $\tilde{\rho}$ was determined from the density of states calculated according to $\rho(k) = \sum_j \delta(k - k_j)$ from which the mean density $\bar{\rho}(k) = d\bar{N}(k)/dk$ was subtracted.

The absolute square of the Fourier transform of the fluctuating part of the density of resonances $|F(l)|^2$ for the network of the "optical" length 223.6 cm is shown in Figure 26.5. The lengths of the bonds of the network fulfill the following relations: $a < b < c < d < e < f$. Results obtained from the experimental spectrum (solid line) are compared to the results obtained from numerical calculations (broken line). The experimental spectrum included 149 identified eigenfrequencies while the numerical one included 150 eigenfrequencies. In both cases, the frequency range was 0.0001–10.2 GHz. Figure 26.5 shows that the

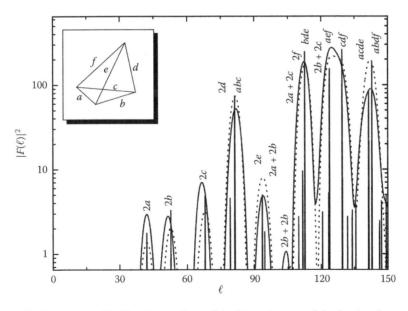

FIGURE 26.5 Absolute square of the Fourier transform of the fluctuating part of the density of resonances of the network of the "optical" length 223.6 cm. Results of the experiment that yielded 149 eigenfrequencies (solid line) are compared with the numerical results, 150 eigenfrequencies (broken line). The more detailed structure of the periodic orbits in the network (sharp peaks) was found on the basis of numerical calculations that yielded 7461 eigenfrequencies. The assignment of peaks of $|F(l)|^2$ to simple periodic orbits (see text) is shown along with the length of the orbits. The "optical" lengths of the bonds of the graph: $a = 21.0$ cm, $b = 26.3$ cm, $c = 34.0$ cm, $d = 39.6$ cm, $e = 46.8$ cm, and $f = 55.9$ cm.

agreement between the experimental and the theoretical results is good. As expected, $|F(l)|^2$ shows pronounced peaks near the lengths of certain periodic orbits. In Figure 26.5, we mark all irreducible periodic orbits [16], i.e., periodic orbits that do not intersect themselves, with the lengths $l < 165$ cm and additionally, for clarity, we show the first repetition of the periodic orbit $2b$ at $l = 2b + 2b = 105.2$ cm and two reducible periodic orbits $abc + 2c$ and $bde + 2a$ at $l = 149.3$ cm and $l = 154.7$ cm, respectively. It should be noted that many peaks for $l > 70$ cm cover several unresolved periodic orbits. The more detailed structure of the periodic orbits in the network (sharp peaks in Figure 26.5) can be found on the basis of numerical calculations, which yielded 7461 eigenvalues in the frequency range 0–500 GHz. However, this number of resolved eigenfrequencies is beyond the scope of this experiment.

26.6 Microwave Networks with Absorption

Microwave networks with absorption are more realistic but also more complicated systems. They have been recently studied in a series of experimental papers as shown in References 28, 29, and 33.

In Reference 28, directed microwave networks with absorption consisting of coaxial cables and Faraday isolators for which the deviation of the level spacing distribution from GOE were observed.

In References 29 and 33, the results of the experimental study of the distribution $P(R)$ of the reflection coefficient R and the distributions of Wigner's reaction matrix [66] (also called the K matrix [67]) for microwave networks with attenuators that correspond to graphs with TRS ($\beta = 1$ symmetry class of random matrix theory [68]) in the presence of absorption were presented. Application of attenuators allowed to vary absorption in the networks in a controlled, quantitative way.

Open quantum graphs (with external leads) have also been analyzed theoretically in References 23 and 24.

26.6.1 Directed Microwave Networks

Microwave networks consisting of coaxial cables and Faraday isolators are examples of simple experimental realization of directed networks (see Figure 26.1b) [28]. A microwave Faraday isolator is a passive device, which transmits the wave moving in one direction while absorbing the wave moving in the opposite direction. Owing to absorption, the introduction of Faraday isolators transforms the problem from the bound system to an open system. In the experiment [28], AerCom 60583 Faraday isolators operating in the frequency range 3.5–7.5 GHz were used. The results of 12 measurements of microwave networks consisting in one of their bonds one Faraday isolator or two Faraday isolators connected in series were averaged to obtain the INNS distribution (solid triangles) in Figure 26.6. The INNS distribution obtained in the same frequency range for microwave networks without Faraday isolators are also shown in Figure 26.6 (empty triangles). Figure 26.6 shows that the INNS distribution for the networks without Faraday isolators is close to the RMT prediction for GOE (solid line) in contrast to the INNS distribution for the networks with the isolators, which follows more closely the RMT prediction for GUE (dashed line). This behavior is especially well seen at small eigenfrequency spacing s, although also for $s > 1$ the experimental results obtained for microwave networks with Faraday isolators lie above the GOE prediction.

One ought to point out that similar deviations of the spectral statistics from GOE to GUE were observed in the experiments with microwave billiards [47–49]. In the experiment [48,49], the deviation from GOE statistics was due to a Faraday isolator connected to a microwave cavity while in the experiment performed by So et al. [47], the transition from GOE to GUE statistics was caused by a piece of magnetized ferrite placed inside a two-dimensional microwave cavity. In Reference 49, the deviation of the level spacing distribution from GOE observed at small spacings in Reference 48 was interpreted as the influence of absorption, which produces a quadratic level repulsion. In the case of microwave networks, Figure 26.6 shows that the influence of Faraday isolators on the INNS distribution is more uniform, including small and higher eigenfrequency spacings, suggesting that in this case we deal with the phenomenon of TRS breaking. This

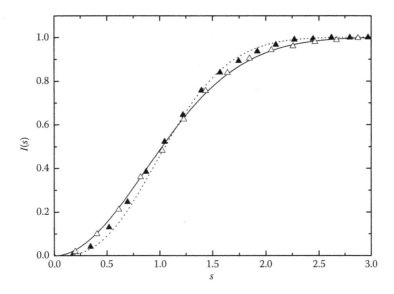

FIGURE 26.6 INNS distribution averaged for 12 realizations of the microwave networks with Faraday isolators (solid triangles) is compared with the averaged results for the microwave networks without the isolators (empty triangles) and theoretical prediction for GOE (solid line) and GUE (dashed line). In both cases, experimental results were obtained for the frequency range 3.5–7.5 GHz.

behavior is in agreement with the numerical results of Reference 21 (see also Reference 19) where it was shown that for the directed graphs without absorption, the GUE level statistics should be observed.

To avoid the problems with strong absorption present in Faraday isolators in further experimental investigations of directed microwave networks, the Faraday isolators were replaced by much less absorbing three-port microwave circulators.

26.6.2 Theoretical Description of Directed Graphs

The directed networks were modeled theoretically in Reference 28. The crucial element of the network— the Faraday isolator in a directed bond—can be described by means of a filter factor, which damps the wave moving in one direction.

The connectivity matrix D of a directed network is not symmetric. With any directed network Γ, one may associate a bidirectional network G with the same number of vertices. Its connectivity matrix C is symmetric,

$$C_{ij} = \max(D_{ij}, D_{ji}). \tag{26.14}$$

The number B of bonds in the graph G is equal to $B = \frac{1}{2}\sum_{ij} C_{ij}$. In order to filter out some waves propagating in one direction while preserving those moving in the opposite direction, a bond scattering technique of analyzing spectra of graphs, similar to the one introduced by Kottos and Smilansky [16] was used.

A plain wave $\Psi_{j'n}(x) = e^{-ikx}$ coming from the vertex j' to the vertex n is scattered into all bonds going out from the vertex n, for which $C_{nj} \neq 0$,

$$\Phi_{nj}(x) = \delta_{jj'} e^{-ikx} + \sigma_{jj'}^{(n)} e^{ikx}. \tag{26.15}$$

The vertex scattering matrix $\sigma_{jj'}^{(n)}$ is completely determined, if Neumann boundary conditions are assumed, which imply

$$\sigma_{jj'}^{(n)} = C_{j'n} C_{nj}(-\delta_{jj'} + 2/v_n). \tag{26.16}$$

Here, v_n denotes the number of bonds meeting at the nth vertex, also called the *valency* of the vertex. Elements of $\sigma_{jj'}^{(n)}$ for all vertex n combine to the entire bond transition matrix of the network G

$$T_{jl,nm} = \delta_{ln}C_{jl}C_{nm}\sigma_{jm}^{(l)}, \tag{26.17}$$

which describe the changes of amplitudes of waves propagating in each bond of the network (in both directions) after one event of scattering on vertices. To take into account the presence of Faraday isolators or microwave circulators, the connectivity matrix D of the directed graph Γ was used in the definition of a diagonal $2B \times 2B$ matrix $\Lambda(k)$

$$\Lambda_{jl,j'l'}(k) = \delta_{jj'}\delta_{ll'}D_{jl}e^{ikL_{jl}}, \tag{26.18}$$

where the phase factor describes the free propagation along the bond (jl) of length L_{jl}. The total evolution of the vector of wave amplitudes of length $2B$ is given by the bond scattering matrix

$$S(k) = \Lambda(k) \cdot T. \tag{26.19}$$

The matrix $S(k)$ is subunitary, since it is obtained by putting to zero some elements of a unitary matrix. Therefore, its eigenvalues $\lambda_j(k)$ are located in the unit circle, $|\lambda_j(k)| \leq 1$. The equation for the eigenmodes of the quantum graph

$$\det(S(k) - 1) = 0 \tag{26.20}$$

may have no real solution.

In microwave network experiments that were discussed so far, the circuits were driven by the continuous wave microwave generators. To find wave vectors k for which the resonant driving of the network will appear, one should analyze the stationary states of the system. In Reference 28, it was showed that the total response function of the network may be approximated by the average enhancement factor

$$r(k) = \frac{1}{2B}\sum_{j=1}^{2B}\frac{1}{1 - \lambda_j(k)}, \tag{26.21}$$

where $\lambda_j(k)$ are the eigenvalues of the bond scattering matrix $S(k)$.

Equation 26.21 shows that the maxima of the response function $r(k)$ occur for the resonant values of the wave vector k for which one of the eigenvalue $\lambda_j(k)$ is close to unity.

Using the approach of the response function, eigenfrequencies of 20 directed graphs in the frequency range 0–20 GHz were calculated [28]. As in the experimental realization [28], only one bond was assumed to be directed. Figure 26.7 shows the INNS distribution averaged for 20 realizations of the directed graphs in the frequency range 0–20 GHz (solid circles). The results for the directed graphs are compared with the numerical data obtained in the same frequency range 0–20 GHz for 20 realizations of standard (bidirectional) graphs (empty circles). Theoretical predictions for GOE and GUE are shown by solid and dashed curves, respectively. Figure 26.7 shows that the INNS spectral statistics for directed graphs deviate at small spacings from the GOE curve and become closer to the GUE predictions. This result also confirms the experimental findings obtained for the microwave directed networks (see Figure 26.6).

26.6.3 Bidirectional Microwave Networks with Absorption

A network with no absorption and no leads to the outside world is a closed system. The presence of absorption and/or leads creates an open system. In the previous section, we discussed the case of directed networks where absorption was introduced by Faraday isolators. Absorption in a bidirectional microwave network can be much more efficiently varied by adding microwave attenuators [33], changing the length of the cables [28], or by changing the coupling to the outside world.

In this section, we would like to present the results of the experimental study of the distribution $P(R)$ of the reflection coefficient R and the distributions of imaginary and real parts of Wigner reaction matrix K [66,67] for microwave networks with absorption that correspond to quantum graphs with TRS. The extended studies of the above problem have been published in Reference 30.

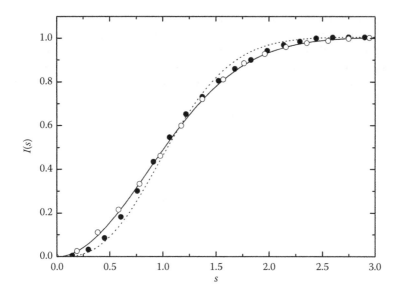

FIGURE 26.7 Numerically calculated INNS distributions averaged for 20 realizations of the directed graphs (solid circles) is compared with the averaged results for the bidirectional graphs (empty circles). Calculations were performed in the frequency range 0–20 GHz. Numerical results for the INNS distributions are compared with theoretical predictions for GOE (solid line) and GUE (dashed line).

26.6.3.1 Distributions of the Reflection Coefficient R and the Wigner's Reaction Matrix K

In the case of a single-channel antenna experiment, the K matrix is related to the scattering matrix S by the following relation:

$$S = \frac{1 - iK}{1 + iK}. \tag{26.22}$$

Equation 26.22 holds for the systems with absorption but without direct processes [67]. It is important to mention that the function $Z = iK$ has a direct physical meaning of the electric impedance that has been recently measured in the microwave cavity experiment [69]. In the one-channel case, the S matrix can be parameterized as

$$S = \sqrt{R}e^{i\theta}, \tag{26.23}$$

where R is the reflection coefficient and θ the phase.

Properties of the statistical distributions of the S matrix with direct processes and imperfect coupling have been studied theoretically in several important papers [70–75]. Recently, the distribution of the S matrix has also been measured experimentally for chaotic microwave cavities with absorption [76]. The distribution $P(R)$ of the reflection coefficient R and the distributions of the imaginary $P(v)$ and the real $P(u)$ parts of the Wigner's reaction K matrix are theoretically known for any dimensionless absorption strength γ [67,77]. In the case of time reversal systems (symmetry index $\beta = 1$), $P(R)$ has been studied experimentally by Méndez-Sánchez et al. [78]. The distributions $P(v)$ and $P(u)$ have been studied for chaotic microwave cavities in References 69 and 79 and for microwave networks for moderate absorption strength $\gamma \leq 7.1$ in References 29 and 33. For systems without TRS ($\beta = 2$) and a single perfectly coupled channel, $P(R)$ was calculated by Beenakker and Brouwer [80] while the exact formulas for the distributions $P(v)$ and $P(u)$ were given by Fyodorov and Savin [67].

From the experimental point of view, absorption of the networks can be changed by the change of the bonds (cables) length [28] or more effectively by the application of microwave attenuators [29,33]. In the numerical calculations, weak absorption inside the cables can be described with the help of the complex wave vector [28] while strong absorption inside an attenuator can be described by a simple optical potential [86]. The corresponding mathematical theory has been developed in Reference 81.

The distribution $P(R)$ of the reflection coefficient R and the distributions of the imaginary and real parts of the Wigner's reaction matrix K for microwave networks with absorption were found in Reference 30 using the impedance approach [33,69,79]. In this approach, the real and imaginary parts of the normalized impedance Z

$$z = \frac{\mathrm{Re}\, Z + i(\mathrm{Im}\, Z - \mathrm{Im}\, Z^r)}{\mathrm{Re}\, Z^r} \tag{26.24}$$

of a chaotic microwave system are measured, with $Z = Z_0(1 + S)/(1 - S)$ and $Z^r = Z_0(1 + S^r)/(1 - S^r)$ being the network and radiation impedance, respectively, expressed by the network (radiation) scattering matrix S (S^r) and Z_0 is the characteristic impedance of the microwave cables. The radiation impedance Z^r is the impedance seen at the output of the coupling structure for the same coupling geometry, but with the vertices of the network removed to infinity. The Wigner's reaction matrix K can be expressed by the normalized impedance as $K = -iz$. The scattering matrix s of a network for the perfect coupling case (no direct processes present) required for the calculation of the reflection coefficient R (see Equation 26.2) can be finally extracted from the formula $s = (1 - z)/(1 + z)$.

Figure 26.8a shows the experimental setup for measuring the single-channel scattering matrix S of fully connected hexagon microwave networks necessary for finding of the impedance Z. We used the Hewlett-Packard 8720A microwave vector network analyzer to measure the scattering matrix S of the networks in the frequency window: 7.5–11.5 GHz. The networks were connected to the vector network analyzer through a lead—an HP 85131-60012 flexible microwave cable—connected to a 6-joint vertex. The other five vertices of the networks were connected by 5 joints. Each bond of the network presented in Figure 26.8a contains a microwave attenuator.

The radiation impedance Z^r was found experimentally by measuring the scattering matrix S^r of the 6-joint connector with 5 joints terminated by 50 Ω loads (see Figure 26.8b).

The experimentally measured fully connected hexagon networks were described in numerical calculations by fully connected quantum hexagon graphs with one lead attached to the 6-joint vertex.

In order to find the distribution $P(R)$ of the reflection coefficient R and the distributions of the imaginary and real parts of the K matrix, we measured the scattering matrix S of 88 and 74 network configurations containing in each bond a single 1 and 2 dB microwave SMA attenuator, respectively. The total optical lengths of the microwave networks containing 1 dB attenuators, including joints and attenuators,

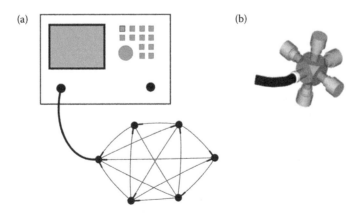

FIGURE 26.8 (a) Scheme of the experimental setup for measurements of the scattering matrix S of the microwave fully connected networks with absorption. Absorption in the networks was varied by the change of the attenuators. (b) Scheme of the setup used to measure the radiation scattering matrix S^r. Instead of a network, five 50 Ω loads were connected to the 6 joints.

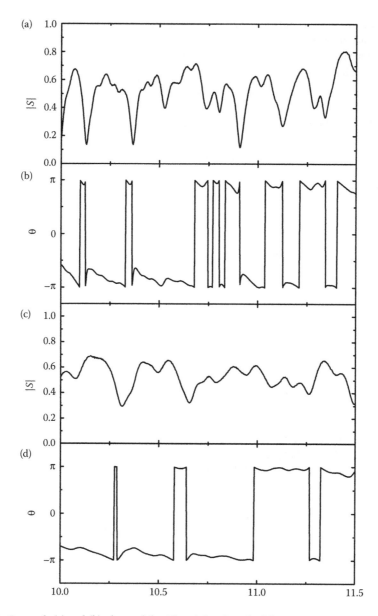

FIGURE 26.9 In panels (a) and (b), the modulus $|S|$ and the phase θ of the scattering matrix S measured for the network with $\gamma = 19.9$ are plotted in the frequency range 10–11.5 GHz. In (c) and (d), $|S|$ and θ of the scattering matrix S are plotted for the network with $\gamma = 47.9$ in the same frequency range. The measurements have been done for the two networks, which in each bond contained: 1 dB attenuator (a and b) and 2 dB attenuator (c and d), respectively. The total "optical" length of the microwave networks including joints and attenuators were 574 and 554 cm, respectively.

were varied from 574 to 656 cm. For the networks with 2 dB attenuators, the optical lengths were varied from 554 to 636 cm. To avoid degeneracy of eigenvalues of the networks, the lengths of the bonds were chosen as incommensurable.

In Figure 26.9, the modulus $|S|$ and the phase θ of the scattering matrix S of the microwave networks with $\gamma = 19.9$ and 47.9, respectively, are presented in the frequency range 10–11.5 GHz. The measurements were done for the two networks containing 1 and 2 dB attenuators, respectively. Their total "optical" lengths including joints and attenuators were 574 and 554 cm, respectively.

For systems with TRS ($\beta = 1$), the explicit analytic expression for the distribution $P(R)$ of the reflection coefficient R is given by [77]

$$P(R) = \frac{2}{(1-R)^2} P_0\left(\frac{1+R}{1-R}\right). \tag{26.25}$$

The probability distribution $P_0(x)$ is given by the expression

$$P_0(x) = -\frac{dW(x)}{dx}, \tag{26.26}$$

where the integrated probability distribution $W(x)$ is expressed by the formula [77]

$$W(x) = \frac{x+1}{4\pi} \Big[f_1(w)g_2(w) + f_2(w)g_1(w) + h_1(w)j_2(w) + h_2(w)j_1(w) \Big]_{w=(x-1)/2}. \tag{26.27}$$

The functions f_1, g_1, h_1, j_1 are defined as follows:

$$f_1(w) = \int_w^\infty dt \frac{\sqrt{t\,|t-w|}\,e^{-\gamma t/2}}{(1+t)^{3/2}} \Big[1 - e^{-\gamma} + \frac{1}{t} \Big], \tag{26.28}$$

$$g_1(w) = \int_w^\infty dt \frac{e^{-\gamma t/2}}{\sqrt{t\,|t-w|}(1+t)^{3/2}}, \tag{26.29}$$

$$h_1(w) = \int_w^\infty dt \frac{\sqrt{|t-w|}\,e^{-\gamma t/2}}{\sqrt{t(1+t)}} \Big[\gamma + (1-e^{-\gamma})(\gamma t - 2) \Big], \tag{26.30}$$

$$j_1(w) = \int_w^\infty dt \frac{e^{-\gamma t/2}}{\sqrt{t|t-w|}(1+t)^{1/2}}. \tag{26.31}$$

Their counterparts with the index 2 are given by the same expressions but the integration is performed in the interval $t \in [0, w]$ instead of $[w, \infty)$.

The distributions of the imaginary and the real parts $P(v)$ and $P(u)$ of the K matrix [67] can also be expressed by the probability distribution $P_0(x)$:

$$P(v) = \frac{\sqrt{2}}{\pi v^{3/2}} \int_0^\infty dq P_0 \Big[q^2 + \frac{1}{2}\Big(v + \frac{1}{v}\Big) \Big], \tag{26.32}$$

and

$$P(u) = \frac{1}{2\pi\sqrt{u^2+1}} \int_0^\infty dq P_0 \Big[\frac{\sqrt{u^2+1}}{2}\Big(q + \frac{1}{q}\Big) \Big], \tag{26.33}$$

where $-v = \mathrm{Im}\,K < 0$ and $u = \mathrm{Re}\,K$ are, respectively, the imaginary and real parts of the K matrix.

Figure 26.10 shows the experimental distributions $P(R)$ (squares) of the reflection coefficient R for the two mean values of the parameter $\bar{\gamma}$, viz., 19.3 and 47.7. The distribution for $\bar{\gamma} = 19.3$ is obtained by averaging over 88 realizations of the microwave networks containing 1 dB attenuators. The distribution for $\bar{\gamma} = 47.7$ is obtained by averaging over 74 realizations of the microwave networks containing 2 dB attenuators. The experimental values of the γ parameter were estimated for each realization of the network by adjusting the theoretical mean reflection coefficient $\langle R \rangle_{\mathrm{th}}$ to the experimental one $\langle R \rangle = \langle ss^\dagger \rangle$, where

$$\langle R \rangle_{\mathrm{th}} = \int_0^1 dR R P(R). \tag{26.34}$$

Figure 26.10 also presents the corresponding distributions $P(R)$ (solid and dashed lines, respectively) evaluated from Equation 26.25. A good overall agreement of the experimental distributions $P(R)$ with their theoretical counterparts is seen.

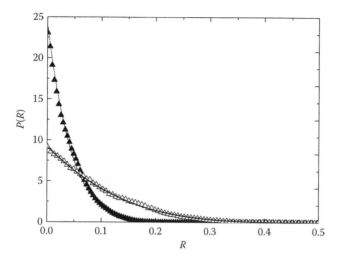

FIGURE 26.10 Experimental distribution $P(R)$ of the reflection coefficient R for the microwave fully connected hexagon networks at $\bar{\gamma} = 19.3$ (empty squares) and $\bar{\gamma} = 47.7$ (full squares). The corresponding theoretical distribution $P(R)$ evaluated from Equation 26.25 is marked by the solid line ($\gamma = 19.3$) and dashed line ($\gamma = 47.7$), respectively.

In Figure 26.11, the experimental distribution $P(v)$ of the imaginary part of the K matrix is shown for the two mean values of the parameter $\bar{\gamma} = 19.3$ and 47.7, respectively. The distribution is the result of averaging over 88 and 74 realizations of the networks with the attenuators 1 and 2 dB, respectively. The experimental results in Figure 26.11 are in general in good agreement with the theoretical ones evaluated from Equation 26.34. However, both experimental distributions are slightly higher than the theoretical ones in the vicinity of their maxima.

Determination of the distribution $P(u)$ of the real part of the Wigner's reaction matrix gives an additional test of the consistency of the γ evaluation. In Figure 26.12, we show this distribution obtained for

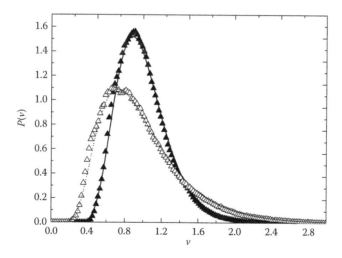

FIGURE 26.11 Experimental distribution $P(v)$ of the imaginary part of the K matrix for the two values of the mean absorption parameter: $\bar{\gamma} = 19.3$ (empty squares) and $\bar{\gamma} = 47.7$ (full squares), respectively. The corresponding theoretical distribution $P(v)$ evaluated from the Equation 26.34 is marked by the solid line ($\gamma = 19.3$) and dashed line ($\gamma = 47.7$), respectively.

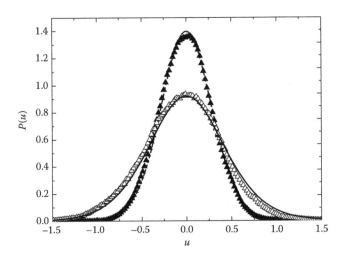

FIGURE 26.12 Experimental distribution $P(u)$ of the real part of the K matrix for the two values of the mean absorption parameter: $\bar{\gamma} = 19.3$ (empty circles) and $\bar{\gamma} = 47.7$ (full circles), respectively. The experimental results are compared to the theoretical distributions $P(u)$ evaluated from Equation 26.33: solid line ($\gamma = 19.3$) and dashed line ($\gamma = 47.7$).

the two values of $\bar{\gamma} = 19.3$ and 47.7, respectively, compared to the theoretical ones evaluated from Equation 26.34. Also, here we observe good overall agreement between the experimental and theoretical results.

In summary, the microwave networks can be successfully used to measure the distribution of the reflection coefficient $P(R)$ and the distributions of the imaginary $P(v)$ and the real $P(u)$ parts of the Wigner's reaction K matrix for microwave networks and graphs in the presence of strong absorption. The application of attenuators in the microwave networks allows for effective change of absorption in the graphs.

26.6.4 Elastic Enhancement Factor

In Reference 82, the results of the experimental study of the two-port scattering matrix $\hat{S}$ elastic enhancement factor $W_{S,\beta}$ [75,83] for microwave networks simulating quantum graphs with preserved TRS (symmetry index $\beta = 1$) and with broken TRS ($\beta = 2$) in the presence of moderate and strong absorption have been presented.

In the case of the two-port scattering matrix

$$\hat{S} = \begin{bmatrix} S_{11} & S_{12} \\ S_{21} & S_{22} \end{bmatrix} \tag{26.35}$$

the elastic enhancement factor $W_{S,\beta}$ is defined by the following relation [75,83]:

$$W_{S,\beta} = \frac{\sqrt{\text{var}(S_{11})\text{var}(S_{22})}}{\text{var}(S_{12})}, \tag{26.36}$$

where $\text{var}(S_{12}) \equiv \langle |S_{12}|^2 \rangle - |\langle S_{12} \rangle|^2$ denotes the variance of the scattering matrix element S_{12}. One of the most important properties of the enhancement factor $W_{S,\beta}$ for $\gamma \gg 1$ is connected with the fact that it should not depend on the direct processes present in the system [83,84]. The reciprocal quantity $\Xi_{S,\beta} = 1/W_{S,\beta}$ was considered theoretically and measured in the function of frequency for a chaotic microwave cavity with TRS [84]. The variances of the fluctuations of multiport impedance parameters that are necessary to estimate the enhancement factor $W_{S,\beta=1}$ were also studied within an electromagnetic scattering theory in Reference 85.

There is a general agreement that for $\gamma \gg 1$, $W_{S,\beta} = 2/\beta$ [75,83,84]. However, calculations performed within RMT indicate that for the other values of the parameter γ, the enhancement factor $W_{S,\beta}$ might

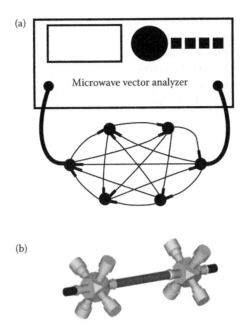

FIGURE 26.13 (a) Scheme of the experimental setup for measuring the two-port scattering matrix $\hat{S}$ of fully connected hexagon microwave networks. The measurements were performed in the frequency window: 4–14 GHz. The microwave networks were connected to the vector network analyzer through the leads connected to the 6-joint vertices. The other four vertices of the networks were the 5 joints. Each bond of the network presented in this figure contains a microwave attenuator. (b) Scheme of the setup used to measure the radiation scattering matrix S_r^{kk} of the 6-joint connector. Instead of a microwave network, five 50 Ω loads were connected to the 6 joints.

depend both on the coupling to the system and the parameter γ itself [84]. In the more particular case of the stochastic environment characterized by a statistically isotropic scattering matrix (occurring in RMT in the case of the perfect coupling when all transmission amplitudes are equal to unity), the enhancement factor should have the universal value $W_{S,\beta=1} = 2$ [85].

The microwave networks with broken TRS were constructed using microwave circulators. A microwave circulator is a nonreciprocal three-port passive device. A wave that enters port 1 of a circulator exits port 2, a wave into port 2 exits port 3, and finally a wave into port 3 exits port 1 (see Figure 26.2 for a schematic representation of a circulator). In the experiment Anritsu PE8403 microwave circulators with the operating frequency range 7–14 GHz and low insertion loss ($i_{12} \simeq i_{23} \simeq i_{31} \simeq 0.4$ dB) were used.

Figure 26.13a shows the experimental setup for measuring the two-port scattering matrix $\hat{S}$ of fully connected hexagon microwave networks. We used Agilent E8364B microwave vector network analyzer to measure the scattering matrix $\hat{S}$ of the networks in the frequency window: 4–14 GHz. The networks were connected to the vector network analyzer through the leads—HP 85133-616 and HP 85133-617 flexible microwave cables—connected to the 6-joint vertices. The 5 joints were the other four vertices of the networks. Each bond of the network presented in Figure 26.13a contains a microwave attenuator.

Figure 26.14 shows the setup for measuring the two-port scattering matrix $\hat{S}$ of microwave networks with broken TRS which contain four Anritsu PE8403 microwave circulators and microwave attenuators. The microwave circulators are enlarged to show the directions of traveling waves. Owing to the operating frequency range of the microwave circulators Anritsu PE8403, the scattering matrix $\hat{S}$ of the networks was measured in the narrower frequency window: 7–14 GHz.

Properties of such networks with microwave circulators but without microwave attenuators were investigated using the integrated nearest-neighbor distribution $I(s)$. The distribution $I(s)$ for microwave networks with broken TRS is shown in Figure 26.15 (full circles). In order to minimize absorption of the networks, the measurements were performed in the narrower frequency range 7–9 GHz. The distribution $I(s)$ was averaged over 20 microwave network configurations. In this way, 1378 eigenfrequencies of the networks were used in the calculations of the distribution $I(s)$. Figure 26.15 shows that the experimental distribution $I(s)$ is close to the theoretical prediction for GUE in RMT (solid line).

In Figure 26.16, the enhancement factor $W_{S,\beta}$ of the two-port scattering matrix $\hat{S}$ of the microwave networks simulating quantum graphs with preserved and broken TRS, respectively, is shown in the function of the parameter γ.

The experimental values of the parameter $\gamma = (\gamma_1 + \gamma_2)/2$ were estimated for each realization of a network by adjusting the theoretical mean reflection coefficients

$$\langle R \rangle_{(k)}^{th} = \int_0^1 dR\, R\, P(R), \tag{26.37}$$

to the experimental ones $\langle R \rangle_{(k)} = \langle s_{kk} s_{kk}^{\dagger} \rangle$ obtained after eliminating the direct processes [30]. Here, the index $k = 1, 2$ denotes the port 1 or 2. The parameters γ_1 and γ_2 were independently measured for the ports 1 and 2, respectively. In the impedance approach [69,79], the scattering matrix s_{kk} of a network for the perfect coupling case (no direct processes present) can be extracted from the formula

$$s_{kk} = (1 - z_{kk})/(1 + z_{kk}), \tag{26.38}$$

where the normalized impedance z_{kk} of a chaotic microwave network is given by

$$z_{kk} = \frac{\operatorname{Re} Z_{kk} + i(\operatorname{Im} Z_{kk} - \operatorname{Im} Z_{kk}^r)}{\operatorname{Re} Z_{kk}^r}. \tag{26.39}$$

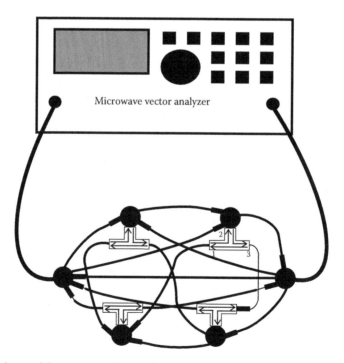

FIGURE 26.14 Scheme of the experimental setup for measuring the two-port scattering matrix $\hat{S}$ of microwave networks with broken TRS. The network additionally to the attenuators contains four microwave circulators. The microwave circulators are enlarged to show the input ports and the directions of traveling waves. The measurements were performed in the frequency window: 7–14 GHz.

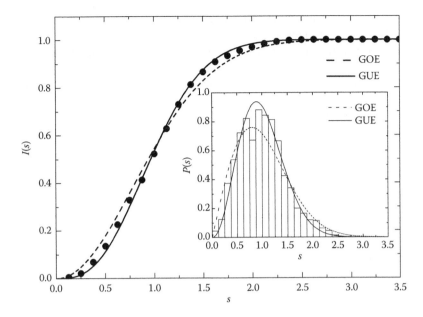

FIGURE 26.15 The integrated nearest-neighbor distribution $I(s)$ obtained for microwave networks with broken TRS (full circles). The measurements were performed in the frequency range 7–9 GHz. The distribution $I(s)$ was averaged over 20 microwave network configurations. The experimental distribution is compared to the theoretical predictions for GOE (broken line) and GUE (solid line).

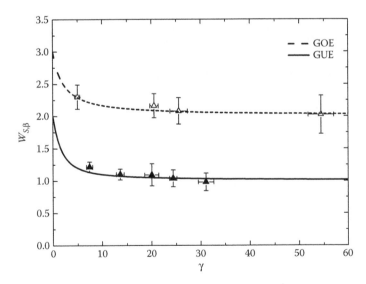

FIGURE 26.16 The enhancement factor $W_{S,\beta}$ of the two-port scattering matrix $\hat{S}$ of the microwave networks simulating quantum graphs with preserved and broken TRS, respectively, in the function of the parameter γ. The experimental results are compared to the theoretical predictions for GOE (broken line) and GUE (solid line).

In the formula 26.39, $Z_{kk} = Z_0(1 + S_{kk})/(1 - S_{kk})$ and $Z^r_{kk} = Z_0(1 + S^r_{kk})/(1 - S^r_{kk})$ are the network and the radiation impedances expressed, respectively, by the network S_{kk} and the radiation S^r_{kk} scattering matrices. Z_0 is the characteristic impedance of the transmission lines feeding the 6-joint vertices. The radiation scattering matrix S^r_{kk} is the scattering matrix measured at the input of the coupling structure for the same coupling geometry, but with the vertices (walls) of the system removed to infinity. The scheme of the setup used to measure the radiation scattering matrix S_{kk}^r of the 6-joint connector is shown in Figure 26.8b. The five 50 Ω loads are connected to the microwave joint to simulate the vertices removed to infinity.

The measurements were done for the networks with preserved TRS containing no attenuators (the smallest value of the parameter $\gamma \simeq 5.0$) as well as for the ones containing fifteen 1 dB attenuators ($\gamma \simeq 20.6$), nine 1 dB and six 2 dB attenuators ($\gamma \simeq 25.6$), and finally fifteen 2 dB attenuators ($\gamma \simeq 54.4$). In all of the cases, our results were averaged over 60 microwave network configurations. The total "optical" length of the networks including joints and attenuators was varied for different network configurations from 538 to 681 cm. The experimental results for the networks with preserved TRS are in general good agreement with the theoretical ones predicted by References 75 and 83. Even for moderate absorption $\gamma \simeq 5$, the experimental result $W_{S,\beta=1} = 2.30 \pm 0.19$ is close to the theoretical one. In order to show the spread of the results obtained for different graph configurations, the assigned error was calculated using the definition of the sample standard deviation. Because of absorption of microwave cables (network bonds), we could not test experimentally predicted by the theory [75,83] increase of the enhancement factor $W_{S,\beta=1} \rightarrow 3$ for small values of the parameter γ. The measurements of the enhancement factor for this range of the parameter γ have been reported in Reference 87. In the experiment, a flat microwave cavity with ohmic losses was used. However, owing to the relatively large spread of the experimental results and their big uncertainties, the limit $\lim_{\gamma \to 0} W_{S,\beta=1} \simeq 3$ still remains to be tested.

The experimental studies of the enhancement factor $W_{S,\beta}$ of the two-port scattering matrix $\hat{S}$ for the systems with broken TRS have not been reported so far. However, Reference 87 has already reported a weak change of the enhancement factor due to partially broken time invariance in microwave cavity. In Figure 26.16, we show the results for the networks with broken TRS for the range of the parameter $7 < \gamma < 32$. The five experimental points were obtained for the networks containing four microwave circulators and different number of microwave attenuators. Beginning from the lowest absorption, the measurements were done for the microwave networks containing no attenuators ($\gamma \simeq 7.4$), seven 1 dB attenuators ($\gamma \simeq 13.6$), fifteen 1 dB attenuators ($\gamma \simeq 20.1$), nine 1 dB, six 2 dB attenuators ($\gamma \simeq 24.4$), and finally for the ones containing fifteen 2 dB attenuators ($\gamma \simeq 31.1$). In all cases, our results were averaged over 80 microwave network configurations. The total "optical" length of the networks, including joints, circulators, and attenuators, was varied for different network configurations from 528 to 699 cm. Also in this case, the experimental results are in good agreement with the theoretical ones $W_{S,\beta=2} \simeq 1$ predicted by References 75, 83, and 84 for moderate and strong absorption.

26.7 Isoscattering Networks

More than 25 years were needed to solve the problem expressed in the famous question "Can one hear the shape of a drum?" posed by Marc Kac in 1966 [88]. It concerns the problem of uniqueness of the spectrum of the Laplace operator on a planar domain with Dirichlet boundary conditions. The theoretical solution, indicating that in general the spectrum is not unique, was found in 1992 [89] and confirmed experimentally, using microwave cavities, 2 years later [90]. Similar problems were also considered in the context of quantum graphs [91]. For example, the question regarding the possibility of the determination of the geometry of a quantum graph in the scattering experiment was, on the basis of the theoretical considerations, recently answered in the negative [92,93]. We present here the results of the first scattering experiment on microwave networks simulating quantum graphs in which these theoretical predictions were tested [11].

The experimental confirmation of the existence of isoscattering networks simulating isoscattering graphs was demonstrated in a recent paper by Hul et al. [11]. Here we consider networks with two most typical physical vertex boundary conditions, the Neumann and Dirichlet ones. The first one imposes the

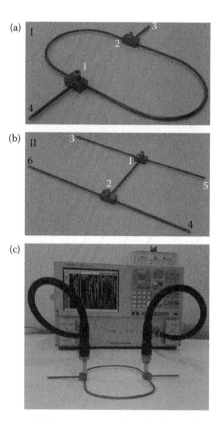

FIGURE 26.17 The isoscattering network No. I consisting of four bonds and four vertices and network No. II consisting of five bonds and six vertices are shown in panels (a) and (b), respectively. Panel (c) shows network I connected to a vector network analyzer. The numbers 1, 2, 3, 5 denote vertices with the Neumann boundary condition. The vertices with the Dirichlet boundary condition are denoted by numbers 4, 6.

continuity of waves propagating in bonds meeting at *i* and vanishing of the sum of their derivatives calculated at a vertex *i*. The latter demands vanishing of the waves at the vertex.

In order to verify experimentally a negative answer to the modified Mark Kac's question, we consider the two microwave networks Figure 26.17 which simulate [11] the two isoscattering graphs. The network No. I consists of four bonds and four vertices, whereas the second one consists of five bonds and six vertices. To obtain the scattering networks, two infinite leads (in the experiment, two elastic microwave cables) were connected to the vertices 1 and 2 of each network. The optical lengths of the microwave networks, which were obtained by rescaling of the physical lengths by the factor $\sqrt{\varepsilon}$, where $\varepsilon \approx 2.08$ is the dielectric constant of a homogeneous material filling the space between the inner and the outer leads of the cables, were the same and had the following value: 1.0504 ± 0.0010 m. The valency of the vertices No. 1 and No. 2, defined by the number of bonds and leads meeting at a given vertex, is four, whereas for the remaining vertices the valency is one. The Neumann boundary conditions (continuity of waves and vanishing of their derivatives) were imposed at all vertices except the vertices No. 4 and No. 6 where the Dirichlet boundary conditions (vanishing of waves) were assumed. Such systems are described by the two-port scattering matrix $\hat{S}(\nu)$:

$$\hat{S}(\nu) = \begin{pmatrix} S_{11}(\nu) & S_{12}(\nu) \\ S_{21}(\nu) & S_{22}(\nu) \end{pmatrix}. \tag{26.40}$$

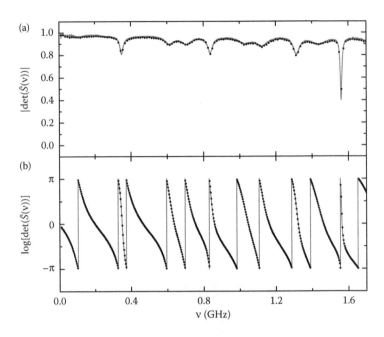

FIGURE 26.18   The amplitude and the phase of the determinant of the scattering matrix obtained for the microwave networks with $n = 4$ (gray solid line) and $n = 6$ (black circles) vertices, panels (a) and (b), respectively.

In the case when the networks are isoscattering and dissipative, the phases

$$log[det(\hat{S}^I(v))] = log[det(\hat{S}^{II}(v))] \tag{26.41}$$

and the modulus of the determinants (amplitude)

$$|det(\hat{S}^I(v))| = |det(\hat{S}^{II}(v))| \tag{26.42}$$

of their scattering matrices should be equal for all values of frequency. In order to measure the two-port scattering matrix $\hat{S}(v)$, the networks were connected to an Agilent E8364B vector network analyzer (VNA) via leads (Figure 26.17).

The moduli and phases of the determinants determined for the microwave network No. I (gray solid line) and No. II (black circles) in the frequency range 0.01–1.7 GHz are shown in panels (a) and (b) of Figure 26.18, respectively. The agreement between the results obtained for both networks is excellent, which confirms that the networks are isoscattering. To test the sensitivity of the spectral properties of the networks to boundary conditions, the measurement of the $\hat{S}(v)$ of network II with the boundary condition of the vertex 5 changed from the Neumann to Dirichlet one was performed.

A comparison of the results obtained for network I (gray solid line) with the ones obtained for network II where the Neumann boundary condition at vertex 5 was replaced by the Dirichlet one (black empty circles), presented in Figure 26.19, clearly shows that such a modification results in destroying isoscattering features. It should be stressed that the results of the measurements are also sensitive to the accuracy of the preparation of all elements of the microwave networks. The uncertainties of the bond lengths, which limit the frequency range of the measurements, are mainly due to the difficulties in the preparation of the Neumann boundary condition at the vertices. The presented results clearly confirm the theoretical prediction on the impossibility of the determination of the geometry of a quantum graph in the scattering experiment.

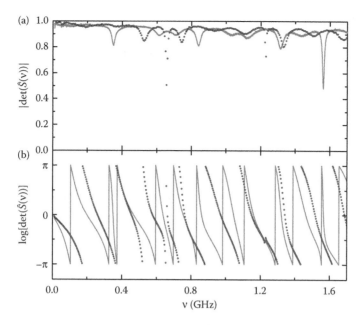

FIGURE 26.19 The amplitude and the phase of the determinant of the scattering matrix, panels (a) and (b), respectively, obtained for the microwave network with $n = 4$ vertices (gray solid line) compared to the results obtained for the microwave network with $n = 6$ vertices where the Neumann boundary condition at the vertex 5 was replaced by the Dirichlet one (black empty circles).

26.8 Conclusions

We show that quantum graphs with Neumann and Dirichlet boundary conditions can be simulated experimentally by microwave networks. Bidirectional microwave networks simulate quantum graphs with TRS (symmetry index $\beta = 1$). The results for the directed microwave networks with Faraday isolators indicate that their certain characteristics such as the INNS distribution can significantly differ from the RMT prediction for GOE, approaching the results characteristic of GUE. Directional microwave networks with microwave circulators simulate quantum graphs with broken TRS (symmetry index $\beta = 2$).

Microwave networks consisting of cables and attenuators are also model examples of experimental systems with absorption. The experimental results for the distribution $P(R)$ of the reflection coefficient R and the distributions $P(v)$ and $P(u)$ of imaginary and real parts of Wigner's reaction matrix obtained for microwave networks with absorption are in overall good agreement with the recent exact theoretical predictions for quantum systems with absorption [77].

Microwave networks can also be used to study the elastic enhancement factor $W_{S,\beta}$ in the presence of moderate and strong absorption. The experimental results for networks with preserved and broken TRS in the presence of strong absorption are, respectively, in good agreement with the theoretical ones $W_{S,\beta=1} \simeq 2$ and $W_{S,\beta=2} \simeq 1$ predicted by References 75, 83, and 84 within the framework of RMT.

An excellent example of the great research potential of quantum simulations based on microwave networks has been demonstrated in the case of isoscattering networks. The presented results [11] clearly confirmed the theoretical prediction on the impossibility of the determination of the geometry of a quantum graph in the scattering experiment.

The results presented in this chapter show that bidirectional and directed microwave networks that simulate bidirectional and directed quantum graphs, respectively, extend substantially the number of microwave systems that can be used to study wave effects predicted by quantum physics.

Acknowledgment

This work was partially supported by the Ministry of Science and Higher Education grant No. N N202 130239.

References

1. L. Pauling, *J. Chem. Phys.* **4**, 673, 1936.
2. H. Kuhn, *Helv. Chim. Acta*, **31**, 1441, 1948.
3. C. Flesia, R. Johnston, and H. Kunz, *Europhys. Lett.* **3**, 497, 1987.
4. R. Mitra and S. W. Lee, *Analytical Techniques in the Theory of Guided Waves* (Macmillan, New York, 1971).
5. E. L. Ivchenko and A. A. Kiselev, *JETP Lett.* **67**, 43, 1998.
6. J. A. Sanchez-Gil, V. Freilikher, I. Yurkevich, and A. A. Maradudin, *Phys. Rev. Lett.* **80**, 948, 1998.
7. Y. Imry, *Introduction to Mesoscopic Systems* (Oxford, New York, 1996).
8. D. Kowal, U. Sivan, O. Entin-Wohlman, and Y. Imry, *Phys. Rev. B* **42**, 9009, 1990.
9. Y. Avishai and J. M. Luck, *Phys. Rev. B* **45**, 1074, 1992.
10. T. Nakayama, K. Yakubo, and R. L. Orbach, *Rev. Mod. Phys.* **66**, 381, 1994.
11. O. Hul, M. Ławniczak, S. Bauch, A. Sawicki, M. Kuś, and L. Sirko, *Phys. Rev. Lett.* **109**, 040402, 2012.
12. S. Gnutzmann and U. Smilansky, *Adv. Phys.* **55**, 527, 2006.
13. K. A. Dick, K. Deppert, M. W. Larsson, T. Märtensson, W. Seifert, L. R. Wallenberg, and L. Samuelson, *Nat. Mater.* **3**, 380, 2004.
14. K. Heo et al., *Nano Lett.* **8**, 4523, 2008.
15. T. Kottos and U. Smilansky, *Phys. Rev. Lett.* **79**, 4794, 1997.
16. T. Kottos and U. Smilansky, *Ann. Phys.* **274**, 76, 1999.
17. T. Kottos and U. Smilansky, *Phys. Rev. Lett.* **85**, 968, 2000.
18. F. Barra and P. Gaspard, *J. Stat. Phys.* **101**, 283, 2000.
19. G. Tanner, *J. Phys. A* **33**, 3567, 2000.
20. G. Tanner, *J. Phys. A* **34**, 8485, 2001.
21. P. Pakoński, K. Życzkowski, and M. Kuś, *J. Phys. A* **34**, 9303, 2001.
22. R. Blümel, Yu Dabaghian, and R. V. Jensen, *Phys. Rev. Lett.* **88**, 044101, 2002.
23. T. Kottos and H. Schanz, *Physica E* **9**, 523, 2003.
24. T. Kottos and U. Smilansky, *J. Phys. A* **36**, 3501, 2003.
25. P. Pakoński, G. Tanner, and K. Życzkowski, *J. Stat. Phys.* **111**, 1331, 2003.
26. T. Kottos, *Acta Phys. Pol. A* **109**, 7, 2005.
27. P. Exner, P. Hejcik, and P. Seba, *Acta Phys. Pol. A* **109**, 23, 2005.
28. O. Hul, Sz. Bauch, P. Pakoński, N. Savytskyy, K. Życzkowski, and L. Sirko, *Phys. Rev. E* **69**, 056205, 2004.
29. O. Hul, O. Tymoshchuk, Sz. Bauch, P. M. Koch, and L. Sirko, *J. Phys. A* **38**, 10489, 2005.
30. M. Ławniczak, O. Hul, S. Bauch, P. Šeba, and L. Sirko, *Phys. Rev. E* **77**, 056210, 2008.
31. M. Ławniczak, S. Bauch, O. Hul, and L. Sirko, *Phys. Scr.* **T135**, 014050, 2009.
32. M. Ławniczak, O. Hul, S. Bauch, and L. Sirko, *Acta Phys. Pol. A* **116**, 749, 2009.
33. O. Hul, S. Bauch, M. Ławniczak, and L. Sirko, *Acta Phys. Pol. A*, **112**, 655, 2007.
34. D. S. Jones, *Theory of Electromagnetism* (Pergamon Press, Oxford, 1964), 254.
35. L. D. Landau and E. M. Lifshitz, *Electrodynamics of Continuous Media* (Pergamon Press, Oxford, 1960).
36. G. Goubau, *Electromagnetic Waveguides and Cavities* (Pergamon Press, Oxford, 1961).
37. R. Blümel, A. Buchleitner, R. Graham, L. Sirko, U. Smilansky, and H. Walther, *Phys. Rev. A* **44**, 4521, 1991.
38. M. Bellermann, T. Bergemann, A. Haffmann, P. M. Koch, and L. Sirko, *Phys. Rev. A* **46**, 5836, 1992.

39. L. Sirko, S. Yoakum, A. Haffmans, and P. M. Koch, *Phys. Rev. A* **47**, R782, 1993.

40. L. Sirko and P. M. Koch, *Appl. Phys. B* **60**, S195, 1995.

41. L. Sirko, A. Haffmans, M. R. W. Bellermann, and P. M. Koch, *Europhys. Lett.* **33**, 181, 1996.

42. L. Sirko, S. A. Zelazny, and P. M. Koch, *Phys. Rev. Lett.* **87**, 043002, 2001.

43. L. Sirko and P. M. Koch, *Phys. Rev. Lett.* **89**, 274101, 2002.

44. H. J. Stöckmann and J. Stein, *Phys. Rev. Lett.* **64**, 2215, 1990.

45. S. Sridhar, *Phys. Rev. Lett.* **67**, 785, 1991.

46. H. Alt, H.-D. Gräf, H. L. Harner, R. Hofferbert, H. Lengeler, A. Richter, P. Schardt, and A. Weidenmüller, *Phys. Rev. Lett.* **74**, 62, 1995.

47. P. So, S. M. Anlage, E. Ott, and R. N. Oerter, *Phys. Rev. Lett.* **74**, 2662, 1995.

48. U. Stoffregen, J. Stein, H.-J. Stöckmann, M. Kuś, and F. Haake, *Phys. Rev. Lett.* **74**, 2666, 1995.

49. F. Haake, M. Kuś, P. Šeba, H.-J. Stöckmann, and U. Stoffregen, *J. Phys. A* **29**, 5745, 1996.

50. L. Sirko, P. M. Koch, and R. Blümel, *Phys. Rev. Lett.* **78**, 2940, 1997.

51. Sz. Bauch, A. Błędowski, L. Sirko, P. M. Koch, and R. Blümel, *Phys. Rev. E* **57**, 304, 1998.

52. N. Savytskyy, A. Kohler, Sz. Bauch, R. Blümel, and L. Sirko, *Phys. Rev. E* **6403**, 6211, 2001.

53. R. Blümel, P. M. Koch, and L. Sirko, *Found. Phys.* **31**, 269, 2001.

54. Y. Hlushchuk, L. Sirko, U. Kuhl, M. Barth, and H.-J. Stöckmann, *Phys. Rev. E* **63**, 046208, 2001.

55. N. Savytskyy, O. Hul, and L. Sirko, *Phys. Rev. E* **70**, 056209, 2004.

56. S. Deus, P. M. Koch, and L. Sirko, *Phys. Rev. E* **52**, 1146, 1995.

57. U. Dörr, H.-J. Stöckmann, M. Barth, and U. Kuhl, *Phys. Rev. Lett.* **80**, 1030, 1998.

58. C. Dembowski, B. Dietz, H.-D. Gräf, A. Heine, T. Papenbrock, A. Richter, and C. Richter, *Phys. Rev. Lett.* **89**, 064101-1, 2002.

59. O. Tymoshchuk, N. Savytskyy, O. Hul, S. Bauch, and L. Sirko, *Phys. Rev. E* **75**, 037202, 2007.

60. P. Kuchment and H. Zeng, *J. Math. Anal. Appl.* **258**, 671, 2001.

61. J. Rubinstein and M. Schatzman, *Arch. Rat. Mech. Anal.* **160**, 271, 2001.

62. E. Doron, U. Smilansky, and A. Frenkel, *Phys. Rev. Lett.* **65**, 3072, 1990.

63. F. Haake, *Quantum Signatures of Chaos* (2nd ed. Springer, Berlin, 2000).

64. H.-J. Stöckmann, *Quantum Chaos—An Introduction* (Cambridge University Press, Cambridge, 1999).

65. M. V. Berry, *Proc. R. Soc. London A* **400**, 229, 1985.

66. G. Akguc and L. E. Reichl, *Phys. Rev. E* **64**, 056221, 2001.

67. Y. V. Fyodorov and D.V. Savin, *JETP Lett.* **80**, 725, 2004.

68. M. L. Mehta, *Random Matrices* (Academic Press, New York, 1991).

69. S. Hemmady, X. Zheng, E. Ott, T. M. Antonsen, and S. M. Anlage, *Phys. Rev. Lett.* **94**, 014102, 2005.

70. G. López, P. A. Mello, and T. H. Seligman, *Z. Phys. A* **302**, 351, 1981.

71. E. Doron and U. Smilansky, *Nucl. Phys. A* **545**, 455, 1992.

72. P. W. Brouwer, *Phys. Rev. B* **51**, 16878, 1995.

73. D. V. Savin, Y. V. Fyodorov, and H.-J. Sommers, *Phys. Rev. E* **63**, 035202, 2001.

74. Y. V. Fyodorov, *JETP Lett.* **78**, 250, 2003.

75. Y. V. Fyodorov, D. V. Savin, and H.-J. Sommers, *J. Phys. A* **38**, 10731, 2005.

76. U. Kuhl, M. Martinez-Mares, R. A. Méndez-Sánchez, and H.-J. Stöckmann, *Phys. Rev. Lett.* **94**, 144101, 2005.

77. D. V. Savin, H.-J. Sommers, and Y. V. Fyodorov, *JETP Lett.* **82**, 544, 2005.

78. R. A. Méndez-Sánchez, U. Kuhl, M. Barth, C. V. Lewenkopf, and H.-J. Stöckmann, *Phys. Rev. Lett.* **91**, 174102-1, 2003.

79. S. Hemmady, X. Zheng, T. M. Antonsen, Jr., E. Ott, and S. M. Anlage, *Acta Phys. Pol. A* **109**, 65, 2006.

80. C. W. J. Beenakker and P. W. Brouwer, *Physica E* **9**, 463, 2001.

81. P. Exner, *Ann. Inst. H. Poincaré: Phys. Theor.* **66**, 359, 1997.

82. M. Ławniczak, S. Bauch, O. Hul, and L. Sirko, *Phys. Rev. E* **81**, 046204, 2010.

83. D. V. Savin, Y. V. Fyodorov, and H.-J. Sommers, *Acta Phys. Pol. A* **109**, 53, 2005.

84. X. Zheng, S. Hemmady, T. M. Antonsen, Jr., S. M. Anlage, and E. Ott, *Phys. Rev. E* **73**, 046208, 2006.

85. B. Michielsen, F. Isaac, I. Junqua, and C. Fiachetti, arxiv:math-ph/0702041v1 13 Feb. 2007.

86. O. Hul, P. Seba, and L. Sirko, *Phys. Rev. E* **79,** 066204, 2009.

87. B. Dietz, T. Friedrich, H. L. Harney, M. Miski-Oglu, A. Richter, F. Schäfer, J. Verbaarschot, and H. A. Weidenmüller, *Phys. Rev. Lett.* **103,** 064101, 2009.

88. M. Kac, *Am. Math. Mon.* **73,** 1, 1966.

89. C. Gordon, D. Webb, and S. Wolpert, *Bull. Am. Math. Soc.* **27,** 134, 1992.

90. S. Sridhar and A. Kudrolli, *Phys. Rev. Lett.* **72,** 2175, 1994.

91. B. Gutkin and U. Smilansky, *J. Phys. A* **34,** 6061, 2001.

92. J. Boman and P. Kurasov, *Adv. Appl. Math.* **35,** 58, 2005.

93. R. Band, A. Sawicki, and U. Smilansky, *J. Phys. A* **43,** 415201, 2010.

VII

Optics and Chaos

27

Optics and Chaos: Chaotic, Rogue, and Noisy Optical Dissipative Solitons

Vladimir L.
Kalashnikov

27.1 Introduction

In the last decade, the concept of a dissipative soliton (DS), which is a strongly localized and stable coherent structure emergent in a nonlinear dissipative system far from the thermodynamic equilibrium, actively developed and became well established. This concept is highly useful in very different fields of science, ranging from field theory and cosmology, optics and condensed-matter physics, to biology and medicine [2,13,15,16,38,52,135,255]. One may paraphrase: It is "apparent that solitons are around us. In the true sense of the word they are absolutely everywhere" [14]. Nonequilibrium character of a system where a DS emerges, requires a well-organized energy exchange of DS with an environment. In turn, this energy flow forms a nontrivial internal structure of a soliton, which provides the energy redistribution inside it (e.g., see [13,15,257]). In this respect, a DS is a primitive analog of a cell.

In particular, a DS can have an inhomogeneous phase-(ϕ-)distribution so that $Q \equiv d^2\phi/dt^2 \neq 0$ (here, t is a coordinate along which a DS is localized). The last value is called a "chirp" and, correspondingly, a DS with a substantially large chirp was named as a "chirped DS" (CDS) [196]. The unique feature of CDS is its capability to accumulate an energy E without stability loss so that $E \propto Q$ [105,114].[*] As a result,

---

[*] Note that the used designations are model-dependent. For instance, E, ϕ, and t are the pulse energy, phase, and local time for a mode-locked laser [13], but are the number of particles (mass of condensate), momentum (wave number), and transverse spatial coordinate for a Bose–Einstein (BE) condensate [136].

CDS is energy (or "mass") scalable [8,105,114,196]. This phenomenon resembles a resonant enhancement of oscillations in environment-coupled systems so that it was proposed to name it as a "DS resonance" (DSR) [40].

This capacity of a DS to accumulate an "energy" (or a "mass," see footnote above) is of interest for a lot of applications. For instance, it provides the energy scaling of ultrashort laser pulses [110] so that more than 10 MW peak powers and, respectively, more than 10^{14} W/cm^2 intensities become reachable directly from a laser operating at over-MHz repetition rates [174]. Such and quite reachable higher pulse powers bring the high-field physics on tabletops of a mid-level university lab [98]. In particular, high-energy ultrashort pulse lasers nowadays allow such experiments as direct gas ionization and high-harmonic generation, pump-probe diffraction experiments with electrons and production of nanometer-scale structures at the surface of transparent materials, characterization and control of electronic dynamics, a variety of biophotonic and biomedical applications, etc. [159,211,214]. Moreover, the over-MHz pulse repetition rates provide a signal rate improvement factor of 10^3–10^4 in comparison with that of classical chirped-pulse amplifiers [98]. As a result, the signal-to-noise ratio enhances essentially, as well. But, besides a direct scientific and practical interest, an "avatar" of DS in the form of an ultrashort laser pulse can be considered as a testbed for exploring the DSs in whole [8]. A rapid progress of modern femtosecond laser technology provides an ideal playground for such an exploration so that the theoretical insights promise to become directly testable and, on the other part, the theory can be urged by new experimental challenges.

As a result of its nonequilibrium character, a DS can demonstrate a highly nontrivial dynamics, including the formation of multi-DS complexes [227], DS explosions [50], noise-like DSs [101], etc. The resulting structures can be very complicated and consist of strongly or weakly interacting solitons (the so-called soliton molecules and gas) [263] as well as the short-range noise-like oscillations inside a larger wave packet [143] or the clusters of strongly interacting dark and gray solitons against a background ("condensate") [238]. The nonlinear dynamics of these structures can cause regular, chaotic-like, and turbulent behavior [103]. Such a behavior can be affected by excitation of internal perturbation modes arising from a nontrivial internal structure of a DS [116]. A perturbed DS resembles a glass of boiling water and it becomes very sensitive to noise (including quantum) influence. This sensitivity of a DS to a noise turns it into a "mesoscopic" quantum object [253], whose properties remain unexplored.

Further contents can be outlined in the following way: (i) a digest of DS theory will be exposed, (ii) different scenarios of DS chaotization will be considered, (iii) noisy and rogue DS will be surveyed, and (iv) some quantum aspects of the soliton theory will be treated.

27.2 Concept of Optical Dissipative Soliton: Physical and Mathematical Aspects

There is a vast amount of literature regarding the theory of DSs. Some preliminary systematization can be found in References 2, 8, 13–16, 105, 110, 201, 204, 236, and 250. However, it is necessary at first to declare the stumbling block of this theory: absence of a unified viewpoint. There exist unbroken walls between the circles of the scientific community exploiting and exploring the concept of a DS: walls between the solid-state and fiber laser representations of the theory, condensed-matter physics, numerical and analytical approaches, solitonic and statistical concepts, etc.

Briefly and conditionally, the theoretical approaches to DS analysis can be divided into (1) numerical, (2) exact analytical, and (3) approximated analytical. The third approach includes the models based on (1) perturbative and (2) adiabatic models (AM) as well as models based on (3) phase-space truncation (i.e., variational approximation [VA] and method of moments [MM]).[*]

As was emphasized, linear and nonlinear dissipation is crucial for DS formation. The simplest and most studied models for such a type of phenomena are based on the different versions of the nonlinear complex

---

[*] The statistical approaches will be only briefly outlined in this chapter.

Ginzburg–Landau equation (NCGLE) (e.g., see [15,21]). With regard to the physics of mode-locked lasers, the cubic NCGLE [94] is known as the Haus master equation. Cubic-quintic as well as nonpolynomial nonlinear extensions of this equation, as was proved, are precise models for description of laser CDSs with different types of mode-locking [15,16,114,147,148,205,206] and are valid even in the case of lumped dynamics [55,261].

The generalized $(1+1)$-dimensional NCGLE can be expressed in the following form [110]:

$$\frac{\partial a\,(z,t)}{\partial z} = \left[-\sigma + \alpha \frac{\partial^2}{\partial t^2} + \mathfrak{F}\left(|a|^2\right) \right] a\,(z,t) + i \left\{ \beta \frac{\partial^2}{\partial t^2} - \gamma |a|^2 - \chi |a|^4 \right\} a\,(z,t). \tag{27.1}$$

Here, for an optical field envelope $a(z,t)$, the slowly varying envelope approximation [184] is assumed, which is valid until $T \gg 1/\omega_0$ and $\Delta \ll \omega_0$ (ω_0 is the optical wave carrier frequency, Δ is the spectral half-width of DS, and T is its width). z is a "propagation distance" (or a time for a Bose–Einstein condensate). In a laser, a DS transits periodically the same elements during its evolution and this period is termed a laser cavity period T_{rep}. For solid-state lasers as a rule, the DS dynamics is slow so that the corresponding evolutional scale exceeds T_{rep}. In this case, z can be scaled on a cavity length and interpreted as a "cavity round-trip number" N. In a fiber laser, all parameters of Equation 27.1 is z-dependent and, thereby, the field evolution within one cavity round-trip has to be taken into consideration, as a rule. $t \equiv [\tau - z(dk/d\omega)|_{\omega=\omega_0}]$ is the local or "reduced" time (τ is a "laboratory" time, ω is a frequency deviation measured from ω_0, and $k(\omega)$ is a wave number). In the Bose–Einstein condensate, a time coordinate has a sense of spatial one.

The energy exchange with an environment can be characterized by a "net-loss" coefficient σ:

$$\frac{\partial a}{\partial z} = -\sigma(E)a = -\ell a + a \times \begin{cases} \dfrac{g_0}{1 + E/E_s} \\[2ex] \dfrac{g_0}{\sqrt{1 + E/E_s}} \end{cases} \tag{27.2}$$

where g_0 is the unsaturated gain defined by a pump (i.e., the gain coefficient for a small signal), $E \equiv \int_0^{T_{rep}} |a|^2 dt$ is the intracavity field energy ($|a|^2$ has a dimension of power), and $E_s \equiv \hbar\omega_0 S/\sigma_g$ is the gain saturation energy (σ_g is the gain cross section and S is the laser beam area). Multiple propagation of the pulse through an active medium during one cavity round-trip, as it takes a place in a thin-disk oscillator, has to be taken into account by a corresponding multiplier before E in Equation 27.2. The power-independent ("linear") loss coefficient is ℓ. The "gain saturation" law is model-dependent (multipliers after brace in Equation 27.2 represent only two possible ones).[*] If a DS is considered in the vicinity of a marginal stability threshold (where $\sigma = 0$ by definition), one may expand σ:

$$\sigma(E) \approx \delta(E - E^*), \tag{27.3}$$

where E^* is the energy of a marginally stable state corresponding to $\sigma = 0$, and $\delta \equiv (d\sigma/dE)|_{E=E^*}$. In a high-energy regime, the time dependence of E (or, in other words, the dynamic gain saturation) begins to play an important role and has to be described by a separate differential equation [110].

The α-term in Equation 27.1 describes a frequency dissipation in a system (in a Bose–Einstein condensate, this term corresponds to a particle velocity dependence of a "leak" from the condensate). If a spectral filter has a profile $\Phi(\omega)$, then its contribution can be described as

$$\frac{\partial a(z,\omega)}{\partial z} = \Phi(\omega)a(z,\omega), \tag{27.4}$$

where $a(z,\omega)$ is the Fourier image of $a(z,\tau)$.

---

[*] It can be very complicated for a fiber laser and is beyond the scope of a simple scalar model (27.1) [215].

Usually, the Lorenzian shape for $\Phi(\omega)$ is assumed:

$$\Phi(\omega) = \frac{g(\omega_0)}{1 + i\omega/\Omega_g}, \tag{27.5}$$

where $g(\omega_0)$ is a gain coefficient or a maximum transmission coefficient (e.g., for an output mirror). It is assumed that spectral filtering is centered at ω_0. Ω_g is a filter (gain) bandwidth and then $\alpha \equiv g(\omega_0)/\Omega_g^2$ in Equation 27.1.

If $\Delta \ll \Omega_g$, one may expand Equation 27.5 and proceed in the time domain

$$\frac{\partial a(z,\tau)}{\partial z} \approx g(\omega_0)\left[1 - \frac{1}{\Omega_g}\frac{\partial}{\partial\tau} + \frac{1}{\Omega_g^2}\frac{\partial^2}{\partial\tau^2} - h.o.t.\right]a(z,\tau), \tag{27.6}$$

where *h.o.t.* means the higher in $\partial/\partial\tau$-order terms, which are negligible as a rule (for some important exceptions, see, e.g., [15,55]).

The $\mathfrak{F}(|a|^2)$-operator describes a so-called self-amplitude modulation (SAM) or a nonlinear gain. In a laser, this term corresponds to some power-discrimination process in a system. Such a process provides an effective gain growth with the power (at least up to some power level), which forms and stabilizes a DS. The mechanisms of SAM are various [188,248], but there are two simplest expressions for the $\mathfrak{F}(|a|^2)$-operator, which are used extensively:

$$\mathfrak{F}(|a|^2) = \begin{cases} \dfrac{\mu\kappa|a|^2}{1 + \kappa|a|^2} \\ \kappa|a|^2(1 - \zeta|a|^2) \end{cases}. \tag{27.7}$$

The first SAM law describes a monotonic decrease (up to zero) of loss (μ is a loss level, or a so-called "loss modulation depth") with the power $|a|^2$. The corresponding nonlinear coefficient κ describes a "strength" of such a "loss saturation."[*] The second SAM law corresponds to a so-called cubic-quintic NCGLE [15] and describes a loss decrease with the power only up to some critical power level ($P_{cr} = 1/2\zeta$) following which a loss begins to increase. The corresponding parameter of "SAM saturation" is defined by ζ. In a BE condensate, the SAM can be connected with the multiparticle interactions resulting in the creation/annihilation of bosons.

The terms within square brackets in Equation 27.1 describe the dissipative factors in a system. The terms within braces are more convenient and correspond to the so-called cubic-quintic nonlinear Schrödinger equation, which describes dispersive (nondissipative) solitons in fiber optics, BE condensate, etc. (e.g., see [7,13,52,185,255]). The remarkable property of cubic nonlinear nondissipative reduction of Equation 27.1 is that it is completely integrable (e.g., see [48,79,170]) and, thereby, is completely traceable mathematically and, to a certain extent, is "linear-like." A "true" (in the sense of "integrability") soliton of such equation corresponds to the case of $\chi = 0$, $\beta < 0$ (so-called "anomalous" group-delay dispersion [GDD]) and $\gamma > 0$ ("self-focusing" nonlinearity or attractive three-bosonic potential).[†]

The imaginary nondissipative linear ("kinetic energy") term in Equation 27.1 describes a GDD [188,248], which is the frequency dependence of the wave-packet propagation constant. Note that the dispersion coefficient β is frequency-dependent, as well. In the simplest case, such a dependence can be expressed by the inclusion of higher-order derivatives into Equation 27.1 [7]. The corresponding terms are called the higher-order dispersions (HODs). As will shown below, the HOD plays an important role in the chaotization of DS dynamics.

The imaginary nonlinear term in Equation 27.1 describes a self-phase modulation (SPM) [188,248], which is the power dependence of the wave-packet phase. As a rule, the cubic term (the corresponding "three-boson interaction" coefficient is γ) is sufficient to take into account the SPM, but the higher-order nonlinear corrections (e.g., χ-coefficient in Equation 27.1) can affect the DS properties substantially [13,

---

[*] It should be noted that μ has to be included in the definition of ℓ for such a form of SAM law.

[†] The situation is completely reversible: a "true" soliton exists for $\beta > 0$ and $\gamma < 0$, as well.

107]. In particular, a strongly "chirped" and, thereby, energy-scalable DS can develop under condition of "anomalous" GDD, when $\gamma > 0$ and $\chi \neq 0$ [40,41,107,158].

27.2.1 Numerical Study of DSs

Extensive numerical study of DSs of the (1+1)-dimensional cubic-quintic NCGLE has been carried out by N.N. Akhmediev with coauthors [8,13–16,18,39,41,80,226]. The simulations have allowed finding the DS stability regions for some two-dimensional projections of NCGLE parametrical space. The summarizing description of the results obtained is presented in [110].

The most impressive results are

    i. Parametric space of DS can have a reduced dimensionality, the corresponding phenomenon was called the dissipative soliton resonance (DSR).

The last term is sound because this phenomenon resembles a resonant enhancement of oscillations in environment-coupled systems. As a result, the DS becomes energy-scalable and can gain the energy without stability loss. This property is illustrated by Figure 27.1, where one can see a swift growth of DS energy along with the dispersion β.

    ii. The next important result is that DSR remains in models with a lumped evolution of DS.

Such models describe a DS evolution in a typical fiber laser, where the DS parameters can change dramatically during the propagation. The existence of DSR in this case is good news from the point of view of soliton energy scalability.

    iii. DSR exists in both normal ($\beta > 0$) and anomalous ($\beta < 0$) regions of GDD.

The calculations demonstrate that the DS scaling properties for an anomalous GDD surpass those for a normal one. An "unusual" (due to the existence of a "true" soliton for $\beta < 0$) anomalous dispersion regime of DS generation has been observed experimentally [62]. Note, that $\chi \neq 0$ is required for such a regime.

In spite of lasting achievements of numerical approaches in the DS exploration, they have some inherent shortcomings: (i) the parametrical space under consideration is not physically relevant owing to the dropping of the gain saturation, (ii) the true dimensionality of DS parametric space is not identified exactly,

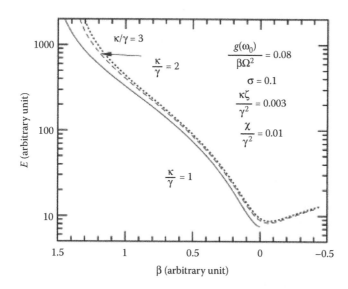

FIGURE 27.1 Soliton energy E versus dispersion β. (Adapted from N. N. Akhmediev, J. M. Soto-Crespo, and Ph. Grelu. *Phys. Lett. A*, 372(17):3124–3128, 2008.)

and (iii) the polarization dynamics does not get taken into account [215]. As has been shown, the inclusion of gain saturation allows a real-world description of the lasers producing DSs as well as it demonstrates a modification of the DSR conditions (e.g., see [43,55,125]). But the patterns revealed cannot be formulated as the quantitative and well-founded laws so far. It is clear that the only advanced and self-consistent analytical theory of DS would provide, in particular, a true representation of dissipative-solitonic parametric space and DSR conditions. A vector extension of the DS model (i.e., taking into account a polarization dynamics) offers supplementary issues as a result of doubling of a phase space dimensionality and taking into account the vector character of field–matter coupling [215].

27.2.2 Analytical Theories of DSs

Below, basic analytical models of DS will be briefly surveyed. For convenience, we divide these models into three groups based on (i) exact DS solution of Equation 27.1, (ii) solutions obtained from the so-called adiabatic approximation, and (iii) approximated solutions obtained from the variational method and the method of moments.

27.2.2.1 Exact DS Solutions of NCGLE

Since Equation 27.1 is nonintegrable, its exact soliton solutions are known only for a few cases. For instance, the cubic-quintic NCGLE in the cubic limit ($\zeta = \chi \equiv 0$) has an exact DS solution in the following form [95,166]:

$$a\left(z,t\right) = a_0 \cosh^{-1+i\psi}\left(\frac{t}{T}\right) \exp\left(iqz\right), \tag{27.8}$$

where a_0 is a soliton amplitude, T is its width, q is a soliton "wave number," and ψ is a so-called chirp, which is a measure of soliton phase inhomogeneity ($\psi \propto QT^2$, see Section 27.1) or its spectral extra-broadening: $\Delta \propto \sqrt{1+\psi^2}/T$ [95] (Δ is a DS spectral half-width). These parameters can be expressed through the parameters of Equation 27.1. There are two disconnected regions of DS existence (such regions are disconnected by virtue of stability condition $\sigma > 0$, which can get broken in the vicinity of $\beta = 0$): (i) anomalous dispersion regime with $\psi \approx 0$ and (ii) normal dispersion regime with $\psi \neq 0$, which corresponds to a CDS [105,110].

Solution (27.8) corresponding to a weakly nonlinear limit of NCGLE does not provide with an in-depth analysis of energy scalable DS. In particular, it is obvious that an absence of the SAM saturation ($\zeta = 0$) enhances the tendency to a collapse-like instability of DS [42]. The general DS solutions of strongly nonlinear versions of Equation 27.1 (e.g., cubic-quintic NCGLE) are unknown. One may hope that they can be revealed on the basis of the algebraic nonperturbative techniques [48,79], which, nevertheless, are not developed sufficiently still. However, one known exact solution of the cubic-quintic NCGLE can provide with some insight into properties of DSs [18,41,80,200,226,240]:

$$a(z,t) = \sqrt{\frac{A}{B + \cosh\left(t/T\right)}} \exp\left[\frac{i\psi}{2} \ln\left(B + \cosh\left(t/T\right)\right) + iqz\right] \tag{27.9}$$

where A, B, T, ψ, and q are the real constants characterizing pulse amplitude, shape, width, chirp, and wave number, respectively. It is important to emphasize that this partial solution exists for only certain algebraic relations imposed on the parameters of Equation 27.1 (i.e., this solution has a so-called "co-dimension one" [236]).

In spite of such a very specific property of Equation 27.9, it allows classifying systematically the DS spectra [200]. The crucial shortcoming of Equation 27.9 is that the strict restrictions are imposed on the NCGLE parameters. As a result, the DS cannot be traced within a broad multidimensional parametric range and the picture obtained is rather sporadic and is of interest only in the close relation with numerical results. Some additional information can be obtained on the basis of perturbation theory [140,164]. This approach provides with a quite accurate approximation for a low-energy CDS, when $\sigma\zeta/\kappa \ll 1$ [107]. The corresponding solution is continuously extendible to Equation 27.8, when ζ and $\chi \to 0$.

As further steps in the development of analytical techniques are required for the DS exploration, two powerful approximate methods have been proposed.

27.2.2.2 Adiabatic Theory of DS

Adiabatic theory of DS has been proposed in References 3 and 196 and can be put in a nutshell in the following way [110]. The theory is developed in three main steps: (i) the condition of $T \gg \sqrt{\beta}$ allows the adiabatic approximation for Equation 27.1; (ii) regularization ("renormalization") procedure is applied to an expression for the soliton frequency deviation $\Omega \equiv d\phi/dt$ (i.e., $Q = d\Omega/dt$), which excludes nonphysical solutions; and (iii) the method of stationary phase for a Fourier image of soliton complex envelope $a(z, \omega)$ is applied, which gives expressions for the soliton spectrum and its energy. The last step requires $|\psi| \gg 1$, i.e., $\beta \gg \alpha$ and $\gamma \gg \kappa$ in Equation 27.1.

This method has been applied for both versions of the SAM law (27.7) as well as for the case of higher-order SPM with $\chi \neq 0$ (see [105,110] for a more technical introduction). The adiabatic theory gives a set of interesting results providing with a deep insight into properties of CDSs.

i. There are two branches of CDSs, which correspond to different soliton energies for a fixed control parameter $C \equiv \alpha\gamma/\beta\kappa$ or different control parameters for a fixed soliton energy.

Figure 27.2 demonstrates the typical power profiles $P(t) = |a(z, t)|^2$ and the corresponding frequency deviations $\Omega(t)$ for these two branches. The so-called positive branch of DS has higher energies and provides the energy scalability by means of pulse width scaling.* A DS profile becomes flat-top in the process of such scaling.

ii. CDS spectra are truncated at some resonance frequency $\pm\Delta$ and have various profiles: flat-top, convex, concave, or concave–convex.

Δ (as well as ω) has a sense of frequency measured from the soliton carrier frequency ω_0. Figures 27.3 and 27.4 demonstrate some typical spectra of CDSs. The spectra can have strongly enhanced ("condensate") spectral components at the edges (at $\pm\Delta$) (Figure 27.4). The nature of this phenomenon is closely connected with DS perturbation and chaotization and will be considered below. When the energy scales and the DS soliton temporal profiles tend to a flat-top one, the spectral profile becomes bell-like ("Lorentzian").

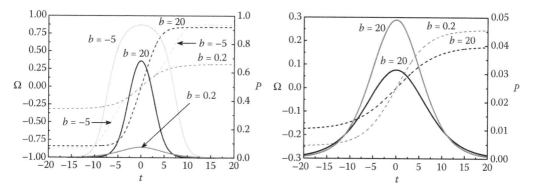

FIGURE 27.2 Left: The positive-branch DS profiles $P(t)$ (solid curves) and frequency deviations $\Omega(t)$ (dashed curves) for the different $b \equiv \zeta\gamma/\chi$. Right: Ditto for the negative-branch DS. $C \equiv \alpha\gamma/\beta\kappa = 1$, $\sigma\zeta/\kappa = 0.01$. (Adapted from V. L. Kalashnikov. *CMSIM*, 1:51–59, 2011.)

---

* The negative branch is quasi-(in the sense of possible stability loss) continuous connected with Equation 27.8.

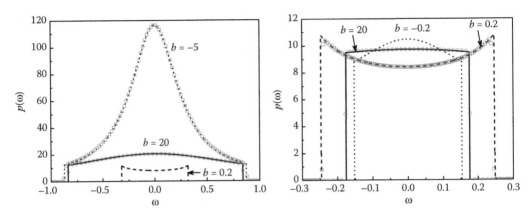

FIGURE 27.3 Power spectra $p(\omega)$ of CDSs shown in Figure 27.2. Points and squares correspond to spectral profiles obtained from numerical simulations. (Adapted from V. L. Kalashnikov. *CMSIM*, 1:51–59, 2011.)

iii. The lasting achievement of adiabatic theory is a disclosure of the structure of DS parametric space and its relevant dimensionality. The corresponding representation of DS parametric space has been named as a "master diagram."

CDS "lives" inside a parametric space with reduced dimensionality (two-dimensional one for $\chi = 0$ and three-dimensional one for $\chi \neq 0$, see Equation 27.1). The structure of this space can be represented in the form of the so-called "master diagram" [124] and is completely characterized by a set of "isogains," that is, a set of curves, where a DS traces some fixed value of the σ-parameter (Equation 27.1). The marginal isogain curve $\sigma = 0$ corresponds to the DS stability border.

iv. Master diagram reveals the energy-scalability properties of DS.

As can be seen from Figure 27.5, the energy scalability of positive and negative branches of CDS differs: the negativec branch requires a substantial decrease of C-parameter (e.g., as a result of the GDD growth) for energy scaling. The positive branch of DS excels the negative one in scalability. For instance, it has a "perfect" scalability for the cubic-quintic NCGLE (left side of Figure 27.5), which means that there exists

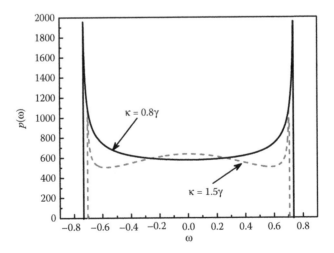

FIGURE 27.4 Positive-branch CDS spectra with enhanced edges. (Adapted from V. L. Kalashnikov. *CMSIM*, 1:51–59, 2011.)

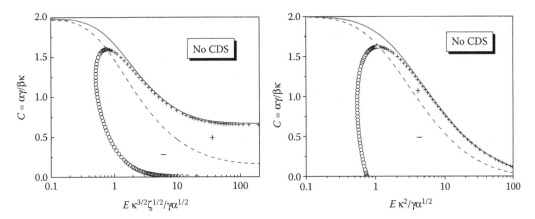

FIGURE 27.5 Left: Master diagram for the CDS of cubic-quintic NCGLE, $\chi = 0$. There exists no soliton above the solid red curve (therein $\sigma < 0$, i.e., the vacuum of Equation 27.1 is unstable). Dashed curve divides the regions, where the positive- and negative-branches of CDS exist. Crosses (circles) correspond to the positive-(negative-)branch for an isogain $\sigma\zeta/\kappa = 0.01$. E is a DS energy. Right: Master diagram for the CDS corresponding to the upper expression for a SAM law in (27.7). Solid curve is the stability threshold. Dashed curve divides the regions of positive and negative branches of DS, which are shown for an isogain $\sigma/\mu = 0.01$ by crosses and circles, respectively. (Adapted from V. L. Kalashnikov. *Solid State Laser*, pages 145–184. InTech, 2012.)

$\lim_{C \to 2/3} E = \infty$ along the marginal isogain. This phenomenon has been named the DSR (see above) and it means that the DS energy scaling for a fiber laser can been achieved by a plain laser lengthening.

As can be seen from the right picture in Figure 27.5, there exists no DSR, when the SAM is defined by the upper expression in Equation 27.7. But the positive branch of CDS is maximally scalable, as well. The corresponding asymptotical energy scalability law can be expressed as $E \approx 18\mu\beta^2/\gamma\alpha^{3/2}$ [110].

v. Simple analytical expressions for the complex spectral amplitude of DS allow developing the perturbation theory in spectral domain.

Such a theory (e.g., see [105,108,130]) proved its usability in spectroscopy and promises a further progress in taking into consideration the higher-order dispersion terms in Equation 27.1.

One may conclude that the adiabatic theory provides with a deep insight into the physics of CDSs. The DS characteristics become easily traceable, and finding a true parametrical space of soliton allows looking at a DS from a unified point of view. Nevertheless, the underlying approximations of (i) strong domination of the nondissipative effects (GDD and SPM) over the dissipative ones (spectral filtering and SAM) and (ii) distributed character of a system impose some restrictions. There are another analytical approaches, which can shed light on some properties of a DS that go beyond the scope of the adiabatic theory.

27.2.2.3 Truncation of Phase Space: Variational Approximation and Method of Moments

This methods are based on a truncation of a space of (unknown) solutions of Equation 27.1 by means of its projection to a subspace of functions corresponding to soliton-like solutions. Such a truncation can be done by some appropriate ansatz into Equation 27.1 with the subsequent integral minimization procedure (for an overview, see [19,20,110,161,194,216,236]). The ansatz, as a rule, is the known analytical solution (27.9) or its representation (reduced and/or phase-modified). As a result, the complex dynamics of DS can be reduced to the comparatively simple one described by a set of ordinary differential equations governing an evolution of the ansatz parameters (pulse amplitude, width, chirp, etc.). Thereby, the problem becomes a semianalytical one [233,239].

The analysis, based on this approach, confirms a reduced dimensionality of DS parametric space. Moreover, in contrast with the adiabatic approximation, the analysis based on truncation of phase space allows exploring a substantially broader range of parameters of Equation 27.1. In particular, this method works in both positive and negative dispersion ranges and allows taking into account a lumped evolution of DS and a DS polarization dynamics in a laser [24,25,27,28,54,56,236].

The example of master diagram obtained from the VA based on the ansatz (27.8) is shown in Figure 27.6. The right solid curve in Figure 27.6 corresponds to the stability threshold obtained from the adiabatic theory: the soliton is unstable on the right of this curve. The positive branch solution with convex spectrum, which is predicted by the adiabatic theory, cannot be obtained from Equation 27.8 so that the solution for the latter is situated on the right of the shown curves corresponding to the different values of γ/κ (Figure 27.6). These curves are the zero-level isogains (i.e., the curves of $\sigma = 0$) for DS solutions of Equation 27.8. Note that the requirement of $\gamma \gg \kappa$ is not essential for the VA. Nevertheless, one may see that all solutions have a single asymptotic (dashed curve) for $C \ll 1$ so that the master diagram is two-dimensional (i.e., it does not depend on the γ/κ value) in this limit. The asymptotical values of the pulse parameters along the dashed curve are [109]:

$$E \approx \frac{17\beta}{\sqrt{\alpha\kappa\zeta}}, \quad T \approx \frac{8}{C}\frac{\gamma}{\kappa}\sqrt{\frac{\alpha\zeta}{\kappa}}. \tag{27.10}$$

The importance of these scaling rules is that they correspond to the flat-top spectrum relevant to Equation 27.8, which has the spectral chirp with a weak frequency dependence within the range of $[-\Delta, \Delta]$. As a result, such a pulse is almost perfectly compressible (has a "maximal fidelity"), that is, the energy loss due to satellite generation in the process of CDS compression is minimal [267].

The prospects of further development of the method of phase space truncation can be shortly characterized as follows: (i) extension of the class of trial ansatzs and advancement of the integration techniques aimed to an adequate description of a broader class of NCGLE; (ii) close cooperation with the adiabatic theory and the numerical simulations aimed to an adequate representation of the DS parametric space and obtaining of the energy scaling laws for the different types of NCGLE; (iii) most drastic advantage of

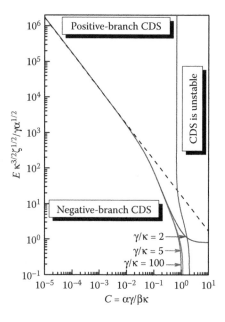

FIGURE 27.6 Master diagram of CDS for the cubic-quintic CNGLE. The right solid curve corresponds to the stability threshold obtained from the adiabatic approximation. Another curves correspond to the thresholds obtained from the variational approximations with the ansatz (27.8) for the different values of γ/κ. The dashed line corresponds to the asymptotic $C \ll 1$. (Adapted from V. L. Kalashnikov. *Solid State Laser*, pages 145–184. InTech, 2012.)

the considered method supposes the extension of NCGLE dimensionality (i.e., $[1+2]$ and $[1+3]$), that is, the inclusion of the transverse spatial coordinates into consideration.*

27.3 Chaotic and Rogue Dissipative Solitons

Nondissipative ("true") solitons of integrable nonlinear systems do not demonstrate a chaotic behavior by definition. The situation changes radically for nonintegrable systems and, especially, for a DS. The latter corresponds to localized structure developing far from equilibrium in the systems with strong energy in/out flows. The scenarios of DS chaotization can be conditionally subdivided into three classes: (i) chaotic single soliton pulsations, (ii) structural soliton chaotization, and (iii) chaotization due to resonant interaction with linear waves.

27.3.1 Chaotic Single DS Pulsations

The chaotic behavior of single DS can be considered in the framework of cubic-quintic CNGLE (27.1) and subdivided into two classes: (i) purely local dynamics with a time-independent σ (there is no dynamical gain saturation) and (ii) nonlocal one with a time-dependent σ (there is a dynamical gain saturation).

The first class of scenarios has been extensively analyzed numerically in References 12 and 224. The revealed types of chaotic DSs are diverse and will be briefly discussed below.

27.3.1.1 Exploding Solitons

For this scenario, the DS envelope becomes increasingly "rippled" in the process of evolution [51,225] (see Figure 27.7). As a result, the soliton acquires a chaotic substructure and cracks into pieces. Then, this chaotic but localized structure "relaxes" and self-collects into a well-shaped DS again. The process repeats forever without a regular period.

The DS explosions can be explained in the following way [223]. DS can be unstable, relatively the slowly growing perturbations. When the amplitude of these perturbations becomes high, the nonlinearity mixes all perturbations that produce the radiative waves. As a result, the dynamics becomes chaotic but the radiative waves "cool" a system that restores a DS after this chaotic stage. Then, the process repeats. At certain values of parameters, the DS explosions can break the time symmetry of Equation 27.1 so that a DS begins to "wander" along the t-coordinate [223]. Such a "wandering" is also possible in the absence of soliton explosion and the corresponding DSs have been named as "creeping solitons" [12]. Their existence can be connected with the asymmetric "internal perturbational modes" of DS (see below).

In some way, the exploding DS belongs to the type of an irregularly repetitive structural soliton chaotization. The structural chaotization will be considered below. But another important feature of an exploding DS is a permanent presence of small (but periodically growing) internal perturbations. Such perturbations (or "internal perturbational modes" [141,193]) can result in the following chaotization scenario.

27.3.1.2 Internal Perturbational Modes of DS

The interesting property of a DS is that it is not completely "solitary" and can have some nontrivial internal filling in the form of permanent nondestructive perturbations, or a solitonic "microflora." Such a "microflora" causes the DS pulsations (both regular and irregular, see Figure 27.8, left) so that the DS dynamics can resemble a strange attractor in low-dimensional systems [224].

The structure of this attractor can be reconstructed on the basis of standard approaches considered in Reference 1. The dimension of a phase space corresponding to a DS dynamics can be estimated on the basis of the "false nearest neighbors" method [1]. The task of this method is to exclude the intersections of trajectories in a phase space, which result from the projection of a true phase space into a low-dimensional one.

---

* The corresponding DS solutions of NCGLE was named as spatiotemporal solitons, or "light bullets" [11,163,202,242,249].

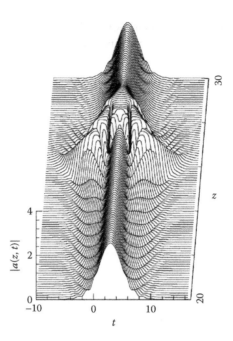

FIGURE 27.7 Exploding DS. (Adapted from J. M. Soto-Crespo and N. Akhmediev. *Math. Comput. Simul.*, 69:526–536, 2005.)

One can see from Figure 27.8 (right) that the phase space is three-dimensional for the GDD of 100 and 150 fs^2. The phase space reconstruction is shown in Figure 27.9, where the propagation lag $z = 35$ is defined from the first minimum of the autocorrelation function of the set of $\max(|a(z,t)|^2)$ [1]. The bifurcation resulting in such long-period pulsations (both regular and irregular) can be classified as the *edge bifurcation* [132,133], which is caused by beatings between the perturbation modes. Such beatings are clearly visible in a DS spectrum evolution and resemble somewhat like the "boiling" of a localized spectrum profile (Figure 27.10).

As one can see from Figure 27.8 (left) and as was found in References 120 and 225, the regions of chaotical behavior can alternate with the region of regular DS pulsations and even coexist with both regularly pulsating and plain DSs. As was conjectured in Reference 116, the excitation of internal perturbational modes can prevent an energy scaling due to chaotization, even destruction of DS, and appearance of turbulence [213]. Also, the region of soliton chaotic pulsations can precede the region of multiple DS generation [26,129] and, thereby, be closely connected with the phenomena of hysteresis and structural soliton chaotization.

27.3.2 Structural Soliton Chaotization and "Rogue" DSs

It was found [42] that an unsaturable SAM, i.e., a cubic nonlinear gain in Equation 27.1, causes a collapse instability of a DS. As a result, a DS becomes nose-like (or "noisy") [101] that is, it transforms into a localized conglomeration of chaotically evolving spikes. Such local chaotization, which does not connect with a global* gain evolution, can be classified as a local structural chaotization.

---

* That is the σ-parameter is t-independent in Equation 27.1 and depends on only DS integral energy $E(z) = \int_0^{T_{rep}} |a(z,t')|^2 \, dt'$ calculated over the current z-slice.

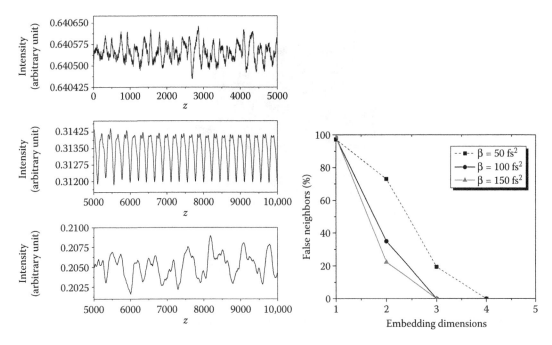

FIGURE 27.8 Left: DS peak-intensity oscillations [116]. Right: Corresponding percentage of false neighbors in dependence on dimension of embedding phase space.

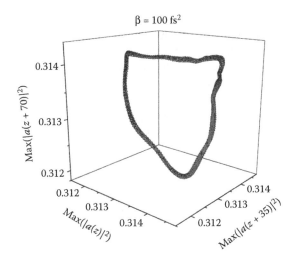

FIGURE 27.9 Phase space reconstruction from the DS peak-power set (5000 points). (Adapted from V. L. Kalashnikov and A. Chernykh. *Phys. Rev. A,* 75:033820, 2007.)

27.3.2.1 Local Structural Chaotization

A structural chaotization can be characterized as a development of femtosecond-scale chaotical structure under a picosecond-scale averaged envelope [101,102,142,144,157,265,266]. One may conjecture that there are two aspects of such structural chaotization.

From the first point of view, a structurally chaotical DS can be treated as an irregularly evolving complex of strongly interacting solitons [129,146,160,264]. An example of such "soliton soup" (or "soliton gas"

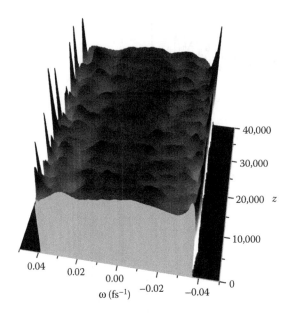

FIGURE 27.10 DS spectral "boiling." (Adapted from V. L. Kalashnikov and A. Chernykh. *Phys. Rev. A*, 75:033820, 2007.)

[171,210]) is shown in Figure 27.11 (left). The chaotical self-emergence and self-annihilation of strongly interacting and bounded solitons can be interpreted as a result of nonresonant interaction with a "vacuum" of Equation 27.1 [109]. A vacuum becomes excited as a result of the growth of DS spectral loss [129] or switching of SAM from the positive to the negative passive feedback regime [142,144,157]. Note, that a "solition gas" (or "solitonic turbulence" [257]) is dynamically nontrivial and can, in particular, evolve in the regular multisoliton complexes (soliton "molecules" or "photon condensate" [32], see Figure 27.11, right) with a possible soliton merging [260] and creation of "supersoliton," which can be interpreted as a "rogue wave" (see below) or "wave-turbulence lasing" [32]. Another possible scenario is an interaction of localized "soliton soup" ("soliton cluster") with a background in the form of the so-called "soliton rains" [44,45]: (i) "soliton cluster" permanently radiates the solitons or (ii) "cluster" permanently absorbs the solitons risen from a background.

From the second point of view, a structural chaotization may be connected with the strong four-wave mixing within a picosecond-scale localized DS (or soliton "condensed phase"). Such a mixing creates chaotic femtosecond-scale intensity beatings causing extremal spectrum broadening (in average): the process, which is akin to the picosecond or noise-driven (highly incoherent) supercontinuum generation [67,207].[*] In the last case, an extremely broadband radiation arises from an initially smooth pulse during its fragmentation and a subsequent power transfer between fragments seeded by noise perturbations [219]. This phenomenon, or generation of *optical rogue waves*, deserves special attention (see below).

Also, the structural chaotization can be interpreted from the point of view of "wave turbulence" [176], when a nonlinear four-wave mixing between the spectral components inside the DS spectrum "leads to a random energy transfer between waves and to enhancement of mode dephasing" [235]. In the time domain, such a turbulence corresponds to the laminar-turbulent transition from a "light condensate" (in the form of a continuous-wave, left picture in Figure 27.12 [238]; or a continuous wave creating a cluster of self-similar evolving DSs, right pictures in Figure 27.12 [113]) to a turbulent cluster of strongly

---

[*] As a rule, the underlying equation for supercontinuum generation is the nonlinear Schrödinger equation supplemented with the HOD and stimulated Raman scattering (SRS) terms [64,218]. The incoherent picosecond supercontinuum generation is characterized by extremal shot-to-shot spectrum variations (see [63] and references therein).

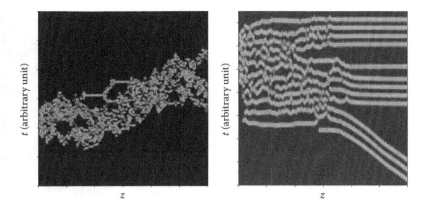

FIGURE 27.11   Contour-plots of $|a(z, t)|^2$. Left: Structurally chaotical complex of DSs ("soliton soup" or "soliton gas"). Right: Self-ordering of structural chaos ("condensation" of "soliton gas"). (Adapted from V. L. Kalashnikov. *CMSIM*, 1:51–59, 2011.)

interacting gray and dark solitons under a background corresponding to an "average" (continuous-wave or DS) envelope.

The interesting feature of DS spectrum following its energy scaling is an appearance of the so-called "spectral condensate" [235], when the main power is concentrated within narrow spectral region at the centrum of the spectrum (Figure 27.3, left; the so-called "finger-like" spectra [196]) or its edges (Figures 27.4 and 27.10 and [130]). A spectral condensation at the centrum of the spectrum can be described in the framework of the adiabatic theory of DS [110], which represents the DS spectrum as a truncated Lorentzian. When the DSR condition $C \to 2/3$ is satisfied, the truncation frequency Δ tends to the constant $\sqrt{\gamma/\beta\zeta}$ while the width of Lorentz spike at the spectrum center tends to zero. In the time domain, such a DS has a broad flat-top envelope. Thus, localization in the spectral domain entails delocalization in the time domain which, in the case of turbulent behavior, leads to a phenomenon of the so-called "spectral incoherent soliton" [172,254] when a coherent-like localization occurs in spectral (not time) domain. The condensation at the spectrum edges can be considered in the frameworks of perturbation theory predicting a resonant enhancement of perturbations with the frequency approaching $\pm\Delta$ [108]. In both cases, the spectral condensation enhances a tendency to chaotization of the DS dynamics.

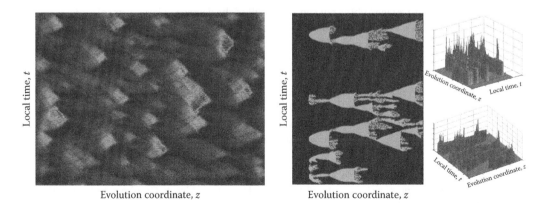

FIGURE 27.12   The laminar-turbulent transitions in a fiber laser. Left: Clustering of continuous-wave light condensate (local power contour-plot, Adapted from E. G. Turitsyna, S. V. Smirnov, S. Sugavanam et al., *Nat. Photon.*, 7:783–786, 2013.) Right: Self-similar clustering of DSs (local power contour-plot and its three-dimensional displays, Adapted from V. L. Kalashnikov. *CMSIM*, 1:29–37, 2014.)

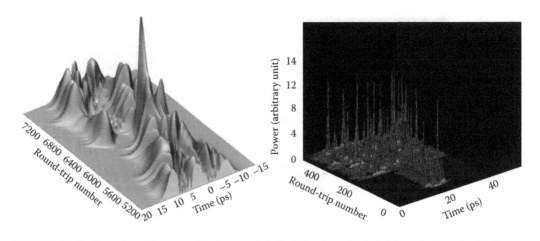

FIGURE 27.13 Examples of the rogue DSs in a mode-locked fiber laser. (Adapted from A. Zavyalov, O. Egorov, R. Iliew, and F. Lederer. *Phys. Rev. A*, 85:013828, 2012 [left] and V. L. Kalashnikov. *CMSIM*, 1:29–37, 2014 [right]). In agreement with the classification of Hammani et al. [83], the left hand picture corresponds to a "persistent and coherent rogue quasi-soliton" whereas the right hand picture corresponds to a "sporadic rogue waves events that emerge from turbulent fluctuations" [182].

27.3.2.2 Optical Rogue Waves

As has been demonstrated above, a local structural chaotization has granularity and turbulence as essential ingredients of phenomenon (compare with the intrafiber propagation analog in Reference 22 and a wave condensation in Reference 83). In this sense, both points of view (i.e., strong soliton interaction/colliding and intensity beatings induced by four-wave mixing) can be considered from a unified viewpoint.

The chaotically emerging DS intensity spikes, which are characterized by *non-Gaussian statistics*[*] of about ten fold (more than twofold by the standard definition [9]) excess of an eventual spike intensity over a stable (averaged) DS one, have been named as optical "rogue/freak-waves" or "extreme events" (e.g., see [149,182,220,228,262] and Figure 27.13).

This term belongs to rather rare but abnormally dangerous event in ocean navigation: an emergence of devastatingly huge wave (with $\approx$30 m height!), which causes damage and even the loss of large ships (22 super carriers between 1969 and 1994) [137]. In a laser, such events cause an extremal spectral broadening, which mimics a CDS in average [149],[†] but can also result in the damage of laser elements.

It should be noted that we did not still consider any noise or HOD (linear wave resonance) contribution to a DS dynamic. Nevertheless, the rogue DS exists under this condition. The linear wave resonance will be considered below, but an absence of noise contribution means that a rogue DS is deterministic by nature [31,149,183,235,262][‡]; therefore, the term "noisy" DS in this context (see above) appears misleading. This statement does not mean that the stochastic processes do not contribute to rogue wave formations; there are examples, when the interplay between stochastic and multistable deterministic dynamics results rogue wave generation [67,195,219]. Noise can play the role of a "starter," or the "switcher," but the nature of rogue DS and, as a consequence, its statistical properties (i.e., extremal increase of the frequency of rare

---

[*] That is, the large spikes appear much more often than they would occur according to Gaussian statistics: a "tail" of the distribution function is extremely stretched [9,203].

[†] In particular, the perturbation starts from a relatively narrow spectral component, then evolves into triangular spectrum shape (which is typical for a chaotic DS), and finishes into a "supercontinuum"-like spectrum with sharp edges [9].

[‡] Nevertheless, rogue wave has an important feature of unpredictability: "it appears from nowhere and disappears without a trace" [9]. This type of rogue waves belongs to "intermittent-like rogue quasi-solitons that appear and disappear erratically" in terms of [83,182] and occur in the upper-left ("low-energy") sector of the "master diagram" in Figure 27.5.

events) are defined by nonlinear factors in a system. Simultaneously, a "deterministic nature" of optical rogue waves allows controlling their statistics and characteristics [65].

As was conjectured, the main mechanism of "rogue soliton" formation is a modulational instability [65,220], which can reveal itself in the formation of the so-called Akhmediev breathers in the framework of the nonlinear Schrödinger equation model [17,138,258]:

$$a\left(z,t\right) = \left[\frac{\left(1-4m\right)\cosh\left(nz\right) + \sqrt{2m}\cos\left(\Omega t\right) + ib\sinh\left(bz\right)}{\sqrt{2m}\cos\left(\Omega t\right) - \cosh\left(nz\right)} \right] \exp\left(iz\right), \qquad (27.11)$$

where Ω is the dimensional modulation frequency, $m = 1/2(1 - \Omega^2/4)$ ($0 < m < 1/2$) determines the frequencies that experience gain, and $n = [8m(1 - 2m)]^{1/2}$ determines the instability growth.

The limiting case of the Akhmediev breather for $m \to 1/2$ corresponds to the so-called Peregrine (or "rational") soliton (see Figure 27.14):

$$a\left(z,t\right) = \left[1 - \frac{4\left(1+2iz\right)}{1+4t^2+4z^2} \right] \exp\left(iz\right). \qquad (27.12)$$

One can see that the Akhmediev breathers demonstrate a high degree of field localization, which is a characteristic feature of rogue solitons [83].

As was shown above, the chaotization of DS is associated with the soliton granulation. In a fiber, such a granulation can be caused by HODs, which produce a resonant interaction of soliton with the linear waves (see below). As a result of granulation, the chaotical collisions between solitons occur (Figure 27.11, left). Such collisions provide an intensive energy exchange so that an extremal "supersoliton" can appear occasionally [78]. This mechanism can be considered as leading in the rogue DS formation. The important feature of both modulational instability and supersoliton scenarios of a rogue soliton formation is that they are very sensitive to a noise so that deterministic and noisy dynamics become strongly interwoven in a real-world system. In particular, such a noise transfer scenario can become leading when a stimulated Raman scattering (SRS) contributes to chaotic and rogue wave-like behavior of DSs (see below and [82]).

The practically important aspect of optical rogue waves is related to modern telecommunication systems. As has been shown [241], a rogue wave can be caused by a "purely linear statistical generation of huge amplitude waves." It means that a "rogue wave" is not something thoroughly established but rather an emerging phenomenon within a broad context.[*] Moreover, "long-range" ("nonlocal") couplings induced, in particular, by SRS can substantially enhance a rogue wave-like behavior [46,82].

27.3.2.3 Nonlocal Structural Chaotization

In a real-world laser, there exist factors that can break a locality of dynamics. That is a current state of the field $a(z, t)$ is influenced by integral history of a system. These factors are, first of all, a dynamic gain (DG) and an SRS.

The first class of chaotical behavior is closely related with the DS frequency shift, which can be caused, for instance, by *DG saturation, SRS,* or *line-width enhancement.*[†]

27.3.2.3.1 DG Saturation: Fine-Grain Aspect

In the framework of this aspect, a global (i.e., z-dependent) evolution of gain is not a decisive factor for DS dynamics. The DG saturation can be taken into account by a replacement of the gain term (i.e., the brace)

---

[*] From a mathematical point of view, the class of underlying statistical models can be very broad and includes effectively linear ones [76].

[†] The frequency shift caused by HOD will be considered below in the framework of the model of resonance interaction with linear waves.

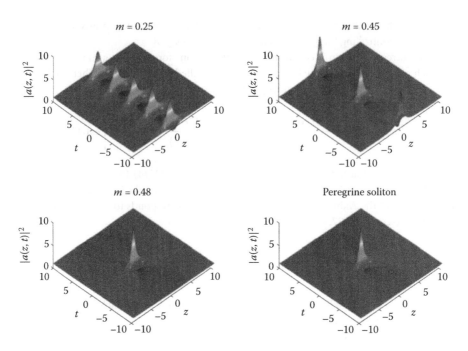

FIGURE 27.14 Examples of the Akhmediev breathers. (Adapted from B. Kibler et al. *Nat. Phys.*, 6:790–795, 2010.)

in Equation 27.2 by $g\left(z,t\right)=g(z,0)\left[1-E_s^{-1}\int_0^t\left|a(z,t'))\right|^2 dt'\right]$ [106,121], where $g(z,0)$ is the gain before a DS and can be expressed by the following map [104]:

$$g(z+1,0)=g(z,0)\exp\left(-\frac{\int_0^{T_{rep}}|a(z,t')|^2 dt'}{E_s}-\frac{T_{rep}}{T_r}-U\right)+\frac{g_m U}{U+\frac{T_{rep}}{T_r}}\left[1-\exp\left(-\frac{T_{rep}}{T_r}-U\right)\right].$$

$$(27.13)$$

Here, T_r is the gain relaxation time and $U=\sigma_a T_{rep}I_p/h\nu_p$ is the dimensionless pump parameter (I_p and ν_p are the pump intensity and wavelength, respectively; σ_a is the pump absorption cross section). $z-$ and $z+1$-arguments mean the current and the next cavity round-trips, respectively. Equation 27.13 presupposes that the DS is well localized, i.e., the temporal window of its localization[*] $\ll T_{rep}$ and the soliton repetition period $\approx T_{rep}$.

As was demonstrated in Reference 121, the pump growth causes a cascade of bifurcations so that regular oscillations of DS change into chaotic ones and vice versa. More careful study based on the numerical modulations in the framework of the model taking into account GDD and exact gain dynamics over the full resonator period [106] has demonstrated that the DG turns a semi-infinite parametric space of DS (Figures 27.5 and 27.6) into an isolated region (Figure 27.15) and, thereby, reduces the soliton energy scalability [55].

Figure 27.15 demonstrates an existence of two borders of DS stability: lower and upper. Lower border corresponds to a generation of satellite before a leading pulse due to preferential amplification of the pulse front caused by DG (D in Figures 27.15 and 27.16). The born satellite interacts intensely with the DS and, as a result, their dynamics becomes chaotical. Further moving off the stability border (i.e., pump or GDD

---

[*] This window defines the limits of integration in $\int_0^t\left|a(z,t'))\right|^2 dt'$, where the lower limit (formally $t=0$) and the ultimate upper limit ($t=t_{max}$) correspond to the borders of this window.

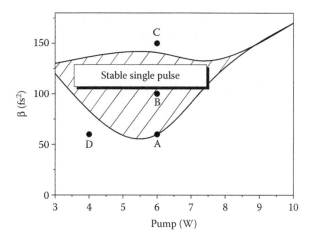

FIGURE 27.15 Region of the DS stability (shaded area) from V. L. Kalashnikov. Chirped-pulse oscillators: An impact of the dynamic gain saturation. Preprint arXiv:0807.1050v1 [physics.optics], available at http://lanl.arxiv.org/abs/0807.1050, 2008. The parameters specified by the points *A*, *B*, *C*, and *D* correspond to the DSs shown in Figure 27.16.

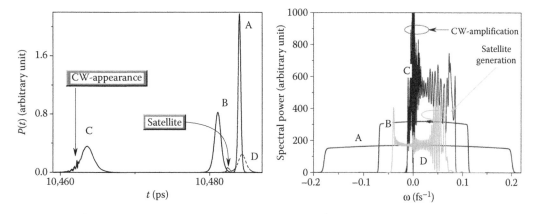

FIGURE 27.16 DS temporal profiles (left) and spectra (right) corresponding to the points in Figure 27.15. (Adapted from V. L. Kalashnikov. Chirped-pulse oscillators: An impact of the dynamic gain saturation. Preprint arXiv:0807.1050v1 [physics.optics], available at http://lanl.arxiv.org/ abs/0807.1050, 2008.)

decrease) can cause either a continuum-wave (CW) generation with a DS "dissolution" or a regular two-soliton generation. For the last scenario, a cascade of further bifurcations with chaotical transits to a pulse multiplication is possible.[*]

The upper stability border is characterized by the appearance of a CW perturbation with subsequent irregular interaction between this perturbation and DS (*C* in Figures 27.15 and 27.16). As a result, the DS becomes chaotical or even disintegrates. Note that the DS spectra are blue-shifted for both *C*- and *D*-regimes in Figure 27.16 because such a shift provides partial compensation (in the normal GDD range) of the time-advance and the preferential amplification of pulse front caused by DG saturation.

---

[*] The pulse multiplication is possible, also, without intermediate chaotization [129].

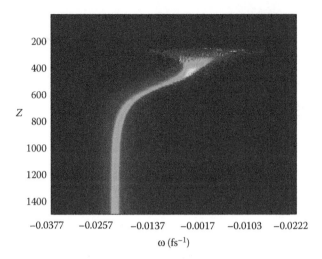

FIGURE 27.17   Spectral evolution of a Dissipative Raman soliton: numerical simulations of an all-normal-dispersion fiber laser with $\kappa = 0.1\gamma$, $\zeta = 0.05\gamma$, $\delta = 0.05$, a 40 nm spectral filter width, and $\beta = 0.2\,\mathrm{ps}^2$ (see Section 27.2). SRS parameters are: $T_1 = 12.2\,\mathrm{fs}$, $T_2 = 32\,\mathrm{fs}$, and $f_R = 0.22$. This parametric point attributed to the master diagram of Figure 27.5 (left) corresponds to $C = 0.029$, $E = 142$ and lies on the boundary of a single Raman DS generation.

27.3.2.3.2 SRS: Raman DS

SRS can become a decisive factor in a high-energy DS dynamics. Its contribution can be taken into account by replacement of SPM terms in Equation 27.1 (i.e., γ- and χ-terms) with [7]:

$$-ia(z,t)\int_0^\infty R(t')\left|a(z,t-t')\right|^2 dt', \tag{27.14}$$

where $R(t) = \gamma(1-f_R)\delta(t) + \gamma f_R h_R(t)$ and $h_R(t) = \left(T_1^2 + T_2^2\right)/T_1 T_2^2 \exp\left(-t/T_2\right)\sin\left(t/T_1\right)$. Here, T_1 is the inverse Raman resonance frequency, T_2 is the damping time of phonon vibrations, and f_R represents the fractional contribution of the Raman response to a nonlinear polarization coefficient γ. Also, a noise source has to be added to Equation 27.1 in order to provide a spontaneous Raman scattering.

   The numerical simulations [111,127] demonstrate that the SRS plays a substantial role in dynamics only for comparatively large energies[*] and enhances a tendency to multipulsing. As a result, the stability boundary of single Raman DS adapted to the master diagram of Figure 27.5 (left) lies substantially lower, that is, the soliton stabilization requires a substantially larger GDD. The cause of destabilization is an enhancement of spectral loss due to SRS. Figure 27.17 demonstrated a single red-shifted Raman DS in the vicinity of the stability border. A distinguishing characteristic of such soliton is a strong perturbation (fragmentation) of its trailing edge caused by the growth of anti-Stokes spectral component (see Figure 27.18). As a result of this perturbation, the Raman DS demonstrates chaotical oscillations of peak power. This effect can be interpreted in the frameworks of concept of "incoherent soliton" [212] when there is enhancement of field perturbations via their long-range (in time and spectral domains) coupling [77]. The reverse side of such a coupling is suppression of wave turbulence in the presence of SRS (Figure 27.19).

27.3.2.3.3 Dynamic Loss Saturation and Linewidth Enhancement

A first law under brace in Equation 27.7 describes a perfectly saturable absorber. As a rule, that is a semi-conductor structure with some finite response time T_r. Then, such a simple law is valid if the DS width $\gg T_r$. If not, the nonlinear loss and phase responses of such structure become nonlocal, i.e., it depends on

---

[*] $E > 20$ in terms of the master diagram presented in Figure 27.5, left.

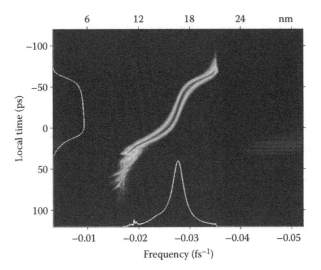

FIGURE 27.18 Wigner function of a chirped DRS. The DRS temporal and spectral profiles are shown as the corresponding axis-projections. The top wavelength shift scale corresponds to an Yb:fiber laser centered around 1070 nm. The parameters are $C = 0.021, E = 111$. (Adapted from V. L. Kalashnikov and E. Sorokin. *Opt. Express*, 22:30118–30126, 2014.)

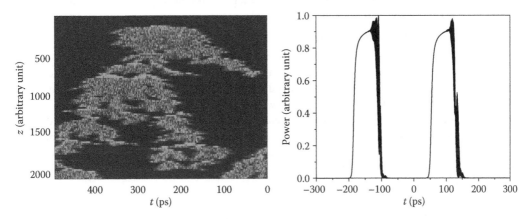

FIGURE 27.19 Left: Evolution of the local field power (contour-plot) for the wave turbulence regime in the absence of SRS ($f_R = 0$). Right: In the presence of SRS ($f_R = 0.22$), two localized DRSs develop. $C = 0.037, E = 111$. (Adapted from V. L. Kalashnikov. Chaotic dissipative Raman solitons. In C. H. Skiadas, editor, *Proceedings of 7th CHAOS Conference*, pages 199–206. ISAST, 2014.)

energy flow (not instant power). Approximately, this effect can be taken into account by replacing Equation 27.7 with [123]:

$$-\mu \left[1 - \left(1 + i\epsilon\right) E_s^{-1} \int_0^t \left|a(z, t')\right|^2 dt'\right] \left[1 + \frac{1}{\Omega_l}\frac{\partial}{\partial t}\right]^{-1}, \tag{27.15}$$

where E_s is a loss saturation energy, Ω_l is an absorption linewidth, ϵ is a so-called linewidth enhancement (or Henry) factor [145], and a loss coefficient μ is not included in ℓ. The differential operator has to be considered as an expansion in the Fourier domain.

As was demonstrated [122,123], such a system has a rich dynamics with multistability, auto-oscillations, and hysteresis. One can conjecture that the chaotical behavior also has to be present here. Moreover, the

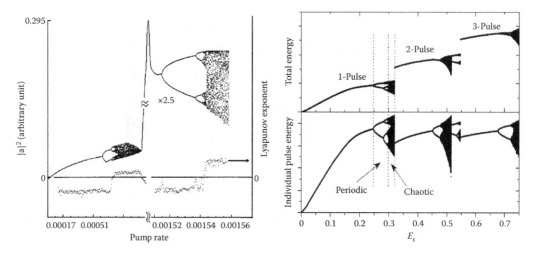

FIGURE 27.20 Iteration map dynamics. (Adapted from V. L. Kalashnikov, I. G. Poloyko, V. P. Mikhailov et al., *J. Opt. Soc. Am. B*, 14(10):2691–2695, 1997 [left] and F. Li, P.K.A. Wai, and J. N. Kutz. *J. Opt. Soc. Am. B*, 27(10):2068–2077, 2010 [right].)

linear term in Equation 27.15 induces HODs and, thereby, generates the dispersive waves, which interact resonantly with a DS and affect its dynamics (see below).

27.3.2.3.4 DG Saturation: Coarse-Grain Aspect

Above, the dynamical aspect of z-dependence of gain was neglected. However, such a dependence can be crucial in many cases. The powerful method for an analysis of this "coarse-grain"* DG is the so-called nonlinear rate equations method, which reduce a dynamic phase-space to some low-dimensional map describing the evolution of DS parameters (intensity, energy, etc.) [104,126,155,173].

Figure 27.20 demonstrates the cascades of bifurcations with alternation of the chaos and regular dynamics in dependence on pump rate (left) or gain saturation energy (right).[†] The left picture demonstrates that a region of DS existence can have an inhomogeneous structure with the slices of chaos and period multiplication. The right picture shows that a transit to multipulsing is accompanied by chaos and period multiplication, as well.

A numerical analysis taking into account a precise dynamics over a whole cavity period demonstrates that the multiple pulse generation can be chaotical, as well (Figure 27.21). The pulses can appear/disappear during T_{rep} without a regular interpulse distance and relative amplitude. Strong interaction between the pulses in this case results from a DG saturation over a resonator period.[‡]

27.3.3 Chaotization Due to Resonant Interaction with Linear Waves

Soliton can be coupled with a linear wave only if the resonant condition is satisfied: a soliton wave number q (see Equations 27.8 and 27.9) has to be equal to a linear wave one. There exists no "true soliton" in this case but a localized soliton with oscillating tail (or tails) can develop (a "quasi-soliton" in terms of References 257 and 259). In the framework of adiabatic theory, the resonant condition defines a CDS spectrum truncation in the absence of HODs: $q = \beta \Delta^2$.[§] The contribution of HODs modifies this condition so that

---

* That means that the time-scale of gain evolution is $\approx T_{rep}$ and is substantially larger than the soliton width. In this case, the DS influence is defined by its full energy (see Equation 27.2).
† Equation 27.2 demonstrates that both parameters are interrelated from the point of view of their contribution to dynamics.
‡ DG saturation plays a role of "long-range force" as opposed to "short-range" forces induced by DS overlapping.
§ This condition is obtained in the framework of the adiabatic theory (see above) and explains the perturbation enhancement ("spectral condensation") along the DS spectrum edges (Figure 27.10 and [130]).

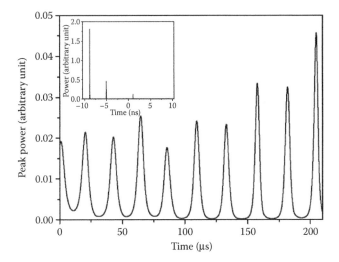

FIGURE 27.21 Evolution of maximum peak power of a multiple DS complex (inset) under a DG action. (Adapted from V. L. Kalashnikov. *CMSIM*, 1:51–59, 2011.)

$q = \beta\omega^2 + \sum_{n=3}^{N} \beta_n \omega^n$ (β_n-terms correspond to HOD contribution in Equation 27.1). If such a resonance condition is satisfied and the corresponding resonance frequency ω_r lies inside the DS spectrum,[*] the generation of dispersive wave begins, that is, the DS interacts resonantly with a continuum [108]. The spectral domain perturbation theory for this phenomena is presented in Reference 108. The numerical simulations demonstrate [117,128,221] that the resonance between a DS and a linear wave causes the strong chaotical perturbations of a soliton and, even, its fragmentation (Figure 27.22), which, nevertheless, preserves a localization of soliton complex (e.g., so-called "optical Newton's cradles" [57]). This phenomenon is closely related to an "optical turbulence" [172,235] and is illustrated by the "wandering" phase trajectory in Figure 27.23 [222]. The dimension of embedding phase-space in Figure 27.23 lies within 2–4 and the attracting manifold has a toroidal shape. These facts are coherent with the concept of chaos induced by the nonlinearly entangled three oscillators [23,192] with the resonant frequencies ω_r (points R_1, R_2, and DW in Figure 27.24c). When the dispersive wave resonance frequency approaches the DS spectrum (point R_2 in Figure 27.24c), the corresponding dispersive wave gets excited. Since the DW position corresponds to the positive group delay, the dispersive wave would propagate slightly faster than the DS and overlap with the pulse. The interference of dispersion wave and DS results in complete change of the whole spectrum and chaotic behavior (Figure 27.24). The spectrum and waveform of DS (Figure 27.24a, bottom and right-hand curves) become strongly modulated and change rapidly (Figure 27.24b). Yet the average spectrum, accumulated over 7000 round-trips, looks stable (Figure 27.24d).

Note that an HOD contribution also results from the limited gain/absorption/transmission spectral line widths (e.g., see Equation 27.15), which are inherent in any real-world optical setup. Therefore, this scenario of DS destabilization is an important factor limiting the soliton energy scalability.

27.4 Noises of Dissipative Solitons

The breakthroughs in optical solitonics shifted the attention from the physics of "power-envelope" to the physics of "spectral subtleties and phase" [50], from an "intensity" to a "field" [33], that is, from the classic

---

[*] For instance, the third-order dispersion β_3 results in a strong resonance between a CDS and a linear wave if $|\omega_r| < \sqrt{\varsigma a_0^2/\beta} \approx |\beta/\beta_3|$.

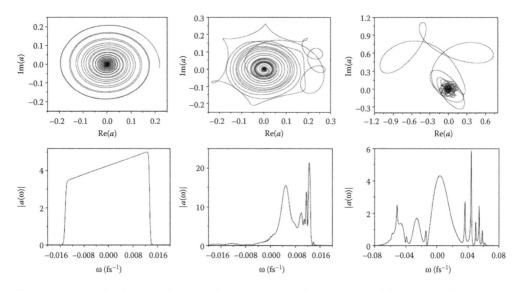

FIGURE 27.22 DS chaotization with a third-order dispersion growth (from left to right): phase portraits and spectra. (Adapted from V. L. Kalashnikov. *Chaos Theory: Modeling, Simulation and Applications*, pages 58–67.World Scientific Publishing Company, London, 2011.)

to the mesoscopic quantum world [253]. It is clear, that a noise can contribute substantially into DS phase and frequency characteristics for this class of phenomena. A system becomes especially sensitive to a noise in the vicinity and within the regions, where a dynamics is chaotic. By physical origin of such contributions that can be roughly divided into those caused by classical and quantum noises.

27.4.1 Classical Noises

Basic theory of the noise characterization has been developed in Reference 243. The theory does not deal with the noise sources but concentrates on the spectral signatures of amplitude fluctuations and timing jitter (i.e., fluctuations of group velocity) of soliton. It was found that these fluctuations result in the different dependence of noise pedestal on the harmonic order in the radio-frequency (RF) spectrum of the pulse train: the amplitude fluctuations contribute to the frequency-independent term, whereas the timing jitter results in the quadratic dependence on RF frequency. Further analysis has revealed a more complicated picture [71]: there exists a linearity in frequency term due to the correlated jitter and energy fluctuations, while pulse width fluctuations contribute to the quadratic term, as well. Further generalization of the model, accounting for the cumulative character of timing fluctuations in a free-running passively mode-locked laser, connected the shape of the power spectrum of the pulse (DS) train with the type of noise (amplitude or timing jitter) and its correlative characteristics [66]. The picture turned out to be more complicated than it looked before. It has been demonstrated that the amplitude-phase coupling in a femtosecond laser [234] affects all aspects of stability, including the carrier-envelope-offset (CEO) stability [97,98] that, for instance, has a crucial significance for the optical frequency comb stabilization.* Thus, an identification of noise sources and an establishment of correlation between them are required for an adequate model of the laser noises.

A physically clear picture of the frequency comb fluctuations induced by different physical mechanisms can be visualized by means of the so-called elastic tape model [30,178,230], where the comb frequencies are symbolized by the equidistant labels on an elastic tape, which is randomly stretched relative to some

---

* Optical frequency comb results from a periodical ($\approx T_{rep}$ or T_{rep}-harmonics) circulation of DSs in a laser. Thereby, a laser DS spectrum is "discretized" and can play a role of metrological standard if the CEO is stabilized. CEO is defined by a slip of the carrier wave, relatively, the DS envelope (such a slip defines a soliton wave number q) [50].

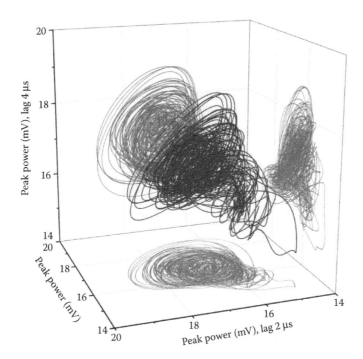

FIGURE 27.23 Phase portrait of the DS peak power evolution. The experimental set corresponds to 210 nJ of intracavity pulse energy of an Cr:ZnSe oscillator, the 2 μs lag corresponds to 290 cavity round-trips. (Adapted from E. Sorokin, N. Tolstik, V.L. Kalashnikov et al., *Opt. Express*, 21:29567–29577, 2013.)

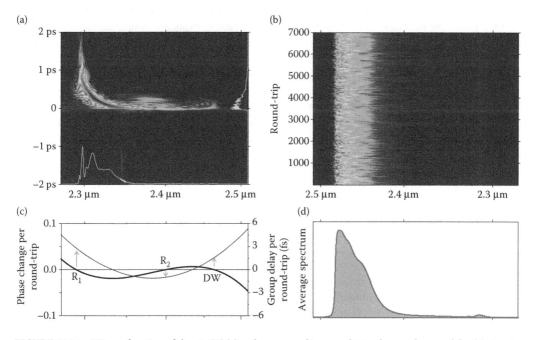

FIGURE 27.24 Wigner function of chaotic DS (a) and corresponding round-trip phase and group-delay (c). Spectra (b) over the 7000 cavity round-trips and their accumulated spectrum (d). (Adapted from E. Sorokin, N. Tolstik, V.L. Kalashnikov et al., *Opt. Express*, 21:29567–29577, 2013.)

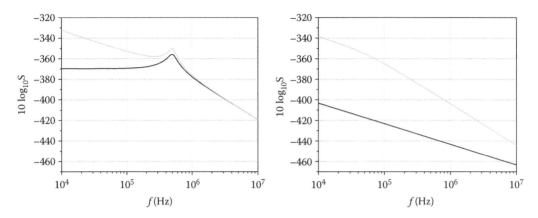

FIGURE 27.25   Left: Timing-jitter noise spectra induced by gain fluctuation in the normal (black curve) and anomalous (gray curve) GDD regimes of a 50 μJ thin-disk Yb:YAG oscillator. Right: Corresponding spectra of the quantum timing noise. (Adapted from V. L. Kalashnikov and A. Apolonski. Microjoule mode-locked oscillators: Issues of stability and noise. In T.Graf, J. I. Mackenzie, H. Jelinkova, G. G. Paulus,V. Bagnoud, and C. Le Blanc, editors, *Proceedings of SPIE: Solid State Lasers and Amplifiers IV, and High-Power Lasers*, vol. 7721, page 77210N SPIE, 2010.)

fixed point. Such a point can be localized in the vicinity of (i) zero frequency (e.g., cavity length fluctuations), (ii) carrier optical frequency (pump fluctuations or quantum noise), or (iii) higher-frequency domain (temperature fluctuations) [178,190]. Knowledge of this pivot frequency (or frequencies) is crucial for an elaboration of successful comb stabilization strategy because the use of stabilizing technique fixing a "wrong" frequency point would even enhance the noise [167,178,190,247]. Here, an integrated representation of the laser noises (both amplitude and phase) is required.

The outlook for such a representation has been outlined in the analytical theory by Paschotta [187, 190,191]. Under the physically sound arguments, the theory allows explicitly connecting the source noise spectra with the temporal and phase noise spectra of a laser irrespective of details of a laser dynamics. The direct generalization of this theory [115] has predicted the timing jitter suppression in a thin-disk multimicrojoule Yb:YAG oscillator operating in the normal GDD regime (Figure 27.25, left).

The situation becomes more intricate for fiber lasers where the stimulated Raman, Brillouin, and Rayleigh scattering as well as the stochastic birefringence can affect the stochastic evolution substantially [96,237,246]. Difference in the time and spatial scales characterizing linear, nonlinear, and stochastic evolution begins playing a crucial role in dynamics and allows applying the multiscale and kinetic theory methods for an analysis of optical phenomena [73,175,180]. As a result of the joint action of multiscale factors, the quite new phenomena can appear: a "stochastic resonance," when a noise enhances a system response to the external periodic perturbation [72,84,156], a "coherence resonance," when a noise creates some coherent-like states in a system [156], or a "stochastic antiresonance," when the fluctuations intensify resonantly within some diapason of system parameters where a switching between the statistical scenarios (e.g., persistent and antipersistent) occurs. Wealth of the phenomena inherent to the incoherent nonlinear wave propagation stimulates developing a new branch of nonlinear optics, the so-called "statistical nonlinear optics" [76], and a concept of "incoherent soliton" [212].

Back on topic of the DS noises, the perturbation theory of the frequency comb generated by a solitonic laser has been developed in Reference 179. The theory based on the MM (see above) has allowed connecting the offset and spacing fluctuations of the frequency comb with the factors governing the laser dynamics. In particular, the transfer functions have been derived, which relate the comb fluctuations with the fluctuations of cavity length and pump power. However, the success of this theory should not be overestimated because data about independence of the noise and the mode-locking regime are controversial (e.g., see [35,189]). Therefore, there is a need for an extensive numerical model that accounts for all factors involved in the laser dynamics and imposes no *a priori* restrictions on the complex field profile $a(z, t)$. An implementation of such modeling has been presented in References 115, 187, 189, and 190. In particular, the numerical simulations of a high-energy thin-disk Yb:YAG oscillator confirmed the timing jitter suppression in the normal GDD regime predicted by the analytical theory (Figure 27.25, left). The details of numerical approach to the noise modeling are more appropriately considered in the following subsection examining the quantum laser noises.

27.4.2 Quantum DS Noises

Unprecedented precision and stability levels required from ultrafast lasers generating DSs compel reducing the noise level down to quantum limit. Therefore, the study of such a limit is urgent for both applications and fundamental physics of DSs.

27.4.2.1 Soliton Quantization

The well-established basis for a quantum theory of noises has been provided by the quantum noise theory of a linear (or linearized) amplifier [47,87,90–92]. It is shown in this theory that the "phase-preserving" character of amplification requires quantum noise for conservation of the field commutators. This means that contribution from additional degrees of freedom is required by the "phase-preservation": these include spontaneous emission, intracavity loss (with the corresponding "baths"), and output coupling of an open oscillator (i.e., contribution from the "input" vacuum fluctuations). The statistical properties of the observables connected with the noise operators can be obtained from the commutator conservation law.

The resulting dynamical equations have a form of Langevin operator equations with the quantum noise sources [29,34,74,75,81,91,99,254]. An implementation of the fluctuation-dissipation theorem and the field uncertainty for an evaluation of the quantum Langevin equations has been reviewed in Reference 99. Thereupon, the operator Langevin equations can be reduced to the stochastic differential ones, where the field and the gain are driven by the integrated (in the Ito or Stratonovich forms) stochastic forces with the diffusion coefficients defined by correlation functions of the noise sources [74,99,153]. There exist different methods for obtaining these coefficients (for an overview, see, for instance, [99]). In particular, the quantum Langevin approach reproduces the Schawlow–Townes limit on the CW laser linewidth [208] (which results also from the commutation relations imposed on the noise sources [252]) as well as provides the powerful methods for laser noise interpretation [85] and its suppression [34,100].

To extend the quantum noise theory to the mode-locked lasers, one would additionally have to consider all nonlinear effects, which come along with the DS propagation. This requires quantization of the dynamically nontrivial solitonic sector of the evolution equations. This approach can be viewed as a part of a more general quantum-field theory [68]. The starting point for an optical soliton quantization is quantization of the nonlinear Schrödinger equation [61,91]. Since the classical nonlinear Schrödinger equation is exactly integrable by the inverse scattering method, it can be exactly quantized, as well [135,152,232]. Formally, the problem is equivalent to the N-body model with a point interaction between bosons [151,231,256].

It is also possible to solve the problem of quantization by linearization of the Heisenberg quantum form of the nonlinear Schrödinger equation around the exact classical soliton solution [91,93,134,150]. This approach (quantum soliton perturbation theory) has allowed further generalization, accounting for the dissipation [197] and the gain [86] contributions to the Langevin noise terms as well as description of a repeater system model of an oscillator. The perturbation theory has been extended to the cubic NCGLE equation governing the dynamics of a passively mode-locked laser [88,89] (an actively mode-locked laser is describable in the framework of the linear Heisenberg–Langevin model [199]). The perturbations of pulse energy, phase, timing, and frequency have been expressed from the characteristics of noise sources: both technical (gain, resonator length, and refractive index fluctuations) and quantum (spontaneous emission). This approach has allowed derivation of the noise spectra and the correlation functions of actively and passively mode-locked lasers.

The noise transfer matrix obtained from the semiclassical perturbation theory allows precise interpretation and its matrix elements are measurable [168,244,245]. As a result, the expressions for noise spectrum and the frequency comb linewidths can be obtained semiphenomenologically for a given laser configuration. The theory limits itself to the case of a dispersion-managed soliton, operating in the anomalous dispersion regime and in the absence of higher-order dispersions and nonlinearities. While a generalization of this approach to the normal dispersion regime and more complicated dispersion and nonlinearity functions seems to be principally possible, this has not been performed yet. Nevertheless, an important feature of this approach is the demonstrated principal measurability of the transfer matrix elements, which provides an independent consistency check route for the model, analytical or numerical.

An important contribution to the theory of noises of mode-locked oscillators is the semiclassical ("phase-preserving") analytical models of References 168, 187, and 191, which reproduce the results of [88,89,199] based on the soliton perturbation theory. The key advantage of the semiclassical model is that it does not restrict itself to the soliton regime and, thereby, its area of applications is broader. In particular, it is applicable for the description of an oscillator operating in the normal-dispersion regime (fiber-based [189] or solid-state [115]), where an ordinary soliton would not exist. The extensive study of CDS formed in the all-normal-dispersion regime has demonstrated that the timing quantum noise can be strongly reduced (see Figure 27.25, right) [115].

The quantization schemes for a CDS have not been developed systematically to date. The main problem is that the relevant nonlinear evolutional equations are not integrable. A possible breakthrough is expected due to further development of the approximated integration techniques [110] and the perturbation methods in the spectral domain [108].

Despite the significant progress achieved on the basis of the soliton-perturbation and semiclassical analytical theories of quantum noises in mode-locked oscillators, these theories are insufficient because they neglect the complicated dynamics of the real-world (especially fiber-based) oscillators and the mutual correlation of various noise mechanisms. Therefore, extensive numerical simulations of the mode-locked oscillators with the inherent noise sources are required.

27.4.2.2 Semiclassical Langevin Methods: Numerical Approach

The numerical model based on the Langevin semiclassical method allows including the spontaneous emission of noise source directly into the classical equations for laser field, population difference, and complex polarization [70,119,131]. The noise results from fluctuation of the dipole moment in an active medium. As an example of results, one can note the demonstration of the "seed-points" inside a mode-locking region with maximum suppression of the noise-induced amplitude fluctuations [53] ("coherence resonance" contributing to a system self-ordering [156]). It was also found that the mode-locking threshold depends on the correlation properties of the noise [118]. The Langevin method also allows easy accounting for technical noises.

The numerical Langevin method based on statistics gathering has recently seen a renaissance [115, 186,187,189] made possible by ever-growing computing resources. In particular, it is becoming feasible to model long pulse trains, which is a necessary condition for the reconstruction of a low-frequency edge of the noise spectrum. The numerical approach has allowed accounting for loss saturation in a semiconductor saturable absorber, the nonlinear dispersion, as well as relatively low-frequency gain relaxation dynamics and related gain dispersion. As a result, the noise coupling mechanisms have been observed, which are not present in the simple analytical models. In particular, these coupling mechanisms cause the excess phase noise resulting in the line broadening above the Schawlow–Towns limit, as well as the linewidth variation across the frequency comb [190] predicted on the basis of the perturbation theory [244]. The different types of the fiber lasers have also been briefly surveyed [189]. It was found that the stretched-pulse mode-locking does not introduce strong excess noise. In the wavebreaking-free fiber lasers (i.e., lasers operating in the all-normal-dispersion regime), the center frequency noise can be effectively coupled to the timing noise that, in combination with the long pulse width, causes a strong excess enhancement of the timing and carrier-envelope offset noises [189,198]. This contradicts the case of comparatively narrow-band thin-disk solid-state laser [115] and the data of References 35 and 209.

An important advantage of the numerical approach is that it allows precise modeling of the intracavity field evolution in the resonators with discrete or strongly inhomogeneous elements. Such evolution is most pronounced in nonsolitonic fiber lasers: DS, similariton, and similariton-soliton types [181,201,250]. These types of fiber lasers are of special interest because they provide the energy scalability of DS.

Some pilot results for the lasers with lumped elements can be obtained analytically on the basis of the VA or the MM (see above). It has been shown, that the dynamics of a system, i.e., the evolution of pulse parameters during a cavity round-trip, can cause significant noise reduction because of noise decoupling [162,177]. On the other hand, the fixed frequency points can shift, as well [178].

27.4.2.3 Quasi-Probability Methods

These methods are based on the quantum phase-space formulation [4–6] of nonlinear quantum optical pulse propagation in the form of the positive-P or Wigner distributions [37,58,60,139,217]. The quantum dynamics of radiation, described in the form of the nonlinear Liouville equation, can be reduced to a set of stochastic differential equations for the classical functions (not operators) describing the P or Wigner quasi-probability operator representations [36,49,59]. In spite of some formal distinctions, this method is equivalent to the linearized stochastic approach above in the soliton quantization scheme [69].

The Wigner representation can be considered as a rigorous substantiation of the semiclassical Langevin methods, described in the previous subsection. However, one has to keep in mind that the Wigner representation (i) is approximated, though practically perfect for large photon numbers, (ii) corresponds to symmetrically (not normally) ordered operator products and has to be corrected to provide a direct interpretation. In particular, the vacuum noise terms have to be included additionally: "shot-noise" as an initial condition, and both gain and loss as complex additive noises [36,49,59]. The advantage of this representation is its reduced dimensionality in comparison with the positive P-representation (e.g., see [165]). At the same time, the latter representation is exact, has no initial and absorbing reservoir vacuum noise terms, and corresponds to the experimentally measurable field averages. Therefore, despite its double dimension, using the P-representation has a big potential.

As an alternative approach to the quasi-probability methods, the modern computational resources allow simulating the dissipative Liouville equation [229] without reduction of quasi-probability representations to a set of stochastic equations. The main advantage of this method is that there is no need to gather the statistics because all required information is contained in the evolving quasi-probability distribution. One may believe that this technique becomes especially effective for the distributed resonators with long-living upper-laser level media, i.e., fiber lasers, where statistics gathering becomes prohibitively expensive computationally. An additional advantage of a direct evaluation of quasi-probability distributions is the possibility to include the measurement process into analysis using the Wigner (or another distribution) function of a measuring device [154].

27.5 Conclusion

The concept of dissipative soliton has become well established over the last two decades. Dissipative solitons, which are localized structures preserving a self-identity during a long evolution, describe an enormously broad range of phenomena ranging from physics to biology, and geophysics to social sciences. An existence far from the equilibrium state of a system results in a substantial nontriviality of dissipative soliton dynamics, which can reveal itself in chaotical, multistable, and extremely noise-sensitive behavior. Thereby, possessing the "energy" (or "mass") scalability (i.e., the definitely "macroscopic" property), a dissipative soliton can be "quantum-sensitive." A dissipative soliton is a "mesoscopic" object and the study of its properties is of fundamental interest.

The impressive progress of modern ultrashort laser pulse technologies provides us with a unique testbed for dissipative soliton exploring. The studies of such optical dissipative solitons advance with seven-league strides. In this short review, only some aspects of optical dissipative soliton dynamics have been considered. The main approaches to dissipative soliton theory have been exposed and the preliminary classification of scenarios of an optical dissipative soliton chaotization has been proposed. Also, the main trends in the studies of noise influence (including quantum one) on an optical dissipative soliton have been reviewed. It should be recalled that the unified theory of optical dissipative solitons remains under way and one may expect new impressive advantages in this field in the near future.

Acknowledgments

I would like to acknowledge Dr. Eugeni Sorokin for contribution to the sections regarding dissipative soliton noises. Supports of the FP7-PEOPLE-2012-IAPP (project GRIFFON, no. 324391) and the Austrian Science Fund (Project no. P24916) are acknowledged.

References

1. H. D. I. Abarbanel, R. Brown, J. J. Sidorovich, and L. Sh. Tsimting. The analysis of observed chaotic data in physical systems. *Rev. Mod. Phys.*, 65(4):1331–1392, 1993.

2. F. Kh. Abdulaev and V. V. Konotop, editors. *Nonlinear Waves: Classical and Quantum Aspects*. Kluwer, Dordrecht, 2004.

3. M. J. Ablowitz and Th. P. Horikis. Solitons in normally dispersive mode-locked lasers. *Phys. Rev. A*, 79(6):063845, 2009.

4. G. S. Agarwal and E. Wolf. Calculus for functions of noncommuting operators and general phase-space methods in quantum mechanics. I. Mapping theorems and ordering of functions of noncommuting operators. *Phys. Rev. D*, 2:2161–2186, 1970.

5. G. S. Agarwal and E. Wolf. Calculus for functions of noncommuting operators and general phase-space methods in Quantum mechanics. II. Quantum mechanics in phase space. *Phys. Rev. D*, 2:2187–2205, 1970.

6. G. S. Agarwal and E. Wolf. Calculus for functions of noncommuting operators and general phase-space methods in quantum mechanics. III. A generalized wick theorem and multitime mapping. *Phys. Rev. D*, 2:2206–2225, 1970.

7. G. Agrawal. *Nonlinear Fiber Optics*. Elsevier, Oxford, 2013.

8. N. Akhmediev and Ph. Grelu. Dissipative solitons for mode-locked lasers. *Nat. Photon.*, 6(1):84–92, 2012.

9. N. Akhmediev and E. Pelinovski. Editorial—Introductory remarks on "discussion & debate: Rogue waves—Towards a unifying concept?" *Eur. Phys. J. Special Topics*, 185:1–4, 2010.

10. N. Akhmediev, J. M. Soto-Crespo, and A. Ankiewicz. Could rogue waves be used as efficient weapons against enemy ships? *Eur. Phys. J. Special Topics*, 185:259–266, 2010.

11. N. Akhmediev, J. M. Soto-Crespo, and Ph. Grelu. Spatiotemporal optical solitons in nonlinear dissipative media: From stationary light bullets to pulsating complexes. *Chaos*, 17(3):037112, 2007.

12. N. Akhmediev, J. M. Soto-Crespo, and G. Town. Pulsating solitons, chaotic solitons, period doubling, and pulse coexistence in mode-locked lasers: Complex Ginzburg–Landau equation approach. *Phys. Rev. E*, 63:056602, 2001.

13. N. N. Akhmediev and A. Ankiewicz. *Solitons: Nonlinear Pulses and Beams*. Chapman & Hall, London, 1997.

14. N. N. Akhmediev and A. Ankiewicz. Solitons around us: Integrable, Hamiltonian and dissipative systems. In K. Porsezian and V. C. Kuriakose, editors, *Optical Solitons: Theoretical and Experimental Challenges*, pages 105–126. Springer, Berlin, 2003.

15. N. N. Akhmediev and A. Ankiewicz, editors. *Dissipative Solitons*. Springer, Berlin, 2005.

16. N. N. Akhmediev and A. Ankiewicz, editors. *Dissipative Solitons: From Optics to Biology and Medicine*. Springer, Berlin, 2008.

17. N. N. Akhmediev and V. I. Korneev. Modulation instability and periodic solutions of the nonlinear Schrödinger equation. *Theor. Math. Phys.*, 69(2):1089–1093, 1986.

18. N. N. Akhmediev, J. M. Soto-Crespo, and Ph. Grelu. Roadmap to ultra-short record high-energy pulses out of laser oscillators. *Phys. Lett. A*, 372(17):3124–3128, 2008.

19. D. Anderson, M. Lisak, and A. Berntson. A variational approach of nonlinear dissipative pulse propagation. *Pramana J. Phys.*, 57(5–6):917–936, 2001.

20. A. Ankiewicz, N. Akhmediev, and N. Devine. Dissipative solitons with a Lagrangian approach. *Opt. Fiber Technol.*, 13(2):91–97, 2007.

21. I. S. Aranson and L. Kramer. The world of the complex Ginzburg–Landau equation. *Rev. Mod. Phys.*, 74:99–143, 2002.

22. F. T. Arecchi, U. Bortolozzo, A. Montina, and S. Residori. Granularity and inhomogeneity are the joint generators of optical rogue waves. *Phys. Rev. Lett.*, 106:153901, 2011.

23. C. Baesens, J. Guckenheimer, S. Kim, and R. S. MacKay. Three-coupled oscillators: Mode-locking, global bifurcations and toroidal chaos. *Physica D*, 49:387–475, 1991.

24. B. G. Bale, S. Boscolo, J. N. Kutz, and S. K. Turitsyn. Intracavity dynamics in high-power mode-locked fiber lasers. *Phys. Rev. A*, 81(3):033828, 2010.

25. B. G. Bale, S. Boscolo, and S. K. Turitsyn. Dissipative dispersion-managed solitons in mode-locked lasers. *Opt. Lett.*, 34(21):3286–3288, 2009.

26. B. G. Bale, K. Kieu, J. N. Kutz, and F. Wise. Transition dynamics for multi-pulsing in mode-locked lasers. *Opt. Express*, 17(25):23137–23146, 2009.

27. B. G. Bale and J. N. Kutz. Variational method for mode-locked lasers. *J. Opt. Soc. Am. B*, 25(7):1193–1202, 2008.

28. B. G. Bale, J. N. Kutz, A. Chong, W. H. Renninger, and F. W. Wise. Spectral filtering for high-energy mode-locking in normal dispersion fiber lasers. *J. Opt. Soc. Am. B*, 25(10):1763–1770, 2008.

29. C. Benkert and M. O. Scully. Role of pumping statistics in laser dynamics: Quantum Langevin approach. *Phys. Rev. A*, 41:2756–2764, 1990.

30. E. Benkler, H. R. Telle, A. Zach, and F. Tauser. Circumvention of noise contributions in fiber laser-based frequency combs. *Opt. Express*, 13:5662–5668, 2005.

31. C. Bonatto, M. Feyereisen, S. Barland, M. Giudici, C. Masoller, J. R. Leite, and J. R. Tredicce. Deterministic optical rogue waves. *Phys. Rev. Lett.*, 107:053901, 2011.

32. U. Bortolozzo, J. Laurie, S. Nazarenko, and S. Residori. Optical wave turbulence and the condensation of light. *J. Opt. Soc. Am. B*, 26(12):2280–2284, 2009.

33. Th. Brabec and F. Krausz. Intense few-cycle laser field: Frontiers of nonlinear optics. *Rev. Mod. Phys.*, 72:545–591, 2000.

34. B. C. Buchler, E. H. Huntington, C. C. Harb, and T. C. Ralph. Feedback control of laser intensity noise. *Phys. Rev. A*, 57:1286–1294, 1998.

35. I. L. Budunoglu, C. Ülgüdür, B. Oktem, and F. Ö. Ilday. Intensity noise of mode-locked fiber lasers. *Opt. Lett.*, 34:2516–2518, 2009.

36. S. J. Carter. Quantum theory of nonlinear fiber optics: Phase-space representations. *Phys. Rev. A*, 51:3274–3301, 1995.

37. S. J. Carter and P. D. Drummond. Squeezed quantum solitons and Raman noise. *Phys. Rev. Lett.*, 67:3757, 1991.

38. M. A. Cazalilla, R. Citro, Th. Giamarchi, E. Orignac, and M. Rigol. One-dimensional bosons: From condensed matter systems to ultracold gases. *Rev. Mod. Phys.*, 83(4):1405–1466, 2011.

39. W. Chang, A. Ankiewicz, J. M. Soto-Crespo, and N. Akhmediev. Dissipative soliton resonance in laser models with parameter management. *J. Opt. Soc. Am. B*, 25(12):1972–1977, 2008.

40. W. Chang, A. Ankiewicz, J. M. Soto-Crespo, and N. Akhmediev. Dissipative soliton resonances. *Phys. Rev. A*, 78(2):023830, 2008.

41. W. Chang, A. Ankiewicz, J. M. Soto-Crespo, and N. Akhmediev. Dissipative soliton resonances in anomalous dispersion regime. *Phys. Rev. A*, 79(2):033840, 2009.

42. A. I. Chernykh and S. K. Turitsyn. Soliton and collapse regimes of pulse generation in passively mode-locking laser systems. *Opt. Lett.*, 20(4):398–400, 1995.

43. A. Chong, W. H. Renninger, and F. W. Wise. Properties of normal-dispersion femtosecond fiber lasers. *J. Opt. Soc. Am. B*, 25(2):140–148, 2008.

44. S. Chouli and Ph. Grelu. Rains of solitons in a fiber laser. *Opt. Express*, 17(14):11776–11781, 2009.

45. S. Chouli and Ph. Grelu. Soliton rains in a fiber laser: An experimental study. *Phys. Rev. A*, 81:063829, 2010.

46. D. V. Churkin, O. A. Gorbunov, and S. V. Smirnov. Extreme value statistics in Raman fiber lasers. *Opt. Lett.*, 36(18):3617–3619, 2011.

47. A. A. Clerk, M. H. Devoret, F. Marquardt, and R. J. Schoelfopf. Introduction to quantum noise, measurement, and amplification. *Rev. Mod. Phys.*, 82:1155–1208, 2010.

48. R. Conte, editor. *The Painlevé Property: One Century Later*. Springer, New York, 1999.

49. J. F. Corney and P. D. Drummond. Quantum noise in optical fibers. II. Raman jitter in soliton communications. *J. Opt. Soc. Am. B*, 18:153–161, 2001.

50. S. T. Cundiff. Soliton dynamics in mode-locked lasers. In N. N. Akhmediev and A. Ankiewicz, editors, *Dissipative Solitons*, pages 183–206. Springer, Berlin, 2005.

51. S. T. Cundiff, J. M. Soto-Crespo, and N. Akhmediev. Experimental evidence for soliton explosions. *Phys. Rev. Lett.*, 88(7):073903, 2002.

52. Th. Dauxois and M. Peyrard *Physics of Solitons*. Cambridge University Press, Cambridge, 2007.

53. M. I. Demchuk, V. L. Kalashnikov, V. P. Kalosha, and V. P. Mikhailov. Mode-locking efficiency improvement of cw solid-state lasers with additional cavity. *Quantum Electron.*, 22:1081–1085, 1992.

54. F. J. Diaz-Otero and P. Chamorro-Posada. Propagation properties of strongly dispersion-managed soliton trains. *Phys. Rev. A*, 285:162–170, 2012.

55. E. Ding and J. N. Kutz. Operating regimes, split-step modeling, and the Haus master mode-locking model. *J. Opt. Soc. Am. B*, 26(12):2290–2300, 2009.

56. E. Ding and J. N. Kutz. Stability analysis of the mode-locking dynamics in a laser cavity with a passive polarizer. *J. Opt. Soc. Am. B*, 26:1400–1411, 2009.

57. R. Driben, B. A. Malomed, A. V. Yulin, and D. V. Skryabin. Newton's cradles in optics: From n-soliton fission to soliton chains. *Phys. Rev. A*, 87:063808, 2013.

58. P. D. Drummond and S. J. Carter. Quantum-field theory of squeezing in solitons. *J. Opt. Soc. Am. B*, 4:1565–1573, 1987.

59. P. D. Drummond and J. F. Corney. Quantum noise in optical fibers. I. Stochastic equations. *J. Opt. Soc. Am. B*, 18:139–152, 2001.

60. P. D. Drummond, C. W. Gardiner, and D. F. Walls. Quasiprobability methods for nonlinear chemical and optical systems. *Phys. Rev. A*, 24:914–926, 1981.

61. P. D. Drummond, R. M. Shelby, S. R. Friberg, and Y. Yamamoto. Quantum solitons in optical fibers. *Nature*, 365:307–313, 1993.

62. L. Duan, X. Liu, D. Mao, L. Wang, and G. Wang. Experimental observation of dissipative soliton resonance in an anomalous-dispersion fiber laser. *J. Opt. Soc. Am. B*, 20(1):265–270, 2012.

63. J. M. Dudley, C. Finot, G. Millot, J. Garnier, G. Genty, D. Agafontsev, and F. Dias. Extreme events in optics: Challenges of the MANUREVA project. *Eur. Phys. J. Special Topics*, 185:125–133, 2010.

64. J. M. Dudley, G. Genty, and S. Coen. Supercontinuum generation in photonic crystal fiber. *Rev. Mod. Phys.*, 78(4):1135–1184, 2006.

65. J. M. Dudley, G. Genty, and B. J. Eggleton. Harnessing and control of optical rogue waves in supercontinuum generation. *Opt. Express*, 16(6):3644–3651, 2008.

66. D. Eliyahu, R. A. Salvatore, and A. Yariv. Effect of noise on the power spectrum of passively mode-locked lasers. *J. Opt. Soc. Am. B*, 14:167–174, 1997.

67. M. Erkintalo, G. Genty, and J. M. Dudley. On the statistical interpretation of optical rogue waves. *Eur. Phys. J. Special Topics*, 185:135–144, 2010.

68. L. D. Faddeev and V. E. Korepin. Quantum theory of solitons. *Phys. Rep.*, 42:1–87, 1978.

69. J. M. Fini, P. L. Hagelstein, and H. A. Haus. Agreement of stochastic soliton formalism with second-quantized and configuration-space methods. *Phys. Rev. A*, 57:4842–4853, 1998.

70. J. A. Fleck. Ultrashort-pulse generation by q-switched lasers. *Phys. Rev. B*, 1:84–100, 1970.

71. I. G. Fuss. An interpretation of the spectral measurement of optical pulse train noise. *IEEE J. Quantum Electron.*, 30:2707–2710, 1994.

72. L. Gammaitoni, P. Hänggi, P. Jung, and F. Marchesoni. Stochastic resonance. *Rev. Mod. Phys.*, 70:223–287, 1998.

73. J. Gao, Y. Cao, W. Tung, and J. Hu. *Multiscale Analysis of Complex Time Series: Integration of Chaos and Random Fractal Theory, and Beyond*. Wiley, New Jersey, 2007.

74. C. W. Gardiner and J. Collet. Input and output in damped quantum systems: Quantum stochastic differential equations and the master equation. *Phys. Rev. A*, 31:3761–3774, 1985.

75. C. W. Gardiner and P. Zoller. *Quantum Noise: A Handbook of Markovian and Non-Markovian Quantum Stochastic Methods with Applications to Quantum Optics*. Springer, Berlin, 2004.

76. J. Garnier, M. Lisak, and A. Picozzi. Toward a wave turbulence formulation of statistical nonlinear optics. *J. Opt. Soc. B*, 29(8):2229–2242, 2012.

77. J. Garnier and A. Picozzi. Unified kinetic formulation of incoherent waves propagating in nonlinear media with noninstantaneous response. *Phys. Rev. A*, 81:033831, 2010.

78. G. Genty, C. M. de Sterke, O. Bang, F. Dias, N. Akhmediev, and J. M. Dudley. Collisions and turbulence in optical rogue wave formation. *Phys. Lett. A*, 374(7):989–996, 2010.

79. A. M. Greco, editor. *Direct and Inverse Methods in Nonlinear Evolution Equations*. Springer, Berlin, 2003.

80. Ph. Grelu, W. Chang, A. Ankiewicz, J. M. Soto-Crespo, and N. Akhmediev. Dissipative soliton resonance as a guideline for high-energy pulse laser oscillators. *J. Opt. Soc. Am. B*, 27(11):2336–2341, 2010.

81. H. Haken. *Light: Volume II - Laser Light Dynamics*. North-Holland Physics Publ., Amsterdam, 1985.

82. K. Hammani, Ch. Finot, J. M. Dudley, and G. Millot. Optical rogue-wave-like extreme value fluctuations in fiber Raman amplifiers. *Opt. Express*, 16(21):16467–16474, 2008.

83. K. Hammani, B. Kibler, Ch. Finot, and A. Picozzi. Emergence of rogue waves from optical turbulence. *Phys. Lett. A*, 374:3585–3589, 2010.

84. P. Hänggi. Escape from a metastable state. *J. Stat. Phys.*, 42:105–148, 1986.

85. C. C. Harb, T. C. Ralph, E. H. Huntington, D. E. McClelland, H.-A. Bachor, and I. Freitag. Intensity-noise dependence of Nd:YAG lasers on their diode-laser pump source. *J. Opt. Soc. Am. B*, 14:2936, 1997.

86. H. A. Haus. Quantum noise in solitonlike repeater system. *J. Opt. Soc. B*, 8:1122–1126, 1991.

87. H. A. Haus and Y. Lai. Quantum theory of soliton squeezing: A linearized approach. *J. Opt. Soc. Am. B*, 7:386, 1990.

88. H. A. Haus, M. Margalit, and C. X. Yu. Quantum noise of a mode-locked laser. *J. Opt. Soc. Am. B*, 17:1240, 2000.

89. H. A. Haus and A. Mecozzi. Noise of mode-locked laser. *IEEE J. Quantum Electron.*, 29:983–996, 1993.

90. H. A. Haus and J. A. Mullen. Quantum noise in linear amplifiers. *Phys. Rev.*, 128:2407, 1962.

91. H. A. Haus and Y. Yamamoto. Quantum noise of an injection-locked laser oscillator. *Phys. Rev. A*, 29:1261–1274, 1984.

92. H. A. Haus and Y. Yamamoto. Theory of feedback-generated squeezed states. *Phys. Rev. A*, 34:270, 1986.

93. H. A. Haus and C. X. Yu. Soliton squeezing and the continuum. *J. Opt. Soc. B*, 17:618–628, 2000.

94. H. A. Haus. Theory of mode locking with a fast saturable absorber. *J. Appl. Phys.*, 46:3049–3058, 1975.

95. H. A. Haus, J. G. Fujimoto, and E. P. Ippen. Structures for additive pulse mode locking. *J. Opt. Soc. Am. B*, 8(10):2068–2076, 1991.

96. C. Headley and G. P. Agrawal. *Raman Amplification in Fiber Optical Communication Systems*. Elsevier, Burlington, 2005.

97. F. W. Helbing, G. Steinmeyer, and U. Keller. Carrier-envelope offset phase-locking with attosecond timing jitter. *IEEE J. Sel. Topics Quantum Electron.*, 9:1030–1040, 2003.

98. F. W. Helbing, G. Steinmeyer, J. Stenger, H. R. Telle, and U. Keller. Carrier-envelope-offset dynamics and stabilization of femtosecond pulses. *Appl. Phys. B*, 74:S35–S42, 2002.

99. C. H. Henry and R. F. Kazarinov. Quantum noise in photonics. *Rev. Mod. Phys.*, 68:801–853, 1996.

100. D. R. Hjelme and A. R. Mickelson. Semiconductor laser stabilization by external optical feedback. *IEEE J. Quantum Electron*, 27:352–372, 1991.

101. M. Horowitz, Y. Barad, and Y. Silberberg. Noiselike pulses with a broadband spectrum generated from an erbium-doped fiber laser. *Opt. Lett.*, 22(11):799–801, 1997.

102. M. Horowitz and Y. Silberberg. Control of noiselike pulse generation in erbium-doped fiber lasers. *IEEE Photon. Technol. Lett.*, 10(10):1389–1391, 1998.

103. F. Ö Ilday. Turbulent times. *Nat. Photon.*, 7:767–769, 2013.

104. J. Jasapara, W. Rudolph, V. L. Kalashnikov, D. O. Krimer, I. G. Poloyko, and M. Lenzner. Automodulations in Kerr-lens mode-locked solid-state lasers. *J. Opt. Soc. Am. B*, 17(2):319–326, 2000.

105. V. Kalashnikov. Chirped dissipative solitons. In L. F. Babichev and V. I. Kuvshinov, editors, *Nonlinear Dynamics and Applications*, pages 58–67. Minsk, 2010.

106. V. L. Kalashnikov. Chirped-pulse oscillators: An impact of the dynamic gain saturation. Preprint arXiv:0807.1050v1 [physics.optics], available at http://lanl.arxiv.org/ abs/0807.1050, 2008.

107. V. L. Kalashnikov. Chirped dissipative solitons of the complex cubic-quintic nonlinear Ginzburg–Landau equation. *Phys. Rev. E*, 80(4):046606, 2009.

108. V. L. Kalashnikov. Dissipative solitons: Perturbations and chaos formation. In Ch. H. Skiadas, I. Dimotikalis, and Ch. SkiadasBabichev, editors, *Chaos Theory: Modeling, Simulation and Applications*, pages 58–67. World Scientific Publishing Company, London, 2011.

109. V. L. Kalashnikov. Dissipative solitons: Structural chaos and chaos of destruction. *CMSIM*, 1:51–59, 2011.

110. V. L. Kalashnikov. Chirped-pulse oscillators: Route to the energy-scalable femtosecond pulses. In A. H. Al-Khursan, editor, *Solid State Laser*, pages 145–184. InTech, Croatia, 2012. ISBN: 978-953-51-0086-7.

111. V. L. Kalashnikov. Chaotic dissipative Raman solitons. *CMSIM*, (4):403–410, 2014.

112. V. L. Kalashnikov. Chaotic dissipative Raman solitons. In C. H. Skiadas, editor, *Proceedings of 7th CHAOS Conference*, Lisbon, Portugal, pages 199–206. ISAST, June 7–10, 2014.

113. V. L. Kalashnikov. Dissipative solitons in presence of quantum noise. *CMSIM*, 1:29–37, 2014.

114. V. L. Kalashnikov and A. Apolonski. Chirped-pulse oscillators: A unified standpoint. *Phys. Rev. A*, 79(4):043829, 2009.

115. V. L. Kalashnikov and A. Apolonski. Microjoule mode-locked oscillators: Issues of stability and noise. In T. Graf, J. I. Mackenzie, H. Jelinkova, G. G. Paulus, V. Bagnoud, and C. Le Blanc, editors, *Proceedings of SPIE: Solid State Lasers and Amplifiers IV, and High-Power Lasers*, Brussels, Belgium, vol. 7721, page 77210N. SPIE, April 12–16, 2010.

116. V. L. Kalashnikov and A. Chernykh. Spectral anomalies and stability of chirped-pulse oscillators. *Phys. Rev. A*, 75:033820, 2007.

117. V. L. Kalashnikov, A. Fernández, and A. A. Apolonski. High-order dispersion in chirped-pulse oscillators. *Opt. Express*, 16(6):4206–4216, 2008.

118. V. L. Kalashnikov, V. P. Kalosha, and V. P. Mikhailov. Self-mode-locking of continuous solid-state lasers with an extra cavity. *J. Appl. Spectrosc.*, 58:234–239, 1993.

119. V. L. Kalashnikov, V. P. Kalosha, V. P. Mikhailov, and I. G. Poloyko. Self-mode locking of four-mirror-cavity solid-state laser by Kerr self-focusing. *J. Opt. Soc. Am. B*, 13:462–467, 1995.

120. V. L. Kalashnikov, V. P. Kalosha, I. G. Poloyko, and V. P. Mikhailov. Structure of the mode-locking zone of a solid-state laser with an antiresonant ring. *Quant. Electron.*, 27(5):424–426, 1997.

121. V. L. Kalashnikov, V. P. Kalosha, I. G. Polyko, and V. P. Mikhailov. Periodic cycles, bifurcation and chaotic behavior of new type optical quasi-solitons in solid-state lasers mode-locked by linear and nonlinear phase shift. In *Proceedings of SPIE, Laser Optics 95*, St. Petersburg, Russia, Vol. 2792, pages 86–93, 27 June–1 July 1995.

122. V. L. Kalashnikov, D. O. Krimer, and I. G. Poloyko. Soliton generation and picosecond collapse in solid-state lasers with semiconductor saturable absorber. *J. Opt. Soc. Am. B*, 17(4):519–523, 2000.

123. V. L. Kalashnikov, D. O. Krimer, I. G. Poloyko, and V. P. Mikhailov. Ultrashort pulse generation in cw solid-state lasers with semiconductor saturable absorber in the presence of the absorption linewidth enhancement. *Opt. Commun.*, 159(4–6):237–242, 1999.

124. V. L. Kalashnikov, E. Podivilov, A. Chernykh, and A. Apolonski. Chirped-pulse oscillators: Theory and experiment. *Appl. Phys. B*, 83(4):503–510, 2006.

125. V. L. Kalashnikov, E. Podivilov, A. Chernykh, S. Naumov, A. Fernandez, R. Graf, and A. Apolonski. Approaching the microjoule frontier with femtosecond laser oscillators: Theory and comparison with experiment. *New J. Phys.*, 7(1):217, 2005.

126. V. L. Kalashnikov, I. G. Poloyko, V. P. Mikhailov, and D. von der Linde. Regular, quasi-periodic, and chaotic behavior in continuous-wave solid-state Kerr-lens mode-locked lasers. *J. Opt. Soc. Am. B*, 14(10):2691–2695, 1997.

127. V. L. Kalashnikov and E. Sorokin. Dissipative Raman solitons. *Opt. Express*, 22:30118–30126, 2014.

128. V. L. Kalashnikov, E. Sorokin, and I. T. Sorokina. Chaotic mode-locking of mid-ir chirped-pulse oscillator. In *CLEO/Europe-EQEC 2011 Conference Digest* (22-26 May 2011, Munich, Germany), pages ca. p. 19–sun, 1996.

129. V. L. Kalashnikov, E. Sorokin, and I. T. Sorokina. Multipulse operation and limits of the Kerr-lens mode-locking stability. *IEEE J. Quantum Electron.*, 39(2):323–336, 2003.

130. V. L. Kalashnikov, E. Sorokin, and I. T. Sorokina. Chirped dissipative soliton absorption spectroscopy. *Opt. Express*, 19(18):17480–17492, 2011.

131. V. P. Kalosha, Müller, M. Herrmann, and J. S. Gatz. Spatio-temporal model of femtosecond pulse generation in Kerr-lens mode-locked solid-state lasers. *J. Opt. Soc. Am. B*, 15:535–550, 1998.

132. T. Kapitula. Stability criterion for bright solitary waves of the perturbed cubic-quintic Schrödinger equation. *Physica D*, 116(1–2):95–120, 1998.

133. T. Kapitula and B. Sandstede. Instability mechanism for bright solitary-wave solutions to the cubic-quintic Ginzburg–Landau equation. *Rev. Mod. Phys*, 15(11):2757–2762, 1998.

134. F. X. Kärtner and L. Boivin. Quantum noise of the fundamental soliton. *Phys. Rev. A*, 53:454, 1996.

135. D. J. Kaup. Exact quantization of the nonlinear Schrödinger equation. *J. Math. Phys.*, 16:2036–2041, 1975.

136. P. G. Kevrekidis, D. J. Frantzeskakis, and R. Carretero-González, editors. *Emergent Nonlinear Phenomena in Bose–Einstein Condensates*. Springer, Berlin, 2008.

137. C. Kharif, E. Pelinovsky, and A. Slunyaev, editors. *Rogue Waves in the Ocean*. Springer, Berlin, 2009.

138. B. Kibler, J. Fatome, C. Finot, G. Millot, F. Dias, G. Genty, N. Akhmediev, and J. M. Dudley. The peregrine soliton in nonlinear fibre optics. *Nat. Phys.*, 6:790–795, 2010.

139. M. Kitagawa and Y. Yamamoto. Number-phase minimum-uncertainty state with reduced number uncertainty in a Kerr nonlinear interferometer. *Phys. Rev. A*, 34:3974–3988, 1986.

140. Yu. S. Kivshar and B. A. Malomed. Dynamics of solitons in nearly integrable systems. *Rev. Mod. Phys.*, 61(4):763–915, 1989.

141. Yu. S. Kivshar, D. E. Pelinovsky, Th. Cretegny, and M. Peyrard. Internal modes of solitary waves. *Phys. Rev. Lett.*, 80(23):5032–5035, 1998.

142. S. Kobtsev, S. Kukarin, S. Smirnov, S. Turitsyn, and A. Latkin. Generation of double-scale femto/pico-second optical lumps in mode-locked fiber lasers. *Opt. Express*, 10(23):20707–20713, 2009.

143. S. Kobtsev, S. Kukarin, S. Smirnov, S. Turitsyn, and A. Latkin. Generation of doublescale femto/pico-second optical lumps in mode-locked fiber lasers. *Opt. Express*, 17:20707–20713, 2009.

144. S. Kobtsev, S. Kukarin, S. Smirnov, S. Turitsyn, and A. Latkin. Different generation regimes of mode-locked all-positive-dispersion all-fiber Yb laser. In K. Tankala and J. W. Dawson, editors, *Proceedings of SPIE*, Vol. 7580, *Fiber Lasers VII: Technology, Systems, and Applications*, San Francisco, California, page 758028. SPIE, March 23, 2010.

145. S. W. Koch and H. Haug. *Quantum Theory of the Optical and Electronic Properties of Semiconductors.* World Scientific, London, 2004.

146. A. Komarov, K. Komarov, and F. Sanchez. Quantization of binding energy of structural solitons in passive mode-locked fiber lasers. *Phys. Rev. A*, 79:033807, 2009.

147. A. Komarov, H. Leblond, and A. Sanchez. Quintic complex Ginzburg–Landau model for ring fiber lasers. *Phys. Rev. E*, 72(2):025604(R), 2005.

148. A. Komarov and A. Sanchez. Structural dissipative solitons in passive mode-locked fiber lasers. *Phys. Rev. E*, 77(6):066201, 2008.

149. M. G. Kovalsky, A. A. Hnilo, and J. R. Tredicce. Extreme events in the Ti:sapphire laser. *Opt. Lett.*, 36(22):4449–4451, 2011.

150. Y. Lai. Quantum theory of soliton propagation: A unified approach based on the linearization approximation. *J. Opt. Soc. B*, 10:475–484, 1993.

151. Y. Lai and H. A. Haus. Quantum theory of solitons in optical fibers. I. Time-dependent Hartree approximation. *Phys. Rev. A*, 40:844, 1989.

152. Y. Lai and H. A. Haus. Quantum theory of solitons in optical fibers. II. Exact solution. *Phys. Rev. A*, 40:854–866, 1989.

153. M. Lax. Fluctuations from the nonequilibrium steady state. *Rev. Mod. Phys.*, 32:25–64, 1960.

154. U. Leonhardt and H. Paul. Measuring the quantum state of light. *Prog. Quantum Electron.*, 19:89–130, 1995.

155. F. Li, P. K. A. Wai, and J. N. Kutz. Geometrical description of the onset of multipulsing in mode-locked laser cavities. *J. Opt. Soc. Am. B*, 27(10):2068–2077, 2010.

156. B. Lindner, J. Garsia-Ojalvo, A. Neimann, and L. Schimansky-Greif. Effects of noise in excitable systems. *Phys. Rep.*, 392:321–424, 2004.

157. X. Liu. Hysteresis phenomena and multipulse formation of a dissipative system in a passively mode-locked fiber laser. *Phys. Rev. A*, 81:023811, 2010.

158. X. Liu. Mechanism of high-energy pulse generation without wave breaking in mode-locked fiber lasers. *Phys. Rev. A*, 82(5):053808, 2010.

159. Y. Liu, S. Tschuch, A. Rudenko, M. Dürr, M. Siegel, U. Morgner, R. Moshammer, and J. Ullrich. Strong-field double ionization of Ar below the recollision threshold. *Phys. Rev. Lett.*, 101(5):053001, 2008.

160. B. A. Malomed. Bound solitons in the nonlinear Schrödinger–Ginzburg–Landau equation. *Phys. Rev. A*, 44(10):6954–6957, 1991.

161. B. A. Malomed. Variational methods in nonlinear fiber optics and related fields. In E. Wolf, editor, *Variational Methods in Nonlinear Fiber Optics and Related Fields*, vol. 43, pages 71–193. Elsevier, Amsterdam, 2002.

162. B. A. Malomed. *Soliton Management in Periodic Systems*. Springer, New York, 2006.

163. B. A. Malomed, D. Mihalache, F. Wise, and L. Torner. Spatio-temporal optical solitons. *J. Opt. B*, 7(5):R53, 2002.

164. B. A. Malomed and A. A. Nepomnyashchy. Kinks and solitons in the generalized Ginzburg–Landau equation. *Phys. Rev. A*, 42(10):6009–6014, 1990.

165. C. Manzoni, J. Moses, F. X. Kärtner, and G. Cerullo. Excess quantum noise in optical parametric chirped-pulse amplification. *Opt. Express*, 19:8357–8366, 2011.

166. O. E. Martinez, R. L. Fork, and J. P. Gordon. Theory of passively mode-locked lasers for the case of a nonlinear complex-propagation coefficient. *J. Opt. Soc. Am. B*, 2(5):753–760, 1985.

167. J. J. McFerran, W. C. Swann, B. R. Washburn, and N. R. Newbury. Suppression of pump-induced frequency noise in fiber-laser frequency combs leading to sub-radian f_{ceo} phase excursion. *Appl. Phys. B*, 86:219, 2007.

168. C. R. Menyuk, J. K. Wahlstrand, J. Willits, R. P. Smith, Th. R. Schibli, and S. T. Cundiff. Pulse dynamics in mode-locked lasers: Relaxation oscillations and frequency pulling. *Opt. Express*, 15:6677–6689, 2007.

169. C. Michel, B. Kibler, and A. Picozzi. Discrete spectral incoherent solitons in nonlinear media with noninstantaneous response. *Phys. Rev. A*, 83:023806, 2011.

170. A. V. Mikhailov, editor. *Integrability*. Springer, Berlin, 2009.

171. F. Mitschke. Compounds of fiber-optic solitons. In N. Akhmediev and A. Ankiewicz, editors, *Dissipative Solitons: From Optics to Biology and Medicine*, pages 175–220. Springer, Berlin, 2008.

172. F. Mitschke, G. Steinmeyer, and A. Schwache. Generation of one-dimensional optical turbulence. *Physica D*, 96:251–258, 1996.

173. S. Namiki, E. P. Ippen, H. A. Haus, and Ch. X. Yu. Energy rate equations for mode-locked lasers. *J. Opt. Soc. Am. B*, 14(8):2099–2111, 1997.

174. S. Naumov, A. Fernandez, R. Graf, P. Dombi, F. Krausz, and A. Apolonski. Approaching the microjoule frontier with femtosecond laser oscillators. *New J. Phys.*, 7:216, 2005.

175. A. Nayfeh. *Perturbation Methods*. Wiley, New York, 1973.

176. S. Nazarenko. *Wave Turbulence*. Springer-Verlag, Berlin, 2011.

177. L. E. Nelson, D. J. Jones, K. Tamura, H. A. Haus, and E. P. Ippen. Ultrashort-pulse fiber ring lasers. *Appl. Phys. B*, 65:277–294, 1997.

178. N. R. Newbury and W. S. Swann. Low-noise fiber-laser frequency combs. *J. Opt. Soc. Am. B*, 24:1756, 2007.

179. N. R. Newbury and B. R. Washburn. Theory of the frequency comb output from a femtosecond fiber laser. *IEEE J. Quantum Electron.*, 41:1388–1402, 2005.

180. A. C. Newell, S. Nazarenko, and L. Biven. Wave turbulence and intermittency. *Physica D*, 152–153:520–550, 2001.

181. B. Oktem, C. Ülgüdür, and F. Ö. Ilday. Soliton-Seimilariton fibre laser. *Nat. Photon.*, 4:307–311, 2010.

182. M. Onorato, S. Residori, U. Bortolozzo, A. Montina, and F. T. Arecchi. Rogue waves and their generating mechanisms in different physical contexts. *Phys. Rep.*, 528:47–89, 2013.

183. A. R. Osborn. *Nonlinear Ocean Waves and the Inverse Scattering Transform*. Academic Press, Burlington, 2010.

184. K. E. Oughstun. *Electromagnetic and Optical Pulse Propagation 2: Temporal Pulse Dynamics in Dispersive, Attenuative Media*. Springer, New York, 2009.

185. X.-F. Pang and Y.-P. Feng. *Quantum Mechanics in Nonlinear Systems*. World Scientific Pub., Singapore, 2005.

186. R. Paschotta. Noise of mode-locked lasers (part I): Numerical model. *Appl. Phys. B*, 79:153, 2004.

187. R. Paschotta. Noise of mode-locked lasers (part II): Timing jitter and other fluctuations. *Appl. Phys. B*, 79:163–173, 2004.

188. R. Paschotta. *Encyclopedia of Laser Physics and Technology*. Wiley, Weinheim, 2008.

189. R. Paschotta. Timing jitter and phase noise of mode-locked fiber lasers. *Opt. Express*, 18:5041–5054, 2010.

190. R. Paschotta, A. Schlatter, S. C. Zeller, H. R. Telle, and U. Keller. Optical phase noise and carrier-envelope offset noise of mode-locked lasers. *Appl. Phys. B*, 82:265–273, 2006.

191. R. Paschotta, H. R. Telle, and U. Keller. Noise of solid-state lasers. In A. Sennaroglu, editor, *Solid-State Lasers and Applications*, pages 259–318. Taylor & Francis Group, Boca Raton, 2007.

192. D. Pazó, E. Sánchez, and M. A. Matías. Transition to high-dimensional chaos through quasiperiodic motion. *Int. J. Bifurc. Chaos*, 11:2683–2688, 2001.

193. D. E. Pelinovsky, Yu. S. Kivshar, and V. V. Afanasjev. Internal modes of envelope solitons. *Physica D*, 116:121–142, 1998.

194. V. M. Perez-Garcia, P. Torres, and G. D. Montesinos. The method of moments for nonlinear Schrödinger equations: Theory and applications. *SIAM J. Appl. Math.*, 67(4):990–1015, 2007.

195. A. N. Pisarchik. Rogue waves in a multistable system. *Phys. Rev. Lett.*, 107:274101, 2011.

196. E. V. Podivilov and V. L. Kalashnikov. Heavily-chirped solitary pulses in the normal dispersion region: New solutions of the cubic-quintic complex Ginzburg–Landau equation. *JETP Lett.*, 82(8):467–471, 2005.

197. M. J. Potasek and Y. Yurke. Dissipative effects on squeezed light generated in systems governed by the nonlinear Schrödinger equation. *Phys. Rev. A*, 38:1335–1348, 1988.

198. O. Prochnov, R. Paschotta, E. Benkler, U. Morgner, J. Neumann, D. Wandt, and D. Kracht. Quantum-limited noise performance of a femtosecond all-fiber ytterbium laser. *Opt. Express*, 17:15525, 2009.

199. F. Rana, R. J. Ram, and H. A. Haus. Quantum noise of actively mode-locked lasers with dispersion and amplitude/phase modulation. *IEEE J. Quantum Electron.*, 40:41–56, 2004.

200. W. H. Renninger, A. Chong, and F. W. Wise. Dissipative solitons in normal-dispersion fiber lasers. *Phys. Rev. A*, 77(2):023814, 2008.

201. W. H. Renninger, A. Chong, and F. W. Wise. Pulse shaping and evolution in normal-dispersion mode-locked fiber lasers. *IEEE J. Sel. Topics Quantum Electron.*, 18(1):389–398, 2012.

202. N. N. Rosanov. Dissipative optical solitons. *J. Opt. Technol.*, 76:187–198, 2009.

203. V. Ruban, Y. Kodama, M. Ruderman, J. Dudley, R. Grimshaw, P. V. E. McClintock, M. Onorato, C. Kharif, E. Pelinovsky, T. Soomere1, G. Lindgren, N. Akhmediev, A. Slunyaev, D. Solli, C. Ropers, B. Jalali, F. Dias, and A. Osborne. Rogue waves—towards a unifying concept? Discussions and debates. *Eur. Phys. J. Special Topics*, 185:5–15, 2010.

204. A. Ruehl, D. Wandt, U. Morgner, and D. Kracht. Normal dispersive ultrafast fiber oscillators. *IEEE J. Sel. Topics Quantum Electron.*, 15(1):170–181, 2009.

205. M. Salhi, A. Haboucha, H. Leblond, and F. Sanchez. Theoretical study of figure-eight all-fiber laser. *Phys. Rev. A*, 77(3):033828, 2008.

206. M. Salhi, H. Leblond, and F. Sanchez. Theoretical study of the erbium-doped fiber laser passively mode-locked by nonlinear polarization rotation. *Phys. Rev. A*, 67(1):013802, 2003.

207. V. G. Savitski, K. V. Yumashev, V. L. Kalashnikov, V. S. Shevandin, and K. V. Dukel'skii. Infrared supercontinuum from a large-mode area PCF under extreme picosecond excitation. *Opt. Quantum Electron.*, 39(15):1297–1309, 2008.

208. A. L. Schawlow and C. H. Towns. Infrared and optical masers. *Phys. Rev.*, 112:1940–1949, 1958.

209. T. R. Schibli, I. Hartl, D. C. Yost, M. J. Martin, A. Marcinkevicuis, M. E. Fermann, and J. Ye. Optical frequency comb with submillihertz linewidth and more than 10 W average power. *Nat. Photon.*, 2:355359, 2008.

210. A. Schwache and F. Mitschke. Properties of an optical soliton gas. *Phys. Rev. E*, 55(6):7720–7725, 1997.

211. G. Sciaini and R. J. D. Miller. Femtosecond electron diffraction: Heralding the era of atomically resolved dynamics. *Rep. Prog. Phys.*, 74(9):096101, 2011.

212. M. Segev and D. N. Christodoulides. Incoherent solitons: Self-trapping of weakly correlated wave packets. *Opt. Photonics News*, pages 70–76, February 2002.

213. V. Seghete, C. R. Menyuk, and B. S. Marks. Solitons in the midst of chaos. *Phys. Rev. A*, 76:043803, 2007.

214. E. Seres, J. Seres, and Ch. Spielmann. Extreme ultraviolet light source based on intracavity high harmonic generation in a mode-locked Ti:sapphire oscillator with 9.4 MHZ repetition rate. *Opt. Express*, 20(6):6185–6190, 2012.

215. S. V. Sergeyev. Fast and slowly evolving vector solitons in mode-locked fibre lasers. *Phil. Trans. R. Soc. A*, 372(2027):20140006, 2014.

216. S. V. Sergeyev, Ch. Mou, E. G. Turitsyna, A. Rozhin, S. K. Turitsyn, and K. Blow. Spiral attractor created by vector solitons. *Phil. Trans. R. Soc. A*, 3:e131, 2014.

217. R. M. Shelby, P. D. Drummond, and S. J. Carter. Phase-noise scaling in quantum soliton propagation. *Phys. Rev. A*, 42:2966–2976, 1990.

218. D. V. Skryabin and A. V. Gorbach. Looking at a soliton through the prism of optical supercontinuum. *Rev. Modern Phys.*, 82:1287–1299, 2010.

219. D. R. Solli, C. Ropers, P. Koonath, and B. Jalali. Optical rogue waves. *Nature*, 450(13):1054–1058, 2007.

220. D. R. Solli, C. Ropers, P. Koonath, and B. Jalali. Optical rogue waves. *Nature*, 450:1054–1057, 2007.

221. E. Sorokin, V. L. Kalashnikov, J. Mandom, G. Guelachvili, N. Picque, and I. T. Sorokina. Cr4+:YAG chirped-pulse oscillator. *New J. Phys.*, 10(8):083022, 2008.

222. E. Sorokin, N. Tolstik, V.L. Kalashnikov, and I.T. Sorokina. Chaotic chirped-pulse oscillators. *Opt. Express*, 21:29567–29577, 2013.

223. J. M. Soto-Crespo and N. Akhmediev. Exploding soliton and front solutions of the complex cubic-quintic Ginzburg–Landau equation. *Math. Comput. Simul.*, 69:526–536, 2005.

224. J. M. Soto-Crespo and N. Akhmediev. Soliton as strange attractor: Nonlinear synchronization and chaos. *Phys. Rev. Lett.*, 95:024101, 2005.

225. J. M. Soto-Crespo, N. Akhmediev, and A. Ankiewicz. Pulsating, creeping, and erupting solitons in dissipative systems. *Phys. Rev. Lett.*, 85(14):2937–2940, 2000.

226. J. M. Soto-Crespo, N. N. Akhmediev, V. V. Afanasjev, and S. Wabnitz. Pulse solutions of the cubic-quintic complex Ginzburg–Landau equation in the case of normal dispersion. *Phys. Rev. E*, 55(4):4783–4796, 1997.

227. J. M. Soto-Crespo and Ph. Grelu. Temporal multi-soliton complexes generated by passively mode-locked lasers. In N. N. Akhmediev and A. Ankiewicz, editors, *Dissipative Solitons*, pages 207–240. Springer, Berlin, 2005.

228. J. M. Soto-Crespo, Ph. Grelu, and N. Akhmediev. Dissipative rogue waves: Extreme pulses generated by passively mode-locked lasers. *Phys. Rev. E*, 84(1):016604, 2011.

229. V. E. Tarasov. *Quantum Mechanics of Non-Hamiltonian and Dissipative Systems*. Elsevier, Amsterdam, 2008.

230. H. R. Telle, B. Lipphardt, and J. Stenger. Kerr-lens, mode-locked lasers as transfer oscillators for optical frequency measurements. *Appl. Phys. B 74*, 9:1–6, 2002.

231. H. B. Thacker. Exact integrability in quantum theory and statistical systems. *Rev. Mod. Phys.*, 53:253, 1981.

232. H. B. Thacker and D. Wilkinson. Inverse scattering transform as an operator method in quantum field theory. *Phys. Rev. D*, 19:3660–3665, 1979.

233. E. N. Tsoy, A. Ankiewicz, and N. Akhmediev. Dynamical models for dissipative localized waves of the complex Ginzburg–Landau equation. *Phys. Rev. E*, 73(3):036621, 2006.

234. H. Tsuchida. Correlation between amplitude and phase noise in a mode-locked Cr:LiSAF laser. *Opt. Lett.*, 23:1686–1688, 1998.

235. S. K. Turitsyn, S. A. Babin, E. G. Turitsyna, G. E. Falkovich, E. V. Podivilov, and D. V. Churkin. Optical wave turbulence. In V. Shrira and S. Nazarenko, editors, *Advances in Wave Turbulence*, pages 113–164. World Scientific, Singapore, 2013.

236. S. K. Turitsyn, B. G. Bale, and M. P. Fedoruk. Dispersion-managed solitons in fibre systems and lasers. *Rep. Prog. Phys.*, 521:135–203, 2012.

237. S. K. Turitsyn, S. A. Babin, A. E. El-Taher, P. Harper, D. V. Churkin, S. I. Kablukov, J. D. Ania-Castanón, V. Karalekas, and E. V. Podivilov. Random distributed feedback fibre laser. *Nat. Photon.*, 4:231–235, 2010.

238. E. G. Turitsyna, S. V. Smirnov, S. Sugavanam, N. Tarasov, X. Shu, S. A. Babin, E. V. Podivilov, D. V. Churkin, G. Falkovich, and S. K. Turitsyn. The laminar-turbulent transition in a fibre laser. *Nat. Photon.*, 7:783–786, 2013.

239. N. G. Usechak and G. P. Agrawal. Semi-analytic technique for analyzing mode-locked lasers. *Opt. Express*, 13(6):2075–2081, 2005.

240. W. van Saarloos and P. C. Hohenberg. Fronts, pulses, sources and sinks in generalized complex Ginzburg–Landau equations. *Physica D*, 56:303–367, 1992.

241. S. Vergeles and S. K. Turitsyn. Optical rogue waves in telecommunication data streams. *Phys. Rev. A*, 83:061801(R), 2011.

242. A. G. Vladimirov, S. V. Fedorov, N. A. Kaliteevskii, G. V. Khodova, and N. N. Rosanov. Numerical investigation of laser localized structures. *J. Opt. B*, 1(1):101–106, 1999.

243. D. von der Linde. Characterization of the noise in continuously operating mode-locked lasers. *Appl. Phys. B*, 39:201–217, 1986.

244. J. K. Wahlstrand, J. T. Willits, C. R. Menyuk, and S. T. Cundiff. The quantum-limited comb lineshape of a mode-locked laser: Fundamental limits on frequency uncertainty. *Opt. Express*, 16:18624, 2008.

245. J. K. Wahlstrand, J. T. Willits, T. R. Schibli, C. R. Menyuk, and S. T. Cundiff. Quantitative measurements of timing and phase dynamics in a mode-locked laser. *Opt. Lett.*, 32:3426–3428, 2007.

246. P. K. A. Wai and C. R. Menyuk. Polarization mode dispersion, decorrelation, and diffusion in optical fibers with randomly varying birefringence. *J. Lightwave Tech.*, 14(2):148–157, 1996.

247. B. R. Washburn, W. C. Swann, and N. R. Newbury. Response dynamics of the frequency comb output from a femtosecond fiber laser. *Opt. Express*, 13:10622–10633, 2005.

248. A. M. Weiner. *Ultrafast Optics*. Wiley, New Jersey, 2009.

249. F. Wise and P. Di Trapani. The hunt for light bullets–spatio-temporal solitons. *Opt. Photon. News*, 13:28–32, 2002.

250. F. W. Wise, A. Chong, and W. H. Renninger. High-energy femtosecond fiber lasers based on pulse propagation at normal dispersion. *Laser Photon. Rev.*, 2(1–2):58–181, 2008.

251. G. Xu, J. Garnier, M. Conforti, and A. Picozzi. Generalized description of spectral incoherent solitons. *Opt. Lett.*, 39(14):4192–4195, 2014.

252. Y. Yamamoto and H. A. Haus. Commutation relations and laser linewidth. *Phys. Rev. A*, 41:5164, 1990.

253. Y. Yamamoto and A. Imamoglu, editors. *Mesoscopic Quantum Optics*. Wiley, New York, 1999.

254. Y. Yamamoto and N. Imoto. Internal and external field fluctuations of a laser oscillator: Part I—Quantum mechanical Langevin treatment. *IEEE J. Quantum Electron.*, 22:2032–2042, 1986.

255. Y. Yang. *Solitons in Field Theory and Nonlinear Analysis*. Springer, New York, 2001.

256. B. Yoon and J. W. Negele. Time-dependent approximation for a one-dimensional system of bosons with attractive δ-function interactions. *Phys. Rev. A*, 16:1451, 1977.

257. V. Zakharov, F. Dias, and A. Pushkarev. One-dimensional wave turbulence. *Phys. Rep.*, 398:1–65, 2004.

258. V. E. Zakharov and A. A. Gelash. Soliton on unstable condensate. Preprint arXiv:1109.0620 [nlin.SI], available at http://arxiv.org/abs/1109.0620, 2011.

259. V. E. Zakharov and E. A. Kuznetsov. Optical solitons and quasisolitons. *JETP*, 86(5):1035–1046, 1998.

260. V. E. Zakharov, V. F. Pushkarev, V. F. Shvets, and V. V. Yan'kov. Soliton turbulence. *JETP Lett.*, 48(2):1035–1046, 1988.

261. A. Zaviyalov, R. Iliew, O. Egorov, and F. Lederer. Lumped versus distributed description of mode-locked fiber lasers. *J. Opt. Soc. Am. B*, 27(11):2313–2321, 2010.

262. A. Zavyalov, O. Egorov, R. Iliew, and F. Lederer. Rogue waves in mode-locked fiber lasers. *Phys. Rev. A*, 85:013828, 2012.

263. A. Zavyalov, R. Iliew, O. Egorov, and F. Lederer. Dissipative soliton molecules with independenly evolving or flipping phases in mode-locked fiber lasers. *Phys. Rev. A*, 80:043829, 2009.

264. A. Zavyalov, R. Iliew, O. Egorov, and F. Lederer. Hysteresis of dissipative soliton molecules in mode-locked fiber laser. *Opt. Lett.*, 34(24):3827–3829, 2009.

265. L. M. Zhao, D. Y. Tang, T. H. Cheng, H. Y. Tam, and C. Lu. Generation of multiple gain-guided solitons in a fiber laser. *Opt. Lett.*, 32(11):1581–1583, 2007.

266. L. M. Zhao, D. Y. Tang, J. Wu, X. Q. Fu, and S. C. Wen. Noise-like pulses in a gain-guided soliton fiber laser. *Opt. Express*, 15(5):2145–2150, 2007.

267. L. Zhu, A. J. Verhoef, K. G. Jespersen, V. L. Kalashnikov, L. Grüner-Nielsen, D. Lorenc, A. Baltuska, and A. Fernandez. Generation of high fidelity 62 fs, 7 JnJ pulses at 1035 nm from a net normal-dispersion Yb-fiber laser with anomalous dispersion higher-order-mode fiber. *Opt. Express*, 21:16255–16262, 2013.

28

Hyperbolic Prism, Poincaré Disk, and Foams

Alberto Tufaile and
Adriana Pedrosa
Biscaia Tufaile

28.1 Introduction

Physically based visualization of foams improves our knowledge of optical systems. As long as the light transport mechanisms for light scattering in foams are understood, some interesting patterns observed can be connected with some concepts involving hyperbolic geometry, and this study mainly involves the classical scattering of light in foams and geometrical optics. This scattering system is an open system, in which the trajectories of light are not confined to a bound region [1]. The trajectories can leave the foam and eventually escape to infinity, as it is shown in Figure 28.1.

The physical system constituted by light interacting with foams can give hints for an understanding of collision processes, chaotic dynamics, the scatterers for wireless communications, and the construction of light traps, just to cite a few areas of interest. Besides these scientific and technological applications, the enthrallment of the interaction of light, mirrors, and curved surfaces can be found in many art works of great painters and writers from medieval to modern times, such as Van Eyck's *Arnolfini Marriage*, *La Reproduction Interdite* by René Magritte, *Soap Bubbles* by Chardin, many works of the Dutch graphic artist M. C. Escher, Lewis Carroll, or Jorge Luis Borges.

A light ray entering a foam can reflect and refract chaotically, because of the geometry of the interface of bubbles. In this study, we present some patterns involving self-similarity and hyperbolic geometry.

28.2 Kaleidoscopes

The branch of optics related to the geometry, the geometric optics, is based on the laws of ray reflection and refraction that are very simple

$$\theta_i = \theta_r,$$
$$n_i \sin \theta_i = n_t \sin \theta_t. \tag{28.1}$$

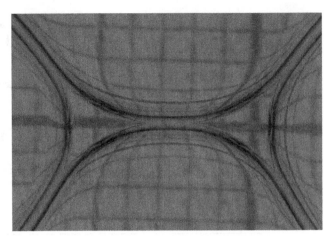

FIGURE 28.1 Some transformations of the image of a square grid behind a foam. The triangles are elements of foams known as Plateau borders. We can observe some contractions and transformations showing that the foam can act as a refractive system.

in which θ_i is the angle of the incident ray in the interface of the foam, θ_r is the reflected ray, θ_t is the angle of the refracted ray, and n_i and n_t are the refractive indices of air and the liquid, respectively. However, the boundary conditions are very difficult to be determined precisely due to the bubble geometry and the paraxial approximation cannot be used for every scattered light ray.

Owing to this aspect, we first explore some features of the pure reflective systems, obtaining some main features that can be compared with the concepts of geometry, and after that, we can apply these concepts to explain some observed patterns.

A kaleidoscope is a physical object made with two or more mirrors side by side, operating in the principle of multiple reflection, as it is shown in Figure 28.2. In general, the mathematics of kaleidoscopes in N dimensions is the study of finite groups of orthogonal $N \times N$ matrices that are generated by reflection matrices. We can observe transformations of such inversions, translations, and rotations. For certain angles of incidence, bubbles reflect the light, and we have conducted some experiments using Christmas balls as reflective spheres to understand the geometry of light scattering in curved surfaces. In Figure 28.3, there is an image of a curved kaleidoscope using three Christmas balls. We can obtain more complex reflections increasing the number of spheres, which resembles some properties of fractal systems, as it is shown in Figure 28.4. These three mirrored spheres could be an analogy to a stereographic projection of a regular kaleidoscope. Such stereographic projection is related to the Poincaré hyperbolic disk illustrated in Figure 28.5e.

28.3 The Poincaré Hyperbolic Disk

According to Needham [2], the Poincaré disk model is a model for hyperbolic geometry, in which a line is represented as an arc of a circle, the ends of which are perpendicular to the disk boundary. What is the definition of parallel rays in this disk? Two arcs that do not meet correspond to parallel rays. In that geometry, arcs that meet orthogonally correspond to perpendicular lines, and arcs that meet on the boundary are a pair of limit rays.

The Poincaré disk is a conformal map of the hyperbolic plane constructed by Beltrami in 1868 and rediscovered 14 years later by Poincaré, which is now universally known as the Poincaré disk.

Using kaleidoscopes, we can see the reflections of the kaleidoscopes with plane mirrors of Figure 28.6a and b in the realm of the Euclidean geometry, while the kaleidoscope with reflective spheres in the realm

FIGURE 28.2 A kaleidoscope can be obtained with two plane mirrors showing some Möbius transformations.

FIGURE 28.3 Image of a curved kaleidoscope using three Christmas balls. We have used two different colors to provide the observation of the sequence of reflections.

of hyperbolic geometry, with the reflections representing Möbius transformations. Considering the disk at the center of the Poincaré disk in Figure 28.6c, we can follow its respective reflections and understand the concept of a new kind of reflective system, known as a hyperbolic kaleidoscope, similar to the kaleidoscopes of Figures 28.3 and 28.4.

The complex pattern observed in the Poincaré disk represents a self-similar pattern, in which there is a fine structure at the border of the disk. This fractal pattern can be observed in the simulation of Figure 28.7a and in the image of the experiment of a hyperbolic kaleidoscope in Figure 28.7b, in which a small sphere is placed at the center of the hyperbolic kaleidoscope. In this way, many physical properties of this system

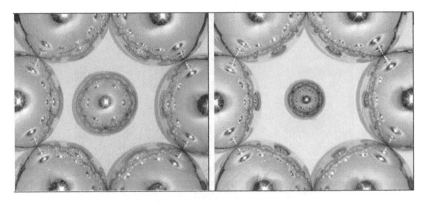

FIGURE 28.4   Image of curved kaleidoscopes using multiple Christmas balls.

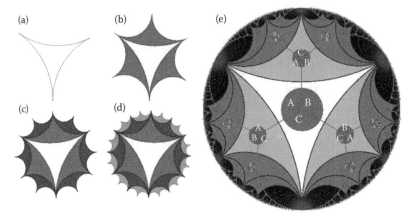

FIGURE 28.5   How to construct the Poincaré hyperbolic disk? From (a) to (d), we present four steps to obtain some reflections and the multiple reflections of an object at the center of the disk.

can be elucidated by the analysis of its geometry, using our sensory experience to guide us in more abstract concepts, such as symmetry groups or the tilling properties.

Besides the connection between geometry and physics, from the point of view of tilling, we can explore the association of the hyperbolic geometry with some features of architecture. For example, Kaplan [3] applied the concepts of non-Euclidean geometry to understand and create ornaments. In this study, patterns with the same properties of the hyperbolic kaleidoscopes are related to polyhedral models, and he presents a geometry-agnostic construction technique to be applied seamlessly to produce Islamic star patterns in the Euclidean plane, hyperbolic plane, and on the sphere.

The array of spheres of Figure 28.8 is made with opaque objects. What happens if one uses transparent objects like bubbles in a foam? The pattern can be observed in Figure 28.9a. The zoom of Figure 28.9b shows multiple reflections and refractions creating a more complex pattern than the case of pure reflective spheres, with the first image of a bubble represented by I, and inside this bubble, we can see the image of other bubbles in II. The process is recurrent, and we can observe the image of three small bubbles in III, in which each bubble acts as a diverging lens and the bubble layers recursively image the optical patterns generated by the bubbles underneath, creating fractal-like patterns. Additional images and computational simulations of this phenomenon can be found in the paper of van der Net et al. [4], in which some beautiful fractal-like patterns of bubble layers arranged in a crystal ordering are presented.

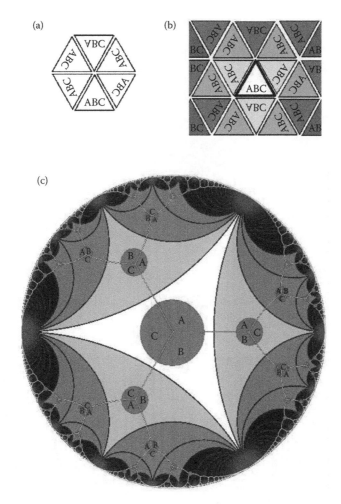

FIGURE 28.6 Comparison between a kaleidoscope with three plane mirrors and the Poincaré disk model.

FIGURE 28.7 In (a), the simulation of a hyperbolic kaleidoscope. In (b), there is an image of the hyperbolic kaleidoscope.

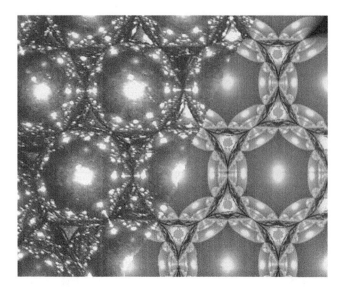

FIGURE 28.8   Experiment and simulation of light scattering of an array of spheres.

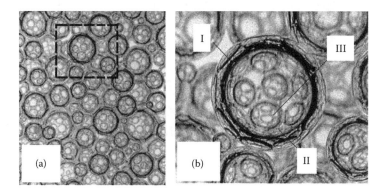

FIGURE 28.9   Image of a foam showing some properties resembling self-similarity.

28.4 Chaotic Scattering

The introduction of the study of chaotic systems in different areas of knowledge represented a paradigm shift, with the change in the perception of events that were already known. In this way, even though the scattering of light in foams is in the realm of classical optics, the approach using dynamical systems revealed some interesting features. For example, we can find the evidence of the butterfly effect in the light scattering in foams, with the sensitivity of initial conditions in the scattering of light, in the interface of bubbles for reflection, and refraction (e.g., see Figure 13 of the paper of van der Net et al. [4]). In Figure 28.10, we present a simulation of the geometrical optics in the Plateau border, with the illustration of some incident light rays (R1) with their refractions (R2) and reflections (R3 and R4). The Plateau border is one of the structures found in foams, and this optical element can present some features observed in chaotic systems because this optical element shares some common features of the hyperbolic kaleidoscopes. In Figure 28.11, we present an image of light scattering of a red laser beam in two bubbles in water.

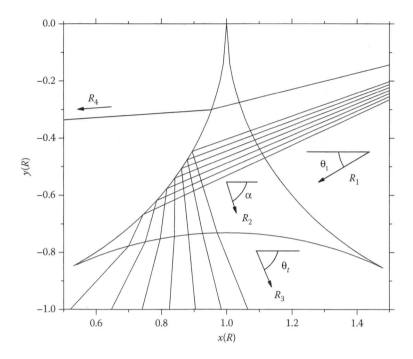

FIGURE 28.10 Light scattering in a Plateau border.

FIGURE 28.11 Butterfly effect in the light scattering with two bubbles.

In this system with just two bubbles, the light scattering presents more possibilities to diverge due to the presence of reflection and refraction at the same time, changing the power of light scattering. In this situation, a light ray can split into two distinct rays: a reflected one and a refracted one, affecting the sensitivity to initial conditions. Analyzing the chaotization mechanisms of light in foams retains some properties of pure reflective systems, but the dynamics of rays passing from one medium to another can acquire

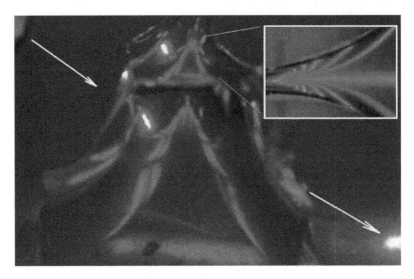

FIGURE 28.12   Cavity hyperbolic prism reflecting the light following the pattern of the Poincaré disk. The inset is the observation of a fractal tree. (Adapted from A. P. B. Tufaile, A. *Journal of Physics: Conference Series* 285, 012006, 2011.)

unusual properties. According to Baryakhtar et al. [5], the law of ray motion itself becomes deterministically chaotic.

On the basis of these assumptions, we have looked for an answer to the following question: How to quantify the chaos of light scattering in foams? To answer this question, we have used the concept of Kolmogorov–Sinai entropy, which is related to the sum of the positive Lyapunov exponents of the system [1]. The Lyapunov exponent quantifies the separation of infinitesimally close trajectories of light rays. We have obtained an empirical formula for the light scattering in foams for the values of Kolmogorov–Sinai entropy in Tufaile et al. [1]. According to our computations, for the case of pure reflective systems, the Kolmogorov–Sinai entropy is 1.98, and for the case of foams, the Kolmogorov–Sinai entropy is 1.82. These values indicate that both the systems have the same degree of chaotization.

To represent the conection between light scattering in foams and the Poincaré disk, we have made a hyperbolic prism and obtained some images with the features of self-similar systems related to the Poincaré disk. In Figure 28.12, we are presenting the cavity of the hyperbolic prism [6,7] with some self-similar structures related to the Poincaré disk.

28.5 Conclusions

The observation of light scattering in foams suggested the existence of some dynamics represented by hyperbolic geometry. Motivated by this representation, due to refraction and reflection at the interfaces, the direction of the rays leaving the interfaces between bubbles can considerably vary for the same incident angle and a small positional offset. A close look at some configurations of the liquid bridges reveals the existence of some triangular patterns surrounded by a complex structure, which bear a resemblance to those observed in some systems involving chaotic scattering and multiple light reflections between spheres. Provided the optical and geometrical properties of the bubbles or sphere surfaces are chosen appropriately, self-similarity is a consequence of multiple scattering of light rays in these cavities. Inspired by the observation of light scattering in foams, we have constructed a hyperbolic prism. The cavity acts as a hyperbolic prism multiplying the scattering of light rays generating patterns related to Poincaré disks.

Acknowledgments

This chapter was supported by Conselho Nacional de Desenvolvimento Científico e Tecnológico (CNPq), Instituto Nacional de Ciência e Tecnologia de Fluidos Complexos (INCT-FCx), and by Fundação de Amparo à Pesquisa do Estado de São Paulo (FAPESP), FAPESP/INCT-FCx/CNPq #573560/2008-0.

References

1. A. P. B. Tufaile, A. Tufaile, and G. Liger-Belair. Hyperbolic kaleidoscopes and chaos in foams in a Hele–Shaw cell. *Journal of Physics: Conference Series* 285, 012006, 2011.
2. T. Needham. *Visual Complex Analysis*, Claredon Press, Oxford, 1997.
3. C. S. Kaplan. Computer graphics and geometric ornamental design. PhD dissertation, University of Washington, 2002.
4. A. van der Net, L. Blondel, A. Saugey, and W. Drenckhan. Simulation and interpretation of images of foams with computational ray tracing techniques. *Colloids and Surfaces A* 309, 159, 2007.
5. V. G. Baryakhtar, V. V. Yanovsky, S. V. Naydenov, and A. V. Kurilo. Chaos in composite billiards. *Journal of the Experimental and Theoretical Physics* 2, 292–301, 2006.
6. A. Tufaile and A. P. B. Tufaile. Hyperbolic prisms and foams in Hele–Shaw cells. *Physics Letters A* 375, 3693–3698, 2011.
7. A. Tufaile, M. V. Freire, and A. P. B. Tufaile. Some aspects of image processing using foams. *Physics Letters A* 378, 3111–3117, 2014.

Parhelic-Like Circle and Chaotic Light Scattering

Adriana Pedrosa
Biscaia Tufaile and
Alberto Tufaile

29.1 Introduction

Scattering problems are at the very heart of physics, from celestial to quantum mechanics, with particles or waves; we are always looking for a target. In our experiment, we are observing the light scattering in foams, and the results can be applied in acoustics, optics, and in spectroscopy. Light through foams presents a complex behavior; for example, though the laws of ray reflection and refraction are simple, the boundary conditions for the scattering of light are very difficult to be determined precisely due to the many awkward technical aspects, such as nonlinearities, or if the thickness of the liquid films is sufficiently close to the wavelength of visible light, there is light interference and thus produces the iridescent colors of soap bubbles.

In this chapter, we discuss chaotic scattering, diffusion, and some aspects of the interface between wave and geometric optics. We have observed that the light-scattering dynamics in foams can present two main processes: a diffusive one related to Gaussian process and another one related to chaotic dynamics, similar to those observed in chaotic saddles, with some rays of light bouncing back and forth for a certain time, and leaving it through one of the several exits. In addition to those behaviors, between geometrics and wave optics, we have also observed the phenomena of the theory of geometrical refraction, with the parlaseric circle.

29.2 The Experimental Apparatus

This experiment involves light scattering in foam. The foam is obtained by shaking a liquid inside a transparent box consisting of two parallel Plexiglas plates separated by a gap ($19.0 \times 19.0 \times 2.0 \, \text{cm}^3$). We can inspect the profiles of light scattered inside the foam and we have used a photographic camera to detect the resulting light patterns. The box contains air and an amount of dishwashing liquid diluted in water ($V = 115 \, \text{cm}^3$). The surfactant used is linear alkylbenzene sulfonate (LAS), with the surface tension being around 25 dyne/cm and the refractive indices equal to 1.333 for the liquid and 1.0 for air. The light sources

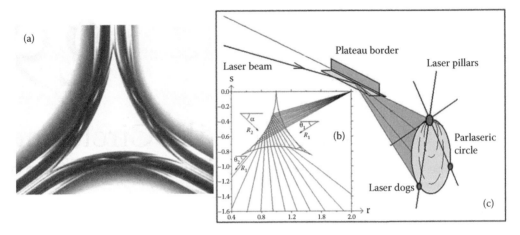

FIGURE 29.1 In (a) the cross section of a Plateau border is shown. An example of ray tracing in a Plateau border (b). The diagram of the experiment is shown in (c).

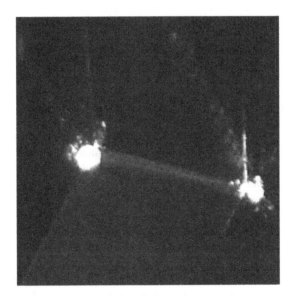

FIGURE 29.2 Image of the laser beam in a Plateau border.

used when photographing the light scattered by the foam were laser diodes with colors red (635 nm), green (532 nm), and blue (405 nm) (Figures 29.1 and 29.2).

We use foams composed of bubbles in a network of surfactant solution and some single structures of foams known as Plateau borders and vertex. The Plateau border is an edge of foam in a junction of three soap films, while the vertex is a place where four Plateau borders meet.

29.3 Diffusion and Chaotic Scattering

Using experiments involving the transport of light in foams, we have observed two main processes of light transport [1]: a diffusive one related to a Gaussian function and another one related to chaotic dynamics, similar to those observed in chaotic saddles, in which a ray of light bounces back and forth for a certain time

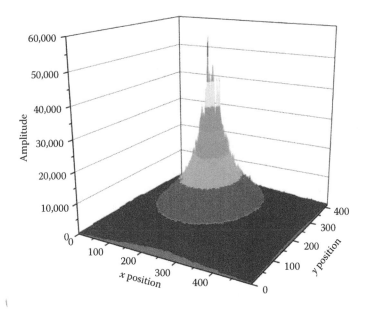

FIGURE 29.3 Light intensity plot of the light scattering in foam.

in the scattering region, and leaves it through one of the several exits, as shown in Figure 29.3, which represents a slice of the profile of light scattering. In this figure, we present the result obtained from the experiment involving light scattering in foams, in which a laser beam is injected into a liquid foam inside a box.

The scattering process spreads the light and limits the depth of light penetration, creating a center glow located just above the interface of the box and the foam. We can represent the overall behavior of the light scattering $g(x, y)$ in foams with the following model:

$$g(x, y) = \frac{1}{2\pi\,\sigma^2} e^{-\left(\frac{x^2+y^2}{2\sigma^2}\right)} + f(x, y), \tag{29.1}$$

in which the first term represents the diffusive process present in all places, while the effects of the chaotic dynamics $f(x, y)$ are mainly present at the center of the foam, around the laser beam direction (Figure 29.4).

29.4 The Parlaseric Circle

Besides these phenomena, we also observed the formation of some caustics. One of these caustics is the light pattern involving the parlaseric circle [2], explained by the theory of geometrical diffraction [3].

The parlaseric circle is a luminous ring generated by light scattering in foams or soap bubbles. In analogy to the atmospheric phenomena known as a parhelic circle, sun dogs, and sun pillars, we have named the features of the patterns observed as a parlaseric circle, laser dogs, and laser pillars. The triangular symmetry of the Plateau borders is analogous to the hexagonal symmetry of ice crystals, which produce these atmospheric phenomena. Working with one Plateau border at a time, we have observed wave optics phenomena that are not used in the explanation of the atmospheric phenomena, such as diffraction and interference (Figures 29.5 and 29.6).

The main features of these patterns are the following: the laser spot and the laser dogs are always inscribed at the circumference of the parlaseric circle, and they are explained by the laws of geometrical optics. The lines crossing the laser spot are explained by the wave optics, and they represent the typical Fraunhofer diffraction of a triangle. The parlaseric circle is explained by the theory of geometrical diffraction [3].

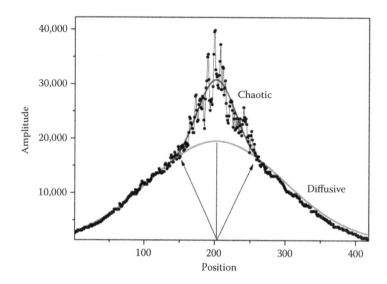

FIGURE 29.4   Chaotic and diffusive light scattering.

FIGURE 29.5   Parlaseric circle, laser dogs, and laser pillars obtained with green light.

According to the geometrical theory of diffraction suggested by Keller [3], when a light beam hits a straight edge obliquely, there is a cone of diffracted rays u_e and the cross section of this cone is a circle given by

$$u_e = Du_i r^{-1/2} e^{ikr}, \tag{29.2}$$

where D is a diffraction coefficient, u_i the incident field, r the distance between the edge and the screen, and $k = 2\pi/\lambda$ is the wave number of the incident field with wavelength λ. This is the case when a laser light hits some structures of foams known as Plateau border, as previously observed by Tufaile and Tufaile [4] (Figure 29.7).

FIGURE 29.6 Parlaseric circle, two laser dogs, and laser pillars obtained with blue light.

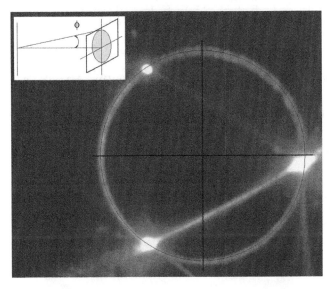

FIGURE 29.7 According to the geometrical theory of diffraction, when a light beam hits a straight edge obliquely, there is a cone of diffracted rays. ϕ is the semi angle of the diffraction cone that is equal to the incident angle measured from the laser beam direction.

Figure 29.8 shows some important aspects of the parlaseric circle pattern, in which *LS* represents the laser spot, *LD1* and *LD2* are the laser dogs, and the laser pillars are represented by the lines of diffraction and interference η_1, η_2, and η_3. Depending on the angle of incidence of light, we can observe two or four laser dogs.

29.5 Vertex Diffraction

In addition to the previous case, we have inspected the case known as vertex-diffracted ray, when light hits a vertex formed by a junction of four Plateau borders, in which the corresponding diffracted waves are

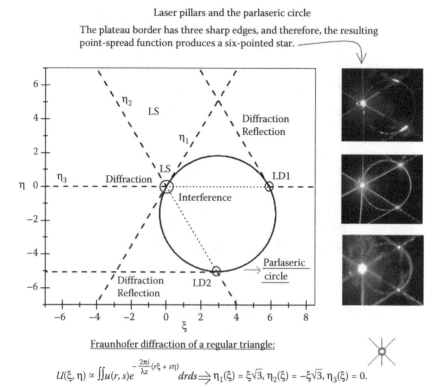

Laser pillars and the parlaseric circle

The plateau border has three sharp edges, and therefore, the resulting
point-spread function produces a six-pointed star.

Fraunhofer diffraction of a regular triangle:

$$U(\xi, \eta) \propto \iint u(r, s) e^{-\frac{2\pi i}{\lambda z}(r\xi + s\eta)} \, dr\, ds \Rightarrow \eta_1(\xi) = \xi\sqrt{3}, \ \eta_2(\xi) = -\xi\sqrt{3}, \ \eta_3(\xi) = 0.$$

FIGURE 29.8 Laser pillars and the parhelic circle pattern diagram. LS is the laser spot, LDs are the laser dogs, and η_n are the laser pillars.

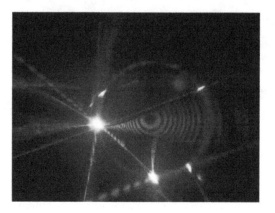

FIGURE 29.9 Vertex diffraction with concentric circles.

spherical with the vertex at their center. Keller suggested that the field on the diffracted ray for this case is
given by

$$u = Cu_i \left(\frac{e^{ikr}}{r} \right). \tag{29.3}$$

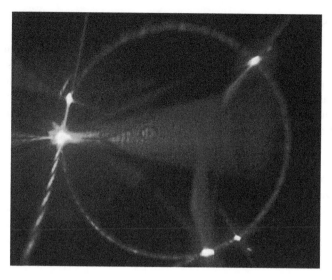

FIGURE 29.10 Vertex diffraction with concentric circles in a larger incident angle.

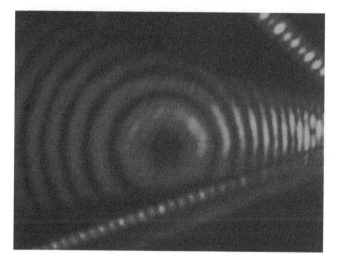

FIGURE 29.11 Another image of the circular fringes at the center of the parlaseric circle.

The images in Figures 29.9 and 29.10 obtained in our experiment that show concentric fringes, could be the vertex diffraction, and in this case, the amplitude varies as r^{-1} since the cross-sectional area of a tube of diffracted rays is proportional to r^2, according to Keller.

However, when we have moved the screen where the images were obtained, the interference pattern remained the same and the width of the fringes change, indicating that this is a spherical wave of diffraction, resembling the interference patterns observed in the case of Newton rings or some patterns obtained in the Michelson interferometer. By increasing the angle of laser beam incidence and the axis of the Plateau border, the size of the fringes decreases.

By inspection of the patterns like those in Figures 29.9–29.11, we are exploring the aspects of thin-film interference that occurs in a curved thin-liquid film and, trying to associate with the case of Pohl interferometer, that is a device based on the amplitude-splitting interference [5].

29.6 Conclusions

We have proposed a combination of two main processes of the light transport in foams: a diffusive one related to Gaussian function, and another one related to chaotic dynamics. Just considering the aspects of the geometrical optics, the curvature of soap film structures causes the incident light to be scattered in different directions, because the direction of the rays leaving the soap film boundaries can vary greatly for the same incident angle with a small positional offset. Besides this, we have obtained some interesting patterns of light diffraction and compared them with the theory of geometrical diffraction suggested by Keller.

The classification of the optical phenomena somehow involves the scale of the scatterer, and the Plateau border is not just a regular prism. The pattern of the parlaseric circle presents some features of the geometrical optics, wave optics, and the theory of geometrical diffraction. All these phenomena are related to the nonlinear effects of light transport in foams, with applications in image processing, construction of new optical elements to generate halos or spherical waves, and with the possibility of improvements in the studies of atmospheric optics. Since all these phenomena can be observed with a tabletop experiment, they can be performed to get a better understanding of some concepts of optics, such as the Huygens principle.

Acknowledgments

This chapter was supported by Conselho Nacional de Desenvolvimento Científico e Tecnológico (CNPq), Instituto Nacional de Ciência e Tecnologia de Fluidos Complexos (INCT-FCx), and by Fundação de Amparo à Pesquisa do Estado de São Paulo (FAPESP), FAPESP/INCT-FCx/CNPq #573560/2008-0.

References

1. A. Tufaile, M.V. Freire, and A.P.B. Tufaile. Some aspects of image processing in foams. *Physics Letters A* **378**, 3111–3117, 2014.
2. E. Conover. Researchers create "laser dogs" with soap bubbles. *Science Magazine*, 2015. http://news.sciencemag.org/physics/2015/02/researchers-create-laser-dogs-soap-bubbles.
3. J. Keller. Geometrical theory of diffraction. *Journal of Optics Society America* **52**, 116–130, 1962.
4. A. Tufaile and A.P.B. Tufaile. Parhelic-like circle from light scattering in Plateau borders. *Physics Letters A* **379**, 529–534, 2015.
5. E. Hecht. *Optics*. Addison-Wesley Longman, England, 1998.

VIII

Chaos Theory in Biology and Medicine

30

Applications of Extreme Value Theory in Dynamical Systems for the Analysis of Blood Pressure Data

Davide Faranda

30.1 Introduction

Over the last few decades, several advancements in experimental techniques have allowed for the measurement of health parameters at a high sampling frequency [3,5]. Although a frequency analysis of the data often reveals major anomalies and helps in the diagnostic process, a great part of the information is hidden in fine properties of the measured time series. Among these properties, extreme fluctuations of health parameters may trigger irreversible processes and result in acute crisis. In the ambit of cardiovascular disease, blood pressure fluctuations may trigger acute hypotensive (hypertensive) episodes and, in some cases, cardiac crisis. For a series of independent and identically distributed (iid) variables, a traditional extreme value analysis straightforwardly gives the probability of observing extremely low (or high) fluctuations of health parameters [13]. However, blood pressure data have internal correlations originating from the quasiperiodic biological processes responsible for blood circulation. Therefore, in order to provide effective warnings against cardiac crisis, the traditional techniques must be accompanied by methods that preserve the dynamical information contained in the data.

In previous studies, the risk of observing acute hypotensive or hypertensive episodes has been assessed either by analyzing blood pressure data averaged over 1 minute [16], or by using neural network multimodels [11], or via spectral techniques [2]. All these techniques usually rely on the identification of a single-threshold pressure value defined by counting the exceedances (hypertensive episodes) or the deficits (hypotensive episodes) with respect to such a threshold. Statistics are always computed under the nongenuine assumption that pressure data are iid. However, when this assumption is satisfied, these methods would only provide information on the tails of the distribution, without inspecting the bulk statistics. This would prevent from providing a global map of the status of the patient, providing the nursing staff with the detailed conditions of the patient. Moreover, they do not associate the probability of extreme pressure events to a time scale of medical interest, e.g., length of a medical treatment or of an operation.

In this chapter we show that, by combining the celebrated theory of Poincaré recurrences [9] with extreme value statistics, one can devise an efficient algorithm to measure the range of expected fluctuations of systolic and diastolic arterial pressures. Such a range can be used to assess the clinical state of a patient and therefore the condition of the patient. The combination of the extreme value statistics with new results from dynamical systems theory allow for a proper treatment of correlations in the data, without the assumption that data are iid. The chapter is structured as follows: Section 30.2 gives an overview on the method and describe the link between recurrences of blood pressure values and rare events. In Section 30.3 we present the multiparameter intelligent monitoring in intensive care (MIMIC) database and describe some general properties of the blood pressure time series. Section 30.4 is dedicated to the data analysis and the definition of an index that directly measure the probability of observing the likelihood for recurrences of pressure values. We conclude by discussing our findings.

30.2 The Method

Here, we briefly give some theoretical elements of the method of recurrences. For a time series $p(t)$ we start by defining the common approaches in the definition of a recurrence of a certain value p^*. Such a description is accompanied by direct visualization of the different methods in Figure 30.1.

- The *exact recurrence* of the value p^* occurs if we have $p(t) = p^*$. This definition is very restrictive as it requires to observe identical values of p^*. In the upper panel of Figure 30.1 exact recurrences are found on the intersections between the orbits and the solid horizontal line.
- The *spatial approach* consists of choosing an interval Δp such that $p(t)$ is a recurrence of $p(t^*)$ if $p^* - \Delta p < p(t) < p^* + \Delta p$. In Figure 30.1, it corresponds to considering all the values falling in the solid rectangle. The spatial approach is equivalent to the so-called peak-over-threshold technique for extremes introduced by Pickands [17]. Extracting the recurrences this way correspond to sampling the statistics of the distances $\text{dist}(p(t), p^*) < \Delta p$, with Δp acting as a threshold value. This statistics will depend on the choice of Δp and instead it will not be sensible on the dynamics. A random reshuffle of the time series will reproduce exactly the same threshold statistics, while destroying all the correlations in the dataset.
- The *time window approach* does not imply the selection of a threshold Δp but rather the introduction of a time scale τ. The recurrence is defined as the closest value of p^* in the interval τ, formally $\min(\text{dist}(p(t), p^*))$ for $t^* < t < t^* + \tau$. We construct the statistics of such recurrences by extracting n values in all intervals of length τ. These points are marked by the circles in the lower panel of Figure 30.1.

We stick on to the latter approach because it preserves the dynamical information contained in the data, i.e., a reshuffle of the data does not produce the same asymptotic distribution as it happens for the Pickands method. In the limit of $n, \tau \to \infty$, a series of minima

$$X_i = \min(g(\text{dist}(p(t), p^*))) \quad t_i < t < t_i + \tau$$

with $g(\cdot)$ a generic observable function, will obey the generalized extreme value (GEV) distribution:

$$F_G(x; \mu, \sigma, \xi) = \exp\left\{ -\left[1 + \xi \left(\frac{x - \mu}{\sigma} \right) \right]^{-1/\xi} \right\}. \tag{30.1}$$

A GEV holds for $1 + \xi(x - \mu)/\sigma > 0$, where $\mu \in \mathbb{R}$ is the location parameter, $\sigma > 0$ the scale parameter, and ξ the shape parameter that discriminates the type of tail behavior: Gumbel law for bulk statistics with

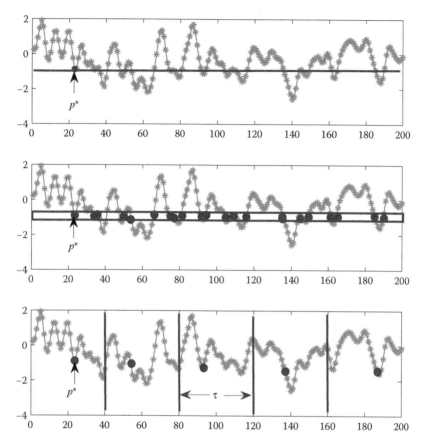

FIGURE 30.1 Example of three different definitions of recurrences of the value p^* of a time series. Upper panel: exact recurrences occur if $p(t) = p^*$ (points on the solid line). Central panel: recurrence of p^* occurs each time $p^* - \Delta p < p(t) < p^* + \Delta p$ (points within the solid rectangle). Lower panel: recurrences are sampled over a time interval τ as the closest values to p^* are measured in each interval.

exponential tails ($\xi = 0$), Fréchet laws for unbounded tails ($\xi > 0$), and Weibull laws for tails bounded from above ($\xi < 0$). As shown by in References 6 through 10, the convergence is given by a combination of the exponential recurrences statistics for chaotic systems and the functional form of the observable function $g(\cdot)$. In particular, letting $g(\cdot) = -\log(\cdot)$, the asymptotic GEV is always a Gumbel law with the shape parameter $\xi = 0$ [9]. The use of the logarithmic function to measure the distances is not arbitrary and it has been justified in several papers as it allows for a correct sampling of the so-called short returns [1,20]. This result is used to answer the following question:

Given a time series representing a chaotic dynamics, and an observation p^*, what is the typical time scale τ such that this observation is recurring in the dynamics?

We can answer by measuring the likelihood of a fit of the X_i's to the Gumbel model, provided that we have extracted the data as described above:

- If the fit succeeds, it means that we are likely to observe a recurrence of p^* in τ. We can repeat the experiment for shorter bin lengths and find the smallest τ such that the fit converges. This defines the shortest convergent recurrence time.
- If the fit fails, the recurrence of p^* in τ is unlikely. We repeat the experiment by increasing the value of τ until the fit succeeds to find the shortest recurrence time.

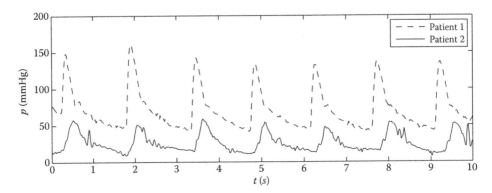

FIGURE 30.2 An example of time series of systemic arterial pressure (dashed line) and pulmonary arterial pressure (solid line) for a patient with unstable post infarction angina.

For the applicability of this method on blood pressure data, first, we have to check the chaotic behaviors of the time series, and then assess the typical time scales involved in the process. For this purpose, in the next section, we perform a first analysis of the blood pressure data, which will clarify these two important issues.

30.3 The Blood Pressure Data

The data have been extracted by the MIMIC II database [19] that includes, for each patient, a time series of systemic and pulmonary arterial pressures sampled at 300 Hz. Data have been collected over a period of about 1 h. The general characteristics of the time series analyzed are presented in Figure 30.2. The series refers to patient 28 of the MIMIC database (*mgh028*), recovered at the Massachusetts General Hospital (MGH) for unstable postinfarction angina. We isolate only 10 s of the time series corresponding roughly to $3 \cdot 10^4$ observations. It is apparent that data are not iid: the periodic signature of the heartbeat is evident, although deviations from the periodic behavior both in time and amplitude can be observed. On longer time scales, ambient conditions, human interactions, and/or pharmacological treatments trigger a chaotic behavior that legitimates the assumptions made in the previous sections. Such chaotic behavior of systemic pressure data has been reported by several authors [21,22]. This first analysis also provides a minimal time scale for the recurrence technique: we will need to consider a period τ longer than the heartbeat oscillation in order to properly sample recurrences of a chaotic system. In the next section, we will start the recurrence analysis by analyzing the effects of τ on the statistics.

30.4 Analysis

Our goal is to provide an accurate information on the status of patients by applying the recurrence techniques described in the previous sections. The final product of such an analysis will be a map of health parameters (here systemic and pulmonary arterial pressures) that inform the nursing staff on the status of the patient and whether he experienced anomalous pressure values with respect to a certain time scale of interest τ. We recall the procedure applied on the data in an algorithmic way:

1. We divide the full time series (the length is typically 1 h) in n intervals of length τ.
2. We fix a reference pressure value p^*.
3. In each interval, we take the closest value of $p(t)$ to the chosen p^*, obtaining n values $p_1, p_2, \ldots, p_n$ as shown in Figure 30.1.
4. We construct a series of logarithmic returns for such values, namely $X_i = -\log(\text{dist}(p_i, p^*))$.
5. We fit the X_i to the GEV model checking the goodness of the Gumbel model.

We repeat the analysis over and over changing the reference points, until all the p^* between $min(p)$ and $max(p)$ have been selected. As mentioned before, if there are enough recurrences of a certain pressure value in τ, the fit will succeed; otherwise it will fail. As a measure of the goodness of the Gumbel model, we use a suitably renormalized likelihood function. This provides a measure of the probability that a patient will experience a pressure value p_{ART} in a time interval τ. The distance between the Gumbel law and the histograms will be quantified by the quantity δ defined as

$$\delta(p_{ART}) = \frac{L(p_{ART})}{L_{max} - L_{min}}$$

Here, $L_{p_{ART}}$ is the negative likelihood function of observing a Gumbel distribution for the recurrences of the value $p^* = p_{ART}$. Since L attains infinity for a *perfect* fit and minus infinity for a completely unreliable fit, then δ takes values in the range $(-1, 1)$, so that

- $\delta(p_{ART}) > 0$ indicates how likely is the recurrence of the value p^* every τ seconds.
- $\delta(p_{ART}) < 0$ indicates how unlikely the value p^* recurs in a τ time interval.

First, we verify the consistency of our method with respect to the usual pressure measurements. The maxima of δ should correspond with the measured systolic and diastolic pressures, respectively p_s and p_d, and they should not sensibly change with τ. This is what is shown in Figure 30.3 for the series of patient *mgh002* and for three different values of τ. This patient was admitted to the hospital for resection and grafting of abdominal aortic aneurysm and his measured systemic pressure parameters were $p_d = 60$ mmHg and $p_s = 130$ mmHg. Effectively, from Figure 30.3 we observe two maxima of δ corresponding to p_s and p_d. However, although the value of p_d is consistent with the one measured at the hospital, the estimate for the systolic pressure is 150 mmHg higher than the one estimated using classical techniques. This is explainable by looking at the shape of the distribution of δ around the p_s value that is skewed toward values smaller than p_s. We recover the value of p_s measured at the hospital by taking a conditional mean of p_{ART} restricted to values higher than 80 mmHg.

These first considerations already point to the problem of measuring only average values of pressure, rather than fully characterizing the probability distributions as we attempt to do here. Moreover, the analysis of δ clearly points to a large range of fluctuations around the p_s-detected value and suggests that this patient could experience hypertensive episodes, underestimated with the classical measurements. We also observe that the parameters slightly depend on τ and that for the shortest time interval $\tau = 1$ s curves of δ are not smooth. This effect can also be detected for other patients and it is explainable with the general argument pointed out in the previous section, i.e., τ must be larger than the heartbeat period in order to get a good convergence for δ. For this reason, in the remaining of this chapter, we will fix $\tau = 4$ s to avoid such problems. We remark that the parameter τ can be varied to correspond to a specific time of interest for medical treatments.

To show the capability of the method, we analyzed several other patients presenting different clinical histories. In Figure 30.4 we present the quantity δ for three particular subjects: patient 1—*mgh002* considered in the previous analysis (left panels); patient 2—*mgh001* presenting a carotid endarterectomy (central panels); and patient 3—*mgh028* showing an unstable postinfarction angina (right panels). In all the cases, the two maxima corresponding to the systolic and diastolic pressures are recognizable. However, their location and the range of fluctuations associated with each of them strongly varies. For patient 2, the hemodynamic data recorded in the database are $p_s = 140$ and $p_d = 50$ mmHg, in agreement with our observations. The analysis also points to a moderate hypertensive risk. For patient 3, we observe a slight deviation from the p_s value reported in the database ($p_s = 130$ and $p_d = 50$ mmHg) again due to the skewed nature of the distribution of the systolic blood pressure already observed in References 14 and 18.

The general medical case of a patient is not given by a single hemodynamic parameter but rather by a combination of them. One of the advantages of the recurrence technique described in this chapter is that it is trivially expandable to multiple hemodynamic parameters. As an example, we consider both the arterious p_{ART} and the pulmonary pressure p_{AP}, and define the joint probability of observing a Gumbel

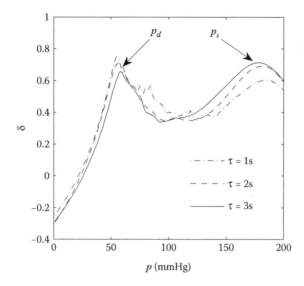

FIGURE 30.3   Normalized likelihood of the Gumbel distribution δ for three different values of $\tau = 1, 2, 3$ s. Maxima indicate the location of the estimated systolic pressure p_s and diastolic pressure p_d.

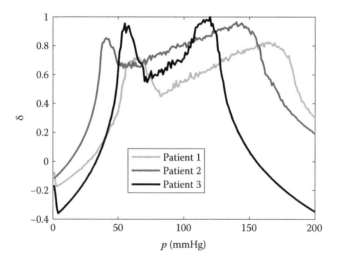

FIGURE 30.4   Normalized likelihood of the Gumbel distribution δ of the systemic arterial pressure data p_{ART} for three different patients: patient 1 admitted to the hospital for resection and grafting of an abdominal aortic aneurysm, patient 2 for carotid endarterectomy, and patient 3 for unstable postinfarction angina. Time interval is $\tau = 4$ s.

distribution for the recurrences of a specific couple of values $p^* = (p_{ART}, p_{AP})$. The structure of the function $\delta(p_{ART}, p_{AP})$ will therefore give an immediate information on the joint risk of experiencing pulmonary and systemic hypertension (or hypotension). For the three patients previously described, the results are reported in Figure 30.5. The left panel refers to patient 1: light gray indicate that the probability of observing a couple of values (p_{ART}, p_{AP}) is high, dark gray that this combination is unlikely. For this patient, there exists a moderate risk of experiencing a combination of pulmonary and systemic hypertensive events. The analysis of patient 2 (central panels) still points to a risk of moderate systemic hypertension. The analysis of pulmonary pressure reveals that this risk is also associated with pulmonary hypertension: the $\delta(p_{ART}, p_{AP})$

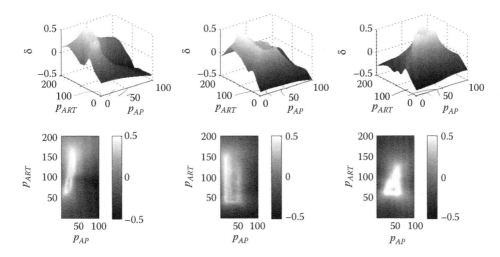

FIGURE 30.5 Normalized likelihood of the Gumbel distribution δ for three different patients obtained by combining systemic arterial p_{ART} and pulmonary arterial p_{AP} pressure data. Time interval is $\tau = 4\,s$. Left panels: patient with resection and grafting of an abdominal aortic aneurysm. Central panels: patient with carotid endarterectomy. Right panels: patient with unstable postinfarction angina. Positive values of δ mean that we are very likely to observe the corresponding values of pressure, negative values that they are unlikely.

profile consists of a sort of ring of probable values, i.e., the bimodal structure observed for the pressure data also repeats for the pulmonary pressure. Finally, $\delta(p_{ART}, p_{AP})$ for patient 3 shows a more complicated pattern associated with a higher risk of hypertensive pulmonary episodes with respect to blood arterial episodes. This is also reported in the logbook of the MGH database.

30.5 Final Remarks

The main achievement of this chapter is to provide a method for the detection of recurrences of pressure values. The recurrence technique shows that the concept of systolic and diastolic blood pressures commonly used in the medical care units can be insufficient for assessing the risk of hypotensive and hypertensive episodes. The main drawback of pure statistical techniques is the misleading assumption that the data are iid, a problem that we overcome by using an approach based on recurrences. Here, we have introduced a parameter δ directly linked to the likelihood of observing a Gumbel statistics expected on a theoretical basis for the recurrences of pressure values. Positive values of δ are associated with higher probabilities and negative values are associated with lower ones. In other words, the profiles of δ provide the following information: for each pair of values p_{ART}, P_{AP}, a positive δ means that, if the conditions of the patient stay stationary, it is likely to observe such values again. This allows to take actions if there is a risk for acute episodes and therefore they provide a useful instrument for operators of medical care units. With respect to the common methods, based on computing the histograms and the probability distribution of pressure values, the study of recurrences provides the following advantages: (i) the probability of observing a certain pressure value is given with respect to a certain time scale τ and condensed in an index with values of $[-1\ +1]$. This means that one can plan medical treatments, operations, or examinations that last for a specific time knowing the risks associated with the patient. (ii) It allows for combining several variables giving the joint probability of observing a combination of health parameters in the time scale of interest τ. (iii) It provides a precise information on the quantity of data needed to assess the Gumbel statistics. This prevents one from overestimating or underestimating the recurrence times. (iv) The technique is easy to implement—a self-written MATLAB code has been used to perform the analysis—and can run in real time. Profiles of δ can be updated with new data, e.g., every minute, giving a map for the changing risk for each patient. This technique can be easily extended to any framework where chaotic time series are available even if they show fractal or multifractal properties and/or power law behavior (see, e.g., [4,12]) as theoretically described in Reference 15. In these cases, traditional statistical approaches are more likely to return biased estimations as they implicitly assume continuous support for the time series as well as independence among the data.

References

1. M Abadi and S Vaienti. Large deviations for short recurrence. *Discrete and Continuous Dynamical Systems—Series A*, 21(3):729–747, 2008.
2. S Axelrod, M Lishner, O Oz, J Bernheim, and M Ravid. Spectral analysis of fluctuations in heart rate: An objective evaluation of autonomic nervous control in chronic renal failure. *Nephron*, 45(3):202–206, 1987.
3. G Beevers, GYH Lip, and E O'Brien. Blood pressure measurement. *British Medical Journal*, 322(7293):1043–1047, 2001.
4. JT Bigger, RC Steinman, LM Rolnitzky, JL Fleiss, P Albrecht, and RJ Cohen. Power law behavior of rr-interval variability in healthy middle-aged persons, patients with recent acute myocardial infarction, and patients with heart transplants. *Circulation*, 93(12):2142–2151, 1996.
5. AV Chobanian, GL Bakris, HR Black, WC Cushman, LA Green, JL Izzo, DW Jones, BJ Materson, S Oparil, and JT Wright. Seventh report of the Joint National Committee on prevention, detection, evaluation, and treatment of high blood pressure. *Hypertension*, 42(6):1206–1252, 2003.
6. D Faranda, X Leoncini, and S Vaienti. Mixing properties in the advection of passive tracers via recurrences and extreme value theory. *Physical Review E*, 89(5):052901, 2014.
7. D Faranda and S Vaienti. A recurrence-based technique for detecting genuine extremes in instrumental temperature records. *Geophysical Research Letters*, 40(21):5782–5786, 2013.
8. D Faranda and S Vaienti. Extreme value laws for dynamical systems under observational noise. *Physica D: Nonlinear Phenomena*, 280:86–94, 2014.
9. ACM Freitas, JM Freitas, and M Todd. Hitting time statistics and extreme value theory. *Probability Theory Related Fields*, 147(3):675–710, 2010.
10. ACM Freitas, JM Freitas, and M Todd. Extreme value laws in dynamical systems for non-smooth observations. *Journal of Statistics Physics*, 142:108–126, 2011.
11. J Henriques and TR Rocha. Prediction of acute hypotensive episodes using neural network multi-models. In *Computers in Cardiology*, p. 549–552. IEEE, Park City, UT, 2009.
12. HV Huikuri, TH Mäkikallio, CK Peng, AL Goldberger, U Hintze, and M Møller. Fractal correlation properties of rr interval dynamics and mortality in patients with depressed left ventricular function after an acute myocardial infarction. *Circulation*, 101(1):47–53, 2000.
13. MR Leadbetter. Extremes and local dependence in stationary sequences. *Z. Wahrsch. Verw. Gebiete*, 65(2):291–306, 1983.
14. Y Li, JG Wang, E Dolan, PJ Gao, HF Guo, T Nawrot, AV Stanton, DL Zhu, E O'Brien, and JA Staessen. Ambulatory arterial stiffness index derived from 24-hour ambulatory blood pressure monitoring. *Hypertension*, 47(3):359–364, 2006.
15. V Lucarini, D Faranda, G Turchetti, and S Vaienti. Extreme value theory for singular measures. *Chaos*, 22(2):023135, 2012.
16. GB Moody and LH Lehman. Predicting acute hypotensive episodes: The 10th annual physionet/computers in cardiology challenge. In *Computers in Cardiology*, p. 541–544. IEEE, Park City, UT, 2009.
17. J Pickands III. Statistical inference using extreme order statistics. *The Annals of Statistics*, 3:119–131, 1975.

18. SC Robinson and M Brucer. Range of normal blood pressure: A statistical and clinical study of 11,383 persons. *Archives of Internal Medicine*, 64(3):409–444, 1939.

19. M Saeed, M Villarroel, AT Reisner, G Clifford, LW Lehman, G Moody, T Heldt, TH Kyaw, B Moody, and RG Mark. Multiparameter intelligent monitoring in intensive care ii (mimic-ii): A public-access intensive care unit database. *Critical Care Medicine*, 39(5):952, 2011.

20. B Saussol. An introduction to quantitative Poincaré recurrence in dynamical systems. *Reviews in Mathematical Physics*, 21(08):949–979, 2009.

21. VK Yeragani, M Mallavarapu, RKA Radhakrishna, M Tancer, and T Uhde. Linear and nonlinear measures of blood pressure variability: Increased chaos of blood pressure time series in patients with panic disorder. *Depression and Anxiety*, 19(2):85–95, 2004.

22. U Zwiener, D Hoyer, R Bauer, B Lüthke, B Walter, K Schmidt, S Hallmeyer, B Kratzsch, and M Eiselt. Deterministic–chaotic and periodic properties of heart rate and arterial pressure fluctuations and their mediation in piglets. *Cardiovascular Research*, 31(3):455–465, 1996.

31

Comb Models for Transport along Spiny Dendrites

Vicenç Méndez and
Alexander Iomin

31.1 Introduction

A comb is a simplified model for various types of natural phenomena that belong to the loopless graphs category. The comb consists of a backbone along the horizontal axis and fingers or teeth along the perpendicular direction (see Figure 31.1 for a two-sided comb).

Comb-like models have been applied to mimic ramified structures as spiny dendrites of neuron cells [27,35] or percolation clusters with dangling bonds [48]. We are interested in the first example, where a comb structure with one-sided teeth of infinite length can be used to describe the movement and binding dynamics of particles inside the spines of dendrites. These spines are small protrusions from many types of neurons located on the surface of a neuronal dendrite. They receive most of the excitatory inputs and their physiological role is still unclear, although most spines are thought to be key elements in neuronal information processing and plasticity [49]. Spines are composed of a head ($\sim 1\ \mu$m) and a thin neck ($\sim 0.1\ \mu$m) attached to the surface of dendrites (see Figure 31.2).

The heads of spines have an active membrane, and as a consequence, they can sustain the propagation of an action potential with a rate that depends on the spatial density of spines [17]. Decreased spine density can result in cognitive disorders, such as autism, mental retardation, and fragile X syndrome [38]. Diffusion over branched smooth dendritic trees is basically determined by classical diffusion and the mean square displacement (MSD) along the dendritic axis grows linearly with time. However, inert particles diffusing along dendrites enter spines and remain there, trapped inside the spine head and then escape through a narrow neck to continue their diffusion along the dendritic axis. Recent experiments together with numerical simulations have shown that the transport of inert particles along spiny dendrites of Purkinje and Pyramidal cells is anomalous with an anomalous exponent that depends on the density of spines [16,44,45]. On the basis of these results, a fractional Nernst–Planck equation and fractional cable equation have been

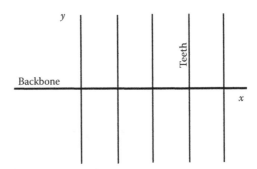

FIGURE 31.1   Two-sided comb.

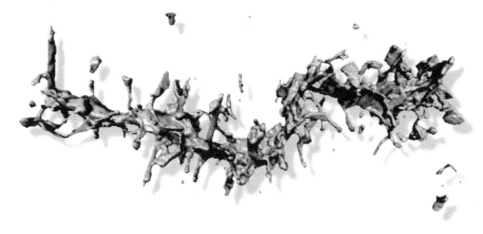

FIGURE 31.2   Electron tomogram of a spiny dendrite. Image taken from Internet (http://www.cacr.caltech.edu/projects/ldviz/results/levelsets/).

proposed for electrodiffusion of ions in spiny dendrites [24]. While many studies have been focused on the coupling between spines and dendrites, they are either phenomenological cable theories [14,24] or microscopic models for a single spine and parent dendrite [8,43]. More recently, a mesoscopic non-Markovian model for spine–dendrite interaction and an extension including reactions in spines and variable residence time have been developed [20,33]. These models predict anomalous diffusion along the dendrite in agreement with the experiments but are not able to relate how the anomalous exponent depends on the density of spines [16,45]. Since these experiments have been conducted with inert particles (i.e., there is reaction inside spines or dendrites), one concludes that the observed anomalous diffusion is exclusively due to the geometric structure of the spiny dendrite. Recent studies on the transport of particles inside spiny dendrites indicate the strong relation between the geometrical structure and anomalous transport exponents [6,10,45]. Therefore, elaboration of such an analytic model that establishes this relation can be helpful for further understanding the transport properties in spiny dendrites. The real distribution of spines along the dendrite, their size, and shapes are completely random [38], and inside spines, the spine necks act as a transport barrier [8]. For these reasons, one may reasonably assume that the diffusion inside the spine is anomalous. In this chapter, we describe some models, based on a comb-like structure that mimics a spiny dendrite; where the backbone is the dendrite and the teeth (lateral branches) are the spines. The models predict (i) anomalous transport inside spiny dendrites, in agreement with the experimental results of Reference 44, (ii) also explain the dependence between the MSD and the density of spines observed in Reference 45, and (iii) the mechanism of translocation wave of CaMKII (Ca^{2+}—calmodulin-dependent protein kinase II, a key regulator of the synaptic function). The chapter is organized as follows. First, we study the statistical properties of combs and explain how to reduce the effect of teeth on the movement along the backbone as a waiting time distribution between consecutive jumps. Second, we justify an employment of a comb-like structure as a paradigm for further exploration of a spiny dendrite. In particular, we show how a comb-like structure can sustain the phenomenon of anomalous diffusion, reaction–diffusion, and Lévy walks. Finally, we illustrate how the same models can also be useful to deal with the mechanism of the translocation wave/translocation waves of CaMKII and its propagation failure. We also present a brief introduction to the fractional integro-differentiation in the appendix at the end of the chapter.

31.2 Random Walks in Combs

The statistical properties of comb-like structures have been widely studied in the last century. The first passage time and the survival probability were studied by some authors [41,48]. More recently, the interest has been centered in the waiting time distribution equivalent to perform a random walk along the teeth [11, 15,33,47], the mean encounter time between two random walkers [1], and the occupation time statistics [40]. In this section, we illustrate two methods to compute the waiting time distribution that mimics the effect of a random walk along a teeth.

We first consider the case of a discrete one-dimensional (1D) chain where the nearest neighbors are separated by a distance a. A random walk, where each walker moves only to one of its nearest neighbors with equal probability after a fixed waiting time τ, is characterized by the following probability distribution functions (pdfs) for the waiting times and jump lengths, respectively:

$$\phi(t) = \delta(t - \tau),$$
$$w(x) = \frac{1}{2}[\delta(x - a) + \delta(x + a)].$$ (31.1)

In this way, systems with discrete time and space can be analyzed in terms of the continuous-time random walk (CTRW). Next, we add to every site of the backbone a secondary branch of length l, to produce a one-sided comb-like structure (see Figure 31.3). On such a structure, a walker that is at a given site of the backbone, can spend a certain amount of time in the secondary branch before jumping to one of the nearest-neighbor sites on the backbone. If we are only interested in the behavior of the system in the direction of the backbone, then, the secondary branches introduce a delay time for jumps between the neighboring sites on the backbone. The random walk on the comb structure can be modeled as a CTRW with Equation 31.1 and a renormalized waiting time pdf $\phi(t)$ that includes the effect of the delay due to the motion along the teeth.

To analytically determine the effect of the teeth, we invoke convolution rules that were introduced in Reference 47 for the case of homogeneous lattices:

  i. Consider a walker that is initially at a certain site within a tooth. If the walker proceeds further into a tooth, i.e., moves away from the backbone, its probability to return to the initial site after time t is a convolution of factors, i.e., a product in the Laplace space.
 ii. The total probability for that walker to return to the initial site is determined by summing over all t from 0 to ∞.
iii. When the walker reaches a crossing, where it can choose between different directions, the total probability is the sum of the probabilities for each possible direction.

The teeth have no other additional crossings, as shown in Figure 31.3. For the sake of generality, we consider the case that when the walker is at a site on the backbone, it can jump to another site on the backbone with probability α, or move onto the secondary branch with probability $1 - \alpha$.

Without loss of generality, we assume that initially the walker is located on the backbone, and we apply rules (i) through (iii) to determine $\phi(t)$. We study three specific cases:

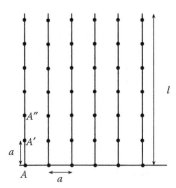

FIGURE 31.3   Sketch of a comb structure with the nearest-neighbor distance a and teeth of length l. The symbols A, A', A''... denote sites on the lattice (see text).

a. *Comb structure with $l = a$.* In this case, there is only one site on the tooth, A'. The walker can only jump in the direction of the backbone with probability α or move onto the branch with probability $1 - \alpha$ and then return to the initial site at the next jump. The time it takes to reach one of the nearest neighbors on the backbone is $t = \tau$ with probability α, $t = 3\tau$ with probability $(1 - \alpha) \times 1 \times \alpha$, $t = 5\tau$ with probability $(1 - \alpha)^2 \times 1^2 \times \alpha$, and so on. We can intuitively write the general form $\phi(t)$ as

$$\phi(t) = \sum_{j=1}^{\infty} \alpha(1 - \alpha)^{j-1} \delta\left[t - (2j - 1)\tau\right]. \tag{31.2}$$

The rules listed above for $\phi(t)$ should reproduce this behavior. For this purpose, we need to work in the Laplace space. Let $\hat{\phi}(s)$ be the Laplace transform of $\phi(t)$. Rules (i) through (iii) lead to the expression

$$\hat{\phi}(s) = \alpha\hat{\phi}_0 \sum_{j=0}^{\infty} \left[(1 - \alpha)\hat{\phi}_0^2\right]^j = \frac{\alpha\hat{\phi}_0}{1 - (1 - \alpha)\hat{\phi}_0^2}, \tag{31.3}$$

where $\hat{\phi}_0$ is the probability distribution for a single jump, $\hat{\phi}_0 = e^{-\tau s}$, which is the Laplace transform of Equation 31.1.

Equation 31.3 is derived as follows. The term $(1 - \alpha)\hat{\phi}_0^2$ in the sum represents, according to rule (i), the probability function for each occurrence of the walker moving onto the secondary branch. This expression must be summed up to infinity, according to rule (ii), to take into account that the walker can move onto the tooth $1, 2, ..., \infty$ times. The factor $\alpha\hat{\phi}_0$ accounts for the final jump to the nearest neighbor on the backbone.

It is easy to see that expression (31.3) may be written as a Taylor series

$$\hat{\phi}(s) = \sum_{j=1}^{\infty} \alpha(1 - \alpha)^{j-1} \left(\hat{\phi}_0\right)^{2j-1}, \tag{31.4}$$

which is the Laplace transform of Equation 31.2. The method for determining $\hat{\phi}(s)$ is shown to be valid in this case.

b. *Comb structure with $l = 2a$.* The secondary branch is two-sites long, A' and A''. Similar to the previous case, we can write the distribution for the time probabilities as

$$\hat{\phi}(s) = \alpha\hat{\phi}_0 \sum_{j=0}^{\infty} \left[\frac{(1 - \alpha)}{2}\hat{\phi}_0^2 \sum_{k=0}^{\infty} \left(\frac{1}{2}\hat{\phi}_0^2\right)^k\right]^j = \frac{\alpha\hat{\phi}_0(2 - \hat{\phi}_0^2)}{2 - (2 - \alpha)\hat{\phi}_0^2}. \tag{31.5}$$

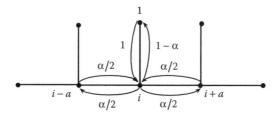

FIGURE 31.4 Comb with a single node along the tooth. In the picture, the probabilities of jumping between nodes of length a are written.

In this equation, a new sum over index k appears, because the walker can move away from the backbone twice. For each such occurrence, we must apply rule (i). We also assume that jumps to the nearest neighbor occur with probability $1/2$ on the linear teeth.

c. *Comb structure with $l \to \infty$.* Each time the walker moves away from the backbone, a new convolution factor appears in $\hat{\phi}(s)$. For the case $l \to \infty$, we have in principle infinitely many convolution factors in the expression for $\hat{\phi}(s)$. Fortunately, we can simplify this situation considerably. Assume that the walker is at the first site on the secondary branch, point A' in Figure 31.3, and moves away from the backbone. Let $\eta_{A'}$ be the probability distribution of returning for the first time to point A' after a time t. Now suppose the same situation but for the initial point A''. It is easy to see that as $l \to \infty$ the limit $\eta_{A''} \to \eta_{A'}$ has to hold, and we can again use rules (i) through (iii) to determine $\eta_{A'}$. Doing so, we obtain the expression

$$\eta_{A'} = \frac{1}{2}\hat{\phi}_0 \sum_{j=0}^{\infty} \left(\frac{1}{2}\hat{\phi}_0 \eta_{A''}\right)^j. \tag{31.6}$$

This expression is equivalent to form (31.3) with $\alpha = 1/2$; on the secondary branch, every jump to a nearest neighbor occurs with probability $1/2$. Introducing the condition $\eta_{A''} = \eta_{A'}$, which is strictly correct for $l = \infty$, and solving Equation 31.6, we find

$$\eta_{A'} = \frac{1 - \sqrt{1 - \hat{\phi}_0^2}}{\hat{\phi}_0}. \tag{31.7}$$

With this result, the distribution $\hat{\phi}(s)$ is obtained straightforwardly from rules (i) through (iii)

$$\hat{\phi}(s) = \alpha\hat{\phi}_0 \sum_{j=0}^{\infty} \left[(1-\alpha)\hat{\phi}_0 \eta_{A'}\right]^j = \frac{\alpha\hat{\phi}_0}{\alpha + (1-\alpha)\sqrt{1 - \hat{\phi}_0^2}}. \tag{31.8}$$

In finding the waiting time pdf associated with the comb structures, it is assumed that we are only interested in the dynamical behavior along the backbone and the rest of the structure is considered as secondary.

The second method consists of finding the master equation for the random walk moving along the backbone. This master equation has to incorporate the movement along the teeth. First, consider the simplest case of a one-sided comb with a single node as shown in Figure 31.4.

Let α be the probability of moving along the backbone, when the walker is at a node of the backbone and $1 - \alpha$ is the probability of jumping to the teeth, if the walker is at the backbone. So, if the movement along the backbone is considered isotropic, the probability of jumping to the right or to the left is $\alpha/2$. The master equation for the probability of finding a walker located at node i of the backbone at time t is

(see Figure 31.4)

$$P(i,t) = \frac{\alpha}{2}P(i+a,t-\tau) + \frac{\alpha}{2}P(i-a,t-\tau) + P(1,t-\tau), \tag{31.9}$$

where we have taken into account that the walker waits a constant time τ at every node between consecutive jumps. The master equation for node 1 of the tooth reads

$$P(1,t) = (1-\alpha)P(i,t-\tau), \tag{31.10}$$

that is coupled to Equation 31.9. Since time is a discrete variable, it is convenient to transform the time coordinate into a new variable z through the transformation

$$P(i,z) = \sum_{t=0}^{\infty} z^t P(i,t).$$

Multiplying Equations 31.9 and 31.10 by z^t and summing from $t=0$ to infinity we have, respectively

$$P(i,z) = \frac{\alpha b}{2}P(i+a,z) + \frac{\alpha b}{2}P(i-a,z) + bP(1,z), \tag{31.11}$$

$$P(1,z) = b(1-\alpha)P(i,z), \tag{31.12}$$

where $b = z^\tau$. Inserting Equation 31.12 into Equation 31.11 and rearranging terms, we obtain the following master equation for the movement along the backbone:

$$\frac{1}{2}P(i+a,z) + \frac{1}{2}P(i-a,z) = P(i,z)\frac{1-(1-\alpha)b^2}{\alpha b}. \tag{31.13}$$

It is convenient to stress that Equation 31.13 is actually the master equation for a walker moving on a comb-like structure constructed by repeating the element depicted in Figure 31.4. Finally, we can write Equation 31.13 in the Laplace space for time by taking into account that $z = e^{-s}$, so that $b = e^{-s\tau}$. Hence, Equation 31.13 becomes

$$\frac{1}{2}\hat{P}(i+a,s) + \frac{1}{2}\hat{P}(i-a,s) = \hat{P}(i,s)\frac{1-(1-\alpha)e^{-2s\tau}}{\alpha e^{-s\tau}}. \tag{31.14}$$

The example consists of generalizing the above structure to a tooth with R nodes or length $l = aR$. Then, we consider a one-sided comb, the basic element of which is given in Figure 31.5. The master equation for the probability of finding a walker at node i of the backbone at time t is, at the z space

$$P(i,z) = \frac{\alpha b}{2}P(i+a,z) + \frac{\alpha b}{2}P(i-a,z) + \frac{b}{2}P(1,z). \tag{31.15}$$

At node R, the master equation is

$$P(R,z) = \frac{b}{2}P(R-1,z), \tag{31.16}$$

and at the nodes $R-1$ and 1 they are

$$P(R-1,z) = \frac{b}{2}P(R-2,z) + bP(R,z), \tag{31.17}$$

$$P(1,z) = b(1-\alpha)P(i,z) + \frac{b}{2}P(2,z). \tag{31.18}$$

At a generic node j of the tooth we obtain

$$P(j,z) = \frac{b}{2}P(j-1,z) + \frac{b}{2}P(j+1,z), \tag{31.19}$$

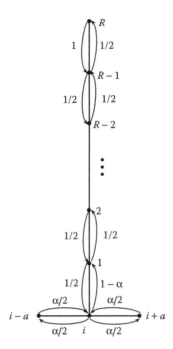

FIGURE 31.5 Comb with R nodes along the teeth. In the picture, the probabilities of jumping between nodes of length a are written.

where $j = 2, \ldots, R - 2$. We may solve Equation 31.19 by proposing the solution $P(j, z) = A\lambda^j$. Inserting this solution, we obtain the characteristic equation $\lambda^2 - 2\lambda/b + 1 = 0$. Hence,

$$P(j, z) = A_1 \lambda_+^j + A_2 \lambda_-^j, \tag{31.20}$$

where $A_{1,2}$ are constant to be determined and

$$\lambda_\pm = \frac{1}{b} \left(1 \pm \sqrt{1 - b^2} \right).$$

On setting $j = 2$ and $j = 3$ into Equation 31.20, we obtain a system of algebraic equations for the constants $A_{1,2}$

$$\begin{aligned} A_1 \lambda_+^2 + A_2 \lambda_-^2 &= P(2, z), \\ A_1 \lambda_+^3 + A_2 \lambda_-^3 &= P(3, z). \end{aligned} \tag{31.21}$$

On the other hand, by inserting Equation 31.20 into Equations 31.16 and 31.17 we obtain the equation relating A_1 and A_2

$$A_1 \left[\lambda_+^{R-2} \left(1 - \frac{3b^2}{4} \right) - \frac{b}{2} \left(1 - \frac{b^2}{2} \right) \lambda_+^{R-3} \right] + A_2 \left[\lambda_-^{R-2} \left(1 - \frac{3b^2}{4} \right) - \frac{b}{2} \left(1 - \frac{b^2}{2} \right) \lambda_-^{R-3} \right] = 0. \tag{31.22}$$

By combining Equations 31.21 and 31.22 we can express $P(2, z)$ in terms of $P(1, z)$ in the form $P(2, z) = h(b)P(1, z)$, where

$$h(b) = \frac{b}{1 - q\sqrt{1 - b^2}} \quad \text{where} \quad q = \frac{\lambda_-^{R-4} - \lambda_+^R}{\lambda_-^{R-4} + \lambda_+^R}. \tag{31.23}$$

Now, we are in a position to get the master equation for the motion along the backbone by substituting $P(2, z) = h(b)P(1, z)$ into Equation 31.18 and the result into Equation 31.15. The final result is

$$\frac{1}{2}P(i+a, z) + \frac{1}{2}P(i-a, z) = \frac{1}{\alpha b}\left[1 - \frac{b^2(1-\alpha)}{2 - bh(b)}\right]P(i, z). \tag{31.24}$$

We now show how to reduce the effect of the teeth on a waiting time distribution for the motion of a walker along the backbone. To this end, we make use of the CTRW, and in particular, of the generalized master equation for finding a walker at point x at time t when it moves in a 1D space

$$\frac{\partial P}{\partial t} = \int_0^t K(t - t')\left[\int P(x - x', t')w(x')dx' - P(x, t')\right]dt'. \tag{31.25}$$

Considering the jump lengths distribution given in Equation 31.1, for jumps between the consecutive nodes of a 1D lattice with spacing a and transforming into the Laplace space in time, Equation 31.25 becomes

$$\frac{1}{2}\hat{P}(i+a, s) + \frac{1}{2}\hat{P}(i-a, s) = \hat{P}(i, s)\left[\frac{s + \hat{K}(s)}{\hat{K}(s)}\right], \tag{31.26}$$

where the memory kernel $K(t)$ is related to the waiting time distribution $\phi(t)$ through their Laplace transforms

$$\hat{K}(s) = \frac{s\hat{\phi}(s)}{1 - \hat{\phi}(s)}. \tag{31.27}$$

Therefore, Equation 31.26 reduces to

$$\frac{1}{2}\hat{P}(i+a, s) + \frac{1}{2}\hat{P}(i-a, s) = \hat{P}(i, s)\frac{1}{\hat{\phi}(s)}. \tag{31.28}$$

The waiting time distribution is obtained by comparing Equations 31.28 and 31.24, to obtain

$$\hat{\phi}(s) = \frac{\alpha b}{1 - (1 - \alpha)\frac{b^2}{2 - bh(b)}}, \tag{31.29}$$

where $b = e^{-s\tau}$. If the one-sided comb has a teeth with one node, $R = 1$, thus from Equation 31.29, the waiting time distribution is

$$\hat{\phi}(s) = \frac{\alpha b}{1 - (1 - \alpha)b^2},$$

which coincides with the result obtained in Equation 31.3. When $R = 2$, then

$$\hat{\phi}(s) = \frac{\alpha b(2 - b^2)}{2 - (2 - \alpha)b^2},$$

which is the same result obtained in Equation 31.5. Finally, let us consider the infinite length of the teeth $R \to \infty$, or $l \gg a$. In this case, Equation 31.29 reduces to

$$\hat{\phi}(s) = \frac{\alpha b}{\alpha + (1 - \alpha)\sqrt{1 - b^2}}, \tag{31.30}$$

that is equal to Equation 31.8. In the limit of large times, $s \to 0$, the waiting time distribution in Equation 31.30 reads $\hat{\phi}(s) \simeq 1 - (1 - \alpha)\sqrt{2s\tau}/\alpha$, which predicts an anomalous diffusion along the backbone. So that, the teeth need to have an infinite length to predict an anomalous diffusion along the whole comb at asymptotically large times.

31.3 Comb-Like Models Mimic Spiny Dendrites

As shown in previous studies, the geometric nature of spiny dendrites plays an essential role in kinetics [6,7,10,16,44,45]. The real distribution of spines along the dendrite, their size, and shapes are completely random [38], and inside spines, not only the spine necks but the spine itself acts as a transport barrier [8, 10,45]. Therefore, a reasonable assumption is a consideration of anomalous diffusion along both the spines and dendrite. So, we propose models based on a comb-like structure that mimics a spiny dendrite, where the backbone is the dendrite and teeth (lateral branches) are the spines (see Figure 31.6) (we distinguish between a smooth and a spiny dendrite). In this case, dynamics inside the teeth corresponds to spines, whereas the backbone describes diffusion along dendrites. Note that the comb model is an analog of a 1D medium where fractional diffusion has been observed and explained in the framework of the CTRW [2,31,36,48] and making use of macroscopic descriptions [50].

Before embarking for the CTRW consideration in the framework of the comb model, let us explain how anomalous diffusion in the comb model relates to the CaMKII transport along the spiny dendrite, and how geometry of the latter relates to the anomalous transport. As admitted above, the spine cavities behave as traps for the contaminant transport. As follows from a general consideration of a Markov process inside a finite region, the pdf of lifetimes inside the cavity with a finite volume and arbitrary form decays exponentially with time t (see, e.g., Reference 41) $\varphi(t) = \frac{1}{\tau} \exp(-\frac{t}{\tau})$. Here τ is a survival time (mean life time), defined by the minimum eigenvalue of the Laplace operator and determined by geometry of the cavity. For example, in References 6 and 10, for spines with a head of volume V and the cylindrical spine neck of length L and radius a, the mean life time is $\tau = LV/\pi a^2 D = L^2/D$, where D is diffusivity of the spine. Therefore, the mean probability to find a particle inside the spine after time t (i.e., the survival probability inside the cavity from 0 to t) averaged over all possible realizations of τ is given by the integral

$$\Psi(t) = \int_t^\infty \int_0^\infty \varphi(t'/\tau) f(\tau) d\tau dt', \qquad (31.31)$$

where $f(\tau)$ is a distribution function of the survival times τ (recall that the size and shape of spines are random [38]). Finally, the waiting time pdf can be easily calculated from Equation 31.31, as follows:

$$\phi(t) = -\partial_t \Psi(t) = \int_0^\infty \varphi(t/\tau) f(\tau) d\tau. \qquad (31.32)$$

In the simplest case, when the distribution is the exponential $f(\tau) = (1/\tau_0) \exp(-\tau/\tau_0)$, one obtains from Equation 31.32 that the general kinetics is not Markovian and the waiting time pdf is a stretched exponential for large times

$$\phi(t) = \frac{1}{\tau_0} \int_0^\infty \frac{e^{-t/\tau}}{\tau} e^{-\tau/\tau_0} d\tau = \frac{2}{\tau_0} K_0 \left(2\sqrt{t/\tau_0} \right) \sim \left(\frac{t}{\tau_0} \right)^{-\frac{1}{4}} \exp(-\sqrt{t/\tau_0}), t/\tau_0 \gg 1.$$

The situation is more interesting when the distribution of the survival times is the power law $f(\tau) \sim 1/\tau^{1+\alpha}$, $(0 < \alpha < 1)$. In this case, the waiting time pdf is the power law as well $\phi(t) \sim 1/t^{1+\alpha}$ that leads to

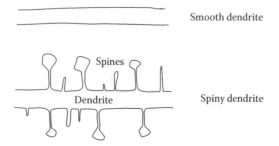

Smooth dendrite

Spines

Dendrite

Spiny dendrite

FIGURE 31.6 Drawing of a smooth and spiny dendrites.

subdiffusion motion along the dendrite. This result follows from the CTRW theory, since all underlying microprocesses are independent Markovian ones with the same distributions [36].

Now, we explain the physical reason of the possible power law distribution $\phi(t)$. At this point, we paraphrase some arguments from Reference 5 with the corresponding adaptation to the present analysis. Let us consider the escape from a spine cavity from a potential point of view, where geometrical parameters of the cavity can be related to a potential U. For example, for the simplest case mentioned above, it is $U = VL/\pi a^2$, which "keeps" a particle inside the cavity, whereas $D\tau_0$ plays a role of the kinetic energy, or of the "Boltzmann temperature." Therefore, an escape probability from the spine cavity/well is described by the Boltzmann factor $\exp(-U/D\tau_0)$. This value is proportional to the inverse waiting, or survival time

$$t \sim \exp\left(\frac{U}{D\tau_0}\right). \tag{31.33}$$

As admitted above, potential U is random and distributed by the Poisson distribution $P(U) = U_0^{-1}\exp(-U/U_0)$, where U_0 is an averaged geometrical spine characteristic. The probability to find the waiting time in the interval $(t, t + dt)$ is equal to the probability to find the trapping potential in the interval $(U, U + dU)$, namely $\phi(t)dt = P(U)dU$. Therefore, from Equation 31.33 one obtains

$$\phi(t) \sim \frac{1}{t^{1+\gamma}}. \tag{31.34}$$

Here $\gamma = \frac{D\tau_0}{U_0} \in (0, 1)$ establishes a relation between geometry of the dendrite spines and subdiffusion observed in References 44 and 45 and supports application of the comb model, which is a convenient implement for analytical exploration of anomalous transport in spiny dendrites in the framework of the CTRW consideration.

31.3.1 Anomalous Diffusion in Spines

Geometry of the comb structure makes it possible to describe anomalous diffusion in spiny dendrites structure in the framework of the comb model.

Usually, anomalous diffusion on the comb is described by the two-dimensional (2D) distribution function $P = P(x, y, t)$, and a special behavior is that the displacement in the x-direction is possible only along the structure backbone (x-axis at $y = 0$). Therefore, diffusion in the x-direction is highly inhomogeneous. Namely, the diffusion coefficient is $D_{xx} = D_x\delta(y)$, whereas the diffusion coefficient in the y-direction (along teeth) is a constant $D_{yy} = D_y$. Owing to this geometrical construction, the flux of particles along the dendrite is

$$J_x = -D_x\delta(y)\frac{\partial P}{\partial x}, \tag{31.35}$$

and the flux along the finger describes the anomalous trapping process that occurs inside the spine

$$J_y = -D_y\frac{\partial^{1-\gamma}}{\partial t^{1-\gamma}}\bigg|_{RL}\frac{\partial P}{\partial y}, \tag{31.36}$$

where $P(x, y, t)$ is the density of particles and

$$\frac{\partial^{1-\gamma}}{\partial t^{1-\gamma}}\bigg|_{RL}f(t) = \frac{\partial}{\partial t}I_t^\gamma f(t) \tag{31.37}$$

is the Riemann–Liouville fractional derivative, where the fractional integration I_t^γ is defined by means of the Laplace transform

$$\hat{\mathcal{L}}\left[I_t^\gamma f(t)\right] = s^{-\gamma}\hat{f}(s). \tag{31.38}$$

So, inside the spine, the transport process is anomalous and $\langle y^2(t) \rangle \sim t^\gamma$, where $\gamma \in (0, 1)$. Making use of the continuity equation for the total number of particles

$$\frac{\partial P}{\partial t} + \text{div} \mathbf{J} = 0, \tag{31.39}$$

where $\mathbf{J} = (J_x, J_y)$, one has the following evolution equation for transport along the spiny dendrite:

$$\frac{\partial P}{\partial t} - D_x \delta(y) \frac{\partial^2 P}{\partial x^2} - D_y \frac{\partial^{1-\gamma}}{\partial t^{1-\gamma}} \bigg|_{RL} \frac{\partial^2 P}{\partial y^2} = 0. \tag{31.40}$$

The Riemann–Liouville fractional derivative in Equation 31.40 is not convenient for the Laplace transform. To ensure feasibility of the Laplace transform, which is a strong machinery for treating fractional equations, one reformulates the problem in a form suitable for the Laplace transform application.

To shed light on this situation, let us consider a comb in the three dimension (3D) [5]. This model is described by the distribution function $P_1(x, y, z, t)$ with the evolution equation given by the equation

$$\frac{\partial P_1}{\partial t} - D_x \delta(y) \delta(z) \frac{\partial^2 P_1}{\partial x^2} - D_y \delta(z) \frac{\partial^2 P_1}{\partial y^2} - \frac{\partial^2 P_1}{\partial z^2} = 0. \tag{31.41}$$

It should be stressed that the z coordinate (do not confuse with the z variable introduced in the previous section) is a supplementary, virtue variable, introduced to describe the fractional motion in spines by means of the Markovian process. Thus, the true distribution is $P(x, y, t) = \int_{-\infty}^{\infty} P_1(x, y, z, t) dz$ with the corresponding evolution equation

$$\frac{\partial P}{\partial t} - D_x \delta(y) \frac{\partial^2 P_1(z=0)}{\partial x^2} - D_y \frac{\partial^2 P_1(z=0)}{\partial y^2} = 0. \tag{31.42}$$

A relation between $P(x, y, t)$ and $P_1(x, y, z = 0, t)$ can be expressed through their Laplace transforms

$$\hat{P}_1(x, y, z = 0, s) = \frac{\sqrt{s}}{2} \hat{P}(x, y, s), \tag{31.43}$$

where $\hat{P}(x, y, s) = \hat{\mathcal{L}}[P(x, y, t)]$ and $\hat{P}_1(x, y, z, s) = \hat{\mathcal{L}}[P_1(x, y, z, t)]$.

Equation 31.43 establishes a relationship between the distributions $P_1(x, y, z = 0, t)$ and $P(x, y, t)$ in the Laplace space. Both distributions are related through the expression

$$P(x, y, t) = \int_{-\infty}^{\infty} P_1(x, y, z, t) dz.$$

If we transform the above equation by Fourier–Laplace, we obtain

$$\hat{P}(k_x, k_y, s) = \hat{P}_1(k_x, k_y, k_z = 0, s). \tag{31.44}$$

Then, Equation 31.43 is nothing but a relation between $\hat{P}_1(k_x, k_y, k_z = 0, s)$ and $\hat{P}_1(k_x, k_y, z = 0, s)$. To find $\hat{P}_1(k_x, k_y, k_z, s)$ we transform Equation 31.41 by Fourier–Laplace and after collecting terms we find

$$\hat{P}_1(k_x, k_y, k_z, s) = \frac{1 - D_x k_x^2 P_1(k_x, y = 0, z = 0, s) - D_y k_y^2 P_1(k_x, k_y, z = 0, s)}{s + k_z^2}, \tag{31.45}$$

where the initial condition has been assumed to be $P_1(x, y, z, t = 0) = \delta(x) \delta(y) \delta(z)$ for simplicity. Setting $k_z = 0$ one gets

$$\hat{P}_1(k_x, k_y, k_z = 0, s) = \frac{1 - D_x k_x^2 P_1(k_x, y = 0, z = 0, s) - D_y k_y^2 P_1(k_x, k_y, z = 0, s)}{s}. \tag{31.46}$$

Inverting Equation 31.45 by Fourier over k_z we obtain

$$\hat{P}_1(k_x, k_y, z, s) = \frac{1 - D_x k_x^2 P_1(k_x, y = 0, z = 0, s) - D_y k_y^2 P_1(k_x, k_y, z = 0, s)}{2\sqrt{s}} e^{-\sqrt{s}|z|}.$$

Then setting $z = 0$, one obtains

$$\hat{P}_1(k_x, k_y, z = 0, s) = \frac{1 - D_x k_x^2 P_1(k_x, y = 0, z = 0, s) - D_y k_y^2 P_1(k_x, k_y, z = 0, s)}{2\sqrt{s}}. \tag{31.47}$$

Combining Equations 31.46 and 31.47, one has

$$\hat{P}_1(k_x, k_y, z = 0, s) = \frac{\sqrt{s}}{2}\hat{P}_1(k_x, k_y, k_z = 0, s),$$

then the Fourier inversion over k_x and k_y yields Equation 31.43. Finally, performing the Laplace transform of Equation 31.42 one obtains

$$s\hat{P}(x, y, s) - D_x \delta(y)\frac{\partial^2 \hat{P}_1(x, y, z = 0, s)}{\partial x^2} - D_y \frac{\partial^2 \hat{P}_1(x, y, z = 0, s)}{\partial y^2} = P(x, y, t = 0) \tag{31.48}$$

and substituting relation (31.43), dividing by $\sqrt{s}$, and then performing the Laplace inversion, one obtains the comb model with the fractional time derivative

$$\frac{\partial^{\frac{1}{2}} P}{\partial t^{\frac{1}{2}}} - D_x \delta(y)\frac{\partial^2 P}{\partial x^2} - D_y \frac{\partial^2 P}{\partial y^2} = 0, \tag{31.49}$$

where $2D_{x,y} \to D_{x,y}$ and the Caputo derivative* $\frac{\partial^\gamma}{\partial t^\gamma}$ can be defined by the Laplace transform for $\gamma \in (0, 1)$ [32]

$$\hat{\mathcal{L}}\left[\frac{\partial^\gamma f}{\partial t^\gamma}\right] = s^\gamma \hat{f}(s) - s^{\gamma-1} f(t = 0). \tag{31.50}$$

The fractional transport takes place in both the dendrite x-direction and the spines y coordinate. To make fractional diffusion normal in dendrites, we add the fractional integration $I_t^{1-\gamma}$ by means of the Laplace transform (31.38), as well as $\hat{\mathcal{L}}\left[I_t^{1-\gamma} f(t)\right] = s^{\gamma-1}\hat{f}(s)$. This yields Equation 31.49, after generalization $\frac{1}{2} \to \gamma \in (0, 1)$

$$\frac{\partial^\gamma P}{\partial t^\gamma} - D_x \delta(y) I_t^{1-\gamma}\frac{\partial^2 P}{\partial x^2} - D_y \frac{\partial^2 P}{\partial y^2} = 0. \tag{31.51}$$

Performing the Fourier–Laplace transform in Equation 31.51 we obtain

$$P(k_x, k_y, s) = \frac{P(k_x, k_y, t = 0) - D_x k_x^2 P(k_x, y = 0, s)}{s + D_y k_y^2 s^{1-\gamma}}, \tag{31.52}$$

where the Fourier–Laplace image of the distribution function is defined by its arguments $\hat{\mathcal{L}}\hat{\mathcal{F}}_x\hat{\mathcal{F}}_y[P(x, y, t)] = P(k_x, k_y, s)$. If $P(x, y, t = 0) = \delta(x)\delta(y)$, inversion by Fourier over y gives

$$P(k_x, y, s) = \frac{1 - D_x k_x^2 P(k_x, y = 0, s)}{s^{(2-\gamma)/2}\sqrt{D_y}} \exp\left(-|y| s^{\gamma/2}/\sqrt{D_y}\right). \tag{31.53}$$

---

* To avoid any confusion between the Riemann–Liouville and the Caputo fractional derivatives, the former one stands in the text with an index RL: $\frac{\partial^\alpha}{\partial t^\alpha}|_{RL}$, while the latter fractional derivative is not indexed $\frac{\partial^\alpha}{\partial t^\alpha}$. Note that it is also convenient to use Equation 31.50 as a definition of the Caputo fractional derivative.

Taking $y = 0$, the above equation provides

$$P(k_x, y = 0, s) = \frac{1}{s^{(2-\gamma)/2}\sqrt{D_y} + D_x k_x^2}, \tag{31.54}$$

which yields after inserting into Equation 31.52

$$P(k_x, k_y, s) = \frac{1}{s + D_y k_y^2 s^{1-\gamma}} \left(1 - \frac{D_x k_x^2}{s^{(2-\gamma)/2}\sqrt{D_y} + D_x k_x^2}\right). \tag{31.55}$$

We can calculate the density of particles at a given point x of the dendrite at time t, namely $P(x, t)$, by integrating over y in the Fourier space

$$P(k_x, s) = P(k_x, k_y = 0, s) = \frac{s^{-\gamma/2}\sqrt{D_y}}{s^{(2-\gamma)/2}\sqrt{D_y} + D_x k_x^2}, \tag{31.56}$$

then

$$\langle x^2(s)\rangle = -\left.\frac{\partial^2}{\partial k_x^2} P(k_x, s)\right|_{k_x=0} = \frac{2D_x}{\sqrt{D_y}}\frac{1}{s^{2-\frac{\gamma}{2}}}, \tag{31.57}$$

so that

$$\langle x^2(t)\rangle = \frac{2D_x}{\sqrt{D_y}} t^{1-\frac{\gamma}{2}}. \tag{31.58}$$

Equation 31.58 predicts subdiffusion along the spiny dendrite that is in agreement with the experimental results reported in Reference 44. It should be noted that this result is counterintuitive. Indeed, subdiffusion in spines or fingers should lead to the slower subdiffusion in dendrites or the backbone with the transport exponent less than in a usual comb, since these two processes are strongly correlated. But this correlation is broken due to the fractional integration $I_t^{1-\gamma}$ in Equation 31.51. On the other hand, if we invert Equation 31.56 by Fourier–Laplace, we obtain the fractional-diffusion equation for $P(x, t)$

$$\frac{\partial^{1-\frac{\gamma}{2}} P}{\partial t^{1-\frac{\gamma}{2}}} = \frac{D_x}{\sqrt{D_y}} \frac{\partial^2 P}{\partial x^2},$$

which is equivalent to the generalized master equation (31.25), in the diffusion limit

$$\frac{\partial P}{\partial t} = \int_0^t K(t - t') \frac{\partial^2 P(x, t')}{\partial x^2} dt', \tag{31.59}$$

if the Laplace transform of the memory kernel is given by $\hat{K}(s) = \frac{D_x}{\sqrt{D_y}} s^{\gamma/2}$, which corresponds to the waiting time pdf in the Laplace space given by

$$\hat{\phi}(s) = \frac{1}{1 + \frac{\sqrt{D_y}}{D_x} s^{1-\frac{\gamma}{2}}}, \tag{31.60}$$

that is $\phi(t) \sim t^{-2+\frac{\gamma}{2}}$ as $t \to \infty$. Let us employ the notation for a dynamical exponent d_w used in References 44 and 45. If $d_w = 4/(2 - \gamma)$ then the MSD grows as t^{2/d_w}. On the other hand, it has been found in experiments that d_w increases with the density of spines ρ_s and the simulations prove that d_w grows linearly with ρ_s. Indeed, the experimental data admit almost any growing dependence of d_w with ρ_s due to the high variance of the data (see Figure 31.6 in Reference 45). Equation 31.58 also establishes a phenomenological relation between the second moment and ρ_s. When the spine density is zero, then $\gamma = 0$, $d_w = 2$, and normal diffusion takes place. If the spine density ρ_s increases, the anomalous exponent of the pdf (31.60) $1 - \gamma/2 = 2/d_w$ must decrease (i.e., the transport is more subdiffusive due to the increase of ρ_s) so that d_w has to increase as well. So, our model qualitatively predicts that d_w increases with ρ_s, in agreement with the experimental results in Reference 45.

31.3.2 Lévy Walks on Fractal Comb

In this section, we consider a fractal comb model [28] to take into account the inhomogeneity of distribution of spines. Here, we consider the comb model for a phenomenological explanation of an experimental situation, where we introduce a control parameter that establishes a relation between diffusion along dendrites and the density of spines. Suggesting a more sophisticated relation between the dynamical exponent and the spine density, we can reasonably suppose that the fractal dimension, due to the box counting of the spine necks, is not an integer: it is embedded in the 1D space, thus the spine fractal dimension is $\nu \in (0,1)$. According to fractal geometry (roughly speaking), the most convenient parameter is the fractal dimension of the spine volume (mass) $\mu_{\mathrm{spine}}(x) \equiv \mu(x) \sim |x|^\nu$. Therefore, following Nigmatulin's idea on construction of a "memory kernel" on a Cantor set in Fourier space $|k|^{1-\nu}$ [37] (and further developing in References 3, 30, and 42), this leads to a convolution integral between the nonlocal density of spines and the pdf $P(x,y,t)$ that can be expressed by means of the inverse Fourier transform [28] $\hat{\mathcal{F}}_x^{-1}\left[|k_x|^{1-\nu}P(k_x,y,t)\right]$. Therefore, the starting mathematical point of the phenomenological consideration is the fractal comb model

$$\frac{\partial^\gamma P}{\partial t^\gamma} - D_x \delta(y) I_t^{1-\gamma} \frac{\partial^2 P}{\partial x^2} - D_y \frac{\partial^2}{\partial y^2} \hat{\mathcal{F}}_{k_x}^{-1}\left[|k_x|^{1-\nu}P(k_x,y,t)\right] = 0. \tag{31.61}$$

Performing the same analysis in the Fourier–Laplace space, presented in the previous section, then Equation 31.56 reads

$$P(k_x,s) = P(k_x,k_y=0,s) = \frac{s^{-\gamma/2}\sqrt{D_y}}{s^{(2-\gamma)/2}\sqrt{D_y} + D_x|k_x|^\beta}, \tag{31.62}$$

where $\beta = 3/2 + \nu/2$.

Contrary to the previous equation, analysis expression (31.57) does not work any more, since superlinear motion is involved in the fractional kinetics. This leads to divergence of the second moment due to the Lévy flights. The latter are described by the distribution $\sim 1/|x|^{1+\beta}$, which is separated from the waiting time probability distribution $\phi(t)$. To overcome this deficiency, we follow the analysis of the Lévy walks suggested in References 52 and 53. We consider our exact result in Equation 31.62 as an approximation obtained from the joint distribution of the waiting times and the Lévy walks. Therefore, a cutoff of the Lévy flights is expected at $|x| = t$. This means that a particle moves at a constant velocity inside dendrites not all times, and this laminar motion is interrupted by localization inside spines distributed in space by the power law.

Performing the inverse Laplace transform, we obtain a solution in the form of the Mittag–Leffler function [4]

$$P(k_x,t) = E_{1-\gamma/2}\left(-D|k|^\beta t^{1-\gamma/2}\right), \tag{31.63}$$

where $D = \frac{D_x}{\sqrt{D_y}}$. For the asymptotic behavior $|k| \to 0$, the argument of the Mittag–Leffler function can be small. Note that in the vicinity of the cutoff $|x| = t$, this corresponds to the large t ($|k| \sim \frac{1}{t} \ll 1$), thus we have [4]

$$E_{1-\gamma/2}\left(-D|k|^\beta t^{1-\gamma/2}\right) \approx \exp\left(-\frac{D|k|^\beta t^{1-\gamma/2}}{\Gamma(2-\gamma/2)}\right). \tag{31.64}$$

Therefore, the inverse Fourier transform yields

$$P(x,t) \approx A_{\gamma,\nu} \frac{Dt^{1-\gamma/2}}{\Gamma(2-\gamma/2)|x|^{(5+\nu)/2}}, \tag{31.65}$$

where $A_{\gamma,\nu}$ is determined from the normalization condition. (The physical plausibility of estimations (31.64) and (31.65) also follows from the plausible finite result of Equation 31.65, which is the normalized distribution $P(x,t) \sim 1/|x|^{(3+\nu+\gamma)/2}$, where $|x| = t$.) Now, the second moment corresponds to integration with the cutoff at $x = t$ that yields

$$\langle x^2(t) \rangle = K_{\gamma,\nu} t^{\frac{3-\gamma-\nu}{2}}, \tag{31.66}$$

where $K_{\gamma,\nu} = \frac{4A_{\gamma,\nu}D_x}{(1-\nu)\Gamma(2-\gamma/2)\sqrt{D_y}}$ is a generalized diffusion coefficient. Transition to the absence of spines means first transition to normal diffusion in teeth with $\gamma = 1$ and then $\nu = 0$ that yields

$$\langle x^2(t) \rangle = K_{1,0} t. \tag{31.67}$$

31.3.3 Fractional Reaction–Diffusion along Spiny Dendrites

Geometrically, spiny dendrites in the 3D space are completely described by a comb structure in the 2D, where the spine density on the cylinder surface is projected on the 1D axis (say the x-axis): $\rho(x, r = \text{const}, \theta) \to \rho(x)$. Here $\rho(x, r = \text{const}, \theta)$ is the spine density, whereas $\rho(x)$ is the density of the comb teeth. In what follows, we consider $\rho(x) = g = \text{const}$, which is, probably, the most realistic case. Fractional diffusion inside the spines is described by fractional diffusion inside the teeth. Therefore, one considers a two-sided comb model as in Figure 31.1, and the starting mathematical point of the phenomenological consideration is the Fokker–Planck equation obtained in Reference 35.

It reads

$$\frac{\partial P}{\partial t} - \delta(y)\frac{\partial^2 P}{\partial x^2} - g\frac{\partial^2 P}{\partial y^2} = 0. \tag{31.68}$$

This equation is obtained by the rescaling with relevant combinations of the comb parameters D_x and D_y, such that the dimensionless time and coordinates are $D_x^3 t/D_y^2 \to t\, D_x/D_y \to x, D_x/D_y \to y/\sqrt{g}$, correspondingly [26], and parameter g can be considered as a constant density of the fingers.

As admitted above, a variety of interactions inside spines leads to correlated noises in dendritic spines [51]. The strong correlations of that leads to anomalous (subdiffusive) motion inside the spines. Following a phenomenological description by the CTRW, this subdiffusion is controlled by a waiting time pdf $\phi(t)$ decaying according to the power law. Therefore, normal diffusion of the contaminant density $P(x, y, t)$, for example, activated CaMKII in spines is replaced by the anomalous transition term

$$g\frac{\partial^2 P}{\partial y^2} \Rightarrow g\int_0^t K(t-t')\frac{\partial^2 P(t')}{\partial y^2}dt', \tag{31.69}$$

where $K(t)$ is again the time memory kernel of the generalized master equation and is defined in Equation 31.27. For subdiffusion, $\phi(t) = \frac{1}{1+t^{1+\gamma}}$ with $0 < \gamma < 1$ that yields [23] $\hat{K}(s) = s^{1-\gamma}$.

One may recognize that Equation 31.69 is a formal expression for the anomalous transport with a very complicated form in the time domain, which, in turn, is very inconvenient for the analytical treatment. Therefore, the comb model may be presented in the Laplace domain. Substituting Equation 31.69 into Equation 31.68, then performing the Laplace transform, and taking into account Equation 31.27, one obtains the comb model in the Laplace domain

$$s\hat{P} = \delta(y)\frac{\partial^2 \hat{P}}{\partial x^2} - gs^{1-\alpha}\frac{\partial^2 \hat{P}}{\partial y^2} + P_0. \tag{31.70}$$

Here $P_0 = P(x, y, t = 0)$ is the initial condition. As admitted, the kernel $\hat{K}(s)$ is problematic for the Laplace inversion, since it leads to the appearance of the initial condition. To overcome this obstacle, one multiplies Equation 31.70 by $s^{\alpha-1}$ and then performs the Laplace inversion that yields

$$\int_0^t (t-t')^{-\gamma}\left[\frac{\partial P(x,y,t')}{\partial t'} - \delta(y)\frac{\partial^2 P(x,y,t')}{\partial x^2}\right]dt' = g\frac{\partial^2 P(x,y,t)}{\partial y^2}. \tag{31.71}$$

Amending this equation by the reaction term, one arrives at the integro-differential equation

$$\int_0^t (t - t')^{-\gamma} \left[\frac{\partial P(x, y, t')}{\partial t'} - \delta(y) \frac{\partial^2 P(x, y, t')}{\partial x^2} \right] dt' = g \frac{\partial^2 P(x, y, t)}{\partial y^2} + g\hat{C}[P(x, y, t)], \tag{31.72}$$

which describes the 2D inhomogeneous reaction–diffusion in the dispersive medium. Here $\hat{C}[P(x, y, t)] \equiv \hat{C}(P)$ is a reaction kinetic term. In particular, to model reaction kinetics inside dendrites, it can be considered either linear $\hat{C}(P) = CP$, or logistic $\hat{C}(P) = CP(x, y, t)[1 - P(x, y, t)]$ [9]. Integration with the power law kernel $t^{-\gamma}$ ensures anomalous diffusion in both the dendrite and spines.

In what follows, we use convenient notations of fractional integro-differentiation given in Equation 31.50 and the text below. Owing to this notation, from Equation 31.72 the equation for $P(x, y, t)$ reads

$$\frac{\partial^\gamma P}{\partial t^\gamma} - \delta(y) I_t^{1-\gamma} \frac{\partial^2 P}{\partial x^2} - g \frac{\partial^2 P}{\partial y^2} = g\hat{C}(P). \tag{31.73}$$

To give a first and brief insight into the problem of front propagation, let us consider the linear reaction and $\gamma = 1$. In this case, one obtains a "simple" solution [25] for the traveling wave along the x-axis (inside dendrites). Introducing the total pdf $P_1(x, t) = \int dy\, P(x, y, t)$, one obtains

$$P_1(x, t) = \frac{\sqrt{2g^{1/2}}}{\pi \sqrt{t^{1/2}}} \exp \left[\frac{x^2}{2\sqrt{gt}} - Cgt \right]. \tag{31.74}$$

This yields the coordinates of the front $x \sim t^{3/4}$ that spreads with the decaying velocity $v \sim t^{-1/4}$. This solution illustrates the asymptotic failure of the reaction-transport front propagation due to the subdiffusion inside spiny dendrites.

31.4 Front Propagation in Combs

Recently, a mechanism of the translocation wave of CaMKII has been suggested [18]. As shown, an activated CaMKII contaminant travels along dendrites with additional translocation inside spines. The process of activation (the conversion of primed CaMKII into its active state) corresponds to the irreversible reaction that, in the absence of spines, is described by the Fisher–Kolmogorov–Petrovskii–Piskunov (FKPP) equation (it also relates to the logistic kinetic term) [18]. Therefore, in the framework of the above-suggested scheme of the dispersive subdiffusive comb (31.73), a nonlinear reaction at subdiffusion in dendrites takes place along the bound x-axis, whereas subdiffusion in fingers describes the translocation inside spines. Therefore, reaction-transport equation (31.73) now reads

$$\frac{\partial^\gamma P}{\partial t^\gamma} - \delta(y) I_t^{1-\gamma} \left[D \frac{\partial^2 P}{\partial x^2} + \hat{C}(P) \right] = g \frac{\partial^2}{\partial y^2} P. \tag{31.75}$$

Here D describes the diffusivity inside dendrites, whereas $\hat{C}(P) = CP(1 - P)$ is the nonlinear reaction term. Again, integrating over y to obtain the kinetic equation for the total distribution $P_1(x, t)$, we have

$$\frac{\partial^\gamma P_1}{\partial t^\gamma} - I_t^{1-\gamma} \left[D \frac{\partial^2 P_0}{\partial x^2} + \hat{C}(P_0) \right] = 0. \tag{31.76}$$

For the sake of brevity, we denoted $P_0 = P(x, y = 0, t)$. Consider the fractional comb model (31.75) without reaction

$$\frac{\partial^\gamma P}{\partial t^\gamma} - \delta(y) I_t^{1-\gamma} D \frac{\partial^2 P}{\partial x^2} = g \frac{\partial^2}{\partial y^2} P \tag{31.77}$$

and perform the Laplace transform, thus one obtains

$$s^\gamma \hat{P} - \delta(y) D s^{\gamma-1} \frac{\partial^2 \hat{P}}{\partial x^2} = g \frac{\partial^2 \hat{P}}{\partial y^2} + s^{\gamma-1} \delta(x)\delta(y), \tag{31.78}$$

where for the initial condition we take $P(t=0) = \delta(x)\delta(y)$. Looking for the solution in the form

$$\hat{P}(x,y,s) = \exp[-\sqrt{s^\gamma/g}|y|] f(x,s), \tag{31.79}$$

one can see that $\hat{P}_0 = \hat{P}(x, y = 0, s) = f(x,s)$ and integrating Equation 31.79 over y yields

$$\hat{P}_0(x,s) = \sqrt{\frac{s^\gamma}{4g}} \hat{P}_1(x,s). \tag{31.80}$$

Substituting Equation 31.80 into Equation 31.76 in the Laplace space, one obtains

$$s\hat{P}_1 - P_1(t=0) - \frac{Ds^{\frac{\gamma}{2}}}{2\sqrt{g}} \frac{\partial^2 \hat{P}_1}{\partial x^2} - \frac{Cs^{\frac{\gamma}{2}}}{2\sqrt{g}} \hat{P}_1 = -C\hat{\mathcal{L}}[P_0^2]. \tag{31.81}$$

Multiplying this equation by e^{st} and using identity $e^{st} s^\alpha f(s) = \frac{\partial}{\partial t} e^{st} s^{\alpha-1} f(s)$, we integrate with the corresponding contour to obtain the inverse Laplace transform. This yields

$$\frac{\partial P_1}{\partial t} - \frac{1}{2\sqrt{g}} \frac{\partial}{\partial t} I_t^{1-\frac{\gamma}{2}} \left[D \frac{\partial^2 P_1}{\partial x^2} + CP_1 \right] = -\frac{C}{4g} \left[\frac{\partial}{\partial t} I_t^{1-\frac{\gamma}{2}} P_1 \right]^2. \tag{31.82}$$

The nonlinear term is obtained by the following chain of transformations:

$$C[P_0^2] = C \left[\hat{\mathcal{L}}^{-1} \hat{P}_0 \right]^2 = \frac{C}{4g} \left[\hat{\mathcal{L}}^{-1} s^{\frac{\gamma}{2}} \hat{P}_1 \right]^2 = \frac{C}{4g} \left[\frac{\partial}{\partial t} I_t^{1-\frac{\gamma}{2}} P_1 \right]^2. \tag{31.83}$$

Note that a specific property of these transformations is an irreversibility with respect to the Laplace transform, since, as well known, the Laplace transform of the Riemann–Liouville fractional derivative involves the (quasi) initial value terms such as $P_1(t=0) = \delta(x)$ [36].

To evaluate the overall velocity of the asymptotic front, let us introduce a small parameter, say ε, at the derivatives with respect to time and space [22]. To this end, we rescale $x \to x/\varepsilon$ and $t \to t/\varepsilon$, and $P_1(x,t) \to P_1^\varepsilon(x,t) = P_1\left(\frac{x}{\varepsilon}, \frac{t}{\varepsilon}\right)$. Therefore, one looks for the asymptotic solution in a form of the Green approximation

$$P_1^\varepsilon(x,t) = \exp\left[-\frac{G^\varepsilon(x,t)}{\varepsilon} \right]. \tag{31.84}$$

The main strategy of implication of this construction is the limit $\varepsilon \to 0$, one has $\exp\left[-\frac{G^\varepsilon(x,t)}{\varepsilon}\right] = 0$, except the condition when $G^\varepsilon(x,t) = 0$. This equation determines the position of the reaction-spreading front (see Equation 31.74). Moreover, we consider the limit $G(x,t) = \lim_{\varepsilon \to 0} G^\varepsilon(x,t)$ as the principal Hamiltonian function [22] that makes it possible to apply the Hamiltonian approach for calculation of the propagation front velocity. In this case, partial derivatives of $G(x,t)$ with respect to time and coordinate have physical senses of the Hamiltonian and momentum:

$$\frac{\partial G(x,t)}{\partial t} = -H \quad \text{and} \quad \frac{\partial G(x,t)}{\partial x} = p. \tag{31.85}$$

Now the method of hyperbolic scaling, explained above, can be applied. Therefore, we have the ansatz (31.84) for the pdf inside dendrites. Inserting expression (31.84) into Equation 31.82, one considers fractional integrations in time. Let us start from the last term in Equation 31.82, which is the reaction term.

We rewrite it in the following convenient form:

$$\frac{\varepsilon}{\Gamma(1-\frac{\gamma}{2})}\frac{\partial}{\partial t}\int_0^{\frac{t}{\varepsilon}} dt'(t')^{-\gamma/2}\exp[-G^{\varepsilon}(t-\varepsilon t',x-\varepsilon x')/\varepsilon]. \tag{31.86}$$

Then performing expansion

$$G^{\varepsilon}(t-\varepsilon t',x-\varepsilon x')\approx G^{\varepsilon}(x,t)-\varepsilon\frac{\partial G^{\varepsilon}(x,t)}{\partial t}t'-\varepsilon\frac{\partial G^{\varepsilon}(x,t)}{\partial x}x',$$

and substituting this in Equation 31.86, one obtains

$$\frac{1}{\Gamma(1-\frac{\gamma}{2})}\left[-\frac{\partial G^{\varepsilon}(x,t)}{\partial t}\right]\exp\left[-\frac{G^{\varepsilon}(x,t)}{\varepsilon}\right]\int_0^{\frac{t}{\varepsilon}}(t')^{-\gamma/2}\exp\left[\frac{\partial G^{\varepsilon}(x,t)}{\partial t}t'+\frac{\partial G^{\varepsilon}(x,t)}{\partial x}x'\right]dt'. \tag{31.87}$$

It should be noted that we neglect differentiation of the upper limit of the integral, since this term is of the order of $O(\varepsilon^{1+\gamma/2})\sim o(\varepsilon)$ that vanishes in the limit $\varepsilon\to 0$. The same procedure of expansion is performed for the diffusion term in Equation 31.82 that yields

$$\frac{\varepsilon^3}{\Gamma(1-\frac{\gamma}{2})}\frac{\partial^3}{\partial t\partial x^2}\exp\left[-\frac{G^{\varepsilon}(x,t)}{\varepsilon}\right]\int_0^{\frac{t}{\varepsilon}}(t')^{-\frac{\gamma}{2}}\exp\left[\frac{\partial G^{\varepsilon}(x,t)}{\partial t}t'+\frac{\partial G^{\varepsilon}(x,t)}{\partial x}x'\right]dt'. \tag{31.88}$$

Differentiating in the limit $\varepsilon\to 0$ and taking into account that the Hamiltonian H and the momentum p in Equation 31.85 are independent of x and t explicitly (that leads to the absence of mixed derivatives), one obtains the Laplace transform of the subdiffusive kernel $t^{-\frac{\gamma}{2}}$. After these procedures in Equations 31.87 and 31.88, the kinetic equation (31.82) becomes a kind of Hamilton–Jacobi equation that establishes a relation between the Hamiltonian and the momentum

$$H=\left[\frac{Dp^2+C}{2\sqrt{g}}\right]^{\frac{2}{2-\gamma}}, \tag{31.89}$$

and the action is $G(x,t)=\int_0^t[p(s)\dot{x}(s)-H(p(s),x(s))]ds$. The rate v at which the front moves is determined at the condition $G(x,t)=0$. Together with the Hamilton equations, this yields

$$v=\dot{x}=\frac{\partial H}{\partial p},\quad v=\frac{H}{p}. \tag{31.90}$$

Note that the first equation in Equation 31.90 reflects the dispersion condition, whereas the second one is a result of the asymptotically free particle dynamics, when the action is $G(x,t)=px-Ht$. Taking into account $x=vt$, one obtains Equation 31.90 (see also details of this chapter, e.g., in References 12 and 34). A combination of these two equations can be replaced by

$$v=\min_{H>0}\frac{H}{p(H)}=\min_{p>0}\frac{H(p)}{p}. \tag{31.91}$$

We also have from the front velocity conditions (31.90) $\frac{\partial}{\partial p}\ln H=1/p$ that, eventually, yields from Equation 31.91

$$v=\left[\left(\frac{4}{g}\right)^{\frac{2}{2-\gamma}}\frac{D}{2-\gamma}\left(\frac{C}{2+\gamma}\right)^{\frac{2+\gamma}{2-\gamma}}\right]^{\frac{1}{2}}. \tag{31.92}$$

To proceed, we first admit that the limiting case of this result with $\gamma=0$ corresponds to the CaMKII propagation along the dendrite only (i.e., there are no teeth). Therefore, Equation 31.92 after rescaling $D/\sqrt{g}\to D$ and $C/\sqrt{g}\to C$ recovers the FKPP scheme for $\gamma=0$ that yields $v=\sqrt{DC}$.

The absence of failure of the activation front propagation should be admitted. It has a simple explanation due to the absence of a reaction "sink" term $-hP$ in Equation 31.75 by neglecting the possibility of spines to accumulate a large amount of Ca^{2+} [29,46], where h is a translocation/accumulation rate [18]. Introducing this term in Equation 31.75, our anticipation is that the hyperbolic scaling for this new equation yields a solution similar to Equation 31.91 with $H = 0$ that corresponds to the failure of front propagation. Moreover, this asymptotic solution for $P_1(x,t)$ always takes place, as one of the possible solutions.

Inserting the sink term in Equation 31.75, one obtains

$$\frac{\partial^\gamma P}{\partial t^\gamma} - \delta(y) I_t^{1-\gamma} \left[D \frac{\partial^2 P}{\partial x^2} + CP(1-P) \right] - g \frac{\partial^2 P}{\partial y^2} - ghP = 0. \tag{31.93}$$

Repeating the same procedures of the Laplace transform and integration over y with definition $\hat{P}_1 = \int_{-\infty}^{\infty} \hat{P}(x,y,s) dy$, and using the substitute

$$P_1(x,s) = 2\sqrt{g/s^\gamma} P(x, y = 0, s),$$

one obtains

$$s\hat{P}_1 - \delta(x) = \frac{D\sqrt{s^\gamma}}{2\sqrt{g}} \frac{\partial \hat{P}_1}{\partial x} + \frac{C\sqrt{s^\gamma}}{2\sqrt{g}} \hat{P}_1 - hg s^{1-\gamma} \hat{P}_1. \tag{31.94}$$

Here, we neglect the nonlinear term, since, as follows from the above analysis, in the further hyperbolic-scaling approximation, this term does not contribute to the Hamilton–Jacobi equation, and we also know how to handle it. Again multiplying this equation by e^{st} and using the same identity $e^{st} s^\alpha f(s) = \frac{\partial}{\partial t} e^{st} s^{\alpha-1} f(s)$, as above, we obtain the inverse Laplace transform. Thus, Equation 31.94 reads

$$\frac{\partial P_1}{\partial t} = \frac{D}{2\sqrt{g}} \frac{\partial}{\partial t} I_t^{1-\frac{\gamma}{2}} \left[\frac{\partial P_1}{\partial x} + \frac{C}{D} P_1 \right] - hg \frac{\partial}{\partial t} I_t^\gamma P_1. \tag{31.95}$$

Application of the hyperbolic scaling with the asymptotic solution (31.84) yields

$$2\sqrt{g}H = \left[Dp^2 H^{\frac{\gamma}{2}} + CH^{\frac{\gamma}{2}} - 2hg^{\frac{3}{2}} H^{1-\gamma} \right]. \tag{31.96}$$

Let us consider a specific case $\gamma = 2/3$ that yields

$$H = \left[\frac{Dp^2 + C - 2hg^{\frac{3}{2}}}{2\sqrt{g}} \right]^{\frac{3}{2}}. \tag{31.97}$$

For $C > 2hg^{\frac{3}{2}}$ there is no failure and the front asymptotically propagates with a constant velocity. For $C \leq 2hg^{\frac{3}{2}}$ the only solution is $H = 0$ and yields $v = 0$. So, $2hg^{\frac{3}{2}}$ is the minimum reaction rate necessary to sustain propagation along the spiny dendrite due to the presence of translocation. Analogously, $(C/2h)^{2/3}$ can also be viewed as the minimum value for the density of spines necessary to have propagation failure. Both results are in agreement with the results obtained from very different models based on the cable model [13].

In the general case, one compares the interplay between the activation $CH^{\frac{\gamma}{2}}$ and the translocation $-2hg^{\frac{3}{2}} H^{1-\gamma}$ terms in Equation 31.96 in the limit $H \to 0$. For $\gamma \in [\frac{2}{3}, 1)$, the translocation term is dominant and leads to the solution with $H = 0$ and the failure of the front propagation, correspondingly. When $0 < \gamma < \frac{2}{3}$, the activation in dendrites can be dominant. This situation is more complicated, and the activation–translocation front can propagate with an asymptotically finite velocity.

Finally, let us consider the linear counterpart of Equation 31.75 with the linear reaction term $\hat{C}(P) = CP$. This analysis will be useful to uncover the behavior of the tail of the total distribution and check if the front accelerates or travels with constant velocity. Rewrite the equation for the total distribution $P_1(x, t)$. As follows from the fractional differentiation of Equation 31.82, this equation reads (see also Reference 35)

$$\frac{\partial^{1-\frac{\gamma}{2}} P_1}{\partial t^{1-\frac{\gamma}{2}}} = \frac{D}{2\sqrt{g}} \left[\frac{\partial^2}{\partial x^2} + \frac{C}{D} \right] P_1 \tag{31.98}$$

with the initial condition $P_1(x, t = 0) = \delta(x)$. After the Fourier transform $\hat{\mathcal{F}}[P_1(x, t)] = \bar{P}_1(k, t)$, one obtains the solution in the form of the Mittag–Leffler function

$$\bar{P}_1(k, t) = E_{1-\frac{\gamma}{2}} \left[A(k) t^{1-\frac{\gamma}{2}} \right], \tag{31.99}$$

where $A(k) = (C - Dk^2)/2\sqrt{g}$. At the asymptotic condition, when $x, t \gg 1$, we have $C \gg Dk^2$ that yields an asymptotic behavior of the Mittag–Leffler function as a growing exponent (for the large positive argument) [4]

$$\bar{P}_1(k, t) \approx \exp \left[\left(\frac{C}{2\sqrt{g}} - \frac{D}{2\sqrt{g}} k^2 \right)^{\frac{2}{2-\gamma}} t \right] \approx \exp \left[\left(\frac{C}{2\sqrt{g}} \right)^{\frac{2}{2-\gamma}} \left(1 - \frac{2}{2-\gamma} \frac{Dk^2}{C} \right) t \right]. \tag{31.100}$$

After the Fourier inversion, one obtains

$$P_1(x, t) = \exp \left[\left(\frac{C}{2\sqrt{g}} \right)^{\frac{2}{2-\gamma}} t - \frac{(2-\gamma)x^2 (2\sqrt{g})^{\frac{2}{2-\gamma}}}{8DC^{\frac{\gamma}{2-\gamma}} t} \right], \tag{31.101}$$

that, finally, yields the nonzero and constant overall velocity of the reaction front propagation. Note that for normal diffusion, $\gamma = 0$, one arrives at the Fisher velocity $v = \sqrt{DC/g} \to \sqrt{DC}$, see the limiting case $\gamma = 0$ in Equation 31.92.

31.5 Conclusion

In this chapter, we show that a comb is a convenient model for analytical exploration of anomalous transport and front propagation phenomena along spiny dendrites. We have studied the properties of a random walk motion along the backbone in the presence of teeth. Teeth are lateral branches crossing the backbone and we have shown here that their effect on the movement along the whole structure can be reduced to a waiting time distribution at the nodes of the backbone during the movement of the random walker. Recent experiments and numerical simulations have predicted anomalous diffusion along spiny dendrites.

We have shown here that this anomalous phenomenon can be explained in the framework of a comb model with infinitely long teeth. Moreover, due to the random distribution of spines along the parent dendrite and the presence of a binding reaction inside spines, one can present a physically reasonable justification of the power law of the waiting time distribution that leads to subdiffusion in both the teeth and the backbone.

We have shown how to predict anomalous diffusion in spines by constructing a fractional-diffusion equation. By using the CTRW formalism, we have computed the MSD for the transport along the whole comb. We presented an illustration of how to take into account the inhomogeneous distribution of spines along the dendrite by using a fractal comb. We have also constructed the corresponding fractional-diffusion equations and computed the MSD. On the other hand, the constructed toy models are simple enough, like the comb model that makes it possible to suggest and understand a variety of reaction-transport schemes, including anomalous transport, by applying a strong machinery of fractional calculus and hyperbolic scaling for asymptotic methods. This approach allows to suggest an analytical description of reaction-transport scenarios in spiny dendrites, where we consider both a linear reaction in spines (see Equations 31.72 and 31.73), and nonlinear reaction along dendrites, considered in the framework of the FKPP scheme [21]. To this end, we suggest a fractional subdiffusive comb model, where we apply a Hamilton–Jacobi approach to estimate the overall velocity of the reaction front propagation. We proposed an alternative approach of a recently suggested mechanism of the translocation wave of CaMKII [18], where the activated CaMKII contaminant travels along dendrites with additional translocation inside spines, and the process of activation corresponds to the irreversible reaction described by the FKPP equation (31.82). One of the main effects, observed in the framework of the considered model, is the failure of the front propagation due to either the reaction inside spines, or interaction of reaction with spines. In the first case, the spines are the source of reactions, whereas in the latter case, the spines are a source of damping, for example, they act as a sink of an activated contaminant (CaMKII). The situation is controlled by three parameters: CaMKII activation C, CaMKII translocation rate h, and the fractional transport exponent γ. The latter reflects the geometrical structure of the transport system: when $0 < \gamma < \frac{2}{3}$, the activation in dendrites can be dominant, and the activation–translocation front can propagate with an asymptotically nonzero and constant velocity. For $\gamma = 2/3$, we have found a criteria for the emergence of propagation failure or for the sustenance of the propagation in terms of the reaction rate, the translocation rate, and the spine's density.

It should be admitted, in conclusion, that the physical arguments suggested above, explain why anomalous transport, namely subdiffusion, of either CaMKII or neutral particles is possible and supports implementation of the comb model. These arguments are based on the geometry of dendritic spines that determines an expression for the transport exponent in Equation 31.34. This situation becomes more sophisticated in the case of a nonlinear FKPP reaction. Indeed, as shown, the power law kernel of the transition probability considered due to the geometry arguments is insensitive to the nonlinear reaction. This consideration completely differs from a mesoscopic non-Markovian approach, developed in References 20 and 33, where spine–dendrite interaction and an extension including reactions in spines have been described in the framework of a variable residence time. This leads to an essential complication of the transition probability due to the nonlinear reaction term [19,33].

Acknowledgments

This chapter was supported by the Ministerio de Economia y Competitividad (Spain) under Grant No. FIS2012-32334.

Appendix: Fractional Integro-Differentiation

The consideration of a non-Markovian process in the framework of kinetic equations leads to the study of the so-called fractional Fokker–Planck equation, where both time and space processes are not local [36]. In this case, derivations are substituted by integrations with the power law kernels. One arrives at the so-called fractional integro-differentiation.

A basic introduction to fractional calculus can be found, e.g., in Reference 39. Fractional integration of the order of α is defined by the operator

$$_aI_t^\alpha f(t) = \frac{1}{\Gamma(\alpha)} \int_a^t f(\tau)(t-\tau)^{\alpha-1} d\tau, \quad (\alpha > 0), \tag{A31.1}$$

where $\Gamma(\alpha)$ is a gamma function. There is no constraint on the limit a. In our consideration, $a = 0$ since this is a natural limit for the time. A fractional derivative is defined as an inverse operator to $_aI_t^\alpha \equiv I_t^\alpha$ as $\frac{d^\alpha}{dt^\alpha} = I_t^{-\alpha} = D_t^\alpha$; correspondingly $I_t^\alpha = \frac{d^{-\alpha}}{dt^{-\alpha}} = D_t^{-\alpha}$. Its explicit form is convolution

$$D_t^\alpha = \frac{1}{\Gamma(-\alpha)} \int_0^t \frac{f(\tau)}{(t-\tau)^{\alpha+1}} d\tau. \tag{A31.2}$$

For arbitrary $\alpha > 0$, this integral is, in general, divergent. As a regularization of the divergent integral, the following two alternative definitions for D_t^α exist [32]:

$$_{RL}D_{(0,t)}^\alpha f(t) \equiv D_{RL}^\alpha f(t) = D^n I^{n-\alpha} f(t) \frac{1}{\Gamma(n-\alpha)} \frac{d^n}{dt^n} \int_0^t \frac{f(\tau) d\tau}{(t-\tau)^{\alpha+1-n}}, \tag{A31.3}$$

$$D_C^\alpha f(t) = I^{n-\alpha} D^n f(t) \frac{1}{\Gamma(n-\alpha)} \int_0^t \frac{f^{(n)} d\tau(\tau)}{(t-\tau)^{\alpha+1-n}}, \tag{A31.4}$$

where $n-1 < \alpha < n$, $n = 1, 2, \ldots$. Equation A31.3 is the Riemann–Liouville derivative, whereas Equation A31.4 is the fractional derivative in the Caputo form [32]. Performing integration by parts in Equation A31.3 and then applying Leibniz's rule for the derivative of an integral and repeating this procedure n times, we obtain

$$D_{RL}^\alpha f(t) = D_C^\alpha f(t) + \sum_{k=0}^{n-1} f^{(k)}(0^+) \frac{t^{k-\alpha}}{\Gamma(k-\alpha+1)}. \tag{A31.5}$$

The Laplace transform can be obtained for Equation A31.4. If $\hat{\mathcal{L}}[f(t)] = \tilde{f}(s)$, then

$$\hat{\mathcal{L}}\left[D_C^\alpha f(t)\right] = s^\alpha \tilde{f}(s) - \sum_{k=0}^{n-1} f^{(k)}(0^+) s^{\alpha-1-k}. \tag{A31.6}$$

The following fractional derivatives are helpful for the present analysis:

$$D_{RL}^\alpha[1] = \frac{t^{-\alpha}}{\Gamma(1-\alpha)}, \quad D_C^\alpha[1] = 0. \tag{A31.7}$$

We also note that

$$D_{RL}^\alpha t^\beta = \frac{t^{\beta-\alpha} \Gamma(\beta+1)}{\Gamma(\beta+1-\alpha)}, \tag{A31.8}$$

where $\beta > -1$ and $\alpha > 0$. The fractional derivative from an exponential function can be simply calculated as well by virtue of the Mittag–Leffler function (see, e.g., References 4 and 39):

$$E_{\gamma,\delta}(z) = \sum_{k=0}^\infty \frac{z^k}{\Gamma(\gamma k+\delta)}. \tag{A31.9}$$

Therefore, we have the following expression:

$$D_{RL}^\alpha e^{\lambda t} = t^{-\alpha} E_{1,1-\alpha}(\lambda t). \tag{A31.10}$$

References

1. E. Agliari, A. Blumen, and D. Cassi. Slow encounters of particle pairs in branched structures. *Phys. Rev. E*, 89:052147, 2014.
2. V.E. Arkhincheev and E.M. Baskin. Anomalous diffusion and drift in a comb model of percolation clusters. *Sov. Phys. JETP*, 73:161, 1991.
3. E. Baskin and A. Iomin. Electrostatics in fractal geometry: Fractional calculus approach. *Chaos Solitons Fractals*, 44(4–5):335–341, 2011.
4. A. Erdélyi and H. Bateman. *Higher Transcendental Functions*, volume 3. McGraw-Hill, New York, 1955.
5. D. ben Avraham and S. Havlin. *Diffusion and Reactions in Fractals and Disordered Systems*. Cambridge University Press, UK, 2005.

6. A.M. Berezhkovskii, A.V. Barzykin, and V.Y. Zitserman. Escape from cavity through narrow tunnel. *J. Chem. Phys.*, 130(24):245104, 2009.

7. A. Biess, E. Korkotian, and D. Holcman. Diffusion in a dendritic spine: The role of geometry. *Phys. Rev. E*, 76:021922, 2007.

8. B.L. Bloodgood and B.L. Sabatini. Neuronal activity regulates diffusion across the neck of dendritic spines. *Science*, 310(5749):866–869, 2005.

9. P.C. Bressloff. Propagation of CaMKII translocation waves in heterogeneous spiny dendrites. *J. Math. Biol.*, 66(7):1499–1525, 2013.

10. M.J. Byrne, M.N. Waxham, and Y. Kubota. The impacts of geometry and binding on CaMKII diffusion and retention in dendritic spines. *J. Comput. Neurosci.*, 31(1):1–12, 2011.

11. D. Campos and V. Méndez. Reaction–diffusion wave fronts on comblike structures. *Phys. Rev. E*, 71(5):051104, 2005.

12. D. Campos, S. Fedotov, and V. Méndez. Anomalous reaction-transport processes: The dynamics beyond the law of mass action. *Phys. Rev. E*, 77:061130, 2008.

13. S. Coombes and P. Bressloff. Saltatory waves in the spike-diffuse-spike model of active dendritic spines. *Phys. Rev. Lett.*, 91:028102, 2003.

14. S. Coombes and P.C. Bressloff. Saltatory waves in the spike-diffuse-spike model of active dendritic spines. *Phys. Rev. Lett.*, 91(2):028102, 2003.

15. L. Dagdug, A.M. Berezhkovskii, Y.A. Makhnovskii, and V.Y. Zitserman. Transient diffusion in a tube with dead ends. *J. Chem. Phys.*, 127(22):224712, 2007.

16. C.I. De Zeeuw and T.M. Hoogland. Anomalous diffusion imposed by dendritic spines (commentary on Santamaria et al.). *Eur. J. Neurosci.*, 34(4):559–560, 2011.

17. B.A. Earnshaw and P.C. Bressloff. A diffusion–activation model of CaMKII translocation waves in dendrites. *J. Comput. Neurosci.*, 28(1):7789, 2009.

18. B.A. Earnshaw and P.C. Bressloff. A diffusion–activation model of CaMKII translocation waves in dendrites. *J. Comput. Neurosci.*, 28(1):77–89, 2010.

19. S. Fedotov. Non-Markovian random walks and nonlinear reactions: Subdiffusion and propagating fronts. *Phys. Rev. E*, 81:011117, 2010.

20. S. Fedotov and V. Méndez. Non-Markovian model for transport and reactions of particles in spiny dendrites. *Phys. Rev. Lett.*, 101(21):218102, 2008.

21. R.A. Fisher. The wave of advance of advantageous genes. *Ann. Eugenics*, 7(4):355–369, 1937.

22. M.I. Freidlin. *Markov Processes and Differential Equations. Asymptotic Problems.* Birkhäuser Basel, Switzerland, 1996.

23. D. Froemberg, H. Schmidt-Martens, I. Sokolov, and F. Sagués. Front propagation in $A + B \rightarrow 2A$ reaction under subdiffusion. *Phys. Rev. E*, 78:011128, 2008.

24. B.I. Henry, T.A.M. Langlands, and S.L. Wearne. Fractional cable models for spiny neuronal dendrites. *Phys. Rev. Lett.*, 100(12):128103, 2008.

25. A. Iomin. A toy model of fractal glioma development under RF electric field treatment. *Eur. Phys. J. E*, 35(6):42, 2012.

26. A. Iomin and E. Baskin. Negative superdiffusion due to inhomogeneous convection. *Phys. Rev. E*, 71:061101, 2005.

27. A. Iomin and V. Méndez. Reaction–subdiffusion front propagation in a comblike model of spiny dendrites. *Phys. Rev. E*, 88(1):012706, 2013.

28. A. Iomin. Subdiffusion on a fractal comb. *Phys. Rev. E*, 83:052106, 2011.

29. E. Korkotian and M. Segal. Spatially confined diffusion of calcium in dendrites of hippocampal neurons revealed by flash photolysis of caged calcium. *Cell Calcium*, 40(5–6):441–449, 2006. Calcium microdomains and the fine control of cell function.

30. A. Le Mehaute, R.R. Nigmatullin, and L. Nivanen. *Fleches du Temps et Geometric Fractale* (Hermes, Paris, 1998).

31. I.A. Lubashevskii and A.A. Zemlyanov. Continuum description of anomalous diffusion on a comb structure. *J. Exp. Theor. Phys.*, 87(4):700–713, 1998.

32. F. Mainardi. Fractional relaxation–oscillation and fractional diffusion-wave phenomena. *Chaos Solitons Fractals*, 7(9):1461–1477, 1996.

33. V. Méndez, S. Fedotov, and W. Horsthemke. *Reaction-Transport Systems: Mesoscopic Foundations, Fronts, and Spatial Instabilities (Springer Series in Synergetics)*. Springer, Berlin, Germany, 1st edition, 2010.

34. V. Méndez, D. Campos, and S. Fedotov. Analysis of fronts in reaction-dispersal processes. *Phys. Rev. E*, 70:066129, 2004.

35. V. Méndez and A. Iomin. Comb-like models for transport along spiny dendrites. *Chaos Solitons Fractals*, 53:46–51, 2013.

36. R. Metzler and J. Klafter. The random walk's guide to anomalous diffusion: A fractional dynamics approach. *Phys. Rep.*, 339(1):1–77, 2000.

37. R.R. Nigmatullin. Fractional integral and its physical interpretation. *Theor. Math. Phys.*, 90(3):242–251, 1992.

38. E.A. Nimchinsky, B.L. Sabatini, and K. Svoboda. Structure and function of dendritic spines. *Annu. Rev. Physiol.*, 64:313–353, 2002.

39. I. Podlubny. *Fractional Differential Equations*. Academic Press, San Diego, 1998.

40. A. Rebenshtok and E. Barkai. Occupation times on a comb with ramified teeth. *Phys. Rev. E*, 88:052126, 2013.

41. S. Redner. *A Guide to First-Passage Processes*. Cambridge University Press, 2001.

42. F.-Y. Ren, J.-R. Liang, X.-T. Wang, and W.-Y. Qiu. Integrals and derivatives on net fractals. *Chaos Solitons Fractals*, 16(1):107–117, 2003.

43. B.L. Sabatini, M. Maravall, and K. Svoboda. Ca(2+) signaling in dendritic spines. *Curr. Opin. Neurobiol.*, 11(3):349–356, 2001.

44. F. Santamaria, S. Wils, E. De Schutter, and G.J. Augustine. Anomalous diffusion in Purkinje cell dendrites caused by spines. *Neuron*, 52(4):635–648, 2006.

45. F. Santamaria, S. Wils, E. De Schutter, and G.J. Augustine. The diffusional properties of dendrites depend on the density of dendritic spines. *Eur. J. Neurosci.*, 34(4):561–568, 2011.

46. M. Segal. Dendritic spines and long-term plasticity. *Nat. Rev. Neurosci.*, 6(4):277–284, 2005.

47. C. Van den Broeck. Waiting times for random walks on regular and fractal lattices. *Phys. Rev. Lett.*, 62:1421–1424, 1989.

48. G.H. Weiss and S. Havlin. Some properties of a random walk on a comb structure. *Physica A*, 134(2):474–482, 1986.

49. R. Yuste. *Dendritic Spines*. MIT Press, Massachusetts, 2010.

50. V.Y. Zaburdaev, P.V. Popov, A.S. Romanov, and K.V. Chukbar. Stochastic transport through complex comb structures. *J. Exp. Theor. Phys.*, 106(5):999–1005, 2008.

51. S. Zeng and W.R. Holmes. The effect of noise on CaMKII activation in a dendritic spine during ltp induction. *J. Neurophysiol.*, 103(4):1798–1808, 2010.

52. G. Zumofen and J. Klafter. Laminar-localized phase coexistence in dynamical systems. *Phys. Rev. E*, 51:1818–1821, 1995.

53. G. Zumofen, J. Klafter, and A. Blumen. Anomalous transport: A one-dimensional stochastic model. *Chem. Phys.*, 146(3):433–444, 1990.

32

Applications of Chaos Theory Methods in Clinical Digital Pathology

Wlodzmierz Klonowski

32.1 Introduction

The aim of this chapter is to provide an insight into applications of nonlinear methods that have been used in chaos theory [1] and in nonlinear dynamics. I shall consider applications of Higuchi's fractal dimension method in the analysis of biosignals and of pathophysiological images [2,3].

A living organism is a complex nonlinear system. So, it is much more appropriate to use nonlinear methods for biomedical applications, despite the fact that for short time intervals, linear methods such as fast Fourier transform (FFT) may work well. Nonlinear methods may be applied to linear cases—for example, one might try to approximate a straight line using a parabolic function only to find that the coefficient of the quadratic term was practically equal to zero. It is the opposite that obviously fails—if one uses linear approximation, then, one will never be able to appropriately approximate a nonlinear function. Similarly, nonlinear dynamical methods may be applied to static signals and images but not *vice versa*.

32.2 Fractal Analysis of Signals

One often forgets that "classical" methods such as FFT work only for stationary signals [4]. But even when analyzing stationary signals, FFT may give quite misleading results. For example, spectral analysis of a signal of frequency v and amplitude A modulated with frequency κ gives two harmonic signals, each of amplitude $A/2$, with frequencies $(v + \kappa)/2$ and $(v - \kappa)/2$, respectively (as illustrated by an example in

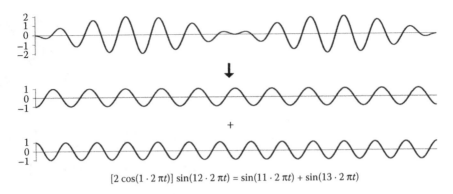

$$[2\cos(1 \cdot 2\,\pi t)]\sin(12 \cdot 2\,\pi t) = \sin(11 \cdot 2\,\pi t) + \sin(13 \cdot 2\,\pi t)$$

FIGURE 32.1 Fourier decomposition of the signal of frequency 12 Hz with the amplitude modulated with frequency 1 Hz (upper) results in two harmonic signals—one with frequency 11 Hz and another with frequency 13 Hz; the basic frequency 12 Hz completely disappears from the spectrum [4].

Figure 32.1), what is rather obvious for any person knowing trigonometric identity:

$$\sin\alpha + \sin\beta = 2\sin\left[\frac{(\alpha+\beta)}{2}\right]\cos\left[\frac{(\alpha-\beta)}{2}\right]$$

Multiplying both sides by $(A/2)$ and putting

$$\left[\frac{(\alpha+\beta)}{2}\right] = 2\pi\nu t;\left[\frac{(\alpha-\beta)}{2}\right] = 2\pi\kappa t$$

that is

$$\alpha = 2\pi\left[\frac{(\nu+\kappa)}{2}\right]t;\beta = 2\pi\left[\frac{(\nu-\kappa)}{2}\right]t$$

where the corresponding angle (phase) at moment t is equal to the frequency multiplied by $2\pi t$, and changing the left- and right-hand sides one obtains

$$[A\cos(2\pi\kappa t)]\sin(2\pi\nu t) = \left(\frac{A}{2}\right)\left[\sin\left(2\pi\left[\frac{(\nu+\kappa)}{2}\right]t + \sin(2\pi)\left[\frac{(\nu-\kappa)}{2}\right]t\right)\right] \qquad (32.1)$$

So, if a simple amplitude-modulated signal is analyzed using FFT, the basic frequency of the analyzed signal ν disappears from the Fourier spectrum (Figure 32.1).

Nonlinear analysis methods, for example, fractal dimension methods, do not have such shortcomings. But there exist several very differently calculated quantities that are called "fractal dimension" and one should always clearly say which of these quantities is to be used. Here, in all our further consideration when the term "fractal dimension" will be used, it denotes the quantity calculated using a very simple algorithm proposed by Higuchi [5] D_f, calculated from the time series directly in the time domain, without the necessity of data embedding in a phase space [2]. D_f should not be misled with fractal dimension of an attractor in the system's phase space.

Running fractal dimension, $D_f(t)$, may be calculated using a sliding window as short as that including 70–100 data points. Higuchi's fractal dimension is always between 1 (Euclidean dimension of a curve) and 2 (Euclidean dimension of a plane). $(D_f - 1)$ shows what "fraction of a plane" is occupied by the graph representing the analyzed signal. It is well seen while analyzing the chirp signal that is a cosine function frequency of which increases linearly in time (Figure 32.2a)—its $D_f(t)$ if calculated in a sliding window of a constant length very quickly increases from 1 at $t = 0$ up to nearly 2 (Figure 32.2b). It is also demonstrated that the greater the signal's frequency is, the greater is its D_f.

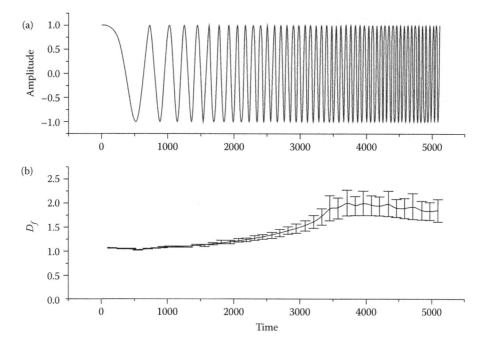

FIGURE 32.2 Chirp signal—a cosine with linearly increasing frequency $\omega(t) = a + b * t$; so, the phase changes as a quadratic function: $\varphi(t) = t * (a + b * t) = a * t + b * t^2$ (a) and its Higuchi' fractal dimension (b). For $D_f(t)$ values close to 1 the error is small; for $D_f(t)$ values close to 2, the absolute value of the error increases considerably (error bars are put on the graph automatically, but the value of D_f does not exceed 2).

Any signal may be characterized locally in time by its $D_f(t)$; it is not necessary to use the surrogate data test before applying Higuchi's fractal dimension method [4] because it does not matter if the analyzed signal is "really chaotic"—it may be deterministic or chaotic, nonstationary, and noisy.

32.3 Fractal Analysis of Images

When speaking about the fractal analysis of two-dimensional (2D) images, most scientists usually have in mind the so-called *box fractal dimension* and its variations [6–8]. We have proposed two methods of fractal analysis that make use of Higuchi's fractal dimension.

The analyzed 2D image is first preprocessed to form one-dimensional (1D) sequences that are subsequently analyzed using Higuchi's fractal dimension algorithm. The first method, *image signature's fractal dimension* (ISF), is based on the analysis of a 1D sequence called the *signature* of the 2D object's contour. The second method, *image landscapes' fractal dimension* (ILF), is based on the analysis of two 1D sequences called the *horizontal and vertical landscapes*.

32.3.1 ISF method

A contour is a closed planar curve specified in any rectangular coordinate system by the set of pairs (x_i, y_i) $(i = 1, \dots, I)$. We transform Cartesian coordinates of the 2D contour to be analyzed into a 1D series r_i that we call the contour's *signature*—the set r_i $(i = 1, \dots, I)$ of r-coordinates of all contour's points in the polar coordinate system centered at (x_0, y_0)

$$r_i = \sqrt{(x_i - x_0)^2 + (y_i - y_0)^2} \tag{32.2}$$

where x_0 and y_0 are the arithmetic averages of the Cartesian coordinates of all points of the contour. The signature is then analyzed using Higuchi's algorithm.

The shortcoming of the ISF method is the necessity of segmentation of the image to obtain contour(s) of the object(s) that one wants to analyze.

32.3.2 ILF Method

The ILF method is based on constructing from the analyzed 2D image two 1D sequences that we call *landscapes*. Digitized grayscale images are stored in the form of matrices where matrix elements can take on values ranging from $g_{min} = 0$ for a black pixel up to $g_{max} = (2^b - 1)$ for a white pixel, where b denotes the number of bits ($g_{max} = 255$ for $b = 8$). Most color images are overlays of three monochrome images. In this chapter, we analyze only grayscale images.

First, the analyzed color image is transformed into a grayscale one. Then stepping through a gray value image length of N pixels and height of M pixels row by row, we calculate the sum of the gray values in each row, $G_m (m = 1, \ldots, M)$, and dividing the so-obtained numbers by the largest of them, G_R, one obtains the *horizontal landscape, hgs,*

$$NGS_m = \frac{G_m}{G_R} \in [0, 1] \quad m = 1, \ldots, M \tag{32.3}$$

where

$$G_m = \sum_{n=1}^{N} g_{mn}; \quad G_R = \max(G_m) \quad m = 1, \ldots, M$$

Similarly, stepping through the same image column by column, we calculate the sum of the gray values in each column, $G_n (n = 1, \ldots, N)$, and dividing the so-obtained numbers by the largest of them, G_C, one obtains the *vertical landscape, vgs*

$$NGS_n = \frac{G_n}{G_C} \in [0, 1] \quad n = 1, \ldots, N \tag{32.4}$$

where

$$G_n = \sum_{m=1}^{M} g_{mn}; \quad G_C = \max(G_n) \quad n = 1, \ldots, N$$

If necessary, other landscapes may be constructed using a similar counting technique, stepping through the same image in different directions, e.g., in diagonal directions, or in some rectangular frames. Normalization in Equations 32.3 and 32.4 is convenient but not essential since Higuchi's fractal dimension is invariant with respect to scaling of the data.

The landscapes are then analyzed using Higuchi fractal dimension in a sliding window of length L moved in each step l points (pixels) to the right. In most cases, we used a sliding window of length $L = 100$ moved in each step $l = 1$ point (pixel) to the right and Higuchi's parameter $k_{max} = 8$ (cf. References 2 and 3). D_f is calculated for each window and assigned to the last (the Lth) pixel of the window. So, we obtain $M - L + 1$ fractal dimension values for the horizontal landscape and their arithmetical mean value is D_h; similarly, we obtain $N - L + 1$ fractal dimension values for the vertical landscape and their arithmetical mean value is D_v. Since the first $L - 1$ points (pixels) are "lost" due to the use of a sliding window, one may repeat the procedure starting from the last pixel on the right side and moving the window l points to the left; in such a case, one would have four $D_f(t)$ graphs and four mean fractal dimension values; two D_h values are then averaged and so are two D_v values.

32.3.2.1 Examples: ILF Analysis of Brodatz Textures

The ILF method was originally proposed for characterizing the surfaces of nanomaterials [9]. We have tested the method by analyzing Brodatz textures (cf. References 10 and 11). The analysis confirmed that (Figures 32.3 and 32.4)

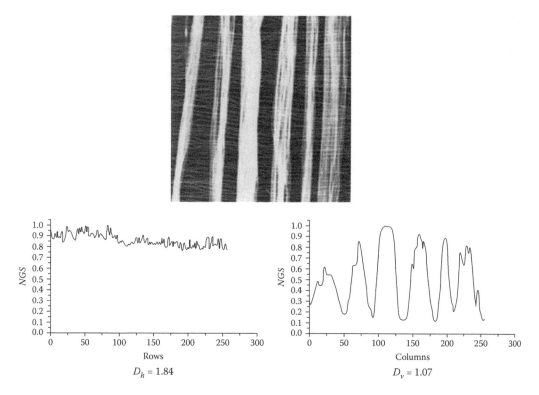

FIGURE 32.3 Example of the image of a surface with anisotropic roughness properties (texture) (cf. Reference 11)—fractal dimension of the horizontal and vertical landscapes differs considerably. The landscapes were analyzed using moving window length of 128 points, moved in each step 1 point forward. On the other hand, fractal dimensions of landscapes for surfaces that show isotropic roughness properties change appropriately with changes in surface properties—the smaller the unevenness of the surface, the greater the fractal dimensions of its landscapes (Figure 32.4).

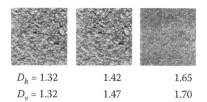

$D_h = 1.32$ 1.42   1.65
$D_v = 1.32$ 1.47   1.70

FIGURE 32.4 Images of surfaces with (nearly) isotropic and uniform properties of different roughness; fractal dimensions of the horizontal landscapes D_h and of the vertical landscapes D_v are nearly equal.

- If the fractal dimension of a landscape fluctuates only slightly, it means that the changes on the analyzed image in the considered direction are smooth; but if the fractal dimension of a landcape shows significant fluctuations, it means that there are nonstationarities/discontinuities in the analyzed image in the considered direction.
- If the values of a fractal dimension of two landscapes are nearly identical, it means that the image and the surface this image represents show nearly isotropic properties (roughness) that may be characterized using just one number, D_f—the average of D_h and D_v; but if D_h and D_v differ considerably, it means that the properties of the image change differently in horizontal and vertical directions and that the surface this image represents shows texture.

32.4　Results: Applications in Clinical Digital Pathology

32.4.1　ISF Method in Differentiation between Malignant and Benign BC

We have also applied the analysis of Higuchi's fractal dimension to the contours of breast masses (cf. Reference 12). We thank Professor R.M.Rangayyan (University of Calgary, Alberta, Canada) for data of contours of breast masses (cf. Reference 13) used for our calculations. Signatures of contours (cf. Equation 32.2) of benign masses show significantly higher values of Higuchi's fractal dimension than those of malignant breast tumors (Figure 32.5).

While the contour of a benign breast mass seems to be more regular than the contour of a malignant breast tumor (Figure 32.5), the fractal dimension of the malignant breast tumors is lower than the fractal dimension of benign breast masses. If the contours are zoomed up, one may observe that these benign

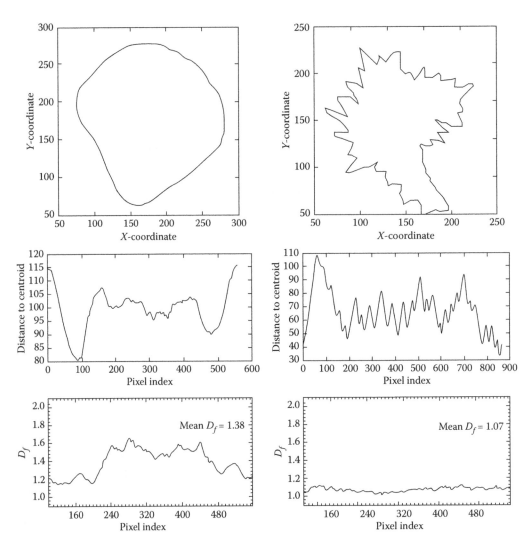

FIGURE 32.5　Contours of a benign mass (upper left) and of a malignant breast tumor (upper right), their signatures (cf. Equation 32.2, middle row), and the signatures associated with Higuchi's fractal dimension (lower row). Signature of a malignant tumor shows a lower fractal dimension than that of a benign mass [12].

masses show much more small irregularities than those of malignant tumors. That is why the signature of a benign mass shows many small "fluctuations" while that of a malignant tumor does not (cf. Figure 32.5), leading to the differences in their fractal dimension. The problem of calculation of length of the coastline considered by Mandelbrot was quite analogous; in fact, fractal dimension turned out to be the best characteristic that actually gives the possibility to compare properties of different coastlines.

We analyzed 37 cases of benign masses and 20 cases of malignant tumors [12]. There exist significant differences in mean values of D_f between signatures of benign mass contours and those of malignant tumors contours—D_f of benign masses are significantly higher; standard deviations of mean D_f values are small and the ranges (mean $D_{f\text{-}benign} \pm SD_{benign}$) and (mean $D_{fmalignant} \pm SD_{malignant}$) do not overlap [12].

However, sensitivity and specificity of the ISF method do not allow to use this method in clinical pathology [14]. Moreover, image segmentation to obtain contours of breast masses is very time consuming.

32.4.2 ILF Analysis of Anatomopathological Images for Cancer Staging

For the ILF method, image segmentation is not necessary. For validation of this method, we used it to differentiate clinical cases of three grades of *anal intraepithelial neoplasia* (AIN) [15] for which today adequate quantitative and nonsubjective screening techniques have not turned out to be satisfactory. AIN is a disease that is characterized by epithelial dysplasia and can lead to anal carcinoma. The presence of dysplastic cells and an abnormal epithelial texture in the anal tissue strongly depends on the presence of human papilloma viruses (HPVs). The number of abnormal cells or abnormal nuclei as well as their distribution throughout the epithelium yields three grades of dysplasia—AIN1, AIN2, and AIN3. AIN3 is the highest grade with the highest risk for invasive anal carcinoma. This classification is widely used but there is considerable interobserver variation in staging of AIN.

We have analyzed 120 histological slices of abnormal anal tissues: 36 of AIN1, 56 of AIN2, and 28 of AIN3. Microscopic images of eosin-stained slices were of a dimension of 749×579 pixels (i.e., 579 rows, 749 columns) each. These color images were transformed into 8-bit grayscale—for the pixel belonging to the mth row and the nth column, the gray value g_{mn} is assigned accordingly

$$g_{mn} \in [0, 255] \tag{32.5}$$

where the value 0 corresponded to a black pixel and the value 255, i.e., $(2^8 - 1)$ corresponded to a white pixel.

From this gray image, the horizontal and vertical landscapes were generated and their Higuchi's fractal dimension was calculated [15] (Figure 32.6). For the calculation of Higuchi's fractal dimension, our longtime experience suggested to make use of a moving window length of 100 points, moved in each step 1 point forward, with $k_{max} = 8$. D_f is not an absolute characteristic of the image and so, for the given grading process, all images should be analyzed using the same parameters of the algorithm.

None of the analyzed slides yielded anisotropic fractal dimension values.

We averaged the values of D_f and calculated standard deviations for each of the three grades of AIN. The results are shown in Figure 32.7.

Finally, we obtained the following values of D_f and standard deviations:

For AIN1: $D_f = 1.205 \pm 0.007$
For AIN2: $D_f = 1.243 \pm 0.008$
For AIN3: $D_f = 1.319 \pm 0.010$

It is important that the mean values of D_f for AIN2 were between those for AIN1 and AIN3. It is obvious that the differences between the grades of AIN are statistically significant. But in single cases, the tissue that had been classified, for example, as being AIN2 had D_f values that would suggest its belonging to another grade—AIN1 if D_f was much smaller or AIN3 if D_f was much greater than the mean value for AIN2.

One should always remember that AIN grading is subjective and that it depends on individual interpretation and may vary among pathologists.

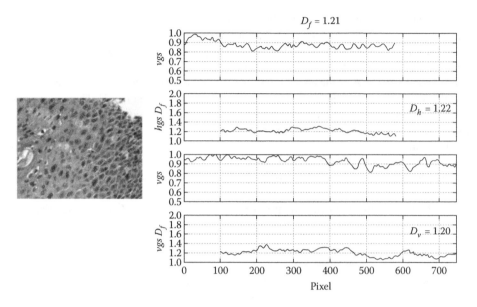

FIGURE 32.6   Example of ILF analysis of one histological slice of AIN.

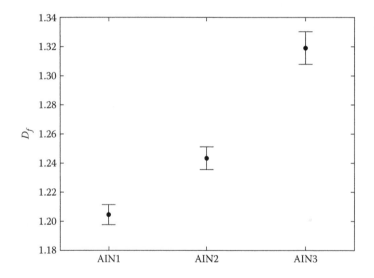

FIGURE 32.7   Differentiation of three AIN grades using ILF methods.

   The ILF method is easy and quick and it may be successfully applied at least as an auxiliary method for differentiation between AIN stages. But ILF can be easily applied to other histological specimen, for example, for the staging of neuromuscular diseases based on magnetic resonance imaging (MRI).

   The major advantage of the method is that image segmentation is not necessary. Usually, image segmentation introduces subjective parameters such as threshold values and, furthermore, fully automatic algorithms are rare and mostly fail because of slide to slide color or brightness variations. What might be a serious problem in other methods, namely that the intensity is different for different images even within the same series, for the ILF method is of no importance. Thus, the time necessary for calculations may be even further reduced. Moreover by virtue of its scale independence, fractal dimension calculated from an image appears to be nearly independent of the orientation of the surface.

32.5 Discussion and General Remarks

Using computers for image analysis, both supervised and unsupervised, is obvious but it raises some questions. The most important question that biomedical researchers have been asked since the introduction of a computer to the laboratories is "Do you really understand what your computer calculates?" Unfortunately, the answer to this question is quite often not positive. It concerns in particular commercial software when the source code is not given to the end users. But even information technology (IT) specialists often have problems to understand the details of the source code written by others, not to speak about biologists and medical doctors. And doctors are especially "conservative" in the sense that they do not want to use methods more advanced than Fourier transform they had learnt and understood.

The second question concerns repeatability of results. One would like that at least classification of the images from a given series would lead to the same result when repeated. But already the problem of choosing the region(s) of interest (ROI) may lead to differences between researchers. It seems that the whole image is as unbiased as one can get, so that any remaining bias originated in how the images were acquired. It was exactly the case in the application of the fractal method in the analysis of textures of anatomopathological slides for the staging of cases of AIN—some original slides showed white irregular "holes" but any attempt to analyze ROIs without holes only worsen the final results.

The third question concerns the very important, but not so well known, the so-called *curse of dimensionality*. To create a classifier, first, one needs descriptors for each object class that can be expressed by numbers. It might seem that to obtain a more accurate classification, one could add more features, maybe even a few hundred of these features. For example, one of the best software packages designed especially for calculation features of digitized images is *MaZda*; it has been under development in the Technical University of Lodz, Poland, since 1998, starting from COST Actions B11 and B21 [16]. But in fact, after a certain point, increasing the dimensionality of the problem by adding new features would actually degrade the performance of the classifier—as the dimensionality increases, the classifier's performance increases until the optimal number of features is reached; further increasing the dimensionality without increasing the number of training samples results in a decrease in classifier performance; this is called *overfitting* (cf. Reference 17). So, with only a small number of images in a database, it is just intractable to check all possible combinations of features in any given method nor to compare different proposed methods of image roughness and texture analysis to decide which combination or which method is the best one. Even the so-called feature extraction methods, such as the *principal component analysis*, that reduce the dimensionality of the classification problem, are not always helpful. Fractal dimension of a surface is invariant with respect to linear transformation of the data and to transformation of scale. So, the normalization in Equations 32.3 and 32.4 is convenient for presentation of the landscapes, but it is not really necessary since it does not change the value of Higuchi's fractal dimension.

32.6 Conclusions

The proposed method may serve for a simple quantitative assessment of surface roughness and texture, not only in the analysis of anatomopathological images and of MRI images, but also for the assessment of nanosurfaces. Our philosophy is that to be applicable, the method should preferably be really simple and easily understandable by nonspecialists in the field. The presented fractal methods are very simple and both of them draw from multiple disciplines and have multidisciplinary application.

Complicated and very specialized methods are good for research studies, but if one wants to propose a texture analysis procedure that could become widely applicable in hospitals and clinics, this should be rather a quite simple method, based on only a few texture parameters (perfectly just one or two) quick and easily understandable by medical doctors.

Acknowledgments

W. Klonowski was supported by the Nalecz Institute of Biocybernetics and Biomedical Engineering, Polish Academy of Sciences, Warsaw, through its statutory project.

References

1. Skiadas C.H., Dimotikalis I., Skiadas C. Eds. 2011. *Chaos Theory: Modeling, Simulation and Applications*. World Scientific. ISBN: 9814350338.
2. Klonowski W. 2007. The metaphor of "chaos." In: Konopka A.K. Ed. *Systems Biology. Principles, Methods and Concepts*. Boca Raton, London, New York: CRC Press/Taylor & Francis, pp. 115–138.
3. Klonowski W. 2007. From conformons to human brains: An informal overview of nonlinear dynamics and its applications in biomedicine. *Nonlinear Biomedical Physics* 1:5. Open access. http://www.nonlinearbiomedphys.com/content/pdf/1753-4631-1-5.pdf.
4. Klonowski W. 2009. Everything you wanted to ask about EEG but were afraid to get the right answer. *Nonlinear Biomedical Physics* 3:2. Open access http://www.nonlinearbiomedphys.com/content/pdf/1753-4631-3-2.pdf.
5. Higuchi T. 1988. Approach to an irregular time series on the basis of the fractal theory. *Physica D* 31:277–283.
6. Ogata Y., Katsura K. 1991. Maximum likelihood estimates of the fractal dimension for random spatial patterns. *Biometrika* 78(3):463–474.
7. Klonowski W. 2000. Signal and image analysis using chaos theory and fractal geometry. *Machine Graphics and Vision* 9(1/2):403–431.
8. Sandu A.L., Rasmussen I.A. Jr., Lundervold A., Kreuder F., Neckelmann G., Hugdahl K., Specht K. 2008. Fractal dimension analysis of MR images reveals grey matter structure irregularities in schizophrenia. *Computerized Medical Imaging Graphics* 32(2):150–158.
9. Klonowski W., Olejarczyk E., Stepien R. 2005. A new simple fractal method for nanomaterials science and nanosensors. *Materials Science—Poland* 23(3):607–612.
10. Brodatz P. 1966. *Textures: A Photographic Album for Artists and Designers*. New York, NY: Dover.
11. Original Brodatz Texture Database http://multibandtexture.recherche.usherbrooke.ca/original_brodatz_more.html.
12. Klonowski W., Stepien R., Stepien P. 2010. Simple fractal method of assessment of histological images for application in medical diagnostics. Open access http://www.nonlinearbiomedphys.com/content/pdf/1753-4631-4-7.pdf.
13. Rangayyan R.M., Nguyen T.M. 2007. Fractal analysis of contours of breast masses in mammograms. *Journal of Digital Imaging* 20(3):223–237.
14. Stepien P., Stepien R. 2010. Analysis of contours of tumor masses in mammograms by Higuchi's fractal dimension. *Biocybernetics and Biomedical Engineering* 30(4):49–56.
15. Klonowski W., Pierzchalski M., Stepien P., Stepien R., Sedivy R., Ahammer H. 2013. Application of Higuchi's fractal dimension in analysis of images of anal intraepithelial neoplasia. *Chaos, Solitons and Fractals* 48:54–60.
16. Szczypinski P.M., Strzelecki A., Materka A., Klepaczko A. 2009. *MaZda*—A software package for image texture analysis. *Computer Methods and Programs in Biomedicine* 94(1):66–76.
17. Spruyt V. 2014. About the curse of dimensionality http://www.datasciencecentral.com/profiles/blogs/about-the-curse-of-dimensionality.

Chaos in Mechanical Sciences

XI

System Augmentation for Detection and Sensing: Theory and Applications

Kiran D'Souza and
Bogdan I. Epureanu

33.1 Introduction: Inverse Problems in Nonlinear Systems

Sometimes nonlinearities present in the dynamics of systems are weak and a linearized model can be used for modeling and analysis. However, this linearized model is only valid over a specified range of parameters. Additionally, many systems have strong nonlinearities that can lead to bifurcations (qualitative changes in the dynamics) and chaotic dynamics.

The reason why linear and linearized systems are often preferred over nonlinear systems is that there are a large set of linear techniques that have already been developed in a wide array of fields, including system identification, structural health monitoring, and sensing. These techniques capitalize on the two features that linear systems must satisfy, namely, superposition and proportionality. A set of features often extracted from linear systems are their modal properties (natural frequencies, mode shapes, damping levels). A whole suite of experimental modal analysis techniques have been developed to capture these properties from a linear system [1]. These methods can be separated into time and frequency domain methods. Each of these categories can be further broken down into single input single output, single input multiple output, and multiple input multiple output (MIMO) approaches. Yang et al. [2] surveyed just the time-based MIMO methods and discussed each of their characteristics. Yang separated the approaches into free and impulse response methods such as the poly-reference complex exponential [3], eigensystem realization algorithm [4], and Ibrahim time domain [5] methods. Additionally, there are a set of forced response methods that include autoregressive moving average vector [6], direct system parameter identification (DSPI) [7], and vector backward autoregressive with exogenous model [8].

In addition to providing insight into the behavior of the dynamics of the system, the modal properties can also be used for applications such as damage detection and sensing. A review of the four general categories for linear nondestructive evaluation is given by Ibrahim [9] and Heylen [10]. The first category are eigenstructure assignment techniques, which were surveyed by Andry et al. [11]. More recently, Jiang et al. [12] used eigenstructure assignment to optimize the controllers to increase the sensitivity of closed loop frequencies to structural changes. The second category consists of optimal matrix update methods [13,14]. These methods use an optimization function with a cost function such as a minimum Frobenius norm update subject to a set of constraints (e.g., maintaining the sparsity pattern of the original system matrices). The third category is sensitivity methods, which determine changes in the system parameters based on their sensitivity to changes in modal properties of the system. Leung et al. [15] presented a new method for determining the parameter changes using sensitivity equations for asymmetric systems. The last category are the minimum rank perturbation methods [16,17]. These methods use the fact that damage often initiates in localized regions of a system, which lead to localized changes in the system matrices, which in turn lead to changes in the system matrices that are of minimum rank. In terms of sensing applications, the shift in resonant frequencies due to changes in structural parameters has been studied in both micro- and macro-scale applications. At the microscale, for instance, chemical and biological detection have been explored using microstructures such as microcantilevers [18–20] and microbeams [21]. At the macro scale, recent methods have optimally placed the resonant frequencies to enhance the sensitivity to changes in the parameters [12,22,23].

In contrast, the field of nonlinear experimental analysis is much less developed and there are many different types of nonlinearities that can cause a broad range of behaviors. Worden and Tomlinson [24] described a way to separate nonlinear analysis methods into three categories. The first category uses linear techniques to examine the nonlinear system and then detects when the behavior of the system does not match a linear system. For example, the early work in this area conducted by Ewins [1] studied the frequency response distortion due to the loss of the amplitude invariance in nonlinear systems. These approaches tend to be qualitative, indicating the presence of a nonlinearity, but they do not provide quantitative insights into the system. The second category tries to capture nonlinear effects by modifying existing linear approaches so they can work for nonlinear systems. For instance, a method based on Volterra series [25] was created by Gifford [26] to obtain the linear and nonlinear parameters of the system from higher-order frequency response functions. The third category consists of discarding the linear theory and generating new techniques to directly address the nonlinearity such as center manifold theory [27,28] and nonlinear normal modes [29–33].

System augmentation was developed so that nonlinear systems undergoing complex dynamics, including chaotic dynamics can be analyzed using linear techniques [34–37,42]. Additionally, it was used to explore the enhanced sensitivity that can be obtained from nonlinear dynamics [38–40]. The enhanced sensitivity of chaotic systems to parametric changes was inspired by the work of Yin and Epureanu [41]. The remainder of this chapter introduces the details behind system augmentation, discusses several linear techniques that have been used with augmented linear systems, explores the results of applying the method to various systems, and provides some concluding remarks.

33.2 System Augmentation

The key idea behind system augmentation is that for many nonlinear systems an augmented (higher-dimensional) system can be constructed to follow the same trajectory as the nonlinear system when the dynamics in the higher-dimensional (augmented) space are projected onto the original (physical) space. The higher dimensionality comes from defining each distinct nonlinearity as an augmented variable. Consider a one degree of freedom (DOF) nonlinear system given by

$$m\ddot{x} + d\dot{x} + kx + k_n x^5 = g(t), \tag{33.1}$$

where m, d, k, k_n, and $g(t)$ are the mass, damping, linear stiffness, nonlinear stiffness, and forcing, respectively, and the augmented variable would be defined as the nonlinearity, i.e., $y = x^5$. In general, each nonlinearity can be defined as an augmented variable and the core requirement is that the augmented variables must be linearly independent from the physical states and the other augmented variables. This requirement enables the use of system augmentation with linear methods such as modal analysis.

After defining the augmented variables of a system, the next step is to define the associated augmented equation. For instance, the augmented system for the nonlinear system described by Equation 33.1 can be described as

$$m\ddot{x} + d\dot{x} + kx + k_n y = g(t),$$
$$m_a \ddot{y} + d_a \dot{y} + k_c x + k_a y = h(t),$$

(33.2)

where m_a, d_a, k_a, and $h(t)$ are the augmented mass, damping, stiffness, and forcing and k_c is the coupled stiffness. The augmented variable y is defined as the nonlinearity and its derivatives $\dot{y}$ and $\ddot{y}$ can be computed by their relation to the state variables or by finite differencing. The parameters of the augmented equation m_a, d_a, k_c, and k_a are chosen by the user to optimally suit their needs. The augmented forcing $h(t)$ is computed to enforce that Equation 33.2 holds, which can be done since the left-hand side of the bottom relation in Equation 33.2 is completely known when x is measured. This augmented forcing is the key to the system augmentation approach because it enforces that the higher-dimensional (augmented) system will project down onto the original (physical) subspace. The augmented system behaves in a linear manner; it just has a complex augmented forcing that makes the system follow a particular trajectory.

In the general case, the nonlinear equations of motion of a structure can be written as

$$\mathbf{M\ddot{x}} + \mathbf{D\dot{x}} + \mathbf{Kx} + \mathbf{f(x, \dot{x}, \ddot{x})} = \mathbf{g}(t),$$

(33.3)

where $\mathbf{M}$, $\mathbf{D}$, and $\mathbf{K}$ are the linear mass, damping, and stiffness matrices, and $\mathbf{f(x, \dot{x}, \ddot{x})}$ contains all nonlinearities. For certain forms of nonlinearities, Equation 33.1 can be rewritten as

$$\mathbf{M\ddot{x}} + \mathbf{D\dot{x}} + \mathbf{Kx} + \mathbf{N}_I \ddot{\mathbf{y}} + \mathbf{N}_D \dot{\mathbf{y}} + \mathbf{N}_S \mathbf{y} = \mathbf{g}(t),$$
$$\mathbf{N}_{AI} \ddot{\mathbf{y}} + \mathbf{N}_{CD} \dot{\mathbf{x}} + \mathbf{N}_{AD} \dot{\mathbf{y}} + \mathbf{N}_{CS} \mathbf{x} + \mathbf{N}_{AS} \mathbf{y} = \mathbf{h}(t),$$

(33.4)

where $\mathbf{N}_I$, $\mathbf{N}_D$, and $\mathbf{N}_S$ are constant matrices that contain (physical) parameters that multiply nonlinearities (e.g., k_n from Equation 33.1), $\mathbf{N}_{AI}$, $\mathbf{N}_{AD}$, and $\mathbf{N}_{AS}$ are constant matrices that contain augmented (fictitious, user chosen) parameters (e.g., m_a, d_a, k_a from Equation 33.1) and $\mathbf{N}_{CD}$ and $\mathbf{N}_{CS}$ are constant augmented (fictitious, user chosen) matrices (e.g., k_c from Equation 33.1) that provide additional coupling between the augmented and physical states. The values chosen for all constant matrices in the augmented equations are chosen by the user to optimally suit their needs. For instance, these parameters have been optimized for sensitivity enhancement of the resonant frequencies of the augmented system [38–40] and been designed to maintain the structural physicality of the system [35].

The system augmentation approach is valid for nonlinear structural systems and also for a wide array of other nonlinear systems where the nonlinearities are associated with the measured states of the system. For example, consider the Lorenz equations given by

$$\dot{x} - \sigma(y - x) = 0,$$
$$\dot{y} - x(\rho - z) + y = 0,$$
$$\dot{z} - xy + \beta z = 0,$$

(33.5)

with states x, y, and z measured. Although this system is a first-order set of differential equations used to model atmospheric convection, the system augmentation methodology works in the same way. The augmented variables would be defined as the nonlinearities, i.e., $a_1 = xz$ and $a_2 = xy$. The augmented equations can then be described by

$$\dot{x} - \sigma(y - x) = 0,$$
$$\dot{y} - \rho x + y + a_1 = 0,$$
$$\dot{z} + \beta z - a_2 = 0,$$
$$\dot{a}_1 + c_{y1}y + k_1 a_1 = h_1(t),$$
$$\dot{a}_2 + c_{z2}z + k_2 a_2 = h_2(t),$$

(33.6)

where constants c_{y1}, c_{z2}, k_1, and k_2 are chosen constants, and $h_1(t)$ and $h_2(t)$ are calculated by using the left side of the bottom two relations in Equation 33.6. In general, there can be more coupling terms in the augmented equations to couple each augmented equation to all the other equations (i.e., c_{x1}, c_{z1}, c_{a1}, c_{x2}, c_{y2}, and c_{a2}).

The system augmentation approach works irrespective of the complexity of the dynamics of the system. Consider, for example, two cases where the parameters for the Lorenz system are given by case 1: $\sigma = 3$, $\beta = 8/3$, and $\rho = 28$, and case 2: $\sigma = 10$, $\beta = 8/3$, and $\rho = 28$. Plots of the phase space for both these systems are shown in Figure 33.1 for the initial condition of $(x, y, z) = (1, 1, 1)$. System augmentation will work the same if the system is approaching a fixed point as in Figure 33.1a or if the dynamics are chaotic Figure 33.1b.

A key step in many identification, detection, or sensing algorithms for structural systems is to extract modal information from the system. With certain adjustments [42], modal analysis techniques such as DSPI [7] and smooth orthogonal decomposition (SOD) [43] can be used to determine the eigenvalues of these augmented nonstructural systems. Selection of the parameters in the augmented equations of course has a direct impact on the eigenvalues of the augmented system. For both case 1 and 2, the parameters used were $c_{y1} = 0$, $c_{z2} = 0$, $k_1 = 5$, and $k_2 = 5$. Combining the augmented parameter information with the response data from Figure 33.1 the eigenvalues for the system can be extracted using DSPI and SOD for both cases. This information is summarized in Table 33.1. The differences in the eigenvalues of the system come from just the change in the parameter σ. Various algorithms have been developed to exploit the shifts in frequencies or other modal properties of systems to relate them to changes in system parameters; some of the results associated with system augmentation are highlighted in Section 33.4.

The reason why the system augmentation approach works is the specialized augmented forcing that enforces that the augmented system follow the path of the nonlinear system when projected onto the physical space. Plots of the augmented forcing for both cases of the Lorenz system are shown in Figure 33.2. In case 2, the complexity of the chaotic motion of the physical nonlinear system enters into the linear augmented system by the complex augmented forcing. This motion directly excites the augmented states, which excites the physical states through their coupling. In contrast, the augmented forcing of the system for case 1 is much simpler and is approaching a static state as the system approaches the fixed point.

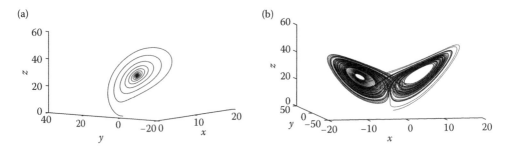

FIGURE 33.1 Plot of the phase space of the Lorenz system for (a) case 1 when the system approaches a fixed point and (b) case 2 when the system dynamics are chaotic.

TABLE 33.1 Augmented Eigenvalues of the System Computed Analytically and by Two Modal Analysis Techniques

	Eigenvalues		
	Exact	SOD	DSPI
Case 1	−2.667	−2.667	−2.665
	−5.000	−5.000	−4.992
	−5.000	−5.000	−5.001
	−7.220	−7.220	−7.220
	−11.22	−11.22	−11.22
Case 2	−2.667	−2.667	−2.660
	−5.000	−5.000	−4.992
	−5.000	−5.000	−5.013
	11.83	11.83	11.86
	−22.83	−22.83	−22.78

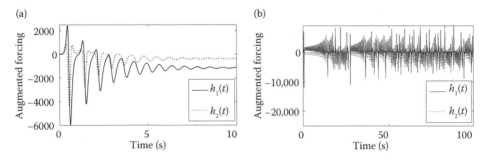

FIGURE 33.2 Plot of the augmented forcing acting on the augmentations of the Lorenz system for (a) case 1 when the system approaches a fixed point and (b) case 2 when the system dynamics are chaotic.

In either case, the method works the same and linear approaches can be used to extract features of the system such as its eigenvalues and eigenvectors.

33.3 Combining System Augmentation with Linear Methods

In this section, some of the linear analysis tools that have been used with system augmentation are discussed.

33.3.1 Modal Analysis

A useful tool for gaining insight into linear systems as well as for further analysis (i.e., sensing and damaged detection) is modal analysis. Extracting augmented modal properties from a nonlinear system requires proper measurement of the states and the use of a MIMO modal analysis approach. The authors have used both DSPI [7] and SOD [43,44] with augmented systems to extract the modal properties; however, other MIMO approaches can be used. The states that must be measured correspond to any states that are required to construct the augmented variable(s). The measured states as well as the constructed augmented variable(s) are included in the MIMO approach as an output. Forcing that is random does not need to be measured as an input into the system, but other forcing should be measured and included as an input along with the augmented forcing. The augmented forcing is directly computed from the augmented variable, the measured states, and the chosen parameters. Although DSPI and SOD were originally developed for second-order structural systems, they can also be used to extract system eigenvalues and eigenvectors of nonstructural systems such as the Lorenz system described in Section 33.2.

33.3.2 Matrix Updating

The modal information extracted from augmented systems can be combined with a variety of model updating methods. An overview of four classes of modal-based matrix updating methods is given in Section 33.1. Most of these methods could be combined with system augmentation for model updating. For example, minimum rank perturbation theory (MRPT) [16,17] has been combined with system augmentation with the main focus being on damage detection. MRPT uses the fact that damage often occurs in localized regions of a structure, which relate to localized regions in the system matrices, which in turn correspond to changes to the mass and/or stiffness matrices that are of minimum rank. MRPT had to be modified before it could be used with augmented systems. MRPT enforces that changes in a system occur symmetrically in the system matrices (of a nominally symmetric system), which is true for general linear structural systems. However, damage to augmented linear systems do not have to lead to symmetric changes in the system matrices. If a change occurs to the nonlinear parameters (in the physical equations) that damage will not be reflected in the augmented parameters (in the augmented equations) because the augmented equations are designed to be exactly satisfied for the nominal system. Furthermore, when nonstructural systems such as the Lorenz system are considered, MRPT, cannot be applied since the nominal system is not symmetric.

To overcome the limitations of MRPT, a generalized MRPT (GMRPT) was developed. This method allows for nominally asymmetric systems as well as asymmetric changes to systems. Three separate methods were developed depending on the application. The first one is for the case where the user has access to both the right and the left eigenvectors of the system [34,42]. The second method is an iterative method [35] that estimates the left eigenvectors. The approach is to invert first the mass matrix through the relation

$$\mathbf{U}_d^T \mathbf{M} \mathbf{V}_d = \mathbf{I}, \text{ so that } \mathbf{U}_d^T = \mathbf{V}_d^{-1} \mathbf{M}^{-1}, \tag{33.7}$$

where $\mathbf{U}_d$ and $\mathbf{V}_d$ correspond to the left and right eigenvectors of the damaged system. Then the model is updated. Next, the stiffness matrix is inverted through the relation

$$\mathbf{U}_d^T \mathbf{K} \mathbf{V}_d = \Lambda_d^2, \text{ so that } \mathbf{U}_d^T = \Lambda_d^2 \mathbf{V}_d^{-1} \mathbf{K}^{-1}, \tag{33.8}$$

where Λ_d are the resonant frequencies of the damaged system. Finally, the model is updated again. The process is then repeated until the process converges and the updates become negligible. Note that this iterative process is only required when there are simultaneous changes to the mass and stiffness matrices. The third method uses the fact that multiple augmentations exist for the same system. Based on this, a method was developed to update the nonlinear system first, then reformulate the augmented matrices in a symmetric fashion, and then use regular MRPT [35]. Several applications of GMRPT are discussed in Section 33.4.1.

Another linear method was developed for the case where the hot spots (areas where damage is first expected to occur) of the system are known in advance and little sensory information is available. This damage detection method is known as damage identification by hot spot projection (DIHSP). It works the same for linear and augmented systems and can be combined with a reduced order health assessment (ROHA) and sensor placement algorithm [37]. The basic idea of ROHA is that the modes most sensitive to changes in the hot spots are identified, and sensors are placed to capture these modes. The damage identification is done in a reduced space by comparing the measured results with a set of basis vectors for each of the damage scenarios. Several applications of ROHA and DIHSP are discussed in Section 33.4.1.

33.3.3 Sensitivity Enhancement

The resonant frequencies of augmented systems have also been used for sensing applications using feedback control for sensitivity enhancement. The idea behind the use of only resonant frequencies (instead of all the modal information) for sensing applications comes from the fact that resonant frequencies are more accurately measured [45] and require less sensory data than mode shapes. There are some key drawbacks however, which include the lack of sensitivity of the resonant frequencies to changes in the system properties [46,47] and the limited number of frequencies that can be accurately measured. These drawbacks can lead an underdetermined problem when solving for changes in the system [48,49]. These limitations have been addressed by several researchers using linear feedback control to enhance the sensitivity of the resonance frequencies to changes in system parameters [12,50–52]. In particular, Jiang et al. [12] optimized the placement of the eigenvalues and eigenvectors of the system using an eigenstructure assignment technique that minimized control effort while maximizing sensitivity enhancement of the resonant frequencies to parameter changes. Using system augmentation that optimization algorithm was extended for nonlinear systems (and linear systems) using nonlinear feedback auxiliary signals (NFAS) [38–40].

Consider a one DOF linear oscillator given by

$$m\ddot{x} + d\dot{x} + kx = g(t). \tag{33.9}$$

Using nonlinear feedback, the system can be modified to

$$m\ddot{x} + d\dot{x} + kx + K_{CL}x + K_{CN}y = g(t), \tag{33.10}$$

where K_{CL} is a linear control gain, K_{CN} is a nonlinear control gain, and y is a chosen nonlinear function of x. By adding the augmented (nonlinear) variable y, the system now increases its dimensionality by one, and the following augmented equation can be added to the system:

$$m_a\ddot{y} + d_a\dot{y} + k_cx + k_ay = h(t). \tag{33.11}$$

For traditional linear feedback, the sensitivity of the single resonant frequency can only be adjusted by changing K_{CL}, whereas for NFAS, $K_{CL}, K_{CN}, m_a, d_a, k_c$, and k_a can all be placed to optimize the sensitivity of the augmented frequencies to changes in the system. The optimization uses an eigenstructure assignment technique to place the eigenvalues and eigenvectors of the system such that $J(\tau)$ is minimized in

$$J(\tau) = C_1/SE + C_2CE + C_3SR, \tag{33.12}$$

where τ are the parameters being optimized, SE is the overall sensitivity enhancement that should be maximized, CE is the control effort that needs to be minimized, SR is a sensitivity reduction term, and C_1, C_2, and C_3 are weighting coefficients. Note that the SE term is an overall amount that the sensitivity of the resonant frequencies increase with respect to the open loop case for the parameters to be detected. The SR term is actually reducing the sensitivity of the system to various changes in the system which are not to be detected, such as uniform changes in mass or temperature that might arise from temperature or humidity changes. The optimization also enforces some constraints such as maintaining stability in the physical linearized system (the augmented system is allowed to become unstable, since it is fictitious), and also maintaining a linearity in the change in frequency with the change in parameter.

Overall, NFAS have significant advantages over traditional linear feedback excitations, including the need for a reduced number of physical actuator points and sensor locations due to the additional augmented states and forcing. Also, NFAS have the added benefit of interrogating the system with physically stable systems characterized with unstable augmentations that often provide added sensitivity. Additionally, the ability to set the parameters in the augmented DOFs without any physical controller effort is an added benefit of NFAS. Several applications of NFAS and traditional linear feedback for sensitivity enhancement are discussed in Section 33.4.2.

33.4 Selected Applications of System Augmentation

In this section, results obtained using system augmentation with a variety of linear techniques are detailed for several systems. For all the systems analyzed, a nominal nonlinear system was considered known. One (or multiple) augmented systems were generated for each of these systems by associating each nonlinearity with an augmented variable. Then, the nonlinear parameters were identified and included in the appropriate nonlinear constant matrices ($\mathbf{N}_I$, $\mathbf{N}_D$, $\mathbf{N}_S$ from Equation 33.4). The augmented matrices ($\mathbf{N}_{AI}$, $\mathbf{N}_{AD}$, $\mathbf{N}_{AS}$, $\mathbf{N}_{CD}$, $\mathbf{N}_{CS}$ from Equation 33.4) were chosen to either maintain the symmetry and keep a uniform scale in the augmented system or to be optimized for sensitivity enhancement of the resonant frequencies of the system. The system augmentation approach was then coupled with various techniques to detect or sense changes in the nominal system by using knowledge of the nominal system and measurements from the damaged/altered system.

33.4.1 Damage Detection and System Identification

A complete set of results using system augmentation for damage detection and/or system identification can be found in the literature [34–37,42]. Consider the systems shown in Figure 33.3 [36]. Figure 33.3a is a 3-bay frame structure that consists of 44 steel beams connected at 16 nodes. Two of the nodes are pinned to the ground and therefore allow no translation in any direction, while two other nodes are connected to the ground with a slider joint that allows translation in one direction. Figure 33.3b is based on the frame structure shown in Figure 33.3a, but includes 12 cubic springs connecting nodes to other nodes. Also, Coulomb friction acts to oppose translation at both slider joints.

Results for a case where the Coulomb friction coefficient is increased by 50% and 100% at points A and B, cubic springs are decreased in stiffness by 30%, 40%, and 35% at points E, F, and G, and linear beams are reduced in stiffness by 20% and 15% at points E and H, respectively, are summarized in Figure 33.4 [36]. For these results, the multiple augmentations GMRPT approach [35,36] was used to detect the changes in both the linear and nonlinear system parameters using the first 10 out of the 96 modes (the complete mode was considered measured). First, the nonlinear parameter changes were identified using 15 distinct augmentations for the system by varying the entries in the augmented matrices ($\mathbf{N}_{AI}$, $\mathbf{N}_{AD}$, $\mathbf{N}_{AS}$, $\mathbf{N}_{CD}$, $\mathbf{N}_{CS}$). The system was then updated, and a symmetric augmentation was formulated. Finally, the changes

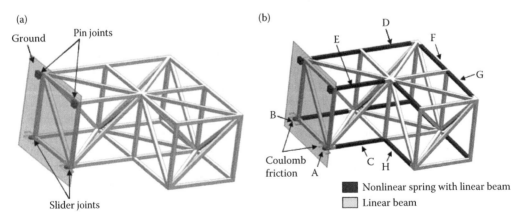

FIGURE 33.3 (a) Schematic representation of a 3-bay frame connected to ground by two pins and two roller joints and (b) nonlinear 3-bay frame with Coulomb friction at the two slider joints and 12 elements with cubic stiffness. (Adapted from K. D'Souza and B. I. Epureanu. Nonlinear model updating based on system augmentation for nonlinear damage detection. In *Proceedings of the Third European Workshop on Structural Health Monitoring*, pages 515–522, Granada, Spain, July 2006.)

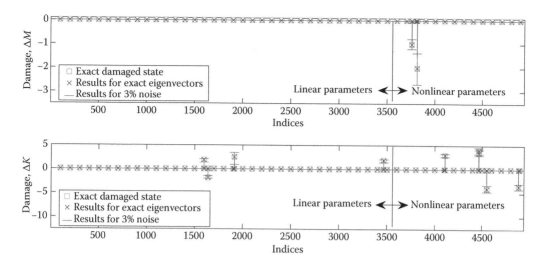

FIGURE 33.4 Predicted damage in the nonlinear 3-bay frame with damage associated with an increase in Coulomb friction at both locations (upper plot), and a reduction of stiffness in three cubic springs and two linear springs (lower plot). (Adapted from K. D'Souza and B. I. Epureanu. Nonlinear model updating based on system augmentation for nonlinear damage detection. In *Proceedings of the Third European Workshop on Structural Health Monitoring*, pages 515–522, Granada, Spain, July 2006.)

in the linear parameters were detected using linear MRPT. The top plot in Figure 33.4 shows element by element the values of the mass perturbation matrix $\Delta\mathbf{M}$, while the bottom plot in Figure 33.4 shows the stiffness perturbation matrix $\Delta\mathbf{K}$. Note that the Coulomb friction nonlinearities were incorporated into the augmented mass matrix so that the discontinuous nonlinearity could be integrated in the augmented equations (as opposed to differentiated like the cubic stiffness nonlinearities contained in the augmented stiffness matrix). The exact results are shown along with the case of having the exact eigenvectors and the case where 3% random noise is added to the eigenvectors. For the case with noisy data, 100 separate calculations were performed, and average and standard deviation error bars are plotted. Figure 33.4 shows that the changes can be identified exactly for the noise-free case and remains very accurate for as much as 3% measurement noise.

To consider the case where there are only a few measurement locations, a ROHA method was developed to place the sensors and expand the partial eigenvector information to the full space [37]. This method was developed to work with DIHSP to determine the location and magnitude of the damages in a system when information about the hot spots are known. The method is implemented the same on both linear and augmented linear systems. Figure 33.5 [37] shows a linear 5-bay frame structure with 2 plates, which are pinned to the ground at their perimeter. Hence, they exhibit stretching induced by bending. Using a one-mode Galerkin approximation for each plate, a linear and cubic stiffness is introduced to the frame structure at the point of attachment. The frame structure consists of 70 steel beams connected at 24 nodes, 4 of which are pinned to the ground. The system has 22 hot spots, one corresponding to each of the transverse (bending) stiffnesses of the 20 horizontal beams, and one for each plate. Each beam was separated into 5 beam elements, except for the horizontal beams. The nonlinear system had 1332 DOFs and the augmented linear system had 1334 DOFs. There were 20 sensors measuring the displacement of the 20 longitudinal beams.

Figure 33.6 shows the results of the method being applied to two cases. The *x*-axis in each plot represents the 22 damage scenarios (i.e., the 20 transverse stiffnesses of the 20 longitudinal beams and the 2 plates). The *y*-axis in each plot represents the percent damage for each scenario. Figure 33.6a shows results for a case with a 15% loss of stiffness in plate *A*, and 20% loss in plate *B* (see Figure 33.5). Standard deviation error bars are plotted for 100 separate numerical simulations, in which 5% random eigenvector noise was

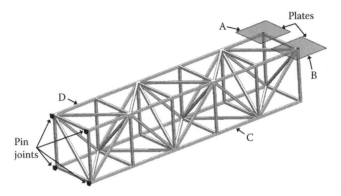

FIGURE 33.5 A linear 5-bay structure with two plates which introduce cubic stiffness nonlinearities. (Adapted from K. D'Souza and B. I. Epureanu. *AIAA Journal*, 46(10):2434–2442, 2008.)

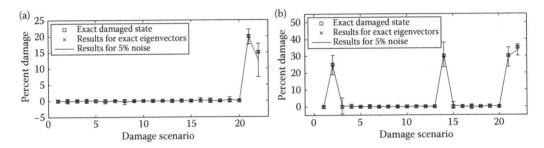

FIGURE 33.6 Predicted damage in the nonlinear 5-bay frame with (a) damage in both cubic stiffnesses and (b) damage in both cubic stiffnesses and two linear hot spots, with 5% random eigenvector noise using 20 sensors. (Adapted from K. D'Souza and B. I. Epureanu. *AIAA Journal*, 46(10):2434–2442, 2008.)

added. When there is zero noise, exact damage is predicted. For 5% noise, the actual damage is predicted very accurately. Figure 33.6b shows the results for a case with a 35% loss of stiffness in plate *A*, 30% loss in plate *B*, 30% loss in beam *C*, and 25% loss in beam *D* (see Figure 33.5). Standard deviation error bars are plotted for 100 separate numerical simulations, in which 5% random eigenvector noise was added. When there is zero noise, exact damage is predicted. For 5% noise, the actual damage is predicted very accurately, with little false damage predicted in other scenarios.

33.4.2 Sensitivity Enhancement and Reduction

A complete set of results of using system augmentation with NFAS for sensitivity enhancement and/or reduction can be found in the literature [38–40]. Consider the linear six DOF system shown in Figure 33.7 [39]. There are controller input actuators at the first and sixth masses. These actuators can change the linear stiffness at their respective locations as well as include a nonlinear feedback excitation. These NFAS were optimized to minimize the control effort while maximizing the sensitivity of the augmented system's resonant frequencies to changes in parameters of the system (in this case the six linear springs). The nonlinearity used to excite the 1st and 6th masses were cubic stiffnesses. Three separate controllers were designed to maximize the sensitivity of the first three resonant frequencies to changes in all six parameters using both NFAS and linear control (without system augmentation).

An interesting property of augmented linear systems was observed with relation to NFAS. Namely, the optimized augmented modal properties of the system are often unstable, although the physical nonlinear system is constrained to be stable in the optimization algorithm. An example of this case is provided in

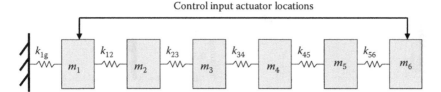

FIGURE 33.7 Linear mass-spring system that has nonlinear control applied at two locations with parameters $m_i = 1$, $i = 1, 2, \ldots, 6$, $k_{1g} = k_{34} = 10^4$, $k_{12} = k_{23} = 1.5 \cdot 10^4$, $k_{45} = 2 \cdot 10^4$, and $k_{56} = 2.5 \cdot 10^4$. (Adapted from K. D'Souza and B. I. Epureanu. *ASME Journal of Vibration and Acoustics*, 132(2):1–9, 2010.)

TABLE 33.2 Eigenvalues of a Physically Stable System with an Unstable Augmented System

Eigenvalues of Physical Linearized System	Exact Eigenvalues of Augmented System	Eigenvalues of Augmented System Computed by DSPI
	0.5042 + 12.9049i	0.4862 + 12.9045i
	0.5042 − 12.9049i	0.4862 − 12.9045i
5.6881i	54.5368i	54.5368i
75.8141i	82.97i	82.9699i
144.43i	145.7084i	145.7087i
179.3784i	179.779i	179.779i
222.2159i	222.3982i	222.3981i
264.0378i	264.0428i	264.0428i

Source: Adapted from K. D'Souza and B. I. Epureanu. *ASME Journal of Vibration and Acoustics*, 132(2):1–9, 2010.

Table 33.2 [39]. The first column in this table has the eigenvalues of the physical linearized system, which are all imaginary numbers. The second column are the exact augmented eigenvalues of this system. Note that there are two additional eigenvalues (since the augmented system is higher dimensional than the physical system), and that two of the eigenvalues have positive real parts, which means the augmented system is unstable.

The fact that the augmented system is unstable is not an issue in this case since DSPI can extract the augmented modes of the system as shown in the third column of Table 33.2. The timeseries response used to extract these modal properties is shown in Figure 33.8 [39]. The reason why a physically stable

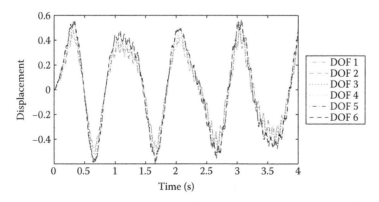

FIGURE 33.8 Response of the six physical DOFs of a physically stable system with an unstable augmentation. (Adapted from K. D'Souza and B. I. Epureanu. *ASME Journal of Vibration and Acoustics*, 132(2):1–9, 2010.)

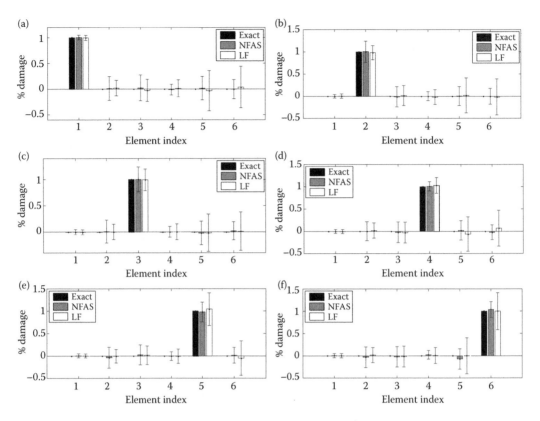

FIGURE 33.9 Detection results for resonant frequencies contaminated with ±0.25 random noise for LF and NFAS for (a) 1st, (b) 2nd, (c) 3rd, (d) 4th, (e) 5th, and (f) 6th elements. (Adapted from K. D'Souza and B. I. Epureanu. *ASME Journal of Vibration and Acoustics*, 132(2):1–9, 2010.)

system can have unstable augmented systems is the specialized augmented forcing. For this type of system, the augmented forcing ensures the response of the system remains bounded even though the augmented system is unstable.

A comparison of the results when using traditional linear feedback and NFAS, where all the six positions are considered known is shown in Figure 33.9 [39]. The x-axis in each plot corresponds to the six different elements that can be damaged, while the y-axis in each plot corresponds to the percent damage. Both methods are shown to be able to detect small changes in the stiffness with noise levels of ±0.25 (approximately 1% of the lowest natural frequency) added to the frequencies of the damaged system. For these noisy cases, 100 separate numerical simulations were run and average and standard deviation error bars are plotted. NFAS have significantly smaller error bars than linear feedback for elements 4, 5, and 6, while linear feedback slightly outperforms NFAS for elements 2 and 3. It should be noted that in addition to NFAS generally outperforming linear feedback in detection, they also have substantially less control effort (4–10 times less).

To show some of the additional advantages of NFAS over traditional linear feedback, the system shown in Figure 33.10 [40] was explored. The beam was discretized into 28 elements with two DOFs per node, so the linear beam had 56 DOFs. The beam was actuated by piezoelectric patches placed in a bimorph configuration. The excitation of the system was modeled as an induced moment on the system due to an applied voltage. Two piezo-sensors were used to measure the response of the system. The sensor information corresponds to voltages that relate to the curvature of the beam that can in turn relate to the rate of change in the slope of the beam at the sensor. This sensory information was fed back into the system to

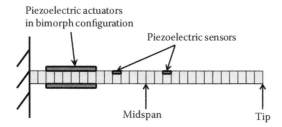

FIGURE 33.10 Linear beam excited by piezo-actuators using nonlinear feedback auxiliary signals and two piezo-sensors. (Adapted from K. D'Souza and B. I. Epureanu. *Journal of Sound and Vibration*, 32913:2463–2476, 2010.)

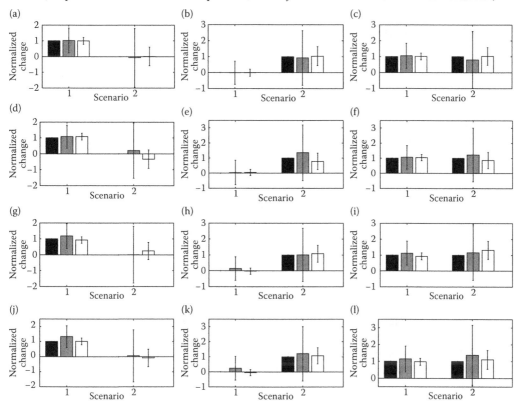

FIGURE 33.11 Sensed mass: (i) by the open loop system (gray), (ii) by a closed loop system designed to be *insensitive* to uniform mass and stiffness changes using two sets of nonlinear feedback auxiliary signals (white), and (iii) the exact changes (black). Scenario 1 represents changes in mass at the tip and Scenario 2 represents changes at the midspan. (a) Scenario 1; (b) Scenario 2; (c) Scenarios 1 and 2; (d) Scenario 1 with uniform change in mass; (e) Scenario 2 with uniform change in mass; (f) Scenarios 1 and 2 with uniform change in mass; (g) Scenario 1 with uniform change in stiffness; (h) Scenario 2 with uniform change in stiffness; (i) Scenarios 1 and 2 with uniform change in stiffness; (j) Scenario 1 with uniform change in mass and stiffness; (k) Scenario 2 with uniform change in mass and stiffness; and (l) Scenarios 1 and 2 with uniform change in mass and stiffness. (Adapted from K. D'Souza and B. I. Epureanu. *Journal of Sound and Vibration*, 32913:2463–2476, 2010.)

apply the NFAS. Four states of the system were known, two corresponding to physical measurements and two corresponding to augmented states of the system.

A set of results were obtained for the case where the NFAS were set up to be sensitive to small changes in the mass (added mass of 0.1% of the beam) at the tip and/or midspan, and insensitive to small global changes such as uniform changes in the mass and/or stiffness of the system, which could occur due to changes in temperature or humidity. The results of the exact changes, the changes predicted for the case where no controller is used, and the use of NFAS are summarized in Figure 33.11 [40]. Note that the linear feedback case was not used in this example; owing to the limited actuation and control effort, a well-conditioned sensitivity matrix could not be obtained for traditional linear feedback. The exact frequencies were calculated for the system and a random noise approximately $\pm 0.5\%$ of the lowest resonant frequency was added to the data. For the noisy case, 100 simulations are plotted and average and standard deviation error bars are plotted. These results show the improvements of NFAS over the uncontrolled system and demonstrate how these signals can be used to enhance sensitivity to the desired parameters while reducing the sensitivity of the resonant frequencies to environmental or operational conditions.

33.5 Conclusions

The system augmentation approach enables the use of a variety of linear tools to be used with nonlinear systems. These systems can be strongly nonlinear whereby the dynamics can include a variety of nonlinear phenomena such as chaos and bifurcations. The critical requirements are that the functional form must be known or identifiable, and the nonlinearity needs to be related to some of the measured states of the system. These augmented systems are higher dimensional than the corresponding linear systems with each distinct nonlinearity corresponding to an additional augmented equation. The augmented equation is defined by the user (and hence it is completely known), and it has a specialized augmented forcing that enforces the system follow a single trajectory of the nonlinear system if the dynamics are projected onto the physical (lower-dimensional) space. In addition to using linear tools for nonlinear systems, system augmentation enables the use of NFAS that can enhance the sensitivity and selectivity of the system to certain changes. These signals have been used with linear and nonlinear systems to enhance the sensitivity of the system.

References

1. D. J. Ewins. *Modal Testing: Theory and Practice*. Research Studies Press, Taunton, 1984.
2. Q. J. Yang, P. Q. Zhang, C. Q. Li, and X. P. Wu. A system theory approach to multi-input multi-output modal parameters identification methods. *Mechanical Systems and Signal Processing*, 8(2):159–174, 1994.
3. H. Vold and G. T. Rocklin. The numerical implementation of a multi-input modal estimation for mini-computers. In *Proceedings of the 1st International Modal Analysis Conference*, pages 542–548, Orlando, Florida, 1982.
4. J. N. Juang and R. S. Pappa. An eigensystem realisation algorithm for modal parameter identification and model reduction. *Journal of Guidance, Control, and Dynamics*, 8(5):620–627, 1985.
5. W. Shong and P. Q. Zhang. MIMO ITD identifying technique for mini-computer. *Journal of University of Science and Technology of China*, 18(2):195–202, 1988.
6. P. Q. Zhang, C. Q. Li, and Q. J. Yang. Identification of structural modal parameters by ARMAV model method. *Journal of Experimental Mechanics*, 4(1):137–145, 1989.
7. J. Leuridan. *Some Direct Parameter Model Identification Methods Applicable for Multiple Modal Analysis*, PhD thesis. Department of Mechanical and Industrial Engineering, University of Cincinnati, 1984.
8. C. Hung, W. Ko, and Y. Peng. Identification of modal parameters from measured input and output data using a vector backward auto-regressive with exogenous model. *Journal of Sound and Vibration*, 276(3-5):1043–1063, 2004.
9. S. R. Ibrahim and A. A. Saafan. Correlation of analysis in modeling and structures, assessment and review. In *Proceedings of the 5th IMAC*, volume 2, pages 1651–1660, London, England, 1987.
10. W. Heylen and P. Sas. Review of model optimization techniques. In *Proceedings of the 5th IMAC*, volume 2, pages 1172–1182, London, England, 1987.

11. A. N. Andry, E. Y. Shapiro, and J. C. Chung. Eigenstructure assignment for linear-systems. *IEEE Transactions on Aerospace and Electronic Systems*, 19(5):711–729, 1983.

12. L. J. Jiang, J. Tang, and K. W. Wang. An optimal sensitivity-enhancing feedback control approach via eigenstructure assignment for structural damage identification. *ASME Journal of Vibration and Acoustics*, 129(6):771–783, 2007.

13. P. L. Liu. Identification and damage detection of trusses using modal data. *Journal of Structural Engineering*, 121(4):599–608, 1995.

14. H. S. Kim and Y. S. Chun. Structural damage assessment of building structures using dynamic experimental data. *Structural Design of Tall and Special Buildings*, 13(1):1–8, 2004.

15. A. Y. T. Leung, L. F. Chen, and W. L. Wang. A linearized procedure for solving inverse sensitivity equations of non-defective systems. *Journal of Sound and Vibration*, 259(3):513–524, 2003.

16. D. C. Zimmerman and M. Kaouk. Structural damage detection using minimum rank update theory. *ASME Journal of Vibration and Acoustics*, 116(2):222–231, 1994.

17. M. Kaouk, D. C. Zimmerman, and T. W. Simmermacher. Assessment of damage affecting all structural properties using experimental modal parameters. *ASME Journal of Vibration and Acoustics*, 122(4):456–463, 2000.

18. T. Thundat, E. A. Wachter, S. L. Sharp, and R. L. Warmack. Detection of mercury-vapor using resonating microcantilevers. *Applied Physics Letters*, 66(13), 1995.

19. G. Y. Chen, T. Thundat, E. A. Wachter, and R. J. Warmack. Adsorption-induced surface stress and its effects on resonance frequency of microcantilevers. *Journal of Applied Physics*, 77(8), 1995.

20. B. Ilic, D. Czaplewski, M. Zalalutdinov, H. G. Craighead, P. Neuzil, C. Campagnolo, and C. Batt. Single cell detection with micromechanical oscillators. *Journal of Vacuum Science and Technology*, 19(6):2825–2828, 2001.

21. D. DeVoe. Piezoelectric thin film micromechanical beam resonators. *Sensors and Actuators A*, 88(2):263–272, 2001.

22. Z. Wang, F. Au and Y. Cheng. Statistical damage detection based on frequencies of sensitivity-enhanced structures. *International Journal of Structural Stability and Dynamics*, 8(2):231–255, 2008.

23. L. J. Jiang and K. W. Wang. An experiment-based frequency sensitivity enhancing control approach for structural damage detection. *Smart Materials and Structures*, 18(6):1–12, 2009.

24. K. Worden and G. R. Tomlinson. Nonlinearity in experimental modal analysis. *Philosophical Transactions of the Royal Society of London: A—Mathematical, Physical, and Engineering Sciences*, 359(1778):113–130, 2001.

25. D. Mirri, G. Iuculano, F. Filicori, G. Pasini, G. Vannini, and G. P. Gualtieri. A modified Volterra series approach for nonlinear dynamic systems modeling. *IEEE Transactions on Circuits and Systems I—Fundamental Theory and Applications*, 49(8):1118–1128, 2002.

26. S. J. Gifford. *Volterra Series Analysis of Nonlinear Structures,* PhD thesis. Department of Mechanical Engineering, Heriot-Watt University, 1989.

27. W. Steiner, A Steindl, and H. Troger. Center manifold approach to the control of a tethered satellite system. *Applied Mathematics and Computation*, 70(2-3):315–327, 1995.

28. M. Q. Xiao and T. Basar. Center manifold of the viscous Moore-Greitzer PDE model. *SIAM Journal on Applied Mathematics*, 61(3):855–869, 2000.

29. S. W. Shaw and C. Pierre. Normal modes for non-linear vibratory systems. *Journal of Sound and Vibration*, 164(1):85–124, 1993.

30. S. J. Levinson. A padé approach for eigenvalue identification in underwater acoustic normal mode computations. *Journal of the Acoustical Society of America*, 99(2):831–835, 1996.

31. X. Ma, M. F. A. Azeez, and A. F. Vakakis. Non-linear normal modes and non-parametric system identification of non-linear oscillators. *Mechanical Systems and Signal Processing*, 14(1):37–48, 2000.

32. E. Pesheck, C. Pierre, and S. W. Shaw. Accurate reduced-order models for a simple rotor blade model using nonlinear normal modes. *Mathematical and Computer Modelling*, 33(10-11):1085–1097, 2001.

33. F. X. Wang, A. K. Bajaj, and K. Kamiya. Nonlinear normal modes and their bifurcations for an inertially coupled nonlinear conservative system. *Nonlinear Dynamics*, 42(3):233–265, 2005.

34. K. D'Souza and B. I. Epureanu. Damage detection in nonlinear systems using system augmentation and generalized minimum rank perturbation theory. *Smart Materials and Structures*, 14(5):989–1000, 2005.

35. K. D'Souza and B. I. Epureanu. Multiple augmentations of nonlinear systems and generalized minimum rank perturbations for damage detection. *Journal of Sound and Vibration*, 316(1-5):101–121, 2008.

36. K. D'Souza and B. I. Epureanu. Nonlinear model updating based on system augmentation for nonlinear damage detection. In *Proceedings of the Third European Workshop on Structural Health Monitoring*, pages 515–522, Granada, Spain, July 2006.

37. K. D'Souza and B. I. Epureanu. Sensor placement for damage detection in nonlinear systems using system augmentations. *AIAA Journal*, 46(10):2434–2442, 2008.

38. K. D'Souza and B. I. Epureanu. Nonlinear feedback auxiliary signals for system interrogation and damage detection. *Philosophical Transactions of the Royal Society of London: A—Mathematical, Physical and Engineering Sciences*, 464(2100):3129–3148, 2008.

39. K. D'Souza and B. I. Epureanu. Damage detection in nonlinear systems using optimal feedback auxiliary signals. *ASME Journal of Vibration and Acoustics*, 132(2):1–9, 2010.

40. K. D'Souza and B. I. Epureanu. Detection of global and local parameter variations using nonlinear feedback auxiliary signals and system augmentation. *Journal of Sound and Vibration*, 32913:2463–2476, 2010.

41. S. Yin and B. I. Epureanu. Structural health monitoring based on sensitivity vector fields and attractor morphing. *Philosophical Transactions of the Royal Society of London: A—Mathematical, Physical and Engineering Sciences*, 364:2515–2538, 2006.

42. A. Sloboda, K. D'Souza, and B. I. Epureanu. Identifying parameter variations in nonlinear systems using system augmentation and modal analysis. *Journal of Sound and Vibration*, under review, 2015.

43. U. Farooq and B. F. Feeny. Smooth orthogonal decomposition for modal analysis of randomly excited systems. *Journal of Sound and Vibration*, 316:137–146, 2008.

44. K. D'Souza and B. I. Epureanu. Noise rejection for two time-based multi-output modal analysis techniques. *Journal of Sound and Vibration*, 3306:1045–1051, 2010.

45. E. Dascotte. Practical application of finite element tuning using experimental data. In *Proceeding of the 8th International Modal Analysis Conference*, pages 1032–1037, Orlando, Florida, 1990.

46. A. S. J. Swamidas and Y. Chen. Monitoring crack growth through change of modal parameters. *Journal of Sound and Vibration*, 186(2):325–343, 1995.

47. R. D. Adams, P. Cawley, C. J. Pye, and B. J. Stone. A vibrational technique for non-destructively assessing the integrity of structures. *Journal of Mechanical Engineering Science*, 20(2):93–100, 1978.

48. N. Stubbs and R. Osegueda. Global non-destructive damage evaluation in solids. *Modal Analysis: The International Journal of Analytical and Experimental Modal Analysis*, 5(2):67–79, 1990.

49. N. Stubbs and R. Osegueda. Global damage detection in solids-experimental verification. *Modal Analysis: The International Journal of Analytical and Experimental Modal Analysis*, 5(2):81–97, 1990.

50. L. R. Ray and S. Marini. Optimization of control laws for damage detection in smart structures. In *SPIE Symposium on Mathematics and Control in Smart Structures*, volume 3984, pages 395–402, Newport Beach, California, March 2000, SPIE.

51. J. N. Juang, K. B. Lim, and J. L. Junkins. Robust eigensystem assignment for flexible structures. *Journal of Guidance, Control, and Dynamics*, 12(3):381–387, 1989.

52. B. H. Koh and L. R. Ray. Feedback controller design for sensitivity-based damage localization. *Journal of Sound and Vibration*, 273(1-2):317–335, 2004.

34

Unveiling Complexity of Church Bells Dynamics Using Experimentally Validated Hybrid Dynamical Model

Piotr Brzeski,
Tomasz Kapitaniak, and
Przemyslaw Perlikowski

34.1 Introduction

Bells are musical instruments that are closely connected with European cultural heritage. Although the design of a bell, its clapper, and a belfry has been developed for centuries, mathematical modeling of their behavior has been encountered recently. It is surprising from an engineering point of view because bells and their supports are structures that are exposed to severe loading conditions during ringing. To ensure that they will work reliably for ages, we have to consider many factors and one of the most important is to ensure safe and effective operation. Moreover, the dynamics of a yoke–bell–clapper system is extremely complex and difficult to analyze due to its nonlinear characteristic, repetitive impacts, and complicated excitation.

The first attempt to describe a bell's behavior using the equation of motion was made by Veltmann in the nineteenth century [18,19]. His work was stimulated by the failure of the famous Emperor's bell in the Cologne Cathedral when the clapper remained always on the middle axis of the bell instead of striking it. Veltmann explained the reason of this phenomena using a simple model developed based on the equations of a double physical pendulum. Heyman and Threlfrall [8] used a similar model to estimate inertia forces induced by a swinging bell. The knowledge of loads induced by ringing bells depends on the mounting layout and is crucial during the design and restoration processes of belfries. In Europe, one can distinguish three different types of swinging bells: Central European, English, and Spanish. In the first of them, bells tilt on their axis and maximum amplitude of oscillation is usually below 90°. In the English system, bells perform nearly a complete rotation (the bell stops close to the upper position), while in the Spanish system, bells rotate continuously in the same direction.

Muller [15] and Steiner [17] analyzed the dynamic interactions between bells and bell towers and described the dynamic forces appearing in bells mounted in the Central European manner. There are also similar studies concerning the English system [9,10] and the Spanish system [3,4]. Ivorra et al. [6,7] showed how the mounting layout affects the dynamic forces induced by bells. The authors proved that in the Spanish system forces transmitted to the supporting structure are significantly lower than in the English and Central European systems. Nevertheless, studies described in References 5 and 13 prove that often a minor modification in the support's design can significantly decrease the probability of damage to the bell tower.

Klemenc et al. contributed with a series of papers [11,12] devoted to the analysis of the clapper-to-bell impacts. The presented results prove that full-scale finite-element model is able to reproduce the effect of collisions but requires long computational times and complex, detailed models. Therefore, it would be difficult to use such models to analyze the dynamics of bells. Because of that, we recently observed the tendency to use hybrid dynamical models that are much simpler and give accurate results with less modeling and computational effort. In Reference 14, the authors propose a lumped parameter model of the bells mounted in the Central European system and prove that with the model we are able to predict the impact acceleration and the bell's period of motion.

This chapter is organized as follows. In Section 34.2, we describe the hybrid dynamical model of the church bell and validate it by comparing the results of numerical simulations with data obtained experimentally. In Section 34.3, we characterize the seven most common working regimes and Section 34.4 investigates how they can be obtained. Finally, in Section 34.5, the conclusions are drawn.

34.2 Model of the System

The model that is presented in this chapter is built based on the analogy between a freely swinging bell and the motion of the equivalent double physical pendulum. The first pendulum has a fixed axis of rotation and models the yoke and the bell that is mounted on it. The clapper is imitated by the second pendulum attached to the first one. The proposed model involves eight physical parameters. The photo of the bell that has been measured to obtain the parameters values is presented in Figure 34.1a–c show the schematic model of the bell indicating the position of the rotation axes of the bell, the clapper, and presenting parameters involved in the model. For simplicity, henceforth, we use the term "bell" with respect to the combination of the bell and its yoke, which we treat as one solid element.

Parameter L describes the distance between the rotation axis of the bell and its center of gravity (point C_b), and l is the distance between the rotation axis of the clapper and its center of gravity (point C_c).

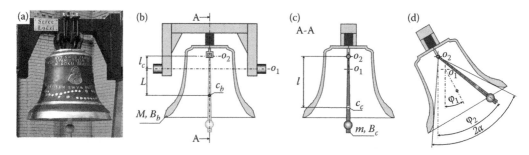

FIGURE 34.1 *The Heart of Lodz*, the biggest bell in the Cathedral Basilica of St. Stanislaus Kostka (a) and its schematic model (b), (c), and (d) along with physical and geometrical quantities involved in the mathematical model of the system.

The distance between the bell's and the clapper's axes of rotation is given by parameter l_c. The mass of the bell is described by parameter M, while parameter B_b characterizes the bell's moment of inertia referred to it axis of rotation. Similarly, parameter m describes the mass of the clapper and B_c stands for the clapper's moment of inertia referred to it axis of rotation.

The considered model has two degrees of freedom. Figure 34.1d presents two generalized coordinates that we use to describe the state of the system: the angle between the bell's axis and the downward vertical is given by φ_1 and the angle between the clapper's axis and downward vertical by φ_2. Parameter α (see Figure 34.1d) is used to describe the clapper to the bell impact condition, which is as follows:

$$|\varphi_1 - \varphi_2| = \alpha \tag{34.1}$$

We use Lagrange equations of the second type and derive two coupled second-order ODEs that describe the motion of the considered system (full derivation can be found in our previous publication [2]):

$$\left(B_b + ml_c^2\right)\ddot{\varphi}_1 + ml_c l\ddot{\varphi}_2 \cos\left(\varphi_2 - \varphi_1\right) - ml_c l\dot{\varphi}_2^2 \sin\left(\varphi_2 - \varphi_1\right)$$
$$+ \left(ML + ml_c\right) g \sin\varphi_1 + D_b\dot{\varphi}_1 - D_c\left(\dot{\varphi}_2 - \dot{\varphi}_1\right) = M_t(\varphi_1) \tag{34.2}$$

$$B_c\ddot{\varphi}_2 + ml_c l\ddot{\varphi}_1 \cos\left(\varphi_2 - \varphi_1\right) + ml_c l\dot{\varphi}_1^2 \sin\left(\varphi_2 - \varphi_1\right)$$
$$+ mgl \sin\varphi_2 + D_c\left(\dot{\varphi}_2 - \dot{\varphi}_1\right) = 0 \tag{34.3}$$

where g stands for gravity and $M_t(\varphi_1)$ describes the effects of the linear motor propulsion. The motor is active—and excites the bell—when its deflection from the vertical position is smaller than $\pi/15$ rad (12°). The generalized momentum generated by the motor $M_t(\varphi_1)$ is given by the piecewise formula:

$$M_t(\varphi_1) = \begin{cases} T\,\text{sgn}(\dot{\varphi}_1)\,\cos\left(7.5\varphi_1\right), & \text{if } |\varphi_1| \leq \dfrac{\pi}{15} \\[2mm] 0, & \text{if } |\varphi_1| > \dfrac{\pi}{15} \end{cases} \tag{34.4}$$

where T is the maximum achieved torque. Although the above expression is not an accurate description of the effects generated by the linear motor, in Reference 2 we prove that it is able to reproduce the characteristics of modern bells' propulsions.

There are 11 parameters involved in the mathematical model presented above. The parameters have the following values: $M = 2633\,\text{kg}$, $m = 57.4\,\text{kg}$, $B_b = 1375\,\text{kgm}^2$, $B_c = 45.15\,\text{kgm}^2$, $L = 0.236\,\text{m}$, $l = 0.739\,\text{m}$, $l_c = -0.1\,\text{m}$ and $\alpha = 30.65° = 0.5349\,\text{rad}$, $D_c = 4.539\,\text{Nms}$, $D_b = 26.68\,\text{Nms}$, $T = 229.6\,\text{Nm}$. As aforementioned, all parameters values have been evaluated specifically for the purpose. For integration of the model described above, we use the fourth–order Runge–Kutta method.

When the condition 34.1 is fulfilled, we stop the integration process. Then, instead of analyzing the collision course, we restart simulation updating the initial conditions of Equations 34.2 and 34.3 by switching the bell's and the clapper's angular velocities from the values before the impact to the ones after the impact. The angular velocities after the impact are obtained taking into account the energy dissipation and the conservation of the system's angular momentum that are expressed by the following formulas:

$$\frac{1}{2}B_c\left(\dot{\varphi}_{2,AI} - \dot{\varphi}_{1,AI}\right)^2 = k\frac{1}{2}B_c\left(\dot{\varphi}_{2,BI} - \dot{\varphi}_{1,BI}\right)^2 \tag{34.5}$$

$$\left[B_b + ml_c^2 + ml_c l \cos\left(\varphi_2 - \varphi_1\right)\right]\dot{\varphi}_{1,BI} + \left[B_c + ml_c l \cos\left(\varphi_2 - \varphi_1\right)\right]\dot{\varphi}_{2,BI}$$
$$= \left[B_b + ml_c^2 + ml_c l \cos\left(\varphi_2 - \varphi_1\right)\right]\dot{\varphi}_{1,AI} + \left[B_c + ml_c l \cos\left(\varphi_2 - \varphi_1\right)\right]\dot{\varphi}_{2,AI} \tag{34.6}$$

where index *AI* stands for "after impact," index *BI* for "before impact" and parameter k is the coefficient of energy restitution. In our simulations, we assume $k = 0.05$ referring to a series of experiments performed by Rupp et al. [16]. In our previous investigation [2], we have analyzed the influence of k on the response of the system and we have proved that system's dynamics barely changes for small alterations of parameter k (±20%). Hence, we claim that there is no need to further adjust the value of k for the considered bell. ODEs 34.2 and 34.3 together with the discreet model of impact create a hybrid dynamical system that can be used to simulate the behavior of church bells.

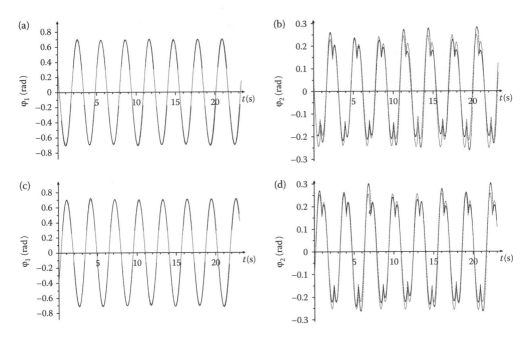

FIGURE 34.2 Comparison between time traces of the bell (a,c) and the clapper (b,c) obtained numerically (gray lines) and experimentally (black dots). Subplots (a,b) correspond to the data obtained from the first recording and (c,d) from the second one.

34.2.1 Validation of the Model

In this section, we investigate the full model and validate it focusing on normal ringing conditions. Our study is based on the bell in the Cathedral Basilica of St. Stanislaus Kostka and we check the reliability of the model by comparing the results of numerical simulations with experimentally obtained time traces of *The Heart of Lodz*.

We launched the linear motor propulsion and after a start-up procedure, when the amplitude of the bell's motion stabilized, we began recording with a high-speed camera (Basler piA640-210gm). We used black-and-white stickers to mark the position of the bell and the clapper and indicate reference length. We used Kinovea software to create—based on the recordings—the data-sheets with markers' abscissa and ordinate depending on time and processed the data in Mathematica software.

In Figure 34.2, we compare the results of numerical simulations and time traces obtained from two separate recordings. In subplots (a,b), we present data collected from the first recording and in subplots (c,d) from the second one. Gray lines in Figure 34.2 represent the periodic attractor obtained numerically and black dots correspond to experimental results. In subplots (a) and (c), we show time traces of the bell while subplots (b) and (d) are devoted to the clapper's behavior. Comparing the trajectories of the bell, one can say that numerical results show remarkable agreement with the experiment. Simultaneously, time traces of the clapper do not show such a convergence as the clapper of the examined bell performed nonperiodic motion. Divergence from the numerically obtained periodic attractor is mainly visible around the moments of impact.

Analyzing Figure 34.2, one can say that the hybrid model introduced in this chapter is able to simulate the bell's behavior with excellent accuracy and precisely determine crucial features of the clapper's motion such as the period of motion, average amplitude of motion, and predict moments of the clapper to the bell collisions. Still, slight divergence between the lines is visible around the moments of impacts. It is caused by the oscillations of the bell's shell, which are not included in the mathematical model of the

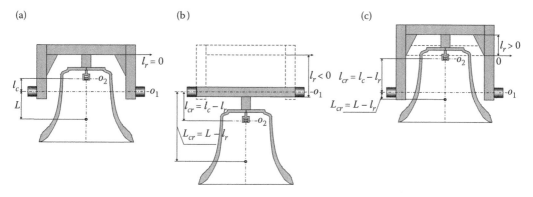

FIGURE 34.3 Description of parameter l_r: (a) reference yoke design $l_r = 0$, (b) $l_r < 0$, (c) $l_r > 0$.

system. In Reference 2, we describe this phenomenon in detail and prove that it does not compromise the reliability and practical significance of the considered model.

34.2.2 Influencing Parameters

Most of the parameters involved in the model are self-dependent. Moreover, we have to remember that we investigate a musical instrument; hence we cannot change some of its features that could affect the sound it generates. The two features that we can safely modify in real applications are the driving motor and the yoke of the bell. Therefore, we analyze how such changes influence the system's dynamics. As a reference, we use values of parameters characteristic for *The Heart of Lodz* and alter them to simulate modifications in the propulsion or mounting layout. We assume the linear motor driving that is described by piecewise function $M_t(\varphi_1)$ (34.4) and modify the output of the motor by changing the maximum generated torque T, which we use as the first controlling parameter. To describe the modifications of the yoke, we introduce the second parameter l_r whose meaning is described in Figure 34.3.

We take *The Heart of Lodz* as a reference yoke for which $l_r = 0$. If the bell's center of gravity is lowered, then $l_r < 0$ and as the value of l_r we take the distance by which the bell's center of gravity is shifted with respect to the reference yoke. Similarly, if we elevate the bell's center of gravity, we assume $l_r > 0$ and take its displacement as the value of l_r. As a maximum considered value of l_r, we take 0.235 m because for $l_r = 0.236$ m rotation axis of the bell goes through its center of mass.

It should be stressed that each change of the l_r value affects other parameters of the model. So, when the value of l_r is changed, the three different parameters should be swapped: L has to be replaced by L_r, l_c by l_{cr}, and finally B_b by B_{br}. New parameters L_r, l_{cr}, and B_{br} are given by the following formulas:

$$L_r = L - l_r$$
$$l_{cr} = l_c - l_r \tag{34.7}$$
$$B_{br} = \left(B_b - ML^2\right) + ML_r^2$$

34.3 Most Common Bells' Working Regimes

In this section, we present and describe periodic attractors that can be considered as proper working regimes and have practical applications. These regimes are often called ringing schemes and can be classified in groups that have common characteristics such as, especially, number of collisions during one period of motion, course of collisions, and time between them. In Figure 34.4 we present six most common working regimes by showing phase portraits of the bell and the clapper and indicating the effects of collisions. Subplots of Figure 34.4 were obtained for different values of parameters T and l_r.

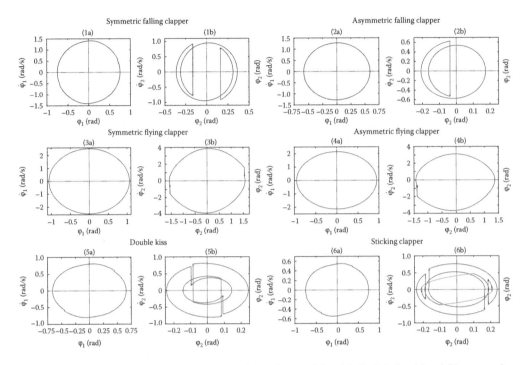

FIGURE 34.4 Presentation of the most common working schemes of church bells. Subplots (a) and (b) present phase portraits of the bell and the clapper, respectively.

We say that the bell works in a "falling clapper" manner if the collisions between the bell and the clapper occur when they perform an antiphase motion (see Figure 34.4 (1a,b) and (2a,b)). This type of behavior is common for bells that are mounted in the European manner. In the "falling clapper" ringing scheme, the amplitude of the clapper's motion is smaller than the bell's, and the clapper's velocity sign changes when collision occurs. We can distinguish a symmetric type of "falling clapper" with two collisions per one period of motion (Figure 34.4 (1a,b) obtained for $T = 350$ Nm and $l_r = 0.05$ m) and its asymmetric version with one impact per period (Figure 34.4 (1a,b) obtained for $T = 150$ Nm and $l_r = -0.03$ m). These ringing schemes differ mainly in the time intervals between the successive impacts.

The second characteristic working regime is called the "flying clapper." In this regime, collisions occur when the bell and the clapper perform in-phase motion. The amplitude of the clapper's motion is larger than the bell's, and the clapper's velocity sign remains the same after the collisions. The collisions have more gentle course than in the "falling clapper" manner, hence sometimes it may be difficult to achieve a nice resounding of the bell. In Figure 34.4 (3a,b) and (4a,b) we present two types of "flying clapper" behavior: a symmetric attractor with two impacts per period obtained for $T = 450$ Nm, $l_r = -0.91$ m and an asymmetric one with only one impact per period that we receive for $T = 325$ Nm and $l_r = -1.21$ m.

Bells mounted in the English manner usually work in the "sticking clapper" regime in which the clapper and the bell remain in contact for a certain amount of time. In the considered system prior to the sliding mode we observe a number of successive impacts (usually three) that have a "falling clapper" course. In Figure 34.4 (6a,b), we show phase portraits of the bell (Figure 34.4 (6a)) and the clapper (Figure 34.4 (6b)) working in the "sticking clapper" manner obtained for $T = 125$ Nm and $l_r = 0.2$ m. The energy amount that is transferred between the bell and the clapper decreases with each subsequent collision. Hence, the sound effects caused by each hit are different and not all collisions may be noticed by the listener.

Rarely, we can observe the so-called "double kiss" working regime in which we observe four impacts per one period of motion (see Figure 34.4 (5a,b) obtained for $T = 175$ Nm and $l_r = 0.16$ m). During one period, the clapper hits each side of the bell's shell twice. The first collision on each side is in the "falling clapper" manner while the second impact has the "flying clapper" course. This behavior is especially attractive for the listeners but it is difficult to achieve.

Apart from what is described above, we can also observe stable periodic attractors with no collisions, but then no sound is produced and the bell cannot work as a musical instrument. Unfortunately, no impacting attractors occur in a wide range of T and l_r values. Moreover, we can distinguish other periodic attractors that can be successfully employed such as asymmetric "flying clapper" behavior with doubled period and four impacts per period that can be easily taken as a typical "flying clapper." Similarly, a quasi-periodic attractor with almost equal time intervals between the subsequent impacts can sound almost like a periodic ringing scheme. Section 34.4 analyzes how the most common working schemes can be achieved by proper designing of the propulsion mechanism and the yoke of the bell.

34.4 Influence of the Yoke Design and Forcing Amplitude on the System's Dynamics

The yoke's design—described by parameter l_r—and the amplitude of forcing—parameter T—define the working regime of the bell. This section analyzes how these parameters influence the response of the system. In Figure 34.5, we show the result of series of numerical simulations obtained using 154 different sets of parameters values from the following ranges: $l_r \in \langle -1.3, 0.23 \rangle$ m and $T \in \langle 100, 625 \rangle$ Nm. For each set of parameters, we start the simulation from zero initial conditions ($\varphi_1 = 0$, $\varphi_2 = 0$, $\dot{\varphi}_1 = 0$, $\dot{\varphi}_2 = 0$) and mark on the plot which ringing scheme we obtain. We include only the solutions the basins of attraction of which contain zero initial conditions ($\varphi_1 = 0$, $\varphi_2 = 0$, $\dot{\varphi}_1 = 0$, $\dot{\varphi}_2 = 0$). Such an approach is practically justified because in most cases, the bell and the clapper start their motion from the hanging-down position with zero velocities.

We consider seven most characteristic types of the behavior of bells that are described in detail in the previous section.

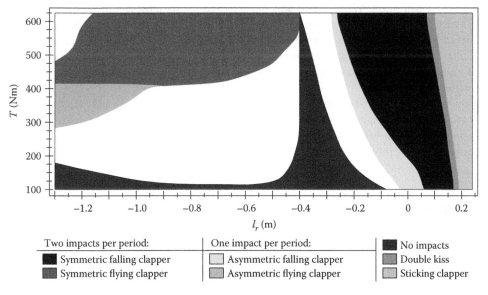

FIGURE 34.5 Two parameter ringing schemes diagram showing the behavior of the system for different values of l_r and T.

Analyzing Figure 34.5, we see that the biggest areas correspond to the three most common ringing schemes: symmetric "falling clapper" and "flying clapper" and "sticking clapper" that is typical for bells mounted in the English manner. Thanks to that, these behaviors are relatively easy to achieve and remain so even after a long period of time when values of some parameters can change a bit (for example, maximum torque generated by the linear motor). We see that, in general, the design of the yoke determines how the bell will operate. For $l_r < -0.4$ m, we can observe "flying clapper" with two or one impacts per period but these ringing schemes can be achieved only when T is bigger than some threshold value which decreases with the decrease of l_r. If T is not sufficient, we will not observe any impacts or the system will reach a different attractor (white)—periodic or nonperiodic one. For $l_r \in (-0.4$ to $-0.284)$ m, we cannot obtain any of the analyzed ringing schemes despite the forcing amplitude. Hence, when the yoke is designed improperly, it may be impossible to force the system to ensure proper operation. For $l_r > -0.284$ m, the system that can work in the following manner: "falling clapper" with one (yellow) or two (black) impacts per period, "double kiss" (light blue) or "sticking clapper" (pink). Each of these ringing schemes can be achieved despite the value of T but the yoke should be designed for the purpose. Hence, for these working regimes, there is no need to use very powerful driving motors and the system that can work more efficiently. In our previous publication [1], we considered transitions between different dynamical states of the yoke—bell—clapper system and analyzd the time that is needed to reach the presumed solution.

34.5 Conclusions

In this chapter, we investigated a plethora of different dynamical behaviors encountered during the analysis of the hybrid dynamical model of the church bell. The model was developed based on the bell in the Cathedral Basilica of Stanislaus Kosta, Lodz, Poland. Its parameter values were determined during the series of measurements and experiments involving the bell [2]. Finally, we validated the model by comparing the results of numerical simulations with experimental data and proved that it provides all crucial information about the system's dynamics, which is beneficial for bell-founders, bell-hangers, and engineers working on bells or bell towers.

In the next part of the chapter, we focused on solutions that can be considered as proper working regimes of the instrument. We presented and characterized the six most common behaviors and presented a method that can be used to determine the conditions under which a given type of behavior can be achieved. We used the amplitude of the forcing T and the yoke design (described by the parameter l_r) as the influencing parameters and developed two parameter ringing scheme diagrams. Such plots provide full information on how the geometry of the yoke and maximum output of the driving motor influence the dynamics of the system.

Ringing scheme charts can be calculated for any bell and used to design its yoke and propulsion. The presented tools can help to improve the working conditions of existing bells as well as the design of mounting layouts for new instruments. Thus, using the numerical analysis of the system's dynamics, we can ensure that the bell will work properly and reliably for ages and regardless of small changes of parameters.

Acknowledgments

This work is funded by the National Science Center, Poland based on the decision number DEC-2013/09/N/ST8/04343.

We would especially like to thank the Parson of Cathedral Basilica of St. Stanislaus Kostka Prelate Ireneusz Kulesza for his support and unlimited access to the bell. We have been able to measure the bell's template thanks to the bell's founder Mr. Zbigniew L. Felczyński. The data on the clapper, the yoke, and the motor have been obtained from Mr. Paweł Szydlak.

References

1. P. Brzeski, T. Kapitaniak, and P. Perlikowski. Analysis of transition between different ringing schemes of the church bell. *International Journal of Impact Engineering*, 85:57–66, 2015.
2. P. Brzeski, T. Kapitaniak, and P. Perlikowski. Experimental verification of a hybrid dynamical model of the church bell. *International Journal of Impact Engineering*, 80:177–184, 2015.
3. S. Ivorra and J. R. Cervera. Analysis of the dynamic actions when bells are swinging on the bell-tower of Bonrepos i Mirambell Church (Valencia, Spain). *Proceedings of the 3rd International Seminar of Historical Constructions*, pages 413–419, 2001.
4. S. Ivorra and F. J. Pallares. Dynamic investigations on a masonry bell tower. *Engineering Structures*, 28(5):660–667, 2006.
5. S. Ivorra, F. J. Pallares, and J. M. Adam. Dynamic behaviour of a modern bell tower – A case study. *Engineering Structures*, 31(5):1085–1092, 2009.
6. S. Ivorra, F. J. Pallarés, and J. M. Adam. Masonry bell towers: Dynamic considerations. *Proceedings of the ICE—Structures and Buildings*, 164:3–12(9), 2011.
7. S. Ivorra, M. J. Palomo, G. Verdudu, and A. Zasso. Dynamic forces produced by swinging bells. *Meccanica*, 41(1):47–62, 2006.
8. B. D. Threlfall and J. Heyman. Inertia forces due to bell-ringing. *International Journal of Mechanical Sciences*, 18:161–164, 1976.
9. A. Selby and J. Wilson. Durhamm cathedral tower vibrations during bell-ringing. In *Proceedings of the Conference Engineering a Cathedral*, pages 77–100. London: Thomas Telford, 1993.
10. A. Selby and J. Wilson. Dynamic behaviour of masonry church bell towers. In *Proceedings of the CCMS Symposium*, pages 189–199. New York: Chicago ASCE, 1997.
11. J. Klemenc, A. Rupp, and M. Fajdiga. A study of the dynamics of a clapper-to-bell impact with the application of a simplified finite-element model. *Engineering with Computers*, 27(3):261–272, 2011.
12. J. Klemenc, A. Rupp, and M. Fajdiga. Dynamics of a clapper-to-bell impact. *International Journal of Impact Engineering*, 44(0):29–39, 2012.
13. M. Lepidi, V. Gattulli, and D. Foti. Swinging-bell resonances and their cancellation identified by dynamical testing in a modern bell tower. *Engineering Structures*, 31(7):1486–1500, 2009.
14. G. Meneghetti and B. Rossi. An analytical model based on lumped parameters for the dynamic analysis of church bells. *Engineering Structures*, 32(10):3363–3376, 2010.
15. F. P. Muller. Dynamische und statische gesichtspunkte beim bau von glockenturmen. Karlssruhe: Badenia Verlag GmbH, pages 201–212, 1986.
16. R. Spielmann, A. Rupp, M. Fajdiga, and B. Aztori. Kirchenglocken-kulturgut, musikinstrumente und hochbeanspruchte komponenten. *Glocken–Lebendige Klangzeugen. Schweizerische Eidgenossenschaft, Bundesamt fur Kultur BAK*, pages 22–39, 2008.
17. J. Steiner. Neukonstruktion und sanierung von glockenturmen nach statischen und dynamischen gesichtspunkten. Karlssruhe: Badenia Verlag GmbH, pages 213–237, 1986.
18. W. Veltmann. Uerber die bewgugn einer glocke. *Dinglers Polytechnisches Journal*, 220:481–494, 1876.
19. W. Veltmann. Die koelner kaiserglocke. enthullungen uber die art und weise wie der koelner dom zu einer miSSrathenen glocke gekommen ist. Hauptmann, Bonn, 1880.

35

Multiple Duffing Problems Based on Hilltop Bifurcation Theory on MFM Models

Ichiro Ario

35.1 Introduction

In fundamental research on nonlinear dynamics, it is important to understand the essential properties for nonlinear mechanics such as bifurcation or chaotic behavior, as these are physical factors in the loss of structural stability in structures. It is very useful to exactly demonstrate the nonlinear dynamics of structures by numerical analysis because such real behavior after a bifurcation point is complex. In addition, if the structural model had several degrees-of-freedom (D.O.F.) as a number of parameters, its behavior would be much more complex. If a real structure had some imperfection as a perturbation of the perfect design, the bifurcation point would be hidden and it might have mode-jumping phenomena to jump to another stable equilibrium path and/or a loss of stiffness, which is inappropriate behavior for the structure. There are also a number of factors that can disrupt the stability of a structure. Problems related to the elastic stability of a structure have been researched by many researchers in this field. Notably, the work of Thompson and Hunt [1] has been identified with the general theory of elastic stability using catastrophe theory in engineering, and Thompson and Stewart [2], who introduced nonlinear dynamics and chaos based upon the instability of a structural model. In particular, it is well known that shallow arches and/or trusses have *unstable-symmetric bifurcation* with regard to elastically geometrical nonlinearity [3–5]. Von Mises truss, which is called "a two-bar truss," has snapthrough behavior, which is one of the structural instabilities in static nonlinear mechanics in a shallow truss without dissipation that is considered in some References 6 and 7. In addition, Cook and Simiu [8] experimentally studied the periodic and chaotic snapthrough motions of the Stoker column [9]. Ario [10] has expressed the formula of the single Duffing equation and its applications under geometrical conditions of a D.O.F. Dynamic systems with snapthrough behavior have been extensively studied because of the important phenomena of structural instability such as jumping, buckling, and snapping; see, for example, References 11–15.

In the fields of chaos or complex systems, it is well known that the Duffing equation, which is one of the nonlinear oscillation problems, is a standard model with a double-well potential system for *the Duffing oscillation* [16–23]. Awrejcewicz [24] has expressed the excellent academic work of "Bifurcation and Chaos in Coupled Oscillators," which is chaos phenomena for a mechanical oscillator's model with two D.O.F.s as a coupled problem. But it is recognized that this model does not have multiple singular points or hilltop bifurcation issues in a static problem. Almost all dynamic Duffing problems, however, have been considered to be limited to a single D.O.F., as their advanced model including the higher-order stiffness terms or a two-parameter model without the multiple D.O.F. and multiple singular point problems.

On the other hand, Holnicki et al. [25,26] have suggested a mechanical concept which contains the component of a micropantographic truss as the multifolding microstructures (MFM) theory. There are several hinge nodal joints as multiple D.O.F. on an MFM model. To analyze the nonlinear behavior of an MFM system subjected to static or dynamic loading, it has been clarified that there are nonlinear equilibrium paths for several different foldings using theoretical bifurcation analysis [27–29]. However, there have been hardly any follow-up surveys on a Duffing oscillator with the multiple D.O.F. of the MFM folding nonlinear system and its oscillation. The reason is because it is hard to compute the multiple D.O.F. of the Duffing system and there are multiple bifurcation points and/or paths on periodic structures in nonlinear statics [29]. Its chaos problem of multiple Duffing equations with a hilltop bifurcation singular point is broken-through for the multiple D.O.F. in nonlinear modeling dynamics of engineering by Reference 30. Furthermore, the imperfection problem of its hilltop bifurcation is described in the advanced paper [31]. In particular, a dynamic system with hilltop bifurcation is assumed to show rare behavioral characteristics, and has not been discussed enough.

An ordinary Duffing oscillation such as the von Mises truss has basically followed a homoclinic orbit in general, and this type of potential system is called a double-well system of total potential energy. However, it has never been applied to the chaos science of the MFM model with multiple D.O.F. using discreted geometrical nonlinear stiffness joints.

Incidentally, a problem that is more complex than the double-well system is the chaotic behavior of a three-well Duffing system, and this has been numerically investigated [32]. The chaotic behavior of a nonlinear damped three-well potential of Φ^6-Van der Pol oscillator under external and parametric excitations is studied in detail by means of Melnikov's analysis [33,34]. A three-well Duffing system with two external forcing terms is investigated [35]. These papers have mainly been proposed to solve a nonlinear equation with the fifth high-order terms of the powered function for the potential energy as a three-well Duffing system. There is, however, no proper model for this structure to understand the mechanism and what happens in this system, even though it has the third-order stiffness of a normal Duffing's equation, without higher-order terms of a three-well potential system. There is, however, an infinite-well system with periodic solutions, for example, in the nonlinear equation for the pendular problem. No suitable theoretical model for hilltop bifurcation analysis has been shown regarding the extended dimensional Duffing problem for a low-dimensional D.O.F. based on the consideration of static bifurcation. This is because it has a much more complex behavior and its mechanism from the viewpoint of static bifurcation analysis is hardly understood. Even if we obtained dynamic solutions for a number of D.O.F. of the system, it might be impossible to deeply comprehend the global and local properties of this complex behavior.

This chapter describes the importance of investigating the strongly nonlinear mechanics and/or dynamics when there is a relationship between the multiple heteroclinic orbits and elastic instability in a basic folding model of a critical problem of an MFM system, which has a bifurcation point. In this system, there is a hilltop-type bifurcation, and it is important to analyze the connections between the nonlinear response and the structural stability under an increasing load and under oscillation of a diamond-shaped truss. We propose that it is suitable to use an MFM model with several D.O.F. to analyze the relationship between strange attractor of the Duffing oscillation and bifurcation points when there are multiple potential wells. It is more useful to use all members in the system when they exhibit elastic behavior rather than plastic behavior because the member might not suffer any damage, allowing the system to be reused as a super elastic one. And the MFM folding system like a pantograph truss might have a contact issue after snapthrough. In this research, the folding system does not have any contact issue to consider and it is a purely ideal behavior for a mathematical model. If we obtained dynamic solutions for a couple of D.O.F. of the system, it would be possible to comprehend the mechanism related between the global and local properties of the complex nonlinear behavior. Although the top point in the system is subject to an impact or cyclical loading, instead of the given impact condition, it might be given the initial condition after the impact loading as equal physical parameters.

In order to observe the large-scale deformations due to the dynamic invariant behavior, we analyzed a diamond-shaped truss model with several D.O.F. and several bifurcation points and/or geometrical nonlinear paths. This research has investigated the theory of the nonlinear dynamics of an elastic multifolding pantographic truss as a simplified MFM model with horizontal displacements holding. We do this from the viewpoint of nonlinear mechanics by conducting a structural analysis of *the dynamic snapthrough, (the hilltop) bifurcations*, and *the homoclinic orbits*. In the static analysis of a nonlinear model with one D.O.F., we may observe snapthrough behavior in structural instability. However, a nonlinear structure with two or more D.O.F. in the interactive parameters has a bifurcation point as one of the singular points. In order to forecast this strange behavior, it is important to analyze the nonlinear dynamics of the bifurcation problem. In particular, we consider this theoretical model of a simplified folding system with the rare hilltop bifurcation point and focus our analysis on the dynamic stability, convergence, static bifurcation, and oscillation of the system.

Through this chapter, we have clearly demonstrated some more interesting models and results in the extended Duffing problem in nonlinear dynamics for a fundamental function of expanding and folding pantographic trusses, which have been going through structural instability of large and/or hierarchical snapthrough behaviors or subject cyclic loadings in the field of engineering science. Apart from the case of a simple truss, it is found that there are several folding patterns to solve the extended Duffing equation for a diamond-shaped truss with a singular problem in geometrical nonlinear truss. This is based on periodic structures or materials such as porous bodies, honeycomb structures, and pore-forming materials in engineering science with multiple bifurcation phenomena. The hilltop bifurcation phenomena in a pantographic truss are considered to be an appropriate chaotic model for determining advanced microstructure issues in multiple-well Duffing problems. Such models are very useful in engineering and basic science for determining the essential and invariant nonlinear phenomena of the extended Duffing oscillator.

35.2 Duffing Equation for a Snapthrough Behavior in a Two-Bar Truss

In this section, a simple and mechanical model, "a two-bar truss" with geometric nonlinearity for nonlinear dynamics of fundamental chaos physics, is introduced [10]. From this section, it becomes a general Duffing's equation based on structural instability of snapthrough behavior in static nonlinear mechanics in a D.O.F. without any bifurcation points.

35.2.1 Formula for the Duffing Equation from a von Mises Truss Model

This section describes how to assemble a formula for the Duffing equation with an ordinary double-well potential from a two-bar elastic truss, which is shallow and it is called a von Mises truss also with an unstable state of snapthrough behavior before the theoretical analysis for the folding system. This truss contains part of a multifolding system and it has geometrically left–right symmetric conditions. We consider the mechanical system of a material point that can slip in a smooth plane shown in Figure 35.1. A mass point m is joined by two dampers c with elastic stiffness EA on the position h. And there is no stress on the dampers and springs in the initial position. It is assumed that the bar is not buckled and is perfectly elastic in an ideal situation.

The energy principle is applied to solve a righting force, taking into account geometrical nonlinearity. The length of a couple of the initiated bar which has an aspect ratio $\gamma_1 = h_1/L$ of the height divided by L is the same for both the left and right sides. It is shown that

$$\ell_1 = L\sqrt{1+\gamma_1^2} = \ell_2, \tag{35.1}$$

and the deformed length $\hat{\ell}_1$ similarly as

$$\hat{\ell}_1 = L\sqrt{1+(\gamma_1-\bar{v}_1)^2} = \hat{\ell}_2. \tag{35.2}$$

Here, it is denoted in a normalized form as $\bar{v}_1 = v_1/L$.

We have several definitions of strain, and the normal strain is used for basic engineering in general. In the appendix of Reference 28, we have suggested the difference and/or similarity by comparison between two kinds of strains: the Green's strain and the normal strain. It is similar to trace essentially nonlinear physical behavior with theoretical approach and it is enough to confirm the nonlinear accuracy. The

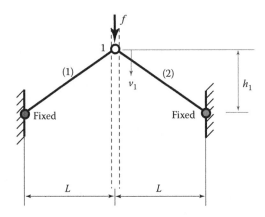

FIGURE 35.1 Theoretical model on a D.O.F. of a two-bar truss. (Adapted from I. Ario, *International Journal of Nonlinear Mechanics*, 39(4), 2004, 605–617.)

Green's strain is more useful to analyze the theoretical confirmation [28]. Now, the axial strain on a member is subsequently defined by the Green's strain in the following:

$$\varepsilon_i \equiv \frac{1}{2}\left\{\left(\frac{\hat{\ell}_1}{\ell_1}\right)^2 - 1\right\} \tag{35.3}$$

$$= \frac{1}{2}\left\{\frac{1+(\gamma_1-\bar{v}_1)^2}{1+\gamma^2} - 1\right\} = \frac{1}{2}\frac{\bar{v}_1^2 - 2\gamma_1\bar{v}_1}{1+\gamma_1^2} \tag{35.4}$$

$$= \frac{1}{2}\frac{(\bar{v}_1 - 2\gamma_1)\bar{v}_1}{1+\gamma_1^2}, \quad i = 1, 2. \tag{35.5}$$

Therefore, the total potential energy $\mathcal{V}$ is the sum of the strain energy saved into two elastic bars (EA) and forced potential energy caused by the unknown force f.

$$\mathcal{V} = \sum_{i=1}^{2} \frac{EA_i\ell_i}{2}(\varepsilon_i)^2 - f\bar{v}_1 L \tag{35.6}$$

$$= 2 \times \frac{EA\ell_1}{2}(\varepsilon_1)^2 - f\bar{v}_1 L = EAL\sqrt{1+\gamma_1^2}\frac{1}{4}\frac{(\bar{v}_1 - 2\gamma_1)^2}{(1+\gamma_1^2)^2}\bar{v}_1^2 - f\bar{v}_1 L \tag{35.7}$$

$$= \frac{\beta L}{4}\bar{v}_1^2(\bar{v}_1 - 2\gamma)^2 - f\bar{v}_1 L. \tag{35.8}$$

Here, $EA_i = EA, \varepsilon_2 = \varepsilon_1, \ell_2 = \ell_1, \beta = EA/(1+\gamma^2)^{3/2}$ from symmetric condition. Applying the principle of minimum potential energy $(\partial\mathcal{V}/\partial v_1 = \partial\mathcal{V}/\partial(\bar{v}_1 L) = 0)$ gives the primary equilibrium path with nonlinear stiffness:

$$f = F(\bar{v}_1) = \beta\bar{v}_1(\bar{v}_1 - \gamma_1)(\bar{v}_1 - 2\gamma_1) \tag{35.9}$$

where $F(\bar{v}_1)$ means that it is a static nonlinear relationship between the loading parameter f and the normalized displacement $\bar{v}_1$. Accordingly, as a function of $\bar{v}_1$; $F(\bar{v}_1) = f$, this nonlinear equilibrium equation is applied to a kinematic equation depending on time t, $m\ddot{\bar{v}}_1(t) + c\dot{\bar{v}}_1(t) + F(\bar{v}_1(t)) = 0$:

$$\ddot{\bar{v}}_1(t) + \bar{c}\dot{\bar{v}}_1(t) + \bar{\beta}\bar{v}_1(t)(\bar{v}_1(t) - \gamma_1)(\bar{v}_1(t) - 2\gamma_1) = 0. \tag{35.10}$$

An ordinary Duffing equation is obtained. Here, $\bar{c} = c/m, \bar{\beta} = \beta/m$ means a damping parameter with unit of a mass.

35.2.2 Stability Near the Singular Point(s)

We are able to rewrite Equation 35.10 in the following:

$$\ddot{\bar{v}}_1 = \bar{\beta}\bar{v}_1(\bar{v}_1 - \gamma_1)(\bar{v}_1 - 2\gamma_1) - \bar{c}\dot{\bar{v}}_1. \tag{35.11}$$

Then, there is the problem of solving this nonlinear ordinary differential equation, which has the singular points $(\bar{v}_1, \dot{\bar{v}}_1) = (0,0), (\gamma_1, 0), (2\gamma_1, 0)$. As we are interested in the behavior near solutions of (35.11), if this equation is linearized, the Jacobian matrix on the field of vectors shows

$$J \equiv \begin{pmatrix} \dfrac{\partial\dot{\bar{v}}_1}{\partial\bar{v}_1} & \dfrac{\partial\dot{\bar{v}}_1}{\partial\dot{\bar{v}}_1} \\ \dfrac{\partial\ddot{\bar{v}}_1}{\partial\bar{v}_1} & \dfrac{\partial\ddot{\bar{v}}_1}{\partial\dot{\bar{v}}_1} \end{pmatrix} = \begin{pmatrix} 0 & 1 \\ \bar{\beta}(2\gamma_1^2 - 6\gamma_1\bar{v}_1 + 3\bar{v}_1^2) & -\bar{c} \end{pmatrix}. \tag{35.12}$$

We can obtain eigenvalues for this matrix J:

$$\lambda_{1,2} = \frac{1}{2}\left(-\bar{c} \pm \sqrt{\bar{c}^2 + 4\bar{\beta}\left(2\gamma_1^2 - 6\gamma_1\bar{v}_1 + 3\bar{v}_1^2\right)}\right). \tag{35.13}$$

For the saddle point $(0,0)$, it becomes

$$\lambda|_{(0,0)} = \frac{1}{2}\left(-\bar{c} \pm \sqrt{\bar{c}^2 + 8\bar{\beta}\gamma_1^2}\right). \tag{35.14}$$

The parameter $\mathrm{Re}\,(\lambda_i)$ has both positive and negative values for $\bar{c}, \bar{\beta}\gamma_1^2 > 0$. The area of instability in this system can be found from the determinant of Jacobian J. On the other hand, eigenvalues on focus point $(\gamma_1, 0)$ is

$$\lambda|_{(\gamma_1,0)} = \frac{1}{2}\left(-\bar{c} \pm \sqrt{\bar{c}^2 - 4\bar{\beta}\gamma_1^2}\right). \tag{35.15}$$

The other eigenvalues on $(2\gamma_1, 0)$ is

$$\lambda|_{(2\gamma_1,0)} = \frac{1}{2}\left(-\bar{c} \pm \sqrt{\bar{c}^2 + 8\bar{\beta}\gamma_1^2}\right). \tag{35.16}$$

When $\bar{\beta}\gamma_1^2, \bar{c} > 0$, it becomes $\mathrm{Re}(\lambda_i) < 0$. The solutions around these points become asymptotically stable based on eigenvalues at the fixed point. This fixed point on $(0,0)$ is called the saddle point, and the focusing points for $\bar{c} > 0$ are on $(\gamma_1, 0), (2\gamma_1, 0)$.

35.2.3 Stability of System

The stability of this system is decided by a determinant of the Jacobian matrix in the following:

$$\det J \begin{cases} > 0 & : \quad \text{stable,} \\ = 0 & : \quad \text{critical,} \\ < 0 & : \quad \text{unstable.} \end{cases} \tag{35.17}$$

Substituting into Equation 35.12, it becomes

$$\det J = \bar{\beta}\left(2\gamma_1^2 - 6\gamma_1\bar{v}_1 + 3\bar{v}_1^2\right) = 0, \tag{35.18}$$

and the singular point $\bar{v}_{cr}$ is

$$\bar{v}_{cr} = \left(1 \pm \frac{1}{\sqrt{3}}\right)\gamma_1, \quad \text{or} \quad \bar{v}_{cr} = \left(1 \pm \frac{1}{\sqrt{3}}\right)\frac{h_1}{L}. \tag{35.19}$$

In a D.O.F. system, $\bar{\beta}$ is related without any dynamic stability. The state of this system is classified as

$$\begin{cases} \bar{v}_1(t) < \left(1 - \frac{1}{\sqrt{3}}\right)\frac{h_1}{L} & : \text{ stable,} \\[2mm] \bar{v}_1(t) = \left(1 - \frac{1}{\sqrt{3}}\right)\frac{h_1}{L} & : \text{ critical,} \\[2mm] \left(1 - \frac{1}{\sqrt{3}}\right)\frac{h_1}{L} < \bar{v}_1(t) < \left(1 + \frac{1}{\sqrt{3}}\right)\frac{h_1}{L} & : \text{ unstable,} \\[2mm] \bar{v}_1(t) = \left(1 + \frac{1}{\sqrt{3}}\right)\frac{h_1}{L} & : \text{ critical,} \\[2mm] \left(1 + \frac{1}{\sqrt{3}}\right)\frac{h_1}{L} < \bar{v}_1(t) & : \text{ stable.} \end{cases} \tag{35.20}$$

Substituting these singular points $\bar{v}_{cr}$ to Equation 35.13,

$$\lambda_{1,2} = \frac{1}{2}\left(-\bar{c} \pm \bar{c}\right)\begin{cases} = 0 \\ = -\bar{c} \end{cases} \tag{35.21}$$

has the eigenvalue of 0. In this case, this state is critical, and this information can be used to consider how to achieve dynamic stability.

35.2.4 A Homoclinic Orbit

Nondamping oscillation has an "∞"-forming a homoclinic orbit. This orbit is the classical standard orbit for damping trajectories, and the orbit has characteristic behavior for a damping system. Now, let us consider the total potential energy of this orbit. At time t, total energy of this system using nonlinear stiffness Equation 35.9 is

$$\mathcal{U} = \frac{(\dot{\bar{v}}_1)^2}{2} + \int F(\bar{v}_1)\, d\bar{v}_1 \tag{35.22}$$

$$= \frac{(\dot{\bar{v}}_1)^2}{2} + \bar{\beta}\frac{\bar{v}_1^2}{2}\left(\frac{\bar{v}_1^2}{2} - \gamma_1^2\right) + C. \tag{35.23}$$

Here, C is the constant of integration, and it is assumed $C = 0$. The first term of the right hand side of Equation 35.23 shows kinetic energy while the second term shows the potential energy of the stiffness. When Equation 35.23 becomes $\mathcal{U} = 0$, a homoclinic orbit is shown as

$$\Gamma_0 \equiv (\dot{\bar{v}}_1)^2 - \bar{\beta}\bar{v}_1^2\left(\gamma_1^2 - \frac{\bar{v}_1^2}{2}\right) = 0. \tag{35.24}$$

Assuming that a given trajectory of $(\bar{v}_1(t), \dot{\bar{v}}_1(t))$ is put to $\Gamma(t)$, if there is any trace inside the homoclinic orbit, it is denoted as $\Gamma(t) < \Gamma_0$; otherwise, if it is outside, it is denoted as $\Gamma(t) > \Gamma_0$. If $\Gamma_0 > 0$, the kinetic energy is $\mathcal{U} > 0$, and if $\Gamma_0 < 0, \mathcal{U} < 0$. The velocity $|\dot{\bar{v}}_1(t)|$ on the orbit Γ_0 is obtained by

$$|\dot{\bar{v}}_1(t)| = \sqrt{\bar{\beta}}\,\bar{v}_1(t)\sqrt{\gamma_1^2 - \frac{(\bar{v}_1(t))^2}{2}}, \quad \text{on } \Gamma_0. \tag{35.25}$$

35.3 Nonlinear Elastic Folding Analysis for Multifolding Models

In this section, we consider the folding mechanism for the pantographic structures subject to a vertical load at the top of the system shown in Figure 35.2a. The system is a pin-jointed elastic truss, and all nodes of the system displace vertically only. No allowance is made for friction or gravity in this nonlinear geometric problem; please refer to the details for the general formula of an MFM system [29].

35.3.1 Theoretical Approach for the Generalization to Multiple Layers: The MFM Model

We consider a structural system that contains several pin-joined diamond trusses with left–right symmetry, as shown in Figure 35.2a. In general, nonlinear equilibrium equations under geometrical conditions are complicated and are more difficult to solve. In particular, if we use the standard strain from engineering to determine the total strain energy, it would be difficult to complete mathematical proofs of the theory. Hence, even though this system has beautiful symmetry, this is not useful for understanding why symmetry-breaking happens. In the case of this model, we have to consider the theoretical bifurcation analysis that includes nonlinear dynamics.

Now, let us consider a theoretical estimation for the multifolding truss model without any imperfection. We assume a periodic height $h = \gamma L$ for the ith layer, as follows:

$$h_i = h = \gamma L, \quad i = 1, \ldots, n,$$

where the width L of the truss is fixed. Therefore, an initial length for each bar in the geometry of the figure is expressed as

$$\ell_i = \sqrt{L^2 + h_i^2} = L\sqrt{1 + \gamma^2}, \quad i = 1, \ldots, n. \tag{35.26}$$

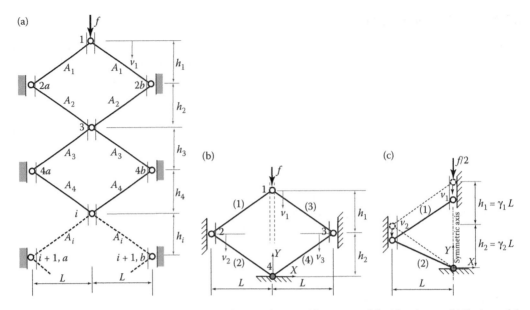

FIGURE 35.2 Theoretical folding model to analyze mechanism (a) MFM model with n layers. (b) Basic model. (c) One half of the model. (Adapted from I. Ario and M. Nakazawa, *International Journal of Non-Linear Mechanics*, 45, 2010, 337–347.)

The deformed length of each bar is denoted as $\hat{\ell}_i$ and is a function of the height and the nodal displacement variables

$$\hat{\ell}_i = L\sqrt{1 + \{\gamma - q_i(\bar{v}_i, \bar{v}_{i+1})\}^2}, \quad i = 1, \ldots, n-1$$

$$\hat{\ell}_n = L\sqrt{1 + \{\gamma - \bar{v}_n\}^2}, \tag{35.27}$$

where $\gamma = h/L > 0$ and $q_i(\bar{v}_i, \bar{v}_{i+1}) = \bar{v}_i - \bar{v}_{i+1}, \bar{v}_i = v_i/L, \quad (i = 1, \ldots, n)$.

Using Equations 35.3 through 35.5 of the Green's expression for strain ε_i for the ith elastic bar, the total energy for half of the system is given by

$$V = \sum_{i=1}^{n} \frac{EA_i \ell_i}{2} (\varepsilon_i)^2 - f^* \bar{v}_1 L$$

$$= \sum_{i=1}^{n} \frac{EA_i L\sqrt{1 + \gamma^2}}{2} \frac{1}{4} \left\{ \frac{1 + (\gamma - q_i)^2}{1 + \gamma^2} - 1 \right\}^2 - \frac{f}{2} \bar{v}_1 L, \tag{35.28}$$

where f^* is half of the load parameter ($f^* = f/2$) and $\bar{v}_{n+1} = 0$. When each layer has the same height, $\gamma = \gamma_i$, and when each member has the same stiffness, $EA = EA_i$. Then, the total potential energy can be described as

$$V = \frac{L}{8} \sum_{i=1}^{n} \beta(q_i)^2 (q_i - 2\gamma)^2 - \frac{f}{2} \bar{v}_1 L, \tag{35.29}$$

where the stiffness parameter $\beta = EA/(1 + \gamma^2)^{3/2}$ is a function of γ. From Equation 35.29, we can obtain the equilibrium equations based on the principle of minimum energy as follows:

$$F_i(\ldots, v_i, \ldots) \equiv \frac{\partial V}{\partial v_i} = \frac{\partial V}{\partial \bar{v}_i} \frac{\partial \bar{v}_i}{\partial v_i} = 0, \quad \text{for } i = 1, \ldots, n. \tag{35.30}$$

Hence, for the first, ith, and nth equilibrium equations

$$F_1\left(\bar{v}_1, \bar{v}_2\right) = \frac{\beta}{2}(\bar{v}_1 - \bar{v}_2)((\bar{v}_1 - \bar{v}_2) - \gamma)((\bar{v}_1 - \bar{v}_2) - 2\gamma) - \frac{f}{2} = 0,$$

$$F_i\left(\bar{v}_{i-1}, \bar{v}_i, \bar{v}_{i+1}\right) = (\bar{v}_{i-1} - \bar{v}_i)((\bar{v}_{i-1} - \bar{v}_i) - \gamma)((\bar{v}_{i-1} - \bar{v}_i) - 2\gamma)$$

$$- (\bar{v}_i - \bar{v}_{i+1})((\bar{v}_i - \bar{v}_{i+1}) - \gamma)((\bar{v}_i - \bar{v}_{i+1}) - 2\gamma) = 0,$$

$$F_n\left(\bar{v}_{n-1}, \bar{v}_n\right) = (\bar{v}_{n-1} - \bar{v}_n)((\bar{v}_{n-1} - \bar{v}_n) - \gamma)((\bar{v}_{n-1} - \bar{v}_n) - 2\gamma)$$

$$- \bar{v}_n(\bar{v}_n - \gamma)(\bar{v}_n - 2\gamma) = 0.$$

Using the implicit function theorem, it is then possible to solve these equilibrium equations for all the variables $\bar{v}_i$, $(i = n, \ldots, 1)$, as follows

$$F_n(\bar{v}_{n-1}, \bar{v}_n) = 0 \rightarrow \bar{v}_n = \mathcal{F}_n(\bar{v}_{n-1}), \tag{35.31}$$

$$F_i(\bar{v}_{i-1}, \bar{v}_i, \bar{v}_{i+1}) = F_i(\bar{v}_{i-1}, \bar{v}_i, \mathcal{F}_{i+1}(\bar{v}_i)) = 0 \quad \rightarrow \bar{v}_i = \mathcal{F}_i(\bar{v}_{i-1}), \tag{35.32}$$

$$F_1(\bar{v}_1, \bar{v}_2) = F_1(\bar{v}_1, \mathcal{F}_2(\bar{v}_1)) = 0, \tag{35.33}$$

where $\mathcal{F}(\cdot)$ denotes a nonlinear function. Thus, we obtain all the solutions for each nonlinear equilibrium path by finding each of the normalized nodal displacements in turn.

The stability of the system is given by a nonzero value for the determinant of the tangent stiffness matrix, Jacobian $J \in \mathbf{R}^{n \times n}$. J is defined as follows:

$$J \equiv (J_{ij}) = \left(\frac{\partial^2 \mathcal{V}}{\partial v_i \partial v_j}\right) = \left(\frac{\partial^2 \mathcal{V}}{\partial \bar{v}_i \partial \bar{v}_j} \frac{\partial \bar{v}_i}{\partial v_i} \frac{\partial \bar{v}_j}{\partial v_j}\right) = \left(\frac{\partial F_i}{\partial \bar{v}_j} \frac{\partial \bar{v}_j}{\partial v_j}\right), \quad \text{for } i, j = 1, \ldots, n, \tag{35.34}$$

and instability is defined as

$$\det J(v_i) = 0. \tag{35.35}$$

It is then possible to use the nonlinear equations during instability to determine the buckling load and the postbuckling shape of the truss at the singular point(s).

35.3.2 Theoretical Approach for a Folding Truss Shaped Like a Diamond

This structural system shown in Figure 35.2b has the diamond truss with right–left symmetry. Hence, in this research, the theoretical bifurcation analysis is limited to considering a collapse with symmetric deformation. By allowing for symmetric models only, we can therefore consider the half model shown in Figure 35.2c for the theoretical MFM analysis.

Now, let us consider a theoretical estimation for the multifolding truss model. We assume a periodic height for each layer of $h_i = \gamma_i L$ where the width L of the truss is fixed. Therefore, the initial length for each bar in the geometry of the figure is expressed as ℓ_1 and ℓ_2 like Equation 35.26. The deformed length of each bar denoted as $\hat{\ell}_i$ is a function of the height and the nodal displacement variables

$$\hat{\ell}_1 = L\sqrt{1 + (\gamma_1 - q_1(\bar{v}_1, \bar{v}_2))^2}, \tag{35.36}$$

$$\hat{\ell}_2 = L\sqrt{1 + (\gamma_2 - q_2(\bar{v}_2))^2}, \tag{35.37}$$

where $\gamma_i = h_i/L > 0$, $\quad q_1(\bar{v}_1, \bar{v}_2) = \bar{v}_1 - \bar{v}_2$, $q_2(\bar{v}_2) = \bar{v}_2$, $\bar{v}_i = v_i/L$, $\quad (i = 1, 2)$.

Using the Green's expression for strain, the total energy of half of the system is given by

$$V = \sum_{i=1}^{2} \frac{EA_i\ell_i}{2} \left(\varepsilon_i\right)^2 - f^*\bar{v}_1 L \tag{35.38}$$

$$= \sum_{i=1}^{2} \frac{EA_iL\sqrt{1+\gamma_i^2}}{2} \frac{1}{4} \left\{ \frac{1+(\gamma_i-q_i)^2}{1+\gamma_i^2} - 1 \right\}^2 - \frac{f}{2}\bar{v}_1 L, \tag{35.39}$$

where f^* is half of load parameter $f^* = f/2$ and $\bar{v}_3 = 0$. When each layer has the same height, $\gamma = \gamma_i$ and each member has the same stiffness, $EA = EA_i$. Then the total potential energy can be described as

$$V = \frac{\beta L}{8} \sum_{i=1}^{2} (q_i)^2 \left(q_i - 2\gamma\right)^2 - \frac{f}{2}\bar{v}_1 L, \tag{35.40}$$

where the stiffness parameter $\beta = EA/(1+\gamma^2)^{3/2}$, which is a function of γ. From Equation 35.40, we can obtain the equilibrium equations based on the principal of minimum energy [1] in the following:

For the first and second equilibrium equations chained

$$F_1 = \frac{\partial V}{\partial v_1} = \frac{\partial V}{\partial q_1} \frac{\partial q_1}{\partial \bar{v}_1} \frac{1}{L} = \frac{\beta}{2} q_1(q_1 - \gamma)(q_1 - 2\gamma) - \frac{f}{2}$$

$$= \frac{\beta}{2}(\bar{v}_1 - \bar{v}_2)((\bar{v}_1 - \bar{v}_2) - \gamma)((\bar{v}_1 - \bar{v}_2) - 2\gamma) - \frac{f}{2} = 0, \tag{35.41}$$

$$F_2 = \frac{\partial V}{\partial v_2} = \sum_{i=1}^{2} \frac{\partial V}{\partial q_i} \frac{\partial q_i}{\partial \bar{v}_2} \frac{1}{L} = \beta \sum_{i=1}^{2} q_i(q_i - \gamma)(q_i - 2\gamma)$$

$$= \beta \left\{ (\bar{v}_1 - \bar{v}_2)((\bar{v}_1 - \bar{v}_2) - \gamma)((\bar{v}_1 - \bar{v}_2) - 2\gamma) + \bar{v}_2(\bar{v}_2 - \gamma)(\bar{v}_2 - 2\gamma) \right\} = 0. \tag{35.42}$$

These equations equal to zero in the limit value problem of total potential energy. It is possible to obtain all variables $\bar{v}_i$ by substituting the obtained solutions into the next limited condition using the theorem of implicit function.

$$F_2(\bar{v}_1, \bar{v}_2) = 0 \rightarrow \bar{v}_2 = \mathcal{F}_2(\bar{v}_1), \tag{35.43}$$

$$F_1(\bar{v}_1, \bar{v}_2) = F_1(\bar{v}_1, \mathcal{F}_2(\bar{v}_1)) = 0, \tag{35.44}$$

where $F(\cdot)$ denotes a function of the nonlinear solutions. Finally, we obtain all solutions completely as nonlinear equilibrium paths.

The components of the Jacobian matrix $J \in \mathbf{R}^{2\times2}$ for this model are obtained as

$$J_{11} = \frac{\partial F_1}{\partial \bar{v}_1} = \frac{\beta}{2} \left(2\gamma^2 - 6\gamma(\bar{v}_1 - \bar{v}_2) + 3(\bar{v}_1 - \bar{v}_2)^2\right),$$

$$J_{12} = \frac{\partial F_1}{\partial \bar{v}_2} = -\frac{\beta}{2} \left(2\gamma^2 - 6\gamma(\bar{v}_1 - \bar{v}_2) + 3(\bar{v}_1 - \bar{v}_2)^2\right),$$

$$J_{21} = \frac{\partial F_2}{\partial \bar{v}_1} = \beta \left(2\gamma^2 - 6\gamma(\bar{v}_1 - \bar{v}_2) + 3(\bar{v}_1 - \bar{v}_2)^2\right),$$

$$J_{22} = \frac{\partial F_2}{\partial \bar{v}_2} = 3\beta(2\gamma - v_1)(v_1 - 2v_2).$$

Then, this determinant is expressed in the following:

$$\det \begin{pmatrix} J_{11} & J_{12} \\ J_{21} & J_{22} \end{pmatrix} = \frac{\beta^2}{2} \left\{2\gamma^2 - 6\gamma(\bar{v}_1 - \bar{v}_2) + 3(\bar{v}_1 - \bar{v}_2)^2\right\} \left(2\gamma^2 - 6\gamma\bar{v}_2 + 3\bar{v}_2^2\right). \tag{35.45}$$

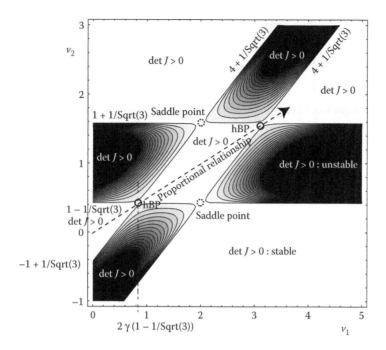

FIGURE 35.3 Distribution of determinate function of $\det(J(\bar{v}_1, \bar{v}_2))$ for $\gamma = 1$.

When the determinant of the Jacobian equals zero, this becomes an equation for the condition of stability:

$$\det \begin{pmatrix} J_{11} & J_{12} \\ J_{21} & J_{22} \end{pmatrix} = 0. \tag{35.46}$$

The buckling load and the postbuckling paths at the singular points can be determined from these nonlinear equations as the boundary of stable and unstable state in this pantographic system. From the condition (35.46), we can obtain the relationship as the stable (or unstable) map between $\bar{v}_1$ and $\bar{v}_2$ shown in Figure 35.3. There is a bound on $\det(J) = 0$ as a critical condition of structural stability. This solution of $\det(J(\bar{v}_1, \bar{v}_2)) = 0$ as the boundary of stable and unstable states is obtained in the following:

$$\bar{v}_2 = \begin{cases} \left(1 \pm \dfrac{1}{\sqrt{3}}\right) \gamma, \\ \left(1 \pm \dfrac{1}{\sqrt{3}}\right) \gamma + \bar{v}_1. \end{cases} \tag{35.47}$$

These solutions are corresponding to the boundary lines on $\det J = 0$. These points where the boundary lines are crossing at, are the singular points for structural instability. It determines the buckling load and postbuckling paths at these singular points to solve the determinant condition of Jacobian as the 0-eigenvalues and/or eigenvectors problem for finding critical state of this diamond truss model.

35.3.3 Bifurcation Analysis for a Diamond Truss Model ($n = 2$)

Let us consider solving equilibrium paths for the basic model shown in Figure 35.2b based on the nonlinear bifurcation theory. The height of each layer is identical, $\gamma = \gamma_i$. In order to find the variable $\bar{v}_i$, we use the implicit function from Equation 35.43, which shows the solutions in the following:

$$\bar{v}_2 = F_2(\bar{v}_1) \begin{cases} = \bar{v}_1/2, & \text{for primary path,} \\ = \left(\bar{v}_1 \pm \sqrt{-8\gamma^2 + 12\gamma\bar{v}_1 - 3\bar{v}_1^2}\right)/2, & \text{for bifurcation paths.} \end{cases} \tag{35.48}$$

These solutions to Equation 35.48 without any imperfection are plotted in Figure 35.4a.

Since we have the expressions $\bar{v}_2 = F_2(\bar{v}_1)$, we can now express the equilibrium equations for the primary and bifurcation paths in terms of variable $\bar{v}_1$ in the following way:

$$f_{\text{pri.}} = \frac{\beta}{8}\bar{v}_1(\bar{v}_1 - 2\gamma)(\bar{v}_1 - 4\gamma), \qquad \text{for primary path,} \tag{35.49}$$

$$f^{\pm}_{\text{bif}(\bar{v}_2=\bar{v}_3)} = \beta\left(6\gamma^3 - 11\gamma^2\bar{v}_1 + 6\gamma\bar{v}_1^2 - \bar{v}_1^3\right)$$
$$= \beta(\gamma - \bar{v}_1)(2\gamma - \bar{v}_1)(3\gamma - \bar{v}_1), \quad \text{as } \bar{v}_2 = \bar{v}_3 \quad \text{for bifurcation path,} \tag{35.50}$$

$$f^{\pm}_{\text{bif}(\bar{v}_2=-\bar{v}_3)} = \frac{\beta}{16}\left(48\gamma^3 - 80\gamma^2\bar{v}_1 + 42\gamma\bar{v}_1^2 - 7\bar{v}_1^3\right)$$
$$= \frac{7\beta}{16}\left(2\gamma\left(1 - \frac{1}{\sqrt{7}}\right) - \bar{v}_1\right)(2\gamma - \bar{v}_1)\left(2\gamma\left(1 + \frac{1}{\sqrt{7}}\right) - \bar{v}_1\right),$$
$$\text{as } \bar{v}_2 = -\bar{v}_3 \quad \text{for bifurcation path.} \tag{35.51}$$

The primary equilibrium path and the secondary paths are obtained by the above equations, and these paths are shown in Figure 35.4b. In addition, we obtain the critical positions based on the condition $df/d\bar{v}_1 = 0$ of the maximum/minimum value for an equilibrium primary path in the following (see Figure 35.5):

$$\bar{v}_1^{\text{BP}} = \left(1 \mp \frac{1}{\sqrt{3}}\right)2\gamma. $$

Finally, the maximum load is given by

$$f_{\max} = f(\bar{v}_1^{\text{BP}}) = \frac{2\beta}{3\sqrt{3}}\gamma^3 = \frac{2}{3\sqrt{3}}\frac{EA}{(1+\gamma^2)^{3/2}}. \tag{35.52}$$

When $\gamma = 1$, the maximum load becomes

$$f_{\max} = \frac{EA}{3\sqrt{6}} = 0.136\,EA. \tag{35.53}$$

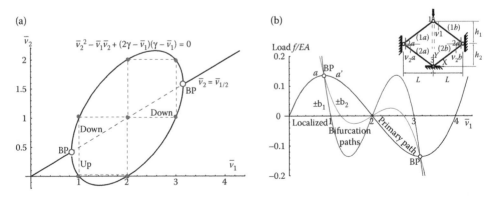

FIGURE 35.4 Nonlinear equilibrium paths for a diamond truss system ($h_1 = h_2 = 1, L = 1$). (a) Relationship between $\bar{v}_1$ and $\bar{v}_2$ from Equation 35.48. (b) Equilibrium paths. (Adapted from I. Ario, *Chaos, Solitons & Fractals*, 51, 2013, 52–63.)

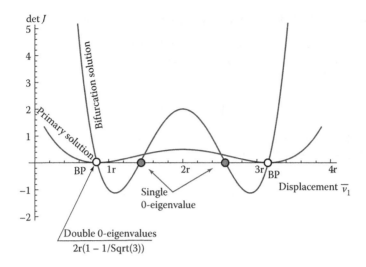

FIGURE 35.5 Determination of Jacobian $\det(J(\bar{v}_1))$.

35.3.4 Structural Stability Using det J

In this section, we consider the function of the determinant J for the primary path and the bifurcation paths from the determinant of Jacobian Equation 35.45 and the relationship between $\bar{v}_1$ and $\bar{v}_2$ of Equation 35.48. At first, substituting the primary solution $\bar{v}_2 = \bar{v}_1/2$ in Equation 35.45 results in the following:

$$\det J|_{\mathrm{pri}} = \frac{\beta^2}{32}\left(8\gamma^2 - 12\gamma\bar{v}_1 + 3\bar{v}_1^2\right)^2. \tag{35.54}$$

From this function $\det J|_{\mathrm{pri}} = 0$ where it is the critical boundary between structural stable and unstable states, which are obtained as follows:

$$\bar{v}_{\mathrm{cr}} = \left\{\left(1 \pm \frac{1}{\sqrt{3}}\right)2\gamma, \ \left(1 \pm \frac{1}{\sqrt{3}}\right)2\gamma\right\}. \tag{35.55}$$

Second, substituting the bifurcation solution $\bar{v}_2 = \left(\bar{v}_1 \pm \sqrt{-8\gamma^2 + 12\gamma\bar{v}_1 - 3\bar{v}_1^2}\right)/2$ in Equation 35.45 results in the following:

$$\det J|_{\mathrm{bif}} = \frac{\beta^2}{2}\left(8\gamma^2 - 12\gamma v_1 + 3v_1^2\right)\left(11\gamma^2 - 12\gamma\bar{v}_1 + 3\bar{v}_1^2\right). \tag{35.56}$$

From this function $\det J|_{\mathrm{bif}} = 0$, we obtain the solutions as follows:

$$\bar{v}_{\mathrm{cr}} = \left\{\left(1 \pm \frac{1}{\sqrt{3}}\right)2\gamma, \ \left(2 \pm \frac{1}{\sqrt{3}}\right)\gamma\right\}. \tag{35.57}$$

35.4 Dynamic Analysis for a Diamond Folding Truss without Contact

Consider kinematic energy without a term of damping from three equilibrium equations (35.49), (35.50), and (35.51).

35.4.1 Homoclinic Orbit for Primary Equilibrium Path

Before this problem, exchange the variable of the top displacement $\bar{v}_1$ of Equation 35.49 to the base point Q_1, which is corresponding to the central coordinated point for the global orbit in the following:

$$\begin{cases} \dot{Q}_1 = \dot{\bar{v}}_1 \\ Q_1 = 2\gamma + \bar{v}_1. \end{cases} \tag{35.58}$$

When the external force potential energy is from the integration of the primary equilibrium path exchanged coordinate (Equation 35.58), the external force potential energy is described as

$$\begin{aligned} \mathcal{V}(Q_1) &= \int f_{\text{pri}}(Q_1)\, L dQ_1 = \frac{\beta L}{8} \int (Q_1 + 2\gamma) Q_1 (Q_1 - 2\gamma)\, dQ_1 \\ &= \frac{\beta L}{32} Q_1^2 \left(Q_1^2 - 8\gamma^2 \right) + C. \end{aligned} \tag{35.59}$$

Here, an integral constant C is zero when $\mathcal{V} = 0$ as $Q_1 = 0$. It is shown as

$$\mathcal{V}(Q_1) = \frac{\beta L}{32} Q_1^2 \left(Q_1^2 - 8\gamma^2 \right). \tag{35.60}$$

Now, a kinematic equation with a mass m of the first integration is given as

$$h = \frac{m}{2} \dot{Q}_1^2 + \mathcal{V}(Q_1). \tag{35.61}$$

Then, it is defined as an orbit Γ_h^{pri} based on the primary equilibrium stiffness in the following:

$$\Gamma_h^{\text{pri}} \equiv \dot{Q}_1^2 + \frac{2}{m} \left(\mathcal{V}(Q_1) - h \right) = 0. \tag{35.62}$$

In particular, if $h = 0$, then it is called the homoclinic orbit as Γ_0^{pri}.

$$\Gamma_0^{\text{pri}} = \dot{Q}_1^2 + \frac{2}{m} \left(\frac{\beta L}{32} Q_1^2 \left(Q_1^2 - 8\gamma^2 \right) \right) = 0. \tag{35.63}$$

Therefore, the velocity is related to the function of the normalized displacement as follows:

$$\begin{aligned} \dot{Q}_1 &= \pm \sqrt{\frac{2}{m} \left(-\mathcal{V}(Q_1) \right)} \\ &= \pm \frac{1}{4} \sqrt{\frac{\beta L}{m}} Q_1 \sqrt{-Q_1^2 + 8\gamma^2} \\ &= \pm \frac{1}{4} \sqrt{\bar{\beta} L} (2\gamma - \bar{v}_1) \sqrt{-(2\gamma - \bar{v}_1)^2 + 8\gamma^2} \quad \text{on the global orbit } \Gamma_0^{\text{pri}}, \end{aligned} \tag{35.64}$$

where $\bar{\beta} = \beta/m$.

35.4.2 Homoclinic Orbit for Bifurcation Path

Exchange the variable of the negative displacement $\bar{v}_1$ of Equation 35.50 to the base point Q_1 to shift the displacement $Q_1 = 2\gamma - \bar{v}_1$ like the same procedure of Equation 35.58.

The external force energy for the bifurcation path is described as

$$V(Q_1) = \int f_{bf1}(Q_1)\, dQ_1 = \int \beta(Q_1 - \gamma)Q_1(Q_1 + \gamma)\, dQ_1$$

$$= \frac{\beta L}{4} Q_1^2 \left(Q_1^2 - 2\gamma^2\right), \quad \text{as } C = 0. \tag{35.65}$$

Here, it is assumed that this integral constant C is also zero, and a kinematic equation of the first integration is given in the same way as Equation 35.61.

$$\dot{Q}_1 = \pm\sqrt{\frac{2}{m}\left(-\frac{\beta L}{4}Q_1^2(Q_1^2 - 2\gamma^2)\right)}$$

$$= \pm\frac{1}{\sqrt{2}}\sqrt{\frac{\beta L}{m}}\, Q_1 \sqrt{-Q_1^2 + 2\gamma^2}$$

$$= \pm\frac{1}{\sqrt{2}}\sqrt{\beta L}\, (2\gamma - \bar{v}_1) \sqrt{-(2\gamma - \bar{v}_1)^2 + 2\gamma^2} \quad \text{on the local orbit } \Gamma_0^{bf1}. \tag{35.66}$$

35.5 Dynamic Analysis for Folding Truss

We begin by assuming the Duffing equations for the primary and bifurcation paths on a perfect model, and we use these as an approximation for decomposing the invariant equations. The dynamics of the folding truss, including mass, damping, and nonlinear stiffness $F(v(t))$, are described by the following kinetic equation:

$$M\ddot{v}(t) + C\dot{v}(t) + F(\bar{v}(t)) = 0, \tag{35.67}$$

where $M = \text{diag}.[m_{11}, m_{22}, \ldots, m_{ii}, \ldots]$ and $C = \text{diag}.[c_{11}, c_{22}, \ldots, c_{ii}, \ldots]$. $M \in \mathbf{R}^{n \times n}$ is the mass matrix; $C \in \mathbf{R}^{n \times n}$ is the damping matrix; $F(\cdot)$ is the nonlinear stiffness vector; $\{\ddot{v}_i(t)\}^T = \ddot{v}(t) \in \mathbf{R}^n$ is normalized acceleration; $\{\dot{v}_i(t)\}^T = \dot{v}(t) \in \mathbf{R}^n$ is the velocity; $\{\bar{v}_i(t)\}^T = \bar{v}(t) \in \mathbf{R}^n$ is the normalized displacement; and n is the total number of D.O.F. in the system. If the mass and damping in this system are given as independent uniform variables, where $m_{ii} = m$ and $c_{ii} = c$, $(i = 1, \ldots, n)$ are the diagonal components of their respective matrices, then from Equation 35.44, we obtain the equation for the nodal variables $\ddot{v}_1(t)$, $\dot{v}_1(t)$, and $\bar{v}_1(t)$. This becomes in the following equation:

$$m\ddot{v}_1(t) + c\dot{v}_1(t) + F_1\left(\bar{v}_1(t), \mathcal{F}_2(\bar{v}_1(t))\right) = 0. \tag{35.68}$$

We solve this equation using numerical analysis with both the incremental load method and the incremental displacement method as follows:

$$m\ddot{v}_1(t) + c\dot{v}_1(t) + (\beta\mathcal{F}_1(\bar{v}_1) - f) = 0, \tag{35.69}$$

dividing each term by m, and this external force f and the displacement $\bar{v}_1$ depend on time t, we obtain the following equation for the primary equilibrium path:

$$\ddot{v}_1(t) + \bar{c}\dot{v}_1(t) + \frac{\bar{\beta}}{8}\bar{v}_1(t)(\bar{v}_1(t) - 2\gamma)(\bar{v}_1(t) - 4\gamma) = \bar{f}(t), \tag{35.70}$$

where $\bar{c} = c/m$, $\bar{\beta} = \beta/m$, and $\bar{f} = f/m$ (and includes both the primary path and the bifurcation loads). On the other hand, the kinetic equation for the bifurcation equilibrium paths are shown from Equations 35.50 and 35.51 in the following:

$$\ddot{v}_1(t) + \bar{c}\dot{v}_1(t) + \bar{\beta}(\gamma - \bar{v}_1(t))(2\gamma - \bar{v}_1(t))(3\gamma - \bar{v}_1(t)) = \bar{f}(t) \tag{35.71}$$

and

$$\ddot{\bar{v}}_1(t) + \bar{c}\dot{\bar{v}}_1(t) + \frac{\bar{\beta}}{16}(48\gamma^3 - 80\gamma^2\bar{v}_1 + 42\gamma\bar{v}_1^2 - 7\bar{v}_1^3) = \bar{f}(t). \tag{35.72}$$

If we did not consider the damping parameter $\bar{c}$ to have a very small effect, it would be a nondamping model, which is useful to analyze the property of a nonlinear dynamic system with the total potential energy remaining constant.

35.5.1 Free Duffing Equation Model without an External Force

If the external force in Equation 35.71 and/or 35.72 does not exist, this is limited to the free Duffing problem. It is expressed as

$$\ddot{\bar{v}}_1(t) + \bar{c}\dot{\bar{v}}_1(t) + \bar{\beta}\mathcal{F}_1(\bar{v}_1(t)) = 0. \tag{35.73}$$

This dynamic equation is generalized so that the stiffness is nonlinear without an external force. For example, in the initial conditions of this equation, they are given the normalized values $m = 1, \beta = 1$, $v_1(0) = 0, \dot{v}_1(0) = 0.6, 0.7, 0.8, \ldots, 1.3$ and two kinds of damping parameters $\bar{c} = 0.02$ and $\bar{c} = 0.1$. Then, the trajectories of the free Duffing equation are shown in Figure 35.6. Figure 35.6a is the case of $\bar{c} = 0.02$ and Figure 35.6b is the case of $\bar{c} = 0.1$. In the figure, the dotted lines are the trajectories for a two-bar truss model where the height is the same. In Figure 35.6a, points A, B, and C are the stable fixed points for a diamond truss, and point B is the saddle point for a two-bar truss. These trajectories have three local orbits in this diamond truss without any contact issue when the cycle is limited.

35.5.2 Model under Cyclic Loading

Let us consider the numerical model with the cyclic loading as the external force of Equation 35.73. It is assumed that the cyclic loading is $\bar{f}(t) = f(t)/m = f_0 \cos \omega t$, and this term depends on time. The forced Duffing equation is expressed as

$$\ddot{\bar{v}}_1(t) + \bar{c}\dot{\bar{v}}_1(t) + \bar{\beta}\mathcal{F}_1(\bar{v}_1(t)) = f_0 \cos \omega t. \tag{35.74}$$

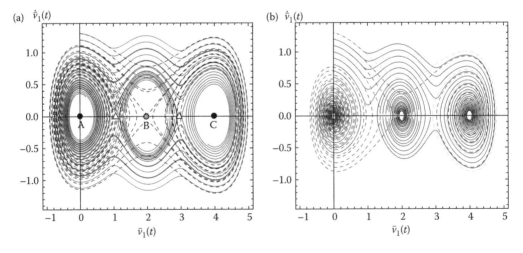

FIGURE 35.6 Trajectories from $(0, \dot{\bar{v}}_1(t))$. (a) $\bar{c} = 0.02$. (b) $\bar{c} = 0.1$. (Adapted from I. Ario, *Chaos, Solitons & Fractals*, 51, 2013, 52–63.)

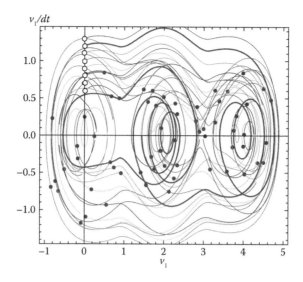

FIGURE 35.7 Parameters depend on the initial velocity; $\bar{c} = 0.02, f_0 = 0.3, \bar{v}_1(0) = 0, \omega = 1$.

It is applied to this equation as an example of the numerical point attractor. We prepare the fixed parameter $\bar{c} = 0.02, f_0 = 0.3, \bar{v}_1 = 0, \omega = 1$ and the initial condition with several velocities $\dot{v}_1(0) = 0.6$, $0.7, 0.8, \ldots, 1.3$. Then, this numerical result is shown in Figure 35.7. In the figure, the dots mean $n\pi$ on the trajectory. It is similar to the shape of the trajectory compared with the orbit of the free Duffing equation. This depends on the amplitude of the parameter f_0. The trajectory of the numerical solution looks like a smooth wave.

Next, let us consider the amplitude of the external cyclic loading, for example, $f_0 = \{0.2, 0.3, 0.5, 0.7\}$, respectively, on the constant damping parameter $\bar{c} = 0.26$ when other fixed parameters are $\omega = 1$, $m = 1, \beta = 1, v_1(0) = 0, \dot{v}_1(0) = 0$. We have obtained interesting trajectories in Figure 35.8. These trajectories depend on different load parameters 0.2, 0.3, 0.5, and 0.7 when the damping parameter is the same as $\bar{c} = 0.26$. If this model was the free Duffing oscillation without any external forces, a trajectory would converge into the stable fixed points at $\{(v_1, \dot{v}_1)_{\text{fix}}|(0,0), (2,0), (4,0)\}$ shown at points A, B, and C in Figure 35.6a. Moreover, we know the saddle fixed points at $\{(v_1, \dot{v}_1)_{\text{saddle}}|(1,0), (3,0)\}$. A trajectory in Figure 35.8a is running around the static fixed point at $(v_1, \dot{v}_1)_{\text{fix}} = (0,0)$ inside the local limit cycle. Hence, this trajectory with the load parameter $f_0 = 0.2$ is not out of the cycle and has related the balance of the external periodic force, and kinematic, stiffness, and damping. Figure 35.8b shows that the trajectory of $f_0 = 0.3$ gets out the first local limit cycle and gets into around $(v_1, \dot{v}_1)_{\text{fix}} = (2,0)$. Both Figure 35.8a and b show the gathering around three static fixed points as the global behavior. And these trajectories are different looping around the different fixed points. Hence, it is shown that the trajectory has been influenced by the stable fixed points and the magnitude of the load parameter in Figure 35.6.

Trajectories depend on different load parameters with the fixed damping parameter $\bar{c} = 0.26$ started from the initial point $(0, \dot{0})$. These numerical results drawn with the local homoclinic orbits are shown in Figure 35.9. The trajectory under the load parameter $f_0 = 0.2$ is shown in Figure 35.9a and it closes in the only left zone "L" where in the partial left of local homoclinic orbits "C". It is like a limit circle on one side of locally multiple-potential wells. The trajectory in case of $f_0 = 0.3$ is shown in Figure 35.9b and it spins around the zone "L" and "C" where in the part of local homoclinic orbits. The trajectory in case of $f_0 = 0.5$ is shown in Figure 35.9c and it circles the central zone "C" and out of local homoclinic orbits. The trajectory in case of $f_0 = 0.6$ is shown in Figure 35.9d and it jumps out of orbits and into the local zone "L" and "R." By drawing trajectories that depend on different load parameters with local bifurcation orbits,

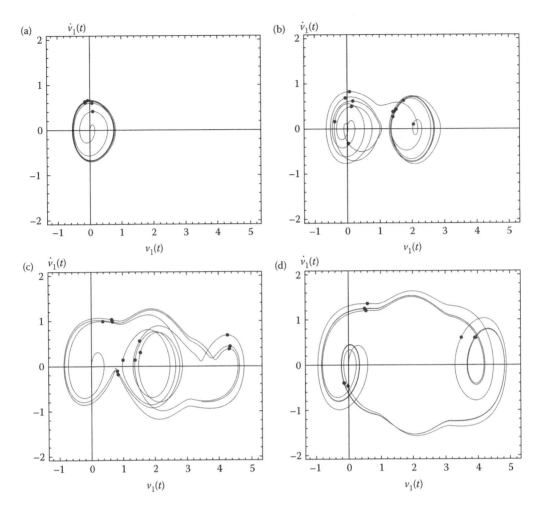

FIGURE 35.8 Trajectories depend on different load parameters (damping parameter is the same as $\bar{c} = 0.26$). (a) $f_0 = 0.2$. (b) $f_0 = 0.3$. (c) $f_0 = 0.5$. (d) $f_0 = 0.7$. (Adapted from I. Ario, *Chaos, Solitons & Fractals*, 51, 2013, 52–63.)

it is (corresponding) recognized that there are dynamic properties or habit of focus points, saddle points, which potential-wells, and their orbits in this model easily, based on geometrically structural condition.

Moreover, except for these phase portraits, it is shown that response of time histories depends on each load parameter f_0 in Figure 35.10. In the figure, it is denoted "•" the periodic time, $t = 2\pi i / \omega_0$. It is found that there are several fixed points as temporarily stationary positions at the displacements $\bar{v}_1 = \{0, 2, 4\}$ when this diamond truss has three-well-potential model, which depends on the D.O.F. of its geometrical structure. When we observe response of time histories that depends on each load parameter f_0, two waves of $f_0 = 0.2$ and 0.7 look regular. On the other hand, the other waves of $f_0 = 0.3$ and 0.5 look irregular responses and jumping to another fixed point with a switchover state. Then, the load parameter has the critical value to take over a closed domain of another well-potential fixed point in the multiple well-potential system with the multiple homoclinic bifurcations.

To get the characteristic property for time response shown in Figure 35.10, we obtain the spectra result using FFT for each parameter, shown in Figure 35.10b. There are the peak points with the characteristic property of frequency response on the figure. Frequency 14 Hz is all peak points for all parameters and 5 Hz is for two parameters $f_0 = 0.7, 0.5$.

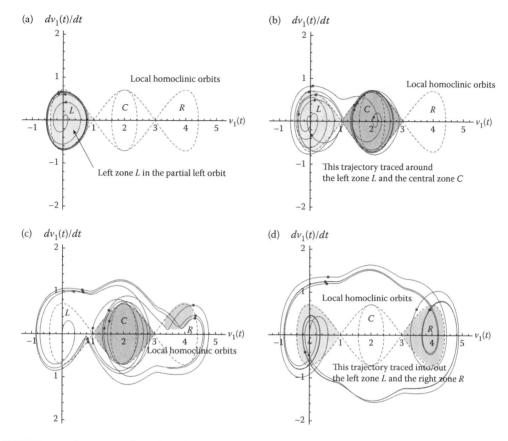

FIGURE 35.9 Trajectories depend on different load parameters with local orbits (damping parameter is same as $\bar{c} = 0.26$). (a) $f_0 = 0.2$. (b) $f_0 = 0.3$. (c) $f_0 = 0.5$. (d) $f_0 = 0.7$.

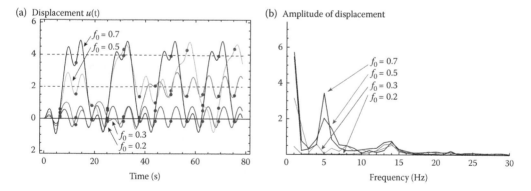

FIGURE 35.10 Time histories and spectra depend on different load parameters (damping parameter is same as $\bar{c} = 0.26$). (a) Time histories. (b) Spectra comparing with the load parameters. (Adapted from I. Ario, *Chaos, Solitons & Fractals*, 51, 2013, 52–63.)

35.5.3 Viewing Global Chaotic Behavior Using Poincaré Map

We have obtained the chaotic attractor that has the periodic solution at $t_i = 2\pi i/\omega_0$ for Equation 35.74. In this numerical research, the discrete time takes from $i = 1$ to 100 as finite time until 630 s. Here, we give

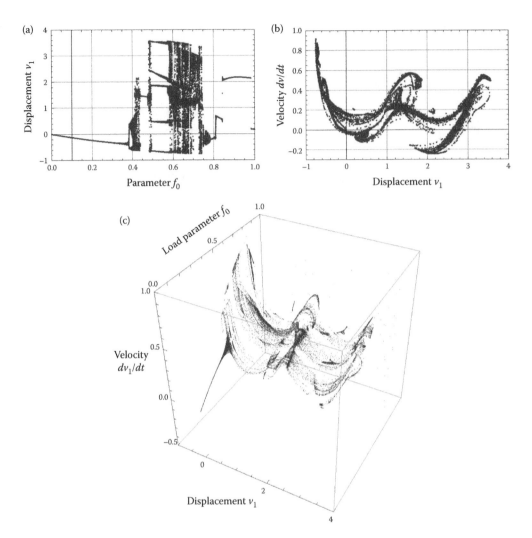

FIGURE 35.11 Chaos attractor (damping parameter $\bar{c} = 0.26$, natural angular frequency $\omega_0 = 1.0$, discrete time $t_i = 2\pi i/\omega_0$). (a) Relationship f_0 and $v_1(t_i)$. (b) $v_1(t_i)$ and $\dot{v}_1(t_i)$. (c) Viewing on three-dimensional space $(v_1, \dot{v}_1, f_0)$. (Adapted from I. Ario, *Chaos, Solitons & Fractals*, 51, 2013, 52–63.)

the numerical condition $m = 1, \bar{c} = 0.26, \beta = 1, \omega_0 = 1, v_1(0) = 0, \dot{v}_1(0) = 0$, and f_0 is increasing in steps of 0.002 as a parameter. Then, the solution is obtained in Figure 35.11. In the figure, replace γ means the load parameter f_0 and v_1 means the vertical displacement at the top node. It looks interesting as a chaos attractor, shown in Figure 35.11a. With the increasing load parameter f_0, the single periodic solution of the vertical displacement v_1 at $f_0 = 0.35$ branches to the double periodic solutions and spreading solutions. At $f_0 = 0.48$, it is divergence of the solutions, from near $f_0 = 0.5$, it shows the five paths with the periodic solutions clearly. The distribution of solutions has a classical bifurcation diagram, which depends on the load parameter f_0. And, we have obtained the phase as numerical result between u and v, which is like a wave that is quite a different attractor in 35.11b. It indicates a bifurcation diagram and the distribution of the periodic solutions $(v_1(t), \dot{v}_1(t)) = (v_1(2\pi i), \dot{v}_1(2\pi i))$ depending on the load parameter under the forced oscillation. The velocity is small when the displacement value is -0.4, 1, or 3. On the other hand, the velocity is high when the position is -0.7 or 3.3. In particular, there are three faces that have high, neutral, and negative velocity at the position 1.7. It seems that this physical meaning of this attractor is

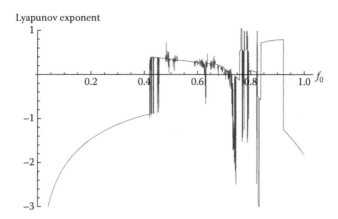

FIGURE 35.12   Lyapunov exponent. (Adapted from I. Ario, *Chaos, Solitons & Fractals*, 51, 2013, 52–63.)

harmonic balance between kinematic energy and the potential energy of nonlinear stiffness for the three-well type of the forced Duffing oscillation.

This behavior looks strange and there is a complex motion of the chaos attractor with a bifurcation point. In order to find out why there is a complex motion under an invariant law, we present the periodic solutions in view of three-dimensional spaces shown in Figure 35.11c. We are able to observe the distribution of point attractors as the periodic solution for a spreading and/or gathering state. The point attractors look like a cloud depending on the load parameter with essential nonlinearity. There might be several bifurcations such as periodic doubling bifurcation or torus bifurcation from Figure 35.11a. It might be seen periodic and 1/5 subharmonic as well as the phase of chaos or quasi-periodic attractor in view of Figure 35.11c. It will be necessary to investigate more details for understanding this relationship well in future. Then, the Lyapunov exponent means the index of dynamic instability and it is shown in Figure 35.12. The bifurcation point on the bifurcation diagram corresponds to the changing point from a negative value (stable) to a positive value (unstable) in Figure 35.12 when the incremental load parameter at $f_0 = 0.42$. From this amplitude, the load parameter keeps dynamic instability as the spreading distribution of chaos attractor.

Furthermore, in order to obviously see "more shaking behavior" as a chaos attractor, when we try to change the damping parameter $\bar{c} = 0.26$ to $\bar{c} = 0.15$ and the increment of the periodic external force amplitude 0.002 to 0.0005 on the physical parameters given by the previous model, let us show the numerical results. Then, we obtain the distribution of the numerical periodic solution. There is a phase portrait between the amplitude of the periodic external force and the periodic displacements $(v_1(t_i),\ t_i = 2\pi i/\omega_0)$, like "shaking in time and space" shown in Figure 35.13 when its amplitude γ is controlled from 0.285 to 0.310.

In the interaction between the displacement and velocity, although the range of the external force is small, the behavior of the point attractors means there is a global trajectory with a complex motion and changeable speed in the whole system. In Figure 35.13, the chaos attractor has look the ends of several waves, similar to a stall behavior shown by the marks B, C, and D. Point A is a stable attractor such as an eye or vortex after the chaos attractor. Figure 35.13a plots the attractor of the period time $t_i = 2\pi i$, and Figure 35.13b shows the periodic time $t_i = \pi i$, including half time on Figure 35.13a. Their solutions as a short section between 0.285 and 0.295 are corresponding to the same points to compare with both the rotated and normal Figure 35.13a. In Figure 35.13b, A', B',..., D' and BP are coupling points from A, B,..., D and BP in Figure 35.13a. BP means a singular bifurcation point that is corresponding to the limit point; this is hilltop bifurcation in statically nonlinear analysis.

This structural model of the chaos attractor does not have a high-order term of more than fifth powered stiffness. However, there is a two-D.O.F. system and an attractor is shown that is different from the strange attractor in the ordinary Duffing oscillator.

(a)   Velocity dv_1/dt

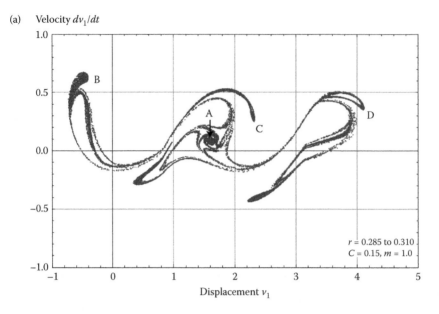

(b)   Velocity dv_1/dt

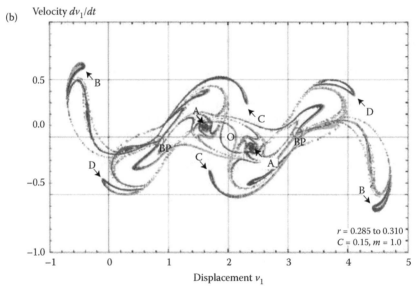

FIGURE 35.13 Chaos attractor (damping parameter $\bar{c} = 0.15$). (a) $v_1(\pi t)$. (b) $\dot{v}_1(\pi t)$. (Adapted from I. Ario, *Chaos, Solitons & Fractals*, 51, 2013, 52–63.)

35.6 Conclusions

In this academic research, the nonlinear behavior of a (multiple) folding truss with static bifurcation has been analyzed from a theoretical basis and by performing numerical simulations.

1. We have obtained the formula of the Duffing equation with the nonlinear stiffness from snapthrough behavior of a two-bar truss structure in statically structural instability.
2. It was found that there is a hierarchical structure that has global and local dynamics in several heteroclinic orbits. This was found by analyzing the (multi-) folding system with a static bifurcation

model. This extends the Duffing oscillation model with a multiwell system, compared with the standard double-well system of a homoclinic orbit in the model of structural mechanics.

3. As a concept for a multifolding model, a quite interesting chaos attractor was discovered, including a bifurcation diagram of a diamond folding model with a bifurcation single point. The phase between v and $\dot{v}$ is like a wave, which is quite a different attractor than in an ordinary Duffing equation.

4. We found that the low-order nonlinear structure without high-order term stiffness has indicated an interesting chaos phenomena in the engineering field. We have also obtained a different and strange attractor using a nonlinear truss model, which was assembled to a diamond shape with the bifurcation point. In this way, we extended the multiple Duffing oscillator with several D.O.F. We will think about more solutions in detail for this multifolding mechanics model to solve the multiple Duffing issue.

5. This model is useful and effective in showing a model of the multiple-well Duffing problem as an expanded two-bar truss as a fundamental model. This field is investigated through fundamental analysis from the viewpoint of structural instability. Hilltop bifurcation phenomena in a pantographic truss should be researched as a proper model to help solve nano- and microstructures as advanced models in the multiple-well Duffing problems.

Acknowledgment

This research on the problem of multiple symmetry-breakings has received Grant-in-Aid to conduct Exploratory Research of the Japan Society for the Promotion of Science (JSPS). I am extremely grateful to them for supporting scientific research from 2010 to 2012.

References

1. J.M.T. Thompson and G.W. Hunt, *A General Theory of Elastic Stability*, Wiley, London, 1973.
2. J.M.T. Thompson and H.B. Stewart, *Nonlinear Dynamics and Chaos; Geometrical Methods for Engineers and Scientists*, John Wiley & Sons Ltd., 1986.
3. J.S. Humphereys and S.R. Bodner, Dynamic buckling of shallow shells under impulsive loading, *Journal of the Engineering Mechanics Division, Proceedings of the ASCE*, 88, EM2, 1962, 17–36.
4. W. Nachbar and N.C. Huang, Dynamic snapthrough of a simple viscoelastic truss, *Quarterly of Applied Mathematics*, 25, 1967, 65–82.
5. N.C. Huang, Axisymmetric dynamic snapthrough of elastic clamped shallow spherical shells, *American Institute of Aeronautics and Astronautics Journal*, 7(2), 1969, 215–220.
6. N.C. Huang, Dynamic buckling of some elastic shallow structure subjected to periodic loading with high frequency, *International Journal of Solids and Structures*, 8, 1972, 315–326.
7. P. Nawrotzki and C. Eller, Numerical stability analysis in structural dynamics, *Computer Methods in Applied Mechanics and Engineering*, 189, 2000, 915–929.
8. R.G. Cook and E. Simiu, Periodic and chaotic oscillations of modified stoker column, *Journal of Engineering Mechanics*, 117(9), 1991, 2049–2064.
9. J.J. Stoker, *Nonlinear Vibrations in Mechanical and Electrical Systems*, Wiley, New York, 1950.
10. I. Ario, Homoclinic bifurcation and chaos attractor in elastic two-bar truss, *International Journal of Non-Linear Mechanics*, 39(4), 2004, 605–617.
11. G.W. Hunt, G.J. Lord, and A.R. Champneys, Homoclinic and heteroclinic orbits underlying the postbuckling of axially-compressed cylindrical shells, *Computer Methods in Applied Mechanics and Engineering*, 170, 1999, 239–251.
12. D.E. Musielak, Z.E. Musielak, and J.W. Benner, Chaos and routes to chaos in coupled Duffing oscillators with multiple D.O.F., *Chaos, Solitons & Fractals*, 24, 2005, 907–922.

13. S. Lenci and G. Rega, Global optimal control and system-dependent solutions in the hardening Helmholtz-Duffing oscillator, *Chaos, Solitons & Fractals*, 21, 2004, 1031–1046.
14. H. Caoa, Y. Jiang, and Y. Shan, Primary resonant optimal control for nested homoclinic and heteroclinic bifurcations in single-dof nonlinear oscillators, *Journal of Sound and Vibration*, 289, 2006, 229–244.
15. S. Lenci and G. Rega, Load carrying capacity of systems within a global safety perspective Part I. Attractor/basin integrity under dynamic excitations, *International Journal of Non-Linear Mechanics*, 46, 2011, 1232–1239.
16. G. Duffing, Erzwungene Schwingungen bei veränderlicher eigenfrequenz, Vieweg, Braunschweig, 1918.
17. A. Lyapunov, Problème Gènèral de la Stabilité de Mouvement, *Annals of Mathematics Studies*, 17, Princeton University Press, 1947.
18. V.I. Arnold, *Geometrical Methods in the Theory of Ordinary Differential Equations*, Springer-Verlag, New York, 1983.
19. S. Winggins, *Introduction to Applied Nonlinear Dynamical Systems and Chaos*, Springer-Verlag New York, 1990.
20. S. Smale, Diffeomorphisms with many periodic points. In *Differential and Combinatorial Topology*, S.S. Cairns (ed.), Princeton University Press, 1963, 63–80.
21. F.C. Moon and P.J. Holmes, A magnetoelastic strange attractor, *Journal of Sound and Vibration*, 65(2), 1979, 275–296.
22. Y. Ueda, Steady motions exhibited by Duffing's equation: A picture book of regular and chaotic motions, In *New Approaches to Nonlinear Problems in Dynamics, SIAM*, 1980, 311–322.
23. J. Awrejcewicz, *Bifurcation and Chaos in Simple Dynamical Systems*, World Scientific, Singapore, 1989.
24. J. Awrejcewicz, *Bifurcation and Chaos in Coupled Oscillators*, World Scientific, Singapore, 1991.
25. J. Holnicki-Szulc, P. Pawłowski, and M. Wiklo, High-performance impact absorbing materials—The concept, design tools and applications, *Smart Materials and Structures*, 12, 2003, 461–467.
26. J. Holnicki-Szulc and P. Pawłowski, The concept of multifolding and its experimental validation, *Proceedings of XXI ICTAM*, Warsaw, 2004.
27. I. Ario and A. Watson, Dynamic folding analysis for multi-folding structures under impact loading, *Journal of Sound and Vibration*, 308/3-5, 2007, 591–598.
28. I. Ario and A. Watson, Structural stability of multi-folding structures with contact problem, *Journal of Sound and Vibration*, 324, 2009, 263–282.
29. I. Ario and M. Nakazawa, Non-linear dynamic behaviour of multi-folding microstructure systems based on origami skill, *International Journal of Non-Linear Mechanics*, 45, 2010, 337–347.
30. I. Ario, Multiple Duffing problem in a folding structure with hill-top bifurcation, *Chaos, Solitons & Fractals*, 51, 2013, 52–63.
31. I. Ario, Multiple Duffing problem in a folding structure with hill-top bifurcation and possible imperfections, *Meccanica*, 49, 2014, 1967–1983.
32. H. Cao, X. Chia, and G. Chen, Suppressing or inducing chaos in a model of robot arms and mechanical manipulators, *Journal of Sound and Vibration*, 271, 2004, 705–724.
33. M.S. Siewe, F.M.M. Kakmeni, C. Tchawoua, and P. Woafo, Bifurcations and chaos in the triple-well Φ^6-Van der Pol oscillator driven by external and parametric excitations, *Physica A*, 357, 2005, 383–396.
34. J. Awrejcewicz and M.M. Holicke, *Smooth and Nonsmooth High-Dimensional Chaos and the Melnikov-Type Methods*, World Scientific Publishing, Singapore, 2007.
35. Z. Jing, J. Huang, and J. Deng, Complex dynamics in three-well duffing system with two external forcings, *Chaos, Solitons & Fractals*, 33, 2007, 795–812.

Chaotic Pattern Recognition

743

36

The Science and Art of Chaotic Pattern Recognition

B. John Oommen,
Ke Qin, and
Dragos Calitoiu

36.1 Introduction

36.1.1 What Is *Chaotic* Pattern Recognition?

Pattern recognition (PR) is the study of how a system can observe the environment, learn to distinguish patterns of interest from their background, and make decisions about their classification or categorization. In general, a pattern can be any entity described with features, where the dimensionality of the feature space can range from being a few to thousands. The four best approaches for PR are template matching, statistical classification, syntactic or structural recognition, and artificial neural networks (ANNs). The latter approach attempts to use some organizational principles such as learning, generalization, adaptivity, fault tolerance, distributed representation, and computation in order to achieve the recognition. The main characteristics of ANNs are that they have the ability to learn complex nonlinear input–output relationships, use sequential training procedures, and adapt themselves to data. Some popular models of ANNs have been shown to be capable of associative memory (AM) and learning. The learning process involves updating the network architecture and modifying the weights between the neurons so that the network can efficiently perform a specific classification/clustering task.

An AM permits its user to specify part of a pattern or key, and to thus retrieve the values associated with that pattern. One limitation of the AM model is its dependency on an external input. As opposed to biological NNs, once an output pattern has been identified, the ANN remains in that state until the arrival of an external input. To be more specific, once a pattern is recalled from a memory location, the brain is not "stuck" to it; it is also capable of recalling other associated memory patterns without being prompted by any additional external stimulus. This ability to "jump" from one memory state to another *in the absence of a stimulus* is one of the hallmarks of the brain, and this is one phenomenon that we want to emulate.

The evidence that indicates the possible relevance of chaos to brain functions was first obtained by Freeman [10] through his clinical work on the large-scale collective behavior of neurons in the perception of olfactory stimuli. Freeman developed a model for an olfactory system having cells in a network connected by both excitatory and inhibitory synapses. He described how a chaotic system state in the neighborhood of a desired attractor can fall on a stable direction when a perturbation is applied to the system's parameter. From this model, he conjectured that the quiescent state of the brain is *chaos*, while during perception, when attention is focused on any sensory stimulus, the brain activity becomes more *periodic*, where the periodic orbits can be interpreted as specific memories. If the patterns stored in memory are identified with an infinite number of unstable periodic attractors which are embedded in an attractor, then the transition from the quiescent state onto an "attention" state can be interpreted as the controlling of chaos. The controlling of chaos gives rise to periodic behavior, culminating in the identification of the sensory stimulus that has been received. Thus, mimicking this identification on an NN can lead to a new model of PR, *which is the goal of this research endeavor.*[*]

The NN models described here are not altogether unique—they are "distant cousins" of the family of Hopfield-like chaotic NNs (CNNs) which have been proposed in the literature. However, their use in generating repetitive/periodic responses to achieve PR is, to our knowledge, new, and the papers which report this phenomenon are surveyed here.

A CNN with nonfixed weights for the connections between the neurons can reside in one of the infinite number of possible states that are allowed by the functions of the network. In the general case, the dimensionality of the possible state space is finite, although the number of states is infinite. When the weights of the CNN are specified, the volume of the state space decreases, but the number of possible states continues to be infinite. During its evolution, a CNN with fixed weights can be in one of the infinite states. In the case when one inserts one of the memorized patterns as an input to the network, we want the network to repetitively generate that pattern at the output, or even resonate with that pattern. The system should then, in this setting, generate *that* pattern repeatedly with, hopefully, a small periodicity, where, if it is truly periodic, the actual period of resonance is not of critical importance. Between two consecutive appearances of the memorized pattern, we would like the network to be in one of the nonmemorized infinite number of states, and in the memorized states with an arbitrary small probability.

The *repetition* with the memorized pattern given as input, and the *transition* through several states from the infinite set of possible states (even when the memorized pattern is inserted as the input) represent the difference between this kind of PR and the classical types which correspond to the strategies associated with statistical, syntactical, and structural PR.

---

[*] Unfortunately, if the external excitation forces the brain out of chaos completely, it can lead to an epileptic seizure. An ambitious goal of research in chaotic brain modeling is to see how these episodes can be anticipated, remedied, and/or prevented. Although some initial results of how this can be achieved are currently available, we will not address this issue here, as it is outside the scope of this chapter.

The dynamic NNs proposed by Adachi and Aihara [1], referred to here as Adachi neural network (AdNN), can store a large number of dynamic spatiotemporal patterns and spontaneously recall these associated patterns on the arrival of a specific external stimulus. In the AdNN, the learning process is solved in a *single* step by the computation of the weights, and this is analogous to how the training is achieved in Hopfield-like NNs. Observe that the testing process, however, is a dynamical one. How to "juggle" between chaos and repetition/periodicity is, indeed, an art, and this is what we attempt to control by design.

36.1.2 Repetition and Periodicity

The salient characterizing phenomenon that a chaotic PR system must exhibit is, on the one hand, chaos for untrained patterns. That is relatively easy to achieve if the system is inherently chaotic. The more difficult phenomenon is the concept of *changing* its behavior in the presence of a trained pattern. The question really is one of determining *how* this change should be manifested. Of course, if the system is said to "recognize" patterns, the pattern that is to be recognized must necessarily be found at the output. Indeed, to make the PR system nontrivial, the least that we require is that this recognized pattern appears *frequently* or *repeatedly* at the output. If augmented with this repeated occurrence, the pattern also appears in a periodic manner, and it would be an added feature. At this juncture of the development of the field, however, we are not going to necessarily demand it. All that we do require is that a pattern to be recognized must appear *frequently* or *repeatedly* at the output.

36.1.3 On Designing Chaotic PR Systems

Traditionally, PR systems work with the following model: given a set of training patterns, a PR system learns the characteristics of the class of the patterns, and retains this information either parametrically or nonparametrically. When a testing sample is presented to the system, a decision of the identity of the sample class is made using the corresponding "discriminant" function, and this class is "proclaimed" by the system as the identity of the pattern.

Compared with such traditional PR systems, the goal of the field of *chaotic* PR systems can be expressed as follows: we do not intend a chaotic PR system to report the identity of a testing pattern with such a "proclamation." Rather, what we want to achieve, on one hand, is to have the chaotic PR system give a strong repetitive, *periodic*, or *more frequent* signal when a pattern is to be recognized. Further, between two consecutive recognized patterns, none of the trained patterns must be recalled. Finally, and most importantly, if an untrained pattern is presented, the system must emit a chaotic signal.

In the training phase, one would present the chaotic PR system with a set of patterns, and it thus "learns" the weights of the CNN. We now visit the issue of how the repetition of a pattern at the output is observed/recognized in the testing phase. To achieve this, Adachi et al. used the Hamming distance as the measure of similarity between the input and the output patterns. To be consistent, and in the interest of simplicity, we have used this same metric throughout this chapter. The Hamming distance is defined as

$$Dist(j) = \sqrt{\sum_{i=1}^{N}(x_i - P_i^j)^2}, \tag{36.1}$$

where N is the number of neurons, x_i is the output of the of ith neuron, P_i^j is the value of ith neuron associated with Pattern j, $j = 1, 2, \cdots, K$, and where there are K patterns being examined. Obviously, if $Dist(j) = 0$, it implies that the output is exactly Pattern j, which scenario is also referred to as a "perfect match" or a "hit." Otherwise, if $Dist(j) = 100$, one can see that the output pattern that is recalled is the *inverse* version of Pattern j.

Using the Hamming distance as a metric, the testing involves detecting a repetition/periodicity in the system, signaling the user that a learned pattern has occurred, and then inferring what the repeated/periodic pattern is. We shall now demonstrate how the latter task is achieved in a *simulation*. In a simulation setting, we are not dealing with a real-life chaotic system (as the brain[*]). Indeed, in this case, the output of the CNN is continuously monitored, and a repetitive/periodic behavior can be observed by studying the frequency spectrum, or by processing, in the time domain, the outputs as they come. Note that the latter is an infeasible task, as the number of distinct outputs could be countably infinite. This is a task which the brain (or, in general, a chaotic system), seems to be able to do, quite easily, and in multiple ways.

More specifically, extracting the chaotic/repetitive information from the NNs output can be done in one of the two ways: by either a frequency-based statistical analysis or by a Fourier-based analysis. We explain both of these below:

1. Frequency-based statistical analysis. In a frequency-based statistical analysis, if during the iterations, the Hamming distance between the output and a certain memorized pattern is 0 (or nearly 0), and if this index appears repetitively or frequently, we can say that this pattern is recognized "successfully." Further, if more than one pattern is recognized during the process (which is a scenario that we would like to avoid), we choose the one that is retrieved most frequently as the recognized pattern. This method can be implemented very easily. However, it is relatively inefficient, especially when there are a number of known patterns.

2. Fourier-based analysis. To achieve a Fourier-based analysis, one must still resort to computing the Hamming distances. It is obvious that most of the time, the Hamming distance between the output and the memorized pattern is relatively large. It is 0 or small only at a few time instances. This is, indeed, quite informative. As a consequence of this observation, we can apply a Fourier analysis to extract these peaks or troughs. Subsequently, one can determine (at least conceptually) whether the network recognizes the given pattern based on this information. Although such a process is conceivable, the question of how the Fourier transform can be used to yield this information is yet unsolved—it is currently being investigated.

Since we have to work with serial machines, to demonstrate the repetition/periodicity, we have opted to compare the output patterns with the various trained patterns. In other words, as a *prima facie* case, we have resorted to a frequency-based statistical analysis to achieve this recognition. Whenever the distance between the output pattern and *any* trained pattern is less than a threshold, we mark that time instant with a distinct marker characterized by the class of that particular pattern. The question of determining the repetition/periodicity of a pattern is now merely one of determining the repetition/periodicity of *these markers*.[†]

36.1.4 The Data Sets Used

Throughout this chapter, we shall present various Chaotic NNs that do, indeed, possess the above-mentioned PR properties. To demonstrate their properties, all the experiments conducted were tested on three data sets, which we shall refer to as the "benchmark" sets. These three benchmark data sets were, namely, the figures used in References 1, 2, and 5, the so-called "LOVE" data set used in Reference 11, and a subset of the numeral data sets used in References 7 and 8. These data sets are given in Figures 36.1 through 36.3, and are referred to as the *Adachi*, *LOVE*, and *numeral* data sets, respectively.

The reader will observe that all these data sets are binary in nature. The reason for this is the following: by virtue of the state/output functions of the various NNs (currently explained), the output of each neuron is forced to converge to either "0" or "1." In other words, every Adachi-like NN is constrained to only work

---

[*] How the brain is being able to record and recognize such a repeated/periodic behavior amidst chaos is yet unexplained [10].

[†] The unique characteristic of such a PR system is that each trained pattern has a unique attractor. When the testing pattern is fed into the system, the system converges to the attractor that characterizes it best. The difficult question, *which is still unsolved*, is one of knowing which attractor it falls on. This is an extremely interesting area for future research.

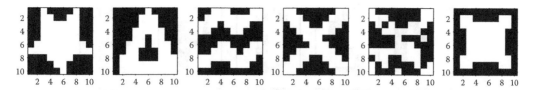

FIGURE 36.1 The set of patterns used in Adachi's and the experiments reported here. The first four patterns constitute the set used by Adachi et al. Pattern 5 is a noisy version of Pattern 4, in which 15% of the pixels have had their pixel values changed from "0" to "1" and vice versa. Pattern 6 is an untrained pattern.

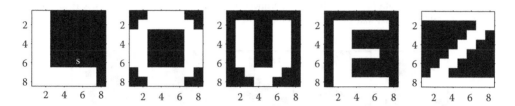

FIGURE 36.2 The set of patterns used in the experiments reported in Reference 11. The first four patterns depict the characters "L," "O," "V," and "E," and were used to train the network. Pattern 5 is an untrained pattern, which represents the character "Z."

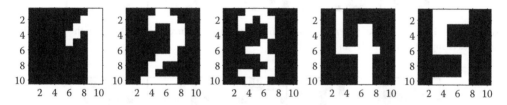

FIGURE 36.3 A subset of the set of patterns used in the experiments reported in References 7 and 8. The first four patterns represent the digits "1," "2," "3," and "4," and were used to train the network. Pattern 5 is an untrained pattern, which represents the digit "5."

with binary signals. The issue of designing NNs which have chaotic and PR properties for nonbinary data sets remains open. Indeed, this has not even been tackled or reported in the literature.

We would like to emphasize that the pattern, from a PR perspective, is not a "shape." Rather, it is a 100-dimensional vector. Every component of this vector can be modified. Thus, since the noise is bit wise, the current PR exercise may not work if the "images" are subject to translation/rotation operations. To be able to also consider translations in an image-processing application, one should preprocess the patterns and thus extract the features that will serve as the above-mentioned bit-wise array, which will be subsequently used to represent the latter vector.

In the case of the *Adachi* data set, the first four patterns constitute the training set, and the last pattern is an untrained pattern. Pattern 5 is a noisy version of Pattern 4 in which 15% of the pixels had their pixel values changed from "0" to "1" and vice versa. Similarly, in the case of the *LOVE* data set used in Reference 11, the first four patterns depict the characters "L," "O," "V," and "E," and were used to train the network. Pattern 5 is an untrained pattern which represents the character "Z." Finally, in the case of the *numeral* data set used in References 7 and 8, the first four patterns represent the digits "1," "2," "3," and "4" which were used to train the network, and Pattern 5 is an untrained pattern representing the digit "5." The more complete *numeral* data set consisted of the numerals from {"0" . . . "9"} (depicted in Figure 36.4 and their

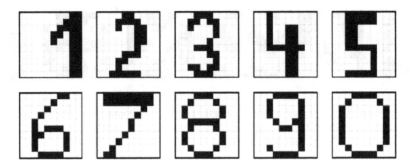

FIGURE 36.4 The set of patterns used in the PR experiments in Reference 8. These were the 10 × 10 bitmaps of the numerals "0" ... "9." The initial state used was randomly chosen.

FIGURE 36.5 The second set of patterns (with 10% noise) used in Reference 8, in which 10% of the pixels had their pixel values changed from "0" to "1" and vice versa.

FIGURE 36.6 The third set of patterns (with 15% noise) used in Reference 8, in which 15% of the pixels had their pixel values changed from "0" to "1" and vice versa.

noisy versions in Figures 36.5 and 36.6), and the results pertaining to these will also be reported in the relevant sections where it has been used.

36.2 The Pioneering Model in This Art: The AdNN

36.2.1 The Design of the AdNN

The AdNN is a network of neurons with weights associated with the edges, a well-defined present-state/next-state function, and a well-defined state/output function. It is composed of N neurons (Adachi et al. set $N = 100$), topologically arranged as a completely connected graph, i.e., each neuron communicates with every other neuron, including itself. It is described by means of three equations relating the two internal states $\eta_i(t)$ and $\xi_i(t)$, $i = 1 \ldots N$, and the outputs, $x_i(t)$, as per

$$x_i(t+1) = f(\eta_i(t+1) + \xi_i(t+1)), \tag{36.2}$$

$$\eta_i(t+1) = k_f \eta_i(t) + \sum_{j=1}^{N} w_{ij} x_j(t), \tag{36.3}$$

$$\xi_i(t+1) = k_r \xi_i(t) - \alpha x_i(t) + a_i. \tag{36.4}$$

In the above, $x_i(t)$ is the output of the neuron i which has an analog value in $[0,1]$ at the discrete time t. The internal states of the neuron i are $\eta_i(t)$ and $\xi_i(t)$, and $f(\cdot)$ is the logistic function with the steepness parameter, ε, satisfying $f(y) = 1/(1 + e^{-y/\varepsilon})$. k_f and k_r are the decay parameters for the feedback inputs and the refractoriness, respectively. $\{w_{ij}\}$ are the synaptic weights from the ith constituent neuron to the jth constituent neuron, and a_i denotes the temporally constant external inputs to the ith neuron.[*] α is the refractory scaling parameter.

While the network dynamics are described by Equations 36.3 and 36.4, the outputs of the neurons are obtained by Equation 36.2. The feedback interconnections are determined according to the following symmetric autoassociative matrix of the p-stored patterns by

$$w_{ij} = \frac{1}{p} \sum_{s=1}^{p} (2x_i^s - 1)(2x_j^s - 1),$$

where x_i^s is the ith component of the sth stored pattern, and p is the number of trained patterns.

The AdNN possesses rich dynamical phenomena such as AM, and *some* PR and chaotic properties as one or more of its parameters change. The details of these phenomena are reported in Reference 17. However, since this is not the primary intent of this chapter (and in the interest of brevity), the details of these properties are not fully listed here. Rather, we crystallize the experimental findings of Adachi et al. as below.[†]

36.2.2 AM Properties of the AdNN

For the data set given in Figure 36.1, under the settings $k_f = 0.9, k_r = 0.2, \varepsilon = 0.015, \alpha = 10$, and $a_i = 2$, the AdNN demonstrates interesting AM properties, as shown in Figure 36.7. From this figure, we can clearly see that the AdNN can dynamically retrieve all the four trained patterns (and their inverse versions) several times during 1000 recorded iterations, which implies that the AdNN does, indeed, possess AM properties.

36.2.3 PR Properties of the AdNN

Adachi et al. also noticed that the AdNN has some properties which can *probably* be used to develop a simplified PR system because it responds with distinct outputs for the various external stimuli. These properties can be cataloged into three cases as below:

1. The external stimulus corresponds to a trained pattern, e.g., Pattern 4. In this case, the output of the AdNN is periodic, as shown in Figure 36.8.

---

[*] A PR system receives the pattern that needs to be recognized as an input, for a limited time (or a number of iterations). This "to-be-recognized pattern" is denoted a_i. From the perspective of system dynamics, the reader must see that this input can be considered as a perturbation (an additional constant introduced into the PR system for a limited time) and the goal of the PR system is to classify/discriminate between these perturbations. One should also see that this perturbation is acting as a control parameter and not as noise.

[†] More details about the AdNN's properties can be found in Reference 1.

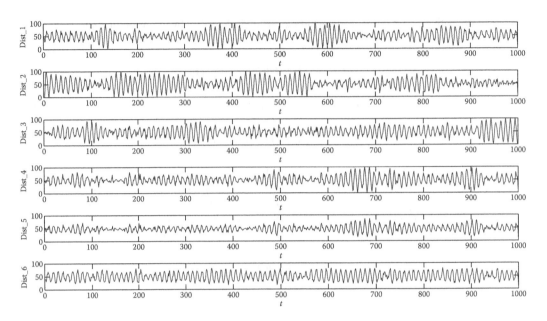

FIGURE 36.7 The AM properties of the AdNN: The Hamming distance between the output pattern of the AdNN and the four stored patterns when the applied external stimulus is a constant, 2.

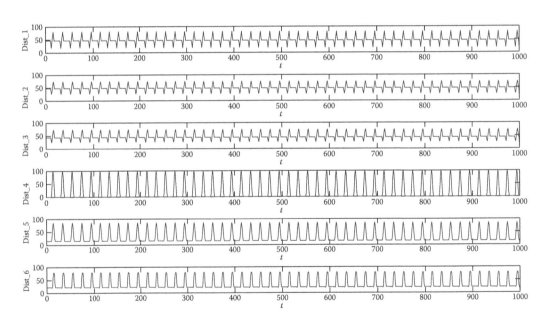

FIGURE 36.8 The PR properties of the AdNN (Case 1): The Hamming distance between the output pattern of the AdNN and the four stored patterns when the external applied stimulus corresponds to a trained pattern, Pattern 4.

2. The external stimulus corresponds to a trained pattern but with some noise, e.g., Pattern 5. In this case, the output of the AdNN is *not* periodic, as shown in Figure 36.9.

3. The external stimulus corresponds to an untrained pattern, e.g., Pattern 6. In this case, the output of the AdNN is *not* periodic, as shown in Figure 36.10.

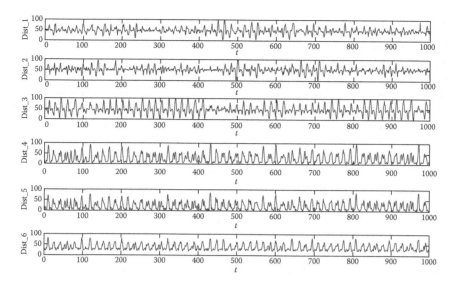

FIGURE 36.9 The PR properties of the AdNN (Case 2): The Hamming distance between the output pattern of the AdNN and the four stored patterns when the external applied stimulus corresponds to a trained pattern with 15% noise, Pattern 5.

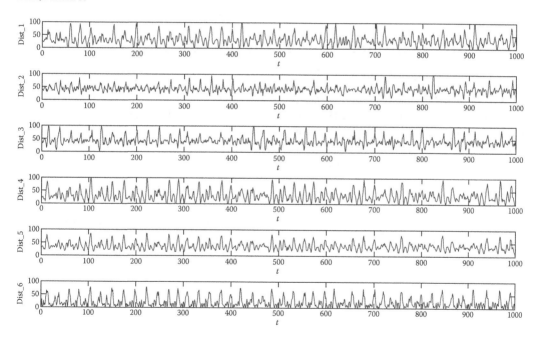

FIGURE 36.10 The PR properties of the AdNN (Case 3): The Hamming distance between the output pattern of the AdNN and the four stored patterns when the external applied stimulus corresponds to an untrained pattern, Pattern 6.

These phenomena almost attain the goal required of a chaotic PR system, but it yet does not quite yield the standards that we require because the AdNN responds to *trained and untrained* patterns with periodic and nonperiodic outputs, respectively. However, it responds to a noisy trained pattern also with nonperiodic outputs, which is a characteristic that we do not want. Instead, we would like a chaotic PR system to respond to both trained and noisy trained patterns with repetitive/periodic outputs. This has motivated the research works that follow.

We summarize below the AdNN's salient features and limitations:

1. Topologically, the AdNN is a completely connected NN. The computational cost for obtaining the outputs of all the neurons is thus $O(N^2)$. For the examples cited in References 1 and 3–8, which use 10×10 pixel arrays, this involves $O(10^4)$ computations *per time step*.
2. Adachi et al. pointed out that when $a_i = 2$, the network yields a periodic response after a long-term (about 21,000 iterations) transient phase. In this phase, the AdNN can behave as a dynamic AM. Thus, it can dynamically recall all the memorized patterns as a consequence to an input which serves as a "trigger."
3. The transient phase is not periodic. Additionally, in this phase, the network is not trapped in any equilibrium point, rendering it more dynamic than conventional associative networks.
4. The AdNN displays a complex dynamic behavior distinct from traditional chaotic systems. Reportedly, even its maximum Lyapunov exponent (LE) is near negative zero, implying an intermittent instability. For example, when $k_r = 0.9$, the maximum LE is approximately -0.0536, and the transient phase is unstable. This is explained in more detail in Section 36.3.2.
5. Under certain fixed parametric conditions, if the external stimulations correspond to trained patterns, the AdNN can behave like a PR system. However, as alluded to above, the PR capabilities of the AdNN are limited.

The first AdNN-like model which enhances this for PR is the modified AdNN (M-AdNN) described below.

36.3 A New CNN Model for PR: The M-AdNN

36.3.1 The Design of the M-AdNN

We now present a CNN model which *modifies* the AdNN to enhance its PR capabilities. This NN, referred to as the M-AdNN, is actually also a Hopfield-like model, and manipulates the internal structure and present-state/next-state equations of the original AdNN, and can be seen to be the pioneering work in designing chaotic PR systems. Structurally, it is also composed of N neurons, topologically arranged as a completely connected graph. Again, each neuron i, $i = 1 \ldots N$, has internal states $\eta_i(t)$ and $\xi_i(t)$, and an output $x_i(t)$. Calitoiu et al. [8] presented a brief rationale for each modification, and these are listed below:

1. The fundamental difference between the AdNN and the M-AdNN in terms of their present-state/next-state equations is that the latter has only a *single* global neuron (and its corresponding two global states) which is used for the state-updating criterion for *all* the neurons. Thus, $\eta_i(t)$ and $\xi_i(t)$ are updated as per

$$x_i(t+1) = f(\eta_i(t+1) + \xi_i(t+1)),$$

$$\eta_i(t+1) = k_f \eta_m(t) + \sum_{j=1}^{N} w_{ij} x_j(t),$$

$$\xi_i(t+1) = k_r \xi_m(t) - \alpha x_i(t) + a_i. \tag{36.5}$$

Observe that at time $t + 1$, the global states are updated based on the values of the states of *the single neuron of index m*, i.e., $\eta_m(t)$ and $\xi_m(t)$. In all the papers that describe the M-AdNN, the value of m is set to be N. This is in contrast to the AdNN in which the updating at time $t+1$ uses the internal state values of *all* the neurons at time t.
2. The weight assignment rule for the M-AdNN is the classical variant $w_{ij} = \frac{1}{p} \sum_{s=1}^{p} x_i^s x_j^s$. This is in contrast to the AdNN which uses $w_{ij} = \frac{1}{p} \sum_{s=1}^{p} (2x_i^s - 1)(2x_j^s - 1)$.

3. Additionally, Calitoiu et al. set $a_i = x_i^5$, as opposed to the AdNN in which $a_i = 2 + 6x_i^5$. Thus, as the researchers who proposed the M-AdNN argued, the M-AdNN will be more "receptive" to external inputs, hopefully, leading to a superior PR system.

In the interest of completeness and continuity, we first list below the M-AdNN's salient features, and go into the relevant details in the subsequent sections:

1. Being a variant of the AdNN, the M-AdNN is topologically a completely connected NN.
2. The main difference between the M-AdNN and the classical auto-AM model is that unlike the latter, the M-AdNN can be seen to be chaotic, this behavior being a consequence of the dynamics of the underlying system. The M-AdNN seems to be more representative of the way by which the brain achieves PR.
3. The most significant difference between the AdNN and the M-AdNN is the way by which the values of the internal state(s) are updated. The M-AdNN uses two global internal states, both of which are associated with a *single* neuron. By using these global states, the transient phase is reduced to be approximately 30 [8], which is in contrast to the AdNN's transient phase—which can be as large as 21,000.
4. The authors of Reference 8 showed that the largest LE of the AdNN is $\lambda = \log k_r < 0$. Thus, when $k_r = 0.9$, it implies that the AdNN is not *really chaotic*. In contrast, the largest LE of the M-AdNN is positive since it is given by $\lambda = \frac{1}{2} \log N + \log k_r$, rendering it *truly chaotic*. Since this chaotic property is fundamental to this chapter, we shall explain this, in greater depth, in Section 36.3.2.
5. The experimental results reported in Reference 8 claimed that the M-AdNN responds periodically for a trained input pattern. These results are presented in Section 36.3.3. However, it turns out that this periodic output occurs for both trained and untrained patterns, which was a phenomenon that was subsequently discovered in Reference 14.

Calitoiu et al. [7] later enhanced the M-AdNN to yield an even more interesting NN, the modified blurring AdNN (Mb-AdNN), which was capable of justifying the process of "blurring" in an NN. Since it is not directly relevant to this chapter, the details of the Mb-AdNN are omitted here. But it is very pertinent to mention that, topologically, this NN is also a completely connected NN, and that it requires a computational cost of $O(N^2)$ for obtaining the outputs of all the neurons.

36.3.2 LE Analysis of the M-AdNN

For a dynamical system, its sensitivity to its initial conditions is quantified by its exponents LEs. For example, consider two trajectories with initial conditions near to an attracting manifold. When an attractor is chaotic, the related trajectories, on average, diverge at an exponential rate characterized by the largest LE. This concept is also generalized for the spectrum of LEs, λ_i ($i = 1, \ldots, 2r$), by considering a small r-dimensional sphere of initial conditions, where r is the number of equations (or, equivalently, the number of state variables) used to describe the system. As time, t, progresses, the sphere evolves into an ellipsoid whose principal axes expand (or contract) at rates given by the LEs. The presence of positive exponents is sufficient for diagnosing chaos, and represents local instability in particular directions.

There are many ways, both numerically and analytically, to compute the LEs of a dynamical system. Generally, for systems with a small dimension, the best way is to analytically compute it using its formal definition. As opposed to this, for systems with a high dimension, it is usually not easy to obtain the entire LE spectrum in an analytic manner. In this case, we have several other alternatives to judge whether a system is chaotic. One of these is to merely determine the largest LE (instead of computing the entire LE spectrum) since the existence of a single positive LE indicates chaotic behavior. In this setting, this is usually achieved by using a numerical scheme as described in Algorithm 36.1. In this algorithm, the basic idea is to follow two orbits that are close to each other, and to calculate their average logarithmic rate of separation [9,19,20].

Algorithm 36.1 Numerical Calculation of the LE

Input: The NN whose LEs have to be determined.
Output: The LEs of the NN.
Method

1. Start with any initial condition in the basin of attraction.
2. Iterate until the orbit is on the attractor.
3. Select a nearby point (separated by d_0, as indicated in Reference 19. In our work, we set $d_0 = 10^{-12}$).
4. Advance both orbits by one iteration and calculate the new separation d_1.
5. Evaluate $\log|d_1/d_0|$ in any convenient base.
6. Readjust one orbit so that its separation is d_0 in the same direction as d_1.
7. Repeat Steps 4–6 many times and calculate the average of Step 5.

End Algorithm Calculation_Lyapunov_Exponent

In practice, this algorithm is both simple and convenient if we have the right to access the equations that govern the system. Furthermore, if it is easy to obtain the partial derivatives of the system, we can also calculate the LE spectrum by QR decomposition [9,18,20]. Moving now to a formal analytic strategy, we consider a discrete time ($t \in Z^+$) dynamical system $A \longmapsto F(A)$ and its Jacobian matrix of partial derivatives $J(A) = D_A F(A)$. Consider also the sequence $\{A_0, A_1, \ldots, A_{k-1}\}$ generated by successive iterations of the initial condition, A_0. For this sequence, we introduce the matrix

$$J_k(A) = J(F^{k-1}(A))J(F^{k-2}(A))\ldots J(F(A))J(A_0). \tag{36.6}$$

The LE are given by the logarithms of the eigenvalues of $\Lambda := [J^T(A)J(A)]^{1/2}$, where $J^T(A)$ denotes the transpose of $J(A) = \lim_{k\to\infty} J_k(A)$.

Theorem 36.1

The M-AdNN, described by the set of Equations 36.5, is locally unstable as demonstrated by its Lyapunov spectrum.

Proof. Consider the dynamical system for the M-AdNN, whose asymptotic dynamical matrix J is given as

$$J(A) = \begin{pmatrix} [J_{ij}^1] & [J_{ij}^2] \\ [J_{ij}^3] & [J_{ij}^4] \end{pmatrix}, \tag{36.7}$$

where each $[J_{ij}^k]$ is an $N \times N$ submatrix of $J(A)$, for $1 \leq k \leq 4$, $1 \leq i \leq N$, and $1 \leq j \leq N$. Each $[J_{ij}^k]$ is a result of the partial derivatives of $\eta_i(t+1)$ and $\xi_i(t+1)$ with regard to $\eta_j(t)$ and $\xi_j(t)$, respectively. This will be clarified currently.

The term $J_{ij}^k(t)$, for $1 \leq k \leq 4$, is the arbitrary element at time "t," and each J_{ij}^k in Equation 36.7 is obtained by taking the limit as $t \longrightarrow \infty$. Therefore, $J_{ij}^k(t)$ takes on the following four distinct forms:

1. $J_{ij}^1(t) = \dfrac{\partial \eta_i(t+1)}{\partial \eta_j(t)}$ when $1 \leq i \leq N$ and $1 \leq j \leq N$;

2. $J_{ij}^2(t) = \dfrac{\partial \eta_i(t+1)}{\partial \xi_j(t)}$ when $1 \leq i \leq N$ and $1 \leq j \leq N$;

3. $J_{ij}^3(t) = \dfrac{\partial \xi_i(t+1)}{\partial \eta_j(t)}$ when $1 \leq i \leq N$ and $1 \leq j \leq N$;

4. $J_{ij}^4(t) = \dfrac{\partial \xi_i(t+1)}{\partial \xi_j(t)}$ when $1 \leq i \leq N$ and $1 \leq j \leq N$.

Because $x_i(t+1) = f(\eta_i(t+1) + \xi_i(t+1))$, where f is the logistic function with the steepness parameter ε satisfying $f(y) = 1/(1 + \exp(-y/\varepsilon))$, and $\dfrac{df(y)}{dy} = \dfrac{d}{dy}\dfrac{1}{1+\exp(-y/\varepsilon)}$ (or $\dfrac{df(y)}{dy} = \dfrac{1}{\varepsilon}x(1-x)$ where $x \equiv f(y)$), the following explicit forms of $J_{ij}(t)$ result:

a. $J_{ij}^1(t) = \dfrac{w_{ij}}{\varepsilon}x_j(t)(1 - x_j(t))$ when $1 \leq i \leq N - 1$ and $1 \leq j \leq N$;

b. $J_{ij}^1(t) = k_f + \dfrac{w_{ij}}{\varepsilon}x_j(t)(1 - x_j(t))$ when $i = N$ and $1 \leq j \leq N$;

c. $J_{ij}^2(t) = \dfrac{w_{ij}}{\varepsilon}x_j(t)(1 - x_j(t))$ when $1 \leq i \leq N$ and $1 \leq j \leq N$;

d. $J_{ij}^3(t) = -\dfrac{\alpha}{\varepsilon}x_j(t)(1 - x_j(t))$ when $\leq i \leq N$ and $1 \leq j \leq N$ and $i = j$;

e. $J_{ij}^3(t) = 0$ when* $1 \leq i \leq N$ and $1 \leq j \leq N$ and $i \neq j$;

f. $J_{ij}^4(t) = -\dfrac{\alpha}{\varepsilon}x_j(t)(1 - x_j(t))$ when $1 \leq i \leq N - 1$ and $1 \leq j \leq N$ and $i = j$;

g. $J_{ij}^4(t) = 0$ when $1 \leq i \leq N - 1$ and $1 \leq j \leq N$ and $i \neq j$;

h. $J_{ij}^4(t) = k_r - \dfrac{\alpha}{\varepsilon}x_j(t)(1 - x_j(t))$ for $i = N$ and $1 \leq j \leq N$.

Since we seek the asymptotic value $J(A)$, we observe that the derivative (i.e., $x_i(t)(1 - x_i(t))$) is always positive and attains the value zero only when $x(t) = 0$ or 1. Thus, the outputs of the neurons converge to the values zero or unity, and the term $x_i(t)(1 - x_i(t))$ has an asymptotic value of zero for all $1 \leq i \leq N$. Thus, to obtain $J(A)$ we enforce the limiting argument to get

$$J(A) = \begin{pmatrix} 0 & \cdots & 0 & k_f & 0 & \cdots & 0 \\ \vdots & \ddots & \vdots & \vdots & \vdots & \ddots & \vdots \\ 0 & \cdots & 0 & k_f & 0 & \cdots & 0 \\ 0 & \cdots & 0 & 0 & \cdots & 0 & k_r \\ \vdots & \ddots & \vdots & \vdots & \ddots & \vdots & \vdots \\ 0 & \cdots & 0 & 0 & \cdots & 0 & k_r \end{pmatrix}. \tag{36.8}$$

Since $\Lambda_A = [J(A)^T J(A)]^{1/2}$ we have

$$\Lambda_A = \begin{pmatrix} 0 & \cdots & 0 & 0 & 0 & \cdots & 0 \\ \vdots & \ddots & \vdots & \vdots & \vdots & \ddots & \vdots \\ 0 & \cdots & 0 & 0 & 0 & \cdots & 0 \\ 0 & \cdots & 0 & N^{1/2}k_f & 0 & \cdots & 0 \\ 0 & \cdots & 0 & 0 & 0 & \cdots & 0 \\ \vdots & \ddots & \vdots & \vdots & \vdots & \ddots & \vdots \\ 0 & \cdots & 0 & 0 & 0 & \cdots & 0 \\ 0 & \cdots & 0 & 0 & \cdots & 0 & N^{1/2}k_r \end{pmatrix}. \tag{36.9}$$

Observe that this matrix has $2N$ eigenvalues. One eigenvalue with multiplicity $2N-2$ is 0, and the others are $N^{1/2}k_f$ and $N^{1/2}k_r$. Thus, $\mu_1 = \ldots = \mu_{2N-2} = 0$; $\mu_{2N-1} = N^{1/2}k_f$; and $\mu_{2N} = N^{1/2}k_r$.

Since the LE are given by the logarithms of the eigenvalues, we have $\lambda_1 = \ldots = \lambda_{2N-2} = -\infty$; $\lambda_{2N-1} = \frac{1}{2}\ln N + \ln(k_f)$; and $\lambda_{2N} = \frac{1}{2}\ln N + \ln(k_r)$.

We know that the LE of a superstable point (or superstable orbit) is $-\infty$, and that the LE of a bifurcation point is 0. Every value between $-\infty$ and 0 implies a stable solution. The M-AdNN has *one* positive LE with

---

* Observe that $J_{ij}^3(t)$ has the value zero; this result is obtained as follows: $\dfrac{\partial \xi_i(t+1)}{\partial \eta_j(t)} = \dfrac{\partial(k_r \xi_i(t) - \alpha x_j(t) + a_j)}{\partial \eta_j(t)} = 0$.

the value $\frac{1}{2}\ln N + \ln(k_f)$ and *another* positive exponent with the value $\frac{1}{2}\ln N + \ln(k_r)$. The presence of the positive exponents proves the theorem.   ∎

Remark 36.1

1. Adachi and Aihara [1] set $k_f = 0.2$, $k_r = 0.9$, and $N = 100$, implying that the AdNN involves a 200×200 matrix. Thus, the AdNN has 100 LE with the value -1.609437, and 100 LE with the value -0.105360, which shows that the AdNN is *not chaotic*.

2. Using the same coefficients and parameters, the M-AdNN has *one* LE with the value 2.197225, another *one* with the value 0.693148, and 198 LE with the value $-\infty$. Thus, the M-AdNN can truly demonstrate chaotic properties.

3. Although fascinating, the reason why the system switches from chaos to repetition/periodicity has been explained in an elegant manner in Reference 8, but is omitted here in the interest of brevity.

4. A comparison of the AdNN and M-AdNN with respect to their PR capabilities can be briefly summarized as follows:

   a. The transitory phase for the AdNN is a few orders of magnitude larger than that of the M-AdNN.

   b. When it concerns the PR capabilities of the AdNN, we observe that the system is periodic, although the output pattern does not always mimic the input. Furthermore, the periodicity of the AdNN is again a few orders of magnitude larger than that of the M-AdNN.

   More details about these are explained in the next subsection.

36.3.3 Experimental Results

In this section, we discuss the AM and PR capabilities of the M-AdNN using the strategy discussed in Section 36.1.3. To present our results in the right perspective, we have tested the schemes for two sets of data described in Section 36.1.4 (see Figures 36.1 and 36.4). The first was the set which Adachi and his coauthors used [1]. The second set is more realistic, and is one which involves the recognition of numerals. We now report the results obtained from each of these two sets.

36.3.3.1 PR with Adachi's Data Set

For the *Adachi* data set presented in Figure 36.1a–d, the underlying M-AdNN had $N = 100$ neurons. The constants used for Equation 36.5 were $\varepsilon = 0.00015$, $k_f = 0.2$, and $k_r = 0.9$. After the training, the system was presented with 10×10 binary-valued arrays which contained "exact" versions of one of these four patterns. The testing was initiated with an initial random pattern given in Figure 36.1e. This *same* initial pattern was used in all the testing experiments.

36.3.3.2 Periodicity-Based Accuracy for the Nonnoisy and Noisy Adachi's Data Set

The M-AdNN trained as described above, demonstrated a periodic response when the nonnoisy external stimuli were applied, after an initial nonperiodic transient phase. It is interesting to note that the transient phase was *very* short—the mean length of the transient phase was 38.5 time units. The actual length of the transient phase in each case is given in Table 36.1. Thereafter, the system resonated sympathetically with the input pattern, with a fairly small periodicity. The periodicity of the response is also tabulated in Table 36.1. Note that the periodicity is "small" (between 18 for Pattern 2, and 39 for Patterns 3 and 4) since N_{Max} (which, in our case, is 1000) itself could be arbitrarily large. Thus, there are enough repetitions in the output of the M-AdNN for the experimenter to observe the periodicity, and to infer the identity of the presented pattern.

It should be observed that since the system was presented with "nonnoisy" patterns, any accuracy (i.e., the jump from chaos to repetition/periodicity) less than 100% would be unacceptable, and this accuracy is, indeed, what was obtained for the M-AdNN.

TABLE 36.1 Transitory Phase and the Periodicity for the
M-AdNN Tested with Adachi's "Noise-Free" Data Set
(Figure 36.1), When the Network Evolved from an Initial
State Given in Figure 36.1e

Pattern	Number of Steps in Transitory Process	Periodicity
1	39	20
2	37	18
3	39	39
4	39	39

TABLE 36.2 Transitory Phase and the Periodicity for the
M-AdNN with Noisy Samples from Adachi's Data Set,
When the Network Evolved from an Initial State Given in
Figure 36.1e

Pattern	Number of Steps in Transitory Process	Periodicity
1	39	39
2	39	39
3	19/39	18
4	19	18

To investigate the power of the M-AdNN for noisy samples, it was also tested for the noisy data set proposed by Adachi et al. The M-AdNN was trained using $N = 100$ neurons, and the constants for Equation 36.5 were set as $\varepsilon = 0.00015$, $k_f = 0.2$, and $k_r = 0.9$. After the training, the system was presented with 10×10 binary-valued arrays which contained noisy versions of one of these four patterns, as explained in Section 36.1.4. As in the "noise-free" case, the testing was initiated with the initial random pattern given in Figure 36.1e.

For each class (i.e., each trained prototype), the Hamming distance $d_p(t)$ between it and the network's output pattern was computed. For a pattern to be recognized, the value of the threshold specified in the algorithm was set to be the same as the level of noise.

The M-AdNN trained as described above demonstrated a periodic response with the noisy external patterns, after an initial nonperiodic transient phase. As before, the transient phase was very short—the mean length of the transient phase was 34 time[*] units. The actual length of the transient in each case is given in Table 36.2. It should be observed that, the system resonated sympathetically with the input pattern, with a fairly small periodicity, also tabulated in Table 36.3. Again, unlike the case of traditional PR systems, the chaotic PR would operate by noticing periodicity, and in this case, there is enough repetition for the M-AdNN to observe the periodicity, and to infer the identity of the presented pattern. Observe that the recognition accuracy (i.e., the moving away from chaos to repetition/periodicity) is 100%, even though the images are significantly degraded.

36.3.3.3 PR with the *Numeral* Data Set

In this subsection, we report the results of training/testing on the *numeral* data set described in Section 36.1.4. The training set had 10 patterns, given in Figure 36.4, and consisted of 10×10-bit maps of the numerals {"0" . . . "9"}.

---

[*] In the case of Pattern 3, the first periodic pattern was visible at time index 19. But the periodicity was observable only after time index 39.

TABLE 36.3 Transitory Phase and the Periodicity for the
M-AdNN, for the Nonnoisy Versions of the Training Set,
Namely, the Numerals

Pattern	Number of Steps in Transitory Process	Periodicity
1	15	7,15
2	24	26
3	24	26
4	24	26
5	24	26
6	24	27
7	24	26
8	24	26
9	24	26
10	24	26

Note: The first pattern has a limit cycle with double periods,
the first with 7 units and the second with 15.

36.3.3.4 Periodicity-Based Accuracy for the Nonnoisy and Noisy Numeral Data Set

The trained M-AdNN again demonstrated a periodic response when the nonnoisy external stimuli were applied, after an initial nonperiodic transient phase. Here too, the transient phase was *very* short—its mean length was 23.1 time units, and most of the transitory phases were of length 24 units. The actual length of the transient phase in each case is given in Table 36.4. As before, the system resonated sympathetically with the input pattern, with a fairly small periodicity. The periodicity of the response is also tabulated in Table 36.4, where the initial starting input pattern was randomly chosen. Again, note that the periodicity was fairly "small" (26 in most cases—the first pattern had a limit cycle with double periods, the first with 7 units, and the second with 15). The accuracy of recognition (i.e., the switching from chaos to repetition/periodicity) was again 100%.

To investigate the power of the M-AdNN for noisy numeral samples, we mention two cases, namely, the case when the noise was 10% (see Figure 36.5) and the case when the noise was 15% (see Figure 36.6). In both cases, after an initial nonperiodic transient phase, the trained M-AdNN demonstrated a periodic response when it was presented with the noisy numerals. As before, the transient phase (see Table 36.4) was rather insignificant (it had an average value of 20.7 when the noise was 10%), but the periodic resonance was significant. The resonance periodicity[*] was consistently "small" (also tabulated in Table 36.4).

The salient characteristics of the PR can be summarized as

1. The system is almost insensitive to the *initial* state, but rather, very sensitive only to the input pattern. Indeed, even if the initial state was any of the *other* stored patterns, the transient and periodicity are essentially unaffected.
2. The identity does not depend on the specific period—many trained patterns can lead to the same periodicity. Such a unique mapping between the trained pattern and the periodicity, unfortunately, does not exist.
3. Numerous additional simulations for the case when the testing patterns are not training patterns were also done. It turns out that no output with a distance smaller than the chosen threshold was obtained.

---

[*] The periodicity (7,15) means that we encounter a "double cycle." Thus, after the transient phase, the training pattern occurs at time instants 7, 22, 29, 44, etc. This is actually because we have an 8-shaped limit cycle with the smaller loop of the "8" having a periodicity of 7, and the larger loop having a periodicity of 15.

TABLE 36.4 Transitory Phase and the Periodicity for the M-AdNN, When the Testing Is Done with Patterns from the Numeral Training Set Containing 10% Noise and Also Patterns Containing 15% Noise

Pattern	Transient (10%)	Periodicity (10%)	Transient (15%)	Periodicity (15%)
1	8	7, 7, 8	24	25
2	24	26	8	7,7,8
3	24	26	8	7,7,8
4	24	26	8	7,7,8
5	24	26	8	7,7,8
6	24	27	8	7,7,8
7	24	26	8	7,7,8
8	24	26	8	7,7,8
9	24	26	8	2,5,7,8
10	7	22	7	22

Note: The first pattern with 10% noise has a limit cycle with triple periods, the first two with a period of 7, and the last with a period of 8. Similar scenarios are observed in the 15% noise case.

4. The threshold was not too crucial. The researchers of Reference 8 opted to use a threshold that is directly related to the acceptable level of noise found in the testing pattern. In the experiments reported, they used noise levels from 0% to 15%. The question of how this "acceptable threshold" varies with the noise is open. If the threshold is too small, a "nonperiodic" signal could possibly be inferred. But as mentioned above, they increased the threshold as the permitted noise level was increased.

5. The recognition accuracy (i.e., the switching from chaos to repetition/periodicity) was 100%, even though the images were quite degraded.

6. The question of why the M-AdNN yields cycles of finite periodicity remains open. However, we should point out that since the number of neurons is so high, there is a possibility that there are other attractors, which were not located, whose periodicities are many orders of magnitude larger (e.g., of the order of 10,000) than the periodicities that were otherwise observed for the trained patterns. But even if that were the case, the large differences in these periodicities would help in discriminating between these two "classes" of attractors: the "useful" ones which achieve PR, and the "nonuseful" ones. But the existence of the latter still seems to be a conjecture.

The reader will observe that the results are quite remarkable, especially when one sees the extremely poor quality of the testing samples.

One can imagine a direct application of this Chaotic NN in optical character recognition (OCR) systems. In this case, the set of training patterns will consist of the characters to be recognized—represented within the bit pattern of the character window. The new pattern that is presented to the OCR system to be recognized, will cause the system to resonate with the output being the ideal character stored in the weights of the network. Since such a PR system will switch from being chaotic to repetition/periodicity with 100% accuracy, we believe that the *overall* system will have a greater accuracy if the actual text is preprocessed appropriately. Observe that we are not required to invoke elaborate segmentation or skeletization procedures—the background can remain relatively noisy.

We summarize the characteristics of the M-AdNN as follows: by decreasing the multiplicity of the eigenvalues of the AdNN's control system, we can effectively drive the system into chaos, and also increase the possibility of periodic behavior. The M-AdNN has the desirable property that it can recognize various input patterns by the system essentially *sympathetically* "resonating" with a finite periodicity whenever *these* samples (or their reasonable resemblances) are presented.

36.4 Ideal Chaotic PR: The Ideal M-AdNN

36.4.1 The Design of the Ideal M-AdNN

The M-AdNN described above is a fascinating NN which has been shown to possess the required repetitive/periodic property desirable for PR applications. However, subsequent research discovered that the PR properties of the M-AdNN are not as powerful as originally reported. Indeed, it resonates periodically for *trained* input patterns. But unfortunately, it also resonates for unknown patterns and produces these unknown patterns at the output periodically. Consequently, we must reconsider how the parameters of the M-AdNN for its weights, steepness, and external inputs, can be specified so as to yield a new NN capable of a truly *chaotic PR* behavior. We shall refer to this NN as the *ideal* M-AdNN.

Quite simply put, the ideal M-AdNN has the same topology of the M-AdNN. But the parameters, the weight assignment rule, and all its parameters have now been modified to yield the properties that we are seeking. All these issues will be discussed in the forthcoming sections.

36.4.2 The Weights of the Ideal M-AdNN

In the interest of clarity, we first argue that the assignment of the weights as per Equation 36.10 for the M-AdNN is not appropriate. We rather advocate another form of the Hebbian rule given by Equation 36.15 instead. The reason for this is explained below.

As we have observed above, the M-AdNN uses a form of the Hebbian rule to determine the weights of the connections in the network. This rule is defined by the following equation:

$$w_{ij} = \frac{1}{p} \sum_{s=1}^{p} P_i^s P_j^s, \tag{36.10}$$

where P_i^s denotes the ith neuron of the sth pattern P^s, and p is the number of training patterns.

At this juncture, we emphasize that the Hebbian rule, Equation 36.10, is founded on one fundamental premise: any pair of learning vectors, P and Q, must be orthogonal, and thus, for all P and Q, if P^T is the transpose of P:

$$P^T Q = \begin{cases} 0, (P \neq Q) \\ N, (P = Q). \end{cases} \tag{36.11}$$

The reasons why the formula given by Equation 36.10 is based on the above, is because of the following: if P denotes an N-by-1 vector (e.g., the $P1$ of Figure 36.1 is a 100-by-1 vector), Equation 36.10 can be rewritten as

$$W = \frac{1}{p} \sum_{s=1}^{p} P^s (P^s)^T. \tag{36.12}$$

Thus, the output for any given input vector P^k is

$$O = W P^k = \frac{1}{p} \sum_{s=1}^{p} P^s (P^s)^T P^k$$

$$= \frac{1}{p} \sum_{s=1}^{p} P^s \left[(P^s)^T P^k \right] = \frac{N}{p} P^k. \tag{36.13}$$

As per Equation 36.11, the output $O = \frac{N}{p}P^k$ is a scalar multiple* of the input P^k. Otherwise, if the learning vectors $\{P\}$ are not orthogonal, Equation 36.13 has the form

$$O = \frac{1}{p}\sum_{s=1}^{p}P^s[(P^s)^T P^k] = \frac{1}{p}\left(NP^k + P^{noise}\right), \tag{36.14}$$

where $P^{noise} = \sum_{s=1,s\neq k}^{p} b_s P^s$ and $b_s = (P^s)^T P^k$. Obviously, $O = \frac{N}{p}P^k$ if and only if $P^{noise} = 0$, explaining the rationale for orthogonality.

The need for orthogonality is not so stringent, because, as per the recorded research, when the number of neurons is much larger than the number of patterns, and the learning vectors are randomly chosen from a large set, the probability of having the learning vectors being orthogonal is very high. Consequently, Equation 36.11 is true, albeit probabilistically.

36.4.2.1 Summary

Based on the above observations, we conclude that to enhance the M-AdNN, we should not use the Hebbian rule as dictated by the form given in Equation 36.10, since the data sets used by both Adachi et al. and Calitoiu et al. are defined on $\{0,1\}^N$, and the output is further restricted to be in $[0,1]$ by virtue of the logistic function. One can easily verify that any pair of the given patterns in Figure 36.3 are *not* orthogonal. In fact, this is why Adachi and Aihara computed the weight by scaling all the patterns to -1 and 1 using the formula given by Equation 36.15 instead of Equation 36.10:

$$w_{ij} = \frac{1}{p}\sum_{s=1}^{p}(2P_i^s - 1)(2P_j^s - 1). \tag{36.15}$$

By virtue of this argument, in designing the ideal M-AdNN, one uses Equation 36.15 to determine the network's weights.

It is pertinent to mention that since the patterns P are scaled to be in the range between -1 and 1, it *does* change the corresponding property of orthogonality. Actually, it is very interesting to compare the inner product of the so-called Adachi data set before and after scaling (see Table 36.5). From it, we see that the inner products of any pair of *scaled* patterns while not being exactly 0, are very close to 0—i.e., these patterns are *almost* orthogonal. In contrast, the inner products of the corresponding *unscaled* patterns are very large. Again, this table confirms that Equation 36.15 is a more suitable (and reasonable) expression than Equation 36.10.

36.4.3 The Steepness, Refractory, and Stimulus Parameters

36.4.3.1 Significance of ε for the AdNN

The next issue that we need to consider concerns the value of the steepness parameter ε of the output function. From Section 36.2, we see that the output function is defined by the logistic function $f(x) = \frac{1}{1+e^{-x/\varepsilon}}$, which is a typical sigmoid function.

One can see that ε controls the steepness of the output. If $\varepsilon = 0.01$, then $f(x)$ is a normal sigmoid function. If ε is too small, say, 0.0001, the logistic function almost degrades to become a unit step function (see Figure 36.11). The question, really, is one of knowing how to set the "optimal" value for ε. To provide a rationale for determining the best value of ε for the ideal M-AdNN, we concentrate on Adachi's neural

---

* Observe that without loss of generality, we can easily force $O = P^k$ instead of $O = \frac{N}{p}P^k$, by virtue of a straightforward normalization.

TABLE 36.5 Inner Products of the Patterns of the Adachi Set

	P1	P2	P3	P4
P1	100	10	10	6
P2	10	100	−4	4
P3	10	−4	100	12
P4	6	4	12	100
P1	49	27	27	26
P2	27	50	24	26
P3	27	24	50	28
P4	26	26	28	50

Note: In the left table, all patterns are scaled to −1 and 1 by $2P-1$. In the table on the right, all the patterns are defined on $\{0, 1\}^N$. The inner products in the table on the left are much smaller and almost 0, implying that the patterns are *almost* orthogonal.

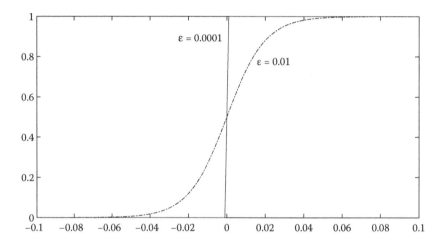

FIGURE 36.11 A graph demonstrating the effect of the parameter ε to change the steepness of the sigmoidal function. The ε for the dashed and solid lines are 0.01 and 0.0001, respectively.

model [1] defined by

$$y(t+1) = ky(t) - \alpha f(y(t)) + a, \tag{36.16}$$

where $f(\cdot)$ is a continuous differentiable function, which as per Adachi and Aihara [1], is the logistic function.*

The properties of Equation 36.16 greatly depend on the parameters $k, \alpha, a,$ and $f(\cdot)$. In order to obtain the full spectrum of the properties represented by Equation 36.16, it is beneficial for us to first consider $f(\cdot)$ in terms of a unit step function, and to work with a fixed point analysis. To do this, we consider a few distinct cases listed below:

1. If the system has only *one* fixed point.
    a. In this case, obviously, $y(t) = 0$ is not a fixed point of Equation 36.16 unless $a = 0$.

---

* Historically, the original form of this equation initially appeared in the paper by Nagumo and Sato [13]. The only difference between the function that these researchers used and the one discussed here is the function $f(\cdot)$, since in Reference 13, they utilized the unit step function instead of the logistic function.

b. If Equation 36.16 has only one positive fixed point, say y_0, then the value of this point can be resolved from Equation 36.16 as $y_0 = \frac{a - \alpha}{1 - k}$. In the paper by Nagumo and Sato [13], the authors set $\alpha = 1, k < 1$, and let a vary from 0 to 1. Apparently, the above value of y_0 is always negative, which contradicts our assumption that y_0 is a positive fixed point. Similarly, we can see that Equation 36.16 does not have any negative fixed points either. In other words, we conclude that Equation 36.16 does not have any fixed point.

2. If the system has period-2 points, say y_1 and y_2 (where $y_1 \neq y_2$), we see that these points must satisfy

$$\begin{cases} y_2 = ky_1 - \alpha f(y_1) + a, \\ y_1 = ky_2 - \alpha f(y_2) + a. \end{cases} \qquad (36.17)$$

a. If 0 is one of the period-2 points (without loss of generality, $y_1 = 0, y_2 \neq 0$), then we can solve the above equations to yield the two solutions as

$$\begin{cases} y_2 = a, \\ y_1 = (k + 1)a - \alpha. \end{cases}$$

From this, we see that the two points exist if and only if $a \neq 0$ and $\alpha = (1 + k)a$.

b. If the system has two positive period-2 points (assume $y_1 > 0, y_2 > 0$), we see that the system can be rewritten as

$$\begin{cases} y_2 = ky_1 - \alpha + a, \\ y_1 = ky_2 - \alpha + a, \end{cases}$$

which is equivalent to implying that $y_2 - y_1 = k(y_1 - y_2)$. Clearly, this equality cannot hold since $y_1 \neq y_2$ and $k > 0$. In an analogous manner, we can conclude that the system does not have two negative period-2 points either.

c. If the system has two period-2 points with different signs (without loss of generality we can assume that $y_1 > 0, y_2 < 0$), we solve Equation 36.17 as

$$\begin{cases} y_1 = \dfrac{a(k + 1) - k\alpha}{1 - k^2}, \\ y_2 = \dfrac{a(k + 1) - \alpha}{1 - k^2}. \end{cases} \qquad (36.18)$$

Since $y_1 > 0$ and $y_2 < 0$, this yields the inequality

$$\frac{k\alpha}{1 + k} < a < \frac{\alpha}{1 + k}. \qquad (36.19)$$

In Nagumo's paper [13], the authors set $k = 0.5$ and $\alpha = 1$, implying that the system has period-2 fixed points only when $1/3 < a < 2/3$.

From the above, we know that the Nagumo–Sato model does not have any fixed points, but that it rather does have period-2 points. The next question that we have to resolve is whether it possesses chaotic properties. To determine this, we have to investigate what happens to the iteration orbits when the initial point is not exactly on, but close to a period-2 point. Indeed, this question can be answered by considering the inequality

$$\left| \frac{\partial g^{(2)}(y)}{\partial y} \right|_{y = y_1} < 1, \qquad (36.20)$$

where $g(y) = ky - \alpha f(y) + a$, and $f(\cdot)$ is still a unit step function. A simple algebraic manipulation displays that the inequality given by Equation 36.20 is always true whenever $k < 1$,

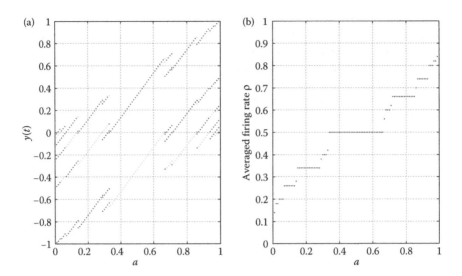

FIGURE 36.12 The period-n orbits (a) and the averaged firing rate (b) of the Nagumo–Sato model, in which a varies from 0 to 1.

implying that whatever the initial condition, after a sufficiently large number of iterations, the orbits converge to the period-2 points. In other words, all the period-2 points are attracting, implying that the system does not demonstrate any chaotic phenomena.

3. If we now consider the period-n orbits, one can see that it is not possible to resolve the set of equations that generalize from Equations 36.18 and 36.19 for n variables. However, the analysis that corresponds to generalizing the *stability* equation is exactly the same as in Equation 36.20, and in every case, one can again show that the inequality is true.[*] This again leads to the conclusion that all the period-n orbits are also attracting. This can be verified by considering the graphs in Figure 36.12.

The situation is, however, completely different if the unit step function is replaced by the sigmoid function. This is a relevant issue because, as stated earlier, the difference between the Nagumo–Sato model and the Adachi model lies in the output function $f(\cdot)$, where the former uses the unit step function and the latter uses a sigmoid function. The issue is further accentuated because the logistic function may degrade to a unit step function if an inappropriate value of ε is utilized, as shown in Figure 36.11. Indeed, when ε is very small, the sigmoid function will "degrade" to the unit step function, leading us to the analysis which we have just concluded.

36.4.3.2 Significance of ε for the M-AdNN

Since the M-AdNN is a modified version of the AdNN (which uses the logistic function as the output function), setting the value of ε is again quite pertinent. As we shall argue now, in order to enable the system to have chaotic properties, the parameter ε should not be close to zero.

To demonstrate the role of the parameter ε of the Adachi model, as before, we analyze the behavior of the model starting from the fixed points or period-n points.

---

[*] This has been done for values of n which equal 3, 4, and 5. The general analysis for arbitrary values of n can also be done using the Schwarz derivative and the period-doubling bifurcation theorem. However, this goes beyond the scope of this chapter and is thus omitted here.

Let us first analyze whether the Adachi neuron has any fixed points, i.e., if

$$y(t+1) = y(t) = ky(t) - \alpha f(y(t)) + a,$$

is ever satisfied. Since this equation is a transcendental equation, obtaining the exact fixed point(s) is far from trivial. Nevertheless, we are still able to achieve some analysis on the bifurcation parameters.

As we know, if a fixed point(s) y^* exists, it should satisfy

$$\alpha f(y^*) = ky^* - y^* + a.$$

Our first task is to see if a fixed point y^* does exist.

To do this, we compose

$$F(y) = -\alpha f(y) + (k-1)y + a, \tag{36.21}$$

which is done so that the root of $F(y)$ corresponds to the fixed point of y.

It is easy to observe the following: for any given negative y, $f(y)$ is almost zero by virtue of its sigmoidal nature. Thus, $F(y) \approx (k-1)y + a > 0$. Similarly, for any given positive y (e.g., $y=1$), $f(y)$ is almost unity, again by virtue of its sigmoidal nature. Thus, $F(y) \approx -\alpha + k - 1 + a < 0$. Since $F(y)$ is a continuous function over $(-\infty, \infty)$, and since it changes its sign, there must exist at least one y^* satisfying $F(y^*) = 0$ implying that this value of y^* is the fixed point.

The next question is how many fixed points does the system have? This question can be answered by verifying the monotonicity of Equation 36.21. Since $f'(y) = -\frac{1}{\varepsilon}f(y)[1 - f(y)]$, we get

$$F'(y) = -\frac{\alpha}{\varepsilon}f(y)[1 - f(y)] + k - 1. \tag{36.22}$$

Again, since we always have $f(y)[1 - f(y)] \approx 0$, it implies that $F'(y) \approx k - 1 < 0$, which, in turn means that $F(y)$ is a decreasing function. As a result of the fact that it is positive on one side of the fixed point, negative on the other side, and simultaneously always has a negative derivative, we conclude that the Adachi model has only *a single* fixed point, y^*.

If y^* is stable, it should satisfy the condition

$$\left| \frac{dy(t+1)}{dy(t)} \right|_{y=y^*} < 1. \tag{36.23}$$

From the dynamical form of $y(t+1)$, this derivative can be rewritten as:

$$\left| k - \frac{\alpha}{\varepsilon}f(y^*)(1 - f(y^*)) \right| < 1, \tag{36.24}$$

which is equivalent to:

$$\begin{cases} \alpha f^2(y) - \alpha f(y) + \varepsilon(k+1) > 0, \\ \alpha f^2(y) - \alpha f(y) + \varepsilon(k-1) < 0. \end{cases} \tag{36.25}$$

Since each neuron's output is between 0 and 1, we see that for all values in $(0,1)$ the inequalities given by Equation 36.25 are always true. This leads us to the discriminants

$$\begin{cases} \Delta_1 = \alpha^2 - 4\alpha\varepsilon(k+1) < 0, \\ \Delta_2 = \alpha^2 - 4\alpha\varepsilon(k-1) > 0. \end{cases} \tag{36.26}$$

Obviously, since $k - 1 < 0$, $\alpha > 0$, and $\varepsilon > 0$, the second inequality is always true. Concentrating now on the first inequality, we can solve for ε and see that it has to satisfy

$$\varepsilon > \frac{\alpha}{4(k+1)}. \tag{36.27}$$

Briefly, this states that if the fixed point is stable, ε should be greater than $\alpha/4(k+1)$; otherwise, it is not stable.

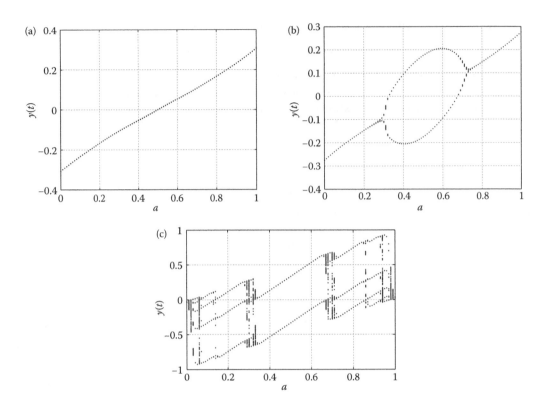

FIGURE 36.13 This figure shows that as the steepness parameter, ε, varies, the dynamics of the Adachi neuron changes significantly. In the three figures, the values of ε are 0.18, 0.15, and 0.015, respectively.

This condition is experimentally verified in Figure 36.13a–c. In our experiment, the parameters were set to be $\alpha = 1$ and $k = 0.5$. In this case, the "tipping point" for ε is $1/6 \approx 0.1667$. As one can very clearly see, if $\varepsilon = 0.18 > 0.1667$, all the fixed points are stable (Figure 36.13a); otherwise, if $\varepsilon = 0.15 < 0.1667$, there exist period-doubling bifurcations[*] (Figure 36.13b). As ε is further decreased, one can observe chaotic windows (Figure 36.13c).

We conclude this section by emphasizing that ε cannot be too small, for if it were, the Adachi neural model would degrade to the Nagumo–Santo model, which does not demonstrate any chaotic behavior. This is also seen from Figure 36.14a and b. In Figure 36.14a, where the parameter ε is 0.015, we can see that the orbit of the iteration is nonperiodic, while in Figure 36.14b, in which ε is 0.00015, the orbit of the trajectory is forced to have a periodicity of 5. In fact, we have calculated the whole LE spectrum of the system given by Equation 36.16 when it concerns the parameters ε, as shown in Figure 36.15a. In order to get a better view, we have amplified the graph in the area $\varepsilon \in [0, 0.04]$ and in the area $\varepsilon \in [-1.5, 0.5]$ in 36.15b. As one can see when $\varepsilon = 0.015$, the LE is positive, and has a value which is approximately 0.3, which indicates chaos. On the other hand, when $\varepsilon = 0.00015$, the LE is negative. Of course, there are some other optional values which also lead to chaos—values such as $\varepsilon = 0.007, 0.008, 0.012$, etc., can also lead to chaos.

Indeed, our arguments show that the value of ε as set in Reference 7 to be $\varepsilon = 0.00015$, is not appropriate. Rather, to develop the ideal M-AdNN, we have opted to use a value of ε which is *two* orders of magnitude larger, i.e., $\varepsilon = 0.015$.

---

[*] In this chapter, we analyze the bifurcation in a rather simple way. The reader must note that this could also be achieved by using the Schwarz derivative and the period-doubling bifurcation theorem.

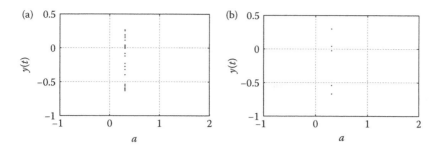

FIGURE 36.14 This figure shows that if the value of ε is too small, it will force the orbit of the trajectory to be periodic (b). The values of ε for (a) and (b) are 0.015 and 0.00015, respectively.

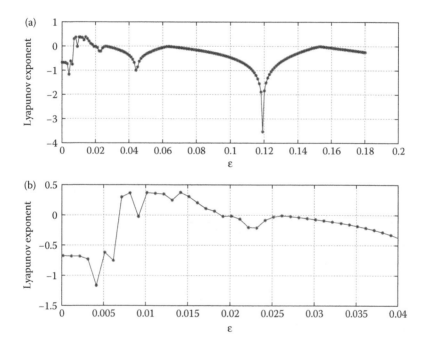

FIGURE 36.15 This figure shows the LE spectrum of the system. Some values between 0.007 and 0.02 may cause the system to be chaotic. The reader should observe that the other parametric settings are: $k = 0.5$, $a = 0.31$. The reason why we have used such settings is because one can observe chaotic windows under these conditions, as shown in Figure 36.13c.

36.4.4 LEs Analysis of the Ideal M-AdNN

We shall now analyze the LEs of the ideal M-AdNN, both from the perspective of a *single* neuron and of the entire network.

To initiate this, we first present the theoretical basis for obtaining the LE spectrum using the QR decomposition. In the interest of compactness, we use the notation that J_t represents the Jacobian matrix at time t, i.e., $J_t = J(t)$, whose transpose is written as J_t^T. Also, x_0 is assumed to be a randomly chosen starting point, and $\{f(x_0), f_2(x_0), f_3(x_0) \dots f_t(x_0)\}$ denotes the trajectory obtained as a consequence of iterating as per the system's dynamical equation. Again, to simplify the representation, we shall use the notation that $x_k = f_k(x_0)$.

As we know, the LEs of a network are defined by the eigenvalues of the limit of the matrix[*]

$$\Lambda = \lim_{t \to \infty} [J_t \cdot J_t^T]^{\frac{1}{2t}}. \tag{36.28}$$

Generally, it is hard to compute the limit of Equation 36.28, which is why one often invokes the QR decomposition. To achieve this, we note that by the chain rule

$$Df_t(x_0) = J(f_{t-1}(x_0)) \dots J(f(x_0)) \cdot J(x_0),$$

where $J(x_i) = Df(x)|_{x=x_i} = J(f_i(x_0))$. As clarified in References 9 and 20, we can rewrite $J_0(x_0)$ using its QR decomposition as $J_0(x_0) = Q_1 R_1$, where Q_1 is orthogonal implying that $Q_1 \cdot Q_1^T = I$. If we define $J_k^* = J(f_{k-1}(x_0))Q_{k-1}$, we can write *its* QR decomposition as $J_k^* = Q_k R_k$. Thus, $J(f_{k-1}(x_0)) \cdot Q_{k-1} = Q_k R_k$. Consequently, we obtain $J(f_{k-1}(x_0)) = Q_k R_k Q_{k-1}^{-1}$. By applying this equation to the chain rule, the differential $Df_t(x_0)$ can be transformed to be

$$\begin{aligned} Df_t(x_0) &= Q_t R_t Q_{t-1}^{-1} \dots Q_{t-1} R_{t-1} Q_{t-2}^{-1} \dots R_1 \\ &= Q_t R_t \cdots R_1. \end{aligned} \tag{36.29}$$

The LEs, $\{\lambda_i\}$, can then be obtained as

$$\lim_{t \to \infty} \frac{1}{t} \ln |v_{ii}(t)| = \lambda_i, \tag{36.30}$$

where $\{v_{ii}(t)\}$ are the diagonal elements of the product $R_{t-1} R_{t-2} \dots R_1$.

Again, for the reader's convenience, the algorithmic details are formally given in Algorithm 36.2.

Algorithm 36.2 Calculation of the LE by QR Decomposition

Input: The NN whose LEs have to be determined.
Output: The LEs of the NN.
Method

1. Choose a starting point x_0 and compute the Jacobian matrix $J_0(x_0) = \frac{df(x)}{dx}$ at x_0.
2. Decompose $J_0(x_0)$ by the QR decomposition: $J_0(x_0) = Q_1 \cdot R_1$, where Q_1 is an orthogonal matrix and R_1 is an upper triangular matrix. Let $\Upsilon = R_1$.
3. Compute the Jacobian matrix at x_1, $J_1(x_1) = \frac{df(x_1)}{dx_1}$, where $x_1 = f(x_0)$.
4. Let $J_2^* = J_1 \cdot Q_1$ and decompose J_2^* using the QR decomposition: $J_2^* = Q_2 \cdots R_2$, where Q_2 is also an orthogonal matrix and R_2 is an upper triangular matrix. Let $\Upsilon = R_2 \cdot \Upsilon$.
5. Repeat Steps 3 and 4 $(t-1)$ times till we finally obtain $\Upsilon = R_t R_{t-1} \dots R_1$.
6. The LE spectrum is defined by the logarithm of the diagonal elements v_{ii} of Υ: $\lambda_i = \frac{1}{t} \log |\Upsilon_{ii}|$.

End Algorithm Calculation_Lyapunov_Exponent_ByQR

36.4.4.1 Numeric Lyapunov Analysis of an ideal M-AdNN

We shall now embark on a Lyapunov analysis of the ideal M-AdNN. To do this, we first undertake the Lyapunov analysis of a single neuron in the interest of simplicity. Indeed, it can be easily proven that a single neuron is chaotic when the parameters are properly set.

---

[*] The conditions for the existence of the limit are given by the Oseledec theorem.

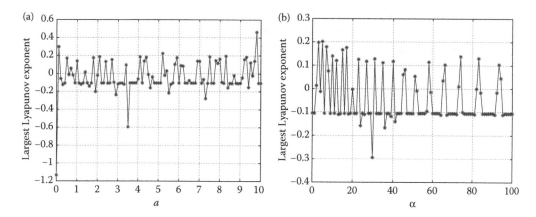

FIGURE 36.16 The variation of the largest LE of a single neuron with the parameters a and α. These two figures are plotted using Algorithm 36.1. In (a), a varies from 0 to 10 with a step size of 0.1. In (b), α varies from 0 to 100 with a step size of 1.

Consider a primitive component of the ideal M-AdNN, where the model of a *single* neuron can be described as[*]:

$$\eta(t+1) = k_f\eta(t) + wx(t), \tag{36.31}$$

$$\xi(t+1) = k_r\xi(t) - \alpha x(t) + a, \tag{36.32}$$

$$x(t+1) = \frac{1}{1 + e^{-(\eta(t+1)+\xi(t+1))/\varepsilon}}. \tag{36.33}$$

As we know from the records of References 1 and 6, the external stimulus a is considered as a constant, i.e., $a = 2$. In References 1 and 6, the authors have shown the LE spectrum of Equation 36.16 while a varies from 0 to 1. However, for a single Adachi neuron, the role of a has not yet been discussed. As a matter of fact, most of the papers which concern the Lyapunov analysis of the AdNN and its variants, focus on the parameters k_f and k_r [1,7,8,15,16]. Indeed, the roles of α and a have never been discussed seriously in the literature.

Figure 36.16a and b shows the largest LE spectrum as obtained by Equations 36.31 through 36.33. We can clearly observe that the largest LE fluctuates sharply with a and α: for some values of a and α, the largest LE is positive and for others it is negative. From this, we can understand that in all the papers [1,7,8,15,16], the values of $a = 2$ and $\alpha = 10$ were really set arbitrarily.

Indeed, these parameters could have been replaced by one of many other values. As a matter of fact, we have verified in an experimental way that the AdNN possesses *more powerful* AM properties when $a = 7$. As we can see from Figure 36.16a and b, when $a = 2$ and $\alpha = 10$, the largest LE is negative which indicates that there is no chaos. This also explains why the AdNN converges after a long transient phase—which is a phenomenon also reported in Reference 1—approximately 21,000 iterations.[†]

36.4.4.2 Lyapunov Analysis of the Ideal M-AdNN

We shall now perform an analysis of the ideal M-AdNN (i.e., the entire network) from the viewpoint of its LEs.

---

[*] It should be observed that $w_{ii} = 1, i = 1, 2, \ldots, N$ for the entire network. For a single neuron, the value of w should be $w = 1$. We should observe that for a primitive neuron, there is no difference between the AdNN, the M-AdNN, and the ideal M-AdNN.

[†] Elsewhere, we have verified that the size of the transient phase varies with the actual initial input [12].

The ideal M-AdNN originates from the AdNN and the M-AdNN. It differs from the M-AdNN in its parameter settings w_{ij}, ε, and a_i, which is quite crucial, because the same network can exhibit completely different phenomena depending on the settings themselves. Consequently, the theoretical Lyapunov analysis of the ideal M-AdNN should be exactly the same as that of the M-AdNN (explained in Section 36.3.2), except that the LEs are evaluated at the *current* parameter settings.

As demonstrated in Reference 8 (see Section 36.3.2), the Jacobian matrix J of the ideal M-AdNN (and the M-AdNN) can be seen to have the form[*]:

$$J = \begin{pmatrix} J_{ij}^1 & J_{ij}^2 \\ J_{ij}^3 & J_{ij}^4 \end{pmatrix} = \begin{pmatrix} 0 & \cdots & 0 & k_f & 0 & \cdots & 0 \\ \vdots & \ddots & \vdots & \vdots & \vdots & \ddots & \vdots \\ 0 & \cdots & 0 & k_f & 0 & \cdots & 0 \\ 0 & \cdots & 0 & 0 & \cdots & 0 & k_r \\ \vdots & \ddots & \vdots & \vdots & \ddots & \vdots & \vdots \\ 0 & \cdots & 0 & 0 & \cdots & 0 & k_r \end{pmatrix} = \begin{cases} k_f, & 0 < i \le N, & j = N \\ k_r, & N < i \le 2N, & j = 2N \\ 0, & otherwise, \end{cases}$$

where J_{ij}^n is an N-by-N matrix, and $J_{ij}^1(t) = \frac{\partial \eta_i(t+1)}{\partial \eta_j(t)}, J_{ij}^2(t) = \frac{\partial \eta_i(t+1)}{\partial \xi_j(t)}, J_{ij}^3(t) = \frac{\partial \xi_i(t+1)}{\partial \eta_j(t)}$, and $J_{ij}^4(t) = \frac{\partial \xi_i(t+1)}{\partial \xi_j(t)}$, respectively.

Generally speaking, it is not easy to compute the limit of Equation 36.28. However, due to the special form of J that we encounter here, we are able to calculate the value of Λ easily. As illustrated above

$$J_t = \begin{pmatrix} 0 & \cdots & 0 & k_f^t & 0 & \cdots & 0 \\ \vdots & \ddots & \vdots & \vdots & \vdots & \ddots & \vdots \\ 0 & \cdots & 0 & k_f^t & 0 & \cdots & 0 \\ 0 & \cdots & 0 & 0 & \cdots & 0 & k_r^t \\ \vdots & \ddots & \vdots & \vdots & \ddots & \vdots & \vdots \\ 0 & \cdots & 0 & 0 & \cdots & 0 & k_r^t \end{pmatrix},$$

whence we get

$$\Lambda = \lim_{t \to \infty} [J_t \cdot (J_t)^T]^{\frac{1}{2}} = \begin{pmatrix} k_f & \cdots & k_f & 0 & \cdots & 0 \\ \vdots & \ddots & \vdots & \vdots & \ddots & \vdots \\ k_f & \cdots & k_f & 0 & \cdots & 0 \\ 0 & \cdots & 0 & k_r & \cdots & k_r \\ \vdots & \ddots & \vdots & \vdots & \ddots & \vdots \\ 0 & \cdots & 0 & k_r & \cdots & k_r \end{pmatrix}. \tag{36.34}$$

By a simple algebraic analysis, we see that Λ has three different eigenvalues: Nk_f, Nk_r, and 0. As a result, the LE are

$$\lambda_1 = \ldots \lambda_{N-1} = -\infty,$$
$$\lambda_N = \log N + \log k_f > 0,$$
$$\lambda_{N+1} = \ldots \lambda_{2N-1} = -\infty,$$
$$\lambda_{2N} = \log N + \log k_r > 0.$$

In conclusion, the ideal M-AdNN has two positive LEs, which indicates that the network is truly a chaotic network!

---

[*] The algebraic details are not repeated from Section 36.3.2. But the reader should note that since we have chosen the Nth neuron to be the *global* neuron, only the Nth column vector of J_{ij}^1 and J_{ij}^4 is nonzero.

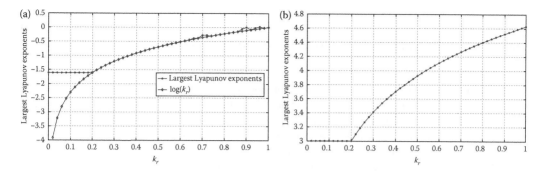

FIGURE 36.17 The spectrum of the largest LEs of (a) the AdNN and of (b) the ideal M-AdNN, in which we vary k_r from 0 to 1 with a step size of 0.02, and where $k_f = 0.2$. Figure 36.17a is calculated using Algorithm 36.1, while Figure 36.17b is calculated using Algorithm 36.2.

At this juncture, we would like to point out that in the work of Reference 8 (given in Section 36.3.2), the authors used another form of Equation 36.28 which can be written as

$$\Lambda = \lim_{t \to \infty} [(J_t)^T \cdot J_t]^{\frac{1}{2}}. \tag{36.35}$$

Consequently, the largest LE is the maximum of the $\{\frac{1}{2}\log N + \log k_f, \frac{1}{2}\log N + \log k_r\}$, which is different from the one derived here. However, from a realistic perspective, the choice of the definition does not matter. Indeed, independent of the definition we use, the result is consistent: the largest LE is positive, which implies that the network is chaotic.

It is very interesting to compare this result with the one presented for the AdNN. Indeed, as we can see from Reference 2, the AdNN has two different LEs: $\log k_f$ and $\log k_r$. Thus, the largest LE is $\max\{\log k_f, \log k_r\}$. One can observe the following from Figure 36.17a: when $k_r < 0.2$, the largest LE is approximately -1.6, while when $k_r > 0.2$, the largest LE is[*] $\log k_r$. As opposed to this, the largest LE of the ideal M-AdNN is always positive, as shown in Figure 36.17b. The difference is that by binding the states of all the neurons to a single *"global"* neuron, we force the ideal M-AdNN to have two positive LEs.

36.4.5 Chaotic and PR Properties of the Ideal M-AdNN

We shall now report the properties of the ideal M-AdNN. The protocol of our experiments is quite simple: if the network output repeats or resonates a known input pattern with *any* periodicity, we deem this pattern to have been recognized from a chaotic PR perspective. Otherwise, we say that it is unrecognizable. These properties have been gleaned as a result of examining the Hamming distance between the input pattern and the patterns that appear at the output. As a comparison, we also list the simulation results of the M-AdNN. By comparing the Hamming distance of the ideal M-AdNN and the M-AdNN, we can then conclude which of the schemes performs better when it concerns the crucial PR properties. In this regard, we mention that the experiments were conducted using two data sets, namely the Adachi data set given in Figure 36.1 and the numeral data sets given in Figure 36.3. In both the cases, the patterns were described by 10×10 pixel images and thus, the networks had 100 neurons.

Before we proceed, we remark that although the experiments were conducted for a variety of scenarios, in the interest of brevity, we present here only a few typical sets of results—essentially, to catalog the overall conclusions of the investigation.

---

[*] Note that due to the fact that the computation is done numerically, there are some numerical errors, and thus, the largest LE is not exactly, but very close to $\log k_r$ at some points.

We discuss the properties of the ideal M-AdNN for two different settings: the NNs AM properties and its PR properties. In all cases, the parameters were set to be $k_f = 0.2$, $k_r = 0.9$, $\varepsilon = 0.015$, and all internal states $\eta_i(0)$ and $\xi_i(0)$ started from 0. Further, we catalog our experimental protocols as follows:

1. AM properties. Although the AM properties are not the main issues of this chapter, it is still interesting to examine whether the ideal M-AdNN possesses any AM-related properties. This is because the ideal M-AdNN is a modified version of the AdNN and the M-AdNN. By comparing the differences between the ideal M-AdNN and the AdNN or the M-AdNN, we can obtain a better understanding of its dynamical properties. In this case, the external stimulus a is a constant, i.e., $a = 2$.

2. PR properties. This investigation is really the primary intent of this chapter. In this case, we investigate whether our new network is able to achieve PR. This is accomplished by checking whether the output can respond correctly to different inputs. In our experiments, we tested the network with known patterns, noisy patterns, and with unknown patterns. In this case, the external stimulus was set to $a = 2 + 6P$, where P is the input pattern.

36.4.5.1 AM Properties

We now examine whether the ideal M-AdNN possesses any AM-related properties for certain scenarios, i.e., if we fix the external input $a_i = 2$ for all neurons. The observation that we report is that during the first 1000 iterations (due to the limitations of the file size, we present here only the first 36 images), the network only repeats black and white images. This can be seen in Figure 36.18.

It is very easy to understand this phenomenon: first of all, we see that all the neurons have the same output, 0, at time t (i.e., $t = 0$). This is true because we start the iteration with all internal states $\eta_i(0) = 0$, $\xi_i(0) = 0$, and $x_i(0) = 0$. As a result, the first image of Figure 36.18 is completely black.[*] At the next time step, $\eta_i(1) = 0$, $\xi_i(1) = a > 0$, which causes the output of all the neurons to be $x_i(1) = f(\eta + \xi) \approx 1$, which is why we see a completely white image. At the third time step, we first computed the summation $\Sigma_{j=1}^{j=N} w_{ij}x_j(1)$, where w_{ij} is defined by Equation 36.15. We can verify that the summation of each line of the matrix w is either -0.5 or 0.5. Thus, $\eta_i(2)$ is either 0.5 or -0.5. Meanwhile, we must note that $\alpha = 10$, which results in a negative value for $\xi_i(2)$ and $\eta_i(2) + \xi_i(2)$, and consequently, $x_i(1) = f(\eta + \xi) \approx 0$. This is the reason why we see the third image of Figure 36.18 to be completely black. If we follow the same arguments, we see that at any time instant, the outputs of all the neurons only switch between 1 and 0 synchronically, which means that the output image of Figure 36.18 switches between black and white, implying that the network possesses *no* AM properties at all.

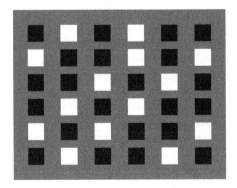

FIGURE 36.18 The visualization of the output of the ideal M-AdNN under the external input $a_i = 2$. We see that the output switches between images which are entirely only black or white.

---

[*] In our visualization, the value 0 means "black," while 1 means "white."

By way of comparison, we mention that the M-AdNN also does not possess any AM-related properties. In this regard, the M-AdNN and the ideal M-AdNN are similar.

36.4.5.2 PR Properties

The PR properties of the ideal M-AdNN are the main concern of this chapter. As illustrated in Section 36.1.3, the goal of a chaotic PR system is the following: the system should respond repetitively/periodically to trained input patterns, while it should respond chaotically (with chaotic outputs) to untrained input patterns. We now confirm that the ideal M-AdNN does, indeed, possess such phenomena.

We now present an in-depth report of the PR properties of the ideal M-AdNNs by using a Hamming distance-based analysis. The parameters that we used were $k_f = 0.2$, $k_r = 0.9$, $\varepsilon = 0.015$, and $a_i = 2 + 6x_i$. The PR-related results of the ideal M-AdNN are reported for the three scenarios, i.e., for trained inputs, for noisy inputs, and for untrained (unknown) inputs, respectively. We report only the results for the setting when the original pattern and the noisy version are related to P4. The results obtained for the other patterns are identical, and are omitted here for brevity.

1. The external input of the network corresponds to a known pattern, say P4.

   To report the results for this scenario, we request the reader to observe Figure 36.19, where we can find that P4 is retrieved periodically as a response to the input pattern. This occurs 391 times in the first 500 iterations. On the other hand, the other three patterns never appear in the output sequence. The phase diagrams of the internal states that correspond to Figure 36.19 are shown in Figure 36.20, where the x-axes are $\eta_{86}(t) + \xi_{86}(t)$ (the neuron with the index 86 has been set to be the global neuron), and the y-axes are $\eta_i(t) + \xi_i(t)$, where the index $i = 80, \ldots, 88$ (i is chosen randomly). From Figures 36.19 and 36.20, we can conclude that the input pattern can be recognized periodically if it is one of the known patterns. Furthermore, from Figure 36.20, we verify that the periodicity is 14, because all the phase plots have exactly 14 points.

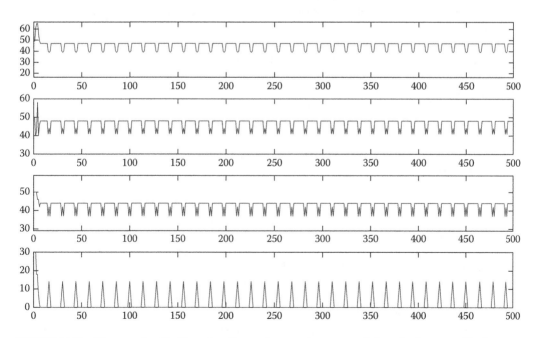

FIGURE 36.19 PR properties: The Hamming distance between the output and the trained patterns. The input was the pattern P4. Note that P4 appears periodically, i.e., the Hamming distance is zero at periodic time instances.

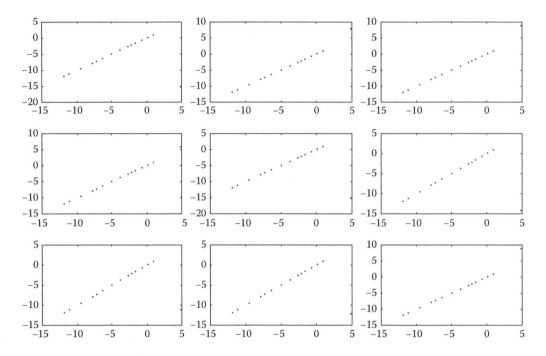

FIGURE 36.20   PR properties: The phase diagrams of the internal states corresponding to Figure 36.19.

As a comparison, we report that the M-AdNN is also able to recognize known patterns under *certain* conditions. However, PR is not a phenomenon that we can "universally" and "religiously" ascribe to the M-AdNN because it fails to pass the other necessary tests!

2. The external input of the network corresponds to a noisy pattern, in this case P5, which is a noisy version of P4.

   Even when the external stimulus is a garbled version of a known pattern (in this case P5 which contains 15% noise), it is interesting to see that *only* the original pattern P4 is recalled periodically. In contrast, the other three known patterns are *never* recalled. This phenomenon can be seen from Figure 36.21. By comparing Figures 36.19 and 36.21, we can draw the conclusion that the ideal M-AdNN can achieve chaotic PR even in the presence of noise and distortion.

   As in the previous case, the phase diagrams of the internal states that correspond to Figure 36.21 are shown in Figure 36.22, where the x-axes are $\eta_{86}(t) + \xi_{86}(t)$ (the neuron with the index 86 has been set to be the global neuron), and the y-axes are $\eta_i(t) + \xi_i(t)$, where the index $i = 80, \ldots, 88$ (i is chosen randomly). As before, from Figure 36.22, we verify that the periodicity is 14, because all the phase plots have exactly 14 points.

   Indeed, even if the external stimulus contains some noise, the ideal M-AdNN is still able to recognize it correctly, by resonating periodically! Furthermore, the NN has also been tested using noisy patterns which contained up to 33% noise. We are pleased to report that it still resonates the input correctly, as shown in Figure 36.23. If the noise is even higher, i.e., more than 34%, the PR properties tend to gradually disappear. In this case, understandably, the simulation results are almost the same as when one does the testing with *unknown* patterns.

3. The external input of the network corresponds to an unknown pattern, P6.

   In this case, we investigate whether the ideal M-AdNN is capable of distinguishing between known and unknown patterns. Thus, we attempt to stimulate the network with a completely unknown pattern. In our experiments, we used the pattern P6 of Figure 36.1 initially used by Adachi et al. From Figure 36.24, we see that neither the known patterns nor the unknown pattern appear.

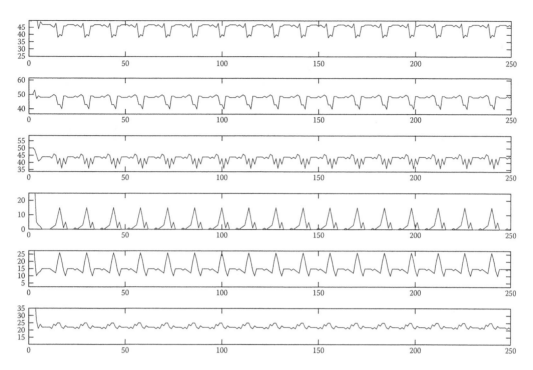

FIGURE 36.21 PR properties: The Hamming distance between the output and the trained patterns. The input was the pattern P5. Note that P4 (not P5) appears periodically, i.e., the Hamming distance is zero at periodic time instances.

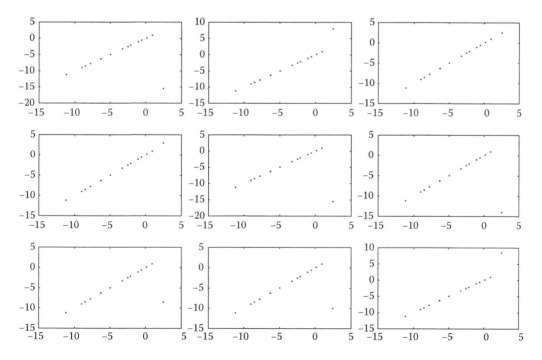

FIGURE 36.22 PR properties: The phase diagrams of the internal states corresponding to Figure 36.21.

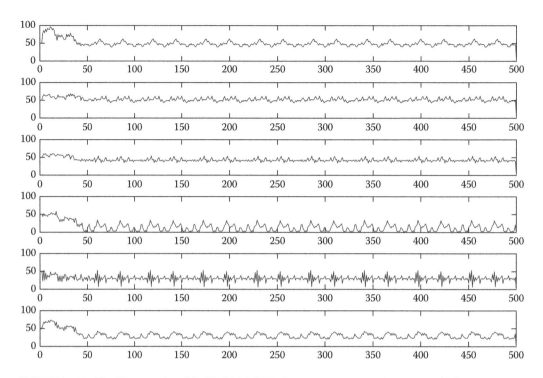

FIGURE 36.23 The PR properties of the ideal M-AdNN when it encounters a noisy pattern which contains up to 33% noise. Observe that the system still resonates the corresponding known pattern correctly.

As in the previous two cases, the phase diagrams of the internal states that correspond to Figure 36.24 are shown in Figure 36.25, where the *x*-axes are $\eta_{86}(t) + \xi_{86}(t)$ (where, as before, the neuron with the index 86 has been set to be the global neuron), and where the *y*-axes are $\eta_i(t) + \xi_i(t)$ where the index $i = 80, \ldots, 88$ (*i* is chosen randomly), respectively. The lack of periodicity can be observed from Figure 36.25, since the plots themselves are dense.

By way of comparison, we can see that the M-AdNN fails to pass this test, as can be seen from Figure 36.26. In this figure, the input is an unknown pattern, P6. We expect the network to respond chaotically to this input. Unfortunately, one sees that it still yields a repetitive/periodic output. In short, no matter what the input is, the M-AdNN always yields the input pattern itself, and that, *periodically*. Obviously, strictly speaking, this property cannot be deemed to represent PR. As opposed to this, the ideal M-AdNN responds intelligently to the various inputs with correspondingly different outputs, each resonating with the input that excites it—which is the crucial "golden" hallmark characteristic of a *chaotic* PR system. Indeed, the switch between "order" (resonance) and "disorder" (chaos) seems to be consistent with Freeman's biological results—which, we believe, is quite fascinating!

36.4.5.3 Summary

We now summarize the contributions of this section. While the AdNN [1,3–6] has properties which are pseudochaotic, it also possesses *limited* PR characteristics. As opposed to this, the M-AdNN proposed by Calitoiu et al. [8], is a fascinating NN which has been shown to possess the required repetitive/periodic property desirable for PR. In this section, we have explained why the PR properties claimed in Reference 8 are not as powerful as originally claimed. Thereafter, we have presented arguments for the basis for setting the parameters of the M-AdNN to lead to the ideal M-AdNN. By appropriately tuning the parameters of

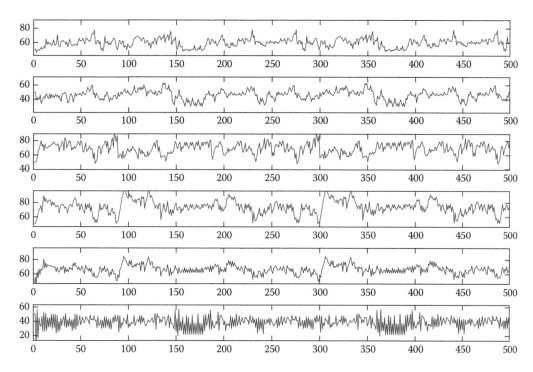

FIGURE 36.24 PR properties: The Hamming distance between the output and the trained patterns. The input pattern was P6.

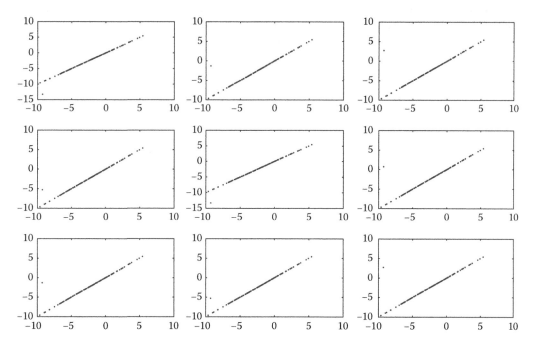

FIGURE 36.25 PR properties: The phase diagrams of the internal states corresponding to Figure 36.24.

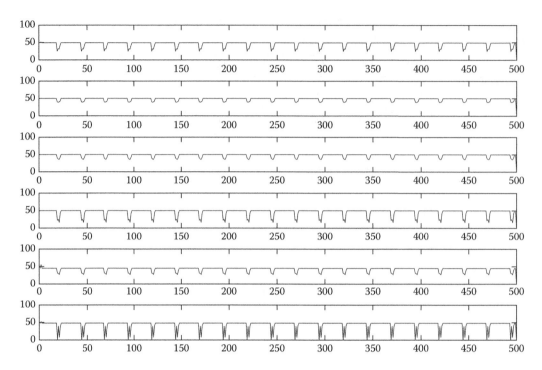

FIGURE 36.26 The Hamming distance between the output and the trained patterns of the M-AdNN. The output repeats the input pattern, P6, periodically.

the M-AdNN for its weights, steepness, and external inputs, we have obtained this new NN, the ideal M-AdNN. Using a rigorous Lyapunov analysis, we have demonstrated the chaotic properties of the ideal M-AdNN. We have also verified that the system is truly chaotic for untrained patterns. But most importantly, we have shown that it is able to *switch to being periodic* whenever it encounters patterns with which it was trained (or noisy versions of the latter).

As the reader can observe, we have not been too concerned about the application domain. Rather, our work has been of an investigatory sort. Thus, although the data sets that we used were rather simplistic, they were used to submit a *prima facie* case. On one hand, if we could achieve our goal with these data sets, we believe that this research direction has the potential of initiating new research avenues in PR and AM. On the other hand, from a practical perspective, we believe that this is also beneficial for other computer science applications such as image searching, content-addressed memory, etc., and this is considered open currently.

36.5 The Linearized AdNN

36.5.1 The Design of the Linearized AdNN

The problem with all the above instantiations and variations of the AdNN is the excessive computational cost. Indeed, in all the works listed above, the number of computations is a quadratic function of the number of neurons. This is essentially because all the neurons are connected to each other, thus rendering the topology to be completely connected. In this section, we shall show how we can reduce the number of computations to an optimal (linear) number, and yield an NN which effectively possesses the same capabilities. This, essentially, is the motivation of this section.

To achieve this goal, in this section, we resort to the following strategy. We first determine a spanning tree of the weighted completely connected graph. Clearly, such a spanning tree has the desired *linear* number of edges. But the problem encountered is the following: if the number of edges is reduced, the training process, which assigns the weights to the edges, becomes ineffective. To thereafter obtain a set of modified weights which is most suitable for this spanning tree, we use a gradient-based search algorithm. The latter attempts to determine the new weights of the edges of this graph to yield a bidirectional spanning tree-based CNN. The consequence of such a gradient-based search algorithm is that unlike the original AdNN, in which the weight matrix is symmetric, in the case of the tree-based NN, the consequent weights may not be symmetric.

Since the topology of the new AdNN-like NN has been changed, the system's properties are also consequently quite different. It turns out that unlike the properties of the AdNN rigorously recorded in Reference 14, the new machine, the linearized AdNN (L-AdNN) only goes through two different phases: chaos and PR, as will be seen in Section 36.5.5.1. However, as opposed to the AdNN, the M-AdNN, and the ideal M-AdNN, on achieving recognition, the L-AdNN does not emit the recognized pattern at the output *periodically*, but rather *repetitively*.

36.5.2 Designing the L-AdNN

The design of every NN involves three steps. First of all, the designer has to decide on the *number* of neurons, and their interconnections. This essentially dictates the topology of the network. The second phase involves determining the weights of the edges associated with the network topology. This is done by training the NN using the input-training samples. The last phase is the testing phase where the accuracy and quality of the NN is estimated. We shall discuss each of these phases in the next subsections.

36.5.2.1 The Topology of the L-AdNN

To minimize the computational burden of the new AdNN, we shall first arrive at a topology with a linear number of edges. To achieve this, we have to understand that the first approximation we have is the one in which all the neurons are connected to each other, i.e., the completely connected graph. Further, this graph has one advantage, which is that the initial weights can be obtained by the traditional one-shot training phase [1,3–8] proposed in the literature.

We arrive at the modified linear version in two steps. First of all, we reduce the number of edges to be linear. To achieve this, we must ensure that all the nodes are connected, implying that we must retain a suitable *spanning tree* structure of the nodes. The second step will involve the computation of the weights associated with this new structure, which we will address subsequently.

We know that every complete graph has a large number of possible spanning trees. The question now is that of determining which is the spanning tree to be used as an initial approximation. To decide on this, we observe that the outputs of the neurons are the weighted combinations of the inputs, where the weights are those associated with the edges. Since we want the initial approximation to maximally approximate the original AdNN, we propose to use the maximum spanning tree (MST) of the AdNN (with the edge weights given by their absolute values) to be the initial linear approximation to the completely connected AdNN.[*] This is formalized by Algorithm Topology_L-AdNN below.

Unlike the AdNN and its variants, this initial approximation of the L-AdNN is tree shaped, since we have pruned the edges which we believe are the most "redundant." In this regard, since the original AdNN has N neurons and $N \times N$ edges, we recommend the use of the Prim algorithm which is much more efficient than the Kruskal algorithm in such a setting.

---

[*] There is no reason to believe that any single MST would be superior to another, and so, currently, we simply choose any one of them. But it should be remembered that this is just an initial approximation. The final weights of the L-AdNN are, all the same, computed after the gradient search optimization described currently. Our experience has shown that the choice of the initial MST is not so crucial.

Algorithm 36.3 Topology_L-AdNN

Input: N, the number of neurons in the NN, and a set of p patterns which it has to "memorize."
Output: The topology and initial weights of the L-AdNN.
Method

1. Create a completely connected graph $\mathcal{G}$ which is to represent the AdNN.
2. Compute the weights of the edges of $\mathcal{G}$, $\{w_{ij}\}$, by the following:
   $w_{ij} = \frac{1}{p}\sum_{s=1}^{p}(2x_i^s - 1)(2x_j^s - 1)$, where x_i^s is the ith component of the sth stored pattern.
3. Create a nonnegative modified set of edge weights for the graph as $w_{ij}' = |w_{ij}| \; \forall i, j$.
4. Compute the MST of $\mathcal{G}$ with the associated weights of $\{w_{ij}'\}$ as the initial approximation for the L-AdNN.

End Algorithm Topology_L-AdNN

36.5.2.2 The Weights of the L-AdNN: Gradient Search

Since we have removed most of the "redundant" edges from the completely connected graph by using an MST, it is clear that the NN at hand will not adequately compare with the original AdNN. Thus, our next task is to determine a new set of weights so as to force the L-AdNN to retain some of its PR properties, namely those corresponding to the trained patterns. We explain below the process for achieving this.

In order to have the L-AdNN "mimic" the original AdNN, we must at least impose the constraint that they work analogously from an input/output perspective. Consider the state and output equations of the AdNN given below, described by means of their two internal states $\eta_i(t)$ and $\xi_i(t)$, $i = 1 \ldots N$, and the output $x_i(t)$:

$$x_i(t+1) = f(\eta_i(t+1) + \xi_i(t+1)), \tag{36.36}$$

$$\eta_i(t+1) = k_f\eta_i(t) + \sum_{j=1}^{N} w_{ij}x_j(t), \tag{36.37}$$

$$\xi_i(t+1) = k_r\xi_i(t) - \alpha x_i(t) + a_i. \tag{36.38}$$

The L-AdNN, on the other hand, is defined by the following equations:

$$x_i^l(t+1) = f(\eta_i^l(t+1) + \xi_i^l(t+1)), \tag{36.39}$$

$$\eta_i^l(t+1) = k_f\eta_i^l(t) + \sum_{e_{ij}\in T} w_{ij}^{l^*} x_j^l(t), \tag{36.40}$$

$$\xi_i^l(t+1) = k_r\xi_i^l(t) - \alpha x_i^l(t) + a_i, \tag{36.41}$$

where $\{w_{ij}^{l^*}\}$, x_i^l, ξ_i^l, and η_i^l are the weights, outputs, and state variables of the L-AdNN, respectively, and have similar meanings to $\{w_{ij}\}$, x_i, ξ_i, and η_i of the AdNN. Before we proceed, we should emphasize that since T is a tree, the summations in Equation 36.40 must be taken only over the edges of the tree, by virtue of the fact that $\{w_{ij}^l\}$ equals 0 if the corresponding edge $\{e_{ij}\}$ is not contained in T. Our aim now is to be able to compute the weights $\{w_{ij}^{l^*}\}$, in such a way that the two NNs have properties which are, as close to possible, analogous from an input/output perspective. We propose to do this by a gradient search algorithm which starts with an initial value of $\{w_{ij}(0)\}$ and converges at the $\{w_{ij}^{l^*}\}$ which minimizes a suitable error criterion described below.

In order to find the optimal values of $\{w_{ij}^{l^*}\}$, we define the square error between the original output of the AdNN and the new output, as generated by the L-AdNN for the set of weights $\{w_{ij}^l(n)\}$ at the nth step of the *update** by

$$E_p = \frac{1}{2} \sum_{i=1}^{N} (x_i^{A,p} - x_i^{L,p}(n))^2, \tag{36.42}$$

where $x_i^{A,p}$ implies the output of the ith neuron when the pth pattern is presented to AdNN network, and $x_i^{L,p}$ implies the output of the ith neuron when the pth pattern is presented to L-AdNN network. The overall global error is defined by

$$E = \sum_{p=1}^{P} E_p, \tag{36.43}$$

where P is the number of trained patterns.

In order to adjust w_{ij}^L to obtain the least global error E, we consider the gradient, Δw_{ij}^L, and move w_{ij}^L by an amount which equals Δw_{ij}^L in the direction where the error is minimized. This can be formalized as below:

$$\Delta w_{ij}^L = -\beta \frac{\partial E}{\partial w_{ij}^L} = -\beta \frac{\partial \sum_{p=1}^{P} E_p}{\partial w_{ij}^L} = -\beta \sum_{p=1}^{P} \frac{\partial E_p}{\partial w_{ij}^L}$$

$$= -\beta \sum_{p=1}^{P} \frac{\partial E_p}{\partial x_i^{L,p}(n)} \frac{\partial x_i^{L,p}(n)}{\partial w_{ij}^L}$$

$$= \beta \sum_{p=1}^{P} (x_i^{A,p} - x_i^{L,p}(n)) \frac{\partial x_i^{L,p}(n)}{\partial \eta_i^{L,p}(n)} \frac{\partial \eta_i^{L,p}(n)}{w_{ij}^L(n)}, \tag{36.44}$$

where β is the learning rate of the gradient search. By the chain rule, our next task is to incorporate the partial derivative, $\frac{dx}{dy}$. Indeed, since $x = f(y)$, where $f(\cdot)$ is the AdNN's and L-AdNN's logistic function, it is easy to verify that $\frac{dx}{dy} = \frac{1}{\varepsilon} \cdot x \cdot (1-x)$. Consequently,

$$\Delta w_{ij}^L = \frac{\beta}{\varepsilon} \sum_{p=1}^{P} (x_i^{A,p} - x_i^{L,p}(n)) \cdot x_i^{L,p}(n) \cdot (1 - x_i^{L,p}(n)) \cdot x_j^{L,p}(n). \tag{36.45}$$

The formal algorithm which achieves the update is given in Algorithm 36.4.

In the interest of continuity, we shall now proceed with explaining the other analytic properties of the L-AdNN before highlighting its actual convergence properties. The experimental results describing its convergence for three benchmark data sets are given in Section 36.5.4.

36.5.3 LE Analysis of the L-AdNN

Since the LE analysis of the L-AdNN is very similar to that of the AdNN and the M-AdNN, in the interest of brevity, we omit the details here—they can be found in Reference 15. Without further comment, we merely submit the results here.

The LE of the L-AdNN are

$$\lambda_1 = \lambda_2 = \cdots = \lambda_N = \log k_f,$$

$$\lambda_{N+1} = \lambda_{N+2} = \cdots = \lambda_{2N} = \log k_r.$$

Choosing the parameters k_r and k_f is a complex and nontrivial issue. While on one hand, we would like to force the network to be *chaotic*, on the other hand, it is not advantageous to have it "*too chaotic.*"

---

* The reader must observe the difference between the time instants "t" and "n." "t" is the time that is actually used in the training and testing of the NNs. The index "n," on the other hand, is merely a conceptual clock used in the gradient search algorithm to lead us to the best weights for the L-AdNN, namely, $\{w_{ij}^{L*}\}$.

Algorithm 36.4 Weights_L-AdNN

Input: The number of neurons, N, a set of P patterns, and the initial weights $\{w_{ij}^L\}$ of the L-AdNN. These initial weights are $\{w_{ij}^A\}$ for the edges in the MST, and are set to *zero* otherwise. The parameters and the setting which we have used are the learning rate $\beta = 0.05$, $\varepsilon = 0.015$, $\alpha = 10$, $k_f = 0.2$, and $k_r = 1.02$.

Output: The weights $\{w_{ij}^{L^*}\}$ of the L-AdNN.

Method

1. Compute the outputs of the L-AdNN corresponding to the P trained inputs.
2. For all edges of the L-AdNN, compute Δw_{ij}^L as per Equation 36.45. Otherwise, set $\Delta w_{ij}^L = 0$.
3. $w_{ij}^L \leftarrow w_{ij}^L + \Delta w_{ij}^L$.
4. Go to Step 1 until E, given by Equation 36.43, is less than a given value or $\Delta w_{ij}^L \approx 0$.

End Algorithm Weights_L-AdNN

Thus, by choosing the values of the coefficient k_r to be close to unity, for example, at $k_r = 1.02$, and by choosing $k_f = 0.2$, we can force the largest Lyapunov exponent (LLE) to be $\lambda_{N+1} = \lambda_{N+2} = \ldots = \lambda_{2N} = \log 1.02 > 0$, rendering the L-AdNN to be chaotic. Although such a choice of parameters also yields the other phenomena that we seek, unfortunately, we cannot, at this juncture, provide any deeper insight into the biological significance of these settings!

36.5.4 Experimental Results: Convergence of the L-AdNN

Having described the various topological, analytic, and Lyapunov properties of the L-AdNN, we shall now describe its convergence, chaotic, and PR properties for the benchmark data sets.

The results of a typical numerical experiment which proceeds along the gradient search formalized in Algorithm Weights_L-AdNN are shown in Figures 36.27 through 36.29 for the *Adachi, LOVE*, and *numeral* data sets, respectively.

A word about the parameter β used is not out of place. Algorithm Weights_L-AdNN essentially uses a hill-climbing method (in the direction of the gradient) to obtain a minimum value of the error function, and the correctness of a hill-climbing method for optimization[*] has been known for more than four decades. The choice of the parameter for any hill-climbing scheme is always by trial and error unless we use the second derivatives. If it is too small, the convergence is sluggish. If it is too large, it may not converge or can oscillate. In all the experiments reported for the *Adachi* data set, the value of β was set to be 0.05 because the scheme converged. The value of β was set at 0.01 for the *LOVE* and *numeral* data set because they yielded the best convergence results.

Consider Figure 36.27 for the *Adachi* data set. From it, the reader will observe that the global error, E, converges very rapidly. Indeed, after merely 60 iterations, the average value of the matrix of Δw_{ij}^L converges to a value arbitrarily close to 0 (from the perspective of the machine's accuracy), and becomes invariant, thereafter. The value of E, the total error, does not actually converge to *zero*. This means that even though the weights can be made to converge by using a gradient search, the actual performance of the two networks will not be exactly identical. Indeed, it turns out that, in general, the value of E can *never* be made arbitrarily close to zero since both the networks are *chaotic*. Clearly, two different CNNs will not have *exactly* the same outputs unless they are subject to rigorous synchronization.

For the *LOVE* data set, the reader can observe from Figure 36.28 that the average error converges to a value arbitrarily close to zero (~ -0.005) after 150 time steps. The figure on the bottom shows the variation of the global error over the same time frame. The analogous results for the *numeral* data set are given in Figure 36.29.

---

[*] Note that the results do not claim that one attains the global optimum, but rather a *local* optimum. Our experience is that this is sufficient for the L-AdNN to operate desirably.

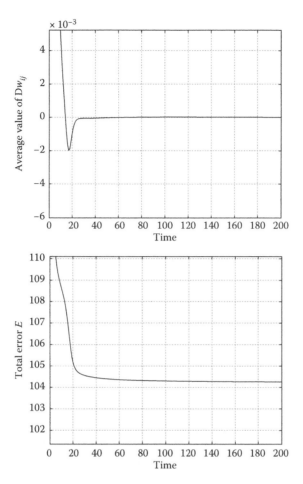

FIGURE 36.27 The figure on the top shows the variation of the average of Δw_{ij}^l (averaged over all values of i and j) over the first 200 iterations of the gradient search scheme for the *Adachi* data set. The average converges to a value arbitrarily close to zero after 60 time steps. The figure on the bottom shows the variation of the global error over the same time frame. Observe that this quantity does not converge to *zero*.

36.5.5 PR Properties of the AdNNs

We shall now report the comparative PR properties of the L-AdNN and AdNN. These properties have been gleaned as a result of examining the Hamming distance between the input pattern and the patterns that appear in the output. By way of a reminder, we observe that as mentioned before, unlike the AdNN, the M-AdNN, and the ideal M-AdNN, on achieving recognition, the L-AdNN does not emit the recognized pattern at the output *periodically*, but rather *repetitively*.

36.5.5.1 Comparative Chaotic Properties

We now report the PR properties of the L-AdNN. In the ideal setting, we would have preferred the L-AdNN to be chaotic when exposed to untrained patterns, and the output to appear repetitively (i.e., more frequently) when exposed to trained patterns. Besides yielding this phenomenon, the L-AdNN also goes through a chaotic phase and a PR phase as some of its parameters change.

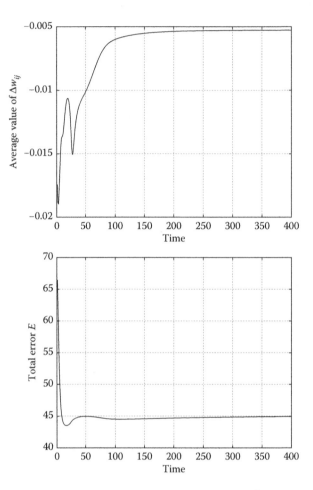

FIGURE 36.28   The figure on the top shows the variation of the average of Δw_{ij}^l (averaged over all values of i and j) over the first 400 iterations of the gradient search scheme for the *LOVE* data set. The average converges to a value arbitrarily close to zero (~ -0.005) after 150 time steps. The figure on the bottom shows the variation of the global error over the same time frame. Observe that this quantity does not converge to *zero*.

We summarize the results[*] for the L-AdNN by stating that by using different settings of α and a_i, the latter demonstrates the following amazing results tabulated in Table 36.6. From this table, we see that if the L-AdNN is presented with a *trained* pattern, the output is the same trained pattern occurring frequently (indicated in the table by the symbol F). But the output is chaotic (indicated in the table by the symbol C) for untrained patterns. While this phenomenon is not observed when $a_i = 2$ (when the L-AdNN is always chaotic), the frequent behavior is noticeable when $a_i = 2 + 6x_i$.

To further clarify the dynamics of the L-AdNN, we observe that for the *Adachi* data set, if $a_i = 2$, the L-AdNN is always chaotic whether the input is a trained or an untrained pattern, as can be seen from Figure 36.30a. The case when $a_i = 2 + 6x_i$, is seen in Figure 36.30b.

Observe that no pattern is retrieved during the system's evolution. Since a finer measure of how closely the output mimics the input is their Hamming distance (which is exactly *zero* if the output is a precise replica of the input), we have also tabulated the frequency of the Hamming distance within the first 1000 time intervals in Table 36.7 for the various input patterns. Table 36.8 demonstrates that this distance is

---

[*] We have used these settings and parameters so as to be consistent with the data and results reported by Adachi et al. (1997)

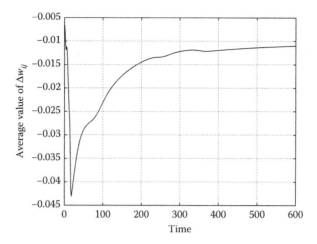

FIGURE 36.29 The figure shows the variation of the average of Δw_{ij}^L (averaged over all values of i and j) over the first 600 iterations of the gradient search scheme for the *numeral* data set. The average converges to a value arbitrarily close to zero (~ -0.011) after 400 time steps.

TABLE 36.6 Chaotic and PR Properties of the L-AdNN Obtained *for All the Data Sets* for Different Values of the Decay Factor, α, and the External Stimulation Coefficient, a_i

a_i	α	**P1**	**P2**	**P3**	**P4**	**P5**
	$\alpha = 4$	C	C	C	C	C
$a_i = 2$	$\alpha = 10$	C	C	C	C	C
	$\alpha = 4$	C	C	C	C	C
$a_i = 2 + 6x_i$	$\alpha = 10$	F	F	F	F	C

Note: The data sets used are the *Adachi*, *LOVE*, and *numeral* data sets (see Figures 36.1 through 36.3), and the values of the parameters are $k_f = 0.2$, $k_r = 1.02$, and $\varepsilon = 0.015$. The legend for the table is the following: C—chaotic; F—frequent.

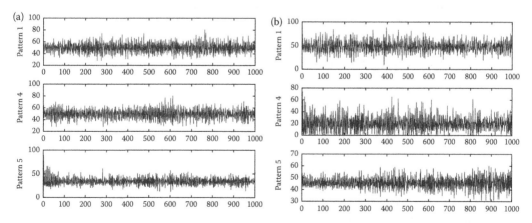

FIGURE 36.30 The Hamming distance between the output and the trained patterns or untrained patterns while using different settings for the *Adachi* data set. The input is Pattern 4 in both (a) and (b). The external stimulation weights are $a_i = 2$ and $a_i = 2 + 6x_i$ in (a) and (b), respectively. The output graphs are displayed for Pattern 1, Pattern 4, and Pattern 5.

TABLE 36.7   Frequency Distribution of the Hamming Distance for the *Adachi* Data Set between the Input and the Output as the Measure Occurs within 10 Intervals, When the External Input Is $a_i = 2$, and the Refractory Factor Is $\alpha = 10$

Input Pattern	[0,5]	[6,19]	[20,29]	[30,39]	[40,49]	[50,59]	[60,69]	[70,79]	[80,89]	[90,100]
					Hamming Distance					
Pattern 1	1	3	25	129	279	397	143	22	1	0
Pattern 2	3	5	33	121	291	374	146	24	3	0
Pattern 3	1	2	15	117	342	358	146	18	1	0
Pattern 4	1	6	12	157	338	339	121	24	2	0

TABLE 36.8   Frequency Distribution of the Hamming Distance for the *LOVE* Data Set between the Input and the Output as the Measure Occurs within 10 intervals, When the External Input Is $a_i = 2$, and the Refractory Factor Is $\alpha = 10$

Input Pattern	[0,5]	[6,19]	[20,29]	[30,39]	[40,49]	[50,59]	[60,69]	[70,79]	[80,89]	[90,100]
					Hamming Distance					
Pattern 1	1	56	231	179	32	0	1	0	0	0
Pattern 2	1	9	120	261	95	13	1	0	0	0
Pattern 3	1	56	275	117	48	2	1	0	0	0
Pattern 4	1	39	275	161	23	0	1	0	0	0

TABLE 36.9   Frequency Distribution of the Hamming Distance for the *Numeral* Data Set between the Input and the Output as the Measure Occurs within 10 Intervals, When the External Input Is $a_i = 2$, and the Refractory Factor Is $\alpha = 10$.

Input Pattern	[0,5]	[6,19]	[20,29]	[30,39]	[40,49]	[50,59]	[60,69]	[70,79]	[80,89]	[90,100]
					Hamming Distance					
Pattern 1	13	112	228	316	151	22	129	28	0	1
Pattern 2	1	149	387	148	88	57	96	35	36	3
Pattern 3	1	164	333	148	107	161	53	22	10	1
Pattern 4	2	141	333	191	84	135	75	30	8	1

close to zero (i.e., between 0 and 5) quite infrequently, implying that under these settings, the L-AdNN is chaotic. It is worth mentioning that most of the Hamming distances are in the interval [40, 59]—i.e., an almost 50% erroneous pattern, which is what we expect from a truly noisy pattern. If the same settings are used in the AdNN, the system behaves as an AM, which has been well illustrated in Reference 1.

36.5.5.2 Comparative PR Properties

If the parameters of the AdNN are set as prescribed in Reference 1, and the AdNN weights the external inputs with $a_i = 2 + 6x_i$, the system becomes closer to mimicking a PR phenomenon. Adachi et al. claim that such an NN yields the stored pattern at the output periodically, with a small periodicity. However, it turns out that if the input is an untrained pattern, the AdNN can still retrieve *it* and *the other four trained patterns* with a relatively lower frequency. This property is more or less one which resembles an AM system instead of a PR system. This can be seen from Table 36.10.

As opposed to this, for the L-AdNN, the most interesting scenario occurs when $\alpha = 10$ and the weight for the external input is $a_i = 2 + 6x_i$. In this case, the system's output is noticeably chaotic. But the input pattern can be retrieved very frequently as long as the input is one of the trained patterns.

TABLE 36.10 AdNN: In This Table, the Input Is The untrained Pattern 5 in Figure 36.1

	Pattern 1	Pattern 2	Pattern 3	Pattern 4	Pattern 5
Frequency	44	0	1	34	110

Note: The numbers in this table are the frequencies of each pattern being retrieved within the first 1000 time intervals.

TABLE 36.11 Frequency Distribution of the Hamming Distance for the *Adachi* Data Set between the Input and the Output as the Measure Occurs within 10 Intervals, When the External Input Is $a_i = 2 + 6x_i$ and the Refractory Factor Is $\alpha = 10$

Input Pattern	Hamming Distance									
	[0,5]	[6,19]	[20,29]	[30,39]	[40,49]	[50,59]	[60,69]	[70,79]	[80,89]	[90,100]
Pattern 1	217	342	226	112	63	28	10	2	0	0
Pattern 2	197	369	222	131	58	21	1	1	0	0
Pattern 3	198	325	251	130	57	32	3	4	0	0
Pattern 4	196	357	235	126	54	25	5	2	0	0
Pattern 5	46	495	334	104	18	2	0	1	0	0

TABLE 36.12 Frequency Distribution of the Hamming Distance for the *LOVE* Data Set between the Input and the Output as the Measure Occurs within 10 Intervals, When the External Input Is $a_i = 2 + 6x_i$ and the Refractory Factor Is $\alpha = 10$

Input Pattern	Hamming Distance									
	[0,5]	[6,19]	[20,29]	[30,39]	[40,49]	[50,59]	[60,69]	[70,79]	[80,89]	[90,100]
Pattern 1	108	293	66	27	5	0	1	0	0	0
Pattern 2	104	326	43	23	3	1	0	0	0	0
Pattern 3	102	290	88	14	3	2	1	0	0	0
Pattern 4	102	302	82	9	3	2	0	0	0	0
Pattern 5	46	396	52	2	3	1	0	0	0	0

We clarify this by explaining Figure 36.30 in greater detail. The input is Pattern 4, and the output is obtained by using the two settings $a_i = 2$ and $a_i = 2 + 6x_i$ in Figure 36.32a and b, respectively. The figure shows the Hamming distance between the output and the various possible inputs, namely Patterns 1–5 (although, in the interest of simplicity, we have only displayed the output for Patterns 1, 4, and 5). The output for all the patterns is chaotic when $a_i = 2$. However, when $a_i = 2 + 6x_i$, the output displays the input (Pattern 4) or a reasonably similar pattern with a very high frequency—as demonstrated by the low Hamming distance. The Hamming distance between the output and all the other patterns—the trained and the untrained pattern—is noticeably very large (i.e., with almost 50% noise).

The corresponding table of Hamming distances is given in Table 36.11. Consider the first column of Table 36.11. The reader will observe that the trained input patterns can be approximately retrieved almost 20% of the time. On the contrary, when encountering an untrained input pattern (Pattern 5 in Figure 36.1), it is observed in the output with a much lower frequency—less than 5%. Adachi et al. did observe this for the AdNN, namely that the system with the external stimulations yielded a stored pattern with a relative higher frequency, and an untrained pattern with a relative lower frequency—which phenomenon can be utilized to achieve PR. The amazing point here is that the L-AdNN displays the same phenomenon even though it merely has a *linear* number of connections, and the fact that the weights used are not directly obtained from the patterns themselves, but by migrating from the latter by using a gradient search

TABLE 36.13 Frequency Distribution of the Hamming Distance for the *Numeral* Data Set between the Input and the Output as the Measure Occurs within 10 Intervals, When the External Input Is $a_i = 2 + 6x_i$ and the Refractory Factor Is $\alpha = 10$

Input Pattern	Hamming Distance									
	[0,5]	[6,19]	[20,29]	[30,39]	[40,49]	[50,59]	[60,69]	[70,79]	[80,89]	[90,100]
Pattern 1	212	425	116	101	80	36	23	5	1	1
Pattern 2	231	397	122	115	76	40	17	1	0	1
Pattern 3	237	396	127	110	73	28	25	3	0	1
Pattern 4	252	380	117	108	80	40	17	5	0	1
Pattern 5	147	509	95	140	57	30	21	0	1	0

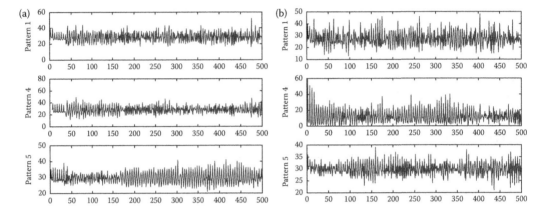

FIGURE 36.31 The Hamming distance between the output and the trained patterns or untrained patterns while using different settings for the *LOVE* data set. The input is Pattern 4 in both (a) and (b). The external stimulation weights are $a_i = 2$ and $a_i = 2 + 6x_i$ in (a) and (b), respectively. The output graphs are displayed for Pattern 1, Pattern 4, and Pattern 5.

algorithm. As far as we know, analogous properties have not been reported earlier for any type of NN (chaotic or otherwise).

Without repeating the same facts, we merely state that the same phenomenon is observed for the *LOVE* and *numeral* data sets as seen in Figures 36.31 and 36.32, respectively, for which the frequency distribution of the Hamming distances are given in Tables 36.12 and 36.13 respectively, when $a_i = 2$, and in Tables 36.12 and 36.13 respectively, when $a_i = 2 + 6x_i$.

36.5.5.3 Summary

We summarize the characteristics of the L-AdNN as follows. By extracting a spanning tree from the original completely connected graph, and then computing the best weights for this spanning tree by means of a gradient-based algorithm, the computational cost of the AdNN has been reduced significantly. The new linear-time AdNN-like network possesses interesting chaotic and PR properties for different settings.

36.6 The Logistic NN

36.6.1 The Design of the Logistic NN

In this section, we present the logistic NN (LNN) in which the primitive building block is the logistic neuron. The behavior of the LNN is analogous to that of the L-AdNN discussed in the previous section.

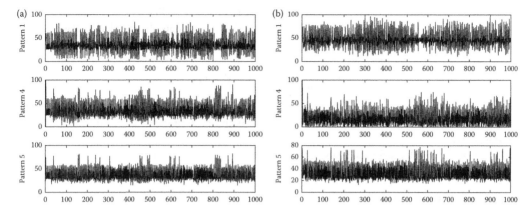

FIGURE 36.32 The Hamming distance between the output and the trained patterns or untrained patterns while using different settings for the *numeral* data set. The input is Pattern 4 in both (a) and (b). The external stimulation weights are $a_i = 2$ and $a_i = 2 + 6x_i$ in (a) and (b), respectively. The output graphs are displayed for Pattern 1, Pattern 4, and Pattern 5.

Unlike the AdNN, the M-AdNN, and the ideal M-AdNN, on achieving recognition, this network too does not emit the recognized pattern at the output *periodically*, but rather *repetitively*.

The logistic map is the simplest chaotic map. Indeed, Inoue and Nagayoshi, in a pioneering attempt, proposed the first CNN based on the logistic map [11]. The LNN described in this section is much "simpler" than the one reported in Reference 11, and it arguably possesses more fascinating properties.

Consider the discrete Hopfield NN model, characterized by the following equations:

$$y_i(t+1) = ky_i(t) + \alpha \left(\sum_j w_{ij}x_j(t) + a_i \right). \tag{36.46}$$

In an attempt to obtain AM and PR properties, we shall modify the structure by introducing a logistic *feedback* component. Therefore, our new network possesses a present-state/next-state function, and a state/output function, which are described by means of the following equations relating the only internal state $\eta_i(t)$ and the output $x_i(t)$ as follows:

$$\eta_i(t+1) = k\eta_i(t) + \alpha \left(\sum_j w_{ij}x_j(t) + a_i \right) - \beta z_i(t)x_i(t), \tag{36.47}$$

$$z_i(t+1) = 4z_i(t)(1 - z_i(t)), \tag{36.48}$$

$$x_i(t+1) = \frac{1}{1 + e^{-\eta_i(t+1)/\varepsilon}}. \tag{36.49}$$

Observe that the new model is composed of N neurons, topologically arranged as a completely connected graph. Each neuron i, $i = 1, 2, \ldots N$, has an internal state $\eta_i(t)$ and an output $x_i(t)$. When $t = 0$, the initial input $x_i(0)$ is assigned a pattern (known or unknown). The output $x_i(1)$ is fed as the *feedback input* at time $t = 1$, and so on. This input must be differentiated from the "eternal stimulus," $\{a_i\}$. As we shall see, to obtain AM properties, the $\{a_i\}$ is set to be a constant, for example, at $a_i = 2$. However, for PR properties, the $\{a_i\}$ correspond to the initial input, $\{x_i(0)\}$.

With regard to the present-state/next-state function, at time instant $t + 1$, the internal state $\eta_i(t+1)$ is determined by the previous internal state $\eta_i(t)$, the external stimulus, and the net input which is obtained via the feedback $x_i(t)$. In the equations, k is the damping factor (of the "nerve" membrane), where $0 < k \leq 1$. Also, since α and β are constants, we have set α to be unity, and used β to be the parameter which is tuned to get the desired performance. Also, while $z_i(t)$ is a chaotic feedback factor characterizing the logistic map given by Equation 36.48, the weights $\{w_{ij}\}$ are the edge weights obtained by the classic definition

$$w_{ij} = \frac{1}{p} \sum_{s=1}^{p} (2x_i^s - 1)(2x_j^s - 1).$$

Observe that this is a one-shot assignment and that it, in and of itself, does not include an additional training phase.

We now proceed with the formal analysis of the LNN described by Equations 36.47 through 36.49.

36.6.2 LE Analysis of the LNN

We shall now analyze the LNN, both from the perspective of a *single* neuron and of the network in its entirety.

36.6.2.1 Lyapunov Analysis of a Single Logistic Neuron

We first undertake the Lyapunov analysis of a single neuron. Indeed, it can be easily proven that a single neuron is chaotic by considering its Jacobian matrix and its QR decomposition [18].

By a straightforward computation, we see that the Jacobian matrix of the neuron is

$$J = \begin{pmatrix} \dfrac{\partial \eta(t+1)}{\partial \eta(t)} & \dfrac{\partial \eta(t+1)}{\partial z(t)} \\ \dfrac{\partial z(t+1)}{\partial \eta(t)} & \dfrac{\partial z(t+1)}{\partial z(t)} \end{pmatrix} = \begin{pmatrix} k + \dfrac{1}{\varepsilon}(\alpha - z(t)) \cdot x(t) \cdot (1 - x(t)) & -\beta x(t) \\ 0 & 4 - 8z(t) \end{pmatrix}.$$

It is very interesting to note from the above Jacobian matrix, that the LEs are only related to $x(t), z(t)$, and the parameters k, ε, and α, and that they are not related to β. Figure 36.33 shows the variation of the LEs of a single neuron with the parameter k. A single neuron has two LEs since it is defined by a two-dimensional discrete system (Equations 36.47 and 36.48). From this figure, we can see that one of the LEs is positive (~ 0.6918), which implies that the behavior of every single neuron is chaotic. We also see from the figure that the positive eigenvalue approaches a constant value,[*] because it is only determined by $z(t)$. As we know, $z(t)$ is a chaotic map possessing the positive LE $\log 2 \approx 0.6932$. Consequently, we claim that a single LN is truly chaotic because of the logistic feedback factor $z(t)$.

36.6.2.2 Lyapunov Analysis of the LNN

We shall now perform an analysis of the LNN (i.e., the entire network) from the viewpoint of its LEs. Consider the Jacobian matrix J of the LNN whose structure, topology, and weights have been determined in Section 36.6.1. J can be seen to have the form

$$J_{ij} = \frac{\partial \eta_i(t+1)}{\partial \eta_j(t)}, \tag{36.50}$$

where $i = 1, 2, \ldots, N$ and $j = 1, 2, \ldots, N$.

---

[*] One should also observe that the other LE is negative. This LE corresponds to the first eigenvalue of the Jacobian matrix.

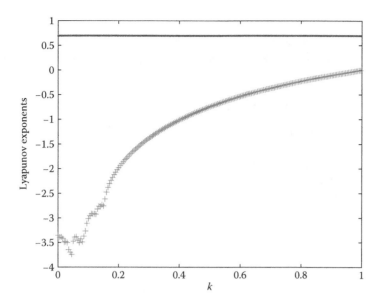

FIGURE 36.33 The variation of the LE of a single neuron with the parameter k.

By manipulating Equations 36.47 through 36.49, and by resorting to some simple algebraic steps, it is easy to show that

$$\frac{\partial \eta_i(t+1)}{\partial \eta_j(t)} = k \cdot \frac{\partial \eta_i(t)}{\partial \eta_j(t)} + \alpha \cdot w_{ij} \cdot \frac{\partial x_j(t)}{\partial \eta_j(t)} - \beta \cdot z(t) \cdot \frac{\partial x_i(t)}{\partial \eta_j(t)}. \tag{36.51}$$

The reader should observe that

$$x = 1/(1 + e^{-\eta/\varepsilon}),$$

which is equivalent to stating that

$$\eta = \frac{1}{\varepsilon}[\ln x - \ln(1 - x)].$$

Thus, the derivative $dx/d\eta = 1/(d\eta/dx) = \frac{1}{\varepsilon} \cdot x \cdot (1 - x)$.

The term $x_i(t) \cdot (1 - x_i(t))$ is always positive and close to the value zero whenever $x(t)$ is close to 0 or 1. Consequently, Equation 36.51 has the form

1. $\frac{\partial \eta_i(t+1)}{\partial \eta_j(t)} = k$ when $i = j, i, j = 1, 2, \ldots N$
2. $\frac{\partial \eta_i(t+1)}{\partial \eta_j(t)} = 0$ when $i \neq j, i, j = 1, 2, \ldots N$.

From the above, we see that the Jacobian matrix has the form

$$J = \begin{pmatrix} k & 0 & \ldots & 0 \\ 0 & k & 0 & 0 \\ \vdots & 0 & \ddots & \vdots \\ 0 & \ldots & \ldots & k \end{pmatrix}.$$

By definition, the LE are given by the logarithms of the eigenvalues of matrix Λ, where $\Lambda = (J^T J)^{\frac{1}{2}}$. Thus, the LNN has the Lyapunov spectrum

$$\lambda_i = \ln k, (i = 1, 2, \ldots, N).$$

Remark 36.2

Although the LEs are negative (as in the case of the AdNN), using the same terminology as Adachi et al., we can still refer to the LNN as being *chaotic*, because, for all practical purposes, there is no observable periodicity.

36.6.3 Chaotic, AM, and PR Properties of the LNN

We shall now report the AM and PR properties of the LNN. These properties have been gleaned as a result of examining the Hamming distance between the input pattern and the patterns that appear at the output. The experiments were conducted using two data sets, namely the figures used by Adachi et al. given in Figure 36.1, and the numeral data sets used by Calitoiu et al. [7,8] given in Figure 36.3. Although the experiments were conducted for a variety of scenarios, in the interest of brevity, we present here only a few typical sets of results—essentially, to catalog the overall conclusions of the investigation.

36.6.3.1 AM Properties

We discuss the properties of the LNN in three different settings. In all the three cases, the parameters were set to be $k = 0.55$, $\beta = 30$, and $z_i(0) = 0.3$. We should also point out that $z_1(0) = z_2(0) = \cdots = z_{100}(0) = 0.3$. In other words, the initial values of all the neurons were forced to be the same. Also, to emphasize, these AM properties were demonstrated by setting each "external stimulus" to be $a_i = 2$.

A word about the settings of the parameters used is not out of place. As per its biological significance, k, the damping factor of the nerve membrane, varies from 0 to 1. Thus, k was set as $k = 0.55$, which is a value relatively close to the mean. β is the input feedback parameter which is set by trial and error to obtain the best performance, as one does in the training phase of any PR problem. $z_i(t)$ is the chaotic logistic state variable constrained between 0 and 1. In fact, one observes that even though $z_i(0)$ for $i = 1, 2, \ldots, 100$ could be different, it does not affect our numerical or theoretical analysis. Therefore, the initial value $z_i(0) = 0.3$ was set rather arbitrarily. We remark that these specific settings ensured that the state variables $\eta_i(t)$ did not increase excessively or even disappear.

We first discuss the AM-related results of the LNN for the three scenarios, i.e., for trained inputs, for noisy inputs, and for untrained (unknown) inputs, respectively. In each case, as mentioned above, we report only a few results, i.e., for the case when the original pattern and the noisy version are related to P4. The results obtained for the other patterns are identical, and omitted here in the interest of brevity.

1. The initial input of the network is a known pattern, say P4.

   The observation that we report is that during the first 1000 iterations, the network can dynamically retrieve all the known patterns. This can be seen in Figure 36.34 in which we plot the Hamming distance[*] of the output pattern with P4.

2. The initial input of the network is a noisy pattern, in this case P5, which is a noisy version of P4.

   In this case, the network can still dynamically retrieve all the known patterns and their inverse versions during the first 1000 iterations. This can be seen in Figure 36.34b. The reader should observe that although the initial input is a *noisy* pattern, the four memorized patterns (and their inverse versions) appear in the output sequence frequently.

   We now magnify the results for a larger time frame, since from Figure 36.34 we see that the known pattern P2 is never recalled because we merely repeated the experiment for only 1000 iterations. Indeed, if the time frame is expanded for 10,000 iterations, we see that the pattern P2 appears 5 times. But even in this expanded time frame, P5 and P6 *never* appear. The plot of the Hamming distance is shown in Figure 36.35.

---

[*] As mentioned earlier, the reader must observe that if the identical pattern is observed, the Hamming distance is 0, and if the inverse version is observed, the Hamming distance is 100.

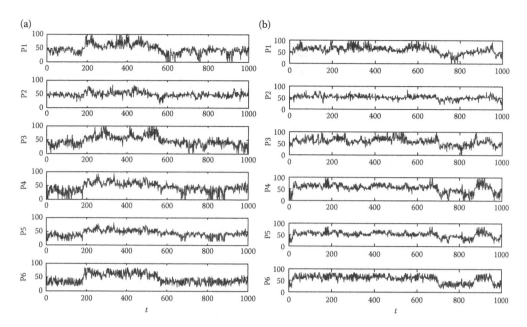

FIGURE 36.34 AM properties: The Hamming distance between the output and the trained patterns. In (a) the input patterns are the fourth pattern from Figure 36.1 and (b) the input patterns are the fifth pattern from Figure 36.1.

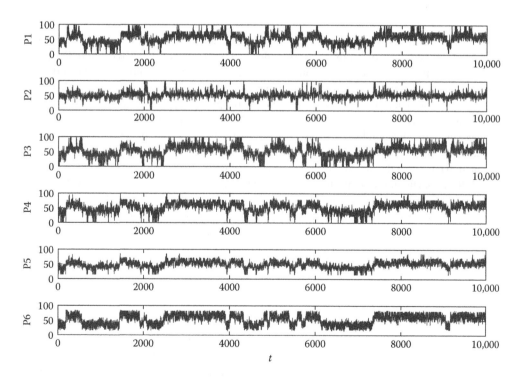

FIGURE 36.35 AM properties: The Hamming distance between the output and the trained patterns for a magnified time frame. The trained pattern is the fourth pattern of Figure 36.1.

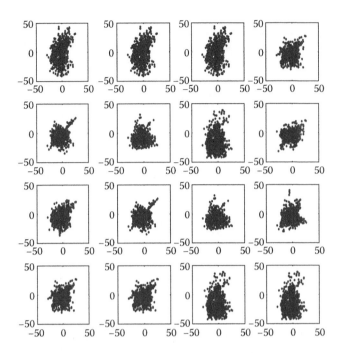

FIGURE 36.36 The 16 phase diagrams of the neurons' internal states. In this figure, from left to right and from top to bottom, the x-axis are $\{\eta_{49}(t), \eta_{50}(t), \ldots, \eta_{64}(t)\}$, respectively. The y-axis in each case is $\eta_{89}(t)$.

   Based on the above observations, we can easily conclude that the four patterns and their inverse versions are the attractors of the network. Indeed, in both these cases, we see that independent of the initial input, the network will be attracted to these attractors, even though the network will not stay at a fixed or periodic state—which is a typical phenomenon of chaotic systems. To demonstrate this chaotic behavior more clearly, we have plotted the phase diagram of the internal states of a few randomly chosen neurons, namely neurons $\{49, 50, \ldots, 64\}$, respectively versus that of neuron 89. These plots are shown in Figure 36.36. They display the trajectories of the internal states, where the chaotic behavior as a chaotic associate memory is clear.

 3. The initial input of the network is an unknown pattern, P6.

   If the initial input is a completely unknown pattern, say, P6, a system possessing AM should still be able to reproduce/retrieve all the memorized patterns. On the other hand, the system should never yield an unknown pattern. These phenomena can be seen from Figure 36.37. Observe that the untrained pattern never appears in the output sequence.

A record of the statistics (frequencies) of all the above three cases is cataloged in Table 36.14.

36.6.3.2 PR Properties

Similar to the results reported in Section 36.6.3, we now present an in-depth report of the LNN's PR properties by using Hamming distance-based analyses. The parameters that were used were $k = 0.55$, $\beta = 30$, $a_i = 2x_i$, and $z_0 = 0.3$. The difference between these experiments and the AM-related ones involves the "external stimulus" $\{a_i\}$, which is not a constant, but a scaled version of the $\{x_i\}$. The PR-related results of the LNN are reported for the three scenarios, i.e., for trained inputs, for noisy inputs, and for untrained (unknown) inputs, respectively. As in the AM case, we report only the results for the setting when the

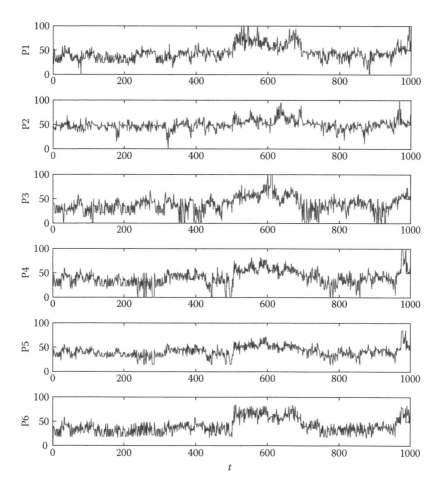

FIGURE 36.37 AM properties: The Hamming distance between the output and the trained patterns. In this figure, the input pattern is the sixth pattern of Figure 36.1.

original pattern and the noisy version are related to P4. The results[*] obtained for the other patterns are identical, and omitted here for brevity.

1. The initial input of the network is a known pattern, say P4.

    To report the results for this scenario, we request the reader to observe Figure 36.38, where Figure 36.38a we can find that P4 is retrieved very frequently as a response to the input pattern. This occurs 197 times (almost 1/5) in the first 1000 iterations. On the other hand, the other three patterns never appear in the output sequence. The phase diagrams of the internal states that correspond to Figure 36.38a are shown in Figure 36.40a, where the x- and y-axes are the same as in Figure 36.36. From Figures 36.38a and 36.40a, we can conclude that the input pattern can be recognized successfully if it is one of the known patterns.

2. The initial input of the network is a noisy pattern, in this case P5, which is a noisy version of P4.

    Even when the external stimulus is a garbled version of a known pattern (in this case P5 which contains 15% noise), it is interesting to see that *only* the original pattern P4 is recalled frequently (as high as 187 times in the first 1000 iterations). In contrast, the other three known patterns are never

---

[*] We again remind the reader that unlike the AdNN, the M-AdNN, and the ideal M-AdNN, on achieving recognition, this network too does not emit the recognized pattern at the output *periodically*, but rather *repetitively*.

TABLE 36.14 Statistics (Frequencies) of the Occurrences of the Various Patterns during the First 10,000 Iterations

		Input Patterns					
		P_1	P_2	P_3	P_4	P_5	P_6
Frequency statistic	P_1	99	96	108	102	92	89
	$*P_1$	187	85	201	197	191	186
	P_2	38	45	32	34	38	31
	$*P_2$	31	19	41	41	31	32
	P_3	190	190	185	174	189	222
	$*P_3$	235	90	219	219	236	218
	P_4	98	98	127	135	84	95
	$*P_4$	40	18	30	30	43	33
	P_5	0	0	0	0	1	0
	$*P_5$	0	0	0	0	0	0
	P_6	0	0	0	0	0	1
	$*P_6$	0	0	0	0	0	0

Note: In the table, $*P_i$ signifies the inverse version of the pattern P_i. From this table, we can see that independent of the identity of the input pattern (trained or untrained), the network can recall the four memorized patterns and their inverse versions dynamically.

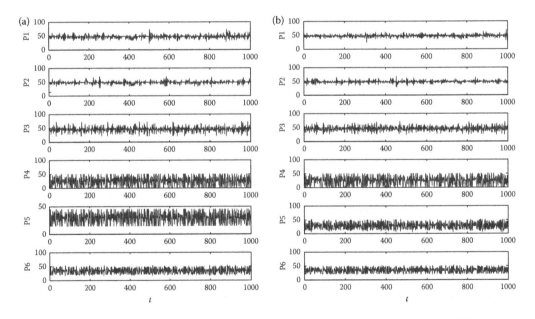

FIGURE 36.38 PR properties: The Hamming distance between the output and the trained patterns. The input patterns are the P4 and P5 patterns of Figure 36.1, respectively. Observe that P5 is a 15%-noisy version of P4.

recalled. These phenomena can be seen from the Figure 36.38b. By comparing (a) and (b) of Figure 36.38, we can draw the conclusion that the LNN can achieve a chaotic PR even in the presence of noise and distortion. Indeed, even if the external stimulus contains some noise, the LNN is still able to recognize it correctly.

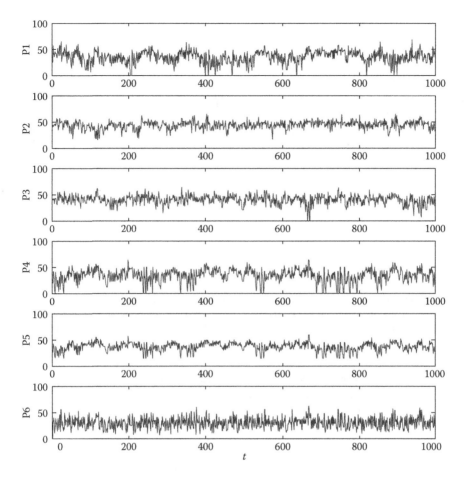

FIGURE 36.39 PR properties: The Hamming distance between the output and the trained patterns. The input pattern is the sixth pattern of Figure 36.1. P6 is an unknown pattern.

3. The initial input of the network is an unknown pattern, P6.

In this case, we investigate whether the LNN is capable of distinguishing between known and unknown patterns. Thus, we attempt to stimulate the network with a completely unknown pattern. In our experiments, we used the pattern P6 of Figure 36.1 initially used by Adachi et al. From Figure 36.39, we see that some of the known patterns (P1, P3, and P4) are retrieved several times. As opposed to this, the noisy pattern P5 and the unknown pattern P6 never appear. The corresponding phase diagram is shown in Figure 36.40b.

It is very interesting to compare the difference between Figure 36.40a and b. From Figure 36.40a, we can intuitively see that the phase diagram is more "ordered"—although it is not periodic. As opposed to this, in Figure 36.40b, it is more "disordered," implying the presence of strange attractors. The only explanation for these phenomena is due to the external stimulus: the LNN is "ordered" when exposed to known patterns, and tends to be "disordered" when exposed to unknown/untrained patterns. In other words, the LNN responds intelligently to the various inputs with different outputs, which is a crucial characteristic of a PR system. Moreover, the switch between "order" and "disorder" seems to be consistent with Freeman's biological results—which, we believe, is quite fascinating!

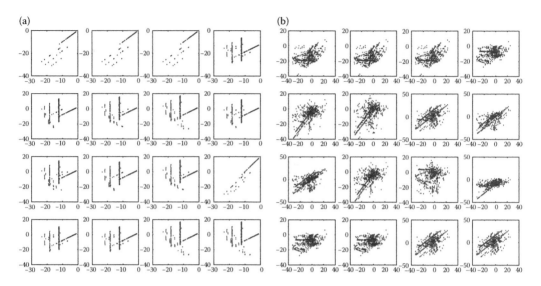

FIGURE 36.40   PR properties: The phase diagrams of the neurons' internal states. In both (a) and (b), from left to right and from top to bottom, the x-axis are $\{\eta_{49}(t), \eta_{50}(t), \ldots, \eta_{64}(t)\}$, respectively. The y-axis is $\eta_{89}(t)$ in every case. While Figure 36.40a corresponds to a known external stimulus, Figure 36.40b corresponds to an unknown external stimulus.

TABLE 36.15   Frequency-Based Statistics for the Adachi Data Set, Which Demonstrate the LNN's PR Capabilities for Figures 36.38 and 36.39

		\multicolumn{6}{c}{Input Patterns}				
		P_1	P_2	P_3	P_4	P_5	P_6
	P_1	188	0	0	0	0	16
	P_2	0	232	0	0	0	0
Frequency	P_3	0	0	204	0	0	3
statistic	P_4	0	0	0	197	187	36
	P_5	0	0	0	0	1	0
	P_6	0	0	0	0	0	1

The statistical frequencies of the Hamming distances for the Adachi data set for all the three cases are listed in Table 36.15. We believe that this is also a convincing proof of the fact that the LNN can achieve a chaotic PR.

To summarize, we catalog our results by stating that the LNN is a novel CNN which possesses good AM and PR properties as follows:

- Case I—AM: When exposed to an external stimulus with $a_i = 2$, it performs as an AM. In other words, it is able to retrieve all the known patterns independent of what the initial input pattern is.
- Case II—PR: When exposed to an external stimulus $a_i = 2x_i$, it achieves PR. Further, in this case
  - If x_i corresponds to a known pattern or a noisy pattern, the LNN can recognize it correctly. In this case, the phase diagrams of the neurons are "ordered."
  - If the x_i corresponds to an unknown pattern, the LNN can retrieve all the known patters (akin, to some degree, to an AM phenomenon), but the frequency with which the patterns occur is much lower than in Case I. In this case, the phase diagram is "disordered."

Observe that like all the other CNNs discussed here, the AM and PR characteristics of the LNN depend on the settings and the inputs. We conclude by stating that, as far as we know, this new model of CNNs has been unreported in the literature, and the results given here are novel.

36.7 Overall Conclusions

In this chapter, we have concentrated on the field of *chaotic* PR. This field, as we have observed, is distinct from the other strategies used in PR such as statistical, syntactic, and structural, because it deals with using chaos and repetition/periodicity to achieve the "recognition." As a benchmark of a chaotic PR system, the chapter has placed the following stipulations: first of all, one must be able to train the system with a set of "training" patterns. Subsequently, as long as there is no testing pattern, the system must be chaotic. Additionally, if the system is presented with an unknown testing pattern, the behavior must ideally be as follows. If the testing pattern is not one of the trained patterns, the system must continue to be chaotic. As opposed to this, if the testing pattern is truly one of the trained patterns (or a noisy version of a trained pattern), the system must emit, at its output, the trained pattern repetitively, and hopefully, periodically. This switch from *chaos* to *order*, i.e., repetition, is truly ambitious, and is precisely an amazing phenomenon that the brain is able to accomplish.

This chapter has presented the various NNs that have been proposed in the literature to attain this goal. In particular, the chapter has used as its starting model the AdNN, and shown how various authors have built on the latter to yield the M-AdNN and the ideal modified Adachi neural network (Ideal-M-AdNN). In all these cases, the output of the NN is *periodic* when a trained pattern appears at its input. The chapter has also described the L-AdNN and the LNN in which the specific trained pattern appears at the output repeatedly, though not periodically.

In summary, this chapter describes, quite exhaustively, the art and science of *chaotic* PR, and is a record of the state-of-the-art results pertaining to this field.

References

1. M. Adachi and K. Aihara. Associative dynamics in a chaotic neural network. *Neural Networks*, 10(1):83–98, 1997.
2. M. Adachi and K. Aihara. An analysis on instantaneous stability of an associative chaotic neural network. *International Journal of Bifurcation and Chaos*, 9(11):2157–2163, 1999.
3. M. Adachi and K. Aihara. Characteristics of associative chaotic neural networks with weighted pattern storage—A pattern is stored stronger than others. In *The 6th International Conference on Neural Information*, vol. 3, pp. 1028–1032, Perth, Australia, 1999.
4. M. Adachi, K. Aihara, and M. Kotani. Pattern dynamics of chaotic neural networks with nearest-neighbor couplings. In *The IEEE International Symposium on Circuits and Systems*, vol. 2, pp. 1180–1183, Westin Stanford and Westin Plaza, Singapore, 1991.
5. M. Adachi, K. Aihara, and M. Kotani. An analysis of associative dynamics in a chaotic neural network with external stimulation. In *International Joint Conference on Neural Networks*, vol. 1, pp. 409–412, Nagoya, Japan, 1993.
6. K. Aihara, T. Takabe, and M. Toyoda. Chaotic neural networks. *Physics Letters A*, 144(6–7):333–340, 1990.
7. D. Calitoiu, B. J. Oommen, and D. Nussbaum. Desynchronizing a chaotic pattern recognition neural network to model inaccurate perception. *IEEE Transactions on Systems Man and Cybernetics Part B—Cybernetics*, 37(3):692–704, 2007a.
8. D. Calitoiu, B. J. Oommen, and D. Nussbaum. Periodicity and stability issues of a chaotic pattern recognition neural network. *Pattern Analysis and Applications*, 10(3):175–188, 2007b.

9. J. P. Eckmann and D. Rulle. Ergodic theory of chaos and strange attractors. *Reviews of Modern Physics*, 57(3):617–656, 1985.

10. W. J. Freeman. Tutorial on neurobiology: From single neurons to brain chaos. *International Journal of Bifurcation and Chaos in Applied Sciences and Engineering*, 2:451–482, 1992.

11. M. Inoue and A. Nagayoshi. A chaos neuro-computer. *Physics Letters A*, 158:373–376, 1991.

12. G. C. Luo, J. S. Ren, and K. Qin. Dynamical associative memory: The properties of the new weighted chaotic Adachi neural network. *IEICE Transactions on Information and Systems*, E95d(8):2158–2162, 2012.

13. J. Nagumo and S. Sato. On a response characteristic of a mathematical neuron model. *Biological Cybernetics*, 10(3):155–164, 1971.

14. K. Qin and B. J. Oommen. Chaotic pattern recognition: The spectrum of properties of the Adachi neural network. In *Lecture Notes in Computer Science*, vol. 5342, pp. 540–550, Florida, USA, 2008.

15. K. Qin and B. J. Oommen. Adachi-like chaotic neural networks requiring linear-time computations by enforcing a tree-shaped topology. *IEEE Transactions on Neural Networks*, 20(11):1797–1809, 2009.

16. K. Qin and B. J. Oommen. Networking logistic neurons can yield chaotic and pattern recognition properties. In *IEEE International Conference on Computational Intelligence for Measure Systems and Applications*, pp. 134–139, Ottawa, Ontario, Canada, 2011.

17. K. Qin and B. J. Oommen. The entire range of chaotic pattern recognition properties possessed by the Adachi neural network. *Intelligent Decision Technologies*, 6(1):27–41, 2012.

18. M. Sandri. Numerical calculation of Lyapunov exponents. *The Mathematica Journal*, 6(3):78–84, 1996.

19. J. C. Sprott. Numerical calculation of largest Lyapunov exponent, 1997. Available at http://sprott. physics.wisc.edu/chaos/lyapexp.htm.

20. A. Wolf, B. J. Swift, L. H. Swinney, and A. J. Vastano. Determining Lyapunov exponent from a time series. *Physica*, 16D:285–317, 1985.

Chaos in Socioeconomic and Human Sciences

XI

Chaos in
Socioeconomic
and Human
Sciences

37

Why Economics Has Not Accomplished What Physics Has?

Marisa Faggini and
Anna Parziale

37.1 Introduction

It has been more than 20 years since ideas from deterministic chaos began appearing in the economics literature. Economists began to look at chaotic analyses of the late 1970s and the 1980s, including important works such as those by Medio (1979), Stutzer (1980), Benhabib and Day (1981), Day (1982), and Grandmont (1985), just to name a few. A common feature of chaos models is that nonlinear dynamics tend to arise as the result of relaxing the assumptions underlying the competitive market general equilibrium approach (Faggini, 2009).

This interdisciplinary spread of ideas was accompanied by expectations that many major problems in economics could be easily solved using chaos-inspired techniques. It is probably true that many early expectations for chaos have not been fulfilled.

An assessment of the impact that chaotic dynamics has had on economics requires an understanding of the paradigm of research dominant in this area, in which the generic method of isolation, of inclusion and exclusion, of focusing on key elements and neutralizing the rest, and of simplification and idealization are applied.

This paradigm is essentially based on the following assumptions:

a. In the absence of exogenous shocks, the economy tends toward a determinate and intrinsically stable equilibrium as its natural end.
b. The economic agents, resumed in the behavior of the representative agent, are described as rational-calculating individuals who maximize their utility or profits on the basis of complete information about the quantities, costs, prices, and demand of their products. They have extraordinary capacities, particularly concerning the area of information processing and computation.

c. Linear models or at least the linearization of models have been traditionally preferred by economists. This is why linear models with one solution or one equilibrium position can be explicitly solved without using numerical procedures.

So described, economics is largely a matter of formalized thin fiction and has little to do with the wonderful richness of the facts of the real world.

The idea that markets are inherently dynamically unstable has always played a minor role in studies of economic phenomena, and this has changed only marginally with the diffusion of chaos theory. This is because chaos theory has stimulated the search for a mechanism that generates observed movements in real economic data and minimizes the role of exogenous shocks. If stochastic models explain many of these sudden fluctuations caused by external random shocks, in a chaotic system, these abrupt fluctuations are considered to be internally generated as part of the deterministic process (Gilmore, 1996). The fluctuations are within the system. They are the result of complex interactions among the system's elements, and although it is difficult to predict the system behavior, the same cannot be said of the process that created it, as it is deterministic. In this sense, chaos theory represents a shift in thinking about methods for studying economic activity and in the explanation of many economic phenomena.

The theory of chaos stresses that the world does not necessarily work as a linear relationship with perfectly defined or with direct relations in terms of expected proportions between causes and effects.

Therefore, researchers in economics and finance have been interested in testing nonlinear dependence and chaos in economic models and data. A wide variety of reasons for this interest have been suggested, including an attempt to improve the forecasting accuracy of linear time series models and to better explain the dynamics of the underlying variables of interest using a richer class of models than that permitted by limiting the set to the linear case.

But chaos theory has not had the same impact in economics as it has in hard sciences like physics. The search for chaos in economics has gradually become less enthusiastic, as no empirical support for the presence of chaotic behaviors in economics has been found. The literature did not provide a solid support for chaos as a consequence of the high noise level that exists in most economic time series, the relatively small sample sizes of data, and the weak robustness of chaos tests for these data.

In economics, data sets are the outcome of a complex process including institutional or structural changes and monetary regime switches, shocks, wars, political crises, etc. The rich nature as well as the impact of these changes reveals interesting features in time series (structural instability and nonlinearity) that needs to be studied by developing new techniques able to filter these complex dynamics (Krystos and Vorlow, 2009).

The following sections contain a survey of basic chaos tests and empirical works on economic and financial data performed to uncover the evidence of chaotic dynamics. The aim of this chapter is to show why chaotic processes in the economic data series have proved problematic and certain nonlinear time series tools and techniques inspired by chaos have thrived.

37.2 Tools for Analyzing Economic Time Series

From a theoretical point of view, researchers efforts have concentrated in showing that even very standard, commonly accepted models of intertemporal economic equilibrium can generate endogenous fluctuations and chaos even if a common feature of these models is that nonlinear and chaotic dynamics tend to arise as the result of relaxing the assumptions underlying the competitive market general equilibrium approach.

Nevertheless, the relevance of addressing chaos in economic models is associated with detecting the presence of chaotic motion in economic data. To show that a mathematical model exhibits chaotic behavior does not prove that chaos is also present in the corresponding experimental data. From an empirical point of view, it is difficult to distinguish between fluctuations provoked by random shocks and endogenous fluctuations determined by the nonlinear nature of the relation between economic aggregates.

For this purpose, chaos tests are developed to investigate the basic features of chaotic phenomena: non-linearity, fractal attractor, and sensitivity to initial conditions.

37.2.1 Correlation Dimension

A necessary but not sufficient condition to define a system as being chaotic is that the strange attractor has a fractal dimension.

The notion of dimension refers to the degree of complexity of a system expressed by the minimum number of variables that is needed to replicate the system (Schwartz and Yousefi, 2003). For example, a cube has three dimensions, a square has two dimensions, and a line has one. The topological dimension is always an integer. A chaotic system has noninteger dimensionality called fractal dimension. The fractal dimension measures the probability that two points chosen at random will be within a certain distance of each other and examines how this probability changes as the distance is increased.

In the literature, there are many methods[*] for calculating the fractal dimension (Hausdorff dimension, the box-counting dimension, the information dimension, and the correlation dimension), which nevertheless do not provide equivalent measures (Hentschel and Procaccia, 1983). Among these different algorithms, the correlation dimension proposed by Grassberger and Procaccia (1983a), based on phase space reconstructions of the process to estimate,[†] has the advantage of being straightforward and quickly implemented.

Let us consider the monodimensional series $\{x_t\}_{t=1}^{n}$ and, from this, the sequence of $N = n - m + 1$ m-dimensional vectors, $X_t = (x_t, x_{t-1}, \ldots, x_{t-m+1})$ that gives us the reconstructed series $\{x_t\}_{t=m}^{n}$.

If the unknown system that generated $\{x_t\}_{t=1}^{n}$ is n-dimensional, and provided that the embedding dimension[‡] is $m \geq 2n+1$,[§] we have that the set of m-histories recreates the dynamics of the data-generating system and can be used to analyze its dynamics (Packard et al., 1980, Takens, 1981).

Let us suppose that $C(N, m, \varepsilon)$ is the number of points separated by a distance less than ε; for a given embedding dimension, the correlation function[¶] is given by

$$C(N, m, \varepsilon) = \frac{1}{N(N-1)} \sum_{m \leq t \neq s \leq N} H(\varepsilon - \|X_t - X_s\|) \, \varepsilon > 0 \tag{37.1}$$

$H(z)$ is the Heaviside function given by $H(z) = 1$ for all $z \geq 0$ and 0; otherwise, ε is the sufficiently small distance between vectors X_t and X_s, and $\|\cdot\|$ is the norm operator.

The correlation function $C(N, m, \varepsilon)$ gives the probability that a randomly selected pair of delay coordinate points is separated by a distance less than ε. It measures the frequency with which temporal patterns are repeated in the data.

To determine the correlation dimension from Equation 37.1, we have to determine how $C(N, m, \varepsilon)$ changes as ε changes. As ε grows, the value of $C(N, m, \varepsilon)$ grows because the number of near points to be included in Equation 37.1 increases. Grassberger and Procaccia (1983b) show that for a sufficiently

---

[*] See Farmer et al. (1983), Barnsley (1988), Falconer (1990), and Cutler (1991).

[†] This procedure is based on the method of delay time coordinates, by Takens (1981) that showed that this type of reconstruction yields a topological equivalent attractor leaving the dynamic parameters invariant.

[‡] Basic elements to reconstruct the time series from the original one are the delay time and the embedding dimension. In the literature, there are some techniques like the false nearest neighbor and the mutual information function to respectively choose the embedding dimension and the delay time.

[§] According to the numerical results provided by Packard et al. (1980), it is possible to get reasonable results with much smaller embedding dimensions. This point is particularly interesting in different economic applications since in such cases, the dimension of the true phase space is often not known *a priori*. Over the years, this insight has widely been adopted in economic literature on chaos where the common practice is to choose m around 10–12 (Schwartz and Yousefi, 2003).

[¶] Example by Barnett and He (2000).

small ε, $C(N, m, \varepsilon)$ grows at rate D_C and can be well approximated by

$$C(N, m, \varepsilon) \approx \varepsilon^{D_C} \tag{37.2}$$

That is, the correlation function is proportional to the same power of D_C that represents the value of the correlation dimension.

More formally, the dimension associated with the reconstructed dynamic is given by

$$D_C = \lim_{\varepsilon \to 0} \frac{\log C(N, m, \varepsilon)}{\log \varepsilon} \tag{37.3}$$

That is, it is given by the slope of the regression of $\log C(N, m, \varepsilon)$ versus $\log \varepsilon$ for small values of ε and depends on the chosen embedding dimension.

If, as m increases D_C continues to rise, then this relationship is symptomatic of a stochastic system. If the data are generated by a chaotic system, D_C will reach a finite limit at some relatively small m (saturation point). The importance of the correlation dimension arises from the fact that the minimum number of variables required to model a chaotic attractor is the smallest integer greater than the correlation dimension itself.

The reliability of implementing this algorithm suffers from some problems. Because it is based on the method of delay time coordinates introduced by Takens (1981), the estimates of the embedding dimension and delay time are so crucial that an unfortunate delay time choice yields misleading results concerning the dimension of well-known attractors.

Other than the problems associated with these estimates, the correlation dimension suffers from two other problems related to the choice of a sufficiently small ε and the norm operator. With the limited length of the data, it will almost always be possible to select a sufficiently small ε so that any two points will not lie within ε of each other (Ramsey and Yuan, 1989).

Regarding the norm operator, while Brock's (1986) theorem gives the conditions under which the correlation function remains independent of the choice of the norm, Kugiumtzis (1997) shows the invalid application of this theorem for a short noisy time series, such as economic and financial series. Therefore, under such circumstances, the most reliable results are obtained by using the Euclidian norm (Schwartz and Yousefi, 2003). The presence of noise in time series not obtained from experiments could further compromise the distinction between stochastic and deterministic behavior. Therefore, a data-filtering procedure (linear and nonlinear) is required to reduce the unwelcome noise level without distorting the original signal[*] to obtain reliable results.

Reliability could also be compromised by using short data sets (Ramsey and Yuan, 1989, 1990). In fact, in the case of high-dimensional chaos, it will be very difficult to make estimates without an enormous amount of data. This suggests that the correlation dimension can only distinguish low-dimensional chaos from high-dimensional stochastic processes, particularly with economic data. Furthermore, if the fractal dimension is found, the correlation dimension, as in all nonparametric methods, does not provide information about the dynamics of the process that generated it because it does not preserve time-ordering data (Gilmore, 1993a,b).

37.2.2 Brock–Dechert–Scheinkman Test

The Brock–Dechert–Scheinkman (BDS) test[†] introduced by Brock et al. (1987, 1996) is a nonparametric method based on the correlation function developed by Grassberger and Procaccia (1983a), defined in

---

[*] Caputo et al. forthcoming in *Chaos, Solitons, and Fractals*.

[†] Subsequent to its introduction, the BBS test was generalized by Savit and Green (1991) and Wu et al. (1993), and more recently, DeLima (1998) introduced an iterative version of the BBS test.

Equation 37.1, and used to test for serial dependence and nonlinear structures[*] in a time series. The BDS test incorporates the embedding dimensions, but it assumes that the delay time equals 1.[†]

Therefore, the BDS test is not considered to be a direct test for chaos; rather, it is used as a model selection tool to obtain some information about what kind of dependency exists after removing nonlinear dependency from the data.

The standardized residuals from an autoregressive conditional heteroskedasticity (ARCH)-type model are extracted and then tested for nonlinear dependence. If there is no dependence, the data are not chaotic because the ARCH-type model has captured all nonlinearities (Hsieh, 1989, 1991); otherwise, the BDS test is applied to residuals to check if the best-fit model for a given time series is a linear or nonlinear model.

The BDS tests the null hypothesis that the variable of interest is independently and identically distributed (IID). Because IID implies randomness, if a series is proved to be IID, it is random (Barnett et al., 1997).

Under the null hypothesis of whiteness, the BDS statistic is obtained by[‡]

$$W(N, m, \varepsilon) = \sqrt{N} \frac{C(N, m, \varepsilon) - C(N, 1, \varepsilon)^m}{\hat{\sigma}(N, m, \varepsilon)} \tag{37.4}$$

The correlation function asymptotically follows standard normal distribution $N(0,1)$: $\lim_{N \to \infty} W(N, m, \varepsilon) \sim N(0, 1)$, $\forall m$, and ε. $\hat{\sigma}(N, m, \varepsilon)$[§] is the standard sample deviation of $C(N, m, \varepsilon) - C(N, 1, \varepsilon)^m$.

Moving from the hypothesis that a time series is IID, the BDS tests the null hypothesis that $C(N, m, \varepsilon) = C(N, 1, \varepsilon)^m$, which is equivalent to the null hypothesis of whiteness against an unspecified alternative.

Hsieh (1991) shows that the BDS test can detect the presence of four types of non-IID behaviors resulting from a nonstationarity of the series, a linear stochastic system (such as autoregressive moving average [ARMA] processes), a nonlinear stochastic system (such as ARCH/GARCH processes), or a nonlinear deterministic system, which could feature low-order chaos.[¶] If the series are IID, in which linear or even conditional heteroskedasticity can describe the relations between data, chaotic tests will not be required. However, if this is not the case, investigating the main properties of chaoticity should not be disregarded.

Moreover, because it is based on the correlation dimension, the BDS test suffers from the same limitations. In particular, its performance depends on the size of data sets (N) and ε,[‖] even though Brock et al. (1991) showed how the statistics of this test are correctly approximated in finite samples if

The number of data N is greater than 500.

ε lies between 0.5 θ and 2 θ, where θ is the standard deviation of the series.

The embedding dimension m is lower than $N/200$.

Moreover, it has been found that the BDS test has low power against certain forms of nonlinearity, such as self-exciting threshold AR processes and neglected asymmetry in volatility (Kuan, 2008).

37.2.3 Lyapunov Exponents

The time series analysis tools described above—the BDS test and the correlation dimension—allow for the distinction between nonlinear systems with a certain degree of complexity and those without, relying on specific features of these systems: nonlinearity for the BDS test and fractal dimension for the correlation dimension. Therefore, considering the fact that the BDS test produces an indirect evidence of nonlinear

---

[*] "There are three particularly well-known tests currently in use for testing for nonlinearity: BDS test, White's neural network test, and the Hinich bispectrum test" (Barnett et al., 1997, p. 8).

[†] See Barnett et al. (1997) and Matilla-García et al. (2004) for the problems when fixing delay time to one. Moreover, we have to consider the BDS-G test suggested by Matilla-García et al. (2004) as a new way for selecting an adequate delay time that allows the obtaining of a good approximation of the correlation dimension.

[‡] Example by Barnett and He (2000).

[§] See Brock et al. (1996).

[¶] See Fillol (2001).

[‖] To deepen this point see Kyrtsou et al. (2001).

dependence, which is a necessary but not sufficient condition for chaos (Barnett et al., 1995, 1997, Barnett and Hinich, 1992), and even though a nonlinear dynamic is low dimensional, it cannot be considered chaotic; thus, we need a more appropriate tool to detect chaotic behavior. The Lyapunov exponent may provide a more useful characterization of chaotic systems because unlike the correlation dimension, which estimates the complexity of a nonlinear system, it indicates a system's level of chaos.

The Lyapunov exponent (L) investigates another different, and perhaps more specific, characteristic of chaotic systems: their sensitivity to initial conditions. Two points with arbitrarily close but unequal initial conditions will diverge at exponential rates. The trajectories remain within a bounded set if the dynamic system is chaotic.

In calculating the divergence of the trajectories, we are interested in identifying what is known as the greatest exponent or Lyapunov characteristic exponent.[*] This exponent measures the average exponential divergence or convergence between trajectories that differ only in having an "infinitesimally small" difference in their initial conditions and remains well defined for noisy systems.

To estimate λ from experimental or observational data, there are two classes of methods, both of which are based on reconstructing the space state by the delay coordinate methods. The direct methods[†] proposed by Wolf et al. (1985) and Rosenstein et al. (1993) are based on the calculation of the growth rate of the difference between two trajectories with an infinitesimal difference in their initial conditions.

In Jacobian methods,[‡] data are used to estimate the Jacobians of underlying processes, and λ is calculated from these. Nychka et al. (1992) proposed a regression method similar in some respects to the Gençay and Dechert test (1992), which involves the use of neural networks to estimate the Jacobians and λ; it is known as the Nychka–Ellner–Gallant–McCaffrey (NEGM) test. Some remarkable advantages of the Jacobian methods over the direct methods are their robustness to the presence of noise and their satisfactory performance in moderate sample sizes (Shintani and Linton, 2004).

The general idea on which these methods are based is to follow two nearby points and calculate their average logarithmic rate of separation.

Consider x_0 and x_0' as two points in the state space with distance $\|x_0 - x_0'\| = \delta_{x_0} \ll 1$.

Here, δ_{xt} is the distance after T iterations between two trajectories emerging from these points; thus,

$$\delta_{xt} \approx \delta_{x_0} e^{\lambda T}$$

where T is the iteration number and λ is the maximal Lyapunov exponent, which measures the average rate of divergence or convergence of two nearby trajectories. This process of averaging is the key to calculating accurate values of λ using small, noisy data sets.

In a system with attracting fixed points of a periodic orbit, the distance $\delta x(x_0, t)$ diminishes asymptotically with time. If the system is unstable, the trajectories diverge exponentially for a while but eventually settle down. If the system is chaotic, $\delta x(x_0, t)$ behaves erratically. Hence, it is better to study the mean exponential rate of the divergence of trajectories from two initially close points using the following algorithm:

$$\lambda = \lim_{T \to \infty} \frac{1}{T} \sum_{t=1}^{T} \ln \frac{|\delta x(x_0, t)|}{|\delta x_0|} \tag{37.5}$$

The exponents can be positive or negative, but at least one exponent must be positive for an attractor to be classified as chaotic. In particular, if $\lambda < 0$, the system converges to a stable fixed point or stable periodic orbits. A negative value of the Lyapunov exponent is characteristic of dissipative or nonconservative

---

[*] "[...] maximal Lyapunov exponent [...] is the inverse of a time scale and quantifies the exponential rate by which two typical nearby trajectories diverge in time. In many situations the computation of this exponent only is completely justified, [...]. However, when a dynamical system is defined as a mathematical object in a given state space, [...] there exist many different Lyapunov exponent as there are space dimensions," (Kantz and Schreiber, 1997, p. 174).

[†] Some limitations of these methods are highlighted in Shintani and Linton (2004).

[‡] To obtain the Lyapunov exponent from observational data, Eckmann and Ruelle (1985) and Eckmann et al. (1987) proposed a method, known as the Jacobian method, which is based on nonparametric regression.

systems. If $\lambda = 0$, the system is conservative and converges to a stable cycle limit. If $\lambda > 0$, the system is unstable and chaotic. Nearby points, no matter how close, will diverge. Therefore, if the system is chaotic, it will at least have a positive Lypunov exponent.[*] In fact, one definition of chaotic systems is based on a positive Lyapunov exponent (Deneckere and Pelikan, 1986, Mayer-Kress, 1986, Wolf et al., 1985). Finally, if $\lambda = \infty$, the system is random.

37.2.4 Topological Methods

The tools described so far, which are based on detecting the metric and dynamical invariants of attractors, are highly sensitive to noise (Barnett and Serletis, 2000). In particular, their applications often require large, clean data sets. Now, if we consider the data sets provided in economics, which are small and noisy, the possibility of detecting chaotic behaviors is very limited. To overcome these limits, the attention of researchers is redirected toward the topological tools that provide the basis for a new way of testing data for chaotic behavior (Mindlin et al., 1990, Mindlin and Gilmore, 1992, Tufillaro et al., 1990).

Topological tools are characterized by the study of the organization of the strange attractor[†] and exploit an essential property of chaotic systems, i.e., the tendency of a time series to nearly, although never exactly, repeat itself over time. This property is known as the recurrence property. Topological methods feature close returns plots (CRPs) (Gilmore, 1993a,b, 1996, 2001, McKenzie, 2001), and recurrence analysis (Eckmann et al., 1987). This latter is based on embedded delayed coordinates, while the former is implemented without embedding.

These techniques make it possible to reveal correlations in data sets that are not possible to detect in the original time series. They do not require assumptions on the stationarity of a time series or the underlying motion equations and have been successfully applied in the sciences to detect chaos in experimental data. Moreover, they are particularly applicable to economic and financial data because they work well on relatively small data sets (Faggini, 2007, 2011) and are robust against noise. They preserve the time ordering of data and provide information about the system that generated them.

Of course, this does not mean that these tools are without problems. First, the threshold term ε, which compares the data, is subjective. This is true even if, unlike the BDS test, this term can be varied without altering the qualitative nature of the observed pattern of close returns (McKenzie, 2001). Second, both metric and topological tests suffer with nonstationary data.[‡] However, the problem of nonstationarity, while a problem for both metric and topological tests, seems to be more stringent for the former because it does not maintain the time ordering of data. Moreover, for recurrence analysis, we must take into account all limitations concerning the procedure of embedding coordinates.

37.2.5 Close Returns Test and Recurrence Analysis

The close returns test consists of two parts: a qualitative component that is a graphical representation of the presence of chaotic behaviors, the CRP, and a quantitative one that tests the null hypothesis that the data are IID against both linear and nonlinear alternatives. It exhibits the same performance as the BDS test. Unlike the methods based on the correlation dimension, this tool detects the recursive behavior of a chaotic time series.

Let $\{x_t\}$ be a time series whose trajectories are orbiting in the face space. If the orbit is one period, the trajectory will return to the neighborhood of x_t after an interval equals 1; if the orbit is two periods, it will return after an interval equals 2, and so on.

---

[*] "[...] the magnitude of the exponent reflecting the time scale on which system dynamics become unpredictable," (Wolf et al., 1985, p. 285).

[†] A strange attractor is the set of points toward which a chaotic system converges.

[‡] "This is especially a concern when testing data for emerging countries, which typically undergo significant regulatory changes in the development period," (McKenzie, 2001, p. 40).

Therefore, if x_t evolves near a periodic orbit for a sufficiently long time, it will return to the neighbourhood of x_t after some interval (T). The criterion of closeness requires that the difference $\|x_i - x_{i+T}\|$ be very small. Computing all differences $\|x_t - x_{t+i}\|$, where $t = (1, \ldots, n)$, $i = (1, \ldots, n-1)$, and n is the length of the sample, the close returns test detects the observations for which $\|x_i - x_{i+T}\|$ is smaller than a threshold value ε. In the plot[*] (Figure 37.1), the horizontal axis indicates the number of observations t, where $t = (1, 2, \ldots, N)$, and the vertical axis is i, where $i = (1, 2, \ldots, N-1)$. N is the observation number. If this difference is smaller, it is coded black; otherwise, it is coded white.

If the data are IID, the distribution of black dots will be random. In the plot, no pattern is evident (Figure 37.1). If the time series is deterministic, it is possible to observe horizontal line segments. More specifically, the short horizontal line is symptomatic of chaotic dynamics (Figure 37.2).

The second part of the test concerns the construction of a histogram that resumes the information of the CRP.

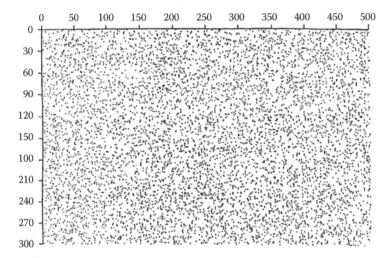

FIGURE 37.1 CRP of random data.

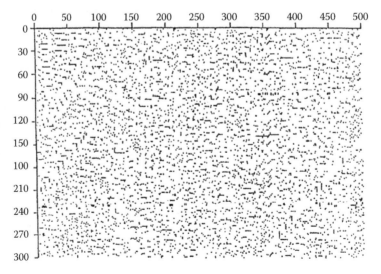

FIGURE 37.2 CRP series from a logistic map.

[*] Examples by Gilmore (1993a,b, 1996, 2001).

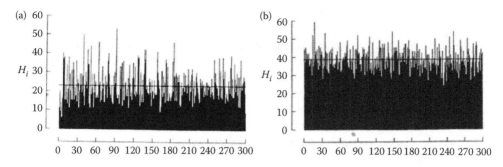

FIGURE 37.3   Histogram of Henon series (a) and random series (b).

The histogram displays the number of close returns for each i, where $H_i = \sum \Theta(\varepsilon - |x_t - x_{t+i}|$ and Θ is the Heaviside function. In the histogram, the chaotic data will show some peaks (Figure 37.3a); otherwise, it will be uniformly distributed around some value $\overline{H}$ (Figure 37.3b).

To define if on average $H_i = \overline{H}$ and to determine whether the null hypothesis of IID can be accepted, we have to calculate the chi-squared statistic:

$$\chi_t^2 = \frac{\sum_{i-1}^{k} \left[H_i - \bar{H} \right]^2}{np} \tag{37.6}$$

In Equation 37.6, $\overline{H} = np$ and i is equal to the number of observations over p, which is estimated in the following way:

$$p = \frac{\text{total number of close returns}}{\text{total area of the plot}}$$

Then the estimated χ_t^2 is compared to the critical test value (χ_c^2) obtained with $(k-1)$ degrees of freedom. If $\chi_c^2 > \chi_t^2$, the null hypothesis that the data are IID is rejected (Gilmore, 1993, 2001, McKenzie, 2001).

37.2.6 Recurrence Analysis

Like the close returns test, recurrence analysis consists of two parts: the recurrence plot (RP) developed by Eckmann et al. (1987), a graphical tool that evaluates the temporal and phase space distance, and recurrence quantification analysis (RQA),[*] a statistical quantification of RP.

Recurrence analysis and the CRP are more similar because they are based on the same methodology but differ in the plot construction. RPs are symmetrical over the main diagonal. Moreover, while the close returns test analyzes the time series directly and fixes a value ε to estimate nearby points, the RP is based on the reconstruction of time series and an estimation of the points that are close. This closeness is measured by a critical radius so that a point is plotted as a colored pixel only if the corresponding distance is below or equal to this radius. From the occurrence of lines parallel to the diagonal in the RP, it can be seen how fast neighbored trajectories diverge in phase space. The line segments parallel to the main diagonal are points that move successively closer to each other in time and would not occur in random as opposed to a deterministic process. Chaotic behavior produces very short diagonals, whereas deterministic behavior produces longer diagonals. Thus, if the analyzed time series is chaotic, then the RP shows short segments parallel to the main diagonal; on the other hand, if the series is white noise, then the RP does not show any kind of structure.

Nevertheless, sometimes, the graphical output of RP is not easy to interpret because the signature of determinism (DET), the set of lines parallel to the main diagonal, might not be so clear. As a consequence,

---

[*] See Zbilut and Webber (1992).

Zbilut and Webber (1998) and Zbilut et al. (2000) proposed a statistical quantification of RP, which is well known as RQA.

RQA defines the measures of the diagonal segments in an RP. These measures are recurrence rate (REC), DET, averaged length of diagonal structures, entropy (ENT), and trend (TREND).

REC is the ratio of all recurrent states (recurrence points percentage) to all possible states and is the probability of recurrence of a special state. REC is simply what is used to compute the correlation dimension of data. An RP can be considered to be a two-dimensional pictorial representation of the points that contribute to Equation 37.1 for a particular value of ε.

DET is the ratio of recurrence points forming diagonal structures to all recurrence points. DET[*] measures the percentage of recurrent points forming line segments that are parallel to the main diagonal. A line segment is a point's sequence, which is equal to or longer than a predetermined threshold. These line segments reveal the existence of deterministic structures, and absence instead of randomness.

Maxline (MAXLINE) represents the average length of diagonal structures and indicates the longest line segments that are parallel to the main diagonal. It is claimed to be proportional to the inverse of the largest positive Lyapunov exponent. A periodic signal produces long line segments, while the *noise* does not produce any segments. Short segments indicate chaos.

ENT (Shannon ENT) measures the distribution of those line segments that are parallel to the main diagonal and reflects the complexity of the deterministic structure in the system. This ratio indicates the time series *structuredness* so that high values of ENT are typical of periodic behaviors, while low values are typical of chaotic behaviors. A high ENT value indicates a large diversity in diagonal line lengths; low values indicate a small diversity in diagonal line lengths.[†] "[. . .] short line max values therefore are indicative of chaotic behaviors."[‡] The value TREND measures the paling of the patterns of RPs away from the main diagonal (used for detecting drift and nonstationarity in a time series).

An extensive survey of different software used to apply these techniques is provided in Belaire-Franch and Contreras (2002).

37.2.7 Empirical Investigations

The application of these tests to economic and financial time series produced controversial results. Investigators found substantial evidence for nonlinearity but relatively weak evidence for chaos per se.

In this section, we will summarize the main results obtained by applying chaos tests to economic and financial data. We will distinguish between the applications to macroeconomic and financial data because of the variation in the results between the two. Of course, this survey is not meant to be exhaustive but only indicative of the state of the art.

37.2.8 Chaos in Economic Data

In applying the BDS test to the residuals of linear models, Sayers (1986) and Frank and Stengos (1987) rejected the presence of chaos in the data of work stoppages and Canadian macroeconomic series, respectively. Brock and Sayers (1988) conducted their analysis using U.S. macroeconomic data[§] and showed the presence of nonlinearity but presented a weak evidence of chaos. Barnett and Chen (1988) demonstrated the presence of chaos in the U.S. Divisia monetary aggregates. The estimated value of the correlation dimension reaches a saturation point between 1.3 and 1.5, indicating the presence of a chaotic attractor in a bidimensional face space. In Frank et al. (1988), the application of the correlation dimension to residuals

---

[*] "This is a crucial point: a recurrence can, in principle, be observed by chance whenever the system explores two nearby points of its state space. On the contrary, the observation of recurrent points consecutive in time (and then forming lines parallel to the main diagonal) is an important signature of deterministic structuring" (Manetti et al., 1999).

[†] See Trulla et al. (1996).

[‡] See Iwanski and Bradley (1998, p. 10) and Atay and Altintas (1999, p. 6595).

[§] Employment 1950-I to 1983-IV, unemployment 1949-I to 1982-IV, monthly postwar industrial production, and pig iron production 1877–1937.

of the AR model and Lyapunov exponent did not show the presence of chaos in macroeconomic data from 1960 to 1988 for West Germany, Italy, Japan, and England. DeCoster and Mitchell (1991), in applying the correlation dimension and the BDS test to weekly monetary variables (Divisia M2 and M3), showed that nonlinear and even chaotic monetary dynamics for U.S. data cannot be dismissed. The same procedure and conclusions were reported by Frank and Stengos (1989). They investigated daily prices from the mid-1970s to the mid-1980s for gold and silver using the correlation dimension and Kolmogorov ENT. They found that the correlation dimension is between 6 and 7 and that the Kolmogorov ENT is about 0.2 for both assets.

Different conclusions were reported by Yang and Brorsen (1992). They did not find chaos in the daily prices of some agricultural commodities or in several future markets, including those of gold and silver. The evidence of nonlinearity is not consistent with chaos because the shuffled data do not satisfy the saturation condition of the correlation dimension. Cromwell and Labys (1993) apply the correlation dimension and the Lypunov exponent to commodity prices for the period from 1960 to 1992. They find chaos in the daily prices of corn but not in those of sugar, coffee, and cacao.

In Chavas and Holt (1991), the application of the correlation dimension, the BDS test, and the Lyapunov exponent provided clear evidence that the dynamic process generating the pork cycle is nonlinear, even if the evidence in favor of chaos is less conclusive. These results are confirmed by Streips (1995), who by applying the same tests on observations of the monthly U.S. hogcorn price ratio for the period 1910–1994, shows that data are chaotic.

Kohzadi and Boyd (1995) tested for the presence of chaos and nonlinear dynamics in monthly cattle prices for the period 1922–1990. The Grassberger and Procaccia, BDS, and Hurst exponent tests showed an evidence of chaos in these data. Bajo-Rubio et al. (1992) presented estimates of the correlation dimension and the largest Lyapunov exponent for daily data regarding the Swedish Krona versus the Deutsche Mark, ECU, U.S. Dollar, and Yen exchange rates. They find indications of deterministic chaos in all exchange rate series. However, because of the limited number of data (1985–1991), the estimates for the largest Lyapunov exponents are not reliable, except in the Swedish Krona–ECU case where they find a low-order chaotic behavior.

The close returns test has been implemented in detecting and visualizing chaotic behavior in a macroeconomic time series using some monetary aggregates. Montoro and Paz (1997) did not find an evidence of chaos in the Divisia M2 series, in contrast with the results obtained in several previous works.

Monthly aggregate air transport service series for over two decades have been analyzed using BDS test by Adrangi et al. (2001a). While a strong evidence of nonlinearity is found in the data, this evidence is not consistent with chaos. The ARCH-type model explains the nonlinearity in the data well.

Daily oil products for the Rotterdam and Mediterranean petroleum markets have been tested for the presence of chaos and nonlinear dynamics by Panas and Ninni (2000). The correlation dimension, BDS test, and Lyapunov exponents show a strong evidence of chaos in a number of oil products: Naphtha, Mogas Prem, Sulfur FO 3.5% e FO 1.0%, and Gasoil e Mogas REG.UNL. Moreover, Panas (2002) investigated the price behavior in the London Metal Exchange (LME) market using the two most attractive nonlinear models: long memory and chaos. The results indicate that the dynamics of the LME market can be attributed to long memory (aluminum and copper), i.e., a persistent process exhibiting self-similarity, short-memory behavior (nickel and lead price returns), antipersistent (or intermediate memory in the case of zinc returns), and a deterministic chaotic process (in the case of tin returns).

Serletis and Shintani (2006) find a statistically significant evidence against low-dimensional chaos in Canadian and U.S. data.

Barkoulas (2008) applied both metric and topological methodologies to test for a deterministic chaotic structure in simple-sum and Divisia monetary aggregates. The results did not satisfy any of the three indications of chaos. The monetary dynamics are not chaotic. Faggini (2007, 2011), starting from the conclusion by Frank et al. (1988) that refused the chaotic hypothesis in macroeconomic data, applied visual recurrence analysis to the same time series and found chaos in gross national product (GNP) data from Japan and United Kingdom.

37.2.9 Chaos in Financial Data

As well-known, economic theories have traditionally been dominated by a linear modeling, based on concepts like Gaussian distributions and random walks, generating the so-called capital market theory built on the assumption of normally distributed returns and the efficient market hypothesis (EMH), by which the markets follow a random walk. As a first consequence, the future would be unrelated to the past or the present, with no possibility of identifying TRENDs or cycles.

The apparent randomness of financial markets led some economists to become interested in chaos theory as a theoretical framework able to explain those fluctuations. One of the first applications of chaos tests to financial data was carried out by Scheinkman and LeBaron (1989). They analyzed the U.S. weekly returns from the Center for Research in Security Prices (CRSPs), applying the BDS test on residuals of linear models. They found a rather strong evidence of nonlinearity and some evidence of chaos. This is because the correlation dimension of the shuffled residuals appeared to be much greater than that of the original residuals. That is, it does not reach a saturation point, indicating the completely random behavior of the shuffled data. If time series data are chaotic, the estimated correlation dimension of residuals is the same as that of the original data. When the data are stochastic, the correlation dimension of the residuals increases as the embedding dimension increases. The residuals are less structured than the original data.

The same conclusion was reached by Hsieh (1991), who uses the BDS test to detect chaos in weekly stock returns data from the CRSPs from 1963 to 1987. All the data were first filtered by an autoregression model. The correlation dimension technique was used by DeCoster et al. (1992) to find evidence of chaos in daily sugar, silver, copper, and coffee future prices for the period 1960–1989.

Mayfield and Mizrach (1992) and Vaidyanathan and Krehbiel (1992) found an evidence of chaos behavior in the S&P 500 index, though this conclusion is based on the results of correlation dimension rather than on Lyapunov exponent estimates.

In Blank (1991), the estimates of correlation dimensions and Lyapunov exponents on soybean and S&P 500 future prices are consistent with the presence of deterministic chaos. Yang and Brorsen (1993) found an evidence of nonlinearity in several future markets, which is consistent with deterministic chaos in about half of the cases. Hsieh (1993) found an evidence of nonlinearity in four currency futures contracts, but found that nonlinearity is the result of predictable conditional variances. In Brorsen and Yang (1994), nonlinear dependence is not removed for the value-weighted index or the S&P 500 stock index. Standardized residuals from the GARCH model are not IID for two of three returns series. The application of BDS test proves that deterministic chaos cannot be dismissed.

Abhyankar et al. (1995) tested for the presence of nonlinear dependence and chaos in real-time returns on the U.K. FTSE-100 index (about 60.000 data points). Their results suggest that GARCH can explain some but not all of the observed nonlinear dependence. The application of Hinich test (1982), BDS test, and Lypunov estimates fail to detect chaos in the data.

Sewell et al. (1996, p. 92) examined weekly changes for the period 1980–1994 in six major stock indices (the United States, Korea, Taiwan, Japan, Singapore, and Hong Kong) and the World Index as well as the corresponding foreign exchange rates between the United States and the other five countries. They concluded that "[t]hese results do not prove the existence of chaos in these markets but are consistent with its existence in some cases." Abhyankar et al. (1997) and Serletis and Shintani (2003) reject the null hypothesis of low-dimensional chaos in the S&P 500 and Dow Jones Industrial Average.

Serletis and Gogas (1997) tested for chaos in seven East European black market exchange rates by applying the BDS test, the NEGM test (Nychka et al., 1992), and the Lyapunov exponent. They found evidence consistent with a chaotic nonlinear generation process in two out of the seven series: the Russian ruble and East German mark. In particular, the BDS test rejects the null hypothesis of IID for three of the seven markets, whereas the Lyapunov exponent estimator proposed by Nychka et al. (1992) supports the hypothesis of chaotic dynamics in two markets. Barkoulas and Travlos (1998) investigated the existence of a deterministic nonlinear structure in the stock returns of the Athens Stock Exchange (an emerging capital market) and found no strong evidence of chaos.

Gao and Wang (1999) examined the daily prices of four future contracts (S&P 500, JPY, DEM, and Eurodollar) and found no evidence of deterministic chaos. Andreou et al. (2000) examined four major currencies against Greek drachma (GRD) and found an evidence of chaos in two out of four cases.

Adrangi et al. (2001b), implementing the BDS test, correlation dimension, and Kolmogrov ENT, investigated the presence of low-dimensional chaos in crude oil, heating oil, and unleaded gasoline future price from the early 1980s. They find a strong evidence of nonlinearity inconsistent with chaos.

Gilmore (2001) implemented the close returns test to examine the presence of chaos in some foreign exchange rates. The results did not support chaotic explanations without excluding other possible forms of the nonlinear structure (Gilmore, 1993a,b). The test was implemented on the residuals of some ARCH-type filters as well, and the results have indicated that the models have captured some, although not all, of the nonlinear dependence among data. Additionally, McKenzie (2001) investigated the presence of chaos in a wide range of major national stock market indices using the close returns test. The results indicate that the data are not chaotic, although considerable nonlinearities are present.

Daily data for the Swedish Krona against the Deutsche Mark, the ECU, the U.S. Dollar, and the Yen exchange rates are tested by Bask (2002) using the Lypunov exponent. In most cases, the null hypothesis that the Lyapunov exponent is zero is rejected in favor of a positive exponent.

Shintant and Linton (2004) used 18.490 daily observations of the Dow Jones Industrial Average from January 1928 to October 2000. They used the Lyapunov exponent estimator proposed by Nychka et al. (1992). The results did not indicate chaotic behaviors in the data. Foreign exchange rates versus IRR (Iranian Rial) have been investigated by Torkamani et al. (2007), who show that the data in this market have complex chaotic behavior with a large degree of freedom. The tests applied are the correlation dimension and the Lyapunov exponent.

Antoniou and Vorlow (2005) investigated the "compass rose" patterns revealed in phase portraits (delay plots) of FTSE-100 stock returns and found a strong nonlinear and possibly deterministic signature in the data-generating processes.

The nonlinearity and chaoticity of the exchange rate time series are investigated by Liu (2009). The BDS test and surrogate data method indicated that the exchange rate time series of Canadian Dollars to United States Dollar (CD/USD), Japanese Yen to United States Dollar (JY/USD), and United States Dollar to British Pound (USD/BP) exhibit nonlinearity, while the exchange rate time series of United States Dollar to EURO (USD/EURO) is linear. The largest positive Lyapunov exponents have provided evidence for the possibility of deterministic chaos in the daily exchange rate time series of CD/USD, JY/USD, and USD/BP.

Das and Das (2007) implemented the Lyapunov exponent and surrogate data method to investigate the chaoticity of foreign exchange rates of several countries. They found an indication of deterministic chaos in all exchange rate series.

Adrangi et al. (2010), employing the daily bilateral exchange rates of the dollar, conducted a battery of tests for the presence of low-dimension chaos (correlation dimension tests, BDS tests, and tests for ENT). The strong evidence of nonlinear dependence in the data is not consistent with chaos. The nonlinear dependencies in the dollar exchange rate returns series arise from the GARCH model rather than from a chaotic structure.

More recently, the rate of returns series for six Indian stock market indices were tested for chaos (Mishra, 2011). The results from the test of independence on filtered residuals suggest that the existence of nonlinear dependence, at least to some extent, can be attributed to the presence of conditional heteroskedasticity. To account for the remaining nonlinearity in the data, the Lyapunov exponent is estimated.[*] The result is a positive value in two out of six cases, indicating that these return series are generated by a chaotic system.

---

[*] It is clearly observable that in two out of seven cases, the point estimates of Lyapunov exponents are positive. This implies that the returns of Bank Nifty and CNX IT exhibit a chaotic nonlinear-generating process, and therefore, the nonlinear structure in these return series is possibly deterministic in nature. Further, a negative Lyapunov exponent (i.e., in the case of CNX NIFTY, BSE SENSEX, BSE 200, and BSE 100) indicates that the nature of the series is consistent with a stochastic process rather than a deterministic low-dimensional chaotic system.

Analyses on financial data are performed by recurrence analysis in Mizrach (1996), McKenzie (2001), Holyst and Zebrowska (2000), Holyst et al. (2001), and Strozzi et al. (2002). Belaire-Franch et al. (2002) analyzed (RP e RQA) the exchange rates of 16 Organization for Cooperation and Development (OECD) countries from 1957 to 1998. After a filtering procedure with the ARMA model, RQA applied on residuals shows the presence of chaotic dynamics in the data.

Daily index prices, consisting of free float-adjusted market capitalization stock indices of developed and emerging markets between January 1995 and December 2009, were analyzed by Bastos and Caiado (2011) using recurrence analysis. The statistical tests suggested that the dynamics of stock prices in emerging markets are characterized by higher values of RQA measures when compared to their developed counterparts.

37.3 Why Economics Has Not Accomplished What Physics Has?

Although the literature on tests for chaos in economic and financial time series is voluminous, there are no uncontroversial results to speak of. Clearly, the review presented above suggests that there is ample evidence of the presence of nonlinearities and some evidence of deterministic chaos.

The difficulty of using chaos theory in economics is a direct consequence of some problems related to the application of these techniques to economic data.

First of all, data quantity and data quality are crucial when applying these techniques, and the main obstacle in empirical economic analysis is addressing short and noisy data sets. Data quantity and data quality in economics constitute a significant obstacle in chaotic-economic theory.

Moreover, testing macroeconomic series is regarded with some suspicion; not only are the gathered data insufficient to perform tests, but the macro time series also involve mixed effects: it is not only the distinction between noise and nonlinearities that must be determined but also the eventual source of nonlinearity; they are usually aggregated and derived from a system whose dynamics and measurement probes may be changing over time. Granger (2001) shows that the aggregation of independent series hides the nonlinear or chaotic signals. Current tests used to detect a chaotic structure often fail to find an evidence of chaos in aggregated data, even if those data are generated by a nonlinear deterministic process.

Little or no evidence for chaos has been found in macroeconomic time series. Investigators have found substantial evidence for nonlinearity but relatively weak evidence for chaos per se. This is due to the small sample sizes and high noise levels for most macroeconomic series. In contrast to laboratory experiments, through which a large number of data points can be easily obtained, most economic time series consist of monthly, quarterly, or annual data, with the exception of some high-frequency financial series. In fact, the analysis of financial time series has led to results that are, as a whole, more reliable than those of macroeconomic series. This is because financial time series data are available in large quantities over many disaggregated time intervals, though this literature is not free of controversial results.

Controversial results also arise from using inappropriate analytical methodologies that are more similar to a standard statistical protocol. To distinguish between chaotic and nonchaotic behaviors, all researchers, before applying chaos tests, filtered the data using either linear or nonlinear models (Frank and Stengos, 1989, Blank, 1991, Cromwell and Labys, 1993, Yang and Brorsen, 1992, 1993), in most cases, using ARCH-type models. When these do not capture all the nonlinearities in the economic and financial data (Hsieh, 1991, Vaidyanathan and Krebhiel, 1992), chaos analysis is conducted on the residuals. This procedure has been used by Frank and Stengos (1989), DeCoster et al. (1992), Chavas and Holt (1991), and Bask (2002), among others.

The filtering procedure was supported by Brock (1986), who stated that before testing for a possible nonlinear dependency among the observations, we need to remove all linear correlations that may cause the null hypothesis to be rejected. He also argued that with an infinite amount of noise-free data, possible nonlinear structures should be unaffected by the implementation of a linear-filtering process. Removing all linear structures is difficult, but a good approximation can be achieved by using an ARMA fit to stationary data. With the assumption that the residuals are filtered for linear dependence, it is reasonable to assert that any resulting dependence found in the residuals must be nonlinear. Then, when nonlinearity is found, ARCH-type models are applied to detect the source. If unexplained nonlinearity remains, chaos tests are applied. In this assertion by Brock (1986), we must highlight the fact that the linear-filtering procedure is irrelevant if the data are infinite, noise-free, and stationary,[*] conditions that are not testable for economic and financial data.

More generally, the open question is whether the chaotic properties of a process are invariant to linear and nonlinear transformations. It has been proved that linear and nonlinear filters can distort potential chaotic structures (Chen, 1993, Wei and Leuthold, 1998) and may affect the dimensionality of the original data (Chen, 1993, Panas and Ninni, 2000, Panas, 2002), providing a false indication of chaos. Chen (1993) shows that the correlation dimension is invariant to filtering by the MA (moving average model) because, in this way, the fractal structure of the dynamics is lost.

Moreover, sometimes, the conclusions both for and against chaos are reached by applying only one type of chaos test. For example, Kohzadi and Boyd (1995) find chaos by using only the BDS test and rescaled range (R/S) analysis.

To produce convincing results, we have to employ all tests for chaos to exploit their different potentials and limits. Few published papers have jointly applied the BDS test, the correlation dimension test, and the test for a positive Lyapunov exponent. Very few use topological tools (Barkoulas, 2008).

The consequence of this resonates in the words of Granger and Terasvirta (1992, p. 195): "Deterministic (chaotic) models are of little relevance in economics and so we will consider only stochastic models." The question was intensified by Jaditz and Sayers (1993), who reviewed a wide variety of data to conclude that there was no evidence for chaos, though they did not deny the indication of nonlinear dynamics of some sort. Moreover, controversies are also produced by the nature of the tests themselves. There may be very little robustness of such tests across variations in sample size, test methods, and data-aggregation methods. It is widely known that problems related to the quality and lack of sufficient amounts of data, the issue of an appropriate level of disaggregation, and the proper definition of methods for detecting white noise create considerable obstacles to constructing a meaningful and coherent statistical theory about the dynamics of economic and financial data. Following the studies discussed so far, we have to admit that no natural deterministic explanation exists for the observed economic fluctuations that are produced by external shocks or by inherent randomness and, consequently, an inherent unpredictability.

Moreover, up until now, we have been interested only in low-complexity chaotic behavior. The failure to detect low-dimensional chaos does not preclude the possibility of there being high-dimensional chaos in these variables (Day, 1994). It is possible that the underlying nonlinear structure of the economy is more complex and that the chaotic dynamics it exhibits are of a higher dimensionality. The algorithms presented above were developed to detect chaos in experimental data. Because physicists can often generate very large samples of high-quality data from laboratory experiments, they find these algorithms to be directly applicable to their research. Consequently, further theoretical advances are required to develop tests that are able to detect more complex forms of chaotic behaviors.

The early excitement about chaos theory centered on the fact that many of the diagnostics used in physics could be applied to any time series, independent of the theories of what causes it to change. But again, most of these tests were designed for the low-noise worlds of experimental physics. They also worked best with what is called "low-dimensional chaos," meaning they were designed for chaotic systems that themselves were not too complicated. Analysis of time series in the social sciences often gave indications of a previously unanticipated structure, but few or no strong statements could be made about the sort of low-dimensional chaos studied in physics.

We have to remember that economics is not an exact science; so, by its nature, economics cannot conform rigorously to the models of the exact sciences. This, of course, does not mean that a social science is impossible, but researchers must adapt to a different way of approaching science because human actions

---

[*] Stationarity is an important property of data because in the stationary time series, the statistical properties do not change over time.

are conditioned by social mechanisms just as the phenomena of nature are governed by mechanisms of their own: "In both spheres the aim of scientific work is to reveal the mechanisms involved and there is every reason to expect such work to be as meaningful and illuminating in the social sphere as it is in the natural domain" (Lawson, 1998, p. 170).

The excessive reliability of social sciences on the paradigms and methods employed in physics have provoked a misrepresentation of what a social system is and what aspects or elements it includes. In physics, different objects of the same type behave quite the same. The details of context or history in which the object finds itself does not usually determine its behavior.

The problems of economic agents are of a different order than the problems of physical objects. In this changeable context, we have to face a difficult task: of finding whether or not there are durable patterns for constructing explicit models. We should have a description of all complex interrelationships and feedback loops that exist among the economic variables. We cannot draw conclusions from individual relationships between particular variables of interest, excluding variables or relationships that have a secondary or indirect effect. This can be dangerous, not so much because it aggregates or simplifies the structure of the economy, but because it may ignore or misrepresent some aspect of the economy's structure that plays an important role in its dynamic behavior.

In human affairs, it is beginning to look as if history and tradition, even individual acts, and decisions, are far more powerful determinants of how a society is organized than the economic and political "forces" that nineteenth-century social theory reduced to social laws.

Social systems are products of not only physical aspects, and living dynamics, but also of symbolic communication. Consequently, the internal structure of a social system is typically far more ordered, that is, that of a purely physical system.

In other words, the effort to understand economies or indeed almost any human social system changes the system! I am not talking about some subtle quantum-observer effect but something much more dramatic.

37.4 Conclusions

After an exuberant flurry of publications, the search for chaos in economics has been gradually becoming less enthusiastic over the last two decades, as no empirical support for the presence of chaotic behaviors in economics has been found. The literature described above does not provide a solid support for chaos as a consequence of the high noise level that exists in most aggregated economic time series, the relatively small sample sizes of data, and the weak robustness of chaos tests for these data.

To determine if a time series was generated by deterministic chaos is not easy. Many common time series problems (such as seasonality and TRENDs) can confuse most of the diagnostic tools that people use. These complications have led to many conflicting results. Building an easy-to-use test that can handle the intricacies of a real-world time series is a tough problem, one which will probably not be solved anytime soon. Moreover, most of the theoretical structure in chaos is based on purely deterministic models that have no noise, or at most just a very small amount of noise, affecting the dynamics of the system. This approach works well in many physical situations, but it does not offer a very good picture of most social situations. It is hard to look at social systems isolated from the environment in the way that one can analyze fluid in a laboratory beaker. Once noise plays a major role in the dynamics, the problems involved in analyzing nonlinear systems become much more difficult.

To summarize, surely there is no reason to suppose that the economy is linear (George and Oxley, 2007); so, we should be able to find an evidence of nonlinearity in economic data.

Therefore, if we prove that the data are chaotic, we will be able to prove that there is a deterministic system that generates them. In fact, chaos theory provides the possibility of discovery of the pattern and complex regulation governing the behavior of such variables and uses them to predict future TRENDs in the short term.

This could be a big step forward in clarifying the "nature" of the economy. The result of such an assumption is that variations of these variables are not predictable.

Chaotic nonlinear systems can endogenize shocks. Chaotic dynamic models allow for the explanation of persistent and irregular fluctuations without stochastic exogenous shocks introduced ad hoc.

Moreover, if the economy is chaotic, then we can create a complete and closed model for it. This development would significantly help forecasting and control efforts in the short run.

Thus, even if we agree with Barnett (2006, p. 255) that "...the economics profession, to date, has provided no dependable empirical evidence of whether or not the economy itself produces chaos, and I do not expect to see any such results in the near future. The methodological obstacles in mathematics, numerical analysis, and statistics are formidable," we do not have the slightest idea of whether or not the economy exhibits chaotic nonlinear dynamics, and hence, we are not justified in excluding the possibility. If there be chaotic process in some macroeconomic series, applying some inadequate and premature policies may lead to disruption and irregularity in the TREND of variables and make the complex conditions governing them substantially more complicated and therefore uncontrollable.

Moreover, part of the controversy and the resulting conclusions are due to the misconception that low-dimensional chaos can be expected to be present generically in all economic phenomena. We cannot assume deterministic chaos for any measured time series.

The results of chaos tests do not prove the existence of chaos in all economic variables but are consistent with its existence; in some cases, this could only mean that some economic phenomena are less complex than others and that the economy of a country or simply a single market of an economy is chaotic, not that an economy as a whole is chaotic.

Given these considerations, studies in this area should grow both in size and importance as a field of their own within economics, although the empirical task of extracting an evidence of chaotic dynamics from economic time series is objectively more difficult than in the natural sciences, like physics.

So, we have to create techniques that acknowledge that a basic difference exists between physics, which is generally an exact science, and economics. To date, the application of chaos theory has been only a mechanical transfer that has not taken into account the specific features of economic systems. Surely, compared with neoclassical theory models, chaos theory in economics allows researchers to interpret phenomena considered noninfluential, exogenous, and stochastic, the result of complex interactions among the system's elements, and although it is difficult to predict system behavior, the same cannot be said of the process that created it, as it is deterministic.

References

Abhyankar A. H., Copeland L. S., and Wong W. 1995. Non-linear dynamics in real-time equity market indices: Evidence from the UK. *Economic Journal*, 105, 864–880.

Abhyankar A. H., Copeland L. S., and Wong W. 1997. Uncovering nonlinear structure in real-time, stock-market indexes: The S&P 500, the DAX, the Nikkei 225, and the FTSE-100. *Journal of Business and Economic Statistics*, 15, 1–14.

Adrangi B., Allender M., Chatrath A., and Raffiee K. 2010. Nonlinear dependencies and chaos in the bilateral exchange rate of the dollar. *International Business and Economics Research Journal*, 9(3), 85–96.

Adrangi B., Chatrath A., and Raffiee K. 2001a. The demand for U.S. air transport service: A chaos and nonlinearity investigation. *Transportation Research Part E*, 37, 337–353.

Adrangi B., Chatrath A., Dhanda K. K., and Raffiee K. 2001b. Chaos in oil prices? Evidence from futures markets. *Energy Economics*, 23, 405–425.

Andreou A. S., Pavlides G., and Karytinos A. 2000. Nonlinear time-series analysis of the Greek exchange-rate market. *International Journal of Bifurcation and Chaos*, 10(7), 1729–1758.

Antoniou A. and Vorlow C. E. 2005. Price clustering and discreteness: Is there chaos behind the noise? *Physica A*, 348, 389–403.

Atay F. M. and Altintas Y. 1999. Recovering smooth dynamics from time series with the aid of recurrence plots. *Physical Review E*, 59, 6593–6598.

Bajo-Rubio O., Fernández-Rodríguez F., and Sosvilla-Rivero S. 1992. Chaotic behavior in exchange-rate series. First result for the Peseta–U.S. Dollar case. *Economics Letters*, 39, 207–211.

Barkoulas J. T. 2008. Testing for deterministic monetary chaos: Metric and topological diagnostics. *Chaos, Solitons and Fractals*, 38, 1013–1024.

Barkoulas J. and Travlos N. 1998. Chaos in an emerging capital market? The case of the Athens stock exchange. *Applied Financial Economics*, 8(3), 231–243.

Barnett W. A. 2006. Comments on chaotic monetary dynamics with confidence. *Journal of Macro-economics*, 28, 253–255.

Barnett W. A. and Serletis A. 2000. Martingales, nonlinearity, and chaos. *Journal of Economic Dynamics and Control*, 24, 703–724.

Barnett W. and He Y. 2000. Unsolved econometric problems in nonlinearity, chaos, and bifurcation. Economics Working Paper Archive at WUSTL, n. 09–06.

Barnett W. A. and Hinich M. J. 1992. Empirical chaotic dynamics in economics. *Annals of Operations Research*, 37, 1–15.

Barnett W. A. and Chen P. 1988. The aggregation theoretic monetary aggregates are chaotic and have strange attractors: An econometric application of mathematical of chaos, in Barnett W. A., Berndt E., and White H. (eds.), *Dynamic Econometric Modeling, Proceedings of the 3rd International Symposium on Economic Theory and Econometrics*, Cambridge University Press, USA.

Barnett W. A., Gallant A. R., Hinich M. J., Jungeilges J., Kaplan D., and Jensen M. J. 1997. A single-blind controlled competition between tests for nonlinearity and chaos. *Journal of Econometrics*, 82, 157–192.

Barnett W. A., Gallant A. R., Hinich M. J., Jungeilges J., Kaplan D., and Jensen M. J. 1995. Robustness of nonlinearity and chaos test to measurement error, inference method, and sample size. *Journal of Economic Behavior and Organization*, 27, 301–320.

Barnsley M. 1988. *Fractals Everywhere*, Boston Academic Press, USA.

Bask M. 2002. A positive Lyapunov exponent in Swedish exchange rates. *Chaos, Solitons and Fractals*, 14(8), 1295–1304.

Bastos J. A. and Caiado J. 2011. Recurrence quantification analysis of global stock markets. *Physica A: Statistical Mechanics and Its Applications*, 390, 1315–1325.

Belaire-Franch J. and Contreras D. 2002. Recurrence plots in nonlinear time series analysis: Free software. *Journal of Statistical Software*, 7(9), http:// aeser.anaeco.uv.es/pdf/dt/dt01-01.pdf.

Belaire-Franch J., Contreras-Bayarri D., and Tordera-Lledó L. 2002. Assessing non-linear structures in real exchange rates using recurrence-plot strategies. *Physica D: Nonlinear Phenomena*, 171(4), 249–264.

Benhabib J. and Day R. 1981. Rational choice and erratic behaviour. *Review of Economic Studies*, 48, 459–472.

Blank S. C. 1991. Chaos in futures markets: A nonlinear dynamically analysis. *The Journal of Futures Markets*, 11, 711–728.

Brock W. A. 1986. Distinguishing random and deterministic system. *Journal of Economic Theory*, 40, 68–195.

Brock W. A., Dechert W. D., and Scheinkman J. 1996. A test for independence based on the correlation dimension. *Econometric Reviews*, 15, 197–235.

Brock W. A., Hsieh D. A., and Le Baron B. 1991. *Non Linear Dynamics, Chaos, and Instability: Statistical Theory and Economic Evidence*, MIT Press, Cambridge, MA.

Brock W. A., Dechert W. D., and Scheinkman J. 1987. A Test for Independence Based on the Correlation Dimension, Department of Economics, University of Wisconsin, Madison, SSRI Working Paper 8702.

Brock W. A. and Sayers C. 1988. Is the business cycle characterized by deterministic chaos? *Journal of Monetary Economics*, 22.1, 71–90.

Brorsen B. W. and Yang S. 1994. Nonlinear dynamics and the distribution of daily stock index returns. *The Journal of Financial Research*, 17, 187–203.

Caputo E., Sello S., and Marcolongo V. 1994. Nonlinear analysis of experimental noisy time series in fluidized bed systems. Submitted to: *Chaos Solitons and Fractals*.

Chavas J. and Holt M. 1991. On nonlinear dynamic: The case of pork cycles. *American Journal of Agricultural Economics*, 73, 819–828.

Chen P. 1993. Searching for economic chaos: A challenge to econometric practice and nonlinear tests, in: Day R. H. and Chen P. (eds.), *Nonlinear Dynamics and Evolutionary Economics*, Oxford University Press, Oxford, New York, pp. 217–233.

Cromwell J. B. and Labys W. C. 1993. Testing for Nonlinear Dynamics and Chaos in Agricultural Commodity Prices, Memo Institute for Labor Study, West Virginia University.

Cutler C. D. 1991. Some results on the behaviour and estimation of the fractal dimensions of distributions on attractors. *Journal of Statistical Physics*, 62, 651–708.

Das A. and Das P. 2007. Chaotic analysis of the foreign exchange rates. *Applied Mathematics and Computation*, 185, 388–396.

Day R. H. 1982. Irregular growth cycles. *American Economic Review*, 72, 406–414.

Day R. H. 1994. Complex economic dynamics. *An Introduction to Dynamical Systems and Market Mechanisms*, vol. 1, MIT Press, Cambridge, MA.

DeCoster G. P., Labys W. C., and Mitchell D. W. 1992. Evidence of chaos in commodity futures prices. *The Journal of Futures Markets*, 12, 291–305.

DeCoster G. P. and Mitchell D. W. 1991. Nonlinear monetary dynamics. *Journal of Business and Economic Statistics*, 9, 455–461.

DeLima P. J. F. 1998. Nonlinearities and nonstationarities in stock returns. *Journal of Business Economic Statistics*, 16, 227–236.

Deneckere R. and Pelikan S. 1986. Competitive chaos. *Journal of Economic Theory*, 30, 13–25.

Eckmann J. P. and Ruelle D. 1985., Ergodic theory and strange attractor. *Review of Modern Physics*, 57, 617–656.

Eckmann J. P., Kamphorst S. O., and Ruelle D. 1987. Recurrence plots of dynamical systems. *Europhysics Letters*, 4(9), 973–977.

Faggini M. 2007. Visual recurrence analysis: An application to economic time series, in: Salzano M. and Colander E. D. (eds.), *Complexity Hints for Economic Policy*, Springer, Milan, Italy.

Faggini M. 2009. Chaos and chaotic dynamics in economics. *Nonlinear Dynamics, Psychology and Life Sciences*, 13(3), 327–340.

Faggini M. 2011. Chaos detection in economics. Metric versus topological tools. MPRA Paper No. 30928.

Falconer K. 1990. *Fractal Geometry*, Chicester, Wiley.

Farmer J. D., Ott E., and Yorke J. A. 1983. The dimension of chaotic attractor. *Physica D*, 7, 153–180.

Fillol J. 2001. Limits of the Tools for Detection of Chaos in Economy. Application to the Stock Returns MODEM, University of Paris X-Nanterre, http://www.univorleans. fr/DEG/GDR ecomofi/Activ/fillolpau.pdf.

Frank M., Gencay R., and Stengos T. 1988. International chaos? *European Economic Review*, 32, 1569–1584.

Frank M. and Stengos T. 1987. Some evidence concerning macroeconomic chaos. Discussion Paper n. 1987-2, University of Guelph.

Frank M. and Stengos T. 1989. Measuring the strangeness of gold and silver rates of return. *Review of Economic Studies*, 56, 553–567.

Gao A. H. and Wang G. H. K. 1999. Modeling nonlinear dynamics of daily futures price changes. *The Journal of Futures Markets*, 19(3), 325–351.

Gençay R. and Dechert W. D. 1992. An algorithm for the n Lyapunov exponents of an n dimensional unknown dynamical system. *Physica D*, 59, 142–157.

George D. and Oxley L. T. 2007. Economics on the edge of chaos. *Environmental Modelling and Software*, 22, 580–589.

Gilmore C. G. 1993a. A new test for chaos. *Journal of Economic Behaviour Organisations*, 22, 209–237.

Gilmore C. G. 1993b. A new approach to testing for chaos, with applications in finance and economics. *International Journal of Bifurcation Chaos*, 3(3), 583–587.

Gilmore C. G. 1996. Detecting linear and nonlinear dependence in stock returns: New methods derived from chaos theory. *Journal of Business and Accounting*, 29(9) and (10), 1357–1377.

Gilmore C. G. 2001. An examination of nonlinear dependence in exchange rates, using recent methods from chaos theory. *Global Finance Journal*, 12, 139–151.

Grandmont J. M. 1985. On endogenous competitive business cycles. *Econometrica*, 50, 1345–1370.

Granger C. W. J. and Terasvirta T. 1992. Experiments in modeling nonlinear relationships between time series, in: Martin C. and Stephen E. (eds.), *Nonlinear Modeling and Forecasting, Diagnostic Testing for Nonlinearity, Chaos, and General Dependence in Time Series Data*, Addison-Wesley, Redwood City, CA, pp. 189–198.

Granger C. W. J. 2001. Overview of nonlinear macroeconometric empirical models. *Macroeconomic Dynamics*, 5, 466–481.

Grassberger P. and Procaccia I. 1983a. Characterization of strange attractors. *Physical Review Letters* 50, 346–349.

Grassberger P. and Procaccia I. 1983b. Measuring the strangeness of strange attractors. *Physica D*, 9, 189–208.

Hentschel H. G. E. and Procaccia I. 1983. The infinite number of generalized dimensions of fractals and strange attractors. *Physica D*, 8, 435–444.

Hinich M. 1982. Testing for Gaussianity and linearity of a stationary time series. *Journal of Time Series Analysis*, 3(3), 169–176.

Holyst A. and Zebrowska M. 2000. Recurrence plots and Hurst exponents for financial markets and foreign exchange data. *Internal Journal of Theoretical Applied Finance*, 3, 419.

Holyst A., Zebrowska M., and Urbanowicz K. 2001. Observations of deterministic chaos in financial time series by recurrence plots: Can one control chaotic economy. *The European Physical Journal B*, 20, 531–535.

Hsieh D. 1989. Testing for non-linearity in daily foreign-exchange rate changes. *Journal of Business*, 62, 339–368.

Hsieh D. 1991. Chaos and nonlinear dynamics: Applications to financial markets. *Journal of Finance*, 46, 1839–1877.

Hsieh D. 1993. Using non-linear models to search for risk premia in currency futures. *Journal of International Economics*, 35, 113–132.

Iwanski J. S. and Bradley E. 1998. Recurrence plots of experimental data: To embed or not to embed? *Chaos*, 8, 861–871.

Jaditz T. and Sayers C. L. 1993. Is chaos generic in economic data?, *International Journal of Bifurcation and Chaos*, 3, 745–755.

Kantz H. and Schreiber T. 1997. *Non Linear Time Series Analysis*, Cambridge University Press, UK.

Kohzadi N. and Boyd M. K. 1995. Testing for chaos and nonlinear dynamics in cattle prices. *Canadian Journal of Agricultural Economics*, 43, 475–484.

Kuan G. M. 2008. Lecture on Time Series Diagnostic Tests, Institute of Economics Academia Sinica, Taiwan, February 17.

Kugiumtzis D. 1997. Assessing different norms in nonlinear analysis of noisy time series. *Physica D*, 105, 62–78.

Kyrtsou C. and Vorlow C. 2009. Modelling non-linear comovements between time series, *Journal of Macroeconomics*, 31(1), 200–211.

Kyrtsou C., Labys W. C., and Terraza M. 2001. Noisy Chaotic Dynamics in Commodity Markets, Division of Resource Management Working Paper RESMWP-03-01.

Lawson T. 1998. The predictive science of economics, in *Foundations of Research in Economics: How do Economists Do Economics?*, Medema S. G. and Samuels W. J. (eds), Edward Elgar, UK.

Liu L. 2009. Testing for nonlinearity and chaoticity in exchange rate time series. SEI Online.

Manetti C., Ceruso M., Giuliani A., Webber J. C., and Zbilut J. P. 1999. Recurrence quantification analysis as a tool for characterization of molecular dynamics simulations. *Physical Review E*, 59, 992–998.

Matilla-García M., Queral R., Sanz P., and FZquez J. V. 2004. A generalized BDS statistic. *Computational Economics*, 24, 277–300.

Mayer-Kress G. 1986. *Dimension and Entropies in Chaotic Systems*, Springer-Verlag, Berlin.

Mayfield E. S. and Mizrach B. 1992. On determining the dimension of real-time stock-price data. *Journal of Business and Economic Statistics*, 10(3), 367–374.

McKenzie M. D. 2001. Chaotic behavior in national stock market indices. *Global Finance Journal*, 12(1), 35–53.

Medio A. 1979. *Teoria Non-Lineare del Ciclo Economico*, Il Mulino, Bologna.

Mindlin G. B., Hou X. J., Solari H. G., Gilmore R., and Tufillaro N. B. 1990. Classification of strange attractors by integers. *Physical Review Letters*, 64, 2350–2353.

Mindlin G. B. and Gilmore R. 1992. Topological analysis and synthesis on chaotic time series. *Physica D*, 58, 229–242.

Mishra S. K. 2011. Volatility in Indian stock markets. Available at SSRN: http://ssrn.com/abstract=1788306.

Mizrach B. 1996. Determining delay time for phase space reconstruction with application to the FFDm exchange rate. *Journal of Economic Behaviour and Organisations*, 30, 369–381.

Montoro J. D. and Paz J. V. 1997. Detecting macroeconomic chaos. *Innovation in Mathematics: Proceedings of the Second International Mathematica Symposium*, Southampton, pp. 353–360.

Nychka D. W., Daniel F. M., Stephen P. E., and Ronald G. A. 1992. Estimating the Lyapunov exponent of a chaotic system with nonparametric regression. *Journal of the American Statistical Association*, 87, 682–695.

Packard N. H., Crutcfield J. P., Farmer J. D., and Shaw R. S. 1980. Geometry from a time series. *Physical Review Letters*, 45, 712–716.

Panas E. 2002. Long memory and chaotic models of prices on the London metal exchange. *Resources Policy*, 27, 235–246.

Panas U. E. and Ninni V. 2000. Are oil markets chaotic? A non-linear dynamic analysis. *Energy Economics*, 22, 549–568.

Ramsey J. B. and Yuan H. J. 1989. Bias and error bars in dimension calculations and their evaluation in some simple models. *Physics Letters*, A134, 397–398.

Ramsey J. B. and Yuan H. J. 1990. The statistical properties of dimension calculations using small data sets. *Nonlinearity*, 3, 155–176.

Ramsey J. B. and Yuan H. J. 1989. Bias and error bars in dimension calculations and their evaluation in some simple models. *Physics Letters A*, 134, 397–398.

Rosenstein M. T., Collins J. J., and De Luca C. J. 1993. Reconstruction expansion as a geometry-based framework for choosing proper delay times. *Physica D*, 65, 117–134.

Savit R. and Grenn M. 1991. Time series and dependent variables. *Physica D*, 50, 95–116.

Sayers C. L. 1986. *Work Stoppage: Exploring the Nonlinear Dynamics*, Department of Economics, University of Houston.

Scheinkman J. and LeBaron B. 1989. Nonlinear dynamics and stock returns. *Journal of Business*, 62(3), 311–337.

Schwartz B. and Yousefi S. 2003. On complex behavior and exchange rate dynamics. *Chaos Solutions and Fractals*, 18, 503–523.

Serletis A. and Gogas P. 1997. Chaos in East European black-market exchange rates. *Research in Economics*, 51, 359–385.

Serletis A. and Shintani M. 2003. No evidence of chaos but some evidence of dependence in the U.S. stock market. *Chaos, Solitons and Fractals*, 17, 449–454.

Serletis A. and Shintani M. 2006. Chaotic monetary dynamics with confidence. *Journal of Macroeconomics*, 28, 228–252.

Sewell S. P. et al. 1996. Using chaos measures to examine international capital market integration. *Applied Financial Economics*, 6(2), 91–101.

Shintani M. and Linton O. 2004. Nonparametric neural network estimation of Lyapunov exponents and a direct test for chaos. *Journal of Econometrics*, 120, 1–33.

Streips M. A. 1995. The problem of the persistent hog price cycle: A chaotic solution. *American Journal of Agricultural Economics*, 77, 1397–1403.

Strozzi F., Zaldìvar J. M., and Zbilut J. P. 2002. Application of nonlinear time series analysis techniques to high-frequency currency exchange data. *Physica A*, 312, 520–538.

Stutzer M. J. 1980. Chaotic dynamics and bifurcation in a macro model. Staff Report 55, Federal Reserve Bank of Minneapolis.

Takens F. 1981. Detecting strange attractors in turbulence, in: Rand D. A. and Young L. S. (eds.), *Dynamical Systems and Turbulence*, vol. 898, Lecture Notes in Mathematics, Springer, Berlin.

Torkamani M. A., Mahmoodzadeh S., Pourroostaei S., and Lucas C. 2007. Chaos theory and application in foreign exchange rates vs. IRR (Iranian Rial), *World Academy of Science, Engineering and Technology*, 30, 328–332.

Trulla L. L., Giuliani A., Zbilut J. P., and Webber C. L. Jr. 1996. Recurrence quantification analysis of the logistic equation with transients. *Physics Letters A*, 223, 255–260.

Tufillaro N. B., Solari H. G., and Gilmore R. 1990. Relative rotation rates: Fingerprints for strange attractors. *Physical Review A*, 41, 5717–5720.

Vaidyanathan R. and Krehbiel T. 1992. Does the S&P 500 futures mispricing series exhibit nonlinear dependence across time? *The Journal of Futures Markets*, 12, 659–677.

Wei A. and Leuthold R. M. 1998. Long agricultural futures prices: ARCH, long memory, or chaos processes? OFOR Paper, n. May 03.

Wolf A., Swift J. B., Swinney H. L., and Vastano J. A. 1985. Determining Lyapunov exponents from time series. *Physica D*, 16, 285–317.

Wu K., Savit R., and Brock W. 1993. Statistical tests for deterministic effects in broad band time series. *Physica D*, 69, 172–188.

Yang S. R. and Brorsen B. W. 1992. Nonlinear dynamics of daily cash prices. *American Journal of Agricultural Economics*, 74, 706–715.

Yang S. R. and Brorsen B. W. 1993. Nonlinear dynamics of daily futures prices: Conditional heteroskedasticity or chaos? *The Journal of Futures Markets*, 13, 175–191.

Zbilut J. P., Giuliani A., and Webber C. L. Jr. 2000. Recurrence quantification analysis as an empirical test to distinguish relatively short deterministic versus random number series. *Physics Letters A*, 267, 174–178.

Zbilut J. P. and Webber C. L. Jr. 1998. Detecting deterministic signals in exceptionally noisy environments using cross-recurrence quantification. *Physics Letters A*, 246, 122–128.

Zbilut J. P. and Webber C. L. Jr. 1992. Embeddings and delays as derived from quantification of recurrence plots. *Physics Letters A*, 171, 199–203.

38

Human Fuzzy Rationality as a Novel Mechanism of Emergent Phenomena

Ihor Lubashevsky

38.1 Introduction

38.1.1 Complex Systems with Cooperative Behavior

There is a wide variety of systems consisting of many elements that may be classified as *complex systems with cooperative behavior* because their behavior as whole entities is a result of cumulative contribution of many elements partly synchronized in individual dynamics. This variety contains objects of different nature, including physical and chemical systems, one on the side, and ecological, economic, and social systems, on the other side. Two characteristic examples of phenomena observed in such complex systems are presented in Figure 38.1.

The left picture shows frost patterns, ice flowers, that usually grow on window panes in the winter time. Such dendritic crystals emerge through a growth instability occurring when their growth rate is limited, for example, by the rearrangement of water molecules near the crystal interface and the dissipation of heat generated by this rearrangement. For geometric reasons, any small protuberance forming randomly on

FIGURE 38.1 Examples of objects emerging through self-organization processes: ice flowers formed on a window pane in the winter time (left picture) and a flock of birds moving as a whole (right picture). (Left picture from Schnobby. Photo: Ice-ferns. http://en.wikipedia.org/wiki/File:Frost_patterns_4.jpg (licensed under the terms of the CC-BY-SA 3.0). Right picture from Rae, Alastair. Photo: Red-billed quelea flocking at waterhole. http://www.flickr.com/photos/merula/342898722/in/set-72157594423311516 (licensed under the terms of the CC-BY-SA-2.0).)

the interface locally enhances the heat diffusion from the interface, whereas the surface tension tends to flatten the protuberance. As a result, if the tip radius of this protuberance is not too small, it will grow. This amplification of small interface perturbations occurs again and again until a dendrite is produced. So the formation of ice flowers is governed by the cooperative interaction of many water molecules rather than their individual properties directly.

The right picture in Figure 38.1 shows a flock of birds. In its flight, every bird tries to repeat the maneuvers performed by the neighboring birds, which synchronizes the individual motion of birds. It enables us to regard their flock as a certain object with some properties ascribed to it as a whole rather than the birds individually. In particular, it becomes possible to single out certain relatively sharp boundaries of the region occupied by the birds.

The two examples illustrate the distinctive features and phenomena that can be summarized within the notions of *emergence* and *self-organization*. Following O'Connor and Wong (2012) and De Wolf and Holvoet (2005), we may say that

> A phenomenon, some property, or structure and pattern can be categorized as *emergent* at the *macro-level* if it dynamically arises from the interaction between many constituent elements of a given complex system at the *micro-level*. Such emergents are "novel" or "irreducible" with respect to the individual elements.

The notion of *self-organization* describes another characteristic feature of many complex systems; following De Wolf and Holvoet (2005), we state that

> Self-organization is a dynamical and adaptive process where systems acquire and maintain structures themselves, without external control. The "structure" term refers to spatial patterns forming in such systems, temporal patterns representing time variations of certain quantities, or both.

According to the modern state of the art in understanding the formation and evolution of the spatiotemporal patterns (see, e.g., Meyers 2009, p. VI), emergent phenomena are considered to be a result of self-organization processes. Figure 38.2 illustrates the details of this relationship. At the micro level, the system description deals with individual constituent elements and their interaction with one another. Emergent phenomena are ascribed to the macro level of such a system when its structureless state, i.e., its

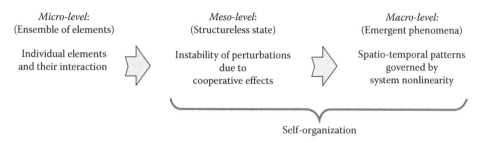

FIGURE 38.2 Illustration of emergence governed by self-organization processes in multielement systems.

homogeneous equilibrium becomes unstable under certain conditions. It is essential that this instability is a cumulative effect of the element interaction, meaning that only perturbations comprising many constituent elements can be grown.

38.1.1.1 Stationary Point Instability and Self-Organization Phenomena

In order to explain this mechanism of emergence in mathematical terms, let us consider some hypothetical system of N elements whose individual state is quantified by a variable q_i (here i is the element index, q_i may be a vector). The collection $\mathbb{Q} = \{q_i\}_{i=1}^{i=N}$ of all these variables specifies a state of the given system as a whole; further $\mathbb{Q}$ will be referred to as the phase space of the system. Without loss of generality, the equation governing the system dynamics can be written as

$$\frac{dq_i}{dt} = \mathcal{F}_i(\mathbf{q}, t), \tag{38.1}$$

where $\mathcal{F}_i(\mathbf{q}, t)$ is a certain function of all the variables $\mathbf{q} = \{q_1, q_2, \dots, q_N\}$ and, maybe, the time t. Below it will be referred to as a "force" acting on element i from the other elements. The assumed direct dependence of the force $\mathcal{F}_i$ on the time t enables us to take into account possible random factors within the same notation. The notion of phase space has been used to underline the fact that if we know the "position" $\mathbf{q} = \{q_i\}$ of the system in the space $\mathbb{Q}$ and, naturally, the forces $\{\mathcal{F}_i\}$ are given, then its dynamics is completely determined, at least, in principle. Without uncontrollable factors acting on the system, which are usually treated as some structureless noise, self-organization phenomena are related to intrinsic causes rather than extrinsic ones. It allows us to consider the forces $\{\mathcal{F}_i\}$ or, strictly speaking, their regular components not to contain the time t in the list of their arguments, i.e., we may set $\mathcal{F}_i = \mathcal{F}_i(\mathbf{q})$ for all i.

Let us assume that the system at hand is currently in a certain equilibrium state specified by a stationary point $\mathbf{q}^{st} = \{q_i^{st}\}$ meeting, by definition, the condition

$$\mathcal{F}_i(\mathbf{q}^{st}) = 0 \quad \text{for all } i = 1, 2, \dots, N. \tag{38.2}$$

Based on the governing equation (38.1), the stability analysis of the system dynamics in the vicinity of the stationary point $\mathbf{q}^{st}$ is usually reduced to the linearized version of Equation 38.1,

$$\frac{d\delta q_i}{dt} = \sum_{j=1}^{N} \widehat{\mathcal{F}}_{ij}^{(1)} \cdot \delta q_j, \tag{38.3}$$

with respect to small deviations $\delta q_i(t) = q_i(t) - q_i^{st}$ of the system from the stationary point. This deviation could be caused, for example, by weak random external factors acting on the system. If the matrix $\|\widehat{\mathcal{F}}_{ij}^{(1)}\|$ possesses an eigenvector $\{\psi_i^\lambda\}$ with an eigenvalue λ such that its real part is positive, $\mathrm{Re}\,\lambda > 0$, then the equilibrium state $\mathbf{q}^{st}$ becomes unstable. It means that small perturbations of the form

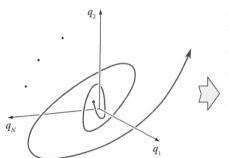

FIGURE 38.3 Explanation of emergence arising via a system instability. The left fragment schematically shows the system dynamics near the origin of the phase space $\mathbb{Q} = \{q_i\}_{i=1}^{i=N}$ regarded as an unstable stationary point. The right fragment illustrates the formation of spatial patterns, e.g., caused by the Kelvin–Helmholtz instability. (After Grahamuk. Photo: Wave clouds forming over Mount Duval, Australia. http://en.wikipedia.org/wiki/Image:Wavecloudsduval.jpg [licensed under the terms of the CC-BY-SA 3.0].)

$$\delta q_i(t) = a(t)\psi_i^\lambda$$

will grow with time as $a(t) \propto \exp\{\text{Re } \lambda \cdot t\}$. To study the further nonlinear dynamics of the system instability, in particular, to specify the type of bifurcation, the terms of higher order in the expansion

$$\mathcal{F}_i\{\mathbf{q}\} = \sum_{j=1}^{N} \widehat{\mathcal{F}}_{ij}^{(1)} \cdot \delta q_j + \text{higher-order terms in } \{\delta q_j\}$$

must be taken into account. Nevertheless, the analysis of the stationary point answers a question such as whether a system hand can undergo instability giving rise to self-organization processes and emergent phenomena. Only special situations like the dynamics of Hamiltonian systems or systems becoming unstable only with respect to *finite* perturbations require a more sophisticated analysis.

Summarizing the aforesaid, it should be underlined once more that in describing emergent phenomena in multielement ensembles, the notion of the stationary point and its stability is one of the cornerstones of modern theory. The corresponding scenario of emergent phenomena arising via a system instability is illustrated in Figure 38.3. The given mechanism of emergence appeals, explicitly or implicitly, to the basic concepts of Newtonian mechanics and is based on the mathematical formalism of statistical physics.

As far as systems where humans play a crucial role are concerned, it should be noted that during the last few decades there has been considerable progress in modeling such objects based on this approach (e.g., Chakrabarti et al. 2006; Galam 2008; Castellano et al. 2009; Nowak and Strawinska 2009; Stauffer and Solomon 2009; Slanina 2009; Helbing 2012; Galam 2012). In particular, the notion of energy functional (Hamiltonian) and the corresponding master equation were employed to simulate opinion dynamics, and the dynamics of culture and languages (e.g., Castellano et al. 2009; Slanina 2009; Stauffer 2009); the social force model inheriting the basic concepts from Newtonian mechanics was used to simulate traffic flow, pedestrian motion, the motion of bird flocks, fish schools, and swarms of social insects (e.g., Castellano et al. 2009; Helbing and Johansson 2009; Schadschneider et al. 2009). A detailed review of other techniques based on kinetic theory, fluid dynamics, the Ginsburg–Landau equations, etc. applied to traffic flow and similar problems can also be found in Helbing (2001), Chowdhury et al. 2000; Nagatani (2002). Continuing the list of examples, we note the application of the Lotka–Volterra model and the related reaction-diffusion systems to stock market, income distribution, and population dynamics (Yaari et al. 2009). The replicator equations developed initially in the theory of species evolution were applied to the moral dynamics (Hegselmann 2009). The notion of a fixed-point attractor as a stable equilibrium point in the system dynamics that corresponds to some local minimum in a certain potential relief is widely met in social psychology (Vallacher 2009). The latter is extended even to collections of such fixed point attractors forming a basin. Besides, social psychology uses the notion of latent attractors (i.e., invisible ones under equilibrium and whose presence affects strong perturbations), periodic attractors representing limit cycles, and deterministic chaos. In addition, the concept of synchronization of interacting oscillators was used to model social coordination (Oullier and Kelso 2009).

In spite of these achievements, we have to note that the mathematical theory of social systems is currently at its initial stage of development. Indeed, animate beings and objects of the inanimate world are highly different in their basic features, in particular, such notions as willingness, learning, prediction, motives for action, moral norms, personal and cultural values are just inapplicable to inanimate objects. Moreover, as far as the organization of life is concerned, questions as to how emergence arises have received only rather vague answers. In particular, whether it is possible to describe, at least in principle, human mind and mental processes appealing to neural, philological, or biochemical subsystems of the human body is rather far from being understood up to now.

This situation prompts us to pose a question as to what *individual* physical notions and mathematical formalism should be developed to describe social systems in addition to the available ones inherited from modern physics. For example, Kerner's hypothesis about the *continuous* multitude of metastable states representing the synchronized phase of traffic flow, on the one hand, stimulated developing the three-phase traffic model explaining a number of observed phenomena in congested traffic flow (Kerner 2004, 2009). On the other hand, a *microscopic* mechanism enabling the coexistence of many different metastable states actually at the same point of the corresponding phase space is a challenging problem.

The present chapter discusses one of such notions, namely, the fuzzy rationality, which can be regarded as a specific implementation of the bounded capacity of human cognition (Dompere 2009). Its particular goal is to demonstrate that the fuzzy rationality can be responsible for complex emergent phenomena in social systems. However, before passing directly to the main subject, let us present some physiological and psychological evidence for the use of the fuzzy rationality concept in modeling human behavior.

38.1.2 Physiological and Psychological Background

The pivot point of the theory to be constructed below is based on two fundamental features of human cognition and behavior. The first one can be categorized as the bounded capacity of human cognition and its specific implementations in motor control over mechanical systems. The second one is related to basic human cognitive biases characterizing systematic deviations of humans in their behavior or judgment from rationality optimal or good actions and strategies.

38.1.2.1 Human Intermittent Control

During the last few decades there has been a great deal of research on physiological mechanisms governing human motor actions such as postural control in quiet standing, balancing an inverted stick (real or virtual one), controlling the steering wheel while driving or the sail while yachting and windsurfing (see, e.g., Milton et al. 2009b; Loram et al. 2009, 2011; Balasubramaniam 2013, and references therein). As a result, a large body of evidence suggests that such human actions can be described within the concept of intermittent (discontinuous) control. This type control is characterized by its implementation via a sequence of discrete actions separated by some time intervals. Craik and Vince (Craik 1947, 1948; Vince 1948) seem to be the first who demonstrated that unlike the majority of engineering systems human motor control proceeds via a sequence of discrete movements with maximum frequency 2–3 actions per second, at least, in pursuing targets with unpredictable dynamics. Recent noninvasive measurements also demonstrated that during postural sway the muscle movements for balance control are not continuous but are intermittent and pulsatile (Loram and Lakie 2002; Loram et al. 2006). Similar conclusions have been obtained from studies of stick balancing at the fingertip (Cabrera et al. 2006; Hosaka et al. 2006). It is well established that peripheral reflexes are continuous in nature and, in particular, peripheral servo-mechanisms can drive muscle length at high frequency up to 8 Hz (Evans et al. 1983; Rack et al. 1983). It enables us to distinguish between this high bandwidth mechanism driving muscles directly and the low-frequency core of human balance control. Generalizing the available experimental data, it is possible to say that there is a certain similarity between voluntary manual control and normal standing, besides, the observed variability of their basic characteristics may indicate a significant role of intentional mechanisms in human movement control (Loram et al. 2009). It enables us to pose a question about a general model for such human actions that appeals to the fundamental features of human behavior. As far as the current state of the art in this field (Loram et al. 2011) is concerned, it is possible to say that human intermittent control appears to be a natural biological strategy based on opening the sensory feedback loop while executing control action. Moreover, flexible, refined control by the higher nervous system relies on inhibition of lower continuous feedback system and attenuation of sensation during movement (Bays et al. 2006; Voss et al. 2006).

Among the explanations of human control intermittency proposed so far are the interplay between noise and delays in sensorimotor system (Cabrera and Milton 2002), the clock-driven ("act-and-wait") (Craik 1947; Loram and Lakie 2002; Insperger 2011), and the event-driven ("drift-and-act") control. The latter hypothesis, referred to also as the event-driven approach, has become the most widely employed recently (Milton et al. 2009a; Gawthrop et al. 2011; Kowalczyk et al. 2012; Milton 2013). The event-driven models build up on the fact that human operators cannot detect small deviations of the controlled system from the goal state. Therefore, the control is switched off as long as the deviation remains below a certain threshold value. Whenever the deviation exceeds the threshold, the control is switched on so that the system is driven back to the goal state. Some studies hypothesize that human operators may also exploit the dynamical properties of the controlled system. For instance, in quiet standing, humans may ignore the large angular deviation of the body if the body already moves toward the upright position due to inertia (Bottaro et al. 2008; Asai et al. 2009; Suzuki et al. 2012; Asai et al. 2013). Even so, the latter mechanism is assumed to operate not on its own, but jointly with the threshold-based control.

The importance of the threshold notion in human control is thus widely acknowledged. Still, virtually all current models of event-driven human control utilize the very basic, even somewhat naive version of the threshold mechanism. In essence, they presume that the threshold represents the fixed, precise boundary of the sensory deadzone, and the control is triggered at the very moment the deviation from the goal state crosses this threshold. A notable exception is the model of quiet standing proposed by Bottaro et al. (2008). It employs the stochastic mechanism of generating control bursts: the probability of control activation increases with deviation of the system from the reference state. However, this model still remains one of a kind, while the models based on the standard threshold concept are used ubiquitously, possibly because despite being relatively simple they may explain many features of the experimentally observed dynamics (Boulet et al. 2010; Gawthrop et al. 2011; Kowalczyk et al. 2012; Milton 2013). Recurrently (Zgonnikov et al. 2014), a new approach arguing that control activation in humans may not be threshold-driven, but instead intrinsically stochastic, noise-driven. Specifically, it was suggested that control activation stems from stochastic interplay between the operators need to keep the controlled system near the goal state, on the one hand, and the tendency to postpone interrupting the system dynamics, on the other hand.

38.1.2.2 Status Quo Bias

In psychology and behavioral economics, there have been accumulated a large body of evidence for various cognitive biases in human behavior, i.e., human tendencies to think in certain ways, which can lead to systematic deviations from a standard of rationality or good judgment. One of them is the "status quo bias," a preference for the current state of affairs. The status quo bias leads people to prefer that things remain the same, or that they change as little as possible, if they absolutely must be altered. Status quo bias interacts with other nonrational cognitive processes such as loss aversion, existence bias, endowment effect, longevity, mere exposure, and regret avoidance (Wikipedia 2014).

There are several routes to status quo bias (for a recent review see, e.g., Eidelman and Crandall 2012). Choice could be difficult (Iyengar and Lepper 2000; Schwartz 2000), and decision makers may prefer to do nothing (Ritov and Baron 1990) because for them to maintain their current course of action may be easier (Samuelson and Zeckhauser 1988). As evidence, decision makers are more likely to postpone making a decision as alternatives are added (Tversky and Shafir 1992), besides status quo alternatives may require less mental effort to maintain (Eidelman and Crandall 2009). According to Samuelson and Zeckhauser (1988), the status quo bias fuses decision-makers preferences to do nothing (i.e., the omission bias, Ritov and Baron 1992; Spranca et al. 1991) with a preference to maintain the current choice.

In addition to these cognitive limitations imposed by choice, there are also informational limitations because decision outcomes often are not certain. So while previous decisions are "good enough," sticking with what worked in the past is a safe option that makes for a smart choice (Eidelman and Crandall 2012).

38.1.3 Concept of Dynamical Traps: Stationary Point Generalization

The concept of dynamical traps is actually some generalization of the notion of stationary points for social systems and systems governed by humans. In its building, this concept takes into account the main features of human intermittent control and the gist of cognitive biases discussed above.

In general, the concept of dynamical traps assumes that humans (operators) governing the dynamics of a certain system try to follow an optimal strategy in controlling the system motion but fail to do this perfectly because similar actions are indistinguishable for them. So when, for example, two actions are close to each other in quality from the standpoint of a human operator making a decision, their choice may be random because he ought to consider them equivalent. In particular, dealing with a dynamical system in a phase space $\mathbb{R}_{xy}$ its stationary point $\{x^{st}, y^{st}\}$ being initially stable is replaced by a certain neighborhood $\mathbb{Q}_{tr}$, called the dynamical trap region, such that when the system goes into $\mathbb{Q}_{tr}$, its dynamics is stagnated. This mimics vain actions of an operator in directing the system motion toward the point $\{x^{st}, y^{st}\}$ precisely. Indeed, when the system under the operator control gets any point in $\mathbb{Q}_{tr}$, the operator may consider the current situation perfect because he just does not "see" the point $\{x^{st}, y^{st}\}$ and until the system leaves $\mathbb{Q}_{tr}$, he has no reason to keep the control active. In other words, if the operator does not know what to do, he does nothing to change the current situation. A similar region of dynamical traps can be introduced with respect to the optimal strategy of operator actions.

In this chapter, the main attention is focused on systems where the optimal dynamics implies the stability of a certain equilibrium point in the corresponding phase space. Section 38.2 is devoted to constructing models for dynamical traps appealing to the basic features caused by the bounded capacity of human cognition and a certain freedom in taking the appropriate actions for governing the corresponding system. In Section 38.3, it will be demonstrated that complex cooperative phenomena can emerge in ensembles of many elements due to the dynamical trap effects in their individual behavior. It should be underlined that such emergent phenomena have to be regarded as a novel class of self-organization processes because their origin is not the instability of some stationary points.

38.2 Dynamical Traps Caused by the Bounded Capacity of Human Cognition

To elucidate the introduction of dynamical traps, let us consider the required modification of the dynamical equation formalism in two steps. At the first step, the notion of stationary point will be modified to take into account the bounded capacity of human cognition treated as a certain fuzziness in human actions. Then, the phase space of a system governed by human operator will be extended to allow for human active behavior and the freedom in taking the appropriate actions.

38.2.1 Dynamical Traps and Human Fuzzy Rationality

Let us consider a simple hypothetical system governed by a human operator, for example, car driving. The dynamics of the given system is represented as the motion of a point $\{x, y\}$ on a phase plane $\mathbb{R}_{xy}$.

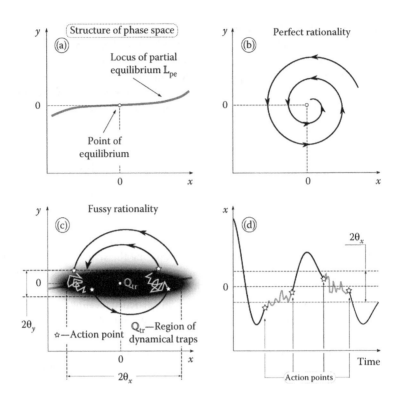

FIGURE 38.4 The presumed structure of the phase space $\mathbb{R}_{xy}$ of the system considered in Section 38.2.1, model (38.4) (a); a schematic illustration of its dynamics near the equilibrium point $\{0, 0\}$ (stable stationary point) in the cases of the perfect rationality (b) and the fuzzy rationality (c,d). Not to overload the drawings, the frame origin is shifted from the point $\{0, 0\}$.

38.2.1.1 The Limit of Perfect Rationality

We presume that if the operator was able to govern the system perfectly following a certain optimal strategy, then its dynamics would be described by the coupled equations:

$$\tau\frac{dx}{dt} = F_x(x, y), \quad \tau\frac{dy}{dt} = F_y(x, y). \tag{38.4}$$

Here, τ is a time scale characterizing the operator perception delay; the "forces" $F_x(x, y)$ and $F_y(x, y)$ are determined by *both* the physical regularities of the system mechanics and the active behavior of the operator in controlling the system motion. The origin $\{0, 0\}$ of the coordinate frame is placed at the equilibrium point of system (38.4), i.e., the equalities

$$F_x\big|_{\substack{x=0\\y=0}} = 0, \quad F_y\big|_{\substack{x=0\\y=0}} = 0 \tag{38.5}$$

are assumed to hold. In this context, the perfect rationality of the operator means his ability (i) to locate precisely the current position of the system on the phase plane $\mathbb{R}_{xy}$, (ii) to predict strictly its further motion, and then (iii) to correct the current motion continuously. Exactly in this case, it is possible to consider that the operator orders the strategies of behavior according to their preference and then chooses the optimal one. As a result, the equilibrium point $\{0, 0\}$ must be stable when the aim of operator actions is to keep the system in close vicinity to this point (Figure 38.4a,b).

The motion of the given system has been presumed to be a *cumulative* effect of the physical regularities and the operator actions. The notion of *partial equilibrium* implements this feature, namely, the operator is considered to be able to halt the system motion at a certain multitude $\mathbb{L}_{pe}$ of points in the phase space $\mathbb{R}_{xy}$ to be called the locus of partial equilibrium and treated here as some smooth curve. So, in the case of perfect rationality, a gradual system motion toward the equilibrium point $\{0,0\} \in \mathbb{L}_{pe}$ is due to the intelligent actions of the operator, which can locate this point on the plane $\mathbb{R}_{xy}$ precisely. The coordinate frame under consideration has been chosen in such a manner that the x-axis be tangent to the partial equilibrium locus $\mathbb{L}_{pe}$ at the point $\{0,0\}$ (Figure 38.4a).

Let us touch on the dynamics of a car following a lead car moving ahead with a fixed velocity V to exemplify these constructions. The motion of the following car is usually described in terms of the headway distance h and its velocity v whose time variations are governed by the social force model generally written as

$$\frac{dh}{dt} = V - v, \quad \frac{dv}{dt} = F_v(v, h, V), \tag{38.6}$$

where $F_v(v, h, V)$ is a certain function.[*] In specifying the function $F_v(v, h, V)$, a driver is typically assumed to respond to the combined effect of two stimuli. One of them is to keep the speed of his car equal, on the average, to the lead car speed V and can be quantified in terms of the relative velocity $u = v - V$. The other is to maintain an optimal headway distance $h_{opt}(v, V)$ generally determined by the values of v and V. This stimulus similarly can be quantified by the difference $h - h_{opt}(v, V)$. So $\{h_{opt}, V\}$ is the equilibrium point of the car following in the phase space $\mathbb{R}_{hv}$ and the partial equilibrium locus $\mathbb{L}_{pe}$ is the line $v = V$. Indeed, for simple kinematic reasons, any point on this line corresponds to a steady-state arrangement of the two cars, which can be frozen by the driver just fixing the car velocity, whereas keeping the headway distance equal to its optimal value $h_{opt}(V, V)$ is due to the driver intelligent action.

It is worthwhile to note that the social force model (38.6) matches actually the perfect rationality in driver behavior. In fact, the *detailed* description of driver actions requires a certain extension of the phase space, including, at least, the car acceleration a as an individual phase variable. The matter is that, on the one hand, by physical reasons the driver cannot change freely the position x and velocity v of his car, he is able to affect the car dynamics via changing the acceleration a only. On the other hand, the acceleration on its own contributes to the driver perception of the car motion quality. So, in the approximation of perfect rationality, the description of the car following is reduced to the problem of minimizing a certain cost functional whose integrand, a cost function, depends on the headway distance h, the velocity v, and also the acceleration a (Lubashevsky et al. 2003b). As a result, the governing equation contains $\ddot{a}$ in the leading order and, thus, does not meet the Newtonian mechanics paradigm. However, since the rational driver can perfectly predict the car motion, the final governing equation is reduced to the social force model (38.6), where the "force" $F_v(v, h, V)$ depends not only on the current values of the headway distance h and the car velocity v but also on the parameters of the equilibrium point $\{h_{opt}(V, V), V\}$ the attaining of which is the goal of the driver actions. Naturally, beyond the perfect rationality approximation, the car dynamics cannot be described in the frameworks of Newtonian mechanics (Lubashevsky et al. 2003a).

38.2.1.2 Case of Fuzzy Rationality

The perfect rationality cannot be implemented in the reality because of the bounded capacity of human cognition. As far as the system at hand is concerned, this limitation manifests itself in the fact that the operator is not able to order states of the system motion by their preference when they are close to one another in quality. In this case, the rational behavior becomes physically impossible for the operator and model (38.4) cannot pretend to describe the system dynamics. To tackle this problem, let us note the following before modifying the governing equations (38.4).

Pursuing two individual succeeding goals can be singled out in the operator actions. The first one is to halt the system motion by driving it to any point $g_t \in \mathbb{L}_{pe}$ of the partial equilibrium locus because it

---

[*] Model (38.6) also admits a generalization relating the current acceleration $a(t) = dv/dt$ to the headway distance $h(t - \bar{\tau})$ and the car velocity $v(t - \bar{\tau})$ taken at the previous moment of time with some time shift $\bar{\tau}$ (for a review, see, e.g., Helbing 2001).

could be tough governing fast motion of the system affected not only by the operator intentions but also the physical regularities. The second one is to drive the system toward the equilibrium position {0, 0}, for example, within close proximity to $\mathbb{L}_{pe}$. As a result, the mechanisms governing the system motion along the x- and y-axes are different; let us discuss them separately.

In the chosen coordinate frame, the curve $\mathbb{L}_{pe}$ is tangent to the x-axis at the equilibrium point {0, 0}, thereby, to simplify the further constructions, we may confine our consideration to its certain neighborhood and regard the partial equilibrium locus, $\mathbb{L}_{pe}$ as the x-axis. Outside the partial equilibrium locus, the system state varies in time under any action taken by the operator. However, if after driving the system to some point $g_t \in \mathbb{L}_{pe}$ the operator fixes the variable y, then there will be no "forces" causing the system motion along $\mathbb{L}_{pe}$; any point of $\mathbb{L}_{pe}$ is steady state. It enables us to approximate the "force" $F_x(x, y)$ by a linear function $F_x(x, y) = y \cdot f(x)$ and regard the cofactor $f(x)$ to be mainly determined by the system mechanics. In other words, the fact that the operator behavior is not perfect seems not to influence substantially the rate of system motion along the x-axis and, thus, there is no necessity to modify the first equation of system (38.4). Moreover, in the case under consideration, the function $f(x)$ can be regarded as some constant $f(x) = f$ without loss of generality.

The situation is just opposite with respect to the motion along the y-axis; the operator actions are to affect it directly. The operator is able to take any reasonable course of actions in order to drive the system toward the partial equilibrium locus and after getting some point $g_t \in \mathbb{L}_{pe}$, he can just freeze the system motion along the y-axis to halt the system motion as a whole. The point g_t is not necessary to be the point $g \in \mathbb{L}_{pe}$ that the operator intended to get initially because reaching any point at the partial equilibrium locus is acceptable to halt the system motion. The point g in turn is not mandatory to coincide with the equilibrium one because of the bounded capacity of operator cognition. At the next step of governing the system dynamics, the operator has no necessity to be in "hurry"; now it is possible for him to draw a decision on taking actions for reaching the currently desired point g during a relatively long time interval. In pursuing the latter goal, the operator can drive the system either in close vicinity to the partial equilibrium locus $\mathbb{L}_{pr}$ or deviating the system from $\mathbb{L}_{pr}$ considerably to enable fast motion.

Therefore, to go beyond the frameworks of the perfect rationality in constructing a model for the system motion, the following should be taken into account. First, the characteristic time scale τ_{tr} of system dynamics in close vicinity to the partial equilibrium locus $\mathbb{L}_{pe}$ must exceed essentially the corresponding time scale far from it, i.e., the inequality $\tau_{tr} \gg \tau$ should hold.

Second, the cognition limitations make it impossible for the operator to locate precisely not only the equilibrium point {0, 0} at $\mathbb{L}_{pe}$ but also the position of the partial equilibrium locus $\mathbb{L}_{pe}$ itself. In order to specify this uncertainty, let us introduce the perception thresholds, {θ_x, θ_y}, that characterize the dimensions of the neighborhood $\mathbb{Q}_{tr}$ of the equilibrium point {0, 0} within which this point as well as the corresponding fragment of $\mathbb{L}_{pe}$ can be located by the operator with high probability. Since the control over the variable y is of prime priority in governing the system dynamics, the threshold θ_y can be treated as a small parameter. Therefore, the region $\mathbb{Q}_{tr}$ is actually some narrow neighborhood of the partial equilibrium locus $\mathbb{L}_{pe}$ or, more rigorously, its certain fragment containing the equilibrium point. Any point of $\mathbb{Q}_{tr}$ is regarded by the operator as the equilibrium state with high probability.

Third, the point $g \in \mathbb{L}_{pe}$ characterizing the course of actions chosen by the operator at a given moment of time is not fixed, it can migrate inside the region $\mathbb{Q}_{tr}$ as time goes on. The movement of this point has to be rather irregular until it remains inside the domain $|g| \lesssim \theta_x$.

It is worthy of notice that the standard concept of stability is inapplicable to the system motion near the partial equilibrium locus $\mathbb{L}_{pe}$. Indeed, although the system has not reached the desired equilibrium point {0, 0}, the operator freezes its motion near some other point $g_t \in \mathbb{L}_{pe}$ and then keeps the system near $\mathbb{L}_{pe}$ until he makes a decision about driving the system toward {0, 0}. So, before making this decision, the system motion near $\mathbb{L}_{pe}$ looks like stable fluctuations near $\mathbb{L}_{pe}$, after that it has to be classified as unstable.

Let us discuss a model based on (38.4) that captures the key aspects of such operator behavior. The operator chooses a point g on the x-axis to which he is going to drive the system. While his control over the system motion is active, the system dynamics is governed by the equations

$$\tau \frac{dx}{dt} = f \cdot y, \tag{38.7}$$

$$\tau \frac{dy}{dt} = \Omega \cdot F(x - g, y) \tag{38.8}$$

with the cofactor Ω equal to unity, $\Omega = 1$. Here, the subscript y is omitted at the "force" $F(x - g, y)$. The two equations actually describe the system dynamics outside the region $\mathbb{Q}_{tr}$. When, roughly speaking, the system enters the region $\mathbb{Q}_{tr}$, the operator regarding its any point as an acceptable destination just freezes the system motion along the y-axis to such a degree that real variations in the variable y become imperceptible to him and, thus, are not controllable. This action is described by the stepwise transition

$$\{\Omega = 1\} \Rightarrow \{\Omega = \Delta_r\} \quad \text{with a probability rate} \quad \frac{1}{\tau} P\left(\frac{x}{\theta_x}, \frac{y}{\theta_y}\right). \tag{38.9}$$

Here, $\Delta_r = \Delta_r(t)$ is some small random value, $|\Delta_r| \ll 1$, which, in addition, can change in time also in an uncontrollable way, $P(z_x, z_y) \leq 1$ is a certain function of two arguments such that

$$P(z_x, z_y) \approx 1 \quad \text{for } |z_x| \lesssim 1 \quad \text{and} \quad |z_y| \lesssim 1,$$
$$P(z_x, z_y) \ll 1 \quad \text{for } |z_x| \gtrsim 1 \quad \text{or} \quad |z_y| \gtrsim 1.$$

When the system leaves the region $\mathbb{Q}_{tr}$ under the actions of uncontrollable factors or the operator gets a decision to correct the system location along the x-axis, he resumes governing the system dynamics, which is represented as the stepwise transition

$$\{\Omega = \Delta_r\} \Rightarrow \{\Omega = 1\} \quad \text{with the probability rate} \quad \frac{1}{\tau}\left[1 - P\left(\frac{x}{\theta_x}, \frac{y}{\theta_y}\right)\right]. \tag{38.10}$$

Finally, the given model should be completed by an equation describing the operator perception of the desired destination point inside $\mathbb{Q}_{tr}$. In this chapter, we write it in a symbolic form

$$\frac{dg}{dt} = \widehat{R}(x, y, t, g | \theta_x, \theta_y), \tag{38.11}$$

where the presence of the time t in the list of arguments allows for random factors in the dynamics of the variable g.

It should be noted that the formulated model of the system dynamics enables us to specify some general features of the "force" $F(x - g, y)$. In fact, the existence of the partial equilibrium locus $\mathbb{L}_{pe}$ has allowed us to single out two stimuli in governing the system motion, which determine the operator actions. They may be reformulated as follows. The goal of the first one is to keep up the variable y in close proximity to the partial equilibrium locus $\mathbb{L}_{pe}$ in order to depress the fast system motion. This stimulus can be quantified in terms of the variable y. As a result, the component $F_I(x - g, y)$ of the "force" $F(x - g, y)$ caused by the first stimulus may be written as $F_I(x - g, y) = -\sigma y$, where $\sigma > 0$ is some kinetic coefficient. The second one is related to the operator actions of driving the system toward the desired point $g \in \mathbb{L}_{pe}$. The corresponding component $F_{II}(x - g, y)$ does not change its sign in crossing the x-axis and in the simplest case may be represented in the form $F_{II}(x - g, y) = -\beta(x - g)$, where β is another kinetic coefficient. Therefore, the expression

$$F(x - g, y) = -\beta(x - g) - \sigma y \tag{38.12}$$

is the simplest ansatz catching the basic features of such systems. The sign of β depends on the sign of the kinetic coefficient f entering Equation 38.7; if $f > 0$, then the coefficient $\beta > 0$ too, otherwise, $\beta < 0$. Indeed, only for the combinations meeting the inequality $\beta f > 0$, the operator action causes the system motion toward the desired destination g. Since the two cases can be mapped on each other by the replacement $y \to -y$, we may assume that $f > 0$ and $\beta > 0$ without loss of generality. The form of the governing equations (38.7 through 38.11) within approximation (38.12) allows us to call the given system the oscillator with dynamical traps.

Following Todosiev, 1963, the time moments when the operator suspends or resumes the control over the system motion will be referred to as action points. Besides, the neighborhood $\mathbb{Q}_{tr}$ of the partial equilibrium locus $\mathbb{L}_{pe}$ will be called the region of dynamical traps because after transition (38.9) the system can reside inside it for a long time. It should be pointed out that a similar notion of dynamical traps was also introduced for relaxation oscillations in systems with singular kinetic coefficients (Lubashevsky et al. 1998) and congested traffic flow (Lubashevsky et al. 2002). Besides, the concept of dynamical traps is met in describing Hamiltonian systems with complex dynamics (for a review, see, e.g., Zaslavsky 2008) that denote some regions in the corresponding phase space with anomalously long residence time, however, the nature of the latter traps is different.

The stated concept of human behavior combines the principles of the perfect rationality and the characteristic features exhibited by the bounded capacity of human cognition in ordering possible actions, events, etc. by their preference. As a result, actions or events similar in properties are treated as equivalent and their ordering becomes possible only on scales exceeding some perception threshold. Naturally, this threshold should be regarded in the "weak" sense; it characterizes the probability of classifying actions or events as equivalent rather than it determining a certain strict boundary between them. To underline these aspects, it will be called the fuzzy rationality. In the case under consideration, the fuzzy rationality reflects itself in two effects: the stagnation of the system motion in the region of dynamical traps and probabilistic nature of the system dynamics in this region.

38.2.1.3 Continuous Model for the Stagnation Effect Caused by Dynamical Traps

To simulate the probabilistic on–off transitions in the operator actions when the system goes in or out of the trap region in terms of differential equations, we need a model that mimics the effects of dynamical traps in the frameworks of regular system motion with stagnation in the region $\mathbb{Q}_{tr}$. It is illustrated in Figure 38.5, where the left fragment depicts the three characteristic events occurring when the system goes through the region of dynamical traps $\mathbb{Q}_{tr}$, namely, halting the system motion when entering $\mathbb{Q}_{tr}$, the motion in $\mathbb{Q}_{tr}$ under influence of random factors, and resuming the governing of motion. The latter can be caused by two factors, the system leaving the region $\mathbb{Q}_{tr}$ or the operator just getting the decision of driving the system toward the desired point $g \in \mathbb{L}_{pe}$. Both of them determine some cumulative mean lifetime $\tau_{tr} \gg \tau$ of the

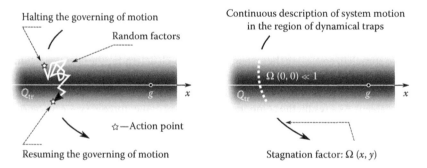

FIGURE 38.5 The *left fragment* illustrates the characteristic sequence of events in going through the region of dynamical traps $\mathbb{Q}_{tr}$, namely, halting the governing of the system motion in entering $\mathbb{Q}_{tr}$, then, random motion of the system in $\mathbb{Q}_{tr}$, and, finally, resuming the governing of motion due to the system leaving the region $\mathbb{Q}_{tr}$ or the operator decision of driving the system toward the desired position g. The *right fragment* represents the corresponding effective continuous description of regular system motion in the region $\mathbb{Q}_{tr}$ based on introduction of the stagnation factor $\Omega(x, y)$ being continuous function of its arguments meeting the inequality $\Omega(0, 0) \ll 1$.

system inside $\mathbb{Q}_{tr}$. The right fragment of Figure 38.5 exhibits an effective model mimicking this behavior, at least, semiquantitatively. It assumes the system motion along the y-axis to be governed by the regular force $\Omega(x, y)F(x - g, y)$, where the continuous function $\Omega(x, y)$, the stagnation factor, takes a small value $\Delta \sim \tau/\tau_{tr} \ll 1$ at the central points of the region $\mathbb{Q}_{tr}$, in other words, $\Omega(0, 0) = \Delta$. As the point $\{x, y\}$ goes away from the central points of $\mathbb{Q}_{tr}$ and leaves it, the stagnation factor exhibits gradual growth up to unity, i.e., $\Omega(x, y) \approx 1$ for $|x| \gtrsim \theta_x$ or $|y| \gtrsim \theta_y$. Let us make use of the following ansatz,

$$\Omega(x, y) = \frac{\Delta + (x/\theta_x)^2 + (y/\theta_y)^2}{1 + (x/\theta_x)^2 + (y/\theta_y)^2}. \tag{38.13}$$

In addition, in the frameworks of the given model, irregular variations of the final destination point g in the "mind" of the operator should be also ignored. Within this approximation, model (38.7 through 38.11) reads

$$\tau\frac{dx}{dt} = f \cdot y, \tag{38.14}$$

$$\tau\frac{dy}{dt} = \Omega(x, y) \cdot F(x, y) + \epsilon\xi(t). \tag{38.15}$$

Here, the additive noise term, the Langevin force $\epsilon\xi(t)$, is added to take into account random factors in the region of dynamical traps that are not related to the operator cognitive processes, $\xi(t)$ is white noise with unit amplitude,

$$\langle\xi(t)\rangle = 0, \quad \langle\xi(t)\xi(t')\rangle = \delta(t - t'), \tag{38.16}$$

and ϵ is the intensity of the Langevin forces.

The next section demonstrates the fact that the dynamical trap effect together with noise can induce new type of nonequilibrium phase transition.

38.2.1.4 Oscillator with Dynamical Traps: Noise-Induced Oscillations

Let us consider a particular example of model (38.14, 38.15) that can be interpreted as the oscillator with dynamical traps, namely,

$$\frac{dx}{dt} = v, \tag{38.17}$$

$$\frac{dv}{dt} = -\omega_0^2\Omega(v)\left[x + \frac{\sigma}{\omega_0}v\right] + \epsilon_0\xi_v(t). \tag{38.18}$$

Here, x and v are the dynamical variables usually treated as a coordinate and velocity of a certain particle, ω_0 is the circular frequency of oscillations provided the system is not affected by other factors, σ is the damping decrement, and the term $\epsilon_0\xi_v(t)$ in Equation 38.18 is a random Langevin "force" of intensity ϵ_0 proportional to the white noise $\xi_v(t)$ with unit amplitude. The function $\Omega(v)$ describes the dynamical trap effect arising in the vicinity of zero-value velocity. For this function, the following simple *ansatz* (cf. (38.13)) is used:

$$\Omega(v) = \frac{v^2 + \Delta^2\vartheta_{th}^2}{v^2 + \vartheta_{th}^2}, \tag{38.19}$$

where the parameter ϑ_{th} characterizes the thickness of the trap region and the parameter $\Delta \leq 1$ measures the trapping efficacy. When $\Delta = 1$ the dynamical trap effect is ignorable, for $\Delta = 0$ it is most effective.

The characteristic features of the given system are illustrated in Figure 38.6. The shadowed domain shows the trap region where the regular "force," the former term in Equation 38.18, is depressed. The latter is described by the factor $\Omega(v)$ taking small values in the trap region (for $\Delta \ll 1$). Inside the trap region, the system is mainly governed by the random Langevin "force." Outside the trap region, it is approximately harmonic.

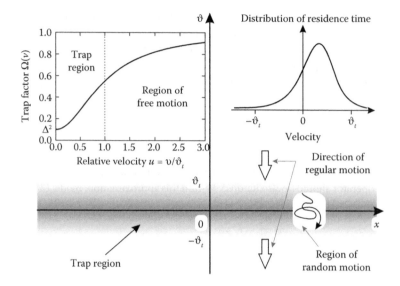

FIGURE 38.6 Characteristic structure of the phase space $\{x, v\}$. The shadowed domain represents the trap region where the regular "force" is depressed and the system motion is random. The regular "force" depression is described by the factor $\Omega(v)$ illustrated in the left window. The essence of the trap effect on the system dynamics is shown in the right window. Outside the trap region, the system dynamics is mainly regular.

In order to analyze the system dynamics, a dimensionless time t and the dynamical variables η and u are used. Namely, the time t is measured in units of $1/\omega_0$, i.e., $t \to t/\omega_0$ and the units of the coordinate x and the velocity v are ϑ_{th}/ω_0 and ϑ_{th}, respectively. So, by introducing the new variables

$$\eta = \frac{x\omega_0}{\vartheta_{th}} \quad \text{and} \quad u = \frac{v}{\vartheta_{th}},$$

the dynamical equations (38.17), Equation 38.18 reads (for the dimensionless time t)

$$\frac{d\eta}{dt} = u, \quad \frac{du}{dt} = -\Omega[u]\left(\eta + \sigma u\right) + \epsilon\xi(t), \tag{38.20}$$

where the noise $\xi(t)$ obeys conditions like equalities (38.16), the parameter $\epsilon = \epsilon_0/(\sqrt{\omega_0}\vartheta_{th})$, and the function

$$\Omega[u] = \frac{u^2 + \Delta^2}{u^2 + 1}.$$

Without noise, this system has only one stationary point $\{\eta = 0, u = 0\}$ being stable because it possesses a Liapunov function,

$$\mathcal{H}(\eta, u) = \frac{\eta^2}{2} + \frac{u^2}{2} + \frac{1 - \Delta^2}{2} \ln\left(\frac{u^2 + \Delta^2}{\Delta^2}\right). \tag{38.21}$$

This Liapunov function attains the absolute minimum at the point $\{\eta = 0, u = 0\}$ and obeys the inequality,

$$\frac{d\mathcal{H}(\eta, u)}{dt} = -\sigma u^2 < 0 \quad \text{for} \quad u \neq 0. \tag{38.22}$$

In particular, if $\sigma = 0$ and $\epsilon = 0$, then function (38.21) is the first integral of the system. In what follows, the values σ and ϵ will be treated as small parameters.

The present section demonstrates the fact that the noise $\xi(t)$ can cause a phase transition in the given system. It manifests itself in that the distribution function $\mathcal{P}(\eta, u)$ changes from a unimodal to a bimodal one. The dynamics of system (38.20) was analyzed numerically using a high-order stochastic Runge–Kutta

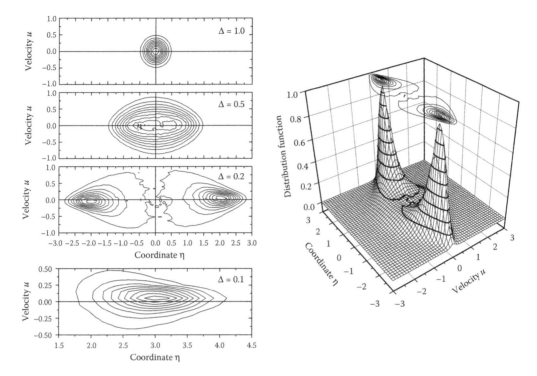

FIGURE 38.7 Evolution of the distribution function $\mathcal{P}(\eta, u)$ as the parameter Δ decreases. Left fragment: the contour plots for different values of Δ, the lower window depicts only one maximum of the distribution function. Right fragment: the form of the distribution function $\mathcal{P}(\eta, u)$ for $\Delta = 0.2$. In numerical calculations, the values $\sigma = 0.1$ and $\epsilon = 0.1$ were used. (After Lubashevsky, I. et al., *Eur. Phys. J. B-Condensed Matter Complex Syst.*, 36, 115–118, 2003).

method (Burrage 1999) (see also Burrage and Burrage 1998). The distribution function $\mathcal{P}(\eta, u)$ was calculated numerically by finding the cumulative time during which the system is located inside a given mesh on the (η, u)-plane for a path of a sufficiently long time of motion, $t \approx 500,000$. The size of the mesh was chosen to be about 1% of the dimension characterizing the system location on the (η, u)-plane.

The evolution of the distribution function $\mathcal{P}(\eta, u)$ is shown in Figure 38.7 in the form of the level contours dividing the variation scale into 10 equal parts. The upper window corresponds to the case of $\Delta = 1$ where the trap effect is absent and the distribution function is unimodal. The third window illustrates the case when the distribution function has the well-pronounced bimodal shape shown in Figure 38.7. Comparing the three upper windows in Figure 38.7, it becomes evident that there is a certain relation $\Phi_c(\Delta, \sigma, \epsilon) = 0$ between the parameters Δ, σ, and ϵ when the system undergoes a second-order phase transition, which manifests itself in the change of the shape of the phase space density $\mathcal{P}(\eta, u)$ from unimodal to bimodal. In particular, for $\sigma = 0.1$ and $\epsilon = 0.1$, the critical value of the parameter Δ is $\Delta_c(\sigma, \epsilon) \approx 0.5$ as seen in the second window.

38.2.1.5 Mechanism of the Phase Transition

To understand the mechanism of the noise-induced phase transition observed numerically in the given system, consider a typical fragment of the system motion through the trap region for $\Delta \ll 1$ that is shown in Figure 38.8. When it goes into the trap region $\mathcal{Q}_t$, $-\vartheta_t \ll v \ll \vartheta_t$, the regular "force" $\Omega[u] (\eta + \sigma u)$ containing the trap factor $\Omega[u]$ and governing the regular motion becomes small. So, inside this region, the system dynamics becomes random due to the remaining weak Langevin "force" $\epsilon\xi(t)$. However, the boundaries $\partial_+\mathcal{Q}_t$ (where $v \sim \vartheta_t$) and $\partial_-\mathcal{Q}_t$ (where $v \sim -\vartheta_t$) are not identical in properties with respect

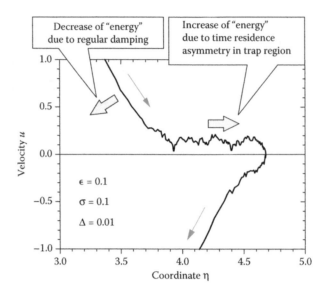

FIGURE 38.8 A typical fragment of the system path going through the trap region. The parameters $\sigma = 0.1, \epsilon = 0.1$, and $\Delta = 0.01$ were used in numerical simulations in order to make the trap effect more pronounced.

to the system motion. At the boundary $\partial_+ \mathcal{Q}_t$, the regular "force" leads the system inwards the trap region $\mathcal{Q}_t$, whereas at the boundary $\partial_- \mathcal{Q}_t$ it causes the system to leave the region $\mathcal{Q}_t$. Outside the trap region $\mathcal{Q}_t$, the regular "force" is dominant. Thereby, from the standpoint of the system motion inside the region $\mathcal{Q}_t$, the boundary $\partial_+ \mathcal{Q}_t$ is "reflecting" whereas the boundary $\partial_- \mathcal{Q}_t$ is "absorbing."

As a result, the distribution of the residence time at different points of the region $\mathcal{Q}_t$ should be asymmetric, as schematically shown in Figure 38.6 (the right window). This asymmetry is also seen in the distribution function $\mathcal{P}(\eta, u)$ obtained numerically. Its maxima are located at the points with nonzero values of the velocity, which is clearly visible in the lower window of Figure 38.7 (left column). Therefore, during location inside the trap region, the mean velocity of the system must be positive and it tends to go away from the origin. This effect gives rise to an increase in the "energy" $\mathcal{H}(\eta, u)$. Outside the trap region, the "energy" $\mathcal{H}(\eta, u)$ decreases according to expression (38.22). So, when the former effect becomes sufficiently strong, i.e., the random "force" intensity ϵ exceeds a certain critical value, $\epsilon > \epsilon_c(\Delta, \sigma)$, the distribution function $\mathcal{P}(\eta, u)$ becomes bimodal.

The system location with respect to the velocity v is due to the regular "force" being sufficiently strong outside the trap region, so the system spends the main time inside this region. Its location with respect to the coordinate x is caused by the fact that the region where the Langevin "force" mainly affects the system dynamics decreases in thickness as the coordinate x increases. The latter tendency takes place because the regular "force," the first term in Equation 38.18, is proportional to x. In fact, the thickness $U(\eta)$ of the trap region in the vicinity of the point $\{\eta, u = 0\}$ can be estimated using the condition of the equality of the characteristic times, $\tilde{t}_s$ and $\tilde{t}_d$, during which the system crosses the trap region under action of the regular "force" and the random Langevin "force." So

$$\tilde{t}_s \sim \frac{U}{\Omega[U]\eta} \sim \tilde{t}_d \sim \frac{U^2}{\epsilon^2}.$$

and setting for the sake of simplicity $\Delta = 0$, we get the estimate

$$U(\eta) \sim \epsilon^{2/3} \eta^{-1/3}.$$

Moreover, let the mean velocity in the trap region caused by the residence time asymmetry be about $U(\eta)$. then, the characteristic increase $\delta\eta$ of the coordinate η obtained by the system when crossing the trap

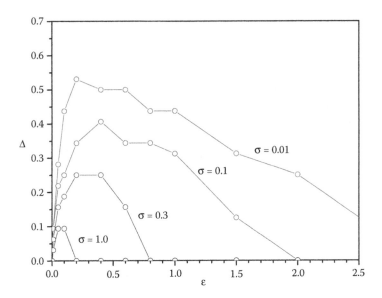

FIGURE 38.9 The region on the plane $\{\varepsilon, \Delta\}$, where the distribution $\mathcal{P}(\eta, u)$ takes the bimodal form for different values of the parameter σ. (Based on Lubashevsky, I.A. et al., *Bull. Lebedev Physics Institute*, 1, 15–20, 2007; Hajimahmoodzadeh, M., Dynamical states in condensed systems with anomalous kinetic coefficients (supervised by Lubashevsky I.), PhD thesis, Moscow State University, Moscow, Russia, 2005.)

region is estimated as $\delta\eta \sim U\tilde{t}_s \sim 1/\eta$. Thereby the dynamical trap effect becomes weaker as the "energy" $\mathcal{H}(\eta, u)$ increases. By contrast, according to Exp. (38.22), the higher the "energy," the stronger its dissipation caused by the regular "force."

As a general *conclusion* to this section, we note that in contrast to the classical phase transitions, the position of the new noise-induced phases is not specified by the zero values of regular "forces" even approximately. The cause of the observed phase transition is the asymmetry of the residence time distribution inside the trap region. This asymmetry is due to the cooperative effect of the regular "force" outside the trap region and the random Langevin "force" inside it. The regular "force" does not change the direction when crossing the trap region, inside this region it is only depressed. As a result, for the motion inside the trap region, one of its boundaries is "reflecting," whereas the other is "absorbing," which induces the residence time asymmetry. The latter gives rise to increase in the system "energy." Outside the trap region, the regular "force" causes the "energy" to decrease. The constructive role of noise is illustrated in Figure 38.9 showing the region on the parameter plane $\{\Delta, \varepsilon\}$ where the bimodal form of the distribution $\mathcal{P}(\eta, u)$ arises as a result of the noise-induced phase transition. Without noise, $\varepsilon = 0$, in the case of single oscillator with dynamical traps, the origin, $u = 0$ and $\eta = 0$, is always stable and for finite values of the trap intensity, $\Delta > 0$, rather small noise, $\varepsilon \ll 1$, cannot give rise to the bimodal distribution $\mathcal{P}(\eta, u)$.

38.2.2 Action Dynamical Traps

The model of oscillator with dynamical traps considered above admits the following interpretation. Let some object with inertia be driven by a human operator in one-dimensional space x. The goal of the operator is to maintain the object at the desired state, the origin $x = 0$, by implementing an optimal in some sense control strategy such that the object acceleration $a = d^2x/dt^2$ becomes a certain function of the current position x and the velocity $v = dx/dt$ of this object, $a = a_{\text{opt}}(x, v)$. However, if the object currently resides in some vicinity of the desired state, the operator prefers to halt active control over the system. The equations describing the system dynamics under the operator control are given by a model similar to system (38.17, 38.18) without noise

$$\dot{x} = v, \quad \dot{v} = \Omega(x, v) \cdot a_{\text{opt}}(x, v).$$

The cofactor $\Omega(x, v)$ is some function such that $\Omega(x, v) \approx 1$ for all the values (x, v) that are far enough from the origin and $\Omega(x, v) \ll 1$ in a certain neighborhood $\mathbb{Q}_{\text{tr}}$ of the origin in the space $\{x, v\}$. The cofactor $\Omega(x, v)$ describes the fuzzy perception of the operator. When the current state is far from the origin, the operator perfectly follows the optimal action strategy $a_{\text{opt}}(x, v)$. If the current position is recognized as "good enough," i.e., $(x, v) \in \mathbb{Q}_{\text{tr}}$, the operator halts active control over the system. So, during a considerable period of time, the system is affected only by random factors of a small amplitude; in other words, the system is "trapped" in a vicinity of the desired position. One may notice that in the case of linear feedback strategy

$$a_{\text{opt}} \propto -(x + \sigma v),$$

where $\sigma > 0$ is a constant damping parameter; the given system under human control is analogous to the physical system of a damped harmonic oscillator. This model captures the basic behavior properties of the fuzzy rational operator, i.e., the operator who does not react to small deviations from the desired phase space position. When the system deviates significantly from the goal state, the operator decides to start controlling the system in order to return it to an acceptable state. This can be achieved by varying the control parameter, namely, the acceleration, in a way that is optimal in some sense. The latter, however, contradicts the gist of the human fuzzy perception. Human operators have no capacity of controlling precisely not only the object position and its velocity but also following strictly the optimal strategy of behavior.

To understand how to cope with this problem, let us turn to the example of car following lead car, which should demonstrate characteristic features of dynamical systems governed by human operators with fuzzy rationality (Lubashevsky et al. 2003a). Car drivers are unable to continuously keep perfect awareness of the surrounding situation, so they usually set the acceleration to some constant value based on the current circumstances. Once fixed, the value of acceleration is changed only when the driver realizes that the deviation from some "optimal" acceleration value has become too large to be ignored. In other words, considerable deviations of the current acceleration a from the optimal value a_{opt} cause the operator to start active control over the car motion. However, when the difference $a - a_{\text{opt}}$ is rather small, there are no stimuli for the driver to act, i.e., to change the acceleration. Thus, one may imagine a certain region around the optimal strategy $a_{\text{opt}}(x, v)$, wherein each strategy is regarded as acceptable (Figure 38.10). Instead of precisely following the optimal strategy, the operator just keeps the actually implemented strategy inside this region, making some corrections only when the mismatch $a - a_{\text{opt}}$ exceeds some fuzzy threshold. For this reason, the region of acceptable strategies around a_{opt} will be called the action dynamical trap. The

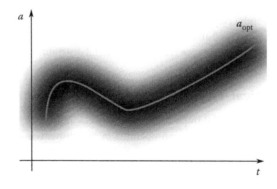

FIGURE 38.10 Action dynamical trap: the vicinity of the optimal strategy $a_{\text{opt}}(t)$ in the space of all action strategies $\{a(t)\}$.

"thickness" of the action dynamical trap is determined by the capacity of the operator perception and levels of concentration and motivation to follow the optimal control strategy. The action dynamical trap model is proposed to capture the discussed effects of fuzzy rationality in choosing and implementing the action strategies in human-controlled dynamical processes.

38.2.2.1 Oscillator with Action Dynamical Traps

The pivot point of the proposed approach is that we regard human actions as an independent component of the system rather than some predetermined function of its physical state. We extend the physical phase space $\{x, v\}$ by introducing a new phase variable, in the given case, the system acceleration a, i.e.,

$$\{x, v\} \rightarrow \{x, v, a\}.$$

It enables us to ascribe to the system an additional degree of freedom corresponding to the operator actions. Now, the model capturing the dynamical trap effect in controlling the deviation $a - a_{\text{opt}}$ is written as

$$
\begin{aligned}
\dot{x} &= v, \\
\dot{v} &= a, \\
\tau \dot{a} &= -\Omega_a \big[a - a_{\text{opt}}(x, v)\big] \cdot \big[a - a_{\text{opt}}(x, v)\big],
\end{aligned}
\tag{38.23}
$$

where τ is the operator reaction time parameter, and functions $a_{\text{opt}}(x, v)$ and $\Omega_a(a - a_{\text{opt}})$ are to be specified.

We define the operator control strategy as a linear feedback aimed at maintaining the system at the origin: $a_{\text{opt}}(x, v) = -\omega^2(x + \frac{\sigma}{\omega} v)$, where ω and σ are nonnegative constant coefficients. However, as the operator is fuzzy rational, the optimal control strategy should incorporate the dynamical trap effect in correcting the velocity variations:

$$a_{\text{opt}}(x, v) = -\Omega_v(x, v)\omega^2(x + \frac{\sigma}{\omega} v).$$

Thus, the control strategy a_{opt} is optimal from the standpoint of fuzzy rational human operator. Here, the dynamical trap cofactor $\Omega_v(x, v)$ is claimed not to depend on x. It reflects the assumption that the control over the system velocity v is of prior importance for the operator comparing to the control over the coordinate x. The desired effect can be mimicked by any function $\Omega(v)$ such that $\Omega \ll 1$ if $v \approx 0$ and $\Omega \approx 1$ otherwise. Without loss of generality, we use the ansatz

$$\Omega(x, v) := \Omega_v(v) = \frac{\Delta_v v_{\text{th}}^2 + v^2}{v_{\text{th}}^2 + v^2},
\tag{38.24}$$

where $v_{\text{th}} > 0$ is the threshold value of velocity and $\Delta_v \in [0, 1]$ is the dynamical trap intensity coefficient. When Δ_v equals unity, there is no dynamical trap effect—the operator is strictly rational and reacts even to the tiniest deviations. The case $\Delta_v = 0$ matches the situation when the operator ignores the small deviations but engages actively in the control over the system when the deviation becomes large enough.

One may notice that if we set $\Omega_a(a - a_{\text{opt}}(x, v)) \equiv 1$, system (38.23) describes following the optimal action strategy a_{opt} precisely by the operator whose reaction time is τ. As the human operator is not capable of doing so in the case of small deviations $a - a_{\text{opt}}$, we write

$$\Omega_a(a - a_{\text{opt}}) = \frac{\Delta_a a_{\text{th}}^2 + (a_{\text{opt}} - a)^2}{a_{\text{th}}^2 + (a_{\text{opt}} - a)^2},$$

where, in analogy to Equation 38.24, $\Delta_a \in [0, 1]$ indicates the presence of the action dynamical trap and a_{th} is the threshold in perceiving acceleration deviations from the optimal value.

In order to reduce the number of system parameters, we change the time and spatial scales as follows:

$$t \to t\frac{1}{\omega}, \quad x \to x\frac{a_{th}}{\omega^2}.$$

It is easy to check that in these dimensionless units parameters, ω and a_{th} are both equal to unity. Thus, the above expressions for Ω_a and a_{opt} take the form

$$\Omega_a(a - a_{opt}) = \frac{\Delta_a + (a_{opt} - a)^2}{1 + (a_{opt} - a)^2},$$

$$a_{opt}(x, v) = -\Omega_v(v)(x + \sigma v).$$
(38.25)

38.2.2.2 Oscillations Caused Solely by Action Dynamical Traps

System (38.23 through 38.25) possesses the only equilibrium point at the origin. Linear stability analysis reveals that this equilibrium is stable for all the values of the system parameters σ, τ, and Δ_a such that

$$\frac{\tau}{\sigma} < \Delta_a.$$
(38.26)

If the effect of the action dynamical trap is absent, $\Delta_a = 1$, the system is stable for $\tau < \sigma$, i.e., when the operator reaction time τ is relatively small and (or) the capability of suppressing the velocity deviations σ is relatively high.

When the action dynamical trap effect comes into play, $\Delta_a \ll 1$, system (38.23 through 38.25) is stable only if $\tau \ll \sigma$. This may be interpreted in such sense that the operator cannot precisely maintain the desired state of the system, unless the operator's reaction is almost immediate ($\tau \ll 1$) or the velocity feedback gain σ is extremely large. Moreover, when Δ_a reaches zero, the system governed by Equations 38.23 through 38.25 becomes unstable at the origin regardless the values of the other parameters. It is notable that the system stability does not depend on the parameters Δ_v and v_{th} quantifying the intensity of the velocity dynamical trap and the velocity perception threshold, respectively.

In this section, we focus on the operator affected by both velocity and acceleration dynamical traps: $\Delta_v = \Delta_a = 0$. We also comment briefly on the case when the operator is perfectly rational in controlling either velocity ($\Delta_v = 1$) or acceleration ($\Delta_a = 1$) deviations. The intermediate values of the parameters $\Delta_{a,v}$ far from the boundary values have in fact little physical meaning, corresponding to the hypothetical case when the operator stays focused on controlling the small deviations, but at the same time applies reduced effort in doing so. For this reason, although the below results hold for any $\Delta_{a,v}$, we refrain from the detailed analysis of the system dynamics in case of $0 < \Delta_{a,v} < 1$.

We analyzed the behavior of system (38.23 through 38.25) numerically under the adopted assumptions for various values of system parameters. The absolute and relative error tolerance parameters of the routine used for the numerical simulations were chosen in a way that varying them 10-fold could not affect the results of the simulations. The initial conditions for simulations were formed by assigning small random values to the phase variables.

We observed two major patterns of the system dynamics depending on the parameters σ, τ, and v_{th}. The system either performs periodic oscillations or becomes uncontrollable by the operator, with all phase variables exhibiting unbounded growth. Generally, the periodic behavior can be observed when the operator response latency τ is in some sense small and (or) the feedback gain σ is relatively large. The form of the found limit cycle almost does not depend on the particular values of these parameters. Figure 38.11 represents the example of the limit cycle found for $\sigma = 1$, $\tau = 0.9$, $v_{th} = 0.2$. The fragment of the acceleration time pattern $a(t)$ corresponding to this phase portrait is depicted in the top frame of Figure 38.11, as well as is the evolution of the optimal action strategy $a_{opt}(t)$. As clearly seen, the implemented action strategy remains in the vicinity of the optimal one. When the difference between these two strategies becomes sufficiently small, the acceleration growth ratio is also small. It reflects the fact that under this condition the

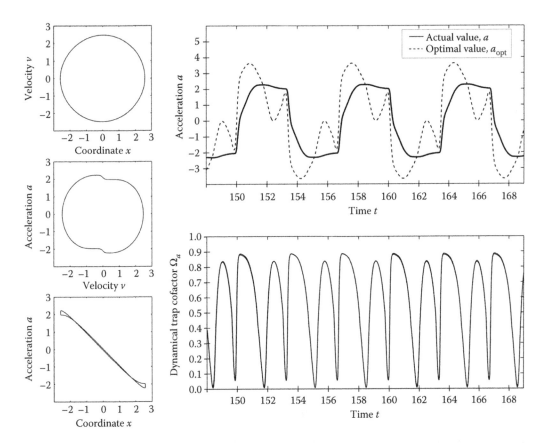

FIGURE 38.11 The projections of the limit cycle formed by the phase trajectory of system (38.23 through 38.25) (left fragment) and the corresponding time patterns of the actual and optimal acceleration (right fragment, top frame) as well as dynamical trap cofactor (right fragment, bottom frame). The values of parameters used for simulation are $\sigma = 1$, $\tau = 0.9$, $v_{th} = 0.2$, $\Delta_a = \Delta_v = 0$. (Based on Zgonnikov, A. and I. Lubashevsky, *Progr. Theor. Exp. Phys.*, 2014, 033J02, 2014.)

operator almost does not change the control variable, a, for a certain period of time. However, when the deviation from the optimal action strategy becomes large, the operator behavior turns to be active and the actual acceleration changes fast. This is also reflected in the bottom frame of Figure 38.11, where the time pattern of the dynamical trap cofactor Ω_a is represented. The values of Ω_a near unity correspond to the periods of the acceleration active growth or decrease, while the stagnation of a is characterized by values of Ω_a close to zero. When $a - a_{opt}(t)$ becomes large, Ω_a "switches on," and the operator starts to actively control the system. Occasionally, only little effort is needed to adjust the current control strategy to the optimal one (see, for instance, $t \approx 156$ in Figure 38.11); however, sometimes the operator has to correct the actions substantially (e.g., $t \in [160,161]$).

As τ grows and (or) σ decreases, the amplitude of the oscillations increases and eventually the periodic pattern evolves to the uncontrolled motion. One can note that for any fixed value of τ the oscillation magnitude monotonically decays with σ. Indeed, the better the operator can handle the velocity deviations, the closer the system is to the desired state. Similarly, the limit cycle shrinks as τ decreases: as the operator reaction becomes faster, it becomes easier to keep the system near the desired position. On the contrary, the operators with large τ and small σ are not capable of controlling the system. Such operators destabilize the system by unintelligent actions, so the phase variables reach infinite values. There is a boundary in the parameter space (τ, σ) at which the transition from the periodic behavior pattern to the unbounded

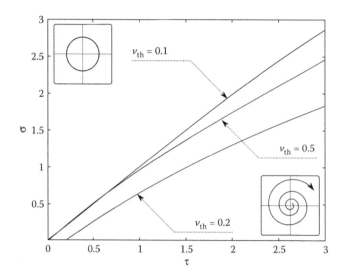

FIGURE 38.12 Phase diagram of the system (38.23 through 38.25). The curves separating the areas of periodic and unbounded motion were reconstructed numerically for three various values of v_{th}. The system motion has been identified as unbounded since the oscillations amplitude had been consistently increasing over a large period of time (10,000 units). (After Zgonnikov, A. and I. Lubashevsky, *Progr. Theor. Exp. Phys.*, 2014, 033J02, 2014.)

growth occurs. We found that this boundary depends essentially on the parameter v_{th}. Figure 38.12 illustrates the numerically obtained boundaries for $v_{th} = \{0.1, 0.5, 2\}$. One can see that increasing values of the velocity tolerance v_{th} paradoxically lead to a larger area of the parameter space corresponding to the periodic motion around the origin. This may be explained in the following way. The operator characterized by the relatively large response delay τ may destabilize the system by the delayed and therefore improper actions when trying to act continuously, i.e., to compensate for the tiniest deviations (v_{th} close to zero). However, the operator neglecting small velocity deviations (v_{th} of order unity) can handle maintaining the bounded motion of the system, even though by cost of increased motion amplitude.

38.3 Emergent Phenomena in Ensembles of Elements with Dynamical Traps

The goal of this section is to demonstrate that dynamical traps caused by the bounded capacity of human cognition can be responsible for emergent phenomena in multielement systems. We will consider three examples; the first two are a system of particles moving along a straight line and a chain of beads simulating string oscillations. The first one mimics motion of some transport flow along a certain path. The second one focuses on general features of oscillations affected by dynamical traps on their own. In both of them, the dynamical traps are interpreted in the sense shown in Section 38.2.1. The third system is used to analyze possible effects caused by the human fuzzy rationality in following some optimal strategy of behavior in the frameworks presented in Section 38.2.2.

38.3.1 Ensemble of "Lazy" Particle Moving on a Straight Path

Keeping in mind the aforementioned discussion about human behavior, we consider a one-dimensional ensemble of "lazy" particles. These particles are characterized by their positions and velocities $\{x_i, v_i\}$ as well as possessing some motives for active behavior. Particle i "wishes" to get the "optimal" middle position between the nearest neighbors. So one of the stimuli for it to accelerate or decelerate is the difference

$\eta_i = x_i - \frac{1}{2}(x_{i-1} + x_{i+1})$ provided its relative velocity $\vartheta_i = v_i - \frac{1}{2}(v_{i-1} + v_{i+1})$ with respect to the pair of the nearest neighbors is sufficiently low. Otherwise, especially if particle i is currently located near the optimal position, it has to eliminate the relative velocity ϑ_i, being the other stimulus for particle i to change its state of motion. Since a particle cannot predict the dynamics of its neighbors, it has to regard them as moving uniformly with the current velocities. So both the stimuli determine directly its acceleration dv_i/dt. The model to be formulated in the next section combines both of these stimuli within a linear approximation similar to $(\eta_i + \sigma\vartheta_i)$, where σ is the relative weight of the second stimulus.

When, however, the relative velocity ϑ_i attains sufficiently low values, the current situation for particle i cannot become worse, at least, rather fast. So, in this case, particle i "prefers" not to change the state of motion and to retard the correction of its relative position. This assumption leads to the appearance of some common cofactor $\Omega(\vartheta_i)$ in the governing equation like this

$$\frac{dv_i}{dt} \propto -\Omega(\vartheta_i)(\eta_i + \sigma\vartheta_i).$$

The cofactor $\Omega(\vartheta)$ has to meet the inequality $\Omega(\vartheta) \ll 1$ for $\vartheta \ll \vartheta_c$ and $\Omega(\vartheta) \approx 1$ when $\vartheta \gg \vartheta_c$, where ϑ_c is a certain critical value quantifying the particle "perception" of speed. The inclusion of such a factor is the implementation of the dynamical trap effect. Now let us specify the model.

38.3.1.1 Model

The following linear chain of N point-like particles is considered (Figure 38.13). Each internal particle $i \neq 1, N$ can freely move along the x-axis interacting with the nearest neighbors, namely, particles $i-1$ and $i+1$ via ideal elastic springs with some quasi-viscous friction. The dynamics of this particle ensemble is governed by the collection of coupled equations

$$\frac{dx_i}{dt} = v_i, \tag{38.27}$$

$$\frac{dv_i}{dt} = -\Omega(\vartheta_i, h_i)[\eta_i + \sigma\vartheta_i + \sigma_0 v_i] + \epsilon\xi_i(t). \tag{38.28}$$

Here, for $i = 2, 3, \ldots, N-1$, the variables η_i and ϑ_i to be called the symmetry distortion and the distortion rate, respectively, are specified as

$$\eta_i = x_i - \frac{1}{2}(x_{i-1} + x_{i+1}), \tag{38.29}$$

$$\vartheta_i = v_i - \frac{1}{2}(v_{i-1} + v_{i+1}), \tag{38.30}$$

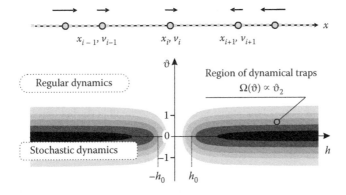

FIGURE 38.13 The particle ensemble under consideration and the structure of the phase space. The darkened region depicts the points where the dynamical trap effect is pronounced. For the relationship between the variables x_i, v_i, h_i, and ϑ_i, see formulae (38.30) and (38.31).

the mean distance h_i between the particles at the point x_i, by definition, is

$$h_i = \frac{1}{2}(x_{i+1} - x_{i-1}), \tag{38.31}$$

and $\{\xi_i(t)\}$ is the collection of mutually independent white noise sources of unit amplitude, i.e.,

$$\langle \xi_i(t) \rangle = 0, \quad \langle \xi_i(t)\xi_{i'}(t') \rangle = \delta_{ii'}\delta(t - t'). \tag{38.32}$$

In addition, the parameter ϵ is the noise amplitude, σ is the viscous friction coefficient of the springs, and σ_0 is a small parameter that can be treated as a certain viscous friction related to the particle motion with respect to the given physical frame. It is introduced to prevent the system motion as a whole reaching infinitely high velocity. The symbol $\langle \cdots \rangle$ denotes averaging over all the noise realizations, and $\delta_{ii'}$ and $\delta(t - t')$ are the Kronecker symbol and the Dirac δ-function. The factor $\Omega(\vartheta_i, h_i)$ is due to the effect of dynamical traps, and following the previous sections the *ansatz*

$$\Omega(\vartheta, h) = \frac{\vartheta^2 + \Delta^2(h)}{\vartheta^2 + 1} \tag{38.33}$$

with a function $\Delta(h)$ such that

$$\Delta^2(h) = \Delta^2 + \left(1 - \Delta^2\right)\frac{h_0^2}{h^2 + h_0^2} \tag{38.34}$$

is used. The parameter $\Delta \in [0.1]$ quantifies the dynamical trap influence, and the spatial scale h_0 specifies the small distances within which the trap effect is to be depressed, i.e., for $h \ll h_0$ the value $\Delta(h) \approx 1$, whereas when $h \gg h_0/\Delta$, the value $\Delta(h) \approx \Delta$. If the parameter $\Delta = 1$, the dynamical traps do not exist at all, in the opposite case, $\Delta \ll 1$, their influence is pronounced inside a certain neighborhood of the h-axis (trap region) whose thickness is about unity (Figure 38.13). The temporal and spatial scales have been chosen so that the thickness of the trap region is about unity as well as the oscillation circular frequency is equal to unity outside the trap region. The terminal particles, $i = 1$ and $i = N$, are assumed to be fixed, i.e.,

$$x_1(t) = 0, \quad x_N(t) = (N - 1)l, \tag{38.35}$$

where l is the particle spacing in the homogeneous chain. The particles are treated as mutually impermeable ones. So when the coordinate x_i and x_{i+1} of an internal particle pair become identical, an absolutely elastic collision is assumed to happen, i.e., if $x_i(t) = x_{i+i}(t)$ at a certain time t, then the timeless velocity exchange

$$v_i(t + 0) = v_{i+1}(t - 0), \quad v_{i+1}(t + 0) = v_i(t - 0) \tag{38.36}$$

comes into being. Multiparticle collisions are ignored.

The system of Equations 38.27 through 38.36 forms the model under consideration.

The stationary point $x_i^{\text{st}} = (i - 1)l$ is stable with respect to small perturbations. It stems from the linear stability analysis with respect to perturbations of the form

$$\delta x_i(t) \propto \exp\{\gamma t + \mathbf{i}kl(i - 1)\}, \tag{38.37}$$

where γ is the instability increment, k is the wave number, and the symbol $\mathbf{i}$ denotes the imaginary unit. The boundary conditions (38.35) are fulfilled by assuming the wave number k to take the values $k_m = \pi m/[(N - 1)l]$ for $m = \pm 1, \pm 2, \ldots, \pm(N - 2)$. For large values of the particle number N, the parameter k can be treated as a continuous variable. Using the standard technique, the system of

Equations 38.27 and 38.28, for perturbation (38.37) leads us to the following relation of the instability increment $\gamma(k)$ and the wave number k:

$$\gamma = -\Omega_0 \left[\frac{1}{2}\sigma_0 + \sigma \sin^2 \left(\frac{kl}{2} \right) \right] + \mathbf{i} \sqrt{2\Omega_0 \sin^2 \left(\frac{kl}{2} \right) - \Omega_0^2 \left[\frac{1}{2}\sigma_0 + \sigma \sin^2 \left(\frac{kl}{2} \right) \right]^2}. \qquad (38.38)$$

In deriving expression (38.38), *ansatz* (38.33) has been used, enabling us to set $\Omega_0 = \Omega(0, l) = \Delta^2(l)$. Whence it follows that $\mathrm{Re}\,\gamma(k) > 0$ for $k > 0$, so the homogeneous state of the chain is stable with respect to infinitely small perturbations of the particle arrangement.

38.3.1.2 Nonlinear Dynamics

The nonlinear dynamics of the given system has been analyzed numerically. The integration of the stochastic differential equations (38.27), (38.28) was performed using the E2 high-order stochastic Runge–Kutta method (Burrage and Burrage 1998; Burrage 1999). Particle collisions were implemented analyzing a linear approximation of the system dynamics within *one* elementary step of the numerical procedure and finding the time at which a collision has happened. Then, this step treated as a complex one was repeated. The integration time step of 0.02 was used, and the obtained results were checked to be stable with respect to decreasing the integration time step. The ensemble of 1000 particles was studied in order to make the statistics sufficient and to avoid a strong effect of the boundary conditions. The integration time T was chosen from 5000 to 8000 time units in order to make calculated distributions stable. At the initial stage, all the particles were distributed uniformly in space, whereas their velocities were randomly and uniformly distributed within the unit interval.

The results of numerical simulation were used to evaluate the following partial distributions:

$$\mathcal{P}(z) = \frac{1}{(N - 2M)(T - T_0)} \sum_{i=M}^{N-M} \int_{T_0}^{T} dt\, \delta(z - z_i(t)), \qquad (38.39)$$

where the time dependence $z_i(t)$ describes the dynamics of one of the variables $\eta_i(t)$, $\vartheta_i(t)$, and $v_i(t)$ ascribed to particle i, and z is a given point of the space $\mathbb{R}_z$ describing the symmetry distortion η, the distortion rate ϑ, and the particle velocity v, respectively. The variables $\{\eta, \vartheta, v\}$ enable one to represent the system dynamics portrait within the space $\mathbb{R}_\eta \times \mathbb{R}_\vartheta \times \mathbb{R}_v$ or its subspace, N is the total number of particles in the ensemble, and M is the number of particles located near each of its boundaries. They are excluded from the consideration in order to weaken a possible effect of the specific boundary conditions. The same is true for the lower boundary of time integration T_0; its value is chosen to eliminate the effect of the specific initial conditions.

The numerical implementation of the integration over time in expression (38.39) was related to the direct summation of the obtained time series, and the partition of the corresponding space $\mathbb{R}_z$ was chosen so that the results be practically independent of the cell size. The value of M was also chosen using the result stability with respect to the double increase in M. Typically, the value $M \sim 50$ was chosen for $N = 1000$; for $N = 3$, naturally, $M = 1$, and $T_0 \sim 500\text{–}1000$.

38.3.1.3 Three-Particle Ensemble

The given oscillator chain made of three particles is actually the system studied in Section 38.2.1. In this case, only the middle particle is movable and the variables $\eta := \eta_2$ and $\vartheta := \vartheta_2$ are its coordinate and velocity. Here, we also present the results for the three-particle ensemble in order to have a feasibility of distinguishing characteristics of local nature from many particle effects.

Figure 38.14 compares the distribution functions $\mathcal{P}(\eta)$ and $\mathcal{P}(\vartheta)$ obtained in the cases where the dynamical trap effect is absent ($\Delta = 1$) and when the dynamical traps affect the particle motion substantially ($\Delta \ll 1$). The upper windows correspond to the system with weak dissipation, $\sigma + \sigma_0 = 0.1$, whereas the lower ones are related to the case of strong dissipation, $\sigma + \sigma_0 = 1.0$.

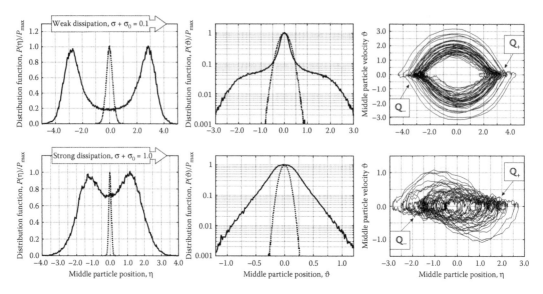

FIGURE 38.14 The distribution functions of the coordinate η and the velocity ϑ of the movable particle in the three-particle ensemble. These distributions were obtained by averaging the simulation results over a time interval of 100,000. Solid lines correspond to the case of strong trap effect, $\triangle = 0.1$; dotted line match the absence of dynamical traps, $\triangle = 1.0$. The right windows depict some path fragment formed by the movable particle during a time interval of 1000 time units. Other parameters used are $\epsilon = 0.1$ and two values of the dissipation rate $\sigma + \sigma_0 = 0.1$ (upper row) and $\sigma + \sigma_0 = 1.0$ (lower row). (After Lubashevsky, I. et al., *Eur. Phys. J. B-Condensed Matter Complex Syst.*, 44, 63–70, 2005.)

In agreement with the previous results, it is seen that the decrease of the parameter $\triangle$, i.e., the dynamical trap intensification induces the conversion of the function $\mathcal{P}(\eta)$ from the unimodal form to the bimodal one, with the dissipation no more than weakening this effect. A new result is the essential dependence of the velocity distribution on the dissipation rate. In the case of weak dissipation, the movable particle performs alternately fast motions outside the trap region and slow motion inside it. The fast motion paths connect the neighborhoods $\mathbb{Q}_-$, $\mathbb{Q}_+$ of the $\mathcal{P}(\eta)$-function maxima, whereas the slow motion arises when the particle wanders inside these regions. This feature is visualized in Figure 38.14: the right upper window shows a fragment of the particle path of duration about 1000 time units. Therefore, the obtained distribution function $\mathcal{P}(\vartheta)$ as seen in Figure 38.14 (middle upper window) is actually made of two monoscale components. For the case of strong dissipation, the two neighborhoods $\mathbb{Q}_-$ and $\mathbb{Q}_+$ are not directly connected by the fast motion paths (Figure 38.14, right lower window). Now, they rather uniformly spread over a certain domain on the $\{\eta, \vartheta\}$-plane; previously, they were located inside a sufficiently narrow layer. As a result, the velocity distribution converts into a monoscale function having a quasi-cusp form $\propto \exp\{-|\vartheta|\}$. We relate the cusp formation to the properties of the system dynamics near the trap region. If the dynamical trap effect is absent, $\triangle = 1$, all these distributions, as must be the case, are of the Gaussian form shown in Figure 38.14 with dotted lines.

38.3.1.4 Multi Particle Ensemble

To analyze cooperative phenomena arising in such systems, the dynamics of 1000–particle ensembles was implemented. Let us, first, consider local properties exhibited by these ensembles. The term "local" means that the corresponding state variable can take practically independent values when the particle index i changes by one or two. The variable η_i (expression (38.29)) may be regarded in such a manner. It describes the symmetry of particle arrangement in space, when $\eta_i = 0$ particle i takes the middle position between the nearest neighbors, particles $i - 1$ and $i + 1$. A nonzero value of η_i denotes its deviation from this position, in other words, a local distortion of the ensemble symmetry. The latter was the reason for the used name of the variables η_i as well as the variables $\vartheta_i = d\eta_i/dt$.

Figure 38.15 exhibits the distribution of the variables η and ϑ depending on the dissipation rate σ and the initial distance l between particles, i.e., their mean density. Comparison of Figure 38.14 and Figure 38.15 shows us that in this case of weak dissipation the distribution functions of the symmetry distortion $\mathcal{P}(\eta)$ and the distortion rate $\mathcal{P}(\vartheta)$ are qualitatively similar to those of the corresponding three-particle ensemble. Only a few new features appear. First, for the system with high particle density ($l = 5$), a small spike is visible at the center, $\eta = 0$, of the distribution function $\mathcal{P}(\eta)$, which is pronounced in the case of strong dissipation. It corresponds to the symmetrical state of the particle ensemble being stable without dynamical traps and is destroyed for the three-particle ensemble. In the given case, "many-particle" effects seem to reconstruct it in part. So in the given case, the particle arrangement is characterized by three states, two of them match the extrema of the distribution function $\mathcal{P}(\eta)$ and the symmetrical state singled out to some degree.

As for the three-particle ensemble, the distortion rate distribution is again composed of two monoscale components, narrow and wide ones. Previously, we have related them to the fast and slow motions. Figure 38.15 (upper second window) also justifies this. The narrow component is due to the particle motion inside the trap region and should be practically independent of the mean distance between particles. By contrast, the wide one depends remarkably on the particle density because it matches the fast motion of particles outside the trap region and, thus, has to be affected by their relative dynamics. Exactly, this effect is demonstrated in Figure 38.15 visualizing also the corresponding properties of the particle paths.

For the 1000-particle ensemble with strong dissipation, $\sigma \approx 1.0$, the situation changes dramatically, although the characteristic scales of the corresponding distributions turn out to be of the same order in magnitude. In the given case, the distribution function $\mathcal{P}(\eta)$ of the symmetry distortion has only one maximum at $\eta = 0$; however, its form is characterized by two scales. In other words, it looks like a sum of two monoscale components. One of them is sufficiently wide; its thickness is about the same value that is obtained for the corresponding particle ensemble with weak dissipation. This component exhibits a remarkable dependence on the particle density, enabling us to relate it to the particle motion outside the trap region. The other is characterized by an extremely narrow and sharp form shown in detail in the inner window in Figure 38.15 for the dense particle ensemble. Its sharpness leads us to the assumption that "many-particle" effects in such systems with dynamical traps cause the symmetrical state to be singled out from the other possible states in the system properties.

By contrast, the distortion rate behaves similarly to the previous case except for some details. When the mean particle density is high ($l = 5$), the wide component of the distortion rate distribution disappears and only the narrow one remains, with the latter having a quasi-cusp form $\propto \exp\{-|\vartheta|\}$. For the system with low density, the peak of the distortion rate distribution splits into two small spikes.

These features can be explained by referring to the row right windows in Figure 38.15, which exhibit typical path fragments formed by motion of a single particle on the $\{\eta\vartheta\}$-plane. Roughly speaking, now three motion types can be singled out: some stagnation inside a narrow neighborhood of the origin $\{\eta = 0, \vartheta = 0\}$ (clearly visible in the right window), slow wandering inside the trap region that, on average, follows a line with a finite positive slope (visible in the left window), and the fast motion outside the trap region (visible again in the left window). The fast motion fragments typically stem from an arbitrary point of the low motion region and lead to a certain neighborhood of the origin. It seems that the systems with low-density particles have the possibility to go sufficiently far from the origin that during the fast motion come into the stagnation region rarely. As a result, first, the distortion rate distribution function is of a two scale form and contains two spikes on the peak. In the case of high density, the fast motion is depressed substantially and the system migrates mainly in the slow motion region entering the stagnation region many times. So the distortion rate distribution converts into a single-scale function and the symmetric state occurs often, giving rise to a significant sharp component of the distortion distribution located near the point $\eta = 0$.

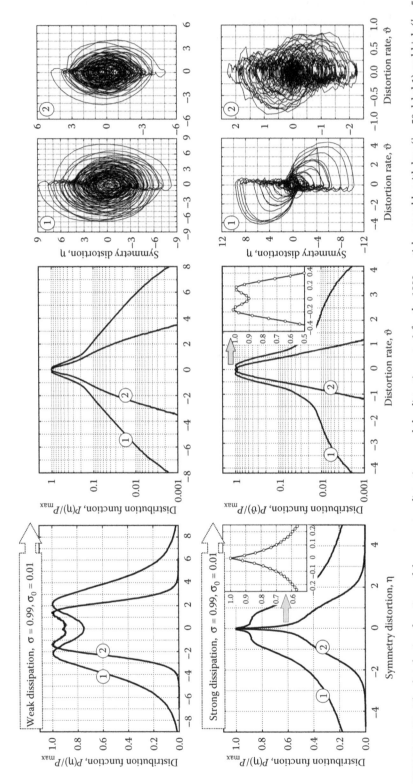

FIGURE 38.15 The distribution functions of the symmetry distortion η and the distortion rate ϑ for the 1000-particle ensemble with low ($l = 50$, label 1) and high ($l = 5$, label 2) density and weak ($\sigma \approx 0.1$) and strong ($\sigma \approx 1.0$) dissipation. The right four windows depict characteristic path fragments of duration of 1000 time units formed by a single particle with index $i = 500$ on the phase plane (η, ϑ), which was chosen due to its middle in the given ensemble. The other parameters used are the noise amplitude $\epsilon = 0.1$, the trap effect measure $\Delta = 0.1$, the small regularization friction coefficient $\sigma_0 = 0.01$, and the regularization spatial scale $h_0 = 0.25$. The time interval within which the data were averaged changed from 2000 to 5000 in order to make the obtained distributions stable. (After Lubashevsky, I. et al., *Eur. Phys. J. B-Condensed Matter Complex Syst.*, 44, 63–70, 2005.)

Now, we discuss the nonlocal characteristics of the 100-particle ensembles. Figure 38.16 depicts the velocity distributions. As is seen, it depends essentially on both the parameters, the mean particle density and the dissipation rate. When the mean particle density is low and the dissipation is weak ($l = 50$ and $\sigma \approx 0.1$), the velocity distribution is practically of Gaussian form; however, its width reaches extremely large values about 10. We recall that without dynamical traps the width of the corresponding distribution does not exceed 0.5 (Figure 38.14). The 10-fold increase of the particle density, $l : 50 \mapsto 5$, shrinks the velocity distribution to the same order and its scale reaches values similar to that of the distortion rate distribution in magnitude. However, in this case, the form of the velocity distribution is a monoscale function of the well-pronounced cusp form $\propto \exp\{-|v|\}$. In the case of strong dissipation ($\sigma \approx 1.0$), the situation is opposite. The system with low density ($l = 50$), as previously, is characterized by an extremely wide velocity distribution, its width is about 10. However, now its form deviates substantially from the Gaussian one. For the corresponding ensemble with high density ($l = 5$), the velocity distribution is Gaussian with width about 1. The latter, nevertheless, is much larger than the same width in the absence of dynamical traps.

These features of the velocity distribution characterizes the cooperative behavior of particles rather than their individual dynamics. In other words, there should be strong correlations in the motion of not only neighboring particles but also distant ones. Therefore, the velocity variations responsible for the formation of such distributions describe in fact the motion of multiparticle clusters. To justify this, we refer to the right column windows in Figure 38.16. They demonstrate some typical fragments of the time patterns formed by the velocities of individual particles. When the mean particle density is low ($l = 50$), these patterns

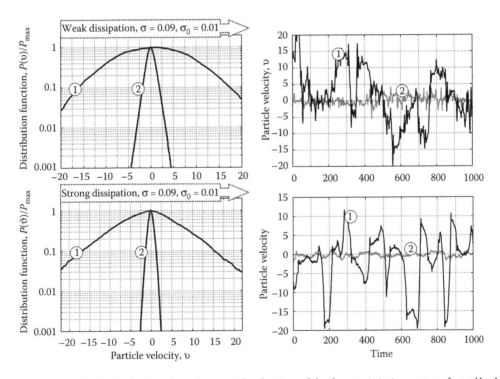

FIGURE 38.16 The distribution functions of the particle velocities and the characteristic time patterns formed by the velocity variations of the 500th particle. Dynamics of the 1000-particle ensemble with low ($l = 50$, label 1) and high ($l = 5$, label 2) mean density and weak ($\sigma \approx 0.1$) and strong ($\sigma \approx 1.0$) dissipation was implemented for the calculation time up to 8000 time units to make the obtained distributions stable with respect to time increase. The other parameters used are the noise amplitude $\epsilon = 0.1$, the trap effect measure $\triangle = 0.1$, the small regularization friction coefficient $\sigma_0 = 0.01$, and the regularization spatial scale $h_0 = 0.25$. (Based on Lubashevsky, I. et al., *Eur. Phys. J. B-Condensed Matter Complex Syst.*, 44, 63–70, 2005.)

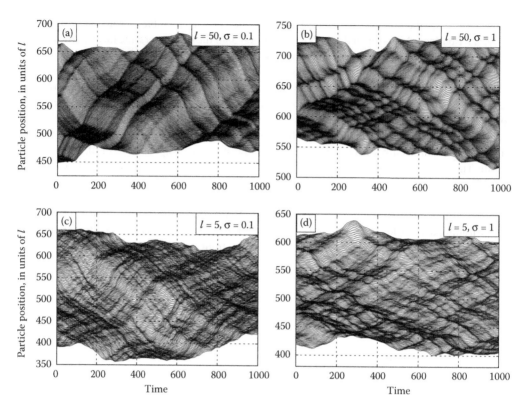

FIGURE 38.17 The time patterns formed by 200 paths of particle motion during 1000 time units and chosen in the middle of the given ensemble. Here, the curve thickness has been chosen so that the apparent color can depict local variations in the path spacing due to changes either in the particle density or in the velocities of cooperative particle motion (in this way the different long-lived states of the given particle ensemble become apparent). The parameters used are the noise amplitude $\epsilon = 0.1$, the trap effect measure $\triangle = 0.1$, the small regularization friction coefficient $\sigma_0 = 0.01$, and the regularization spatial scale $h_0 = 0.25$, the other system parameters are shown in the frames. (Based on Lubashevsky, I. et al., *Eur. Phys. J. B-Condensed Matter Complex Syst.*, 44, 63–70, 2005; Lavrenov, A.Y. et al., *Bull. Lebedev Phyisics Institute*, 2, 1–8, 2005.)

look like a sequence of fragments $\{v_\alpha\}$ inside which the particle velocity varies in the vicinity of some level v_α. The values $\{v_\alpha\}$ are rather randomly distributed inside a certain region of thickness $V \sim 10$ in the vicinity of $v = 0$. The continuous transitions between these fragments occur via sharp jumps. The typical duration of these fragments is about $T \sim 100$, which enables us to regard them as long-lived states because the temporal scales of individual particle dynamics are about several units. Moreover, these long-lived states can persist only if a group of many particles moves as a whole because the characteristic distance L individually traveled by a particle involved in such state is about $L \sim VT \sim 1000 \gg l$.

The spatial structure of these cooperative states is visualized in Figures 38.17 and 38.18. Figure 38.17 depicts time patterns formed by paths $\{x_i(t)\}$ of 200 particles of duration about 1000 time units. These particles were chosen in the middle part of the 1000-particle ensembles with low density. For high-density ensembles, such patterns also develop but are not so pronounced. As is seen, a large number of different mesoscopic states formed in these systems. They differ from one another in size, the direction of motion, the speed, the life time, etc. Moreover, the life time of such a state can be much longer than the characteristic time interval during which particles forming it currently will belong to this state individually. Besides, the patterns found could be classified as hierarchical structures. Some relatively small domains formed by cooperative motion of individual particles in their turn make up together larger superstructures. In other

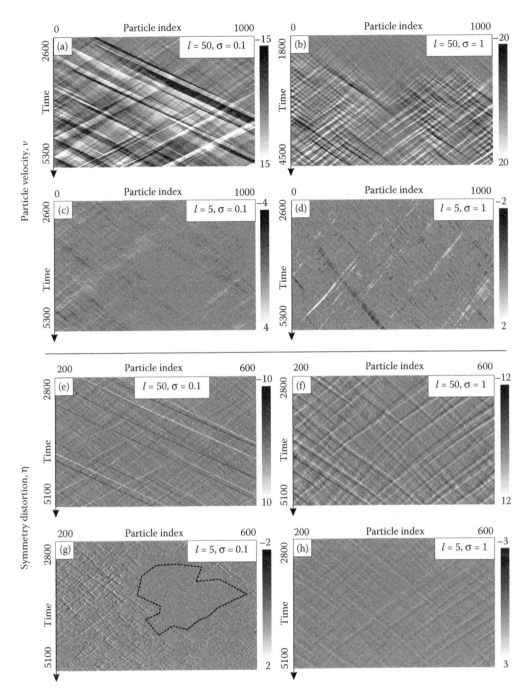

FIGURE 38.18 The time patterns formed by particle motion during 1000 time units: (a–d) the particle velocity patterns, (e–h) the symmetry distortion patterns. The parameters used are the noise amplitude $\epsilon = 0.1$, the trap effect measure $\triangle = 0.1$, the small regularization friction coefficient $\sigma_0 = 0.01$, and the regularization spatial scale $h_0 = 0.25$, the other system characteristics are shown in the frames. The plot is based on results by Lavrenov et al. (2005).

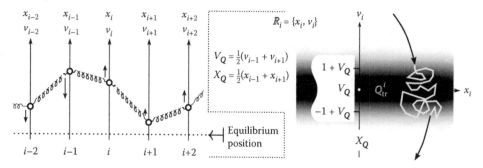

FIGURE 38.19 The chain of N beads under consideration and the structure of their individual phase space $\mathbb{R}_i = \{x_i, v_i\}$ ($i = 1, 2, \ldots, N$). The formal initial $i = 0$ and terminal $i = N + 1$ beads are assumed to be fixed, specifying the equilibrium bead position. The springs are drawn with broken lines to underline that the bead interaction is only mimicked by elastic springs to some degree rather the system under consideration is of some kind of bead-spring models.

words, the observed long-lived cooperative states have their "own" life independent, in some sense, of the individual particle dynamics. The latter properties are the reason for regarding them as certain dynamical phases arising in the systems under consideration due to the dynamical traps affecting the individual particle motion. The term "dynamical" has been used to underline that the complex cooperative motion of particles is responsible for these long-lived states, without the continuous particle motion such states cannot exist. These statements are also justified in Figure 38.18 showing the time patterns of the particle velocities and the symmetry distortion.

38.3.2 String Composed of "Lazy" Beads

Now, let us consider a chain of N "lazy" beads (Figure 38.19). Each of these beads can move in the vertical direction and its dynamics is described in terms of the deviation $x_i(t)$ from the equilibrium position and the motion velocity $v_i(t) = dx_i/dt$ depending on time t; here, the bead index i runs from 1 to N. The equilibrium position $x_i = 0$ is specified assuming the formal initial ($i = 0$) and terminal ($i = N + 1$) beads to be fixed. Each bead i "wishes" to get the "optimal" middle position with respect to its nearest neighbors. So one of the stimuli for it to accelerate or decelerate is the difference

$$\eta_i = x_i - \frac{1}{2}(x_{i-1} + x_{i+1})$$

provided its relative velocity

$$\vartheta_i = v_i - \frac{1}{2}(v_{i-1} + v_{i+1})$$

with respect to the pair of the nearest beads is sufficiently low. Otherwise, especially if bead i is currently located near the optimal position, it has to eliminate the relative velocity ϑ_i, representing the other stimulus for bead i to change its state of motion. The model to be formulated below combines both of these stimuli within one cumulative impetus $\propto (\eta_i + \sigma\vartheta_i)$, where σ is the relative weight of the second stimulus. Actually, this ansatz coincides with approximation (38.12) provided the system variables are measured in units where the kinetic coefficient $\beta = 1$.

When, however, the relative velocity ϑ_i becomes less than a threshold θ, i.e., $|\vartheta_i| \lesssim \theta$, bead i is not able to recognize its motion with respect to its nearest neighbors. Since a bead cannot "predict" the dynamics of its neighbors, it has to regard them as moving uniformly with the current velocities. So from its point of view, under such conditions, the current situation cannot become worse, at least, rather fast. In this case, bead i just "allows" itself to do nothing, i.e., not to change the state of motion and to retard the correction of its relative position. This feature is the reason why such beads are called "lazy." below, we will use the dimensionless units in which the perception threshold is equal to unity $\theta = 1$ as well as in the later expression for cumulative impetus the required proportionality factor is equal to unity too.

Under these conditions, the equation governing the system dynamics is written in the following form:

$$\frac{dv_i}{dt} = -\Omega(\vartheta_i)[\eta_i + \sigma\vartheta_i + \sigma_0 v_i]. \tag{38.40}$$

If the cofactor $\Omega(\vartheta_i)$ was equal to unity, the given system would be no more than a chain of beads connected by elastic springs characterized by the friction coefficient σ. The term $\sigma_0 v_i$ with the coefficient $\sigma_0 \ll 1$ that can be treated as a certain viscous friction of the bead motion with respect to the given physical frame has been introduced to prevent the system motion as a whole reaching extremely high velocities. The factor $\Omega(\vartheta_i)$ is due to the effect of dynamical traps and following the general ansatz (38.13), we write

$$\Omega(\vartheta) = \frac{\Delta + \vartheta^2}{1 + \vartheta^2}, \tag{38.41}$$

where, as before, the parameter $\Delta \in [0, 1]$ quantifies the intensity of dynamical traps. If $\Delta = 1$, the dynamical traps do not exist at all; in the opposite case, $\Delta \ll 1$, their influence is pronounced inside the neighborhood $\mathbb{Q}_{tr}^i$ of the axis $v_i = (v_{i-1} - v_{i+1})/2$ (the trap region) whose thickness is about unity (Figure 38.19). For the terminal fixed beads, $i = 0$ and $i = N + 1$, we set

$$x_0(t) = 0, \qquad x_{N+1}(t) = 0, \tag{38.42}$$

which play the role of "boundary" conditions for Equation 38.40.

38.3.2.1 Numerical Results

The system of Equation 38.40 was solved numerically using the standard explicit Runge–Kutta algorithm of fourth order with fixed time step. Initially, all the beads were located at the equilibrium positions $\{x_i|_{t=0} = 0\}$, and perturbations were introduced into the system via ascribing random independent values to their velocities. The time step dt of numerical integration was chosen in such a way that its decrease or increase by several times have no considerable effects. The system dynamics was found to depend remarkably on the intensity of "dissipation" quantified by the parameter σ. We remind that the parameter σ specifies the relative weight of the stimuli to take the middle "optimal" position and to eliminate the relative velocity; the larger the parameter σ, the more significant the latter stimulus. So let us discuss the obtained results for the cases of "strong," "intermediate," and "weak" dissipation individually.

It should be noted beforehand that, first, all the results of numerical simulation to be presented below were obtained for the dynamical traps of high intensity, namely, for $\Delta = 10^{-4}$. Emergent phenomena in such systems for different values of the dynamical trap intensity as well as the influence of stochastic factors are worthy of individual analysis. Second, the parameter σ_0 quantifying additional friction introduced to depress extremely high values of the bead velocities $\{v_i\}$ was set equal to $\sigma_0 = 0.01$. Third, in plotting a collection of phase portraits of bead motion, e.g., $\{x_i(t), v_i(t)\}_{i=1}^N$, the bead coordinates $\{x_i\}$ are shown with some individual shifts, namely, $x_i \rightarrow x_i + 50 \cdot i$ to simplify the portrait visualization.

38.3.2.2 "Strong" Dissipation

The system dynamics with "strong" dissipation is exemplified by numerical data obtained using the kinetic coefficient $\sigma = 3$. In this case, the system instability was detected numerically only for the chains with the number of beads $N \geq 3$. Figure 38.20a depicts the found limit cycles of the bead oscillations for the chain with $N = 3$. The corresponding time patterns $\{x_2(t)\}, \{\vartheta_2(t)\}$ of the middle bead $i = 2$ showing time variations of its position and relative velocity are exhibited in Figure 38.20b. As seen, the periodic motion of these beads looks like relaxation oscillations with the "slow" motion fragments matching $\vartheta_i = 0$, i.e., the synchronized motion of neighboring beads. It is worthwhile to note that the given bead periodic motion is

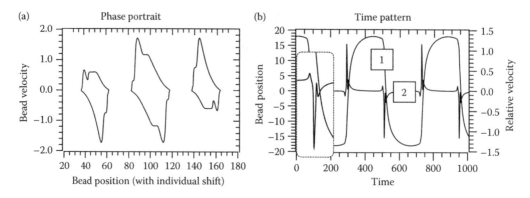

FIGURE 38.20 The phase portraits of the individual bead motion on the phase plane $\mathbb{R}_{xv}$ for the chain of three beads with "strong" dissipation (a) and the corresponding time patterns of the middle bead motion (b), namely, the time variations in the bead position (1) and the relative velocity (2). (After Lubashevsky, I., *Adv. Complex Syst.*, 15, 1250045, 2012.)

not the standard relaxation oscillations related to alternative stepwise transitions between two quasistable states directly specified by system properties. In fact, for the given system, only the states $\{\vartheta_i = 0\}$ are singled out in properties and the time pattern $\{\vartheta_2(t)\}$ exhibits considerable spikewise variations only within the "fast" motion fragments, whereas at the other moments of time it is located near the point $\vartheta = 0$. In addition, it should be pointed out that the state $\{x_i = 0, v_i = 0\}$ is metastable, i.e., stable with respect to small perturbations. Moreover, not all large perturbations in the bead velocities were found numerically to give rise to the limit cycle formation; some of them faded away. However, when the instability was initiated, the steady-state oscillations appeared usually after the time interval $T \gtrsim 10^4$ exceeding the period of these oscillations 10-fold.

For the given bead chains, evolution of the phase portraits as the number of beads N increases is illustrated in Figure 38.21. While their size is not too large, namely, $N \sim 10$, the regular periodic motion of the beads remains stable; however, various patterns of limit cycles can be formed depending on the initial perturbations. Two found examples are shown in Figure 38.21a,b. As seen in Figure 38.21b, a limit cycle can have its own complex structure, which is a property of the system dynamics rather than a numerical artifact; it was verified by decreasing the integration time step by several times. As the number of beads increases, the system dynamics becomes irregular (Figure 38.21c,e); at least, on time scales about $T \sim 10^5$, no periodic bead motion was found for $N \sim 50$. However, the irregularity of individual trajectories seems to grow gradually with the number of beads N. The latter feature is demonstrated in Figure 38.21d,f; the structure of the shown trajectories is visually more regular for the chain of 50 beads in comparison with the chain of 100 beads.

38.3.2.3 "Intermediate" Dissipation

The chains of beads with the kinetic coefficient $\sigma = 1$ are treated as characteristic examples of such systems with "intermediate" dissipation. In this case, the instability was detected in the system of two beads, which is the minimal number of beads when the instability caused by dynamical traps without noise can appear in principle. As noted before, for one oscillator with dynamical traps, noise must be present for the instability to arise.

Following the presentation of the previous subsection, Figure 38.22 depicts the phase portraits of the two-bead chain dynamics Figure 38.22a and the time patterns $\{x_1(t)\}$, $\{\vartheta_1(t)\}$ of the first bead (Figure 38.22b). As previously, the periodic motion of these beads looks like relaxation oscillations with the anomalous behavior discussed above and again not all the perturbations give rise to the instability onset. However, in the case of "intermediate" dissipation, the regular periodic motion of beads finally arises for

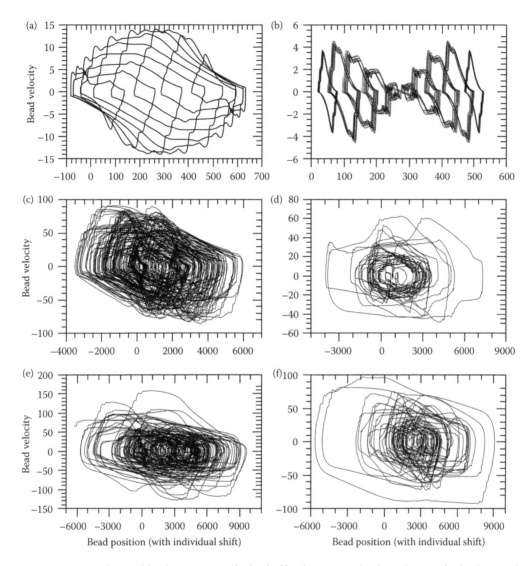

FIGURE 38.21 Evolution of the phase portraits of individual bead motion on the phase plane $\mathbb{R}_{xv}$ for the chains with "strong" dissipation as the number of beads N increases. The frames (a) and (b) depict the data for $N = 10$. The frame (c) exhibits the phase portraits of equidistant 10 beads for the chain of 50 beads and a fragment of the corresponding phase trajectory of a middle bead is shown in the frame (d). The frames (e) and (f) demonstrate actually the same data for the chain of 100 beads. (After Lubashevsky, I., *Adv. Complex Syst.*, 15, 1250045, 2012.)

the chains of many beads and only one type of limit cycle patterns was detected numerically. In particular, Figure 38.23a exhibits the stable limit cycles developed in the chain of 50 beads; here, 11 equidistant beads are shown. The corresponding spatial profile of the bead ensemble treated as a certain "beaded string" as well as its velocity profile are exemplified in Figure 38.23b. Namely, it exhibits the characteristic spatial form of the distribution $\{x_i(t)\}_{i=1}^{N}$ describing the deviation of the "beaded string" from the equilibrium position as well as the distribution of the bead velocities $\{v_i(t)\}_{i=1}^{N}$ taken at a certain moment of time t. We point out that although the found spatial form of the "beaded string" oscillations looks like the fundamental mode of elastic string vibration, the system dynamics has nothing in common with vibrations of elastic stretched strings. It becomes visible explicitly in the form of the velocity distribution $\{v_i(t)\}_{i=1}^{N}$ whose dynamics can

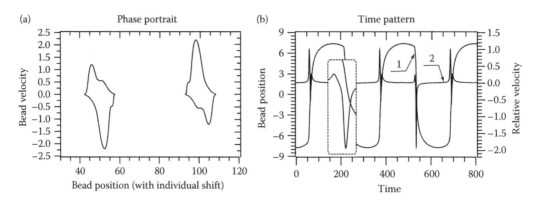

FIGURE 38.22 The phase portraits of the individual bead motion on the phase plane $\mathbb{R}_{xv}$ for the chain of two beads with "intermediate" dissipation (a) and the corresponding time patterns of the motion of the first bead (b), namely, the time variations in the bead position (1) and the relative velocity (2). (After Lubashevsky, I., *Adv. Complex Syst.*, 15, 1250045, 2012.)

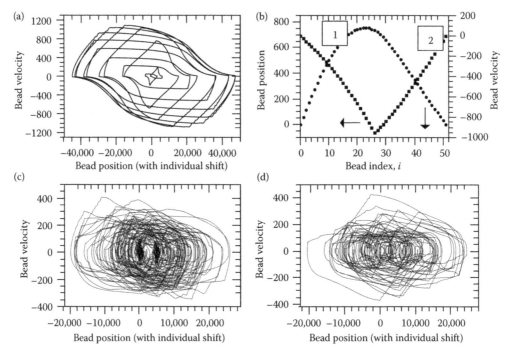

FIGURE 38.23 The phase portraits of the individual bead motion on the phase plane $\mathbb{R}_{xv}$ for the chains of 50 and 100 beads with "intermediate" dissipation, the frames (a) and (c), respectively. Here, 11 equidistant beads are shown. The frame (d) depicts the phase portrait of the middle bead motion for the 100-bead chain. For the regular bead motion, the frame (b) illustrates the characteristic spatial form of the distribution $\{x_i(t)\}_{i=1}^{N}$ (1) as well as the corresponding distribution $\{v_i(t)\}_{i=1}^{N}$ (2) of the bead velocities fixed at a certain time moment. Here, arrows show the current direction of motion of the points $\{x_i(t)\}$ and $\{v_i(t)\}$ on the phase plane $\mathbb{R}_{xv}$. For the irregular motion, the shown fragments are of duration 3×10^3 (c) and 2×10^4 (d), the total simulation time was 10^6. (After Lubashevsky, I., *Adv. Complex Syst.*, 15, 1250045, 2012.)

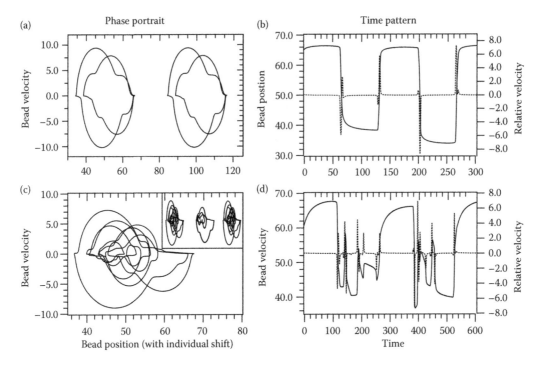

FIGURE 38.24 The phase portraits of the individual bead motion on the phase plane $\mathbb{R}_{xv}$ for the chains of two and three beads with "weak" dissipation (a,c) and the corresponding time patterns of the motion of the first and middle beads (b,d), namely, the time variations in the bead position (solid line) and the relative velocity (dotted line). The inset in the frame (c) depicts the full collection of three phase portraits, whereas its main part exhibits the middle phase portrait in detail. (After Lubashevsky, I., *Adv. Complex Syst.*, 15, 1250045, 2012.)

be represented, at least, qualitatively as the propagation of a certain cusp along the "beaded string." As the number N of beads in the chains increases, the periodic bead motion either becomes unattainable for the majority of initial perturbations or requires extremely long time to arise; at least, for the chain of 100 beads, the system dynamics remained irregular for all the simulations of duration $T \sim 10^6$, which is illustrated in Figure. 38.23c,d.

38.3.2.4 "Weak" Dissipation

In the case of "weak" dissipation, the system dynamics turns out to be more complex in properties, which, by way of example, was analyzed for the bead ensembles with $\sigma = 0.1$. In particular, Figure 38.24 exhibits the phase portraits and the corresponding time patterns for the chains of two and three beads. In both the systems, the periodic bead motion is stable; it occurs each time finite amplitude perturbations of the equilibrium state become unstable. The corresponding phase portraits and the time pattern are illustrated in Figure 38.24a,c and Figure 38.24b,d, respectively. However, already for the chain of three beads the limit cycles can have a rather complex form as does the corresponding time patterns of the velocity variations (Figure 38.24c,d). To make certain that this limit cycle form is a property of the given system, its reproducibility was verified changing the time step in numerical integration or introducing additional small random Langevin forces into Equation 38.40.

As the number N of beads increases, the complexity of the system dynamics does not grow gradually, which is exemplified in Figure 38.25. For the three-bead ensemble, two additional types of limit cycles were fixed (Figure 38.25a,b) whose structure is rather simple in comparison with one shown in Figure 38.24c. For the four-bead chain (Figure 38.25c), only one type of limit cycles was found numerically; it is similar

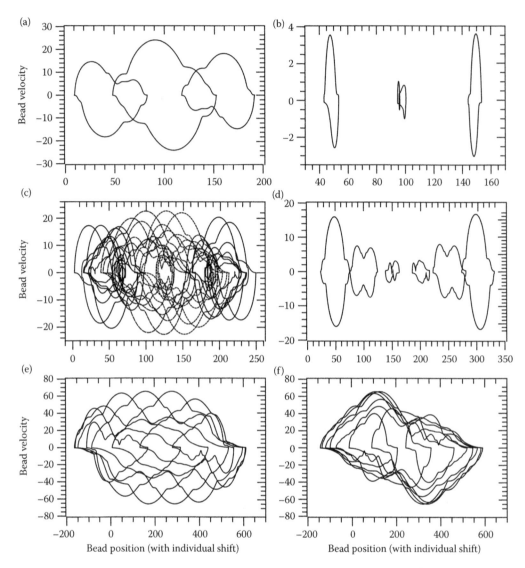

FIGURE 38.25 The phase portraits of the individual bead motion on the phase plane $\mathbb{R}_{xv}$ for the chains of beads with "weak" dissipation. The frames (a,b) depict the limit cycles for the three-bead ensemble found numerically in addition to one shown in Figure 38.24c, the frame (c) exhibits limit cycles for the four-bead chain where the motion trajectories of the second and third beads are plotted out with dashed lines, the plot (d) matches the six-bead chain, and the frames (e,f) show the limit cycle patterns found for the eight bead chain. (After Lubashevsky, I., *Adv. Complex Syst.*, 15, 1250045, 2012.)

to the one shown in Figure 38.24c and matches a rather complex periodic motion of individual beads with a relatively large amplitude. To make it clear, the limit cycles of the second and third beads are plotted out here with dashed lines. The dynamics of five-bead chain is similar in properties. The six-bead chain exhibits the opposite behavior illustrated in Figure 38.25d. The only one stable periodic motion found numerically is of a rather simple geometry and its amplitude is relatively small. The dynamics of seven and eight bead chains is of the same type of complexity, in particular, Figures 38.25e,f plot the found limit cycle patterns for the eight-bead ensemble, which can be treated as derivatives of the pattern shown in Figure 38.25a.

As the number of beads increases, a new feature of the system dynamics was fixed for the 12-bead ensemble. In addition to the periodic oscillations represented by patterns similar to ones plotted out in Figure 38.25e,f, a limit cycle collection actually of the same form as shown in Figure 38.25b,d was found (Figure 38.26a). The given pattern again was verified to be stable with respect to changing the integration time step dt and introducing additional small Langevin forces. Attempts to find a similar periodic motion for the ensembles of 11 or 13 beads were not successful. Moreover, for the 12-bead ensemble, only a few of the generated initial perturbations give rise to it. A more detailed analysis demonstrated the fact that this type of bead motion is actually an intermediate stage of the instability development for the majority of the generated initial perturbations even for the 12-bead chain. For example, Figure 38.26b exemplifies the usual geometry of bead trajectories when, on the one hand, the initially induced uncorrelated motion of beads has faded away and, on the other hand, the periodic stable motion has not arisen yet. As seen, these trajectories together make up some region looking like the pattern in Figure 38.26a scattered by some noise. For many-bead ensembles, only the stable periodic motion of the type shown in Figure 38.25f survives; however, the transient processes of the instability development go through this stage, which is demonstrated in Figures 38.26c,d. Namely, Figure 38.26c shows the motion trajectories of seven equidistant beads within the 30-bead ensemble; the shown fragments of duration of 10^3 match the simulation time $T \sim 7 \times 10^5$. Figure 38.26d exhibits the bead trajectories of the same system after an additional time interval $\Delta T \sim 2 \times 10^4$ when the regular periodic motion of the beads was fixed to start its formation explicitly (in the given figure, this moment is pointed out with gray arrows). For large ensembles of beads, for example, the 80-bead chain, the periodic motion was not fixed, at least, on time scales $t \lesssim 10^6$, maybe, because of a fast growth of the required waiting time as the number of beads increases. It poses a question as to whether noise with an extremely small amplitude can cause a stochastic dynamics of such systems.

As the "dissipation" parameter σ becomes smaller, on the one hand, the motion complexity should be met even in the dynamics of ensembles with a few beads. On the other hand, the finite size of such ensembles has to manifest itself in its properties. It is justified by Figure 38.27 demonstrating the phase portrait of the motion of a middle bead in the four-bead ensemble with $\sigma = 0.03$. Figure 38.27a depicts the bead trajectory as a whole whereas Figure 38.27b exhibits its central part. Changing the integration time step and the integration time T it was justified that the bound pattern is stable and is not an artifact. As seen, this phase portrait does have a complex multiscale structure.

38.3.2.5 Effects of Weak Noise

In the previous section, the emergence of spatial patterns was studied in the systems without noise. Here, we consider effects caused by weak noise, additively introduced into the governing Equation 38.40, namely,

$$\frac{dv_i}{dt} = -\Omega(\vartheta_i)[\eta_i + \sigma\vartheta_i + \sigma_0 v_i] + \epsilon\xi_i(t). \tag{38.43}$$

where, as previously, $\xi_i(t)$ is white noise of unit amplitude and ϵ is the intensity of random Langevin forces.

This system was studied numerically. Initially, all the beads were located at the equilibrium positions $\{x_i|_{t=0} = 0\}$ and perturbations were introduced into the system via ascribing random independent values to their velocities. Equation 38.43 was integrated using the E2 high-order stochastic Runge–Kutta method (Burrage 1999; Burrage and Burrage 1998). The integration time step of 0.001 was used; the obtained results were checked to be stable with respect to decreasing the integration time step 10-fold. The integration time was equal to 10^5–10^6, which enabled us to deal with the steady-state dynamics. The other parameters used in simulation were taken to be equal to $\Delta = 10^{-3}$ and $\sigma_0 = 0.01$. Besides, to simplify the data visualization, the bead coordinates are shown with some individual shifts, namely, $x_i \rightarrow x_i + 50 \cdot i$.

In the previous sections, it has been found that the dynamics of this system without noise depends on the intensity of "dissipation" quantified by the parameter σ. We remind that the parameter σ specifies the relative weight of the stimuli to take the middle "optimal" position and to eliminate the relative velocity; the larger the parameter σ, the more significant the latter stimulus. When the parameter σ is not too small, the system tends to get the regime of regular dynamics represented by a collection of limit

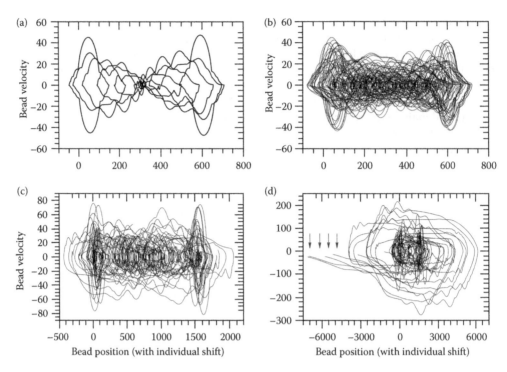

FIGURE 38.26 The phase portraits of the individual bead motion on the phase plane $\mathbb{R}_{xv}$ for the chains of beads with "weak" dissipation. The frame (a) depicts a special type of the limit cycle patterns fixed for the 12-bead chain as a rare event, in contrast, the frame (b) exhibits an intermediate stage of the instability development observed usually for these chains with the number of beads $N \gtrsim 10$ before limit cycle patterns similar to one shown in Figure 38.25d appear. Here, the bead trajectories of duration of 10^3 after the simulation time $T \sim 6 \times 10^5$ are presented. The frames (c,d) exhibit the transient processes for the 30-bead chain when (c) the bead trajectories are located in the vicinity of the limit cycle pattern similar to one shown in the frame (a) and at the moment (d) corresponding to the explicit formation of the stable periodic motion (pointed out by gray arrows). Here, trajectories of duration of 10^3 are shown after the simulation time $T \sim 7 \times 10^5$ and after an additional time interval $\Delta T \sim 2 \times 10^4$. (After Lubashevsky, I., *Adv. Complex Syst.*, 15, 1250045, 2012.)

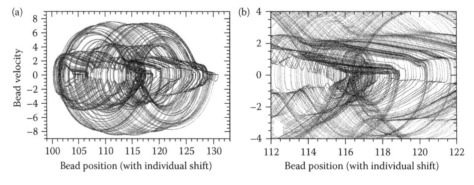

FIGURE 38.27 The phase portrait of the motion of a middle bead on the phase plane $\mathbb{R}_{xv}$ for the four-bead ensemble with $\sigma = 0.03$. The frame (a) shows the whole trajectory, whereas the frame (b) exhibits its central part. In the plot, the bead trajectory is visualized as a sequence of dots separated by the time interval 0.01. The shown trajectory fragment is of duration of 2×10^3 and 4×10^3 for the plots (a) and (b), respectively. In numerical simulation, the integration time step $dt = 5 \times 10^{-3}$ was used and the integration time was $T \sim 5 \times 10^5$. (After Lubashevsky, I., *Adv. Complex Syst.*, 15, 1250045, 2012.)

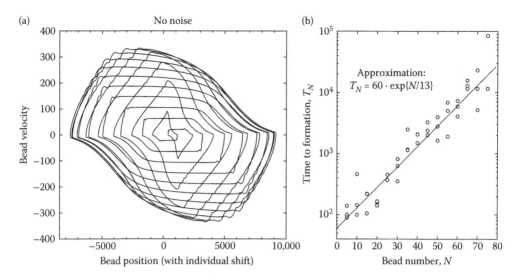

FIGURE 38.28 The characteristic phase portrait of the steady-state dynamics exhibited by systems without noise and not too weak "dissipation" (a). The chain of 30 beads with $\sigma = 1$ was used in constructing the shown pattern where the limit cycles of each second bead are visualized. The frame (b) depicts the characteristic time T_N required for such a system to get the steady-state dynamics vs the number N of beads. The scattered points are the data obtained for each value of N on three trials, $\sigma = 1$ was used in simulation. (After Lubashevsky, I. and D. Parfenov, *CMSIM*, 1, 31–38, 2013.)

cycles of individual bead motion. It should be noted that these limit cycles could be of complex form when the number of beads is not too large, namely, $N \lesssim 10$. Nevertheless, for systems with a large number of beads, the resulting phase portrait takes a rather universal form shown in Figure 38.28a; actually here the plot of Figure 38.23a is reproduced. However, the "time to formation" T_N, i.e., the mean time required for a given bead chain to get the steady-state regular dynamics grows exponentially as the number of beads increases. For example, for beads with $\sigma = 1$ this time can be approximated by the function (Figure 38.28b).

$$T_N \approx T_c \cdot \exp\{N/N_c\} \quad \text{with } T_c \sim 60 \quad \text{and} \quad N_c \sim 13 \tag{38.44}$$

On the one hand, this strong dependence explains that for chains of oscillators with not too weak "dissipation," only chaotic motion was found when the number of beads becomes sufficiently large, $N \gtrsim 100$. On the other hand, it enables us to pose a question regarding the chaotic dynamics of such systems for $N \to \infty$ as a certain phase state.

In the case of weak "dissipation," the system dynamics exhibits sharp transition to a stable chaotic regime as the coefficient σ decreases. It is demonstrated in Figure 38.29 showing the transition from the regular dynamics for $\sigma = 0.1$ to a chaotic motion when $\sigma = 0.09$. As seen in Figure 38.29, the chaotic portrait can be conceived of as a highly chaotic kernel surrounded by fragments of the regular limit cycle destroyed by instability.

Noise forces these systems to undergo two-phase transitions as its intensity ϵ increases. The first one can be categorized as the transition from the regular bead motion to a cooperative chaotic bead motion. The latter means that the beads correlate substantially with one another in motion but individual trajectories are rather irregular and the magnitude of this irregularity cannot be due to the present noise only. The second transition is determined by the formation of highly irregular mutually independent oscillations in the bead position. To illustrate the first-phase transition, Figure 38.30 depicts two-phase portraits of the middle bead motion for different values of ϵ. As seen, for $\epsilon = 0.01$, the phase portrait looks like a regular limit cycle disturbed by small noise. In contrast, when the noise intensity increases by two times,

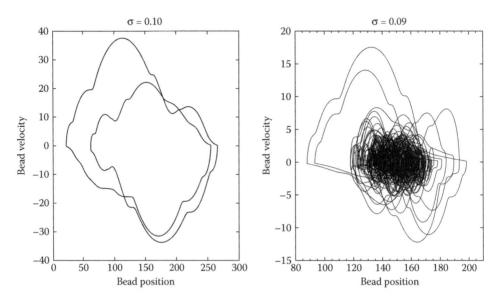

FIGURE 38.29 The phase portraits of the middle bead motion of the five-bead chain for the "dissipation" parameter σ taking the values 0.1 (left frame) and 0.09 (right frame). The period of the shown limit cycle is about 200; the chaotic phase portrait was obtained by visualizing the system motion within time interval about 5×10^5. (After Lubashevsky, I. and D. Parfenov, *CMSIM*, 1, 31–38, 2013.)

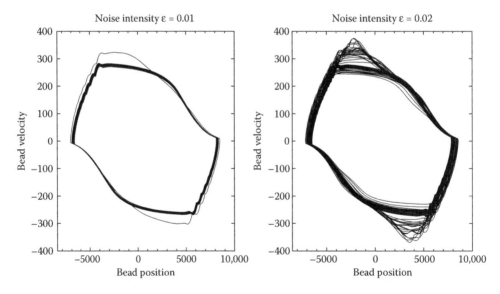

FIGURE 38.30 The phase portraits of the middle bead motion of the 30-bead chain with $\sigma = 1$ for two values of the noise intensity $\varepsilon = 0.01$ and 0.02. In plotting these portraits, bead trajectories of motion during time interval about 2×10^4 were used. (After Lubashevsky, I. and D. Parfenov, *CMSIM*, 1, 31–38, 2013.)

i.e., $\epsilon = 0.02$, the corresponding phase portrait becomes rather complex in form and the volume of the phase space layer containing the shown trajectory as a whole grows sharply. The two features has exactly enabled us to classify the found effect as a phase transition. It should be noted that this phase transition from regular motion to stochastic chaos, in contrast to the second transition to highly irregular motion, does not manifest itself in the one-particle distributions of all the variables x, v, η, ϑ ascribed to the beads individually, so it could be categorized as a "weak" phase transition.

38.3.3 Chain of Beads with Action Dynamical Traps

As demonstrated in the previous sections, in chains of particles governed by Newtonian-type laws outside the region of dynamical traps, affecting the particle motion only near the stationary points, complex cooperative phenomena (except for long-term transient processes) arise due to the cumulative effects of dynamical traps and noise. However, the bounded capacity of human cognition hinders the operators from following the strictly optimal strategy in governing their systems. To allow for such effects, in Section 38.2.2, the concept of action dynamical traps was developed; it extends the physical phase space via introducing additional phase variable quantifying the operator behavior. The present section demonstrates that in systems with elements whose individual behavior is described by the concept of action dynamical traps, complex emergent phenomena can be caused by such dynamical traps on their own.

38.3.3.1 Model

The analyzed system is actually a rather natural modification of the model studied in Section 38.3.2, the string composed of "lazy" beads, illustrated in Figure 38.19. Namely, it is a chain of $N+2$ beads moving along parallel vertical axes; the motion of each bead is characterized by its coordinate x_i, velocity v_i, and acceleration a_i. Each bead tends to minimize the absolute values of its relative coordinate and velocity with respect to its neighbors, namely, $\eta_i = x_i - \frac{1}{2}(x_{i-1} + x_{i+1})$ and $\vartheta_i = v_i - \frac{1}{2}(v_{i-1} + v_{i+1})$. Two terminal particles are assumed to be fixed: $x_0(t) \equiv x_{N+1}(t) \equiv 0$. Following the concept of action dynamical traps (Section 38.2.2), the dynamics of such system could be described by the following equations:

$$\frac{dx_i}{dt} = v_i,$$

$$\frac{dv_i}{dt} = a_i, \tag{38.45}$$

$$\frac{da_i}{dt} = \Omega_a \big[a_i, a_{\mathrm{opt}}(\eta_i, \vartheta_i, v_i) \big] \cdot \big[a_{\mathrm{opt}}(\eta_i, \vartheta_i, v_i) - a_i \big],$$

for $i = 1, \ldots, N$. Here,

$$a_{\mathrm{opt}}(\eta, \vartheta, v) = -\Omega_\vartheta(\vartheta)(\eta + \sigma\vartheta + \sigma_0 v) \tag{38.46}$$

is the optimal strategy of the operator behavior, which is considered to depend mainly on the current values of the relative position η and velocity ϑ. σ could be treated as a relative weight of the velocity variations as a stimulus causing operator actions (with respect to the first stimulus η_i); $\sigma_0 v_i$ stands for the friction force, which characterizes the physical properties of the environment where the system is placed ($\sigma_0 \ll 1$). The dynamical trap effect in system (38.45), (38.46) is modeled by cofactors Ω_ϑ and Ω_a defined as follows:

$$\Omega_\vartheta(\vartheta) = \frac{\Delta_\vartheta + \vartheta^2}{1 + \vartheta^2}, \quad \Omega_a(a, a_{\mathrm{opt}}) = \frac{\Delta_a + (a_{\mathrm{opt}} - a)^2}{1 + (a_{\mathrm{opt}} - a)^2}, \tag{38.47}$$

where parameters $0 \leq \Delta_\vartheta, \Delta_a \leq 1$ determine the intensity of dynamical traps: the lesser these parameters, the stronger the effect of corresponding dynamical traps.

It should be pointed out that we assume the former dynamical trap cofactor Ω_ϑ not to depend on particle coordinate; it could be explained in such a manner that the control over system relative velocity ϑ is of prior importance for the operator comparing to the control over position η. Thus, if the relative velocity becomes sufficiently small, the operator prefers to retard the correction of the coordinate in order not to make the velocity variations take undesirably large values.

The cofactor Ω_a in Equation 38.45 stands for the effect of action dynamical traps. Indeed, assuming $\Omega_a = 1$, one could easily see that the last equation in Equation 38.45 in fact implies the equality

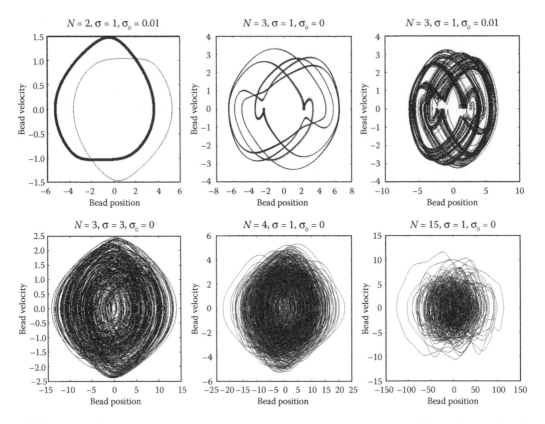

FIGURE 38.31 The phase trajectory projections of system (38.45 through 38.47) for bead chains of various length ($N = 2, 3, 4, 15$). The corresponding parameters used for simulation are shown at the plots, respectively. It should be reminded that the total number of beads in the ensemble is $N + 2$ in the notions used in this section. (Based on Zgonnikov, A. and I. Lubashevsky, *CMSIM*, 1, 60–66, 2013.)

$a_i = a_{\mathrm{opt}}(\cdots)$. However, we consider that the operator, first, is hardly able to precisely implement the strategy a_{opt} defined by Equation 38.46, and, second, cannot distinguish between the strategies that are close in some sense to the optimal one. Therefore, one may think of a certain neighborhood of the optimal strategy in the space of all possible strategies, such that each strategy from this region is treated as the optimal one by the operator. So in case the operator feels that current control regime is optimal, he just keeps maintaining the current value of the control effort constant so that $da/dt \approx 0$. When the operator realizes that the current strategy is far from the optimal one, she starts adjusting it to the desired value which means that $da/dt \sim (a_{\mathrm{opt}} - a)$.

These speculations led us to the system (38.45 through 38.47) as a model that may reflect some of the mentioned properties of human bounded rationality.

38.3.3.2 Complex Oscillations Caused by Active Dynamical Traps

We analyze numerically the collective behavior of the bead chain by solving Equations 38.45 through 38.47 using the standard (4, 5)-Runge–Kutta algorithm. Owing to the fact that the behavior of the studied system significantly varies depending on the number of interacting particles, the below analysis is divided into three parts according to the cases (1) $N = 2$; (2) $N = 3$; (3) $N \geq 4$. We should specify that all of the following results were obtained for small values of parameters Δ_{ϑ} and Δ_a, namely, 0.001, which correspond to the strong effect of dynamical trap. Below, all phase space portraits depict projections of three-dimensional phase trajectories on the "coordinate-velocity" plane generated by the system motion during the time interval of $T = 10^4$ given small randomly assigned initial disturbances. In case of multibead chains, the middle particles trajectories are represented; particle motion structure is similar for all particles in the given ensemble; however, particles in the center of the chain have slightly larger fluctuation amplitude.

The case of the single particle oscillating between two fixed neighbors was actually studied in Section 38.2.2. It turns out that the chain of two interacting beads exhibits similar behavior patterns [the phase portrait of one bead is shown in Figure 38.31(upper left plot)], except for the asymmetry of phase trajectories caused by the motion of the second bead (cf. Figure 38.11). In both cases, the structure of the limit cycles is stable with respect to variations of the system parameters. Namely, the found pattern remains for the following values of system parameters: $\sigma = 1, 3$; $\sigma_0 = 0, 0.01, 0.1$.

From Figure 38.31 it could be seen that the dynamical trap effect causes the instability of the single bead motion; the limit cycle emerges. Similar phenomena could be observed in almost the same form for each bead in the pair of coupled oscillators. The situation dramatically changes when the ensemble of three beads is taken into consideration. Adding just one more oscillator to the system causes the anomalous cooperative phenomena to emerge, particularly, a complex three-dimensional attractor arises in the system phase space (see Figure 38.31).

Notably, unlike the previous cases ($N = 1, 2$), introducing the external friction force ($\sigma_0 \neq 0$) causes the attractor to become significantly blurred, while increasing the relative weight of the bead velocity as the stimulus for the operator actions makes the particle dynamics to take form of chaotic oscillations (Figure 38.31(lower plots)).

In case of the relatively large number of interacting elements, the system dynamics becomes highly irregular. The chain of four beads demonstrates the oscillatory behavior. It is worth underlining that the well-defined attractor (Figure 38.31(upper left plot)) could be destructed just by adding one bead to the ensemble without changing any of the system parameters.

The system motion trajectories for $N = 15$ are of even greater irregularity due to the increased number of bead and corresponding cooperative effect. For larger N, the system motion exhibits the patterns of similar structure, but the amplitude of the fluctuations increases with N.

38.4 Conclusion

A new mechanism of emergent phenomena in social systems or systems governed by cumulative action of humans and physical regularities has been discussed. It is the fuzzy rationality caused by the bounded capacity of human cognition and manifesting itself in the limited capability of humans in ordering events, actions, strategies of behavior, etc. according to their preference perfectly. This is most pronounced when, for example, an individual should make a choice between several possible actions similar in quality. As a result, he has to consider them equivalent, thereby, their choice becomes random and practically independent of the real action quality "hidden" for the individual. Only in the case where two actions at hand are characterized by a significant difference in quality, the choice is determined by the preference relation. When the control over the system dynamics is concerned, the fuzzy rationality affects the choice between the "hidden" optimal strategy of behavior and actions in its proximity. In this case, the optimal strategy becomes unattainable and individuals consider a whole multitude of possible actions "optimal" with high probability. As a result, the dynamics of a given system as well as the control by its elements (individuals) is stagnated until the system motion goes rather far from the optimal one, which was expected to induce a system instability.

38.4.1 Concept of Dynamical Traps

First, to capture these features of human behavior, the concept of dynamical traps caused by human fuzzy rationality has been considered; it can be regarded as a certain generalization of the notion of stationary point in dynamical systems. The proposed model focused the attention on the systems where the optimal dynamics implies the stability of a certain stationary point in the corresponding phase space. In such systems, the bounded capacity of human cognition gives rise to some neighborhood of the equilibrium point, the region of dynamical traps, wherein each point is regarded as an equilibrium one by the operators. So when a system enters this region and while it is located in it, maybe for a long time, the operator control is suspended. It is demonstrated that under rather general assumptions the description of such human behavior is reduced to the model of oscillator with dynamical traps.

As shown in the frameworks of this model, the cumulative effect of the dynamical trap and noise can induce nonequilibrium phase transitions, of a new type. In contrast to the classical phase transitions, the self-formation of new noise-induced phases is not determined by the zero values of regular "forces" even approximately. The cause of the found phase transition is the asymmetry of the residence time distribution inside the trap region. This asymmetry is due to the cooperative effect of the regular "force" outside the trap region and the random Langevin "force" inside it.

Second, a human operator controlling a dynamical system is usually not capable of selecting or calculating the optimal action strategy that allows it to reach and maintain the desired end-state or goal. However, during the control process, the operator is able to realize that the currently implemented strategy deviates from the optimal one if this deviation becomes large enough. Once aware of the mismatch, the operator can adjust the actions until he feels that the current value of the control parameter is acceptable. In order to capture this feature of human cognition, the extended phase space of the dynamical system under human control with the control parameter as an independent phase variable has been introduced. Namely, the standard "coordinate-velocity" phase space inherited from the Newtonian mechanics is extended by the acceleration variable. This enables us to introduce a certain region alongside the optimal strategy in the space of all action strategies; each strategy within this region is treated as acceptable by the operator. The latter region is called the action dynamical trap. As demonstrated, the action dynamical traps on their own can be responsible for self-sustained periodic oscillations around the stationary point.

38.4.2 Emergent Phenomena in Multielement Systems with Dynamical Traps

The self-organization of spatiotemporal patterns in the chain of oscillators with dynamical traps was studied in detail. This model assumes the individual element behavior to be governed by two stimuli: one of them is to optimize the spatial arrangement of a given element with respect to its nearest neighbors, the other is to minimize their relative velocities. However, when the relative velocity becomes rather small, an element being "lazy" loses motives for active behavior in correcting the current situation because in this case it cannot become worthy. This suspension of activity is regarded as the effect of dynamical traps.

The cooperative phenomena arising in these systems have been studied by analyzing the velocity distributions and visualizing some time patterns formed by the element dynamics. The time patterns of the one element velocity dynamics have demonstrated the presence of long-lived states. These patterns look like a sequence of fragments within which the element velocity varies in the vicinity of some level continuously joined by sharp jumps in the element velocity. The velocity levels are rather uniformly distributed inside a wide interval and the life time of these fragments exceeds essentially the time scales of the individual dynamics of the elements. It has been shown that such long-lived states can persist if only multielement clusters moving as a whole are formed. The visualized multielement patterns have also demonstrated the presence of such cooperative structures. Moreover, it has become clear that these long-lived states persist independently in some sense of the individual dynamics of elements forming them currently. In other words, the life time of such a state can exceed substantially the time interval during which the element forming it at a current time belongs to it. Keeping the latter in mind, we refer to them as to dynamical states. These states in turn can form superstructures, so the observed patterns are classified as hierarchical structures.

In the chains of oscillators with dynamical traps, noise plays a constructive role. Without noise only regular spatio-temporal patterns can emerge; nevertheless, transient processes could be highly long-term and complex in form. In particular, the mean time required for regular structures to form in such systems is found to grow exponentially with the number of elements. It enables us to pose a question as to whether noise with an extremely small amplitude can affect substantially the properties exhibited by systems of many elements with fuzzy rational behavior and be responsible for emergent phenomena with anomalous properties. For example, it has been found that the transition between the regimes of regular and cooperative chaotic bead motion manifests itself only the sharp growth of the volume of the phase space layer containing the element trajectories, whereas all the one-particle distribution functions do not change their forms remarkably.

In the chain of oscillators with action dynamical traps, complex structures with highly irregular behavior have been found to emerge without noise, i.e., due to the action dynamical traps on their own. The various complex patterns of the system motion are shown to arise depending on the system parameters. In particular, with the increasing number of elements the system motion becomes significantly irregular, for large numbers of elements exhibiting chaotic oscillations.

Acknowledgments

This work was supported in part by the JSPS "Grants-in-Aid for Scientific Research" Program, Grant No. 245404100001. Sections 38.2.2 and 38.3.3 were written in collaboration with A. Zgonnikov.

References

Asai, Y. et al. 2009. A model of postural control in quiet standing: Robust compensation of delay-induced instability using intermittent activation of feedback control. *PLoS One* 4.7, e6169.

Asai, Y., S. Tateyama, and T. Nomura. 2013. Learning an intermittent control strategy for postural balancing using an EMG-based human–computer interface. *PLoS One* 8.5, e62956 (1–19).

Balasubramaniam, R. 2013. On the control of unstable objects: The dynamics of human stick balancing. In: *Progress in Motor Control: Neural, Computational and Dynamic Approaches*. Ed. by M. J. Richardson, M. A. Riley, and K. Shockley. New York: Springer Science + Business Media, pp. 149–168.

Bays, P. M., J. R. Flanagan, and D. M. Wolpert. 2006. Attenuation of selfgenerated tactile sensations is predictive, not postdictive. *PLoS Biology* 4.2, e28(1–4).

Bottaro, A. et al. 2008. Bounded stability of the quiet standing posture: An intermittent control model. *Human Movement Science* 27.3, 473–495.

Boulet, J. et al. 2010. Stochastic two-delay differential model of delayed visual feedback effects on postural dynamics. *Philosophical Transactions of the Royal Society A: Mathematical, Physical and Engineering Sciences* 368.1911, 423–438.

Burrage, K. and P. M. Burrage. 1998. General order conditions for stochastic Runge–Kutta methods for both commuting and non-commuting stochastic ordinary differential equation systems. *Applied Numerical Mathematics* 28.2, 161–177.

Burrage, P. M. 1999. Runge–Kutta methods for stochastic differential equations. PhD thesis. University of Queensland.

Cabrera, J. L., C. Luciani, and J. Milton. 2006. Neural control on multiple time scales: Insights from human stick balancing. *Condensed Matter Physics* 9.2, 373–383.

Cabrera, J. L. and J. G. Milton. 2002. On–off intermittency in a human balancing task. *Physical Review Letters* 89.15, 158702(1–4).

Castellano, C., S. Fortunato, and V. Loreto. 2009. Statistical physics of social dynamics. *Reviews of Modern Physics* 81.2, 591–646.

Chakrabarti, B. K., A. Chakraborti, and A. Chatterjee. 2006. *Econophysics and Sociophysics: Trends and Perspectives*. Weinhaim: Wiley-VCH Verlag GmbH & Co. KGaA, p. 622. ISBN: 9783527406708.

Chowdhury, D., L. Santen, and A. Schadschneider. 2000. Statistical physics of vehicular traffc and some related systems. *Physics Reports* 329.4, 199–329.

Craik, K. J. W. 1947. Theory of the human operator in control systems. I. The operator as an engineering system. *British Journal of Psychology. General Section* 38.2, 56–61.

Craik, K. J. W. 1948. Theory of the human operator in control systems. II. Man as an element in a control system. *British Journal of Psychology. General Section* 38.3, 142–148.

De Wolf, T. and T. Holvoet. 2005. Emergence versus self-organisation: Different concepts but promising when combined. In: *Engineering Self-Organising Systems: Methodologies and Applications*. Ed. by S. A. Brueckner et al. Vol. 3464. Lecture Notes in Computer Science. Berlin: Springer-Verlag, pp. 11–15.

Dompere, K. K. 2009. Fuzzy Rationality. Berlin: Springer-Verlag.

Eidelman, S. and C. S. Crandall. 2009. A psychological advantage for the status quo. In: *Social and Psychological Bases of Ideology and System Justication*. Ed. by J. T. Jost, A. C. Kay, and H. Thorisdottir. Oxford: Oxford University Press, pp. 85–106.

Eidelman, S. and C. S. Crandall. 2012. Bias in favor of the status quo. *Social and Personality Psychology Compass* 6.3, 270–281.

Evans, C. M. et al. 1983. Response of the normal human ankle joint to imposed sinusoidal movements. *The Journal of Physiology* 344.1, 483–502.

Galam, S. 2008. Sociophysics: A review of Galam models. *International Journal of Modern Physics C* 19.03, 409–440.

Galam, S. 2012. *Sociophysics: A Physicist's Modeling of Psycho-Political Phenomena*. New York: Springer.

Gawthrop, P. et al. 2011. Intermittent control: A computational theory of human control. *Biological Cybernetics* 104.1–2, 31–51.

Grahamuk. *Photo: Wave clouds forming over Mount Duval, Australia*. http://en.wikipedia.org/wiki/Image:Wavecloudsduval.jpg (licensed under the terms of the CC-BY-SA 3.0).

Hajimahmoodzadeh, M. 2005. Dynamical states in condensed systems with anomalous kinetic coefficients. (supervised by Lubashevsky I.) PhD thesis. Moscow State University, Moscow, Russia.

Hegselmann, R. 2009. Moral dynamics. In: *Encyclopedia of Complexity and Systems Science*. Ed. by R. A. Meyers. New York: Springer Science+Business Media, LLC., pp. 5677–5692. ISBN: 978-0-387-30440-3.

Helbing, D. 2001. Traffic and related self-driven many-particle systems. *Reviews of Modern Physics* 73.4, 1067–1141.

Helbing, D. ed. 2012. *Social Self-Organization*. Berlin: Springer-Verlag.

Helbing, D. and A. Johansson. 2009. Pedestrian, crowd and evacuation dynamics. In: *Encyclopedia of Complexity and Systems Science*. Ed. by R. A. Meyers. New York: Springer Science+Business Media, LLC., pp. 6476–6495. ISBN: 978-0-387-30440-3.

Hosaka, T. et al. 2006. Balancing with noise and delay. *Progress of Theoretical Physics Supplement* 161, 314–319.

Insperger, T. 2011. Stick balancing with reflex delay in case of parametric forcing. *Communications in Nonlinear Science and Numerical Simulation* 16.4, 2160–2168.

Iyengar, S. S. and M. R. Lepper. 2000. When choice is demotivating: Can one desire too much of a good thing? *Journal of Personality and Social Psychology* 79.6, 995–1006.

Kerner, B. S. 2004. *The Physics of Traffic: Empirical Freeway Pattern Features, Engineering Applications, and Theory*. Berlin: Springer-Verlag.

Kerner, B. S. 2009. Traffic congestion, modeling approaches to. In: *Encyclopedia of Complexity and Systems Science*. Ed. by R. A. Meyers. New York: Springer Science+Business Media, LLC., pp. 9355–9411. ISBN: 978-0-387-30440-3.

Kowalczyk, P. et al. 2012. Modelling human balance using switched systems with linear feedback control. *Journal of The Royal Society Interface* 9.67, 234–245.

Lavrenov, A. Y. et al. 2005. The structures of the states of the oscillator chain with dynamical traps. *Bulletin of the Lebedev Phyisics Institute* 2, 1–8.

Loram, I. D., P. J. Gawthrop, and M. Lakie. 2006. The frequency of human, manual adjustments in balancing an inverted pendulum is constrained by intrinsic physiological factors. *The Journal of Physiology* 577.1, 417–432.

Loram, I. D. and M. Lakie. 2002. Human balancing of an inverted pendulum: Position control by small, ballistic-like, throw and catch movements. *The Journal of Physiology* 540.3, 1111–1124.

Loram, I. D., M. Lakie, and P. J. Gawthrop. 2009. Visual control of stable and unstable loads: What is the feedback delay and extent of linear time-invariant control? *The Journal of Physiology* 587.6, 1343–1365.

Loram, I. D. et al. 2011. Human control of an inverted pendulum: Is continuous control necessary? Is intermittent control effective? Is intermittent control physiological? *The Journal of Physiology* 589.2, 307–324.

Lubashevsky, I. 2012. Dynamical traps caused by fuzzy rationality as a new emergence mechanism. *Advances in Complex Systems* 15.08, 1250045.

Lubashevsky, I., P. Wagner, and R. Mahnke. 2003a. Bounded rational driver models. *The European Physical Journal B—Condensed Matter and Complex Systems* 32.2, 243–247.

Lubashevsky, I. et al. 2003. Noise-induced phase transition in an oscillatory system with dynamical traps. *The European Physical Journal B-Condensed Matter and Complex Systems* 36.1, 115–118.

Lubashevsky, I. et al. 2005. Long-lived states of oscillator chains with dynamical traps. *The European Physical Journal B-Condensed Matter and Complex Systems* 44.1, 63–70.

Lubashevsky, I. A., V. V. Gaychuk, and A. V. Demchuk. 1998. Anomalous relaxation oscillations due to dynamical traps. *Physica A: Statistical Mechanics and its Applications* 255.3, 406–414.

Lubashevsky, I. A. et al. 2007. Phase transition in a harmonic oscillator with dynamical traps. *Bulletin of the Lebedev Physics Institute* 1, 15–20.

Lubashevsky, I. and D. Parfenov. 2013. Complex dynamics and phase transitions caused by fuzzy rationality. *Chaotic Modeling and Simulation* 1, 31–38.

Lubashevsky, I., P. Wagner, and R. Mahnke. 2003b. Rational-driver approximation in car-following theory. *Physical Review E* 68.5, 056109.

Lubashevsky, I. et al. 2002. Long-lived states in synchronized traffic flow: Empirical prompt and dynamical trap model. *Physical Review E* 66.1, 016117.

Meyers, R. A. 2009. Preface. In: *Encyclopedia of Complexity and Systems Science*. Ed. by R. A. Meyers. New York: Springer Science+Business Media, LLC., pp. VI–VIII. ISBN: 978-0-387-30440-3.

Milton, J. et al. 2009a. Balancing with positive feedback: The case for discontinuous control. *Philosophical Transactions of the Royal Society A: Mathematical, Physical and Engineering Sciences* 367.1891, 1181–1193.

Milton, J. et al. 2009b. The time-delayed inverted pendulum: Implications for human balance control. *Chaos: An Interdisciplinary Journal of Nonlinear Science* 19.2, 026110–026110.

Milton, J. G. 2013. Intermittent motor control: The "drift-and-act" hypothesis. In: *Progress in Motor Control Neural, Computational and Dynamic Approaches*. Ed. by M. J. Richardson, M. A. Riley, and K. Shockley. New York: Springer Science+Business Media, pp. 169–193.

Nagatani, T. 2002. The physics of traffic jams. *Reports on Progress in Physics* 65.9, 1331–1386.

Nowak, A. and U. Strawinska. 2009. Applications of physics and mathematics to social science, Introduction to. In: *Encyclopedia of Complexity and Systems Science*. Ed. by R. A. Meyers. New York: Springer Science+Business Media, LLC., pp. 322–326. ISBN: 978-0-387-30440-3.

O'Connor, T. and H. Y. Wong. 2012. Emergent properties. In: *The Stanford Encyclopedia of Philosophy*. Ed. by E. N. Zalta. Spring 2012. http://plato.stanford.edu/archives/spr2012/entries/properties-emergent/.

Oullier, O. and J. A. S. Kelso. 2009. Social coordination, from the perspective of coordination dynamics. In: *Encyclopedia of Complexity and Systems Science*. Ed. by R. A. Meyers. New York: Springer Science+Business Media, LLC., pp. 8198–8213. ISBN: 978-0-387-30440-3.

Rack, P. M. et al. 1983. Reflex responses at the human ankle: The importance of tendon compliance. *The Journal of Physiology* 344.1, 503–524.

Rae, A. *Photo: Red-billed quelea flocking at waterhole*. http://www.flickr.com/photos/merula/342898722/in/set-72157594423311516 (licensed under the terms of the CC-BY-SA-2.0).

Ritov, I. and J. Baron. 1990. Reluctance to vaccinate: Omission bias and ambiguity. *Journal of Behavioral Decision Making* 3.4, 263–277.

Ritov, I. and J. Baron. 1992. Status-quo and omission biases. *Journal of Risk and Uncertainty* 5.1, 49–61.

Samuelson, W. and R. Zeckhauser. 1988. Status quo bias in decision making. *Journal of Risk and Uncertainty* 1.1, 7–59.

Schadschneider, A. et al. 2009. Evacuation dynamics: Empirical results, modeling and applications. In: *Encyclopedia of Complexity and Systems Science*. Ed. by R. A. Meyers. New York: Springer Science+Business Media, LLC., pp. 3142–3176. ISBN:978-0-387-30440-3.

Schnobby. *Photo: Ice-ferns*. http://en.wikipedia.org/wiki/File:Frost_patterns_4.jpg (licensed under the terms of the CC-BY-SA 3.0).

Schwartz, B. 2000. Self-determination: The tyranny of freedom. *American Psychologist* 55.1, 79–88.

Slanina, F. 2009. Social processes, physical models of. In: *Encyclopedia of Complexity and Systems Science*. Ed. by R. A. Meyers. New York: Springer Science+Business Media, LLC., pp. 8379–8405. ISBN: 978-0-387-30440-3.

Spranca, M., E. Minsk, and J. Baron. 1991. Omission and commission in judgment and choice. *Journal of Experimental Social Psychology* 27.1, 76–105.

Stauffer, D. 2009. Opinion dynamics and sociophysics. In: *Encyclopedia of Complexity and Systems Science*. Ed. by R. A. Meyers. New York: Springer Science+Business Media, LLC., pp. 6380–6388. ISBN: 978-0-387-30440-3.

Stauffer, D. and S. Solomon. 2009. Physics and mathematics applications in social science. In: *Encyclopedia of Complexity and Systems Science*. Ed. by R. A. Meyers. New York: Springer Science+Business Media, LLC., pp. 6804–6810. ISBN: 978-0-387-30440-3.

Suzuki, Y. et al. 2012. Intermittent control with ankle, hip, and mixed strategies during quiet standing: A theoretical proposal based on a double inverted pendulum model. *Journal of Theoretical Biology* 310, 55–79.

Todosiev, E. P. 1963. *The action point model of the driver-vehicle system*. Tech. rep. 202A-3. (Ph.D. Dissertation, Ohio State University, 1963). The Ohio State University.

Tversky, A. and E. Shar. 1992. Choice under conflict: The dynamics of deferred decision. *Psychological Science* 3.6, 358–361.

Vallacher, R. R. 2009. Social psychology, applications of complexity to. In: *Encyclopedia of Complexity and Systems Science*. Ed. by R. A. Meyers. New York: Springer Science+Business Media, LLC., pp. 8405–8420. ISBN: 978-0-387-30440-3.

Vince, M. A. 1948. The intermittency of control movements and the psychological refractory period. *British Journal of Psychology. General Section* 38.3, 149–157.

Voss, M. et al. 2006. Sensorimotor attenuation by central command signals in the absence of movement. *Nature Neuroscience* 9.1, 26–27.

Wikipedia. 2014. *Status quo bias—Wikipedia, The Free Encyclopedia*. [Online; accessed 13-November-2014]. URL: http://en.wikipedia.org/w/index.php?title=Status_quo_bias&oldid=633114317.

Yaari, G., D. Stauffer, and S. Solomon. 2009. Intermittency and localization. In: *Encyclopedia of Complexity and Systems Science*. Ed. by R. A. Meyers. New York: Springer Science+Business Media, LLC., pp. 4920–4930. ISBN: 978-0-387-30440-3.

Zaslavsky, G. M. 2008. *Hamiltonian Chaos and Fractional Dynamics*. Oxford: Oxford University Press.

Zgonnikov, A. and I. Lubashevsky. 2013. Complex dynamics of multiparticle system governed by bounded rationality. *Chaotic Modeling and Simulation* 1, 60–66.

Zgonnikov, A. and I. Lubashevsky. 2014. Extended phase space description of human-controlled systems dynamics. *Progress of Theoretical and Experimental Physics* 2014.3, 033J02.

Zgonnikov, A. et al. 2014. To react or not to react? Intrinsic stochasticity of human control in virtual stick balancing. *Journal of The Royal Society Interface* 11, 20140636.

39

Chaos in Monolingual and Bilingual Speech

Elena Babatsouli

39.1 Introduction

> Don't you think so, *reader*, *rather*,
> Saying lather, bather, father?
> Finally, which rhymes with enough,
> Though, through, bough, cough, hough, sough, tough?

Reading the above extract from Nolst Trenité's poem, *The Chaos*, may cause perplexity to second-language speakers of English depending on their fluency levels. The poem, originally published in the author's 1920 *Drop Your Foreign Accent: Engelsche Uitspraakoefeningen* textbook for Dutch learners of English, amasses 800 examples of irregularity in English pronunciation and spelling. Its title signifies confusion at the apparent lack of order, randomness, and unpredictability of grammatical rules and, whimsically, denotes the presence of *complexity* in oral and written language. Purposefully chosen to introduce this chapter on *chaos* in monolingual and bilingual speech, it makes additional allusions to the complexity involved in language learning.

A question that may naturally arise is: how does *chaos theory*, an essentially mathematical field, apply to the scientific study of language? Chaos theory originates in the 1960s in Edward Lorenz's work on meteorology and its maxims have been employed to the understanding and scientific explanation of various complex natural phenomena. One such phenomenon is human language, which in its raw, primal form is called *speech*. The smallest units of speech, i.e., phonemes, and how they are organized in language(s) is the focus of phonology in linguistics. Humans learn to speak naturally as a means of communication, while writing is learned subsequently in instructional settings. Learning to speak, or language acquisition, involves being able to *perceive* and then *produce* the speech sounds of a particular language. Language acquisition occurs in both monolingual and bilingual (or multilingual) settings. Bilingualism refers to the regular use of a second language alongside a first (Grosjean, 2013), irrespective of the enacting conditions that led to gaining competence in each of the languages (Babatsouli and Ingram, 2015). In a broader sense, bilingualism will be assumed here to include second-language acquisition, which examines how a second (third, fourth, etc.), typically foreign, language is learned naturally or through instruction after the native language has been acquired.

Mirroring the trend across scientific disciplines (Gleick, 1997), recent approaches to language processing and learning have recognized the need for a holistic, multiplanar viewpoint and turned to chaos theory for answers. "Now that science is looking, chaos seems to be everywhere" (Gleick, 1997:3). Tenets of chaos theory are interspersed in various theoretical approaches of the study of language and its learning, although not specifically stated so (Ellis, 2008). Seeds in phonology are found in Mohanan's work (1992, 1993). The overwhelming amount of work, however, that explicitly involves chaos theory in language is found in applied linguistics with regard to second-language and bilingual acquisition (Larsen-Freeman, 1997; De Bot, 2008; De Bot et al., 2007; Schmid and Lowie, 2011). For speech development in a first language, the work of Babatsouli et al. (2014) quantitatively shows the developmental path of the acquisition of words in a child's speech with the existence of a major strange attractor from the age of 2 years and 9 months (2;9) to the age of 3 years and 4 months (3;4), and the existence of a double-logistic growth thereafter until complete acquisition at the age of 4 years. This chapter reviews the aspects of chaos theory, as these apply to our understanding of language and language learning. The chapter consists of seven parts: (1) introduction, (2) chaos theory, (3) language and its learning as complex, dynamic, and adaptive systems, (4) chaos in phonological development, (5) quantifiable speech variables, (6) sound development in a child's speech, and (7) conclusions.

39.2 Chaos Theory

Hesiod's *Theogony* (1000-700 BC) advocates that *chaos*, from Greek χάος/'xaos/, was the beginning of all creation and that from it sprung "matter," "time," and "creation." The term literally denotes abyss, chasm, and disorder. In the field of mathematics, *chaos theory* uses mathematical information models to study how dynamical systems behave (Skiadas and Skiadas, 2009). The theory on *dynamical systems* (Devaney, 2003) maintains that any point (state), represented by real numbers, in a geometrical space (state space) is dependent on time. This makes for a system. Small changes in the state of a system lead to small changes in the numbers. Small changes in the numbers bear extensively diverging outcomes for dynamical systems (Lorenz, 1972), a phenomenon popularized as the butterfly effect (Gleick, 1997).

The paradox with dynamical systems is that change (their evolution) depends on a fixed rule that deterministically makes their future behavior fully dependent on initial conditions; however, even trivial changes in these initial conditions inhibit long-term prediction of future conduct (Kellert, 1993). Predictability depends on accurate measurements of the current state, the timescale of the dynamics examined, and on how much uncertainty can be tolerated in the prediction. So, although the behavior of dynamical systems is predictable, in principle, this is not always the case (Kautz, 2011). Lorenz is claimed by Danford (2013, online citation) to define chaos as follows: "When the present determines the future, but the approximate present does not approximately determine the future." A dynamical system is chaotic when it shows sensitivity to initial conditions, the regions of its space overlap, and it shows dense periodic orbits (Hasselblatt and Katok, 2003). Dynamical systems are not necessarily chaotic everywhere; they may only exhibit chaotic behavior in certain parts of the state space, sometimes converging toward specific self-instigated states, "attractor" states, rather in the opposite direction, "repeller" states. The aspect of unpredictability in the behavior of dynamical systems makes for their randomness, nonlinearity, and complexity.

Complexity theory, or complex systems theory (Sorin and Eran, 2003), is associated with chaos theory and investigates how the parts of a system interact with each other and the environment determining the system's collective behavior. Some characteristics of complex systems are evolution and adaptation, self-organization, interdependence, collective behavior, emergence, pattern formation, nonlinearity, and networking. Although dynamical systems are found in natural phenomena, such as the climate, swinging of a clock's pendulum, water flowing in a pipe, etc., complex systems are physically undefined; they are only represented in mathematical information models based on statistics, information theory, and non-linear dynamics. Owing to their dynamics and complexity, such systems are difficult to formally model and simulate.

Discussions on the tenets of chaos theory viewed from the perspective of language may be found in a number of publications (Mohanan, 1992, 1993; Van Geert, 1994; Larsen-Freeman, 1997, 2002; De Bot et al., 2005; De Bot, 2008; Jessner, 2008, Larsen-Freeman and Cameron, 2008a,b; Schmid and Lowie, 2011; Verspoor et al., 2011; De Bot et al., 2013). In many of these publications, the theory of chaos is known as chaos/complexity theory (C/CT) or dynamic systems theory (DST), and their complementarity is acknowledged (Larsen-Freeman, 2007). A short catalog of the basic C/CT and DST principles follows: nonlinearity; complete interconnectedness; sensitivity to initial conditions; ongoing dependence on own components, resources, and the environment; steady self-reorganization; and change; change is iterative: each state being a developmental function of the previous one; paths of change vary (both linear and fluctuating) and may be chaotic; transitory paths are cyclical; novel properties emerge out of fundamental properties; there is scale invariance; phenomena have fractal nature, which may also be self-similar; and final states are not infinitely steady.

39.3 Language and Its Learning as Complex, Dynamic, and Adaptive Systems

That human language is a complex system of communication needs no reference. To begin with, complexity is an inherent feature of abstract linguistic systems per se; natural languages have their own sets of structural rules, some of which are inherently opaque. Human language in general, and abstract linguistic systems in particular are also *dynamic* and prone to change; the set of rules that make up the grammar of a particular language is not fixed ad infinitum—languages are found to evolve in a process of diachronic change (Morten and Kirby, 2003). For any specific targeted language, communication depends on learning and appropriately using the complex sets of rules that govern how phonemes (smallest units of speech) and/or graphemes (smallest units of text) are used. Oral competence in the targeted language is acquired seemingly effortlessly before children reach school age. Nonetheless, the ability to speak involves an amazing array of human faculties that simultaneously interact to achieve speech, such as cognition, perception, and articulation. Despite the assumed automaticity, language is an emergent system that is learned in stages and by induction; various theories, such as DST (Port and van Gelder, 1995), neural networks modeling (Fausett, 1994), and optimality theory (OT) (Tesar and Smolensky, 1998) support that a linguistic structure emerges from interacting constraints (MacWhinney, 1999). To sum up, language is complex and dynamic, both as an abstract entity and a quotient of human capacity.

When two or more (rather than one) languages are learned, either simultaneously or successively, there is an interaction between them and the resulting competence in any one of them is often compromised by the very presence of the other. While the bilingual speaker has been shown to maintain independent grammatical systems or, more specifically, phonologies for each language (Keshavarz and Ingram, 2002), there is an evidence of both competition (Marian and Spivey, 2003) and interaction (Poplack, 1980) between them. In other words, linguistic capacity is further affected system-internally by neuropsycholinguistic factors, what De Bot et al. (2007) have pinpointed as a "cognitive ecosystem" (12) with "limited resources" and "carrying capacity." The languages of a bilingual speaker fluctuate along a "language mode continuum" (Grosjean, 2001) that is the result of cerebral activation (Green, 1986). *Mental lexicon* in linguistics refers to an abstract system of logical structures in human cognition that processes information (auditory, visual, contextual, etc.), stores it, and has the capacity to retrieve it for comprehension and production. Attempts to understand the cognitive processes involved in speech have evolved from those in structural frameworks (Levelt, 1989; De Bot, 1992) to recent ones adopting tenets of chaos theory (Lowie and Verspoor, 2011; Schwieter and Tokowicz, 2015) and computational modeling (Li and Zhao, 2015).

Language learning is instigated by the self-organization of the system (i.e., by the interaction of various system-internal factors) as stated, and also by adaptation to the environment. Alongside considerations of inherent, system-internal linguistic mechanisms, one should not disregard system-external mechanisms, such as *input*, the *timing of exposure to language*, and what "The Five Graces Group" refer to as "inter-related patterns of experience" and "social interaction" (Beckner et al., 2009:2). Language starts with linguistic input. Skinner (1938) claimed that verbal behavior is the result of some external "originating force" (p. 51): "a discriminative stimulus acting prior to the response" (pp. 178–9). A major ambiguity in language acquisition, known as Platos' problem (Hornstein and Lightfoot, 1981), deliberates on how humans acquire complex knowledge that considerably exceeds the input they are exposed to. There is no language acquisition in the absence of linguistic input. Hearing impairment leads to speech impairment. It is also known that where two (or more) languages are involved, the quality and quantity of linguistic input (Müller and Hulk, 2001; Piske and Young-Scholten, 2009; Bosch and Ramon-Casas, 2011) are important factors in determining ultimate attainment and native-like competence.

Input and timing of exposure are interacting factors; linguistic input alone does not safeguard linguistic capacity. There is a "critical period" (Penfield and Roberts, 1959) in the acquisition of language beyond which humans progressively lose the capacity for speech. Studies on feral children who received no or limited linguistic input in toddlerhood support this (Curtiss, 1977). The presence of *accent*, i.e., mispronunciation, in second-language speech is another effect of the "critical period," since people are usually exposed to the novel linguistic system of a second language after the critical period has passed. Consequently, the evolution of linguistic capacity is unquestionably dependent on *initial conditions, continuous self-organization*, and *environmental adaptation*.

Language emergence occurs through numerous processes of *change*. The process of change in language involves evolution. Evolution, however, may have a positive direction (growth), as in "language development" or a negative direction (deterioration), as in "language attrition." Language attrition (Köpke et al., 2007) refers to the gradual deterioration and even loss of capacity in a particular language, usually because another language is used more. Nevertheless, language evolution also involves periods of stagnation, as in "language fossilization" (Selinker, 1972), when the process of learning halts prematurely. The changes, or the dynamics, of speech are evident in variability during production. Variability refers to both inter-variability in processing and production during speech acquisition (Dinnsen, 1992; Bosch and Ramon-Casas, 2011; Van Dijk et al., 2011; Lowie and De Bot, 2015) and to intravariability or individual variation (Goad and Ingram, 1987; De Bot et al., 2007). The source of variability is located in biological constraints, environmental effects, and linguistic complexity (Ingram, 1981a) and is typical of both monolingual and bilingual scenarios. Interestingly, the dynamics of speech do not cease when the learning process ceases or stagnates. Speech variability is also evidenced in idiolects (individual speaker specificities) (Kuhl, 2003), dialects (social variation in speech), and accents (mispronunciation of languages). In other words, linguistic performance is never static, but is constantly affected by the terms of usage and the environment.

39.4 Chaos Theory in Phonological Development

Phonological development examines the acquisition of phonology based on cognitive and articulatory resources. Development is best viewed "in terms of a multidimensional descriptive space" (58) that is "unfolding" and "unwrapping" (Van Geert, 1994:61). This assumes that there is an initial state (genetics), a seed upon which growth occurs autocatalytically when facilitating conditions (environment) allow (Chomsky, 2007). Phonological development is a nonlinear, dynamic "morphogenesis" (the creation of form) in the sense of "pattern formation and adaptation" (Mohanan, 1992), in which iteration is a basic developmental mechanism (Van Geert, 1994). Any state in the developmental sequence is "a developmental function of the previous developmental state" (ibid, p. 20). In other words, principles inherent in chaos theory are involved in its generic themes: innateness and the environment; iteration/cyclicity; nonlinearity; developmental paths; etc.

39.4.1 Nonlinearity

Nonlinearity, contrast, and opposition is in the very essence of phonology. Although it is not the purpose here to review phonological theory, a rudimentary introduction may give a flavor. Speech segments, or phonemes in their abstract form, are the smallest units of speech that differentiate the meaning of words in a language and are assessed in terms of sets of "distinctive features" that catalog all possible human speech in an intricate network of articulatory and acoustic correlates (Jakobson, 1941/1968). For instance, *t* and *k* differentiate *tough* and *cough*, as in Trenité's poem in the introduction. Although *t* and *k* share articulatory characteristics in terms of the *manner of articulation*, i.e., both of them are *stops* (there is obstruction in the mouth and airflow ceases) and *voicing*, i.e., both of them are *voiceless* (vocal cords do not vibrate), their *place of articulation* differs. That is, *t* is *alveolar* (produced with the tip of the tongue on the alveolar ridge) and *k* is *velar* (produced with the body of the tongue touching the velum, the soft palate at the back of the mouth). Every segment may only have one or the other property and there is, inherently, a binary opposition between any two segments: e.g., *t* and *k* are ±*velar*. Thus, the phonological system of a language is essentially multilayered and the hierarchy of these layers is universal and invariable across languages (Trubetzkoy, 1939/1958).

To complicate matters further, differences between the *underlying form* (theoretical and abstract characteristics) of a speech segment and a *surface form* (its actual production) also arise within a language and across different languages. For instance, English *t* in *tough* is aspirated (a burst of air follows the obstruction), while *t* in *ate* is not. Moreover, while *t* is found in many languages, these languages do not necessarily have the aspirated form (e.g., in Greek and Spanish). The various combinations of phonemes and their surface forms make for the difference in the phonological systems between languages. So, while both *d* in *reader* and *th* in *rather* (see Trenité's poem in the introduction) are found in English, only *d* is found in Dutch. This and the distance, or sometimes opaqueness, between the underlying and surface forms inhibits learning in both monolingual and bilingual acquisition of phonology. As *th* is lacking in Dutch (to take an example from second-language speech), a Dutch speaker of English will tend to substitute *th* in *rather* with *d* as in *reader*, which explains the pairing of *reader* with *rather* as in Trenité's poem. Such "errors" in a second language are evidence of speech learning in development. Similar "errors" are present in child speech during phonological development. Rounding up, it is unanimously agreed that phonological development is nonlinear both in terms of particular phonological theorizing and language acquisition theory (Mohanan, 1992; Rice and Avery, 1995; Gierut, 1996; Bernhardt and Stemberger, 1998; Larsen-Freeman, 1997; Gierut, 2007; De Bot, 2008).

39.4.2 Developmental Paths

Nonlinearity in phonological development has also been seen in terms of developmental patterns. When language learning is involved, the term *development* is preferred over *acquisition* (Verspoor et al., 2011) because of the semantic connotations of progress in the word. Progress can only be observed in relation to *time*. Development in monolingual and bilingual acquisition of phonology takes place over a long period of time, that is, years. Deep-rooted theoretical controversies in phonological development that relate to timescales are "what is a stage," "what is/are the developmental path(s)," and "whether development is continuous or discontinuous." The term "*stage*" may refer to a phase, a period, a step, a point, a juncture, as well as time. Owing to the multiple semantic references involved in it, *stage* has been arbitrarily used in both language acquisition and phonological development to refer to similar, but not the same, entities. However, clearly defining what a *stage* is will help determine the presence or absence of stages in phonological development that will ultimately provide substantial evidence for a theory of language acquisition (Bernhardt and Stemberger, 1998:6–7).

With regard to developmental paths, a pattern that has been identified qualitatively in speech development (Stemberger et al., 1999; Stoel-Gammon, 2011) is the U-shape: it traces speech productions that first appear (progress), then disappear (regress), and then apparently reappear (progress again) over time.

A "fluctuation" phenomenon (Dodd et al., 2003:622), whereby speech sounds seem acquired for a period and then they disappear until a later stage is known as "reversal" (Wellman et al., 1931). This is historically supported by the experimental study of memory in the work of Ebbinghaus (1913), who pioneered the idea of a forgetting and a learning S-curve, what is often referred to as the logistics curve (or function) of growth of some population P, originally due to P. F. Verhulst (Bacaër, 2011). Bills (1934) elaborated on this with a more detailed description of learning curves identifying the three properties: *negative acceleration, positive acceleration*, and *plateaus*. In economics, development refers to a system-learning process with varying rates of progression and the S-curve has different appearances depending on the timescale of observation. Stemberger et al. (1999:1) state that "U-shaped learning appears to be especially prevalent in the development of phonology, where it is known as *regression*." Regression, despite an evidence of progression, may also be found in an undeveloped "frozen form" (Ferguson and Farewell, 1975) or in a "phonological idiom" (Moskowitz, 1971) that persists over time and does not uniformly improve across the board.

Predominant in the question of developmental path(s) is the continuity/discontinuity hypotheses. Proponents of the continuity hypothesis, mostly tackled in terms of child speech development, argue that the child gradually and naturally goes from the initial to the end state (Tesar and Smolensky, 1998). In discontinuity arguments (Ferguson and Farewell, 1975; Vihman, 1992), a stage is a compulsory step in the developmental sequence with clearly defined, noninterrelating steps that are associated with distinct maturational advances in the biology of humans (Jakobson, 1941/1968). However, how we view time in the study of developmental phenomena affects the outcomes because "our notions of continuity and discontinuity depend on the time scale we actually contemplate" (Van Geert, 1994:6), as well as, on defining what it is that is being continued/discontinued, i.e., defining exact variables that can be quantified and traced along the time span of development.

39.5 Quantifiable Speech Variables

Overall, chaos theory is often used in the study of language in a metaphorical sense. However, chaos theory is an essentially mathematical field. Studying development in terms of chaos encourages viewing growth as a "quantifiable variable (i.e., in the sense of numbers)" (Van Geert, 1994). The primary quantifiable variable in speech is the produced phoneme (speech sound or phone), which is represented either in text by a phonetic transcriptional system or in acoustics by the numerical values of acoustic correlates. Other quantifiable linguistic variables are the syllable, the word, the sentence, etc.

Speech data are collected in digital recordings and may be subject to phonetic transcription and/or acoustical analysis for the study of speech. The most widely used transcriptional system is the International Phonetic Association (IPA) chart (graphemes, diacritics, and suprasegmentals). So, *tough* and *cough* are [tʰʌf], [kʰʌf] in IPA transcription, respectively, while *through* and *bough* are [θɹuː] and [baʊ]. It is obvious that in spite of the similarity in spelling, there is variability in production. Phonetic transcription of a word may also vary depending on whether the utterance is articulated in running speech or independently; for instance, *don't* [doʊnt] and *you* [ju] become [doʊntʃə] in running speech. Phonetic transcription, however, is only a representation of actual speech sounds and may not portray minute production differences that are present even in a single speaker's repetitions. An acoustical analysis of speech data (e.g., power spectra, pitch, intensity, formants, etc.) is the reply to such discrepancies and is common in the study of speech (Flipsen et al., 1999; Escudero et al., 2009). Praat (Boersma and Weenink, 2014) is a widely used computer program that facilitates acoustical analyses, as well as the statistics (e.g., multidimensional scaling, principal component, and discriminant analyses) on those analyses.

Measuring acquisition in phonological development has been of much interest. Quantitative approaches in phonological development involve statistical frequencies on the extent to which a phoneme is adequately proven to be consistently produced as in targeted speech (Ingram, 1981b; Shriberg et al., 1997; Macleod et al., 2011; Freedman and Barlow, 2012). Children's "productions of sounds in word contexts are usually examined in terms of degree of production accuracy and the percentage of children in an age group who reached the level of accuracy in phoneme production" (Dodd et al., 2006:26). Despite the abundance of speech data, phonological development experiences a shortage of dense longitudinal data of many individual data sets that are also quantifiable, which is true in both monolingual and bilingual acquisition (Bernhardt and Stemberger, 1998; Van Dijk et al., 2011). The longest-spanning case studies of a child's phonological development, for instance, are qualitative in nature (Leopold, 1949), while the norms established through cross-sectional studies (Prather et al., 1975; PAL, 1995) are based on child group (rather than on individual) data collected longitudinally in spaced-out patches and averaged out. Moreover despite the use of statistical methods in applied linguistics, there is an evidence of the need for better foundations on them (Lazaraton, 2012), as well as some progress in that direction (Larson-Hall, 2009; Lowie and Seton, 2013). These may be some of the reasons that modeling and simulating phonological development during its entire span and on mathematical grounds is lagging behind. "Only if [. . .] models are transformed into calculus, however simple its form, will deductive inferences become possible" (Van Geert, 1994:19).

39.6 A Sound's Development in a Child's Speech

Here, data collected and processed by the author of the chapter are considered in order to obtain the growth pattern of sounds in child speech development. In particular, a female child's speech during her phonological development in simultaneous Greek/English bilingualism was recorded on an almost daily basis, longitudinally from age 2 years and 6 months (2;6) until 3 years and 10 months (3;10). The data in both languages were entered in orthographic and IPA phonetic transcription and time aligned to sound in a CLAN (MacWhinney, 2000) database. The main focus has been on the development of English, being the weaker language in the child's bilingualism. This chapter, in particular, investigates the acquisition path of a specific phoneme, the voiced interdental fricative /ð/, as in the word *this*. The targeted /ð/ words in the child's English speech and the age of their first production are *brother, father, further, other, that, the, there, these, they, this, together* (2;6); *another, bothers, them, then* (2;7); *bothering, mother, there's, those* (2;8); *bother, either, that's* (2;9); *others* (2;11); *lather, themselves* (3;3); *without* (3;4); *than* (3;5); *themselves* (3;7); *though* (3;8); *neither* (3;9); *breathe* (3;10). In Figure 39.1, the total number of the child's targeted /ð/ as it changes weekly is shown. We can see that the weekly change is chaotic.

During development, the child cannot predominantly and correctly produce the sound [ð] and she substitutes it with sounds that she can produce correctly (adult like) by this age, the sounds such as [l] and [d]. The developmental path of the acquisition of /ð/ is shown in Figure 39.2 as the percentage of produced [ð] to targeted /ð/.

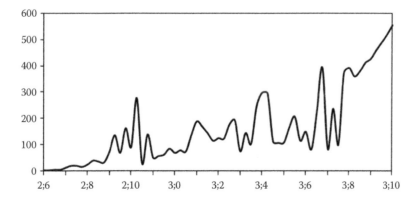

FIGURE 39.1 The number of the child's targeted /ð/ versus age.

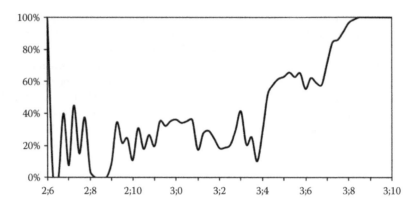

FIGURE 39.2 The acquisition level of /ð/ in development.

We can see that there are basically two stages of development, a cyclic and a progressive. From age 2;6 to age 3;4, the acquisition level of /ð/ exhibits a disordered cyclicity, a strange attractor. From age 3;4, there is growth in a double-logistic-like manner with a strange attractor connecting two single logistic-like growth patterns from age 3;5 to about age 3;7. At age 3;9, the child completely acquires this sound. The two developmental stages of Figure 39.2 may be associated with the second and third parts of a U-shape. U-shaped phonological and linguistic learning and development has been discussed previously in the literature in a qualitative manner (Bills, 1934; Werker et al., 2004) but, in this chapter, a quantitative description is provided both for the depth and length of U, which are 0.75 and 10 months, respectively. The depth of U may be seen more clearly in Figure 39.3 where the map of the acquisition level of /ð/ between succeeding and preceding ages in increments of 1 week is shown.

The two strange attractors in the child's acquisition of /ð/ during development are clearly visible. The major strange attractor in the cyclic stage (the bottom of the U-shaped pattern) is centered at the 25% acquisition level, whereas the minor strange attractor is slowing down for 2 months, and the double-sigmoid-like progressive stage is centered at the 62% acquisition level of /ð/.

It is worth noting that the child's development of other speech sounds exhibits similar paths even though the acquisition levels at different ages differ from sound to sound depending on the place and manner of articulation of the speech sound. The point here was to quantitatively demonstrate the developmental paths of a particular speech sound; providing a full picture of the development of all phonemes in her speech would be outside the scope of this chapter.

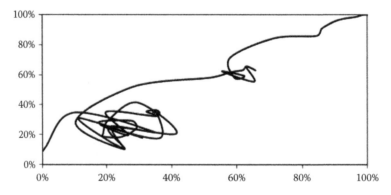

FIGURE 39.3 The map of the acquisition level of /ð/ between succeeding (vertical axis) and preceding (horizontal axis) ages in increments of 1 week.

39.7 Conclusions

As seen, scientific explanations on speech and its acquisition/learning are interspersed by the tenets of chaos theory that include the following: nonlinearity, complexity, self-reorganization, change, cyclicity, iteration, scale invariance, etc. Quantifying language by using variables such as speech segments allows mathematical computations that in turn provide empirical and quantitative support to impressionistic approaches of chaos theory in the understanding of language in all its variations: first, second, bilingual, dialectal, child, or adult. Dense and longitudinal speech data facilitate detailed investigations in terms of chaos theory in language. To this end, this chapter utilized the data of a Greek/English bilingual child case study in phonological development. Specifically, the chapter quantitatively demonstrated chaotic patterns such as the existence of disordered cylicity, two strange attractors of variable duration, as well as a double-logistic-like growth pattern in the developmental acquisition of the child's voiced interdental fricative /ð/ in English from age 2;6 to 3;10. Detailed and longitudinal investigations of changes (i.e., variability) of speech in all its possible variations will further substantiate chaos in monolingual and bilingual speech.

References

E. Babatsouli and D. Ingram. What bilingualism tells us about phonological acquisition. In: Bahr R. H. and Silliman E. R. (eds.) *Routledge Handbook of Communication Disorders*. Routledge: Taylor & Francis, pp. 173–182, 2015.

E. Babatsouli, D. Ingram, and D. A. Sotiropoulos. Phonological word proximity in child speech development. *Chaotic Modeling and Simulation* 4 (3):295–313, 2014.

N. Bacaër. Verhulst and the logistic equation (1838), in Bacaër N. (ed.) *A Short History of Mathematical Population Dynamics*, London: Springer-Verlag London, pp. 35–39, 2011.

C. Beckner, R. Blythe, J. Bybee, M. H. Christiansen, W. Croft, J. Holland, D. Larsen-Freeman, J. Ke, and T. Schoenemann. Language is a complex adaptive system: Position paper, The "Five Graces Group." *Language Learning* 59 (1):1–26, 2009.

B. M. Bernhardt and J. Stemberger. *Handbook of Phonological Development: From a Nonlinear Constraints-Based Perspective*. San Diego: Academic Press, 1998.

A. G. Bills. *General Experimental Psychology, Longmans Psychology Series*. New York, NY: Longmans, Green & Co, 1934.

P. Boersma and D. Weenink. *Praat: Doing Phonetics by Computer* [Computer program], Version 5.4.04, http://www.praat.org/, 2014.

L. Bosch and M. Ramon-Casas. Variability in vowel production by bilingual speakers: Can input properties hinder the early stabilization of contrastive categories? *Journal of Phonetics* 39 (4):514–526, 2011.

N. Chomsky. Approaching UG from below. In: Uli S. and Gärtner H. M. (eds.) *Interfaces+Recursion=Language?* New York: Mouton de Gruyter, pp. 1–29, 2007.

S. Curtiss. *Genie: A Psychological Study of a Modern-Day "Wild Child."* New York: Academic Press, 1977.

C. M. Danforth. Chaos in an atmosphere hanging on a wall. *Mathematics of Planet Earth 2013*, http://mpe2013.org/2013/03/17/chaos-in-an-atmosphere-hanging-on-a-wall/, 2013. Retrieved online on January 2016.

K. De Bot. A bilingual production model: Levelt's 'speaking' model adapted. *Applied Linguistics* 13:1–24, 1992.

K. De Bot. Introduction: Second language development as a dynamic process. *The Modern Language Journal* 92:166–178, 2008.

K. De Bot, W. Lowie, S. L. Thorne, and M. H. Verspoor. DST as a comprehensive theory of second language development. In: del Pilar M. G. M., Mangado M. J. G., and Martínez A. M. (eds.) *Contemporary Approaches to Second Language Acquisition*. Amsterdam: John Benjamins, pp. 199–220, 2013.

K. De Bot, W. Lowie, and M. H. Verspoor. *Second Language Acquisition: An Advanced Resource Book*. London: Routledge, 2005.

K. De Bot, W. Lowie, and M. H. Verspoor. A dynamic systems theory approach to second language acquisition. *Bilingualism: Language and Cognition* 10:7–21, 2007.

R. L. Devaney. *An Introduction to Chaotic Dynamic Systems* (2nd ed.), Boulder, CO: Westview Press, 2003.

D. A. Dinnsen. Variation in developing and fully developed phonologies. In Ferguson C. A., Menn L., and Stoel-Gammon C. (eds.), *Phonological Development: Models, Research, Implications*. Timonium, MD: York Press, pp. 191–210, 1992.

B. Dodd, A. Holm, S. Crosbie, and J. Bloomfield. English phonology: Acquisition and disorder. In: Hua Z. and Dodd B. (eds.) *Phonological Development and Disorders: A Multilingual Perspective*. Clevedon, England: Multilingual Matters, pp. 25–54, 2006.

B. Dodd, A. Holm, Z. Hua, and S. Crosbie. Phonological development: A normative study of British speaking children. *Clinical Linguistics and Phonetics* 17 (8):617–643, 2003.

H. Ebbinghaus. *A Contribution to Experimental Psychology*. New York: Teachers College Press, 1913.

R. Ellis. *The Study of Second Language Acquisition*. Oxford: Oxford University Press, 2008.

P. Escudero, P. Boersma, A. S. Rauber, and R. Bion. A cross-dialect acoustic description of vowels: Brazilian and European Portuguese. *Journal of the Acoustical Society of America* 126:1379–1393, 2009.

L. Fausett. *Fundamentals of Neural Networks*. Englewood Cliffs, NJ: Prentice-Hall, 1994.

C. Ferguson and C. Farewell. Words and sounds in early language acquisition: Initial consonants in the first fifty words. *Language* 51:419–439, 1975.

P. Jr. Flipsen, L. Shriberg, G. Weismer, H. Karlsson, and J. McSweeny. Acoustic characteristics of /s/ in adolescents. *Journal of Speech, Language, and Hearing Research* 42 (3):663–677, 1999.

S. E. Freedman and J. A. Barlow. Using whole-word production measures to determine the influence of phonotactic probability and neighborhood density on bilingual speech production. *International Journal of Bilingualism* 16 (4):369–387, 2012.

J. A. Gierut. Featural categories in English phonemic acquisition. In: Bernhardt B. M., Ingram D., and Gilbert J. (eds.) *Proceedings of the UBC International Conference on Phonological Acquisition*. Somerville, MA: Cascadilla Press, pp. 42–52, 1996.

J. A. Gierut. Phonological complexity and language learnerability. *American Journal of Speech Language Pathology* 16(1):6–17, 2007.

J. Gleick. *Chaos: Making a New Science*. New York: Random House, 1997.

H. Goad and D. Ingram. Individual variation and its relevance to a theory of phonological acquisition. *Journal of Child Language* 14:419–432, 1987.

D. W. Green. Control, activation, and resource. *Brain and Language*, 27:210–223, 1986.

F. Grosjean. The bilingual's language modes. In: Nicol J. (ed.), *One Mind, Two Languages: Bilingual Language Processing*, Oxford: Blackwell, pp. 1–22, 2001.

F. Grosjean. Bilingualism: A short introduction. In: Grosjean F. and Li P. (eds.) *The Psycholinguistics of Bilingualism*. Oxford, Malden, MA: Wiley-Blackwell, pp. 5–26, 2013.

B. Hasselblatt and A. Katok. *A First Course in Dynamics: With a Panorama of Recent Developments*. Cambridge, UK: Cambridge University Press, 2003.

N. Hornstein and D. Lightfoot. *Explanation in Linguistics: The Logical Problem of Language Acquisition*. London and New York: Longman, 1981.

D. Ingram. On variant patterns of language acquisition. Presentation at the *Biennial Meeting for the Society of Research on Child Development*, Boston, MA, 1981a.

D. Ingram. *Procedures for the Phonological Analysis of Children's Language*. Baltimore, MD: University Park Press, 1981b.

R. Jakobson. *Child Language, Phonological Universals and Aphasia*. The Hague: Mouton, 1941/1968.

U. Jessner. A DST model of multilingualism and the role of multilinguistic awareness. *Modern Language Journal* 92:270–283, 2008.

R. Kautz. *Chaos: The Science of Predictable Motion*. Oxford: Oxford University Press, 2011.

S. H. Kellert. *In the Wake of Chaos: Unpredictable Order in Dynamical Systems*. Chicago: University of Chicago Press, 1993.

M. H. Keshavarz and D. Ingram. The early phonological development of a Farsi-English bilingual child. *International Journal of Bilingualism* 6 (3):255–269, 2002.

B. Köpke, M. Schmid, M. Keijzer, and S. Dostert (eds.) *Language Attrition: Theoretical Perspectives.* Amsterdam: John Benjamins, 2007.

J. W. Kuhl. *The Idiolect, Chaos, and Language Custom Far from Equilibrium: Conversations in Morocco,* PhD thesis, The University of Georgia, 2003.

D. Larsen-Freeman. Chaos/complexity science and second language acquisition. *Applied Linguistics* 18 (2):141–165, 1997.

D. Larsen-Freeman. A chaos/complexity theory perspective. In: Kramsch C. (ed.) *Language Acquisition and Language Socialization: An Ecological Perspective.* London: Continuum Press, pp. 33–46, 2002.

D. Larsen-Freeman. On the complementarity of chaos/complexity theory and dynamic systems theory in understanding the second language acquisition process. *Bilingualism: Language and Cognition* 10 (1):35–37, 2007.

D. Larsen-Freeman and L. Cameron. *Complex Systems and Applied Linguistics.* Oxford: Oxford University Press, 2008a.

D. Larsen-Freeman and L. Cameron. Research methodology on language development from a complex systems perspective. *The Modern Language Journal* 92:200–213, 2008b.

J. Larson-Hall. *A Guide to Doing Statistics in Second Language Research Using SPSS.* London: Routledge, 2009.

A. Lazaraton. Current trends in research methodology and statistics in applied linguistics. *TESOL Quarterly* 34 (1):175–181, 2012.

W. J. M. Levelt. *Speaking: From Intention to Articulation.* Cambridge, MA: MIT Press, 1989.

P. Li and X. Zhao. Computational modeling of bilingual language acquisition and processing: Conceptual and methodological considerations. In: Schwieter J. W. (ed.) *The Cambridge Handbook of Bilingual Processing.* Cambridge: Cambridge University Press, 2015.

E. N. Lorenz. Predictability: Does the flap of a butterfly's wings in Brazil set off a tornado in Texas? Paper presented at *139th Meeting of the American Association for the Advancement of Science,* Boston, MA: December 29, 1972.

W. Lowie and K. De Bot. Variability in bilingual processing: A dynamic approach. In: Schwieter J.W. (ed.) *The Cambridge Handbook of Bilingual Processing.* Cambridge: Cambridge University Press, 2015.

W. Lowie and B. J. Seton. *Essential Statistics for Applied Linguistics.* Basingstoke: Palgrave MacMillan, 2013.

W. Lowie and M. H. Verspoor. The dynamics of multilingualism: Levelt's speaking model revisited. In: Schmid M. S. and Lowie W. (eds.) *Modeling Bilingualism: From Structure to Chaos.* Amsterdam: John Benjamins, pp. 26–288, 2011.

A. A. Macleod, K. Laukys, and S. Rvachew. The impact of bilingual language learning on whole-word complexity and segmental accuracy among children aged 18 and 36 months. *International Journal of Speech and Language Pathology* 13:490–499, 2011.

B. MacWhinney. *The Emergence of Language.* Mahwah, NJ: Lawrence Erlbaum, 1999.

B. MacWhinney. *The CHILDES Project: Tools for Analyzing Talk.* Mahwah, NJ: Lawrence Erlbaum, 2000.

V. Marian and M. Spivey. 2003. Competing activation in bilingual language processing: Within- and between language competition. *Bilingualism: Language and Cognition,* 6 (2):97–115, 2003.

K. P. Mohanan. Emergence of complexity in phonological development. In: Ferguson C. A., Menn, L. and Stoel-Gammon C. (eds.) *Phonological Development: Models, Research, Implications.* Maryland, MD: York Press, pp. 635–662, 1992.

K. P. Mohanan. Fields of attraction in phonology. In: Goldsmith J. (ed.) *The Last Phonological Rule: Reflections on Constraints and Derivations.* London: The University of Chicago Press Ltd., pp. 61–116, 1993.

C. H. Morten and S. Kirby. *Language Evolution.* Oxford: Oxford University Press, 2003.

A. I. Moskowitz. *Acquisition of Phonology.* Unpublished PhD dissertation, University of California, Berkeley, USA, 1971.

N. Müller and A. Hulk. Crosslinguistic influence in bilingual acquisition: Italian and French as recipient languages. *Bilingualism: Language and Cognition* 4:1–21, 2001.

PAL (Panhellenic Association of Logopaedics). *Assessment of Phonetic and Phonological Development.* Athens: PAL (in Greek), 1995.

W. Penfield and L. Roberts. *Speech and Brain Mechanisms.* Princeton: Princeton University Press, 1959.

T. Piske and M. Young-Scholten (eds.). *Input Matters in SLA.* Clevedon: Multilingual Matters, 2009.

S. Poplack. Sometimes I'll start a sentence in Spanish Y TERMINO EN ESPAÑOL: Towards a typology of code-switching. *Linguistics*, 18:581–618, 1980.

R. F. Port and T. van Gelder (eds.). *Mind as Motion.* Cambridge, MA: MIT Press, 1995.

E. M. Prather, D. L. Hedrick, and C. A. Kern. Articulation development in children aged two to four years. *Journal of Speech and Hearing Disorders* 40:179–191, 1975.

K. Rice and P. Avery. Variability in a deterministic model of language acquisition: A theory of segmental elaboration. In: Archibald J. A. (ed.) *Phonological Acquisition and Phonological Theory.* Hillsdale, NJ: Lawrence Erlbaum, pp. 23–42, 1995.

M. S. Schmid and W. Lowie. *Modeling Bilingualism from Structure to Chaos: In Honor of Kees de Bot.* Amsterdam: John Benjamins, 2011.

L. Selinker. Interlanguage. *International Review of Applied Linguistics* 10:209–231, 1972.

L. D. Shriberg, D. Austin, B. A. Lewis, J. L. McSweeny, and D. L. Wilson. The percentage of consonants correct (PCC) metric: Extensions and reliability data. *Journal of Speech, Language and Hearing Research*, 40:708–722, 1997.

C. H. Skiadas and C. Skiadas. *Chaotic Modelling and Simulation: Analysis of Chaotic Models, Attractors and Forms.* New York, NY: Taylor & Francis/CRC Press, 2009.

B. F. Skinner. *The Behavior of Organisms: An Experimental Analysis.* Cambridge, MA: B. F. Skinner Foundation, 1938.

S. Sorin and S. Eran. Complexity: A science at 30. *Europhysics News* 34 (2):54–57, 2003.

J. P. Stemberger, B. M. Bernhardt, and C. E. Johnson. U-shaped learning in phonological development. Presentation at the *6th International Child Language Congress.* Online: http://roa.rutgers.edu/files/471-1101/471-1101-stemberger-0-0.pdf, 1999.

J. W. Schwieter and N. Tokowicz. Bilingual processing: A dynamic and rapidly changing. In: Schwieter J. W. (ed.) *The Cambridge Handbook of Bilingual Processing.* Cambridge: Cambridge University Press, 2015.

C. Stoel-Gammon. Relationships between lexical and phonological development in young children. *Journal of Child Language* 38:1–34, 2011.

B. Tesar and P. Smolensky. Learnability in optimality theory. *Linguistic Inquiry* 29:229–268, 1998.

N. S. Trubetzkoy. *Fundamentals of Phonology.* Travaux du Cercle Linguistique, 1939/1958.

N. S. Trubetzkoy. Grundzügeder Phonologie. Travaux du Cercle Linguistique de Prague, 7. C. Baltaxe (trans.). *Principles of Phonology.* Berkeley: University of California Press, 1939/1969.

M. Van Dijk, M. H. Verspoor, and W. Lowie. Variability and DST. In: Verspoor M., De Bot K., and Lowie W. (eds.) *A Dynamic Systems Approach to Second Language Development: Methods and Techniques.* Amsterdam, Philadelphia: John Benjamins, pp. 55–84, 2011.

P. Van Geert. *Dynamic Systems of Development: Change between Complexity and Chaos.* New York: Harvester Wheatsheaf, 1994.

M. H. Verspoor, K. De Bot, and W. Lowie. *A Dynamic Approach to Second Language Development: Methods and Techniques.* Amsterdam/Philadelphia: John Benjamins, 2011.

M. M. Vihman. Early syllables and the construction of phonology. In: Ferguson C. A., Menn L., and Stoel-Gammon C. (eds.) *Phonological Development: Models, Research, Implications.* Timonium, MD: York Press, pp. 393–422, 1992.

B. L. Wellman, I. M. Case, I. G. Mengert, and D. E. Bradbury. Speech sounds of young children. *University of Iowa Studies in Child Welfare*, 2 (5). Iowa city: The Iowa Child Welfare Station: University of Iowa, 1931.

J. F. Werker, D. G. Hall, and L. Fais. Reconstruing U-shaped functions. *Journal of Cognition and Development* 5 (1):147–151, 2004.

Chaos In Music

40

Composers and Chaos: A Survey of Applications of Chaos Theory in Musical Arts and Research

Scott Mc Laughlin

40.1 Introduction

40.1.1 Chaos and Music

The general case for linking music and chaos theory can be made on several levels, from the metaphoric to the procedural. At the most basic level is the irresistible connection between the visible patterns of chaotic equations and the motion and contours of organic systems themselves; it is not for nothing that Conway referred to his early cellular automata system as the *Game of Life*. This organic quality is a point of connection with music, which has a long and rich relationship with metaphors of nature. As long as people have spoken about music they have referred to nature as a simile, music sounds "fluid" or "wild": notwithstanding when music is referred to anthropomorphically, as an avatar of persons or personae, "human" and "nature" can both be subsumed in music. Many composers who work with chaos theory, this writer included, describe this connection as their own gateway to using non linear dynamic systems as a compositional tool or subject. Composers have a tendency to see the world in terms of music, or at least in terms of the aspects of music they resonate with, and for certain composers, chaos theory can be seen to have musical qualities in relation to variation and transformation through iteration. In his essay "Music and Fractals," composer Charles Wuorinen (1938–) describes what for him are fractal characteristics of music:

> [...] in traditional western diatonic-tonal music, there is a strong tendency for similar structures to appear on different time-scales. Thus the same harmonic progression may

determine the course of a whole movement, of a sizable section of it, or of a single short phrase. Traditional terminology shows, by its indifference to scale, how deeply imbedded self-affine structures in compositions are: "C major" can refer to a single sonority, the key of a phrase, a movement-section, a whole movement, or the key of an entire work. And in post-tonal music similar structures on differing time-scales are certainly present.

Wuorinen goes on to describe a general method of "fractal composition" he developed based on the principle of nesting. By selecting a group of pitch intervals that will be important on a local level of notes and phrases. These same intervals would also be "flung over the duration of the whole work or movement, [...] used to divide a large span of time into sections whose lengths are directly proportional to the interval-sizes of the original succession." Wuorinen goes on to point out that this process is not the composition itself, only a preparation.

> It should be clear that what I have sketched above is not really 'composing' in the sense we might historically apply to the familiar figures of the western tradition. It is rather the preparation for the composing, [...]. Having made such preparation, then, I have found it possible to compose with a kind of intuitive freedom which still assures macrostructural coherence. Those who try for this coherence without structural underpinnings usually fail. [19]

For Wuorinen, the act of composition requires a level of intuitive "play," the site of composition for him remains at this level, with the process being relegated to a preparatory act: "play" here is a particularly appropriate term; Johan Huizinga describes "play" as a primary condition for the generation of culture, and music as something that arises from this [6]. Composers move between composition as intuitive play—in the sense of exploration that Huizinga expands upon—and composition as a rigorous process, which in itself can arise as an idea concretized and extended from play, or can be a formula of the composer's personal practice or handed down through pedagogy or genre.

James Gleick's popular science book *Chaos: Making a New Science* was an entry point to non linear dynamics for many artists, explaining the concepts in a way that sparked ideas without requiring specialist knowledge. In it, he referred to "[t]he first chaos theorists, the scientists who set the discipline in motion, shared certain sensibilities. They had an eye for pattern, especially pattern that appeared on different scales at the same time. They had a taste for randomness and complexity, for jagged edges and sudden leaps" [5]. This passage could just as easily be referring to modernist composers as all of these are attributes that could describe particular flavors of twentieth-century music, and without too much of a stretch could even describe Beethoven or other older composers. Judy Lochead draws our attention to the philosopher of science Katherine Hayles, who makes the connection between chaos theory and the post-structural thought that has redefined humanities in the last 50 years, citing Roland Barthes' "extolling the virtues of noisy interpretations of literature." Lochead goes on to describe Hayles, vision of chaos theory's intersection with artistic culture:

> For Hayles, the broader framework is a new cultural paradigm, which she refers to as "chaotics," affecting not simply scientific practices but social and intellectual ones as well. Bound up with this new paradigm is postmodern philosophical and critical scholarship, which has called into question distinctions fundamental to knowledge itself: distinctions between order and disorder, unity and disunity, depth and surface, rational and irrational, presence and absence. "Chaotics" in the sense that Hayles articulates it, comprises one slice of the intellectual and cultural shift that has been the subject of much recent scholarship in the humanities and the sciences. [9]

This idea is echoed in a wide-ranging essay by Kenneth McLeod on chaos theory and critical theory where he describes chaos theory as a possible bridge to reconcile overly simplistic and positivistic readings of culture that perpetuate the dualism of science as objective and art as subjective [10]. This idea that chaos theory connects to the noisy interpretation of music resonates strongly with composers, who are always caught between opposing Apollonian and Dionysian impulses, the intellect and the passions. Depending on who you listen to, music as an art form is wholly derived from one of these impulses: we valorize the towering intellect of the "genius" of Beethoven, while simultaneously valorizing the "raw energy" of Jimi Hendrix, while Mozart is a "prodigy," an ontological curiosity, part genius, part primitive.

This perceived incompatibility between the objective and subjective impulses is slowly eroded by post-structural thought in the twentieth century and even today. Most composers would agree that their practice is a mixture, to varying levels, of rigorous process and intuitive decision-taking. Peter Paul Nash, discussing the work of Charles Wuorinen, says of chaos theory that it "could be the most radical way of reconsidering the world since the great modernist and reductive theories of the early twentieth century; since it is so anti-reductive, and admits the irregularities, proportionalities, hierarchies and sheer messiness of 'real life.' The implications for music are obvious. Indeed a succession of 'proto-fractal' composers can be traced back into the past: Sibelius, Beethoven, J. S. Bach, and even further back" [19].

40.1.2 Parameterization of Music

Explaining the relationship of chaos theory to various aspects of musical composition requires a grounding in the basic tools and procedures used by composers of Western art music and a brief narrative on the history of Western music and composition in the twentieth and twenty-first centuries.

Musical composition in the twenty-first century involves a dizzying profusion of styles, sounds, and aesthetic positions, many of which are mutually contradictory, and fashions that change from generation to generation. There are almost no objective truths in composition, as composer Jan Beran puts it, "In music, there is usually no clear definition of optimality or, if there is, no unique optimal answer exists" [2]. The most interesting work is often done by composers who break the rules of the previous generation or by composers who find new ways to think about problems that seemed already solved. In demonstrating the links between composition and chaos, arguably, the most important point to make about the multiplicity of individual voices and styles is that musical composition is a highly personal balance of intuition and process. Every composer finds their own way of working with materials, and for composers working with chaos theory, each has their own method and reasoning. The act of composition is multivalent, involving overlapping and interfering influences and materials, with each composer and each work involving multiple possible sites of compositional acts. At the extreme edges of the spectrum, some composers such as Tom Johnson (1939–) work almost exclusively with bare process. The site of composition for Johnson is the point at which he decides on what process to use in the piece, and how this process should be mapped in a musically meaningful way. As an example, his piece *The Chord Catalogue* (1986) consists of every chord possible in the space of a single octave (there are 8178), played one by one in ascending order of size and starting pitch: first all the two-note chords, then all the three-note chords, etc. Once this is decided, there is no room for intuition or taste, the composition simply unfolds as this set of sounds in time. Another example of Johnson's is his *Rational Melodies II* that uses 384 iterations of the "Dragon Pattern" Lindenmayer system as a process. The piece simply replaces the left and right folding operations of the L-system with steps up and down through a musical scale: conceivably, any scale could be inserted into the process to create variations on this piece. This process piece seeks to highlight the process itself as the music, with minimum artistic (or interpretive) intervention. At the opposing end of this notional spectrum are composers who write completely by intuition, where the act of composition is contingent on the sounding material and is situated in decisions taken reflexively during the moment-to-moment rehearsal of the material (sounds, melodies, concepts, etc.), either via the composer's inner ear or by near at an instrument and feeling their way through the musical possibilities to decide what should happen next; what the musical consequences of the preceding musical decisions imply for them in that moment. Of course, the majority of composers have a personal blend of both possibilities, the site of the compositional act varies and may change from piece to piece. For example, composition may involve the strict application of a process that is later manipulated intuitively according to taste, or the application of small local processes to a material, the results of which are ordered according to what "sounds best." It is commonplace even to examine the works of the old masters of Western music (Beethoven, Mozart, Bach, etc.) to find passages that sound fluid and improvised to the casual ear, but which are actually composed by the application of common musical formulae. Process and intuition exist hand in hand. Composition is more than simply the ordering of notes on a page. Notes have histories and their meanings change with respect to the context of the listener. Composers have contexts, influences, objectives (from the concrete/professional to the abstract and personal). Composition is a form of communication by structuring materials into meaningful relationships:"meaningful to whom" is an entirely different question.

Inherent in all composition is the existence, implicit or not, of abstract musical structures, patterns of sound in memory. These may be anything from complex mathematically derived relationships between disparate materials, to the most simple of antecedent–consequent relationships (such as nursery rhymes), but all of these are multiply nested hierarchical structures of difference. Music proceeds by the exposition of material and by the subsequent development of that material by a variety of means; such as exploration of its internal relationships and structures as in the baroque fugue, or its relationships with itself as in the classical sonata where the main melodies are deconstructed, extended, and distorted, or even with external materials such as other musics or extra-musical concepts. Beran notes that this development is often algorithmic in nature: "Whether we listen to a sonata, a symphony, or an Indian raga, 'logical' construction is an inherent part of music. Standard musical techniques, such as retrograde, inversion, arpeggio, or augmentation are mathematical functions" [2]. This development of the material requires some balance between the most basic of musical concepts, repetition, and variation. To demonstrate with recourse to extremes, the spectrum extends from total repetition, the repetition of a single sound for all infinity, to total variation, the constant and extreme random variation of sound in all possible parameters and dimensions. Most music consists of materials that is both repeated and varied, with occasional sharp contrasts of new materials that are themselves repeated and varied before returning to the previous material or another variation on it. The balance of repetition and variation provides novelty and change without departing too far from the normative musical behavior implied by a genre.

This is where the connection to chaos can be made and the behavior of dynamical systems often demonstrates these same characteristics of repetition and variation across multiple overlapping scales. Jeff Pressing notes that phenomena such as fixed points, limit cycles, bifurcations, chaos, and strange attractors, can be viewed as similar to "the transformation or distortion of a simple entity (a [musical] motive), often followed by some sort of return to the original motive": an example of this is given below in the section on Pressing's work [15]. To show how composers make the connection between dynamical systems and music requires a brief exposition of compositional methods post-1900.

The compositional act is often reduced in the public mind to either the micro-choices, the moment-to-moment choice of notes/rhythms/instruments/etc, or the macro-choices, the structures and forms that are then filled by the composer. The former image feeds into the Romantic notion of composer as "genius," notes springing from his mind and thrust onto paper the, pieces erupting spontaneously from "his" passions. The latter image feeds into the opposing but complementary composer as architect, the meticulous craftsman of forms, symphonies, cycles, etc., where all the elements of the music are carefully related to one another, like a vast puzzle of sound. This latter idea has been fostered largely in the twentieth century through the valorization of mathematical approaches to musical composition, which was engendered by the increasing trend toward parametrization of music, a trend that arguably began with Arnold Schoenberg's "The Method of Composing with Twelve Notes Related Only to One Another" (1920–1922). Schoenberg's technique was later referred to as *serialism* (also twelve-tone music, or dodecaphonism), this method was developed to afford each of the twelve pitches a statistically even weighting within a piece, to avoid the gravitational pull that the tonic note exerted within tonal music, and to allow all notes to relate to each other equally, to "emancipate dissonance" as Schoenberg explained. However, it also derived from the idea that a work of art should have a single seed, an idea, in this case it is the *series*, the idea that musical continuity could be maintained in a piece of music by providing a strict serial ordering for the twelve possible chromatic pitches. This ordering persists throughout the work in one form or another but is varied through transposition, inversion, retrograding, and variation in other parameters of music, rhythm, register, instrumentation, phrasing, etc.

The serial system itself does not generate music, it simply provides an ordering of the twelve pitches, but this acts as a foil for the composer's creativity. The strict ordering can be observed or subverted, with an infinity of possible points along this scale that allows varying levels of perceivable order and disorder. In the generations that followed Schoenberg, the system rapidly became popular among a certain school of composers of art and concert music, and individual composers were quick to grasp the basic concept of the system and alter it to create their own individual compositional language. Later European composers such as Pierre Boulez (1925–) and Karlheinz Stockhausen (1928–2007) expanded the serial approach beyond pitches to include all other parameters that could conceivably be performed on a numerically defined scale, including note duration, dynamics, register, etc. The short but dramatic period of *total serialism* in the 1950–1960s is characterized by works where musical parameter was controlled in relation to a single series, the numbers 1–12 in a specific ordering. The next level of mathematical manipulation was developed largely by American composers such as Milton Babbitt (1916–2011) and others in tandem with the development of computation in the United States. Thus, the parametrization of music is extended via numerical manipulation with the development of musical theories from set theory, such as musical combinatoriality: a development the musicologist Richard Steinitz described as "a subterranean world of software research in which [...] intuitive artistic impulses easily get submerged" [17]. Ultimately, however, these mathematical constructs still served a single purpose, to create structures of musical relationships in the context of free twelve-note harmony. These mathematical structures were seen as a necessity for modern composers, to allow them to continue the compositional traditions that came before them. Composers of the late Romantic period such as Brahms or Mahler were able to create large-scale musical forms because of their common musical language, *tonality*. This system affords a great range of flexible musical structures based on the hierarchy of notes in a key, and the attraction that they have on each other. It is this audible hierarchy that allows the listener to make predictions, allowing the composer to potentially thwart those predictions, thus leading to a game of cat and mouse between the composer and the listener, the give-and-take, push-and-pull, tease-and-satisfy structure of almost all music in the world. Composers in the twentieth century and beyond who choose not to use this tonal language accept the challenge of finding another way to create musical structures that would (in theory) be perceivable, to have this same flexibility to bring the listener on a musical journey. The parametrization of music is one strategy that can achieve this, by allowing mathematical structures to be applied to musical information.

40.2 Problem of Mapping

In considering music simultaneously as a personal, social, and cultural phenomenon, it is clear that there have always been connections between music and external concepts. It is a common understanding of music that it represents mental/cultural/social ideas in patterns of sound: where sound is a semiotic substrate, a medium for meaning. This can be seen in the way we commonly think of pieces of music in terms of emotional states (sad and happy songs) or as a metonym of social concepts such as nationhood, localities, or other sub cultures (anthems, football chants), or even as avatars of current and historical personages or events in our own lives. These are forms of mapping; in these cases the music is mapped to a representation in the form of metaphor or metonymy. "Classical" music is not immune to this. With the rise of romanticism and humanistic individualism in the eighteenth and nineteenth centuries, there was a trend toward mapping pieces of music to a "program," an overarching idea or narrative that the music represented through sound: even the so-called "absolute" music (formally structured symphonies and sonatas, the supposed opposite of such narrativizing) can be considered metaphorically as the enlightenment model of "reason" in musical form.

In addition to these representational mappings, we can also consider literal mapping, where a numerical sequence is mapped directly to a musical parameter: for example, the notes G, A, B, C, D can be assigned the numbers 1–5, so the sequence 1–2–1–4–3, 1–2–1–5–4 generates the pitch sequence G–A–G–C–B, G–A–G–D–C, the melody of *Happy Birthday*. Musical parametrization, in tandem with the rise of electronic and computer musics, afforded abstraction of music into categories that could be considered numerically as discrete sets and scales, and thus manipulated with mathematical tools. For example, musical dynamics, the softness or loudness of a sound, is traditionally notated according to a scale approximately stretching from an intimate and quiet *pianissimo* sound (*ppp*) to a full orchestral blast of *fortissimo* sound (*fff*), with about eight degrees of intensity–*ppp, pp, p, mp, mf, f, ff, fff*. Some composers in the 1960s expanded this to twelve degrees (commensurate with the twelve degrees of the chromatic musical scale) and treated it not as a performative quality to be judged by the musical context (by the composer or the performer), but as a parameter to which numerical processes could be applied to increase the perceived order or disorder of a musical line. In information-theoretic terms, the amount of information (variation) in the musical line is increased with the addition of further parameters that are altered to greater degrees in a shorter span of time, increasing the density of musical information, and also simultaneously increasing both the potential for musical connections and reducing the possibility for any individual connection to stand out among the disorder.

Once the leap has been made to consider musical parameters as elements of a numerical set, mathematical structures can then be applied to any aspect of music in order to generate repetition and variation on different scales. Beran notes that many composers in the twentieth century made explicit use of mathematical ideas and structures in their music. Beran also notes that "perhaps it was psychologically not too clever to admit this publicly. The general audience tends to have a more romantic view of music and wants to relax. A sober explanation of the logical construction is likely to stigmatize a composition as purely intellectual" [2]. These developments can be situated alongside attempts by theorists to develop general mathematical foundations of music; an idea that has roots as far back as Pythagoras but that resists strong formulation to this day. In a 1975 paper in *Nature*, Voss and Clarke [22] presented their analysis of generalized recorded music as $1/f$ noise. Beran shows this claim to be over generalized, but acknowledges the implication that music may have a fractal dimension [2].

There are several aesthetic issues that arise with literal mapping. Whereas musical metaphor is clearly a question of representation, with literal mapping, it is not so clear. In Beethoven's Symphony No. 3 "The Pastoral," there are clear musical references to the countryside, such as the oboe melody that sounds like a cuckoo. This case is clearly representational; the listener will never confuse the oboe with an actual cuckoo. In the case of literal mapping, the representation is not as obvious. The contour of a mountain range can be sketched on graph paper and the relative dimensions scaled and applied to a melodic pattern. Does this melody represent the mountain? Does the composer intend the listener to associate the melody with the mountain? Or is it simply that the contour registered in the composer's mind as a suitable melodic shape? Iannis Xenakis (1922–2001) was a composer and architect who worked closely with mathematical systems in his music, most notably using stochastic systems to create structures of musical probability. His biographer Nouritza Matossian has this to say about Xenakis' approach to sonification:

> Xenakis never claimed that a rigorous mathematical basis is sufficient to produce a well-formed piece of music. Those who are partially informed about the mathematical theory expect the music to be a mirror of mathematical processes and equations. *Pithoprakta* [Xenakis' composition of 1956] is no more a translation of probability theory than an artichoke or a celery is a *translation* [Matossians italics] of the Fibonacci series, or a flowing river is a translation of random functions. [...] Even though underlying structures are shared, particularity of musical resources ensures uniqueness in each one. [23]

One problem with mapping that is particularly noticeable in the case of chaos theory is that the structures of chaos often require visualization to become clear, as composer Rolf Wallin (1957–) puts it:

> One of the first serious turn-downs that appear in an attempt to find viable ways of using Chaos in music is of course the fact that music is serial information, flowing into our ears in

one dimension, degree of air pressure, in relation to another dimension, time. The Mandelbrot pictures also have two dimensions, horizontal and vertical, but these two dimensions are much more closely combined and caught by the eye in one glimpse. One could represent the vertical dimension in the Mandelbrot set with frequency and the horizontal with time, but the aural information would be almost as meaningless as to scan Mona Lisa and put the information into a synthesizer: visual and aural information are seldom directly compatible. [20]

This problem of mapping is exemplified by media stories that purport to show "The Sound of the Big Bang" or "Listening to Atoms," when what they really mean is that some data have been sonified. The data manipulation that takes place in the act of sonification is often ignored in these reports, when this is often the process most salient to the final "sound," In sonification, the data are mapped onto musical parameters by scaling to a range of musically meaningful values and applying these either to audio synthesis or musical instrument digital interface (MIDI) sequencing. The character of the music is largely dependent on the choices made in mapping. The site of composition here is in the decisions taken about what data are used, how the data are filtered and scaled, whether quantization is applied (to pitch, rhythm, dynamics), what type of sounds is chosen, etc. The vacuum of space ensures that there is no actual "sound of the Big Bang," and the oscillations of the microwave radiation that generated the data are not technically "sound" as it is outside the range of human hearing, so the sonification is a representation in metaphorical terms, but also a literal mapping of the data into musical sound; it is not the actual sound of the Big Bang, but it can tell us something about the Big Bang, even if we have no basis of comparison for these sounds that can be meaningful: perhaps the same sonification applied to other radiation sources would provide a meaningful comparison.

Of course, there will be some invariant factors, and ideally this is what sonification can show if it is done sensitively. Scientific sonification is most successful when considered as a form of functional communication and that the choice of mapping tools should reflect whatever it is in the data that the researcher wishes to highlight. For composers, this functional level is not a constraining factor and is often not present at all as the compositional choices are subject only to artistic consideration.

40.3 Case Studies of Composers

What follows is a selection of case studies to show some of the different ways in which composers have taken chaos theory as a fundamental tool or material in their work, the different mapping strategies and variation between metaphoric and literal approaches.

40.3.1 Iannis Xenakis

While it is true that Xenakis did not work directly with chaos theory, in the public mind at least, he is the modernist composer most often linked with the direct marriage of mathematics and music. Before rising to fame as a composer, he trained with Le Corbusier as an architect, contributing to several well-known projects in France such as the Dominican priory of Sainte Marie de La Tourette near Lyon and the Philips Pavilion at *Expo 58* in Brussels. His first recognized composition *Metastaseis* is seminal for its use of spatial concepts in sound masses (building on the work of Edgard Varese), the movement and transformation of stochastically defined masses: music as the architecture of time. The publication of his book *Formalized Music: Thought and Mathematics in Composition* in 1963 sealed his formalist reputation. This perceived connection between mathematics and music is underlined by the often unremitting brutality of Xenakis' music, though as we have seen already in relation to the mapping of data to musical parameters, there is no reason for earthquake magnitude data to sound musically like an earthquake unless the composer deliberately chooses to make such a metaphorical link by using earthquake-like sounds. Xenakis was interested in how mathematics could be used to structure sound. His early writings attacked the excesses of total serialism as amounting to "an irrational and fortuitous dispersion of sounds over the whole extent of the sonic spectrum" [23], but this attack was a lead-in to his own compositions that applied the mathematics of probability theory, stochastics, Markovian theory, and game theory, to musical material and structures.

Xenakis' compositional development pre-dates the development and popularization of chaos theory, but there are a few examples worth exploring briefly. In the late 1980s, Xenakis worked with a programmer to write GENDYN, a system in BASIC that would synthesize electronic music by using multiple layers of stochastic processes controlled by an initial set of variables input by the user. Later analysis of this program by Hoffman characterized the systems as "a dynamic system where the dynamic behaviour of the system is fractal in nature (like all random walks) and, further, that the cascading of random walks means that the sound is governed by strange attractors" [3]. In Xenakis' acoustic music, the closest he came to use dynamic systems was the use of cellular automata (CA) in some of his later pieces, such as *Horos* and *Ata*. Musicologist Makis Solomos has analyzed the extent and specific application of CA in Xenakis' late works, finding that

> [...]of course, we will see that—as it happens with all his "theories"—in his concrete use of cellular automata, Xenakis takes liberties with his model, and introduces "licences", "gaps" (*écarts* in French), manual interventions; in other terms, his use of cellular automata is mediated through *bricolage*. [16]

In *Horos*, Xenakis uses the CA designated with Wolfram's code 4200410 as a filter on existing pitch sets (derived from sieve theory). The CA that he used generated expanding and contracting patterns, including variations and symmetries. The CA also had values of 0–4 for each cell, 0 meaning that pitch would not sound, and 1–4 (excluding 3, which has no mapping) meaning that pitch is assigned to each of the orchestral groups. In addition to generating a repeating and varying set of harmonic progressions, these progressions could be automatically assigned to instrumental timbres, generating a subtle variation. Solomos explains that orchestral timbre and color were of particular interest to Xenakis in this work; the CA allowed him to generate a series of chords made of constantly fluctuating blocks of orchestral color. Xenakis' use of the CA only goes as far as generating the chords and the instrumental families to which each pitch of the chord was assigned, he was calculated the CA on a pocket calculator and appears to have used this process in only 24 bars of a work probably exceeding 200 bars. Once this information was calculated, the remaining details of the musical implementation were carried out intuitively according to practical considerations; register, specific instrumentation (the process only defines families of instruments), articulation, rhythmic and dynamic character, etc. Solomos makes the interesting point that Xenakis' interest in CA may have been instigated by a 1985 article by Stephen Wolfram showing how CA could be used to model turbulence in fluids [16]. Xenakis was interested in the chaotic and generative properties of these mathematics; had he been born a generation later, we can speculate that he may have applied dynamical systems equations to his music. However, Solomos makes a strong case for Xenakis as a composer for whom process "has not to be applied mechanically," that the principle of manual interventions by the composer on the output of the process is an important site of composition. Solomos quotes composer Horacio Vaggione: "Science, regardless of its deductive or empirical nature, tends at least ideally towards an equivalence of process and result. Music shows no tendency of this kind, for the rigour of the generative process does not guarantee the music coherence of the work." Solomos concludes that while mathematical constructions of many varieties often for the backbone of Xenakis' musical thought, ultimately the composer uses these structures as a creative foil, a source for *bricolage*, "mediation, [...] opposed to rational thought" arising out of the very way that the composer works with his tools, in this case, the tools are mathematical. Solomos continues:

> Indeed, with Xenakis implementations of cellular automata, we learn more about the way he is working (the way he makes *bricolage*) than about cellular automata! [16]

40.3.2 György Ligeti

40.3.2.1 Aesthetic

Ligeti (1923–2006) belongs to the same generation as Xenakis and many of the post-War total serialists. Like Xenakis, he eschewed the rigorous serial approach, beginning his mature works with a series of "cloud" pieces (*Atmospheres* and *Lux Aeterna* are good examples), but without the strict stochastic calculation of Xenakis; Ligeti worked instead with the principles of Renaissance polyphony reworked for a modernist post-tonal idiom, and his clouds were made of multitudes of shifting and interacting musical lines that he described as "micro-polyphony." Ligeti had a great interest in mathematics, and made use of rule-based processes (again derived from Renaissance contrapuntal principles) in his music, but these are always subservient to his dramatic instinct in structuring the music. Composer Rolf Wallin discusses Ligeti's aesthetic:

> [...] after a lengthy description of the scientific background of the Mandelbrot set, he cuts the whole thing off by declaring his distance to the 'scientific' ("Naturwissenschaftlige") music of Xenakis, and that he rather utilizes the underlying idea of chaos theory as an inspiration for his compositional techniques. He points out that *the idea of small, simple 'cells' that form complex and interesting patterns when iterated many times, reoccurs in many kinds of music.* The idea of a transition from order to chaos and back again is also very relevant to Ligeti's own music, at least from the 60s and 70s. [20]

Ligeti made it clear that ideas in mathematics were influential in music, given Ligeti's choice of titles and program notes it would be fair to say that Ligeti wanted to enthuse listeners about the ideas that influenced music, but without implying that music was "cerebral." Ligeti's public image balanced modernist–formalist credentials with a sense of surreal mischief (see his references to M.C. Escher, Hieronymus Bosch, and *Alice in Wonderland*). This distancing of Ligeti from the perceived intellectual "coldness" of much post-War modernist music is seen in writings about him also. While it may say more about the demographic of the publication than about the author or composer, Richard Steinitz's article in *The Musical Times* begins with two paragraphs of near-apologia that the reader should not fear the mathematics in Ligeti's music, that:

> The inspiration [Ligeti] has drawn from the amazing revelations of recent [mathematical] research is more poetic and procedural than precisely computed. We are dealing more with kinship than with calculation. [...] Those who faltered at the first algebraic hurdle in Xenakis's *Formalised Music* need not take fright. [17]

However, this is not to criticize the composer or the musicologist; it merely situates the aesthetic concerns of Ligeti as outside, or possibly even opposing the formalized processes of Xenakis. Steinitz goes on to characterize Ligeti's engagement with chaos theory as "impressively knowledgeable at a technical level [...] far deeper than a mere superficial flirtation with popular science," noting that Ligeti sustained an ongoing dialog with Heinz-Otto Peitgen, a pioneering researcher in fractal geometry and visualization of dynamical systems. Ligeti's interest in dynamical systems is in parallel with his musical concerns, the balance of order and chaos, the mechanisms of repetition and variation, and the generation of complex structures from simple relationships: Steinitz points to an early piece of Ligeti's *Clocks and Clouds* (1973) as taking its cue from the Karl Popper lecture of the same name, which "anticipates chaos theory in its comparison of both deterministic and indeterministic interpretations of the universe" [17]. The composer's application of chaos theory to his music is metaphorical in nature; there is no calculation of equations and no data are sonified, but through metaphor he finds powerful ways to communicate ideas that are central to dynamical systems in general. Ligeti's pieces resonate with ideas that go into them, and leave the listener perhaps understanding more about dynamical systems than could be communicated through literal mappings.

40.3.2.2 Technical

The pieces that best exemplify Ligeti's fascination with chaos theory all come from the mid-1980s, a time when Ligeti rediscovered the piano, and wrote a series of 17 virtuosic piano studies and a piano concerto, redefining the instrument for a generation of composers. Three of the studies—"Vertige," "L'escalier du diable" (alluding to self-similarity in Cantor sets), and "Coloana Infinită"—approach the idea of self-similarity (and infinitely ascending spirals and vortices) in three different ways. Steinitz describes them as

> [...] initially deterministic systems containing hidden variables [...which] may be more intuitive than calculated [...], in which the amplification of error leads to more-or-less dramatic developments. [...] All three contain recursively overlapping and proliferating scales, which not only multiply but are progressively magnified. [17]

Ligeti takes the idea of sensitivity to initial conditions and applies it to his compositional technique. His strategy is analogous rather than metaphorical. Ligeti's processes exhibit the same type of variation as seen in chaotic systems but he does not actually calculate these values in advance. This is mimesis, achieving a state similar to a chaotic equation but in the discrete world of musical pitches and rhythmic durations, mimicking the disorder for representation in musical language and as part of a musical structure that requires a certain level of dramatic structuring and dialectical oppositions.

Two examples from Ligeti's "Piano Studies" demonstrate how he achieves the transitions to chaos in a discrete domain by a combination of musical manipulation and perceptual sleight of hand so that the cascading of seemingly insignificant variations into complexity occurs almost unnoticed by the listener. "Vertige" consists almost entirely of descending chromatic scales, so fast that, as the score says, "the individual notes—even without pedal—almost melt into continuous lines" [8]. Two scales repeat and overlap each other; as they repeat, the gaps between scales become smaller and the number of repeating scales rapidly doubles from two to four (period doubling). As this happens, the length of each scale starts to increase, from 16 notes at the start to 21 or more within the first 30 seconds of music; as Steinitz puts it, "the semitonal material—the musical liquid—remains consistent, but no two waves disturbing its surface are ever the same" [17]. At this point, the process is put in the background as a slowly descending melody takes the foreground, Ligeti's composerly instinct is to juxtapose the process with a more dramatic material. The process and melody continue to alternate the foreground for the remainder of the piece.

In "Désordre," a similar process is achieved with two melodies (left hand and right hand) beginning the piece in rhythmic synchronization, each melody using different pitches but following the same rhythm and contour. However, as the piece progresses, the melodies become out of phase, rapidly becoming chaotic: Figure 40.1 shows the left- and right-hand melodies (graphed as pitch against time), highlighting the phrase beginnings and the phase offset. The graph also highlights self-similar aspects as the melody iterates through successive phrases. As with "Vertige", Ligeti achieves this quasi-chaotic motion with gradual shortening of rhythmic phrase lengths; Steinitz describes the 'resizing of the same shape [...] through continuous iteration [as] as fractal characteristic" [18], relating it to the Koch curve. While this method is similar to that of "Vertige", the crucial difference here is that there is a sense of melodic identity from the start. In "Vertige," the material is simply rushing scales that overlap and blur each other, in 'Désordre', the melodic character allows clear perception of two melodic objects which go out of phase, eventually dissolving into a chaos of non identity, the defining characteristics of the melodic patterns are obliterated by the process. The melody is defined by its pitch contour, and its phrase lengths that are marked by loud accents. The piece starts with both melodies sounding as one, rhythmically together. As they fall out of phase, there is a perceptual shift when it becomes apparent that they are different melodies, now desynchronized, their accents make the patterning of both melodies clear. As the compositional process reduces the phrase lengths, the accents become closer together, and because musical durations are discrete, this rapidly reaches a point where the unaccented notes are removed entirely, leaving only a constant stream

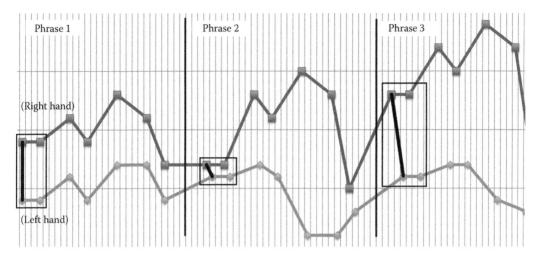

FIGURE 40.1 Melodic contours in Ligeti's "Disorder" (score page 1).

of accents. Without the differentiation provided by the accent, there is no melodic contour, no phrase, and no melodic identity, chaos.

40.3.3 Jeff Pressing

Jeff Pressing (1946–2002) was a polymath working in a wide range of disciplines from physical chemistry to ethnomusicology. A keyboard player and improvisor, he wrote several papers on cognitive aspects of improvised performance based on research into cognitive models of non linearity in movement [14]. He also experimented with using discrete nonlinear maps as tools for generating musical materials and structures. His 1988 paper "Nonlinear Maps as Generators of Musical Design" [15] provides a detailed discussion of his strategies for using the logistic map and similar structures to generate numerical material, which he can then map onto various musical parameters. Pressing discusses and provides audio examples for the following structures: logistic map, cube root logistic map, Metz map (modified), predator prey map (and with reversed variables), complex quadratic map (and with barrier), quaternion quadratic map with barrier, quaternion logistic map, and symmetrized quaternion with barrier. All examples in the following section are drawn from Pressing's paper.

40.3.3.1 Aesthetic

Pressing takes an aesthetically neutral approach to the generation of musical material from nonlinear equations, regarding the usefulness of the results as a matter of individual "judgement," describing the results as "idiosyncratic, but [having] a listenable degree of structural consistency," and as making "musical sense only from a twentieth-century musical perspective." He draws parallels between the outcomes of his experiments and "free" types of music, such as "certain kinds of folk music, free jazz, and European 'cloud' music (Xenakis, Ligeti, Penderecki)": these are all musics where musical ideas transform fluidly, and structures can be unpredictable but tend to oscillate around a few identifiable motives. Pressing also suggests that the use of mathematics to generate material can be considered as a "found process," in the same aesthetic light as "found objects" in all art forms.

40.3.3.2 Technical

Pressing chooses specific regions of the logistic map (equation below), often using sections of a-values just before the onset of a new quasi-stable cycle in the period-doubling cascade; he describes these sections as being "of considerable interest for the concept of musical variation" [15].

$$x_{n+1} = ax_n(1 - x_n), 0 < a \leq 4 \tag{40.1}$$

Pressing uses sections where a repeating pattern destabilizes, then stabilizes again, either returning to the previous pattern, or moving to a new pattern. These regions of the logistic map move between chaotic and quasi-periodic states, and this motion suits a musical variation very well, allowing the listener some time to absorb a musical idea, then altering it—or in this case allowing the process to alter it—then either returning to its home state or arriving at a new idea. Pressing notes that "when this is translated into musical terms, the result can be something rather like some twentieth-century variation techniques" [15]. For Pressing, one site of composition is in the fine-tuning of the *a*-value to achieve the right kind of variation for the parameter at the moment in the musical argument, selecting the data from a point in the dynamical system process where a stable line begins to become unstable, this affords a type of musical tension.

The data from the selected sections of the dynamical system are then mapped to musical parameters, for which Pressing discusses a variety of strategies. Often, each parameter is mapped to a different section of *a*-values, depending on the level of variation required. In some cases, he may use a canonic strategy, applying the same section of values to the pitch of different musical voices but temporally offset, and possibly offset in register to create voices in bass, treble, etc. He also suggests applying several different parameters in the same voice, all sharing a single value but offset by some amount: of course the relationships here may not be perceivable as different parameters scale in different ways. Here are three examples of Pressing's mappings with an increasing number of *a*-values used in each: the parameters used here are for audio syntheses, *interonset time* is the duration between events, *attack time* is the duration between the event beginning in silence and its reaching full volume and, *dynamics* is the level of "full volume" for that event:

- 3 parameters: pitch (3.93744), interonset time (3.93740), attack time (3.97776)
- 4 parameters: pitch (3.6785), interonset time (3.99943), attack time (3.99943), dynamics (3.99943)
- 5 parameters: two-voice rhythmic heterophony, 5th value is interonset time for the 2nd voice, which shares all other parameters (3.00, 2.5, 3.9999, 3.84, 3.80)

Pressing's choice of scaling reflects both his aesthetic preferences and the medium for which he is writing. As his examples are intended to be used in computer audio synthesis, he can forego translating the data into values that will work in musical notation. The composer scales the data to a range of three octaves (two octaves below middle C [64 Hz] to an octave above [512 Hz]) to keep the melodic patterns within normal musical range, but notes that while "quantization to tempered (or other) tuning norms could be applied, [...] for the most part this was not done, as it seemed to yield less interesting results" [15]. As many computer music systems use data in the range 0–1, scaling was often not required except to make sectional changes; for example, interonset time and dynamics can be scaled as 0–1 in relation to a variable "beat duration" or tempo.

Figure 40.2 presents a sample of Pressing's data from his 1988 article shown as a graph, where the repetition and variation are clearly seen; both the subtle changes to the cyclic values and the sudden chaotic deviations. Figure 40.3 shows the same data in musical notation, where the periodic 3-note pattern is clearly visible, the subtle changes can be seen in microtonal pitch deviations, and the chaotic changes are seen as changes in phrase length as well as pitch value.

Pressing goes on in the article to discuss his exploration of two-dimensional maps using complex values for *x* and *a*, and using quaternions with reflecting barriers (to reduce the problem of infinities and divergences). His mapping strategies do not change significantly with these equations, he is mainly interested in generating other shapes that may be suitable for musical applications: for example, he notes that "the cube root logistic map produces a jazzlike melody and walking bass-line, rather surprisingly. The apparent effect of introducing such fractional exponents into the equation is to 'smooth out' chaotic effects" [15]. He makes several useful suggestions for further development that take into account more strictly "musical" composition procedures, in some cases reducing the agency of the process.

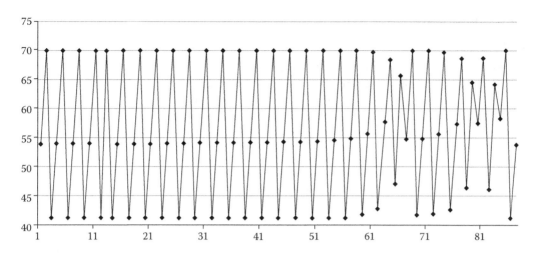

FIGURE 40.2 Pressing's logistic map data example: Expressed here as floating-point MIDI numbers.

FIGURE 40.3 Pressing's logistic map example frequencies in musical notation: $+/-$ numbers indicate cent deviations, where 100 cent = 1 semitone.

Further development could take such forms as generalizing the above procedures by changing a- or c-values (for equations with complex values) in a coordinated fashion during the course of a piece, quantizing all variables to promote traditionalist music perception, linking parameters between voices in a more meticulous fashion to clarify contrapuntal effects, using functional inverses to effect time reversal, and pursuing mathematical extensions such as continuous variables, fractional exponents, and rational functions of polynomials.

40.3.4 Rolf Wallin

40.3.4.1 Aesthetic

The Norwegian composer Rolf Wallin (1957–) writes mostly acoustic music for concert performance, and used fractals, as he describes it, "an important component in the technical shell" of his work between 1989 and 1995 [21]. Wallin is one of the few composers who has written relatively extensively on his relationship with chaos theory. He worked with one particular equation on a number of pieces in this period, using it both as a structural tool and to generate local-level values, but like Wuorinen uses it only to create a framework:

> These fascinating mathematical phenomena, simple equations from which strangely organic structures emanated, had provided a means to generate a quite detailed skeleton on which I could sculpt the music.
>
> [...] there was a constant collaboration between my subjective artistic choice and the computer: I gave the computer specific tasks and chose the "best one" out of different solutions offered. When a satisfactory skeleton had been erected, I could let my imagination interact with it to shape the final score. [21]

In various articles, and in an interview with the author, Wallin expounds on the organic quality of chaos theory as important to his work. The complexity of the system generated is like an "ecosystem," where all the parts have interlocking relationships, and the patterns generated by the chaotic functions are not "complicated, cumbersome, [or] difficult, but rather something that has a many faceted quality, a world within the world to explore" [21]. In working with these equations, he makes a point under explored by other composers, that a level of immersion in the chaotic system itself is an important part of generating interesting musical material from it. Jeff Pressing explained above how he used specific a-values of the equations to generate specific types of chaotic contours; Wallin describes the process of immersion in the system as almost like getting to know an unfamiliar animal. The variables of the system can be tuned to generate different outputs, different environments and climates for the composer to explore. This is an essentially fractal quality of the systems that interests composers, the ability to explore an idea to infinite depth. Wallin argues that some equations are not musically interesting (at least from his perspective), taking the logistic map to task as follows:

> [...] quite entertaining for a while, until we understood that the self similarity in this system is so self similar that it soon gets boring. The same three motives bite one another's tail *ad infinitum*. No new motives appear when "diving" into the graph by enlarging smaller regions. [20]

This shows as much as anything else that an equation that one composer can find endlessly fascinating can be barren for another composer, the choice of mapping strategies has a fundamental effect on how useful the system is for the composer.

The balance between the process and the intuitive composition is also addressed by Wallin. He describes the chaotic equations as a "ghost in the machine," where the system can generate material that would not have been thought of the composer alone, but also that the process-driven system can be used as a foil against our own intuition and its tendency to write what we already know; use of process in this way adds noise to the system of human creativity. The composer points to a specific example in one of his pieces:

> In fact, the last third of Stonewave is one long linear sweep through a microscopic jungle of numbers arranging themselves in less and less predictable patterns, with a "pocket" of extreme repetitiveness before exploding into the last chaotic bars. [21]

Wallin points to other composers who do not work with chaos theory but who also referred to rich external systems as sources that have an alien quality, something outside our experience but being recognizably a communication, an intelligence. Messiaen's use of birdsong and Indian classical music, Cage's reflection on the *i-ching* and Buddha, all of these things, and chaos theory, point to an intelligence outside human thought and interference. However, these are again simply starting points and scaffolding for the music, as Wallin says, "listenable music, not the system itself, is the ultimate goal." [21]

> The music is neither more nor less fractal than African drum rhythms or a Bartók string quartet. After all, the concept of complexity from the interaction of simple particles and of the parts resembling the whole can be traced back to the early history of mankind, because it is so obvious to us that the world is constructed that way. Chaos theory therefore brings nothing strictly new to our perception of the world, it only confirms it scientifically in a shockingly simple way, and it also for the first time gives us a possibility to simulate some of Nature's own structures by mathematical means. [...] This is of course no "artistic proof" against Ligeti's music, as the aesthetical value lies in the music itself, not in its origin. [...] there is something defeatistic in denying the possibilities of actually using fractals in music while admiring the Peitgen pictures for their artistic value, well knowing that they are made in a much more "scientific" way than Xenakis makes his music. [20]

40.3.4.2 Technical

Wallin's work explores one particular equation, a "symmetrically coupled nonlinear system" introduced to him by Jan Frøyland at the University of Oslo. Here, it is presented as a three-dimensional version where population values (x, y, z) mutually feed back into each other, their evolution governed by the two parameters (c, d):

$$xn + 1 = 2 * d * (yn + zn) + 2 * xn * (c + xn) \tag{40.2}$$

$$yn + 1 = 2 * d * (xn + zn) + 2 * yn * (c + yn) \tag{40.3}$$

$$zn + 1 = 2 * d * (xn + yn) + 2 * zn * (c + zn) \tag{40.4}$$

His pieces from this period such as *Onda di Ghiaccio* and *Stonewave* use two formulas, "one to generate the formal structure, the other, the detail ("melodic") content, providing an 'event skeleton' on which I could compose the whole piece."[*] With this skeleton in place, the musical detail of the work can be intuitively composed.

40.3.5 Eleri Pound

40.3.5.1 Aesthetic

Pound (1982–) mixes a background in mathematics with concerns from experimental music. Much of her use of mathematics involves calculation of microtonal tunings with specific properties: she collaborated with mathematician Robert Sturman on research into using Arnold Tongues to derive new sets of pitch intervals that are not generally found in conventional tuning [13]. She has also worked with chaos theory to produce pieces that "perform" non linear dynamics, such as *Circles 9* for voices that visualizes and sonifies the principle of sensitivity to initial conditions. The piece is in the form of a a very literal mapping, the score is generated by using the initial sung pitches as inputs to the function, computed in real time. The piece is a sonification of the resultant data, but rather than simply playing the computed pitches back as synthesized sound, the pitches are interpreted aurally by singers and performed in real time, adding a human and performative dimension.

---

[*] Email correspondence with the author, April 3, 2013.

40.3.5.2 Technical

$$[x < h]f(x) = x/h \quad [x \geq h]f(x) = (1-x)/(1-h) \tag{40.5}$$

Circles 9 uses the "witches hat" function (above) as the starting point for a slowly unfolding vocal piece for 3–5 singers, with accompanying visuals of the process data. The piece "performs" the function by calculating a score on-the-fly as a product of discrepancies in the performed tuning of piece's opening note. Three singers begin the piece by singing the same pitch together, the computer analyzes the three sung pitches and uses these frequency values as separate inputs to the function. Although to the listener the three singers will sound in-tune, it is practically impossible for all the singers to produce precisely the same Hz value. Given these three different starting values, the computer then iterates the function 180 times independently for each voice, producing a score that will vary in each performance, with each voice having its own unique path through the function. The audio module of the computer patch then iterates through the score; once every N seconds, the singers are sent the next frequency value via ear-pieces and the singers reproduce this: so the three different routes through the function are heard via the singers. The whole piece is then structured as three sections, each having a different value for h: in the initial calculation, the computer alters the value of h every 60 iterations: the values are 0.113, 0.507, and 0.907.

Figures 40.4 and 40.5 show the visualization of the computer score at two different points in the piece. The large window shows the current detail of the three different voices and the lower small window shows the progress through an idealized version of the entire piece assuming an input value of 330 Hz.

40.3.6 Jonathan Dawe

Jonathan Dawe's (1965–) work plots a new path that can be related to some of the strategies outlined above. The composer applies fractal geometry (among other ideas) to the pre-existing musical material,

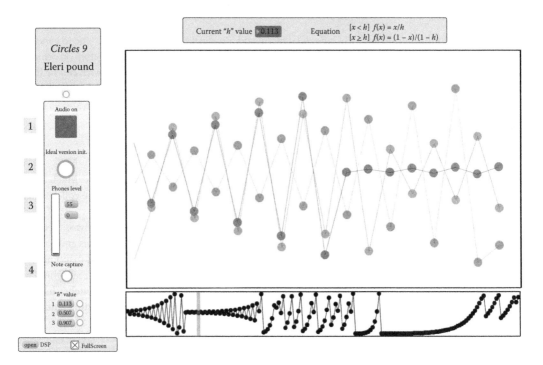

FIGURE 40.4   Pound's *Circles 9* at $h = 0.113$.

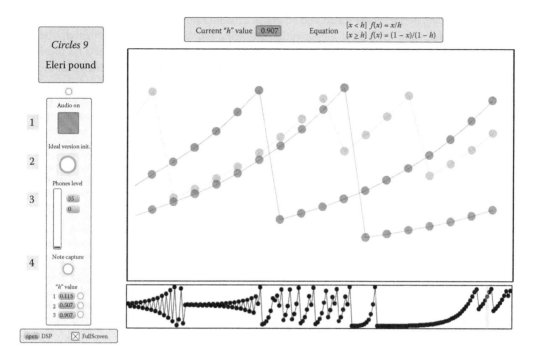

FIGURE 40.5 Pound's *Circles 9* at $h = 0.907$.

often from long-dead composers such as Vivaldi and Charpentier. Rather than using equations to generate the musical material as a sonification of that process, Dawe treats the musical material (melodies, harmonic progressions, spectral information, etc.) as shapes and contours to be manipulated by the fractal processes, making his procedure more "geometric" in a sense than sonification. He describes his process and applications in different compositions thus:

> I've focused mostly on growth patterns that are self-replicating, spawning multiple copies of the original material after some number of generations of growth. *Piano Concerto* (2002) is a large-scale work that engineers Cellular Automata on local, middle-ground, and structural levels, drawing its original shapes from the 1610 *Vespers* of Claudio Monteverdi [4].
>
> In *Symphony of Imaginary Numbers* (1998) the musical domains of Mozart's 'Hafner' Symphony and Johann Christian Bach's Symphony (the inspiration for Mozart's piece) are added, subtracted, divided and multiplied as complex numbers. From the union of these two works a new piece is created [4].
>
> *The Horn Trio* a single-movement work, [which] systematically presents and re-presents its musical ideas. In the drama of this work, new levels of structure are created as older previous material is reinforced through pitch, and rhythmic instrumental doublings, fashioning the foundation for off-shoot melodies. These new filigrees, always derived from the music beneath them, then, in turn, become the stable platform for still fresher recursions. The first original structures thus becoming deep pillars of design. The beginning material from which the Horn Trio builds its dynamic design comes from a comes from a restructuring of the twelve-tone series in Karlheinz Stockhausen's early work, *Sonatine for Violin and Piano* (1951). [7]

This continual transformation of familiar material provides the music with a hauntology of sorts, a sense of hearing something familiar in a new way, what Dawe describes as a "fragmentation of references" [1]; this relates strongly to modernistic processes from all domains in post-nineteenth-century art, such as

James Joyce, Luciano Berio, Pablo Picasso, and others. In a manner similar to Ligeti's "Désordre" discussed above, Dawe's manipulation is a quasi-topographical distorting and varying of the older material, treating the older material as objects, and to a certain extent relying on their perception as objects for the distortion to be effective. To perceive the material as new, the distortion needs to carry the listener through an aesthetic process whereby the material is recognized as familiar (notionally familiar in style is enough, not necessarily that the listener knows the specific melody), then varied just enough to maintain the connection to its source, until the transformation passes a threshold where the original material is an imprint but the material is now Dawe's own. Fractal processes are useful here as they can be "tuned" to make small but unpredictable alterations, moving just far enough in each iteration that the listener maintains a connection to the previous iteration.

As with Xenakis and Wallin above, Dawe's rigorous use of process can be separated from more intuitive or traditional compositional techniques. In email correspondence with the author, Dawe describes his "local" working as rigorous, the note-to-note application of fractal mathematics to fragments of music, he says the "material to be grown (usually) ranges from 2 to 7 pitches, but can be more." The large-scale structures of the pieces occasionally have mathematical input, or can, as the composer describes, "develop as outgrowths of the 'local' fractal procedures," but are mostly constructed with more conventional compositional techniques.[*] Figures 40.6 and 40.7 show chordal and melodic strategies respectively for mapping the CA to music: both examples are taken from *Mozart Super Fractals* (2007) for piano and clarinet, provided by the composer.[†] In the chordal example, the CA object is read from left to right with live cells read as vertical pitch groupings (chords), where the numbers represent pitch classes 0–11: 0 = C, 1 = C-sharp, 2 = D, … 11 = B, etc. In the melodic example, the same pitch-class mapping is used but here the verticals of the CA are read as consecutive (descending) pitches rather than simultaneous pitches, with each beat of the group subdivided into as many smaller beats as there are cells in the vertical: so the first vertical has one live cell and takes up a single quaver/eighth note, whereas while the second vertical has three live cells so the single quaver/eighth note is subdivided into three triplet (three notes in the time of two) semiquaver/sixteenth notes. Dawe takes his initial material (usually, melodic fragments taken from

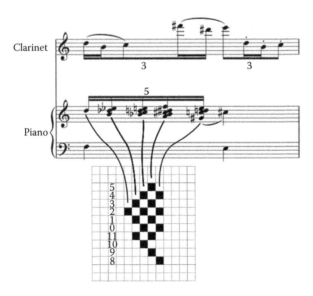

FIGURE 40.6   Jonathan Dawe's chordal mapping of Fredkin CA.

---

[*]   Quotes drawn from email correspondence with the author, May 10, 2013.

[†]   Some aspects of the notation have been altered for clarity of the figure, e.g., pitch spellings have been enharmonically altered, accents have been omitted.

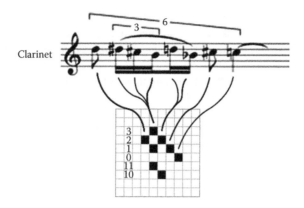

FIGURE 40.7   Jonathan Dawe's melodic mapping of Fredkin CA.

older musics), and maps it as cells on a CA grid, by reversing the procedures discussed above. The nature of the CA means that there may be some aspects of repetition or fractal character as the CA iterates and develops, which can act as a musical development and variation.

Dawe's earlier music uses serial techniques and rotational arrays (after Stravinsky), which the composer chose for their recursive properties. This serial approach is conjoined with the parametrization of music that allows the pitch and other aspects to be treated and manipulated numerically. As seen above, Dawe is interested in recursion and iterative processes to transform material. The basic strategy of treating musical elements as pitches in a series, or cells on graph paper, allows any mathematical structure to be considered. Dawe says that he almost always maps in two dimensions x = time and y = pitch/frequency, and within the limitation that simple mapping he utilizes an impressive array of structures: cellular automata in the *Piano Concerto* and *Zipoli Automata* (Pascal's triangle), fractal and affine geometries in *Royal Water Music*, *Armide*, and the *Clarinet Quintet* among others, fibonacci patterns in *Cracked Orlando*, Farey series, complex numbers. There is a close correspondence between the composer's practical application of these functions, and his ability to see in these functions a musical possibility. The composer's job is to see music in everything, and then find a way to get at that musical possibility and translate it.

40.4 Conclusions

In this overview, I hope to draw attention to the ways in which composers relate to, and work with, chaos theory and nonlinear dynamical systems. The tension between composition as process driven or intuition driven is highlighted in the way that different composers draw on chaos theory as a resource. The varying sites of composition for these composers show the multidimensionality of composition as a practice, and how chaos theory can be situated as a similarly anti reductive mode of thought.

Many excellent composers and researchers have been omitted from this chapter, partly to keep the themes to manageable amount. Eleri Pound is the only representative of experimental and post-Cagean composition included here, but interested readers should also seek out the writings and music of David Dunn, David Burraston, and Alan Lamb for their contributions to non linear systems of environmental sound, noise, and noise music. An overview of my own works and relationship to chaos theory can be found in my 2011 article, "Dynamical Systems, Mimesis, and Analogy in Experimental Music" [11].

References

1. American Composers' Orchestra, Playing It Unsafe: Jonathan Dawe's Journey, https://www .youtube.com/watch?v=UQ-YPtAfkW4, accessed 20/05/2013.

2. Beran, J., Music: Chaos, fractals, and information, *Chance*, Vol. 17, No. 4, pp. 7–16, 2004.

3. Brown, A.E., Andrew, R., and Jenkins, G., The interactive dynamic stochastic synthesizer, in Norris, Michael. (eds). *Proceedings of the Australasian Computer Music Conference*, Wellington, New Zealand, pp. 18–22, 2004.

4. Dawe, J., *Fractals*, http://www.jonathandawe.com/projects/fractals.html, accessed 20/05/2013.

5. Gleick, J., *Chaos: Making a New Science*, New York: Viking 1988.

6. Huizinga, J., *Homo Ludens; A Study of the Play-Element in Culture*. Boston: Beacon Press, 1955.

7. Innova Records, Jonathan Dawe, A Noise Did Rise: Chamber Works 1993–1999, http://www.innova.mu/sites/www.innova.mu/files/liner-notes/316.htm, accessed 20/05/2013.

8. Ligeti, G., *Études pour piano: 2ème livre, Mainz: Schott*, 1998.

9. Lochhead, J., Hearing chaos, *American Music*, Vol. 19, No. 2, pp. 210–246, 2001.

10. McLeod, K., Interpreting chaos: The paradigm of chaotics and new critical theory, *College Music Symposium*, Vol. 45, pp. 42–56, 2005.

11. Mc Laughlin, S., Dynamical systems, mimesis, and analogy in experimental music, *Chaotic Modeling and Simulation*, Vol. 1: pp. 127–137, October 2011.

12. Matossian, N., *Xenakis*. London: Kahn & Averill, 1986.

13. Pound, E.A., *Chaos as compositional order*, in Skiadas, C.H., Dimotikalis, I., Skiadas, C (eds), *Chaos Theory: Modelling, Simulation and Applications*, World Scientific Publishing Co., 2011.

14. Pressing, J., Improvisation: Methods and models in: *Generative Processes in Music* J., Sloboda (ed) Oxford University Press, 1987.

15. Pressing, J., Nonlinear maps as generators of musical design, *Computer Music Journal*, Vol. 12, No. 2, pp. 35–46, 1988.

16. Solomos, M., Cellular automata in xenakis music: Theory and practice, in Solomos M., Georgaki A., Zervos, G (eds), *Definitive Proceedings of the International Symposium Iannis Xenakis*, Athens, May 2005.

17. Steinitz, R., Music, maths & chaos, *The Musical Times*, Vol. 137, No. 1837, pp. 14–20, 1996.

18. Steinitz, R., The dynamics of disorder, *The Musical Times*, Vol. 137, No. 1839, pp. 7–14, 1996.

19. Nash, P.P., *Music of Charles Wuorinen*, http://classes.yale.edu/fractals/panorama/music/wuorinen/wuorinen.html, accessed 02/05/2013.

20. Wallin, R., Fractal music—red herring or promised land? or just another of those boring papers on chaos, Lecture given at the *Nordic Symposium for Computer Assisted Composition Stockholm*, 1989.

21. Wallin, R., Lobster soup, *Nordic Sounds*, Vol. 1, 1998.

22. Voss, R.F. and Clarke, J., 1/f noise in music and speech. *Nature*, Vol. 258, pp. 317–318, 1975.

23. Xenakis, I., 1971. *Formalized Music: Thought and Mathematics in Composition*, London: Indiana University Press, 1971.

Index

913

Printed and bound by CPI Group (UK) Ltd, Croydon, CR0 4YY

24/10/2024

01778295-0017